THE RIBOSOME

STRUCTURE, FUNCTION, & EVOLUTION

THE RIBOSOME

STRUCTURE, FUNCTION, & EVOLUTION

Edited by

Walter E. Hill
University of Montana
Missoula

Albert Dahlberg
Brown University
Providence, Rhode Island

Roger A. Garrett
University of Copenhagen
Copenhagen, Denmark

Peter B. Moore
Yale University
New Haven, Connecticut

David Schlessinger
Washington University School of Medicine
St. Louis, Missouri

Jonathan R. Warner
Albert Einstein College of Medicine
Bronx, New York

AMERICAN SOCIETY FOR MICROBIOLOGY
WASHINGTON, D.C.

1325 Massachusetts Avenue, N.W.
Washington, DC 20005

Library of Congress Cataloging-in-Publication Data

The Ribosome : structure, function, and evolution / edited by Walter E. Hill . . . [et al.].
p. cm.
Papers presented at a conference held on Aug. 6–11, 1989 in East Glacier Park, Montana.
Includes index.
ISBN 1-55581-020-9
1. Ribosomes—Congresses. I. Hill, Walter E. (Walter Ensign),
1937– . II. American Society of Microbiology.
QH603.R5R54 1990
574.87′34—dc20 90-757
CIP

Cover photograph: "Two-Piece Reclining Figure No. 2," Henry Moore, 1959. From the permanent collection of the Saint Louis Art Museum, on loan to the Missouri Botanical Garden, St. Louis, Mo. Photograph by Gregg Bogosian. Published by permission of the Saint Louis Art Museum and Missouri Botanical Garden.

To the Memory of

HEINZ-GÜNTER WITTMANN

DAVID ELSON

CONTENTS

III. Probing rRNA Function

IV. Initiation

V. Elongation

VI. Termination

VII. Ribosome Formation

VIII. Antibiotic Mechanisms and Probes

IX. Translational Fidelity

X. Evolution of Ribosomes

CONTRIBUTORS

Patrick N. Allen ■ Thimann Laboratories, University of California, Santa Cruz, Santa Cruz, California 95064 (3)

R. Amils ■ Centro de Biología Molecular, U.A.M.-C.S.I.C., Universidad Autonoma de Madrid, Canto Blanco, Madrid 28049, Spain (56)

Evelyn Arndt ■ Max-Planck-Institut für Molekulare Genetik, Ihnestrasse 73, D-1000 Berlin-Dahlem, Federal Republic of Germany (53)

Thomas Atha ■ Molecular Biology Institute and Department of Biology, University of California at Los Angeles, Los Angeles, California 90024 (12)

John Atkins ■ Howard Hughes Medical Institute and Department of Human Genetics, University of Utah Medical Center, Salt Lake City, Utah 84132 (46)

J. Auer ■ Lehrstuhl für Mikrobiologie der Universität, Maria-Ward-Strasse 1a, D-8000 Munich 19, Federal Republic of Germany (54)

J. P. G. Ballesta ■ Centro de Biología Molecular, Canto Blanco, Madrid 28049, Spain (43)

Diane M. Baronas-Lowell ■ Department of Cell Biology, Albert Einstein College of Medicine, 1300 Morris Park Avenue, Bronx, New York 10461 (38)

Andrea Barta ■ Institut für Biochemie der Universität Wien, Währingerstrasse 17, A-1090 Vienna, Austria (28)

Bonnie Bartel ■ Department of Biology, Massachusetts Institute of Technology, Cambridge, Massachusetts 02139 (55)

Florence Baudin ■ Laboratoire de Biochimie, Institut de Biologie Moléculaire et Cellulaire du Centre National de la Recherche Scientifique, 15 rue René Descartes, 67084 Strasbourg Cedex, France (9)

W. Bennett ■ Max Planck Research Unit for Structural Molecular Biology, D-2000 Hamburg 52, and Max Planck Institute for Molecular Genetics, D-1000 Berlin 33, Federal Republic of Germany (8)

N. Bilgin ■ Institute of Molecular Biology, University of Uppsala Biomedical Center, Box 590, S-751 24 Uppsala, Sweden (30, 44)

A. Böck ■ Lehrstuhl für Mikrobiologie der Universität, Maria-Ward-Strasse 1a, D-8000 Munich 19, Federal Republic of Germany (54)

Gregg Bogosian ■ Animal Sciences Division, Monsanto Corporation, Mail Zone BB3M, 700 Chesterfield Village Parkway, Chesterfield, Missouri 63198 (48)

L. Bosch ■ Department of Biochemistry, Leiden University, Gorlaeus Laboratories, 2333 AL Leiden, The Netherlands (34)

M. Boublik ■ Roche Institute of Molecular Biology, Roche Research Center, Nutley, New Jersey 07110 (6)

D. Brechemier-Baey ■ Institut de Biologie Physico-Chimique, Paris, France (47)

Richard Brimacombe ■ Max-Planck-Institut für Molekulare Genetik, Abteilung Wittmann, Ihnestrasse 73, D-1000 Berlin-Dahlem, Federal Republic of Germany (4)

Chris M. Brown ■ Department of Biochemistry, University of Otago, P.O. Box 56, Dunedin, New Zealand (32)

R. H. Buckingham ■ Institut de Biologie Physico-Chimique, Paris, France (47)

Alex B. Burgin ■ Department of Biology and Institute for Molecular and Cellular Biology, Indiana University, Bloomington, Indiana 47405 (35)
Maria A. Canonaco ■ Max-Planck-Institut für Molekulare Genetik, Abteilung Wittmann, Ihnestrasse 73, D-1000 Berlin-Dahlem, Federal Republic of Germany (21)
Yuen-Ling Chan ■ Department of Biochemistry and Molecular Biology, The University of Chicago, Chicago, Illinois 60637 (14)
F. Claesens ■ Institute of Molecular Biology, University of Uppsala Biomedical Center, Box 590, S-751 24 Uppsala, Sweden (30)
Barry S. Cooperman ■ Department of Chemistry, University of Pennsylvania, Philadelphia, Pennsylvania 19104 (42)
E. Coppin-Raynal ■ Laboratoire de Génétique de l'Université Paris-Sud, 91405 Orsay, France (47)
Michael R. Culbertson ■ Laboratory of Molecular Biology, University of Wisconsin, 1525 Linden Drive, Madison, Wisconsin 53706 (49)
Eric Cundliffe ■ Department of Biochemistry and Leicester Biocentre, University of Leicester, Leicester LE1 7RH, United Kingdom (41)
Philip R. Cunningham ■ Roche Institute of Molecular Biology, Roche Research Center, Nutley, New Jersey 07110 (17)
John Czworkowski ■ Clayton Foundation Biochemical Institute, Department of Chemistry and Biochemistry, University of Texas at Austin, Austin, Texas 78712 (29)
Albert E. Dahlberg ■ Brown University, Providence, Rhode Island 02912 (16, 33)
Elizabeth De Stasio ■ Lawrence University, Appleton, Wisconsin 54912 (16)
I. Diaz ■ Institute of Molecular Biology, University of Uppsala Biomedical Center, Box 590, S-751 24 Uppsala, Sweden (30)
Daniel B. Dix ■ Molecular, Cellular, and Developmental Biology, University of Colorado, Boulder, Colorado 80309 (45)
David E. Draper ■ Department of Chemistry, Johns Hopkins University, Baltimore, Maryland 21218 (10)
Diane Dunn ■ Howard Hughes Medical Institute and Department of Human Genetics, University of Utah Medical Center, Salt Lake City, Utah 84132 (46)
Jean-Pierre Ebel ■ Laboratoire de Biochimie, Institut de Biologie Moléculaire et Cellulaire du Centre National de la Recherche Scientifique, 15 rue René Descartes, 67084 Strasbourg Cedex, France (9)
Jan Egebjerg ■ Biostructural Chemistry, Chemistry Institute, Aarhus University, DK-8000 Aarhus C, Denmark (11)
M. Ehrenberg ■ Institute of Molecular Biology, University of Uppsala Biomedical Center, Box 590, S-751 24 Uppsala, Sweden (30, 44)
Bernard Ehresmann ■ Laboratoire de Biochimie, Institut de Biologie Moléculaire et Cellulaire du Centre National de la Recherche Scientifique, 15 rue René Descartes, 67084 Strasbourg Cedex, France (9)
Chantal Ehresmann ■ Laboratoire de Biochimie, Institut de Biologie Moléculaire et Cellulaire du Centre National de la Recherche Scientifique, 15 rue René Descartes, 67084 Strasbourg Cedex, France (9)
Yaeta Endo ■ Department of Biochemistry, Yamanashi Medical College, Yamanashi 409-38, Japan (14)
Francis J. Eng ■ Department of Cell Biology, Albert Einstein College of Medicine, 1300 Morris Park Avenue, Bronx, New York 10461 (38)
Carmen L. Fernández ■ Department of Chemistry, University of Pennsylvania, Philadelphia, Pennsylvania 19104 (42)
N. Figueroa ■ Centre de Génétique Moleculaire du Centre National de la Recherche Scientifique, 91190 Gif-sur-Yvette, France (47)
Daniel Finley ■ Department of Cellular and Molecular Physiology, Harvard Medical School, Boston, Massachusetts 02115 (55)
J. Frank ■ Wadsworth Center for Laboratories and Research, New York State Department of Health, Albany, New York 12201-0509, and School of Public Health, State University of

New York at Albany, Albany, New York 12222 (5)
Vincent Frick ■ Roche Institute of Molecular Biology, Roche Research Center, Nutley, New Jersey 07110 (17)
Roger A. Garrett ■ Institute for Biological Chemistry B, University of Copenhagen, Sølvgade 83, DK-1307 Copenhagen K, Denmark (11)
Ute Geigenmüller ■ Washington University School of Medicine, 660 South Euclid Avenue, St. Louis, Missouri 63110-1093 (25)
Susan A. Gerbi ■ Division of Biology and Medicine, Brown University, Providence, Rhode Island 02912 (39)
Ray Gesteland ■ Howard Hughes Medical Institute and Department of Human Genetics, University of Utah Medical Center, Salt Lake City, Utah 84132 (46)
Klaus Giese ■ Max-Planck-Institut für Molekulare Genetik, Abteilung Wittmann, Ihnestrasse 73, D-1000 Berlin-Dahlem, Federal Republic of Germany (57)
Anton Glück ■ Department of Biochemistry and Molecular Biology, The University of Chicago, Chicago, Illinois 60637 (14)
Thomas Gluick ■ Division of Biological Sciences, University of Montana, Missoula, Montana 59812 (18)
Andreas Gnirke ■ Washington University School of Medicine, 660 South Euclid Avenue, St. Louis, Missouri 63110-1093 (25)
Larry Gold ■ Department of Molecular, Cellular, and Developmental Biology, University of Colorado, Boulder, Colorado 80309 (20)
H. U. Göringer ■ Max-Planck-Institut für Molekulare Genetik, Berlin, Federal Republic of Germany (16)
Manolo Gouy ■ Laboratoire de Biométrie, Université Lyon I, 69622 Villeurbanne Cedex, France (50)
Michael W. Gray ■ Department of Biochemistry, Dalhousie University, Halifax, Nova Scotia B3H 4H7, Canada (52)
Barbara Greuer ■ Max-Planck-Institut für Molekulare Genetik, Abteilung Wittmann, Ihnestrasse 73, D-1000 Berlin-Dahlem, Federal Republic of Germany (4)
Claudio O. Gualerzi ■ Max-Planck-Institut für Molekulare Genetik, Abteilung Wittmann, Berlin, Federal Republic of Germany, and Laboratory of Genetics, Department of Cell Biology, University of Camerino, Camerino, Italy (21)
Boyd Hardesty ■ Clayton Foundation Biochemical Institute, Department of Chemistry and Biochemistry, University of Texas at Austin, Austin, Texas 78712 (29)
Dieter Hartz ■ Department of Molecular, Cellular, and Developmental Biology, University of Colorado, Boulder, Colorado 80309 (20)
Tomomitsu Hatakeyama ■ Max-Planck-Institut für Molekulare Genetik, Ihnestrasse 73, D-1000 Berlin-Dahlem, Federal Republic of Germany (53)
Thomas-Peter Hausner ■ Department of Molecular Biophysics and Biochemistry, Yale University, 333 Cedar Street, New Haven, Connecticut 06510-8024 (25)
Kathryn A. Hijazi ■ Department of Molecular Genetics, University of Texas M. D. Anderson Cancer Center, Houston, Texas 77030 (33, 47)
Walter E. Hill ■ Division of Biological Sciences, University of Montana, Missoula, Montana 59812 (18)
Allan Jacobson ■ Department of Molecular Genetics and Microbiology, University of Massachusetts Medical School, Worcester, Massachusetts 01655 (23)
Claus Jeppesen ■ Division of Biology and Medicine, Brown University, Providence, Rhode Island 02912 (39)
Stewart P. Johnson ■ Department of Pathology, Duke University Medical Center, Durham, North Carolina 27710 (38)
F. Jörgensen ■ Institute of Molecular Biology, University of Uppsala Biomedical Center, Box 590, S-751 24 Uppsala, Sweden (44)
Qida Ju ■ Department of Cell Biology, Albert Einstein College of Medicine, 1300 Morris Park Avenue, Bronx, New York 10461 (38)
Patricia E. Jung ■ Animal Sciences Division, Monsanto Corporation, Mail Zone BB3M, 700

Chesterfield Village Parkway, Chesterfield, Missouri 63198 (**48**)
James F. Kane ■ Animal Sciences Division, Monsanto Corporation, Mail Zone BB3M, 700 Chesterfield Village Parkway, Chesterfield, Missouri 63198 (**48**)
Berthold Kastner ■ Department of Biochemistry, University of Otago, P.O. Box 56, Dunedin, New Zealand (**32**)
Arati Khanna-Gupta ■ Department of Biology and Center for Molecular Bioscience and Biotechnology, Lehigh University, Bethlehem, Pennsylvania 18015 (**40**)
Andreas K. E. Köpke ■ Max-Planck-Institut für Molekulare Genetik, Ihnestrasse 73, D-1000 Berlin-Dahlem, Federal Republic of Germany (**53**)
Wolfgang Krömer ■ Max-Planck-Institut für Molekulare Genetik, Ihnestrasse 73, D-1000 Berlin-Dahlem, Federal Republic of Germany (**53**)
Ernst Kuechler ■ Institut für Biochemie der Universität Wien, Währingerstrasse 17, A-1090 Vienna, Austria (**28**)
C. G. Kurland ■ Institute of Molecular Biology, University of Uppsala Biomedical Center, Box 590, S-751 24 Uppsala, Sweden (**30, 44**)
Anna La Teana ■ Max-Planck-Institut für Molekulare Genetik, Abteilung Wittmann, Berlin, Federal Republic of Germany, and Laboratory of Genetics, Department of Cell Biology, University of Camerino, Camerino, Italy (**21**)
James A. Lake ■ Molecular Biology Institute and Department of Biology, University of California at Los Angeles, Los Angeles, California 90024 (**12, 51**)
Niels Larsen ■ Biostructural Chemistry, Chemistry Institute, Aarhus University, DK-8000 Aarhus C, Denmark (**11**)
E. Lazaro ■ Centro de Biología Molecular, Canto Blanco, Madrid 28049, Spain (**43**)
R. Leberman ■ European Molecular Biology Laboratory, Outstation, Grenoble, France (**13**)
Peter Leeds ■ Laboratory of Molecular Biology, University of Wisconsin, 1525 Linden Drive, Madison, Wisconsin 53706 (**49**)
Wen-Hsiung Li ■ Center for Demographic and Population Genetics, University of Texas, P.O. Box 20334, Houston, Texas 77225 (**50**)
Anders Liljas ■ Molecular Biophysics, Chemical Center, Lund University, Box 124, S-221 00 Lund, Sweden (**24**)
Roland Lill ■ Institut für Physiologische Chemie, Universität München, D-8000 Munich 2, Federal Republic of Germany (**27**)
V. Mandiyan ■ Roche Institute of Molecular Biology, Roche Research Center, Nutley, New Jersey 07110 (**6**)
Richard T. Marconi ■ Division of Biological Sciences, University of Montana, Missoula, Montana 59812 (**18**)
I. Marin ■ Centro de Biología Molecular, U.A.M.-C.S.I.C., Universidad Autonoma de Madrid, Canto Blanco, Madrid 28049, Spain (**56**)
A. T. Matheson ■ Department of Biochemistry and Microbiology, University of Victoria, Victoria, British Columbia V8W 2Y2, Canada (**54**)
David S. McPheeters ■ Division of Biology, 147-75 California Institute of Technology, Pasadena, California 91125 (**20**)
William C. Merrick ■ Department of Biochemistry, School of Medicine, Case Western Reserve University, Cleveland, Ohio 44106 (**22**)
Chuck Merryman ■ Division of Biological Sciences, University of Montana, Missoula, Montana 59812 (**18**)
B. H. Mims ■ Department of Molecular Genetics, University of Texas M. D. Anderson Cancer Center, Houston, Texas 77030 (**47**)
Philip Mitchell ■ Max-Planck-Institut für Molekulare Genetik, Abteilung Wittmann, Ihnestrasse 73, D-1000 Berlin-Dahlem, Federal Republic of Germany (**4**)
Danesh Moazed ■ Thimann Laboratories, University of California, Santa Cruz, Santa Cruz, California 95064 (**3**)
Wim Möller ■ Department of Medical Biochemistry, Sylvius Laboratory, University of Leiden, P.O. Box 9503, 2300 RA Leiden, The Netherlands (**31**)
Peter B. Moore ■ Department of Chemistry, Yale University, New Haven, Connecticut 06511

Bernice E. Morrow ■ Department of Cell Biology, Albert Einstein College of Medicine, 1300 Morris Park Avenue, Bronx, New York 10461 (38)
Marylène Mougel ■ Laboratoire de Biochimie, Institut de Biologie Moléculaire et Cellulaire du Centre National de la Recherche Scientifique, 15 rue René Descartes, 67084 Strasbourg Cedex, France (9)
David Munroe ■ Center for Cancer Research, Massachusetts Institute of Technology, 77 Massachusetts Avenue, Cambridge, Massachusetts 02139 (23)
Emanuel J. Murgola ■ Department of Molecular Genetics, University of Texas M. D. Anderson Cancer Center, Houston, Texas 77030 (33, 47)
Wouter Musters ■ Biochemisch Laboratorium, Vrije Universiteit, de Boelelaan 1083, 1081 HV Amsterdam, The Netherlands (15, 37)
Didier Nègre ■ Roche Institute of Molecular Biology, Roche Research Center, Nutley, New Jersey 07110 (17)
Knud H. Nierhaus ■ Max-Planck-Institut für Molekulare Genetik, Abteilung Wittmann, Ihnestrasse 73, D-1000 Berlin-Dahlem, Federal Republic of Germany (25)
L. Nilsson ■ Department of Biochemistry, Leiden University, Gorlaeus Laboratories, 2333 AL Leiden, The Netherlands (34)
Harry F. Noller ■ Thimann Laboratories, University of California, Santa Cruz, Santa Cruz, California 95064 (3)
Masayasu Nomura ■ Department of Biological Chemistry, University of California, Irvine, Irvine, California 92717 (1)
Kelvin Nurse ■ Roche Institute of Molecular Biology, Roche Research Center, Nutley, New Jersey 07110 (17)
Melanie I. Oakes ■ Molecular Biology Institute and Department of Biology, University of California at Los Angeles, Los Angeles, California 90024 (12)
O. W. Odom ■ Clayton Foundation Biochemical Institute, Department of Chemistry and Biochemistry, University of Texas at Austin, Austin, Texas 78712 (29)
James Ofengand ■ Roche Institute of Molecular Biology, Roche Research Center, Nutley, New Jersey 07110 (17)
Monika Osswald ■ Max-Planck-Institut für Molekulare Genetik, Abteilung Wittmann, Ihnestrasse 73, D-1000 Berlin-Dahlem, Federal Republic of Germany (4)
Y. M. Ostanevich ■ Joint Institute of Nuclear Physics, Dubna, USSR (13)
Norman R. Pace ■ Department of Biology and Institute for Molecular and Cellular Biology, Indiana University, Bloomington, Indiana 47405 (35)
F. T. Pagel ■ Department of Molecular Genetics, University of Texas M. D. Anderson Cancer Center, Houston, Texas 77030 (47)
M. Y. Pavlov ■ Institute of Protein Research, Academy of Sciences of the USSR, Pushchino, Moscow Region, USSR (13)
A. G. Pisabarro ■ Centro de Biología Molecular, U.A.M.-C.S.I.C., Universidad Autonoma de Madrid, Canto Blanco, Madrid 28049, Spain (56)
Rudi J. Planta ■ Biochemisch Laboratorium, Vrije Universiteit, de Boelelaan 1083, 1081 HV Amsterdam, The Netherlands (15, 37)
Cynthia L. Pon ■ Max-Planck-Institut für Molekulare Genetik, Abteilung Wittmann, Berlin, Federal Republic of Germany (21)
Ted Powers ■ Thimann Laboratories, University of California, Santa Cruz, Santa Cruz, California 95064 (3)
Catherine Prescott ■ Brown University, Providence, Rhode Island 12912 (16)
M. Radermacher ■ Wadsworth Center for Laboratories and Research, New York State Department of Health, Albany, New York 12201-0509 (5)
C. Ramírez ■ Department of Biochemistry and Microbiology, University of Victoria, Victoria, British Columbia V8W 2Y2, Canada (54)
L. Ramirez ■ Centro de Biología Molecular, U.A.M.-C.S.I.C., Universidad Autonoma de Madrid, Canto Blanco, Madrid 28049, Spain (56)
Hendrik A. Raué ■ Biochemisch Laboratorium, Vrije Universiteit, de Boelelaan 1083, 1081 HV Amsterdam, The Netherlands (15, 37)

Jaanus Remme ■ Institute of Chemical Physics and Biophysics of the Academy of Sciences of Estonian SSR, Jakobistreet 2, Tartu, USSR (25)
Hans-Jörg Rheinberger ■ Max-Planck-Institut für Molekulare Genetik, Abteilung Wittmann, Ihnestrasse 73, D-1000 Berlin-Dahlem, Federal Republic of Germany (25)
A. Richter ■ Institute of Molecular Biology, University of Uppsala Biomedical Center, Box 590, S-751 24 Uppsala, Sweden (44)
Jutta Rinke-Appel ■ Max-Planck-Institut für Molekulare Genetik, Abteilung Wittmann, Ihnestrasse 73, D-1000 Berlin-Dahlem, Federal Republic of Germany (4)
Rafael Rivera-León ■ Division of Biology and Medicine, Brown University, Providence, Rhode Island 02912 (39)
James M. Robertson ■ Applied Biosystems, Inc., Foster City, California 94403 (3, 27)
A.-M. Rojas ■ Institute of Molecular Biology, University of Uppsala Biomedical Center, Box 590, S-751 24 Uppsala, Sweden (30, 44)
Pascale Romby ■ Laboratoire de Biochimie, Institut de Biologie Moléculaire et Cellulaire du Centre National de la Recherche Scientifique, 15 rue René Descartes, 67084 Strasbourg Cedex, France
I. N. Rublevskaya ■ Institute of Protein Research, Academy of Sciences of the USSR, Pushchino, Moscow Region, USSR (13)
Carla A. Rutgers ■ Biochemisch Laboratorium, Vrije Universiteit, de Boelelaan 1083, 1081 HV Amsterdam, The Netherlands (15)
E. Sanchez ■ Centro de Biología Molecular, U.A.M.-C.S.I.C., Universidad Autonoma de Madrid, Canto Blanco, Madrid 28049, Spain (56)
Mark G. Sandbaken ■ Laboratory of Molecular Biology, University of Wisconsin, 1525 Linden Drive, Madison, Wisconsin 53706 (49)
J. L. Sanz ■ Centro de Biología Molecular, U.A.M.-C.S.I.C., Universidad Autonoma de Madrid, Canto Blanco, Madrid 28049, Spain (56)
Haruo Saruyama ■ Hokkaido Green-Bio Institute, Naganuma-cho, Higashi 5 Kita 15, Hokkaido 069-13, Japan (25)
Rocco Savino ■ Division of Biology and Medicine, Brown University, Providence, Rhode Island 02912 (39)
Andrew Scheinman ■ Molecular Biology Institute and Department of Biology, University of California at Los Angeles, Los Angeles, California 90024 (12)
David Schlessinger ■ Department of Molecular Microbiology, Washington University School of Medicine, St. Louis, Missouri 63110 (36)
Murray N. Schnare ■ Department of Biochemistry, Dalhousie University, Halifax, Nova Scotia B3H 4H7, Canada (52)
Dierk Schüler ■ Max-Planck-Institut für Molekulare Genetik, Abteilung Wittmann, Ihnestrasse 73, D-1000 Berlin-Dahlem, Federal Republic of Germany (4)
I. N. Serdyuk ■ Institute of Protein Research, Academy of Sciences of the USSR, Pushchino, Moscow Region, USSR (13)
Manuela Severini ■ Max-Planck-Institut für Molekulare Genetik, Abteilung Wittmann, Berlin, Federal Republic of Germany, and Laboratory of Genetics, Department of Cell Biology, University of Camerino, Camerino, Italy (21)
Gary Shankweiler ■ Molecular Biology Institute and Department of Biology, University of California at Los Angeles, Los Angeles, California 90024 (12)
Joseph G. Sinning ■ Roche Institute of Molecular Biology, Roche Research Center, Nutley, New Jersey 07110 (17)
Peter M. Smooker ■ Max-Planck-Institut für Molekulare Genetik, Abteilung Wittmann, Ihnestrasse 73, D-1000 Berlin-Dahlem, Federal Republic of Germany (57)
P. Sorensen ■ Institut de Biologie Physico-Chimique, Paris, France (47)
Alexander S. Spirin ■ Institute of Protein Research, Academy of Sciences of the USSR, Pushchino, Moscow Region, USSR (2)
Roberto Spurio ■ Laboratory of Genetics, Department of Cell Biology, University of Camerino, Camerino, Italy (21)
Anand K. Srivastava ■ Department of Molecular Microbiology, Washington University School

of Medicine, St. Louis, Missouri 63110 (36)

Katrin Stade ■ Max-Planck-Institut für Molekulare Genetik, Abteilung Wittmann, Ihnestrasse 73, D-1000 Berlin-Dahlem, Federal Republic of Germany (4)

Barbara Stebbins-Boaz ■ Division of Biology and Medicine, Brown University, Providence, Rhode Island 02912 (39)

Günter Steiner ■ Institut für Biochemie der Universität Wien, Währingerstrasse 17, A-1090 Vienna, Austria (28)

Seth Stern ■ Thimann Laboratories, University of California, Santa Cruz, Santa Cruz, California 95064 (3)

Georg Stöffler ■ Institut für Mikrobiologie, Medizinische Fakultät, Universität Innsbruck, Fritz-Pregl-Strasse 3, A-6020 Innsbruck, Austria (7)

Marina Stöffler-Meilicke ■ Institut für Klinische und Experimentelle Virologie, Freie Universität Berlin, Hindenburgdamm 27, D-1000 Berlin 45, Federal Republic of Germany (7)

Alap R. Subramanian ■ Max-Planck-Institut für Molekulare Genetik, Abteilung Wittmann, Ihnestrasse 73, D-1000 Berlin-Dahlem, Federal Republic of Germany (57)

William E. Tapprich ■ Brown University, Providence, Rhode Island 02912 (16, 18)

Anchalee Tassanakajohn ■ Division of Biological Sciences, University of Montana, Missoula, Montana 59812 (18)

Warren P. Tate ■ Department of Biochemistry, University of Otago, P.O. Box 56, Dunedin, New Zealand (32)

Cheryl L. Thomas ■ Department of Biochemistry and Program in Molecular and Cellular Biology, University of Massachusetts, Amherst, Massachusetts 01003 (26)

Linda K. Thomas ■ Molecular, Cellular, and Developmental Biology, University of Colorado, Boulder, Colorado 80309 (45)

Robert C. Thompson ■ Molecular, Cellular, and Developmental Biology, University of Colorado, Boulder, Colorado 80309 (45)

Amelia A. Tiedeman ■ Department of Molecular Genetics, University of Texas M. D. Anderson Cancer Center, Houston, Texas 77030 (33)

Kathleen Triman ■ Thimann Laboratories, University of California, Santa Cruz, Santa Cruz, California 95064 (3)

S. Tumminia ■ Roche Institute of Molecular Biology, Roche Research Center, Nutley, New Jersey 07110 (6)

D. Ureña ■ Centro de Biología Molecular, U.A.M.-C.S.I.C., Universidad Autonoma de Madrid, Canto Blanco, Madrid 28049, Spain (56)

A. Vanet ■ Department of Biochemistry, Leiden University, Gorlaeus Laboratories, 2333 AL Leiden, The Netherlands (34)

H. van Heerikhuizen ■ Biochemisch Laboratorium, Vrije Universiteit, de Boelelaan 1083, 1081 HV Amsterdam, The Netherlands (37)

P. H. Van Knippenberg ■ Department of Biochemistry, Gorlaeus Laboratories, Leiden University, Einsteinweg 5, 2333 CA Leiden, The Netherlands (19)

Jan Van 't Riet ■ Biochemisch Laboratorium, Vrije Universiteit, de Boelelaan 1083, 1081 HV Amsterdam, The Netherlands (15)

Alexander Varshavsky ■ Department of Biology, Massachusetts Institute of Technology, Cambridge, Massachusetts 02139 (55)

H. Verbeek ■ Department of Biochemistry, Leiden University, Gorlaeus Laboratories, 2333 AL Leiden, The Netherlands (34)

A. Verschoor ■ Wadsworth Center for Laboratories and Research, New York State Department of Health, Albany, New York 12201-0509 (5)

E. Vijgenboom ■ Department of Biochemistry, Leiden University, Gorlaeus Laboratories, 2333 AL Leiden, The Netherlands (34)

Bernard N. Violand ■ Animal Sciences Division, Monsanto Corporation, Mail Zone BB3M, 700 Chesterfield Village Parkway, Chesterfield, Missouri 63198 (48)

T. Wagenknecht ■ Wadsworth Center for Laboratories and Research, New York State Department of Health, Albany, New York 12201-0509, and School of Public Health, State University of New York at Albany, Albany, New York 12222 (5)

Vassie C. Ware ■ Department of Biology and Center for Molecular Bioscience and Biotechnology, Lehigh University, Bethlehem, Pennsylvania 18015 (**40**)

Jonathan R. Warner ■ Department of Cell Biology, Albert Einstein College of Medicine, 1300 Morris Park Avenue, Bronx, New York 10461 (**38**)

S. Weinstein ■ Department of Structural Chemistry, Weizmann Institute, Rehovot, Israel, and Max Planck Institute for Molecular Genetics, D-1000 Berlin 33, Federal Republic of Germany (**8**)

Bryn Weiser ■ Thimann Laboratories, University of California, Santa Cruz, Santa Cruz, California 95064 (**3**)

J. Weiser ■ Institute of Microbiology, Czechoslovak Academy of Sciences, Videnska 1083, 14220 Prague 4, Czechoslovakia (**30**)

Robert Weiss ■ Howard Hughes Medical Institute and Department of Human Genetics, University of Utah Medical Center, Salt Lake City, Utah 84132 (**46**)

Carl J. Weitzmann ■ Roche Institute of Molecular Biology, Roche Research Center, Nutley, New Jersey 07110 (**17, 42**)

Jennifer Weller ■ Division of Biological Sciences, University of Montana, Missoula, Montana 59812 (**18**)

Eric Westhof ■ Laboratoire de Biochimie, Institut de Biologie Moléculaire et Cellulaire du Centre National de la Recherche Scientifique, 15 rue René Descartes, 67084 Strasbourg Cedex, France (**9**)

Patricia G. Wilson ■ Laboratory of Molecular Biology, University of Wisconsin, 1525 Linden Drive, Madison, Wisconsin 53706 (**49**)

Wolfgang Wintermeyer ■ Institut für Molekularbiologie, Universität Witten/Herdecke, D-5810 Witten, Federal Republic of Germany (**27**)

Heinz-Günter Wittmann ■ Max-Planck-Institut für Molekulare Genetik, Abteilung Wittmann, Ihnestrasse 73, D-1000 Berlin-Dahlem, Federal Republic of Germany (**8, 53**)

Brigitte Wittmann-Liebold ■ Max-Planck-Institut für Molekulare Genetik, Ihnestrasse 73, D-1000 Berlin-Dahlem, Federal Republic of Germany (**53**)

Ira G. Wool ■ Department of Biochemistry and Molecular Biology, The University of Chicago, Chicago, Illinois 60637 (**14**)

Jacek Wower ■ Department of Biochemistry and Program in Molecular and Cellular Biology, University of Massachusetts, Amherst, Massachusetts 01003 (**26**)

A. Yonath ■ Department of Structural Chemistry, Weizmann Institute, Rehovot, Israel, and Max Planck Research Unit for Structural Molecular Biology, D-2000 Hamburg 52, Federal Republic of Germany (**8**)

G. Zaccai ■ Institut Laue-Langevin, Grenoble, France (**13**)

Robert A. Zimmermann ■ Department of Biochemistry and Program in Molecular and Cellular Biology, University of Massachusetts, Amherst, Massachusetts 01003 (**26**)

PREFACE

This volume originates primarily from presentations made in a conference held 6–11 August 1989 in the pristine air of scenic East Glacier Park, Montana. The conference dealt with the structure, function, and evolution of ribosomes and attracted over 250 investigators in the field. It was clear from the animated discussions that there has been a resurgence of interest in the ribosome, as new and refined techniques have been used to explore the ribosome structure and function.

A key to the conference was its sense of history. Ribosomes were first isolated and analyzed just over 30 years ago. Since that time many workshops and meetings have been held to stimulate and coordinate the activity in the field. Several workshops sponsored by the European Molecular Biology Organization (EMBO) are notable in this regard. In addition, in 1972, a Cold Spring Harbor Symposium on ribosomes was held, which resulted in a symposium volume. In 1979, a Steenbock Symposium on ribosomes was held at Madison, Wisconsin. The proceedings of this conference appeared in a "Red Book" which has been an important reference on ribosomes ever since. In 1985, a conference was held at Port Aransas, Texas, which led to the publication of a "Green Book" which has likewise become a valuable reference.

It is hoped that this "Blue Book" will serve the same purpose. This book is not a laboratory manual, nor does it include details of experimental data or techniques. An excellent volume of *Methods in Enzymology* (vol. 164) has provided much of that information recently. This book is designed instead to give an overview of the state of knowledge about the structure, function, and evolution of the ribosome as of the summer of 1989.

The historical aspects of the field were covered ably by Drs. Nomura and Spirin, who were asked to present historical overviews from their perspective. Their contributions constitute the first section of the book. Drs. Elson and Wittmann had also been asked to contribute in this portion of the conference, but the untimely death of Dr. Elson and the illness of Dr. Wittmann, which ultimately caused his death, prevented them from so doing. They will be greatly missed.

The remainder of the book is organized into sections that deal with the aspects of ribosome structure and function that are under active investigation today. Although the majority of the work described has been performed on the procaryotic systems, notably *Escherichia coli*, we have chosen to interdigitate procaryotic and eucaryotic studies rather than divide them according to their nuclear membrane.

Each author was asked to provide a short review not only of his or her own work, but of the entire area in which they are active. It is hoped that this will make the book more understandable to those outside the field. Although the chapters have some overlap, the overlap is useful to the extent that it will enable the reader to understand the topic from different perspectives. We recognize that there are many who have made important contributions to this field whom we were unable to ask to provide chapters. We regret that there was not sufficient space to do so.

I am very grateful to the many who made the conference and this volume possible. The editors have performed superlatively and ASM has helped enormously. Special thanks are due to Pnina Spitnik-Elson and the late David Elson, who encouraged me to undertake this assignment during the last conference in Texas. We also acknowledge the financial support of the National Science Foundation, the National Institutes of Health, the University of Montana, Gene-Trak, Monsanto, Hoffmann-LaRoche, Life Technologies, and Pharmacia/LKB. Finally, a note of appreciation is given to all those in the ribosome group at the University of Montana who contributed unselfishly of their time and energy to make the conference possible.

Walter E. Hill

Introduction

Comments on the 1989 International Conference on Ribosomes

PETER B. MOORE

The organizers of scientific conferences often ask one of their colleagues to close the meeting by making some comments that capture its spirit and summarize the state of the field. Entrusted with this task, I am placed in a position that resembles that of the corpse at an Irish wake. Its participation is vital, but the congregation will get upset if it says too much. In this case, it would be inappropriate to summarize the talks given. The papers in this volume render that function redundant. Instead, I prefer to make a few specific observations.

ON THE CIVILITY OF MEETING

A prominent characteristic of this meeting was the unemotional tone of most of the presentations and discussions. Veterans of earlier conventions of the Amalgamated Ribosome Workers know that it has not always been thus. Ten or fifteen years ago, there were lots of violent disagreements about facts, techniques, personal ancestry, and much else of substance.

The fact that we no longer experience such conflicts is evidence of the maturity of the ribosome field. We agree about most of the basic facts. Furthermore, tensions have been alleviated by the tendency of mature fields to spread out and to diversify. Today there are lots of different questions one can address in the ribosome system, all of importance. In this environment, each participant can find a niche where he or she can pursue science relatively unperturbed by competitive pressures. While this development has increased our individual psychological comfort levels, it is hazardous. The danger is that the field may become so diffuse that no one cares intensely any more about anything. That would be tragic.

It is interesting that the one issue that produced heat was evolution. As everyone knows, rRNA sequences are a rich data base from which to extract inferences about the evolutionary relationships of the species. Furthermore, evolution is so central to biology that no biologist can ignore it. The deep phylogenetic branchings discussed at this meeting are particularly fascinating since they speak to the origins of life on this planet.

Evolutionary questions, especially ones that deal with deep branchings, may rouse passions forever because they are likely to be unresolvable. All methods that I am aware of for deriving taxonomic relationships from the sequences of functionally homologous molecules assume that a monotonic relationship exists between the accumulation of mutations and the passage of time. The bigger the difference in sequence, however that difference is defined, the longer the period of time since the species in question diverged. Although this assumption is highly plausible and is supported by comparisons of sequences of RNAs obtained from organisms whose taxonomic relationships are known on nonmolecular grounds, it is not directly testable as such. No one was there to witness what happened.

Implicit in the assumption that sequence differences build up progressively with time is the larger assumption that evolution at the molecular level is rational. Evolution is history, however, not physics or chemistry, and history is not necessarily rational, as human experience in the 20th century eloquently attests. Accidents happen. Might it be possible that a set of deep evolutionary branchings could be proposed on the basis of sequence analysis upon which all agreed (which is not the case today) but which was nevertheless wrong?

ON THE DOMINANCE OF THE RNA BIOCHEMISTS

Anyone perusing the table of contents of this volume will be impressed by the degree to which

Peter B. Moore ■ Department of Chemistry, Yale University, New Haven, Connecticut 06511.

RNA dominates the discourse in the ribosome field in 1989. Less than 10% of the talks given at this meeting dealt directly with ribosomal proteins. This is another huge change from the past. Twenty years ago, the portion of the ribosome on which everyone focused was its protein. Before I go further, let me declare that I too believe in RNA; RNA is the focus of much of what is going on in my own laboratory. I nevertheless feel the pendulum has swung too far.

Many have portrayed this shift in emphasis as having been driven by scientific considerations only. I submit that technological considerations played at least as large a role. In the late 1960s we could work with ribosomal proteins but were unable to study the large rRNAs effectively. So we concentrated on ribosomal proteins. Starting in the late 1970s, molecular biologists devised techniques that revolutionized our ability to work with nucleic acids. We can do experiments on rRNAs today that are far more powerful than anything we ever attempted with ribosomal proteins in the past. So today we work on rRNA; we would be crazy not to.

Inspiring progress is being made today in investigations of the structure and function of the large rRNAs. There is every reason to believe that much of value will emerge from RNA studies in the next several years. But we should bear in mind that we have not learned all there is to know about the ribosomal proteins, and we should all hope that technical developments will revive protein studies soon.

ON THE IMPACT OF MOLECULAR GENETICS

Anyone reading the conference volumes on ribosomes that precede this one would have (correctly) concluded that ribosome biogenesis and protein initiation, especially in eucaryotes, were poorly understood and that both areas were all but dead for lack of means for pursuing them effectively. Both have revived in the last few years as a result of the imaginative application of genetics and genetic engineering. We look forward to the insight such investigations should yield. There is reason to expect that other areas we have ignored for technical reasons will benefit from similar approaches.

ON THE IMPORTANCE OF RIBOSOME ENZYMOLOGY

The study of the enzymology of ribosomes has always been difficult. On the one hand, the events that occur on the ribosome during protein synthesis are very complex; on the other hand, they are dependent to varying degrees on a whole host of extraribosomal factors and enzymes as well as on ionic conditions, which makes the task of setting up a realistic, well-controlled, protein-synthesizing system in vitro very difficult. Studies of this kind have also been plagued by the difficulty of making ribosomal preparations that are simultaneously free of contaminating activities and highly active.

It is gratifying that several groups are continuing to work in this area despite the difficulties. The information that they are harvesting is more reliable and more penetrating than any we have had in the past. Workers in the more glamorous area of ribosome structure should support these efforts enthusiastically. They are rapidly defining the chemical properties of the ribosome which structural investigations must ultimately rationalize. To put it another way, knowledge of the structure of the ribosome will be of little value if it is not accompanied by a sophisticated understanding of its functional properties.

As the proceedings of this conference make clear, the two-site model for the ribosome, which the world has accepted for a generation, is dead. The existence of a third site for tRNA binding, the exit site, is now established beyond reasonable doubt. This is unquestionably the most significant advance in our understanding of the ribosomal events of protein synthesis in many years. Biochemistry textbooks must now be rewritten.

ON THE STATE OF STRUCTURAL STUDIES OF THE RIBOSOME

Our understanding of ribosome structure at low resolution has improved considerably recently as a result of the merger of data on rRNA structure, protein-binding sites, and protein locations. Several molecular models of the small ribosomal subunit from *Escherichia coli* are included in this volume. However approximate these models ultimately turn out to be, there is no doubt that they correlate structure with function in a useful way and suggest experiments that should extend our understanding of the ribosome.

Some have complained about the limitations of protein distribution models that depict individual proteins as spheres. Current protein models are indeed limited by our inability to represent individual proteins more accurately, and there is similarly no doubt that the rRNA tertiary structure models based on current protein maps are less accurate than they might be for the same reason. Limited maps of this kind are (unhappily) the best we can do for the moment as far as ribosomal proteins are concerned.

I used to hope that low-resolution structural maps would suffice to elucidate many of the most important aspects of the relationship between structure and function in the ribosome. I am less confident today, and the fact that some find the primitive protein maps now available unsatisfactory reenforces my pessimism. We are not going to be satisfied with anything less than an atomic resolution structure for the ribosome. For that reason, those pursuing ribosome crystallography are urged to proceed, and new groups should be encouraged to get involved.

The technique that has contributed the most to recent progress in ribosome research has been chemical probing of rRNA. There seems no end to the valuable information that can be extracted from such experiments. There is an aspect of these studies that is bothersome, however. It has to do with the presentation of such data in terms of explicit, three-dimensional molecular models.

Detailed, atomic resolution pictures of biological macromolecules were introduced into biochemistry by the X-ray crystallographers several decades ago as a means for reporting the results of their investigations. Even though their data do not always extend to atomic resolution, and even though they do not always interprete their data correctly, these pictures were (are) appropriate because most crystallographic data can be interpreted only in terms of molecular models at atomic resolution. As a result of the efforts of crystallographers, a whole host of computer programs have been developed enabling one to manipulate molecular images, compute molecular energies, and make pictures of the result.

Unfortunately, these crystallographic programs are now widely used by scientists who do not have atomic resolution data but publish molecular pictures to represent their results anyway. Their pictures are often largely the products of theoretical computations supported by an amount of data that is small in comparison with what is needed to support a high-resolution structure. The problem with these pictures is that they look just like the crystallographic ones, even though their status with respect to the truth is very different. At the very least, it creates a pedogogical problem when it becomes difficult to distinguish an experimentally determined structure from something that is intended only as a heuristic model.

It should be added that the computer codes available today for calculating molecular conformations are only approximately correct. Even the relatively simple task of using experimental data on base pair interactions to calculate free energies for RNA structures yields results correct only to first order. We have all learned not to place too much confidence in RNA secondary structures proposed solely on the basis of such computations. The more general codes that attempt atom-by-atom energy computations (e.g., AMBER and CHARM) are more derivative than that. They use pairwise potentials of mean force to compute the interactions between different atoms within the system of interest. These potentials cannot be perfectly accurate because they are linear approximations to a physical reality which is fundamentally nonlinear. Moreover, it has been known for years that three-body and higher-order interactions contribute measurably to molecular interactions (Maitland et al., 1987); they are not taken into account in an explicit way. Furthermore, all such codes have problems dealing with the effects of solvation and are weak in their treatment of coulombic interactions. Thus, the properties of these programs make them less than ideal for computing the structures of polyelectrolytes like RNA or ribonucleoproteins in water solution. *Caveat emptor*!

The people who compute molecular structures from nuclear magnetic resonance data have a problem not unlike the one under discussion here: how to depict uncertainties in models for macromolecular structures. They have found a solution that might be useful. Each structure is presented as a superposition of a whole set of structures, each of which meets the restraints of the experimental data. Examples of such pictures are now commonplace in the literature (e.g., Wüthrich, 1986). In pictures such as these, it is easy to distinguish features of a molecular model that are well established from those that are not.

ORGANIZATION OF THE 1989 MEETING

Scientific meetings such as the one that led to this volume come to pass only when a single individual takes it upon himself to do what needs to be done. The ribosome field owes Professor Walter Hill of the University of Montana an immense debt of gratitude for organizing the 1989 meeting. Had he not conceived of running a meeting in East Glacier, Montana, and then invested a huge amount of time and energy in the enterprise, it would never have come to pass. It is not often that segments of the international scientific community gather in the State of Montana. Those who got to enjoy the magnificent scenery of Glacier National Park and the hospitality of the locals during this conference learned how much we have been missing.

REFERENCES

Maitland, C. G., M. Rigby, E. B. Smith, and W. A. Wakeham. 1987. *Intermolecular Forces*. Oxford University Press, Oxford.

Wüthrich, K. 1986. *NMR of Proteins and Nucleic Acids*. John Wiley & Sons, Inc., New York.

I. HISTORICAL

I. HISTORICAL

Systematic study of the structure and function of ribosomes started over 30 years ago when it was discovered that these complex macromolecules were the sites where proteins were synthesized. As techniques were developed to isolate and characterize ribosomes, the number of workers in the field increased dramatically. However, the problem was difficult, and the excitement that was present in the 1960s, when the protein synthesis pathway was being worked out and the genetic code was established, waned in the '70s. The field has been revived in recent years by the invention of a host of new techniques which enable researchers to extract more and more detailed information about the structure and function of this extremely complex macromolecule.

At the conference in East Glacier, we chose to honor those who had worked in the field over many years. We asked two of the most distinguished members, Masayasu Nomura and Alexander Spirin, to address us specifically concerning the history of the study of ribosomes as they viewed it. The following two chapters reflect this history as seen by these two outstanding scientists.

Chapter 1

History of Ribosome Research: a Personal Account

MASAYASU NOMURA

EARLY DAYS OF RIBOSOME RESEARCH

It was the spring of 1958 when I met Jim Watson for the first time. I had come from Japan to work as a postdoctoral fellow in Sol Spiegelman's laboratory at the University of Illinois. Jim came to the campus to give a series of lectures sponsored by Salvador Luria, Jim's former mentor and then a professor at Illinois working in the same department as Sol Spiegelman. In addition to giving formal lectures to students and faculty members on the campus, Jim spent time in Spiegelman's laboratory discussing with Sol the nature of the ribosomes seen in bacterial extracts by analytical ultracentrifugation. The specific question was whether ribosomes consist of a single species (70S) or of two species (30S and 50S subunits). People who were interested in ribosomes (then variously called by such terms as "microsomal particles" or "ribonucleoprotein particles of the microsome fraction") had sometimes observed a single peak and sometimes two (or more) peaks in chlieren patterns obtained with the analytical ultracentrifuge. Jim and Sol agreed that these differences in the patterns were due to the solution conditions, specifically Mg^{2+} concentrations, used to prepare bacterial extracts and isolate the particles, as suspected from the work of Chao on yeast ribosomes which had just been published (Chao, 1957). The exciting question was the significance of this unequal subunit structure. Of course, the "subunit structure" of DNA, the double-helical structure, was the key to revealing the secret of self-replication of DNA. Ribosomal particles present in various living cells apparently all consisted of two nonidentical subunits, small and large. What does this mean? Sol asked me this question one evening, and this question remained with me for a long time.

The story of ribosomes is, of course, intimately connected to the story of protein synthesis. Thus, the history of ribosome research is also intimately connected to the history of research on protein synthesis. Before coming to the United States as a postdoctoral fellow, I was interested in protein synthesis, randomly reading research papers on protein synthesis published in several journals which, though only in limited kinds, were accessible to scientists in Japan at that time. From these readings, I started to learn about ribonucleoprotein particles in the microsome fraction (which contained endoplasmic reticulum membrane fragments) in mammalian cells as the possible site of protein synthesis. For example, Zamecnik and his collaborators published several papers on studies using intact animals, tissues, and tissue homogenates which demonstrated that radioactive amino acids administered to animals or added to in vitro systems appeared first in protein attached to ribonucleoprotein particles in the microsome fractions, followed by gradual incorporation into protein in the soluble protein fraction (Littlefield et al., 1955). However, when I came to the United States in late 1957, protein synthesis in cell extracts from bacterial cells had not been clearly demonstrated, even though Zamecnik and Keller (1954) had already developed a cell-free system from rat liver and people such as Sol Spiegelman were struggling to develop a similar system from bacteria. Nevertheless, we believed in the unity of fundamental principles in biology and thus in the importance of ribosomal particles seen in bacterial extracts in connection with protein synthesis.

During Jim Watson's visit to Urbana, I arranged to spend 3 months in his laboratory at Harvard the following summer. Just a half year before, I had come to the United States and was struggling to develop an in vitro system in which information transfer from DNA to protein could be studied. However, as a postdoctoral fellow I was eager to learn as many

Masayasu Nomura ■ Department of Biological Chemistry, University of California, Irvine, Irvine, California 92717.

different approaches as possible, and working in Jim Watson's laboratory was certainly appealing.

At Harvard, Jim Watson, together with Alfred Tissières, had started to isolate and characterize ribosomes from *Escherichia coli*. Other people in the Watson laboratory included David Schlessinger and Charles G. Kurland. They were busy estimating the molecular weights of ribosomes and rRNAs and studying other physicochemical and chemical properties of ribosomes, and Jim and Alfred were just preparing their first paper on the characterization of *E. coli* ribosomes (Tissières and Watson, 1958). With respect to my research, Jim suggested that I study the effects of chloramphenicol on ribosomes. This was my formal introduction to ribosome research. I discovered that extracts prepared from *E. coli* treated with chloramphenicol showed patterns very different from those of normal extracts in the analytical ultracentrifuge, indicating formation of abnormal particles. By repeated differential centrifugation, I isolated the particles, named chloromycetin particles (later called chloramphenicol particles), and showed that they were deficient in proteins (Nomura and Watson, 1959; Kurland et al., 1962). As I will mention later, although chloramphenicol particles were once suggested as artifacts (Yoshida and Osawa, 1968), results of later studies did not agree with this suggestion (e.g., Lefkovits and Di Girolamo, 1969; Sykes et al., 1977), and I now think that the formation of these protein-deficient particles in *E. coli* under these conditions is real and reflects certain features of the regulation and assembly of ribosomes.

Although my own chloramphenicol particle project was concerned with the subject of ribosome biosynthesis, the mission of the ribosome research of the Watson laboratory and for several other laboratories was the quest for the Holy Grail of biology at that time, namely, the mechanism of information transfer from gene to protein. Francis Crick had just formulated the central dogma and specifically proposed that RNAs contained in the ribosomes are the information carriers between DNA and protein. Although Crick's article (Crick, 1958) had not been published yet, the concept of the central dogma was already known, and there was a sense of excitement and urgency. In fact, only a few days after my arrival at Harvard (in June 1958), the first Nucleic Acid Gordon Conference was held at New Hampton, N.H., with Crick from England and other major players on this subject participating. Attending the conference, I had the chance to meet and talk with these people, gradually learning ideas and facts in the mainstream of the emerging new biology, molecular biology as we now call it, and the significance of ribosome research in this context.

In addition to the proposed role of rRNAs as the information carrier from gene to protein, Crick also speculated on the structure of ribosomes in his 1958 article on the central dogma. The favored working model was a structure similar to that of spherical viruses with many identical protein subunits (Crick, 1958). Of course, these early speculations were soon destined to be challenged by the discovery of mRNA and by the work of Jean-Pierre Waller on the heterogeneity of ribosomal proteins (r-proteins), as will be described below.

Although many people working on protein synthesis were involved in the studies of ribosomes, there were only a handful of laboratories working on ribosomes per se with a full commitment. Among them, one of the most active was the group at the Carnegie Institution of Washington. Dick Roberts, Roy Britten, Ellis Bolton, and Dean Cowie (and others who joined the group later, such as Brian McCarthy) were intensively studying the kinetics of flow of radioactive amino acids and bases into protein and RNA, respectively, in *E. coli*, devising various techniques and strategies for this purpose. Thus, they devised and started to extensively use sucrose density gradient centrifugation to separate ribosomal particles (Britten and Roberts, 1960), and this quickly became an essential technique for research on ribosomes. Dick Roberts also organized a session on ribosomes in the Biophysical Society meeting at the Massachusetts Institute of Technology, Cambridge, in 1958 and published a slim book entitled *Microsomal Particles and Protein Synthesis*, which contained papers presented in this symposium (Roberts, 1958).

In the introduction to this book, Roberts discussed the confusion and inadequacy of various phrases such as "microsomal particles" used during the meeting and proposed the name "ribosome" to "designate ribonucleoprotein particles in the size range 20 to 100S," stating that "it has a pleasant sound" (Roberts, 1958). Thus, 1958 was a memorable year for the history of ribosome research.

The concept that rRNAs in the ribosomes are the information carriers soon started to encounter experimental observations that were difficult to explain in simple terms. For example, purified *E. coli* 16S and 23S rRNAs appeared to be too homogeneous in size to function as a template for thousands of different *E. coli* proteins with different sizes. Thus, when Arthur Pardee announced the results of his experiments conducted at the Pasteur Institute, Paris, that the expression of the β-galactosidase gene (*lacZ*$^+$) takes place immediately after its entry from male bacterial cells to *lacZ* female cells in bacterial conjugation (Pardee, 1958; later published in Pardee et al., 1959), some people interpreted the results to mean that metabolically stable rRNAs are not the intermediates and hence that DNA might be the direct template for protein synthesis (see the discussion in Pardee, 1958). There was even a suggestion that bacterial ribosomes serve a function different from protein synthesis and perhaps represent a storage substance. In fact, in contrast to mammalian cell extracts, reliable cell-free protein-synthesizing activity had not been demonstrated with *E. coli* extracts until around 1960 (Lamborg and Zamecnik, 1960; Tissières et al., 1960; however, there was a paper published in 1959 by Zillig's group in Munich, Federal Republic of Germany [see Schachtschabel and Zillig, 1959]). As late as 1959, the central dogma was still a hypothesis, and the mechanism of information transfer was in a confused state.

Like many investigators at that time, I was very much concerned with this question and carried out experiments to study the question of whether RNA synthesized after bacteriophage T2 infection, which would be the intermediate information carrier between the phage DNA and phage-specific proteins, is identical to the rRNAs, as the hypothesis of rRNA being the information carrier would predict. A few years earlier, Volkin and Astrachan had discovered a small amount of RNA synthesized in *E. coli* after phage T2 infection by using a [^{32}P]phosphate labeling technique (Volkin and Astrachan, 1956), but they were much more concerned with the observed quantitative conversion of this RNA to phage DNA and considered a hypothesis that RNA might be a precursor for phage T2 DNA (Astrachan and Volkin, 1958). I remember the discussion that I had with Jim Watson on this subject at Harvard in the early summer of 1959 and the actual hard experimental work that I did with Benjamin Hall during the following 3 months in Sol Spiegelman's laboratory at Illinois. We discovered that the "T2-specific RNA" (the name we gave to the mRNA synthesized after T2 infection) was associated with ribosomes in the presence of a high concentration of Mg^{2+}, as judged by electrophoretic analysis, but was released from ribosomes as free RNA upon decreasing the Mg^{2+} concentration. (However, from one sucrose gradient centrifugation analysis, we incorrectly thought that the RNA sedimented mainly as 30S-like particles.) We also found that T2-specific RNA had sedimentation coefficients of about 8 to 10S and could be physically separated from 16S and 23S rRNAs (Nomura et al., 1960). However, I left Spiegelman's laboratory without completing the work (to work in Seymour Benzer's laboratory, then at Purdue, on phage genetics), and the experimental results were published without solving the crucial question of whether T2-specific RNA was a new type of RNA functioning together with preexisting ribosomes or whether it was "special" rRNA contained in "special" ribosomes synthesized after phage infection. As is well known, about a year later, brilliant density transfer experiments using heavy isotopes were done by Sydney Brenner, Francois Jacob, and Matt Meselson (1961) at the California Institute of Technology, Pasadena, proving that the former is correct. The name "messenger RNA" was coined by Jacob and Monod for this new type of RNA.

I should note that in the work on T2-specific RNA we observed, in control uninfected *E. coli* cells labeled with ^{32}P for 2 min, the heterogeneous RNA sedimenting at 8 to 10S, which was similar to phage-specific mRNA and was distinct from 16S and 23S rRNAs that were also labeled with ^{32}P. This was a clear demonstration of the presence of a new type of RNA (mRNA) distinct from rRNAs in normal cells, but, again, we failed to make a definitive conclusion in this regard. The correct and explicit interpretation was given a year later in the paper (Gros et al., 1961) published by Watson's group which appeared in *Nature* side by side with the paper by Brenner, Jacob, and Meselson mentioned above. For some time, I regretted that I did not have time to do more experiments or time to think more seriously about the results so that we could develop some clear concepts of mRNA by ourselves. However, to do so would have been difficult in any event. We were unaware of the new developments, both experimental and conceptual, that were taking place in Cambridge, England, as well as in Paris (for a historical account of mRNA discovery, see Judson, 1978). In addition, many simple basic facts about ribosomes were still lacking in 1959. For example, it had not been firmly established that the observed two major species of rRNAs, 16S and 23S rRNAs in the case of bacterial rRNAs (studied by several workers, especially Kurland [1960] in Watson's laboratory), are really two distinct species, each representing covalently linked unique and homogeneous molecules without any subunit structures. In fact, some workers explicitly suggested the possibility that these high-molecular-weight rRNAs consisted of discrete subunits of smaller sizes that are nonconvalently linked and that the smaller-size RNA detected by pulse-labeling might be precursors for rRNAs (e.g., see Aronson and McCarthy, 1961; McCarthy and Aronson, 1961; McCarthy et al., 1962). Thus, the discovery of the new type of heterogeneous RNA with sizes smaller than rRNAs alone could not distinguish between the two possibilities: one that the new RNA was mRNA and the other that it was rRNA precursors. In fact, after the discovery of mRNA, there was still a period of confusion. Some people studied mRNA assuming that pulse-labeled RNA is mostly mRNA, whereas others studied the synthesis of ribosomes assuming that it is mostly rRNA precursors. Of course, we now know that pulse-labeled RNA in *E. coli* consists of both stable RNAs (rRNAs and tRNAs) and mRNAs, that their proportions vary depending on growth conditions, and that there are certainly no subunits for *E. coli* rRNAs. Thanks to the progress made during the past decade, our current knowledge of the primary and secondary structures of rRNAs as well as the structure of rRNA transcription units (operons) is on a very firm basis, and most of the confusion and debate about the results in these early days can now be easily resolved.

The discovery of mRNA clarified the role of ribosomes in protein synthesis and, in addition, helped in the development of reliable cell-free protein-synthesizing systems which were needed to study ribosomes. The role of ribosomes as the site of protein synthesis had been firmly established by that time even in bacterial systems. By carrying out pulse-chase experiments using radioactive amino acids, McQuillen, Roberts, and Britten (1959) demonstrated that in growing *E. coli*, the radioactive proteins appeared first in the 70S ribosomes and then chased into other cellular fractions. In addition, incorporation of radioactive amino acids into protein was also demonstrated in cell-free extracts from *E. coli* around that time (Lamborg and Zamecnik, 1960; Tissières et al., 1960). However, it was the in vitro experiments of Nirenberg and Matthaei (1961), using externally added synthetic as well as natural polynucleotides as mRNA, that really opened up the new era in the field of protein synthesis, enabling studies to solve the genetic code as well as studies of the mechanism of protein synthesis in vitro.

IN VITRO PROTEIN SYNTHESIS STUDIES IN THE 1960s

The 1960s were no doubt the golden age for research in protein synthesis. With the discovery of poly(U)-directed polyphenylalanine synthesis by Nirenberg, solving the genetic code became suddenly a real possibility, and many molecular biologists and biochemists joined the race to solve the code. I heard the news of Nirenberg's discovery in Japan from Jim Watson. It was in the early fall of 1961, and Jim had just attended the International Congress of Biochemistry at Moscow and was on the way back to the United States. I remember the excited way he told me the news and the prospect of solving the genetic code, but I do not think that I expected to see the code solved so soon, that is, only 5 years thereafter.

In November of the previous year, 1960, after 3 years in the United States as a postdoctoral fellow, I returned to Japan and took an assistant professorship at Osaka University. At the beginning, I thought about continuing the work on mRNA that I had started in Spiegelman's laboratory. However, I soon realized that I would not be able to compete in such an isolated and poorly equipped place; in addition to the original groups that participated in the discovery of mRNA, many competent molecular biologists and biochemists had already started working on the same subjects, and general conditions in Japan at that time were far from satisfactory, especially for people who wanted to do science in a newly emerging and competitive field. The news of Nirenberg's discovery, though it was so exciting, did not affect my thoughts, and I decided not to work on mRNA or protein synthesis per se. Instead, I worked on the mode of action of colicins, although as side projects I continued to work on some aspects of mRNA metabolism and ribosome biosynthesis. Thus, for several years I was not in the mainstream of research on protein synthesis, mostly watching the exciting developments in this field from the sidelines.

Because of the development of a reliable and easy in vitro protein-synthesizing system using poly(U), new information on ribosome function in protein synthesis started to appear rapidly in the early 1960s. For example, Wally Gilbert, a new convert from theoretical physics, and working in Jim Watson's laboratory at that time, started to analyze the poly(U)-polyphenylalanine system in a systematic way (Gilbert, 1963a, 1963b). He demonstrated that the 50S subunit carries a tRNA-binding site and that the tRNA carrying growing polyphenylalanine is bound to the 50S subunits (Gilbert, 1963b), whereas the mRNA, poly(U), is apparently bound to 30S subunits (Takanami and Okamoto, 1963). The presence of polysomes was also demonstrated in this in vitro system (Gilbert, 1963a), confirming the independent discovery of polysomes in reticulocytes synthesizing hemoglobin by Alex Rich and co-workers (Warner et al., 1962; polysomes were also independently discovered by Gierer [1963] and Wettstein et al. [1963]). Partial reactions of protein synthesis were also demonstrated by using this system with synthetic polynucleotides as mRNA, and whenever possible, the partial reactions were tested with separated subunits: binding of mRNA to ribosomes (70S or 30S; e.g., Takanami and Okamoto, 1963), mRNA-dependent binding of tRNA to ribosomes (70S or 30S; e.g., Kaji and Kaji, 1964), and peptidyltransferase reaction (70S or 50S; Traut and Monro, 1964; Monro, 1967). The basic concept of polypeptide chain elongation, two sites (A and P sites) on ribosomes for tRNA binding, was described by Jim Watson in his 1964 article (Watson, 1964). Of course, more detailed biochemical mechanisms involved in protein synthesis had to await identification and purification of elongation factors and detailed biochemical analysis of their interactions with ribosomes. These jobs were done by many competent biochemists such as Richard Schweet at the University of Kentucky; Joan Ravel and Boyd Hardesty, both at the University of Texas; Kivie Moldave at the University of Pittsburgh; Jean Lucas-Lenard, Ann-Lise Haenni, and others in Fritz Lipmann's laboratory at Rockefeller University; Herb Weissbach and his co-workers, first at the National Institutes of Health (NIH) and then at the Roche Institute of Molecular Biology, Nutley, N.J.; Peter Lengyel and his co-workers at Yale; and Yoshito Kaziro and his co-workers at the University of Tokyo.

Another subject studied with the in vitro system, which proved to be important and useful for later ribosome research, was the mode of action of various antibiotics that specifically inhibit protein synthesis. The most dramatic experiments were the demonstration of misreading of the genetic code induced by streptomycin in the in vitro system carried out by Julian Davies, Walter Gilbert, and Luigi Gorini (Davies et al., 1964). These experiments, together with Gorini's previous observations on streptomycin-induced suppression of arginine-requiring mutations of *E. coli* (Gorini et al., 1961; Gorini and Kataja, 1964), demonstrated the important role of ribosomes in maintaining the fidelity of translation of the genetic code, heralding the long and extensive research on this subject (Kurland, 1980; Gallant and Foley, 1980; Ehrenberg et al., 1986). For example, some important general concepts such as the kinetic proofreading model proposed by Hopfield (1974) arose as an outcome of studies on the fidelity of translation. Luigi Gorini, who died in 1978, will be remembered as the pioneer of this subject.

The site of action of a number of different antibiotics was localized to the ribosomes by using the in vitro system. For example, the site of action of streptomycin was originally proposed by Spotts and Stanier (1961) to be ribosomes on the basis of indirect arguments. In vitro poly(U)-dependent polyphenylalanine synthesis experiments using 70S ribosomes (or a combination of 30S and 50S subunits) derived from strains of *E. coli* either resistant or sensitive to streptomycin established that the site of action of the drug is, in fact, ribosomes (Speyer et al., 1962), specifically 30S subunits (Davies, 1964; Cox et al., 1964).

Although the use of synthetic polynucleotides as mRNA in in vitro protein-synthesizing systems was so convenient and powerful in the study of ribosomes and protein synthesis, especially in solving the genetic code, the use of natural mRNA such as RNA from RNA phage f2 (or R17) in the in vitro system (Nathans et al., 1962) was also important, especially for studies of the initiation and termination steps and their regulation. There were many important discoveries and events related to this subject. For example, regarding the initiation step, one of the most important was proving that protein synthesis in *E. coli* starts with formylmethionine at the unique codon, AUG (or GUG), on mRNA. Starting from the discovery of the existence of fMet-tRNA by Marcker and Sanger in 1964 (Marcker and Sanger, 1964), several early workers made important contributions to establishing this conclusion (e.g., Webster et al., 1966; Adams and Capecchi, 1966; Clark and Marcker, 1966). The discovery of initiation factors essential for the initiation of natural mRNA (Stanley et al., 1966; Revel and Gros, 1966; Brawerman and Eisenstadt, 1966) was also important in revealing the complexity of the biochemistry of initiation of protein synthesis. I do not describe these events here, nor do I describe studies carried out on chain termination. (The mechanism of polypeptide chain termination was studied mainly by Mario Capecchi at Harvard and Tom Caskey and Marshal Nirenberg and their co-workers at NIH.) I shall just describe briefly the discovery that initiation starts on the small subunits in vitro, in which I participated during the late 1960s. At that time, I was already back in the United States and working on partial reconstitution of ribosomes (as well as on the mode of action of colicin) at the University of Wisconsin, Madison.

RIBOSOME SUBUNIT EXCHANGES AND INITIATION OF PROTEIN SYNTHESIS

As mentioned earlier, one of the fundamental discoveries in early studies on ribosome structure was the unequal bipartite structure, that is, the presence of small and large subunits. At that time, we wondered whether the unequal bipartite structure was necessary for ribosome function, or perhaps for the mechanism of ribosome synthesis. For example, it was formally possible to think about a mechanism in which the 30S subunits are assembled on the 50S subunits and vice versa. These kinds of questions stayed with me and influenced my thinking on the initiation of protein synthesis when we (and other investigators) discovered in 1967 that initiation starts on the small subunits in vitro (Nomura and Lowry, 1967). There must have been similar influences on other investigators who considered and demonstrated cyclic dissociation and reassociation of subunits during ribosome function. In fact, at the time of these discoveries we (Nomura and Lowry, 1967; Nomura, 1968) as well as Kaempfer, Meselson, and Raskas (1968) thought that we had found the reason for the existence of this bipartite structure, although it is still not entirely clear why the bipartite structure evolved as the only ribosome structure in all living cells.

Of course, as mentioned above, people had been asking how ribosome functions are distributed between the two subunits. For example, it had been shown that the 30S subunit had mRNA-binding function and the 50S subunit alone could carry out the peptidyltransferase reaction. Nevertheless, these studies were carried out in a way analogous to the studies of active centers of enzymes without considering the significance of the dissociation and reassociation of ribosomes. Many people used synthetic polynucleotides as mRNA and 70S ribosomes for in vitro protein synthesis; 70S ribosomes could attach to mRNA and initiate protein synthesis under optimized reaction conditions (usually in the presence of 10 to 20 mM Mg^{2+}, which prevents 70S ribosome dissociation). Thus, most people did not think about the possibility that initiation starts on the small subunits. As will be related later, I was working on the reconstitution of ribosomes at that time and was eager to demonstrate some partial reactions carried out by the 30S subunits alone or by the 50S subunits alone that could be used to assess the functional capacity of reconstituted subunits. Thus, in early 1967, we discovered stimulation of fMet-tRNA binding to 30S subunits by phage f2 RNA (Nomura and Lowry, 1967). On the basis of this and other experimental results, we proposed that the initiation of protein synthesis starts on 30S subunits rather than 70S ribosomes. I recall that my short talk on this subject at the Nucleic Acid Gordon Conference in the same year left many people skeptical. However, several other people thought that the new model was reasonable. For example, Robert Thach, who was then at Harvard University, had observed that although the AUG-dependent fMet-tRNA-binding reaction (in the presence of initiation factors IF1 and IF2) itself had an optimum Mg^{2+} concentration of about 10 mM, preincubation of his 70S ribosome preparations at a lower Mg^{2+} concentration (0.5 mM) caused a large stimulation of the subsequent binding reaction (Ohta et al., 1967). Such observations could be best explained on the basis of dissociation of 70S ribosomes during preincubation caused by the lowered Mg^{2+} concentration. Binding of fMet-tRNA to 30S subunits directed by synthetic or natural mRNAs followed by joining of 50S subunits was quickly confirmed in several laboratories (Hille et al., 1967; Ghosh and Khorana, 1967).

Of course, as already mentioned, my laboratory was not the only one that seriously considered the importance of ribosomal subunits in protein synthesis at that time. David Schlessinger and his co-workers at Washington University were studying polysomes in *E. coli* extracts prepared by "gentle" methods. They observed that the extracts contained only polysomes and free subunits, not single 70S ribosomes, and suggested that single ribosomes do not exist in vivo (Mangiarotti and Schlessinger, 1966), implying that protein synthesis starts with free ribosomal subunits. Although other investigators, such as Bernard Davis, observed the presence of single ribosomes in similar studies and disagreed with Schlessinger's conclusion (see, e.g., Kohler et al., 1968; for reviews, see Davis, 1974, and Kaempfer, 1974), Schlessinger and his co-workers showed that the "native" 30S and 50S subunits found in crude extracts can form 70S ribosomes only in the presence of mRNA and tRNA and that mRNA containing the AUG sequence is more effective than poly(U) in promoting association (Schlessinger et al., 1967).

Another laboratory that made an important contribution to this subject was that of Matthew Meselson at Harvard. Using ribosomes labeled with heavy isotopes, Raymond Kaempfer, Meselson, and Herschel Raskas demonstrated that during growth, *E. coli* ribosomes undergo subunit exchange (Kaempfer et al., 1968) and that subunit exchange takes place in crude bacterial extracts only under conditions of protein synthesis (Kaempfer, 1968; Kaempfer and Meselson, 1969). These experiments were entirely consistent with the 30S subunit initiation model and gave strong support to the model. In addition, Kaempfer demonstrated a similar subunit exchange in a eucaryotic system (Kaempfer, 1969), suggesting that the basic concept of translation initiation on small ribosomal subunits is correct in eucaryotic systems too. (Of course, later work by Marilyn Kozak [1978] extended this earlier conclusion and led to the formulation of the 40S subunit scanning model for the initiation of protein synthesis in eucaryotes.)

Finally, Christine Guthrie, then a graduate student in my laboratory, and I carried out further experiments to test the 30S subunit initiation model. Using 70S ribosomes labeled with heavy isotopes and carrying out tRNA binding experiments in the presence of excess light 50S subunits, we were able to conclude that the binding of fMet-tRNA to 70S ribosomes does not take place directly but proceeds through the formation of 30S · mRNA · fMet-tRNA as an intermediate (Guthrie and Nomura, 1968).

In the original formulation of the model entertained by us, Meselson's group, and David Schlessinger's group, it was assumed that only 30S and 50S subunits exist in the cellular pool and that 70S ribosomes can be formed only after the formation of the 30S initiation complex on mRNA. In addition, there was evidence that 30S subunits are able to form the initiation complex at the initiator AUG codon but do not respond to other codons efficiently, whereas 70S ribosomes have the ability to respond to all other codons and to bind the corresponding aminoacyl-tRNAs. Thus, we argued that the bipartite structure is fundamentally important to prevent nonspecific initiation of protein synthesis on mRNA. However, after considerable debate, it was subsequently concluded that free 70S ribosomes do exist in bacterial cells in vivo (Davis, 1974; Kaempfer, 1974). In eucaryotic cells, the presence of large amounts of nonfunctioning 80S ribosomes is often observed unequivocally, and these 80S ribosomes are prevented from entering translation by other mechanisms. In fact, we could conceive (hypothetical) mechanisms allowing free 70S ribosomes to start only at correct initiation sites without dissociation into subunits. Therefore, it is still not entirely clear why Nature has selected the bipartite ribosome structure and a mode of translation involving cyclic dissociation and reassociation.

PARTIAL IN VITRO DISSOCIATION OF r-PROTEINS AND REASSEMBLY OF RIBOSOMES

The discovery of reconstitution of ribosomes is in a way connected to the discovery of mRNA described earlier. In October 1960, after finishing my postdoctoral work with Seymour Benzer, I visited Pasadena on the way back to Japan and gave a seminar on T4 *rII* genetics at the California Institute of Technology. Jacob and Brenner had just finished the experiments on mRNA mentioned earlier and had left Pasadena to return to Europe. I met Matt Meselson and learned about their experiments with a sense of excitement mixed with a feeling of slight regret for the lost opportunity of the discovery of mRNA. Back in Osaka, I started to work mainly on the mode of action of colicins, as mentioned above, but maintained a serious interest in mRNA and ribosomes. Thus, when the paper on mRNA by Brenner, Jacob, and Meselson appeared in *Nature* in 1961 (Brenner et al., 1961), I noticed their peculiar observation immediately. When they centrifuged bacterial extracts in CsCl to separate ribosomes by densities, they observed two bands containing ribosomal particles. The lighter band (the B band), corresponding to a density of 1.61, contained mRNA as well as growing polypeptide chains; the heavier band (the A band), corresponding to a density of 1.65, did not contain either. The presence of the A band was not relevant to the main theme of the paper, but it was stated that 30S and 50S subunits corresponded to the A band and the 70S ribosomes corresponded to the B band. We knew then that the 30S and 50S subunits and their aggregates, the 70S ribosomes, have the same chemical composition, and I wondered why their densities should be different.

In the summer of 1962, I had a chance to visit Meselson's laboratory at Harvard to examine this question. I followed the protocol of the original experiments, recovered the particles from the two bands, and analyzed them by sucrose gradient centrifugation. We found that the B band contained undegraded ribosomal subunits, whereas the A band contained smaller 40S and 23S "core" particles, which were derived from the 50S and 30S subunits by dissociation of a fraction of the proteins during density gradient centrifugation (Meselson et al., 1964).

Upon returning to Osaka, I continued experiments with Robert K. Fujimura to characterize these core particles. At the beginning, we prepared the core particles by centrifuging 50S (or 30S) subunits in 5 M CsCl for 36 h at a high speed and recovering the particles from the middle of centrifuge tubes. However, this method was not convenient for large-scale preparations, and we simply mixed 50S (or 30S) subunits in CsCl and kept them in a cold room for 36 h, thinking that conversion of the subunits into the core particles would take place by irreversible splitting of proteins by high concentrations of CsCl. Thus, we were very much surprised when we found intact 50S (or 30S) subunits, rather than core particles, upon removal of CsCl. Why no splitting of ribosomal proteins? We immediately realized that splitting of ribosomal proteins is reversible and that the reaction can be pushed in the direction of protein splitting only by separating split protein products from the core particles by centrifugation. This reasoning was easily confirmed by mixing 40S (or 23S) core particles with split proteins from 50S (or 30S) subunits and analyzing the products by sucrose gradient centrifugation. In this way, we discovered partial reconstitution of ribosomal subunits. This was late in 1962. However, measurement of functional activities of the reconstituted particles was not easy in Japan at that time because of difficulties in obtaining (or preparing) various reagents and synthetic polynucleotide mRNA necessary for in vitro protein synthesis assays. In the meantime, I left Japan, in the fall of

1963, to take a position at the University of Wisconsin. Thus, formal proof that reconstituted particles are completely active in protein synthesis could be made in the new laboratory at Madison in 1965, that is, 3 years after the initial discovery, and Keiichi Hosokawa participated in completing this project (Hosokawa et al., 1966).

At about the same time, Theophil Staehelin and Matt Meselson, who were taking similar approaches to study the structure and function of ribosomes, independently succeeded in showing reconstitution of ribosomes from core particles and split proteins (Staehelin and Meselson, 1966).

In connection with these partial reconstitution studies, I should also mention studies on the structure of ribosomes that Alexander Spirin and his co-workers started at the A. N. Bakh Institute of Biochemistry in Moscow in the early 1960s. One approach they took was to make systematic alterations (or partial degradation) of ribosomes and analyze the altered structures by using various physical methods. Thus, in 1963, they discovered that a mixture of 50S and 30S subunits, after treatment with a high concentration of NH_4Cl followed by centrifugation and resuspension in a low-ionic-strength buffer without Mg^{2+}, is converted to a mixture of more asymmetric particles with lower sedimentation coefficients; they called this process unfolding (Spirin et al., 1963). They showed that most r-proteins stay with rRNAs and the unfolding process is reversible, as judged by sedimentation analysis. From these and other studies, particularly electron microscopic analysis of the structures of ribosomes and chloramphenicol particles (which accumulate in chloramphenicol-treated *E. coli* cells, as I mentioned earlier), Spirin proposed a model for ribosome structure which postulated a folding of a ribonucleoprotein strand into a compact structure (Spirin, 1964a, 1964b). In retrospect, the chloramphenicol particle preparations were probably contaminated with other materials, most likely GroE protein particles, and gave incorrect morphological information. Nonspecific binding of r-proteins to rRNA in the unfolded particles was not excluded either. Thus, the model was not based on solid experimental evidence. Nonetheless, this model was perhaps the first specific and detailed model proposed for assembly of ribosomes and stimulated thought on ribosome structure.

Although scientific exchanges between the East and the West at that time were not easy, Spirin maintained good contacts and interactions with scientists in the Western countries by attending meetings and traveling. For example, his success in the unfolding of ribosomes was aided by such an interaction. In earlier studies of ribosomes, people observed degradation of rRNA under conditions of decreased Mg^{2+}. David Elson (1958), at the Weizmann Institute of Science, Rehovot, Israel, first noted the presence of (nonspecifically) bound RNase I in the usual ribosome preparations. Under low Mg^{2+}, ribosomes unfold and the exposed rRNA becomes sensitive to RNase I. Spirin could unfold ribosomes successfully because he removed RNase I by washing ribosomes with 0.5 M NH_4Cl, the unpublished technique developed by Robert Bock at the University of Wisconsin, Madison (acknowledged by Spirin in his paper [1964b]). (Unfolding and refolding studies were also carried out later by Ray Gesteland in Watson's laboratory at Harvard, using an RNase I-deficient strain that he isolated [Gesteland, 1966], and by us under conditions of reconstitution [Traub and Nomura, 1969]). In addition, the MRE600 strain lacking RNase I became available around this time [Cammack and Wade, 1965], and the two classes of strains, Q13 derivatives originated from Gesteland's mutant and MRE600 derivatives, became the favorite strains for biochemical studies of *E. coli* ribosomes.) Conversely, the earlier work done by Spirin's group in Moscow was also well known to Western scientists through Spirin's well-written monograph entitled *Macromolecular Structure of Ribonucleic Acids*, published in English in 1964. Spirin also had an interest in chloramphenicol particles and attempted conversion of chloramphenicol particles to normal 50S and 30S subunits in vitro (Spirin, 1963). Therefore, it must have been natural for Spirin and his co-workers to extend their interest to the newly discovered r-protein-splitting reaction induced by centrifugation in CsCl. Thus, in 1966, Spirin and his co-workers also described their experiments showing that protein splitting under these conditions is reversible (Lerman et al., 1966; Spirin and Belitsina, 1966).

PURIFICATION AND CHARACTERIZATION OF r-PROTEINS AND rRNAs

The success in establishing the partial reconstitution system in 1965 gave me a chance to change my major effort from colicin research back to ribosome research, and I started to think more seriously about ribosome problems. At the very beginning of the reconstitution work, we could easily demonstrate that r-proteins split from 30S subunits were different from those split from 50S subunits in their ability to reconstitute functionally active ribosomal subunits (Hosokawa et al., 1966). Although by this time Waller had demonstrated, by using a starch gel electrophoresis technique, that there are many different r-proteins in ribosomes and that r-proteins in the 30S subunits are different from those in the 50S subunits (Waller and Harris, 1961; Waller, 1964), there was still skepticism and rumors of contrary results. Therefore, it was very satisfying to confirm Waller's conclusion by demonstrating functional differences between the two split protein fractions. Peter Traub, who joined my laboratory in 1965, then started to separate and purify individual r-proteins in the split protein fractions and was soon able to purify five 30S split proteins, demonstrating that they are chemically and functionally different (Traub et al., 1967).

After the pioneering work by Waller, several laboratories started major efforts to purify and characterize all of the r-proteins in *E. coli* ribosomes. Alfred Tissières, who had left Watson's laboratory and returned to Switzerland, created a new center of ribosome research in Geneva and, together with several excellent postdoctoral fellows from the United States, such as Robert Traut, Peter Moore, and Harry Noller, started to publish a series of papers characterizing r-proteins and demonstrating their chemical heterogeneity (Traut et al., 1967; Moore et al., 1968).

Charles Kurland, who worked in the Watson laboratory as a graduate student, came to Madison as a faculty member in the Department of Zoology at the University of Wisconsin. He organized a ribosome research group and, together with Gary Craven's group in the Laboratory of Molecular Biology on the same campus, started purification and characterization of r-proteins (Hardy et al., 1969; Craven et al., 1969). After the success in complete reconstitution of 30S subunits to be described below, we also started to purify all of the r-proteins from 30S subunits. (I should note that on the Madison campus in the 1960s there were also several other laboratories actively working on various aspects of ribosomes and protein synthesis. Robert Bock, in the Department of Biochemistry, who was one of the few biochemists who had started to work on ribosomes in the late 1950s, discovered that high-salt washing removes RNase from ribosomes, as mentioned above, and was then studying rRNA in addition to tRNA. Both Julian Davies in the Department of Biochemistry and Bernard Weisblum in the Department of Pharmacology started to work on ribosomes in connection with the mode of action of various antibiotics. In the Laboratory of Biophysics, Walter Hill was estimating the shape and size of ribosomes, using low-angle X-ray-scattering techniques in John Anderegg's laboratory. Of course, Gobind Khorana and his co-workers were busy synthesizing polynucleotides and solving the genetic code; in 1968 the campus celebrated his winning the Nobel prize for this work. I enjoyed the interactions with all of these people and remember those days at Madison fondly. Madison in the 1960s was certainly a very active center for research on ribosomes and protein synthesis.)

Another major group which started r-protein fractionation was that of Heinz-Günter Wittmann and his co-workers in Berlin. Wittmann, after having performed pioneering work on the characterization of amino acid alterations in various tobacco mosaic virus (TMV) mutant coat proteins, decided to work on r-proteins and ribosomes. I remember reading a few sentences at the end of his 1966 Cold Spring Harbor Symposium paper on TMV proteins, announcing that he was starting r-protein research (Wittmann and Wittmann-Liebold, 1966). Papers describing purification and characterization of many r-proteins started to be published in 1967 (Kaltschmidt et al., 1967), and Wittmann's Abteilung in the Max Plank Institute in Berlin was soon to become one of the largest and most active centers in ribosome research.

Several other laboratories were also engaged in purification and characterization of r-proteins at about the same time. They included Paul Sypherd's laboratory, then at the University of Illinois, Wim Möller's laboratory at the University of Leiden, and Shozo Osawa's laboratory, then at Hiroshima University. With so many proteins in ribosomes, the task of purifying and characterizing all of them was certainly not an easy one. Yet all of the techniques necessary for protein separation and determination of amino acid sequences were available, and this was probably one reason why many researchers started to undertake this enormous task. (In contrast, sequencing large RNA molecules such as rRNA was beyond any practical hope in the mid-1960s.)

Initially, one problem encountered in dealing with r-proteins was their poor solubility in ordinary buffers used for physiological experiments. Two methods were used to extract r-proteins from ribosomes. One was the use of acetic acid, which was originally employed by Fraenkel-Conrat to extract proteins from TMV (Fraenkel-Conrat, 1957). The second was the LiCl-urea method developed by Pnina Spitnik-Elson (1965) at the Weizmann Institute. Column chromatographic separation was carried out mostly in the presence of urea to maintain solubility of r-proteins. Thus, we were originally concerned with the denatured state of r-proteins. Therefore, it was a great relief for me to see that r-proteins purified in the presence of urea were able to restore the functional activities of ribosomes when added together to inactive core particles in the partial reconstitution experiments. This observation also strengthened our determination to attempt complete reconstitution of ribosomes by using rRNA and a mixture of r-proteins isolated in the presence of urea.

One serious debate on r-proteins which took place in the late 1960s was the stoichiometry of r-proteins in ribosomal particles. Accurate measurements of the stoichiometry were technically not easy at that time (and perhaps even now), especially with the dilemma that ribosomes prepared without high-salt washing contained many non-r-proteins (as exemplified by RNase I and initiation factors mentioned earlier), and ribosome preparations obtained after salt washing tended to lose some r-proteins. (In addition, estimated molecular weights of r-proteins were sometimes in considerable error compared with the values from their amino acid sequences, as we now know.) Although the original suggestions were that most or all r-proteins exist in a single copy per ribosome (Moore et al., 1968), Kurland and his co-workers obtained data indicating that quite a few r-proteins exist in fractional amounts and suggested that ribosomes are physically and functionally heterogeneous in vivo (Kurland et al., 1969). Although this was an intriguing possibility, subsequent studies led to the general agreement that ribosomes are largely homogeneous and that all r-proteins exist in a single copy except for L7/L12, which is present in four copies per particle (e.g., Subramanian, 1975; Hardy, 1975; Marquis and Fahnestock, 1978). Demonstration of stoichiometric requirements of essential r-proteins in reconstitution of 30S subunits (Held et al., 1973) was also consistent with this general conclusion.

In connection with the purification of r-proteins, I should note the great technical contribution of the two-dimensional gel electrophoresis system developed by Kaltschmidt and Wittmann in 1970 (Kaltschmidt and Wittmann, 1970). With over 50 different r-proteins purified and studied independently by several groups and called by different names, it was difficult to compare the results obtained by different groups. The two-dimensional gel method, which enabled most of the r-proteins to be separated in a single gel, made the comparison much easier. Most important, the major groups engaged in r-protein purification agreed to adopt the nomenclature used by the Berlin group based on the position of a protein in this gel system (Wittman et al., 1971). At the beginning, it was very inconvenient for our group to change the names of proteins, as it must have been for some other groups. By that time, we had constructed the 30S ribosome assembly map and carried out many reconstitution experiments. In fact, I had just written a comprehensive review on bacterial ribosomes discussing all of these new results with our nomenclature (Nomura, 1970). Nevertheless, changing to a common nomenclature was, of course, a very good decision for science in the long run.

I should also mention that most of us who were engaged in r-protein purification or analysis using the two-dimensional gel methods saw minor peaks in column chromatography or occasional faint unidentified spots on stained gels. Without any necessity to look for (hypothetical) missing components required for ribosome function and viewing them as minor components, these proteins were mostly disregarded as non-r-proteins or modified or degraded r-protein derivatives, and the consensus was reached that *E. coli* ribosomes contain 53 proteins (counting L7/L12 as two different proteins). However, Wada at Kyoto University worked persistently on these unidentified proteins seen on stained gels and recently showed that at least two (called A and B) of the four he studied are probably genuine r-proteins (Wada and Sako, 1987). Thus, enumeration of the number of *E. coli* r-proteins done around 1970 was not complete, and the final listing may yet have to be awaited. Nevertheless, completion of primary sequence determination of all of the formally listed r-proteins, which was done mostly in Berlin and completed by the early 1980s (Wittmann-Liebold, 1984), was a real achievement. (I should add that before completion of primary sequence determination of r-proteins by conventional protein sequence techniques, many of the genes for these r-proteins were isolated and sequenced by our laboratory as well as by others. Amino acid sequences deduced from nucleotide sequences of the genes helped completion of the sequence work.)

In contrast to r-proteins, enumeration of rRNA molecules in *E. coli* ribosomes was simpler. In addition to the two major rRNA molecules, 16S and 23S, initially characterized (Kurland, 1960; Stanley and Bock, 1965), 5S RNA was discovered in 1963 by Rosset and Monier (1963). The debate over whether the large 16S rRNA (or 23S rRNA) molecule represents a single covalently linked RNA molecule or an aggregate consisting of smaller subunits (see above) was also largely settled by the mid-1960s. However, considering that 16S and 23S rRNAs were so large and knowing how much work was required to determine the primary structure of alanine tRNA (Holley et al., 1965), it seemed to be almost hopeless to determine their primary structures in the early 1960s. But in 1965, Fred Sanger at the Medical Research Council laboratory, Cambridge, England, introduced rapid sequencing methods, the use of ^{32}P-labeled RNA and fingerprinting techniques (Sanger et al., 1965), and the power of the techniques was in fact demonstrated by the subsequent complete sequence determination of 5S RNA by Sanger and his co-workers (Brownlee et al., 1968). Thus, heroic efforts to sequence two large rRNA molecules were initiated in the late 1960s, originally by Peter Fellner in Fred Sanger's laboratory (Fellner and Sanger, 1968) and then by Jean Pierre Ebel's group in Strasbourg, France, where Peter Fellner also moved from Cambridge. Although the Strasbourg group completed the sequencing of 16S rRNA by RNA fingerprinting methods in 1978 (Carbon et al., 1978), it was really the development of gene cloning methods and DNA sequencing techniques (again, thanks to Fred Sanger's ingenious dideoxy method as well as Maxam and Gilbert's original chemical method) that has almost completely changed the nature of rRNA sequencing work. As is now well known, starting with the sequencing of *E. coli* 16S rRNA (encoded by the *rrnB* operon) in 1978 (Brosius et al., 1978) and *E. coli* 23S rRNA in 1980 (Brosius et al., 1980) by Noller and his co-workers, sequences of many rRNA molecules of various origins were determined, and much useful information has started to come forth from rRNA structural work.

COMPLETE RECONSTITUTION OF RIBOSOMES FROM rRNA AND r-PROTEINS

There was every reason for attempting to reconstitute ribosomes from RNA and proteins. We wanted to know the functional and structural roles of many components we found in ribosomes and how ribosomes are assembled from these components. The striking demonstration of reconstitution of infectious TMV particles from RNA and protein, achieved in 1955 by Fraenkel-Conrat and Williams (1955), was an obvious model to follow. Of course, by the mid-1960s it had become clear that the structure of ribosomes is fundamentally different from and much more complicated than that of TMV, and there was no guarantee of success. In fact, several people attempted reconstitution of ribosomes in the early 1960s but without success. Nevertheless, soon after we demonstrated restoration of protein-synthesizing activity in partial reconstitution experiments as mentioned above, Peter Traub and I started serious efforts to reconstitute 30S subunits from protein-free 16S rRNA and a mixture of r-proteins.

We first assumed that it would be better to do reconstitution in two steps, first making 23S core particles from 16S rRNA and proteins and then making complete 30S subunits from 23S cores and the split proteins. As was the usual custom for enzyme purification, we originally thought that it was essential to protect r-proteins and any intermediate from heat inactivation and carried out the mixing of RNA and proteins in the cold. In fact, in the case of partial reconstitution, all of the procedures, that is, preparation of cores and split proteins, protein separations, and mixing proteins and cores to reconstitute ribosomes, had been done in the cold; hence, the operation in the cold was justifiable. However, after our initial failures, we decided to try to mimic conditions inside growing *E. coli* cells. Thus, we used 37°C for incubation. In addition, we increased the ionic strength and also added various potential cofactors such as spermine and spermidine, which were thought to be bound to ribosomes in vivo. I found that the estimated values for ionic strength inside bacterial cells available in the literature at that time varied widely, ranging from about 0.2 to 1.0, without any consensus. However, it was certainly much higher than the ionic strength we used for our initial experiments. These two factors, temperature and ionic strength, turned out to be crucial for our success. By the end of 1967, we were able to demonstrate reconstitution reproducibly. The unnecessary components were omitted, and two protein fractions were soon replaced by a mixture of r-proteins extracted directly from 30S subunits. Various parameters were systematically altered, and optimum conditions were defined. The first report on the reconstitution was published in *Proceedings of the National Academy of Sciences* in 1968 (Traub and Nomura, 1968a). Details on various parameters, together with other information such as kinetics of reconstitution and intermediate particles identified in the reaction, were published later (Traub and Nomura, 1969).

Self-assembly of complex biological structures was certainly not what one would readily expect. I knew then the work on cortex structures of *Paramecium* sp. done by Tracy Sonneborn and his school and Sonneborn's belief that the generation of supramolecular cell structures requires more than self-assembly; structural information could be maintained by preexisting structures (e.g., Beisson and Sonneborn, 1965). (After our work on ribosome reconstitution, I had a chance to meet Sonneborn, a great scientist and the founding father of ciliate genetics, when I visited the University of Indiana, Bloomington. He was disappointed with the demonstration of self-assembly of ribosomes but was nevertheless eager in persuading me to think about more complex structures on the basis of his "structural guidance" concept.) In fact, we and other workers had considered the possibility that ribosome assembly might take place only with the aid of some structure as a scaffold. Therefore, when we succeeded in reconstitution, I was very impressed with this remarkable self-assembly and was happy to be able to write in the first paper that "The efficient reconstitution . . . indicates that the entire information for the correct assembly of the ribosomal particles is contained in the structure of their molecular components, and not in some other non-ribosomal factors" (Traub and Nomura, 1968a). Nevertheless, as is well known, ribosome assembly in vivo is not exactly the same as in vitro, and the presence of nonribosomal factors that might stimulate assembly in vivo is still a real possibility. Such factors have been looked for by many investigators (including ourselves) without any clear success so far. Recent demonstration of the presence of proteins ("chaperones") that stimulate assembly of oligomeric enzymes such as ribulose carboxylase-oxygenase (Ellis, 1987; Hemmingsen et al., 1988) may rekindle the search for possible assembly factors involved in the biosynthesis of ribosomes in vivo. (In fact, Varshavsky and his co-workers, who had been working on ubiquitin, just recently discovered that three of the four yeast ubiquitin genes encode hybrid proteins that are cleaved to yield ubiquitin and r-proteins and obtained experimental results suggesting that the transient association between ubiquitin and these r-proteins promotes their incorporation into nascent ribosomes. Thus, ubiquitin appears to function as a chaperone in this eucaryotic system [Finley et al., 1989].)

Before describing the development of new research directions using the 30S reconstitution system and their outcomes, I should mention reconstitution of 50S subunits. After the success of reconstitution of 30S subunits, we obviously attempted to reconstitute 50S subunits from *E. coli* by using the same procedures. However, particles recovered after incubation of reconstitution mixtures did not sediment like 50S subunits, nor did they show protein-synthesizing activity. We then tried to use 50S subunits from *Bacillus stearothermophilus*. The choice was made because we had already used this bacterium for heterologous 30S reconstitution to study RNA-protein interactions in ribosome assembly and to study the heat stability of ribosomes from this thermophilic bacterium (Nomura et al., 1968). We also knew the strong temperature dependency of the 30S reconstitution reaction and thought that 50S reconstitution might require much higher incubation temperatures to overcome kinetic barriers and to achieve a measurable rate of assembly in vitro. This inference turned out to be correct. By mixing 23S rRNA, 5S RNA, and 50S r-proteins derived from *B. stearothermophilus* and incubating the mixtures at 60°C, Volker Erdmann and I could, in fact, demonstrate in vitro reconstitution of 50S subunits (Nomura and Erdmann, 1970). The system was then refined (Erdmann et al., 1971; Fahnestock et al., 1973) and used for various studies; for example, the requirement of 5S RNA for 50S functions was demonstrated by using this system (Erdmann et al., 1971). Stephen Fahnestock, who worked on this system as a postdoctoral fellow with me, took a faculty position at Pennsylvania State University at University Park and continued to work on this system. Volker Erdmann, who moved from Madison to Berlin and joined Wittmann's Abteilung, also continued to study 5S RNA by using this reconstitution system. Jeff Cohlberg, a postdoctoral fellow in my laboratory, then carried out protein purification and, together with Fahnestock's later work (e.g., Auron and Fahnestock, 1981), demonstrated reconstitution with defined molecular components (Cohlberg and Nomura, 1976). Nevertheless, reconstitution of *E. coli* 50S subunits was obviously desirable because of extensive biochemical and genetic studies carried out on *E. coli* ribosomes.

Although there was one report of success in reconstitution of *E. coli* 50S subunits in 1971 (Maruta et al., 1971), the experiments were difficult to reproduce, and the real accomplishment had to await the experiments of Knud Nierhaus and Ferdinand Dohme carried out in Wittmann's Abteilung in Berlin. They devised, in 1974, a two-step incubation method to achieve successful reconstitution (Nierhaus and Dohme, 1974). (Subsequently, a similar method was devised by Cantor and his co-workers [Amils et al., 1978].) A variety of studies using the 30S reconstitution system were already well under way by that time, and Nierhaus and his co-workers started to do similar studies with the newly established 50S reconstitution system (Nierhaus, 1980, 1982).

IN VITRO RIBOSOME ASSEMBLY AND RNA-PROTEIN INTERACTIONS

When the 30S reconstitution system was established in early 1968, many interesting experiments suddenly became possible. We started to work on several projects right away. Basically, there were two goals which were interrelated; one was to study the assembly reaction to get some insights into ribosome structures as well as the mechanism and regulation of ribosome synthesis in vivo, and the other was to study functional roles of each of the molecular components in protein synthesis.

Regarding the first goal, we carried out two different types of experiments without waiting for all of the proteins to be purified. The first was to study the kinetics of reconstitution and identification of the rate-limiting step (Traub and Nomura, 1969), and the results were well publicized and recognized. The second was to study the cooperativity of the assembly reaction, which involved 16S rRNA and some 20 different protein molecules. We asked how many sites existed on 16S rRNA that bind r-proteins independently without cooperativity, that is, how many "nucleation sites" exist in 30S ribosome assembly. I reasoned that if ribosome assembly starts from the formation of a string-and-bead-like structure, as imagined in the original model suggested by Spirin (1964a, 1964b), and if r-proteins bind to individual sites on rRNA that are largely independent of each other, the fraction of rRNA with a complete set of proteins would be very small in the presence of excess rRNA. On the other hand, if assembly is completely cooperative, that is, if all of the protein components interact with RNA as one group, forming active ribosomes utilizing all of the available protein sets and leaving excess RNA as free RNA, then we would find predominantly 30S subunits and free rRNA. Simple reconstitution experiments were carried out with decreasing amounts of the protein mixture and a constant amount of 16S rRNA. I hoped to see complete cooperativity, that is, a single nucleation site for ribosome assembly. However, the assembly was not completely cooperative. The data were more consistent with the presence of two to three independent nucleation sites (Nomura et al., 1969b). I presented this result (together with other experimental results) at the Oak Ridge National Laboratory symposium in the early summer of 1969. I emphasized that there were certainly not many independent r-protein-binding sites on rRNA, and the assembly was highly cooperative. The actual experimental data were published in a special (symposium) issue of the *Journal of Cellular Physiology*, a journal that most molecular biologists and biochemists probably rarely read. Perhaps partly because of this fact, the phenomenon of the breakdown of cooperativity under conditions of excess rRNA was left untouched for many years.

For my part, we embarked on the purification of 30S r-proteins and started to study the basis of the cooperative nature of the assembly reaction, which led to the construction of an assembly map (Mizushima and Nomura, 1970; Held et al., 1974a). In addition, we were busy with many experiments related to the functional role of molecular components. In the meantime, I changed the research direction of my laboratory and did not have the chance to go back to study initial nucleation events in ribosome assembly any further. When Peter Moore and his co-workers recently finished three-dimensional mapping of 30S r-proteins by the neutron-scattering technique, they compared the published assembly map with their model and pointed out that assembly starts from two nucleation sites that are distinctly separated in each of two domains, one starting with S7 and the other starting with one of other several rRNA-binding r-proteins (Moore et al., 1986; Moore, 1987). Also, Nowotny and Nierhaus (1988) have just published a paper suggesting that the binding of S4 and S7 to rRNA may represent two independent nucleation events in ribosome assembly. Of course, I was pleased to see that the original experiments, which were done some 20 years ago, have now been extended and rationalized. Recently, I also started to think about the breakdown of cooperativity under conditions of excess rRNA in relation to regulation of ribosome synthesis in vivo. In fact, as will be explained later, I now think that this breakdown of cooperativity might be responsible for accumulation of protein-deficient precursor particles, such as the chloramphenicol particles mentioned earlier, and protein-deficient particles accumulated at low temperatures in some cold-sensitive mutants of *E. coli.*

While constructing the in vitro assembly map and studying the mechanism of ribosome assembly in vitro, I wished to show that in vitro assembly reaction is relevant to the assembly in vivo. I thought that the best approach would be to isolate conditionally lethal mutants with defects in ribosome assembly. I reasoned that since 30S ribosome assembly was strongly temperature dependent, any small defects affecting the assembly might be manifested more readily at low temperatures, and hence such assembly-defective mutants might be isolated as cold-sensitive mutants. Thus, Christine Guthrie, then a graduate student in my laboratory, screened many cold-sensitive mutants and, in fact, found quite a few mutants apparently defective in ribosome assembly (Guthrie et al., 1969). We showed that there were at least three different types of mutants accumulating protein-deficient particles of different kinds. Unfortunately, both genetic and biochemical analyses were not so easy, and only one mutant, isolated as a spectinomycin-resistant, cold-sensitive mutant, was characterized in detail. As expected, the altered r-protein was shown to be S5 encoded by the *spc* gene. The proteins found in the 21S particles accumulated in this mutant were all shown to be in the upstream region of the assembly map, giving support for the relevance of the in vitro studies to the in vivo situation (Nashimoto and Nomura, 1970; Nashimoto et al., 1971). Unfortunately, when 30S reconstitution was studied by using components from this mutant, which included the mutationally altered S5, we failed to find any significant differences in kinetics of reconstitution between this system and the control using components from the parent strain (Nashimoto et al., 1971). We now know that the observed defects in ribosome assembly seen in vivo were indirect consequences of the primary defect, which is almost certainly a defect in a ribosome function, the initiation of protein synthesis (our unpublished experiments, cited in Nomura, 1987), as I will explain later.

At about the same time, John Ingraham and his co-workers at the University of California, Davis, also discovered that a significant fraction of cold-sensitive mutants of *Salmonella typhimurium* isolated in complex media is defective in ribosome assembly at low temperatures (Tai et al., 1969). It was unfortunate that neither Ingraham's group nor our group could really exploit these many interesting mutants to analyze the ribosome assembly reaction. If the current gene cloning techniques had been available at that time, it would have been easy to clone the gene responsible for the defects. It is very likely that some of the mutants were genuine assembly-defective mutants and might have even shown defects in in vitro assembly. Unfortunately, most of the mutants isolated at that time were lost and cannot be analyzed now. I should also mention that several temperature-sensitive assembly mutants were also subsequently isolated by other workers, e.g., Alex Bollen and his co-workers at the University of Brussels in Belgium (Cabezón et al., 1977; Herzog et al., 1979). Altered r-proteins were identified, but again the defects were studied only in vivo, not in vitro.

One major research direction that was stimulated after the discovery of ribosome reconstitution was the study of the interaction of r-proteins and rRNA. At the beginning, five to seven r-proteins were shown to form a stable RNA-protein complex individually under conditions of in vitro reconstitution. Since most r-proteins are basic and bind to RNA nonspecifically (in low-ionic-strength buffers), there was no guarantee that the complexes observed even under these reconstitution conditions were specific complexes. In our initial studies, we showed the specificity of S4-16S RNA and S8-16S RNA complexes by using radioactive S4 and S8 proteins and showing quantitative conversion of the complexes to ribosomes without dissociation (Mizushima and Nomura, 1970). Subsequently, Robert Zimmermann, first in the Tissières laboratory in Geneva and later at the University of Massachusetts, defined several useful operational criteria for specific rRNA-r-protein interactions observed with individual r-proteins (Zimmermann, 1974), and a consensus was reached for six proteins (S4, S7, S8, S15, S17, and S20) in the case of the 30S assembly reaction. The binding sites on rRNA for these initial binding proteins were studied extensively through the 1970s and 1980s by several groups. These groups included, in addition to Zimmermann's group, Charles Kurland's group, first at the University of Wisconsin and then at the University of Uppsala, Sweden; Harry Noller's group at the University of California, Santa Cruz; Roger Garrett, first in the Wittmann Abteilung in Berlin and then at Aarhus University, Aarhus, Denmark; and people such as Bernard Ehresmann and Chantal Ehresmann in Jean Pierre Ebel's group at Strasbourg. In the very early stages of the studies, the approach used was to treat an RNA-protein complex with RNase, isolate RNA fragments protected by the r-protein, and characterize the fragments by RNA sequencing techniques. Thus, the work was laborious and the progress was slow. Nevertheless, the information obtained in these early studies was the basis for the more recent rapid progress made after the establishment of the secondary structures of rRNAs, which utilized more powerful modern technologies, as exemplified by the studies carried out by Harry Noller and his co-workers using a rapid footprinting method (Stern et al., 1989).

In fact, it is very impressive to see the recent achievements made by Noller and his co-workers, who used various probes to measure accessibility and reactivity of each nucleotide in 16S rRNA, not only before and after binding of the six primary binding proteins but also during the whole process of 30S assembly, by adding one protein at a time, following the assembly map (Stern et al., 1989). They obtained not only information about the location of each protein relative to the structure of 16S rRNA but also evidence for protein-dependent conformational changes in 16S rRNA as an important basis for the cooperativity of ribosome assembly. For someone like me, who participated in the very early days of ribosome research, the recent progress is very impressive and satisfying.

FUNCTIONAL ANALYSIS OF MOLECULAR COMPONENTS

Search for the Functional Role of 16S rRNA

One of the obvious features recognized right after the start of ribosome research in the 1950s was the fact that as much as two-thirds of ribosomal mass is composed of rRNA. In addition, at the very beginning of ribosome research, the ribonucleoprotein particles that had been known and well characterized were plant RNA virus particles; hence, these RNA virus particles served as models for ribosome research. Therefore, it was natural to ascribe an important role to rRNA; people imagined the ribosome as a particle consisting of functionally important RNA and a few different kinds of proteins whose sole function was a structural one, holding RNA in a proper configuration. This tendency continued even after the discovery of mRNA that eliminated the functional role of rRNAs as templates for proteins synthesized on ribosomes. (This was perhaps one of the reasons why the heterogeneity of r-proteins, though discovered as early as 1961, was not immediately accepted, as mentioned earlier.) Thus, contrary to the statements frequently made in the current literature, the earlier workers seriously considered the functional importance of rRNA. Only when the presence of many different r-proteins was established did people start to think about the possibility of these r-proteins participating directly in ribosome functions, with the hope of explaining the ribosome structures containing so many different proteins which were very different from the structure of the model nucleoprotein particles, virus particles. Even so, it was reasonable to think about direct interaction between mRNA (or tRNA) and rRNA by mechanisms such as base pairing or base-phosphate interactions. This possibility was first suggested by Watson in 1964 (Watson, 1964), and experimental studies were carried out in his laboratory. For example, Peter Moore (1966) carried out perhaps the first systematic chemical modification experiments in order to examine the role of rRNA (and r-proteins) in mRNA binding.

When we started functional studies using the 30S reconstitution system in 1968, the first question we asked was whether 16S rRNA is important. Thus, in addition to demonstrating the specific (and absolute) requirement of rRNA for reconstitution, we carried out experiments to examine the effects of chemical modification of 16S rRNA on ribosome reconstitution. We used nitrous acid (Nomura et al., 1968), monoperphthalic acid, and water-soluble carbodiimide (unpublished results cited in Nomura, 1970), and we found that 16S rRNA was very sensitive to these reagents and lost its ability to reconstitute functional 30S subunits. The kinetics of inactivation was studied, and an inference of the number and kinds of base modifications corresponding to a single hit was made (Nomura et al., 1968) as was done in the earlier chemical modification studies of infectious TMV RNA (Schuster and Wilhelm, 1963). Unfortunately, since inactive "30S" particles reconstituted from chemically modified 16S rRNA usually showed somewhat altered (and heterogeneous) sedimentation patterns, it was difficult to prove direct involvement of rRNA in the translation function of ribosomes. In addition, treating such a large molecule (and the structure was unknown at that time) as 16S rRNA with chemical reagents, it was obviously difficult to make uniform modifications at specific sites. The use of reagents and conditions allowing reactions only with specific bases in a particular state, as we attempted with, for example, monoperphthalic acid, which reacted with adenine and cytosine residues in single-stranded regions, was perhaps the best we could do at that time. (Of course, that was in 1968. Now we have the genes for rRNA, we know the complete sequences and have very reliable models for the secondary structures of rRNA, and we have in vitro mutagenesis techniques to alter, as we wish, any specific base residues. Thus, the chemical modification experiments we dreamed of in 1968 can now be done. In fact, Jim Ofengand and his co-workers at the Roche Institute of Molecular Biology are now carrying out precisely these types of experiments; they transcribe various mutant 16S rRNA genes fused to the T7 promoter with T7 RNA polymerase in vitro, use the 16S RNA products carrying the desired base alterations for in vitro reconstitution, and in this way analyze functional roles of specific base sequences or secondary structures in various regions of the molecules [e.g., Denman et al., 1989a; Denman et al., 1989b]. Similarly, in vitro mutagenesis techniques are also now used to analyze the functional roles of rRNA structures in vivo [e.g., Jacobs et al., 1987; Hui and deBoer, 1987].)

Perhaps the first clear indication of the importance of rRNA in ribosome function came from the analysis of colicin E3-inactivated ribosomes carried out in our laboratory in collaboration with Jim Dahlberg's laboratory in the Department of Physiological Chemistry, also on the Madison campus. As mentioned earlier, I had initiated the work on the mode of action of colicins but without any expectation that it would become related to the ribosome research. By that time, we had discovered that one of the colicins we studied, colicin E3, specifically inhibits protein synthesis in *E. coli* (Nomura, 1963) and that ribosomes, specifically 30S subunits, isolated from E3-treated cells were almost completely inactive in in vitro protein synthesis (Konisky and Nomura, 1967). Such inactive ribosomes were apparently physically intact, as judged by the sedimentation analysis of ribosomes and extracted rRNA as well as by gel electrophoretic patterns of r-protein components. Thus, the basis for colicin-mediated inactivation was initially a mystery. This earlier E3 work was done by Jordan Konisky, the first graduate student who worked with me at Madison. Mike Bowman, a Ph.D.-M.D. student, then carried out reconstitution analysis of the E3-inactivated 30S subunits and convincingly demonstrated that it is RNA that is inactive and that the protein fraction is fully functional (Bowman et al., 1971). With the help of Jim Dahlberg's group, the presence of a single nucleolytic cleavage at a position about 50 nucleotides from the 3′ end was demonstrated (Bowman et al., 1971; see also Senior and Holland, 1971).

At that time, I was somewhat concerned with the unsettled question of whether the inactivation was due to the cleavage itself or the loss of the fragment. However, the reconstitution experiments on the colicin E3-cleaved 30S subunits were clear-cut and informative regardless of the answers to this unsettled question. Jordan Konisky had shown that E3 inactivates both poly(U)-dependent Phe-tRNA binding and poly(AUG)-dependent fMet-tRNA binding (Konisky and Nomura, 1967; Konisky, 1968). Mike Bowman showed that the cleavage reaction was most likely between A-1493 and G-1494 (according to the current nomenclature; Bowman et al., 1971; Nomura et al., 1974). In view of the current knowledge on the importance of the 3′ end of 16S rRNA which is complementary to the Shine-Dalgarno sequence of mRNA and that on the decoding site, which involves bases near C-1400 interacting with bases near A-1500 (Noller et al., 1986), removal of G-1494 and the distal part from the rest of 16S rRNA molecule is surely expected to abolish the decoding as well as the initiation function of 30S subunits. But for full appreciation of the results, we had to wait for later studies on the function of 16S rRNA.

Another reconstitution experiment that led to the suggestion of the importance of 16S rRNA was the heterologous reconstitution using components from *E. coli* and *B. stearothermophilus*. By repeating many reconstitution experiments, Bill Held, then a postdoctoral fellow in my laboratory, concluded that substitution of *Bacillus* 16S rRNA (singly or together with S12) for *E. coli* 16S rRNA (or together with S12) decreased R17 RNA translation activity without decreasing poly(U) translation activity. We suggested that direct interaction of some parts of 16S rRNA with mRNA was involved in recognition of initiation signals on natural mRNA (Held et al., 1974b).

Of course, soon afterwards, the paper by Shine and Dalgarno (1974) appeared in *Proceedings of the National Academy of Sciences* proposing a more specific initiation mechanism that would become one of the most celebrated models related to ribosome function. This was in the spring of 1974. As I will describe later, my personal effort by this time had shifted from structure and function to genetics and regulation. I remember reading the paper by Shine and Dalgarno and thinking the model to be very ingenious and interesting, but I did not think about testing the model myself. Thus, a year later, when the paper by Joan Steitz and Karen Jakes (1975) appeared providing experimental support for the Shine-Dalgarno model, I was very impressed with Joan's persistence and decisiveness as well as the elegance of their experiments. They used the colicin E3 cleavage reaction to demonstrate the interaction of the 3′ end of 16S rRNA and mRNA. Of course, Joan Steitz was the first to isolate the ribosome-binding site from natural mRNA and determined the sequences surrounding the AUG start codon (Steitz, 1969), and she maintained her interest in the initiation of protein synthesis and continued the work on this subject. Nevertheless, we were also working on the initiation of protein synthesis. In addition, we were the ones who discovered and studied the E3 cleavage reaction, and we had just suggested the interaction of mRNA and 16S rRNA. Yet I did not think about carrying out these experiments. While admiring Joan's beautiful experiments, I concluded that I should concentrate on the genetics and regulation of synthesis of ribosomes. Around that time, we had just succeeded in isolating the λ transducing phages carrying many r-protein genes located at the *str-spc* region of the *E. coli* chromosome and were ready to embark on the task of identifying and characterizing these many genes. However, before starting to describe the story of genetics and regulation, I still have to describe the work on the functional roles of r-proteins and more recent work on the fine structure of ribosomes.

Search for the Functional Roles of Individual r-Proteins

As already mentioned, the major efforts in ribosome research in the late 1960s were directed towards purification of r-proteins, and we also purified individual 30S r-proteins. Makoto Ozaki and Shoji Mizushima, both postdoctoral fellows from Japan, carried out the job and took over the reconstitution studies that had been done by Peter Traub, who left us in 1969. With the complete reconstitution system available, it was natural to define the components of the system and to analyze their roles in ribosome assembly, structure, and function. Many reconstitution experiments, such as single-component omission experiments, were carried out not because we believed that proteins and not RNA were responsible for ribosome function but because the questions were well defined and the analysis was technically feasible. Of course, it was also true that with many r-proteins just discovered, we expected that each of these different protein molecules played a unique role. As mentioned above, this expectation was confirmed with respect to their roles in the assembly reaction. In addition, direct participation of at least some r-proteins in ribosomal function was also expected, because reconstitution experiments with the unfractionated 30S protein mixture and 16S rRNA had already indicated that the alteration caused by an interesting mutation, streptomycin resistance (Str-R), affects a protein and not 16S rRNA (Traub and Nomura, 1968b). The effects of Str-R mutations on translational fidelity had been well documented on the basis of Luigi Gorini's work. Thus, identification of the protein controlled by the *str* locus (S12) and its characterizations were the first major reconstitution experiments that we carried out using the total reconstitution system; in fact, we obtained results indicating that at least this protein, S12, is either directly or indirectly involved in translation (Ozaki et al., 1969).

However, with many other proteins, single-component omission affected assembly, producing particles that had reduced sedimentation coefficients and were deficient not only in the protein omitted but also in certain other proteins (Nomura et al., 1969a; Mizushima and Nomura, 1970). In these instances, it was difficult to decide whether the large decrease observed in the activity of the reconstituted particles was caused by the absence of the protein omitted or by the observed defects in assembly. Only with a few proteins did omission affect function without causing measurable effects on physical assembly or on the physical properties of the assembled particle. In addition, the decrease in activity in these instances was only partial or insignificant (except for S12, as mentioned above). Regarding the partial decrease, I had already asked the question, in connection with some experiments using the partial reconstitution system, as to whether the partial activity of the reconstituted particles in such instances is due to a partial activity of all the reconstituted particles or whether reconstituted particles take two (or more) different conformations and only a portion of the particles with the correct conformation show a near full activity. The experimental results were in support of the latter possibility (Traub et al., 1968), implying that proteins omitted were important for maintaining the correct conformation but were dispensable for ribosomal function in protein synthesis.

Thus, in the early 1970s, I was somewhat disappointed in the failure to define functional roles in protein synthesis for each of the 30S r-proteins. We could make a definitive conclusion in this respect only when omission of a protein (or replacement of a protein with a heterologous or modified protein) affected preferentially one function (e.g., initiation) without affecting another (e.g., elongation), as in the case of replacement of *E. coli* 16S rRNA (and/or S12) with the corresponding component from *B. stearothermophilus* as mentioned above. With single-component omission experiments as originally designed, this situation was relatively rare and, when observed, the degree of the observed preferential effects was variable in repeated experiments. Thus, despite a large number of experiments carried out by Shoji Mizushima, and subsequently by William Held, we could not make definitive documentation of the results and conclusions therefrom regarding functional roles of individual proteins, and we did not publish many of these data except for the discussion of the initial results in the 1969 Cold Spring symposium article (Nomura et al., 1969a).

In fact, one important conclusion that came from our reconstitution studies in the late 1960s and early 1970s was that some r-proteins are clearly required for facilitating the assembly reaction or maintaining the stability (or a correct conformation) of the assembled particles but are not involved directly in protein synthesis. This conclusion was presented together with direct supporting experimental evidence for some r-proteins in 1975 (Held and Nomura, 1975). Around that time, we stopped the work on ribosome reconstitution more or less completely. (Similar conclusions were later made for some *E. coli* 50S r-proteins by Nierhaus and his co-workers [Spillmann and Nierhaus, 1978; Nowotny and Nierhaus, 1980].)

Although we stopped working on the functional aspects of ribosomes by the mid-1970s, other investigators continued to work on these subjects by using various approaches. Nierhaus and his co-workers continued to work on *E. coli* 50S reconstitution in order to define the minimum r-proteins required for peptidyltransferase activity. Since peptide bond formation was considered an enzymatic reaction and all of the known enzymes were proteins, the assumption was that this catalytic reaction was carried out by a protein(s) and not RNA. Also, it was assumed that the functions of 5S and 23S rRNAs were probably related to the binding of substrates, i.e., peptidyl-tRNA and incoming aminoacyl-tRNA, at their correct positions, and/or the maintenance of the correct structure so that the presumed "peptidyltransferase protein(s)" could function in a proper way. Thus, the search for the crucial r-protein in the 50S subunit pursued by Staehelin and his co-workers in the late 1960s (Staehelin et al., 1969) continued in the 1970s. After much effort to identify the peptidyltransferase proteins, Nierhaus and his co-workers concluded that L2, L3, L4, L15, and L16, together with 23S rRNA, constitute the peptidyltransferase center (Nierhaus, 1980; Hampl et al., 1981; Schulze and Nierhaus, 1982). However, many other proteins were also found to be strongly stimulatory, and it was not obvious why so many proteins must participate in the catalysis of a single biochemical reaction. In addition, subsequent genetic studies have demonstrated that one of these essential proteins, L15, is completely dispensable for cell growth (Lotti et al., 1983; Ito et al., 1984). In addition, reconstitution studies of *B. stearothermophilus* 50S subunits by Auron and Fahnestock (1981) led to identification of seven r-proteins most directly involved in the peptidyltransferase reaction, but these seven did not include the homolog of *E. coli* L15.

In addition to reconstitution studies, many other experiments such as affinity labeling experiments were carried out in the 1970s in the hope of identifying the r-proteins directly involved in ribosome functions in protein synthesis (for reviews, see Cantor et al., 1974; Cooperman, 1980). I should note that even in the case of affinity labeling experiments, the emphasis was on protein rather than RNA. People occasionally mentioned labeling of rRNA in these earlier experiments (e.g., Pellegrini et al., 1972) but did not pursue it because of the technical difficulty of dissecting and identifying the labeled region of rRNA. With protein portions labeled, at least one could ask and identify which proteins carried the radioactive affinity probes. I remember that Steve Fahnestock carried out affinity labeling of *E. coli* 50S subunits as well as *B. stearothermophilus* 50S subunits with radioactive bromoacetylphenylalanyl-tRNA in my laboratory in 1972. We were surprised to see that the majority of radioactivity was in the RNA fraction, with only a small fraction in the protein fraction. Steve eventually identified the cross-linked *E. coli* protein as L2, as was found by Cantor and his co-workers (Oen et al., 1973), but could not do anything further with the rRNA fraction (S. Fahnestock, unpublished experiments). Thus, the work on the functional analysis of ribosomal components carried out in the 1970s was dominated by the analysis of r-proteins. Gradual acquisition of knowledge on localization of r-proteins within ribosomes, to be described below, also encouraged people to interpret their experiments on ribosome functions in terms of individual r-proteins.

The real change of emphasis in connection with ribosomal functions from r-proteins to rRNA started around 1980. In the summer of 1979, the second major ribosome meeting took place at the University of Wisconsin, Madison. Carl Woese, who had been working on the origin of the genetic code and evolution at the University of Illinois, gave a talk entitled "Just So Stories and Rube Goldberg Machines: Speculations on the Origin of the Protein Synthetic Machinery," which later appeared as an article in the symposium book (Woese, 1980). With the unique title given and the way he criticized the general trend of research on the functional aspects of ribosomes, his talk and the published article must have attracted the attention of many participants. I knew about the early papers of Crick (1968) and Orgel (1968) suggesting that the primitive ribosome could have been made entirely of RNA, and I myself had suggested the importance of 16S rRNA in 30S functions, as mentioned earlier. Yet I did not fully appreciate the importance of evolutionary arguments while working on ribosome reconstitution in the 1960s and early 1970s. Of course, by 1975 I had stopped working (and thinking) on this subject, switching my efforts to genetics and regulation of ribosome synthesis. I remember reading Carl Woese's article and following his arguments; I had to agree with his contention that rRNA must be important for ribosome functions. He did not specifically discuss peptidyltransferase, but by implication it was obvious that even such enzymatic reactions could have been carried out by RNA (at least in the primitive ribosome).

In addition to the results obtained from the reconstitution studies indicating the dispensability of many r-proteins for in vitro protein synthesis, isolation of mutants with r-proteins missing from the ribosomes (Dabbs, 1979; reviewed in Dabbs, 1986) also started to influence our thoughts on the functional role of r-protein, supporting Carl Woese's contention that rRNA is crucially involved in translation function. Another factor that influenced the course of ribosome research was the isolation of mutations affecting rRNA. Because of the presence of seven copies of rRNA operons, it was previously difficult to isolate discrete rRNA mutations (other than those affecting base modifications). Most of the ribosomal mutations, such as antibiotic resistance mutations or conditionally lethal mutations, were in r-protein genes, supporting the importance of r-proteins. However, after isolation of rRNA genes on plasmids combined with advances in techniques for genetic manipulation, it became possible to isolate rRNA gene mutations by using such plasmids, as first pioneered by Edward Morgan and his co-workers (Sigmund and Morgan, 1982). The isolation of rRNA gene mutations in organelle ribosomes also helped this trend (e.g., Sor and Fukuhara, 1984; Montandon et al., 1985).

Of course, by the late 1970s, rRNA genes were cloned, the primary structures of rRNAs were elucidated, and it became technically easier to dissect large rRNA molecules. Thus, for example, labeling of rRNAs with radioactive affinity probes was seriously recognized, and the sites of labeling could be studied. In this way, the importance of the C-1400 region of

16S rRNA as the decoding site was demonstrated by cross-linking of the 5′ anticodon base of a tRNA at the P site to C-1400 of 16S rRNA by Jim Ofengand, Robert Zimmermann, and their co-workers (Ofengand et al., 1979; Prince et al., 1982). Finally, the discovery of the ability of RNA to function as a catalyst (Kruger et al., 1982; Guerrier-Takada et al., 1983) had, of course, a decisive impact on our thoughts on this subject, and people started to think that even peptidyltransferase activity could be catalyzed by 23S rRNA. Thus, from the sites of affinity labeling with peptidyl-tRNA analogs (Barta et al., 1984) and mutations conferring resistance to antibiotics inhibitory to peptidyltransferase (obtained by using multicopy plasmids), Noller and his co-workers suggested a model for the functional organization of the peptidyltransferase site that involves interaction between two widely separated domains of 23S rRNA (Barta et al., 1984; Douthwaite et al., 1985; Noller et al., 1986). If 23S rRNA is in fact the peptidyltransferase, the function of the various proteins required for reconstitution of activity in vitro might be to fold 23S rRNA in the proper way so that these separate domains of the rRNA molecule are brought together to form the catalytic center. Such a model provides an elegant explanation for the requirement for so many different r-proteins to generate a single catalytic activity. Of course, discussion of this subject is not history. The question of how ribosomal components, 23S rRNA, and/or several r-proteins previously implicated really perform the catalytic function to form peptide bonds surely represents one of the major subjects left for future ribosome research.

I should also add that the role of several r-proteins such as S12 (encoded by the *str* locus), which were originally thought to be more directly involved in ribosome function, also became interpreted on the basis of their indirect role in ribosome function. An example might be the maintenance of a local RNA structure (an active center) that is perhaps directly involved in the interaction with mRNA or tRNA. This view is still consistent with the observations that omission of a single protein in the in vitro reconstitution system or mutational alterations affecting that protein give more drastic or specific effects on ribosome function without affecting the assembly or global conformation of rRNA. Actual experimental results supporting this view were recently obtained by Harry Noller's group (Allen and Noller, 1989).

Regarding the possibility of a functionally important interaction of S12 with 16S rRNA, it should be noted that streptomycin may in fact bind to 16S rRNA as originally suggested by Biswas and Gorini (1972). In the original studies, even though S12 was identified as the determinant of streptomycin sensitivity, we showed that S12 (from sensitive *E. coli*) itself did not bind streptomycin (Ozaki et al., 1969). More recently, it was found that Str-R mutations in *Euglena gracilis* chloroplasts are due to an alteration in 16S rRNA (Montandon et al., 1985). By introducing the same base substitution at the equivalent position in *E. coli* 16S rRNA (C-912 to U) by in vitro mutagenesis, Montandon and his co-workers (Montandon et al., 1986) then demonstrated that this RNA alteration in fact renders *E. coli* cells (and ribosomes) streptomycin resistant. Furthermore, Moazed and Noller recently showed that streptomycin protects nucleotides 912 to 915 (of sensitive ribosomes) from attack by chemical probes (Moazed and Noller, 1987). These observations support the suggestion mentioned above, namely, that S12 interacts with a specific region of 16S rRNA and influences its conformation. (It is well known that streptomycin binds nonspecifically to negatively charged nucleic acids, and this property was used extensively by biochemists to remove nucleic acids from crude extracts for enzyme purification. In addition, previous studies had shown that only one molecule of streptomycin binds to a ribosome [see Chang and Flaks, 1972, and studies cited therein], whereas two molecules of streptomycin bound to one molecule of 16S rRNA in the experiments of Biswas and Gorini [1972]. This was part of the reason why the conclusion of Biswas and Gorini was not accepted immediately at that time, and even now this conclusion needs more rigorous experimental proof.)

I have not yet discussed studies on temperature-sensitive r-protein mutants whose ribosomes showed temperature sensitivity when assayed in vitro (e.g., Champney, 1980). Clearly, mutational alterations in these r-proteins affected ribosomal functions (directly or indirectly) in protein synthesis without affecting assembly. There were also many biochemical studies on r-proteins suggesting direct participation of r-proteins in ribosomal functions. For example, acidic protein L7/L12, first detected and then extensively studied by Möller and his co-workers at the University of Leiden (Möller and Maassen, 1986), constitutes a unique stalk structure as visualized by electron microscopy (Strycharz et al., 1978) and may have a direct role in factor-dependent GTP hydrolysis (Möller and Maassen, 1986). Extensive studies carried out on S1 by several laboratories indicate that this protein almost certainly functions directly in translation, probably acting as an mRNA-binding protein (reviewed in Subramanian, 1983). Thus, even though the essential elements in ribosome function are almost certainly rRNAs, r-proteins may surely contribute to both efficiency and flexibility (e.g., regulation) in translation. With many modern technologies together with solid information on the three-dimensional arrangements of r-proteins now available, we expect that studies will soon define precise roles of r-proteins in protein synthesis.

TOWARD ELUCIDATION OF THE FINE STRUCTURE OF RIBOSOMES

In the spring of 1973, Jim Lake proposed that we collaborate to study ribosome structure by using immunoelectron microscopy. Both Jim and David Sabatini had been working in George Palade's laboratory at the Rockefeller University in New York and just recently moved together to the New York University School of Medicine. George Palade was, of course, one of the first who recognized, together with Albert Claud and Keith Porter in the early 1950s, the presence of many spherical granules in cytoplasm under the electron microscope and demonstrated, in collaboration with Philip Siekevitz, that these granules were rich in RNA, which together with other observations by biochemists and microbiologists led to the discovery of ribosomes (for historical accounts of these earlier studies, see Palade, 1958; Tissières, 1974). In fact, David Sabatini and others had started to work on the structure of eucaryotic ribosomes by using electron microscopy in the Palade laboratory before moving to the New York University School of Medicine (Nonomura et al., 1971).

Electron microscopy had been one of the obvious tools to study the structure of ribosomes, and several papers had been published on the size and morphology of *E. coli* ribosomes. Although the earlier studies (e.g., Hall and Slayter, 1959; Huxley and Zubay, 1960) did not reveal any detailed fine structures, the results obtained were useful to give reasonable estimates for the sizes and shapes of ribosomes and their subunits in conjunction with the other physical measurements such as hydrodynamic studies and small-angle X-ray-scattering studies (reviewed in Van Holde and Hill, 1974). In addition, with improved techniques, more recent studies were beginning to show some morphological features reproducibly (e.g., Lubin, 1968; Nonomura et al., 1971).

In the early 1970s, the major goal of ribosome structure studies was to determine three-dimensional arrangements of all proteins within each of the ribosomal subunits. The general conviction was that a knowledge of the three-dimensional organization of ribosomes would be essential for understanding the mechanism by which the ribosomes function in protein synthesis. The specific goal to find protein arrangements was set because almost all of the r-proteins in the ribosomes had been purified and characterized by that time, and it was therefore easier to dissect and analyze the structure by using each of these many proteins as a landmark. With each of the rRNA molecules being so large and their sequences unknown, it was difficult to dissect RNA portions and study their three-dimensional arrangements. I also had an interest in spatial arrangements of r-proteins. When we first constructed the assembly map of 30S subunits, which indicated interdependence of r-proteins during assembly, we suggested that the observed interdependence might reflect physical proximity (Mizushima and Nomura, 1970). However, there was no experimental basis for this suggestion (the alternative was "conformational alterations at a distance" to explain the interdependence), and I wished to see whether this suggestion was correct.

Regarding immunoelectron microscopy, I knew of an impressive study visualizing antibody molecules with electron microscopy (Valentine and Green, 1967) as well as an ingenious study to localize several specific proteins on the phage T4 structure by this technique (Yanagida and Ahmad-Zadeh, 1970). In my own laboratory, Larry Kahan, then a postdoctoral fellow who had training in immunochemistry as a graduate student, had prepared antibodies against most of the *E. coli* 30S r-proteins. We were using these antibodies to carry out comparative studies of *B. stearothermophilus* r-proteins in connection with heterologous reconstitution mentioned earlier (Higo et al., 1973) and, more important, to detect radioactive r-proteins synthesized in vitro in the genetic and regulation studies we were just initiating (Kaltschmidt et al., 1974). Although my main interest was already in genetics and regulation, I agreed to carry out a serious collaboration with Jim Lake, especially because Larry Kahan was very much interested in this project. Thus, we purified our antibodies or prepared *B. stearothermophilus-E. coli* hybrid 30S subunits for use as controls and sent them to Jim Lake first at New York and later at the University of California, Los Angeles. He started to use electron microscopy to localize r-proteins. The first results of this joint venture were published in late 1974 (Lake et al., 1974). At about the same time, George Stöffler in Wittmann's laboratory, who had also prepared antibodies against many purified r-proteins, started similar immunoelectron microscopic studies to localize r-proteins within ribosome structures (Tischendorf et al., 1974). Larry Kahan continued the collaboration with Jim Lake after he joined the Department of Physiological Chemistry at Wisconsin. Although another postdoctoral associate, Bill Strycharz, in my laboratory also started collaboration with Jim Lake for mapping of 50S r-proteins, I was too busy with genetics and regulation studies and was largely detached from the joint project.

Mapping of r-proteins by specific antibodies was straightforward in principle, but getting reliable results was another matter. Thus, the mapping locations obtained by the two groups quite often showed serious discrepancies. This was partly due to technical problems, such as the presence of contaminating r-proteins in r-protein preparations used to raise polyclonal antibodies. "Specific" antibodies occasionally reacted with other r-proteins in addition to the nominal r-protein one was analyzing. In addition, the general low resolution in electron micrographs with respect to the morphologies of ribosomal particles, especially 30S subunits, caused errors in interpretation of the ribosome structure seen in electron micrographs and hence confusion in protein localization. However, after Jim Lake (and others, such as Vladimir Vasiliev at the Institute of Protein Research, Poushchino, USSR; see Vasiliev, 1974) documented an asymmetric nature of the 30S subunit (Lake, 1976; Leonard and Lake, 1979), discrepancies became less serious. The models showing morphological features of ribosomal subunits used by various investigators also became very similar to those of Jim Lake, providing a reliable basis for later structural and functional studies.

Around 1980, when Bill Strycharz left our laboratory, we stopped the collaboration with Jim Lake on r-protein mapping. Soon thereafter, Jim started to devise a method to map specific regions of rRNA by using synthetic biotinylated oligodeoxynucleotides as probes (Oakes et al., 1986a), and the progress in r-protein mapping by the Lake-Kahan group became slower (Oakes et al., 1986b). Stöffler and his coworkers continued their efforts and have now localized a majority of r-proteins (Stöffler and Stöffler-Meilicke, 1986; Stöffler-Meilicke and Stöffler, 1987; see also Walleczek et al., 1988).

Immunoelectron microscopy was used not only for r-protein mapping but also to localize other critical regions within the ribosome structures. For example, 5′ and 3′ ends of rRNAs (Mochalova et al., 1982; Shatsky et al., 1979; Shatsky et al., 1980; Olson and Glitz, 1979), the location of unique dimethyladenine residues in 16S rRNA (Politz and Glitz, 1977), and the protein exit site (Bernabeu and Lake, 1982) of ribosomes were determined by this technique by several investigators, and sites interacting with mRNA and tRNAs were inferred, giving visual three-dimensional models for functioning ribosomes.

In addition to the immunoelectron microscopic approach, there were several other approaches to elucidate spatial arrangements of r-proteins. One popular approach used in the early 1970s by several investigators was the cross-linking of neighboring r-proteins by using bifunctional reagents. Active investigators included people who had been engaged in the purification of r-proteins: Robert R. Traut, who had moved from Geneva to the University of California at Davis; Charles G. Kurland, who had moved from Madison to the University of Uppsala; and Gary Craven at the University of Wisconsin. In addition, people such as Joel Flaks at the University of Pennsylvania and Donald Hayes at the Institute of Physical Chemistry in Paris were also among the cross-linking study groups. (The early work was reviewed by Traut [1974] and Traut et al. [1980].) This approach started to generate large amounts of data, but the significance of the results was sometimes difficult to evaluate because the yield of cross-linked products was generally low and identification of cross-linked proteins was often not unambiguous. Because our laboratory had antibodies, Larry Kahan and then Bill Strycharz quite often helped other investigators, especially Traut's group, in identifying cross-linked proteins.

Two other, more sophisticated approaches were also initiated in the early 1970s. One was the use of a fluorescence energy transfer technique carried out by Charles Cantor and his co-workers at Columbia University to estimate the distance between two r-proteins that were labeled with different fluorescent dye molecules. In the initial phase of their study, we helped Cantor's group by providing purified r-proteins and reconstituting 30S subunits containing fluorescent dye-labeled r-proteins. The second was the use of a neutron-scattering technique to measure the distance between two deuterated r-proteins in a reconstituted ribosomal subunit. With both approaches, the plan was to construct a three-dimensional map of r-proteins from measurements of distances of sufficient numbers of protein pairs. The second approach, the neutron-scattering approach, was designed and initiated by Peter Moore and Donald Engelman at Yale University. The feasibility of this approach was first published in 1972 (Engelman and Moore, 1972), and the first results appeared in print in 1975 (Engelman et al., 1975). I remember the occasion when I learned for the first time about Peter Moore's plan to construct a three-dimensional r-protein map of the 30S subunit. Peter Moore had just finished his postdoctoral time in Europe and taken a faculty position at Yale. I was very impressed with the plan, but without background knowledge about the scattering technique I could not guess how difficult the project would be, how long it would take to be completed, and in fact whether it would ever be completed.

Although Cantor's group obtained a partial map of protein localization (Huang et al., 1975), they did not continue the project through to completion. This was unfortunate, since continuation would have produced results that could have been compared with the neutron-scattering results. As for the latter approach, Peter Moore, Donald Engelman, and their co-workers continued their efforts persistently and, as is well known, finally completed the monumental work of localizing all 21 30S r-proteins in 1987 (Capel et al., 1987; Moore, 1987). It took 15 years after their initial announcement to complete the project, but it was so fruitful. They compared their model with the models obtained by the immunoelectron microscopic studies and with the large amounts of data obtained by cross-linking studies and presented a reasonable interpretation for these data. They also compared their model with the assembly map that we constructed many years before and concluded that the assembly dependence reflects physical proximity of proteins; I was pleased to see this conclusion. Thus, their three-dimensional model for r-protein localization within the 30S subunit, together with ribosome structure models obtained by immunoelectron microscopy as well as secondary structure models of rRNAs to be mentioned below, now appears to give a solid foundation for interpreting many previous experimental observations as well as for guiding future experimentation to refine and probe further details of 30S subunit structure.

Localization of r-proteins in the 50S subunits by using the same neutron-scattering approach was also initiated in Wittman's Abteilung in Berlin (Nowotny et al., 1986). I should note that during the course of intensive studies on ribosome structures, new strategies and technical innovation and improvements were developed, as exemplified by the neutron-scattering studies. Thus, these strategies and the experience obtained during these many years of studying ribosome structures have also been (and will continue to be) useful to the studies of other complex biological structures.

I have already mentioned the elucidation of the primary structure of *E. coli* 16S and 23S rRNAs. The achievements were almost immediately followed by a dramatic development of construction of secondary structure models for these rRNAs. The presence of secondary structure in RNA (and specifically in rRNA from rat liver ribosomes) was first demonstrated by Paul Doty and his co-workers at Harvard in 1959, and a model for rRNA structure containing hairpinlike helices connected by flexible single-stranded regions was proposed (Doty et al., 1959; Fresco et al., 1960). An essentially similar model was also proposed by Spirin and his co-workers in Moscow (see Spirin, 1964a, for a summary of their early studies). Subsequent studies on the secondary structures of isolated rRNA and rRNA in situ were essentially an extension and elaboration of the initial model and indicated the presence of extensive base-paired helical regions in rRNAs. With the progress of sequencing of *E. coli* 16S rRNA (by RNA fingerprinting techniques), specific secondary structure models were also proposed (e.g., Fellner, 1974). However, the first reliable models were those proposed by Woese, Noller, and their co-workers on the basis of extensive phylogenetic evidence together with supporting evidence based on reactivity of rRNAs to chemical and enzymatic probes (Woese et al., 1980; Noller et al., 1981). Carl Woese had been carrying out extensive comparative studies on (partial) sequences of rRNAs from many different bacterial species, with the conviction that the structures of functionally important regions of 16S rRNA must be conserved during evolution (the initial results were published in 1975 [Woese et al., 1975]). When the primary sequence of *E. coli* 16S rRNA was completed by Noller's group, Woese and Noller used the enormous amounts of data accumulated by that time and presented forcible arguments for the power of the phylogenetic approach to elucidate the secondary structure of rRNA (Woese et al., 1980; Noller and Woese, 1981; Woese et al., 1983). I believe that the impact of the studies must have been strong not only to investigators studying ribosomes but also to the people outside the ribosome field. Later, when I studied the structure of mRNA (the target site for certain repressor r-proteins), we cloned (as DNA) and sequenced the corresponding regions from other bacterial species and used a similar argument to deduce secondary structures (e.g., Cerretti et al., 1988). In addition, construction of compensatory mutations by using in vitro recombinant techniques also became a standard experimental approach to test the validity of hypothetical secondary structures of a variety of RNAs.

Of course, secondary structures of rRNAs were also studied by other groups, notably Chantal Ehresmann, Jean Pierre Ebel, and their co-workers at Strasbourg (e.g., Stiegler et al., 1981) and Brimacombe and his co-workers in Wittmann's Abteilung in Berlin (e.g., Glotz et al., 1981), and with ever-increasing amounts of data, we now have very reliable models for the secondary structures of rRNAs. In addition, as I have already mentioned, Noller's group localized the sites in 16S rRNA involved in the interaction with each of 20 30S r-proteins during assembly and used this information, together with Peter Moore's three-dimensional map of r-proteins, to construct a preliminary model for the three-dimensional folding of 16S rRNA in the 30S subunits (Noller et al., 1987). Other approaches, such as

RNA-RNA cross-linking approaches (Brimacombe et al., 1986) and DNA hybridization microscopy (Oakes et al., 1986a), are also currently being used to elucidate the three-dimensional structure of rRNAs. These studies as well as other approaches to study fine structures of ribosomes, such as X-ray crystallographic analysis of ribosome crystals initiated by Yonath, Wittmann, and their co-workers (Yonath et al., 1986), are being pursued intensively.

In view of the striking progress in this field as well as powerful new technologies available, there is no doubt that we shall soon be able to obtain reliable three-dimensional structural models for both 30S and 50S subunits, displaying folding of rRNA chains as well as locations of all r-proteins, and that these models will continue to be refined in resolution and accuracy. With such models in hand, we should be able to discuss experimental results on structure and function at the level of individual amino acids and nucleotides in their three-dimensional positions. Compared with the time when I started to work on ribosomes in the late 1950s, the progress in this field is truly remarkable.

The fundamental goal of ribosome research is to understand the functions of ribosomes from their structures. I have described how people first tried to define functions of ribosomes and how they then struggled to understand mechanisms involved in some of the ribosome functions thus defined. With impressive progress in our knowledge on structures, it should become possible to understand how the ribosome really works in protein synthesis. There may be two directions in ribosome research in this respect. One is to simplify the complexity of ribosomes and to identify key elements responsible for carrying out a basic biochemical reaction involved in ribosome functions. Identification of key segments (or nucleotide residues) of 23S rRNA or r-proteins or both that are involved in the peptidyltransferase reaction, as mentioned above, is an example. As was done for r-proteins by using reconstitution techniques, recombinant DNA technologies can now dissect rRNAs and delete or modify rRNAs (and r-proteins), enabling us to design and construct artificial (and simpler) ribosomes or parts of ribosomes. Understanding of the mechanics of protein synthesis in terms of behavior of amino acids and nucleotides may be achieved in this way.

The other direction is to understand the significance of complexity of ribosome structures. Even though some r-proteins were demonstrated to be dispensable for protein synthesis (and cell growth) by reconstitution experiments or mutational analysis, these r-proteins must be present in ribosomes for some significant reason, and the complexity of ribosome structure must have evolved for the advantage of organisms. In addition to protein synthesis per se, ribosomes are expected to interact with other cellular components and structures for some other functions. One clear example is the interaction of ribosomes with the protein export machinery, including specific cytoplasmic components (such as signal recognition particles in eucaryotes [Walter and Blobel, 1981] or trigger factor in *E. coli* [Lill et al., 1988]) as well as the cell membrane or the endoplasmic reticulum membrane (e.g., Hortsch et al., 1986; for a review of protein export, see Randall et al., 1987). Some specialized structures of ribosomes may play a role in such interactions, contributing to the complexity of ribosome structure. (It is interesting to note in this regard that the *secY* [*prlA*] gene, which encodes an integral membrane protein essential for protein export [Ito et al., 1983], is in the *spc* r-protein operon [Shultz et al., 1982; Cerretti et al., 1983].)

Translation is an important step at which efficient and sophisticated regulation of gene expression can take place. Even in procaryotes, where all of the regulation of gene expression was once thought to be transcriptional, translational regulation plays a major role, as exemplified by its extensive use in r-protein synthesis in *E. coli*, as I will relate later. In eucaryotes, where translational regulation is a logical way to regulate gene expression efficiently, its dramatic use started to be recognized only recently (see, e.g., articles in Ilan, 1987). It is certain that translational regulation is extensively used in eucaryotes in a variety of ways. Thus, the complexity of ribosome structures may also be related to the abundance and varieties of translational regulation in many gene expression systems, and studies on individual regulatory systems might reveal the significance of ribosome structures in this context.

Protein synthesis is no doubt the major synthetic reaction in growing cells. For example, in exponentially growing *E. coli* cells (in very rich media), more than 90% of the energy produced is consumed for making proteins (Ingraham et al., 1983). Thus, we expect that some important regulatory mechanisms operate to regulate the global rate of protein synthesis. There are abundant examples of such global regulation in eucaryotic cells, and this must also be the case for *E. coli* cells. In addition, protein synthesis is central in cellular growth, and it is expected to interact with other important regulatory mechanisms controlling cellular growth. Thus, the ribosomes as the machinery of translation may receive a variety of regulatory signals and also generate other regulatory signals for proper growth. An example in *E. coli* is its participation in stringent control, responding to amino acid deficiency in the environment and generating a signal molecule, guanosine tetraphosphate (ppGpp), to redirect many cellular activities to adjust to the starvation conditions. At least one r-protein of *E. coli*, L11, was shown to be important in the synthesis of ppGpp (Friesen et al., 1974), even though it is apparently dispensable for protein synthesis and cell growth (Dabbs, 1979). Thus, studies of structure and function of ribosomes will inevitably overlap with studies of regulation of translation and cellular growth. It is known that regulation of total rate of protein synthesis, that is, regulation of growth rate, is intimately related to regulation of ribosome synthesis in *E. coli*, and probably in eucaryotic cells too. In the following sections, I will trace the history of research carried out on the synthesis of ribosomes and its regulation, which started independently of structure-function studies but nevertheless has been and will be interconnected to these studies.

BEGINNING OF RESEARCH ON RIBOSOME SYNTHESIS

The origin of the major questions in the current studies on the regulation of ribosome synthesis can be traced back to the systematic analysis of bacterial growth initiated by Ole Maaløe, Niels Ole Kjeldgaard, and their associates at the State Serum Institute in Copenhagen in the mid-1950s. In 1958, they published two classical papers that described the results of their analysis of cell size and of DNA and RNA content in *S. typhimurium* during balanced growth in several different media and during transition between two different media that supported cell growth at different rates (Schaechter et al., 1958; Kjeldgaard et al., 1958). They discovered that the growth rate, and hence the rate of protein synthesis, is roughly proportional to the RNA content of the cell in a variety of media they tested and that the rate of RNA accumulation increases immediately upon nutritional shiftup. Similar studies were carried out by Boris Magasanik and Fred Neidhardt, then at Harvard Medical School in Boston, who reached the same conclusion (Neidhardt and Magasanik, 1960). The concept of constant efficiency of ribosome action under various nutritional conditions was proposed. This concept would lead to many later studies by bacterial physiologists and molecular biologists on the regulation of ribosome synthesis and, eventually, to our current understanding of the mechanism of growth rate control, as I will elaborate later.

Of course, right after the discovery of ribosomes in the 1950s, several independent studies were initiated on the biosynthesis of ribosomes with different goals in mind. For example, many investigators were interested in actual pathways and biochemical mechanisms used for the synthesis of such a discrete supramolecular structure in vivo. Again perhaps because of the analogy to RNA virus particles, some wondered whether ribosomes self-replicate. A related question was whether ribosome synthesis involved dissociation of the rRNA and r-proteins in preexistent ribosomes. One approach to this kind of question that was new in the late 1950s was to carry out density transfer experiments. The famous Meselson-Stahl density transfer experiments to prove semiconservative DNA replication had just been successfully carried out in 1957 (Meselson and Stahl, 1958). Jim Watson was also interested in using this approach to study ribosome synthesis. David Schlessinger, then a graduate student in Jim's laboratory, spent the summer of 1959 in Matt Meselson's laboratory, then at the California Institute of Technology, Pasadena, to try density transfer experiments but without definitive success because of uncontrolled degradation of ribosomes in the CsCl density gradients. As I have already mentioned, Sydney Brenner and Francois Jacob came to the California Institute of Technology the following summer and succeeded in finding conditions (Mg^{2+} concentration as a crucial factor) to band ribosomes in CsCl, but observed two bands, A and B, in the gradients. Two years later, after he had moved to Harvard, Matt Meselson and I showed that the A band contains 42S and 23S core particles derived from 50S and 30S subunits, respectively, as I mentioned earlier. During my stay at Harvard that summer, we also carried out density transfer experiments successfully and demonstrated that at least core particles (containing rRNA and the major portion of r-proteins) are conserved during bacterial growth. There was no separation of rRNA and r-proteins (Meselson et al., 1964). The project was completed several years later by Kaempfer, Meselson, and Raskas, who used sucrose gradient centrifugation to separate ribosomal subunits labeled with heavy isotopes from those labeled with light isotopes, thus avoiding the problem of protein splitting during CsCl equilibrium centrifugation (Kaempfer et al., 1968). Conservation of the whole bacterial ribosome during bacterial growth was demonstrated, but because of the limited sensitivity of the technique, the possibility of exchange of a few r-proteins during bacterial growth could not be excluded.

Regarding autonomous replication of ribosomes, i.e., rRNA synthesis from RNA template, such a possibility was first suggested to be unlikely by Cyrus Levinthal from his experiments using actinomycin D (Levinthal et al., 1962) and then by the demonstration of the presence of genes (DNA) encoding both 16S rRNA and 23S rRNA using RNA-DNA hybridization by Yankofsky and Spiegelman (1962a, 1962b). For the current ribosome researchers, it may perhaps be difficult to imagine that the proof of the existence of rRNA genes was so important at that time, but that was in fact the case. Thus, soon after the demonstration of RNA-DNA hybridization by Benjamin Hall and Sol Spiegelman at the University of Illinois (Hall and Spiegelman, 1961), Spiegelman made efforts to prove this to be the case for *E. coli*; then, together with F. M. Ritossa, he was able to obtain strong evidence that the nucleolar organizer in *Drosophila melanogaster* represents an rRNA gene cluster (Ritossa and Spiegelman, 1965), as suspected from the work of Don Brown and John Gurdon on *Xenopus laevis* carried out just a year before (Brown and Gurdon, 1964).

I have mentioned the pioneering work done by the group at the Carnegie Institution of Washington on ribosome biosynthesis initiated in the late 1950s and the confusion on the nature of pulse-labeled RNA. Because they did not distinguish pulse-labeled mRNA from precursor rRNA, some of their conclusions obtained from kinetic analysis had to be modified. Nevertheless, as will be mentioned below, their approaches were followed and their major conclusion was confirmed by later workers in the 1960s, namely, that ribosome assembly in vivo is probably a stepwise process.

STUDIES ON RIBOSOME BIOSYNTHESIS IN THE 1960s

In the 1960s, one of the dominant activities in studies of ribosome synthesis was essentially a continuation of the type of experiments initiated by the Carnegie group, although other studies I have mentioned, such as growth rate control and identification and mapping of rRNA and r-protein genes, were also actively pursued. In addition, after the discovery of the *relA* gene (originally called the RC locus) by Gunther Stent and Sydney Brenner in 1961, which defined the phenomenon of stringent control clearly (Stent and Brenner, 1961), much effort was devoted to elucidating the mechanism involved in this stringent control and attempting to formulate unified models to explain both stringent control and growth rate-dependent control, as I will describe later. I should also mention here that studies on rRNA synthesis during amino acid starvation in *relA* mutant strains led to the discovery of protein-deficient particles (relaxed particles) similar to chloramphenicol particles (Nakada et al., 1964). Thus, many experiments were carried out on these relaxed particles (and also on chloramphenicol particles) as possible intermediates in the process of ribosome assembly in vivo.

Regarding the studies of in vivo ribosome assembly pathways, three strategies were thought to be useful by analogy to the strategies used to elucidate metabolic pathways of low-molecular-weight compounds in the 1940s and 1950s. The first was kinetic analysis of the flow of radioactive RNA precursors into mature ribosomes as initiated by the Carnegie group; the second was the use of inhibitors in the hope of causing accumulation of intermediate particles, as was presumed to be the case for chloramphenicol (or relaxed) particles; and the third was genetic, exemplified by isolation of cold-sensitive mutants with defects in ribosome assembly. This research was carried out in the late 1960s, as I mentioned earlier in connection with in vitro assembly studies.

Among several groups that followed the kinetic studies initiated by the Carnegie group, two were the most active; one was Shozo Osawa's group at Hiroshima University, Japan, and another was David Schlessinger's group at Washington University, St. Louis, Mo. These groups used conditions in which radioactive pulse-labeled RNA was thought to be mainly rRNA (rather than mRNA) and detected several particles that behaved like precursors to mature ribosomal subunits. In this way, two different particles, 21S and 26S, were proposed as precursors to 30S particles (some workers detected only one precursor particle); similarly, 32S and 43S particles were proposed as precursors to 50S particles. Although some of these particles were isolated and r-proteins found in the particles were analyzed by column chromatography (earlier studies were reviewed by Osawa [1968]), most of these earlier results were ambiguous, mainly because the structures and protein compositions of mature ribosomes had not been well characterized at that time. The main conclusion was that ribosome assembly was a stepwise process and "intermediate particles" probably exist, reflecting the presence of kinetic holdup points. As mentioned earlier, the kinetics of ribosome assembly was later studied in vitro by using reconstitution systems, and the similarity (or differences) between intermediate particles detected in vitro and those detected in vivo was examined (e.g., Nierhaus et al., 1973). Although the observed similarity was satisfying, characterization of the presumed in vivo intermediate particles was never carried out quantitatively, mainly because of technical difficulties (e.g., the steady-state amounts of these particles found in exponentially growing cells were very small). The important question of whether each of the proposed intermediate particles represents a homogeneous molecular species or a collection of heterogeneous species (as in the case of chloramphenicol particles [Hosokawa and Nomura, 1965] or 21S particles accumulated in spectinomycin-resistant, cold-sensitive mutants mentioned earlier [Nashimoto et al., 1971]) was never answered. Therefore, in my opinion, the proof of presumed "normal intermediate particles" studied in those days as real intermediate particles in the normal ribosome assembly pathways is yet to be done.

I should also mention that a similar kinetic approach was used in the 1960s through early 1970s to study the pathways in the synthesis of ribosomes in eucaryotic systems. These studies were at least more productive than those done in *E. coli*. Short pulse-labeling detected large precursor forms (e.g., 45S RNA in human cells) of rRNA molecules, followed by their processing to mature rRNAs (28S, 18S, and 5.8S rRNAs in human cells), and gross pathways involved in maturation were elucidated. Both Bob Perry's laboratory (Perry, 1962) and Jim Darnell's laboratory (Scherrer et al., 1963) pioneered in this approach, followed by many investigators, such as S. Penman, G. Attardi, B. E. H. Maden, and J. R. Warner. Thus, together with the work on isolated rRNA genes from *Xenopus* cells in 1968 (Birnstiel et al., 1968; Brown and Weber, 1968), linkage of 28S and 18S rRNA genes as well as their cotranscription was firmly established by the end of the 1960s.

In the case of *E. coli*, detection of such a large precursor RNA was difficult in earlier studies. In fact, as already mentioned, the initial studies of the Carnegie group suggested the opposite: the joining of small (hypothetical)-subunit RNAs to form larger rRNA molecules. Although the presence of a transcription unit consisting of 16S, 23S, and 5S rRNA genes was suggested around 1970 from indirect experiments using rifampin (e.g., Doolittle and Pace, 1971), the real proof for this conclusion came only when Dunn and Studier at Brookhaven National Laboratory and David Schlessinger and his co-workers at Washington University detected a large rRNA precursor (30S precursor RNA) by pulse-labeling in a mutant of *E. coli* defective in RNase III (Dunn and Studier, 1973; Nikolaev et al., 1973). David Schlessinger continued his studies on the pathways involved in ribosome assembly in vivo, analyzing details of rRNA processing from the primary transcript to produce mature rRNAs. In *Bacillus subtilis*, physical linkage of the three (16S, 23S, and 5S) rRNA genes was shown earlier by Michio Oishi and his co-workers, then at the Public Health Research Institute of the City of New York (Colli and Oishi, 1969; Colli et al., 1971), using RNA-DNA hybridization techniques combined with CsCl equilibrium centrifugation methods to separate DNA-RNA hybrids from DNA.

Particles similar to chloramphenicol particles and relaxed particles were detected after treatment of *E. coli* cells with a variety of inhibitors or under certain harmful conditions, and different names were given to these particles, depending on the particular conditions used (such as "streptomycin particles" for particles accumulated after streptomycin treatment). I remember reading many papers published on this subject and feeling a sense of frustration because of redundancy and confusion. Although we originally suggested that chloramphenicol particles might be related to precursors for normal ribosomes (Nomura and Watson, 1959; Kurland and Maaløe, 1962), and in fact density transfer experiments that I did with Keiichi Hosokawa at Osaka University in the early 1960s indicated direct conversion of at least a portion of the particles to normal ribosomes after removal of the drug (Nomura and Hosokawa, 1965), we also found that during prolonged chloramphenicol treatment, preexistent ribosomes break down (Nomura and Watson, 1959). In addition, we found that the usual chloramphenicol particle preparations were contaminated with proteins that could be separated by electrophoresis but not easily by centrifugation techniques. We also showed that the particles were heterogeneous (Hosokawa and Nomura, 1965). Thus, the system was complex, and analytical techniques had not been sufficiently developed for carrying out clean and definitive experiments. Although there was nothing wrong in the papers that we had published on chloramphenicol particles, and I was no longer working on this subject at the University of Wisconsin, I felt a sense of guilt because I was the one who first worked on such particles. I remember that I became very reluctant to accept conclusions derived from such complex systems and even became reluctant to do experiments using intact cells for analysis of ribosome synthesis.

Studies on relaxed particles and nascent rRNA isolated from relaxed particles (or chloramphenicol particles) in the mid-1960s led to some excitement and debate. Around 1965, several people suggested that nascent rRNA served as mRNA for r-proteins and that the messenger function stopped immediately after a single translation event because of inhibition by the r-proteins produced or by methylation of the nascent (submethylate) rRNA (Roberts, 1965; Nakada, 1965). This concept was even described in one of the most popular textbooks on molecular biology at that time (Watson, 1965). The model was attractive because it could explain coordinated synthesis of rRNA and r-proteins. Perhaps for this reason, several papers appeared claiming experimental evidence for the model, that nascent rRNA (isolated from relaxed particles or chloramphenicol particles) stimulated the synthesis of r-proteins in vitro (e.g., Otaka et al., 1964; Nakada, 1965; Muto, 1968). There was even a claim that relaxed particles were by themselves able to synthesize r-proteins that they lacked without participation of complete mature ribosomes (Nakada, 1965). Before the discovery of mRNA, rRNA was thought to be the information carrier between genes and proteins. Therefore, the model that nascent rRNA might be the message for r-proteins was a natural extension of the ideas prevalent at that time. In addition, the analogy of ribosomes to RNA virus particles might also have contributed to the plausibility of such a claim. It had already been known that the genomes of RNA phages such as f2 functioned as mRNA and that this messenger function ceased when the intracellular concentration of one of its products, the phage coat protein, increased, leading to assembly of mature phage particles.

Of course, even then the coding capacity of rRNA appeared to be insufficient to produce all of the r-proteins. However, the possibility existed that rRNA genes present in multiple copies on the *E. coli* genome were heterogeneous and that some of the multiple bands seen by Waller on starch gels represented essentially the same r-proteins with minor sequence differences due to heterogeneity of rRNA. I have already mentioned the general reluctance to accept Waller's demonstration of heterogeneity of r-proteins in the mid-1960s. Even accepting the presence of many r-proteins, it was not easy to exclude completely a modified form of the model, namely, that rRNA codes for some, but not all, r-proteins. Perhaps because of its attractiveness, the model had supporters even after more cautious investigators repeated the published experiments and reached conclusions contrary to the original experimental results (Manor and Haselkorn, 1967; Sypherd, 1967). I remember the very interesting talk given by Gunther Stent at the Rutgers symposium in the fall of 1966. He combined this model with his major thesis at that time, obligatory coupling of transcription with translation (Stent, 1964), and explained almost everything known at that time on the regulation of rRNA synthesis, stringent control, growth rate-dependent control, nutritional shiftup, and the ability of chloramphenicol to relax RNA regulation (Stent, 1967). I enjoyed listening to Gunther Stent's talk, but I was not tempted by regulation studies. In the same symposium, I presented the reconstitution experiments and was happy that they were generating solid data that did not require much speculation. Nevertheless, the model and the talk given (and the accompanying article) by Gunther Stent were very stimulating to me. I (and undoubtedly others) thought about mapping (and isolating) rRNA genes and r-proteins genes to test the model. The feasibility of such experiments had just been demonstrated for *B. subtilis* by Michio Oishi and Noboru Sueoka, then at Princeton University, as well as Julius Marmur and his colleagues at Albert Einstein College of Medicine (Oishi and Sueoka, 1965; Dubnau et al., 1965). These workers mapped rRNA genes by monitoring their replication in a synchronized culture upon transfer from light to heavy media; in fact, both the majority of rRNA genes and several putative r-protein genes were shown to be clustered near the origin of chromosomal replication. This rough mapping was at least consistent with the hypothesis that rRNA codes for r-proteins. In *E. coli*, however, rRNA genes had not been mapped, and I became involved in this task years later. In the mid-1970s, we isolated transducing phages carrying many r-protein genes without carrying rRNA genes; conversely, rRNA genes isolated by us as well as by others did not detectably encode r-proteins. Although the hypothesis had already lost credibility by that time, I was glad to see definitive evidence to exclude it. Nevertheless, I should point out that, as I shall describe later, we now know that r-protein gene expression is feedback regulated by r-proteins mostly at the level of translation and thereby is coupled with ribosome assembly; so there is, in a sense, a similarity between the current accepted model and this old (incorrect) hypothesis.

Regarding stringent control, when Stent and Brenner demonstrated the presence of the *relA* gene in 1961, they suggested a model, in analogy to the operon theory of Jacob and Monod, that uncharged tRNA is a repressor of transcription of rRNA and amino acids are inducers (Stent and Brenner, 1961). The proposal was made on the basis of their realization that starvation for any 1 of 20 amino acids caused inhibition of RNA synthesis and that a single mutation (*relA*) abolished this regulation, as in the case of mutations in repressors or operators in "conventional" operons. Simultaneously, a similar proposal was made by Kurland and Maaløe (1962) on the basis of the effects of chloramphenicol on RNA synthesis. Stimulation of RNA synthesis by chloramphenicol was explained by an increase in the degree of charging of tRNA caused by the inhibition of protein synthesis by the drug. Soon after these proposals, the model was tested in vitro, and initial experiments gave positive results; uncharged tRNA inhibited RNA synthesis in vitro, and charged tRNA gave only a weak inhibition (Tissières et al., 1963).

Furthermore, participation of uncharged tRNA in stringent control was convincingly demonstrated by Fred Neidhardt and his co-workers, then at Purdue University (Fangman and Neidhardt, 1964; Eidlic and Neidhardt, 1965). Thus, the model of uncharged tRNA as repressor for the regulation of RNA synthesis appeared to be plausible. In their publicized monograph published in 1966, Maaløe and Kjeldgaard (1966) explained growth rate-dependent control of rRNA synthesis on this basis, arguing that cellular concentrations of amino acids (relative to their acceptor tRNAs) are the crucial factors responsible for growth rate dependency of rRNA synthesis. However, subsequent in vitro studies on the effects of tRNA on RNA synthesis gave variable results and eventually led to the conclusion that inhibition of RNA synthesis during amino acid starvation was probably not due to direct inhibition of transcription by uncharged tRNA (e.g., Bremer et al., 1966). It was around this time that Gunther Stent formulated another explanation for stringent control (and growth rate-dependent control) on the basis of the notion of obligatory coupling of transcription to translation, as mentioned above.

Of course, as we now know, stringent control acts on rRNA and tRNA synthesis, and the synthesis of most mRNA is not inhibited during stringent conditions; in fact, synthesis of some mRNAs, e.g., *his* operon mRNA, is stimulated (e.g., Stephens et al., 1975). Therefore, many in vitro RNA synthesis experiments using *E. coli* DNA or phage T4 DNA as templates carried out at that time were probably not relevant to the question. The explanation of stringent control (and growth rate-dependent control) that Gunther Stent suggested was also mostly based on the presumption that stringent control acts on mRNA synthesis. Only later, around 1968, did experimental results contrary to this presumption start to appear. Then, it soon became dogma that mRNA synthesis in general is not under stringent control (the earlier studies on stringent control were reviewed by Edlin and Broda [1968]).

Thus, although the 1960s was the golden age for the field of protein synthesis, the field of biosynthesis of ribosomes and its regulation was, in general, in a confused state. There were various interesting ideas and models but not enough definitive experiments to prove or disprove these various models, at least partly because of the lack of solid background information on the structural components and their genes. The field had to await isolation of rRNA and r-protein genes as well as the development of various new technologies, including gene manipulation.

ISOLATION OF r-PROTEIN GENES FOR REGULATION STUDIES

In 1971, I spent about 6 months in the Department of Molecular Biology at Aarhus University, Denmark. At that time, my laboratory was very busy studying the structure and function of ribosomes as well as the mechanism of the ribosome assembly reaction, using the recently developed reconstitution systems. However, I decided to take a sabbatical leave in order to think about my long-range research plans. Kjeld Marcker, together with Alan Smith, had just recently demonstrated at Cambridge, England, that protein synthesis in eucaryotes, as in procaryotes, initiates with a unique methionine tRNA (Smith and Marcker, 1970; Brown and Smith, 1970), and then moved to the newly created Department at Aarhus University. He and his co-workers, including Alan Smith, were starting several projects related to gene expression in eucaryotic systems. Switching to eucaryotic systems, as many other molecular biologists did around that time, was a tempting possibility for me, and I was interested in learning the necessary techniques as well as questions and systems for future research.

However, my short sabbatical stay in Aarhus affected my subsequent research course differently than I had originally anticipated. This was because of the influence of daily contacts with Niels Ole Kjeldgaard and his colleagues. Niels Ole had also just moved from Copenhagen, where he, together with Ole Maaløe, had spent many years systematically analyzing the biosynthesis of macromolecules in relation to bacterial growth and its control, as I have described earlier. Around that time, one exciting development in the field of regulation was the discovery of ppGpp and pppGpp in 1969 by Mike Cashel and Jon Gallant at the University of Washington, Seattle. While studying nucleotide metabolism during amino acid starvation, they discovered the formation of two new compounds, which they called magic spots I and II, and suggested that they are involved in the inhibition of the synthesis of RNA (Cashel and Gallant, 1969). These two compounds were subsequently identified as ppGpp and pppGpp, respectively (Cashel and Kalbacher, 1970). With this discovery, many investigators started to study the role of these compounds in the regulation of rRNA synthesis as well as in stringent control in general. Kjeldgaard's group was also studying the relationship between ppGpp and rRNA regulation extensively. Talking with the people in Kjeldgaard's laboratory, my interest in the biosynthesis of ribosomes was rekindled. I realized that the problems of the regulation of ribosome synthesis were still there, by then more clearly defined yet unsolved. I learned about the article just published by Ole Maaløe, "An Analysis of Bacterial Growth" (Maaløe, 1969), which students in the laboratory were reading as gospel. In this article, Maaløe reappraised the question of growth rate-dependent control of rRNA synthesis and abandoned his previous position, that is, the unified explanation of stringent control and growth rate-dependent control on the basis of availability of amino acid relative to uncharged tRNA acting as repressor. To explain the apparent growth rate dependency of r-protein (and rRNA) synthesis rates, Maaløe proposed the passive control model. He hypothesized that r-protein promoters are not directly controlled, but passively transcribed; under nutritionally rich conditions, many biosynthetic operons are repressed; hence, more RNA polymerase becomes available for transcription of r-protein genes, and the opposite situation takes place under nutritionally poor conditions. In addition, to explain coordination of rRNA and r-protein synthesis, he suggested the possibility that one of the r-proteins is an inducer of rRNA synthesis and that this inducer is constantly synthesized and, after functioning as the inducer, removed by incorporation into new ribosomes. I should mention that a few years earlier, Robert Schleif, then at the University of California, Berkeley, had demonstrated that relative synthesis rates of r-proteins (measured as bulk r-proteins) are growth rate dependent and hence are apparently coordinated with the synthesis of rRNA (Schleif, 1967). To formulate the passive control model, Maaløe used Schleif's observation and proposed that growth rate-dependent (passive) control acts primarily on r-protein synthesis and acts only as a secondary consequence on rRNA synthesis. As will be described later, this is opposite to what we now know about regulation. In any event, quite often I discussed these models and suggestions with Kjeldgaard. Thus, by the end of my short stay in Aarhus, I had more or less decided not to switch to eucaryotic systems but to stay with *E. coli* ribosome research. I thought that Maaløe's model was ingenious but too speculative. In reviewing the previous studies in this field, I was convinced that the regulation problems should be studied by using a cell-free gene expression system, and for that purpose one needed ribosomal genes in the form of isolated DNA. In addition, such DNA would make several new experiments possible, e.g., direct measurements of specific r-protein mRNA and physical mapping of r-protein genes (and rRNA genes) and their promoters.

After I returned to Madison from Denmark, we gradually started to do genetics and regulation work, although the major activity of my laboratory continued to be the studies of the structure and function of ribosomes as well as in vitro ribosome assembly reactions. The isolation of r-protein genes was not easy and took a long time. Recombinant DNA techniques were not yet available. Thus, we tried to use classical approaches to isolate genes by incorporating them into λ transducing phages. Although the phenomenon of specialized transduction was discovered in 1956 (Morse et al., 1956), isolation of bacterial genes as transducing phages was limited for a long time only to genes located close to the attachment sites of certain temperate phages such as λ and $\phi 80$. However, two major developments had taken place by the early 1970s, allowing isolation of other genes as λ or $\phi 80$ transducing phages. The first was the development of a technique to transpose bacterial genes to other chromosomal locations by using temperature-sensitive F′ factors. With this technique, Beckwith and Signer (1966) succeeded in the isolation of $\phi 80$ transducing phages carrying the *lac* operon genes. The second was the discovery of secondary attachment sites for λ, reported in 1972, which allowed isolation of λ transducing phages carrying genes located close to various secondary attachment sites if the primary λ attachment site was removed (Shimada et al., 1972).

By the time we decided to isolate r-protein genes, evidence had been accumulated to suggest that there was an r-protein gene cluster in the *str-spc* region of the *E. coli* chromosome (72 min on the *E. coli* genetic map). Several antibiotic resistance mutations were mapped in this region, thanks to the efforts made by many geneticists and microbiologists such as Julian Davies at the University of Wisconsin, and some of the gene products had been identified as r-proteins, first by reconstitution techniques and then by direct amino acid sequence analysis of suspected proteins, as mentioned earlier. In addition, following the pioneering work by Joel Flaks and his co-workers at the University of Pennsylvania, who mapped one r-protein gene (later identified as the gene for S7) in this region by using gel mobility differences between *E. coli* K-12 and B strains ("K-character"; Leboy et al., 1964), several other r-protein genes had also been mapped in this region by the same approach, using hybrids between *E. coli* and other enteric bacterial species such as *Salmonella typhosa*. These studies were done mainly by two groups, Shozo Osawa's group at Hiroshima University and Paul Sypherd's group at the University of Illinois (reviewed by Sypherd and Osawa, 1974). Therefore, it was this *str-spc* region that we attempted to isolate as a transducing phage or as part of an F′ plasmid.

After we tried several strategies without success, one day in late 1973 I was reading an article by Beckwith, Signer, and Epstein published in the 1966

thesis could be explained on the basis of gene dosage effects. It had been known for some time that in faster-growing bacterial cells, the rate of initiation of DNA replication increases without an increase in elongation rate, leading to the formation of multiple DNA replication forks and hence an increase in copy numbers of genes located close to the origin of replication. As I shall describe later, we now know that the total rRNA synthesis rate is largely gene dosage independent and therefore that the above argument was theoretically incorrect. In addition, when the mapping of all of the rRNA operons in *E. coli* was eventually completed, seven rRNA operons were found to be localized in various chromosomal locations well separated from each other (Ellwood and Nomura, 1982). By calculation, we were able to show that the above argument cannot be correct either simply because, according to the correct mapping positions of rRNA genes, the calculated increase in gene dosage with increased growth rate is too small and cannot account for the increase in rRNA synthesis. As in the case of r-protein genes, exact mapping and detailed characterization of rRNA genes had to await physical isolation of rRNA genes.

The first pioneering work for the physical characterization of rRNA operons was done by Norman Davidson and his co-workers at the California Institute of Technology in the early 1970s. They studied F′ factor F14, which covered a large chromosomal segment including the *ilv* region and was therefore predicted to carry most of the rRNA genes according to Atwood's original experiments (Deonier et al., 1974). They also examined the ϕ80d3 ilv^+ $su7^+$ transducing phage isolated by Larry Soll at Harvard Medical School (Ohtsubo et al., 1974). By using electron microscopic heteroduplex techniques, they demonstrated that F14 plasmid carried only two copies of rRNA operons, which were separated from each other by more than 100 kilobases, and that ϕ80d3 ilv^+ $su7^+$ contained a single-copy rRNA operon. In their extensive studies to physically characterize biologically interesting DNA molecules, Norman Davidson and his co-workers had refined and improved the method to visualize DNA-DNA heteroduplexes by using electron microscopy. It was very impressive to see, in the papers they published in late 1974, heteroduplex DNA molecules displaying segments corresponding to the 16S rRNA coding region, the 23S rRNA coding region, and a bubble corresponding to the spacer region in between.

Just around that time, we were busy isolating deletion and insertion derivatives of the λ transducing phages carrying r-protein genes, as mentioned above. Therefore, I decided to learn the heteroduplex techniques myself, and I spent a week in Norman Davidson's laboratory in late 1974 learning the techniques and analyzing DNA samples of mutant phages that we had isolated. This visit turned out to be useful. We immediately started to characterize the structures of various mutant phages and plasmid DNA molecules. In fact, soon after I came back from Pasadena, we discovered, by chance, the presence of rRNA genes (*rrnB* operon) on the λ $rif^{d}18$ transducing phage that we had been studying (Lindahl et al., 1975) and embarked on a serious study of rRNA operons. I wished to know the detailed gene organization in an individual rRNA operon, especially the structure of its promoter(s). I also wished to know exact chromosomal locations of all rRNA operons in *E. coli* to assess their physiological significance. Thus, when I learned that Louise Clarke and John Carbon at the University of California, Santa Barbara, had constructed the first plasmid bank containing randomly sheared *E. coli* chromosomal DNA fragments (Clarke and Carbon, 1976), I decided to screen these recombinant plasmids for the presence of rRNA genes. We obtained the plasmid bank from Clarke and Carbon, and Marian Ellwood Kenerley, then a graduate student, started this massive (we thought at that time) effort with the help of several other members of the laboratory. It was just before we learned about Southern transfer techniques developed by E. M. Southern at the University of Edinburgh, and we carried out hybridization with radioactive rRNA probes in agarose gels without transfer to nitrocellulose filters. In the meantime, several transducing phages carrying rRNA operons from different chromosomal regions were isolated by Niels Fiil and his collaborators in Copenhagen as well as by Masayuki Yamamoto in our laboratory. After we identified 16 plasmids carrying rRNA genes among 2,000 Clarke-Carbon plasmids, we classified these plasmids, together with transducing phages carrying rRNA operons, by using mainly heteroduplex techniques and concluded that we had altogether isolated seven different rRNA operons, five of which we could localize on the chromosome (Kenerley et al., 1977). At about the same time, Pal Venetianer and his co-workers, working at the Hungarian Academy of Sciences in Szeged, published a short paper describing simple Southern hybridization experiments on total *E. coli* DNA digested by several different enzymes and demonstrating that the total number of rRNA operons is in fact seven (Kiss et al., 1977). Although it is now so common, this methodology for estimating the number of gene copies was just beginning to be used at that time, and I was impressed with the elegance of the experiments. I was also relieved that we had isolated all seven operons.

In thinking back about the time in the mid-

1970s, when we worked very hard mostly isolating genes for rRNA and r-proteins, I often wonder whether we should have concentrated more on regulation problems per se rather than generating so much data on the structures of genes. With the rapid development of recombinant DNA techniques combined with the invention of DNA sequencing techniques by Maxam and Gilbert (1977) and Fred Sanger and his co-workers (Sanger et al., 1977), the study of gene structure became relatively easy, and solid new information was constantly generated from our work. For example, regarding rRNA operons, we discovered, in collaboration with Elsebet Lund and Jim Dahlberg on the same campus, the presence of genes for tRNA in the spacer region of all rRNA operons (Lund et al., 1976) as well as at the distal ends of some operons (Morgan et al., 1978). We expended considerable effort to characterize all of the tRNA genes associated with rRNA operons but failed to find physiological significance in this association. (Interestingly, the presence of similar spacer tRNA genes in chloroplast rRNA operons, later discovered by chloroplast molecular biologists, supported the theory of the procaryotic origin of chloroplasts in plants.) Occasionally I thought about starting to devote our entire effort to studies of regulation rather than to studies of structures of genes and operons. But, as I have already mentioned, I had had enough experience watching many interesting experiments in the 1960s that asked important biological questions but made incorrect conclusions because of the absence of solid experimental information on the systems. Therefore, I justified, correctly or incorrectly, our activities as solid groundwork required for the future crucial experiments to solve problems of regulation.

The mapping of all rRNA operons in *E. coli* was finally completed in 1981 (Ellwood and Nomura, 1982). It certainly did not give answers about the regulation of rRNA synthesis, but it was a useful achievement. I have mentioned the calculation of changes in rRNA gene dosage with growth rate on the basis of their map locations. In addition, we confirmed the observation we had already made in the course of mapping studies, that the direction of transcription of all rRNA operons (and r-protein operons in the *str-spc* and *rif* regions) is identical to that of DNA replication; we suggested that this striking chromosomal organization may have evolved to prevent head-on collisions of the DNA replication machinery with actively transcribing RNA polymerase molecules on the rRNA operons (Nomura et al., 1977; Ellwood and Nomura, 1982). Recently, Bonita Brewer at the University of Washington extended our initial observations by reviewing many other observations found in the literature and made compelling arguments to support the conclusion that collisions between RNA polymerase and the replication complex are avoided by the appropriate orientation of transcription units on the *E. coli* chromosome (Brewer, 1988). Furthermore, Brewer and Walton Fangman (1988), working on DNA replication in yeast cells, noticed that in tandemly repeated rRNA genes in yeast cells such a collision can in fact be observed. I was glad to see that our efforts of mapping RNA operons helped establish such a general principle outside the ribosome research area per se.

REGULATION OF THE SYNTHESIS OF r-PROTEINS: INITIAL CONFUSION AND LATER DISCOVERY OF TRANSLATIONAL FEEDBACK REGULATION

Despite my (biased) opinion on (and reluctance to carry out) experiments using intact cells in the mid-1960s, we started such experiments in the mid-1970s, which were in essence similar to those used by the Danish school. This was partly because of the progress in our knowledge on ribosome structures as well as in several analytical techniques that took place in the intervening time; for example, with the development of two-dimensional gel electrophoresis (Kaltschmidt and Wittmann, 1970), it became possible to separate most r-proteins and measure synthesis rates of individual r-proteins. Success in the isolation of transducing phages carrying large clusters of r-protein genes, such as λ *spc*1 and λ *fus*2, also made direct analysis of r-protein mRNA possible for the first time. Fortunately, just around that time, in the summer of 1973, Pat Dennis joined our group. Before coming to Madison, he had already been studying growth rate-dependent control of ribosome synthesis in Hans Bremer's laboratory at the University of Texas in Dallas by measuring synthesis rates of bulk rRNA and r-proteins and other parameters during steady-state growth as well as during nutritional shiftup. With experience and background, Pat Dennis started to work very efficiently and productively. We carried out many measurements to answer several questions related to basic phenomenology. For example, as I have mentioned earlier, growth rate-dependent control of the bulk r-proteins was already known, but it was not known whether the synthesis rates of all of the individual r-proteins are coordinately regulated. Pat Dennis's experiments done in 1974 (Dennis, 1974) demonstrated that their synthesis rates are indeed coordinated and growth rate dependent. Apparently, the synthesis rates of almost all r-proteins are balanced with the amounts used for production of ribosomes. The results led to a clear formulation of the question which was going to preoccupy me (and others in the field) for several years to follow: What mechanisms ensure the coordinate and stoichiometric synthesis of so many r-proteins? We also carried out experiments to determine whether r-protein synthesis is also under stringent control. By using conditions causing partial inhibition of tRNA charging, and by measuring r-protein synthesis rates relative to total protein synthesis rates in *relA*$^+$ and *relA* cells, we were able to demonstrate that r-protein synthesis is also under stringent control (Dennis and Nomura, 1974).

We then used transducing phages carrying the *spc* and α r-protein operons, which had just been isolated and characterized, to measure the amounts of r-protein mRNA under these various conditions. We found that the amount of r-protein mRNA changes in parallel to changes in r-protein synthesis rates. In exponentially growing cells, the amount of r-protein mRNA (for the *spc* and α operon r-proteins) per unit amount of protein increased roughly in proportion to the square of the growth rate, as was found for the r-protein synthesis rate (Dennis and Nomura, 1975a). Changes in the amount of r-protein mRNA during nutritional shiftup were also demonstrated directly. Similarly, during deprivation of charged tRNA, the amount of r-protein mRNA decreased in a *relA*$^+$ strain and increased in *relA* strains (Dennis and Nomura, 1975b). Since accurate measurements of the synthesis rates of r-protein mRNA and comparison of synthesis rates under various conditions were not easy (e.g., because of differences in nucleotide pool sizes), we published the data on the amounts of r-protein mRNA without measuring mRNA synthesis rates. Although we knew the problems of half-life of mRNA and carefully phrased conclusions in the published papers, we nonetheless suggested transcriptional control (which would determine the amounts of mRNA). Obviously, we did not doubt the appropriateness of the suggestion, which was consistent with the general dogma of regulation of gene expression in procaryotes at that time.

In addition, after the discovery of stringent control of r-protein synthesis, Lasse Lindahl carried out in vitro experiments using transducing phage DNA as template. We wanted to examine whether ppGpp, the presumed direct effector molecule responsible for stringent control, inhibited the synthesis of r-proteins, and we found positive results (Lindahl et al., 1976) that were consistent with the suggestion we had made from in vivo results. (However, later experiments using a genetically engineered strain, which carried a mutation at the target site for the L11 operon-specific translational repressor, demonstrated that the synthesis of L11 and L1, which were also analyzed in this in vitro study, is not stringently controlled in this strain [Cole and Nomura, 1986a], which suggested that the inhibitory effects observed in this experiment might not reflect the in vivo situation, as I will describe later.)

Analysis of r-protein mRNA at various growth rates was also done around the same time by Kirsten Gausing in Ole Maaløe's laboratory in Copenhagen. She carried out perhaps the most comprehensive analysis in this respect; she measured not only the amounts of r-protein mRNA but also the synthesis rates of total RNA, rRNA, r-protein mRNA, and (by subtracting rRNA and tRNA synthesis rates from total RNA synthesis rates) total mRNA (Gausing, 1977). To my surprise, Gausing found that whereas rRNA synthesis rates increased roughly in proportion to the square of growth rates as expected, r-protein mRNA synthesis rates did not increase in the same way; the increase was close to a linear increase with growth rate. In other words, r-protein mRNA synthesis rates did not increase in the same way as the increase in r-protein synthesis rates with growth rate. Although Kirsten Gausing concluded correctly that control of transcription of r-protein genes is different from that of rRNA genes, she also concluded that r-protein synthesis is transcriptionally regulated; she argued that the calculated ratio of r-protein mRNA synthesis rate to total mRNA synthesis rate increased with growth rate in a way similar to the differential synthesis rate of r-proteins, and hence, transcriptional activities determine differential r-protein synthesis rates. I found this logic difficult to accept. I read Gausing's 1977 paper many times (it was not an easy paper for me to read) and tried to visualize the regulation. Although I felt that these difficult measurements and calculated values might involve significant experimental errors, I had to agree with the large observed difference in growth rate dependency of rRNA and r-protein mRNA synthesis rates, which would correspond to a large difference between r-protein and r-protein mRNA synthesis rates with respect to growth rate dependency. In this respect, the data suggested to me posttranscriptional control of r-protein synthesis. However, her measurements of the amounts of r-protein mRNA agreed with our measurements suggesting transcriptional control of r-protein synthesis, which was our previous suggestion as well as Gausing's conclusion. Although, with the publication of Gausing's paper (together with our previous publications), the notion that growth rate-dependent control of r-protein synthesis takes place at the level of transcription became generally accepted, I occasionally thought about Gausing's data and felt uneasy about the conclusion.

Regarding the apparent coordination of rRNA and r-protein synthesis, investigators in the field, including us, considered three possibilities. The first was that rRNA synthesis was the primary target of regulatory mechanisms and that the regulation of r-protein synthesis was a secondary consequence of the regulation of rRNA synthesis. The second possibility was opposite to the first, namely, that r-protein synthesis was regulated (either directly or passively, as proposed by Maaløe) and that the regulation of rRNA synthesis was a secondary consequence of the regulation of r-protein synthesis. The r-protein inducer hypothesis suggested by Maaløe (1969), as I mentioned above, was one such specific example. The third possibility was that both rRNA and r-protein syntheses were regulated, and exact coordination was achieved either by a balance of transcriptional and translational efficiencies inherent in the DNA and mRNA structures themselves, or by degradation of products synthesized in excess, or by both. Gausing's conclusion that rRNA genes and r-protein genes are transcribed independently was thought to support the third possibility. In addition, Maaløe's passive control model was still considered applicable to the regulation of r-protein gene transcription in this case. Simply put, a regulatory mechanism for rRNA gene transcription had to be found in some way, and the apparent excess synthesis of rRNA followed by degradation observed in Gausing's experiments was emphasized in her paper.

Among the three possibilities given above, the first was related to the old (incorrect) hypothesis of rRNA functioning as mRNA for r-proteins, as I mentioned earlier. Even though the old hypothesis was incorrect, this first possibility was attractive to me, and we occasionally suggested r-proteins as negative-feedback repressors to explain certain observations (e.g., Dennis and Nomura, 1975b), but feedback repression was thought to act at the level of transcription. Regarding the second possibility, I did not find any experimental evidence or attractive theoretical reason to support it, although this was originally convenient for formulation of Maaløe's passive control model. Regarding the third possibility, we (as well as other investigators) thought about a formal model to explain the coordinated and stoichiometric synthesis for most r-proteins. This model (called "simple model" in some of the articles we published around 1980) assumed that r-protein promoters have the same strength and respond to the same regulatory signals, and all r-protein genes in a given operon are translated with equal efficiency (except the gene for L7/L12). This model seemed to be reasonable. In fact, Pat Dennis, who had established his own laboratory at the University of British Columbia in 1976, reported that the transcriptional activity of r-protein promoters in the *rif* region and those in the *spc* region are identical (Dennis, 1977). In addition, the possibility was also considered that the promoters for rRNA operons and r-protein operons might share operator sites; e.g., they might both respond to the same signal, but the strength of the rRNA promoters would be higher than the strength of r-protein promoters (because the rRNA gene products are not amplified by translation), so that approximate balancing of production of rRNA and r-proteins would be achieved. Although exact molecular mechanisms were not known at that time, there were examples for coordinated regulation of unlinked genes, such as genes involved in arginine biosynthesis or genes involved in SOS functions. In these systems, it was suspected (correctly) that unlinked genes share target sequences and would bind common repressor proteins. Thus, we thought that if this simple model was correct, one might expect some DNA sequence similarities (common box sequences) among r-protein promoters (and rRNA promoters).

After identification of six transcription units (operons) in the two isolated r-protein gene clusters, Leonard Post devoted himself to sequencing these promoters. We originally noted some features common among these promoters, such as rotationally symmetric G+C-rich regions surrounding the −10 Pribnow boxes, and mentioned the possibility that they were target sites for regulation (Post et al., 1978). However, after completion of the sequence determination of all six promoters, we thought that the similarities were not compelling, and we eventually abandoned this simple model (Post et al., 1980; discussed in Nomura and Post, 1980). Around this time, DNA sequences of promoter regions of four rRNA operons became available as the result of work done by both Joan Steitz's laboratory (Young and Steitz, 1979) and our laboratory (deBoer et al., 1979). Although the major P1 promoters of all four rRNA operons shared some common unique sequences not found in other promoters, they were not found among r-protein promoters. However, Andrew Travers at the Medical Research Council in Cambridge, England, subsequently compared these published sequences and proposed a consensus sequence in the G+C-rich region near the promoter as the site that would respond to stringent control signals (Travers, 1980a, 1984). Although we were not convinced by the proposed sequence similarity, Travers obtained some experimental results to support his model (Travers, 1980b; Travers et al., 1986; the studies were carried out by using the gene for tyrosine tRNA) and became a major proponent of coordinated regulation of rRNA and r-protein gene expression through transcriptional control mediated by ppGpp.

As mentioned above, I was interested in the possibility of some kind of feedback mechanism to explain coordination of r-protein and rRNA synthesis. Thus, while promoter sequencing was being carried out, we initiated systematic studies on gene dosage effects when Ann Fallon joined our group in 1976, with later collaboration by Sue Jinks-Robertson and Geneva Strycharz.

Our first approach to test the feedback model was very indirect; we did not yet try to test it directly by using in vitro systems. We examined the effects of increased gene dosages on r-protein synthesis rates and on mRNA synthesis rates in lysogenic strains merodiploid for a group of r-protein genes (in the *str-spc* region) or in strains carrying these extra r-protein genes on a multicopy plasmid. We also measured the amounts of r-protein mRNA. We found that the rates of transcription of the r-protein genes increased in proportion to the increase in gene dosage but that the rates of r-protein synthesis did not increase relative to the synthesis rates of other r-proteins whose genes existed in a single copy. We also observed that the amounts of r-protein mRNA were only slightly elevated relative to the control levels. After the initial results were obtained, I immediately thought about a translational feedback model, that is, a model which assumed that some r-proteins are feedback inhibitors and inactivate (and degrade) their own mRNA, thus coupling r-protein synthesis with ribosome assembly. In this way, I could explain coordination and balancing of r-protein synthesis and rRNA synthesis. In addition, I was gradually able to understand why we (and Gausing) observed that the amounts of r-protein mRNA were proportional to the square of growth rate, as in the case of rRNA and r-protein synthesis, giving us the impression of transcriptional control, and yet Gausing observed that the growth rate dependency of r-protein mRNA synthesis was clearly different from the growth rate dependency of rRNA and r-protein synthesis; r-protein mRNA is synthesized in excess, especially in slow growth, and feedback-inhibited excess mRNA is degraded rapidly.

Around that time, August Böck, then at the University of Regensburg in the Federal Republic of Germany, published a paper reporting the absence of gene dosage effects on r-protein synthesis in strains carrying F′ plasmids (Geyl and Böck, 1977). However, they did not analyze the effects on mRNA synthesis, nor did they emphasize any specific model. Their results were also largely consistent with the feedback model. I thought that the translational feedback model was attractive in its simplicity. However, the model was different from the general belief at that time (and our previous suggestion in several published papers, as mentioned above), i.e., that r-protein genes are transcriptionally controlled. Therefore, we repeated somewhat tedious experiments many times to confirm our experimental results. Finally, I became confident of our data and the model and together with my co-workers published a paper proposing the translational feedback model (Fallon et al., 1979). As mentioned above, the failure of our promoter sequence work to find clues of transcriptional regulation made it easier for us to publish this paper. I should also note that by this time I had realized the existence of a similar system, the regulation of phage T4 gene 32, which encodes a single-stranded DNA-binding protein. On the basis of considerable experimental evidence, Larry Gold and his co-workers at the University of Colorado, Boulder, had recently proposed that the gene 32 protein represses the translation of its own mRNA and that repression takes place only after the titration of gene 32 protein onto available intracellular single-stranded DNA (Gold et al., 1976; Lemaire et al., 1978). I thought that the model we were proposing was strikingly analogous to the gene 32 system except that the r-protein system, involving so many proteins and genes, would be surely more complex in the actual mechanisms involved in regulation.

At about the same time as we published our paper, Pat Dennis and Niels Fiil also published a paper describing the results of similar gene dosage experiments done with a strain carrying extra r-protein genes in the *rif* region on a plasmid and suggested the presence of translational regulation, although they did not discuss how such regulation could be achieved (Dennis and Fiil, 1979).

The proof of the translational feedback model came from in vitro experiments. Since we had almost all r-proteins purified and various DNA molecules carrying r-protein genes and, in addition, we had been using an in vitro system to identify r-protein genes, it was not difficult to demonstrate specific inhibitory effects by some specific r-proteins. John Yates, then a graduate student, carried out many in vitro experiments and quickly observed that some r-proteins had in fact specific inhibitory effects on the synthesis of some r-proteins. The inhibited proteins were always in the same operon as the added r-protein; the specificity was clear and dramatic. The effects of these inhibitory r-proteins were demonstrated to be at the level of translation, rather than transcription, as predicted from the gene dosage experiments. Ribosomal proteins S4, S8, and L1 were immediately identified as translational repressors in this way (Yates et al., 1980), followed by identification of L4 and S7 (Yates and Nomura, 1980; Dean et al., 1981a). Independently, Ryuji Fukuda at Kyoto University in Japan (Fukuda, 1980) and Herb Weissbach and his co-workers at the Roche Institute of Molecular Biology (Brot et al., 1980) showed that L10 inhibits its own synthesis from λ *rif*^{d}18 transducing phage DNA in vitro.

The feedback regulation model proved by the in vitro experiments was confirmed by other kinds of experiments. One powerful approach was used by Lasse Lindahl and Janice Zengel, both of whom had left my laboratory by then and settled at the University of Rochester. They fused several r-protein genes to the *lac* operon promoter so that they could overproduce an excess of a particular protein. They found that overproduction of r-protein L4 inhibits the synthesis of a group of r-proteins (Lindahl and Zengel, 1979; Zengel et al., 1980). Dennis Dean, a postdoctoral fellow in my laboratory, used a similar approach and demonstrated that the repressor r-proteins identified in vitro (S4, S8, and L1) in fact function as (translational) repressors in vivo (Dean and Nomura, 1980; Dean et al., 1981b). However, Lindahl and Zengel initially concluded that in vivo repression of the synthesis of the S10 operon proteins by overproduction of L4 was at the level of transcription, whereas in vitro experiments carried out in our laboratory showed that L4, like all other repressor r-proteins identified, acts as a translational repressor (Yates and Nomura, 1980). Both Lasse Lindahl and I felt uneasy about this discrepancy in conclusions between these two types of experimental approaches. However, later work by Lasse Lindahl's group showed that feedback regulation of the S10 operon involves both translational repression by L4, as seen in vitro by us, and transcriptional repression (through an attenuation mechanism; Lindahl et al., 1983) by the same repressor (Freedman et al., 1987).

The discovery of translational feedback regulation was a breakthrough. As mentioned above, some of the previous (apparently complex) observations became understandable or amenable to experimental reexamination. Many specific (and obvious) questions were posed which could be answered by experiments. One of the first such questions we studied was how a single repressor r-protein inhibits translation of more than one protein from a single polycistronic mRNA. Using a simple bicistronic operon, the L11 operon, which consists of the genes for L11 and L1, we demonstrated that the translation of the L1 gene is very tightly coupled with translation of the preceding L11 gene and that the L1 repressor inhibits translation of L11 directly by acting at a single target site located near the L11 translation initiation site and inhibits translation of L1 indirectly through this tight coupling (Yates and Nomura, 1981; Baughman and Nomura, 1983). When we first obtained evidence to indicate this mechanism, we called the phenomenon of coupling of L11 and L1 translation "sequential translation" and then found that Charles Yanofsky had just discovered a similar phenomenon with the *trp* operon and called it translational coupling (Oppenheim and Yanofsky, 1980). In any event, translational coupling was shown to be prevalent in r-protein operons (shown also to be operating in the L10, *spc*, α, and S10 operons) and to constitute the essential basis of (translational) coregulation of several linked r-protein genes.

Regarding the reason why people in the field did not consider translational regulation, it is very likely that people were influenced by the operon model originally proposed by Jacob and Monod and studied so well in the *lac* operon and similar systems, and consequently they accepted the dogma of transcriptional control for the regulation of r-protein synthesis. In fact, all of the gene expression systems in *E. coli* studied at that time were known to be regulated transcriptionally. However, it should be noted that Jacob and Monod, when they formulated the operon theory to explain coregulation of linked genes (Jacob and Monod, 1961), considered two alternatives to explain phenotypes of various regulatory mutants in the *lac* operon system: one was to postulate the repressor acting at the operator which "is and remains attached to the gene themselves (Fig. 6, I)," and the second was to postulate the repressor acting at the operator which "is and remains attached to the cytoplasmic messenger of the linked *z* and *y* genes which must then be assumed to form a single, integral, particle corresponding to the structure of the whole *ozy* segment and functioning as a whole (Fig. 6, II)." They showed these two models in Fig. 6 of their celebrated 1961 paper. Without deciding which model was correct, they first defined the concept of operon as a genetic unit of coordinate expression, and then they discussed information available at that time and favored the first alternative, that is, the genetic model, rather than the cytoplasmic operator model. Of course, they were correct for the *lac* (and many other) operon(s). However, in reading their paper, we now find that the reason for their having favored a transcriptional model was not conclusive. In fact, they stated in one place that their argument "cannot be considered to eliminate the cytoplasmic operator model . . . ." Therefore, even the original operon theory did not eliminate the possibility of coregulation of linked genes at the level of translation, and we should have been open to the possibility of coregulation of linked genes by translational repressors.

Perhaps a more important difference between the r-protein regulation systems and the conventional operon theory is the means by which several unlinked operons are coregulated. As I mentioned earlier, there were known examples of coordinated regulation of unlinked genes (or operons), but in most cases the unlinked genes shared some common target site (operator or target site for activator) of similar DNA sequences to which a single regulatory molecule (repressor or activator) bound, regulating the various unlinked genes. In contrast, the several unlinked r-protein operons studied are coregulated, not because of the use of a common regulatory (repressor) protein but by coupling the translation of all of these unlinked mRNAs with a single major reaction, ribosome assembly. In this respect, the operon theory influenced us, encouraging us to look for a common target site among ribosomal promoters and perhaps causing a delay in the discovery of translational feedback repression. Of course, the presence of a common regulatory target site in unlinked r-protein operons is still a possibility, as hypothesized by Andrew Travers (see above).

In connection with the coupling of mRNA translation with ribosome assembly, I thought about the concept of competition between structurally similar regions of rRNA and mRNA for the binding of repressor r-proteins. We published a paper on this concept (Nomura et al., 1980) before we actually identified mRNA target sites and studied their structures. After the initial identification of several r-protein repressors, we were struck that most of them, such as S4, S7, S8, L1, and L4, were among the primary rRNA-binding proteins, and the binding sites on rRNA for some of them (e.g., S8 and L1) were being intensively studied. Thus, the study of a key step in regulation of r-protein genes, interactions of r-proteins with mRNA, became intertwined with structural studies of ribosomes, i.e., studies of interactions of r-proteins with rRNA. Although speculative structures of mRNA target sites presented in the original paper (Nomura et al., 1980) turned out to be mostly incorrect, the basic concept was demonstrated to be correct for r-proteins L1 (e.g., Thomas and Nomura, 1987; Said et al., 1988) and S8 (Cerretti et al., 1988; Gregory et al., 1988). Interaction of repressor r-protein with mRNA has been studied (and is still being studied) by several laboratories in addition to our own, including the laboratories that have been engaged in studies of r-protein–rRNA interaction. The principal players have been Niels Fiil and Morton Johnson at the Microbiology Institute in Copenhagen, Jim Friesen at the University of Toronto, Robert Zimmermann at the University of Massachusetts, and David Draper at Johns Hopkins University.

GROWTH RATE-DEPENDENT CONTROL AND STRINGENT CONTROL OF r-PROTEIN SYNTHESIS

I have already described some historical events that led to the discovery of translational feedback regulation of r-protein synthesis. However, even after this discovery, some people maintained, by citing Gausing's measurements and arguments, that growth rate-dependent changes of r-protein synthesis were due mostly to transcriptional control and that translational control was used only for minor adjustment. The alternative view which we entertained was that r-protein mRNA is usually synthesized in excess under most growth conditions and growth rate dependency of r-protein synthesis is achieved as a consequence of growth rate-dependent control of rRNA synthesis. As already explained, we argued that according to Gausing's data, transcriptional activities of r-protein genes do not show the typical growth rate dependency, but the degree of translational repression changes with growth rate, leading to changes in mRNA stability and the typical growth rate dependency in the amounts of r-protein mRNA, which was observed by us as well as by Gausing. To confirm the validity of this view, we carried out several different kinds of experiments in the following years. First, we demonstrated that an increase in the degree of translational repression in fact leads to an increase in the rate of degradation of r-protein mRNA (e.g., Cole and Nomura, 1986b). Second, by gene fusion experiments, we showed that the differential synthesis rates of galactokinase protein from the *galK* gene under the control of r-protein promoters do not show growth rate dependency (Miura et al., 1981). Finally, Jim Cole, then a graduate student, constructed a strain carrying a known L1 target site mutation that abolished translational feedback regulation and demonstrated that the syntheses of L11 and L1, which are normally regulated by L1, did not show growth rate dependency (Cole and Nomura, 1986a). Therefore, although the synthesis of some r-proteins may not be feedback regulated (Wikström and Björk, 1988) and detailed mechanisms involved in feedback regulation may also be different for some r-protein genes from the standard mechanism originally postulated (e.g., the S10 operon mentioned above and L14 and L24 genes, which are regulated by "retroregulation" by a translational repressor as recently shown in my laboratory [Mattheakis et al., 1989]), it now appears that growth rate-dependent control of the synthesis of most r-proteins is largely a secondary consequence of the control of rRNA synthesis.

Stringent control of r-protein synthesis as originally observed in our laboratory (Dennis and Nomura, 1974) also became the subject of two different interpretations. Because of the discovery of translational feedback regulation, stringent control of r-protein synthesis can be explained as a secondary consequence of stringent control of rRNA synthesis. On the other hand, it is also conceivable that stringent control acts directly on transcription of r-protein genes. Experiments done by Jim Cole using the mutant mentioned above favored the first possibility for L11 and L1 synthesis (Cole and Nomura, 1986a), whereas the experiments done by Lasse Lindahl and his co-workers favored the second possibility for r-proteins in the S10 operon (Freedman et al., 1985). Thus, the question has not been resolved. Of course, it is possible that different r-protein promoters respond differently to various regulatory signals. Undoubtedly, research in this field will continue, and new aspects of regulation may be revealed. Because of the complexity of the ribosome structure involving rRNAs and many r-proteins as well as the central importance of ribosome functions in bacterial growth, the presence of multiple, interconnected feedback loops as well as interactions with other regulatory systems and signals are expected. Therefore, studies of the regulation of r-protein synthesis should continue to be a rich source of new and informative regulatory mechanisms.

GROWTH RATE-DEPENDENT CONTROL OF rRNA SYNTHESIS

In their initial studies on the cellular content of RNA at various growth rates, the Copenhagen school did not analyze cells growing very slowly, and hence the data obtained showed a strict proportionality between RNA content and growth rate. Therefore, the Copenhagen school emphasized the notion of constant efficiency of ribosomes. Consequently, people quite often suggested that the rate of bacterial growth (and total protein synthesis) is in fact limited by the number of ribosomal particles present in the cell. Later measurements by several investigators showed that the strict proportionality does not hold between RNA content and growth rate, especially at slow growth conditions (e.g., Neidhardt and Magasanik, 1960; Rosset et al., 1966). However, it was Arthur Koch and his co-workers at Indiana University who really demonstrated the presence of excess functional ribosomes under slow growth conditions and explicitly concluded that bacterial growth is not limited by the amount of ribosomes (Koch, 1970, 1971). Therefore, the major question of growth rate-dependent control was how bacterial cells adjust their ribosome synthesis in relation to synthesis of other cellular components so that the optimal growth rate is attained; that is, what mechanism(s) prevents wasteful overproduction of ribosomes yet ensures sufficient production to attain the maximum growth rate under most growth conditions?

I have mentioned earlier the initial model to explain growth rate-dependent synthesis of ribosomes on the basis of the degree of charging of tRNA, which was proposed in the early 1960s (Stent and Brenner, 1961; Kurland and Maaløe, 1962; Maaløe and Kjeldgaard, 1966). I also explained Ole Maaløe's passive control model published in 1969. Maaløe basically maintained and expanded this model, incorporating new information and revising details along with the progress of the field (Maaløe, 1979; Ingraham et al., 1983). As I shall mention later, when I started to seriously think about the question of growth rate-dependent control in the late 1970s, I was not certain whether these models could actually accomplish the optimum adjustment of ribosome synthesis in the sense that I explained in the previous paragraph.

Perhaps the most extensive studies related to the regulation of rRNA synthesis carried out in the 1970s were on the possible role of ppGpp as the key effector molecule in this regulation. The discovery of ppGpp (and pppGpp) was made in 1969 by Cashel and Gallant as I have already described; since then, many investigators have studied this question by using both in vivo and in vitro approaches. One important finding that took place soon after the discovery of ppGpp was that this compound was accumulated during nutritional shiftdown in both *relA*$^+$ and *relA* cells and that the cellular concentrations of ppGpp vary inversely with growth rate and RNA content of cells in both *relA*$^+$ and *relA* cells (Lazzarini et al., 1971). When the model of uncharged tRNA as repressor for RNA synthesis was proposed in the early 1960s, it was an attempt to explain both stringent control and growth rate-dependent control on a unified basis. Subsequently, Fred Neidhardt discovered that *relA* strains, which lost stringent control of RNA synthesis during amino acid starvation, still maintained growth rate-dependent control of RNA synthesis and exhibited normal responses upon nutritional shiftup and shiftdown (Neidhardt, 1963). This finding caused problems for the unitary hypothesis; in fact, Maaløe explicitly stated that this was one of the reasons why he gave up the original tRNA repressor model and switched to the passive control model (Maaløe, 1969). With the new finding mentioned above, it was once more realized that both growth rate-dependent control and stringent control could be explained in a unified way on the basis of regulation by ppGpp, the magic compound. It was inferred that there were two systems responsible for the synthesis of ppGpp; one requires the product encoded by the *relA*$^{+}$ gene and responds to amino acid starvation, and the other does not require the *relA*$^{+}$ gene product and responds to carbon and energy shiftdown. This view was certainly aided by the initial announcement of inhibition of rRNA synthesis by ppGpp in vitro by Andrew Travers, then at Harvard, and his co-workers (Travers et al., 1970, 1971). Since isolated rRNA genes were unavailable at that time, the initial experiments were done by using *E. coli* chromosomal DNA as a template, and the analysis of rRNA products depended on indirect RNA-DNA hybridation methods; perhaps partly for this reason, the reproducibility of the experiments was questioned by some workers.

The ensuing years brought a flood of papers on the relationship between ppGpp and rRNA synthesis. Regarding the in vivo experiments, people analyzed the cellular concentrations of ppGpp and rRNA synthesis rates under a variety of conditions, and a rough inverse correlation was observed between the two in many studies. However, people soon started to find some exceptional situations in which this inverse relationship apparently breaks down, for example, during temperature shiftup (Gallant et al., 1977). Since such exceptional cases and warnings against the simple model of ppGpp as the key effector came from several investigators, including John Gallant, who had discovered ppGpp with Mike Cashel, and even some alternative effector compound was suggested (Pao and Gallant, 1979), the situation became confusing in the late 1970s. However, Hans Bremer and his co-workers at the University of Texas, Dallas, subsequently carried out extensive studies arguing that one should compare ppGpp concentrations with relative differential synthesis rates of stable RNA (rRNA plus tRNA) rather than the absolute synthcsis rate of stable RNA; they demonstrated that a strict correlation exists between ppGpp concentrations and differential synthesis rates of stable RNA under all conditions studied, including those that had been thought to be exceptional (e.g., Ryals et al., 1982a, 1982b).

Testing the role of ppGpp by in vitro experiments also created a confusing situation in the 1970s. Some investigators reported specific inhibition of stable RNA synthesis by ppGpp, whereas others observed little or no effect. After the isolation of rRNA (and tRNA) genes in the mid-1970s, DNA templates for in vitro experiments became defined, but the controversy continued. In addition to ppGpp, several in vitro conditions and the presence of other effectors, such as EF-Tu, fMet-tRNA, and IF2, were claimed to have selective inhibitory or stimulating effects on rRNA synthesis in vitro. After having carried out many of these in vitro experiments, Andrew Travers proposed that ppGpp influences the selectivity of RNA polymerase for stable RNA relative to other mRNA promoters. According to this model, only a particular form (most likely a dimer) of RNA polymerase is capable of transcription from stable promoters, and ppGpp would inhibit the necessary conformational changes (dimerization) of RNA polymerase, thereby reducing binding of polymerase to stable RNA promoters (Travers, 1976; Travers et al., 1980; Travers et al., 1982). Essentially the same model was used by Hans Bremer and his co-workers to explain the inverse relationship between ppGpp concentrations and differential synthesis rates of stable RNA observed in his in vivo studies as mentioned above (Ryals et al., 1982a).

In the late 1970s, some in vitro experiments were also carried out in my laboratory using isolated rRNA genes as DNA templates. I remember the frustration of being unable to reproduce some of the published experiments in our own laboratory. In addition, although we observed inhibition of transcription from rRNA promoters by ppGpp in vitro, the degree of inhibition varied depending on experimental conditions, such as relative concentrations of each of the substrate nucleoside triphosphates, and we were not certain which conditions were really physiological. Thus, although I originally planned to study regulation in vitro, I became less enthusiastic about the in vitro approach in this case.

In vitro experiments with the use of isolated rRNA genes, however, started to produce some solid useful information. After isolation of rRNA operons, we sequenced the regions upstream from the structural genes and carried out in vitro transcription experiments to determine transcription initiation sites and thereby to define promoters. We found that there were two start sites for each of the two rRNA operons we studied, and each start site was preceded by typical *E. coli* promoter consensus sequences. We called these two promoters P1 and P2 (Gilbert et al., 1979; deBoer et al., 1979). The same conclusions were obtained by Richard Young and Joan Steitz at Yale University for two other operons that they studied (Young and Steitz, 1979). Gad Glaser and Mike Cashel at NIH discovered the two start sites (Glaser and Cashel, 1979) for yet another rRNA operon. The use of these two start sites in vivo was soon demonstrated experimentally by us (deBoer and Nomura, 1979) as well as by Lund and Dahlberg

(1979) on the Madison campus. In addition, under the growth conditions used in these two studies, it was shown that P1 is the major promoter and P2 is the minor promoter. This was in 1979, as we were just seeing the beginning of breakthroughs in the study of regulation of r-protein synthesis with isolated r-protein genes. Thus, it was obvious that serious regulation studies on rRNA operons were soon to be initiated by using isolated rRNA operons and modern recombinant DNA techniques.

The discovery of two tandem promoters on each rRNA operon posed the question of whether transcription initiated from both promoters is subject to stringent or growth rate-dependent control or both. Mike Cashel and his co-workers at NIH were the first to give the answer to this question. By constructing a plasmid carrying an rRNA operon with a large internal deletion and carrying out measurements of short transcripts in vivo, they demonstrated that the P1 promoter is under both growth rate-dependent and stringent control but the P2 promoter is not (Sarmientos and Cashel, 1983; Sarmientos et al., 1983). The conclusion regarding growth rate-dependent control was later confirmed by Rick Gourse, then a postdoctoral fellow in my laboratory, by carrying out a systematic deletion analysis of the promoter region (Gourse et al., 1986). This discovery gave a neat explanation for the old observation that the proportionality between ribosome concentration and growth rate holds only in the medium- and fast-growth-rate range but not under slow-growth conditions. The P2 promoter is probably responsible for making excess rRNA and hence excess ribosomes in slowly growing cells, which may be beneficial in order for *E. coli* to adapt quickly to improved nutritional conditions, as pointed out by Arthur Koch (1970, 1971). It was a pleasure to see this explanation. The remaining question was how regulation of the P1 promoter takes place. In vitro experiments carried out by Glaser, Sarmientos, and Cashel (1983) indicated that indeed ppGpp might be the direct effector responsible for inhibition. However, in view of the history of in vitro experiments done on this subject, I would like to see this conclusion be established by more rigorous experiments.

Before I describe another development in rRNA regulation studies in the 1980s, I should mention the discovery of antitermination and upstream activation related to rRNA synthesis. Since rRNA genes are transcribed without translation, people wondered why premature transcription termination would not take place. (As mentioned earlier, Gunther Stent used the concept then [in 1966] available, that rRNA functions as mRNA for r-protein, to explain rRNA transcription without premature termination.) The presence of a suspected antitermination function was first shown in 1980 by Edward Morgan (Morgan, 1980), who had worked on rRNA operons, identifying various spacer tRNAs and distal tRNAs in my laboratory in the late 1970s, and subsequently established his own laboratory at the State University of New York at Buffalo. Here, the phage λ N antitermination system, which had been extensively studied and characterized, served as a paradigm. In fact, the presence of a sequence in rRNA operons that is similar to the important consensus sequence (box A) found in the phage *nut* sites was first pointed out by David Friedman, a phage geneticist at the University of Michigan, and E. R. Olson (Friedman and Olson, 1983). In addition to Morgan's group, Cathy Squires and her co-workers at Columbia University initiated serious studies on this subject, defining those DNA sequence elements, including the box A-like sequence, required for antitermination function (Li et al., 1984). However, deletion analysis later carried out by Rick Gourse and his co-workers in my laboratory indicated that the antitermination mechanism is not used for growth rate-dependent control of rRNA transcription (Gourse et al., 1986). In the same deletion analysis, we also demonstrated the presence of an A-rich region upstream from the P1 promoter (upstream activator), which was required for high levels of transcription from P1. The presence of a similar upstream activator region had also been demonstrated for the tyrosine tRNA gene by Lamond and Travers (1983). But again, this activation mechanism was shown not to be used for growth rate-dependent control (Gourse et al., 1986). These subjects, antitermination and upstream activation, are under current investigation by several groups.

Soon after the model of translational feedback regulation of r-protein synthesis was proved to be essentially correct, I started to think about a similar feedback model for growth rate-dependent control of rRNA synthesis. One striking feature of growth rate-dependent control, which had been well known and about which I thought frequently, was that no matter what the nutritional conditions and growth rate (except for very slow growth rates), the cell makes appropriate amounts of ribosomes so that the protein synthetic capacity is just sufficient to maintain that growth rate whereas the nontranslating ribosome concentration is minimal. I thought that the simplest way to achieve this remarkable feature of regulation is to use a feedback mechanism; perhaps cells always synthesize ribosomes in excess, but the excess ribosomes generate negative-feedback signals, thereby preventing further overproduction. Even if some effector molecules, such as uncharged tRNA or ppGpp, might act directly on template or RNA polymerase, either alone or in conjunction with a hypothetical repressor, to regulate rRNA synthesis, the concentrations of such effector molecules would have to be precisely the one(s) that would give the proper level of rRNA transcription, leading to the accumulation of proper amounts of ribosomes. To achieve this would be difficult if the synthesis of the effector molecules was done in direct response to environmental conditions or nutritional states of cells, without using some kind of feedback mechanism.

I discussed this idea with Sue Jinks-Robertson, who had been working on regulation of r-protein synthesis as a graduate student, and she agreed to carry out the experiments that we designed to test the feedback model. This was in late 1981. Rick Gourse, who joined our group later as a postdoctoral fellow, also participated in this project. Gene dosage experiments gave results clearly in support of the original idea. We found that the total rRNA synthesis rate was essentially gene dosage independent when rRNA gene dosage was increased two- to threefold with plasmid-borne rRNA operons; that is, expression of individual rRNA operons was two- to threefold repressed in the presence of plasmid-borne rRNA operons. Control experiments using strains carrying extra rRNA operons which had deletions demonstrated that cells have the capacity to make more rRNA in a given nutritional condition, but rRNA synthesis is normally limited by a feedback mechanism that senses intact rRNA (Jinks-Robertson et al., 1983). We were very happy with these experimental results, which supported our original ideas. The paper describing these experiments and the model, which we called the ribosome feedback model, appeared in 1983.

At the time we proposed this model, I had some difficulty explaining why ribosome feedback is inefficient at slow growth. However, as I mentioned above, Mike Cashel's demonstration of the P2 promoter as a constitutive promoter gave a reasonable explanation: under medium- to fast-growth conditions, rRNA synthesis is mostly from the major P1 promoter and is subject to feedback regulation, but in slow-growth conditions, the minor P2 promoter, which escapes feedback repression, continues to function.

We continued to work on the regulation of rRNA synthesis. Various experiments were carried out to test the model further, and the results were largely consistent with the basic conclusion of the model. We could show that for feedback to take place, overproduction of intact rRNA is not enough, but overproduction of intact ribosomes is necessary. But how did overproduced ribosomes give a feedback signal(s)?

Jim Cole in my laboratory carried out experiments to elucidate the next step involved in the processing of the feedback signal. Chris Olsson from John Hershey's laboratory at the University of California, Davis, also participated in the project. We thought about two possibilities for the next step. The first was that the concentration of intact, nontranslating ribosomes inhibits the synthesis rate of rRNA either directly or indirectly through a mechanism that does not include any steps involved in the main function of ribosomes, translation. The other possibility was that excess ribosomes exert their feedback on rRNA synthesis through their translational activities; that is, excess ribosomes cause a small excess in translation (or possibly just translation initiation), which in turn generates a signal leading to an eventual decrease in rRNA synthesis. We obtained the gene for initiation factor IF2 from Marianne Grunberg-Manago, and Jim constructed a bacterial strain in which the expression of the IF2 gene is under *lac* promoter-operator control; hence, the cellular concentration of IF2 can be varied by changing the concentration of the *lac* operon inducer IPTG. The two possible pathways mentioned above made opposite predictions for changes in rRNA synthesis rate upon decreased initiation by limiting IF2, and the results clearly demonstrated that the second possibility is correct (Cole et al., 1987). Other types of experiments were also carried out to discriminate between the two possibilities, and the results gave further support to the above conclusion (Yamagishi et al., 1987). Therefore, cells monitor the concentration of ribosomes through their translational activities. Excess translation must generate some signal molecule(s). The question whether it is in fact ppGpp or some other compounds is not yet answered and will undoubtedly be one of the major subjects for future studies in this field.

Since the growth rate-dependent control of ribosome synthesis was clearly demonstrated in 1958, many people participated in the study of this fundamental problem, and massive amounts of data have been accumulated. Although molecular mechanisms involved in this control are still unknown, I feel that we now know some basic principles used in this control system on a firm experimental basis. Various old observations can also be explained in light of our new knowledge. For example, stimulation of rRNA synthesis by chloramphenicol (or other protein synthesis inhibitors) observed by many investigators is expected because excess translation that generates negative-feedback signal(s) is inhibited. Stimulation is expected to be of greater magnitude in slow-growth conditions in which feedback repression of rRNA synthesis is greater, as observed in earlier studies. Surely not all that we now believe will turn out to be correct, and our studies on this subject will continue. In addition, the principles that we have discovered in *E. coli* may not apply to other organisms, especially to multicellular eucaryotic organisms. However, it is satisfying for me to see the progress achieved in this field and to be able to understand the regulatory behavior of *E. coli* cells in making ribosomes under various physiological conditions.

POSTSCRIPTS

When Walter Hill asked me to write a chapter on the history of ribosome research, I was initially hesitant. I felt that I should be more interested in current research and future development. However, I conceded that I had been engaged in ribosome research for a long time and had touched upon many different aspects of this field. I agreed, with the condition that it would be a personal essay and certainly not a comprehensive study of the history of ribosome research. Thus, I have covered mostly the subjects related to my own past research activities. Undoubtedly I must have omitted some important achievements and events, for which I apologize. In particular, I regret that I did not cover ribosome research related to eucaryotic ribosomes (except for a few occasions). My own laboratory has recently started research on ribosome synthesis in yeast cells. A history related to research on eucaryotic ribosomes will be written someday. It remains to be seen whether my laboratory can produce something significant to contribute to the field of eucaryotic ribosomes.

I believe that productivity in ribosome research should be judged in terms of its (eventual) contribution to biology in general. Ribosome research used to be interrelated with the central questions in biology and enjoyed productivity in this regard. Because of the progress and diversification of research in biology and biochemistry, the situation may be different now. Yet I hope that future ribosome research will continue to be exciting and productive and that new generations will continue to open up novel aspects of research on this fundamentally important biological structure, the ribosome.

Our research described in this chapter was carried out by a collaborative effort, and I thank my present as well as previous coworkers, some of whose names are mentioned in this chapter. I also thank S. M. Arfin, S. Fahnestock, L. Kahan, and J. Keener for reading the manuscript and L. Hill for help in preparation of the manuscript.

Our research has been continuously supported by grants from the National Institutes of Health (currently Public Health Service grant 2R37GM35949) and the National Science Foundation (currently grant DMB-8904131).

REFERENCES

Adams, J. M., and M. R. Capecchi. 1966. N-formylmethionyl-sRNA as the initiator of protein synthesis. *Proc. Natl. Acad. Sci. USA* **55**:147–155.

Allen, P. N., and H. F. Noller. 1989. Mutations in ribosomal proteins S4 and S12 influence the higher order structure of 16S ribosomal RNA. *J. Mol. Biol.* **208**:457–468.

Amils, R., E. A. Matthews, and C. R. Cantor. 1978. An efficient in vitro total reconstitution of the *E. coli* 50S ribosomal subunit. *Nucleic Acids Res.* **5**:2455–2470.

Aronson, A., and B. McCarthy. 1961. Studies of *E. coli* ribosomal RNA and its degradation products. *Biophys. J.* **1**:215–226.

Astrachan, L., and E. Volkin. 1958. Properties of ribonucleic acid turnover in T_2-infected *E. coli. Biochim. Biophys. Acta* **29**: 536–544.

Auron, P. E., and S. R. Fahnestock. 1981. Functional organization of the large ribosomal subunit of *Bacillus stearothermophilus. J. Biol. Chem.* **256**:10105–10110.

Barta, A., G. Steiner, J. Brosius, H. Noller, and E. Kuechler. 1984. Identification of a site on 23S ribosomal RNA located at the peptidyltransferase center. *Proc. Natl. Acad. Sci. USA* **81**: 3607–3611.

Baughman, G., and M. Nomura. 1983. Localization of the target site for translational regulation of the L11 operon and direct evidence for translational coupling in *E. coli. Cell* **34**:979–988.

Beckwith, J. R., and E. R. Signer. 1966. Transposition of the *lac* region of *E. coli.* I. Inversion of the *lac* operon and transduction of *lac* by ϕ80. *J. Mol. Biol.* **19**:254–265.

Beckwith, J. R., E. R. Signer, and W. Epstein. 1966. Transposition of the *lac* region of *E. coli. Cold Spring Harbor Symp. Quant. Biol.* **31**:393–401.

Beisson, J., and T. M. Sonneborn. 1965. Cytoplasmic inheritance of the organization of the cell cortex in *Paramecium aurelia. Proc. Natl. Acad. Sci. USA* **53**:275–282.

Bernabeu, C., and J. A. Lake. 1982. Nascent polypeptide chains emerge from the exit domain of the large ribosomal subunit: immune mapping of the nascent chain. *Proc. Natl. Acad. Sci. USA* **79**:3111–3115.

Birnstiel, M., J. Speirs, I. Purdom, and K. Jones. 1968. Properties and composition of the isolated ribosomal DNA satellite of *Xenopus laevis. Nature* (London) **219**:454–463.

Biswas, D. K., and L. Gorini. 1972. The attachment site of streptomycin to the 30S ribosomal subunit. *Proc. Natl. Acad. Sci. USA* **69**:2141–2144.

Bowman, C. M., J. E. Dahlberg, T. Ikemura, J. Konisky, and M. Nomura. 1971. Specific inactivation of 16S ribosomal RNA induced by colicin E3 *in vivo. Proc. Natl. Acad. Sci. USA* **68**:964–968.

Brawerman, G., and J. M. Eisenstadt. 1966. A factor from *E. coli* concerned with the stimulation of cell-free polypeptide synthesis by exogenous ribonucleic acid. II. Characteristics of the reaction promoted by the stimulation factor. *Biochemistry* **5**:2784–2789.

Bremer, H., C. Yegian, and M. Konrad. 1966. Inactivation of purified *E. coli* RNA polymerase by transfer RNA. *J. Mol. Biol.* **16**:94–103.

Brenner, S., F. Jacob, and M. Meselson. 1961. An unstable intermediate carrying information from genes to ribosomes for protein synthesis. *Nature* (London) **190**:576–581.

Brewer, B. J. 1988. When polymerases collide: replication and the transcriptional organization of the *E. coli* chromosome. *Cell* **53**:679–686.

Brewer, B. J., and W. L. Fangman. 1988. A replication fork barrier at the 3′ end of yeast ribosomal RNA genes. *Cell* **55:**637–643.

Brimacombe, R., J. Atmadja, J. Kyriatsoulis, and W. Stiege. 1986. RNA structure and RNA-protein neighborhoods in the ribosome, p. 184–202. *In* B. Hardesty and G. Kramer (ed.), *Structure, Function, and Genetics of Ribosomes.* Springer-Verlag, New York.

Britten, R. J., and R. B. Roberts. 1960. High-resolution density gradient sedimentation analysis. *Science* **131:**32–33.

Brosius, J., T. Dull, and H. F. Noller. 1980. Complete nucleotide sequence of a 23S ribosomal RNA gene from *E. coli. Proc. Natl. Acad. Sci. USA* **77:**201–204.

Brosius, J., M. Palmer, P. J. Kennedy, and H. F. Noller. 1978. Complete nucleotide sequence of a 16S ribosomal RNA gene from *E. coli. Proc. Natl. Acad. Sci. USA* **75:**4801–4805.

Brot, N., P. Caldwell, and H. Weissbach. 1980. Autogenous control of *E. coli* ribosomal protein L10 synthesis in vitro. *Proc. Natl. Acad. Sci. USA* **77:**2592–2595.

Brown, D. D., and J. M. Gurdon. 1964. Absence of ribosomal RNA synthesis in the anucleolate mutant of *Xenopus laevis. Proc. Natl. Acad. Sci. USA* **51:**139–146.

Brown, D. D., and C. S. Weber. 1968. Gene linkage by RNA-DNA hybridization. I. Unique DNA sequences homologous to 4S RNA, 5S RNA and ribosomal RNA. *J. Mol. Biol.* **34:**661–680.

Brown, J. C., and A. E. Smith. 1970. Initiator codons in eukaryotes. *Nature* (London) **226:**610–612.

Brownlee, G. G., F. Sanger, and B. G. Barrell. 1968. The sequence of 5S ribosomal ribonucleic acid. *J. Mol. Biol.* **34:**379–412.

Cabezón, T., A. Herzog, J. Petre, M. Yaguchi, and A. Bollen. 1977. Ribosomal assembly deficiency in an *E. coli* thermosensitive mutant having an altered L24 ribosomal protein. *J. Mol. Biol.* **116:**361–374.

Cammack, K. A., and H. E. Wade. 1965. The sedimentation behavior of ribonuclease-active and inactive ribosomes from bacteria. *Biochem. J.* **96:**671–680.

Cantor, C. R., M. Pellegrini, and H. Oen. 1974. Affinity labeling techniques for examining functional sites of ribosomes, p. 573–586. *In* M. Nomura, A. Tissières, and P. Lengyel (ed.), *Ribosomes.* Cold Spring Harbor Laboratory, Cold Spring Harbor, N.Y.

Capel, M. S., D. M. Engleman, B. R. Freeborn, M. Kjeldgaard, J. A. Langer, V. Ramakrishnan, D. G. Schindler, D. K. Schneider, B. P. Schoenborn, I.-Y. Sillers, S. Yabuki, and P. Moore. 1987. A complete mapping of the proteins in the small ribosomal subunit of *E. coli. Science* **238:**1403–1406.

Carbon, P., C. Ehresmann, B. Ehresmann, and J. P. Ebel. 1978. The sequence of *E. coli* ribosomal 16S RNA determined by new rapid gel methods. *FEBS Lett.* **94:**152–156.

Cashel, M., and J. Gallant. 1969. Two compounds implicated in the function of the RC gene of *E. coli. Nature* (London) **221:**838–841.

Cashel, M., and B. Kalbacher. 1970. The control of ribonucleic acid synthesis in *E. coli.* V. Characterization of a nucleotide associated with the stringent response. *J. Biol. Chem.* **245:**2309–2318.

Cerretti, D. P., D. Dean, G. R. Davis, D. M. Bedwell, and M. Nomura. 1983. The *spc* ribosomal protein operon of *E. coli*: sequence and cotranscription of the ribosomal protein genes and a protein export gene. *Nucleic Acids Res.* **11:**2599–2616.

Cerretti, D. P., L. C. Mattheakis, K. R. Kearney, L. Vu, and M. Nomura. 1988. Translational regulation of the *spc* operon in *E. coli*: identification and structural analysis of the target site for S8 repressor protein. *J. Mol. Biol.* **204:**309–329.

Champney, W. S. 1980. Protein synthesis defects in temperature-sensitive mutants of *E. coli* with altered ribosomal proteins. *Biochim. Biophys. Acta* **609:**464–474.

Chang, F. N., and J. G. Flaks. 1972. Binding of dihydrostreptomycin to *Escherichia coli* ribosomes: characteristics and equilibrium of the reaction. *Antimicrob. Agents Chemother.* **2:**294–307.

Chao, F. C. 1957. Dissociation of macromolecular ribonucleoprotein of yeast. *Arch. Biochem. Biophys.* **70:**426–431.

Clark, B. F. C., and K. A. Marcker. 1966. The roles of N-formylmethionyl-sRNA in protein biosynthesis. *J. Mol. Biol.* **17:**394–406.

Clarke, L., and J. Carbon. 1976. A colony bank containing synthetic Col El hybrid plasmids representative of the entire *E. coli* genome. *Cell* **9:**91–99.

Cohlberg, J. A., and M. Nomura. 1976. Reconstitution of *Bacillus stearothermophilus* 50S ribosomal subunits from purified molecular components. *J. Biol. Chem.* **251:**209–221.

Cole, J. R., and M. Nomura. 1986a. Translational regulation is responsible for growth-rate-dependent and stringent control of the synthesis of ribosomal proteins L11 and L1 in *E. coli. Proc. Natl. Acad. Sci. USA* **83:**4129–4133.

Cole, J. R., and M. Nomura. 1986b. Changes in the half-life of ribosomal protein messenger RNA caused by translational repression. *J. Mol. Biol.* **188:**383–392.

Cole, J. R., C. L. Olsson, J. W. B. Hershey, M. Grunberg-Mango, and M. Nomura. 1987. Feedback regulation of rRNA synthesis in *E. coli*: requirement for initiation factor IF2. *J. Mol. Biol.* **198:**383–392.

Colli, W., and M. Oishi. 1969. Ribosomal RNA genes in bacteria: evidence for the nature of the physical linkage between 16S and 23S RNA genes in *Bacillus subtilis. Proc. Natl. Acad. Sci. USA* **64:**642–649.

Colli, W., I. Smith, and M. Oishi. 1971. Physical linkage between 5S, 16S, and 23S ribosomal RNA genes in *Bacillus subtilis. J. Mol. Biol.* **56:**117–127.

Cooperman, B. S. 1980. Functional sites on the *Escherichia coli* ribosome as defined by affinity labeling, p. 531–554. *In* G. Chambliss, G. R. Craven, J. Davies, K. Davis, L. Kahan, and M. Nomura (ed.), *Ribosomes. Structure, Function, and Genetics.* University Park Press, Baltimore.

Cox, E. D., J. R. White, and J. G. Flaks. 1964. Streptomycin action and the ribosome. *Proc. Natl. Acad. Sci. USA* **51:**703–709.

Craven, G. R., P. Voynow, S. J. S. Hardy, and C. G. Kurland. 1969. The ribosomal protein of *E. coli.* II. Chemical and physical characterization of the 30S ribosomal proteins. *Biochemistry* **8:**2906–2915.

Crick, F. H. C. 1958. On protein synthesis. *Symp. Soc. Exp. Biol.* **12:**138–163.

Crick, F. H. C. 1968. The origin of the genetic code. *J. Mol. Biol.* **38:**367–379.

Dabbs, E. R. 1979. Selection of *Escherichia coli* mutants with proteins missing from the ribosome. *J. Bacteriol.* **140:**734–737.

Dabbs, E. R. 1986. Mutant studies on the prokaryotic ribosome, p. 733–748. *In* B. Hardesty and G. Kramer (ed.), *Structure, Function, and Genetics of Ribosomes.* Springer-Verlag, New York.

Dabbs, E. R., and H. G. Wittmann. 1976. A strain of *E. coli* which gives rise to mutations in a large number of ribosomal proteins. *Mol. Gen. Genet.* **149:**303–309.

Davies, J. E. 1964. Studies on the ribosomes of streptomycin-sensitive and resistant strains of *E. coli. Proc. Natl. Acad. Sci. USA* **51:**659–664.

Davies, J. E., W. Gilbert, and L. Gorini. 1964. Streptomycin suppression, and the code. *Proc. Natl. Acad. Sci. USA* **51:**883–890.

Davis, B. D. 1974. Alternative views on the ribosome cycle, p. 705–710. *In* M. Nomura, A. Tissières, and P. Lengyel (ed.), *Ribosomes.* Cold Spring Harbor Laboratory, Cold Spring Harbor, N.Y.

Dean, D., and M. Nomura. 1980. Feedback regulation of ribosomal protein gene expression in *E. coli. Proc. Natl. Acad. Sci. USA* **77:**3590–3594.

Dean, D., J. L. Yates, and M. Nomura. 1981a. Identification of ribosomal protein S7 as a repressor of translation within the *str* operon of *E. coli. Cell* **24:**413–419.

Dean, D., J. L. Yates, and M. Nomura. 1981b. *E. coli* ribosomal protein S8 feedback regulates part of *spc* operon. *Nature* (London) **289:**89–91.

deBoer, H., S. F. Gilbert, and M. Nomura. 1979. DNA sequences of promoter regions for rRNA operons *rrnE* and *rrnA* in *E. coli. Cell* **17:**201–209.

deBoer, H., and M. Nomura. 1979. *In vivo* transcription of rRNA operons in *E. coli* initiates with purine nucleoside triphosphate at the first promoter and with CTP at the second promoter. *J. Biol. Chem.* **254:**5609–5612.

Denman, R., D. Negre, P. Cunningham, K. Nurse, J. Colgan, C. Weitzmann, and J. Ofengand. 1989a. Effect of point mutations in the decoding site (C1400) region of 16S ribosomal RNA on the ability of ribosomes to carry out individual steps of protein synthesis. *Biochemistry* **28:**1012–1019.

Denman, R., C. Weitzmann, P. R. Cunningham, D. Negre, K. Nurse, J. Colgan, Y. Pan, M. Miedel, and J. Ofengand. 1989b. *In vitro* assembly of 30S and 70S bacterial ribosomes from 16S RNA containing single base substitutions, insertions, and deletions around the decoding site (C1400). *Biochemistry* **28:**1002–1011.

Dennis, P. P. 1974. *In vivo* stability, maturation and relative differential synthesis rates of individual ribosomal proteins in *E. coli* B/r. *J. Mol. Biol.* **88:**25–41.

Dennis, P. P. 1977. Transcription patterns of adjacent segments on the chromosome of *E. coli* containing genes coding for four 50S ribosomal proteins and the β and β' subunits of RNA polymerase. *J. Mol. Biol.* **115:**603–625.

Dennis, P. P., and N. P. Fiil. 1979. Transcriptional and posttranscriptional control of RNA polymerase and ribosomal protein genes cloned on composite ColE1 plasmids in the bacterium *E. coli. J. Biol. Chem.* **254:**7540–7547.

Dennis, P. P., and M. Nomura. 1974. Stringent control of ribosomal protein gene expression in *E. coli. Proc. Natl. Acad. Sci. USA* **71:**3819–3823.

Dennis, P. P., and M. Nomura. 1975a. Regulation of the expression of ribosomal protein genes in *E. coli. J. Mol. Biol.* **97:** 61–76.

Dennis, P. P., and M. Nomura. 1975b. Stringent control of the transcriptional activities of ribosomal protein genes in *E. coli. Nature* (London) **255:**460–465.

Deonier, R. C., E. Ohtsubo, H. J. Lee, and N. Davidson. 1974. Electron microscope heteroduplex studies of sequence relations among plasmids in *E. coli*. VII. Mapping the ribosomal RNA genes of plasmid F14. *J. Mol. Biol.* **89:**619–629.

Doolittle, W. F., and N. R. Pace. 1971. Transcriptional organization of the ribosomal RNA cistrons in *E. coli. Proc. Natl. Acad. Sci. USA* **68:**1786–1790.

Doty, P., H. Boedtker, J. R. Fresco, R. Haselkorn, and M. Litt. 1959. Secondary structure in ribonucleic acids. *Proc. Natl. Acad. Sci. USA* **45:**482–499.

Douthwaite, S., J. B. Prince, and H. F. Noller. 1985. Evidence for functional interaction between domains II and V of 23S ribosomal RNA from an erythromycin-resistant mutant. *Proc. Natl. Acad. Sci. USA* **82:**8330–8334.

Dubnau, D., I. Smith, and J. Marmur. 1965. Gene conservation in *Bacillus* species. II. The location of genes concerned with the synthesis of ribosomal components and soluble RNA. *Proc. Natl. Acad. Sci. USA* **54:**724–730.

Dunn, J. J., and F. W. Studier. 1973. T7 early RNAs and *E. coli* ribosomal RNAs are cut from large precursor RNAs *in vivo* by ribonuclease III. *Proc. Natl. Acad. Sci. USA* **70:**3296–3300.

Edlin, G., and P. Broda. 1968. Physiology and genetics of the "ribonucleic acid control" locus in *E. coli. Bacteriol. Rev.* **32:**206–226.

Ehrenberg, M., D. Andersson, K. Bohman, P. Jelenc, T. Ruusala, and C. G. Kurland. 1986. Ribosomal proteins tune rate and accuracy in translation, p. 573–585. *In* B. Hardesty and G. Kramer (ed.), *Structure, Function, and Genetics of Ribosomes.* Springer-Verlag, New York.

Eidlic, L., and F. C. Neidhardt. 1965. Protein and nucleic acid synthesis in two mutants of *Escherichia coli* with temperature-sensitive aminoacyl ribonucleic acid synthetase. *J. Bacteriol.* **89:**706–711.

Ellis, R. J. 1987. Proteins as molecular chaperones. *Nature* (London) **328:**378–379.

Ellwood, M., and M. Nomura. 1982. Chromosomal locations of the genes for rRNA in *Escherichia coli* K-12. *J. Bacteriol.* **149:**458–468.

Elson, D. 1958. Latent ribonuclease activity in a ribonucleoprotein. *Biochim. Biophys. Acta* **77:**216–217.

Engelman, D., and P. Moore. 1972. A new method for the determination of biological quarternary structure by neutron scattering. *Proc. Natl. Acad. Sci. USA* **69:**1997–1999.

Engelman, D., P. Moore, and B. Schoenborn. 1975. Neutron scattering measurements of separation and shape of proteins in 30S ribosomal subunit *E. coli*: S2-S5, S5-S8, S3-S7. *Proc. Natl. Acad. Sci. USA* **72:**3888–3892.

Erdmann, V. A., S. Fahnestock, K. Higo, and M. Nomura. 1971. Role of 5S RNA in the functions of 50S ribosomal subunits. *Proc. Natl. Acad. Sci. USA* **68:**2932–2936.

Fahnestock, S., V. Erdmann, and M. Nomura. 1973. Reconstitution of 50S ribosomal subunits from protein-free ribonucleic acid. *Biochemistry* **12:**220–224.

Fallon, A. M., C. S. Jinks, G. D. Strycharz, and M. Nomura. 1979. Regulation of ribosomal protein synthesis in *E. coli* by selective mRNA inactivation. *Proc. Natl. Acad. Sci. USA* **76:**3411–3415.

Fangman, W. L., and F. C. Neidhardt. 1964. Protein and ribonucleic acid synthesis in a mutant of *E. coli* with an altered aminoacyl ribonucleic acid synthetase. *J. Biol. Chem.* **239:** 1844–1847.

Fellner, P. 1974. Structure of the 16S and 23 S ribosomal RNAs, p. 169–192. *In* M. Nomura, A. Tissières, and P. Lengyel (ed.), *Ribosomes.* Cold Spring Harbor Laboratory, Cold Spring Harbor, N.Y.

Fellner, P., and F. Sanger. 1968. Sequence analysis of specific areas of the 16S and 23S ribosomal RNAs. *Nature* (London) **219:** 236–238.

Fiil, N., and J. D. Friesen. 1968. Isolation of "relaxed" mutants of *Escherichia coli. J. Bacteriol.* **95:**729–731.

Finley, D., B. Bartel, and A. Varshavsky. 1989. The tails of ubiquitin precursors are ribosomal proteins whose fusion to ubiquitin facilitates ribosome biogenesis. *Nature* (London) **338:** 394–400.

Fraenkel-Conrat, H. 1957. Degradation of tobacco mosaic virus with acetic acid. *Virology* **4:**1–4.

Fraenkel-Conrat, H., and R. C. Williams. 1955. Reconstitution of active tobacco mosaic virus from its inactive protein and nucleic acid components. *Proc. Natl. Acad. Sci. USA* **41:**690–698.

Freedman, L. P., J. M. Zengel, R. H. Acher, and L. Lindahl. 1987. Autogenous control of the S10 ribosomal protein operon of *E. coli*: genetic dissection of transcriptional and posttranscriptional regulation. *Proc. Natl. Acad. Sci. USA* **84:**6516–6520.

Freedman, L. P., J. M. Zengel, and L. Lindahl. 1985. Genetic dissection of stringent control and nutritional shift-up response of the *E. coli* S10 ribosomal protein operon. *J. Mol. Biol.*

185:701–712.

Fresco, J. R., B. M. Alberts, and P. Doty. 1960. Some molecular details of the secondary structure of ribonucleic acid. *Nature* (London) **188:**98–101.

Friedman, D. I., and E. R. Olson. 1983. Evidence that a nucleotide sequence, "box A", is involved in the action of the NusA protein. *Cell* **34:**143–149.

Friesen, J. D., N. P. Fiil, J. M. Parker, and W. A. Haseltine. 1974. A new relaxed mutant of *E. coli* with an altered 50S ribosomal subunit. *Proc. Natl. Acad. Sci. USA* **71:**3465–3469.

Fukuda, R. 1980. Autogenous regulation of the synthesis of ribosomal proteins, L10 and L7/12, in *E. coli. Mol. Gen. Genet.* **178:**483–486.

Gallant, J., and D. Foley. 1980. On the causes and prevention of mistranslation, p. 615–638. *In* G. Chambliss, G. R. Craven, J. Davies, K. Davis, L. Kahan, and M. Nomura (ed.), *Ribosomes. Structure, Function, and Genetics.* University Park Press, Baltimore.

Gallant, J., L. Palmer, and C. C. Pao. 1977. Anomalous synthesis of ppGpp in growing cells. *Cell* **11:**181–185.

Gausing, K. 1977. Regulation of ribosome production in *E. coli*: synthesis and stability of ribosomal RNA and of ribosomal protein messenger RNA at different growth rates. *J. Mol. Biol.* **115:**335–354.

Gesteland, R. F. 1966. Unfolding of E. coli ribosomes by removal of magnesium. *J. Mol. Biol.* **18:**356–371.

Geyl, D., and A. Böck. 1977. Synthesis of ribosomal proteins in merodiploid strains and in minicells of *E. coli. Mol. Gen. Genet.* **154:**327–334.

Ghosh, H. P., and H. G. Khorana. 1967. Studies on polynucleotides. LXXXIV. On the role of ribosomal subunits in protein synthesis. *Proc. Natl. Acad. Sci. USA* **58:**2455–2461.

Gierer, A. 1963. Function of aggregated reticulocyte ribosomes in protein synthesis. *J. Mol. Biol.* **6:**148–157.

Gilbert, S. F., H. A. deBoer, and N. Nomura. 1979. Identification of initiation sites for the in vitro transcription of rRNA operons *rrnE* and *rrnA* in *E. coli. Cell* **17:**211–224.

Gilbert, W. 1963a. Polypeptide synthesis in *E. coli.* I. Ribosomes and the active complex. *J. Mol. Biol.* **6:**374–388.

Gilbert, W. 1963b. Polypeptide synthesis in *E. coli.* II. The polypeptide chain and S-RNA. *J. Mol. Biol.* **6:**389–403.

Glaser, G., and M. Cashel. 1979. *In vitro* transcripts from the *rrn* B ribosomal RNA cistron originate from two tandem promoters. *Cell* **16:**111–121.

Glaser, G., P. Sarmientos, and M. Cashel. 1983. Functional interrelationship between two tandem *E. coli* ribosomal RNA promoters. *Nature* (London) **302:**74–76.

Glotz, C., C. Zwieb, and R. Brimacombe. 1981. Secondary structure of the large subunit ribosomal RNA from *E. coli*, Zea mays chloroplast, and human and mouse mitochondrial ribosomes. *Nucleic Acids Res.* **9:**3287–3306.

Gold, L., P. Z. O'Farrell, and M. Russel. 1976. Regulation of gene 32 expression during bacteriophage T4 infection of *E. coli. J. Biol. Chem.* **251:**7251–7262.

Gorini, L., W. Gundersen, and M. Burger. 1961. Genetics of regulation of enzyme synthesis in the arginine biosynthetic pathway of *E. coli. Cold Spring Harbor Symp. Quant. Biol.* **26:**173–182.

Gorini, L., and E. Kataja. 1964. Phenotypic repair by streptomycin of defective genotypes in *E. coli. Proc. Natl. Acad. Sci. USA* **51:**487–493.

Gourse, R. L., H. A. deBoer, and M. Nomura. 1986. DNA determinants of rRNA synthesis in *E. coli*: growth-rate-dependent regulation, feedback inhibition, upstream activation and antitermination. *Cell* **44:**197–205.

Gregory, R. J., P. B. F. Cahill, D. F. Thurlow, and R. A. Zimmermann. 1988. Interaction of *E. coli* ribosomal protein S8 with its binding sites in ribosomal RNA and messenger RNA. *J. Mol. Biol.* **204:**295–307.

Gros, F., H. Hiatt, W. Gilbert, C. G. Kurland, R. W. Risebrough, and J. D. Watson. 1961. Unstable ribonucleic acid revealed by pulse labeling of *E. coli. Nature* (London) **190:**581–585.

Guerrier-Takada, C., K. Gardiner, T. Marsh, N. Pace, and S. Altman. 1983. The RNA moiety of ribonuclease P is the catalytic subunit of enzyme. *Cell* **35:**849–857.

Guthrie, C., H. Nashimoto, and M. Nomura. 1969. Structure and function of *E. coli* ribosomes. VIII. Cold-sensitive mutants defective in ribosome assembly. *Proc. Natl. Acad. Sci. USA* **63:**384–391.

Guthrie, C., and M. Nomura. 1968. Initiation of protein synthesis: a critical test of the 30S subunit model. *Nature* (London) **219:**232–235.

Hall, B. D., and S. Spiegelman. 1961. Sequence complementarity of T2-DNA and T2-specific RNA. *Proc. Natl. Acad. Sci. USA* **47:**137–146.

Hall, C. E., and H. S. Slayter. 1959. Electron microscopy of ribonucleoprotein particles from *E. coli. J. Mol. Biol.* **1:**329–332.

Hampl, H., H. Schulze, and K. H. Nierhaus. 1981. Ribosomal components from *E. coli* 50S subunits involved in the reconstitution of peptidyltransferase activity. *J. Biol. Chem.* **256:**2284–2288.

Hardy, S. J. S. 1975. The stoichiometry of the ribosomal proteins of *E. coli. Mol. Gen. Genet.* **140:**253–274.

Hardy, S. J. S., C. G. Kurland, P. Voynow, and G. Mora. 1969. The ribosomal proteins of *E. coli.* I. Purification of the 30S ribosomal proteins. *Biochemistry* **8:**2897–2905.

Haseltine, W. A., R. Block, K. Weber, and W. Gilbert. 1972. MSI and MSII made on the ribosome in idling step of protein synthesis. *Nature* (London) **238:**381–385.

Held, W. A., B. Ballou, S. Mizushima, and M. Nomura. 1974a. Assembly mapping of 30S ribosomal proteins from *E. coli*: further studies. *J. Biol. Chem.* **249:**3103–3111.

Held, W. A., W. R. Gette, and M. Nomura. 1974b. Role of 16S ribosomal ribonucleic acid and the 30S ribosomal protein S12 in the initiation of the natural messenger ribonucleic acid translation. *Biochemistry* **13:**2115–2122.

Held, W. A., S. Mizushima, and M. Nomura. 1973. Reconstitution of *E. coli* 30S ribosomal subunits from purified molecular components. *J. Biol. Chem.* **248:**5720–5730.

Held, W. A., and M. Nomura. 1975. *E. coli* 30S ribosomal proteins uniquely required for assembly. *J. Biol. Chem.* **250:** 3179–3184.

Hemmingsen, S. M., C. Woolford, S. M. van der Vies, K. Tilly, D. Dennis, C. P. Georgopoulos, R. Hendrix, and R. J. Ellis. 1988. Homologous plant and bacterial proteins chaperone oligomeric protein assembly. *Nature* (London) **333:**330–334.

Herzog, A., M. Yaguchi, T. Cabezón, M.-C. Corchuelo, J. Petre, and A. Bollen. 1979. A missense mutation in the gene coding for ribosomal protein S17 (*rpsQ*) leading to ribosomal assembly defectivity in *E. coli. Mol. Gen. Genet.* **171:**15–22.

Higo, K., W. Held, L. Kahan, and M. Nomura. 1973. Functional correspondence between 30S ribosomal proteins of *E. coli* and *Bacillus stearothermophilus. Proc. Natl. Acad. Sci. USA* **70:** 944–948.

Hille, M. B., M. J. Miller, K. Iwasaki, and A. J. Wahba. 1967. Translation of the genetic message. VI. The role of ribosomal subunits in binding formylmethionyl-tRNA and its reaction with puromycin. *Proc. Natl. Acad. Sci. USA* **58:**1652–1654.

Holley, R. W., J. Apgar, G. A. Everett, J. T. Madison, S. H. Merrill, J. R. Penswick, and A. Zamir. 1965. Structure of a ribonucleic acid. *Science* **147:**1462–1465.

Hopfield, J. J. 1974. Kinetic proofreading: a new mechanism for reducing errors in biosynthetic processes requiring high specificity. *Proc. Natl. Acad. Sci. USA* **71**:4135–4139.

Hortsch, M., D. Avossa, and D. I. Meyer. 1986. Characterization of secretory protein translocation: ribosome-membrane interaction in endoplasmic reticulum. *J. Cell Biol.* **103**:241–253.

Hosokawa, K., R. K. Fujimura, and M. Nomura. 1966. Reconstitution of functionally active ribosomes from inactive subparticles and proteins. *Proc. Natl. Acad. Sci. USA* **55**:198–204.

Hosokawa, K., and M. Nomura. 1965. Incomplete ribosomes produced in chloramphenicol and puromycin-inhibited *E. coli. J. Mol. Biol.* **12**:225–241.

Huang, K. H., R. H. Fairclough, and C. R. Cantor. 1975. Singlet energy transfer studies of the arrangement of proteins in the 30S *E. coli* ribosome. *J. Mol. Biol.* **97**:443–470.

Hui, A., and H. deBoer. 1987. Specialized ribosome system: preferential translation of a single mRNA species by a subpopulation of mutated ribosomes in *E. coli. Proc. Natl. Acad. Sci. USA* **84**:4762–4766.

Huxley, H. E., and G. Zubay. 1960. Electron microscope observations on the structure of microsomal particles from *E. coli. J. Mol. Biol.* **2**:10–18.

Ikemura, T. 1981. Correlation between the abundance of *E. coli* transfer RNAs and the occurrence of the respective codons in its protein genes. *J. Mol. Biol.* **146**:1–21.

Ilan, J. (ed.). 1987. *Translational Regulation of Gene Expression.* Plenum Publishing Corp., New York.

Ingraham, J. L., O. Maaløe, and F. C. Neidhardt. 1983. *Growth of the Bacterial Cell.* Sinauer Associates, Inc., Sunderland, Mass.

Isono, K. 1980. Genetics of ribosomal proteins and their modifying and processing enzymes in *Escherichia coli*, p. 641–669. *In* G. Chambliss, G. R. Craven, J. Davies, K. Davis, L. Kahan, and M. Nomura (ed.), *Ribosomes. Structure, Function, and Genetics.* University Park Press, Baltimore.

Isono, K., J. Krauss, and Y. Hirota. 1976. Isolation and characterization of temperature-sensitive mutants of *E. coli* with altered ribosomal proteins. *Mol. Gen. Genet.* **149**:297–302.

Ito, K., D. P. Cerretti, H. Nashimoto, and M. Nomura. 1984. Characterization of an amber mutation in the structural gene for ribosomal protein L15 which impairs the expression of the protein export gene, secY, in *E. coli. EMBO J.* **3**:2319–2324.

Ito, K., M. Wittekind, M. Nomura, K. Shiba, T. Yura, A. Miura, and H. Nashimoto. 1983. A temperature-sensitive mutant of *E. coli* exhibiting slow processing of exported proteins. *Cell* **32**: 789–797.

Jacob, F., and J. Monod. 1961. Genetic regulatory mechanisms in the synthesis of proteins. *J. Mol. Biol.* **3**:318–356.

Jacobs, W. F., M. Santer, and A. Dahlberg. 1987. A single base change in the Shine-Dalgarno region of 16S rRNA of *E. coli* affects translation of many proteins. *Proc. Natl. Acad. Sci. USA* **84**:4757–4761.

Jaskunas, S. R., R. R. Burgess, and M. Nomura. 1975a. Identification of a gene for the α-subunit of RNA polymerase at the *str-spc* region of the *E. coli* chromosome. *Proc. Natl. Acad. Sci. USA* **72**:5036–5040.

Jaskunas, S. R., A. M. Fallon, and M. Nomura. 1977. Identification and organization of ribosomal protein gene of *E. coli* carried by λ*fus*2 transducing phage. *J. Biol. Chem.* **252**:7323–7336.

Jaskunas, S. R., L. Lindahl, and M. Nomura. 1975b. Specialized transducing phages for ribosomal protein genes of *E. coli. Proc. Natl. Acad. Sci. USA* **72**:6–10.

Jaskunas, S. R., L. Lindahl, M. Nomura, and R. R. Burgess. 1975c. Identification of two copies of the gene for the elongation factor EF-Tu in *E. coli. Nature* (London) **257**:458–462.

Jinks-Robertson, S., R. L. Gourse, and M. Nomura. 1983. Expression of rRNA and tRNA genes in *E. coli*: evidence for feedback regulation by products of rRNA operons. *Cell* **33**:865–876.

Judson, H. F. 1978. *The Eighth Day of Creation.* Simon and Shuster, New York.

Kaempfer, R. 1968. Ribosomal subunit exchange during protein synthesis. *Proc. Natl. Acad. Sci. USA* **61**:106–113.

Kaempfer, R. 1969. Ribosomal subunit exchange in the cytoplasm of a eukaryote. *Nature* (London) **222**:950–953.

Kaempfer, R. 1974. The ribosome cycle, p. 679–704. *In* M. Nomura, A. Tissières, and P. Lengyel (ed.), *Ribosomes.* Cold Spring Harbor Laboratory, Cold Spring Harbor, N.Y.

Kaempfer, R., and M. Meselson. 1969. Studies of ribosomal subunit exchange. *Cold Spring Harbor Symp. Quant. Biol.* **34**:209–222.

Kaempfer, R., M. Meselson, and H. Raskas. 1968. Cyclic dissociation into stable subunits and re-formation of ribosomes during bacterial growth. *J. Mol. Biol.* **31**:277–289.

Kaji, H., and A. Kaji. 1964. Specific binding of sRNA with the template-ribosome complex. *Proc. Natl. Acad. Sci. USA* **52**: 1541–1547.

Kaltschmidt, E., M. Dzionara, D. Donner, and H. G. Wittmann. 1967. Ribosomal proteins. I. Isolation, amino acid composition, molecular weights, and peptide mapping of proteins from *E. coli* ribosomes. *Mol. Gen. Genet.* **100**:364–373.

Kaltschmidt, E., L. Kahan, and M. Nomura. 1974. *In vitro* synthesis of ribosomal proteins directed by *E. coli* DNA. *Proc. Natl. Acad. Sci. USA* **71**:446–450.

Kaltschmidt, E., and H. G. Wittmann. 1970. Ribosomal proteins. XII. Number of proteins in small and large ribosomal subunits in *E. coli* as determined by two-dimensional gel electrophoresis. *Proc. Natl. Acad. Sci. USA* **67**:1276–1282.

Kenerley, M. E., E. A. Morgan, L. Post, L. Lindahl, and M. Nomura. 1977. Characterization of hybrid plasmids carrying individual ribosomal ribonucleic acid transcription units of *Escherichia coli. J. Bacteriol.* **132**:931–949.

Kiss, A., B. Sain, and P. Venetianer. 1977. The number of rRNA genes in *E. coli. FEBS Lett.* **79**:77–79.

Kjeldgaard, N. O., O. Maaløe, and M. Schaechter. 1958. The transition between different physiological states during balanced growth of *Salmonella typhimurium. J. Gen. Microbiol.* **19**: 607–616.

Koch, A. L. 1970. Overall controls on the biosynthesis of ribosomes in growing bacteria. *J. Theor. Biol.* **28**:203–231.

Koch, A. L. 1971. The adaptive responses of *E. coli* to a feast and famine existence. *Adv. Microb. Physiol.* **6**:147–217.

Kohler, R., E. Ron, and B. Davis. 1968. Significance of the free 70S ribosomes in *E. coli* extracts. *J. Mol. Biol.* **36**:71–82.

Konisky, J. 1968. Biochemical effects of colicins on the bacterium *E. coli.* Ph.D. thesis, University of Wisconsin, Madison.

Konisky, J., and M. Nomura. 1967. Interaction of colicins with bacterial cells. II. Specific alteration of *E. coli* ribosomes induced by colicin E3 *in vivo. J. Mol. Biol.* **26**:181–195.

Kozak, M. 1978. How do eucaryotic ribosomes select initiation regions in messenger RNA? *Cell* **15**:1109–1123.

Kruger, K., P. J. Grabowski, A. J. Zaug, J. Sands, D. E. Gottschling, and T. R. Cech. 1982. Self-splicing RNA: autoexcision and autocyclization of the ribosomal RNA intervening sequence of tetrahymena. *Cell* **31**:147–157.

Kurland, C. G. 1960. Molecular characterization of ribonucleic acid from *E. coli* ribosomes. I. Isolation and molecular weights. *J. Mol. Biol.* **2**:83–91.

Kurland, C. G. 1980. On the accuracy of elongation, p. 597–614. *In* G. Chambliss, G. R. Craven, J. Davies, K. Davis, L. Kahan, and M. Nomura (ed.), *Ribosomes. Structure, Function, and Genetics.* University Park Press, Baltimore.

Kurland, C. G., and O. Maaløe. 1962. Regulation of ribosomal and transfer RNA synthesis. *J. Mol. Biol.* **4**:193–210.

Kurland, C. G., M. Nomura, and J. D. Watson. 1962. The physical properties of the chloromycetin particles. *J. Mol. Biol.* **4:** 388–394.

Kurland, C. G., P. Voynow, S. J. S. Hardy, L. Randall, and L. Lutter. 1969. Physical and functional heterogeneity of *E. coli* ribosomes. *Cold Spring Harbor Symp. Quant. Biol.* **34:**17–24.

Lake, J. A. 1976. Ribosome structure determined by electron microscopy of *E. coli* small subunits, large subunits and monomeric ribosomes. *J. Mol. Biol.* **105:**131–159.

Lake, J. A., M. Pendergast, L. Kahan, and M. Nomura. 1974. Localization of *E. coli* ribosomal proteins S4 and S14 by electron microscopy of antibody-labeled subunits. *Proc. Natl. Acad. Sci. USA* **71:**4688–4692.

Lamborg, M. R., and P. C. Zamecnik. 1960. Amino acid incorporation into protein by extracts of *E. coli. Biochim. Biophys. Acta* **42:**206–211.

Lamond, A. I., and A. A. Travers. 1983. Requirement for an upstream element for optimal transcription of a bacterial tRNA gene. *Nature* (London) **305:**248–250.

Lazzarini, R. A., J. Cashel, and J. Gallant. 1971. On the regulation of guanosine tetraphosphate levels in stringent and relaxed strains of *E. coli. J. Biol. Chem.* **246:**4381–4385.

Leboy, P. S., and E. C. Cox, and J. G. Flaks. 1964. The chromosomal site specifying a ribosomal protein in *E. coli. Proc. Natl. Acad. Sci. USA* **52:**1367–1374.

Lefkovits, I., and M. Di Girolamo. 1969. Properties of ribonucleoprotein particles in chloramphenicol-treated cells of *E. coli* B. *Biochim. Biophys. Acta* **174:**561–565.

Lemaire, G., L. Gold, and M. Yarus. 1978. Autogenous translational repression of bacteriophage T4 gene 32 expression in vitro. *J. Mol. Biol.* **126:**73–90.

Leonard, K. R., and J. A. Lake. 1979. Ribosome structure: hand determination by electron microscopy of 30S subunits. *J. Mol. Biol.* **129:**155–163.

Lerman, M. I., A. S. Spirin, L. P. Gavrilova, and V. F. Golov. 1966. Studies on the structure of ribosomes. II. Stepwise dissociation of protein from ribosomes by caesium chloride and the reassembly of ribosome-like particles. *J. Mol. Biol.* **15:**268–281.

Levinthal, C., A. Keynan, and A. Higa. 1962. Messenger RNA turnover and protein synthesis in *B. subtilis* inhibited by actinomycin D. *Proc. Natl. Acad. Sci. USA* **48:**1613–1638.

Li, S. C., C. L. Squires, and C. Squires. 1984. Antitermination of *E. coli* rRNA transcription is caused by a control region segment containing lambda *nut*-like sequences. *Cell* **38:**851–860.

Lill, R., E. Crooke, B. Guthrie, and W. Wickner. 1988. The "trigger factor cycle" includes ribosomes, presecretory proteins, and the plasma membrane. *Cell* **54:**1013–1018.

Lindahl, L., R. Archer, and J. M. Zengel. 1983. Transcription of the S10 ribosomal protein operon is regulated by an attenuator in the leader. *Cell* **33:**241–248.

Lindahl, L., S. R. Jaskunas, P. P. Dennis, and M. Nomura. 1975. Cluster of genes in *E. coli* for ribosomal proteins, ribosomal RNA, and RNA polymerase subunits. *Proc. Natl. Acad. Sci. USA* **72:**2743–2747.

Lindahl, L., L. Post, and M. Nomura. 1976. DNA-dependent in vitro synthesis of ribosomal proteins, protein elongation factors, and RNA polymerase subunit α: inhibition by ppGpp. *Cell* **9:**439–448.

Lindahl, L., and J. M. Zengel. 1979. Operon-specific regulation of ribosomal protein synthesis in *E. coli. Proc. Natl. Acad. Sci. USA* **76:**6542–6546.

Linn, T., and J. Scaife. 1978. Identification of single promoter in *E. coli* for *rplJ*, *rplL* and *rpoBC*. *Nature* (London) **276:**33–37.

Littlefield, J. W., E. B. Keller, J. Gros, and P. C. Zamecnik. 1955. Studies on cytoplasmic ribonucleoprotein particles from the liver of the rat. *J. Biol. Chem.* **217:**111–123.

Lotti, M., E. R. Dabbs, R. Hasenbank, M. Stöffler-Meilicke, and G. Stöffler. 1983. Characterization of a mutant from *E. coli* lacking protein L15 and localization of protein L15 by immunoelectron microscopy. *Mol. Gen. Genet.* **192:**295–300.

Lubin, M. 1968. Observations on the shape of the 50S ribosomal subunit. *Proc. Natl. Acad. Sci. USA* **61:**1454–1461.

Lund, E., and J. E. Dahlberg. 1979. Initiation of *E. coli* ribosomal RNA synthesis *in vivo*. *Proc. Natl. Acad. Sci. USA* **76:**5480–5484.

Lund, E., J. E. Dahlberg, L. Lindahl, S. R. Jaskunas, P. P. Dennis, and M. Nomura. 1976. Transfer RNA genes between 16S and 23S rRNA genes in rRNA transcription units of *E. coli. Cell* **7:**165–177.

Maaløe, O. 1969. An analysis of bacterial growth. *Dev. Biol. Suppl.* **3:**33–58.

Maaløe, O. 1979. Regulation of the protein-synthesizing machinery-ribosomes, tRNA, factors and so on, p. 487–542. *In* R. F. Goldberger (ed.), *Biological Regulation and Development*, vol. 1. Plenum Publishing Corp., New York.

Maaløe, O., and N. O. Kjeldgaard. 1966. *Control of Macromolecular Synthesis*. W. A. Benjamin, Inc., New York.

Mangiarotti, G., and D. Schlessinger. 1966. Polyribosome metabolism in *E. coli*. I. Extraction of polyribosomes and ribosomal subunits from fragile, growing *E. coli. J. Mol. Biol.* **20:**123–143.

Manor, H., and R. Haselkorn. 1967. Properties of ribonucleic acid components in ribonucleoprotein particle preparations obtained from an *E. coli* $RC^{relaxed}$ strain. *J. Mol. Biol.* **24:**269–288.

Marcker, K., and F. Sanger. 1964. N-formyl-methionyl-sRNA. *J. Mol. Biol.* **8:**835–840.

Marquis, D., and S. R. Fahnestock. 1978. A complex of acidic ribosomal proteins. Evidence of a four-to-one complex of proteins in the *Bacillus stearothermophilus* ribosome. *J. Mol. Biol.* **119:**557–567.

Maruta, H., T. Tsuchiya, and D. Mizuno. 1971. *In vitro* reassembly of functionally active 50S ribosomal particles from ribosomal proteins and RNAs of *E. coli. J. Mol. Biol.* **61:**123–134.

Mattheakis, L., L. Vu, F. Sor, and M. Nomura. 1989. Retroregulation of the synthesis of ribosomal proteins L14 and L24 by feedback repressor S8 in *E. coli. Proc. Natl. Acad. Sci. USA* **86:**448–452.

Maxam, A. M., and W. Gilbert. 1977. A new method for sequencing DNA. *Proc. Natl. Acad. Sci. USA* **74:**560–564.

McCarthy, B., and A. Aronson. 1961. The kinetics of the synthesis of ribosomal RNA in *E. coli. Biophys. J.* **1:**227–245.

McCarthy, B. J., R. J. Britten, and R. B. Roberts. 1962. The synthesis of ribosomes in *E. coli*. III. Synthesis of ribosomal RNA. *Biophys. J.* **2:**57–82.

McQuillen, K., R. B. Roberts, and R. J. Britten. 1959. Synthesis of nascent protein by ribosomes in *E. coli. Proc. Natl. Acad. Sci. USA* **45:**1437–1447.

Meselson, M., M. Nomura, S. Brenner, C. Davern, and D. Schlessinger. 1964. Conservation of ribosomes during bacterial growth. *J. Mol. Biol.* **9:**696–711.

Meselson, M., and F. W. Stahl. 1958. The replication of DNA in *E. coli. Proc. Natl. Acad. Sci. USA* **44:**671–682.

Miura, A., J. H. Krueger, S. Itoh, H. A. deBoer, and M. Nomura. 1981. Growth-rate-dependent regulation of ribosome synthesis in *E. coli*: expression of the *lacZ* and *galK* genes fused to ribosomal promoters. *Cell* **25:**773–782.

Mizushima, S., and M. Nomura. 1970. Assembly mapping of 30S ribosomal proteins from E. coli. *Nature* (London) **226:**1214–1218.

Moazed, D., and H. F. Noller. 1987. Interaction of antibiotics with functional sites in 16S ribosomal RNA. *Nature* (London) **327:** 389–394.

Mochalova, L. V., I. N. Shatsky, A. A. Bogdanov, and V. D.

Vasiliev. 1982. Topography of RNA in the ribosome: localization of the 16S RNA 5′ end by immune electron microscopy. *J. Mol. Biol.* **159:**637–650.

Möller, W., and J. A. Maassen. 1986. On the structure, function and dynamics of L7/L12 from *Escherichia coli* ribosomes, p. 309–325. *In* B. Hardesty and G. Kramer (ed.), *Structure, Function, and Genetics of Ribosomes.* Springer-Verlag, New York.

Monro, R. 1967. Catalysis of peptide bond formation by 50S ribosomal subunits from E. coli. *J. Mol. Biol.* **26:**147–151.

Montandon, P.-E., P. Nicolas, P. Schurmann, and E. Stutz. 1985. Streptomycin-resistance of *Euglena gracilis* chloroplasts: identification of a point mutation in the 16S rRNA gene in an invariant position. *Nucleic Acids Res.* **13:**4299–4310.

Montandon, P.-E., R. Wagner, and E. Stutz. 1986. *E. coli* ribosomes with a C912 to U base change in the 16S rRNA are streptomycin resistant. *EMBO J.* **5:**3705–3708.

Moore, P. B. 1966. Studies on the mechanism of messenger ribonucleic acid attachment to ribosomes. *J. Mol. Biol.* **22:** 145–163.

Moore, P. B. 1987. On the modus operandi of the ribosome. *Cold Spring Harbor Symp. Quant. Biol.* **52:**721–728.

Moore, P. B., M. Capel, M. Kjeldgaard, and D. M. Engelman. 1986. A 19 protein map of the 30S ribosomal subunit of *Escherichia coli*, p. 87–100. *In* B. Hardesty and G. Kramer (ed.), *Structure, Function, and Genetics of Ribosomes.* Springer-Verlag, New York.

Moore, P. B., R. R. Traut, H. Noller, P. Pearson, and H. Delius. 1968. Ribosomal proteins of *E. coli.* II. Proteins from the 30S subunit. *J. Mol. Biol.* **31:**441–461.

Morgan, E. A. 1980. Insertions of TN10 into an *E. coli* ribosomal RNA operon are incompletely polar. *Cell* **21:**257–265.

Morgan, E. A., T. Ikemura, L. Lindahl, A. M. Fallon, and M. Nomura. 1978. Some rRNA operons in *E. coli* have tRNA genes at their distal ends. *Cell* **13:**335–344.

Morse, M. L., E. M. Lederberg, and J. Lederberg. 1956. Transduction in *E. coli* K12. *Genetics* **41:**142–156.

Muto, A. 1968. Messenger activity of nascent ribosomal RNA. *J. Mol. Biol.* **36:**1–14.

Nakada, D. 1965. Formation of ribosomes by a "relaxed" mutant of *E. coli. J. Mol. Biol.* **12:**695–725.

Nakada, D., I. A. C. Anderson, and B. Magasanik. 1964. Fate of the ribosomal RNA produced by a "relaxed" mutant of *E. coli. J. Mol. Biol.* **9:**472–488.

Nashimoto, H., W. Held, E. Kaltschmidt, and M. Nomura. 1971. Structure and function of bacterial ribosomes. XII. Accumulation of 21S particles by some cold-sensitive mutants of *E. coli. J. Mol. Biol.* **62:**121–138.

Nashimoto, H., and M. Nomura. 1970. Structure and function of bacterial ribosomes. XI. Dependence of 50S ribosomal assembly on simultaneous assembly of 30S subunits. *Proc. Natl. Acad. Sci. USA* **67:**1440–1447.

Nathans, D., G. Notani, J. H. Schwartz, and N. D. Zinder. 1962. Biosynthesis of the coat protein of coliphage *f2* by *E. coli* extracts. *Proc. Natl. Acad. Sci. USA* **48:**1424–1431.

Neidhardt, F. C. 1963. Properties of a bacterial mutant lacking amino acid control of RNA synthesis. *Biochim. Biophys. Acta* **68:**365–379.

Neidhardt, F. C., and B. Magasanik. 1960. Studies on the role of ribonucleic acid in the growth of bacteria. *Biochim. Biophys. Acta* **42:**99–116.

Nierhaus, K. H. 1980. Analysis of the assembly and function of the 50S subunit from *Escherichia coli* ribosomes by reconstitution, p. 267–294. *In* G. Chambliss, G. R. Craven, J. Davies, K. Davis, L. Kahan, and M. Nomura (ed.), *Ribosomes. Structure, Function, and Genetics.* University Park Press, Baltimore.

Nierhaus, K. H. 1982. Structure, assembly, and function of ribosomes. *Curr. Top. Microbiol. Immunol.* **97:**81–155.

Nierhaus, K. H., K. Bordasch, and H. E. Homann. 1973. Ribosomal proteins. XLIII. *In vivo* assembly of *E. coli* ribosomal proteins. *J. Mol. Biol.* **74:**587–597.

Nierhaus, K. H., and F. Dohme. 1974. Total reconstitution of functionally active 50S ribosomal subunits from *E. coli. Proc. Natl. Acad. Sci. USA* **71:**4713–4717.

Nikolaev, N., L. Silengo, and D. Schlessinger. 1973. Synthesis of a large precursor to ribosomal RNA in a mutant of *E. coli. Proc. Natl. Acad. Sci. USA* **70:**3361–3365.

Nirenberg, M. W., and J. H. Matthaei. 1961. The dependence of cell-free protein synthesis in *E. coli* upon naturally occurring or synthetic polyribonucleotides. *Proc. Natl. Acad. Sci. USA* **47:** 1588–1602.

Noller, H. F., M. Asire, A. Barta, S. Douthwaite, T. Goldstein, R. R. Gutell, D. Moazed, J. Normanly, J. B. Prince, S. Stern, K. Triman, S. Turner, B. Van Stolk, V. Wheaton, B. Weiser, and C. R. Woese. 1986. Studies on the structure and function of ribosomal RNA, p. 143–163. *In* B. Hardesty and G. Kramer (ed.), *Structure, Function, and Genetics of Ribosomes.* Springer-Verlag, New York.

Noller, H. F., J. Kop, V. Wheaton, J. Brosius, R. R. Gutell, A. M. Kopylov, F. Dohme, W. Herr, D. A. Stahl, R. Gupta, and C. R. Woese. 1981. Secondary structure model for 23S ribosomal RNA. *Nucleic Acids Res.* **9:**6167–6189.

Noller, H. F., S. Stern, D. Moazed, T. Power, P. Svensson, and L.-M. Changchien. 1987. Studies on the architecture and function of 16S rRNA. *Cold Spring Harbor Symp. Quant. Biol.* **52:**695–708.

Noller, H. F., and C. R. Woese. 1981. Secondary structure of 16S ribosomal RNA. *Science* **212:**403–411.

Nomura, M. 1963. Mode of action of colicines. *Cold Spring Harbor Symp. Quant. Biol.* **28:**315–324.

Nomura, M. 1968. The role of 30S ribosomal subunits in initiation of protein synthesis, p. 50–58. *In* H. Wittmann (ed.), *Molecular Genetics. 4. Wissenschaftliche Konferenz der Gesellschaft Deutscher Naturforscher und Arzte.* Springer-Verlag, KG, Berlin.

Nomura, M. 1970. Bacterial ribosome. *Bacteriol. Rev.* **34:**228–277.

Nomura, M. 1987. Role of RNA and protein in ribosome function. *Cold Spring Harbor Symp. Quant. Biol.* **52:**653–658.

Nomura, M., and V. A. Erdmann. 1970. Reconstitution of 50S ribosomal subunits from dissociated molecular components. *Nature* (London) **228:**774–748.

Nomura, M., B. D. Hall, and S. Spiegelman. 1960. Characterization of RNA synthesized in *E. coli* after bacteriophage T2 infection. *J. Mol. Biol.* **2:**306–326.

Nomura, M., and K. Hosokawa. 1965. Biosynthesis of ribosomes: fate of chloramphenicol particles and pulse-labeled RNA in *E. coli. J. Mol. Biol.* **12:**242–265.

Nomura, M., and C. V. Lowry. 1967. Phage F2 RNA-directed binding of formylmethionyl-tRNA to ribosomes and the role of 30S ribosomal subunits in initiation of protein synthesis. *Proc. Natl. Acad. Sci. USA* **58:**946–953.

Nomura, M., S. Mizushima, M. Ozaki, P. Traub, and C. V. Lowry. 1969a. Structure and function of ribosomes and their molecular components. *Cold Spring Harbor Symp. Quant. Biol.* **34:**49–61.

Nomura, M., E. A. Morgan, and S. R. Jaskunas. 1977. Genetics of bacterial ribosomes. *Annu. Rev. Genet.* **11:**297–347.

Nomura, M., and L. E. Post. 1980. Organization of ribosomal genes and regulation of their expression in *Escherichia coli*, p. 671–691. *In* G. Chambliss, G. R. Craven, J. Davies, K. Davis, L. Kahan, and M. Nomura (ed.), *Ribosomes. Structure, Function, and Genetics.* University Park Press, Baltimore.

Nomura, M., J. Sidikaro, K. Jakes, and N. Zinder. 1974. Effects of colicin E3 on bacterial ribosomes, p. 805–814. *In* M. Nomura, A. Tissières, and P. Lengyel (ed.), *Ribosomes*. Cold Spring Harbor Laboratory, Cold Spring Harbor, N.Y.

Nomura, M., P. Traub, and H. Bechmann. 1968. Hybrid 30S ribosomal particles reconstituted from components of different bacterial origins. *Nature* (London) **219:**793–799.

Nomura, M., P. Traub, C. Guthrie, and H. Nashimoto. 1969b. The assembly of ribosomes. *J. Cell. Physiol.* **74:**241–251.

Nomura, M., and J. D. Watson. 1959. Ribonucleoprotein particles within chloromycetin-inhibited *E. coli*. *J. Mol. Biol.* **1:**204–217.

Nomura, M., J. L. Yates, D. Dean, and L. E. Post. 1980. Feedback regulation of ribosomal protein gene expression in *E. coli*: structural homology of ribosomal RNA and ribosomal protein mRNA. *Proc. Natl. Acad. Sci. USA* **77:**7084–7088.

Nonomura, Y., G. Blobel, and D. Sabatini. 1971. Structure of liver ribosomes studied by negative staining. *J. Mol. Biol.* **60:**303–323.

Nowotny, V., R. P. May, and K. H. Nierhaus. 1986. Neutron-scattering analysis of structural and functional aspects of the ribosome: the strategy of the glassy ribosomes, p. 101–111. *In* B. Hardesty and G. Kramer (ed.), *Structure, Function, and Genetics of Ribosomes*. Springer-Verlag, New York.

Nowotny, V., and K. H. Nierhaus. 1980. Protein L20 from the large subunit of *E. coli* ribosomes is an assembly protein. *J. Mol. Biol.* **137:**391–399.

Nowotny, V., and K. H. Nierhaus. 1988. Assembly of the 30S subunit from *E. coli* ribosomes occurs via two assembly domains which are initiated by S4 and S7. *Biochemistry* **27:**7051–7055.

Oakes, M., M. Clark, E. Henderson, and J. Lake. 1986a. DNA hybridization electron microscopy: ribosomal RNA nucleotides 1392–1407 are exposed in the cleft of the small subunit. *Proc. Natl. Acad. Sci. USA* **83:**275–279.

Oakes, M., E. Henderson, A. Scheinman, M. Clark, and J. Lake. 1986b. Ribosome structure, function, and evolution: mapping ribosomal RNA, proteins, and functional sites in three dimensions, p. 47–67. *In* B. Hardesty and G. Kramer (ed.), *Structure, Function, and Genetics of Ribosomes*. Springer-Verlag, New York.

Oen, H., M. Pellegrini, D. Eilat, and C. Cantor. 1973. Identification of 50S proteins at the peptidyl-tRNA binding site of *E. coli* ribosomes. *Proc. Natl. Acad. Sci. USA* **70:**2799–2803.

Ofengand, J., R. Liou, J. Kohut III, I. Schwartz, and R. A. Zimmermann. 1979. Covalent cross-linking of transfer ribonucleic acid to the ribosomal P site. Mechanism and site of reaction in transfer ribonucleic acid. *Biochemistry* **18:**4322–4332.

Ohta, T. S., S. Sarkar, and R. E. Thach. 1967. The role of guanosine-5′-triphosphate in the initiation of peptide synthesis. III. Binding of formylmethionyl-tRNA to ribosomes. *Proc. Natl. Acad. Sci. USA* **58:**1639–1644.

Ohtsubo, E., L. Solls, R. C. Deonier, H. J. Lee, and N. Davidson. 1974. Electron microscope heteroduplex studies of sequence relations among plasmids of *E. coli*. VIII. The structure of bacteriophage ϕ80d$_3$*ilv*$^+$su$^+$7, including the mapping of the ribosomal RNA genes. *J. Mol. Biol.* **89:**631–646.

Oishi, M., and N. Sueoka. 1965. Location of genetic loci of ribosomal RNA on *Bacillus subtilis* chromosome. *Proc. Natl. Acad. Sci. USA* **54:**483–491.

Olson, H. M., and D. G. Glitz. 1979. Ribosome structure: localization of 3′ end of RNA in small subunit by immunoelectronmicroscopy. *Proc. Natl. Acad. Sci. USA* **76:**3769–3773.

Oppenheim, D. S., and C. Yanofsky. 1980. Translational coupling during expression of the tryptophan operon of *E. coli*. *Genetics* **95:**785–795.

Orgel, L. E. 1968. Evolution of the genetic apparatus. *J. Mol. Biol.* **38:**381–393.

Osawa, S. 1968. Ribosome formation and structure. *Annu. Rev. Biochem.* **37:**109–130.

Otaka, E., S. Osawa, and A. Sibatani. 1964. Stimulation of ^{14}C-leucine incorporation into protein in vitro by ribosomal RNA of *E. coli*. *Biochem. Biophys. Res. Commun.* **15:**568–574.

Ozaki, M., S. Mizushima, and M. Nomura. 1969. Identification and functional characterization of the protein controlled by the streptomycin-resistant locus in *E. coli*. *Nature* (London) **222:**333–339.

Palade, G. E. 1958. Microsomes and ribonucleoprotein particles, p. 36–61. *In* R. B. Roberts (ed.), *Microsomal Particles and Protein Synthesis*. Pergamon Press, New York.

Pao, C. C., and J. Gallant. 1979. A new nucleotide involved in the stringent response in *E. coli*. *J. Biol. Chem.* **254:**688–692.

Pardee, A. B. 1958. Experiments on the transfer of information from DNA to enzymes. *Exp. Cell Res. Suppl.* **6:**142–151.

Pardee, A. B., F. Jacob, and J. Monod. 1959. The genetic control and cytoplasmic expression of "inducibility" in the synthesis of β-galactosidase by *E. coli*. *J. Mol. Biol.* **1:**165–178.

Pellegrini, M., H. Oen, and C. Cantor. 1972. Covalent attachment of a peptidyl-transfer RNA analog to the 50S subunit of *E. coli* ribosomes. *Proc. Natl. Acad. Sci. USA* **69:**837–841.

Perry, R. P. 1962. The cellular sites of synthesis of ribosomal and 4S RNA. *Proc. Natl. Acad. Sci. USA* **48:**2179–2186.

Politz, S., and D. G. Glitz. 1977. Ribosome structure: localization of N^6,N^6-dimethyladenosine by electron microscopy of a ribosome-antibody complex. *Proc. Natl. Acad. Sci. USA* **74:**1468–1472.

Post, L. E., A. E. Arfsten, G. R. Davis, and M. Nomura. 1980. DNA sequence of the promoter region for the α ribosomal protein operon in *E. coli*. *J. Biol. Chem.* **255:**4653–4659.

Post, L. E., A. E. Arfsten, F. Reusser, and M. Nomura. 1978. DNA sequences of promoter regions for the *str* and *spc* ribosomal protein operons in *E. coli*. *Cell* **15:**215–229.

Post, L. E., and M. Nomura. 1980. DNA sequences from the *str* operon of *E. coli*. *J. Biol. Chem.* **255:**4660–4666.

Post, L. E., G. D. Strycharz, M. Nomura, H. Lewis, and P. P. Dennis. 1979. Nucleotide sequence of the ribosomal protein gene cluster adjacent to the gene for RNA polymerase subunit β in *E. coli*. *Proc. Natl. Acad. Sci. USA* **76:**1697–1701.

Prince, J. B., B. H. Taylor, D. L. Thurlow, J. Ofengand, and R. A. Zimmermann. 1982. Covalent cross-linking of tRNA$_1^{Val}$ to 16S RNA at the ribosomal P site: identification of cross-linked residues. *Proc. Natl. Acad. Sci. USA* **79:**5450–5454.

Randall, L. L., S. J. S. Hardy, and J. R. Thom. 1987. Export of protein: a biochemical view. *Annu. Rev. Microbiol.* **41:**507–541.

Revel, M., and F. Gros. 1966. A factor from *E. coli* required for the translation of natural messenger RNA. *Biochem. Biophys. Res. Commun.* **25:**124–132.

Ritossa, F. M., and S. Spiegelman. 1965. Localization of DNA complementary to ribosomal RNA in the nucleolus organizer region of *Drosophila melanogaster*. *Proc. Natl. Acad. Sci. USA* **53:**737–745.

Roberts, R. B. 1965. The synthesis of ribosomal protein. *J. Theor. Biol.* **8:**49–53.

Roberts, R. B. (ed.). 1958. *Microsomal Particles and Protein Synthesis*. Pergamon Press, New York.

Rosset, R., J. Julien, and R. Monier. 1966. Ribonucleic acid composition of bacteria as a function of growth rate. *J. Mol. Biol.* **18:**308–320.

Rosset, R., and R. Monier. 1963. A propos de la presence d'acide ribonucleique de faible poids moleculaire dans les ribosomes d'*E. coli*. *Biochim. Biophys. Acta* **68:**653–656.

Ryals, J., R. Little, and H. Bremer. 1982a. Control of rRNA and tRNA syntheses in *Escherichia coli* by guanosine tetraphosphate. *J. Bacteriol.* **151:**1261–1268.

Ryals, J., R. Little, and H. Bremer. 1982b. Control of RNA synthesis in *E. coli* after a shift to higher temperature. *J. Bacteriol.* **151**:1425–1432.

Said, B., J. R. Cole, and M. Nomura. 1988. Mutational analysis of the L1 binding site of 23S rRNA in *E. coli. Nucleic Acids Res.* **16**:10529–10545.

Sanger, F., G. G. Brownlee, and B. G. Barrell. 1965. A two-dimensional fractionation procedure for radioactive nucleotides. *J. Mol. Biol.* **13**:373–398.

Sanger, F., S. Nicklen, and A. R. Coulson. 1977. DNA sequencing with chain-terminating inhibitors. *Proc. Natl. Acad. Sci. USA* **74**:5463–5467.

Sarmientos, P., and M. Cashel. 1983. Carbon starvation and growth rate-dependent regulation of the *E. coli* ribosomal RNA promoters: differential control of dual promoters. *Proc. Natl. Acad. Sci. USA* **80**:7010–7013.

Sarmientos, P., J. E. Sylvester, S. Contente, and M. Cashel. 1983. Differential stringent control of the tandem *E. coli* ribosomal RNA promoters from the *rrnA* operon expressed *in vivo* in multicopy plasmids. *Cell* **32**:1337–1346.

Schachtschabel, D., and W. Zillig. 1959. Untersuchungen zur Biosynthese der Proteine. I. Ueber den Einbau ^{14}C-markierter Aminosauren ins Protein zellfreier Nucleoproteid-Enzym-Systeme aus *E. coli* B. *Hoppe-Seyler's Z. Physiol Chem.* **314**:262.

Schaechter, M., O. Maaløe, and N. O. Kjeldgaard. 1958. Dependency on medium and temperature of cell size and chemical composition during balanced growth of *Salmonella typhimurium. J. Gen. Microbiol.* **19**:592–606.

Scherrer, K., H. Latham, and J. E. Darnell. 1963. Demonstration of an unstable RNA and of a precursor to ribosomal RNA in HeLa cells. *Proc. Natl. Acad. Sci. USA* **49**:240–248.

Schleif, R. 1967. Control of production of ribosomal protein. *J. Mol. Biol.* **27**:41–55.

Schlessinger, D., G. Mangiarotti, and D. Apirion. 1967. The formation and stabilization of 30S and 50S ribosome couples in *E. coli. Proc. Natl. Acad. Sci. USA* **58**:1782–1789.

Schulze, H., and K. H. Nierhaus. 1982. Minimal set of ribosomal components for reconstitution of the peptidyltransferase activity. *EMBO J.* **1**:609–613.

Schuster, H., and R. Wilhelm. 1963. Reaction differences between tobacco mosaic virus and its free ribonucleic acid with nitrous acid. *Biochim. Biophys. Acta* **68**:554–560.

Senior, B. W., and I. B. Holland. 1971. Effect of colicin E3 upon 30S ribosomal subunit of *E. coli. Proc. Natl. Acad. Sci. USA* **68**:959–963.

Shatsky, I. N., A. G. Evstafieva, T. F. Bystrova, A. A. Bogdanov, and V. D. Vasiliev. 1980. Topography of RNA in the ribosome: localization of the 3′-end of the 23S RNA on the surface of the 50S ribosomal subunit by immune electron microscopy. *FEBS Lett.* **122**:251–255.

Shatsky, I. N., L. V. Mochalova, M. S. Kojouharova, A. A. Bogdanov, and V. D. Vasiliev. 1979. Localization of the 3′ end of *E. coli* 16S RNA by electron microscopy of antibody-labelled subunits. *J. Mol. Biol.* **133**:501–515.

Shimada, K., R. A. Weisberg, and M. E. Gottesman. 1972. Prophage lambda at unusual chromosomal locations. I. Location of the secondary attachment sites and the properties of the lysogens. *J. Mol. Biol.* **63**:483–503.

Shine, J., and L. Dalgarno. 1974. The 3′-terminal sequence of *E. coli* 16S ribosomal RNA: complementarity to nonsense triplets and ribosome binding sites. *Proc. Natl. Acad. Sci. USA* **71**:1342–1346.

Shultz, J., T. J. Silhavy, M. L. Berman, N. Fiil, and S. D. Emr. 1982. A previously unidentified gene in the *spc* operon of *E. coli* K12 specifies a component of the protein export machinery. *Cell* **31**:227–235.

Sigmund, C. D., and E. A. Morgan. 1982. Erythromycin resistance due to a mutation in a ribosomal RNA operon of *E. coli. Proc. Natl. Acad. Sci. USA* **79**:5602–5606.

Smith, A. E., and K. A. Marcker. 1970. Cytoplasmic methionine transfer RNAs from eukaryotes. *Nature* (London) **226**:607–610.

Sor, F., and H. Fukuhara. 1984. Erythromycin and spiramycin resistance mutations of yeast mitochondria: nature of the rib2 locus in the large ribosomal RNA gene. *Nucleic Acids Res.* **12**:8313–8318.

Speyer, J. F., P. Lengyel, and C. Basilio. 1962. Ribosomal localization of streptomycin sensitivity. *Proc. Natl. Acad. Sci. USA* **48**:684–686.

Spillmann, S., and K. H. Nierhaus. 1978. The ribosomal protein L24 of *E. coli* is an assembly protein. *J. Biol. Chem.* **253**:7047–7050.

Spirin, A. S. 1963. *In vitro* formation of ribosome-like particles from CM-particles and protein. *Cold Spring Harbor Symp. Quant. Biol.* **28**:267–268.

Spirin, A. S. 1964a. *Macromolecular Structure of Ribonucleic Acids.* Reinhold Publishing Co., New York.

Spirin, A. S. 1964b. On the structure of ribosomes, p. 163–176. *In Structure and Function of the Genetic Material.* Akademie-Verlag, Berlin.

Spirin, A. S., and N. V. Belitsina. 1966. Biological activity of the reassembled ribosome-like particles. *J. Mol. Biol.* **15**:282–283.

Spirin, A. S., N. A. Kiselev, R. S. Shakulov, and A. A. Bogdanov. 1963. On the structure of ribosomes: reversible unfolding of the ribosomal particles into ribonucleoprotein strands and possible model of packing. *Biokhimiya* **28**:920–930.

Spitnik-Elson, P. 1965. The preparation of ribosomal protein from *E. coli* with lithium chloride and urea. *Biochem. Biophys. Res. Commun.* **18**:557–562.

Spotts, C. R., and R. Y. Stanier. 1961. Mechanism of streptomycin action on bacteria: a unitary hypothesis. *Nature* (London) **192**:633–637.

Staehelin, T., D. Maglott, and R. E. Monro. 1969. On the catalytic center of peptidyl transfer: a part of the 50S ribosome structure. *Cold Spring Harbor Symp. Quant. Biol.* **34**:39–48.

Staehelin, T., and M. Meselson. 1966. *In vitro* recovery of ribosomes and of synthetic activity from synthetically inactive ribosomal subunits. *J. Mol. Biol.* **15**:245–249.

Stanley, W. M., and R. M. Bock. 1965. Isolation and physical properties of the ribosomal ribonucleic acid of *E. coli. Biochemistry* **4**:1302–1311.

Stanley, W. M., M. Salas, Jr., A. J. Wahba, and S. Ochoa. 1966. Translation of the genetic message: factors involved in the initiation of protein synthesis. *Proc. Natl. Acad. Sci. USA* **56**:290–295.

Steitz, J. A. 1969. Polypeptide chain initiation: nucleotide sequences of the three ribosomal binding sites in bacteriophage R17 RNA. *Nature* (London) **224**:957–964.

Steitz, J. A., and K. Jakes. 1975. How ribosomes select initiator regions in mRNA: base pair formation between the 3′ terminus of 16S rRNA and the mRNA during initiation of protein synthesis in *E. coli. Proc. Natl. Acad. Sci. USA* **72**:4734–4738.

Stent, G. S. 1964. The operon: on its third anniversary. *Science* **144**:816–820.

Stent, G. S. 1967. Coupled regulation of bacterial RNA and protein synthesis, p. 99–109. *In* H. Vogel, J. O. Lampen, and V. Bryson (ed.), *Organizational Biosynthesis.* Academic Press, Inc., New York.

Stent, G. S., and S. Brenner. 1961. A genetic locus for the regulation of ribonucleic acid synthesis. *Proc. Natl. Acad. Sci. USA* **47**:2005–2014.

Stephens, J. C., S. W. Artz, and B. N. Ames. 1975. Guanosine 5′-diphosphate 3′-diphosphate (ppGpp): positive effector for histidine operon transcription and general signal for amino-acid deficiency. *Proc. Natl. Acad. Sci. USA* **72**:4389–4393.

Stern, S., T. Powers, L.-M. Changchien, and H. Noller. 1989. RNA-protein interactions in 30S ribosomal subunits: folding and function of 16S rRNA. *Science* **244**:783–790.

Stiegler, P., P. Carbon, M. Zuker, J.-P. Ebel, and C. Ehresmann. 1981. Structural organization of the 16S ribosomal RNA from *E. coli*. Topography and secondary structure. *Nucleic Acids Res.* **9**:2153–2172.

Stöffler, G., and M. Stöffler-Meilicke. 1986. Immuno electron microscopy on *Escherichia coli* ribosomes, p. 28–46. *In* B. Hardesty and G. Kramer (ed.), *Structure, Function, and Genetics of Ribosomes*. Springer-Verlag, New York.

Stöffler-Meilicke, M., and G. Stöffler. 1987. The topography of ribosomal proteins on the surface of 30S subunits of *E. coli*. *Biochimie* **69**:1049–1064.

Strycharz, W. A., M. Nomura, and J. A. Lake. 1978. Ribosomal proteins L7/L12 localized at a single region of the large subunit by immune electron microscopy. *J. Mol. Biol.* **126**:123–140.

Subramanian, A. R. 1975. Copies of proteins L7 and L12 and heterogeneity of the large subunit of *E. coli* ribosome. *J. Mol. Biol.* **95**:1–8.

Subramanian, A. R. 1983. Structure and functions of ribosomal protein S1. *Prog. Nucleic Acid Res. Mol. Biol.* **28**:101–143.

Sykes, J., E. Metcalf, and J. D. Pickering. 1977. The nature of the proteins in 'chloramphenicol particles' from *E. coli* A19 (Hfr *rel met rns*). *J. Gen. Microbiol.* **91**:1–16.

Sypherd, P. S. 1967. Message activity of RNA derived from immature ribosomes. *J. Mol. Biol.* **24**:329–332.

Sypherd, P. S., and S. Osawa. 1974. Ribosome genetics revealed by hybrid bacteria, p. 669–678. *In* M. Nomura, A. Tissières, and P. Lengyel (ed.), *Ribosomes*. Cold Spring Harbor Laboratory, Cold Spring Harbor, N.Y.

Tai, P., D. P. Kessler, and J. Ingraham. 1969. Cold-sensitive mutations in *Salmonella typhimurium* which affect ribosome synthesis. *J. Bacteriol.* **97**:1298–1304.

Takanami, M., and T. Okamoto. 1963. Interactions of ribosomes and synthetic polynucleotides. *J. Mol. Biol.* **7**:323–333.

Thomas, M. S., and M. Nomura. 1987. Translational regulation of the L11 ribosomal protein operon of *E. coli*: mutations that define the target site for repression of L1. *Nucleic Acids Res.* **15**:3085–3096.

Tischendorf, G. W., H. Zeichhardt, and G. Stöffler. 1974. Determination of the location of proteins L14, L17, L18, L19, L22 and L23 on the surface of the 50S ribosomal subunit of *E. coli* by immune electron microscopy. *Mol. Gen. Genet.* **134**:187–208.

Tissières, A. 1974. Ribosome research: historical background, p. 3–12. *In* M. Nomura, A. Tissières, and P. Lengyel (ed.), *Ribosomes*. Cold Spring Harbor Laboratory, Cold Spring Harbor, N.Y.

Tissières, A., S. Bourgeois, and F. Gros. 1963. Inhibition of RNA polymerase by RNA. *J. Mol. Biol.* **7**:100–103.

Tissières, A., D. Schlessinger, and F. Gros. 1960. Amino acid incorporation into proteins by *E. coli* ribosomes. *Proc. Natl. Acad. Sci. USA* **46**:1450–1463.

Tissières, A., and J. D. Watson. 1958. Ribonucleoprotein particles from *E. coli*. *Nature* (London) **182**:778–780.

Traub, P., K. Hosokawa, G. R. Craven, and M. Nomura. 1967. Structure and function of *E. coli* ribosome. IV. Isolation and characterization of functionally active ribosomal proteins. *Proc. Natl. Acad. Sci. USA* **58**:2430–2436.

Traub, P., and M. Nomura. 1968a. Structure and function of *E. coli* ribosomes. V. Reconstitution of functionally active 30S ribosomal particles from RNA and proteins. *Proc. Natl. Acad. Sci. USA* **59**:777–784.

Traub, P., and M. Nomura. 1968b. Streptomycin resistance mutation in *E. coli*: altered ribosomal protein. *Science* **160**:198–199.

Traub, P., and M. Nomura. 1969. Structure and function of *E. coli* ribosomes. VI. Mechanism of assembly of 30S ribosomes studied *in vivo*. *J. Mol. Biol.* **40**:391–413.

Traub, P., D. Söll, and M. Nomura. 1968. Structure and function of *E. coli* ribosomes. II. Translational fidelity and efficiency of protein synthesis of a protein-deficient subribosomal particle. *J. Mol. Biol.* **34**:595–608.

Traut, R. R. 1974. Protein topography of ribosomal subunits from *E. coli*, 271–308. *In* M. Nomura, A. Tissières, and P. Lengyel (ed.), *Ribosomes*. Cold Spring Harbor Laboratory, Cold Spring Harbor, N.Y.

Traut, R. R., J. M. Lambert, G. Boileau, and J. W. Kenny. 1980. Protein topography of *Escherichia coli* ribosomal subunits as inferred from protein cross-linking, p. 89–110. *In* G. Chambliss, G. R. Craven, J. Davies, K. Davis, L. Kahan, and M. Nomura (ed.), *Ribosomes. Structure, Function, and Genetics*. University Park Press, Baltimore.

Traut, R. R., and R. E. Monro. 1964. The puromycin reaction and its relation to protein synthesis. *J. Mol. Biol.* **10**:63–72.

Traut, R. R., P. B. Moore, H. Delius, H. Noller, and A. Tissières. 1967. Ribosomal proteins of *E. coli*. I. Demonstration of different primary structures. *Proc. Natl. Acad. Sci. USA* **57**: 1294–1301.

Travers, A. A. 1976. Modulation of RNA polymerase specificity by ppGpp. *Mol. Gen. Genet.* **147**:225–232.

Travers, A. A. 1980a. Promoter sequence for stringent control of bacterial ribonucleic acid synthesis. *J. Bacteriol.* **141**:973–976.

Travers, A. A. 1980b. A $tRNA^{Tyr}$ promoter with an altered in vitro response to ppGpp. *J. Mol. Biol.* **141**:91–97.

Travers, A. A. 1984. Conserved features of coordinately regulated *E. coli* promoters. *Nucleic Acids Res.* **12**:2605–2618.

Travers, A. A., R. Buckland, and P. G. Debenham. 1980. Functional heterogeneity of *E. coli* ribonucleic acid polymerase holoenzyme. *Biochemistry* **19**:1656–1662.

Travers, A. A., R. Kamen, and M. Cashel. 1971. The in vitro synthesis of ribosomal RNA. *Cold Spring Harbor Symp. Quant. Biol.* **35**:415–418.

Travers, A. A., R. Kamen, and R. F. Schlief. 1970. Factor necessary for ribosomal RNA synthesis. *Nature* (London) **228**: 748–751.

Travers, A. A., A. I. Lamond, and H. A. F. Mace. 1982. ppGpp regulates the binding of two RNA polymerase molecules to the *tyrT* promoter. *Nucleic Acids Res.* **10**:5043–5057.

Travers, A. A., A. I. Lamond, and J. R. Weeks. 1986. Alteration of the growth-rate-dependent regulation of *E. coli tyrT* expression by promoter mutations. *J. Mol. Biol.* **189**:251–255.

Valentine, R. C., and N. M. Green. 1967. Electron microscopy of an antibody-hapten complex. *J. Mol. Biol.* **27**:615–617.

Van Holde, K. E., and W. E. Hill. 1974. General physical properties of ribsomes, p. 53–92. *In* M. Nomura, A. Tissières, and P. Lengyel (ed.), *Ribosomes*. Cold Spring Harbor Laboratory, Cold Spring Harbor, N.Y.

Vasiliev, V. D. 1974. Morphology of the ribosomal 30S subparticle according to electron microscopic data. *Acta Biol. Med. Germ.* **33**:779–793.

Volkin, E., and L. Astrachan. 1956. Phosphorus incorporation in *E. coli* ribonucleic acid after infection with bacteriophage T2. *Virology* **2**:149–161.

Wada, A., and T. Sako. 1987. Primary structures of and genes for new ribosomal proteins A and B in *E. coli*. *J. Biochem.* **101**: 817–820.

Walleczek, J., D. Schuler, M. Stöffler-Meilicke, R. Brimacombe, and G. Stöffler. 1988. A model for the spatial arrangement of the proteins in the large subunit of the *E. coli* ribosomes. *EMBO J.*

7:3571–3576.

Waller, J. P. 1964. Fractionation of the ribosomal protein from *E. coli*. *J. Mol. Biol.* **10:**319–336.

Waller, J. P., and J. I. Harris. 1961. Studies on the composition of the protein from *E. coli* ribosomes. *Proc. Natl. Acad. Sci. USA* **47:**18–23.

Walter, P., and G. Blobel. 1981. Translocation of proteins across the endoplasmic reticulum. II. Signal recognition protein (srp) mediates the selective binding to microsomal membranes of in vitro-assembled polysomes synthesizing secretory protein. *J. Cell Biol.* **91:**551–556.

Warner, J. R., A. Rich, and C. E. Hall. 1962. Electron microscope studies of ribosomal clusters synthesizing hemoglobin. *Science* **138:**1399–1403.

Watson, J. D. 1964. The synthesis of proteins upon ribosomes. *Bull. Soc. Chim. Biol.* **46:**1399–1425.

Watson, J. D. 1965. *Molecular Biology of the Gene*. W. A. Benjamin, Inc., New York.

Watson, R. J., J. Parker, N. Fiil, J. Flaks, and J. D. Friesen. 1975. New chromosomal location for structural genes for ribosomal proteins. *Proc. Natl. Acad. Sci. USA* 72:2765–2769.

Webster, R. E., D. L. Engelhardt, and N. D. Zinder. 1966. *In vitro* protein synthesis: chain initiation. *Proc. Natl. Acad. Sci. USA* **55:**155–161.

Wettstein, F. O., T. Staehelin, and H. Noll. 1963. Ribosomal aggregate engaged in protein synthesis: characterization of the ergosome. *Nature* (London) **197:**430–435.

Wikström, P. M., and G. R. Björk. 1988. Noncoordinate translation-level regulation of ribosomal and nonribosomal protein genes in the *Escherichia coli trmD* operon. *J. Bacteriol.* **170:** 3025–3031.

Wittmann, H. G., G. Stöffler, I. Hindennach, C. G. Kurland, L. Randall-Hazelbauer, E. A. Birge, M. Nomura, E. Kaltschmidt, S. Mizushima, R. R. Traut, and T. A. Bickle. 1971. Correlation of 30S ribosomal proteins of *E. coli* isolated in different laboratories. *Mol. Gen. Genet.* **111:**327–333.

Wittmann, H. G., and G. Wittman-Liebold. 1966. Protein chemical studies of two RNA viruses and their mutants. *Cold Spring Harbor Symp. Quant. Biol.* **31:**163–172.

Wittmann-Liebold, B. 1984. Primary structure of *E. coli* ribosomal proteins. *Adv. Protein Chem.* **36:**56–78.

Woese, C. R. 1980. Just so stories and Rube Goldberg machines: speculations on the origin of the protein synthetic machinery, p. 357–376. *In* G. Chambliss, G. R. Craven, J. Davies, K. Davis, L. Kahan, and M. Nomura (ed.), *Ribosomes. Structure, Function, and Genetics*. University Park Press, Baltimore.

Woese, C. R., G. E. Fox, L. Zablen, T. Uchida, L. Bonen, K. Pechman, B. J. Lewis, and D. Stahl. 1975. Conservation of primary structure in 16S ribosomal RNA. *Nature* (London) **254:**83–86.

Woese, C. R., R. Gutell, R. Gupta, and H. F. Noller. 1983. Detailed analysis of the higher-order structure of 16S-like ribosomal ribonucleic acids. *Microbiol. Rev.* **47:**621–669.

Woese, C. R., L. J. Magrum, R. Gupta, R. B. Siegel, D. A. Stahl, J. Kop, N. Crawford, J. Brosius, R. Gutell, J. J. Hogan, and H. F. Noller. 1980. Secondary structure model for bacterial 16S ribosomal RNA: phylogenetic, enzymatic and chemical evidence. *Nucleic Acids Res.* **8:**2275–2293.

Yamagishi, M., H. A. deBoer, and M. Nomura. 1987. Feedback regulation of rRNA synthesis. A mutational alteration in the anti-shine-dalgarno region of the 16S rRNA gene abolishes regulation. *J. Mol. Biol.* **198:**547–550.

Yamamoto, M., and M. Nomura. 1978. Cotranscription of genes for RNA polymerase subunits β and β' with genes for ribosomal proteins in *E. coli*. *Proc. Natl. Acad. Sci. USA* **75:**3891–3895.

Yanagida, M., and C. Ahmad-Zadeh. 1970. Determination of gene product positions in bacteriophage T4 by specific antibody association. *J. Mol. Biol.* **51:**411–421.

Yankofsky, S. A., and S. Spiegelman. 1962a. The identification of the ribosomal RNA cistron by sequence complementarity. I. Specificity of complex formation. *Proc. Natl. Acad. Sci. USA* **48:**1069–1078.

Yankofsky, S. A., and S. Spiegelman. 1962b. The identification of the ribosomal RNA cistron by sequence complementarity. II. Saturation of and competitive interaction at the RNA cistron. *Proc. Natl. Acad. Sci. USA* **48:**1466–1472.

Yates, J. L., A. E. Afrsten, and M. Nomura. 1980. *In vitro* expression of *E. coli* ribosomal protein genes: autogenous inhibition of translation. *Proc. Natl. Acad. Sci. USA* 77:1837–1841.

Yates, J. L., and M. Nomura. 1980. *E. coli* ribosomal protein L4 is a feedback regulatory protein. *Cell* **21:**517–522.

Yates, J. L., and M. Nomura. 1981. Feedback regulation of ribosomal protein synthesis in *E. coli*: localization of the mRNA target sites for repressor action of ribosomal protein L1. *Cell* **24:**243–249.

Yonath, A., M. A. Saper, and H. G. Wittmann. 1986. Studies on crystals of intact bacterial ribosomal particles, p. 112–127. *In* B. Hardesty and G. Kramer (ed.), *Structure, Function, and Genetics of Ribosomes*. Springer-Verlag, New York.

Yoshida, K., and S. Osawa. 1968. Origin of the protein component of chloramphenicol particles in *E. coli*. *J. Mol. Biol.* **33:**559–569.

Young, R. A., and J. A. Steitz. 1979. Tandem promoters direct *E. coli* ribosomal RNA synthesis. *Cell* **17:**225–234.

Yu, M. T., C. W. Vermeulen, and K. C. Atwood. 1970. Location of the genes for 16S and 23S ribosomal RNA in the genetic map of *E. coli*. *Proc. Natl. Acad. Sci. USA* **67:**26–31.

Zamecnik, P. C., and E. B. Keller. 1954. Relation between phosphate energy donors and incorporation of labeled amino acids into proteins. *J. Biol. Chem.* **209:**337–354.

Zengel, J. M., D. Mueckl, and L. Lindahl. 1980. Protein L4 of the *E. coli* ribosome regulates an eleven gene r-protein operon. *Cell* **21:**523–535.

Zimmermann, R. A. 1974. RNA-protein interactions in the ribosome, p. 225–270. *In* M. Nomura, A. Tissières, and P. Lengyel (ed.), *Ribosomes*. Cold Spring Harbor Laboratory, Cold Spring Harbor, N.Y.

Chapter 2

Ribosome Preparation and Cell-Free Protein Synthesis

ALEXANDER S. SPIRIN

The molecular biology of ribosomes began in the 1950s, when ribonucleoprotein particles were visualized and identified in cells (Palade, 1955), isolated from cells, and then studied with respect to their physicochemical properties (Chao and Schachman, 1956; Ts'o et al., 1956; Peterman and Hamilton, 1957; Tissières and Watson, 1958; see also papers in Roberts, 1958). Simultaneously, the protein-synthesizing ability of the particles was demonstrated (Littlefield et al., 1955; Littlefield and Keller, 1957; McQuillen et al., 1959).

The word "ribosome" was proposed in 1958 to designate these protein-synthesizing ribonucleoprotein particles. In his introduction to the volume of the symposium on microsomal particles and protein synthesis, R. B. Roberts (1958) wrote, "This seems a very satisfactory name, and it has a pleasant sound."

The 1960s were the years of the most intensive and extensive studies of ribosomes and the molecular mechanism of protein biosynthesis. That was the time of many discoveries in this field. Cold Spring Harbor symposium volumes on cellular regulatory mechanisms (volume 26, 1961), synthesis and structure of macromolecules (volume 28, 1963), the genetic code (volume 31, 1966), and especially the mechanism of protein synthesis (volume 34, 1969) recorded the progress made. Later development was smoother and slower, but important contributions were made to the techniques of ribosome preparation, leading finally to crystallization, to the characterization of ribosome structure, including the complete primary structures of rRNAs and ribosomal proteins, and to cell-free protein synthesis, ending with the creation of preparative systems. A number of principles of ribosome function were also discovered.

Progress in ribosome preparation and in cell-free protein synthesis is the topic of this chapter. This survey, however, does not aim to be comprehensive. Only the steps that seem to be of principal significance will be mentioned and discussed. Of course, the choice of these steps is subjective, but I believe that an author should have the privilege of expressing his own viewpoint as well as illustrating the text with his own results.

RIBOSOME PREPARATION

In the end, all progress in structural and in vitro functional studies of ribosomes depends on progress in ribosome preparation. That is why I shall consider first the development of the principles and techniques that underlie modern procedures for stabilization, isolation, and purification of intact ribosomes. The following subheadings reflect the main advances made in ribosome preparation during the last three decades.

Magnesium Ions Are Required for Particle Stability

Decisive success in stabilization of particles in cell extracts and their isolation was achieved by the end of the 1950s as the result of an important discovery: a heat-stable, dialyzable factor required for the stability of ribosomes. The factor proved to be Mg^{2+} (see papers in Roberts, 1958). The decrease of Mg^{2+} concentration in the medium below a critical level was shown to induce the dissociation of 80S or 70S ribosomes into two unequal subunits, 60S and 40S or 50S and 30S, respectively (Chao, 1957; Tissières and Watson, 1958). Further depletion of Mg^{2+} from ribosomes led to fragmentation and destruction of the ribonucleoproteins, mainly due to RNase-induced degradation of rRNA.

Later, it was demonstrated that depletion of Mg^{2+} from ribosomal subunits results in unfolding of the compact particles into ribonucleoprotein strands that are very sensitive to RNase attack (Spirin et al.,

Alexander S. Spirin ■ Institute of Protein Research, Academy of Sciences of the USSR, Pushchino, Moscow Region, USSR.

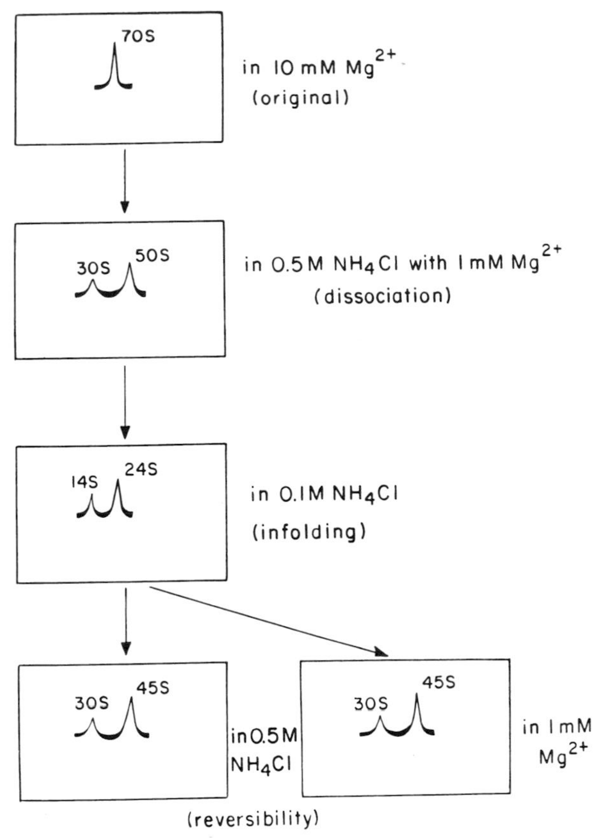

Figure 1. Effect of Mg^{2+} removal and ionic strength on bacterial (*E. coli*) ribosomes: dissociation of 70S ribosomes into 30S and 50S ribosomal subunits (step 1), unfolding of the subunits into 15S and 25S ribonucleoprotein strands (step 2), and refolding of the strand into the compact particles (step 3). (After Spirin et al., 1963, and Spirin, 1964.)

1963; Gavrilova et al., 1966; Gesteland, 1966). Unfolding (Fig. 1) seems to be caused by an increasing electrostatic repulsion of RNA phosphate groups which "explodes" ("melts") the compact RNA structure (Spirin, 1964, 1974). It is important to note that most of the ribosomal proteins remain associated with rRNA in the unfolded state. (The latter fact, in particular, has suggested a backbone role of rRNA in ribosomal ribonucleoprotein assembly and organization [Spirin, 1964].)

Sedimentation in an Ultracentrifuge Provides a Way To Isolate Particles

It was understood rather early that ribosomal particles are an abundant component of cell extracts; at the same time, they are of a uniform size, significantly smaller than cellular organelles and larger than most soluble proteins. These properties made them easily visible in cell extracts as a major sharp component with a unique sedimentation coefficient in an analytical ultracentrifuge. Using this fact, preparative differential ultracentrifugation procedures were developed to isolate ribosomes from cell homogenates (see papers in Roberts, 1958).

Thus, since the end of the 1950s, the preparative ultracentrifuge has been the main tool for the isolation of ribosomes from cell extracts. Pelleting the ribosomes in an angle rotor and then resuspending and repelleting (e.g., Tissières et al., 1959) is still the normal procedure. Pelleting ribosomes through a sucrose cushion of intermediate density (e.g., 1 M or 30% sucrose) or into a sucrose cushion of high density (e.g., 2 M or 70% sucrose) instead of spinning down to the tube bottom directly (either in an angle rotor or in a bucket) was a modification that improved the quality of ribosome preparations (Staehelin et al., 1969; Staehelin and Maglott, 1971; Brown et al., 1974; Bloemendal et al., 1974).

Zonal centrifugation of ribosomal particles in a sucrose gradient at low Mg^{2+} concentration was at first used mainly to separate ribosomal subunits by using swinging-bucket rotors (Britten and Roberts, 1960); later, large-scale zonal rotors were used for the same purpose (Eikenberry et al., 1970). Soon after, zonal centrifugation in Mg^{2+}-containing sucrose gradients, especially using a large-scale zonal rotor, was found to be a very effective way to prepare 70S particles of high quality for both structural and functional studies (Noll et al., 1973a; Noll et al., 1973b).

Alternative techniques have appeared from time to time, but they have not found wide use and have a rather auxiliary significance for preparation of ribosomes. Nevertheless, two groups of methods should be mentioned. First, precipitation of ribosomes from crude extracts with either high Mg^{2+} (Takanami, 1960), streptomycin (Takata and Osawa, 1957; Barbu et al., 1959), or $(NH_4)_2SO_4$ (Elson, 1958; Kurland, 1966, 1971) was proposed. The latter was found to be extremely useful for purification, concentration, and conservation (for months) of biologically active ribosomes and their subunits (e.g., Gavrilova and Smolyaninov, 1971; Gavrilova and Spirin, 1974). Second, gel filtration and chromatographic techniques using Sephacryl (Jelenc, 1980) or Sepharose (Kirillov et al., 1978) were developed; active ribosomes or their subunits were reported to be obtained in this way. Earlier, the polyethylene glycol-dextran aqueous two-phase partition system was successfully used to separate crude cell extracts into ribosomes and soluble protein fractions. Active ribosomes could be recovered from the bottom (dextran) phase (Gordon, 1971).

Absorbed Proteins Are Washed with High Monovalent Salts from the Particles

A very important technical achievement was the development of the procedure of washing ribosomes

Table 1. Composition of eucaryotic monoribosomes and polyribosomes

Constituent	Content/ribosome		
	Size (kilodaltons)	%	
		In monoribosomes	In polyribosomes
rRNA (28S + 18S + 5.8S + 5S)	2,300	50	41
mRNA	~30		0.5
mRNA-bound proteins (messenger ribonucleoproteins)	≦100		<2
Ribosomal proteins	1,700–1,800	40	31
Adsorbed proteins of monoribosomes	500	10	
Adsorbed proteins of polyribosomes	1,500		27

with high monovalent salts, primarily NH_4Cl or KCl. To my knowledge, W. Stanley, Jr., and R. Bock were the first to propose the use of 0.5 M NH_4Cl for the removal of latent RNases from ribosomes (W. Stanley and R. Bock, personal communication; Stanley, 1959; Stanley and Wahba, 1967). This development led to ribosome preparations that were purer than could be prepared before, were more stable in solution because of the absence of RNase activity, and did not fragment under unfolding conditions (Spirin et al., 1963; Salas et al., 1965; Takanami, 1967). (The isolation of bacterial strains, whether mutant or natural, deprived of RNase I, e.g., *Escherichia coli* A19, Q13, or MRE600 [Gesteland, 1966a; Cammack and Wade, 1965] also provided a source of stable ribosomes.)

Later, the procedure of repeated washing of crude bacterial ribosomes with 0.5 to 1 M NH_4Cl containing $MgCl_2$ was widely used to remove from the particles all adsorbed translation factors and other nonribosomal proteins (Stanley et al., 1966; Pestka, 1968; Erbe et al., 1969; Gordon et al., 1971; Dubnoff and Maitra, 1971; Ravel and Shorey, 1971; Gavrilova and Spirin, 1974). To attain ribosomal preparation of high purity, many protocols include 0.5 to 1 M NH_4Cl, KCl, or CsCl in the sucrose cushion for pelleting through the particles (Staehelin et al., 1969; Staehelin and Maglott, 1971; Noll et al., 1973a; Gogia et al., 1986).

Eucaryotic ribosomes and polyribosomes contain particularly large amounts of adsorbed proteins. Also, 0.5 M NH_4Cl was successfully used for washing eucaryotic ribosomes (Moldave and Skogerson, 1967; Moldave and Sadnik, 1979). The 0.5 M KCl wash of eucaryotic ribosomes (including native ribosomal subunits) is the classical source of initiation and translation factors (e.g., Miller and Schweet, 1968; Brown et al., 1974; Heywood and Rourke, 1974; Safer et al., 1976; Schreier et al., 1977; Benne et al., 1978). Elongation factors, aminoacyl-tRNA synthetases, some protein kinases, and many other proteins were found to be associated with eucaryotic monoribosomes and polyribosomes (for reviews, see Spirin and Ajtkhozhin, 1985, and Ryazanov et al., 1987). Most of them can be washed off by high salts. It has been estimated (Minich and Ovchinnikov, 1985) that nonribosomal proteins associated with monoribosomes and polyribosomes of mammalian cell extracts consist of about 0.5×10^6 and 1.5×10^6 daltons per particle, respectively (Table 1). The association of free proteins of the translation machinery with eucaryotic ribosomes may have a biological purpose; it may be a means of dynamic compartmentalization of these proteins on the protein-synthesizing structures of the cell.

It should be mentioned that according to new determinations of buoyant density of washed mammalian monoribosomes and polyribosomes (1.59 g/cm^3 instead of the previous figure of 1.50 g/cm^3), pure ribosomes without accompanying nonribosomal proteins contain 43% protein by weight rather than 50% (Minich and Ovchinnikov, 1985). In some cases, it may be important to have a highly pure preparation of eucaryotic ribosomes for structural and functional studies.

Active Particle Fractions Can Be Isolated from the Total Ribosome Population

In vivo, functionally active ribosomes form polyribosomes. During preparation of the cell extract, some of the ribosomes complete the translation of the polyribosomal mRNA to which they have been bound and thus become free runoff monoribosomes. Another portion of the monoribosomes in cell extracts may be the result of fragmentation of polyribosomes. These particles retain a fragment of mRNA and nascent peptide. Some ribosomes become damaged in the course of extract preparation. There also seem to be preexisting silent (blocked) monoribosomes that do not participate in polyribosome formation in vivo. All of this leads to a complicated, heterogeneous ribosome population in cell extracts and in the isolated total ribosome preparation (Fig. 2).

In most cases, the aim of preparing ribosomes is

to obtain structurally intact and functionally active particles. The choice of an adequate, mild methodology for growing cells, preparing extracts, and isolating ribosomes is critical. Nevertheless, the problem of heterogeneity of ribosome preparations arises every time. One way to solve the problem would be to isolate just the polyribosome fraction, using the difference in sedimentation between monoribosomes and polyribosomes. For most purposes, however, monoribosomes are required, which is why fractionation procedures had to be developed. One way to obtain fractions from total monoribosomes is based on the different stabilities of the different particles to lowering of the Mg^{2+} concentration: ribosomes bearing mRNA fragments and nascent peptides (complexed monoribosomes) are very stable and dissociate into subunits only at very low Mg^{2+}; intact vacant monoribosomes are less complexed and dissociate at intermediate Mg^{2+}; defective particles are the least stable (Tissières et al., 1960; Schlessinger and Gros, 1963; Ron et al., 1968; Spirin et al., 1970; Noll et al., 1973a). Complexed ribosomes, therefore, can be isolated as 70S or 80S particles by centrifugation at low Mg^{2+} concentrations when all other types of monoribosomes are dissociated.

For use in cell-free translation systems with exogenous messengers, the main goal is to have vacant ribosomes. Most polyribosomes and complexed monoribosomes can be converted into vacant ribosomes either by runoff translation in the absence of initiation in vivo (Ron et al., 1968) or in vitro

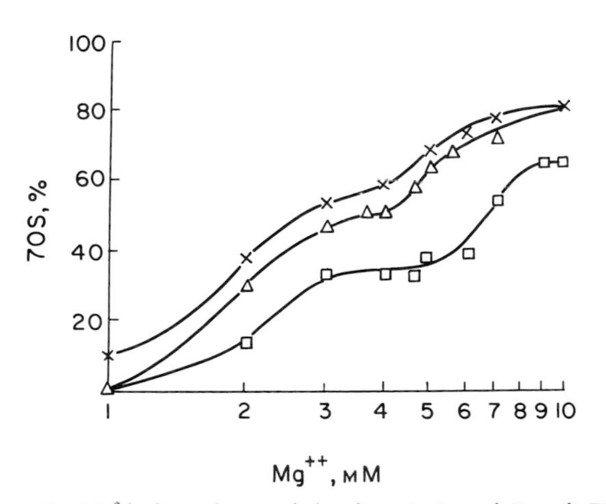

Figure 2. Mg^{2+} dependence of the dissociation of *E. coli* 70S ribosomes for three ribosome preparations. At least four fractions of ribosomal particles differing in dissociation behavior can be distinguished in the total ribosome population: 70S ribosomes resistant to lowering of the Mg^{2+} concentration to 1 mM (seen in one of the preparations); ribosomes dissociating between 3 and 1 mM Mg^{2+}; ribosomes dissociating between 7 and 4 mM Mg^{2+}; and particles dissociating at as high as 10 mM Mg^{2+}. (From Spirin et al., 1970.)

(Staehelin and Falvey, 1971) or by puromycin treatment and high-salt washing off in vitro (applied especially to eucaryotic ribosomes; Lawford, 1969; Blobel and Sabatini, 1971; Brown et al., 1974). Noll and co-workers proposed the use of zonal centrifugation under the conditions (6 mM Mg^{2+} at 100 mM NH_4Cl) in which intact vacant ribosomes were in the form of 70S couples but defective particles were dissociated; in this way, the active ribosome fraction (vacant tight couples) was obtained (Noll et al., 1973a; Noll et al., 1973b; Noll and Noll, 1976).

Quite a different approach to isolate active particles from the total ribosome population was invented in our laboratory. In 1973, a solid-phase translation system was introduced whereby ribosomes were reading a template polynucleotide covalently linked to a cellulose or dextran carrier (Belitsina et al., 1973; see also Belitsina and Spirin, 1979). Only active ribosomes were capable of firmly binding to the carrier-linked polynucleotide under conditions of translation; all inactive and defective particles were easily washed off. When an oligonucleotide template of definite size [e.g., $poly(U)_{100}$] had been linked to the carrier through a cleavable disulfide bridge, the active ribosomes together with their short templates could be eluted by using a sulfhydryl compound (e.g., dithiothreitol), thus yielding homogeneous translating particles (Fig. 3) (Belitsina et al., 1975; Baranov et al., 1979). This technique of preparing 100% translationally active ribosomes also allows one to obtain all of the particles in one or another functional state (e.g., either pretranslocational or posttranslocational) and hence to study them and compare the functional states by physical methods (Spirin et al., 1987).

Intact Ribosomal Particles Can Be Crystallized

Although ordered arrays of ribosomal particles in vivo have been known for a long time (Byers, 1966, 1967; Lake and Slayter, 1972; Taddei, 1972; Unwin, 1977; O'Brien et al., 1980), crystallization of isolated ribosomes and their subunits was achieved only during the last decade. In 1979, Lake and co-workers reported that both ordered (helical) arrays of *E. coli* ribosomal subunits (30S [Clark et al., 1979] and 50S [Lake, 1980]) were produced in vitro. Later, much better crystalline arrays of 50S subunits were obtained by this group (Clark et al., 1982).

Real three-dimensional crystals of bacterial ribosomal particles, however, were first reported by Yonath and Wittmann and associates. Those were microcrystals of the 50S ribosomal subunits from *Bacillus stearothermophilus* (Yonath et al., 1980; Yonath et al., 1982; Yonath et al., 1983). Later, they

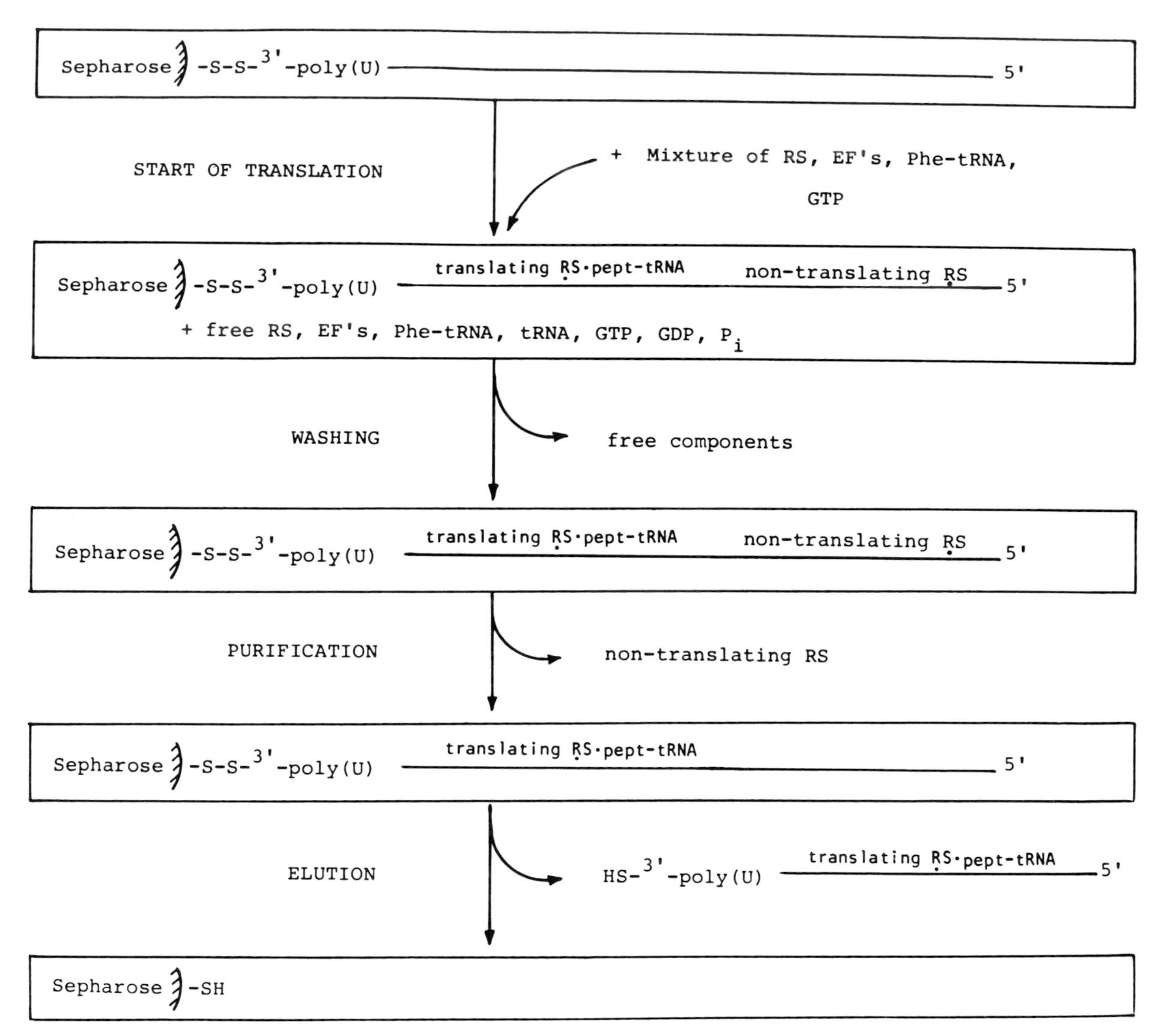

Figure 3. Scheme for isolation of translationally active ribosomes, using columns with a Sepharose-S-S-coupled polynucleotide template: binding the ribosomes to the carrier-linked poly(U) and starting the translation (step 1), washing off the unbound particles and other free components from the column (step 2), eluting the bound but nontranslating particles from the carrier-linked poly(U) (step 3), and eluting the translating particles together with their templates from the column (step 4). (After Belitsina et al., 1975, and Baranov et al., 1979.)

succeeded in growing the crystals of *B. stearothermophilus* 50S subunits to a large size (1 by 0.2 by 0.1 mm) suitable for X-ray studies, and diffraction to 15- to 18-Å (1.5- to 1.8-nm) resolution was demonstrated (Yonath et al., 1986). Large crystals (0.6 by 0.6 by 0.1 mm) were obtained from 50S ribosomal subunits of *Halobacterium marismortui*, for which the diffraction was at 6-Å resolution (Makowski et al., 1987).

Crystallization of 30S ribosomal subunits was first achieved by using the extremely thermophilic bacterium *Thermus thermophilus* as a source of material (Yusupov et al., 1987; Trakhanov et al., 1987). In the same year, microcrystals of 30S ribosomal subunits were reported from this organism as well as from *E. coli* and *H. marismortui* (Glotz et al., 1987). Several crystalline forms of 30S ribosomal subunits from *T. thermophilus* were later produced; one of them is represented by isometric crystals suitable for X-ray analysis and diffraction to 12-Å resolution (Yusupov et al., 1988). Diffraction to 8 Å for the *T. thermophilus* 30S particle crystals has been reported by Yonath et al. (1988).

An important achievement was the crystallization of full 70S ribosomes of bacteria. Microcrystals of 70S ribosomes of *E. coli* (Wittmann et al., 1982), *T. thermophilus* (Karpova et al., 1986; Glotz et al., 1987), and *H. marismortui* (Glotz et al., 1987) were reported. Only recently, large isometric crystals of 70S ribosomes from *T. thermophilus* (up to 0.7 by 0.7 by 0.5 mm; Fig. 4) were produced that gave diffraction to better than 20 Å (Trakhanov et al., 1987; Trakhanov et al., 1989).

CELL-FREE PROTEIN SYNTHESIS

Ribosomes in Cell Extracts Are Capable of Reading Endogenous mRNAs

The first question in the history of cell-free systems was whether disrupted cells or their isolated fractions are capable of synthesizing proteins. A positive answer was given by several groups as early as the beginning of the 1950s (Winnick, 1950a, 1950b; Borsook, 1950; Siekevitz and Zamecnik, 1951; Siekevitz, 1952; Peterson and Greenberg, 1952; Khesin, 1953; Gale and Folkes, 1954). In the second half of that decade, Zamecnik and his colleagues made a really ribosomal system of protein synthesis based on mitchondria-free cytoplasmic extracts of animal cell (Littlefield et al., 1955; Keller and Zamecnik, 1956; Littlefield and Keller, 1957). Zillig's group was the first to succeed in producing a bacterial cell-free translation system (Schachtschabel and Zillig, 1959). Independently, bacterial cell-free translation systems were made by American groups (Lamborg and Zamecnik, 1960; Tissières et al., 1960).

Ribosomes in all of those systems were programmed by endogenous mRNA; they were simply reading the templates to which they had been already attached at the time of cell disruption. Nevertheless, the significance of these systems was great, since they opened the door for studies of molecular mechanisms of protein biosynthesis, including activation of amino acids, involvement of tRNA, GTP requirement, ribosome functions, and participation of soluble translation factors (Zamecnik, 1969).

Figure 4. Photograph of crystals of the *T. thermophilus* 70S ribosomes. The largest has dimensions of 0.7 by 0.7 by 0.5 mm; the diffraction pattern extends to better than 20 Å (Trakhanov et al., 1989). (Courtesy of M. Garber, S. Trakhanov, and M. Yusupov, Institute of Protein Research, Pushchino, USSR.)

Ribosomes in Cell Extracts Can Be Programmed with Exogenous Template Polynucleotides

A revolutionary step in the development of cell-free translation systems was the introduction of exogenous templates. This was first done by Nirenberg and Matthaei in 1961 with a bacterial system (Nirenberg and Matthaei, 1961). Preincubation of the cell extract at physiological temperature was sufficient to remove the endogenous mRNA from the ribosomes. The vacant ribosomes in the extract were found to accept both exogenous natural mRNAs and synthetic polynucleotides as templates. The Nirenberg system became classical (Table 2). The preincubation procedure could be applied to some eucaryotic systems as well.

In more recent versions of the system, the runoff ribosomes, the fraction of active vacant couples, or the salt-washed ribosomes were used in combination with the soluble enzyme or factor fraction freed from any RNA (certainly total tRNA should be added) (Table 2).

At the same time, it was found that wheat germ extract could be directly used for expression of exogenous templates because of the intrinsically low levels of endogenous messengers (Roberts and Paterson, 1973; Marcus et al., 1974; Anderson et al., 1983). Micrococcal Ca^{2+}-dependent RNase treatment was demonstrated to be useful for the removal of endogenous mRNA from reticulocyte lysates (Pelham and Jackson, 1976; Merrick, 1983) as well as from other animal cell extracts (Henshaw and Panniers, 1983). The avoidance of endogenous messenger activity and the expression of exogenous templates became the main way in the use of cell-free translation systems.

Transcription in a Cell-Free System Can Immediately Provide Messages for Ribosomes

Again, Nirenberg and Matthaei seem to be the first to report dependence of the bacterial cell-free system with endogenous templates on the presence of DNA (Matthaei and Nirenberg, 1961). This dependence was seen especially at the later stages of incubation of the system, when preexisting RNA templates had been presumably read out. Some time later, coupled transcription-translation systems were developed by using exogenous bacteriophage DNAs (Wood and Berg, 1962; Byrne et al., 1964).

These systems, however, were poorly active with cellular exogenous DNAs as well as with a number of viral DNAs; in addition, the background due to endogenous polypeptide synthesis was significant. Coupled transcription-translation systems came into

Table 2. Systems developed to prepare ribosomes[a]

System	Components	Comments	References
Cell-free translation			
Crude	DNase-treated supernatant fraction S-30 (including ribosomes, tRNAs, aminoacyl-tRNA synthetases, and translation factors) Synthetic template or natural mRNA Amino acids ATP, GTP, PEP, pyruvate kinase SH compound (mercaptoethanol or dithiothreitol) Mg^{2+}; K^+ or NH_4^+	Fraction S-30 must be preincubated with amino acids, ATP, PEP, and PEP kinase (35 to 37°C, 40 to 80 min) to deprive it of endogenous mRNA activities and then dialyzed.	Nirenberg and Matthaei, 1961; Nirenberg, 1963
Partially purified	Ribosomes Synthetic template or natural mRNA Total tRNA Supernatant fraction S-100 deprived of all nucleic acids (includes aminoacyl-tRNA synthetases and translation factors) Amino acids (Folate) ATP, GTP, PEP, pyruvate kinase SH compound (mercaptoethanol or dithiothreitol) Mg^{2+}; K^+ or NH_4^+		
Transcription-translation cell free			
Crude	Supernatant fraction S-30 (including RNA polymerase, ribosomes, tRNAs, aminoacyl-tRNA synthetases, and translation factors) DNA Amino acids ATP, GTP, UTP, CTP, PEP, pyruvate kinase Folate SH compounds Mg^{2+} or Ca^{2+}; K^+ or NH_4^+	Preincubated to deprive of endogenous DNA and RNA templates	Lederman and Zubay, 1967; DeVries and Zubay, 1967; Zubay, 1973
Partially purified	Ribosomes DNA Total tRNA Supernatant fraction S-100 deprived of nucleic acids (contains RNA polymerase, aminoacyl-tRNA synthetases, and translation factors) Amino acids ATP, GTP, UTP, CTP, PEP, pyruvate kinase Folate SH compounds Mg^{2+} or Ca^{2+}; K^+ or NH_4^+		Gold and Schweiger, 1969, 1971; Schweiger and Gold, 1969
Pure	Ribosomes DNA Total tRNA RNA polymerase Aminoacyl-tRNA synthetases (20 individual proteins) Formyltetrahydrofolate Met-tRNA$_f$ transformylase IF1, IF2, IF3 EF-Tu, EF-Ts, EF-G RF-1, RF-2, RRF Amino acids ATP, GTP, UTP, CTP, PEP, pyruvate kinase $N^{5,10}$-methenyltetrahydrofolate Dithiothreitol Mg^{2+}; K^+ or NH_4^+; spermidine	Several additional protein factors called L, L_α, L_β, L_γ, and L_δ may be also required	Kung et al., 1977; Kung et al., 1979; Zarucki-Schulz et al., 1979

Continued on following page

Table 2.—*Continued.*

System	Components	Comments	References
Pure poly(U)-directed translation	Ribosomes Poly(U) Phe-tRNA EF-Tu, EF-Ts (or EF-1, EF-2) EF-G GTP, PEP, pyruvate kinase SH compounds Mg^{2+}; K^+ or NH_4^+		
Pure bacterial mRNA-directed cell-free translation			
With preaminoacylated tRNA	Ribosomes mRNA Full set of aminoacyl-tRNAs (including fMet-tRNA) IF1, IF2, IF3 EF-Tu, EF-Ts, EF-G RF-1 or RF-2 GTP Dithiothreitol Mg^{2+}, NH_4^+	Additional protein factors (rescue, EF-P, and W) are also required.	Ganoza et al., 1985; Green et al., 1985
With aminoacylation of tRNA	Ribosomes mRNA Total tRNA Aminoacyl-tRNA synthetases (20 individual proteins) Formyltetrahydrofolate Met-tRNA transformylase IF1, IF2, IF3 EF-Tu, EF-Ts, EF-G RF-1, RF-2, RRF Amino acids ATP, GTP, PEP, pyruvate kinase $N^{5,10}$-methenyltetrahydrofolate Dithiothreitol Mg^{2+}; K^+ or NH_4^+		Kung et al., 1978
Factor-free (nonenzymatic) translation	Ribosomes (carefully washed) Poly(U), poly(A), poly(U,C), or poly(U,I) Phe-tRNA or a set of aminoacyl-tRNAs (*p*-Chloromercuribenzoate) Mg^{2+}; K^+ or NH_4^+	Some covalent modifications of ribosomal protein S12 or its absence from ribosomes enhance the factor-free translation.	Gavrilova et al., 1974; Gavrilova et al., 1976; Rutkevitch and Gavrilova, 1982
Template-free elongation	Ribosomes (carefully washed) Lys-tRNA, Ser-tRNA, Thr-tRNA, or Asp-tRNA EF-Tu, EF-G GTP, PEP, pyruvate kinase Dithiothreitol Mg^{2+}, NH_4^+		Belitsina et al., 1981, 1982; Yusupova et al., 1986

[a] Abbreviations: PEP, phosphoenolpyruvate; RF-1 and -2, release factors 1 and 2; RRF, ribosome release factor.

wide use after some major improvements were made by two groups (Lederman and Zubay, 1967; DeVries and Zubay, 1967; Zubay, 1973; Gold and Schweiger, 1969, 1971; Schweiger and Gold, 1969). In the Zubay system (Table 2), the bacterial crude extract was used after prolonged incubation to degrade endogenous RNA and DNA by cellular nucleases; concentrations of the components of the system were optimized. The system became very popular because of the simplicity of its preparation, the stability of the extracts during storage, and its high activity.

The Gold-Schweiger system (Table 2) consists of

the isolated ribosomes and a supernatant protein fraction specially purified of endogenous amino acids and nucleic acids by ion-exchange chromatography. This provides a very low background due to endogenous synthesis and better controlled conditions, but it is more complicated to prepare.

Purified, Modified, and Simplified Peptide-Synthesizing Systems Can Be Composed of Pure Individual Components

In all previous cases, the cell-free systems were based either on the crude cell extract, including ribosomes and all soluble enzymes, factors, and tRNAs (S-30 fraction), or on the ribosome-free extract (S-100 fraction) combined with isolated ribosomes. In cases when the ribosome-free extract had been freed from polynucleotides, total tRNA was added to the mixture. Since major progress was achieved in the isolation and purification of different translation factors as well as of individual aminoacyl-tRNA synthetases and tRNAs, the cell-free systems could be composed from a set of pure components.

The poly(U)-directed cell-free translation system (Table 2) is the easiest to assemble in this way. This is a very good model system for studies of elongation (initiation and termination steps of translation are absent from the system) and for testing ribosome activity. It was exploited intensively by many workers, and much useful information concerning ribosomes, elongation factors, energetics, and other features was obtained.

Special attention should be paid to optimization of the ionic composition in the system. To attain a high rate and high accuracy in the poly(U)-directed system, a "polymix" was proposed that includes Mg^{2+} (5 mM), Ca^{2+} (0.5 mM), K^+ (100 mM), NH_4^+ (5 mM), putrescine (8 mM), and spermidine (1 mM) (Jelenc and Kurland, 1979). At the same time, however, the poly(U)-directed system with purified elongation factors and precharged phenylalanyl-tRNA was found to give comparable activity (5 to 25 pmol of phenylalanine per min per 40 μg of ribosomes) and accuracy (less than 0.1% leucine incorporation) in 6 mM Mg^{2+} and 100 mM K^+, without Ca^{2+}, NH_4^+, putrescine, and spermidine (Gavrilova et al., 1981).

Using natural mRNA as a template, the purified system must be supplemented by initiation factors (three proteins in procaryotic systems or a dozen proteins in eucaryotic systems), termination factors (two or three proteins in procaryotic and one in eucaryotic systems), and a full set of aminoacyl-tRNAs. Such systems were used for studies of initiation and termination phases of translation and the role of individual factors in these processes. A nice example is the bacterial (*E. coli*) system for translation of phage RNAs (MS2 and f2) (Ganoza et al., 1985; Green et al., 1985); several additional protein factors such as rescue factor, elongation factor EF-P, and W were also added and shown to be required for effective translation (Table 2).

A more sophisticated pure system that continuously reacylates tRNA during translation can be formed. Instead of including the set of aminoacylated tRNAs in the system, the full set of tRNAs, all individual aminoacyl-tRNA synthetases (20 proteins), and formyltetrahydrofolate·Met-$tRNA_f$ transformylase should be introduced, along with amino acids, GTP, ATP, an ATP-regenerating system (phosphoenolpyruvate and pyruvate kinase), and a formyl group donor (Table 2) (Kung et al., 1978). Thirty-three individual proteins were used to construct a pure coupled transcription-translation system for gene-dependent synthesis of β-galactosidase (Kung et al., 1977) that includes RNA polymerase, formyltetrahydrofolate·Met-$tRNA_f$ transformylase, 20 aminoacyl-tRNA synthetases, three initiation factors, three elongation factors, two termination factors, and several additional factors (Table 2). This system was later improved (Kung et al., 1979) and also used for synthesis of the proteins of the transcriptional and translational machinery, such as ribosomal proteins L10 and L12, EF-Tu and EF-G, and RNA polymerase subunits (Zarucki-Schulz et al., 1979). These systems can be used for studies of regulation mechanisms of protein synthesis and the role of individual protein factors, as well as in the search for new factors required for transcription and translation.

On the other hand, the possibility of making cell-free systems from pure components opens ways to simplify and modify the natural process. For example, regulation of gene expression in the coupled transcription-translation system can be studied in a highly simplified version of the above-mentioned system wherein only five protein factors are present: RNA polymerase, three initiation factors, and EF-Tu (Robakis et al., 1982). Here only initial dipeptides are formed, and so the system requires a limited set of aminoacyl-tRNAs as well. Identification of dipeptides gives the information about active genes. If necessary for better determination, tripeptides can be formed as a result of adding EF-G and the corresponding aminoacyl-tRNA.

With poly(U) as a template in a cell-free translation system, all of the elongation factors can be omitted. Hence, the system consists of just ribosomes (carefully washed), poly(U), and phenylalanyl-tRNA (Table 2). It was found that in such a factor-free system, binding of aminoacyl-tRNA and transloca-

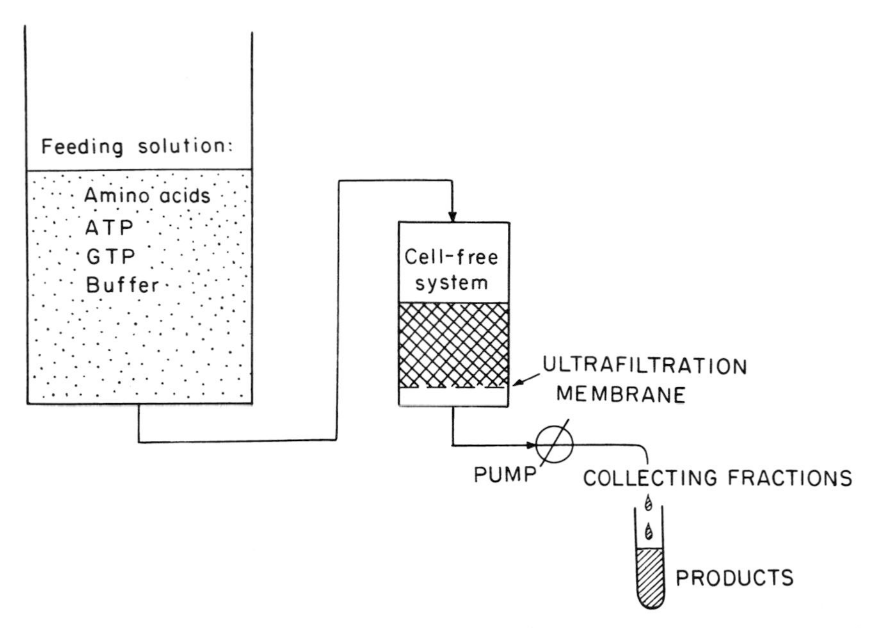

Figure 5. Schematic representation of the device for the continuous-flow cell-free translation system as it was used in the first experiments (Spirin et al., 1988).

tion are accomplished spontaneously (nonenzymatically), thus resulting in slow translation of the template into polyphenylalanine (Pestka, 1969, 1974; Gavrilova and Spirin, 1974; Gavrilova et al., 1976). Factor-free translation could be significantly enhanced as a result of some covalent modifications (SH group blocking), mutational alterations, or complete absence of ribosomal protein S12 from the ribosomes (Gavrilova and Spirin, 1974; Gavrilova et al., 1974; Asatryan and Spirin, 1975). Factor-free translation of poly(A) into oligolysines (Koteliansky and Spirin, 1975) and of poly(U-C) into copolymers of phenylalanine, leucine, serine, and proline (Rutkevitch and Gavrilova, 1982) was also demonstrated. This system is the simplest of all. It has been used for studies of energetics of elongation and accuracy of elongation and, after addition of just one of the elongation factors (either EF-Tu or EF-G), for studies of the contribution of the elongation factors to the rate and accuracy of translation (e.g., Gavrilova et al., 1981).

One more simplified system of peptide synthesis deserves attention: the template-free system for ribosomal synthesis of some polypeptides from aminoacyl-tRNA (Table 2). It was observed that in a system consisting of ribosomes, elongation factors, lysyl-tRNA, and GTP, without any template, oligolysines up to six to seven residues long were synthesized; synthesis depended entirely on ribosome functions and strictly required EF-Tu, EF-G, and GTP (Belitsina et al., 1981). Phenylalanyl-tRNA was found to be incapable of serving as a substrate in this kind of template-free, ribosome-catalyzed elongation. At the same time, when $tRNA^{Lys}$ was misacylated by phenylalanine, the system produced polyphenylalanine from Phe-$tRNA^{Lys}$ (Yusupova et al., 1986). Hence, it is a property of aminoacylated $tRNA^{Lys}$ that allows its participation in the ribosomal elongation cycle without a template. Some other aminoacyl-tRNAs were also found to be capable of serving as substrates for template-free elongation on ribosomes. Among 16 aminoacyl-tRNAs tested, lysyl-, seryl-, threonyl-, and aspartyl-tRNAs proved to be the best, whereas phenylalanyl-, asparaginyl-, methionyl-, isoleucyl-, and some other tRNAs could not serve as substrates in the absence of a messenger (Yusupova et al., 1986). It is interesting that not only was the substrate-binding reaction factor dependent, but the translocation of peptidyl-tRNAs during synthesis was normally catalyzed by EF-G and GTP. In my opinion, the potential of this system for research has not yet been utilized fully.

Cell-Free Systems Are Capable of Continuously Working in the Flow Which Supplies Substrates and Removes Products

All of the cell-free systems described above were made as static incubation mixtures with a constant volume. The situation in a living cell is quite different because consumable substrates are supplied continuously and products are constantly removed. The flux principle in translation was discussed by Kurland as early as 1978, specifically concerning the role of the flux of GTP for providing the rate and fidelity of protein synthesis (Kurland, 1978). It seems that this

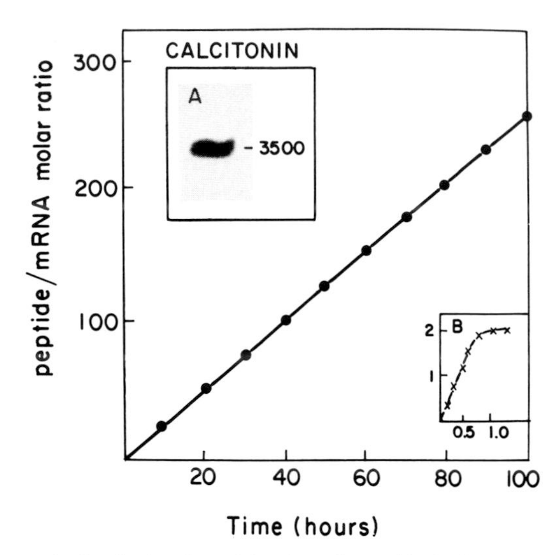

Figure 6. Synthesis of a calcitonin polypeptide in the continuous cell-free translation system based on wheat germ extract at 27°C. Polyribonucleotide transcribed from a synthetic human calcitonin gene by SP6 phage polymerase is used as a template (the composition of the incubation mixture and other conditions are as previously reported by Spirin et al. [1988]). Flow rate is 1 ml/h. Insets: A, electrophoretic pattern of the translation product; B, calcitonin polypeptide synthesis in the standard (static) cell-free translation system of the same composition and volume (given for comparison).

principle has a wider applicability. Not only the continuous regeneration of aminoacyl-tRNAs and GTP but also the removal of the polypeptide product from a polyribosomal compartment may be decisive for providing steady-state kinetics, longevity, and high yield of the protein synthesis process.

Figure 5 is a sketch in which we (Spirin et al., 1988) have attempted to realize the flux principle in a cell-free translation system in a purely mechanical way, i.e., by continuously passing a buffer with substrates (feeding solution) through the incubation mixture. The products are continuously removed from the reaction mixture through an ultrafiltration membrane at the exit of the flow. It has been shown that the flow cell-free translation system is characterized by unusual longevity, linear kinetics of peptide synthesis during its lifetime, and a preparative yield of polypeptide product. If an individual mRNA is used as a template, no purification of the product is required: only the polypeptide synthesized is present as a macromolecular compound in the efflux.

Figure 6 demonstrates an example of the longevity and kinetics in the flow cell-free system. The incubation mixture was based on wheat germ extract and synthetic calcitonin mRNA. It worked linearly during 100 h at 27°C and produced 16 nmol, or more than 50 μg of the polypeptide from 1 ml of incubation mixture. The yields of viral coat proteins in the flow systems based on either bacterial extract and MS2 phage RNA or wheat germ extract and brome mosaic virus RNA were 6 nmol (about 100 μg) and 10 nmol (about 200 μg), respectively, from 1 ml after 20 h of incubation (Spirin et al., 1988). The highest yield was achieved in the flow system designed for globin synthesis in a reticulocyte extract supplemented with purified globin mRNA: 2 mg of the protein was produced in 100 h at 30°C from 0.5 ml of incubation mixture (Ryabova et al., 1989).

Several properties of the continuous-flow translation systems were unexpected, and questions arose as to the supramolecular organization of the protein-synthesizing machinery. First, we did not expect such a strong effect of the flow on the longevity of the cell-free translation. The regeneration of GTP and aminoacyl-tRNAs seemed insufficient to provide such an effect. Our hypothesis is that newly synthesized polypeptides may be inhibitory for ribosomes and that the flowthrough removes this feedback inhibition.

Second, despite the limiting amount of mRNA in the systems, no mRNA degradation was observed during many hours of incubation at 37, 30, or 27°C. In the control experiments, wherein protein synthesis was stopped, mRNA was degraded. Our explanation is that the mRNA molecules involved in intensive translation are well protected by other components of the protein-synthesizing machinery.

Third, through the use of ultrafiltration membranes of different pore sizes, it became clear that no leakage of small protein factors and tRNAs through the membranes takes place during incubation of the reaction mixture in the flow. This fact is especially difficult to explain. We speculate that in both the eucaryotic and procaryotic systems, the actively functioning protein-synthesizing machinery is superorganized in a dynamic complex that retards the removal of its components and thus prevents their leakage.

The bottleneck in the preparative synthesis of polypeptides and proteins in flow cell-free translation systems is in the availability of respective individual mRNAs. About 1 nmol of an individual mRNA should be prepared for the synthesis of 100 to 300 nmol (several milligrams) of a protein. This is feasible but still expensive. That is why it was very tempting to apply the same flow technique to coupled transcription-translation systems. Realization of this simple idea has been accomplished recently, resulting in preparative gene expression under cell-free conditions (Baranov et al., 1989). In the experiments performed, a plasmid DNA was introduced into a bacterial extract freed from endogenous DNAs and mRNAs. The feeding solution contained, in addition to amino acids, ATP, and GTP, also CTP, UTP, and

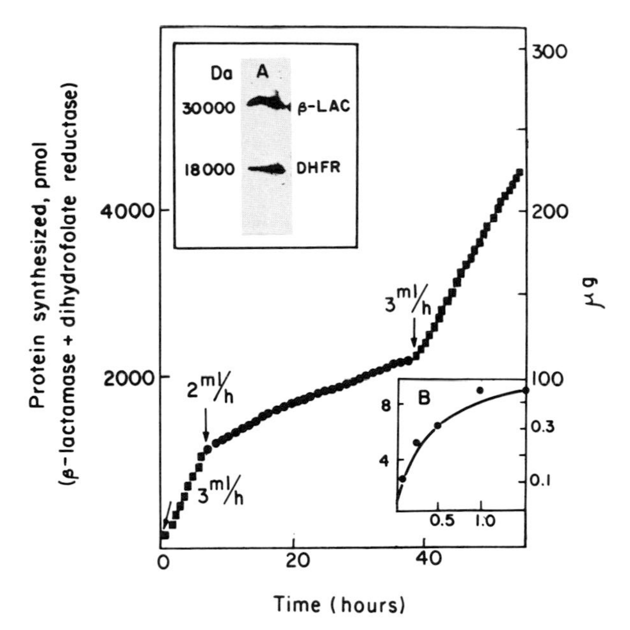

Figure 7. Synthesis of β-lactamase and dihydrofolate reductase (DHFR) in the continuous coupled transcription-translation cell-free system based on bacterial (*E. coli*) extract at 37°C. Plasmid pDF34 (pUD18) is used as a DNA template (see Baranov et al., 1989). Flow rate is changed at the points indicated by arrows from 3 ml/h to 2 ml/h and again to 3 ml/h. Insets: A, electrophoretic pattern of the translation products; B, protein synthesis in the standard (static) coupled transcription-translation system of the same composition and volume (given for comparison). Da, Daltons.

some other supplementary substances. The system was capable of working in a linear manner for tens of hours, responded to the flow rate, and produced pure protein with the same yield as in the case of the flow translation system. Figure 7 represents an example of preparative gene expression in the cell-free system.

In conclusion, I would like to mention possible applications of the preparative cell-free systems of protein synthesis in science and biotechnology: (i) preparative biosynthesis of unstable, cytotoxic, and poorly expressed polypeptides and proteins; (ii) generation of predicted gene products, including translation of open reading frames into polypeptides and proteins not yet discovered; (iii) studies of transcriptional and translational regulations as well as pausing during elongation; (iv) synthesis and study of intermediates of protein folding and modifications; (v) production of poorly available antigen proteins and polypeptides; and (vi) fast protein engineering. Either the longevity of the continuous-flow cell-free systems or the preparative yield or both may contribute to the success of these developments.

I am very grateful to all of my colleagues, especially Vladimir Baranov, Larisa Rozhanskaya, Lubov's Ryabova, and Marat Yusupov, who helped me in writing this chapter. I thank also Alexander Girshovich and Anatoly Gudkov for comments and discussions.

REFERENCES

Anderson, C. W., J. W. Straus, and B. S. Dudock. 1983. Preparation of cell-free protein-synthesizing system from wheat germ. *Methods Enzymol.* **101:**635–644.

Asatryan, L. S., and A. S. Spirin. 1975. Non-enzymatic translocation in ribosomes from streptomycin-resistant mutants of *Escherichia coli*. *Mol. Gen. Genet.* **138:**315–321.

Baranov, V. I., N. V. Belitsina, and A. S. Spirin. 1979. The use of columns with matrix-bound polyuridylic acid for isolation of translating ribosomes. *Methods Enzymol.* **59:**382–397.

Baranov, V. I., I. Y. Morozov, S. A. Ortlepp, and A. S. Spirin. 1989. Preparative gene expression in a cell-free system. *Gene* **84:**463–466.

Barbu, E., J. Panijel, P. Cayeux, and R. Wahl. 1959. Characterization immunochimique de ribonucleoproteines bacteriennes. *C.R. Acad. Sci.* **249:**338–340.

Belitsina, N. V., S. M. Elizarov, M. A. Glukhova, A. S. Spirin, A. S. Butorin, and S. K. Vasilenko. 1975. Isolation of translating ribosomes with a resin-bound polyU-column. *FEBS Lett.* **57:**262–266.

Belitsina, N. V., A. S. Girshovich, and A. S. Spirin. 1973. Translation of resin-bound polynucleotide. *Dokl. Akad. Nauk SSSR* **210:**214–227.

Belitsina, N. V., and A. S. Spirin. 1979. Translation of matrix-bound polyuridylic acid by *Escherichia coli* ribosomes (solid-phase translation system). *Methods Enzymol.* **60:**745–760.

Belitsina, N. V., G. Z. Tnalina, and A. S. Spirin. 1981. Template-free ribosomal synthesis of polylysine from lysyl-tRNA. *FEBS Lett.* **131:**289–292.

Belitsina, N. V., G. Z. Tnalina, and A. S. Spirin. 1982. Template-free ribosomal synthesis of polypeptides from aminoacyl-tRNAs. *BioSystems* **15:**233–241.

Benne, R., M. Brown-Luedi, and J. W. B. Hershey. 1978. Purification and characterization of protein synthesis initiation factors eIF-1, eIF-4C, eIF-4D, and eIF-5 from rabbit reticulocytes. *J. Biol. Chem.* **253:**3070–3077.

Blobel, G., and D. D. Sabatini. 1971. Dissociation of mammalian polyribosomes into subunits by puromycin. *Proc. Natl. Acad. Sci. USA* **68:**390–394.

Bloemendal, H., E. L. Benedetti, and W. S. Bont. 1974. Preparation and characterization of free and membrane-bound polysomes. *Methods Enzymol.* **30:**313–327.

Borsook, H. 1950. Protein turnover and incorporation of labeled amino acids into tissue proteins *in vivo* and *in vitro*. *Physiol. Rev.* **30:**206–219.

Britten, R. J., and R. B. Roberts. 1960. High-resolution density gradient sedimentation analysis. *Science* **131:**32–33.

Brown, G. E., A. J. Kolb, and W. M. Stanley. 1974. A general procedure for the preparation of highly active eukaryotic ribosomes and ribosomal subunits. *Methods Enzymol.* **30:**368–387.

Byers, B. 1966. Ribosome crystallization induced in chick embryo tissues by hypothermia. *J. Cell Biol.* **30:**C1–C6.

Byers, B. 1967. Structure and formation of ribosome crystals in hypothermic chick embryo cells. *J. Mol. Biol.* **26:**155–167.

Byrne, R., J. G. Levin, H. A. Bladen, and M. W. Nirenberg. 1964. The *in vitro* formation of a DNA-ribosome complex. *Proc. Natl. Acad. Sci. USA* **52:**140–148.

Cammack, K. A., and H. E. Wade. 1965. The sedimentation behavior of ribonuclease-active and -inactive ribosomes from bacteria. *Biochem. J.* **96:**671–680.

Chao, F.-C. 1957. Dissociation of macromolecular ribonucleoprotein of yeast. *Arch. Biochem. Biophys.* **70:**426–443.

Chao, F.-C., and H. K. Schachman. 1956. The isolation and characterization of a macromolecular ribonucleoprotein from yeast. *Arch. Biochem. Biophys.* **61**:220–230.

Clark, M. W., M. Hammons, J. A. Langer, and J. A. Lake. 1979. Helical arrays of *Escherichia coli* small ribosomal subunits produced *in vitro*. *J. Mol. Biol.* **135**:507–512.

Clark, M. W., K. Leonard, and J. A. Lake. 1982. Ribosomal crystalline arrays of large subunits from *Escherichia coli*. *Science* **216**:999–1001.

DeVries, J. K., and G. Zubay. 1967. DNA-directed peptide synthesis. II. The synthesis of the x-fragment of the enzyme β-galactosidase. *Proc. Natl. Acad. Sci. USA* **57**:1010–1012.

Dubnoff, J. S., and U. Maitra. 1971. Isolation and properties of protein factors involved in polypeptide chain initiation in *Escherichia coli*. *Methods Enzymol.* **20**:248–260.

Eikenberry, E. F., T. A. Bickle, R. R. Traut, and C. A. Price. 1970. Separation of large quantities of ribosomal subunits by zonal ultracentrifugation. *Eur. J. Biochem.* **12**:113–116.

Elson, D. 1958. Evidence for hydrogen bonds in a ribonucleoprotein. *Biochim. Biophys. Acta* **27**:207–208.

Erbe, R. W., M. M. Nau, and P. Leder. 1969. Translation and translocation of defined RNA messengers. *J. Mol. Biol.* **38**:441–460.

Gale, E. F., and J. P. Folkes. 1954. Effect of nucleic acids protein synthesis and amino-acid incorporation in disrupted staphylococcal cells. *Nature* (London) **173**:1223–1227.

Ganoza, M. C., C. Cuningham, and R. M. Green. 1985. Isolation and point of action of a factor from *Escherichia coli* required to reconstruct translation. *Proc. Natl. Acad. Sci. USA* **82**:1648–1652.

Gavrilova, L. P., D. A. Ivanov, and A. S. Spirin. 1966. Studies on the structure of ribosomes. III. Stepwise unfolding of 50 S particles without loss of protein. *J. Mol. Biol.* **16**:473–489.

Gavrilova, L. P., O. E. Kostiashkina, V. E. Koteliansky, N. M. Rutkevitch, and A. S. Spirin. 1976. Factor-free ("non-enzymic") and factor-dependent systems of translation of polyuridylic acid by *Escherichia coli* ribosomes. *J. Mol. Biol.* **101**:537–552.

Gavrilova, L. P., V. E. Koteliansky, and A. S. Spirin. 1974. Ribosomal protein S12 and "non-enzymatic" translocation. *FEBS Lett.* **45**:324–328.

Gavrilova, L. P., I. N. Perminova, and A. S. Spirin. 1981. Elongation factor Tu can reduce translation errors in poly (U)-directed cell-free system. *J. Mol. Biol.* **16**:67–84.

Gavrilova, L. P., and V. V. Smolyaninov. 1971. Studies on mechanism of translocation in ribosomes. I. Synthesis of polyphenylalanine in *Escherichia coli* ribosomes in the absence of GTP and transfer protein factors. *Mol. Biol.* (Moscow) **8**: 883–891.

Gavrilova, L. P., and A. S. Spirin. 1974. "Nonenzymatic" translation. *Methods Enzymol.* **30**:452–462.

Gesteland, R. F. 1966a. Isolation and characterization of ribonuclease I mutants of *Escherichia coli*. *J. Mol. Biol.* **16**:67–84.

Gesteland, R. F. 1966b. Unfolding of *Escherichia coli* ribosomes by removal of magnesium. *J. Mol. Biol.* **18**:356–371.

Glotz, C., J. Mussig, H. S. Gewitz, I. Makowski, T. Arad, A. Yonath, and H. G. Wittmann. 1987. Three-dimensional crystals of ribosomes and their subunits from eu- and archaebacteria. *Biochem. Int.* **15**:953–960.

Gogia, Z. V., M. M. Yusupov, and T. N. Spirina. 1986. Structure of *Thermus thermophilus* ribosomes. I. Method of isolation and purification of ribosomes. *Mol. Biol.* (Moscow) **20**:519–526.

Gold, L. M., and M. Schweiger. 1969. Synthesis of phage-specific x- and β-glucosyl transferases directed by T-even DNA in vitro. *Proc. Natl. Acad. Sci. USA* **62**:892–898.

Gold, L. M., and M. Schweiger. 1971. Synthesis of bacteriophage-specific enzymes directed by DNA in vitro. *Methods Enzymol.* **20**:537–542.

Gordon, J. 1971. Chain elongation factors, p. 177–199. *In* J. A. Last and A. I. Laskin (ed.), *Protein Biosynthesis in Bacterial Systems*. Marcel Dekker, Inc., New York.

Gordon, J., J. Lucas-Lenard, and F. Lipmann. 1971. Isolation of bacterial chain elongation factors. *Methods Enzymol.* **20**:281–291.

Green, R. H., B. R. Glick, and M. C. Ganoza. 1985. Requirements for *in vitro* reconstruction of protein synthesis. *Biochem. Biophys. Res. Commun.* **126**:792–798.

Henshaw, E. C., and R. Panniers. 1983. Translation systems prepared from the Ehrlich ascites tumor cell. *Methods Enzymol.* **101**:616–629.

Heywood, S. M., and A. W. Rourke. 1974. Cell-free synthesis of myosin. *Methods Enzymol.* **30**:669–674.

Jelenc, P. C. 1980. Rapid purification of high active ribosomes from *Escherichia coli*. *Anal. Biochem.* **105**:369–374.

Jelenc, P. C., and C. G. Kurland. 1979. Nucleoside triphosphate regeneration decreases the frequency of translation errors. *Proc. Natl. Acad. Sci. USA* **76**:3174–3178.

Karpova, E. A., I. N. Serdyuk, Y. V. Tarhovsky, E. V. Orlova, and V. L. Borovyagin. 1986. Crystallization of ribosomes from *Thermus thermophilus*. *Dokl. Akad. Nauk SSSR* **289**:1263–1266.

Keller, E. B., and P. C. Zamecnik. 1956. The effect of guanosine diphosphate and triphosphate on the incorporation of labeled amino acids in protein. *J. Biol. Chem.* **221**:45–59.

Khesin, R. B. 1953. Formation of amylase by cytoplasmic granules isolated from pancreas cells. *Biokhimiya* **18**:462–474.

Kirillov, S. V., V. I. Makhno, N. N. Peshin, and Y. P. Semenkov. 1978. Separation of ribosomal subunits of *Escherichia coli* by Sepharose chromatography using reverse salt gradient. *Nucleic Acids Res.* **5**:4305–4315.

Koteliansky, V. E., and A. S. Spirin. 1975. "Non-enzymatic" translocation in ribosomes using polyadenylic acid as a template. *Dokl. Akad. Nauk SSSR* **221**:477–480.

Kung, H.-F., F. Chu, P. Caldwell, C. Spears, B. V. Treadwell, B. Eskin, N. Brot, and H. Weissbach. 1978. The mRNA-directed synthesis of the x-peptide of β-galactosidase, ribosomal protein L12 and L10, and elongation factor Tu, using purified translational factors. *Arch. Biochem. Biophys.* **187**:457–463.

Kung, H.-F., B. Redfield, B. V. Treadwell, B. Eskin, C. Spears, and H. Weissbach. 1977. DNA-directed *in vitro* synthesis of β-galactosidase. *J. Biol. Chem.* **252**:6889–6894.

Kung, H.-F., B. Redfield, and H. Weissbach. 1979. DNA-directed *in vitro* synthesis of β-galactosidase. *J. Biol. Chem.* **254**:8404–8408.

Kurland, C. G. 1966. The requirements for specific sRNA binding by ribosomes. *J. Mol. Biol.* **18**:90–108.

Kurland, C. G. 1971. Purification of ribosomes from *Escherichia coli*. *Methods Enzymol.* **20**:379–381.

Kurland, C. G. 1978. The role of guanine nucleotides in protein biosynthesis. *Biophys. J.* **22**:373–392.

Lake, J. A. 1980. Ribosome structure and functional sites, p. 207–236. *In* G. Chambliss, G. R. Craven, J. Davies, K. Davis, L. Kahan, and M. Nomura (ed.), *Ribosomes. Structure, Function, and Genetics*. University Park Press, Baltimore.

Lake, J. A., and H. S. Slayter. 1972. Three-dimensional structure of the chromatoid body helix of *Entamoeba invadens*. *J. Mol. Biol.* **66**:271–282.

Lamborg, H., and P. C. Zamecnik. 1960. Amino acid incorporation into protein by extracts of *E. coli*. *Biochim. Biophys. Acta* **42**:206–211.

Lawford, G. R. 1969. The effect of incubation with puromycin on the dissociation of rat liver ribosomes into active subunits. *Biochem. Biophys. Res. Commun.* **37**:143–150.

Lederman, M., and G. Zubay. 1967. DNA-directed peptide synthesis. I. A comparison of T_2 and *Escherichia coli* DNA-directed peptide synthesis in two cell-free systems. *Biochim. Biophys. Acta* **149:**253–258.

Littlefield, J. W., and E. B. Keller. 1957. Incorporation of C^{14} amino acids into ribonucleoprotein particles from the Ehrlich mouse ascites tumor. *J. Biol. Chem.* **224:**13–30.

Littlefield, J. W., E. B. Keller, J. Gross, and P. C. Zamecnik. 1955. Studies on cytoplasmic ribonucleoprotein particles from the liver of the rat. *J. Biol. Chem.* **217:**111–123.

Makowski, I., F. Frolow, M. Saper, M. Shoham, H. G. Wittmann, and A. Yonath. 1987. Single crystals of large ribosomal particles from *Halobacterium marismortui* diffract to 6 Å. *J. Mol. Biol.* **193:**819–822.

Marcus, A., D. Efron, and D. P. Weeks. 1974. The wheat embryo cell-free system. *Methods Enzymol.* **30:**749–754.

Matthaei, J. H., and M. W. Nirenberg. 1961. Characteristics and stabilization of DNAase-sensitive protein synthesis in *E. coli* extracts. *Proc. Natl. Acad. Sci. USA* **47:**1580–1588.

McQuillen, K., R. B. Roberts, and R. J. Britten. 1959. Synthesis of nascent protein by ribosomes in *E. coli. Proc. Natl. Acad. Sci. USA* **45:**1437–1447.

Merrick, W. C. 1983. Translation of exogenous mRNAs in reticulocyte lysates. *Methods Enzymol.* **101:**606–615.

Miller, R. L., and R. Schweet. 1968. Isolation of a protein fraction from reticulocyte ribosomes required for *de novo* synthesis of hemoglobin. *Arch. Biochem. Biophys.* **125:**632–646.

Minich, W. B., and L. P. Ovchinnikov. 1985. Free RNA-binding cytoplasmic proteins are in labile association with polyribosomes. *Biokhimiya* **50:**604–612.

Moldave, K., and I. Sadnik. 1979. Preparation of derived and native ribosomal subunits from rat liver. *Methods Enzymol.* **59:**402–410.

Moldave, K., and L. Skogerson. 1967. Purification of mammalian ribosomes. *Methods Enzymol.* **12:**478–491.

Nirenberg, M. W. 1963. Cell free protein synthesis directed by messenger RNA. *Methods Enzymol.* **6:**17–23.

Nirenberg, M. W., and J. H. Matthaei. 1961. The dependence of cell-free protein synthesis in *E. coli* upon naturally occurring or synthetic polynucleotides. *Proc. Natl. Acad. Sci. USA* **47:**1588–1602.

Noll, M., B. Hapke, and H. Noll. 1973a. Structural dynamics of bacterial ribosomes. II. Preparation and characterization of ribosomes and subunits active in the translation of natural messenger RNA. *J. Mol. Biol.* **80:**519–529.

Noll, M., B. Hapke, M. H. Schreier, and H. Noll. 1973b. Structural dynamics of bacterial ribosomes. I. Characterization of vacant couples and their relation to complexed ribosomes. *J. Mol. Biol.* **75:**281–294.

Noll, M., and H. Noll. 1976. Structural dynamics of bacterial ribosomes. V. Magnesium-dependent dissociation of tight couples into subunits: measurements of dissociate constants and exchange rates. *J. Mol. Biol.* **105:**111–130.

O'Brien, L., K. Shelley, J. Towfighi, and A. McPherson. 1980. Crystalline ribosomes are present in brains from senile humans. *Proc. Natl. Acad. Sci. USA* **77:**2260–2264.

Palade, G. E. 1955. A small particulate component of the cytoplasm. *J. Biophys. Biochem. Cytol.* **1:**59–68.

Pelham, H. R. B., and R. J. Jackson. 1976. An efficient mRNA-dependent translation system from reticulocyte lysates. *Eur. J. Biochem.* **67:**247–256.

Pestka, S. 1968. Studies on the formation of transfer ribonucleic acid-ribosome complexes. III. The formation of peptide bonds by ribosomes in the absence of supernatant enzymes. *J. Biol. Chem.* **243:**2810–2820.

Pestka, S. 1969. Studies on the formation of transfer ribonucleic acid-ribosome complex. VI. Oligopeptide synthesis and translocation on ribosomes in the presence and absence of soluble transfer factors. *J. Biol. Chem.* **244:**1533–1539.

Pestka, S. 1974. Ribonuclease sensitivity of aminoacyl-tRNA: an assay for codon recognition and interaction of aminoacyl-tRNA with 50S subunits. *Methods Enzymol.* **20:**306–316.

Peterman, M. L., and M. G. Hamilton. 1957. The purification and properties of cytoplasmic ribonucleoprotein from rat liver. *J. Biol. Chem.* **224:**725–736.

Peterson, E. A., and D. M. Greenberg. 1952. Characteristics of the amino acid-incorporating system of liver homogenates. *J. Biol. Chem.* **194:**359–375.

Ravel, J. M., and R.-A. L. Shorey. 1971. GTP-dependent binding of aminoacyl-tRNA to *Escherichia coli* ribosomes. *Methods Enzymol.* **20:**306–316.

Robakis, N., Y. Cenatiempo, S. Peacock, N. Brot, and H. Weissbach. 1982. Use of dipeptide synthesis to study the *in vitro* expression of the L10 (B) operon, p. 129–146. *In Interaction of Translational and Transcriptional Controls in the Regulation of Gene Expression.* Elsevier, Amsterdam.

Roberts, B. E., and B. M. Paterson. 1973. Efficient translation of tobacco mosaic virus RNA and rabbit globin 9S RNA in a cell-free system from commercial wheat germ. *Proc. Natl. Acad. Sci. USA* **70:**2330–2334.

Roberts, R. B. (ed.). 1958. *Microsomal Particles and Protein Synthesis.* Pergamon Press, New York.

Ron, E. Z., R. E. Kohler, and B. D. Davis. 1968. Magnesium ion dependence of free and polysomal ribosomes from *Escherichia coli. J. Mol. Biol.* **36:**83–89.

Rutkevitch, N. M., and L. P. Gavrilova. 1982. Factor-free and one-factor-promoted poly (U, C)-dependent synthesis of polypeptides in cell-free systems from *Escherichia coli. FEBS. Lett.* **143:**115–118.

Ryabova, L. A., S. A. Ortlepp, and V. I. Baranov. 1989. Preparative synthesis of globin in a continuous cell-free translation system from rabbit reticulocytes. *Nucleic Acids Res.* **17:**4412.

Ryazanov, A. G., L. P. Ovchinnikov, and A. S. Spirin. 1987. Development of structural organization of protein-synthesizing machinery from prokaryotes to eukaryotes. *BioSystems* **20:**275–288.

Safer, S., S. L. Adams, W. M. Kemper, K. W. Berry, M. Lloyd, and W. C. Merrick. 1976. Purification and characterization of two initiation factors required for maximal activity of a highly fractionated globin mRNA translation system. *Proc. Natl. Acad. Sci. USA* **73:**2584–2688.

Salas, M., M. A. Smith, W. M. Stanley, A. J. Wahba, and S. Ochoa. 1965. Direction of reading of the genetic message. *J. Biol. Chem.* **240:**3988–3995.

Schachtschabel, D., and W. Zillig. 1959. Untersuchungen zur Biosynthese der Proteine. I. Uber den Einbau C^{14}-markierter Aminosauren ins Protein zellfreier Nucleoproteind-Enzyme-Systeme aus *E. coli* B. *Hoppe-Seyler's Z. Physiol. Chem.* **314:** 262–275.

Schlessinger, D., and F. Gros. 1963. Structure and properties of active ribosomes of *Escherichia coli. J. Mol. Biol.* **7:**350–359.

Schreier, M. H., B. Erni, and T. Staehelin. 1977. Initiation of mammalian protein synthesis. I. Purification and characterization of seven initiation factors. *J. Mol. Biol.* **116:**727–753.

Schweiger, M., and L. M. Gold. 1969. Bacteriophage T4 DNA-dependent *in vitro* synthesis of lysozyme. *Proc. Natl. Acad. Sci. USA* **63:**1351–1358.

Siekevitz, P. 1952. Uptake of radioactive alanine *in vitro* into proteins of rat liver fractions. *J. Biol. Chem.* **195:**549–565.

Siekevitz, P., and P. C. Zamecnik. 1951. In vitro incorporation of 1-C^{14}-DL-alanine into proteins of rat-liver granular fractions. *Fed. Proc.* **10:**246–247.

Spirin, A. S. 1964. *Macromolecular Structure and Ribonucleic Acids.* Reinhold, New York.

Spirin, A. S. 1974. Structural transformations of ribosomes (dissociation, unfolding, and disassembly). *FEBS Lett.* **40:**S38–S47.

Spirin, A. S., and M. A. Ajtkhozhin. 1985. Informosomes and polyribosome-associated proteins in eukaryotes. *Trends Biochem. Sci.* **10:**162–165.

Spirin, A. S., V. I. Baranov, G. S. Polubesov, I. N. Serdyuk, and R. I. May. 1987. Translocation makes the ribosome less compact. *J. Mol. Biol.* **194:**119–128.

Spirin, A. S., V. I. Baranov, L. A. Ryabova, S. Y. Ovodov, and Y. B. Alakhov. 1988. A continuous cell-free translation system capable of producing polypeptides in high yield. *Science* **242:** 1162–1164.

Spirin, A. S., N. A. Kisselev, R. S. Shakulov, and A. A. Bogdanov. 1963. On the structure of ribosomes: reversible unfolding of the ribosomal particles into ribonucleoprotein strands and possible model of packing. *Biokhimiya* **28:**920–930.

Spirin, A. S., M. Y. Sofronova, and B. Sabo. 1970. Dissociation on *Escherichia coli* monoribosomes upon lowering the magnesium ion concentration in the medium. *Mol. Biol.* (Moscow) **4:** 618–627.

Staehelin, T., and A. K. Falvey. 1971. Isolation of mammalian ribosomal subunits active on polypeptide synthesis. *Methods Enzymol.* **20:**433–446.

Staehelin, T., and D. R. Maglott. 1971. Preparation of *Escherichia coli* ribosomal subunits active in polypeptide synthesis. *Methods Enzymol.* **20:**449–456.

Staehelin, T., D. Maglott, and R. E. Monro. 1969. On the catalytic center of peptidyl transferase: a part of the 50S ribosome structure. *Cold Spring Harbor Symp. Quant. Biol.* **34:**39–48.

Stanley, W. M. 1959. Ph.D. thesis. University of Wisconsin, Madison.

Stanley, W. M., M. Salas, A. J. Wahba, and S. Ochoa. 1966. Translation of genetic message: factors involved in the initiation of protein synthesis. *Proc. Natl. Acad. Sci. USA* **56:**290–295.

Stanley, W. M., and A. J. Wahba. 1967. Chromatographic purification of ribosomes. *Methods Enzymol.* **12:**524–526.

Taddei, C. 1972. Ribosome arrangement during oogenesis of *Lacerta sicula. Exp. Cell Res.* **70:**285–292.

Takanami, M. 1960. A stable ribonucleoprotein for amino acid incorporation. *Biochim. Biophys. Acta* **39:**318–326.

Takanami, M. 1967. Preparation of ribosomes and their subunits from *Escherichia coli. Methods Enzymol.* **12:**491–494.

Takata, K., and S. Osawa. 1957. Ribonucleoprotein from rabbit appendix microsomes. *Biochim. Biophys. Acta* **24:**207–209.

Tissières, A., D. Schlessinger, and F. Gros. 1960. Amino acid incorporation into proteins by *E. coli* ribosomes. *Proc. Natl. Acad. Sci. USA* **46:**1450–1463.

Tissières, A., and J. D. Watson. 1958. Ribonucleoprotein particles from *E. coli. Nature* (London) **182:**778–780.

Tissières, A., J. D. Watson, D. Schlessinger, and B. R. Hollingworth. 1959. Ribonucleoprotein particles from *Escherichia coli. J. Mol. Biol.* **1:**221–233.

Trakhanov, S., M. Yusupov, V. Shirokov, M. B. Garber, A. Mitschler, M. Ruff, J.-C. Thierry, and D. Moras. 1989. Preliminary X-ray investigation of 70S ribosomes crystals from *Thermus thermophilus. J. Mol. Biol.* **209:**327–328.

Trakhanov, S. D., M. M. Yusupov, S. C. Agalarov, M. B. Garber, S. N. Ryazantsev, S. V. Tischenko, and V. A. Shirokov. 1987. Crystallization of 70S ribosomes and 30S ribosomal subunits from *Thermus thermophilus. FEBS Lett.* **220:**319–322.

Ts'o, P. O. P., J. Bonner, and J. Vinograd. 1956. Microsomal nucleoprotein particles from pea seedlings. *J. Biophys. Biochem. Cytol.* **2:**451–465.

Unwin, P. N. T. 1977. Three-dimensional model of membrane-bound ribosomes obtained by electron microscope. *Nature* (London) **269:**118–122.

Winnick, T. 1950a. Incorporation of labeled amino acids into protein of embryonic and tumor tissue homogenates. *Fed. Proc.* **9:**247.

Winnick, T. 1950b. Studies on the mechanism of protein synthesis in embryonic and tumor tissues. II. Inactivation of fetal rat liver homogenates by dialyses and reactivation by the adenylic acids system. *Arch. Biochem.* **28:**338–347.

Wittmann, H. G., J. Mussig, J. Piefke, H. S. Gewitz, H. J. Rheinberger, and A. Yonath. 1982. Crystallization of *Escherichia coli* ribosomes. *FEBS Lett.* **146:**217–220.

Wood, W. B., and P. Berg. 1962. The effect of enzymatically synthesized ribonucleic acid on amino acid incorporation by a soluble protein-ribosome system from *Escherichia coli. Proc. Natl. Acad. Sci. USA* **48:**94–104.

Yonath, A., C. Glotz, H. L. Gevity, K. T. Bartels, K. vonBochler, J. Mashoski, and H. G. Wittmann. 1988. Characterization of crystals of small ribosomal subunits. *J. Mol. Biol.* **203:**831–834.

Yonath, A., G. Khawitch, B. Tesche, J. Mussig, S. Lorenz, V. A. Erdmann, and H. G. Wittmann. 1982. The nucleation of crystals of the large ribosomal subunits from *Bacillus stearothermophilus. Biochem. Int.* **5:**629–636.

Yonath, A. E., J. Mussig, B. Tesche, S. Lorenz, V. A. Erdmann, and H. G. Wittmann. 1980. Crystallization of the large ribosomal subunits from *Bacillus stearothermophilus. Biochem. Int.* **1:**428–435.

Yonath, A., M. A. Saper, I. Makowski, J. Müssig, J. Piefke, H. D. Bartunik, K. S. Bartels, and H. G. Wittmann. 1986. Characterization of single crystals of the large ribosomal particles from *Bacillus stearothermophilus. J. Mol. Biol.* **187:**633–636.

Yonath, A., B. Tesche, S. Lorenz, J. Mussig, V. A. Erdmann, and H. G. Wittmann. 1983. Several crystal forms of the *Bacillus stearothermophilus* 50S ribosomal particles. *FEBS Lett.* **154:** 15–20.

Yusupov, M. M., S. V. Tischenko, S. D. Trakhanov, S. N. Ryazantsev, and M. B. Garber. 1988. A new crystalline form of 30S ribosomal subunits from *Thermus thermophilus. FEBS Lett.* **238:**113–115.

Yusupov, M. M., S. D. Trakhanov, V. V. Barynin, V. L. Borovyagin, M. B. Garber, S. E. Sedelnikova, O. M. Selivanova, S. V. Tischenko, V. A. Shirokov, and Y. N. Edintsov. 1987. Crystallization of 30S ribosomal subunits from *Thermus thermophilus. Dokl. Akad. Nauk SSSR* **292:**1271–1274.

Yusupova, G. Z., N. V. Belitsina, and A. S. Spirin. 1986. Template-free ribosomal synthesis of polypeptides from aminoacyl-tRNA. Polyphenylalanine synthesis from phenylalanyl-$tRNA^{Lys}$. *FEBS Lett.* **206:**142–146.

Zamecnik, P. C. 1969. An historical account of protein synthesis, with current overtones—a personalized view. *Cold Spring Harbor Symp. Quant. Biol.* **34:**1–16.

Zarucki-Schulz, T., C. Jeres, G. Goldberg, H.-F. Kung, K. H. Huang, N. Brot, and H. Weissbach. 1979. DNA-directed synthesis of proteins involved in bacterial transcription and translation. *Proc. Natl. Acad. Sci. USA* **76:**6115–6119.

Zubay, G. 1973. *In vitro* synthesis of protein in microbial systems. *Annu. Rev. Genet.* **7:**267–287.

II. STRUCTURE OF RIBOSOMES AND rRNA

II. STRUCTURE OF RIBOSOMES AND rRNA

The size and complexity of the ribosome are such that the study of its structure has occupied hundreds of worker-years of research and involved virtually every extant biophysical technique. Although the gross morphology determined by electron microscopy is now widely accepted as a useful, low-resolution model, the structure at high resolution remains elusive. However, remarkable progress has been made, especially in elucidating the position of the individual proteins and the structure of rRNA.

In the following chapters, detailed models of the ribosome are presented which concentrate heavily on the placement of the rRNA in situ. For the first time the ribonucleoprotein mass is beginning to take on a detailed form. Reviewed here are the structures obtained by means of new electron microscopy techniques, results which are giving more and more clearly defined pictures of the gross morphology of the ribosome. The exciting progress in crystallography of ribosomal subunits, giving substantial promise of detailed structural information in the future, is also presented.

The structure of rRNA is being studied in detail, using chemical and biological agents as well as electron microscopy and neutron scattering. While these studies do not purport to give a detailed map of the entire ribosome, considerable progress has been and is being made in these areas.

Chapter 3

Structure of rRNA and Its Functional Interactions in Translation

HARRY F. NOLLER, DANESH MOAZED, SETH STERN, TED POWERS, PATRICK N. ALLEN, JAMES M. ROBERTSON, BRYN WEISER, and KATHLEEN TRIMAN

rRNA is one of the most highly conserved classes of macromolecules found in nature (Gutell et al., 1985). This is most likely a consequence of its central importance in protein synthesis. The notion that rRNA is simply a scaffold for the assembly of ribosomal proteins (Fellner, 1974) has given way to a growing body of evidence which supports the opposite viewpoint: that it is rRNA, rather than the ribosomal proteins, that defines ribosome function (Noller and Woese, 1981). In this view, the fundamental mechanism of translation is based on interactions between mRNA, tRNA, and rRNA. Such a mechanism is consistent with the idea that translation evolved from preexisting RNA-based systems.

In its simplest, conceptual form, the mechanism of translation must account for three fundamental events: codon-anticodon recognition, peptide bond formation, and movement of tRNA and mRNA relative to the ribosome. The first event presents the problem of translational accuracy and its thermodynamic opponent, speed. Codon-anticodon interaction depends on only three base pairs (one of which, the wobble pair, is often very weak) and is insufficiently strong (and consequently extremely inaccurate) in the absence of ribosomes (Grosjean et al., 1978). Such a system could be made more accurate by using more base pairs, but then codon-anticodon pairing would be so strong that translation would be intolerably slow. The ribosome solves this problem, but no one understands precisely how. The second problem, transfer of a peptidyl group between two tRNAs, is unacceptably slow in the absence of ribosomes. The ribosome solves this problem with an activity, peptidyltransferase, that is an integral part of the large subunit (Monro, 1967). Almost nothing is known about the structure or mechanism of this important catalytic function. Finally, the tRNAs and mRNA must move, since the peptidyl- and aminoacyl-tRNAs are continually changing roles, and therefore their stereochemical relationships must change. Woese (1970) proposed an elegant solution to this problem, involving only changes in the conformation of the anticodon loops of tRNA and the codons in mRNA. In the ribosome, however, it appears that tRNAs actually move physically (translocate) from one distinct site to another, as will be discussed below, and this is perhaps the most spectacular problem solved by ribosomes. Again, little is known about how translocation actually happens. In spite of this apparently bleak outlook, recent studies on the structure and function of rRNA give cause for optimism. In this chapter, we focus primarily on results from our own laboratory, which deal with the higher-order structure of rRNA and its interactions with functional ligands.

PROTEIN-RNA INTERACTIONS IN THE SMALL SUBUNIT

Twenty-one different proteins assemble with 16S rRNA to form the *Escherichia coli* small (30S) ribosomal subunit. Some of the main questions concerning this assembly process are: How many of the proteins interact with 16S rRNA? Do the proteins change the conformation of the RNA? Which specific features of the RNA are recognized by the various proteins? What is the basis of the cooperativity of assembly evident in the assembly map of Held et al. (1974)? A major motivation for our studies on 16S

Harry F. Noller, Danesh Moazed, Seth Stern, Ted Powers, Patrick N. Allen, Bryn Weiser, and Kathleen Triman ■ Thimann Laboratories, University of California, Santa Cruz, Santa Cruz, California 95064. **James M. Robertson** ■ Applied Biosystems, Inc., Foster City, California 94403.

rRNA-protein interactions was also to learn about the three-dimensional folding of the RNA, as described in the next section.

Our approach to these questions was made possible by development of a method that permits rapid inspection of the accessibility of each base in a large RNA molecule (Moazed et al., 1986; Stern et al., 1988b). By using a set of chemical probes (kethoxal, dimethyl sulfate, and a carbodiimide reagent), all four bases can be probed. Their susceptibility to chemical modification is then assessed by primer extension, using the modified RNA as template and synthetic DNA oligomers as primers. Reverse transcriptase pauses or stops at the sites of modification, causing premature termination. The sites are identified as bands on sequencing gels. Intramolecular or intermolecular interactions involving the sites of chemical attack (N1 of purines and N3 of pyrimidines) protect the involved bases from the chemical probes, and this is seen as the absence or decrease in intensity of a band on the gels.

We took two approaches to studying the protein-16S rRNA interactions. In one approach, we built up the 30S subunit in a stepwise fashion, beginning with a single primary binding protein, following the assembly map of Nomura and coworkers (Held et al., 1974). The other type of experiment, also borrowed from Nomura et al. (1969), was the single-protein-omission experiment, in which particles were reconstituted but lacking a single protein. In both cases, the effect of assembly of a single protein on the reactivity of bases in 16S rRNA is monitored. Generally, we found the two approaches to be complementary and mutually consistent.

Figure 1 summarizes the effects of assembly of the individual proteins (Stern et al., 1986; Powers et al., 1988a; Powers et al., 1988b; Svensson et al., 1988). All 20 proteins (protein S1 was not tested) produced effects on 16S rRNA. This result shows that each protein interacts, either directly or indirectly, with 16S rRNA. The effects of assembly are complex; most of the proteins produce both protections and enhancements. We interpret protection as being due either to contact between a protein and the protected base or to a conformational change caused by assembly of the protein; enhancements are assigned to protein-induced conformational changes. It is evident that the effects of the primary and secondary binding proteins (e.g., S4, S7, S8, S9, S12, S19, and S20) are generally more extensive than those of the tertiary binding proteins (e.g., S10, S14, and S21). This suggests that the later-assembling proteins may rely relatively less on protein-RNA interactions and more on protein-protein interactions. The majority of the bases protected by the primary binding proteins S4, S7, S15, S17, and S20 fall within their classical "binding-site fragments" (Zimmermann, 1979), as defined by nuclease resistance of specific ribonucleoprotein complexes. Moreover, the protected bases occupy only a relatively small fraction of the structures contained by the classical fragments. This suggests that the proteins actually recognize and bind to compact regions of the rRNA and that the large sizes of some of the binding-site fragments may be due to inherent nuclease resistance of the RNA structure.

Recognition of 16S rRNA by proteins appears to be complex. Recognition of base sequences in helical sections, as is common among DNA-binding proteins, can probably be ruled out. This follows from the fact that sequences in base-paired regions are variable, yet active 30S subunits can be reconstituted from heterologous combinations of RNA and protein from distantly related bacteria (Nomura et al., 1968). Accordingly, recognition must involve more conserved bases, as well as characteristic three-dimensional structural features.

What is the basis of the cooperativity implicit in the assembly map? One can invoke two extreme classes of models to account for the dependence of a late-assembling protein on prior assembly events. In one type of model, the early protein binds to the RNA, and the late protein binds to the early protein. In the other type of model, the early protein induces a conformational rearrangement in the RNA, which creates an RNA-binding site for the late protein. Although no definite conclusions can be made as yet, the available evidence is consistent with the involvement of both classes of mechanism in the assembly process.

Neutron diffraction studies (Capel et al., 1987) have shown that many of the proteins that are linked in the assembly map are near neighbors. Together with the isolation of specific protein-protein complexes (Dijk et al., 1977), it seems likely that protein-protein interactions are an important part of ribosome assembly. Our probing results suggest that protein-dependent conformational changes in 16S rRNA may also play a role in the observed cooperativity.

Assembly of proteins S5 and S12 shows a strong dependency on S16, and S16 assembly depends in turn on S4 (Held et al., 1974). Here, the probing data (Stern et al., 1986; Stern et al., 1988a; Stern et al., 1988c) show a striking parallel to the assembly pathway. S4 causes enhancement of nucleotides 361 to 364, which are protected by assembly of S16 (Fig. 1). S16 produces enhancements at positions 21, 26, 563, 887, and 894, which are in turn protected by assembly of S5 and S12. A second example involves

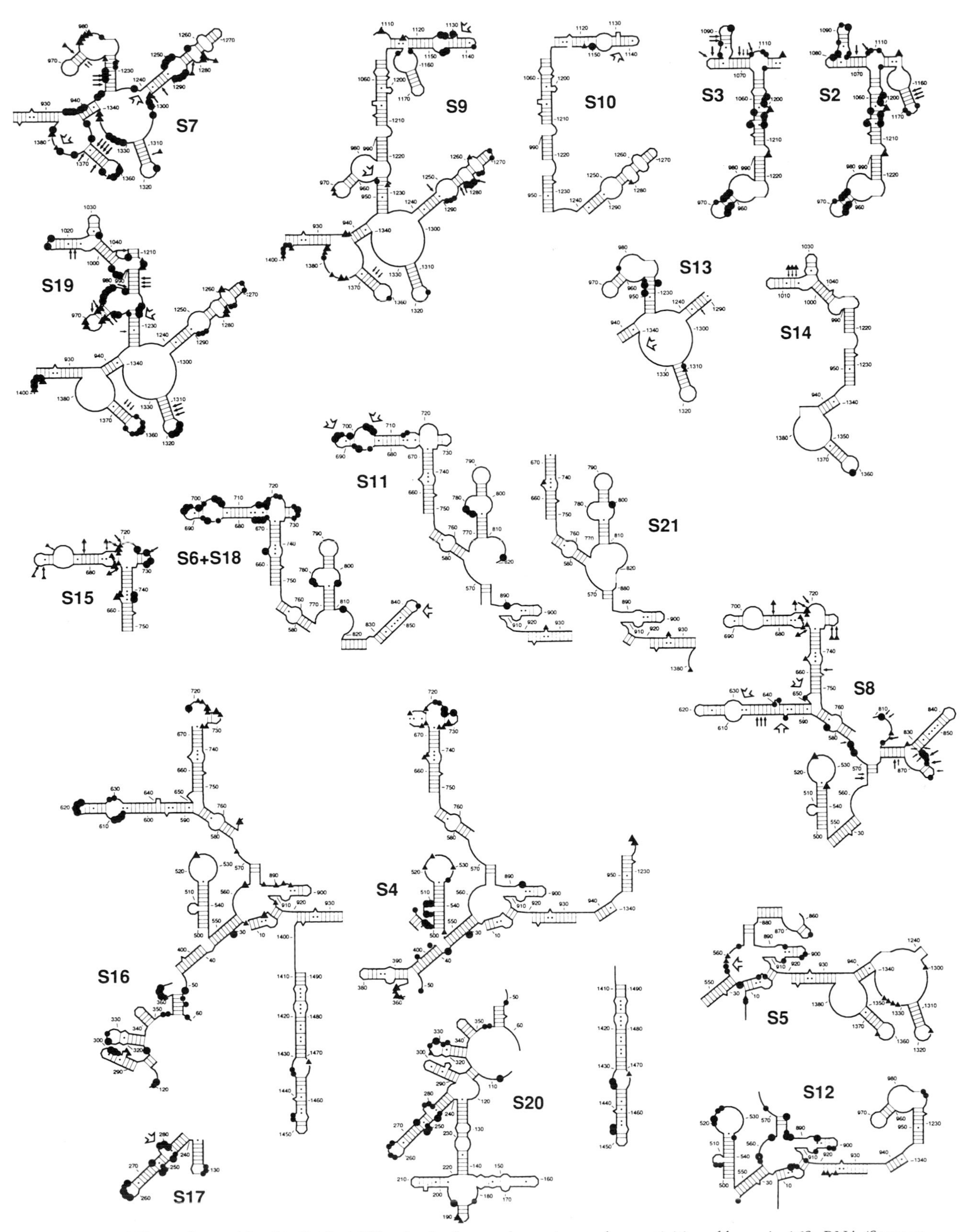

Figure 1. Effects of assembly of individual 30S subunit ribosomal proteins on the reactivities of bases in 16S rRNA (Stern et al., 1986; Stern et al., 1988a; Stern et al., 1988b; Stern et al., 1988c, 1989; Powers et al., 1988a; Powers et al., 1988b; Svensson et al., 1988). Protein-dependent protection from and enhancement of attack by chemical probes are shown by circles and triangles, respectively. Sites where individual proteins have been cross-linked to RNA (reviewed in Brimacombe and Stiege, 1985; Brimacombe et al., 1988) are shown by open arrows.

the dependence of S6 + S18 on S15 (Svensson et al., 1988). S15 causes enhancement of nucleotides 664, 673 to 674, and 717 to 719, which are then protected by assembly of proteins S6 + S18 (Fig. 1). A third example is the dependence of S9 and S19 on S7 (Powers et al., 1988a). S7 causes enhancement of bases in the 980 region, which are then protected by S19. Binding of S7 also causes enhanced reactivity of bases in the 1280 loop, where S9-dependent protections are observed (Fig. 1).

Another phenomenon bearing on the question of cooperativity of assembly is what we have termed polyspecific effects, in which different proteins are found independently to protect the same bases. In this case, protection cannot be due to direct contact between the protected bases and the different proteins. We have proposed that polyspecific effects are caused by conformational changes in the RNA that are stabilized by the binding of proteins to adjacent regions of the RNA structure (Stern et al., 1988c). A likely model is that the proteins bind preferentially to the altered RNA conformer, driving assembly forward in a cooperative manner. Striking examples of polyspecific protection are the effects of S2 and S3 in the 960 and 1050/1200 regions, S5 and S12 in the 900 and adjacent regions, and S11 and S6 + S18 in the 700 region (Fig. 1).

Finally, certain nucleotides are protected only when specific combinations of proteins are bound. We term these cooperative effects. These are distinct from cooperativity of binding, as for example with proteins S6 + S18, which show little capacity for assembly in the absence of each other. An example of cooperative protection is seen with proteins S2 and S3 (Powers et al., 1988b). Each protein produces changes in the probing pattern independently of the other, but when both proteins are present, a new set of effects is seen involving positions 1094, 1104, 1108, and 1111.

Earlier studies on ribosome structure and function implicated certain ribosomal proteins in specific functional roles. Our findings raise the question as to whether the observed functional effects of these proteins are in fact due to their influence on the conformation of functional sites in rRNA. The assembly data for two sets of proteins are particularly striking. One set of proteins, S6, S11, S18, and S21, is located around the cleft of the 30S subunit, in keeping with their implication in mRNA and tRNA binding (Oakes et al., 1986a; Stöffler and Stöffler-Meilicke, 1986). Our probing results (Fig. 1) show that assembly of these proteins specifically affects the reactivities of bases in the 690, 790, and 926 regions (Svensson et al., 1988; Stern et al., 1988c), three of the main sites that are strongly protected by P-site binding of tRNA (Moazed and Noller, 1986, 1990), as described below.

Another set of proteins, S4, S5, and S12, is located at the opposite side of the 30S subset from the cleft and has generally been implicated in translational accuracy (Gorini, 1974). Mutations in protein S12 confer resistance to or dependence on streptomycin, a drug that increases the translational error frequency. S12 alleles conferring streptomycin dependence give rise to ribosomes that are hyperaccurate in the absence of streptomycin. Dependence can be suppressed by certain S4 and S5 (*ram*) alleles, providing evidence for functional interaction between these three proteins. Ribosomes from *ram* strains show a high error frequency. These observations imply that translational accuracy is somehow held in balance at a low level of misreading by a mechanism that is perturbed by certain mutations in S4, S5, or S12.

These same three proteins show striking interactive assembly effects (Fig. 1), the most intriguing of which involve the 530 and 900 loop regions (Stern et al., 1986; Stern et al., 1988a; Stern et al., 1988c), two regions where tRNA-dependent conformational changes are believed to occur (Moazed and Noller, 1986, 1987a), as described below. Proteins S4 and S12 have antagonistic effects on the 530 loop; bases in this region are mainly enhanced by S4 and are protected by S12. S5 and S12 produce several common polyspecific effects in the 900 loop region and in the flanking 20 and 560 regions, at the junction of the three major domains. Thus, both S4 and S5 produce assembly effects which overlap those of S12, but in distinctly different regions of the structure.

Independent support for the notion that the effects of these three proteins on translational accuracy somehow involve 16S rRNA comes from the study of certain protein and RNA mutations. Chemical probing of ribosomes from Sm^d, Sm^r, and *ram* strains shows that the conformation of 16S rRNA is indeed perturbed by mutations in S4 and S12 (Allen and Noller, 1989). In ribosomes from S4 *ram* strains, A-8 and A-26 have enhanced reactivity. In ribosomes from S12 Sm^d mutants, A-908, A-909, A-1413, and G-1487 are all less reactive to our probes (Fig. 2). An interesting correlation is found between the reactivity of A-908 and the miscoding level of Sm^r, Sm^d, Sm^p (streptomycin pseudodependent), and wild-type ribosomes; the reactivity of A-908 increases as the translational error frequency of the ribosomes increases. In the case of *ram* ribosomes, the reactivity of A-908 resembles that of wild type unless tRNA is bound, in which case it becomes hyperreactive. Similarly, the reactivity of A-908 in wild-type ribosomes is enhanced by streptomycin, but only when tRNA is

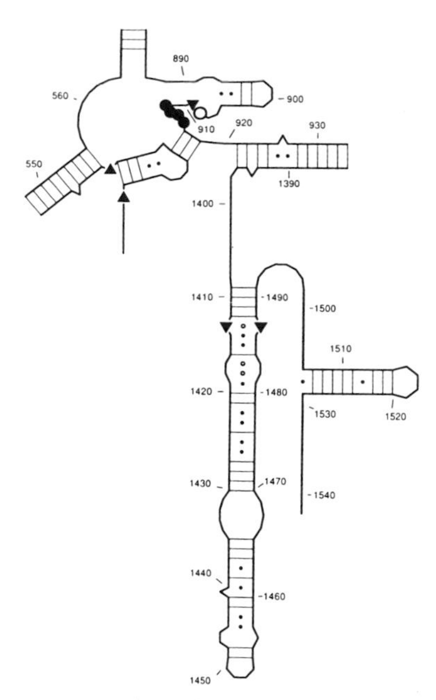

Figure 2. Nucleotides affected by mutations in proteins S4 and S12 (Allen and Noller, 1989). Symbols: ▲, nucleotides enhanced in *ram* ribosomes; ▼, class III nucleotides with reduced activity in Sm^{d} ribosomes; ○, A-908, whose reactivity follows the miscoding potential of mutant ribosomes; ●, bases protected by streptomycin (Moazed and Noller, 1987a).

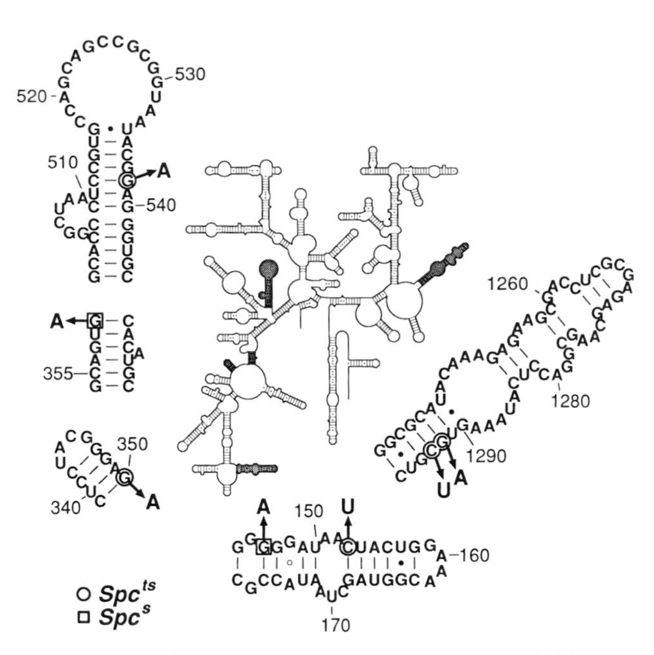

Figure 3. Sites of temperature-sensitive mutations (circled bases) and down mutations (boxed bases) in 16S rRNA (Triman et al., 1989).

bound. Finally, streptomycin causes the reactivity of A-908 in Sm^{d} and Sm^{p} ribosomes to increase to near-wild-type levels. We have proposed a model in which reactivity of A-908 is a measure of the state of equilibrium between two ribosomal conformers (Allen and Noller, 1989). Mutations in S12 would generally cause a shift toward the unreactive (hyperaccurate) conformer, whereas mutations in S4 would shift the equilibrium in favor of the reactive (error-prone) conformer.

A C-to-U mutation at position 912 of 16S rRNA has been reported to confer streptomycin resistance (Montandon et al., 1986), and streptomycin itself protects bases in the 911–915 region (Moazed and Noller, 1987a). In addition, mutations at positions 523 and 525 have recently been shown to confer streptomycin resistance (Melançon et al., 1988; Fromm et al., 1989). Thus, both of the regions of 16S rRNA where protein S12 shows assembly effects in common with S4 and S5, and which are also believed to undergo tRNA-induced conformational changes, are somehow involved in the interaction of streptomycin with ribosomes. In this connection, it is interesting that the neomycin class of antibiotics, which appear to induce miscoding by a mechanism distinct from that of streptomycin (Cundliffe, 1981), cause enhanced reactivity of C-525 (Moazed and Noller, 1987a).

TERTIARY FOLDING OF 16S rRNA

Comparative sequence analysis provided the basis for establishing physiologically authentic secondary structures for 16S and 23S rRNA (Gutell et al., 1985; Noller and Woese, 1981; Woese et al., 1980; Noller et al., 1981). These structures have survived testing not only by hundreds of newly derived rRNA sequences but also by a wide range of biochemical approaches (Moazed et al., 1986; Noller, 1984; Brimacombe and Stiege, 1985). More recently, genetic results have provided further evidence. Temperature-sensitive mutations in 16S rRNA were obtained by a novel selection scheme using random mutagenesis (Triman et al., 1989). The sites of mutation of several of these have been mapped and sequenced, and they have been found to cause mismatches at base-paired positions of the structure (Fig. 3). The next level of structural organization, that of tertiary folding, has proven to be much more resistant to comparative sequence analysis. This is probably due to several factors. First, tertiary base-base interactions are undoubtedly much rarer than secondary base pairs. Second, if tRNA is a reliable indication, the bases involved in tertiary interactions are likely to be much more conserved, and thus the phylogenetic variation required to detect them will be difficult to find. Third, whereas secondary structure is, for the most part, constrained by Watson-Crick pairing rules, tertiary interactions may, as far as we know, involve any combination of bases. Finally, long-range tertiary interactions, such as domain-domain interac-

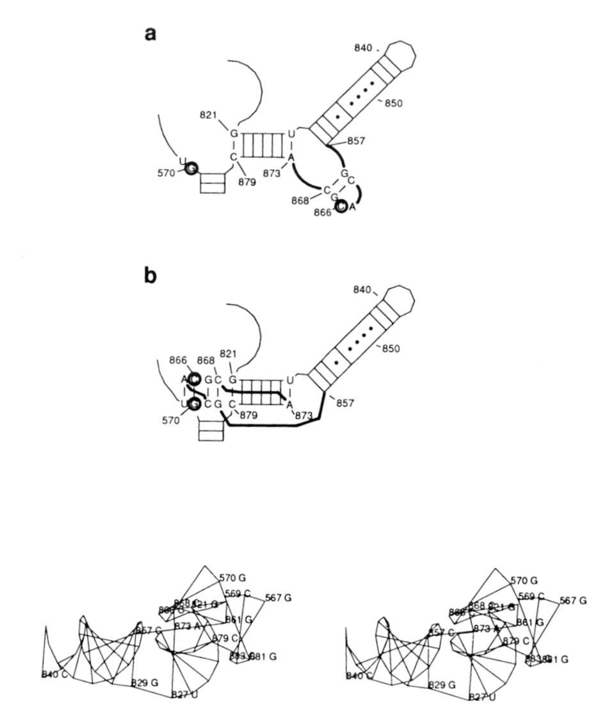

Figure 4. The 570–866 tertiary interaction (Gutell et al., 1986), as modeled by Stern et al. (1988d).

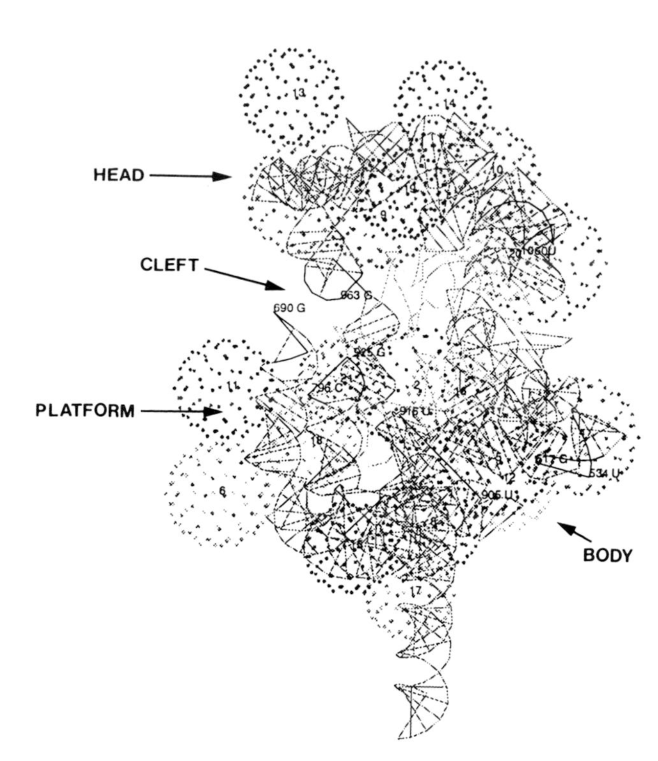

Figure 5. Model for the three-dimensional folding of 16S rRNA (Stern et al., 1988d). The 30S protein positions are from neutron diffraction studies and are shown as dotted spheres (Capel et al., 1987).

tions, have failed to materialize in studies on 16S rRNA, and there is some cause to believe that they may not exist, although several examples have been found in 23S rRNA (Gutell and Fox, 1989; Leffers et al., 1987). Nevertheless, two interesting sets of base-base tertiary interactions have been found in 16S rRNA by the comparative approach. One of these involves interactions between the 510 and 530 loops (Woese and Gutell, 1989), and the other is between positions 570 and 866 (Gutell et al., 1986). Both sets of interactions invoke unusual pseudoknotlike structures. We have suggested a plausible folding scheme for the structure involving the 570–866 interaction (Fig. 4; Stern et al., 1988d).

Without benefit of a crystal structure, a different approach to three-dimensional folding is clearly required. Several groups have undertaken model building, the traditional response of molecular biologists to such problems (Stern et al., 1988d; Expert-Bezançon and Wollenzien, 1985; Brimacombe et al., 1988; Nagano and Harel, 1987; Noller and Lake, 1984). The problem of folding 16S rRNA can be visualized in two stages; the first is to arrange the helical elements of secondary structure folding in space in their correct orientations, and the second is to deduce the folding of the nonhelical elements. Almost all of the model building has so far concentrated on the first stage, which, however, results in an overall general sense of the path of the RNA and is probably the most desirable information to have in hand at the present time.

Our approach to modeling the three-dimensional folding of 16S rRNA is based on its secondary structure (as deduced from comparative sequence analysis), the positions of the centers of mass of the ribosomal proteins (as obtained by neutron diffraction), and chemical probing data on protein-RNA interactions (Fig. 1), which provide a link between the three-dimensional coordinate system defined by the protein positions and the helical elements of the secondary structure. We also used certain protein-RNA and RNA-RNA cross-links obtained using intact 30S subunits by Brimacombe and co-workers (summarized in Brimacombe et al., 1988). The modeling exercise was carried out by computer graphics methods (Stern et al., 1988d).

Figure 5 is a diagram of the model, superimposed on the neutron map (Capel et al., 1987) of the protein positions. About 75% of the RNA is sufficiently constrained to provide a useful model; this contains most of the universally conserved core (Gutell et al., 1985) of the molecule. In all but a few instances, protected and cross-linked sites can be placed within or very close to the spheres representing the anhydrous mass of each protein, while obeying stereochemical rules. The overall shape of the model and locations of specific regions of the RNA correspond well to data derived from electron micro-

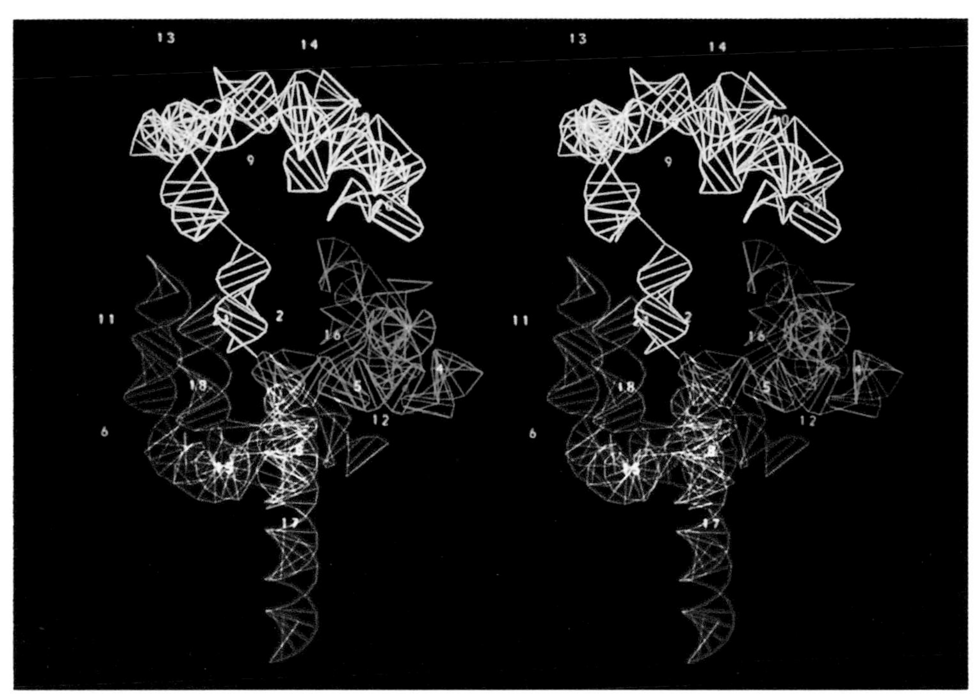

Figure 6. Model for folding of 16S rRNA (Stern et al., 1988d) showing the 5′ domain (blue), central domain (red), and 3′ domain (yellow).

graphs of 30S subunits, although such data were not used to construct the model. Phylogenetic variations in the structure are easily accommodated.

The three major secondary structure domains appear to exist as largely autonomous structural units in three dimensions (Fig. 6). There is extensive contact between the 5′ and central domains, whereas the 3′ major domain has apparently little contact with the rest of the RNA structure. The 5′, central, and 3′ major domains form structures that resemble the body, platform, and head, respectively, as seen in electron micrographs of 30S subunits (Oakes et al., 1986a; Stöffler and Stöffler-Meilicke, 1986).

In certain cases, the locations of specific proteins relative to certain structural features of the RNA suggest how they may play a role in ribosome assembly. The primary binding proteins S4, S7, and S8 are all located at junction regions, at the convergence of several helices. In each case, the relative orientation of these helices determines important overall features of the folding of their three respective domains (S4 in the 5′, S8 in the central, and S7 in the 3′ domain), and, to some extent, the orientation of the three domains relative to one another. A reasonable possibility is that these three proteins help to stabilize the correct relative geometry at these helix junctions. If assembly of the secondary and tertiary binding proteins depends on this geometry, such a role for these primary binders could help to explain why much of 30S assembly depends on them. Indeed, Vasiliev et al. (1977) have obtained electron micrograph images of a complex of 16S rRNA with proteins S4, S7, S8, and S15 which bear close resemblance to complete 30S particles.

Some proteins appear to be positioned at major bends in the RNA. For example, protein S19 is located at a 90° bend that occurs in the internal loop around position 1213 (Fig. 7). Proteins S6 and S18 flank the 90° turn centered at the 720 internal loop, and S15 is positioned where the 650/750 stem emerges at an acute angle from the 590/650 stem. Thus, another possible role for proteins (in this case, both primary and secondary binding proteins) in assembly may be to stabilize bends in the RNA structure.

Our structural model carries a number of strong implications for ribosome function. These are most easily discussed in the context of assignment of functional sites, presented below.

SITES OF INTERACTION OF tRNA WITH rRNA

The ribosome can be thought of as an enzyme ("polypeptide polymerase," as Moore [1985] has suggested) whose "substrate" is tRNA. The "active sites" of the ribosome might then be considered to be those parts of its structure that interact with tRNA. We know of at least two: the site of codon-anticodon

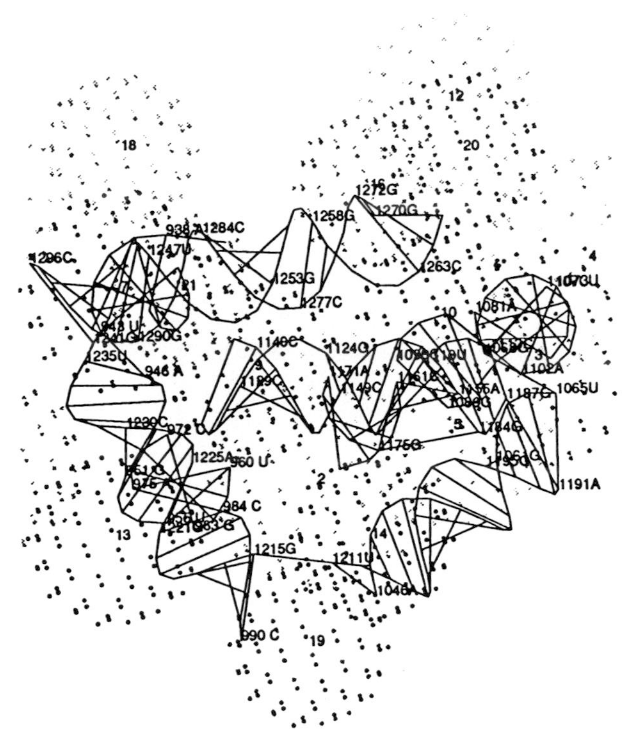

Figure 7. View of the 3′ domain (head) region of the 16S rRNA model, viewed from the top of the subunit (Stern et al., 1988d). Protein S19 is at the bottom.

interaction, or decoding site, and the peptidyltransferase region. The sites of interaction with protein factors, such as elongation factors EF-Tu and EF-G, might thus be akin to regulatory sites and, if the analogy holds, could be involved in allosteric effects that modulate the activity of the tRNA-interactive regions of the ribosome.

What is the nature of a ribosome active site? Several years ago, we showed that chemical modification of a relatively small number of bases in 16S rRNA abolishes the ability of 30S subunits to participate in protein synthesis (Noller and Chaires, 1972). The loss of activity was shown to specifically involve the tRNA-binding function, and bound tRNA protected the subunits from chemical inactivation. Reconstitution experiments clearly showed that it was the RNA, not the protein moiety, that had been functionally modified. We inferred that the subset of modified bases that was protected by bound tRNA is important for tRNA binding and that this important ribosomal active site is made up, at least in part, of RNA. More recently, we have used two different methods to identify bases in rRNA that are protected by tRNA (Brow and Noller, 1983; Moazed and Noller, 1986, 1989a, 1990). The results from the two methods are in close agreement; we discuss the more recent approach, for which the data are more comprehensive.

These experiments are based on the same method (Moazed et al., 1986; Stern et al., 1988b) used in mapping the sites of protein interactions with 16S rRNA, as described above. Ribosomes are subjected to chemical probing, with or without bound tRNA, and the sites and extent of chemical modification in 16S and 23S rRNA are assessed by primer extension. Specific, defined complexes are formed in vitro, according to conventional procedures, to fill the ribosomal P, A, and E sites, respectively. Under the appropriate conditions (10 mM Mg^{2+}, 100 mM NH_4Cl, 37°C, 20 min, using equimolar concentrations of tRNA and ribosomes), *N*-acetyl-Phe-tRNA binds in a poly(U)-dependent fashion to the ribosomal P site. In our hands, about 60% binding occurs under these conditions, and the bound tRNA is fully puromycin reactive, confirming its P-site location. For A-site binding, we first fill the P site with deacylated tRNA and then introduce tRNA into the A site by "enzymatic" binding, as an aminoacyl-tRNA · EF-Tu · GTP ternary complex. E-site binding is defined by its requirement for deacylated tRNA bearing an intact CCA 3′ terminus (Grajevskaya et al., 1982).

Figure 8 summarizes the sites protected by A- and P-site-bound tRNA in 16S rRNA (Moazed and Noller, 1986, 1990), and Fig. 9 shows the sites protected in 23S rRNA by tRNA in the A, P, and E sites. These bases interact, either directly or indirectly, with tRNA, and they must include the bases whose modification causes inactivation of tRNA binding, as concluded from our earlier work (Noller and Chaires, 1972). They are found in regions of the RNA where the surrounding secondary structure is virtually identical in all ribosomes (Gutell et al., 1985). Furthermore, the protected bases are either universally conserved or nearly so across the phylogenetic spectrum. This is the degree of conservation expected for components of active sites of molecules that are involved in an essential, universal function such as translation. Not surprisingly, it is, to our knowledge, matched only by the conservation of structural and sequence elements in tRNA.

All of the same 16S rRNA protections are obtained when, instead of tRNA, its anticodon stem-loop (a 15-nucleotide fragment; Rose et al., 1983) is bound to the ribosome. This shows that at least with respect to 16S rRNA, all of the contacts with the 30S subunit involve only this region of the tRNA. Similarly, most of the P-site protections in 23S rRNA are obtained when only the 3′-terminal UACCA(Ac-Leu) fragment is bound in the presence of sparsomycin and methanol (D. Moazed and H. Noller, unpublished data). Thus, the aminoacyl extremity of tRNA supplies the main interactions with the P site in 23S rRNA. These findings are in accord with what is

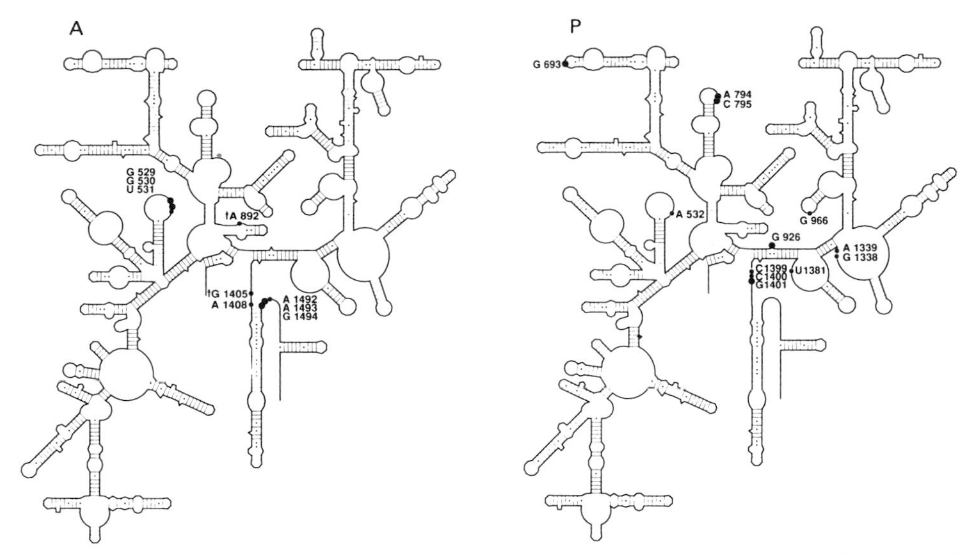

Figure 8. Sites of protection of 16S rRNA by A- and P-site bound tRNA (Moazed and Noller, 1986, 1990).

known about the distribution of function between the two ribosomal subunits. Anticodon-codon recognition requires only the small subunit (Nirenberg and Leder, 1964), aside from proofreading, and peptide bond formation can be catalyzed by isolated 50S subunits (Monro, 1967). Whether or not the "elbow" of tRNA interacts with ribosomes at all remains to be seen. Cross-links to protein S19 from this region of tRNA (Lin et al., 1984) suggest that if it does interact, it may be with proteins.

P-site residues in 16S rRNA seem widely distributed throughout the secondary structure in several regions of the molecule (Fig. 8). At least four of these sites (693, 794–795, 926, and 966) are close to each other in our structural model, however, because of the three-dimensional folding of 16S rRNA (Fig. 10). The helices carrying these four sites have the form of a thumb and three fingers arranged such that another helix (such as the P-site tRNA anticodon stem-loop) can be grasped between them (Stern et al., 1988d).

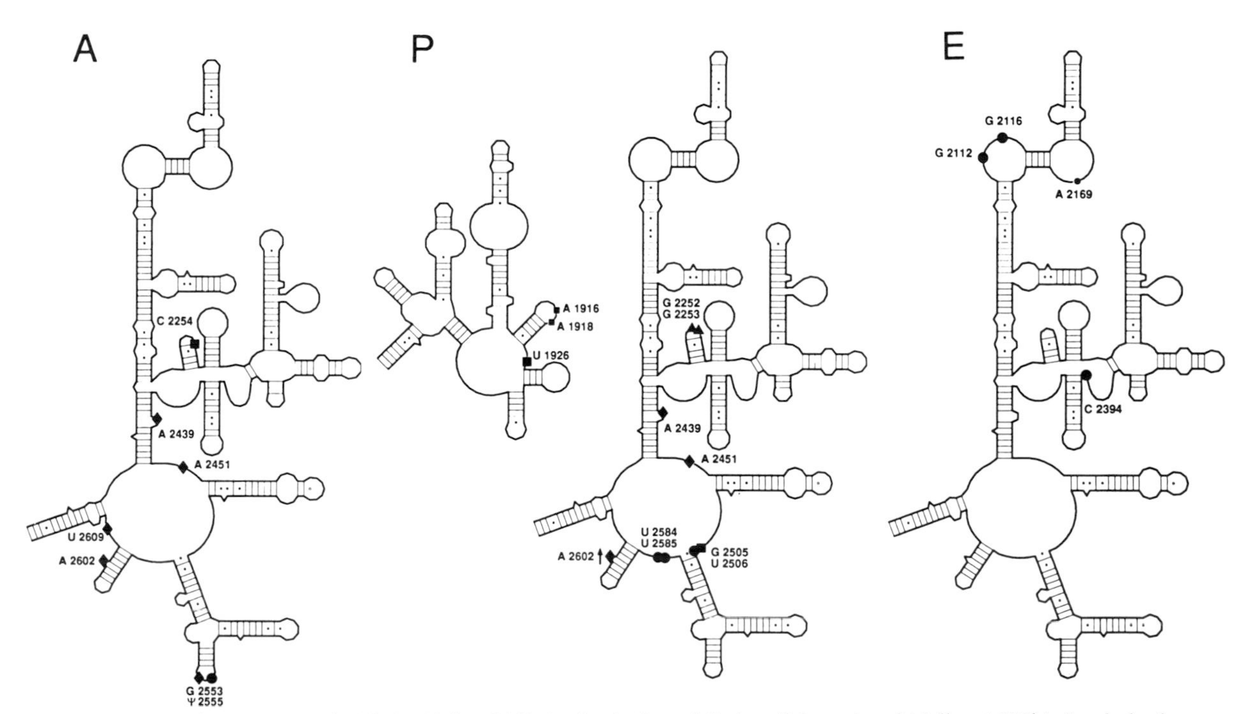

Figure 9. Sites of protection of 23S rRNA by tRNA in the A, P, and E sites (Moazed and Noller, 1989b). Symbols show dependence of protection on the acyl moiety (◆), the 3′-terminal A (●), the 3′-terminal CA (▲), and the rest of the tRNA molecule (■).

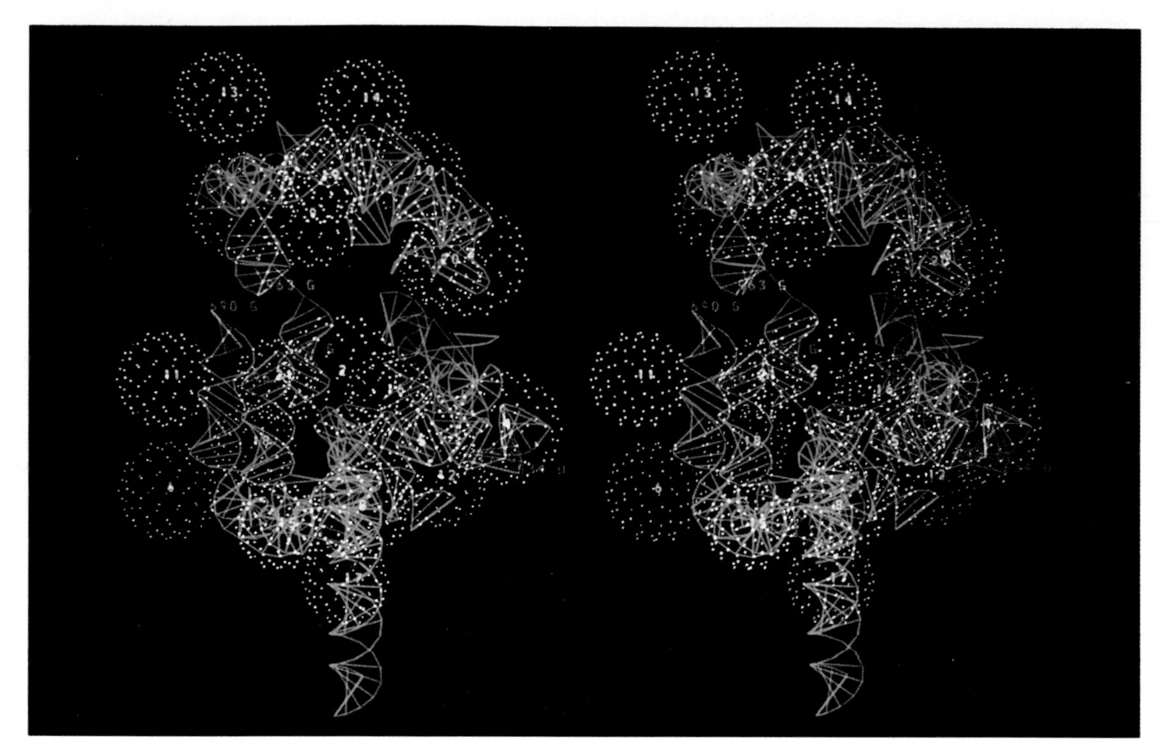

Figure 10. Model for the folding of 16S rRNA showing five of the tRNA P-site protections (G-693, A-794, G-795, G-926, and G-966) in red at the left. The 530 region stem is at the far right (Stern et al., 1988d).

Two of the helices, carrying G-693 and A-794–C-795, respectively, help to form the platform, whereas G-926 protrudes into the cleft from the neck joining the head to the body, and G-966 hangs down from the underside of the head. The apparently extensive interactions between 16S rRNA and P-site tRNA are consistent with the relatively tight binding of tRNA to the P site and its mRNA-independent binding at higher Mg^{2+} concentrations. Another cluster of P-site protections includes C-1399, C-1400, and G-1401 (Fig. 8). Although this region of the molecule is not shown explicitly in our three-dimensional model, it has been localized to the cleft region by a combination of affinity labeling and immunoelectron microscopy (Gornicki et al., 1984; Oakes et al., 1986b). Most likely, it is immediately below the lower three P-site protections shown in Fig. 10, inside the cleft, since the opposite strand of the helix containing G-926 exits at A-1396 at the bottom, such that continuing the sense of the helix would place C-1400 inside the cleft, below and to the left of the 926 helix.

A-site-bound tRNA, in contrast, protects only two clusters of bases, in the 530 loop and in the 1400/1500 region (Fig. 8). Moreover, these two regions are too distant (70 to 100 Å [7 to 10 nm]), judging from our three-dimensional model (Fig. 10) as well as from direct electron microscopy measurements (Trempe et al., 1982), to be simultaneously in contact with the anticodon stem-loop structure, which has a maximum dimension of 25 Å. Thus, any direct contact with A-site tRNA must be reflected by protections in the 1400–1500 region, the site of codon-anticodon interaction (Prince et al., 1982). We are forced to conclude that the effects in the 530 loop are the result of a conformational change resulting from interactions between the tRNA anticodon stem-loop and the 30S subunit in the decoding site. The 530 loop is located on the side of the 70S ribosome where elongation factors EF-Tu and EF-G have been mapped. Since EF-Tu is involved in A-site binding in vivo, it seems reasonable to expect that the 530 loop is in some way involved in its interactions with the ribosome. However, we have so far failed to detect any effects of EF-Tu here by using chemical probing methods.

Most of the bases protected by tRNA in 23S rRNA (Moazed and Noller, 1989a) are found in domain V, but some are also found in domains II and IV (Fig. 9). Several of the domain V bases are located around the central loop that has been identified with the peptidyltransferase function. In fact, some of the protected bases have been labeled by tRNAs bearing photochemically reactive groups on their acyl moieties (Barta et al., 1984) and have been identified as the site of mutations conferring resistance to the peptidyltransferase inhibitors chloramphenicol and anisomycin (Sigmund et al., 1984; Hummel and

Böck, 1987). Two of the protected bases, A-2451 and G-2505, are also protected by chloramphenicol itself as well as by carbomycin, another peptidyltransferase inhibitor (Moazed and Noller, 1987b) (see Fig. 13).

As mentioned above, most of the protections by P-site tRNA are due to interactions involving the 3′-terminal CCA end and its attached acyl moiety. We have further analyzed the structural requirements in tRNA by removing the acyl group, the 3′ A, and the 3′ penultimate C, respectively (Moazed and Noller, 1989a). Specific protections are lost in a discrete, stepwise fashion with the removal of each of these groups, allowing assignment of protected bases to individual structural features of the tRNA (Fig. 9). This provides further evidence that the main interactions between tRNA and 23S rRNA involve only the acyl moiety and the 3′ CCA end of tRNA.

This theme is reinforced by E-site binding, which is absolutely dependent on an intact CCA terminus (Grajevskaya et al., 1982). E-site protections are found only in 23S rRNA (Moazed and Noller, 1989a), in support of the suggestion that E-site binding involves only the large subunit. Three of the E-site effects are found in the upper region of domain V, in the binding site for protein L1 (Branlant et al., 1976). The other site (C-2394) is in a part of domain V that can be placed very close to protein L27 (Wower et al., 1981). From the immunoelectron microscopy locations of these proteins, the E-site-protected nucleotides can be located between the central protuberance and the L1 ridge of the 50S subunit, just to the left of the presumed location of the peptidyltransferase center (Oakes et al., 1986a; Stöffler and Stöffler-Meilicke, 1986), as viewed from the small subunit side. This suggests that the direction of tRNA translocation from A to P to E is from right to left; several independent lines of reasoning lead to this conclusion, as discussed below.

A-site bases in 23S rRNA are not protected when aminoacyl-tRNA is bound as a ternary complex in the presence of kirromycin, an antibiotic that prevents dissociation of the EF-Tu · GDP complex (Moazed and Noller, 1989a). However, A-site protections are found in 16S rRNA under these same conditions. These results provide direct evidence that binding of aminoacyl-tRNA occurs in two steps. During the first step, the anticodon stem-loop interacts with mRNA and the 16S rRNA A site on the small subunit, whereas the EF-Tu moiety (but not tRNA) interacts with 23S rRNA on the large subunit (see below). We refer to this mode of tRNA binding as the A/T state. After GTP hydrolysis, EF-Tu · GDP is released; in the second step, binding of aminoacyl-tRNA is completed when its aminoacyl end interacts with 23S rRNA in the peptidyltransferase region.

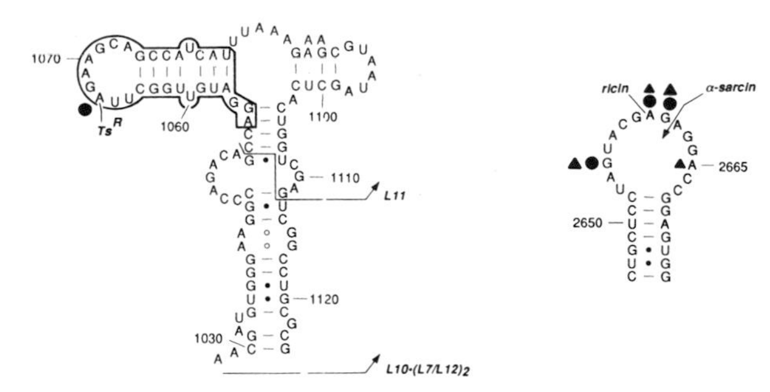

Figure 11. Protection of sites in 23S rRNA by EF-G (●) and EF-Tu (▲) (Moazed et al., 1988). The sites of attack by ricin and α-sarcin (Endo and Wool, 1982; Endo et al., 1987), site of methylation conferring thiostrepton resistance (Thompson et al., 1982), and fragments protected by proteins L10, L7/L12, and L11 are indicated. The fragment cross-linked to EF-G is boxed (Sköld, 1983).

Our findings account for the inability of aminoacyl-tRNA to participate in peptide bond formation when release of EF-Tu is prevented; this is an essential feature of kinetic proofreading models (Hopfield, 1974; Ninio, 1974), in which the accuracy of protein synthesis is amplified by a two-step pathway for the binding of aminoacyl-tRNA.

INTERACTIONS OF EF-Tu AND EF-G WITH 23S rRNA

Elongation factors EF-Tu and EF-G compete with each other for binding to ribosomes (Cabreu et al., 1972; Miller, 1972), and the actions of both factors involve GTP hydrolysis. EF-G has been cross-linked to 23S rRNA (Sköld, 1983) around position 1067, the site of a ribose methylation that confers resistance to thiostrepton, an inhibitor of EF-G–ribosome interaction (Thompson et al., 1982).

Using chemical probing methods, we searched for protection of bases in 16S and 23S rRNA in ribosomes complexed with EF-Tu or EF-G (Moazed et al., 1988). Complexes containing EF-G were stabilized by fusidic acid or by use of the nonhydrolyzable GTP analog GMPPCP or GMPPNP. Complexes between ribosomes and EF-Tu · tRNA · GTP were stabilized by kirromycin. Both factors caused characteristic protection of bases in 23S but not 16S rRNA. The results are summarized in Fig. 11. EF-G protects A-1067 and, more weakly, A-1069, in agreement with the cross-linking and thiostrepton results. It also protects bases in the highly conserved loop at position 2660, the site of action of the cytotoxins α-sarcin and ricin (Fig. 11; Endo and Wool, 1982; Endo et al., 1987). We did not detect protection by EF-Tu in the 1067 region, but found protection in the 2660 loop that overlapped the sites protected by EF-G (Fig. 11). The latter result may account for the mutually exclu-

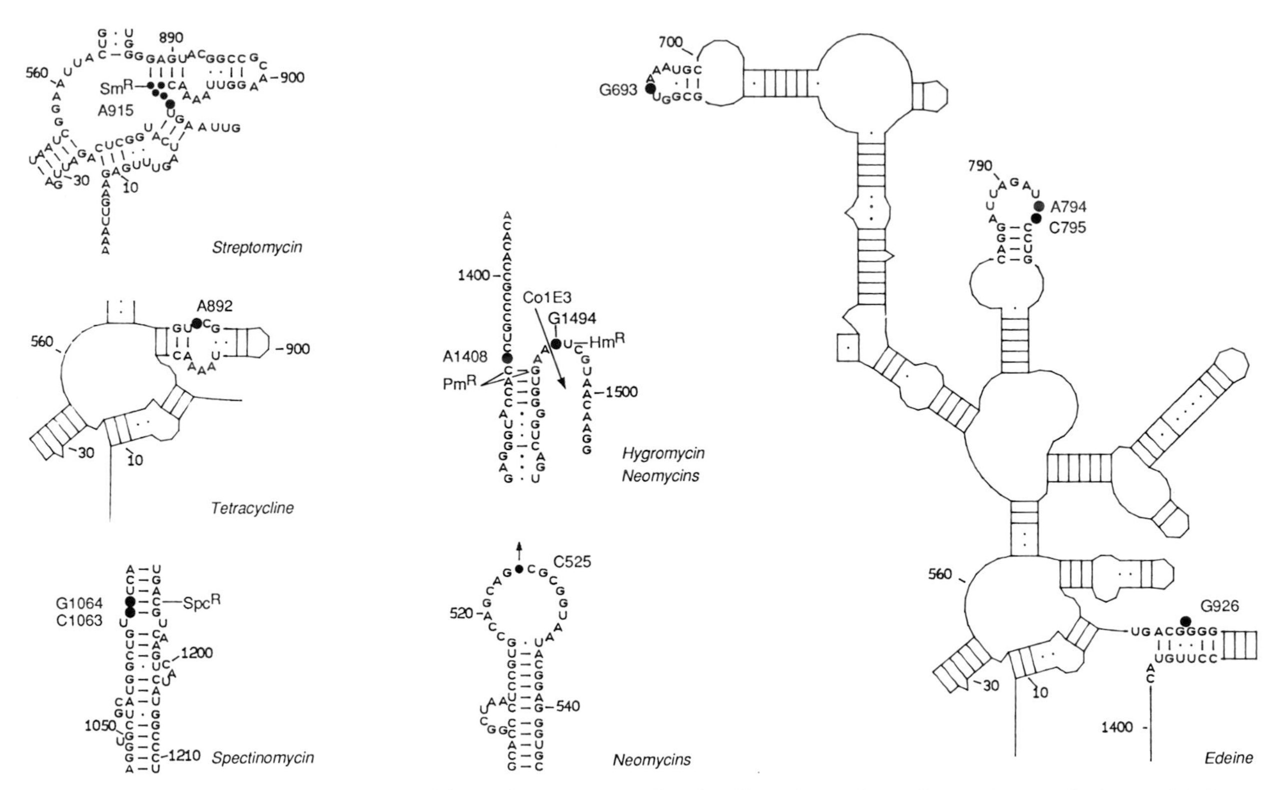

Figure 12. Bases in 16S rRNA protected by antibiotics (Moazed and Noller, 1987a). Sites of mutations conferring antibiotic resistance and the site of colicin E3 cleavage are shown.

sive binding of the two factors. Since both factors are involved in ribosomal GTPase-related events, it is conceivable that the α-sarcin loop itself may somehow be involved in the GTPase function.

INTERACTIONS OF ANTIBIOTICS WITH rRNA

Besides mRNA, tRNA, and protein factors, another important class of ribosomal ligands is the antibiotics (reviewed in Cundliffe, 1981). Their importance to the study of ribosome structure and function is that they block specific steps of the translation process and thus have great potential in helping to identify intermediate states of the mechanism of translation, or simply to arrest all ribosomes at a specific step, as we have done with fusidic acid and kirromycin, for example (see above). Antibiotics can thus be considered to be exquisitely specific enzyme inhibitors, and they probably interact with active sites in the ribosome. If ribosomal active sites are truly composed of RNA, we should expect ribosome-directed antibiotics to interact with specific regions of rRNA. In fact, we have observed discrete protection of rRNA from chemical probing by every such antibiotic that we have so far tested. In almost every instance the protected bases are highly conserved, and most of them have also been implicated by genetic or biochemical studies in functions related to the known mode of action of the antibiotic in question.

Figure 12 summarizes the bases protected in 16S rRNA by several well-known antibiotics (Moazed and Noller, 1987a). Streptomycin, probably the most extensively studied of this group, gives a concise footprint at positions 911 to 915. As mentioned above, this coincides with the site of a mutation conferring streptomycin resistance (position 912) and is in a region where numerous effects on assembly of protein S12 are seen, mutations in which also confer resistance to streptomycin. Tetracycline, a drug that is also believed to affect A-site tRNA binding, though by a different mechanism, protects the nearby residue A-892 and also causes enhancement of U-1052 and C-1054, which are believed to be some distance away in the structure. Spectinomycin protects the N7 position of G-1064, which is base paired to C-1192, mutation of which results in spectinomycin resistance (Sigmund et al., 1984). Similarly, hygromycin and the neomycin group (including paromomycin) affect A-1408 and G-1494; mutation of position 1495 gives hygromycin resistance (Spangler and Blackburn, 1985), whereas mutation of position 1409 or 1491 gives paromomycin resistance (Li et al., 1982). Fi-

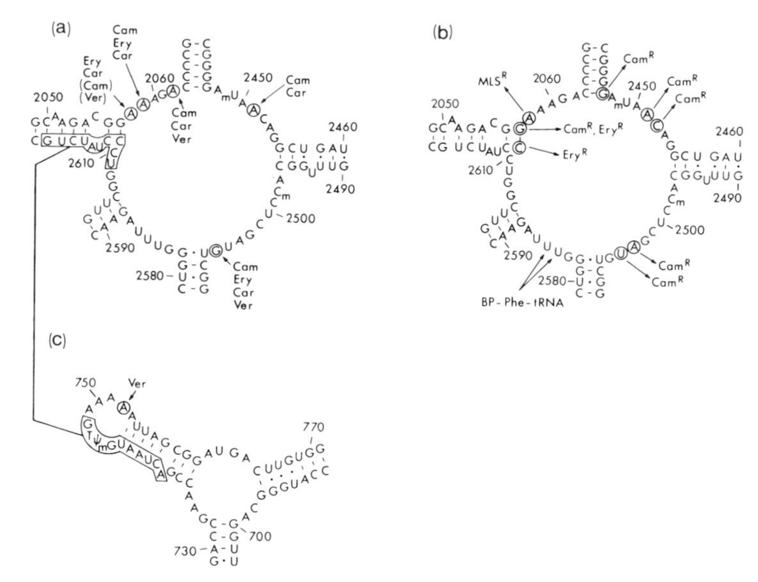

Figure 13. (a and c), Bases protected by antibiotics in 23S rRNA (Moazed and Noller, 1987b). (b) Sites of mutations conferring antibiotic resistance and site of cross-linking of benzophenone-derivatized Phe-tRNA to 23S rRNA (Barta et al., 1984).

nally, edeine, an inhibitor of P-site tRNA binding, protects four of the same sites protected by P-site tRNA (G-693, A-794, A-795, and G-926; Fig. 10).

Similar experiments were carried out on antibiotics whose functional target is the large ribosomal subunit (Moazed and Noller, 1987b). Figure 13 shows the results of probing ribosomal complexes containing chloramphenicol, carbomycin, erythromycin, and vernamycin B. Nearly all of the effects are localized to the central loop of domain V of 23S rRNA, the peptidyltransferase region. Figure 13b shows the positions of mutations conferring resistance to antibiotics and sites of cross-linking of a photoreactive peptidyl-tRNA analog to 23S rRNA. The close agreement between these data and the results of probing the antibiotic complexes (Fig. 13a) is striking. Vernamycin B also protects A-752, in domain II, implying that this remote region of the secondary structure is, in fact, very close to the central loop of domain V. This is confirmed by direct cross-linking of these same two regions of the RNA in experiments by Stiege et al. (1983).

CLASS III SITES

There is a set of bases in 16S rRNA whose reactivity is decreased by tRNA, 50S subunits, or certain antibiotics. These bases, which we term class III sites (Moazed and Noller, 1987a), include A-790, G-791, A-909, A-1394, A-1413, and G-1487 (Fig. 14). Since they are protected by three different kinds of ligands, which can all bind simultaneously to 30S subunits (and indeed potentiate each other's binding), we infer that their protection is the result of ligand-induced conformational changes rather than direct contact with the protected bases. The neomycin class of antibiotics (neomycin, kanamycin, gentamicin, and paromomycin) protect all six bases, as do tRNA and 50S subunits (Moazed and Noller, 1987a). Streptomycin protects only three of them

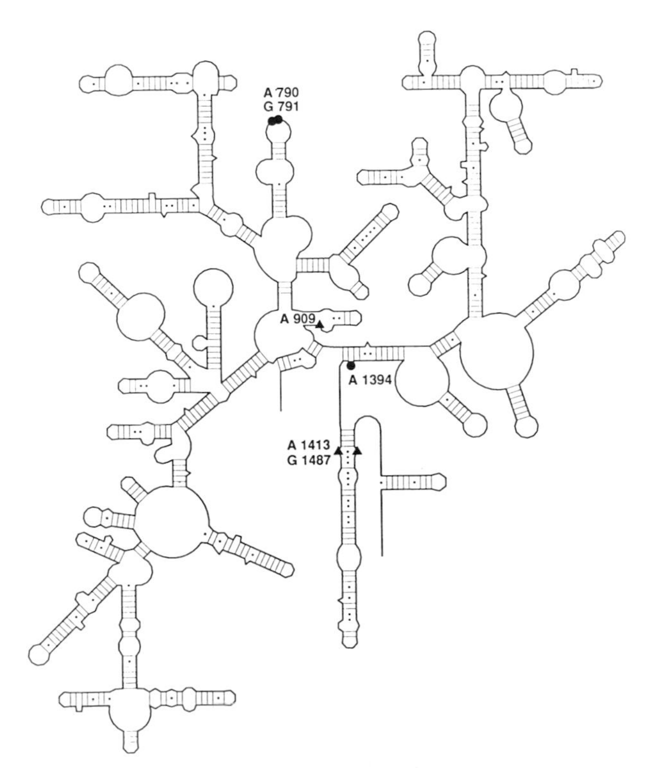

Figure 14. Class III sites (Moazed and Noller, 1987a). All six are protected by tRNA, 50S subunits, and neomycin-class antibiotics. Three (909, 1413, and 1487) are also protected by streptomycin, and the other three (790, 791, and 1394) are protected by edeine.

(A-909, A-1413, and G-1487), however; edeine protects the other three (A-790, G-791, and A-1394).

We consider the class III phenomenon as additional evidence for the involvement of conformational changes in rRNA in the mechanism of translation. It also carries with it the implication that tRNA binding and subunit-subunit interaction are somehow interrelated. This theme is reinforced by the observation that bases protected by tRNA are often located very near bases that are protected by subunit association, both in 16S and in 23S rRNA. The possible significance of this interrelationship is raised in the next section.

INTERMEDIATE STATES IN TRANSLOCATION

Assignment of protected bases to A-, P-, and E-site binding of tRNA provides a direct and independent assay for the location of tRNA on the ribosome. We therefore probed a series of in vitro complexes containing tRNA bound to ribosomes in different states to examine in detail the interactions between tRNA and rRNA during a translational cycle. Specifically, we ask, what are the consequences of peptide bond formation and EF-G-mediated translocation on the state of the ribosome-bound tRNA? According to the classical model of protein synthesis, no change in the binding of the two tRNAs is predicted after peptide bond formation; EF-G-dependent translocation should then move the tRNAs to give a vacant A site. Using the extended three-site model, deacylated tRNA should then occupy the E site. These expectations were verified experimentally in earlier experiments insofar as the movement of mRNA is concerned (Gupta et al., 1971).

In our experiments, three different kinds of in vitro constructs were used to probe tRNA-ribosome complexes in the pre- and post-peptidyl transfer stages of the translational cycle (Moazed and Noller, 1989b). In the first experiment, we examine the binding state of P-site-bound tRNA (*N*-acetyl-Phe-tRNA) before and after peptidyltransferase-catalyzed reaction with puromycin to form an *N*-acetyl-Phe-puromycin peptide bond. Unexpectedly, the P-site footprint on 23S rRNA, but not 16S rRNA, is abolished upon peptide bond formation. Furthermore, an E-site footprint is now observed in 23S rRNA. We conclude that the deacylated tRNA is bound in different "sites" in the two subunits after peptide bond formation; it remains in the P site on the small subunit but has shifted to the E site on the large subunit. We refer to this state of binding as the P/E state. It is remarkable that this movement occurs independently of added factors or GTP.

In the second type of experiment, we reacted the Phe-tRNA · EF-Tu · GTP ternary complex with *N*-acetyl-Phe-tRNA bound to the ribosomal P site, a reaction that more closely resembles the physiological event. In this case, we observed P- and E-site footprints on 23S rRNA and A- and P-site footprints on 16S rRNA; again, after peptide bond formation, the tRNAs appear to occupy different sites relative to 16S and 23S rRNA. We infer that the deacylated tRNA is again in the P/E state, whereas the *N*-acetyl-Phe-Phe-tRNA occupies the A site with respect to 16S rRNA and the P site with respect to 23S rRNA, which we term the A/P state.

When this post-peptidyl transfer complex is subjected to EF-G–plus–GTP-dependent translocation, the 16S A-site footprint is lost, in accord with earlier studies on translocation of mRNA (Gupta et al., 1971). We infer that the EF-G-dependent event involves movement of peptidyl-tRNA with respect to the 30S subunit, vacating the A site. The result is a transition from the A/P state to the P/P state, in which peptidyl-tRNA is bound in a state equivalent to the classical "P-site" binding. At the same time, the deacylated tRNA must vacate the 16S P site, and so it most likely undergoes a shift from the P/E state to the E state but is not released from the ribosome. EF-G thus catalyzes movement of both tRNAs relative to the small subunit, a result that is confirmed in the following experiment.

Pretranslocation complexes have classically been constructed by binding peptidyl-tRNA analogs to ribosomes in the presence of excess deacylated tRNA at high Mg^{2+} concentrations (deGroot et al., 1971). The failure of the peptidyl moiety to react with puromycin in such complexes has usually been taken to mean that it occupies the A site. In such complexes, we find a pattern of tRNA-dependent protection that is identical to the post-peptidyl transfer state, i.e., A- and P-site footprints on 16S rRNA and P- and E-site footprints on 23S rRNA, implying that the tRNAs are in the A/P and P/E states, respectively. EF-G-dependent translocation again results in loss of the 16S A-site footprint, and we conclude that the *N*-acetyl-Phe-tRNA has undergone an A/P-to-P/P transition, whereas the deacylated tRNA has shifted from P/E to the E state. The possibility that the P-site footprint attributed to *N*-acetyl-Phe-tRNA in the pretranslocation state might be due to deacylated tRNA is excluded by another set of experiments in which we substitute tRNA lacking its 3′-terminal CA residues for deacylated tRNA. tRNA (-CA) is unable to give most of the 23S P-site protections, yet when this complex is formed, the standard 23S P-site footprint is again observed, confirming that it is, in fact, due to the *N*-acetyl-Phe-tRNA (Fig. 9).

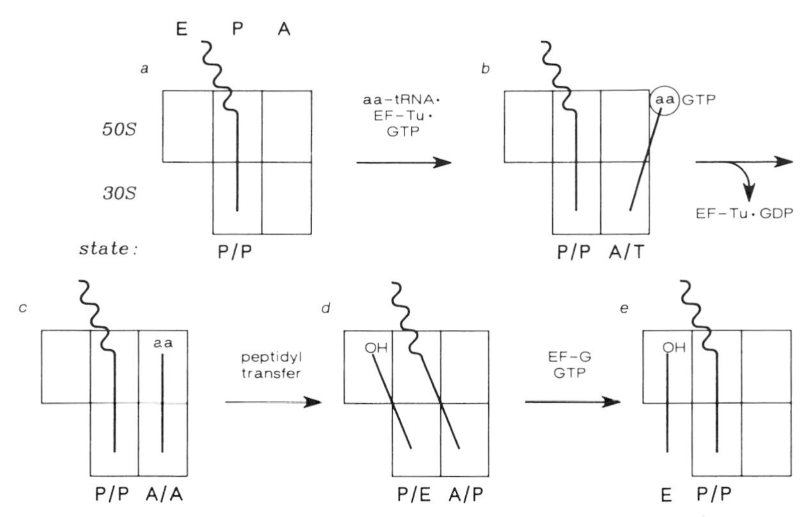

Figure 15. Hybrid site model for the movement of tRNA during translation (Moazed and Noller, 1989b). The binding states (P/P, A/T, etc.) are shown at the bottom of each tRNA. See text for details.

Small but significant differences in the 16S and 23S rRNA footprints give further evidence that the pre- and post-peptidyl transfer states differ (Moazed and Noller, 1989b). In 16S rRNA, A-702 is enhanced, and G-1338 is more strongly protected, as a result of the P/P-to-P/E transition. In 23S rRNA, A-2602 is protected in the A/P state but enhanced in the P/P state.

Our findings can be interpreted in terms of a simple model that is consistent with the results of previous studies on the mechanism of protein synthesis. The diagram shown in Fig. 15 represents the five known sites of interaction of tRNA with the ribosome: A and P on the small subunit, and A, P, and E on the large subunit (the minimum number of sites required to account for the available data). Beginning with a peptidyl-tRNA in the P/P state (Fig. 15a), an aminoacyl-tRNA is brought in as a ternary complex and binds initially in the A/T state (Fig. 15b). After departure of the EF-Tu · GDP complex, the two tRNAs are in P/P and A/A states, respectively, which resemble the classical pre-peptidyl transfer state (Fig. 15c). After peptide bond formation, they shift from P/P to P/E and A/A to A/P, respectively (Fig. 15d), in which both tRNAs move with respect to the large subunit but maintain their locations on the small subunit. Finally, the EF-G-catalyzed step moves the anticodon ends of both tRNAs, together with their associated mRNA, relative to the small subunit (Fig. 15e).

Most remarkable is that tRNA movement occurs independently on the two ribosomal subunits. Thus, from a given binding site on the small subunit, a tRNA may interact with more than one site on the large subunit, and conversely. One important implication is that one end of the tRNA is always anchored to the ribosome during translocation. Another is that the peptidyl moiety remains in a fixed position relative to the structure of the ribosome, explaining why only a single peptidyl-tRNA may bind to the ribosome at a time (Rheinberger et al., 1981).

We conclude that there are at least six distinguishable binding states for tRNA: A/T, A/A, A/P, P/P, P/E, and E (Fig. 15). The newly observed hybrid states escaped detection previously largely because of the inherent limitations of traditional methods. By reformulating the operational definitions of tRNA-binding sites in terms of rRNA footprints, we substitute direct structural correlates for traditional criteria such as filter binding and puromycin reactivity. An example of the limitations of traditional criteria is the previous failure to detect the A/P state. By puromycin reactivity, the A/P and A/A states are indistinguishable. For reasons not yet understood, tRNA in the A/P state is not puromycin reactive, although it affords a 23S footprint which closely resembles that of tRNA bound in the puromycin-reactive P/P state.

Hardesty and co-workers have observed, after peptide bond formation, changes in fluorescence quantum yield and fluorescence anisotropy of fluorescent probes attached to tRNA (Hardesty et al., 1986; Odom and Hardesty, 1987). Their spectroscopic results are very likely a physical manifestation of the P/P-to-P/E transition.

THE MOLECULAR BASIS OF tRNA MOVEMENT

By the very nature of the translation process, mRNA and tRNA must undergo movement relative

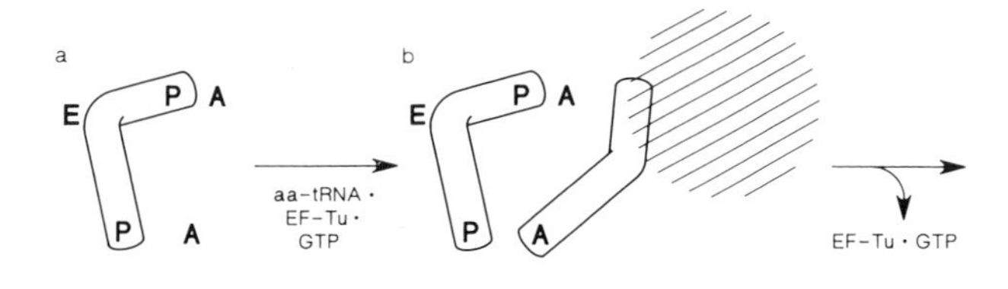

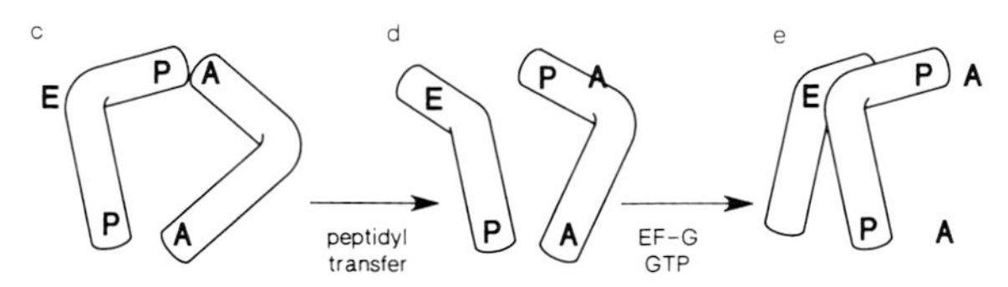

Figure 18. Structural representation of the movement of tRNA. Steps a through e correspond to Fig. 15, with the ribosomal sites located as in Fig. 17.

emerges from the ribosome at the left side. A similar model has been proposed recently by Wower et al. (1989). One consequence of this model is that it places the tRNA molecules between the subunits, very near the sites of subunit-subunit contact, as inferred from electron microscopy data. Further insight into the mechanism of translocation may require understanding the basis of the apparent interrelationship between tRNA binding and subunit association.

REFERENCES

Allen, P. N., and H. F. Noller. 1989. Mutations in ribosomal proteins S4 and S12 influence the higher order structure of 16S ribosomal RNA. *J. Mol. Biol.* **208:**457–468.

Barta, A., G. Steiner, J. Brosius, H. F. Noller, and E. Kuechler. 1984. Identification of a site on 23S ribosomal RNA located at the peptidyl transferase center. *Proc. Natl. Acad. Sci. USA* **81:**3607–3611.

Branlant, C., A. Krol, J. SriWidada, J. P. Ebel, P. Sloof, and R. A. Garrett. 1976. The binding site for protein L1 on 23S ribosomal RNA of *Escherichia coli. Eur. J. Biochem.* **70:**457–469.

Bretscher, M. S. 1968. Translocation in protein synthesis: a hybrid structure model. *Nature* (London) **218:**675–677.

Brimacombe, R., J. Atmadja, W. Stiege, and D. Schüler. 1988. A detailed model of the three-dimensional structure of *Escherichia coli* 16S ribosomal RNA in situ in the 30S subunit. *J. Mol. Biol.* **199:**115–136.

Brimacombe, R., and W. Stiege. 1985. Structure and function of ribosomal RNA. *Biochem. J.* **229:**1–17.

Brow, D. A., and H. F. Noller. 1983. Protection of ribosomal RNA from kethoxal in polyribosomes. Implication of specific sites in ribosome function. *J. Mol. Biol.* **163:**27–46.

Cabreu, B., D. Vazquez, and J. Modolell. 1972. Inhibition by elongation factor G of aminoacyl-tRNA binding to ribosomes. *Proc. Natl. Acad. Sci. USA* **69:**733–736.

Capel, M. S., D. M. Engelman, B. R. Freeborn, M. Kjeldgaard, J. A. Langer, V. Ramakrishnan, D. G. Schindler, D. K. Schneider, B. P. Schoenborn, I.-Y. Sillers, S. Yabuki, and P. B. Moore. 1987. A complete mapping of the proteins in the small ribosomal subunit of *Escherichia coli. Science* **238:**1403–1406.

Cundliffe, E. 1981. Antibiotic inhibitors of ribosome function, p. 402–457. *In* E. F. Gates, E. Cundliffe, P. E. Reynolds, M. H. Richmond, and M. J. Waring (ed.), *The Molecular Basis of Antibiotic Action.* John Wiley & Sons, Inc., New York.

deGroot, N., A. Panet, and Y. Lapidot. 1971. The binding of purified Phe-$tRNA^{Phe}$ and peptidyl-$tRNA^{Phe}$ to *Escherichia coli* ribosomes. *Eur. J. Biochem.* **23:**523–527.

Dijk, J., J. Littlechild, and R. A. Garrett. 1977. The RNA binding properties of "native" protein-protein complexes isolated from the *Escherichia coli* ribosome. *FEBS Lett.* **77:**295–300.

Endo, Y., K. Mitsui, M. Motizuki, and K. Tsurugi. 1987. The mechanism of action of ricin and related toxic lectins on eukaryotic ribosomes. *J. Biol. Chem.* **262:**5908–5912.

Endo, Y., and I. Wool. 1982. The site of action of α-sarcin on eukaryotic ribosomes. *J. Biol. Chem.* **257:**9054–9060.

Expert-Bezançon, A., and P. Wollenzien. 1985. Three-dimensional arrangement of the *E. coli* 16S ribosomal RNA. *J. Mol. Biol.* **184:**53–66.

Fellner, P. 1974. Structure of the 16S and 23S ribosomal RNAs, p. 169–191. *In* M. Nomura, A. Tissières, and P. Lengyel (ed.), *Ribosomes.* Cold Spring Harbor Laboratory, Cold Spring Harbor, N.Y.

Fromm, H., E. Galun, and M. Edelman. 1989. A novel site for streptomycin resistance in the 530 loop of chloroplast 16S ribosomal RNA. *Plant Mol. Biol.* **12:**499–505.

Gavrilova, L. P., O. E. Kostiashkina, V. E. Koteliansky, N. M. Rutkevitch, and A. S. Spirin. 1976. Factor-free (non-enzymic) and factor-dependent systems of translation of polyuridylic acid by *Escherichia coli* ribosomes. *J. Mol. Biol.* **101:**537–552.

Gorini, L. 1974. Streptomycin and misreading of the genetic code, p. 791–803. *In* M. Nomura, A. Tissières, and P. Lengyel (ed.), *Ribosomes.* Cold Spring Harbor Laboratory, Cold Spring Harbor, N.Y.

Gornicki, P., K. Nurse, W. Hellmann, M. Boublik, and J. Ofengand. 1984. High resolution localization of the tRNA anticodon interaction site on *Escherichia coli* 30S ribosomal subunit. *J. Biol. Chem.* **259:**10493–10498.

Grajevskaya, R. A., Y. V. Ivanov, and E. G. Saminsky. 1982. 70S ribosomes of *E. coli* have an additional site for deacylated tRNA binding. *Eur. J. Biochem.* **128:**47–52.

Grosjean, H. J., S. de Henau, and D. M. Crothers. 1978. On the physical basis for ambiguity in genetic coding interactions. *Proc. Natl. Acad. Sci. USA* **75:**610–614.

Gupta, S. L., J. Waterson, M. L. Sopori, S. M. Weissman, and P. Lengyel. 1971. Movement of the ribosome along the messenger ribonucleic acid during protein synthesis. *Biochemistry* **10:** 4410–4421.

Gutell, R. R., and G. E. Fox. 1989. Compilation of large subunit RNA sequences presented in a structural format. *Nucleic Acids Res.* **16:**175–269.

Gutell, R. R., H. F. Noller, and C. R. Woese. 1986. Higher order structure in ribosomal RNA. *EMBO J.* **5:**1111–1113.

Gutell, R. R., B. Weiser, C. R. Woese, and H. F. Noller. 1985. Comparative anatomy of 16S-like ribosomal RNA. *Prog. Nucleic Acid Res. Mol. Biol.* **32:**156–216.

Hardesty, B., O. W. Odom, and H.-Y. Deng. 1986. The movement of tRNA through ribosomes during peptide elongation: the displacement reaction model, p. 495–508. *In* B. Hardesty and G. Kramer (ed.), *Structure, Function, and Genetics of Ribosomes.* Springer-Verlag, New York.

Held, W. A., B. Ballou, S. Mizushima, and M. Nomura. 1974. Assembly mapping of 30S ribosomal proteins from *Escherichia coli. J. Biol. Chem.* **249:**3103–3111.

Hopfield, J. J. 1974. Kinetic proofreading: a new mechanism for reducing errors in biosynthetic processes requiring high specificity. *Proc. Natl. Acad. Sci. USA* **71:**4135–4139.

Hummel, H., and A. Böck. 1987. 23S ribosomal RNA mutations in halobacteria conferring resistance to the anti-80S ribosome targeted antibiotic anisomycin. *Nucleic Acids Res.* **15**:2431–2443.

Leffers, H., J. Kjems, L. Ostergaard, N. Larsen, and R. A. Garrett. 1987. Evolutionary relationships among archaebacteria. A comparative study of 23S ribosomal RNAs of a sulphur-dependent extreme thermophile, an extreme halophile, and a thermophilic methanogen. *J. Mol. Biol.* **195**:43–61.

Li, M., A. Tzagoloff, K. Underbrink-Lyon, and N. C. Martin. 1982. Identification of the paromomycin-resistance mutation in the 15S rRNA gene of yeast mitochondria. *J. Biol. Chem.* **257**:5921–5928.

Lin, F.-L., L. Kahan, and J. Ofengand. 1984. Crosslinking of phenylalanyl-tRNA to the ribosomal A site via a photoaffinity probe attached to the 4-thiouridine residue is exclusively to ribosomal protein S19. *J. Mol. Biol.* **172**:77–86.

McDonald, J. J., and R. Rein. 1987. A stereochemical model of the transpeptidation complex. *J. Biomol. Struct. Dynam.* **4**:729–744.

Melançon, P., C. Lemieux, and L. Brakier-Gingras. 1988. A mutation in the 530 loop of *Escherichia coli* 16S ribosomal RNA causes resistance to streptomycin. *Nucleic Acids Res.* **16**:9631–9639.

Miller, D. L. 1972. Elongation factors EF-Tu and EF-G interact at related sites on ribosomes. *Proc. Natl. Acad. Sci. USA* **69**:753–755.

Moazed, D., and H. F. Noller. 1986. Transfer RNA shields specific nucleotides in 16S ribosomal RNA from attack by chemical probes. *Cell* **47**:985–994.

Moazed, D., and H. F. Noller. 1987a. Interaction of antibiotics with functional sites in 16S ribosomal RNA. *Nature* (London) **327**:389–394.

Moazed, D., and H. F. Noller. 1987b. Chloramphenicol, erythromycin, carbomycin and vernamycin B protect overlapping sites in the peptidyl transferase region of 23S ribosomal RNA. *Biochimie* **69**:879–884.

Moazed, D., and H. F. Noller. 1989a. Interaction of tRNA with 23S rRNA in the ribosomal A, P and E sites. *Cell* **57**:585–597.

Moazed, D., and H. F. Noller. 1989b. Intermediate states in the movement of tRNA in the ribosome. *Nature* (London) **342**:142–148.

Moazed, D., and H. F. Noller. 1990. Binding of tRNA to the ribosomal A and P sites protects two distinct sets of nucleotides in 16S rRNA. *J. Mol. Biol.* **211**:135–145.

Moazed, D., J. M. Robertson, and H. F. Noller. 1988. Interaction of elongation factors EF-G and EF-Tu with a conserved loop in 23S rRNA. *Nature* (London) **334**:362–364.

Moazed, D., S. Stern, and H. F. Noller. 1986. Rapid chemical probing of conformation in 16S ribosomal RNA and 30S ribosomal subunits using primer extension. *J. Mol. Biol.* **187**:399–416.

Monro, R. E. 1967. Catalysis of peptide bond formation by 50S ribosomal subunits from *Escherichia coli. J. Mol. Biol.* **26**:147–151.

Montandon, P. E., R. Wagner, and E. Stutz. 1986. *E. coli* ribosomes with a C912 to U base change in the 16S rRNA are streptomycin resistant. *EMBO J.* **5**:3705–3708.

Moore, P. B. 1985. Polypeptide polymerase: the structure and function of the ribosome in 1985. *Proc. Robert A. Welch Found. Conf. Chem. Res.* **29**:185–215.

Nagano, K., and M. Harel. 1987. Approaches to a three-dimensional model of the *E. coli* ribosome. *Prog. Biophys. Mol. Biol.* **48**:67–101.

Ninio, J. 1974. A semi-quantitative treatment of missense and nonsense suppression in the *strA* and *ram* ribosomal mutants of *Escherichia coli. J. Mol. Biol.* **84**:297–313.

Nirenberg, M., and P. Leder. 1964. RNA codewords and protein synthesis. The effect of trinucleotides upon the binding of sRNA to the ribosomes. *Science* **145**:1399–1407.

Noller, H. F. 1984. Structure of ribosomal RNA. *Annu. Rev. Biochem.* **53**:119–162.

Noller, H. F., and J. B. Chaires. 1972. Functional modification of 16S ribosomal RNA by kethoxal. *Proc. Natl. Acad. Sci. USA* **69**:3115–3118.

Noller, H. F., J. Kop, V. Wheaton, J. Brosius, R. R. Gutell, A. Kopylov, F. Dohme, W. Herr, D. A. Stahl, R. Gupta, and C. R. Woese. 1981. Secondary structure model for 23S ribosomal RNA. *Nucleic Acids Res.* **9**:6167–6189.

Noller, H. F., and J. A. Lake. 1984. Ribosome structure and function: localization of rRNA, p. 217–297. *In* E. Bittar (ed.), *Membrane Structure and Function.* John Wiley & Sons, Inc., New York.

Noller, H. F., and C. R. Woese. 1981. Secondary structure of 16S ribosomal RNA. *Science* **212**:403–411.

Nomura, M., S. Mizushima, M. Ozaki, P. Traub, and C. V. Lowry. 1969. Structure and function of ribosomes and their molecular components. *Cold Spring Harbor Symp. Quant. Biol.* **34**:49–61.

Nomura, M., P. Traub, and H. Bechmann. 1968. Hybrid 30S ribosomal particles reconstituted from components of different bacterial origins. *Nature* (London) **219**:793–799.

Oakes, M., E. Henderson, A. Scheinman, M. Clark, and J. A. Lake. 1986a. Ribosome structure, function and evolution: mapping ribosomal RNA, proteins and functional sites in three dimensions, p. 47–67. *In* B. Hardesty and G. Kramer (ed.), *Structure, Function, and Genetics of Ribosomes.* Springer-Verlag, New York.

Oakes, M. I., M. W. Clark, E. Henderson, and J. A. Lake. 1986b. DNA hybridization electron microscopy: ribosomal RNA nucleotides 1392–1407 are exposed in the cleft of the small subunit. *Proc. Natl. Acad. Sci. USA* **83**:275–279.

Odom, O. W., and B. Hardesty. 1987. An apparent conformational change in $tRNA^{Phe}$ that is associated with peptidyl transferase reaction. *Biochimie* **69**:925–938.

Olson, H. M., L. S. Lasater, P. A. Cann, and D. G. Glitz. 1988. Messenger RNA orientation on the ribosome. Placement by electron microscopy of antibody-complementary oligodeoxynucleotide complexes. *J. Biol. Chem.* **263**:15196–15204.

Pestka, S. 1968. Studies on the formation of transfer ribonucleic acid-ribosome complexes. III. The formation of peptide bonds by ribosomes in the absence of supernatant enzymes. *J. Biol. Chem.* **243**:2810–2820.

Powers, T., L.-M. Changchien, G. R. Craven, and H. F. Noller. 1988a. Probing the assembly of the 3′ major domain of 16S ribosomal RNA. Quaternary actions involving ribosomal proteins S7, S9 and S19. *J. Mol. Biol.* **200**:309–319.

Powers, T., S. Stern, L.-M. Changchien, and H. F. Noller. 1988b. Probing the assembly of the 3′ major domain of 16S rRNA. Interactions involving ribosomal proteins S2, S3, S10, S13, and S14. *J. Mol. Biol.* **201**:697–716.

Prince, J. B., B. H. Taylor, D. L. Thurlow, J. Ofengand, and R. A. Zimmermann. 1982. Covalent crosslinking of $tRNA^{Val}$ to 16S RNA at the ribosomal P site: identification of crosslinked residues. *Proc. Natl. Acad. Sci. USA* **79**:5450–5454.

Rheinberger, H.-J., H. Sternbach, and K. Nierhaus. 1981. Three tRNA binding sites on *Escherichia coli* ribosomes. *Proc. Natl. Acad. Sci. USA* **78**:5310–5314.

Rich, A. 1974. How transfer RNA may move inside the ribosome, p. 871–874. *In* M. Nomura, L. Tissières, and P. Lengyel (ed.), *Ribosomes.* Cold Spring Harbor Laboratory, Cold Spring Harbor, N.Y.

Rose, S. J., P. T. Lowry, and O. C. Uhlenbeck. 1983. Binding of

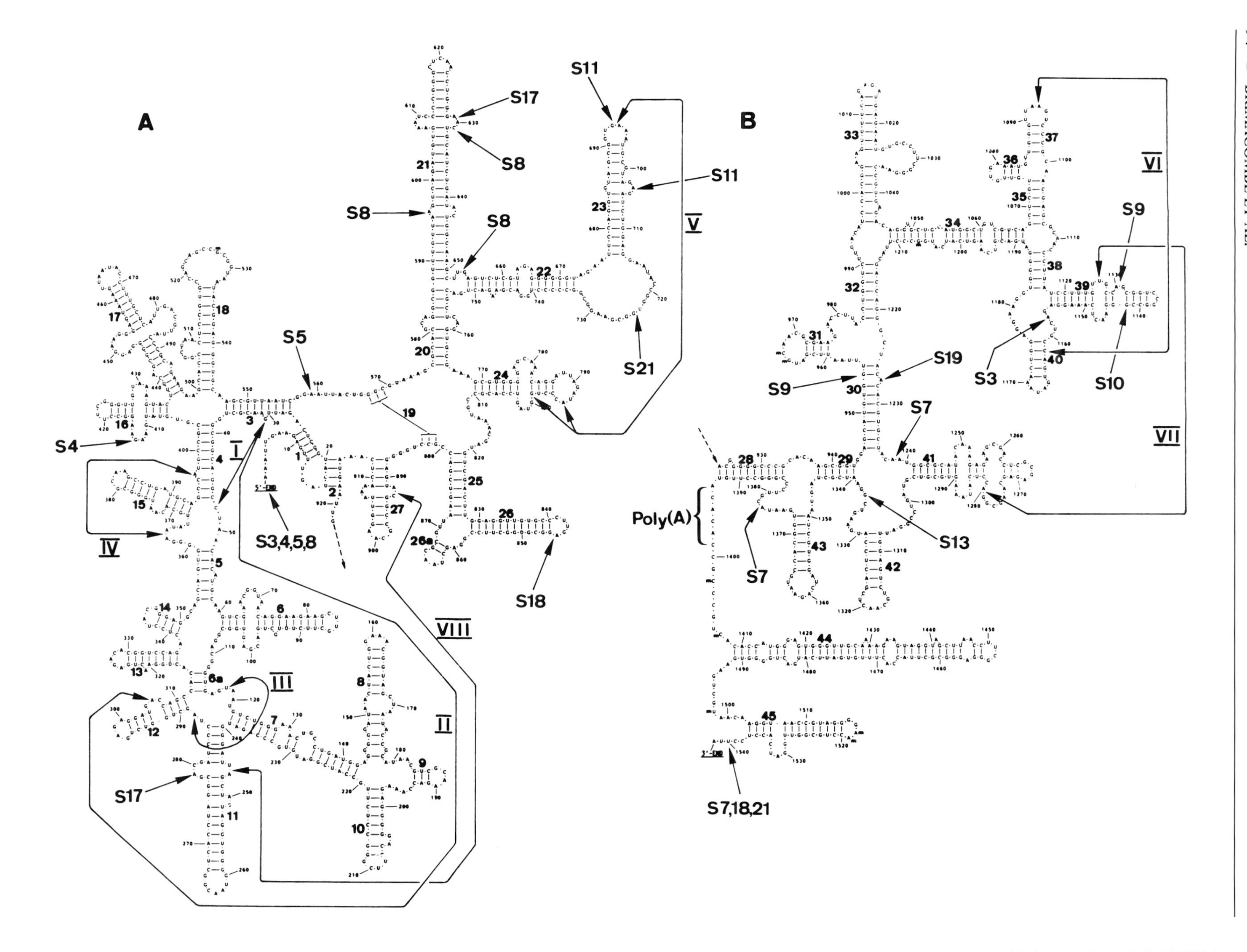
A
B
S17
S8
S11
S11
S8
S8
V
S5
S21
S4
S3,4,5,8
IV
VIII
III
II
S17
S18
S9
S19
S9
S3
S10
VI
VII
S7
Poly(A)
S7
S13
S7,18,21

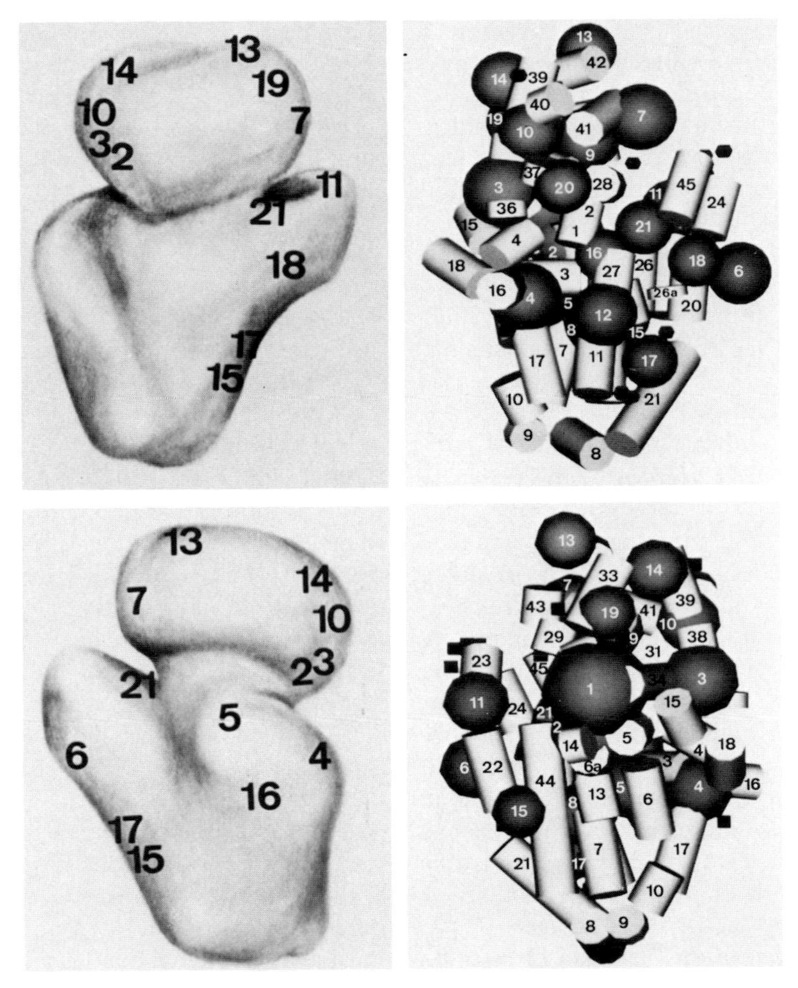

Figure 2. Two views of the computer-generated model of the 30S subunit (Schüler and Brimacombe, 1988) compared with the corresponding electron microscopically derived model, including IEM sites to the ribosomal proteins (Stöffler and Stöffler-Meilicke, 1986). In the computer model, the proteins are represented by dark spheres with white numbers, and the double-helical elements of the RNA (see Fig. 1) are represented by cylinders with black numbers. The small black polygons represent the positions of RNA-protein cross-link sites in the model.

containing the IEM sites for the various proteins (Stöffler and Stöffler-Meilicke, 1986).

Since Stern et al. (1988) have not published any coordinates for the RNA helices in their model, it is not possible for us to make a detailed comparison between their model and ours. Nevertheless, two general points can be made. First, the size, shape, and general appearance of our model (Fig. 2) bear a much closer resemblance to the electron microscopic observations. In particular, all of the proteins are in contact with the RNA, and the least-squares fitting procedure has the effect of "tilting" the entire neutron map such that protein S17 is turned slightly to one side, rather than being the "South Pole" of the protein arrangement (cf. Stern et al., 1988), and it can be seen from Fig. 2 that this leads to a very satisfactory fit with the IEM data for the proteins (with the exception of protein S19, as has already been noted [Schüler and Brimacombe, 1988]).

The second point concerns the type of computer program used to present the models. Our program, in which the helical elements of the 16S RNA are represented as cylinders, was especially developed for this purpose (D. Schüler, unpublished data) and has, in our opinion, distinct advantages over the FRODO type of program preferred by Stern et al. (1988).

Figure 1. Secondary structure of *E. coli* 16S RNA, showing the locations of the tertiary intra-RNA cross-links (Roman numerals) and RNA-protein cross-link sites used to build the three-dimensional model (Brimacombe et al., 1988a). Intra-RNA cross-link site VIII and the cross-link to a poly(A) messenger template are taken from Stiege et al. (1988a) and Stiege et al. (1988b), respectively.

First, the use of simple geometric objects to represent the ribosomal components enables a number of physical parameters to be calculated very easily and compared with experimentally measured values. Thus, we have already shown (Schüler and Brimacombe, 1988) that the displacement between the overall mass centers of the protein and RNA moieties in our model (ca. 20 Å [ca. 2.0 nm]) is in good agreement with the experimental value of 25 Å found by neutron scattering (Ramakrishnan, 1986). Similarly, an estimate for the radius of gyration of the helical elements can readily be calculated (by regarding each helix as a point mass at the center of the appropriate cylinder), and this gives a value of 60.3 Å, which is again in good agreement with the experimental value of 61 Å for the RNA (Koch and Stuhrmann, 1979). Next, it must be remembered that the detailed conformations of the usually irregular RNA helices are not known (see below); furthermore, in the computer models, the protein and RNA moieties often overlap so that part of a helix and part of a protein share the same space. In this situation, the FRODO type of program gives an unjustifiably detailed presentation, whereas the use of simple geometric shapes reflects the limited state of our experimental knowledge more realistically. Last but not least, the latter type of program has a certain three-dimensional depth of its own (Fig. 2), which obviates the need for a stereo viewer, and the model can readily be displayed in this form on most computers, using the simple coordinate system that we have published (Brimacombe et al., 1988a).

LIMITATIONS OF THE MODEL-BUILDING STUDIES

Regardless of whether RNA-protein cross-linking or protein footprinting data are used, it is the fitting of the RNA structure to the neutron map of the proteins that presents the most serious limitation to the resolution of this type of model. On the RNA side, both types of data (cross-linking and footprinting) define precise points on the RNA chain that can be construed as being in close contact with the protein concerned. In contrast, on the protein side, only the position of the mass center of that protein is known. The observed contact with the RNA can thus be to any point on the surface of the protein, and even if the proteins are regarded as perfect spheres (which they surely are not), this introduces a large element of uncertainty in the placement of the RNA helices at each protein cross-link or footprint site. Until a complete crystallographic analysis of the subunit becomes available, there seems to be very little chance of improving this level of resolution, since even if individual crystallographic structures were known for all of the proteins, the problems of orienting each protein structure within the 30S subunit and of correlating these structures with the RNA-protein cross-link or footprint data would still remain. This limitation becomes particularly acute in the thin dimension of the 30S subunit (the plane of the paper in Fig. 2), and it is in this dimension that, according to Stern et al. (1988), the principal discrepancies between their model and ours are to be found.

A further type of uncertainty in the models concerns the conformation of the RNA helices themselves. The great majority of the helical regions in both 16S and 23S RNA contain interruptions in the form of loops and bulges, and the effect of these on the helical conformation is not known. Up to now, for lack of more detailed information, the models have tended to assume that these interruptions do not markedly disturb the helical conformation. Although this may be true in general, we have intra-RNA cross-linking evidence which indicates that in one case in the 23S RNA, a long helical system (comprising helices 76 to 78; Fig. 3 [see below]) actually folds back on itself so that the loop end of helix 78 cross-links to the base of helix 79.

These considerations suggest that the most profitable approach to improve the resolution and reliability of the current RNA models would be to make an intensive search for new intra-RNA cross-links, since these provide direct information with regard to the RNA structure and are independent of the protein arrangement. The construction of a comprehensive network of intra-RNA cross-links was indeed the underlying philosophy in an earlier three-dimensional model of the 16S RNA (Expert-Bezançon and Wollenzien, 1985), although their structure lacked credibility as a result of the imprecise localization (by electron microscopy) of most of the cross-links within the RNA, as well as by their use of isolated 16S RNA rather than intact 30S subunits as the substrate for the majority of the cross-linking studies. In fact, the intra-RNA cross-links that we have already identified in the 16S RNA (Fig. 1) provide the most rigorous set of constraints on the tertiary folding of the RNA. Thus, cross-link I very precisely defines the mutual orientations of helices 3 and 4 (thereby invalidating a previous hypothesis that these two helices should be coaxial [Woese et al., 1983]), cross-link V constrains helices 20, 22, 23, and 24 into a parallelogram that delineates the platform region of the 30S subunit, and so on (see Brimacombe et al., 1988a, for a detailed discussion of the individual cross-links).

In consequence, a considerable part of our current research effort is being devoted to improving techniques for the detection and analysis of intra-RNA cross-links. We expect that in future these cross-links will provide the precise constraints on the RNA structure, whereas the RNA-protein cross-links (or protein footprints) should serve to define the general locations of RNA regions in relation to the proteins concerned.

IMPROVEMENTS IN THE CROSS-LINKING METHODOLOGY

The primer extension method, which has proved so elegantly useful in footprinting and higher-order structure analyses with rRNA (Stern et al., 1988; Ehresmann et al., 1987), is unfortunately not in general applicable to our cross-linking analyses. This is because the whole purpose of our approach is to identify the pairs of structural elements that are participating in the individual cross-links, and the primer extension method cannot provide this information. Instead, some kind of partial digestion procedure must be applied so that complexes containing the various cross-linked regions of the rRNA can be isolated (by gel electrophoresis) and subsequently analyzed. Finding suitable partial digestion conditions has always been a serious problem. In the past, we have used mild digestions with RNase T_1 or, more recently, cobra venom nuclease (Vassilenko et al., 1981) to generate the rRNA fragments containing the cross-links, but these procedures have never been very satisfactory; the spectrum of fragments produced is always heterogeneous, and each class of enzyme inevitably shows a strong selectivity for particular types of structure in the RNA. As a result, many of the cross-links that have been generated fail to appear on the gels in complexes of a suitable size for the subsequent analysis.

This problem has now been finally overcome by using RNase H to generate the RNA fragments (Donis-Keller, 1979). Sets of decadeoxynucleotides have been synthesized which are complementary to the *E. coli* rRNA, spaced at intervals of 30 to 50 nucleotides along the entire lengths of the 16S or 23S molecules. After introducing the cross-links into the ribosomal subunits, non-cross-linked protein is removed, and the RNA is hybridized to a suitable set of decanucleotides and digested with RNase H. The procedure has been described in detail (Brimacombe et al., 1990) and is currently being applied to the analysis of both intra-RNA and RNA-protein cross-links. The method has the obvious advantage that the pattern of digested fragments is tailor made and can be varied at will simply by changing the composition of the deoxynucleotide mixture applied. Furthermore, since any given cross-linked complex now appears on the gel in a single spot (or at the most 2 or 3 spots), as opposed to a heterogeneous series of maybe 10 or more complexes, the yield of each cross-linked complex isolated is often increased by up to an order of magnitude. Yet another (and unexpected) advantage of the RNase H method is that aggregation of the digested complexes, which was previously a common problem in this type of experiment, is virtually eliminated.

Depending on the cross-linking reagent being used, the two-dimensional gel electrophoresis patterns of cross-linked complexes, particularly those derived from the 50S subunit, are often very complicated. In the case of intra-RNA cross-link analyses, the gel patterns can be simplified considerably by selecting for complexes containing certain regions of the RNA. For this purpose, a series of M13 clones has been constructed that contain *E. coli* ribosomal DNA inserts of ca. 40 to 250 nucleotides in length. The DNA from appropriate clones is immobilized on cellulose and is then hybridized with the RNase H digest of the cross-linked RNA complexes. Clearly, only those complexes that contain RNA sequences complementary to the DNA insert will hybridize, and these can subsequently be eluted and separated by gel electrophoresis. The application of this method not only reduces the complexity of the gel patterns but has the added advantage that the cross-link analysis can be focused on regions of the RNA that are of particular interest. Cross-link VIII (Fig. 1) was identified by using this procedure (Stiege et al., 1988a) and was the first tertiary intra-RNA cross-link to be identified in the 16S RNA after the publication of our model (Brimacombe et al., 1988a). As we have already demonstrated (Stiege et al., 1988a), this new cross-link is compatible with the model, although some minor adjustments to the positions of helices 11 and 27 are required (Fig. 2); this cross-link thus represents exactly the type of result that we are looking for in order to refine and improve the 16S model.

In the case of RNA-protein cross-linking experiments, we use a different strategy. Here a selection step is introduced after the two-dimensional gel electrophoretic separation, making use of antibodies to the ribosomal proteins attached to a second antibody, which is in turn immobilized on agarose (Gulle et al., 1988). This technique serves simultaneously to identify the protein contained in the RNA-protein cross-linked complex and to separate the complex from possible contaminants in the same eluted gel fraction. In one case, as many as three different RNA-protein cross-link sites could be identified from

Figure 3. Secondary structure of *E. coli* 23S RNA (modified from Brimacombe and Stiege, 1985), showing locations of intra-RNA and RNA-protein cross-link sites. Both secondary and tertiary intra-RNA cross-links are shown, the latter being designated with Roman numerals (as in Fig. 1). Dotted lines indicate that the RNA region (ca. 30 to 80 bases) containing the cross-link site has been identified, but not the precise position of the site within that region. See text for further details.

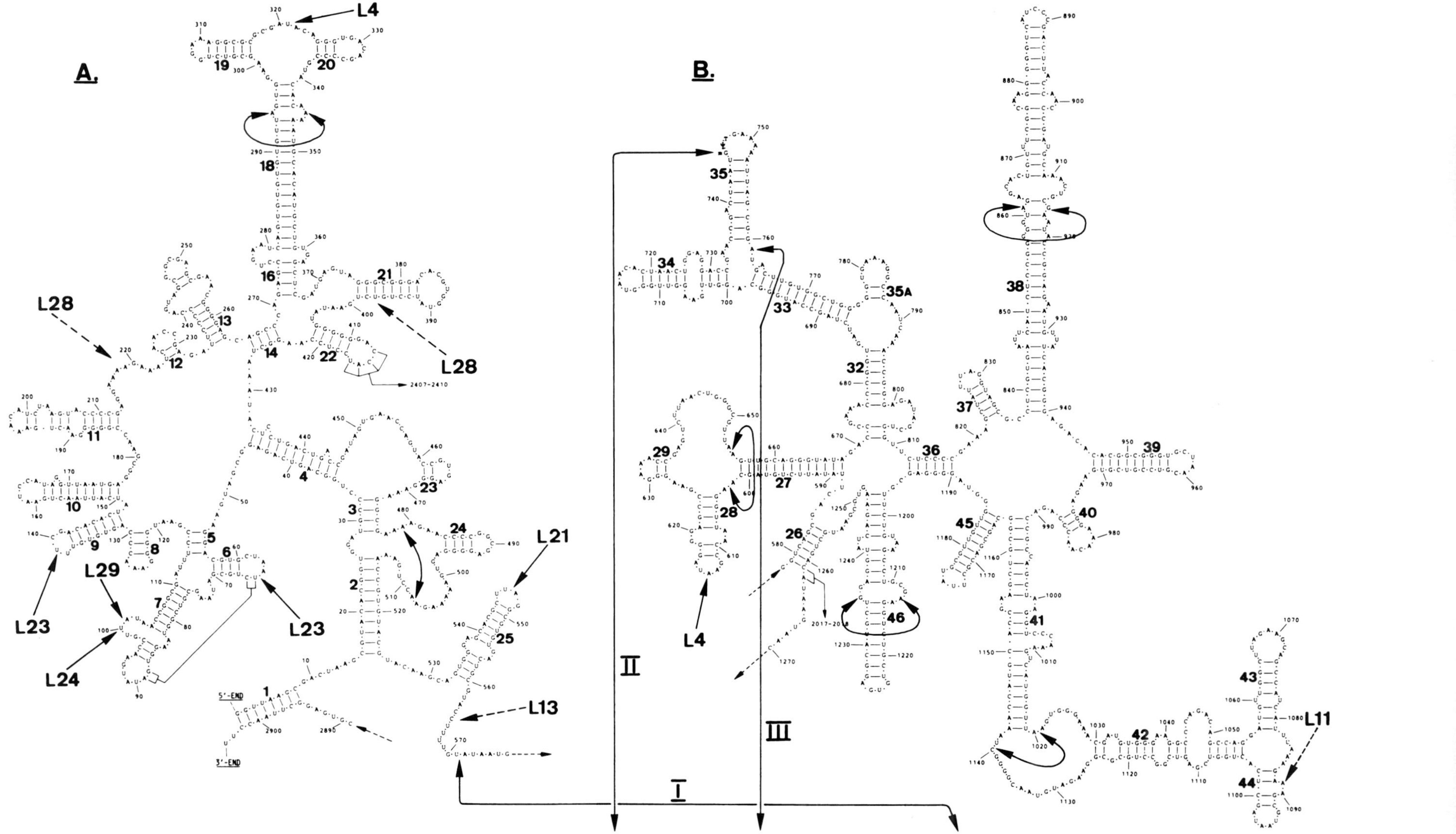

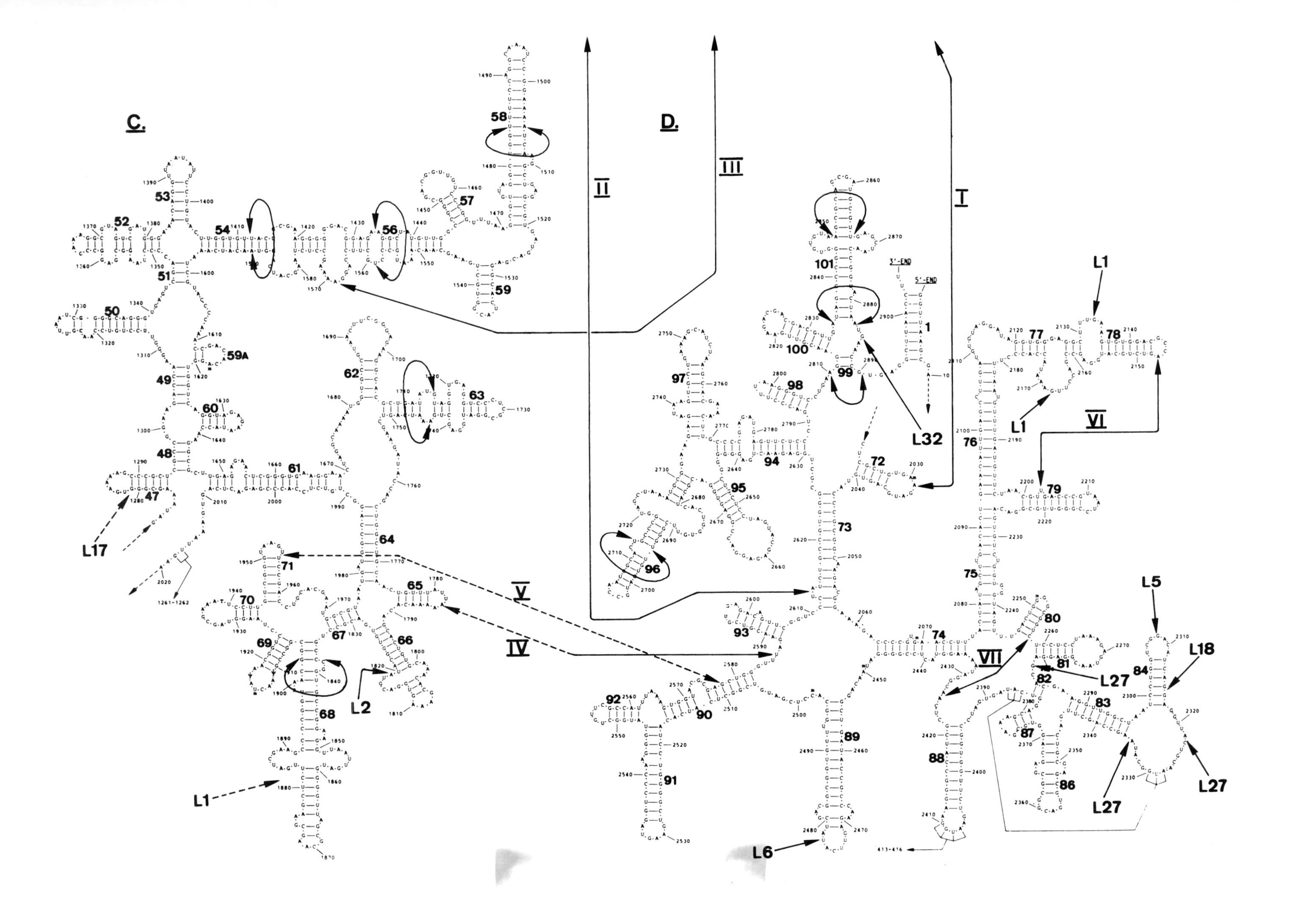

C.
D.
I
II
III
IV
V
VI
VII
L1
L2
L5
L6
L17
L18
L27
L32
3'-END
5'-END

a single gel eluate fraction (Gulle et al., 1988), and the method has led to the localization of many new RNA-protein cross-link sites on the 23S RNA (Fig. 3). As with the M13 clone approach (see above), it is of course possible with this method to focus attention on any desired protein or group of proteins of particular interest.

We have often discussed the reasons that compel us to use oligonucleotide fingerprinting rather than end-label sequencing technology for the analysis of cross-link sites on the RNA (e.g., Brimacombe et al., 1988b). One commonly encountered disadvantage in the fingerprint method, particularly in the case of intra-RNA cross-link studies, is that the cross-linked oligonucleotide remains at the starting point of the two-dimensional chromatogram, where its composition may be obscured by other large oligonucleotides or undigested material. Alternative conditions for the chromatography have, however, been developed (Brimacombe et al., 1990), so that in such cases the larger oligonucleotides (or the cross-linked oligonucleotide) can be clearly separated. This has enabled us to finalize the analysis of a number of intra-RNA cross-links in the 23S RNA whose identities were previously either tentative or ambiguous.

At the time of writing, all of the improved techniques just described are being applied to both the 30S and the 50S subunit, using our full range of intra-RNA and RNA-protein cross-linking agents (Brimacombe et al., 1988a; Brimacombe et al., 1988b). It will take us some time to complete these analyses; the current status of the data that we have obtained for the 23S RNA is summarized in the following section.

CROSS-LINK SITES IN 23S RNA

The cross-linking data set so far available for the 23S RNA is shown in Fig. 3. In this figure, our original secondary structure (Glotz et al., 1981; Brimacombe and Stiege, 1985) has been revised to take account of the latest phylogenetic comparisons of Gutell and Fox (1988) and, with the exception of a few base pairs, is now identical with the latter model. The helices in the secondary structure are numbered according to the system used by Leffers et al. (1987) in their study of archaebacterial 23S RNA, and the phylogenetically conserved tertiary interactions described by these authors are also included in the figure. The tertiary interactions will obviously play a decisive role in building a three-dimensional model of the 23S RNA, the strong interaction between the loop ends of helices 22 and 88 being a particularly important example.

The steadily accumulating sets of RNA-protein cross-link sites in the 23S RNA (Fig. 3) already show a number of topographically interesting features that will also be useful for model building. Thus, protein L4 has cross-link sites to two widely separated regions of the RNA (in helices 20 and 28), and L1 is cross-linked to a region in helix 68 as well as to points within its established binding site (helices 76 to 78 [Branlant et al., 1976a]). L23 has two distinct cross-link sites in helices 6 and 9, although the binding site of this protein is elsewhere (helices 51 to 54 [Vester and Garrett, 1984]). In contrast, protein L11 has a cross-link to a position within its binding site (Schmidt et al., 1981). Proteins L5 and L18 both cross-link to helix 84, which is within a region where they have been observed in a ribonucleoprotein fragment (Branlant et al., 1976b), and we have also found these proteins cross-linked to 5S RNA in situ in the 50S subunit. Furthermore, if Fig. 3 is compared with our preliminary model for the spatial arrangement of the 50S proteins (Walleczek et al., 1988), a number of correlations can be found, such as the proximity of proteins L23 and L29, of L13 and L21, or of L5, L18, and L27. On the other hand, large areas of the 23S molecule are still devoid of RNA-protein cross-linking data, a situation that most probably reflects the selective nature of our older cross-linking methodology. As stressed above, our cross-link analyses using the newer methodology are by no means complete at this time, and many more RNA-protein cross-links remain to be identified.

In the case of the intra-RNA cross-links in the 23S RNA (Fig. 3), we have concentrated on the analysis of cross-links induced by direct UV irradiation (either in vivo [Stiege et al., 1986] or in vitro [Stiege et al., 1983]) in order to develop the improved techniques outlined above. A number of new cross-links have been found, of which the most interesting is that between helices 78 and 79 (cross-link VI in Fig. 3), which implies a "U turn" within a single long helical system, as noted in the foregoing discussion. The long-range tertiary cross-link from helix 35 to helix 73 (cross-link II) has furthermore been localized precisely in that the 3′ component of the cross-link site (which was previously located only within an oligonucleotide 10 residues long [Stiege et al., 1983]) has now been pinpointed. In contrast, the partially localized cross-link IV between helices 65 and 93 from the latter publication has still not been confirmed, but it has nonetheless become clear that at least one other cross-link is formed between these two general regions of the RNA (cross-link V).

The reagent bis(2-chloroethyl)methylamine (nitrogen mustard), which has already proved very useful as an intra-RNA cross-linking agent in the 30S subunit (Atmadja et al., 1986), has not yet been exploited in the 50S subunit beyond our original publication on the subject (Stiege et al., 1982). This is because the gel patterns of cross-linked fragments obtained with this reagent are extremely complex and were thus for the most part beyond the scope of the older methodology. Application of the nitrogen mustard reagent to the 50S subunit by using the new methodology has now become a higher priority, and we are hopeful that this reagent will provide us with sufficient new data to allow the construction of a complete 50S model. In this context, it is also worth mentioning that nitrogen mustard has for a long time been known to induce 16S-23S RNA cross-links at the subunit interface (Zwieb et al., 1978). In the past, all of the available data pertaining to RNA sequences at the subunit interface have come from indirect protection-type experiments in which the accessibility of individual bases to nucleases or chemical reagents in 70S ribosomes is compared in some way with the corresponding behavior of separated subunits (e.g., Vassilenko et al., 1981). Allosteric effects are well known to play a significant role in this type of approach (see below), and it has for instance been demonstrated that certain bases that are protected against kethoxal treatment in 70S ribosomes (Brow and Noller, 1983) show an enhanced reactivity to dimethyl sulfate (Meier and Wagner, 1985). It follows that the successful localization of a direct intra-RNA cross-link (or cross-links) at the subunit interface would be very important for the model-building studies.

FUNCTIONAL IMPLICATIONS OF THE 30S MODEL

The rRNA molecules from different classes of organism have large deletions or expansion segments in relation to the 16S RNA from *E. coli*, and inspection of the 16S RNA model has shown that these regions are clustered into distinct domains in the three-dimensional structure (Brimacombe et al., 1988a). Thus, the expansion segments found in eucaryotic 18S RNA are located in the lower part of the subunit (Fig. 2), and the corresponding deletions in chloroplast 16S RNA relative to the *E. coli* RNA are also concentrated in this region. Similarly, in the small mammalian mitochondrial 12S RNA, there is a large deletion domain in the lower part of the subunit as well as a second one in the head region. In contrast, the majority of the secondary structural elements that form the conserved core of the small-subunit RNA (Brimacombe et al., 1983) are concentrated in the central part of the subunit, an area that

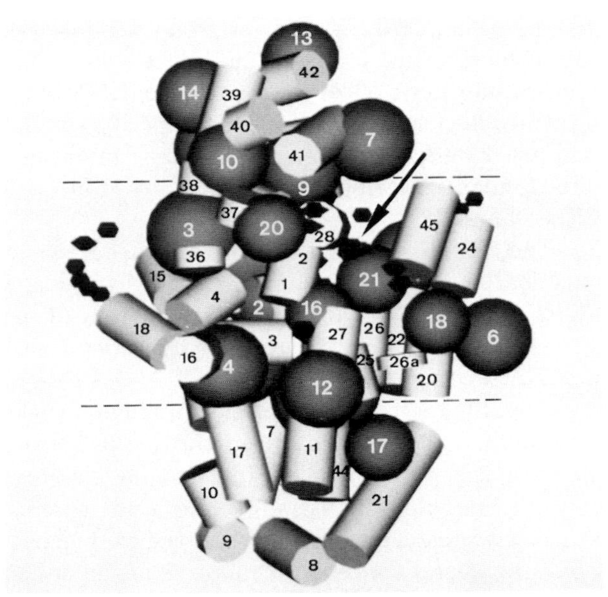

Figure 4. Computer-generated model of the 30S subunit (cf. Fig. 2), showing the positions of functionally important sites (indicated by the small black polygons) in the 16S RNA (cf. Stiege et al., 1988b). The dotted lines delineate the approximate boundaries of the conserved central belt of the subunit, and the broad arrow indicates the position of the decoding site, as evidenced by the positions of the cross-links to tRNA and poly(A) (see text). The group of functional sites lying outside the model on the left are located in the large single-stranded loop at the end of helix 18 (see Fig. 1).

includes the platform and cleft regions (Fig. 4). Further, a number of functionally important positions have been identified in the *E. coli* 16S RNA, including sites of footprinting to tRNA (Moazed and Noller, 1986), sites of footprinting or resistance to antibiotics (Moazed and Noller, 1987), and sites of cross-linking to tRNA (Prince et al., 1982; see below) or a poly(A) messenger analog (Stiege et al., 1988b). These sites, although they are widely distributed throughout the primary and secondary structure of the 16S RNA, are all clustered into the same central area of the 30S model as the conserved core of secondary structure just mentioned, and the modified bases in the 16S RNA itself are also located in this central belt (Stiege et al., 1988b).

A closer look at the model (Fig. 4) reveals that the functional sites are divided into two groups, a large group in or near the cleft (on the right in Fig. 4) and a smaller group on the opposite side of the 30S subunit. The former group includes the sites of cross-linking to tRNA and mRNA and thus represents the decoding site in accordance with electron microscopic observations (e.g., Gornicki et al., 1984), whereas the latter group lies in the immediate vicinity of the electron microscopically located methyl group at position G-527 in the 16S RNA

(Trempe et al., 1982). The latter group also includes sites of footprinting to tRNA, and these presumably represent allosteric effects caused by the tRNA binding. Thus, the whole central region of the 30S subunit (and particularly the cleft area) is clearly implicated in the fundamental ribosomal functions of tRNA and mRNA binding.

The footprinting studies with tRNA have been extended to the 50S subunit, where Moazed and Noller (1989) have made an elegant analysis of the corresponding sites on 23S RNA that are implicated in tRNA binding to the ribosomal A, P, and E sites. However, the footprinting method, although it yields large amounts of detailed information, suffers from two serious limitations. The first (already apparent from the foregoing discussion) is that in the absence of corroborative data, direct effects due to the binding of the ligand concerned cannot be distinguished from allosteric effects; the second is that the method gives no information relating to the orientation of the bound ligand. Particularly in the case of the binding of tRNA and mRNA to the ribosome, this latter question is of crucial importance and calls for alternative types of experimental approach. One such approach, which is being increasingly used in higher-resolution studies with both mRNA and tRNA, is the application of site-directed cross-linking techniques.

SITE-DIRECTED CROSS-LINKING OF tRNA

The anticodon of A-site- or P-site-bound tRNA can be cross-linked to position C-1400 of the 16S RNA (Ofengand et al., 1986; see above), and an affinity label attached to the amino acid in aminoacyl-tRNA has been shown to react with positions 2584 to 2585 or 2451 to 2452 of the 23S RNA when the tRNA is bound to the A site or P site, respectively (Steiner et al., 1988). The two extremities of a bound tRNA molecule are thus already located on the ribosome. In the site-directed cross-linking approach, a photoreactive ligand is introduced at a specific position within the tRNA sequence, and the ribosomal components that subsequently become cross-linked to this ligand are identified. Two variants of the method are currently being applied. In the first, natural modifications in the tRNA are exploited in order to attach the photoreactive ligand; this was the method used for cross-linking the anticodon loop of A-site-bound tRNA just mentioned as well as for cross-linking a modified thiouridine residue at position 8 in *E. coli* $tRNA^{Phe}$ to protein S19 (Ofengand et al., 1986). More recently, a similar method has been used to attach labels to a modified base in the anticodon loop of yeast $tRNA^{Met}$, which, when bound to the P site of *E. coli* ribosomes, reacted with proteins S7 in the 30S subunit and L1 in the 50S subunit (Podkowinski and Gornicki, 1989).

In the second variant of the method, the photolabel is introduced into the tRNA by sequence manipulation techniques; in this way, Wower et al. (1989) have labeled yeast $tRNA^{Phe}$ at positions 73 and 76 with both 2-azidoadenosine and 8-azidoadenosine. In all cases, protein L27 was the primary protein target of cross-linking; in addition, in the case of the 2-azidoadenosine label at position 73, G-1945 in the 23S RNA could be identified as a target. If the tRNA cross-linking data relating to the 50S subunit are compared with Fig. 3 and with the footprinting data for tRNA on the 23S RNA (Moazed and Noller, 1989), a pattern for the orientation of the tRNA begins to emerge. Thus, the latter authors found a footprint for E-site-bound tRNA within the binding site of protein L1 (helices 76 to 78 in Fig. 3), and as just mentioned, protein L1 is the target of a photolabel placed in the anticodon loop of tRNA. Protein L1 also has a cross-link site in helix 68 of the 23S RNA (Fig. 3), and Moazed and Noller (1989) found tRNA footprints in the adjacent helix 69, very close to position G-1945 at the base of helix 71, which is the target of the azidoadenosine placed at position 73 of tRNA (Wower et al., 1989). Position 73 of the tRNA is in turn very close to the amino acid-carrying 3′ terminus of the latter, and the data of Steiner et al. (1988) locate the aminoacyl residue within the ring enclosed by helices 73, 74, 89, 90, and 93 (Fig. 3), where Moazed and Noller (1989) also found footprint sites. Figure 3 shows that this ring is indeed in the neighborhood of helix 71, as evidenced by our intra-RNA cross-links IV and V. The RNA regions containing the cross-link sites to protein L27 (helices 81 to 84) must presumably also fold in to make contact with the 3′-terminal domain of the tRNA.

SITE-DIRECTED CROSS-LINKING OF mRNA

Our own experiments are currently concerned with the site-directed cross-linking of mRNA to the ribosome. As discussed above, the cleft region of the 30S subunit has been clearly implicated in mRNA binding by electron microscopic studies, and the cross-link sites to the anticodon of tRNA at position 1400 of the 16S RNA or to a poly(A) messenger at positions 1394 to 1399 (Fig. 1) are located in the cleft of the 30S model (Fig. 4). Olson et al. (1988) have defined the orientation of the anti-Shine-Dalgarno sequence of the 16S RNA on the 30S subunit by electron microscopy and have deduced that the outgoing message leaves the cleft in a "northerly" direc-

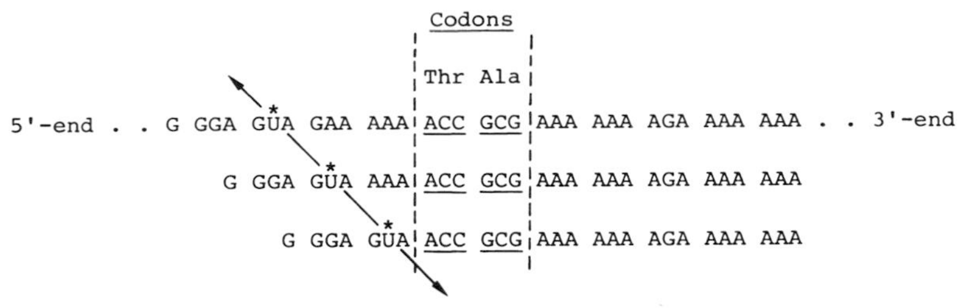

Figure 5. Synthetic mRNA sequences for site-directed cross-linking to the ribosome. Positions of the codons for $tRNA^{Thr}$ and $tRNA^{Ala}$ are indicated, U denoting the thiouridine residue substituted with 4-azidophenacyl bromide (see text).

tion. In another electron microscopic study, Evstafieva et al. (1983) showed that a poly(U) messenger molecule makes a U turn in the ribosome, with entry and exit sites in the same region of the 30S subunit, at the back of the cleft (Fig. 4).

A number of ribosomal proteins have been identified as the targets of cross-linking to mRNA or to mRNA analogs carrying an affinity label at one end (e.g., Babkina et al., 1986). Our own experiments involving direct UV cross-linking of homopolymeric messages identified proteins S1 and S18 as targets in the case of poly(U) and S19 in the case of poly(A), with proteins S3 and S7 reacting to a lesser extent in both instances (our unpublished observations). However, none of these experiments has so far been able to address the central questions in mRNA binding, namely, in which direction the message passes through the cleft and whether or not it is wrapped around the neck of the 30S subunit. In this context, it is also important to remember the rather rigid constraints on the three-dimensional interaction between an mRNA molecule and two tRNA molecules at the A and P sites during transpeptidation; the stereochemical studies of Spirin and Lim (1986) indicate that in this area of contact there is a marked kink between adjacent codons and that the mRNA lies in an orientation such that its 5′ end leads out in the direction of the A site and its 3′ end leads out in the direction of the P site, which is the reverse of the overall direction of flow (from A site to P site in a 3′-to-5′ direction). This implies that the path of the message through the ribosome must be rather convoluted, at least in the immediate vicinity of the decoding site, and elucidation of this path clearly calls for some kind of site-directed cross-linking approach.

Accordingly, we have started a systematic site-directed cross-linking program, using synthetic mRNA molecules that contain one or more unique codons together with a photoreactive residue at a defined position either upstream or downstream from these codons. The principle of the approach (at this preliminary stage) is simply to bind a message to the ribosomal P site by using an uncharged tRNA corresponding to the appropriate codon, to activate the photolabel, and then to analyze the ribosomal proteins or sites on the rRNA that are cross-linked.

The messenger molecules are transcribed with T7 polymerase from DNA templates prepared on an oligonucleotide synthesizer, using the method of Lowary et al. (1986); one set of mRNA molecules currently under test is illustrated in Fig. 5. The messenger sequences have the following properties. The transcribed region of the DNA template sequence contains a single adenine residue, and the transcription reaction is carried out with 4-thio-UTP in place of normal UTP, so that the resulting transcript contains a single 4-thiouridine residue. After isolation and purification of the mRNA, the thiouridine residue can be quantitatively substituted with 4-azidophenacyl bromide (Hixson and Hixson, 1975). The message sequences carry an oligo(A) sequence at their 3′ termini, to enable the cross-linked complexes to be isolated subsequently by binding to oligo(dT)-cellulose (Stiege et al., 1988b); this oligo(A) sequence is interrupted by a G residue to prevent stuttering of the T7 polymerase (O. Uhlenbeck, personal communication). With these features in the message sequence, the number of possible coding triplets that can be unambiguously incorporated is already severely limited, and we elected to use a tandem ACC-GCG codon pair (Fig. 5), coding for threonine and alanine, respectively.

The mRNA is labeled with ^{32}P, and the 70S-tRNA-mRNA complexes are isolated on sucrose gradients. Up to 1 mol of message is bound per ribosome in the presence of the appropriate $tRNA^{Ala}$ or $tRNA^{Thr}$, the binding being unaffected by the presence of the photolabel moiety; in the absence of tRNA, the binding is lower by a factor of 3 to 5. The isolated 70S-tRNA-mRNA complexes are irradiated with UV light, using a filter with a cutoff for wavelengths below 280 nm to avoid the concomitant formation of intra-RNA cross-links within the rRNA, and are then subjected to a second sucrose gradient centrifugation to separate the ribosomal subunits. In all messages that have been tested so far, the cross-linking takes place exclusively to the 30S subunit, and the 30S subunit fractions are accordingly separated into RNA and proteins by a final centrifugation in the presence of sodium dodecyl sulfate; reproducible cross-linking to both the protein and the RNA moieties is observed.

Proteins that have become cross-linked to the ^{32}P-labeled message are identified directly from the supernatant fractions of the sodium dodecyl sulfate-sucrose gradients, using the agarose-immobilized antibody method (Gulle et al., 1988) described above. To make the corresponding identification of the sites on the 16S RNA that have reacted with the photolabel, the cross-linked mRNA-16S RNA complexes are first separated from free 16S RNA by affinity chromatography on oligo(dT)-cellulose. These purified cross-linked fractions (in contrast to our usual mixtures of intra-RNA or RNA-protein complexes; see above) are suitable substrates for analysis by the primer extension method with reverse transcriptase, and they are screened accordingly, using the set of oligodeoxynucleotide primers described by Moazed et al. (1986). A preliminary identification of the cross-linked region is first made by observing the distribution of the ^{32}P-labeled message in RNase H-generated fragments of the 16S RNA.

In the presence of $tRNA^{Ala}$ or $tRNA^{Thr}$, all three messages illustrated in Fig. 5 show a strong and reproducible cross-linking reaction with protein S7, together with a low level of reaction with protein S21. When tRNA is omitted in control experiments, then only the reaction with S21 is observed, allowing us to conclude that the cross-link to S7 represents the tRNA-dependent contribution. As noted above, Olson et al. (1988) have deduced from the position of the Shine-Dalgarno sequence (see helix 45 of the 16S RNA; Fig. 4) that the mRNA moves out of the cleft in a northerly direction. The cross-links to S7 (which lies to the north of helix 45) clearly support this conclusion, since in these messages the photolabel is on the 5′ side of the sequence with respect to the coding triplets. Furthermore, all of the messages in Fig. 5 have the usual T7 leader sequence GGGAG at their 5′ ends (Lowary et al., 1986), and the photolabel is 3′ adjacent to this sequence. Since GGGAG is at the same time a good Shine-Dalgarno sequence, it would be reasonable to expect that in the absence of tRNA, the messages of Fig. 5 are bound via a Shine-Dalgarno interaction. In this case, a cross-linking reaction near the southerly end of helix 45 should occur with these messages; the cross-link to protein S21 observed in the absence of tRNA is consistent with this expectation (Fig. 4).

Of more interest to us are the corresponding sites of cross-linking to the 16S RNA, which should offer a higher level of resolution. At the time of writing, the analyses of the 16S sites cross-linked to the messages in Fig. 5 (and those to an additional set of messages in which the photolabel is on the 3′ side of the sequence in relation to the coding triplets) are still in progress. It is nonetheless already clear that the analytical method works in this system, and preliminary localizations of several cross-link sites have been established. Since many regions of the 16S RNA converge in the cleft region of the 30S subunit (Fig. 2 and 4), the results of such cross-link site analyses with a comprehensive range of different messenger RNA sequences (Fig. 5) will at the same time provide a very searching test of the accuracy of the 16S model.

We are grateful to H. G. Wittmann for his continued support and encouragement and to G. Stöffler and M. Stöffler-Meilicke for providing antibodies.

REFERENCES

Atmadja, J., W. Stiege, M. Zobawa, B. Greuer, M. Oßwald, and R. Brimacombe. 1986. The tertiary folding of *E. coli* 16S RNA, as studied by *in situ* intra-RNA cross-linking of 30S ribosomal subunits with *bis*-(2-chloroethyl)-methylamine. *Nucleic Acids Res.* **14:**659–673.

Babkina, G. T., A. G. Veniaminova, S. N. Vladimirov, G. G. Karpova, V. I. Yamkovoy, V. A. Berzin, E. J. Gren, and I. E. Cielens. 1986. Affinity labelling of *E. coli* ribosomes with a benzylidene derivative of $AUGU_6$ within initiation and pretranslocational complexes. *FEBS Lett.* **202:**340–344.

Branlant, C., V. Korobko, and J. P. Ebel. 1976a. The binding site of protein L1 on 23S ribosomal RNA from *E. coli*. *Eur. J. Biochem.* **70:**471–482.

Branlant, C., A. Krol, J. Sriwidada, and R. Brimacombe. 1976b. RNA sequences associated with proteins L1, L9 and L5, L18, L25 in ribonucleoprotein fragments isolated from the 50S subunit of *E. coli* ribosomes. *Eur. J. Biochem.* **70:**483–492.

Brimacombe, R., J. Atmadja, W. Stiege, and D. Schüler. 1988a. A detailed model of the three-dimensional structure of *E. coli* 16S ribosomal RNA *in situ* in the 30S subunit. *J. Mol. Biol.* **199:**115–136.

Brimacombe, R., B. Greuer, H. Gulle, M. Kosack, P. Mitchell, M. Osswald, K. Stade, and W. Stiege. 1990. New techniques for the analysis of intra-RNA and RNA-protein cross-linking data from ribosomes, p. 131–159. *In* G. Spedding (ed.), *Ribosomes and Protein Synthesis, a Practical Approach.* Oxford University Press, Oxford.

Brimacombe, R., P. Maly, and C. Zwieb. 1983. The structure of ribosomal RNA and its organization relative to ribosomal protein. *Prog. Nucleic Acid Res. Mol. Biol.* **28:**1–48.

Brimacombe, R., and W. Stiege. 1985. Structure and function of ribosomal RNA. *Biochem. J.* **229:**1–17.

Brimacombe, R., W. Stiege, A. Kyriatsoulis, and P. Maly. 1988b. Intra-RNA and RNA-protein cross-linking techniques in *E. coli* ribosomes. *Methods Enzymol.* **164:**287–309.

Brow, D. A., and H. F. Noller. 1983. Protection of ribosomal RNA from kethoxal in polyribosomes. Implication of specific sites in ribosomal function. *J. Mol. Biol.* **163:**27–46.

Capel, M. S., M. Kjeldgaard, D. M. Engelman, and P. Moore. 1988. Positions of S2, S13, S16, S17, S19 and S21 in the 30S ribosomal subunit of *E. coli*. *J. Mol. Biol.* **200:**65–87.

Donis-Keller, H. 1979. Site-specific enzymatic cleavage of RNA. *Nucleic Acids Res.* **7:**179–192.

Ehresmann, C., F. Baudin, M. Mougel, P. Romby, J. P. Ebel, and

B. Ehresmann. 1987. Probing the structure of RNAs in solution. *Nucleic Acids Res.* **15**:9109–9128.

Evstafieva, A. G., I. N. Shatsky, A. A. Bogdanov, Y. P. Semenkov, and V. D. Vasiliev. 1983. Localization of 5′- and 3′-ends of the ribosome-bound segment of template polynucleotides by immune electron microscopy. *EMBO J.* **2**:799–804.

Expert-Bezançon, A., and P. L. Wollenzien. 1985. Three-dimensional arrangement of the *E. coli* 16S ribosomal RNA. *J. Mol. Biol.* **184**:53–66.

Glotz, C., C. Zwieb, R. Brimacombe, K. Edwards, and H. Kössel. 1981. Secondary structure of the large subunit ribosomal RNA from *E. coli, Z. mays* chloroplast, and human and mouse mitochondrial ribosomes. *Nucleic Acids Res.* **9**:3287–3306.

Gornicki, P., K. Nurse, W. Hellman, M. Boublik, and J. Ofengand. 1984. High resolution localization of the tRNA anticodon interaction site on the *E. coli* 30S ribosomal subunit. *J. Biol. Chem.* **259**:10493–10498.

Gulle, H., E. Hoppe, M. Oßwald, B. Greuer, R. Brimacombe, and G. Stöffler. 1988. RNA-protein cross-linking in *E. coli* 50S ribosomal subunits; determination of sites on 23S RNA that are cross-linked to proteins L2, L4, L24 and L27 by treatment with 2-iminothiolane. *Nucleic Acids Res.* **16**:815–832.

Gutell, R. R., and G. E. Fox. 1988. A compilation of large subunit RNA sequences presented in a structural format. *Nucleic Acids Res.* **16**:r175–r269.

Hixson, S. H., and S. S. Hixson. 1975. p-Azidophenacyl bromide, a versatile photolabile bifunctional reagent. Reaction with glyceraldehyde-3-phosphate dehydrogenase. *Biochemistry* **14**: 4251–4254.

Koch, M. H. J., and H. B. Stuhrmann. 1979. Neutron scattering studies of ribosomes. *Methods Enzymol.* **59**:670–706.

Leffers, H., J. Kjems, L. Ostergaard, N. Larsen, and R. A. Garrett. 1987. Evolutionary relationships amongst archaebacteria. A comparative study of 23S ribosomal RNAs of a sulphur-dependent extreme thermophile, an extreme halophile and a thermophilic methanogen. *J. Mol. Biol.* **195**:43–61.

Lowary, P., J. Sampson, J. Milligan, D. Groebe, and O. C. Uhlenbeck. 1986. A better way to make RNA for physical studies, p. 69–76. *In* P. H. van Knippenberg and C. W. Hilbers (ed.), *Structure and Dynamics of RNA.* NATO ASI Series. Plenum Publishing Corp., New York.

Meier, N., and R. Wagner. 1985. Effects of the ribosomal subunit association on the chemical modification of the 16S and 23S RNAs from *E. coli. Eur. J. Biochem.* **146**:83–87.

Moazed, D., and H. F. Noller. 1986. Transfer RNA shields specific nucleotides in 16S ribosomal RNA from attack by chemical probes. *Cell* **47**:985–994.

Moazed, D., and H. F. Noller. 1987. Interaction of antibiotics with functional sites in 16S ribosomal RNA. *Nature* (London) **327**: 389–394.

Moazed, D., and H. F. Noller. 1989. Interaction of tRNA with 23S RNA in the ribosomal A, P, and E sites. *Cell* **57**:585–597.

Moazed, D., S. Stern, and H. F. Noller. 1986. Rapid chemical probing of conformation in 16S ribosomal RNA and 30S ribosomal subunits using primer extension. *J. Mol. Biol.* **187**: 399–416.

Noller, H. F. 1984. Structure of ribosomal RNA. *Annu. Rev. Biochem.* **53**:119–162.

Ofengand, J., J. Ciesiolka, R. Denman, and K. Nurse. 1986. Structural and functional interactions of the tRNA-ribosome complex, p. 473–494. *In* B. Hardesty and G. Kramer (ed.), *Structure, Function, and Genetics of Ribosomes.* Springer-Verlag, New York.

Olson, H. M., L. S. Lasater, P. A. Cann, and D. G. Glitz. 1988. Messenger RNA orientation on the ribosome. Placement by electron microscopy of antibody-complementary oligodeoxynucleotide complexes. *J. Biol. Chem.* **263**:15196–15204.

Podkowinski, J., and P. Gornicki. 1989. Ribosomal proteins S7 and L1 are located close to the decoding site of *E. coli* ribosomes; affinity labelling studies with modified tRNAs carrying photoreactive probes attached adjacent to the 3′-end of the anticodon. *Nucleic Acids Res.* **17**:8767–8782.

Prince, J. B., B. H. Taylor, D. L. Thurlow, J. Ofengand, and R. A. Zimmermann. 1982. Covalent cross-linking of $tRNA^{Val}$ to 16S RNA at the ribosomal P-site; identification of cross-linked residues. *Proc. Natl. Acad. Sci. USA* **79**:5450–5454.

Ramakrishnan, V. 1986. Distribution of protein and RNA in the 30S ribosomal subunit. *Science* **231**:1562–1564.

Schmidt, F. J., J. Thompson, K. Lee, J. Dijk, and E. Cundliffe. 1981. The binding site of protein L11 within 23S ribosomal RNA of *E. coli. J. Biol. Chem.* **256**:12301–12305.

Schüler, D., and R. Brimacombe. 1988. The *E. coli* 30S ribosomal subunit; an optimized three-dimensional fit between the ribosomal proteins and the 16S RNA. *EMBO J.* **7**:1509–1513.

Spirin, A. S., and V. I. Lim. 1986. Stereochemical analysis of ribosomal transpeptidation, translocation, and nascent peptide folding, p. 556–572. *In* B. Hardesty and G. Kramer (ed.), *Structure, Function, and Genetics of Ribosomes.* Springer-Verlag, New York.

Steiner, G., E. Kuechler, and A. Barta. 1988. Photo-affinity labelling at the peptidyl transferase centre reveals two different positions for the A- and P-sites in domain V of 23S rRNA. *EMBO J.* **7**:3949–3955.

Stern, S., B. Weiser, and H. F. Noller. 1988. Model for the three-dimensional folding of 16S ribosomal RNA. *J. Mol. Biol.* **204**:447–481.

Stiege, W., J. Atmadja, M. Zobawa, and R. Brimacombe. 1986. Investigation of the tertiary folding of *E. coli* ribosomal RNA by intra-RNA cross-linking *in vivo. J. Mol. Biol.* **191**:135–138.

Stiege, W., C. Glotz, and R. Brimacombe. 1983. Localization of a series of intra-RNA cross-links in the secondary and tertiary structure of 23S RNA, induced by ultraviolet irradiation of *E. coli* 50S subunits. *Nucleic Acids Res.* **11**:1687–1706.

Stiege, W., M. Kosack, K. Stade, and R. Brimacombe. 1988a. Intra-RNA cross-linking in *E. coli* 30S ribosomal subunits; selective isolation of cross-linked products by hybridization to specific cDNA fragments. *Nucleic Acids Res.* **16**:4315–4329.

Stiege, W., K. Stade, D. Schüler, and R. Brimacombe. 1988b. Covalent cross-linking of poly(A) to *E. coli* ribosomes, and localization of the cross-link site within the 16S RNA. *Nucleic Acids Res.* **16**:2369-2388.

Stiege, W., C. Zwieb, and R. Brimacombe. 1982. Precise localization of three intra-RNA cross-links in 23S RNA, and one in 5S RNA, induced by treatment of *E. coli* 50S ribosomal subunits with *bis*-(2-chloroethyl)-methylamine. *Nucleic Acids Res.* **10**: 7211–7229.

Stöffler, G., and M. Stöffler-Meilicke. 1986. Immuno electron microscopy on *E. coli* ribosomes, p. 28–46. *In* B. Hardesty and G. Kramer (ed.), *Structure, Function, and Genetics of Ribosomes.* Springer-Verlag, New York.

Trempe, M. R., K. Oghi, and D. G. Glitz. 1982. Localization of 7-methylguanosine in the small subunits of *E. coli* and chloroplast ribosomes by immuno electron microscopy. *J. Biol. Chem.* **257**:9822–9829.

Vassilenko, S. K., P. Carbon, J. P. Ebel, and C. Ehresmann. 1981. Topography of 16S RNA in 30S subunits and 70S ribosomes; accessibility to cobra venom nuclease. *J. Mol. Biol.* **152**:699–721.

Vester, B., and R. A. Garrett. 1984. Structure of a protein L23-RNA complex located at the A-site domain of the ribosomal peptidyl transferase centre. *J. Mol. Biol.* **179**:431–452.

Walleczek, J., D. Schüler, M. Stöffler-Meilicke, R. Brimacombe,

and G. Stöffler. 1988. A model for the spatial arrangement of the proteins in the large subunit of the *E. coli* ribosome. *EMBO J.* 7:3571–3576.

Woese, C. R., R. R. Gutell, R. Gupta, and H. F. Noller. 1983. Detailed analysis of the higher-order structure of 16S-like ribosomal RNAs. *Microbiol. Rev.* **47**:621–669.

Wower, J., S. S. Hixson, and R. A. Zimmermann. 1989. Labelling the peptidyl transferase center of the *E. coli* ribosome with photoreactive $tRNA^{Phe}$ derivatives containing azidoadenosine at the 3′-end of the acceptor arm; a new model of the tRNA-ribosome complex. *Proc. Natl. Acad. Sci. USA* **86**:5232–5236.

Zwieb, C., A. Ross, J. Rinke, M. Meinke, and R. Brimacombe. 1978. Evidence of RNA-RNA cross-link formation in *E. coli* ribosomes. *Nucleic Acids Res.* 5:2705–2720.

Chapter 5

Morphologies of Eubacterial and Eucaryotic Ribosomes as Determined by Three-Dimensional Electron Microscopy

J. FRANK, A. VERSCHOOR, M. RADERMACHER, and T. WAGENKNECHT

THE NONCRYSTALLOGRAPHIC METHOD OF RECONSTRUCTION

Progress in the investigation of ribosome quaternary structure has been slow because it has been difficult to grow ribosomal crystals that are sufficiently well ordered and large for electron microscopic or X-ray analysis. Even when suitable crystals are available (e.g., Unwin, 1977; Yonath and Wittmann, 1988), the large size of the asymmetric unit implies that the crystallographic approach is very tedious and time consuming. While awaiting advances in this important area of research, which will ultimately yield information at the atomic level, we are making rapid progress in characterizing the architecture of a variety of ribosomes and ribosomal subunits at a medium resolution level, using a radically different approach. Thus, the first quantitative three-dimensional images of ribosomes are now beginning to emerge, and the feasibility of three-dimensional mapping of ligands has been clearly established.

This breakthrough has been achieved by recent advances in noncrystallographic image processing and three-dimensional reconstruction techniques (Radermacher et al., 1987a; Frank et al., 1988a; Frank et al., 1988b), which allow a quantitative description of ribosomal structure to be derived from images of single-particle specimens. Even though the reproducible resolution obtained thus far in these electron microscopic studies has been in the range 30 to 45 Å (3.0 to 4.5 nm), we can for the first time achieve a matching of prominent, functionally important features seen in the eubacterial and eucaryotic ribosomes. Our results indicate a marked degree of morphological resemblance between ribosomes of the two kingdoms. When reconstructions of the two types of ribosome are presented in roughly equivalent orientations, corresponding features can be identified.

The methods by which the reconstructions are derived have been detailed elsewhere (Radermacher et al., 1987a, 1987b; Radermacher, 1988; Frank et al., 1988a). The specimens are negatively stained and prepared with a double-carbon layer method, which results in uniform staining of the ribosomal particles. Low-dose techniques (approximately 10 $e/Å^2$) are used for imaging in the electron microscope. The reconstruction technique combines a large number of projections (typically in the range of 200 to 500) to form a three-dimensional representation of the particle, using mathematical methods essentially the same as those employed in tomography.

It is important to understand that each reconstruction is computed from a set of ribosomal particles that all face the specimen grid in the same orientation, so that the forces exerted on the particle in the course of the electron microscopic preparation, usually perpendicular to the plane of the specimen grid, lead to a compression of each particle always in the same direction. For the relatively open structure of the monomeric ribosome, in which the subunits are partially separated by an appreciable gap, the degree of the resulting deformation depends critically on the initial orientation of the particle on the specimen grid (e.g., Carazo et al., 1989). Since the various ribosomal subunits and ribosomes assume different preferred orientations, the directions of deformation vary widely, leading to difficulties in attempts to precisely match the structures.

J. Frank and T. Wagenknecht ■ Wadsworth Center for Laboratories and Research, New York State Department of Health, Albany, New York 12201-0509, and School of Public Health, State University of New York at Albany, Albany, New York 12222. ■ **A. Vershoor and M. Radermacher** ■ Wadsworth Center for Laboratories and Research, New York State Department of Health, Albany, New York 12201-0509.

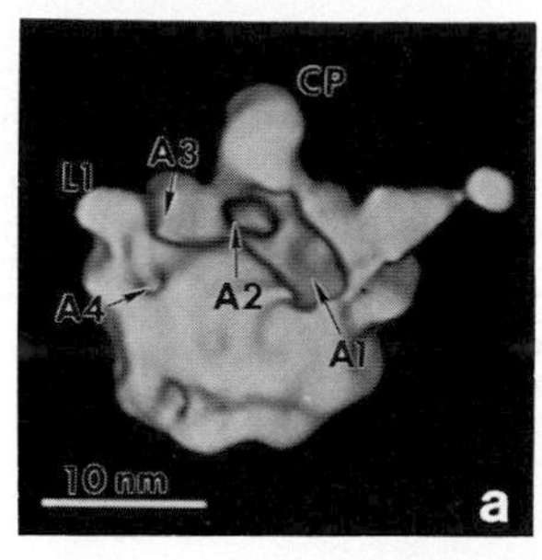

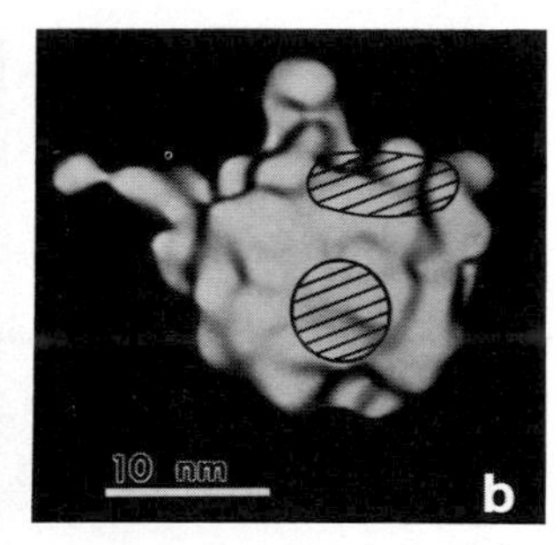

Figure 1. Three-dimensional reconstruction of the 50S ribosomal subunit from *E. coli* determined by the single-exposure random conical tilt series method (Radermacher et al., 1987a). The enclosed volume is 2.1×10^6 Å^3. (a) View from side that binds the 30S subunit (front view). (b) Back view. Abbreviations: CP, central protuberance; L1, L1 ridge; A1, A2, A3, and A4, pockets which together form the interface canyon. In panel b, the two cross-hatched regions denote the regions where nascent polypeptide is exposed in immunoelectron microscopy studies by Ryabova et al. (1988).

E. COLI RIBOSOME

Large (50S) Subunit

The first ribosomal component that was reconstructed three-dimensionally by our noncrystallographic method was the large (50S) ribosomal subunit of *Escherichia coli* (Radermacher et al., 1987b). The preponderance of a single, stable orientation of the subunit on the specimen grid, the so-called crown orientation, made this specimen ideal for reconstruction. Indeed, the resolution attained for the 50S subunit, 30 Å as determined by phase residual analysis (corresponding to a point resolution of 18 Å), has not yet been matched in any of the other reconstructions that we have subsequently obtained. Surface representations of the reconstructed subunit as viewed from the side that binds the small ribosomal subunit (interface side) and from the opposite (solvent) side are shown in Fig. 1.

The reconstructed 50S subunit appears as a refined version of the models that have been derived previously by visual analysis of electron micrographs (Wittmann, 1983). Well-established morphological landmarks such as the central protuberance, L1 ridge, and L7/L12 stalk are unambiguously identifiable (Fig. 1); we note that such is not the case for a reconstruction that has been determined from a crystalline specimen of the *Bacillus stearothermophilus* 50S subunit (Yonath et al., 1987).

Most interesting are the numerous structural features that were either not present or not well defined in the visually derived models. The most notable of these is a feature that we have termed the interface canyon, a deep trough on the interface side of the subunit that is located just below the central protuberance and extends across the subunit from the L1 ridge to the base of the L7/L12 stalk. A number of functional sites have been mapped by immunoelectron microscopy to locations that appear to fall in the vicinity of the interface canyon. These include the binding sites for the elongation factors near the base of the stalk (Girshovich et al., 1981; Girshovich et al., 1986), as well as the peptidyltransferase center near the base of the central protuberance (Olson et al., 1982; Stöffler and Stöffler-Meilicke, 1984; Oakes et al., 1987). The identification of the interface canyon as a true topological feature of the subunit (as opposed, for example, to being an artifact due to positive staining of exposed rRNA) is supported by recent structural studies using unstained, frozen-hydrated specimens (Wagenknecht et al., 1988b).

The interface canyon appears to contain at least three subregions or pockets; the middle one of these is located close to where the peptidyltransferase center is thought to reside (A2 in Fig. 1 and 2). Intriguingly, this pocket is actually a hole which leads to the back of the subunit and which we previously hypothesized to represent the site where the nascent polypeptide initially exits from the peptidyltrans-

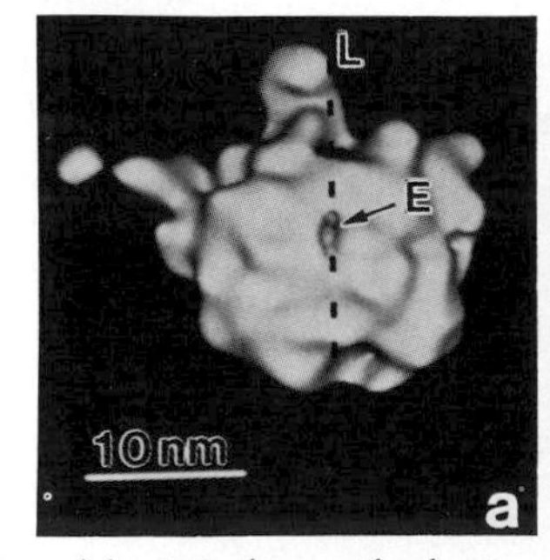

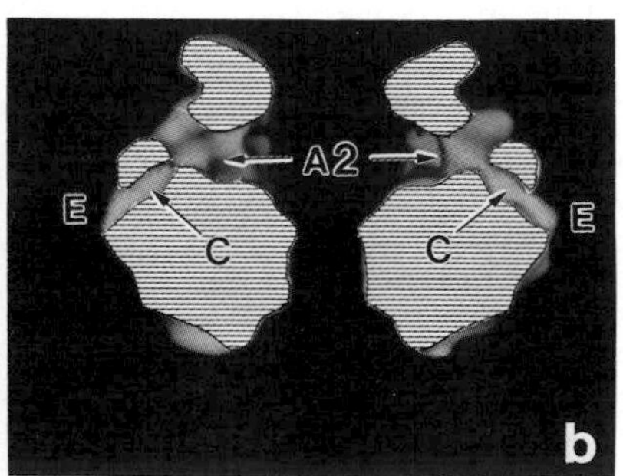

Figure 2. Reconstruction of the 50S ribosomal subunit calculated at a slightly higher threshold than was used for Fig. 1 (0.170 versus 0.137). (a) View of the back of the subunit showing the putative final exit site of the nascent polypeptide. (b) Two halves of the subunit after slicing open along line L in panel a. (Texture is used to indicate the cut surfaces.) C, Channel connecting pocket A2 (near or at the peptidyltransferase center) to the final exit site (E).

ferase center (Radermacher et al., 1987b). An earlier immunoelectron microscopy study (Bernabeu and Lake, 1982) indicated that nascent polypeptide was found lower down on the back of the subunit, a result that did not exclude the possibility that the nascent peptide is exposed at more than one site. A more recent immunoelectron microscopy study by Ryabova et al. (1988) has shown that nascent polypeptide is indeed accessible to antibodies at two sites on the back of the subunit. One of the sites is in agreement with that found in the original study, and the other is located just below the central protuberance and displaced slightly to the L1 side, remarkably consistent with the location that we suggested from the three-dimensional reconstruction on morphological grounds (Fig. 2a).

To account for the observation that no antibody binding occurred between the two sites detected, Ryabova and coauthors (1988) suggested that the nascent chain lies in a groove or channel connecting the two sites and that here the chain is inaccessible to antibody binding. Upon reexamining our reconstruction with a slightly different display threshold value (Radermacher et al., 1988), we were able to locate a channel of low density that leads from the vicinity of the A2 hole toward an indentation on the back, the location of which corresponds to the upper boundary of the second region of antibody binding (Fig. 1b). At the threshold level at which this peripheral channel appears, no other internal channel is found. When the model is cut open along a plane containing the hole A2 and the indentation, representations of the two halves show the channel connecting the front and back (Fig. 2b). Neither our three-dimensional reconstruction nor the results of Ryabova and coworkers (1988) support the idea of a channel running through the middle of the subunit as the site of the nascent polypeptide, as has been proposed elsewhere (Yonath et al., 1987).

70S Monomeric Ribosome

The complete 70S ribosome from *E. coli* has recently been reconstructed in three dimensions (Wagenknecht et al., 1989; Carazo et al., 1989). The analysis was complicated by the tendency of the ribosome to assume a range of orientations on the specimen grid. Consequently, a multivariate statistical analysis (Frank et al., 1988a) was required to identify subsets of particles appearing in a narrow orientation range. Surface representations of the reconstructed ribosome at 40-Å resolution are shown in Fig. 3.

Again, the overall shape is in basic agreement with the visually derived models, and the major

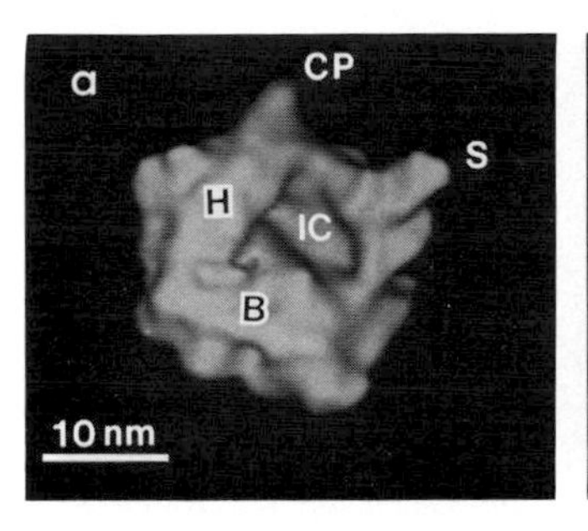

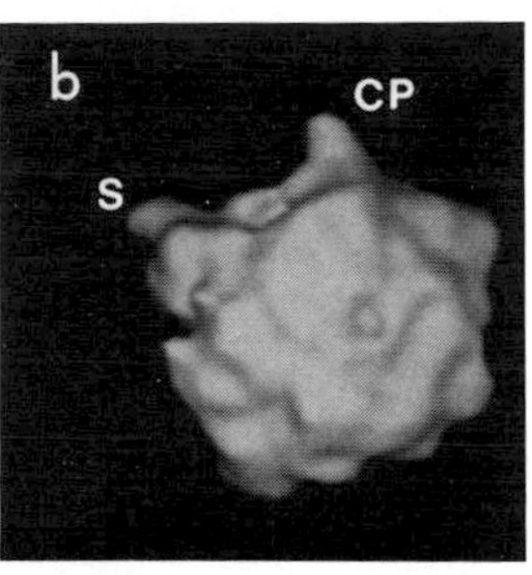

Figure 3. Surface representations of the reconstructed *E. coli* ribosome in two orientations related by a 180° rotation about the vertical axis (Wagenknecht et al., 1989). (A) View of the ribosome in which the 30S subunit comprises the large, essentially triangular mass on the left and overlies the 50S subunit. (B) View showing features of the back of the 50S subunit. The enclosed volume is 3.7 × 10^6 Å^3. Abbreviations: H, head of 30S subunit; B, body of 30S subunit; CP, S, and IC, central protuberance, L7/L12 stalk, and interface canyon, respectively, of the 50S subunit.

structural features are apparent (Fig. 3). Note that the orientational relationship between the small (30S) subunit and the 50S subunit is such that, despite the overlap, a large portion of the interface canyon of the 50S structure remains exposed between the L7/L12 stalk and the central protuberance. It is perhaps significant that in the reconstruction, the region of the peptidyltransferase center is most accessible from the interface canyon on the right-hand side of the 30S subunit (Fig. 3a), implying that perhaps tRNA molecules enter and leave the ribosome from this side. However, it should be appreciated that any flattening that has occurred during drying of the specimen will be normal to the direction of view in Fig. 3 and will have the effect of reducing the space (intersubunit gap) between the small and large subunits.

It would now appear feasible to map the locations of the tRNA molecules on the ribosome directly by three-dimensional reconstruction, provided that we can obtain ribosomes bearing tRNAs bound at specific sites and in a state such that the complexes remain structurally intact during microscopy (Wagenknecht et al., 1988a).

MAMMALIAN CYTOPLASMIC RIBOSOME

Although the 80S ribosome from a higher eucaryote is considerably larger and more complex than the eubacterial ribosome, it evidently carries out very similar functions during the translational process (e.g., Bielka, 1982). Because of the involvement of, for instance, a much larger number of initiation factors than for *E. coli*, additional functional sites must exist on the eucaryotic ribosome, but it is likely that the basic, functionally homologous sites will be similarly located and similar in form. The extent of the apparent morphological resemblance that we see

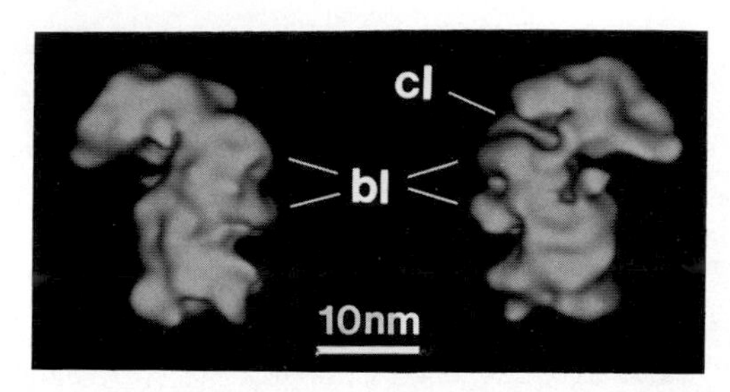

Figure 4. Reconstruction of the 40S small subunit from rabbit reticulocyte ribosome. Marked are the cleft (cl) and the back lobes (bl). (Reproduced with permission from Verschoor et al., 1989.)

between the two ribosomal structures will be discussed in a later section.

The 80S ribosome, as compared with the *E. coli* ribosome, has a larger number of proteins (about 70, versus 55) and an additional small rRNA. Although the large rRNAs in the two subunits are larger (18S versus 16S; 28S versus 23S), the principal increase appears to be in the protein content of the 60S subunit. This increase is sufficient to change the relative proportion of ribosomal protein versus rRNA in the ribosome from 30 to 35% for *E. coli* to 50% for a mammal (Bielka, 1982). Ultimately, we are interested in knowing where on the ribosome the additions of mass are located.

Thus far, we have obtained reconstructions of the monomeric 80S ribosome and of the isolated 40S subunit. We thus have two independent determinations of the structure of the small subunit. A reconstruction of the isolated 60S subunit will be more difficult to achieve because of the more variable behavior of this particle in an electron microscopic preparation. Further, its size and morphology vary among lower and higher eucaryotes, with the result that it appears to assume different preferred orientations among different species. It may prove possible to reconstruct this particle from the elliptically shaped views most characteristic of higher eucaryotes (e.g., Boublik et al., 1982; Montesano and Glitz, 1988).

Small (40S) Subunit

The isolated 40S subunit from rabbit reticulocyte ribosomes was reconstructed to 38.5 Å from one of its two strongly preferred lateral views, the so-called left-featured lateral view. The two sides of the particle are shown in Fig. 4; the view on the right shows the surface of the particle that adsorbs to the carbon. (The other, right-featured lateral view is related to this by a rotation of the particle by approximately 160° [Verschoor et al., 1989].) The basic form familiar from micrographs is evident: a large, beaked head is joined to a roughly cylindrical body by a constricted neck. The upper body is characterized by two noncoplanar back lobes protruding in directions roughly opposite to the direction in which the beak extends. The lower body is distinguished by two basal lobes, or feet.

One of the most distinctive features of the 40S particle is the beak (Frank et al., 1981), which is a complexly shaped structure in three dimensions. Its appearance differs strongly on the two sides of the particle. On the left-featured side (Fig. 4, right) it is seen as an elongate cylindrical protrusion distinct from the head proper; on the opposite side it appears more as a continuous taper of the front of the head.

Several recent functional-site determinations made for the 40S subunit appear to implicate the neck and upper-body regions. The upper of the two back lobes may be considered the analog of the rim of the platform of the 30S subunit. Within the cleft formed between this rim and the neck of the subunit is thought to be the site of exposure of the 3′ end of the rRNA. A recent mapping of the 18S rRNA sequence homologous to the *E. coli* 1400-region sequence representing the site of the codon-anticodon interaction places this site in the cleft region (Oakes et al., 1987).

The binding sites for two of the major initiation factors have also been determined. One of these, eIF-3, a large complex which prevents association of the subunits, has been mapped to bind across the back lobes (Lutsch et al., 1985). A mapping (Bommer et al., 1988) for eIF-2, which forms a quaternary complex with the 40S subunit, the initiator aminoacyl-tRNA, and GTP, places the binding site for this factor along the subunit neck, evidently on its interface aspect.

80S Monomeric Ribosome

The 80S ribosome isolated from rabbit reticulocyte polysomes was reconstructed to 37 Å (Verschoor and Frank, 1990) from the left-featured frontal view (Nonomura et al., 1971), in which the two subunits were interpreted to lie almost side by side. Indeed, we see that this original interpretation was correct; there is minimal overlap of the 40S and 60S structures. The ribosome adsorbs to the carbon support by its more open aspect (Fig. 5, right); in this view, we can see that the separation between the 40S subunit on the left and the 60S subunit on the right extends nearly down to the base of the ribosome, with only a small region of tight contact at the subunit bases. The head of the 40S subunit appears to bridge to the central protuberance of the 60S subunit by a thin, linear connecting structure that is continuous if the reconstruction is visualized at a slightly lower density threshold than the one used in Fig. 5. The mutual orientation of the two subunits can for the first time be determined: the head-to-base axis of the 40S subunit is inclined, so that the subunit tips away from the viewer, whereas the head-to-base axis of the 60S subunit is directed roughly vertically.

The opposite side of the ribosome (Fig. 5, left) shows a strikingly different appearance of the intersubunit space. The entire lower half of the ribosome shows the subunits fused, at least at this resolution; the separation is restricted to the region of the necks of the subunits. This view reveals some of the features of the exit domain, which contains the membrane attachment site as well as the exit site. The latter has been mapped to the lower back of the large subunit in eucaryotes as well as in eubacteria (Bernabeu et al., 1983). Although it is not evident from this view, a further (axial) rotation of the 80S structure by 20 to 40° reveals that the 60S subunit has a strongly ellipsoidal shape, as indeed is evident from micrographs of isolated 60S subunits (e.g., Montesano and Glitz, 1988). This form is markedly different from the rounded form of the *E. coli* 50S subunit and suggests that the extra proteins (and perhaps the 28S rRNA expansion segments as well) in the 60S subunit may form one or more novel domains that give rise to the difference in morphology.

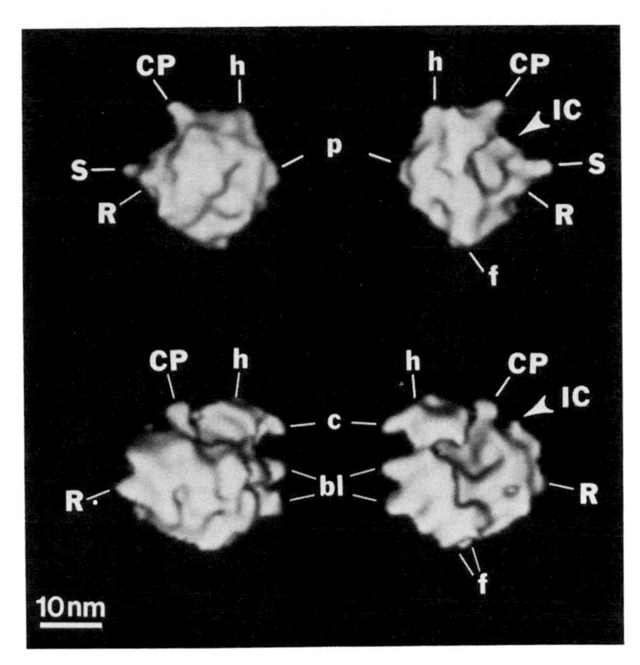

Figure 6. Comparison of 70S *E. coli* ribosome structure (top) with 80S rabbit reticulocyte ribosome structure (bottom). For both structures, the right and left parts of the figure depict views elicited by 180° rotation around a vertical axis. The two reconstructions have been brought into equivalent orientations so that corresponding features can be identified. Small-subunit features: h, head; p, platform; f, feet or basal lobes; c, crest; bl, basal lobes. Large-subunit features: CP, central protuberance; IC, interface canyon; S, stalk; R, ridge.

MORPHOLOGICAL COMPARISON BETWEEN 70S AND 80S RIBOSOMES

The *E. coli* ribosome, as described earlier, was reconstructed by Wagenknecht et al. (1989) and Carazo et al. (1989) using particles exhibiting a view in which the 50S subunit almost directly overlaps the 30S subunit (Lake, 1976; Verschoor et al., 1986). Consequently, as a result of compression, the intersubunit gap, which is seen clearly in certain other views of the ribosome, has all but disappeared in this reconstruction from overlap views (Fig. 3 and 6). However, the existence and original width of this gap can be appreciated from an electron crystallographic reconstruction of the *B. stearothermophilus* 70S ribosome by Arad et al. (1987).

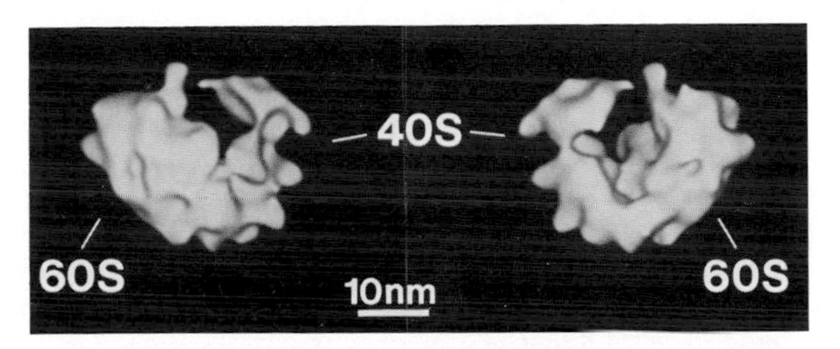

Figure 5. Reconstruction of the 80S cytoplasmic ribosome from rabbit reticulocytes. The two subunits are marked 40S and 60S in the two 180° related views shown.

In contrast to the overlap view of the 70S ribosome, the most distinctive view of the 80S ribosome, which was used for the reconstruction by Verschoor and Frank (in press), shows the subunits lying essentially side by side, so that the intersubunit gap is largely unaffected by compression (Fig. 5). Because the ribosomes in the crystalline sheet of Arad et al. (1987) lay in a side-by-side orientation very similar to that of the 80S ribosomes, we are able to see the striking similarity in overall form of the ribosome in our 80S reconstruction and in their 70S reconstruction. There appears to be good agreement in the shape of the intersubunit space, defined by the mutual fit of the two subunits.

To compare more closely the eucaryotic and eubacterial monosome structures that were obtained in our laboratory from single-particle specimens, we must present them in roughly equivalent orientations (Fig. 6). In the two 180°-related views shown for each, a number of common features of the large subunit are visible: the central protuberance, the interface canyon, and the so-called stalk-base ridge feature. The most striking difference, aside from a general increase in the width or ellipticity of the subunit, arises from the absence of an extended stalk in the eucaryotic structure. Since a stalklike feature is visible in some projections of the isolated 60S subunit (e.g., Montesano and Glitz, 1988), we believe that the stalk may be folded back in random positions when the monosome adsorbs to the carbon support in the orientation that gives rise to the view used in our reconstruction.

For the small-subunit portion of the ribosome, the eucaryotic structure shows a broader and more complex morphology of the foot or basal lobe region, and more distinctly separate back lobes have replaced the notched but continuous platform structure of the 30S subunit. The 40S subunit has several distinctive features that appear to lack eubacterial counterparts, such as the crest and beak features, as well as the additional basal lobe. It appears thus far that such novel features tend to be peripherally situated; a general conservation of the form of the translational domain centered on the upper interface region might be expected in light of the evident strong conservation of function (e.g., Bielka, 1982).

OUTLOOK: REFINEMENT OF STRUCTURAL ANALYSIS AND FUNCTIONAL MAPPING

The use of the negative staining technique limits the information derived from the reconstructions essentially to the surface features. A further disadvantage of this technique is that the uranyl acetate used may stain exposed portions of the RNA preferentially, so that some caution must be exercised in interpreting topological features. To overcome these limitations, we are currently conducting experiments using the technique of unstained, frozen hydrated electron microscopy (Lepault et al., 1983). With this technique, the ribonucleoprotein mass of the particle gives rise to image contrast, and interior detail may be visualized.

A first success has been achieved in characterizing the 50S ribosomal subunit in its crown view (Wagenknecht et al., 1988b). The agreement (thus far in two dimensions) with the results from the stained preparation is remarkable, indicating that the effects of positive staining are not of major importance at the current level of resolution. Eventually, we hope to be able to refine the structures determined for all of the ribosomal particles presented in this review, based on new data obtained with frozen hydrated electron microscopy.

Despite the limited resolution, our new reconstruction technique opens up the possibility of three-dimensional functional mapping, because relatively small differences of mass produced by binding of factors or labels can be reliably detected in three-dimensional difference maps. The successful two-dimensional localization of one of the tRNA-binding sites, the P site, on the 30S subunit by Wagenknecht et al. (1988a) can be seen as a first step in this direction. The progress of such three-dimensional mapping studies obviously depends on the success in obtaining functional complexes, in stoichiometric quantities, that are stable and suitable for electron microscopic preparation.

This work was supported by Public Health Service grant R01-29169 from the National Institutes of Health and by National Science Foundation grant 8313405.

We thank Robert Grassucci for technical support.

REFERENCES

Arad, T., J. Piefke, S. Weinstein, H.-S. Gewitz, A. Yonath, and H. G. Wittmann. 1987. Three-dimensional image reconstruction from ordered arrays of 70S ribosomes. *Biochemie* **69:**1001–1006.

Bernabeu, C., and J. A. Lake. 1982. Nascent polypeptide chains emerge from the exit domain of the large ribosomal subunit: immune mapping of the nascent chain. *Proc. Natl. Acad. Sci. USA* **79:**3111–3115.

Bernabeu, C., E. M. Tobin, A. Fowler, I. Zabin, and J. A. Lake. 1983. Nascent polypeptide chains exit the ribosome in the same relative position in both eucaryotes and procaryotes. *J. Cell Biol.* **96:**1471–1474.

Bielka, H. (ed.). 1982. *The Eukaryotic Ribosome.* Springer-Verlag, Berlin.

Bommer, U.-A., G. Lutsch, J. Behlke, J. Stahl, N. Nesytova, A. Henske, and H. Bielka. 1988. Shape and location of eukaryotic initiation factor eIF-2 on the 40S ribosomal subunit of rat liver. Immunoelectron-microscopic and hydrodynamic investigations. *Eur. J. Biochem.* **172:**653–662.

Boublik, M., W. Hellmann, and F. Jenkins. 1982. Structural homology of ribosomes by electron microscopy. *Proc. 10th Int. Congr. Electron Microsc.* **3:**95–96.

Carazo, J. M., T. Wagenknecht, and J. Frank. 1989. Variations of the three-dimensional structure of the *Escherichia coli* ribosome in the range of overlap views. An application of the method of multicone and local single-cone three-dimensional reconstruction. *Biophys. J.* **55:**465–477.

Carazo, J. M., T. Wagenknecht, M. Radermacher, V. Mandiyan, M. Boublik, and J. Frank. 1988. Three-dimensional structure of the 50S *Escherichia coli* ribosomal subunits depleted of proteins L7/L12. *J. Mol. Biol.* **201:**393–404.

Frank, J., M. Radermacher, T. Wagenknecht, and A. Verschoor. 1988a. Methods for studying ribosome structure by electron microscopy and computer image processing. *Methods Enzymol.* **164:**3–35.

Frank, J., A. Verschoor, and M. Boublik. 1981. Computer averaging of electron micrographs of 40S ribosomal subunits. *Science* **214:**1353–1355.

Frank, J., A. Verschoor, T. Wagenknecht, M. Radermacher, and J. M. Carazo. 1988b. A new non-crystallographic image-processing technique reveals the architecture of ribosomes. *Trends Biochem. Sci.* **13:**123–127.

Girshovich, A. S., E. S. Bochkareva, and V. D. Vasiliev. 1986. Localization of elongation factor Tu on the ribosome. *FEBS Lett.* **197:**192–198.

Girshovich, A. S., T. V. Kurtskhalia, Y. A. Ovchinnikov, and V. D. Vasiliev. 1981. Localization of the elongation factor G on *Escherichia coli* ribosome. *FEBS Lett.* **130**:54–59.

Lake, J. A. 1976. Ribosome structure determined by electron microscopy of *Escherichia coli* small subunits, large subunits and monomeric ribosomes. *J. Mol. Biol.* **105**:131–150.

Lepault, J., F. P. Booy, and J. Dubochet. 1983. Electron microscopy of frozen biological suspensions. *J. Microsc.* **129**:89–102.

Lutsch, G., R. Benndorf, P. Westermann, J. Behlke, U.-A. Bommer, and H. Bielka. 1985. On the structure of native small ribosomal subunits and initiation factor eIF-3 isolated from rat liver. *Biomed. Biochim. Acta* **44**:K1–K7.

Montesano, L., and D. G. Glitz. 1988. Wheat germ cytoplasmic ribosomes. Structure of ribosomal subunits and localization of N^6,N^6-dimethyladenosine by immunoelectron microscopy. *J. Biol. Chem.* **263**:4932–4938.

Nonomura, Y., G. Blobel, and D. Sabatini. 1971. Structure of liver ribosomes studied by negative staining. *J. Mol. Biol.* **60**:303–323.

Oakes, M., A. Scheinman, M. Rivera, D. Souper, G. Shankweiler, and J. Lake. 1987. Evolving ribosome structure and function: rRNA and the translation mechanism. *Cold Spring Harbor Symp. Quant. Biol.* **52**:675–685.

Olson, H. M., B. S. Grant, B. S. Cooperman, and D. G. Glitz. 1982. Immunoelectron microscopic localization of puromycin binding on large subunit of the *Escherichia coli* ribosome. *J. Biol. Chem.* **257**:2649–2656.

Radermacher, M. 1988. Three-dimensional reconstruction of single particles from random and non-random tilt series. *J. Electron Microsc. Techn.* **9**:359–394.

Radermacher, M., J. Frank, and T. Wagenknecht. 1988. The probable exit site of the polypeptide in the ribosome: analysis of densities in three-dimensional reconstruction. *Proc. 9th Eur. Congr. Electron Microsc.* **3**:323–324.

Radermacher, M., T. Wagenknecht, A. Verschoor, and J. Frank. 1987a. Three-dimensional reconstruction from a single-exposure, random conical tilt series applied to the 50S ribosomal subunit of *Escherichia coli*. *J. Microsc.* **146**:113–136.

Radermacher, M., T. Wagenknecht, A. Verschoor, and J. Frank. 1987b. Three-dimensional structure of the large ribosomal subunit from *Escherichia coli*. *EMBO J.* **6**:1107–1114.

Ryabova, L. A., O. M. Selivanova, V. I. Baranov, V. D. Vasiliev, and A. S. Spirin. 1988. Does the channel for nascent polypeptide exist in the ribosome? Immune electron microscopy study. *FEBS Lett.* **226**:255–260.

Stöffler, G., and M. Stöffler-Meilicke. 1984. Immunoelectron microscopy of ribosomes. *Annu. Rev. Biophys. Biophys. Eng.* **13**:303–330.

Unwin, P. N. T. 1977. Three-dimensional model of membrane-bound ribosomes obtained by electron microscopy. *Nature* (London) **269**:118–122.

Verschoor, A., and J. Frank. Three-dimensional structure of the mammalian cytoplasmic ribosome. *J. Mol. Biol.*, in press.

Verschoor, A., J. Frank, T. Wagenknecht, and M. Boublik. 1986. Computer-averaged views of the 70S monosome from *Escherichia coli*. *J. Mol. Biol.* **187**:581–590.

Verschoor, A., N.-Y. Zhang, T. Wagenknecht, T. Obrig, M. Radermacher, and J. Frank. 1989. Three-dimensional reconstruction of mammalian 40S ribosomal subunit. *J. Mol. Biol.* **209**:115–126.

Wagenknecht, T., J. M. Carazo, M. Radermacher, and J. Frank. 1989. Three-dimensional reconstruction of the ribosome from *Escherichia coli*. *Biophys. J.* **55**:465–477.

Wagenknecht, T., J. Frank, M. Boublik, K. Nurse, and J. Ofengand. 1988a. Direct localization of the tRNA-anticodon interaction site on the 30S ribosomal subunit by electron microscopy and computerized image averaging. *J. Mol. Biol.* **203**:753–760.

Wagenknecht, T., R. Grassucci, and J. Frank. 1988b. Electron microscopy and computer image averaging of ice-embedded large ribosomal subunits from *Escherichia coli*. *J. Mol. Biol.* **199**:137–147.

Wittmann, H. G. 1983. Architecture of prokaryotic ribosomes. *Annu. Rev. Biochem.* **52**:35–65.

Yonath, A., K. R. Leonard, and H. G. Wittmann. 1987. A tunnel in the large ribosomal subunit revealed by three-dimensional reconstruction. *Science* **236**:813–816.

Yonath, A., and H. G. Wittmann. 1988. Approaching the molecular structure of ribosomes. *Biophys. Chem.* **29**:17–29.

Chapter 6

Potential of Electron Microscopic Techniques for Structural Analysis of Ribosomes

M. BOUBLIK, V. MANDIYAN, and S. TUMMINIA

Conventional transmission electron microscopy (TEM) played a key role in the discovery of ribosomes (Palade, 1955) and is still the major technique for direct high-resolution imaging of ribosome morphology and topography of its components (Boublik, 1990). Ribosomes are challenging objects for high-resolution TEM imaging. In addition to the high degree of hydration, low contrast, and sensitivity to radiation damage common to all biological specimens, ribosomes have an intricate structure originating from the high number of their components and from the lack of any regularity or symmetry in their structure. The potential of conventional TEM for structural studies of ribosomes can be considerably enhanced by specimen tilt and computer analysis of images taken from various angles. This combination provides a basis for three-dimensional reconstruction of ribosomal particles (Frank et al., 1988).

Application of dedicated scanning TEM (STEM) to the study of biological specimens (Wall, 1979) marks a new trend in electron microscopic imaging, which can be characterized by the transition from a mere photographic record of the specimen to quantitative spectroscopic image analysis (Wall and Hainfeld, 1986). Separation of resolution- and contrast-affecting components in dark-field STEM and optimized placement of detectors for elastic (contrast-forming) electrons make it possible to detect almost 100% of the available elastically scattered electrons. The efficiency of dark-field TEM with a single imaging channel and limited acceptance angle to minimize lens aberrations is only 5%. To compensate for the loss, the radiation dose in TEM has to be about 20 times higher than in STEM to obtain the same signal-to-noise ratio in the dark-field mode used for unstained specimens. The high contrast and superior signal-to-noise ratio associated with the STEM annular detector make it possible to visualize unstained freeze-dried ribosomal particles and rRNAs at extremely low radiation doses of 1 to 50 e/Å (1 Å = 0.1 nm), depending on the magnification (×40,000 to ×250,000). Specimens prepared in this way are free of the main resolution-limiting conditions of TEM, i.e., staining, air drying, and, to a considerable extent, radiation damage. With the elimination of staining, it becomes possible to relate image intensity to the local projected mass of the specimen and thus obtain quantitative data on the molecular mass, mass distribution, and radius of gyration (R_g) of any selected ribosomal particle. Elimination of staining also allows use of heavy metals as high-resolution markers for topographical mapping of ribosomal components and functional sites.

TRANSMISSION ELECTRON MICROSCOPY

Morphology of Ribosomes and Ribosomal Subunits

TEM techniques for conventional imaging of ribosomes used by various groups are more or less standardized (see Hardesty and Kramer, 1986, and Noller and Moldave, 1988, for references). In principle, ribosomal particles at a concentration of about 0.1 A_{260}/ml of buffer solution are adsorbed to a thin (~20- to 30-Å) carbon film on a copper support grid, stained with aqueous uranyl acetate, blotted, and air dried. An example of an electron micrograph obtained by this procedure (described in detail in Boublik, 1990) is given in Fig. 1a, which shows an overall field of 50S ribosomal subunits of *Escherichia coli*. Images such as this are used to evaluate the shape and characteristic dimensions, axial ratio, extent of homology, and degree of preferential orientation of

M. Boublik, V. Mandiyan, and S. Tumminia ■ Roche Institute of Molecular Biology, Roche Research Center, Nutley, New Jersey 07110.

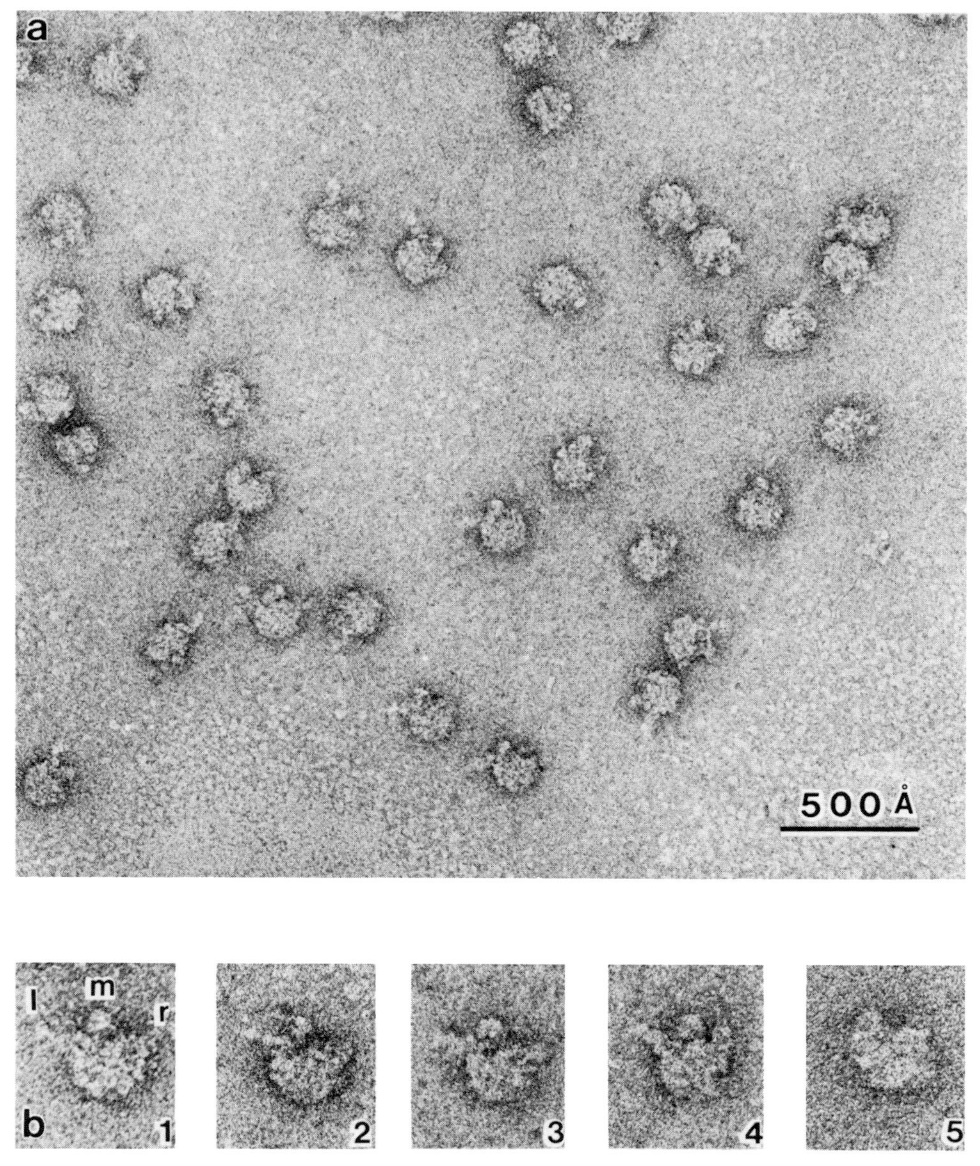

Figure 1. Electron micrographs (TEM) of 50S *E. coli* ribosomal subunits deposited from buffer A (see Table 1, footnote *a*), stained with 0.5% aqueous uranyl acetate, and air dried. (a) Overall field; (b) gallery of selected images at higher magnification. l, m, and r, Left, middle, and right crest in the crown. Bar, 0.05 μm.

ribosomal particles. Enlarged images of individual 50S particles (Fig. 1b) selected from various views are analyzed for fine structural features and can serve as a basis for the proposal of a three-dimensional model. With this approach, it is possible to resolve structural differences between procaryotic and eucaryotic ribosomal monosomes and their respective subunits (Boublik and Hellmann, 1978). However, one cannot reliably resolve structural differences among small ribosomal subunits from various eucaryotes, nor can one resolve ribosomes from the same species but from different stages of development with established differences in protein composition (Boublik and Ramagopal, 1980).

Topographical Mapping of Ribosomal Components and Functional Sites In Situ

Localization of ribosomal constituents in situ by using antibodies (immunoglobulin G [IgG]) directed against a particular protein or rRNA segment considerably broadened our knowledge of the architecture and structure-function relationships of ribosomes. This technique, called immunoelectron microscopy (IEM), has been used to localize ribosomal proteins, rRNAs, and functional sites (reviewed in Lake, 1978; Stöffler and Stöffler-Meilicke, 1986; Stöffler-Meilicke and Stöffler, 1988; and Glitz et al., 1988).

Success in labeling of ribosomal components by

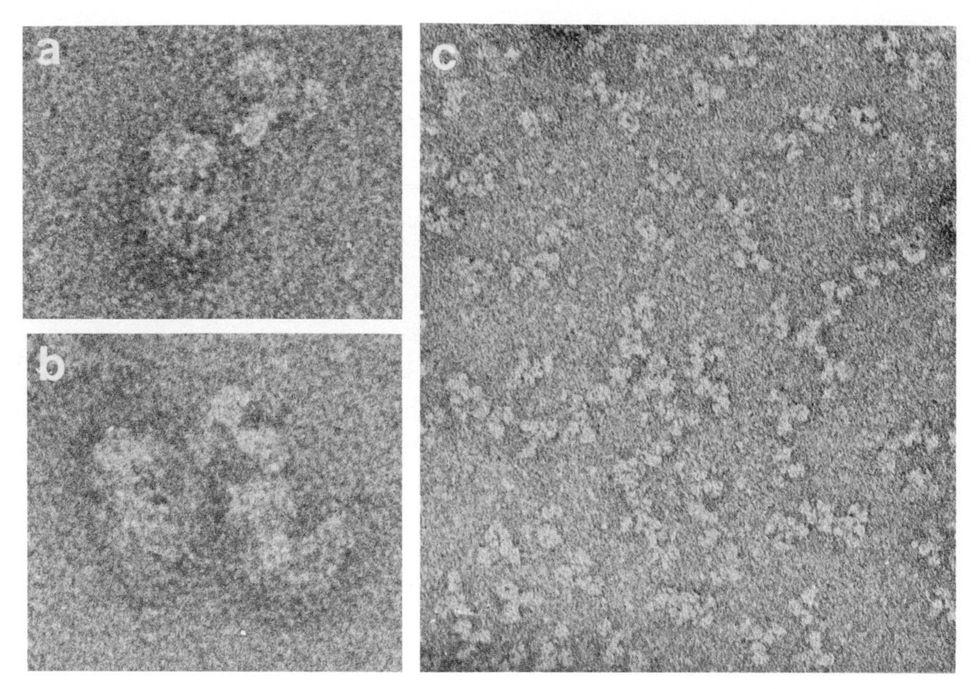

Figure 2. Electron micrographs (TEM) of 30S *E. coli* ribosomal subunits with antidinitrophenol antibody (IgG) attached to the 3′ end of the tRNA (Keren-Zur et al., 1979) (a) and 30S-IgG 30S dimer (b). (c) Example of structurally preserved IgG for IEM labeling.

IEM depends on the purity, structure, and avidity of the antibody (IgG) and accessibility of the antigenic determinant in situ. Antibody can be resolved without additional markers because of its characteristic Y shape. The examples in Fig. 2a and b demonstrate localization of the decoding region of the *E. coli* ribosome to the cleft of the small subunit obtained by a combination of IEM and affinity labeling (Keren-Zur et al., 1979).

Although IEM has promoted rapid advancement of our understanding of the architecture of the ribosome (*E. coli* in particular), it is not the ultimate solution of topographical mapping. Although lack of specificity has been eliminated by using monoclonal antibodies (for references, see Stöffler-Meilicke and Stöffler, 1988), the resolution of IEM remains limited by the size of the IgG marker, specifically by the width of the Fab fragment (~30 to 35 Å). Furthermore, the attachment site of the antibody (the Fab fragment) is not always clearly resolvable and can be hidden by the ribosomal particle. In some cases, the structure of the antibody is distorted and difficult to recognize. An example of well-preserved structure of IgG is given in Fig. 2c. Modifications of IEM using antibody probes to reagents applied in site-specific modifications of rRNAs and to synthetic oligodeoxynucleotides complementing specific rRNA sequences (Glitz et al., 1988) improve the specificity of labeling but not the resolution. The same criticism applies to DNA hybridization electron microscopy using a biotin-avidin complex as a topographical marker (Oakes et al., 1986).

SCANNING TRANSMISSION ELECTRON MICROSCOPY

Molecular Parameters of Ribosomes and Their Subunits

Unstained, freeze-dried specimens of ribosomal particles visualized by STEM appear structurally less expressive than stained air-dried particles in images from conventional TEM. This is obvious from a comparison of electron micrographs of 50S *E. coli* subunits obtained by TEM (Fig. 1) and STEM (Fig. 3). A similar visual impression can be obtained from a comparison of electron micrographs of polysomes obtained by these two techniques. In the TEM image (Fig. 4a), one can resolve the two ribosomal subunits because of the heavy stain deposition in the interface. Polysomes in the STEM image (Fig. 4b) appear as a chain of structureless fluffy balls seemingly following a helical arrangement; mRNA cannot be resolved in either image. Despite the impression, directly digitized STEM images of ribosomes offer substantially more information than do TEM images. Apart from morphological parameters, STEM images yield quantitative data on molecular mass, mass distribution within any selected single particle, location of the center of gravity, and values of the apparent R_g

(Oostergetel et al., 1985). Since the molecular masses of the 30S and 50S subunits and R_g values have been established by hydrodynamic and spectroscopic techniques in which the specimens were investigated in a fully hydrated state (for references, see Hardesty and Kramer, 1986), the validity of STEM data can be rigorously tested. Mass determinations of the 30S and 50S *E. coli* subunits by STEM, 880 ± 40 and 1,680 ± 80 kilodaltons (kDa) (unpublished data), respectively, are in agreement with their theoretical values, 900 and 1,600 kDa, respectively. R_g values of the 30S and 50S subunits calculated from STEM (68 ± 1.4 and 69 ± 2 Å, respectively) (Mandiyan et al., 1988; Mandiyan et al., 1989) fall into the range of 65 to 70 Å obtained by various physicochemical techniques (Hill et al., 1969; Serdyuk et al., 1983; Vasiliev et al., 1986).

Conformation of Ribosomal Components

Ribosomal proteins with an average molecular mass of about ~15 kDa are too small for direct visualization by STEM. However, introduction of a new technique for rRNA deposition under nondenaturing conditions (Boublik et al., 1986b; Boublik et al., 1988) established procedures for direct imaging of free rRNAs in solutions of various ionic strengths and opened the field for imaging rRNA-protein interactions and, ultimately, visualization of the assembly process of ribosomal subunits from the RNAs and individual ribosomal proteins.

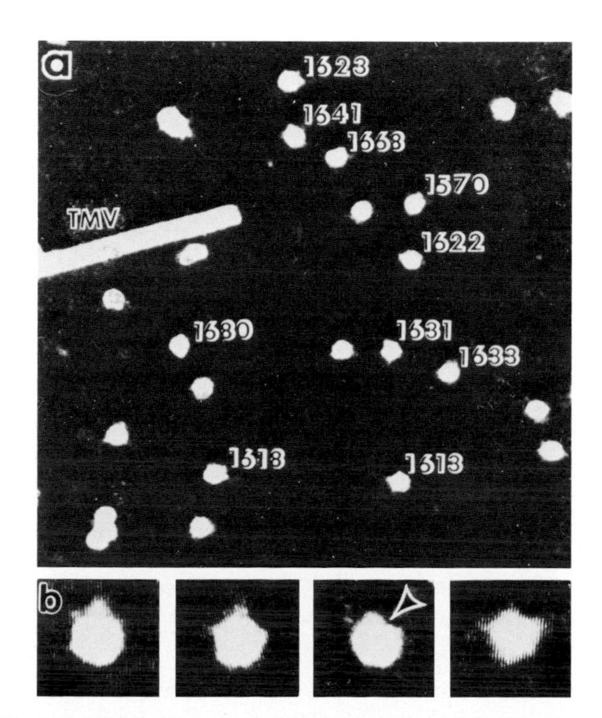

Figure 3. STEM images of unstained freeze-dried 50S *E. coli* subunits. (a) Overall view of particles and their molecular masses (in kilodaltons). (b) Gallery of enlarged images of the subunits; arrowhead points to the crown view.

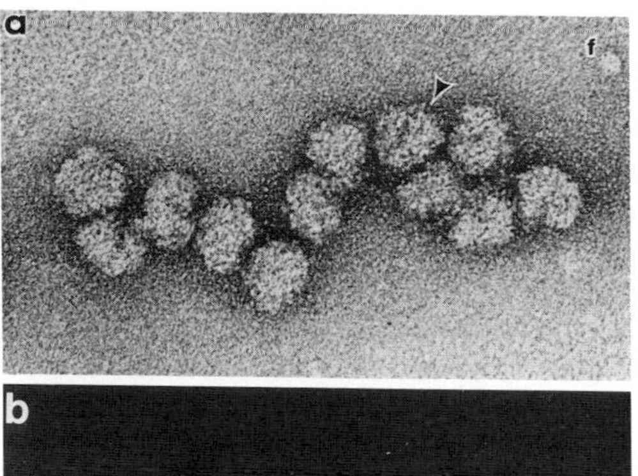

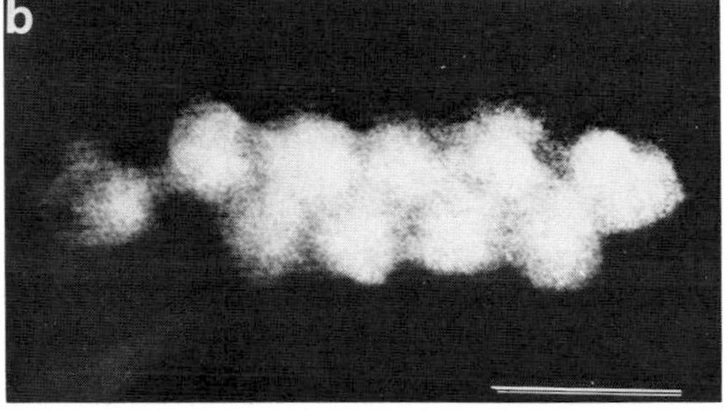

Figure 4. Comparison of images of eucaryotic polysomes as obtained by TEM, stained and air dried (a), and by STEM, unstained and freeze-dried (b). Arrowhead in panel a points to the interface. f, A particle that might be a protein factor. Bar, 0.05 μm.

Figure 5 represents a comparison of 16S rRNA images obtained by STEM (Mandiyan et al., 1988) and by TEM, using the routine spreading techniques with benzylalkylammonium chloride (Sogo et al., 1979). The difference in length of 16S rRNA molecules in STEM (Fig. 5a) and in TEM (Fig. 5b) images reflects the mode of specimen deposition on the support carbon film. 16S rRNA spread on a water hypophase and stretched under denaturing conditions appears as a single strand ~5,000 Å long (Fig. 5b), in agreement with the number of nucleotides (1,542) and the internucleotide distance derived from 2.8-Å axial rise per residue for A-form RNA (Saenger, 1984). The length of the 16S rRNA molecules in STEM images (Fig. 5a) obtained in the absence of denaturing agents and without any stretching forces is ~1,200 Å, only about one-fourth of that in the fully extended state. The apparently reduced length and the high average linear density (M/L) of the 16S rRNA molecules are consistent with the existence of hairpins and loops in the rRNA native secondary structure as established by a variety of techniques (for references, see Noller and Moldave, 1988). The difference in filament thickness is caused by deposition of benzylalkylammonium chloride and tungsten on the RNA molecules. This step obscures the fine structures resolvable on STEM images of unstained freeze-dried RNA molecules (Fig. 5a) and excludes TEM from the possibility for

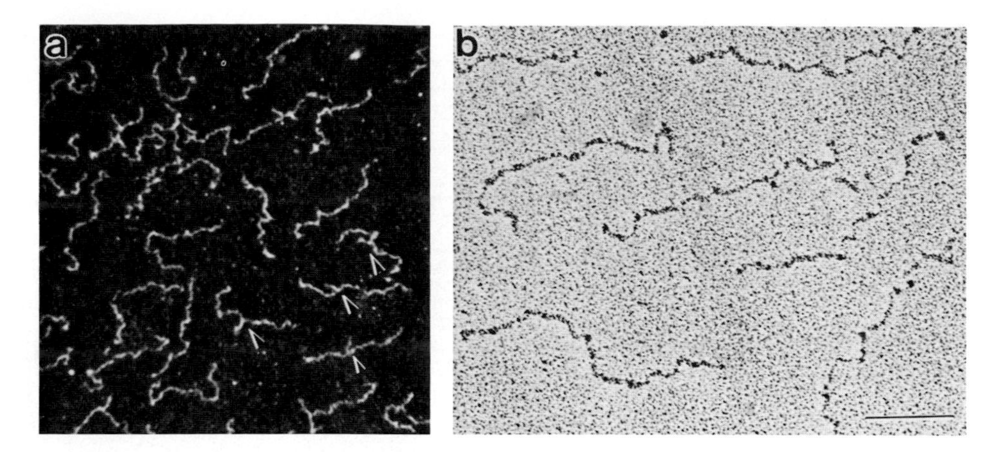

Figure 5. Electron micrographs of *E. coli* 16S rRNA. (a) Obtained by STEM, unstained and freeze-dried. Arrowheads point to the protrusion at ca. two-fifths of the RNA molecule. (b) Obtained by TEM and the conventional benzylalkylammonium dichloride-monolayer technique (Sogo et al., 1979), denatured, air-dried, and shadowed with PtPd. Deposition in both cases is from distilled water. Bar, 0.1 μm.

quantitative mass determinations. The total molecular mass of 16S rRNA as determined by STEM is 551 ± 22 kDa, in excellent agreement with the theoretical data (550 kDa). The average M/L for the 16S RNA molecules in Fig. 5a is about 480 Da/Å. This value is about four times higher than that expected for single-stranded RNA in the denatured extended form and indicates a close association of four single-stranded polynucleotide strands in the main backbone of the observed molecules. The linear density appears to be evenly distributed along the whole molecule except for a short bifurcation seen at the ends of some molecules and a small protrusion ~80 Å long with a mass of ~60 kDa (arrowhead in Fig. 5a). On the basis of the M/L value (750 Da/Å), this protrusion could be composed of six polynucleotide strands. The asymmetric location of the protrusion at about two-fifths of the length of the 16S RNA molecule can be used to determine the polarity of the molecule. However, identification of the 3′ and 5′ ends would require linking of a marker to either end of the 16S rRNA molecule.

Mass and R_g determination proved to be crucial for the characterization of ribosomal subunits, RNA molecules, and the changes in RNA due to the variations of ambient ionic strength. The strong ionic dependence of 16S rRNA is reflected in the conformational states visualized by STEM (Fig. 6). Molecules of 16S rRNA in a solution of very low ionic strength [1.7 μM HEPES (*N*-2-hydroxyethylpiperazine-N′-2-ethanesulfonic acid) · KOH (pH 7.5), 10 μM KCl, 0.34 μM magnesium acetate [$Mg(OAc)_2$]; buffer B, Fig. 6a] appear similar to those molecules deposited in distilled water (Fig. 5a). The molecules are shorter by about 10% but remain extended, their branching pattern becomes more complex, and the R_g value is reduced by about 30% (Table 1). Increasing ionic strength induces further RNA coiling (Fig. 6b) and reduction of R_g. Molecules of 16S rRNA deposited in reconstitution buffer R [30 mM HEPES · KOH (pH 7.5), 330 mM KCl, 20 mM $Mg(OAc)_2$, 1 mM dithiothreitol] appear very tightly coiled (Fig. 6c). However, even under these conditions their R_g is still substantially larger (80 ± 10 Å) than that of the native 30S subunits (68 ± 1.4 Å). These results are in agreement with spectroscopic data (Hill et al., 1969; Tam et al., 1981) indicating that the applied procedure of specimen deposition and freeze-drying did not cause any appreciable structural distortion of 16S rRNA. The value of R_g for 16S rRNA determined by STEM in buffer R (Table 1) is close to the values obtained by neutron scattering (84 Å) and X-ray scattering (86 Å) under similar buffer conditions (Vasiliev et al., 1986). Apart from being tightly coiled, the 16S rRNA molecules in buffer R do not display any characteristic features that would facilitate direct structural comparison with the proposed models for the tertiary structure of 16S rRNA (Brimacombe et al., 1988; Stern et al., 1989), nor do they show any resemblance to the 30S subunits (Fig. 6d). Instead, they appear as tightly packed cores with protruding filaments with no obvious symmetry and no resemblance to the V or Y shapes observed in TEM of shadowed and specially treated specimens (Vasiliev et al., 1986; Shatsky and Vasiliev, 1988).

Interactions of rRNAs with Ribosomal Proteins: Subunit Assembly

Success in protein-free deposition of rRNAs in selected buffers under nondenaturing conditions

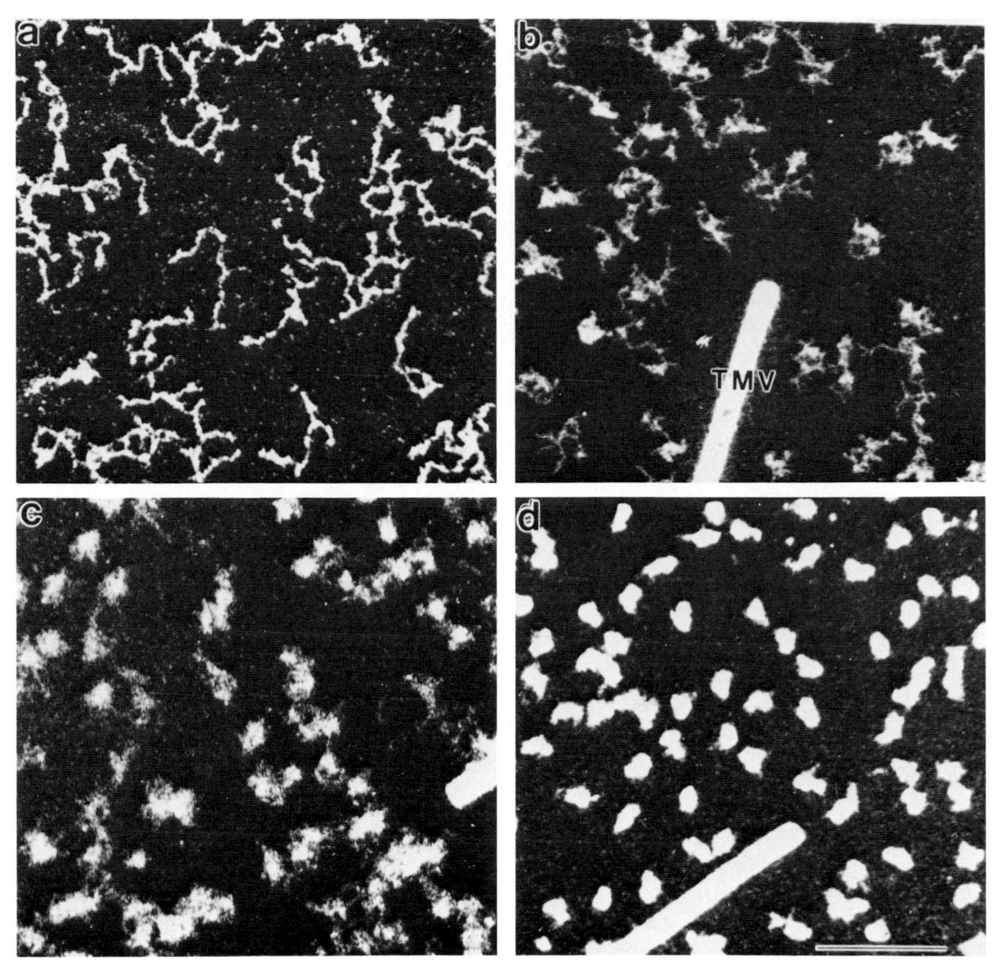

Figure 6. STEM images of 16S rRNA in buffer B (a), buffer A (ribosomal) (b), and buffer R (reconstitution) (c); (d) 30S subunits in buffer A. Tobacco mosaic virus (TMV) was used as the standard for mass determination. Bar, 0.1 μm.

opened the field of electron microscopy to high-resolution imaging of rRNAs and to interactions of rRNA with other biological macromolecules, ribosomal proteins in particular. This approach makes it possible to visualize and quantitatively analyze the conformational changes induced in 16S rRNA by interactions with ribosomal proteins S4, S8, S15, S20, S17, and S7, the six proteins known to bind to 16S rRNA in the initial assembly steps (Nomura and Held, 1974). The reconstituted core particles are characterized by their mass, morphology, R_g values, and the extent and stability of their 16S rRNA secondary structure, which is monitored by circular dichroism spectroscopy (Mandiyan et al., 1989). The stepwise binding of S4, S8, and S15 leads to a corresponding increase of mass and is accompanied by increased folding of 16S rRNA in the core particles, as is evident from the electron micrographs (Fig. 7) and from the decrease of R_g values from 114 to 91 Å (Table 2). Although the binding of S20, S17, and S7 continues the trend of mass increase, the R_g values of these core particles exhibit a variable trend. Addition of S7 to the core particles leads to the formation of a globular mass cluster with a diameter of about 115 Å and a mass of about 300 kDa. The rest of the mass (about 330 kDa) remains loosely coiled, giving the core particle a "medusalike" appearance (Fig. 7f). The morphology of the 16S rRNA-protein core particles, even those containing all of the primary binding proteins, does not resemble that of the native 30S

Table 1. R_g values of *E. coli* 16S and 23S rRNAs in various solutions (calculated from STEM data)

Solution[a]	R_g (Å) ± SD	
	16S rRNA	23S rRNA
Water	305 ± 50	566 ± 67
Buffer B	213 ± 33	371 ± 73
Buffer A	114 ± 20	142 ± 22
Buffer R	80 ± 10	115 ± 16
Buffer A + 0.5 mM Mg^{2+}	136 ± 20	
Buffer A + 10 mM Mg^{2+}	99 ± 10	

[a] Buffers B and R are defined in the text; buffer A is 6,000× buffer B.

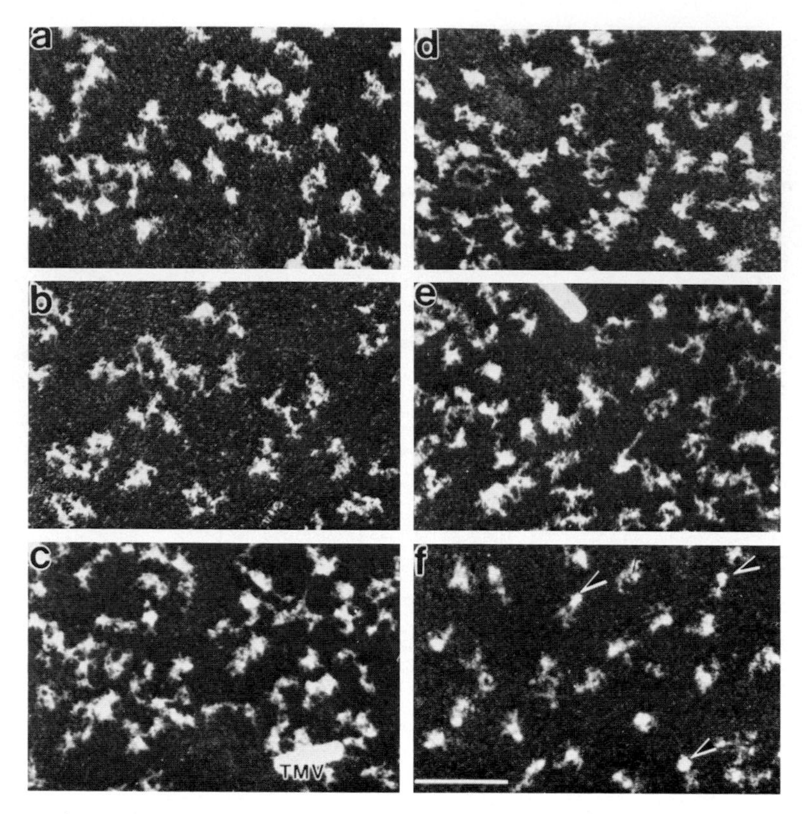

Figure 7. STEM images of 16S rRNA associated sequentially with ribosomal primary binding proteins. (a) S4; (b) S4 and S8; (c) S4, S8, and S15; (d) S4, S8, S15, and S20; (e) S4, S8, S15, S20, and S17; (f) S4, S8, S15, S20, S17, and S7, unstained, freeze-dried, and deposited from buffer A. Arrows in panel f point to the globular mass center of the medusalike coil particle. Tobacco mosaic virus (TMV) was added as an internal standard for mass measurements. Bar, 0.1 μm.

subunit, contrary to what has been reported by others (Vasiliev et al., 1986).

Experiments are in progress with the full set of ribosomal proteins (V. Mandiyan et al., unpublished data). The results so far obtained provide direct evidence that 16S rRNA undergoes significant structural reorganization during the 30S subunit assembly and that perhaps the full complement of proteins is essential for proper folding of the 16S rRNA into a functional 30S subunit.

Table 2. Molecular masses and R_g values of *E. coli* rRNA preparations in buffer A

Prepn	Mass (kDa)[a]	Mass ± SD (kDa)	R_g ± SD (Å)
16S rRNA	550	551 ± 22	114 ± 20
+ S4	573	574 ± 25	102 ± 13
+ S4 + S8	587	586 ± 21	97 ± 14
+ S4 + S8 + S15	597	592 ± 26	91 ± 14
+ S4 + S8 + S15 + S20	606	615 ± 24	94 ± 13
+ S4 + S8 + S15 + S20 + S17	615	621 ± 24	94 ± 12
+ S4 + S8 + S15 + S20 + S17 + S7	635	625 ± 25	108 ± 11
30S *E. coli*	900	880 ± 40	68 ± 1.4

[a] Theoretical value, assuming stoichiometric binding of the proteins.

PERSPECTIVES

In future studies of ribosomes, the emphasis will be on maximizing resolution of electron microscopic imaging and quantitative analysis of the data. This trend should be pursued at the level of instrumentation, specimen preservation, and computer image processing.

The resolution should be improved by reducing the radiation dose and by broadening the spectrum of high-resolution electron spectroscopic techniques, e.g., electron energy loss spectroscopy, which provides highly specific and quantitative information on mass and element distribution on the ribosome (Boublik et al., 1986a). By taking advantage of the unique presence of phosphorus atoms in the structural backbone, as well as the potential of STEM to resolve the difference between the elastic and inelastic scattering (Z contrast), it is possible to localize rRNA in ribosomal subunits (unpublished data).

A major effort will be directed toward preservation of the native structure of ribosomes and their components. The most promising and equally challenging approach seems to be imaging of frozen hydrated specimens embedded in vitreous ice. In topographical mapping of ribosomal proteins, rRNAs, and functional domains, substantial progress in resolution and specificity is expected from the application of cluster compounds, i.e., small organic complexes with regularly spaced heavy atoms (Au, Pt, and W) attached to specific sites on the ribosome (Wall et al., 1982).

Continuous progress in the development of computer programs for analysis of high-resolution electron micrographs of biological specimens is reflected in three-dimensional reconstruction of ribosomal particles from their TEM images from a wide range of projections (Frank et al., 1988; Frank et al., this volume). Success of X-ray crystallography, the only direct technique introduced recently for determination of the complete three-dimensional structure of the ribosome (Yonath and Wittmann, 1988), will depend primarily on the availability of large well-ordered crystals of ribosomes or ribosomal subunits. Important improvements in the resolution of this demanding technique include the use of synchrotron radiation and the prevention of crystal radiation damage by low temperatures (Hope et al., 1989).

Application of the aforementioned approaches should clarify the existing controversies in the structure of ribosomes, both procaryotic and eucaryotic, and provide deeper insight into ribosome involvement in the process of protein synthesis. Detailed knowledge of ribosome architecture will be instrumental for establishing objective criteria for use of ribosomes and their components as phylogenetic probes.

We gratefully acknowledge the excellent assistance of W. Hellmann, G. Shiue, M. Simon, and F. Kito. The expertise of J. S. Wall and J. F. Hainfeld from the Brookhaven STEM Biotechnology Resource was crucial for the advancement of electron microscopic imaging of ribosomes and their components to a quantitative analytical base.

REFERENCES

Boublik, M. 1990. Electron microscopy of ribosome. *In* G. Spedding (ed.), *Ribosomes and Protein Synthesis: a Practical Approach.* IRL Press, Oxford.

Boublik, M., and W. Hellmann. 1978. Comparison of *Artemia salina* and *Escherichia coli* ribosome structure by electron microscopy. *Proc. Natl. Acad. Sci. USA* 75:2829–2833.

Boublik, M., G. T. Oostergetel, D. C. Joy, J. S. Wall, J. F. Hainfeld, B. Frankland, and P. F. Ottensmeyer. 1986a. *In situ* localization of ribonucleic acids in biological specimens by electron energy loss spectroscopy. *Ann. N.Y. Acad. Sci.* **463:** 168–170.

Boublik, M., G. T. Oostergetel, V. Mandiyan, J. F. Hainfeld, and J. S. Wall. 1988. Structural analysis of ribosomes by scanning transmission electron microscopy. *Methods Enzymol.* **164:**49–63.

Boublik, M., G. T. Oostergetel, J. S. Wall, J. F. Hainfeld, M. Radermacher, T. Wagenknecht, A. Verschoor, and J. Frank. 1986b. Structure of ribosomes and their components by advanced techniques of electron microscopy and computer image analysis, p. 68–86. *In* B. Hardesty and G. Kramer (ed.), *Structure, Function, and Genetics of Ribosomes.* Springer-Verlag, New York.

Boublik, M., and S. Ramagopal. 1980. Conformation of ribosomes from the vegetative amoebae and spores of *Dictyostelium discoideum. Mol. Gen. Genet.* **179:**483–488.

Brimacombe, R., J. Atmadja, W. Stiege, and D. Schüller. 1988. A detailed model of the three-dimensional structure of *Escherichia coli* 16S ribosomal RNA *in situ* in the subunit. *J. Mol. Biol.* **199:**115–136.

Frank, J., M. Radermacher, T. Wagenknecht, and A. Verschoor. 1988. Studying ribosome structure by electron microscopy and computer-image processing. *Methods Enzymol.* **164:**3–35.

Glitz, D. G., P. A. Cann, L. S. Lasater, and H. McKuskie Olson. 1988. Antibody probes of ribosomal RNA. *Methods Enzymol.* **164:**493–503.

Hardesty, B., and G. Kramer (ed.). 1986. *Structure, Function, and Genetics of Ribosomes.* Springer-Verlag, New York.

Hill, W. E., S. D. Thompson, and S. W. Anderegg. 1969. X-ray scattering study of ribosomes from *Escherichia coli. J. Mol. Biol.* **44:**89–102.

Hope, H., F. Frolow, K. von Böhlen, I. Makowski, C. Kratky, Y. Halfon, H. Danz, P. Webster, K. S. Bartels, H. G. Wittmann, and A. Yonath. 1989. Cryocrystallography of ribosomal particles. *Acta Crystallogr.* **845:**190–199.

Keren-Zur, M., M. Boublik, and J. Ofengand. 1979. Localization of the decoding region of the 30S *Escherichia coli* ribosomal subunit by affinity immunoelectron microscopy. *Proc. Natl. Acad. Sci. USA* **76:**1054–1058.

Lake, J. A. 1978. Electron microscopy of specific proteins: three dimensional mapping of ribosomal proteins using antibody labels in advanced techniques, p. 173–211. *In* S. K. Koehler (ed.), *Biological Electron Microscopy*, vol. 2. Springer-Verlag KG, Berlin.

Mandiyan, V., S. Tumminia, J. S. Wall, J. F. Hainfeld, and M. Boublik. 1989. Protein-induced conformational changes in 16S rRNA during the initial assembly steps of *Escherichia coli* 30S ribosomal subunit. *J. Mol. Biol.* **210:**323–336.

Mandiyan, V., J. S. Wall, J. F. Hainfeld, and M. Boublik. 1988. Conformation analysis of 16S ribosomal RNA from *Escherichia coli* by scanning transmission electron microscopy. *FEBS Lett.* **236:**340–344.

Noller, H. F., and K. Moldave (ed.). 1988. Ribosomes. *Methods Enzymol.* **164.**

Nomura, M., and W. A. Held. 1974. Reconstitution of ribosomes: studies of ribosome structure, function and assembly, p. 193–224. *In* M. Nomura, A. Tissières, and P. Lengyel (ed.), *Ribosomes.* Cold Spring Harbor Laboratory, Cold Spring Harbor, N.Y.

Oakes, M. I., M. W. Clark, E. Henderson, and J. A. Lake. 1986. DNA hybridization electron microscopy: ribosomal RNA molecules 1392–1407 are exposed in the cleft of the small subunit. *Proc. Natl. Acad. Sci. USA* **83:**275–279.

Oostergetel, G. T., J. S. Wall, J. F. Hainfeld, and M. Boublik. 1985. Quantitative structural analysis of eukaryotic ribosomal RNA by scanning transmission electron microscopy. *Proc. Natl. Acad. Sci. USA* **82:**5598–5602.

Palade, G. E. 1955. A small particulate component of the cytoplasm. *J. Biophys. Biochem. Cytol.* **1:**59–68.

Saenger, W. 1984. *Principles of Nucleic Acid Structure.* Springer-

Verlag, New York.

Serdyuk, I. N., S. C. Agalarov, S. E. Sedelnikova, A. S. Spirin, and R. P. May. 1983. Shape and compactness of the isolated ribosomal 16S RNA and its complexes with ribosomal proteins. *J. Mol. Biol.* **169:**409–425.

Shatsky, I. N., and V. D. Vasiliev. 1988. Electron microscopy studies of ribosomal RNA. *Methods Enzymol.* **164:**76–91.

Sogo, J. M., P. Rodenio, T. Koller, E. Vinuela, and M. Salas. 1979. Comparison of the A-T rich regions and the *Bacillus subtilis* RNA polymerase binding sites in phage ϕ29 DNA. *Nucleic Acids Res.* 7:107–120.

Stern, S., T. Powers, L.-M. Changchien, and H. F. Noller. 1989. RNA-protein interactions in 30S ribosomal subunits. Folding and function of 16S rRNA. *Science* **244:**783–790.

Stöffler, G., and M. Stöffler-Meilicke. 1986. Immunoelectron microscopy on *Escherichia coli* ribosomes, p. 28–45. *In* B. Hardesty and G. Kramer (ed.), *Structure, Function, and Genetics of Ribosomes.* Springer-Verlag, New York.

Stöffler-Meilicke, M., and G. Stöffler. 1988. Localization of ribosomal proteins on the surface of ribosomal subunits from *Escherichia coli* using immunoelectron microscopy. *Methods Enzymol.* **164:**503–520.

Tam, M. F., J. A. Dodd, and W. E. Hill. 1981. Physical characteristics of 16S rRNA under reconstitution conditions. *J. Biol. Chem.* **256:**6430–6434.

Vasiliev, V. D., I. N. Serdyuk, A. T. Gudkov, and A. S. Spirin. 1986. Self-organization of ribosomal RNA, p. 128–142. *In* B. Hardesty and G. Kramer (ed.), *Structure, Function, and Genetics of Ribosomes.* Springer-Verlag, New York.

Wall, J. S. 1979. Biological scanning transmission electron microscopy, p. 333–342. *In* J. J. Hren, J. I. Goldstein, and D. C. Joy (ed.), *Introduction to Analytical Electron Microscopy.* Plenum Publishing Corp., New York.

Wall, J. S., and J. F. Hainfeld. 1986. Mass mapping with the scanning transmission electron microscope. *Annu. Rev. Biophys. Biophys. Chem.* **15:**355–376.

Wall, J. S., J. F. Hainfeld, P. A. Bartlett, and S. J. Singer. 1982. Observation of an undecagold cluster compound in the scanning transmission electron microscope. *Ultramicroscopy* **8:**397–402.

Yonath, A., and H. G. Wittmann. 1988. Crystallographic and image reconstruction studies on ribosomal particles from bacterial sources. *Methods Enzymol.* **164:**95–116.

Chapter 7

Topography of the Ribosomal Proteins from *Escherichia coli* within the Intact Subunits as Determined by Immunoelectron Microscopy and Protein-Protein Cross-Linking†

MARINA STÖFFLER-MEILICKE and GEORG STÖFFLER

When in 1971 antisera were raised for the first time against each of the 53 individual proteins of the 70S ribosome of *Escherichia coli*, no immunological cross-reaction could be detected among the 21 proteins of the 30S subunit (Stöffler and Wittmann, 1971a). The proteins of the 50S subunit were also distinct, with the exception of proteins L7 and L12 (Stöffler and Wittmann, 1971b). Apart from proteins S20 and L26, which were shown to be immunologically identical, no other ribosomal protein of the 30S subunit was related to any protein of the 50S subunit or vice versa. Already at that time we concluded that the ribosomal proteins from *E. coli* had distinct primary structures, with the exception of the two protein pairs L7/L12 and S20/L26, for which similar if not identical primary structures were anticipated. Elucidation of the complete amino acid sequences of all 53 proteins of the *E. coli* ribosome by Wittmann-Liebold and co-workers during the following decade confirmed the immunological results (Wittmann-Liebold, 1986).

Since then, immunochemical methods have yielded considerable information on the structure and function of ribosomes. Antibodies have been used to identify the members of cross-linked protein complexes (e.g., Lutter et al., 1972) and to identify ribosomal proteins either cross-linked to specific RNA sites (Gulle et al., 1987) or attached to ligands such as mRNA, tRNA, protein synthesis factors, or antibiotics that interact with the ribosome during protein biosynthesis (e.g., Hauptmann et al., 1974). The contribution that individual ribosomal proteins make to the various steps of ribosome function during initiation, elongation, and termination has been determined by inhibition studies using protein-specific antibodies (Lelong et al., 1974; Lelong et al., 1979). A wealth of data has come from immunoelectron microscopy (IEM). This technique has yielded information on the topography of ribosomal proteins and specific RNA sites on the ribosomal surface (for references, see Stöffler and Stöffler-Meilicke, 1984), on the arrangement of the ribosomal subunits within the 70S monosome (Kastner et al., 1981; Lake, 1982), and on the location of functional domains (for an overview, see Stöffler and Stöffler-Meilicke, 1984). In this chapter, we review the information obtained by both IEM and immunological identification of protein-protein cross-links with regard to the location of ribosomal proteins of *E. coli* in situ.

IEM

Information on the topography of ribosomal proteins on the surface of ribosomal subunits has been obtained almost exclusively from IEM. A prerequisite for the localization of antibody-binding sites on the ribosomal surface is knowledge of the three-dimensional ribosome structure, which has been derived from the various projections of the ribosomal subunits as seen on electron micrographs. Three-dimensional models of the *E. coli* ribosome have been proposed by different research groups, and the differences among the various models have been discussed extensively (Wittmann, 1986). There is now fairly

† Dedicated to Heinz-Günter Wittmann for his contribution to the understanding of ribosome structure and function.

Marina Stöffler-Meilicke ■ Institut für Klinische und Experimentelle Virologie, Freie Universität Berlin, Hindenburgdamm 27, D-1000 Berlin 45, Federal Republic of Germany. **Georg Stöffler** ■ Institut für Mikrobiologie, Medizinische Fakultät, Universität Innsbruck, Fritz-Pregl-Strasse 3, A-6020 Innsbruck, Austria.

good agreement as to the gross shape of the *E. coli* ribosome and its subunits; taking into account the limitations intrinsic to electron microscopy, the differences among the models are not major and should therefore not be overemphasized.

A further prerequisite for IEM localization of a ribosomal protein is that the protein concerned have at least one antigenic determinant accessible for antibody binding in the intact ribosomal subunit. As long ago as 1973, it was shown by several methods that most proteins of the 30S and 50S subunits have antigenic determinants exposed at the ribosomal surface (Stöffler et al., 1973; Morrison et al., 1973). These results invalidated the proposal that many ribosomal proteins were part of a nucleoprotein core buried within the ribosomal subunit.

Sucrose gradient centrifugation was the most practicable method to separate the immunocomplexes from excess immunoglobulin G and unreacted subunits, and these complexes were stable enough to be used for specimen preparation. The first ribosomal proteins were localized on the surface of the 50S subunit of *E. coli* in our laboratory by 1974 (Wabl, 1974). Shortly thereafter, a number of ribosomal proteins of both the 30S and 50S subunits were localized by us (Tischendorf et al., 1974a, 1974b) and by Lake and co-workers (1974). In these first experiments, epitopes widely separated on the ribosomal surface were described for some ribosomal proteins, e.g., protein S4, and these results were interpreted in terms of ribosomal proteins being elongated in situ (Lake et al., 1974; Tischendorf and Stöffler, 1975). However, it soon became clear that these results were artifacts due to contaminating antibodies. Since then, more sophisticated control experiments have been developed that not only demonstrate the specificity of the reactive antibody for the protein to be localized but also exclude cross-reactivity of the antibody with other ribosomal proteins (Dabbs et al., 1981; Stöffler and Stöffler-Meilicke, 1984).

TOPOGRAPHY OF THE RIBOSOMAL PROTEINS OF THE 30S SUBUNIT AS DETERMINED BY IEM

Electron micrographs of small ribosomal subunits of *E. coli*, reacted with anti-S17, are shown in Fig. 1 to illustrate the procedure of localizing an antibody-binding site in three dimensions. Altogether, 19 of the 21 proteins of the 30S subunit of *E. coli* have been localized on the ribosomal surface by IEM, and 18 of these have been localized in our laboratory (Stöffler-Meilicke and Stöffler, 1984, 1987; Walleczek, 1988; Fig. 2a to d). Except in the case of protein S8, we have performed all of the control experiments necessary to prove the specificity of the antibody used. Most of the locations have been derived by using polyclonal antibodies. In a comparative study, proteins S3 and S7 were localized by using monoclonal and polyclonal antibodies; the antibody-binding sites determined with the two monoclonal antibodies were found to lie within the same area as those obtained with the polyclonal antibodies (Breitenreuter et al., 1984). Protein S2 has been mapped exclusively with monoclonal antibody (Schwedler-Breitenreuter et al., 1985).

In another approach, proteins S4 and S17 were specifically labeled with fluorescein, reconstituted into intact subunits, and subsequently localized with hapten-specific antibodies at the same sites as with protein-specific antibody (Stöffler-Meilicke et al., 1983a; Stöffler-Meilicke et al., 1984). This approach, which excludes errors due to antibody contamination from the outset, was later used by two other groups for the localization of protein S19 (Lin et al., 1984; Olson et al., 1988).

To prevent dissociation of S1 from the ribosome during the localization procedure, the protein was covalently attached to the 30S subunit, using dithiobis(succinimidylpropionate) as a cross-linking reagent. Using fragment-specific antibodies, it was shown that epitopes within the N-terminal part of S1 (amino acids 1 to 193) were located at the large lobe of the 30S subunit, close to the one-third/two-thirds partition, on the side that faces the cytoplasm in the 70S ribosome (Walleczek, 1988; Fig. 2a). Results obtained with monovalent Fab fragments specific for the C-terminal part of S1 (amino acids 332 to 547) suggested that the latter portion of the protein may form a lassolike structure extending at least 10 nm away from the lobe region where the corresponding N-terminal portion has been localized (J. Walleczek, R. Albrecht-Ehrlich, G. Stöffler, and M. Stöffler-Meilicke, submitted for publication). Further experiments using monoclonal antibodies against defined epitopes of S1 will be necessary to determine precisely how antigenic determinants of this large protein are arranged on the surface of the small ribosomal subunit.

Data on the IEM localization of ribosomal proteins from the 30S subunit of *E. coli* have also been obtained by other research groups. Lake and co-workers have localized 11 proteins of the 30S subunit: S3, S4, S5, S6, S8, S10, S11, S12, S13, S14, and S19 (Lake et al., 1974; Lake and Kahan, 1975; Kahan et al., 1981; Winkelmann et al., 1982). The agreement between the two sets of data is good except for the location of S19, which Lake and

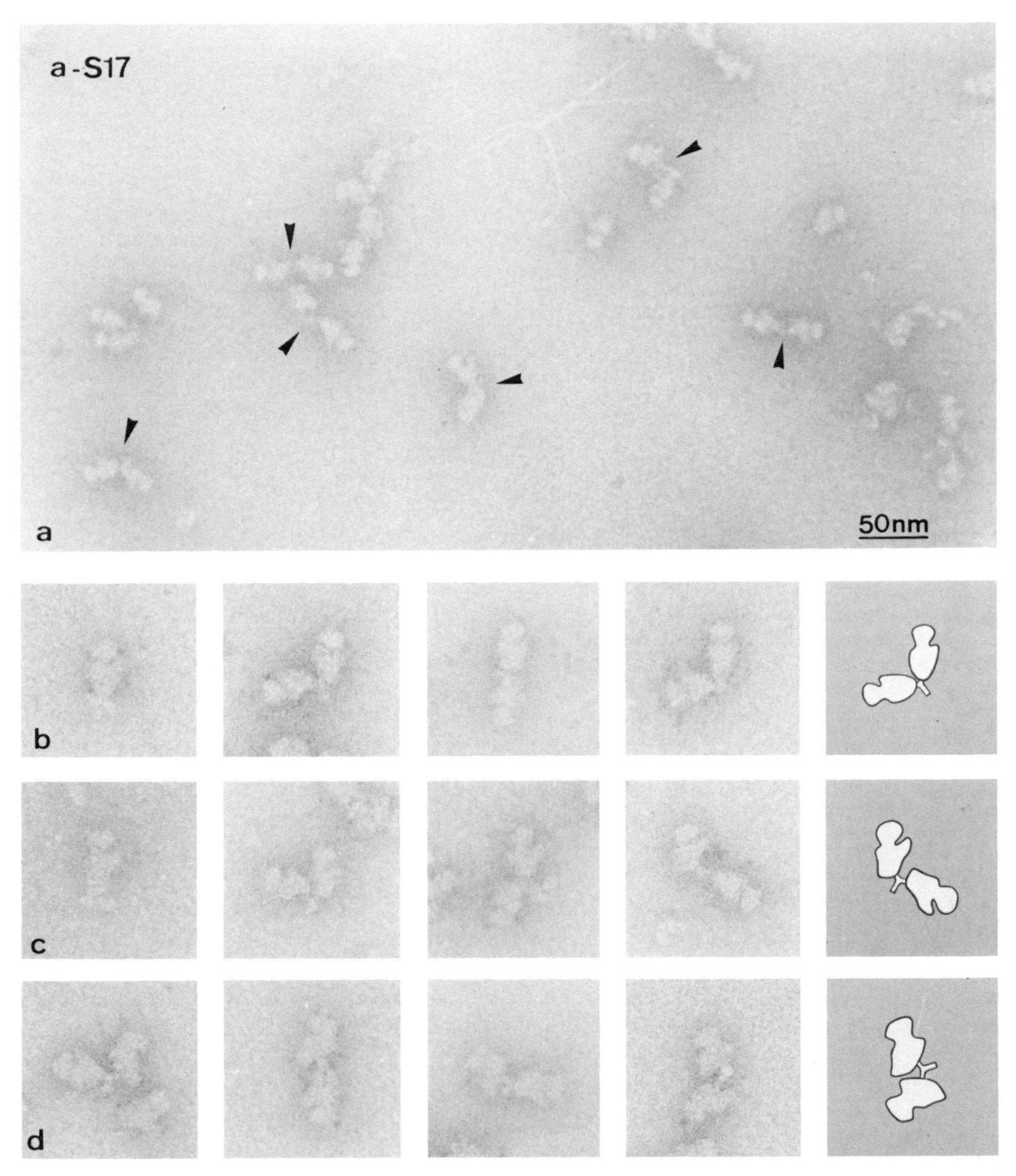

Figure 1. Electron micrographs of 30S subunits from *E. coli*, reacted with anti-S17. (a) General field; (b to d) selected views. Each interpretative scheme relates to the micrograph on the immediate left and illustrates antibody binding to the quasisymmetric (b), the cloven asymmetric (c), and the angled asymmetric (d) projections.

Kahan (1975) mapped at two sites on top of the head of the 30S subunit. One of these sites was close to protein S14 and the other was close to S13, implying that the protein, which consists of 91 amino acids, has to span a distance of almost 80 Å (8 nm). We localize protein S19 at a single site on the interface side of the 30S subunit (Fig. 2d). Support for a location of S19 at this same site also comes from the work of Lin et al. (1984) and Olson et al. (1988). The electron micrographs published by Lin and co-workers, who labeled protein S19 in situ, using Phe-tRNA that had been modified with a dinitrophenyl group as a photoaffinity probe, are indistinguishable from ours (compare Fig. 5 of Lin et al. [1984] and Fig. 4 of Stöffler-Meilicke and Stöffler [1987]). Olson et al. (1988) labeled protein S19 at its amino-terminal proline residue with a dinitrophenyl group, reconstituted the modified protein into 30S subunits, and localized the protein with hapten-specific antibodies. Although their electron micrographs differ slightly from ours, these authors come to a very similar conclusion concerning the three-dimensional location of protein S19, namely, that the protein is located at a single site on the interface side of the 30S

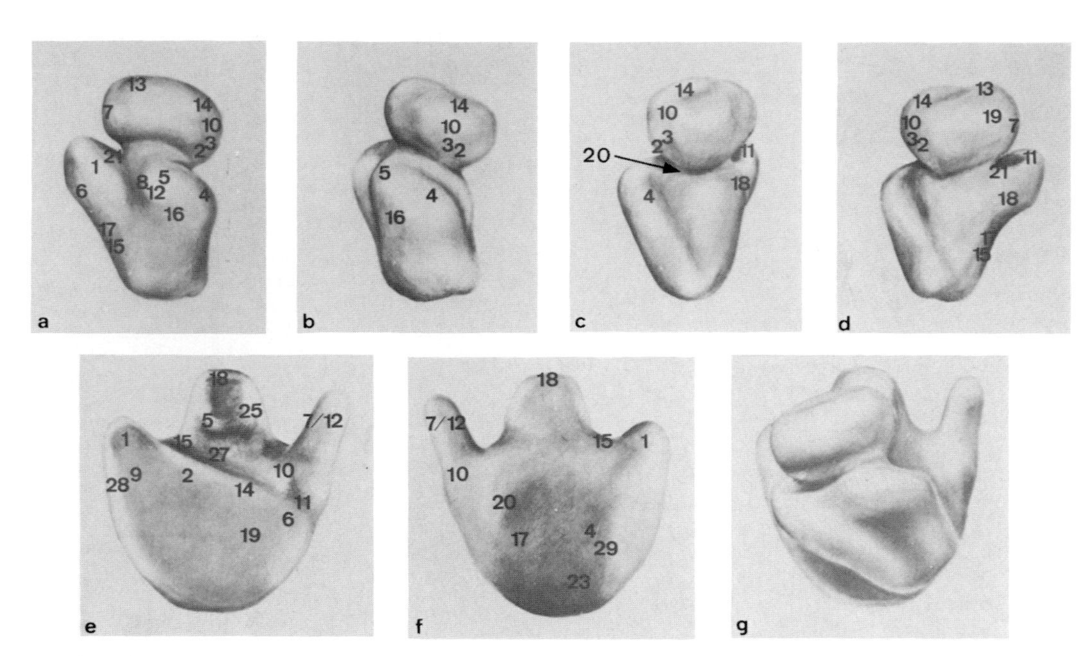

Figure 2. Three-dimensional model of the 30S (a to d), 50S (e and f), and 70S (g) ribosomal particles of *E. coli*. Numbers give locations of the centers of antibody-binding sites for individual ribosomal proteins (Dabbs et al., 1981; Lotti et al., 1983; Lotti et al., 1987; Lotti et al., 1989; Stöffler-Meilicke et al., 1983a; Stöffler-Meilicke et al., 1983b, Stöffler-Meilicke et al., 1984; Stöffler-Meilicke et al., 1985; Stöffler et al., 1984a; Stöffler et al., 1984b; Tate et al., 1984; Breitenreuter et al., 1984; Stöffler-Meilicke and Stöffler, 1984, 1987; Schwedler-Breitenreuter et al., 1985). For protein S1, the location of the N-terminal 193 amino acids is given (Walleczek, 1988). The locations given for proteins S8 and L28 are preliminary, since the necessary control experiments have not been performed. The location of protein S12 is taken from Winkelmann et al. (1982). The arrow indicates the location of protein S20, as deduced from neutron scattering (Capel et al., 1987). According to protein-protein cross-linking (see Table 1) and neutron scattering (Capel et al., 1987), protein S9 is located within the head of the 30S subunit.

subunit head, which in the 70S ribosome is close to the peptidyltransferase center (Fig. 2).

Moore and collaborators have used neutron scattering to measure distances between pairs of deuterated proteins and have placed all 21 proteins of the small ribosomal subunit in a three-dimensional matrix (Capel et al., 1987). In general, there is very good agreement between their results and the 19 proteins localized by IEM. There are some minor discrepancies on the locations of proteins S1, S2, S16, and S19, but these discrepancies could easily be explained by the differences inherent in the principles of the two methods.

Chemical cross-linking of proteins within the large and the small subunits from *E. coli* has been used for many years as a tool to determine protein-protein neighborhoods within the ribosome (Bickle et al., 1972; Lutter et al., 1972), and most of the cross-links reported for both the 30S and 50S subunits have been obtained by using 2-iminothiolane as cross-linking reagent (Traut et al., 1980; Traut et al., 1986). Moore and collaborators compared both their neutron map and our IEM map with the cross-links obtained by Traut and co-workers and (because of the many discrepancies between the two sets of data) came to the conclusion that cross-linking data "are best viewed with caution" (Moore et al., 1986). It seemed probable to us that many of the discrepancies could have arisen from misidentification of the members of the cross-linked protein complexes, and we have accordingly reinvestigated the cross-links formed by 2-iminothiolane and dimethylsuberimidate within the 30S ribosomal subunit of *E. coli*. For an unambiguous identification of the members of the cross-linked protein complexes, immunoblotting techniques were used. The principle of this identification and quantitation of the yield of cross-linking have been described in detail elsewhere (Stöffler et al., 1988; Walleczek et al., 1989b). Examples of immunoblots made to identify the members of the cross-linked protein pairs are shown in Fig. 3 and 4.

The cross-links identified in the 30S subunit treated with 2-iminothiolane or dimethylsuberimidate are summarized in Table 1. The agreement of our cross-linking data with results for both IEM and neutron scattering is very good; proteins that were cross-linked to each other are also close neighbors in both the neutron-scattering and IEM maps (compare Table 1 and Fig. 2a to d). This finding reestablishes the cross-linking method as a powerful technique for obtaining information on neighborhoods among ribosomal proteins, provided that the members of the cross-linked pairs are analyzed by sufficiently reliable methods. In this context, the validity of the conclusions from many cross-linking studies performed with procaryotic as well as with eucaryotic ribosomes (e.g., Uchiumi et al., 1985; Yeh et al., 1986) must be questioned, because all of these studies used only the electrophoretic properties of the ribosomal proteins to identify the members of a cross-link.

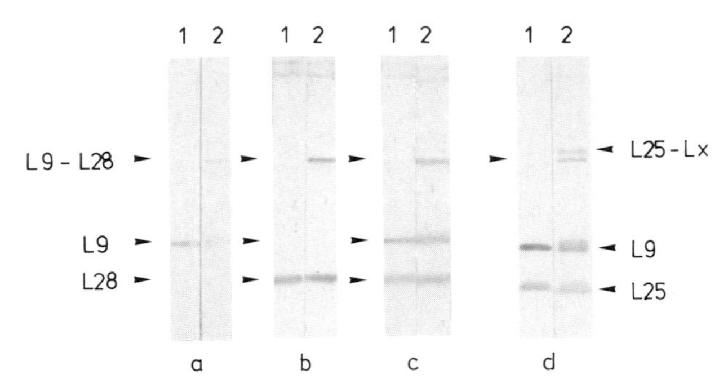

Figure 4. Identification of the cross-link L9-L28. Immunoblots obtained from one-dimensional sodium dodecyl sulfate-polyacrylamide gels were incubated with anti-L9 (a), anti-L28 (b), anti-L9 and anti-L28 simultaneously (c), and anti-L9 and anti-L25 simultaneously (d). Lanes 1, TP50 extracted from untreated 50S subunits; 2, TP50 extracted from subunits cross-linked with dimethylsuberimidate. The immunoblots obtained for the two members of one cross-linked complex reveal high-M_r protein bands with identical M_r values (compare lanes 2 in panels a and b). Incubation of a blot with the two antisera simultaneously (c) gives a single high-M_r protein band and thus confirms the assumption that the two proteins, L9 and L18, have been cross-linked to each other. In contrast, incubation of an immunoblot with anti-L9 and anti-L25 simultaneously (d) gives two distinct cross-link bands, thus excluding the possibility that L25 had been cross-linked to L9.

TOPOGRAPHY OF THE RIBOSOMAL PROTEINS OF THE 50S SUBUNIT AS DETERMINED BY IEM

Figure 5 shows electron micrographs of 50S subunits from *E. coli* reacted with anti-L6 to illustrate how a ribosomal protein is localized in three dimensions on the surface of the 50S subunit. Altogether, 20 ribosomal proteins have been localized on the surface of bacterial 50S subunits (Fig. 2e and f). Antibodies specific for the recently identified proteins L35 and L36 were not available. Nineteen of the locations have been obtained on 50S subunit from *E. coli*, and for all of them the necessary controls of antibody specificity have been performed.

None of the various antisera specific for *E. coli* L2 and L14 gave stable immunocomplexes with intact subunits, but antisera specific for the counterparts of these two proteins in *Bacillus stearothermophilus* did react with 50S particles. Accordingly, proteins L2 and L14 were localized on the surface of 50S subunits from *B. stearothermophilus*, and their locations were transferred onto the *E. coli* model (Hackl et al., 1988; W. Hackl and M. Stöffler-Meilicke, unpublished data; Fig. 2e). Such a transfer of protein locations from the *B. stearothermophilus* ribosome to that of *E. coli* was legitimate, since proteins S4, S5, L1, L6, L9, L23, and L29 from *B. stearothermophilus* have all been localized in the same region as their *E. coli* counterparts (Stöffler-Meilicke et al., 1984; Stöffler and Stöffler-Meilicke, 1986; Hackl and Stöffler-Meilicke, 1988). From these results, we were able to conclude that the architecture of the ribosome is the same in both

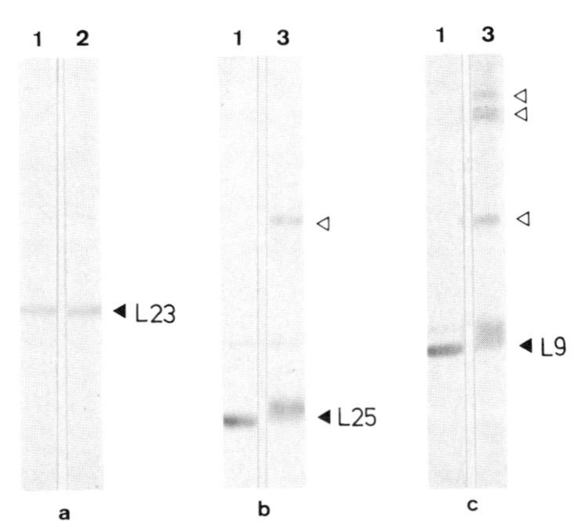

Figure 3. Screening for cross-links by immunoblotting of one-dimensional sodium dodecyl sulfate-polyacrylamide gels. The immunoblots were incubated with anti-L23 (a), anti-L25 (b), and anti-L9 (c). Lanes: 1, 5 μg of TP50 from *E. coli*; 2, 5 μg of TP50 extracted from subunits that had been cross-linked with dimethylsuberimidate; 3, 5 μg of TP50 extracted from subunits that had been cross-linked with 2-iminothiolane. Untreated TP50 gives a single band, corresponding to the unmodified protein (lanes 1). If a certain protein has not been cross-linked to another protein, the immunoblot obtained from TP50 extracted from cross-linked subunits looks the same as that from the control TP50 (compare lanes 1 and 2 in panel a). A cross-linked protein pair will give rise to a new band with a molecular weight higher than that of the non-cross-linked protein (compare lanes 1 and 3 in panel b). The occurrence of more than one high-molecular-weight cross-link band is an indication that the respective protein has been cross-linked to more than one other protein (lane 3 in panel c).

Table 1. Cross-links identified within the 30S subunit from *E. coli*[a]

Cross-link	Cross-link
S2-S3[b,c]	S7-S9[b,c]
S3-S4[b,c]	S7-S13[b]
S3-S10[c]	S11-S21[b,c]
S4-S5[b,c]	S13-S19[b]
S5-S8[b,c]	S15-S17[b,c]
S6-S11[c]	S18-S21[b,c]
S6-S18[b,c]	

[a] From B. Redl, F. Waidmann, and G. Stöffler, unpublished data.
[b] Obtained with 2-iminothiolane.
[c] Obtained with dimethylsuberimidate.

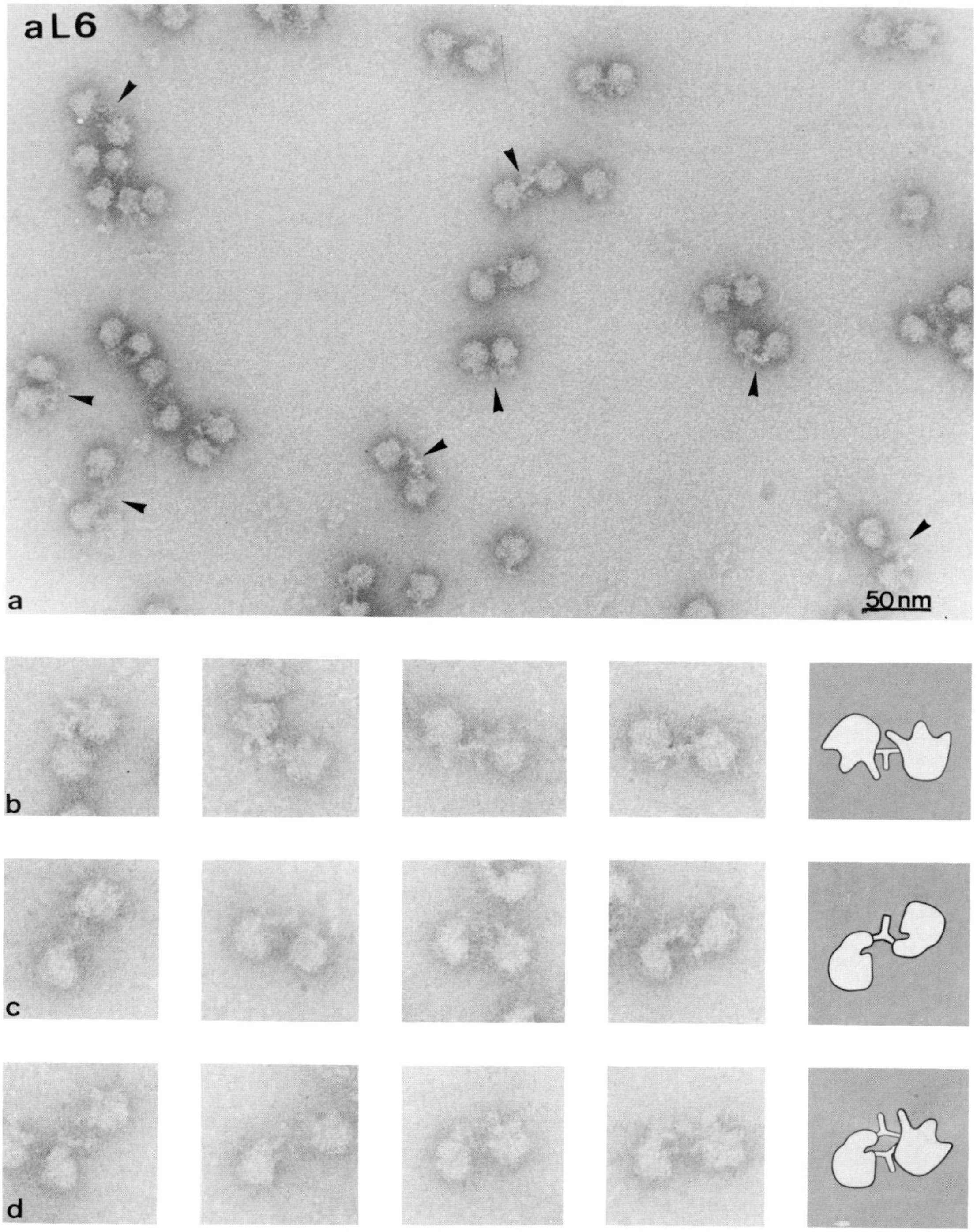

Figure 5. Electron micrographs of 50S subunits from *E. coli*, reacted with anti-L6 (Lotti, 1985). (a) General field; (b to d) selected views. Each interpretative scheme relates to the micrograph on the immediate left and illustrates antibody binding to the crown view (b and d) and the kidney view (c and d).

bacteria. This finding is of general importance, and structural data obtained on ribosomes from any eubacterium could in principle be incorporated into a consensus ribosome model.

For the 50S subunit, comparison of our data with those of others is limited by the fact that only six ribosomal proteins (L1, L5, L7/12, L17, and L27) have been localized by other research groups (Boublik et al., 1976; Strycharz et al., 1978; Lake and Strycharz, 1981; Olson et al., 1986; Nag et al., 1987). The locations obtained for proteins L1 and L17 (Lake and Strycharz, 1981) are identical to those obtained in our study, and those obtained for proteins L27 (Lake and Strycharz, 1981) and L5 (Nag et al., 1987) are very similar to our locations.

There is general agreement that proteins L7/12, which are present in four copies per ribosome, have antigenic determinants located at the rodlike appendage known as the stalk. When we mapped proteins L7/12 with Fab fragments prepared from a polyvalent antiserum, we observed antibody binding at sites ranging from the tip to the base of the stalk but nowhere else (Schaber et al., 1984). On the basis of a study with two monoclonal antibodies, Glitz and co-workers suggested that one L7/12 dimer is sufficient for stalk formation and that a second dimer exists in a folded conformation on the subunit body (Olson et al., 1986), a possibility that cannot be excluded from our own experiments. Their model would be in agreement with the conclusions drawn from energy transfer measurements, namely, that the two L7/12 dimers have separate binding sites on the ribosome, of which only one is present in the stalk (Thielen et al., 1984). Because of the presence of four copies of L7/12 per 50S subunit, each epitope is present four times on the ribosome. Even if monoclonal antibodies were used, it would be impossible to determine by IEM to which of the four identical epitopes an antibody was bound; IEM will therefore be of limited further value for refinement of the arrangement of proteins L7/12 within the 50S particle.

PROTEIN-PROTEIN CROSS-LINKING OF THE RIBOSOMAL PROTEINS OF THE 50S SUBUNIT

Since for the proteins of the 30S subunit there was good agreement between the IEM and our own cross-linking results (see above), we have carried out an extensive cross-linking study to obtain information on the topography of those ribosomal proteins of the 50S subunit not yet mapped. Six cross-linking reagents of different lengths and amino acid specificities have been used, and in each case the members of the cross-linked protein complexes have been identified by immunoblotting (Walleczek et al., 1989a; Walleczek et al., 1989b; Redl et al., 1989).

Altogether, 24 cross-linked protein complexes have been identified (Table 2). Of these cross-links, nine contain protein pairs for which both members have been localized by IEM in comprehensive studies. They fit well with the IEM model, the members of each cross-linked pair being in close neighborhood in the latter (compare Fig. 2e and f with Table 2). The cross-link L5-L7/12 can be explained only if (i) one dimer of L7/12 is not located in the stalk (see above) or (ii) the stalk is flexible (Cowgill et al., 1984).

Table 2. Cross-links identified within the 50S ribosomal subunit from *E. coli*[a]

Cross-link	Cross-link	Cross-link
L1-L33[b–d]	L7/12-L11[d]	L17-L32[b–f]
L2-L9[b,c,e,f]	L9-L28[b–f]	L18-L22[d]
L2-L9-L28[b,c]	L10-L11[d,g]	L19-L25[b–d]
L3-L13[e,f]	L13-L20[d–f]	L20-L21[b,c,e,f]
L3-L19[b,c,e–g]	L13-L21[b,c,g]	L22-L32[b]
L5-L7/12[d]	L14-L19[b,c,e–g]	L23-L29[f]
L6-L19[d]	L16-L27[b–f]	L23-L34[b,c]
L7/12-L10[d]	L17-L30[b]	L27-L33[d]

[a] Proteins that have been localized by IEM on the surface of the 50S ribosomal subunit (Stöffler and Stöffler-Meilicke, 1986; Hackl et al., 1988) are underlined.
[b] Formed by 2-iminothiolane (Walleczek et al., 1989b).
[c] Formed by dithiobis(succinimidylpropionate) (Walleczek et al., 1989a).
[d] Formed by dimethylsuberimidate (Redl et al., 1989).
[e] Formed by *p*-phenylenedimaleimide (Walleczek et al., 1989a).
[f] Formed by *o*-phenylenedimaleimide (Walleczek et al., 1989a).
[g] Formed by diepoxybutane (Walleczek et al., 1989a).

Table 2 includes a number of cross-links for which only one member has been localized by IEM. These cross-links have enabled us to place proteins L3, L13, L16, L21, L22, L28, L30, L32, L33, and L34 in the 50S map.

MODEL FOR THE SPATIAL ARRANGEMENT OF THE PROTEINS OF THE 50S SUBUNIT AS DETERMINED BY COMPUTER GRAPHICS

Combination of our cross-linking results with the IEM data allowed us to generate a three-dimensional model of the protein topography for 29 of the 33 proteins within the 50S ribosomal subunit by using interactive computer graphics (Walleczek et al., 1988). The model was generated in two consecutive steps. In the first step, the coordinates of the surface shape and the protein epitopes on the surface were measured from the IEM model of the 50S ribosomal subunit (Meisenberger et al., 1984; Stöffler and Stöffler-Meilicke, 1986; Hackl et al., 1988) and were fed into the computer graphics system. Since only little is known about the shape, dimensions, or orientation of ribosomal proteins in situ, each protein was represented by a sphere with a volume proportional to its molecular weight. Hydration was assumed to be 0.2 g/g. In the case of proteins L7/12, all four copies of these proteins were represented by a single ellipsoid, whose spatial dimensions corresponded to the dimensions of the L7/12 stalk on the 50S ribosomal subunit. The protein spheres were positioned in such a way that one region of each sphere was in contact with its respective epitope on the surface of the model, and the geometric center of each sphere was placed beneath the surface of the model.

In the second step, the three-dimensional positions of the individual ribosomal proteins were de-

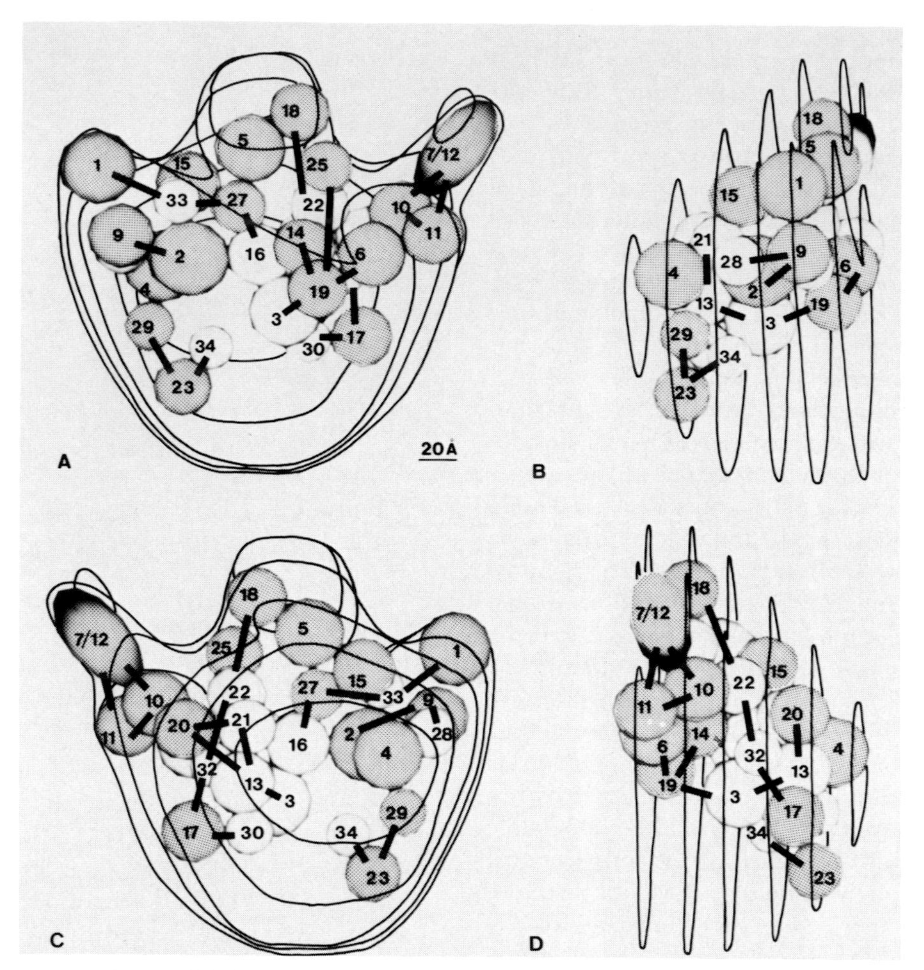

Figure 6. Computer graphics model of the protein topography of the 50S ribosomal subunit (Walleczek et al., 1988). The proteins that have been localized on the surface of the 50S subunit by IEM are shaded; the individual members of cross-linked protein complexes determined in our laboratory are connected by bars except for the cross-link L5-L7/12, which was not included in the modeling process.

rived by incorporating interactively into the model the geometric information of the protein-protein cross-linking results obtained in our laboratory (Table 2). One cross-link not obtained in our laboratory, L14-L32 (Pohl, 1988), was also used in the model-building process. Five proteins, L3, L13, L22, L32, and L33, could be positioned in the model at clearly defined positions, since each of these proteins has been found cross-linked to two or three other proteins whose locations in the ribosome have been clearly established by IEM. For a second group of five proteins, L16, L21, L28, L30, and L34, the geometric constraints were not as rigorous as those for the first group, since each of these five proteins has been cross-linked to only one protein localized by IEM. These latter proteins were positioned so as to be as close as possible to the geometric center of the model, taking into account steric hindrance by other proteins.

Four views of the model are shown in Fig. 6: crown projection with the interface side facing the viewer, kidneylike projection with the L1 protuberance facing the viewer, crown projection with the back of the subunit facing the viewer, and kidneylike projection with the L7/12 stalk facing the viewer. The crown projections correspond exactly to the crown views seen in the electron microscope, whereas the kidneylike projections differ slightly in orientation from the kidney views seen on electron micrographs.

Data in support of our model come from the preliminary location of protein L28, which is located close to protein L9 in both models (compare Fig. 2e and 6A). The location of protein L22 on the back of the 50S subunit, as published by Tischendorf et al. (1974a), also fits well with our model.

At the interface side of the model there are two clusters of proteins, one in the region of the peptidyltransferase center comprising proteins L2, L15, L16, and L27 and the other in the region where elongation factor EF-G-dependent GTP hydrolysis takes place, comprising proteins L6, L7/12, L10, L11, and L14. The lower part of the interface region is free of proteins (Fig. 6B and C). It is of interest that the lower part of the 30S subunit body, which in the 70S ribosome is close to this lower part of the 50S subunit, is also protein free. The neighborhood relationships in our protein model agree extremely well with the functional interrelationships suggested by the assembly map (Herold and Nierhaus, 1987); a detailed comparison of the two sets of data has been given previously (Walleczek et al., 1988).

The protein model of the 50S subunit now provides a useful framework for incorporating data on the three-dimensional structure of 23S RNA; data on the fine structure of 5S RNA and on its location within the 50S ribosomal subunit are already available (Stöffler-Meilicke et al., 1981; Evstafieva et al., 1985; Hancock and Wagner, 1982; Pieler and Erdmann, 1982).

Future studies on ribosomal proteins should concentrate on localizing those proteins for which no topographical data are yet available, namely, L24, L26 (= S20), and L31. One possibility is to localize ribosomal proteins from bacterial species other than *E. coli*. This approach has already been used successfully for the localization of proteins L2 and L14 (see above). The use of monoclonal antibodies that have been selected for high affinity to the ribosome, or haptenization of individual ribosomal proteins and the subsequent use of hapten-specific antibodies, could also be helpful for localizing some of the remaining proteins. Alternatively, antibodies specific for hydrophilic amino acid sequences, which are likely to be exposed on the ribosomal surface, or anti-idiotypic antibodies could be used. For some ribosomal proteins, it may also be necessary to prevent their dissociation from the ribosome by cross-linking, as was necessary for protein S1. Combinations of one or the other of these approaches should eventually lead to a complete map of the ribosomal proteins of the 50S subunit.

Most of this work was carried out at the Max-Planck-Institut für Molekulare Genetik, Berlin. We thank H. G. Wittmann for his constant interest and support and R. Brimacombe and W. P. Tate for critically reading the manuscript. We are highly indebted to R. Albrecht-Ehrlich, R. Hasenbank, and K.-H. Rak for their loyal cooperation over the last 20 years.

REFERENCES

Bickle, T. A., J. W. B. Hershey, and R. R. Traut. 1972. Spatial arrangement of ribosomal proteins: reaction of the *Escherichia coli* 30S subunit with bis-imidoesters. *Proc. Natl. Acad. Sci. USA* **69**:1327–1331.

Boublik, M., W. Hellmann, and H. E. Roth. 1976. Localization of ribosomal proteins L7/L12 in the 50S subunit of *Escherichia coli* ribosomes by electron microscopy. *J. Mol. Biol.* **107**:479–490.

Breitenreuter, G., M. Lotti, M. Stöffler-Meilicke, and G. Stöffler. 1984. Comparative electron microscopic study on the location of ribosomal proteins S3 and S7 on the surface of the *Escherichia coli* 30S subunit using monoclonal and conventional antibody. *Mol. Gen. Genet.* **197**:189–195.

Capel, M. S., D. M. Engelmann, B. R. Freeborn, M. Kjeldgaard, J. A. Langer, V. Ramakrishnan, D. G. Schindler, D. K. Schneider, B. P. Schoenborn, I.-Y. Sillers, S. Yabuki, and P. B. Moore. 1987. A complete mapping of the proteins in the small ribosomal subunit of *Escherichia coli*. *Science* **238**:1403–1406.

Cowgill, C. A., B. G. Nichols, J. W. Kenny, P. Butler, E. M. Bradbury, and R. R. Traut. 1984. Mobile domains in ribosomes revealed by proton nuclear magnetic resonance. *J. Biol. Chem.* **259**:15257–15263.

Dabbs, E. R., R. Ehrlich, R. Hasenbank, H.-G. Schroeter, M. Stöffler-Meilicke, and G. Stöffler. 1981. Mutants of *Escherichia coli* lacking ribosomal protein L1. *J. Mol. Biol.* **149**:553–578.

Evstafieva, A. G., I. N. Shatsky, A. A. Bogdanov, and V. D. Vasiliev. 1985. Topography of RNA in the ribosome: location of the 5S RNA residues A_{39} and U_{40} on the central protuberance of the 50S subunit. *FEBS Lett.* **185**:57–62.

Gulle, H., R. Brimacombe, M. Stöffler-Meilicke, and G. Stöffler. 1987. A rapid immunological spot test for the identification of proteins in covalently linked protein-nucleic-acid complexes. *J. Immunol. Methods* **102**:183–186.

Hackl, W., and M. Stöffler-Meilicke. 1988. Immunoelectron microscopic localisation of ribosomal proteins from *Bacillus stearothermophilus* that are homologous to *Escherichia coli* L1, L6, L23 and L29. *Eur. J. Biochem.* **174**:431–435.

Hackl, W., M. Stöffler-Meilicke, and G. Stöffler. 1988. Three-dimensional location of ribosomal protein BL2 from *Bacillus stearothermophilus*, a key component of the peptidyl transferase center. *FEBS Lett.* **233**:119–123.

Hancock, J., and R. Wagner. 1982. A structural model of 5S RNA from *E. coli* based on intramolecular crosslinking evidence. *Nucleic Acids Res.* **10**:1257–1269.

Hauptmann, R., A. P. Czernilofsky, H. O. Voorma, G. Stöffler, and E. Küchler. 1974. Identification of a protein at the ribosomal donor-site by affinity labeling. *Biochem. Biophys. Res. Commun.* **56**:331–338.

Herold, M., and K. H. Nierhaus. 1987. Incorporation of six additional proteins to complete the assembly map of the 50S subunit from *Escherichia coli* ribosomes. *J. Biol. Chem.* **262**: 8862–8833.

Kahan, L., D. A. Winkelmann, and J. A. Lake. 1981. Ribosomal proteins S3, S6, S8 and S10 of *Escherichia coli* localized on the external surface of the small subunit by immune electron microscopy. *J. Mol. Biol.* **145**:193–214.

Kastner, B., M. Stöffler-Meilicke, and G. Stöffler. 1981. Arrangement of the subunits in the ribosome of *Escherichia coli*: demonstration by immunoelectron microscopy. *Proc. Natl. Acad. Sci. USA* **78**:6652–6656.

Lake, J. A. 1982. Ribosomal subunit orientations determined in the monomeric ribosome by single and by double-labeling immune electron microscopy. *J. Mol. Biol.* **161**:89–106.

Lake, J. A., and L. Kahan. 1975. Ribosomal proteins S5, S11, S13 and S19 localized by electron microscopy of antibody-labeled subunits. *J. Mol. Biol.* **99**:631–664.

Lake, J. A., M. Pendergast, L. Kahan, and M. Nomura. 1974. Localization of *Escherichia coli* ribosomal proteins S4 and S14 by electron microscopy of antibody-labeled subunits. *Proc. Natl. Acad. Sci. USA* **71**:4688–4692.

Lake, J. A., and W. A. Strycharz. 1981. Ribosomal proteins L1, L17 and L27 from *Escherichia coli* localized at single sites on the large subunit by immune electron microscopy. *J. Mol. Biol.* **153**:979–992.

Lelong, J. C., D. Gros, F. Gros, A. Bollen, R. Maschler, and G. Stöffler. 1974. Function of individual 30S subunit proteins of *Escherichia coli*. Effect of specific immunoglobulin fragments

(Fab) on activities of ribosomal decoding sites. *Proc. Natl. Acad. Sci. USA* **71**:248–252.

Lelong, J. C., R. Maschler, M. Crépin, C. Jeantet, G. W. Tischendorf, and F. Gros. 1979. Function of individual *E. coli* 30S ribosomal proteins as determined by *in situ* immunospecific neutralization: a tentative classification. *Biochimie* **61**:881–889.

Lin, F.-L., M. Boublik, and J. Ofengand. 1984. Immunoelectron microscopic localization of the S19 site on the 30S ribosomal subunit which is crosslinked to a site bound transfer RNA. *J. Mol. Biol.* **172**:41–55.

Lotti, M. 1985. Immunological and electron microscopic study on the structure of the *Escherichia coli* ribosome. Ph.D. thesis. Freie Universität Berlin, Berlin, Federal Republic of Germany.

Lotti, M., E. R. Dabbs, R. Hasenbank, M. Stöffler-Meilicke, and G. Stöffler. 1983. Characterisation of a mutant from *Escherichia coli* lacking protein L15 and localisation of protein L15 by immuno-electron microscopy. *Mol. Gen. Genet.* **192**:295–300.

Lotti, M., M. Noah, M. Stöffler-Meilicke, and G. Stöffler. 1989. Localization of proteins L4, L5, L20 and L25 on the ribosomal surface by immuno-electron microscopy. *Mol. Gen. Genet.* **216**:245–253.

Lotti, M., M. Stöffler-Meilicke, and G. Stöffler. 1987. Localization of ribosomal protein L27 at the peptidyl transferase centre of the 50S subunit, as determined by immuno-electron microscopy. *Mol. Gen. Genet.* **210**:498–503.

Lutter, L. C., H. Zeichhardt, C. G. Kurland, and G. Stöffler. 1972. Ribosomal protein neighborhoods. I. S18 and S21 as well as S5 and S8 are neighbors. *Mol. Gen. Genet.* **119**:357–366.

Meisenberger, O., I. Pilz, M. Stöffler-Meilicke, and G. Stöffler. 1984. Small-angle X-ray study of the 50S ribosomal subunit of *Escherichia coli*. A comparison of different models. *Biochim. Biophys. Acta* **781**:225–233.

Moore, P. B., M. Capel, M. Kjelgaard, and D. M. Engelman. 1986. A 19 protein map of the 30S ribosomal subunit of *Escherichia coli*, p. 87–100. *In* B. Hardesty and G. Kramer (ed.), *Structure, Function, and Genetics of Ribosomes*. Springer-Verlag, New York.

Morrison, C. A., R. A. Garrett, H. Zeichhardt, and G. Stöffler. 1973. Proteins occurring at, or near, the subunit interface of *E. coli* ribosomes. *Mol. Gen. Genet.* **127**:359–368.

Nag, B., D. S. Tewari, A. Sommer, H. M. Olson, D. G. Glitz, and R. R. Traut. 1987. Probing ribosome function and the location of *Escherichia coli* ribosomal protein L5 with a monoclonal antibody. *J. Biol. Chem.* **262**:9681–9687.

Olson, H. M., T. V. Olah, B. S. Cooperman, and D. G. Glitz. 1988. Immune electron microscopic localization of dinitrophenyl-modified ribosomal protein S19 in reconstituted *Escherichia coli* 30S subunits using antibodies to dinitrophenol. *J. Biol. Chem.* **263**:4801–4806.

Olson, H. M., A. Sommer, D. S. Tewari, R. R. Traut, and D. G. Glitz. 1986. Localization of two epitopes of protein L7/L12 to both the body and stalk of the large ribosomal subunit. *J. Biol. Chem.* **261**:6924–6932.

Pieler, T., and V. A. Erdmann. 1982. Three-dimensional structural model of eubacterial 5S RNA that has functional implications. *Proc. Natl. Acad. Sci. USA* **79**:4599–4603.

Pohl, T. 1988. Quervernetzungsstudien an Ribosomen aus Prokaryonten. Ph.D. thesis. Freie Universität Berlin, Berlin, Federal Republic of Germany.

Redl, B., J. Walleczek, M. Stöffler-Meilicke, and G. Stöffler. 1989. Immunoblotting analysis of protein-protein crosslinks within the 50S ribosomal subunit of *Escherichia coli*. *Eur. J. Biochem.* **181**:351–356.

Schaber, E., B. Kastner, M. Stöffler-Meilicke, and G. Stöffler. 1984. Decoration of ribosomal subunits from *Escherichia coli* with monovalent Fab fragment specific for proteins L7/L12. *Proc. 8th Eur. Congr. Electron Microsc.* **2**:1555–1556.

Schwedler-Breitenreuter, G., M. Lotti, M. Stöffler-Meilicke, and G. Stöffler. 1985. Localization of ribosomal protein S2 on the surface of the 30S subunit from *Escherichia coli*, using monoclonal antibodies. *EMBO J.* **4**:2109–2112.

Stöffler, G., R. Hasenbank, M. Lütgehaus, R. Maschler, C. A. Morrison, H. Zeichhardt, and R. A. Garrett. 1973. The accessibility of proteins of the *E. coli* 30S ribosomal subunit to antibody binding. *Mol. Gen. Genet.* **127**:89–110.

Stöffler, G., M. Lotti, M. Noah, and M. Stöffler-Meilicke. 1984a. The localization of proteins L5 and L10 on the surface of 50S subunits of *Escherichia coli*. *Proc. 8th Eur. Congr. Electron Microsc.* **2**:1551–1552.

Stöffler, G., M. Noah, M. Stöffler-Meilicke, and E. R. Dabbs. 1984b. The localization of protein L19 on the surface of 50S subunits of *Escherichia coli* aided by the use of mutants lacking protein L19. *J. Biol. Chem.* **259**:4521–4526.

Stöffler, G., B. Redl, J. Walleczek, and M. Stöffler-Meilicke. 1988. Identification of protein-protein cross-links within the *Escherichia coli* ribosome by immunoblotting techniques. *Methods Enzymol.* **164**:119–123.

Stöffler, G., and M. Stöffler-Meilicke. 1984. Immunoelectron microscopy of ribosomes. *Annu. Rev. Biophys. Bioeng.* **13**:303–330.

Stöffler, G., and M. Stöffler-Meilicke. 1986. Immuno electron microscopy on *Escherichia coli* ribosomes, p. 28–46. *In* B. Hardesty and G. Kramer (ed.), *Structure, Function, and Genetics of Ribosomes*. Springer-Verlag, New York.

Stöffler, G., and H. G. Wittmann. 1971a. Sequence differences of *Escherichia coli* 30S ribosomal proteins as determined by immunochemical methods. *Proc. Natl. Acad. Sci. USA* **68**:2283–2287.

Stöffler, G., and H. G. Wittmann. 1971b. Ribosomal proteins. XXV. Immunological studies on *Escherichia coli* ribosomal proteins. *J. Mol. Biol.* **62**:407–409.

Stöffler-Meilicke, M., E. R. Dabbs, R. Albrecht-Ehrlich, and G. Stöffler. 1985. A mutant from *Escherichia coli* which lacks ribosomal proteins S17 and L29 used to localize these two proteins on the ribosomal surface. *Eur. J. Biochem.* **150**:485–490.

Stöffler-Meilicke, M., B. Epe, K. G. Steinhäuser, P. Woolley, and G. Stöffler. 1983a. Immunoelectron microscopy of ribosomes carrying a fluorescence label in a defined position. *FEBS Lett.* **163**:94–98.

Stöffler-Meilicke, M., B. Epe, P. Woolley, M. Lotti, J. Littlechild, and G. Stöffler. 1984. Location of protein S4 on the small ribosomal subunit of *E. coli* and *B. stearothermophilus* with protein- and hapten-specific antibodies *Mol. Gen. Genet.* **197**:8–18.

Stöffler-Meilicke, M., M. Noah, and G. Stöffler. 1983b. Location of eight ribosomal proteins on the surface of the 50S subunit from *Escherichia coli*. *Proc. Natl. Acad. Sci. USA* **80**:6780–6784.

Stöffler-Meilicke, M., and G. Stöffler. 1984. The distribution of ribosomal proteins on the surface of the 30S subunit of *Escherichia coli*. *Proc. 8th Eur. Congr. Electron Microsc.* **2**:1525–1536.

Stöffler-Meilicke, M., and G. Stöffler. 1987. The topography of ribosomal proteins on the surface of the 30S subunit of *Escherichia coli*. *Biochimie* **69**:1040–1064.

Stöffler-Meilicke, M., G. Stöffler, O. W. Odom, A. Zinn, G. Kramer, and B. Hardesty. 1981. Localization of 3′ ends of 5S and 23S rRNAs in reconstituted subunits of *Escherichia coli* ribosomes. *Proc. Natl. Acad. Sci. USA* **78**:5538–5542.

Strycharz, W. A., M. Nomura, and J. A. Lake. 1978. Ribosomal proteins L7/L12 localized at a single region of the large subunit by immune electron microscopy. *J. Mol. Biol.* **126**:123–140.

Tate, W. P., M. J. Dognin, M. Noah, M. Stöffler-Meilicke, and G. Stöffler. 1984. The NH_2-terminal domain of *Escherichia coli* ribosomal protein L11. Its three-dimensional location and its role in the binding of release factors 1 and 2. *J. Biol. Chem.* **259**:7317–7324.

Thielen, A. P. G. M., J. A. Maassen, J. Kriek, and W. Möller. 1984. Mutual orientation of the two L7/L12 dimers on the 50S ribosome of *Escherichia coli* as measured by energy transfer between covalently bound probes. *Biochemistry* **23**:3317–3322.

Tischendorf, G. W., and G. Stöffler. 1975. Localization of *Escherichia coli* ribosomal protein S4 on the surface of the 30S ribosomal subunit by immune electron microscopy. *Mol. Gen. Genet.* **142**:193–208.

Tischendorf, G. W., H. Zeichhardt, and G. Stöffler. 1974a. Determination of the location of proteins L14, L17, L18, L19, L22 and L23 on the surface of the 50S ribosomal subunit of *Escherichia coli* by immune electron microscopy. *Mol. Gen. Genet.* **134**:187–208.

Tischendorf, G. W., H. Zeichhardt, and G. Stöffler. 1974b. Location of proteins S5, S13 and S14 on the surface of the 30S ribosomal subunit from *Escherichia coli* as determined by immune electron microscopy. *Mol. Gen. Genet.* **134**:209–223.

Traut, R. R., J. M. Lambert, G. Boileau, and J. W. Kenny. 1980. Protein topography on *Escherichia coli* ribosomal subunits as inferred from protein crosslinking, p. 89–110. *In* G. Chambliss, G. R. Craven, J. Davies, K. Davis, L. Kahan, and M. Nomura (ed.), *Ribosomes. Structure, Function, and Genetics.* University Park Press, Baltimore.

Traut, R. R., D. S. Tewari, A. Sommer, G. R. Gavino, H. M. Olson, and D. G. Glitz. 1986. Protein topography of functional domains: effects of monoclonal antibodies to different epitopes in *Escherichia coli* protein L7/L12 on ribosome function and structure, p. 286–308. *In* B. Hardesty and G. Kramer (ed.), *Structure, Function, and Genetics of Ribosomes.* Springer-Verlag, New York.

Uchiumi, T., M. Kikuchi, K. Terao, and O. Kikuo. 1985. Cross-linking study on protein topography of rat liver 60S ribosomal subunits with 2-iminothiolane. *J. Biol. Chem.* **260**:5675–5682.

Wabl, M. R. 1974. Electron microscopic localization of two proteins on the surface of the 50S ribosomal subunit of *Escherichia coli* using specific antibody markers. *J. Mol. Biol.* **84**: 241–247.

Walleczek, J. 1988. Untersuchungen zur Protein-Topographie des Ribosoms von *Escherichia coli* mittels chemischer Quervernetzung und Immunelektronenmikroskopie. Ph.D. thesis. Leopold-Franzens-Universität, Innsbruck, Austria.

Walleczek, J., T. Martin, B. Redl, M. Stöffler-Meilicke, and G. Stöffler. 1989a. Comparative cross-linking study on the 50S ribosomal subunit from *Escherichia coli. Biochemistry* **28**: 4099–4105.

Walleczek, J., B. Redl, M. Stöffler-Meilicke, and G. Stöffler. 1989b. Protein-protein cross-linking of the 50S ribosomal subunit of *Escherichia coli* using 2-iminothiolane. *J. Biol. Chem.* **264**:4231–4237.

Walleczek, J., D. Schüler, M. Stöffler-Meilicke, R. Brimacombe, and G. Stöffler. 1988. A model for the spatial arrangement of the proteins in the large subunit of the *Escherichia coli* ribosome. *EMBO J.* **7**:3571–3576.

Winkelmann, D. A., L. Kahan, and J. A. Lake. 1982. Ribosomal protein S4 is an internal protein: localization by immunoelectron microscopy on protein-deficient subribosomal particles. *Proc. Natl. Acad. Sci. USA* **79**:5184–5188.

Wittmann, H. G. 1986. Structure of ribosomes, p. 1–27. *In* B. Hardesty and G. Kramer (ed.), *Structure, Function, and Genetics of Ribosomes.* Springer-Verlag, New York.

Wittmann-Liebold, B. 1986. Ribosomal proteins: their structure and evolution, p. 326–361. *In* B. Hardesty and G. Kramer (ed.), *Structure, Function, and Genetics of Ribosomes.* Springer-Verlag, New York.

Yeh, Y.-C., R. R. Traut, and J. C. Lee. 1986. Protein topography of the 40S ribosomal subunit from *Saccharomyces cerevisiae* as shown by chemical cross-linking. *J. Biol. Chem.* **261**:14148–14153.

Chapter 8

Crystallography and Image Reconstructions of Ribosomes

A. YONATH, W. BENNETT, S. WEINSTEIN, and H. G. WITTMANN

In this chapter, we describe results of the application of X-ray crystallography and image reconstruction to intact ribosomal particles. Our initial attempts at crystallization, the first and most crucial step in these studies, were summarized previously (Yonath et al., 1986c). Here we emphasize recent advances in the crystallographic work and future prospects.

CRYSTALLIZATION AND X-RAY STRUCTURE ANALYSIS OF RIBOSOMAL PARTICLES

Earlier reports from our laboratories have concentrated on our efforts initially to obtain and later to improve the quality of diffraction from crystals of native ribosomal particles. Thanks largely to improvements in the techniques used for growing crystals (Yonath et al., 1988a; Yonath and Wittmann, 1989a) and for collecting crystallographic data from shock-cooled crystals (Hope et al., 1989), we have recently begun to obtain single-crystal diffraction pattern information approaching atomic resolution from some of the crystal forms described earlier. Although our efforts to improve the existing crystal forms continue, these recent successes have encouraged us to devote more effort to the problem of determining initial phases and to the related problem of obtaining crystals of functionally interesting complexes of ribosomal particles with other components of protein synthesis.

Table 1 summarizes the crystals of ribosomal particles grown in our laboratories to date. It is clear from the table that we have been able to obtain crystals from a mutant and a number of chemically modified particles, in addition to native particles. One can also see that some of our efforts to cocrystallize ribosomal particles with other components necessary for protein biosynthesis have already succeeded, in particular several involving complexes with tRNA. Two factors seem to be important in the crystallization of ribosomal particles and their complexes. One is the use of functionally active preparations. We consistently obtain the best crystals from the most active preparations. In addition, we routinely dissolve samples of crystals to verify that the crystalline particles have retained their activity; in all cases, the resolubilized material retains its integrity and biological activity, even when it has been in the crystalline form for periods of several months. A second factor that may be of importance is that the crystals most suitable for crystallographic study are of ribosomal particles from thermophilic or halophilic bacteria; presumably the ribosomes from these organisms are more stable than those from eubacteria during isolation and crystallization.

The column "Resolution" in Table 1 refers to the highest resolution for which diffraction spots could be consistently observed on films. Since there remains considerable variability in the resolution of the diffraction data obtained from different crystals, even among crystals from the same batch, we believe that the best resolution listed in the table is only a temporary upper limit that is likely to improve as we continue to identify and control the sources of the current variability. This is particularly true of the best-diffracting crystals to date, those of the 50S subunit from *Halobacterium marismortui*; these crystals are thin plates that are evidently very susceptible to the mechanical stresses of shock cooling, and

A. Yonath ■ Department of Structural Chemistry, Weizmann Institute, Rehovot, Israel, and Max Planck Research Unit for Structural Molecular Biology, D-2000 Hamburg 52, Federal Republic of Germany. **W. Bennett** ■ Max Planck Research Unit for Structural Molecular Biology, D-2000 Hamburg 52, and Max Planck Institut für Molekulare Genetik, D-1000 Berlin 33, Federal Republic of Germany. **S. Weinstein** ■ Department of Structural Chemistry, Weizmann Institute, Rehovot, Israel, and Max Planck Institut für Molekulare Genetik, Abteilung Wittmann, D-1000 Berlin 33, Federal Republic of Germany. **H. G. Wittmann** ■ Max Planck Institut für Molekulare Genetik, Abteilung Wittmann, D-1000 Berlin 33, Federal Republic of Germany.

Table 1. Characterized three-dimensional crystals of ribosomal particles

Source	Crystal form[a]	Cell dimensions (Å) determined by: Electron microscopy	Cell dimensions (Å) determined by: X-ray crystallography	Resolution (Å)	Comments[b]
70S *E. coli*	A	340 × 340 × 590; $P6_3$			
T. thermophilus					
70S	M		524 × 524 × 306; $P4_12_12$	19	
30S	M		407 × 407 × 170; $P42_12$	7.3	N, H
50S *H. marismortui*	1, P	310 × 350; 105°			
	2, P	148 × 186; 95°	147 × 181; 97°	13	
	3,[c] P	170 × 180; 75°	210 × 300 × 581; $C222_1$	4.5	N, H
50S *B. stearothermophilus*	1, A	130 × 254; 95°			
	2, A	156 × 288; 97°			
	3, A	260 × 288; 105°			
	4, A	405 × 405 × 256; 120°			
	5, A	213 × 235 × 315; 120°			
	6,[c,d] A	330 × 670 × 850; 90°	360 × 680 × 920; $P2_12_12$	18	N
	7,[c,d] P		308 × 562 × 395; 114° C2	11	N, H

[a] Crystals were grown by vapor diffusion from low-molecular-weight alcohols (A), MPD (M), or polyethylene glycol (P).
[b] Crystallographic data were collected from native (N) and derivatized (H) crystals.
[c] Same form and parameters for crystals of a complex of 50S subunits plus tRNA and a segment (18- to 20-mers) of a nascent polypeptide chain.
[d] Same form and parameters for crystals of large ribosomal subunits of a mutant (deficient in L11 protein) of the same source.

we anticipate that the intrinsic order of the crystals is even better than the resolution of 4.5 Å (0.45 nm) listed in the table, which already begins to approach atomic resolution (Fig. 1).

Despite extensive attempts to crystallize small ribosomal subunits, crystals of these particles were obtained only recently (Glotz et al., 1987; Trakhanov et al., 1987; Yusupov et al., 1988; Yonath et al., 1988b), long after the crystallization procedures for several crystal forms containing 50S subunits were well established. The difficulty we and others have encountered in obtaining crystals of 30S subunits is probably related to the relative instability of these particles compared with the 50S subunits from the same organism. Evidence for this point of view has recently been obtained in our laboratories by exposing 70S ribosomes from *Echerichia coli* to a preparation of proteolytic enzymes from *Aspergillus oryzae*. Large variations in the resistance of the two subunits were observed; the 50S subunits remained intact, whereas the 30S subunit completely disintegrated (U. Evers and H. S. Gewitz, unpublished data).

The first microcrystals of 70S ribosomes from *E. coli* were obtained 7 years ago (Wittmann et al., 1982). These crystals were well ordered but too small for crystallographic analysis. More recently, we and

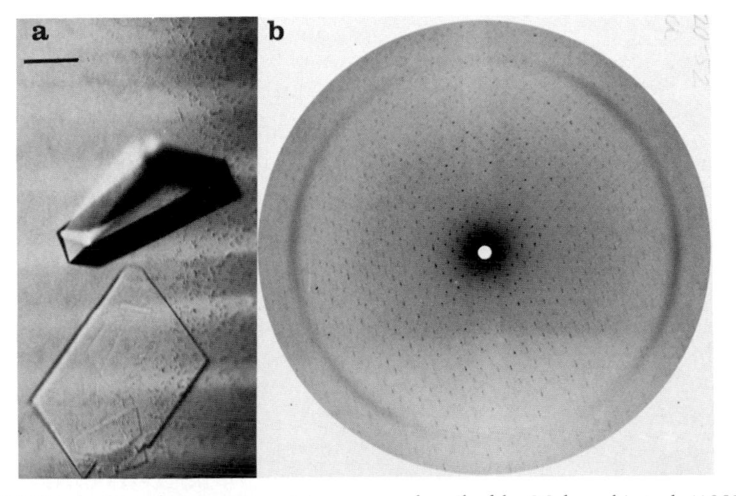

Figure 1. (a) Crystals of 50S subunits of *H. marismortui* grown as described by Makowski et al. (1987). Bar, 0.2 mm. (b) A 1° rotation pattern (recorded on film) of a crystal similar to the one shown in panel a but soaked in a solution containing the components used for its growth and 18% ethylene glycol. The pattern was obtained after 27 h of irradiation at −180°C with a synchrotron X-ray beam (X11 port at EMBL/DESY). Wave length, 1.488 Å; exposure time, 7 min; crystal-to-film distance, 205 mm; resolution, 4.5 Å.

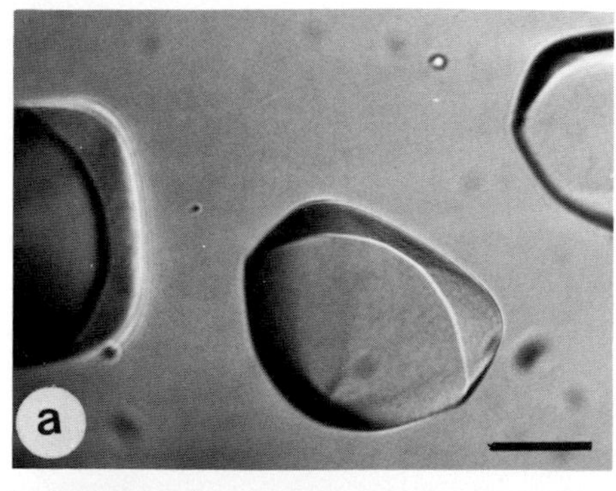

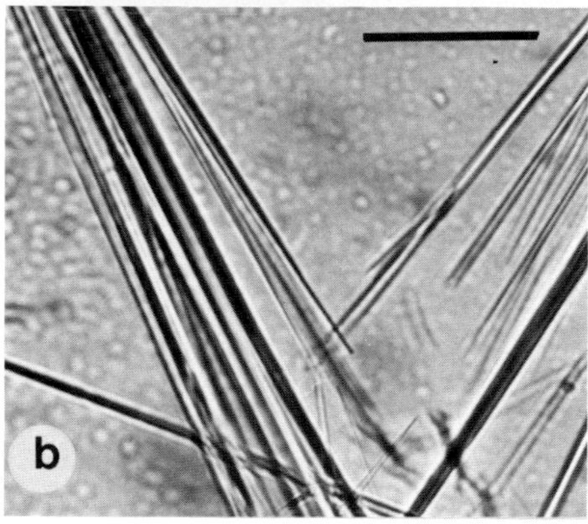

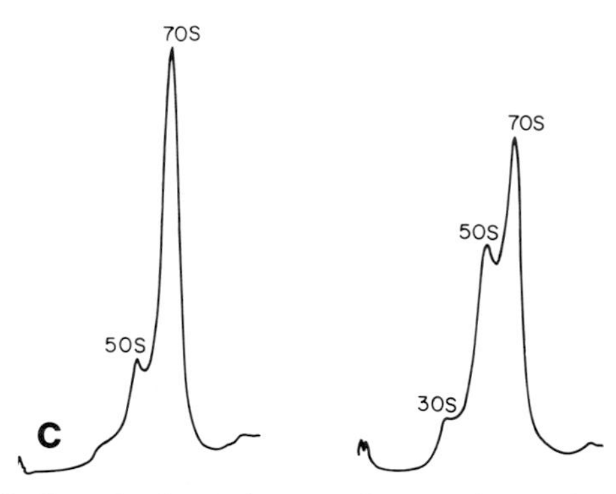

Figure 2. Crystals of 70S ribosomes from *T. thermophilus* (a) and *B. stearothermophilus* (b). Bar, 0.2 mm. (c) Sucrose gradient profiles of the material used for crystallization (left) and of dissolved crystals of *B. stearothermophilus* (right).

others have been able to grow crystals of 70S particles from two thermophilic eubacteria: *Bacillus stearothermophilus* (wild type as well as a mutant) and *Thermus thermophilus* (Glotz et al., 1987; Trakhanov et al., 1987; Berkovitch-Yellin et al., in press [a]). The 70S particles from both sources produce crystals of intermediate size (compared with crystals of other ribosomal particles; Fig. 2), but the crystals consistently diffract to only low resolution: 35 Å for the 70S particles from *B. stearothermophilus* and 19 Å for those from *T. thermophilus*. Given the correlation that we have observed between the activities of samples of 30S or 50S subunits and the quality of the crystals obtained from them, we believe that the poor internal order of the crystals of 70S particles is due to the conformational and functional heterogeneity of the tight couples used for crystallization in our laboratories.

We anticipated that 70S ribosomes obtained by association of purified 50S and 30S subunits will provide a more homogeneous and defined population for obtaining 70S particles. Crystallization experiments have been prepared with several different preparations of high activity but have so far failed to yield crystals. A particularly interesting result was obtained in a similar experiment with highly active hybrids of 50S subunits from *B. stearothermophilus* and 30S subunits from *E. coli*. The crystallization of this hybrid particle was attempted in solutions rich in Mg^{2+}, using a crystallization medium similar to the buffer system used for in vitro poly(U)-programmed polyphenylalanine synthesis. Although large, well-ordered crystals were obtained, sucrose gradients of dissolved crystalline material showed that these crystals contained only 50S particles, despite the presence of 30S subunits in the crystallization mixture (Fig. 3). This result is consistent with our previous observations that large ribosomal subunits crystallize readily under a variety of conditions (Table 1; Yonath et al., 1980; Yonath et al., 1983; Yonath et al., 1984; Yonath et al., 1986a; Yonath et al., 1986b; Yonath et al., 1986c; Müssig et al., 1989). The fact that the 50S subunits crystallized in the presence of heterologous 30S subunits indicates that the interparticle contacts between large subunits in the crystal are stronger than the affinity between the large and small subunits in this hybrid 70S particle, even though the hybrid ribosomes are fully active in vitro. A similar observation has been made for packed two-dimensional sheets of eucaryotic 80S ribosomes. These sheets could be depleted of small subunits and still maintain their packing integrity as a lattice of large subunits (Kühlbrandt and Unwin, 1982).

The crystallization conditions for most of the crystal forms listed in Table 1 are sufficiently refined (and our preparative techniques sufficiently reproducible) that we can obtain crystals from virtually all preparations of active particles. However, some variability in the ribosome preparation evidently still exists, since the exact conditions for the growth of well-ordered and large crystals still must be refined for each ribosomal preparation (Yonath and Wittmann, 1989a). It is interesting that for particles which can be crystallized in more than one form, a preparation that yields large, well-shaped crystals of one crystal form will generally also crystallize well in the other forms. For example, preparations of 50S subunits from *B. stearothermophilus* that produce

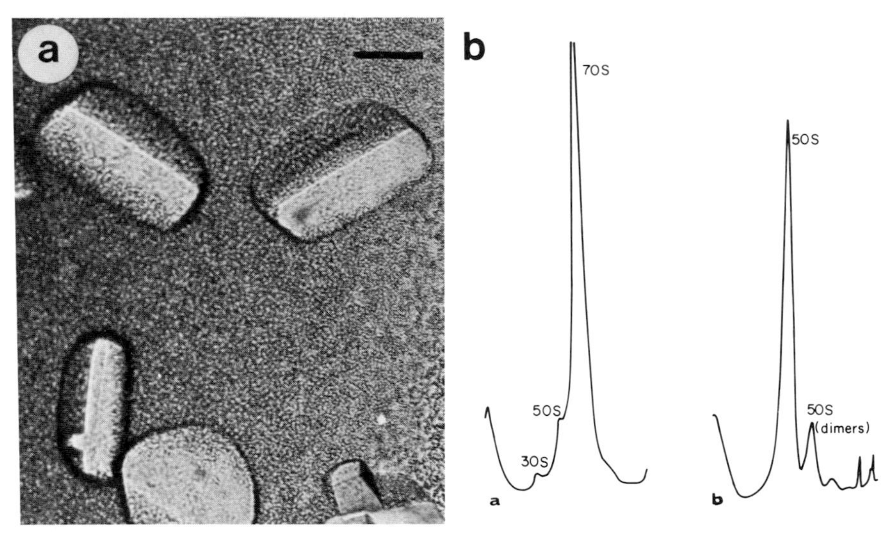

Figure 3. (a) Crystals obtained from an attempt to crystallize a hybrid of 50S from *B. stearothermophilus* and 30S from *E. coli.* Bar, 0.2 mm. (b) Sucrose gradient profile of the material used for crystallization (left) and of the dissolved crystals (right).

good crystals when crystallized from alcohol solutions (form 6 in Table 1) also yield high-quality crystals when polyethylene glycol is used as the crystallizing agent (form 7). Thus, the basic factors governing the ease of crystallization and the yield of large, well-formed ribosomal crystals seem to be related more to the quality of the preparation of the ribosomal particles than to the choice of the crystallization agent.

B. stearothermophilus is the only source from which crystals of all ribosomal particles (70S, 50S, and 30S, including mutated and chemically modified particles) can be grown. The ribosomes and ribosomal subunits of *B. stearothermophilus* have been well characterized by chemical, physical, and immunological methods and thus are particularly attractive for structural study. Furthermore, crystals of the ribosomal particles from *B. stearothermophilus* have been obtained from solutions that resemble the natural environment of the ribosome in the cell, except for the addition of a small amount of polyethylene glycol. For these reasons, we continue our studies of ribosomal crystals from this organism, despite the relatively poor resolution of the diffraction patterns that we have obtained from these crystals to date (Table 1).

The diffracting power of crystalline ribosomal particles is so weak that virtually all of the crystallographic studies have been performed with synchrotron radiation (Bartels et al., 1988). The extremely intense X rays available at synchrotron installations are generated as a by-product of accelerators originally designed for high-energy-particle experiments. However, at temperatures ranging from room temperature to about 4°C, the radiation damage suffered by all crystals of ribosomal particles in an intense synchrotron beam is so rapid and so severe that all reflections beyond about 18 Å resolution decay within a few minutes, a period shorter than the time required to obtain a single X-ray photograph. In our early diffraction studies of crystals of ribosomal particles, which were performed at temperatures above 4°C, the extreme sensitivity of the crystals to radiation damage led us to conclude (incorrectly, as we discovered later) that the diffraction was limited to 15 to 18 Å resolution (Yonath et al., 1984; Yonath et al., 1986a; Yonath et al., 1986b; Yonath et al., 1986c). Even when the problem of radiation damage was evident, the diffraction that we could observe at these temperatures could be seen only in the first X-ray photograph from a crystal. Thus, to measure the diffraction data even to this resolution, a new crystal had to be used for each photograph, and precise alignment of the crystals was impossible. In addition, the exposure of each crystal had to be kept as short as possible to minimize the effect of radiation damage. In a typical attempt to obtain a complete diffraction data set under these conditions, more than 260 individual crystals were used. Typically, however, the combination of randomly oriented crystals and short exposures yielded only partial diffraction data sets which did not contain even a single fully recorded reflection, making evaluation of the data difficult at best.

Since the time of these early diffraction studies, we have been able to overcome the problem of radiation sensitivity for all of the crystals of ribosomal particles by shock freezing the crystals to cryogenic temperatures (for our work, the boiling temperature of liquid nitrogen, about −180°C) before the diffraction experiment. It is generally believed that a major component of the damage to biological samples by ionizing radiation is caused by the diffusion of free radicals, which are produced throughout the region of the sample that is irradiated, including the solvent regions of macromolecular crystals. Small-molecule crystallographers have routinely performed data collection at cryogenic temperatures to protect radiation-sensitive samples from X-ray damage for years, but at the outset of our attempts to apply this approach to ribosomal crystals, the use of low temperatures with macromolecular crystals had generally been less successful, often because the crystals could not be transferred to the cryosolvents thought at the time to be necessary to prevent the solvent in the crystals from freezing. We were able to use the then novel technique of shock freezing in the absence of cryoprotectants to crystals of ribosomal particles (Hope et al., 1989); with this approach, the formation of ice crystals is prevented by lowering the temperature so rapidly that the solvent solidifies as an amorphous glass before crystals can form. It has proven possible to apply the shock-freezing technique to all crystals of ribosomal particles that we have studied so far, with the result that the frozen crystals, if maintained at cryogenic temperatures, suffer no observable radiation damage over the time span needed for collecting several complete diffraction data sets, which allows even the weakest diffraction data available to be measured. We have also recently developed techniques for storing shock-frozen crystals for long periods in solid propane, which solidifies as a glass at a temperature a few degrees above the boiling point of nitrogen. These techniques can also be applied to crystals that have been irradiated, allowing a diffraction experiment to be interrupted and resumed at a later date.

SPECIFIC AND QUANTITATIVE LABELING OF RIBOSOMAL PARTICLES

Determination of the three-dimensional structure of a crystalline compound by X-ray crystallography involves a Fourier summation of the reflections present in the diffraction pattern. Each reflection is a wave characterized by its direction, intensity, and phase. What keeps this summation from being a trivial computational problem is the fact that only the direction and amplitude of a reflection can be measured, whereas the phase cannot be directly determined. For macromolecular crystals, two techniques are commonly used to determine the phases. If an approximate model is available or if the unknown macromolecule is closely related to one whose structure is known, the known structure can often be used to provide preliminary phase information for the unknown molecule with the aid of various search procedures that allow the known structure to be positioned in the crystal lattice of the unknown molecule; this approach is known as molecular replacement. If the crystalline macromolecule is not related to one whose structure is known, its phases are most often determined by the multiple isomorphous replacement (MIR) method. Phase determination by MIR involves the preparation of at least two chemical modifications of the unknown molecule, usually the addition of one or a few electron-dense atoms or groups of atoms to the structure. The modification must be large enough to cause measurable changes in the diffraction pattern of the unknown molecule but not so extensive as to change the structure of the molecule or its crystal lattice. If these conditions are met and the number of added groups is small, the location of the added groups can usually be deduced from the changes in the diffraction pattern, which in turn allows the phases for the unmodified structure to be determined.

Because ribosomal particles are much larger than any macromolecular complex solved by X-ray crystallography to date, one can anticipate that it will be difficult to fulfill all of the conditions of an ideal isomorphous derivative simultaneously; for example, it could be difficult to produce measurable changes in the diffraction pattern of a ribosomal particle with only a few scattering groups. To address at least the foreseeable difficulties of phase determination of such large structures, we have adopted a strategy that combines elements of both the molecular replacement and MIR methods. We have previously reported low-resolution structures derived by image reconstruction from electron micrographs of two-dimensional, ordered arrays of ribosomal particles (Yonath et al., 1987a; Yonath et al., 1987b; Arad et al., 1987b; Yonath and Wittmann, 1989b). These reconstructions are being used as approximate models of the crystalline ribosomal particles for molecular replacement. It is our hope that the low-resolution phases obtained by this approach will be useful for locating additional scattering groups in isomorphous derivatives, perhaps even large numbers of added groups, which will provide higher-resolution MIR phases.

For determination of the phases of proteins with molecular weights of up to about 50,000, the added group of a useful isomorphous derivative typically consists of one or two heavy-metal atoms. An ideal added group for the ribosomal particles would consist of a compact cluster of a proportionately larger number of heavy atoms. Clusters with a core of several heavy-metal atoms linked directly to one another have been synthesized (Jahn, 1989a, 1989b) and are attractive candidates for isomorphous substitutions of ribosomal particles. We are using two such clusters: an undecagold cluster with a total molecular weight of 6,200 and a tetrairidium cluster with a molecular weight of 2,300. As the core of the undecagold cluster is about 8.2 Å in diameter, it can be treated as a single scattering group at low to medium resolution. The tetrairidium cluster has a core diameter of about 5 Å and can thus be treated as a single heavy atom to somewhat higher resolution. Although these clusters have not been used previously for phasing by protein crystallographers, they seem to be ideally suited for crystallographic analysis of ribosomal particles.

In forming isomorphous derivatives of ribosomal particles with these clusters, we are attempting to take advantage of the extensive knowledge of the biochemistry, biophysics, and genetics of the ribosome to prepare singly substituted derivatives that should produce easily interpretable changes in the diffraction patterns even in the absence of approximate molecular replacement phases. Useful derivatives of macromolecules are normally obtained by soaking native crystals in solutions of a heavy-atom compound or by crystallization of the macromolecule from a solution containing the heavy atom. With this approach, the number of heavy atoms that bind to the macromolecule is largely a matter of chance, but the odds of obtaining a useful derivative for a typical macromolecule with these equilibrium techniques are sufficiently high that more sophisticated techniques are rarely needed. However, since ribosomal particles have a much larger surface area than any macromolecule solved by these techniques so far, there is a proportionately larger number of potential binding sites for isomorphous substituents. For this reason, it would clearly be preferable not to leave the binding of the heavy-atom clusters to chance.

An alternative to the equilibrium binding of isomorphous substituents to the crystalline molecule is the formation of a covalent derivative of the molecule at a specific site before crystallization. This approach often requires sophisticated synthetic techniques and time-consuming purification procedures, but it offers us a much better chance of obtaining interpretable changes in the diffraction pattern of a ribosomal particle. Moreover, specific derivatization of selected ribosomal components will ultimately be of considerable value in localizing the components and related functional sites in the three-dimensional structure of the ribosome.

Monofunctional reagents with a maleimido moiety as the reactive group have been prepared from both of the heavy-atom clusters described earlier. Since the clusters are rather bulky, the accessibility of the reactive group was varied by attaching it to the clusters with aliphatic chains of different lengths (Weinstein et al., 1989). The modified clusters were bound to specific sites on the ribosome before crystallization in one of two ways: (i) by direct reaction with chemically active groups on the surface of the ribosome or (ii) by covalent attachment to an isolated ribosomal component that was subsequently reconstituted into the ribosome.

For the first approach, conditions under which a small number of sulfhydryl groups are exposed on the surface of the ribosome were determined. We were able to find conditions under which only one to three sites were exposed on each of the ribosomal particles that could crystallize. For the 50S subunit from *H. marismortui*, for example, we could take advantage of previous studies which showed that the stability, compactness, and biological activity of halophilic ribosomes depend strongly on the concentrations of salts as well as on the delicate equilibrium between the monovalent and divalent ions in the medium (Shevack et al., 1985; H.-S. Gewitz, I. Makowski, S. Weinstein, U. Evers, A. Yonath, and H. G. Wittmann, unpublished data). To obtain an efficient reaction with both clusters, it was necessary to reduce the concentration of the KCl from 3 M, the concentration normally used to stabilize these ribosomes, to 1.2 to 1.5 M. To avoid disintegration of the subunits and depletion of ribosomal proteins at the KCl concentration required for the reaction, the concentration of Mg^{2+} was increased to 20 to 50 mM. Under these conditions, the halophilic ribosomes maintain their integrity for long periods of time and the clusters are bound predominantly to one site.

To limit the reaction of the maleimido group to the sulfhydryl groups of the ribosomal proteins, the binding was conducted at around pH 5.5. The 50S subunits of *H. marismortui* are crystallized and can be safely maintained at this pH. To improve the efficiency of the reaction, it was necessary to attach another aliphatic chain to the −SH groups on the surface of the particle, in addition to the reactive arm of the clusters. It was found that a spacer with a minimum length of about 10 Å between an −SH group on the ribosomal particle and the N atom of the maleimido group of the cluster was needed to bind between 0.5 to 1 equivalent of the gold cluster directly to the 50S subunits of *B. stearothermophilus*. All of the 50S subunits so obtained yielded crystals isomorphous with the native crystals.

Genetic and chemical procedures for obtaining ribosomes of *B. stearothermophilus* in which protein BL11 is missing were developed. The genetic proce-

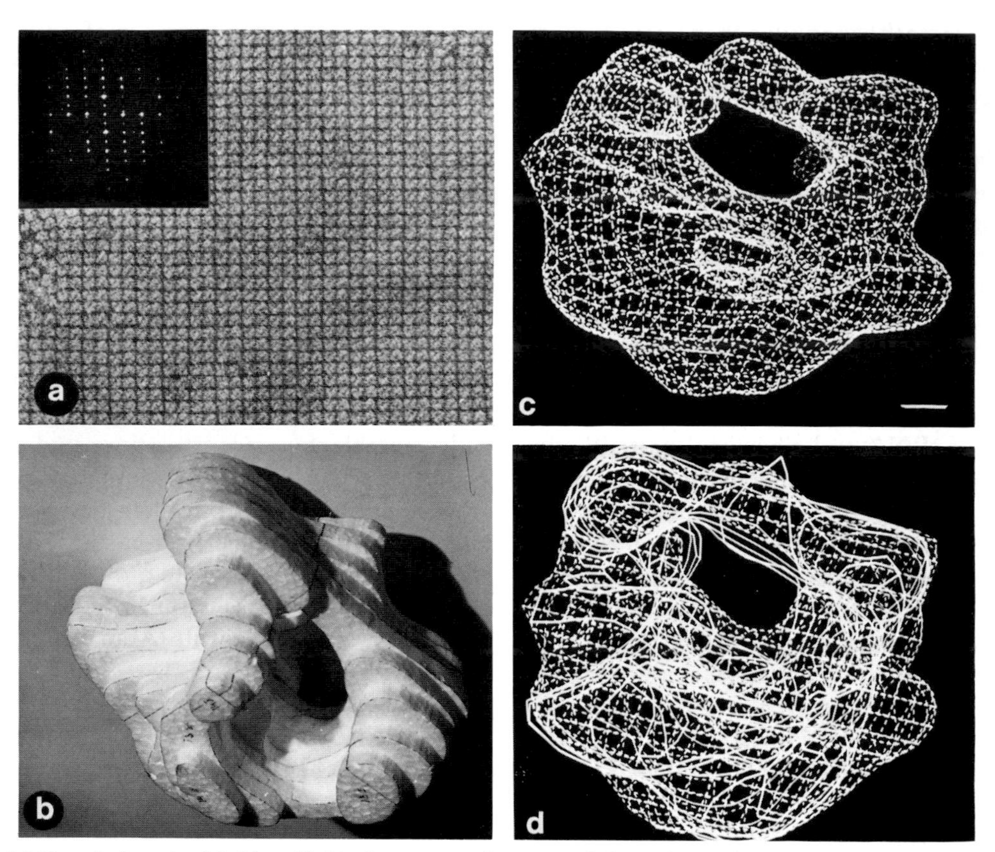

Figure 5. (a) Negatively stained (with gold thioglucose) two-dimensional sheet of 70S ribosomes from *B. stearothermophilus* and its diffraction pattern. (b) Physical model of one of the reconstructed images. (c) Computer graphic representation of another reconstructed model. (d) The images shown in panels b and c superimposed on each other. Both reconstructions were obtained at 47 Å. Bar, 20 Å.

defined in the reconstructions from two-dimensional sheets of 80S ribosomes (Milligan and Unwin, 1986) and from averaged images of single 70S ribosomes (Wagenknecht et al., 1989). It is likely that the poor definition of the empty space in the 80S ribosomes results from the lower resolution at which this reconstruction was performed and, in the single-particle reconstruction of the *E. coli* ribosomes, from shrinkage or collapse of the particles on the grid (Wagenknecht et al., 1989).

Another prominent feature of our reconstructions of negatively stained two-dimensional sheets of 50S subunits from *B. stearothermophilus* is a tunnel of about 100 Å in length and up to 25 Å in diameter through the 50S subunit (Fig. 6). It could also be detected in filtered images of the same sheets viewed unstained in the electron microscope at cryotemperature (Fig. 6; M. Giersig, unpublished data). Our observation of a tunnel in the 50S subunit from *B. stearothermophilus* is consistent with the results of image reconstruction of two-dimensional sheets of 80S eucaryotic ribosomes, which revealed a channel in the large subunit (Milligan and Unwin, 1986). More recently, a similar feature was found in the 50S subunit of *E. coli* ribosomes by Wagenknecht et al. (1989), who derived their model from averaged images of single 70S ribosomes that were selected by statistical criteria. The tunnel is clearly visible in all reconstructions of 50S particles (Yonath et al., 1987b) but is somewhat less well resolved in those of assembled ribosomes (Milligan and Unwin, 1986; Arad et al., 1987b; Wagenknecht et al., 1989), possibly because their tunnels are partially filled with nascent protein chains or because of the lower resolution.

We have tentatively located the large subunit of the 70S ribosome from *B. stearothermophilus* by manually fitting our reconstruction of the large subunit into the reconstruction of the complete ribosome, using interactive computer graphics (Fig. 7). The fitting was based both on similarities in the overall shapes of the models and on orientation of the tunnel. The overall agreement in the shapes of the 50S reconstruction and the part of the 70S reconstruction that was assigned to it is quite striking. However, there are two regions in which the two

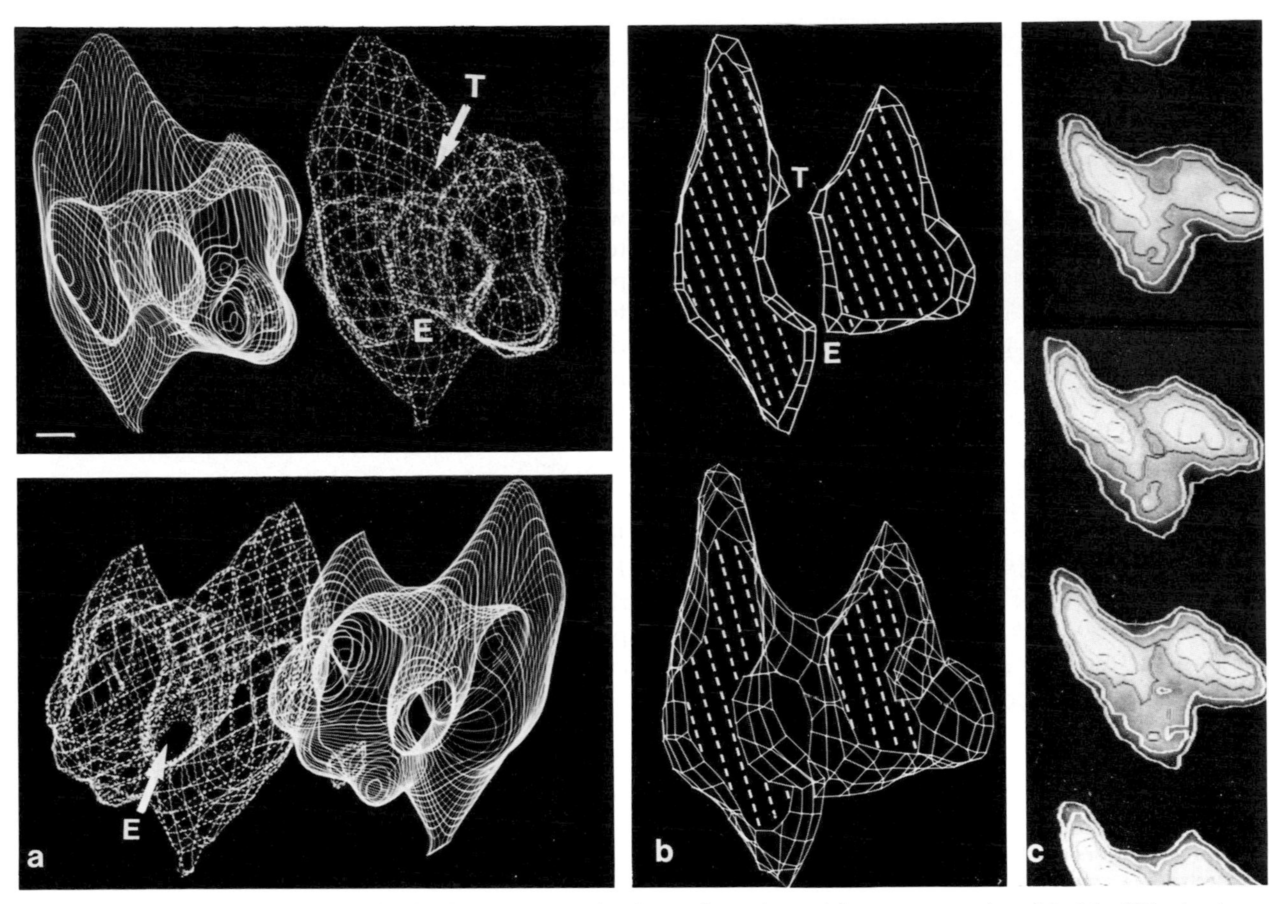

Figure 6. (a) Two computer graphic displays (as a net and in lines) of two views of the reconstructed model of the 50S subunit obtained from negatively stained (with gold thioglucose) two-dimensional sheets. (b) Two sections of thickness of 20 Å (top) and 40 Å (bottom) through the reconstructed model shown in panel a. (c) Filtered images of the unstained sheets of 50S subunits, grown as described by Arad et al. (1987a) and viewed at cryotemperature. Panels a and c were reconstructed or filtered at 28 Å. The entrance to (T) and exit from (E) the tunnel are indicated. Bar, 20 Å.

models differ slightly (Fig. 7). At this stage, it is not clear whether these differences reflect conformational changes occurring upon association of the subunits or whether they simply reflect the differences in the resolution of the two reconstructions.

Since two-dimensional sheets of 30S subunits sufficiently well ordered for diffraction studies have not yet been grown, a reconstruction of the small subunit free from the potential artifacts of single-particle reconstruction techniques is not available. However, after fitting the model of the 50S subunit into that of the 70S ribosomes of *B. stearothermophilus*, we could examine the interface between the two subunits as well as deduce the approximate shape of the small subunit within the 70S particle (Berkovitch-Yellin et al., in press [b]). The subunits are clearly separated by the empty space discussed above (Fig. 7). We will refer to this region as the intersubunit space.

The resulting model for the 30S particle is shown in Fig. 8. There is some similarity between the model of the small subunit so obtained and that observed by visualization of single particles (reviewed in Wittmann, 1983, and Hardesty and Kramer, 1986). However, isolated 30S particles appear somewhat wider in most electron micrographs and in reconstructions from single particles than in our model. Since we can compare our model of the 30S subunit only with models derived with single-particle techniques, artifacts such as flattening of the isolated particles on the grid might contribute to the differences observed. As was the case in our comparison of the reconstructed 50S and 70S particles, these differences may also result from conformational changes in the subunits when they associate to form the 70S ribosome.

More than a decade ago (Malkin and Rich, 1967; Blobel and Sabatini, 1970; Smith et al., 1978), as well as more recently (Rayabova et al., 1988; Yen et al., 1988; Evers and Gewitz, unpublished data), it was observed that the ribosome protects a growing polypeptide chain from enzymatic digestion until the

CONCLUSIONS

From the early stages of this work, it has been clear that the conventional techniques of macromolecular crystallography would not be adequate for determining the structures of ribosomal particles. We have devised an approach that combines phase determination methods of macromolecular crystallography with cryotemperature techniques to obtain higher-resolution data and the exploitation of the extensive information available on the genetic and chemical properties of ribosomes for the rational design of isomorphous derivatives of ribosomal particles by using heavy-atom clusters. These tools have already enabled us to overcome a number of the difficulties associated with crystallographic analysis of ribosomal particles that were once thought to be insurmountable. This broad-based approach, together with recent advances in the instrumentation for X-ray crystallography, leaves no major conceptual obstacle to determination of the three-dimensional structure of the ribosome, although a considerable amount of work remains to be done before this goal is realized. In anticipation of the success of these efforts, we have already begun crystallization studies directed at obtaining crystals in which the ribosome is trapped in different functional states (Bennett and Huber, 1984). In this way, one can use a number of static "snapshots" of a dynamic system such as the ribosome to visualize the sequence of events of the dynamic process and to aid in the interpretation of studies by other techniques.

This work was supported by grant 05 180 MP B0 from the West German Ministry for Research and Technology (BMFT), Public Health Service grant GM 34360 from the National Institutes of Health, grant 85-00381 from the U.S.-Israel Binational Science Foundation, Heinemann grant 4694 81, and Minerva research grants. A.Y. holds the Martin S. Kimmel professorial chair.

REFERENCES

Arad, T., J. Piefke, H. S. Gewitz, B. Hennemann, C. Glotz, J. Müssig, A. Yonath, and H. G. Wittmann. 1987a. The growth of ordered two-dimensional sheets of ribosomal particles from salt-alcohol mixtures. *J. Anal. Biochem.* **167**:113–117.

Arad, T., J. Piefke, S. Weinstein, H. S. Gewitz, A. Yonath, and H. G. Wittmann. 1987b. Three-dimensional image reconstruction from ordered arrays of 70S ribosomes. *Biochimie* **69**: 1001–1005.

Bartels, K. S., G. Weber, S. Weinstein, H. G. Wittmann, and A. Yonath. 1988. Synchrotron light on ribosomes: the development of crystallographic studies of bacterial ribosomal particles. *Top. Curr. Chem.* **147**:57–72.

Bennett, W. S., and R. Huber. 1984. Structural and functional significance of domain motions in proteins. *Crit. Rev. Biochem.* **5**:291–384.

Berkovitch-Yellin, Z., H. A. S. Hansen, W. S. Bennett, R. Sharon, K. von Böhlen, N. Volkmann, J. Piefke, A. Yonath, and H.-G. Wittmann. Crystals of 70S ribosomes from thermolic bacteria are suitable for crystallographic analysis at low resolution. *J. Cryst. Growth*, in press [a].

Berkovitch-Yellin, Z., H.-G. Wittmann, and A. Yonath. Low resolution models for ribosomal particles reconstructed from two dimensional sheets. *Acta Crystallogr.*, in press [b].

Bernabeu, C., and J. A. Lake. 1982. Nascent polypeptide chains emerge from the exit domain of the large ribosomal subunit: immune mapping of the nascent chain. *Proc. Natl. Acad. Sci. USA* **79**:3111–3115.

Blobel, G., and D. D. Sabatini. 1970. Controlled proteolysis of nascent polypeptides in rat liver cell fractions. *J. Cell Biol.* **45**:130–145.

Brimacombe, R., J. Atmadja, W. Stiege, and D. Schüler. 1988. A detailed model of the three-dimensional structure of *E. coli* 16S ribosomal RNA in situ in the 30S subunit. *J. Mol. Biol.* **199**:115–136.

Carazo, J. M., T. Wagenknecht, M. Radermacher, V. Mandiyan, M. Boublik, and J. Frank. 1988. Three-dimensional structure of 50S *E. coli* subunit depleted of protein L7/L12. *J. Mol. Biol.* **201**:393–404.

Clark, W., K. Leonard, and J. Lake. 1982. Ribosomal crystalline arrays of large subunits from *E. coli. Science* **216**:999–1000.

Gewitz, H.-S., C. Glotz, P. Goischke, B. Romberg, J. Müssig, A. Yonath, and H. G. Wittmann. 1987. Reconstitution and crystallization experiments with isolated split proteins from *Bacillus stearothermophilus* ribosomes. *Biochem. Int.* **15**:887–895.

Gewitz, H. S., C. Glotz, J. Piefke, A. Yonath, and H. G. Wittmann. 1988. Two-dimensional crystalline sheets of the large ribosomal subunits containing the nascent protein chain. *Biochimie* **70**: 645–648.

Gilbert, W. 1963. Protein synthesis in E. coli. *Cold Spring Harbor Symp. Quant. Biol.* **28**:287–294.

Glotz, C., J. Müssig, H.-S. Gewitz, I. Makowski, T. Arad, A. Yonath, and H. G. Wittmann. 1987. Three-dimensional crystals of ribosomes and their subunits from eu- and archaebacteria. *Biochem. Int.* **15**:953–960.

Hardesty, B., and G. Kramer (ed.). 1986. *Structure, Function, and Genetics of Ribosomes.* Springer-Verlag, New York.

Hill, W. E., B. E. Tapprich, and B. Tassanakajohn. 1986. Probing ribosomal structure and function, p. 233–252. *In* B. Hardesty and G. Kramer (ed.), *Structure, Function, and Genetics of Ribosomes.* Springer-Verlag, New York.

Hope, H., F. Frolow, K. von Böhlen, I. Makowski, C. Kratky, Y. Halfon, H. Danz, P. Webster, K. Bartels, H. G. Wittmann, and A. Yonath. 1989. Cryocrystallography of ribosomal particles. *Acta Crystallogr.* **45B**:190–199.

Jahn, W. 1989a. Synthesis of water soluble tetrairidium clusters suitable for heavy atom labeling of proteins. *Z. Naturforsch.* **44b**:79–82.

Jahn, W. 1989b. Synthesis of water-soluble undecagold cluster for specific labeling of proteins. *Z. Naturforsch.* **44b**:1313–1322.

Kühlbrandt, W., and P. N. T. Unwin. 1982. Distribution of RNA and proteins in crystalline eukaryotic ribosomes. *J. Mol. Biol.* **156**:611–617.

Kurzchalia, S. V., M. Wiedmann, H. Breter, W. Zimmermann, E. Bauschke, and T. A. Rapoport. 1988. tRNA-mediated labeling of proteins with biotin, a nonradioactive method for the detection of cell-free translation products. *Eur. J.* **172**:663–668.

Lake, J. 1979. Ribosome structural and functional sites, p. 201–236. *In* G. Chambliss, G. R. Craven, J. Davies, K. Davis, L. Kahan, and M. Nomura (ed.), *Ribosomes. Structure, Function, and Genetics.* University Park Press, Baltimore.

Makowski, I., F. Frolow, M. A. Saper, H. G. Wittmann, and A.

Yonath. 1987. Single crystals of large ribosomal particles from *Halobacterium marismortui* diffract to 6 Å. *J. Mol. Biol.* **193:** 819–821.

Malkin, L. I., and A. Rich. 1967. Partial resistance of nascent polypeptide chains to proteolytic digestion due to ribosomal shielding. *J. Mol. Biol.* **26:**329–346.

Milligan, R. A., and P. N. T. Unwin. 1986. Location of the exit channel for nascent proteins in 80S ribosomes. *Nature* (London) **319:**693–696.

Müssig, J., I. Makowski, K. von Böhlen, H. Hansen, K. S. Bartels, H. G. Wittmann, and A. Yonath. 1989. Crystals of wild-type, mutated, derivatized and complexed 50S ribosomal subunits from *Bacillus stearothermophilus* suitable for X-ray analysis. *J. Mol. Biol.* **205:**619–621.

Nierhaus, K. H., R. Brimacombe, and H. G. Wittmann. 1989. Inhibition of protein biosynthesis by antibiotics, p. 29–40. *In* G. G. Jackson, H. D. Schlumberger, and H. J. Zeiler (ed.), *Perspectives in Antiinfective Therapy.* Friedr. Vieweg & Sohn, Braunschweig/Wiesbaden, Federal Republic of Germany.

Piefke, J., T. Arad, H. S. Gewitz, A. Yonath, and H. G. Wittmann. 1986. The growth of ordered two-dimensional sheets of whole ribosomes from *B. stearothermophilus. FEBS Lett.* **209:**104–106.

Oakes, M., E. Henderson, A. Scheiman, M. Clark, and J. Lake. 1986. Ribosome structure, function and evolution: mapping ribosomal RNA, proteins and functional site in three dimensions, p. 47–67. *In* B. Hardesty and G. Kramer (ed.), *Structure, Function, and Genetics of Ribosomes.* Springer-Verlag, New York.

Rayabova, L. A., O. M. Selivanova, V. I. Baranov, V. D. Vasiliev, and A. S. Spirin. 1988. Does the channel for nascent peptide exist inside the ribosome? *FEBS Lett.* **226:**255–260.

Sedelnikova, S. F., S. C. Agalarov, M. B. Garber, and M. M. Yusupov. 1987. Proteins of the *Thermus thermophilus* ribosome: purification of several individual proteins and crystallization of protein TL7. *FEBS Lett.* **220:**227–230.

Shevack, A., H.-S. Gewitz, B. Hennemann, A. Yonath, and H. G. Wittmann. 1985. Characterization and crystallization of ribosomal particles from *Halobacterium marismortui. FEBS Lett.* **184:**68–71.

Smith, W. P., P. C. Tai, and B. D. Davis. 1978. Interaction of secreted nascent chains with surrounding membranes in *Bacillus subtilis. Proc. Natl. Acad. Sci. USA* **75:**5922–5925.

Trakhanov, S. D., M. M. Yusupov, S. C. Agalarov, M. B. Garber, S. N. Rayazantsev, S. V. Tischenko, and V. A. Shirokov. 1987. Crystallization of 70S ribosomes and 30S ribosomal subunits from *Thermus thermophilus. FEBS Lett.* **220:**319–322.

Wagenknecht, T., J. M. Carazo, M. Radermacher, and J. Frank. 1989. Three-dimensional reconstruction of the ribosome from *E. coli. Biophys. J.* **55:**455–464.

Weinstein, S., W. Jahn, H. A. S. Hansen, H. G. Wittmann, and A. Yonath. 1989. Novel procedures for derivatization of ribosomes for crystallographic studies. *J. Biol. Chem.* **264:**19138–19142.

Wittmann, H. G. 1983. Architecture of prokaryotic ribosomes. *Annu. Rev. Biochem.* **52:**35–65.

Wittmann, H. G., J. Müssig, H. S. Gewitz, J. Piefke, H. J. Rheinberger, and A. Yonath. 1982. Crystallization of *E. coli* ribosomes. *FEBS Lett.* **146:**217–220.

Yen, I. J., P. S. Macklin, and D. W. Cleavland. 1988. Autoregulated instability of beta-tubulin mRNAs by recognition of the nascent amino terminus of beta-tubulin. *Nature* (London) **334:** 580–585.

Yonath, A., H. D. Bartunik, K. S. Bartels, and H. G. Wittmann. 1984. Some X-ray diffraction patterns from single crystals of the large ribosomal subunits from *B. stearothermophilus. J. Mol. Biol.* **177:**201–206.

Yonath, A., F. Frolow, M. Shoham, J. Müssig, I. Makowski, C. Glotz, W. Jahn, S. Weinstein, and H. G. Wittmann. 1988a. Crystallography of ribosomes. *J. Crystal Growth* **90:**231–244.

Yonath, A., C. Glotz, H. S. Gewitz, K. S. Bartels, K. von Böhlen, I. Makowski, and H. G. Wittmann. 1988b. Characterization of crystals of small ribosomal subunits. *J. Mol. Biol.* **203:**831–834.

Yonath, A., K. R. Leonard, S. Weinstein, and H. G. Wittman. 1987a. Approaches to the determination of the three-dimensional architecture of ribosomal particles. *Cold Spring Harbor Symp. Quant. Biol.* **52:**729–741.

Yonath, A., K. R. Leonard, and H. G. Wittmann. 1987b. A tunnel in the large ribosomal subunit revealed by three-dimensional image reconstruction. *Science* **236:**813–816.

Yonath, A., J. Müssig, B. Tesche, S. Lorentz, V. A. Erdmann, and H. G. Wittmann. 1980. Crystals of the large ribosomal subunits from *Bacillus stearothermophilus. Biochem. Internat.* **1:**428–435.

Yonath, A., M. A. Saper, I. Makowski, J. Müssig, J. Piefke, H. D. Bartunik, K. S. Bartels, and H. G. Wittmann. 1986a. Characterization of single crystals of the large ribosomal particles from *B. stearothermophilus. J. Mol. Biol.* **187:**633–636.

Yonath, A., M. A. Saper, F. Frolow, I. Makowski, and H. G. Wittmann. 1986b. Characterization of single crystals of large ribosomal particles from a mutant of *Bacillus stearothermophilus. J. Mol. Biol.* **192:**161–162.

Yonath, A., M. A. Saper, and H. G. Wittmann. 1986c. Structural studies on ribosomal particles, p. 112–129. *In* B. Hardesty and G. Kramer (ed.), *Structure, Function, and Genetics of Ribosomes.* Springer-Verlag, New York.

Yonath, A., B. Tesche, S. Lorenz, J. Müssig, V. A. Erdmann, and H. G. Wittmann. 1983. Several crystal forms of the 50S ribosomal particles of *Bacillus stearothermophilus. FEBS Lett.* **154:** 15–20.

Yonath, A., and H. G. Wittmann. 1988. Approaching the molecular structure of ribosomes. *J. Biophys. Chem.* **29:**17–29.

Yonath, A., and H. G. Wittmann. 1989a. Crystallographic and image reconstruction studies on ribosomal particles from bacterial sources. *Methods Enzymol.* **164:**95–117.

Yonath, A., and H. G. Wittmann. 1989b. Challenging the three-dimensional structure of ribosomes. *Trends Biochem. Sci.* **14:** 329–335.

Yusupov, M. M., S. V. Tischenko, S. D. Trakhanov, S. N. Ryazantsev, and M. B. Garber. 1988. A new crystalline form of 30S ribosomal subunits from *Thermus thermophilus. FEBS Lett.* **238:**113–115.

Chapter 9

Detailed Structures of rRNAs: New Approaches

BERNARD EHRESMANN, CHANTAL EHRESMANN, PASCALE ROMBY, MARYLÈNE MOUGEL, FLORENCE BAUDIN, ERIC WESTHOF, and JEAN-PIERRE EBEL

THE CHALLENGE

One of the challenges of these last decades is understanding the detailed conformation of ribosomes and its relation to function. Ribosomes are multimolecular particles with an extraordinary degree of complexity. rRNAs not only are scaffoldings for protein binding but directly participate in ribosomal functions. The conformation of rRNAs and their folding inside the subunits thus represent an essential element for understanding the molecular mechanisms of protein biosynthesis.

The best method for determining the three-dimensional structure of biological molecules is X-ray crystallography. Until now, the crystal structures of only two RNA molecules, yeast $tRNA^{Phe}$ and $tRNA^{Asp}$, have been obtained at high resolution by X-ray diffraction. Although crystals can be obtained from 70S ribosomes, and from 30S and 50S subunits from various bacteria, difficulties in crystal growth remain, and the diffraction does not yield information at high resolution (Yonath and Wittmann, 1989). Moreover, once crystallization and phase problems have been solved, the information will be necessarily static; the system does not lend itself easily to the study of molecular association. Therefore, it is important to devise experimental approaches to study the conformation of RNAs in solution under nearly physiological conditions, to identify sites interacting with ligands, and to monitor possible conformational rearrangements resulting from their association. Biochemical approaches have been extensively developed in recent years, essentially through the use of enzymatic and chemical probes and of RNA-RNA and RNA-protein cross-linking. Furthermore, recent developments in genetic engineering and RNA synthesis by in vitro transcription render it possible to prepare any desired RNA sequence. The challenge is to derive models of structure that integrate experimental data and obey the general rules of thermodynamics and stereochemistry. Computer programs have to be developed for three-dimensional model building. Criteria for judging the validity of these models are rather difficult to establish, and other considerations, such as interactions with other macromolecules and phylogenetic conservation of the structure of RNAs of the same class, must also be taken into account. Most important, site-directed mutagenesis of selected residues, considered critical in the tertiary folding, is an essential tool for testing models.

SECONDARY STRUCTURE OF rRNAs

The information for generating secondary and tertiary structures is contained in the sequence. The secondary folding depends on the ability of bases to form hydrogen bonds. Watson-Crick interactions are the main determinants of the secondary structure of natural RNAs. However, other base-pairing arrangements are possible, the most commonly found being between G and U. Other noncanonical base pairs have also been described (Table 1), inducing some distortion in helices. The RNA chain folds back on itself into a series of hairpins and loops. Helices can be interrupted by interior or bulged loops, separated by interhelical single-stranded regions, or can form more complicated structures such as noncanonical base pairs, branched helices, pseudoknots, and coaxial stacking of helices (Fig. 1).

One criterion that has proven very useful for

Bernard Ehresmann, Chantal Ehresmann, Pascale Romby, Marylène Mougel, Florence Baudin, Eric Westhof, and Jean-Pierre Ebel ■ Laboratoire de Biochimie, Institut de Biologie Moléculaire et Cellulaire du Centre National de la Recherche Scientifique, 15 rue René Descartes, 67084 Strasbourg Cedex, France.

Table 1. Some noncanonical interactions in nucleic acids

Interaction	Hydrogen bond	Observed in:	Reference(s)
A-U	(N7,N6)-(N3,O2)	$tRNA^{Phe}$, $tRNA^{Asp}$: A-14–U-8, A-58–T-54	Quigley et al., 1975; Westhof et al., 1985
A-U	(N1,N6)-(N3,O2)	$tRNA^{Asp}$: A-15–U-48	Westof et al., 1985
A-G	(N1,N6)-(N1,O6)	$tRNA^{Phe}$, $tRNA^{Asp}$: A-44–G-26	Quigley et al., 1975; Westhof et al., 1985
A-G	(N6)-(N7)	$tRNA^{Asp}$: A-46–G-22–Ψ13	Westof et al., 1985
A-A	(N6,N7)-(N7,N6)	$tRNA^{Phe}$, $tRNA^{Asp}$: A-9–A-23–U-12	Quigley et al., 1975; Westhof et al., 1985
G-G	(N3,N2)-(N2,N3)	β-Dodecamer	Wing et al., 1980
G-G	(N2)-(O6)	$tRNA^{Asp}$: G-45–G-10–U-25	Westhof et al., 1985
G-G	(N6)-(O6)	$tRNA^{Phe}$: G-45–G-10–C-25	Quigley et al., 1975
G-G	(N1,N2)-(N7,O6)	$tRNA^{Phe}$: m^7G-46–G-22–C-13	Quigley et al., 1975
U-U	(N3,O4)-(O2,N3)	$tRNA^{Asp}$: U-35–U-35, anticodon-anticodon interaction	Westhof et al., 1985
C-C$^+$	(N4,N3)-(N3,O2)	$tRNA^{Gly}$: C-35–C-35, anticodon-anticodon interaction	Romby et al., 1986
C-C$^+$	(N4,N3,O2)-(O2,N3,N4)	Poly(C)	Cantor and Schimmel, 1980

establishing the secondary structures of rRNAs is phylogenetic conservation. It turns out that molecules have diverged in sequence through evolution while maintaining their structural organization through compensatory base changes. If sequences of RNAs of the same class from different organisms are compared, then a base change in one strand of a putative helix should be compensated by a complementary base change in the other strand. Thus, extensive sequence comparison of 5S rRNAs led to a consensus model for procaryotes and eucaryotes (e.g., Delihas et al., 1984; Wolters and Erdmann, 1988). Secondary structure models for *Escherichia coli* 16S and 23S rRNAs were proposed by groups in Santa Cruz, Strasbourg, and Berlin in the late 1970s. They were based mainly on sequence comparison, accessibility data with RNases and chemicals (essentially kethoxal), RNA fragment interactions, and RNA-RNA cross-links. These models have been refined as the data base has grown (Gutell et al., 1985; Dams et al., 1988; Gutell and Fox, 1988).

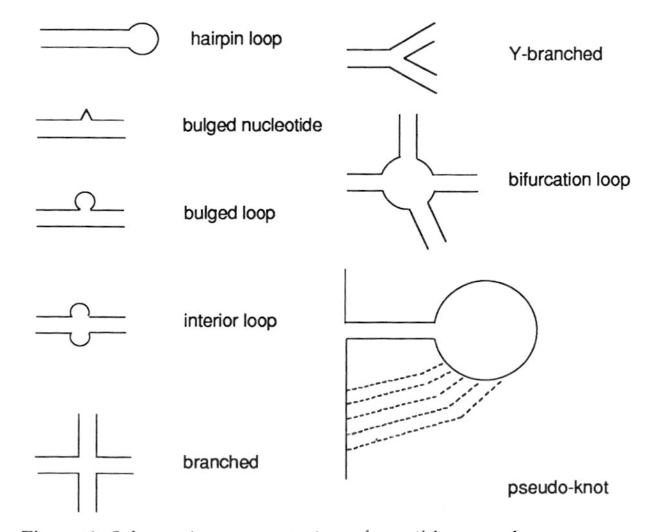

Figure 1. Schematic representation of possible secondary structure motifs in RNAs. Regular double-stranded helices are represented by parallel lines. For the pseudoknot structure, base pairs are denoted by broken lines.

THREE-DIMENSIONAL FOLDING

Our knowledge of the spatial organization of rRNAs inside the subunit has benefited from the improvements in nucleic acid technology. An impressive amount of data has accumulated from various biochemical approaches. The location of proteins on RNAs has been extensively studied by footprinting techniques or by RNA-protein cross-linking (e.g., Stern et al., 1989; Brimacombe et al., 1988). Intra-RNA cross-linking has also enabled RNA regions in close proximity to be identified (Brimacombe et al., 1988). The combination of such data with the spatial location of the proteins in the small ribosomal subunit of *E. coli* by neutron-scattering studies [brought to a conclusion by Capel et al. [1987]) has allowed models of three-dimensional organization of the 30S subunit to emerge (Expert-Bezançon and Wollenzien, 1985; Stern et al., 1988b; Brimacombe et al., 1988). The validity of such models obviously depends on the number and correctness of the constraints imposed on the construction. These low-resolution models provide a first relative orientation of RNA helices inside the subunits, which provides insight into the quaternary structure of the small subunit. However, the RNA structure is far from known at atomic resolution. Yeast $tRNA^{Phe}$ and $tRNA^{Asp}$ are the only biological RNAs of tertiary structure now known at the atomic level (Quigley et al., 1975; Westhof et al., 1985). In light of these data, it appears that the tertiary folding involves numerous hydrogen-bonded interactions (often noncanonical interactions; Table 1), in addition to those involved in normal base pairing. A specific feature of RNA structure is the existence of multiple-stranded interactions and of a

Table 2. Molecular weights and specificities of enzymatic and chemical probes

Probe	Mol wt	Specificity
RNase T_1	11,000	Unpaired G
RNase U_2	12,490	Unpaired A>G
Nuclease S1	32,000	Unpaired N
RNase V_1	15,900	Paired or stacked N
DMS	126	C(N3), A(N1), G(N7)
Diethyl pyrocarbonate	174	A(N7)
CMCT	424	U(N3), G(N1)
Kethoxal	148	G(N1,N2)
ENU	117	Phosphates

specific pattern of chain bending that is typical of protein folding. Another feature of RNA tertiary interactions is the existence of hydrogen bonds between some bases and the ribose-phosphate backbone. These interactions appear to play a crucial role in stabilizing the tertiary folding, especially those involving the 2′-OH group of the ribose. It is relevant that bases involved in hydrogen-bonded tertiary interactions should be either conserved or replaced by compensating bases during evolution.

The elaboration of a detailed tertiary structure model requires a maximum number of data and appropriate computer programs for modeling. This represents a formidable task, given the large number of nucleotides to be handled, especially for large RNAs. Our group has devised a methodological approach based on RNA structure probing, graphic modeling, in vitro RNA synthesis, and mutagenesis. This approach has been applied to defined 16S rRNA regions and 5S rRNAs.

PROBING RNA STRUCTURE IN SOLUTION

Various enzymes and chemicals are available for probing the solution structure of RNAs. The most commonly used are shown in Table 2, together with their molecular weights and their target sites. The mechanisms of action of the probes, the limitations of the technique, and the methods for detection of cuts or modifications are described elsewhere (Ehresmann et al., 1987; Stern et al., 1988a). Probing experiments are generally conducted under different conditions: native (low temperature, in the presence of magnesium and monovalent salt), semidenaturing (low temperature, in the presence of EDTA), and denaturing (high temperature, in the presence of EDTA). This permits an evaluation of the degree of stability of helices. Tertiary interactions and noncanonical interactions that are not very stable are expected to melt under semidenaturing conditions.

Because of their bulky size, nucleases are sensitive to steric hindrance, especially when RNA is tested in the presence of bound proteins. They can also be sterically blocked by particular folding of the RNA. RNases T_1, U_2, and T_2 and nuclease S1 all map accessible unpaired regions. RNase V_1 is particularly useful, since it is the only probe that provides positive information on the existence of a helical structure.

Chemical probes are only weakly sensitive to steric hindrance and yield detailed insight at the atomic level. Most of them specifically modify selected positions of bases. Thus, using dimethyl sulfate (DMS) and 1-cyclohexyl-3-(2-morpholinoethyl) carbodiimide metho-*p*-toluene sulfonate (CMCT), it is possible to test one Watson-Crick position on each of the four bases [C(N3)-A(N1) and U(N3)-G(N1), respectively]. Kethoxal also allows one to map unpaired guanines. The nonreactivity of position N7 of guanine toward DMS, or of adenine toward diethyl pyrocarbonate, reflects the involvement of the base in a noncanonical base pairing or in base stacking. Ethylnitrosourea (ENU) is an *N*-nitroso alkylating reagent that has an affinity for the phosphate group oxygens of nucleic acids, in contrast to other alkylating reagents, which primarily alkylate ring nitrogens. In the case of RNA, the resulting phosphotriesters are unstable and are easily split by mild alkaline treatment. Compared with other chemical probes, ENU is sequence and secondary structure independent. Therefore, ENU is a specific probe of the tertiary structure, mapping phosphates not involved in hydrogen bonds or in cation coordination.

These various probes have been used by several groups to investigate in detail the conformation of rRNAs and to identify the RNA-binding sites of ribosomal proteins or other ribosome ligands (Moazed et al., 1986; Baudin et al., 1987; Baudin et al., 1989; Stern et al., 1989; Egebjerg et al., 1987; Moazed and Noller, 1989; Romby et al., 1988; Romaniuk et al., 1988; Egebjerg et al., 1989). Furthermore, they can be used to detect conformational rearrangements induced by protein binding, by subunit association, or during the various steps of the ribosome cycle. Recently, the use of lead-induced cleavage has been used to probe 16S rRNA (Gornicki et al., 1989). This probe is particularly useful to probe dynamic transitions.

MODELING

The building of a tertiary structure model by computer graphic modeling is based on the knowledge of the secondary structure from solution data. After the elementary motifs constituting the structure are determined, the most appropriate motifs in a specially devised structure bank are chosen, and the secondary structure motifs are assembled into a tertiary structure according to the rules of stereochemistry. The structure bank contains the crystallographically determined nucleic acid structures as well as previously modeled structures. Model building is assisted by several computer programs (Westhof et al., in press). With the program FRAGMENT, one can insert any base sequence on any structural motif existing in the structure bank. Standard RNA helices are used for the base-paired regions and constructed with the program NAHELIX. Structural secondary fragments may consist of several structural motifs. Each fragment is assembled separately from its constituent motifs interactively on the graphic system with the program FRODO (Jones, 1978). The fragment is then subjected to geometrical and stereochemical refinement to ensure proper geometry and prevent bad contacts (Konnert and Hendrickson, 1980; Westhof et al., 1985). The construction of a secondary substructure (for example, a helix with a hairpin loop) is subjected in most cases to geometrical, stereochemical, and energetical constraints, which reduces the number of degrees of freedom. However, assembling those substructures into the whole structure is very often open to numerous possibilities, especially when the links are single-stranded regions. Therefore, the relative spatial orientations between substructures are difficult to assess with confidence on the basis of geometrical and stereochemical constraints alone. Nevertheless, with additional biochemical observations such as the binding sites of proteins and structural study of defined RNA mutants, the possible regions in space can be restricted. At the end of the modeling, as well as during the construction of substructures, a program (ACCESS) based on the algorithm and definition of Richmond (1984) allows one to calculate the accessible surfaces of the probed atoms. The radius of the solvent sphere rolling onto the van der Waals surface is adapted to the chemical probe. The calculated values are then checked against experimental data. A good correlation is generally found between theoretical accessibilities and experimental reactivities. However, the correlation is not straightforward, since the calculation does not take into account the thermodynamic stability or the molecular dynamics in the environment of the tested atom.

Criteria for judging the validity and quality of the modeling are difficult to establish. As guides during the modeling, stereochemical rules are enforced. Yet energetic and stereochemical considerations alone cannot be taken as proofs. The sites of interaction with other macromolecules obtained from footprinting experiments can give a useful indication of the correctness of the model. However, results independent of those used in the modeling should be used, and they are not always available. What appears more convincing is, first, generalization of the model to other molecules of the same class and, second, directed mutagenesis on selected residues considered critical in the tertiary folding.

PROBING THE PHOSPHATES OF *E. COLI* 16S rRNA

Naked RNA

ENU has been used to map phosphates in 16S rRNA engaged in tertiary interactions through hydrogen bonds or in ion coordination (Baudin et al., 1989) (Fig. 2). Seven percent of the phosphates are found involved in such interactions; 57% of them are located in loops or interhelical regions, where they are involved in maintaining local intrinsic structures or long-distance tertiary interactions. The other phosphates (43%) are found in helical regions. These phosphates often occur at the proximity of bulged nucleotides or in irregular helices containing noncanonical base pairs (and bulges) and are assumed to bind cations in order to neutralize negative charges and stabilize unusual phosphate backbone folding.

Effect of 30S Subunit Assembly

Probing the phosphates with ENU within the 30S subunit allowed us to map phosphates in contact with proteins (Fig. 2). For instance, the 3′-terminal domain and the 5′-proximal region are largely exposed to the probe. Results can easily be correlated to footprinting and cross-linking experiments. As a general feature, most of the phosphates that are found protected in the naked RNA are also protected in the subunit, indicating that the conformation of 16S rRNA is not drastically affected by protein binding. However, the loss of protection (for instance, at position 1401) and the enhanced reactivity of several phosphates most likely reflect conformational adjustments or cation displacement. Remarkably, enhanced reactivities are most often observed in or near regions that have been shown to be in contact with proteins (Fig. 2). It is relevant that structural adjustments are observed in two functionally strategic regions: the 1400 region, found to be involved in the decoding process (Ofengand et al., 1986), and region 827–891, which is probably involved in translation initiation, since initiation factor IF3 has been cross-linked in this area (Ehresmann et al., 1986). This emphasizes the role of ribosomal proteins in building the ribosomal functional sites. Similar protein-induced conformational adjustments have also been observed with other probes (Baudin et al., 1987; Gornicki et al., 1989; Stern et al., 1989).

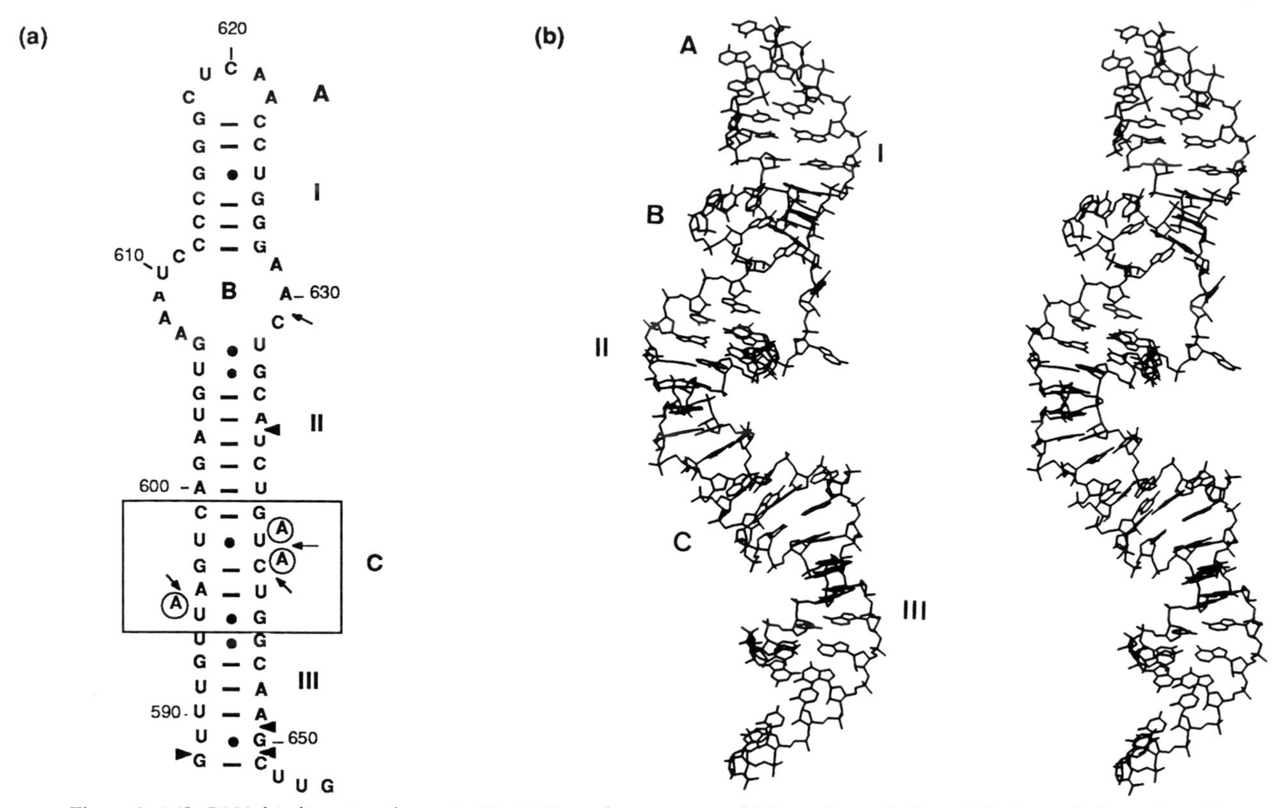

Figure 3. 16S rRNA-binding site of protein S8. (a) Secondary structure folding scheme. I, II, and III denote helices; A and B denote loops; region C is boxed. Protein-induced reactivity changes are indicated: nucleotides protected from DMS are encircled; arrows denote protections from RNases T_1 and U_2 (single strand specific); arrowheads denote protection from RNase V_1 (double strand specific). (b) Stereoscopic view from the three-dimensional model proposed for nucleotides 588 to 654. The various secondary structural elements are indicated. The configuration of UUG-654 is arbitrary. (From Mougel et al., 1987).

recent work has shown that the target site of protein S8 on its mRNA, involved in the translational control of S8, shows striking similarities in both primary and secondary structures, especially in region C (Ceretti et al., 1988; Gregory et al., 1988). These results suggest that the same structural features in both 16S rRNA and the mRNA target site are recognized by S8. Site-directed mutagenesis has further stressed the crucial role of nucleotides in this region (same authors).

RNA-Binding Site of S15

The RNA-binding site of protein S15 is contained within a composite hairpin (nucleotides 655 to 752) interrupted by several interior loops (Fig. 5). One of them, loop C, appears to adopt a complex (and not yet elucidated) conformation in 16S rRNA but is found more reactive in the synthetic fragment. However, the reactivity patterns of the fragment and 16S rRNA become identical in the presence of the protein, suggesting that S15 stabilizes the RNA into a common conformation. Footprinting data suggest that the protein is most likely centered on helix III, in an irregular helical region containing noncanonical A · G base pairs and bulged nucleotides (Fig. 5). Remarkably, S15 also causes enhanced chemical reactivities and enzyme accessibilities, reflecting a local conformational rearrangement induced by protein binding. This conformational adjustment most likely contributes to the cooperative binding of other ribosomal proteins. Indeed, S15 has been reported to be required for the assembly of proteins S6 and S18, which bind in the same region.

CONFORMATION OF 5S rRNA

We have investigated in detail the secondary and tertiary structures of a procaryotic 5S rRNA (the chloroplastic 5S rRNA from spinach [Romby et al., 1988]) and a eucaryotic 5S rRNA (the somatic and oocyte 5S rRNA from *Xenopus laevis* [Romaniuk et al., 1988]). Our results confirm the existence of the five helices proposed in the consensus model of 5S rRNA derived from sequence comparison (Fig. 6b). In addition, the data do not support the existence of tertiary interactions between the three arms of the molecule. In both cases, a detailed atomic model was built by using graphic modeling (Westhof et al., 1989). Both models integrate stereochemical constraints and experimental probing data.

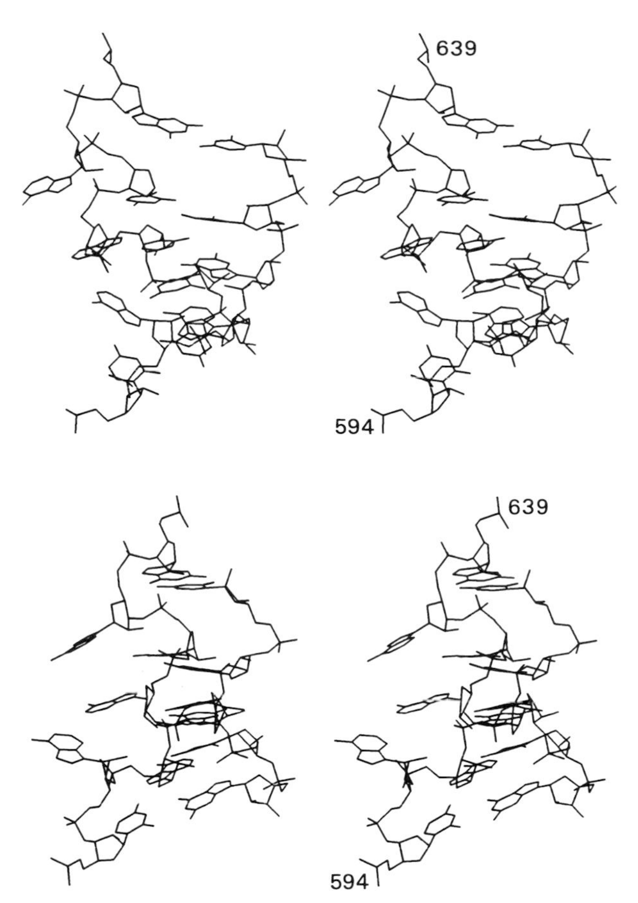

Figure 4. Stereoscopic view of nucleotides 594 to 599 and 639 to 645 (region C). Two orientations are given. The three bulged adenines facing the major groove are clearly seen. (From Mougel et al., 1987.)

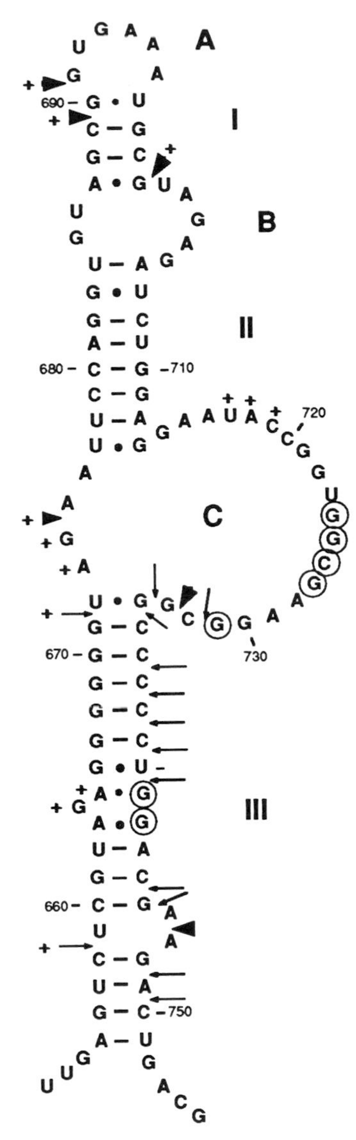

Figure 5. 16S rRNA-binding site of protein S15. The secondary structure is shown, with helices denoted by I to III and loops by A to C. Note that the conformation of internal loop C is not well defined. Protein-induced reactivity changes are indicated: nucleotides protected from DMS or CMCT are encircled; nucleotides displaying an enhanced reactivity are denoted by +; arrows denote protection from RNase V_1; and arrowheads denote protection from RNase T_1.

Chloroplastic 5S rRNA

The model of the 5S RNA adopts a Y-shaped structure, with a short stalk made of stem 1 and the two arms made of stems 2 and 3 (Fig. 7a). Such a Y-shaped configuration is also suggested by Egebjerg et al. (1989) for the *E. coli* 5S rRNA on the basis of probing experiments. In our model, helices B and D are stacked and not far from colinearity. However, the regions depicted as unpaired internal loops appear to fold into complex and organized structures. In particular, loop d (73–79/100–106) displays unusual features: several noncanonical base pairs are formed, involving A(N6,N7)-A(N7,N6) and A(N6, N7)-G(O6,N1), and three residues are bulging out. This results in a completely unwound and extended configuration (Fig. 7a). This particular conformation appears to be stabilized by magnesium, and several phosphates most likely involved in ion coordination have been localized in this region. The question arises as to whether this particular conformation represents a general feature of procaryotic 5S rRNAs. Recent work of Zhang and Moore (1989), based on nuclear magnetic resonance studies, showed that loop d, contained in an *E. coli* RNA fragment (nucleotides 1 to 11 and 69 to 120), does not conform either to the standard phylogenetically derived model (Delihas et al., 1984) or to the spinach chloroplast model (Romby et al., 1988). More detailed information is required for an understanding of the particular folding of this loop in the *E. coli* RNA.

Two ribosomal proteins (CS L12 and CS L13) that bind to the 5S rRNA have been isolated from spinach chloroplast, and their RNA-binding sites have been located by footprinting experiments (Toukifimpa et al., 1989). Protein CS L12 is centered

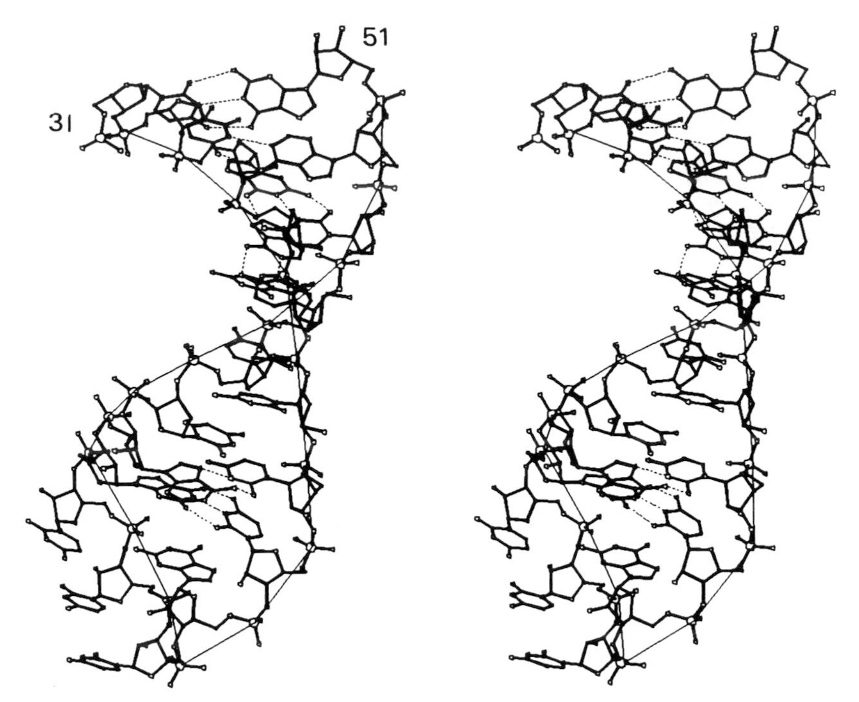

Figure 8. Stereoscopic view of loop c of *X. laevis* 5S rRNA. Nucleotides 31 to 51 are shown. Hydrogen bonds are indicated by dotted lines.

with the "U-33 turn" made at G-39 and residues U-40, C-42, and U-43 playing the role of the anticodon triplet. The validity of the model is now being tested by site-directed mutagenesis. The first results show that mutations in loop c result only in local readjustments and do not affect the rest of the molecule (Romby et al., 1990; Brunel et al., in press). Also, a truncated fragment of RNA corresponding to stem 3 adopts the same conformation as in the entire molecule. These results fully confirm the Y-shaped model and the absence of tertiary interactions.

CONCLUSION

The results presented above show that biochemical approaches, coupled with computer graphic modeling and mutagenesis, represent a powerful approach to investigate the conformation of RNAs at the atomic level. Furthermore, this approach can provide specific information on interactions of RNAs with other macromolecules and may reveal conformational rearrangements. In particular, it has been found that ribosomal proteins induce conformational changes in the rRNA that play an important role in the cooperativity of ribosome assembly and in the construction of functional sites. Dynamic properties of RNAs are probably an essential feature that accounts for the multiple functions in which RNAs are implicated.

REFERENCES

Baudin, F., C. Ehresmann, R. Romby, M. Mougel, J. Colin, L. Lempereur, J. P. Bachellerie, J. P. Ebel, and B. Ehresmann. 1987. Higher-order structure of domain III in *Escherichia coli* ribosomal 16S RNA, 30S subunit and 70S ribosome. *Biochimie* **69:**1081–1096.

Baudin, F., M. Mougel, P. Romby, F. Eyermann, J. P. Ebel, B. Ehresmann, and C. Ehresmann. 1989. Probing the phosphates of the *Escherichia coli* ribosomal 16S RNA in its naked form, in the 30S subunit, and in the 70S ribosome. *Biochemistry* **28:** 5847–5855.

Brimacombe, R., J. Atmadja, W. Stiege, and D. Schüler. 1988. A detailed model of the three-dimensional structure of *Escherichia coli* 16S ribosomal RNA *in situ* in the 30S subunit. *J. Mol. Biol.* **199:**115–136.

Brunel, C., P. Romby, E. Westhof, P. J. Romaniuk, B. Ehresmann, and C. Ehresmann. Effect of mutations in domain 2 on the structural organization of oocyte 5S rRNA from *Xenopus laevis*. *Biochemistry*, in press.

Cantor, C. R., and P. R. Schimmel. 1980. Nucleic acid structural transitions, p. 1146–1160. *In Biophysical Chemistry*, part III. *The Behaviour of Biological Macromolecules*. W. H. Freeman and Co., San Francisco.

Capel, M. S., D. M. Engelman, B. R. Freeborn, M. Kjeldgaard, J. A. Langer, V. R. Ramakrishnan, D. G. Schindler, D. K. Schneider, B. P. Schoenborn, I. Y. Sillers, S. Yabuki, and P. B. Moore. 1987. A complete mapping of the proteins in the small ribosomal subunit of *Escherichia coli*. *Science* **238:**1403–1406.

Ceretti, D. P., L. C. Mattheakis, K. R. Kearney, L. Vu, and M. Nomura. 1988. Translational regulation of the *spc* operon in *Escherichia coli*. Identification and structural analysis of the target site for S8 repressor protein. *J. Mol. Biol.* **204:**309–329.

Christiansen, J., R. S. Brown, B. S. Sproat, and R. A. Garrett. 1987. *Xenopus* transcription factor IIIA binds primarily at junctions between double helical stems and internal loops in oocyte 5S RNA. *EMBO J.* **6**:453–460.

Dams, E., L. Hendricks, Y. Van de Peer, J. M. Neefs, G. Smits, I. Vandenbempt, and R. De Wachter. 1988. Compilation of small ribosomal subunit RNA sequences. *Nucleic Acids Res.* **16**(Suppl.):r87–r173.

Delihas, N., J. Andersen, and R. P. Singhal. 1984. Structure, function and evolution of 5S ribosomal RNAs. *Prog. Nucleic Acid Res. Mol. Biol.* **31**:161–190.

Egebjerg, J., J. Christiansen, R. S. Brown, N. Larsen, and R. A. Garrett. 1989. Protein L18 binds primarily at the junctions of helix II and internal loops A and B in *Escherichia coli* 5S RNA. Implications for 5S RNA structure. *J. Mol. Biol.* **206**:651–668.

Egebjerg, J., H. Leffers, A. Christensen, H. Andersen, and R. A. Garrett. 1987. Structure and accessibility of domain I of *Escherichia coli* 23S RNA in free RNA, in the L24-RNA complex and in 50S subunits. *J. Mol. Biol.* **196**:125–136.

Ehresmann, C., F. Baudin, M. Mougel, P. Romby, J. P. Ebel, and B. Ehresmann. 1987. Probing the structure of RNAs in solution. *Nucleic Acids Res.* **15**:9109–9128.

Ehresmann, C., H. Moine, M. Mougel, J. Dondon, M. Grunberg-Manago, J. P. Ebel, and B. Ehresmann. 1986. Cross-linking of initiation factor IF3 to *Escherichia coli* 30S subunit by trans-diamminedichloroplatinum (II): characterization of two cross-linking sites in 16S rRNA; a possible way of functioning for IF3. *Nucleic Acids Res.* **14**:4803–4821.

Expert-Bezançon, A., and P. L. Wollenzien. 1985. Three-dimensional arrangement of the *Escherichia coli* 16S ribosomal RNA. *J. Mol. Biol.* **184**:53–66.

Gornicki, P., F. Baudin, P. Romby, M. Wiewiorowski, W. Kryzosiak, J. P. Ebel, C. Ehresmann, and B. Ehresmann. 1989. Use of lead(II) to probe the structure of large RNAs. Conformation of the 3′ terminal domain of E. coli 16S rRNA and its involvement in building the tRNA binding sites. *J. Biomol. Struct. Dynam.* **6**:971–984.

Gregory, R. J., P. B. F. Cahill, D. L. Thurlow, and R. A. Zimmermann. 1988. Interaction of *Escherichia coli* ribosomal protein S8 with its binding sites in ribosomal RNA and messenger RNA. *J. Mol. Biol.* **204**:295–307.

Gutell, R. R. and G. E. Fox. 1988. A compilation of large subunit RNA sequences presented in a structural format. *Nucleic Acids Res.* **16**:r175–r269.

Gutell, R. R., B. Weiser, C. R. Woese, and H. F. Noller. 1985. Comparative anatomy of 16S-like ribosomal RNA. *Prog. Nucleic Acid Res. Mol. Biol.* **32**:155–216.

Jones, T. A. 1978. A graphic model building and refinement system for macromolecules. *J. Appl. Crystallogr.* **11**:268–278.

Konnert, J. H., and W. A. Hendrickson. 1980. A restrained parameter thermal-factor refinement procedure. *Acta Crystallogr.* Sect. A **36**:344–349.

Moazed, D., and H. F. Noller. 1989. Interaction of tRNA with 23S rRNA in the ribosomal A, P, and E sites. *Cell* **57**:585–597.

Moazed, D., S. Stern, and H. F. Noller. 1986. Rapid chemical probing of conformation in 16S ribosomal RNA and 30S ribosomal subunits using primer extension. *J. Mol. Biol.* **187**:399–416.

Mougel, M., F. Eyermann, E. Westhof, P. Romby, A. Expert-Bezançon, J. P. Ebel, B. Ehresmann, and C. Ehresmann. 1987. Binding of *Escherichia coli* ribosomal protein S8 to 16S rRNA. A model for the interaction and the tertiary structure of the RNA binding site. *J. Mol. Biol.* **198**:91–107.

Mougel, M., C. Philippe, J. P. Ebel, B. Ehresmann, and C. Ehresmann. 1988. The *E. coli* 16S rRNA binding site of ribosomal protein S15: higher-order structure in the absence and in the presence of the protein. *Nucleic Acids Res.* **16**:2825–2839.

Ofengand, J., J. Ciesiolka, R. Denman, and K. Nurse. 1986. Structural and functional interactions of the tRNA-ribosome complex, p. 473–494. *In* B. Hardesty and G. Kramer (ed.), *Structure, Function, and Genetics of Ribosomes.* Springer-Verlag, New York.

Quigley, G. J., A. Wang, N. C. Seeman, F. L. Suddath, A. Rich, J. L. Sussman, and S. H. Kim. 1975. Hydrogen bonding in yeast $tRNA^{Phe}$. *Proc. Natl. Acad. Sci. USA* **72**:4866–4870.

Richmond, T. J. 1984. Solvent accessible surface area and excluded volumes in proteins. *J. Mol. Biol.* **178**:63–68.

Romaniuk, P. J., I. L. Stevenson, C. Ehresmann, P. Romby, and B. Ehresmann. 1988. A comparison of the solution structures and conformational properties of the somatic and oocyte 5S rRNAs of *Xenopus laevis. Nucleic Acids Res.* **16**:2295–2313.

Romby, P., C. Brunel, E. Westhof, C. Ehresmann, and B. Ehresmann. 1990. Structure of *Xenopus laevis* 5S rRNA as determined by solution data and computer graphic modeling, p. 19–27. *In* D. Vasilescu et al. (ed.), *Water and Ions in Biomolecular Systems.* Birkhäuser Verlag.

Romby, P., E. Westhof, D. Moras, R. Giegé, C. Houssier, and H. Grosjean. 1986. Studies on anticodon-anticodon interactions: hemi-protonation of cytosines induces self-pairing through the GCC anticodon of *E. coli* tRNA-Gly. *J. Biomol. Struct. Dynam.* **4**:193–203.

Romby, P., E. Westhof, R. Toukifimpa, R. Mache, J. P. Ebel, C. Ehresmann, and B. Ehresmann. 1988. Higher-order structure of chloroplastic 5S ribosomal RNA from spinach. *Biochemistry* **27**:4721–4730.

Stern, S., D. Moazed, and H. F. Noller. 1988a. Structural analysis of RNA using chemical and enzymatic probing monitored by primer extension. *Methods Enzymol.* **164**:481–489.

Stern, S., T. Powers, L. M. Changchien, and H. F. Noller. 1989. RNA-protein interactions in 30S ribosomal subunit: folding and function of 16S rRNA. *Science* **244**:783–790.

Stern, S., B. Weiser, and H. F. Noller. 1988b. Model for the three-dimensional folding of 16S ribosomal RNA. *J. Mol. Biol.* **204**:447–481.

Toukifimpa, R., P. Romby, C. Rozier, C. Ehresmann, B. Ehresmann, and R. Mache. 1989. Characterization and footprint analysis of two 5S rRNA binding proteins from spinach chloroplast ribosomes. *Biochemistry* **28**:5840–5846.

Westhof, E., P. Dumas, and D. Moras. 1985. Crystallographic refinement of yeast aspartic acid transfer RNA. *J. Mol. Biol.* **184**:119–145.

Westhof, E., P. Romby, C. Ehresmann, and B. Ehresmann. *In* D. L. Beveridge and R. Lavery (ed.), *Theoretical Biochemistry and Molecular Biophysics. A Comprehensive Survey.* Adenine Press, in press.

Westhof, E., P. Romby, P. J. Romaniuk, J. P. Ebel, C. Ehresmann, and B. Ehresmann. 1989. Computer modeling from solution data of spinach chloroplast and of *Xenopus laevis* somatic and oocyte 5S rRNAs. *J. Mol. Biol.* **207**:417–431.

Wing, R., H. Drew, T. Takano, C. Broca, K. Itakura, and R. E. Dickerson. 1980. Crystal structure analysis of a complete turn of β-DNA. *Nature* (London) **287**:755–758.

Wolters, J., and V. A. Erdmann. 1988. Compilation of 5S rRNA and 5S rRNA gene sequences. *Nucleic Acids Res.* **16**(Suppl.):r1–r70.

Yonath, A., and H. G. Wittmann. 1989. Challenging the three-dimensional structure of ribosomes. *Trends Biochem. Sci.* **14**:329–335.

Zhang, P., and P. B. Moore. 1989. An NMR study of the helix V-loop E region of the 5S RNA from *Escherichia coli. Biochemistry* **28**:4607–4615.

It is striking that the binding reactions are overwhelmingly entropy driven. Specific protein-nucleic acid interactions generally have a strongly favorable entropy of binding (see summary in Mougel et al., 1986), though the binding of the R17 coat protein to an RNA hairpin structure is entirely enthalpy driven (Carey and Uhlenbeck, 1983).

The dependence of the protein-RNA affinity constants on monovalent salt concentration (∂log K/∂log [K^+]) is relatively weak in the few cases studied (Table 1). For proteins binding to an infinite, regular polymer such as DNA, the salt dependence can be interpreted in terms of the number of ionic contacts between nucleic acid phosphates and protein positive charges, and the total electrostatic contribution to the free energy of binding can be estimated by extrapolating the binding constant to 1 M K^+ (Record et al., 1976). The irregular structure of rRNAs precludes a rigorous analysis of this sort. In addition, each of the proteins examined so far shows some affinity for anions (S4 and S8) or multivalent cations (L11); binding of these additional ions alters ∂log K/∂log [K^+] in an unknown way. Thus, only limits can be placed on the contribution of electrostatic interactions to binding: two to five ionic contacts are probably made by the proteins listed, and nonelectrostatic interactions must be substantial. In the case of S8, four phosphates are protected from reaction with ethylnitrosourea (Mougel et al., 1987), consistent with the estimated number of five ionic contacts.

Since ribosomal proteins are basic, one might have predicted extensive ionic contacts between ribosomal proteins and rRNA. While electrostatic interactions in DNA-protein complexes contribute primarily to nonspecific binding, an RNA-binding protein could match a three-dimensional array of positive charges to the conformation of the RNA backbone to obtain specificity; conceivably site-specific binding could be achieved with entirely electrostatic interactions. This extreme strategy is not used by the proteins studied so far, and it seems that electrostatics have a surprisingly minor role to play in the protein-RNA complexes.

In no case do we yet understand the origins of ribosomal protein–RNA complex specificity. Electrostatic interactions probably make only a small contribution; extensive hydrogen bonding would be expected to contribute a favorable enthalpy. Only hydrophobic interactions remain. Measurements of the temperature dependence of ribosomal protein binding affinities are not extensive or accurate enough to quantitatively estimate the hydrophobic stabilization in these complexes, but the weak temperature dependences observed suggest a large hydrophobic contribution. Hydrophobic contacts between a ribosomal protein and the three-dimensional structure of an RNA is a plausible mechanism for achieving specific binding (the wide minor groove of A-form helices and looped or bulged nucleotides all present nonpolar features), and stabilization of protein tertiary structure in an RNA complex could also bury hydrophobic surfaces. The possibility that protein assembly onto rRNA is largely driven by hydrophobic forces deserves serious investigation.

We are accustomed to thinking of protein-nucleic acid binding specificity as the result of a small number of sequence-specific contacts (principally hydrogen bonds) between the two macromolecules, but quite a different kind of complex is possible if there are substantial hydrophobic interactions. The main hydrophobic surface in RNA is the sugar backbone (Alden and Kim, 1979); thus, a protein could recognize an RNA structure in a sequence-independent way, by matching protein and RNA hydrophobic surfaces. A possible example is the TFIIIA protein, which specifically recognizes 5S rRNA but is remarkably indifferent to the RNA sequence (Romaniuk, 1989). Another example is the requirement of the R17 coat protein for a bulged purine residue in its hairpin recognition site. Substitution of many different base derivatives in the bulge position was unable to narrow the specificity to any one hydrogen-bonding position (Wu and Uhlenbeck, 1987). One interpretation of these results is that some aromatic purine carbons contribute to protein binding; the other is that the protein recognizes a distorted helix backbone conformation induced by the purine bulge (binding is insensitive to the sequence of base pairs surrounding the bulge). In either case, hydrophobic interaction between the protein and a surface of the RNA is required to account for the data.

RNA RECOGNITION SITES

The first attempts to define RNA structures recognized by individual proteins were "bind-and-chew" nuclease protection experiments (reviewed in Zimmermann, 1980). In some instances, protected RNA fragments could be isolated and shown to still recognize a protein, a powerful argument that the recognition features are indeed limited to the protected RNA. The availability of well-defined RNA fragments from bacteriophage RNA polymerase transcription of cloned DNA sequences in vitro has allowed these kinds of experiments to be pursued with more ease and precision. Systematic efforts to define the smallest RNA fragment still binding a protein with normal affinity have been carried out for

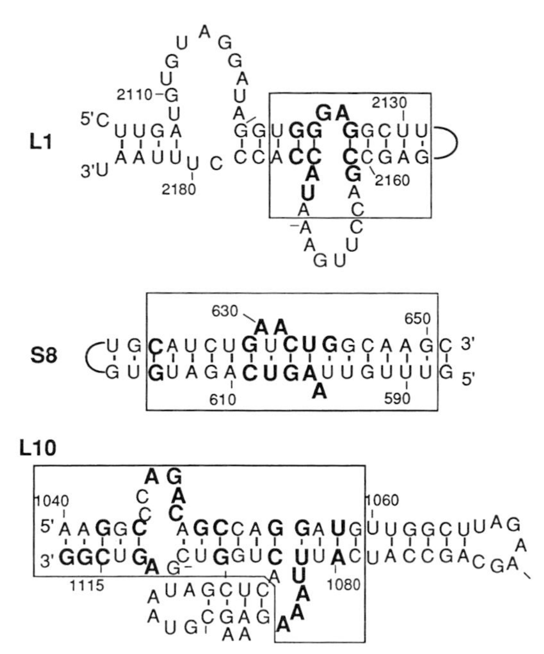

Figure 1. Consensus rRNA binding sites for *E. coli* ribosomal proteins L1, S8, and L10 deduced by comparing the mRNA regulatory target sites for these proteins with the rRNA sites. Boxes indicate RNA segments with secondary structures identical to those of the mRNA binding sites; boldface nucleotides are identical between the mRNA and rRNA sites. The data used to deduce these binding sites are referenced by Draper (1989).

three proteins (Table 1). Besides limiting the region of RNA required for protein recognition, these experiments also define sequences that are able to independently fold into structures resembling fragments of the ribosome.

Binding sites within these RNA fragments can be further delimited by constructing small internal deletions or single-base mutations. This is tedious work. A novel way to rapidly define important structures or nucleotides is available for the set of ribosomal proteins that are translational repressors. These proteins also bind to their own mRNAs, and in several cases where the mRNA- and rRNA-binding sites have been sufficiently well defined, there are striking similarities between the two RNAs (Draper, 1989). Since the mRNA and rRNA have very different functional constraints on their structures, the similar structures and sequences are likely to represent only the essential protein recognition features. The rRNA sites recognizing L1, S8, and L10 are shown in Fig. 1. In each case there is a secondary structure similar to the mRNA (boxed region), within which there are a number of nucleotides identical to the corresponding mRNA positions (shown in boldface). Nuclease protection, site-directed mutagenesis, or other evidence corroborates the importance of the boxed nucleotides for recognition by each protein (reviewed by Draper, 1989).

Two aspects of the RNA structures in Fig. 1 are striking. First, all of the sites are irregular helices with bulges or loops and potentially form unusual structures. For instance, Mougel et al. (1987) propose that the three bulged A's in the S8-binding site interact to form a slightly underwound helix. This suggests that protein recognition takes advantage of three-dimensional aspects of RNA structure. Second, there is a surprising degree of similarity at the sequence level: more than 10 nucleotides are identical between the mRNA and rRNA in each case. For comparison, the R17 coat protein is sensitive to sequence changes at only four positions in a hairpin (Romaniuk et al., 1987), and aminoacyl-tRNA synthetases require as few as two specific nucleotides (Hou and Schimmel, 1988). Some of these similar nucleotides may not be required for protein recognition per se but may be needed to maintain the necessary RNA conformation; the analogous residues in tRNA would be those constant nucleotides (nearly a quarter of a tRNA) that are involved in tertiary interactions. It is also possible that the protein needs to make extensive contacts in order to affect the conformation and activity of the RNA.

Though most of the similar base sequences are in loop or bulge regions, there are a few base pairs whose sequence may be important; these tend to be G · U or G · C base pairs. There is reason to think that G should be the most frequently recognized helical base in RNAs. The minor groove of A-form helices is far more accessible to protein side chains than the major groove, and the 2-amino group of G is the most prominent feature in the minor groove (Alden and Kim, 1979). A G · U pair in a tRNA acceptor stem is a known recognition feature for the alanine tRNA synthetase (Hou and Schimmel, 1988).

The S4 protein is also a translational repressor, but its rRNA- and mRNA-binding sites are much larger (460 and 108 nucleotides, respectively) and more complex structures than those shown in Fig. 1 (Tang and Draper, 1989; Vartikar and Draper, 1989). Clear sequence and structural similarities between the sites are lacking (Draper, 1989). It may be that the S4 recognition site is stabilized by tertiary interactions, so that the similarities between the structures are not obvious in our two-dimensional representations of structure. Tertiary interactions of the sort seen between the D and TψC loops of tRNA could provide a recognition structure similar to those seen in Fig. 1, but spread over several regions of the RNA not obviously related in representations of the secondary structure.

Some information about protein recognition sites can also be gleaned from those ribosomal proteins that recognize RNAs from widely divergent

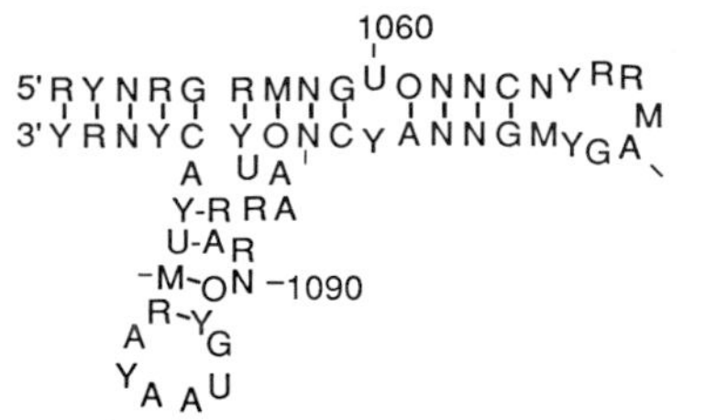

```
                1060
                  |
5'RYNRG RMNGUONNCNYRR
  | | | | |  | | | |   | | | |       M
3'YRNYC YONCYANNGMYGA
   A  UA                  \
  Y-R RA
   U-AR
  -M-ON -1090
  AR-YG
  YA AU
```

Figure 2. The L11-binding site from 23S rRNA. L11 is known to bind a highly conserved RNA structure from eubacterial, archaebacterial, and eucaryotic sources. The sequence shown is conserved among 25 different sequences from these sources. Nucleotides that are not universally conserved are indicated by the following code: R, purine; Y, pyrimidine; O, U or G; M, A or C; N, three or four different nucleotides found.

organisms. Only the aspects of sequence and structure which are conserved between these recognized RNAs can be involved in specific protein interactions. The first such study was a demonstration that *E. coli* L1 binds homologous sequences from *E. coli* and *Dictyostelium discoideum* (Gourse et al., 1981). More extensive studies have since been done with L11, which binds to a conserved 58-nucleotide structure from eubacterial, archaebacterial, and eucaryotic sources (Stark et al., 1980; Beauclerk et al., 1985; El-Baradi et al., 1987a). This structure is shown in Fig. 2, which summarizes conserved bases from 25 different sequences. It is, of course, unlikely that all of the conserved bases are important protein contacts; some bases are probably conserved to maintain other ribosome functions. We have looked at 13 highly conserved nucleotides in bulges or loops so far and find that changes in only 2 have a large effect on L11 binding (P. Ryan and D. E. Draper, unpublished observations). Even so, the conserved structure in Fig. 2 has been a useful means to pare down the number of nucleotides that must be varied in a systematic study of protein recognition features. The binding site for protein L23 has also been defined by comparing rRNA fragments binding the protein from *E. coli* 23S rRNA and yeast 26S rRNA (El-Baradi et al., 1987b).

DO RIBOSOMAL PROTEINS PERTURB rRNA STRUCTURE?

Early physical studies of ribosomal protein-RNA complexes generally suggested that protein binding has little effect on RNA structure. However, changes in the conformations of a few nucleotides in a large RNA may only subtly alter the overall shape, circular dichroism spectrum, or UV hyperchromism of the RNA. Much more sensitive assays for changes in the conformations of individual nucleotides are now available from chemical probing experiments. A number of chemical reagents that react with different positions of pyrimidines and purines are in common use, as well as one reagent reacting with phosphates (Ehresmann et al., 1987). The reactivity of all nucleotides in a large RNA to one of these probes can be determined quickly in one gel electrophoresis run. Several nucleases also provide useful structural information, though their much larger size potentially limits their access to compact RNA structures and their structural specificities are less well understood than are those of chemical reagents. Little quantitative work has been done in this area, but the reactivities of nucleotides toward these reagents are modulated at least an order of magnitude by the structure of the RNA.

At least three factors can cause the reactivities of a reagent toward a particular nucleotide to change when a protein binds. (i) The protein may occlude the nucleotide from solvent and thus prevent reaction. Hydrogen bonding with the reactive position of a base, van der Waals contact with the nucleotide, or even sufficiently close proximity to slow diffusion in the layers of solvent around the base will decrease the reaction rate. (ii) The presence of the protein may locally concentrate the reagent (e.g., by providing a hydrophobic environment for an organic reagent) and enhance the reactivity of nearby residues. (iii) Protein binding may induce changes in the RNA conformation that either enhance or reduce nucleotide reactivity.

In interpreting the footprint data for a large number of ribosomal proteins, Stern et al. (1988a) argue that clusters of strong protections are likely to reflect direct protein contacts, whereas areas with enhanced reactivity have probably changed conformation. They also note situations in which different proteins produce similar changes in the reactivity of a set of nucleotides; these are probably conformational changes as well. Although these principles are useful, it is still difficult to unambiguously interpret an extensive set of enhancements and protections in terms of protein contacts and RNA conformational changes.

Studies of protein complexes with RNA fragments can help interpret these chemical probe experiments. Once an RNA fragment containing all of the protein recognition features is defined, altered chemical reactivities outside this region are likely to be the result of conformational changes induced by protein binding. An example is ribosomal protein S8. The binding site defined by nuclease protection, affinity of RNA fragments, and site-directed mutagenesis is confined to the extended hairpin at positions 588 to 651 (Gregory et al., 1984; Mougel et al., 1987; Gregory et al., 1988); this site is also supported by comparisons with the S8 target site on *spc* mRNA (Fig. 1). S8 does generate some protections against nucleases and chemical reagents within this structure, but extensive protections are found in unrelated areas (565 to 583 and 811 to 879), and some enhancements are clustered at 672 to 732 (Svensson et al., 1988). The footprint alone would make a convincing argument that S8 binds an extensive structure ranging from the 620 hairpin stem to the 840 stem, and including the intervening long-range helices. However, consideration of the overwhelming evidence that all recognition features are located within the 620 stem forces the conclusion that the protein has a significant effect on the RNA structure outside of its actual binding site. Since the protein binds in the middle of a helical segment, it is possible that it alters the RNA twist or bend sufficiently to stabilize tertiary interactions between the 620 stem and other parts of the rRNA. The additional tertiary structure might then afford the extensive nucleotide protections found outside of the 620 stem. In support of tertiary structure in this region, a nucleotide in the 620 hairpin loop is much more reactive in a small fragment than in the intact 16S rRNA (Mougel et al., 1987). Mutations affecting this hairpin loop do not alter S8 binding but are deleterious to ribosome assembly and cell growth (Stark et al., 1984; Gregory et al., 1988). It would be interesting to compare the chemical reactivity patterns of these mutant 16S rRNAs with wild-type sequences to see whether sites of tertiary contact with the 620 loop can be identified.

Another example of this kind of behavior is seen with S4. Studies with RNA fragments show that the region between 39 and 500 binds S4 with the same affinity and salt dependence as intact 16S rRNA. Deletion of a hairpin adjacent to this domain, 500 to 545, has no effect on the measured S4-binding affinity (Vartikar and Draper, 1989), even though it is heavily protected by S4 (Stern et al., 1986). Several other nucleotides located outside the 39–500 domain are also protected by S4. Since the 500–545 hairpin cannot be making any significant contribution to the S4-binding free energy, the only reasonable explanation for its protection is an RNA conformational change induced by S4. In fact, there is substantial evidence that this hairpin responds to binding events in other parts of the rRNA. A good argument can be made that streptomycin binds within a loop of long-range base pairings containing the 900 hairpin (Moazed and Noller, 1987; Gravel et al., 1987), but a mutation within the 500–545 hairpin loop, A-523→C, confers streptomycin sensitivity (Melançon et al., 1988). tRNA bound in the A site protects two clusters of rRNA bases: one set is almost certainly at the site of anticodon-codon interaction (based on UV-induced tRNA-16S rRNA cross-links), and the other is in the 500–545 hairpin loop. These two sites are separated by ~70 Å (7 nm) in any reasonable model of the 16S rRNA folding within the 30S particle, leading Stern et al. (1988b) to conclude that an allosteric conformational change takes place within the ribosome. These observations give some physical basis for thinking about how streptomycin and S4 mutants are able to influence tRNA misreading.

These studies of the S4 and S8 proteins provide evidence for extensive conformation changes taking place in the rRNA surrounding the protein-binding site. Just as important, but perhaps harder to detect, are the effects that proteins might have on RNA structure within their recognition sites. Nuclear magnetic resonance studies of L25 binding to a fragment of 5S rRNA have shown that the resonances of many imino protons within a structured helix-loop region are shifted when L25 binds (Zhang and Moore, 1989). Though these shifts probably reflect subtle changes in RNA conformation, they may nonetheless be functionally significant.

POSSIBLE ROLES OF PROTEINS IN RIBOSOME FUNCTION

The preceding sections have shown that we are starting to discern the RNA features recognized by ribosomal proteins in some detail and that substantial evidence for protein-induced conformation changes in rRNA is accumulating. Much more structural information of this sort will probably be obtained in the next few years. The difficult task ahead is to understand the structural and thermodynamic data available in terms of the functions of protein-rRNA complexes. Since there are very few data suggesting a specific role for any individual protein, I offer here several speculations as to how ribosomal proteins might function in collusion with the rRNAs, with the hope that these ideas will stimulate useful experiments.

Proteins May Stabilize RNA Structures That Might Be Otherwise Difficult To Achieve

Chemical probing experiments with naked 16S rRNA are largely consistent with the phylogenetically conserved structure (Moazed et al., 1986), so it is unlikely that a major function of proteins is to stabilize secondary structure. More subtle kinds of effects are possible: proteins can conceivably bend or twist helical segments to achieve a needed RNA conformation, stabilize weak tertiary interactions, or serve as cross-links holding segments (or domains) of RNA in the correct relative orientation. One result of protein stabilization might be a "stiffening" of structures that might otherwise be too flexible for efficient ribosome functioning.

Although it may be fairly easy to demonstrate that a protein stabilizes a particular rRNA structure (e.g., by showing an increase in the RNA T_m in the protein complex or a decreased flexibility), it will be difficult to show that the stabilization is functionally important. One place where it might be possible to detect a functionally important stabilizing influence of proteins is in the interaction of rRNA with tRNA or mRNA substrates. Careful experiments by Backendorf et al. (1981) showed that a DNA octamer analog of the mRNA Shine-Dalgarno sequence is able to bind to 30S subunits only if S21 is present. The protein either could bind directly to an mRNA-rRNA helical segment or could stabilize a 16S rRNA conformation in which the Shine-Dalgarno sequence is accessible to the mRNA; Backendorf et al. (1981) argue for the latter possibility. In either case, this system provides an excellent opportunity to examine the way in which a ribosomal protein stabilizes a functional RNA-RNA interaction.

Proteins May Facilitate Required RNA Structural Transitions

It seems very likely that RNA structural changes are important for the ribosome cycle. Coupling of GTP hydrolysis to translocation or tRNA binding requires some sort of cooperative or allosteric conformational change in the ribosome, and slow, unimolecular rearrangements take place during initiation complex formation (Wintermeyer and Gualerzi, 1983). It has been proposed that some antibiotics trap RNA in a particular conformation, so that the ribosome becomes "stuck" at a certain point in its cycle (Cundliffe, 1986). Proteins may facilitate such conformational changes, either by lowering the activation energy for switching between structures or by preventing incorrect structures from forming. This kind of role has been proposed to account for the stimulation of protein synthesis and uncoupled GTPase activity by L11 (Cundliffe, 1986).

The cooperative interactions between proteins that take place in ribosome assembly are undoubtedly mediated in part by RNA; some of the patterns of chemical probe enhancement and protection seen in the footprinting experiments of Noller's group, when correlated with the protein assembly map, provide striking evidence for this (see, for example, the discussion of S16 assembly in Stern et al., 1988a). It would be interesting to map the RNA structures comprising these cooperative assembly pathways, since it is likely that the same structures would contribute to any allosteric transitions taking place during protein synthesis. RNA mutations outside of protein-binding sites but blocking assembly of specific proteins would be useful for this purpose.

Proteins May Provide Needed Positive Charge or Other Functional Groups

The RNase P ribozyme provides an instructive analogy for thinking about RNA-protein interactions in ribosomes; it consists of a single RNA and a small, basic protein similar to ribosomal proteins. It is well known that the RNA component of RNase P can recognize and cleave tRNA precursors in the absence of protein, but only at very high salt concentrations (Gardiner et al., 1985). The protein does not alter the substrate specificity or K_m of the RNA enzyme. Comparisons of the holoenzyme with RNA alone in single-turnover experiments suggest that the function of the protein is to provide a controlled amount of positive charge near the tRNA substrate-binding site, so that substrate is able to bind and turn over rapidly at physiological salt concentrations (Reich et al., 1988). Similar roles for ribosomal proteins in aiding the turnover of tRNA or alignment of mRNA on the ribosome can be imagined. Positive charge distributed through the ribosome may also allow the RNA to adopt a more compact conformation and thus promote ribosome function in a nonspecific way.

I am grateful for Public Health Service research career development award CA01081 from the National Cancer Institute. Work in my laboratory was supported by Public Health Service grant GM29048 from the National Institutes of Health.

REFERENCES

Alden, C. J., and S.-H. Kim. 1979. Solvent-accessible surfaces of nucleic acids. *J. Mol. Biol.* **132**:411–434.

Backendorf, C., C. J. C. Ravensberg, J. van der Plas, J. H. van Boom, G. Veeneman, and J. van Duin. 1981. Basepairing potential of the 3′ terminus of 16S RNA: dependence on the functional state of the 30S subunit and the presence of protein S21. *Nucleic Acids Res.* **9**:1425–1443.

Beauclerk, A. A. D., H. Hummel, D. J. Holmes, A. Böck, and E. Cundliffe. 1985. Studies of the GTPase domain of archaebacterial ribosomes. *Eur. J. Biochem.* **151**:245–255.

Carey, J., and O. C. Uhlenbeck. 1983. Kinetic and thermodynamic characterization of the R17 coat protein-ribonucleic acid interaction. *Biochemistry* **22**:2610–2615.

Cech, T. R., and B. L. Bass. 1986. Biological catalysis by RNA. *Annu. Rev. Biochem.* **55**:599–630.

Cundliffe, E. 1986. Involvement of specific portions of ribosomal RNA in defined ribosomal functions: a study utilizing antibiotics, p. 586–604. *In* B. Hardesty and G. Kramer (ed.), *Structure, Function, and Genetics of Ribosomes.* Springer-Verlag, New York.

Deckman, I. C., D. E. Draper, and M. S. Thomas. 1987. S4-α mRNA translation regulation complex. I. Thermodynamics of formation. *J. Mol. Biol.* **196**:313–322.

Donly, B. C., and G. A. Mackie. 1988. Affinities of ribosomal protein S20 and C-terminal deletion mutants for 16S rRNA and

S20 mRNA. *Nucleic Acids Res.* **16**:997–1010.

Draper, D. E. 1989. How do proteins recognize specific RNA sites? New clues from autogenously regulated ribosomal proteins. *Trends Biochem. Sci.* **14**:335–338.

Ehresmann, C., F. Baudin, M. Mougel, P. Romby, J.-P. Ebel, and B. Ehresmann. 1987. Probing the structure of RNAs in solution. *Nucleic Acids Res.* **15**:9109–9128.

El-Baradi, T. T. A. L., V. H. C. F. de Regt, S. W. C. Einerhand, J. Teixido, R. J. Planta, J. P. G. Ballesta, and H. A. Raué. 1987a. Ribosomal proteins EL11 from *Escherichia coli* and L15 from *Saccharomyces cerevisiae* bind to the same site in both yeast 26 S and mouse 28 S rRNA. *J. Mol. Biol.* **195**:909–917.

El-Baradi, T. T. A. L., V. H. C. F. de Regt, R. J. Planta, K. H. Nierhaus, and H. A. Raué. 1987b. Interaction of ribosomal proteins L25 from yeast and EL23 from *E. coli* with yeast 26S and mouse 28S rRNA. *EMBO J.* **4**:2101–2107.

Fellner, P. 1974. Structure of the 16S and 23S ribosomal RNAs. *In* M. Nomura, A. Tissières, and P. Lengyel (ed.), *Ribosomes.* Cold Spring Harbor Laboratory, Cold Spring Harbor, N.Y.

Gardiner, K. J., T. L. Marsh, and N. R. Pace. 1985. Ion dependence of the *Bacillus subtilis* RNase P reaction. *J. Biol. Chem.* **260**:5415–5419.

Gourse, R. L., D. L. Thurlow, S. A. Gerbi, and R. A. Zimmermann. 1981. Specific binding of a prokaryotic ribosomal protein to a eukaryotic ribosomal RNA: implications for evolution and autoregulation. *Proc. Natl. Acad. Sci. USA* **78**:2722–2726.

Gravel, M., P. Melançon, and L. Brakier-Gingras. 1987. Crosslinking of streptomycin to the 16S ribosomal RNA of *Escherichia coli. Biochemistry* **26**:6227–6232.

Gregory, R. J., P. B. F. Cahill, D. L. Thurlow, and R. A. Zimmermann. 1988. Interaction of *Escherichia coli* ribosomal protein S8 with its binding sites in ribosomal RNA and messenger RNA. *J. Mol. Biol.* **204**:295–307.

Gregory, R. J., M. L. Zeller, D. L. Thurlow, R. L. Gourse, M. J. R. Stark, A. E. Dahlberg, and R. A. Zimmermann. 1984. Interaction of ribosomal proteins S6, S8, S15, and S18 with the central domain of 16S ribosomal RNA from *Escherichia coli. J. Mol. Biol.* **178**:287–302.

Hou, Y.-M., and P. Schimmel. 1988. A simple structural feature is a major determinant of the identity of a transfer RNA. *Nature* (London) **333**:140–145.

Melançon, P., C. Lemieux, and L. Brakier-Gingras. 1988. A mutation in the 530 loop of *Escherichia coli* 16S ribosomal RNA causes resistance to streptomycin. *Nucleic Acids Res.* **16**:9631–9639.

Moazed, D., and H. F. Noller. 1987. Interaction of antibiotics with functional sites in 16S ribosomal RNA. *Nature* (London) **327**: 389–394.

Moazed, D., S. Stern, and H. F. Noller. 1986. Rapid chemical probing of conformation in 16 S ribosomal RNA and 30 S ribosomal subunits using primer extension. *J. Mol. Biol.* **187**: 399–416.

Mougel, M., B. Ehresmann, and C. Ehresmann. 1986. Binding of *Escherichia coli* ribosomal protein S8 to 16S rRNA: kinetic and thermodynamic characterization. *Biochemistry* **25**:2756–2765.

Mougel, M., F. Eyermann, E. Westhof, P. Romby, A. Expert-Bezançon, J.-P. Ebel, B. Ehresmann, and C. Ehresmann. 1987. Binding of *Escherichia coli* ribosomal protein S8 to 16 S RNA. A model for the interaction and the tertiary structure of the RNA binding site. *J. Mol. Biol.* **198**:91–107.

Noller, H. F. 1980. Structure and topography of ribosomal RNA, p. 3–22. *In* G. Chambliss, G. R. Craven, J. Davies, K. Davis, L. Kahan, and M. Nomura (ed.), *Ribosomes. Structure, Function, and Genetics.* University Park Press, Baltimore.

Record, M. T., T. M. Lohman, and P. de Haseth. 1976. Ion effects on protein-nucleic acid interactions. *J. Mol. Biol.* **107**:145–158.

Reich, C., G. J. Olsen, B. Pace, and N. R. Pace. 1988. The role of the protein moiety of ribonuclease P, a ribonucleoprotein enzyme. *Science* **239**:178–181.

Romaniuk, P. J. 1989. The role of highly conserved single stranded nucleotides of Xenopus 5S RNA in the binding of transcription factor III A. *Biochemistry* **28**:1388–1395.

Romaniuk, P. J., P. Lowary, H.-N. Wu, G. Stormo, and O. C. Uhlenbeck. 1987. RNA binding site of R17 coat protein. *Biochemistry* **26**:1563–1568.

Ryan, P. C., and D. E. Draper. 1989. Thermodynamics of protein–RNAS recognition in a highly conserved region of the large subunit ribosomal RNA. *Biochemistry* **28**:9949–9956.

Schwarzbauer, J., and G. R. Craven. 1981. Apparent association constants for *E. coli* ribosomal proteins S4, S7, S8, S15, S17 and S20 binding to RNA. *Nucleic Acids Res.* **9**:2223–2237.

Spierer, P., A. A. Bogdanov, and R. A. Zimmermann. 1978. Parameters for the interaction of ribosomal proteins L5, L18, and L25 with 5S RNA from *Escherichia coli. Biochemistry* **17**:5394–5398.

Stark, M. J. R., E. Cundliffe, J. Dijk, and G. Stöffler. 1980. Functional homology between *E. coli* ribosomal protein L11 and *B. megaterium* protein BM-L11. *Mol. Gen. Genet.* **180**: 11–15.

Stark, M. J. R., R. J. Gregory, R. L. Gourse, D. L. Thurlow, C. Zwieb, R. A. Zimmermann, and A. E. Dahlberg. 1984. The effects of site-directed mutations in the central domain of 16 S ribosomal RNA upon ribosomal protein binding, RNA processing, and 30 S subunit assembly. *J. Mol. Biol.* **178**:303–322.

Stern, S., L.-M. Changchien, G. R. Craven, and H. F. Noller. 1988a. Interaction of proteins S16, S17, and S20 with 16 S ribosomal RNA. *J. Mol. Biol.* **200**:291–300.

Stern, S., B. Weiser, and H. F. Noller. 1988b. Model for the three-dimensional folding of 16 S ribosomal RNA. *J. Mol. Biol.* **204**:447–481.

Stern, S., R. C. Wilson, and H. F. Noller. 1986. Localization of the binding site for protein S4 on 16 S ribosomal RNA by chemical and enzymatic probing and primer extension. *J. Mol. Biol.* **192**:101–110.

Svensson, P., L.-M. Changchien, G. R. Craven, and H. F. Noller. 1988. Interaction of ribosomal proteins S6, S8, S15, and S18 with the central domain of 16 S ribosomal RNA. *J. Mol. Biol.* **200**:301–308.

Tang, C. K., and D. E. Draper. 1989. An unusual mRNA pseudoknot structure is recognized by a protein translational repressor. *Cell* **57**:531–536.

Vartikar, J. V., and D. E. Draper. 1989. S4-16 S ribosomal RNA complex: binding constant measurements and specific recognition of a 460 nucleotide region. *J. Mol. Biol.* **209**:221–234.

Wintermeyer, W., and C. Gualerzi. 1983. Effect of *Escherichia coli* initiation factors on the kinetics of N-AcPhe-$tRNA^{Phe}$ binding to 30S ribosomal subunits. A fluorescence stopped-flow study. *Biochemistry* **22**:690–694.

Wu, H.-N., and O. C. Uhlenbeck. 1987. Role of a bulged A residue in a specific RNA-protein interaction. *Biochemistry* **26**:8221–8227.

Zhang, P., and P. B. Moore. 1989. An NMR study of the helix V-loop E region of the 5S RNA from *Escherichia coli. Biochemistry* **28**:4607–4615.

Zimmermann, R. A. 1980. Interactions among protein and RNA components of the ribosome, p. 135–170. *In* G. Chambliss, G. R. Craven, J. Davies, K. Davis, L. Kahan, and M. Nomura (ed.), *Ribosomes. Structure, Function, and Genetics.* University Park Press, Baltimore.

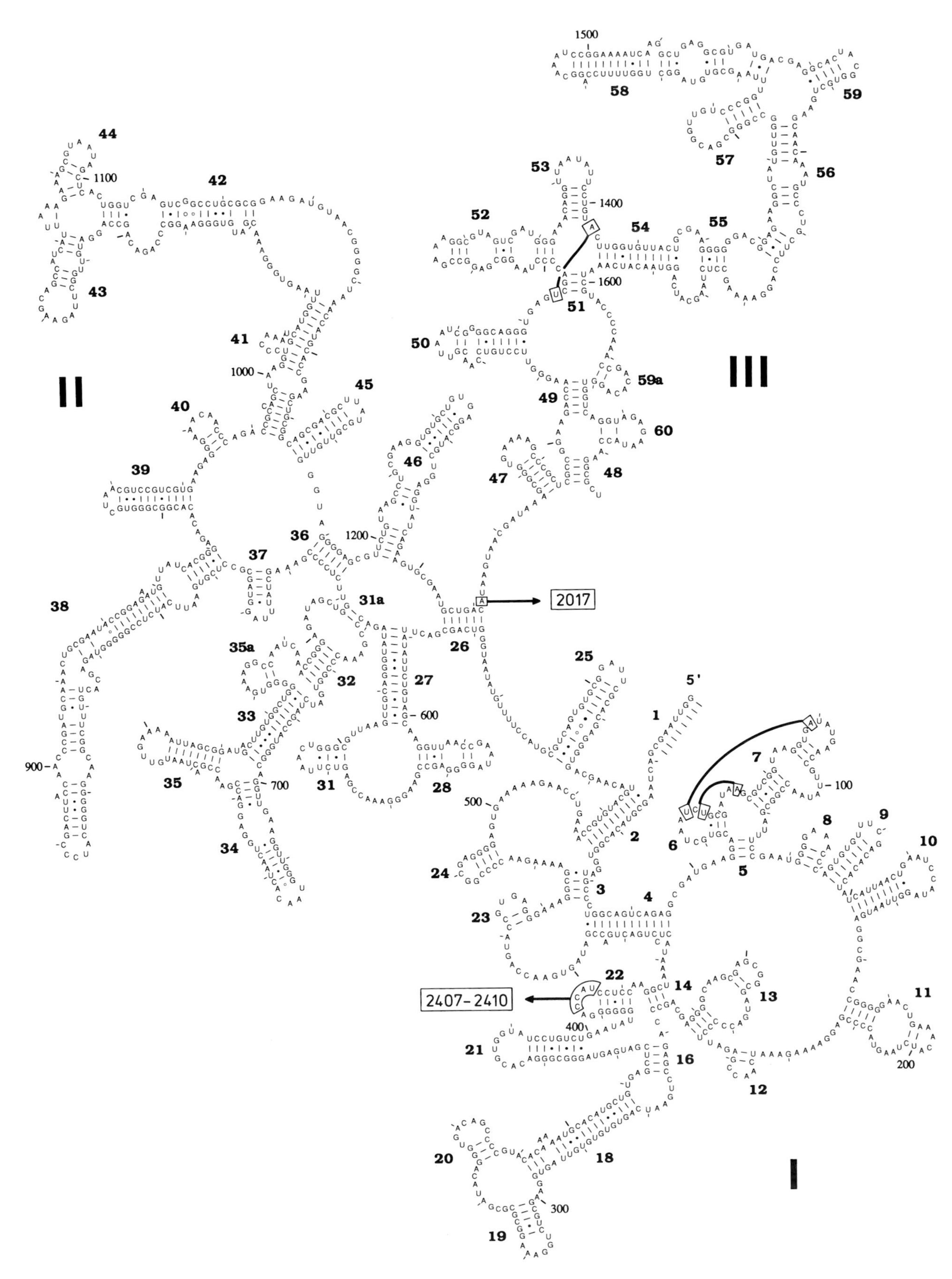

Figure 1. *Part 1.*

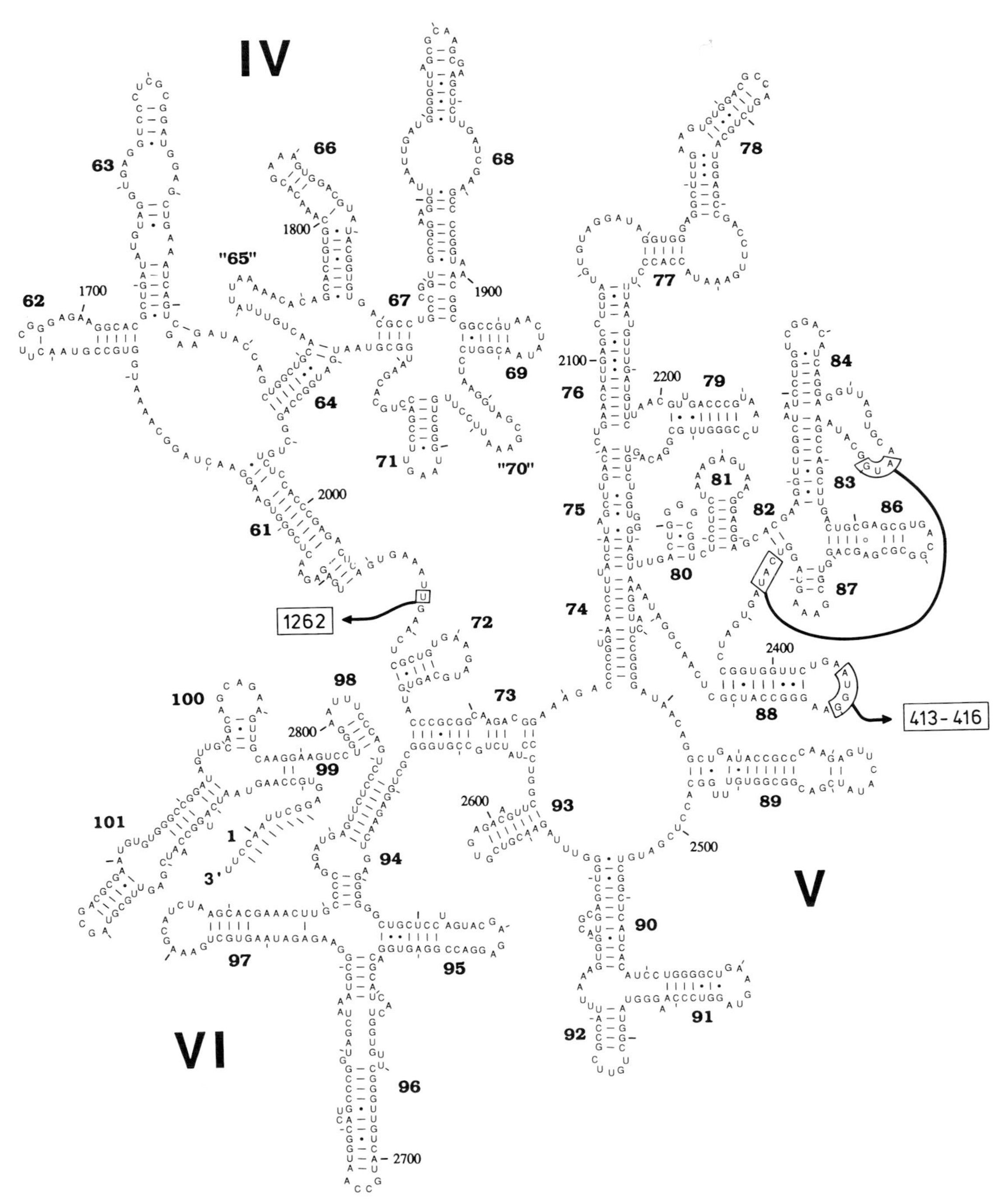

Figure 1. Refined secondary structure model showing the six primary domains (I to VI) of *E. coli* 23S rRNA drawn in the original format of Noller et al. (1981). Phylogenetic comparisons were made of 23S rRNA sequences from 9 eucaryotes, 7 archaebacteria, and 12 eubacteria/chloroplasts (cited in Leffers et al., 1987; Höpfl et al., 1989). Base pairs that are supported by compensating base changes, using the criteria defined earlier (Leffers et al., 1987), or base pairs within a putative double helix for which there is no negative evidence are joined by a line or dot. For those that are simply juxtaposed, the evidence is ambiguous. Helices (two or more base pairs) are numbered as for the model of *D. mobilis* 23S RNA (Leffers et al., 1987). Helices "65" and "70" are uncertain (see text). Every 10th nucleotide from the 5′ end is indicated by a short line; every 50th nucleotide is indicated by a longer line. This figure and Fig. 2 to 4 were produced with EDSTRUC (N. Larsen, unpublished data), converted into PostScript format (PrePS; Larsen, unpublished data), and printed on a Linotronic photosetter. Various modified nucleotides have been characterized in the 23S RNA of *E. coli*, cited by Branlant et al. (1981). They include positions 745 (m^1G), 746 (ψ), 747 (thymidine), 1618 (m^6A), 1911 (ψ), 1915 (m^3U), 1917 (ψ), 1939 (thymidine), 2030 (m^6A), 2069 (m^7G), 2251 (Gm-2′-O), 2449 (unidentified U), and 2498 (Cm-2′-O).

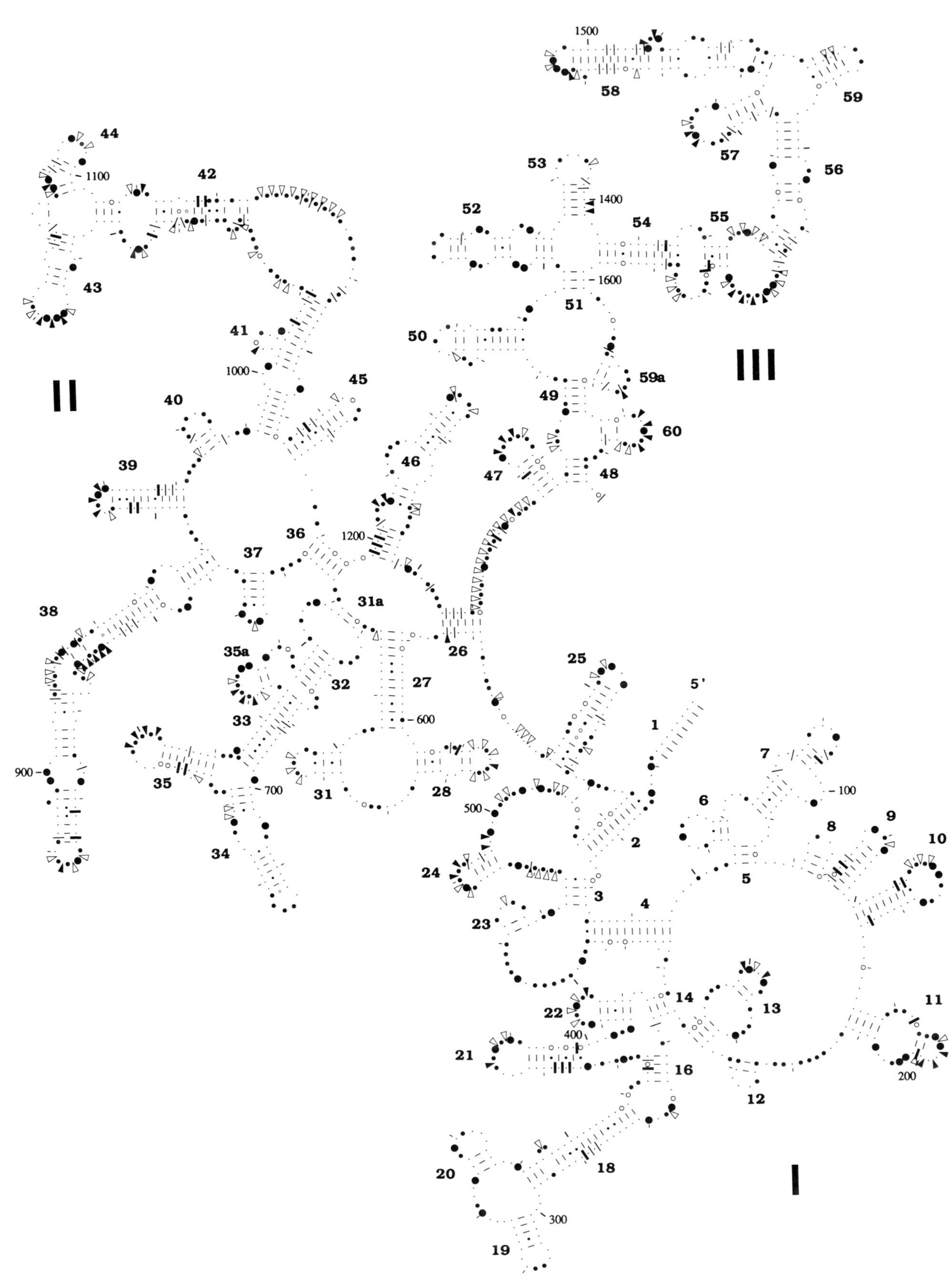

Figure 3. *Part 1.*

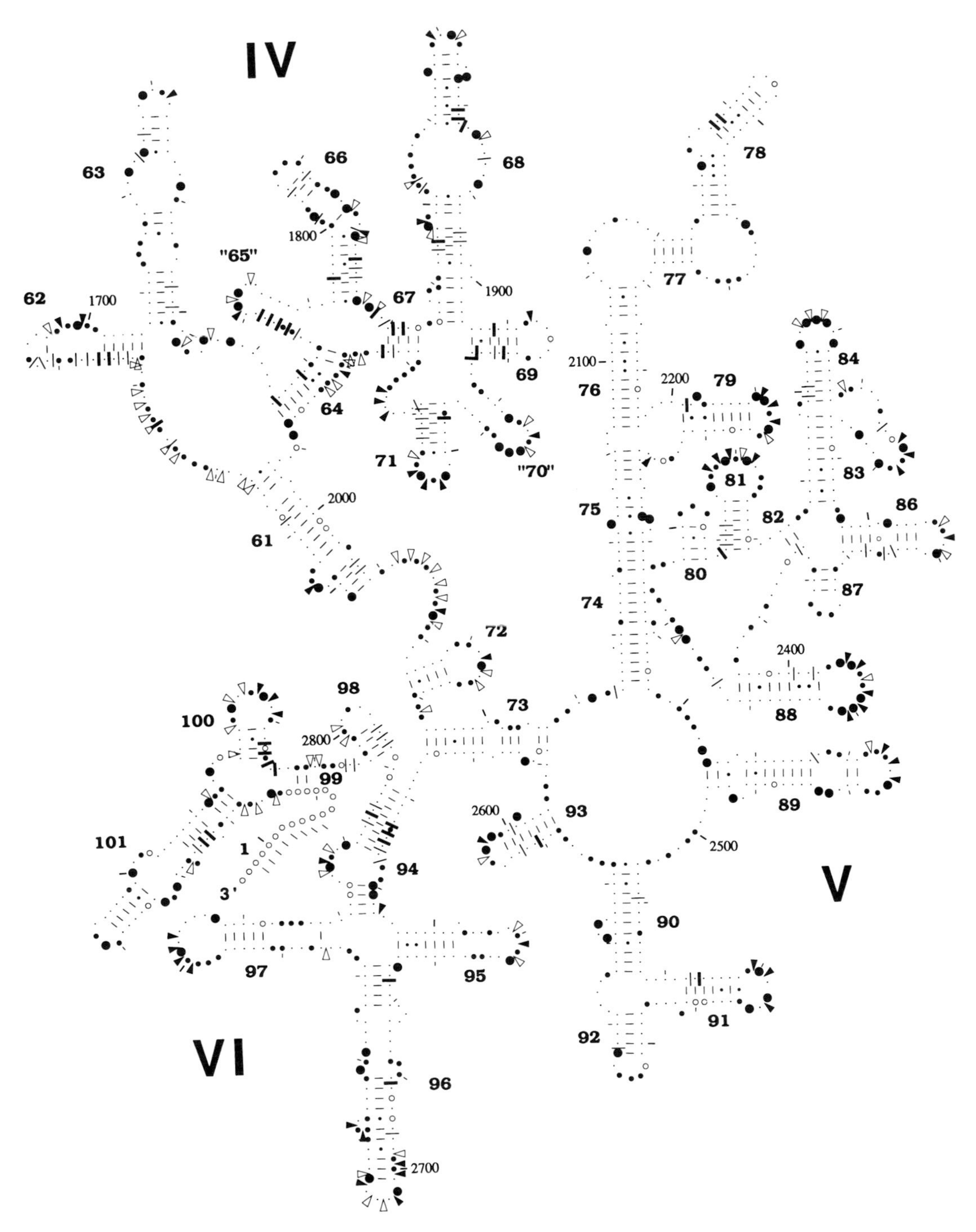

Figure 3. Chemical modification and RNase cutting data superimposed on the secondary structure model. Renatured *E. coli* 23S RNA was treated with the following reagents: dimethyl sulfate, kethoxal, CMCT, RNase T_1, RNase T_2, and RNase CV. Chemical reactivities are classified as either strong (large filled circles) or weak (small filled circles). RNase T_1 and T_2 cuts (single stranded) are classified as either strong (filled arrowheads) or weak (open arrowheads). RNase CV cuts are either strong (thick bars) or weak (thin bars). Open circles reflect the presence of control bands on the autoradiograms or the fact that an oligonucleotide primer was located at the 3′ end of the molecule.

POSSIBLE ARTIFACTUAL HELICES

Some helices exhibit highly conserved base pairs and cannot be proven phylogenetically; such conserved base pairs are included in the model when they appear to fit into a predicted helix. For helices containing several conserved base pairs, such as helices 69, 74, and 93, the phylogenetic evidence is relatively weak. Of these, helix 69 is the only one that is positively supported by its incurring RNase CV

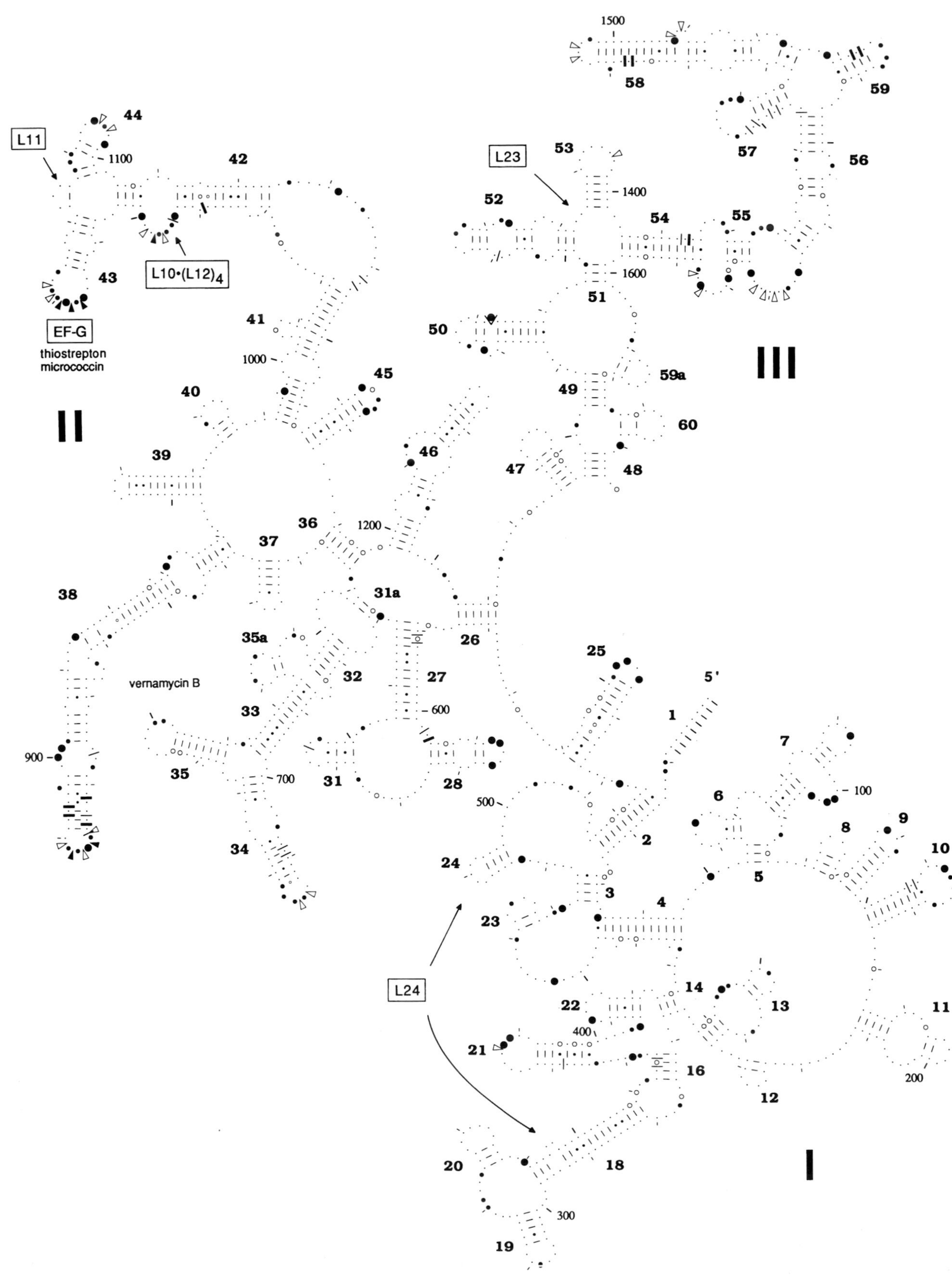

Figure 4. *Part 1.*

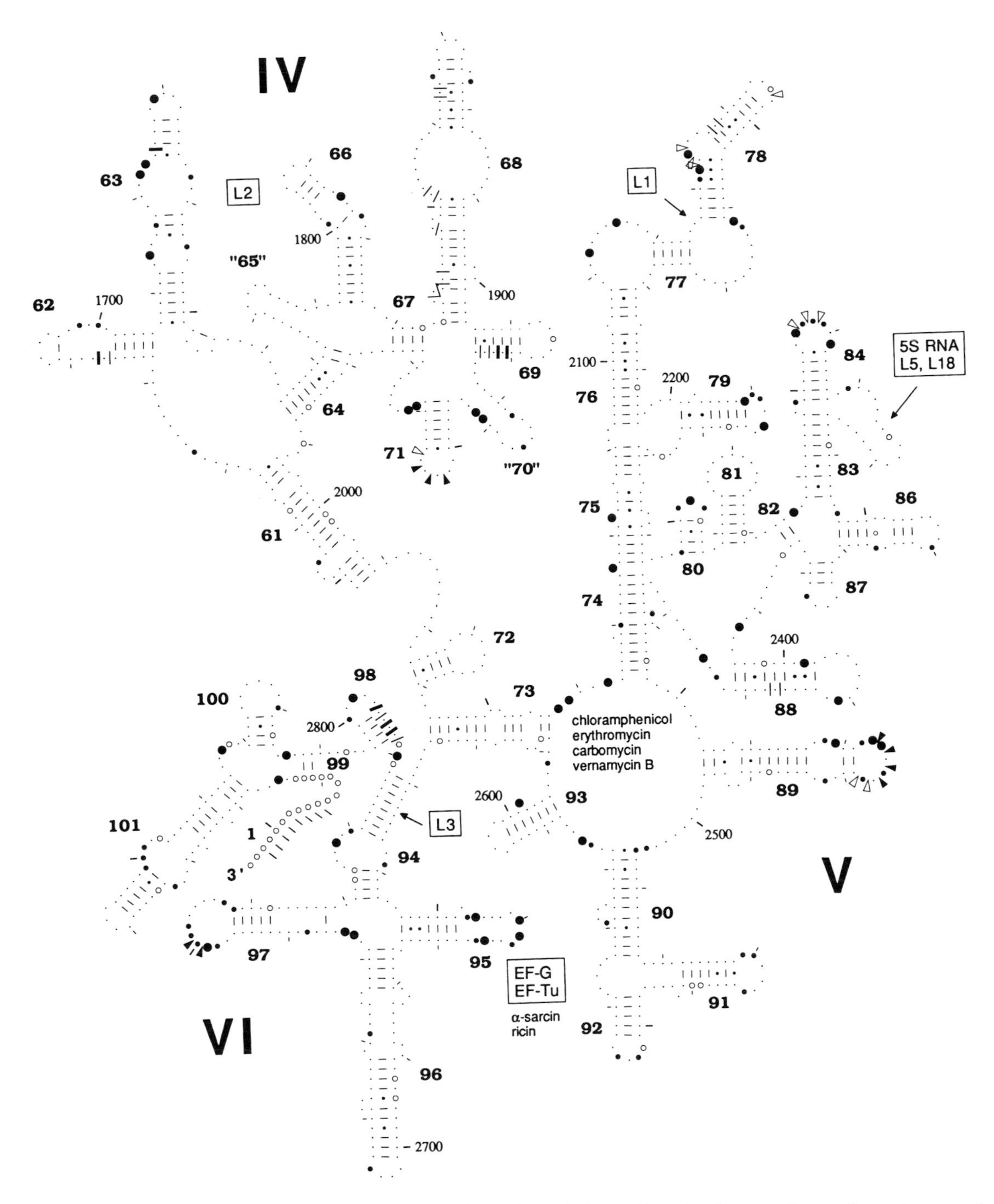

Figure 4. Chemical modification and RNase cutting data obtained from 50S subunits superimposed on the secondary structure model. The same chemical reagents and RNases and the same system of quantification were used as described for 23S RNA in the legend to Fig. 3. The approximate locations of the attachment sites of primary binding proteins are indicated by arrows, as are functional sites and antibiotic-binding sites that have been identified. Literature references are cited in the text.

cuts in 50S subunits. Helices "65" and "70" are extreme examples in that almost all of their "paired" nucleotides are highly conserved. The experimental data are compatible with the formation of both helices in 23S RNA. However, whereas the former is resistant in 50S subunits, nucleotides C–C-1942 within helix "70" are highly reactive and incompatible with base pairing. This is another example of a thermodynamically predicted structure occurring in 23S RNA but not in 50S subunits (see also Moazed et al., 1986). Finally, a new extension to helix 100 was recently suggested that involves the pair A-2813–

U-2833 (Gutell and Fox, 1988), but since this region is strongly reactive in 50S subunits, this pairing, which receives no phylogenetic support, is unlikely.

PUTATIVE TERTIARY INTERACTIONS

Nucleotides involved in the tertiary interactions that were predicted from phylogenetic sequence comparisons (Leffers et al., 1987; Höpfl et al., 1989) are indicated in Fig. 1. They are generally accessible in the free RNA but protected in 50S subunits, which suggests that they are stabilized in the subunit directly or indirectly by proteins. Such protein-dependent stabilization would facilitate the occurrence of base pair mutations and thus simplify their detection by the sequence comparison approach. In contrast, base pairs that are not stabilized by protein would be less amenable to mutations and therefore more likely to be conserved. Consequently, they would tend to escape detection by the sequence comparison approach. Candidates for such tertiary interactions are numerous in 23S RNA, in which the majority of conserved bases are inaccessible (Fig. 2 and 3).

One of the predicted tertiary interactions between domains I and V generates a new helix, 413–416/2407–2410. This is likely to stack coaxially on helix 22 because base complementarity between positions 416 and 2407 of the tertiary interaction correlates with a mismatched pairing between 411 and 416 and vice versa. Such stacking is further supported by the modification data, since nucleotides A-412 and A-2411, which lie at the end of the helix and would not be involved in coaxial stacking, are highly reactive.

ASSEMBLY EFFECTS

A comparison of the data for the 50S subunit with those of the 23S RNA (Fig. 4) reveals that much of the RNA is shielded in the subunit. These shielded regions either are involved in protein binding or are subject to protein-induced conformational changes that could be important for assembly. It is difficult to distinguish between these two possibilities, however, unless all protein attachment sites have been localized.

The approximate binding regions of the protein sites that have been characterized, including L1 (Branlant et al., 1976), L2 (Beauclerk and Cundliffe, 1988), L3 (Leffers et al., 1988), L10 · $(L12)_4$ (Beauclerk et al., 1984), L11 (Schmidt et al., 1981), L23 (Vester and Garrett, 1984), and L24 (Egebjerg et al., 1987), are indicated. Although the data on protein binding are incomplete, a comparison of the subunit and RNA data yields some insight into the assembly mechanism. Thus, the relative inertness of the 5′ part of domain I indicates that it folds up independently of protein; even the putative tertiary interactions (Fig. 1) appear to be stable. Thus, this RNA region may constitute a nucleation site around which the 3′ part of domain I folds, aided by protein L24. The other domains also exhibit a stable RNA region and a protein-stabilized region and probably assemble by a similar mechanism. Thus, in domain III, loops 1308–1312, 1340–1344, and 1602–1610 are structured on binding protein L23 (Vester and Garrett, 1984); similar effects have been observed for protein L2 in domain IV and protein L3 in domain VI (Leffers et al., 1988).

FUNCTIONAL SITES

Functional sites have been identified on the 23S RNA, and some of these, in particular the tRNA-binding sites, are reviewed elsewhere in this volume by Noller and co-workers. In Fig. 4, we pinpoint the following sites on the secondary structure map: the GTPase activity associated with elongation factor EF-G centered on helices 43 and 44 in domain II (Moazed and Noller, 1988; Egebjerg et al., 1989, and references therein), the peptidyltransferase center in domain V (Vester and Garrett, 1988; Wower et al., 1989, and references therein), and elongation factor binding in domain VI at helix 95 and possibly helix 97 (Leffers et al., 1988; Moazed et al., 1988, and references therein). The binding sites of antibiotics, some of which are associated with the preceding functional sites (Moazed and Noller, 1987; Egebjerg et al., 1989), are also indicated. All of the functional sites fall in conserved sequence regions (Fig. 2) that are accessible in the ribosome (Fig. 4); other conserved and accessible sites of unknown function can also be discerned in these figures.

During the course of this work, J.E. was supported by Aarhus University, N.L. received a grant from the Danish Natural Science Research Council Programme for Biomolecular Techniques, and R.A.G. received grants from The Danish Centre for Microbiology.

We appreciate the assistance of Vicka Nissen in preparing the manuscript.

REFERENCES

Andersen, A., N. Larsen, H. Leffers, J. Kjems, and R. A. Garrett. 1986. A domain of 23S rRNA in search of a function, p. 221–237. *In* P. H. Knippenberg and C. W. Hilbers (ed.), *Structure and Dynamics of RNA*. Plenum Publishing Corp., New York.

Beauclerk, A. A. D., and E. Cundliffe. 1988. The binding site for ribosomal protein L2 within 23S RNA of *E. coli*. *EMBO J.* 7:3589–3594.

Beauclerk, A. A. D., E. Cundliffe, and J. Dijk. 1984. The binding site for ribosomal protein complex L8 within 23S rRNA of *E. coli*. *J. Biol. Chem.* **259**:6559–6563.

Branlant, C., A. Krol, M. A. Machette, J. Pouyet, J.-P. Ebel, K. Edwards, and H. Kössel. 1981. Primary and secondary structures of *E. coli* MRE 600 23S rRNA. Comparison with models of secondary structures for maize chloroplast 23S rRNA and for large portions of mouse and human 16S mitochondrial rRNAs. *Nucleic Acids Res.* **9**:4303–4324.

Branlant, C., A. Krol, J. Sriwidada, J. P. Ebel, P. Sloof, and R. A. Garrett. 1976. The binding site of protein L1 on 23S RNA of *E. coli*. Determination of the RNA region contained in the L1-RNP and determination of the order of RNA subfragments within this region. *Eur. J. Biochem.* **70**:457–469.

Christiansen, J., J. Egebjerg, N. Larsen, and R. A. Garrett. 1990. Analysis of rRNA structure. Experimental and theoretical considerations, p. 229–252. *In* G. Spedding (ed.), *Ribosomes and Protein Synthesis: a Practical Approach*. IRL Press, Oxford.

Egebjerg, J., S. Douthwaite, and R. A. Garrett. 1989. Antibiotic interactions at the GTPase-associated centre within *E. coli* 23S rRNA. *EMBO J.* **8**:607–611.

Egebjerg, J., H. Leffers, A. Christensen, H. Andersen, and R. A. Garrett. 1987. Structure and accessibility of domain I of *E. coli* 23S rRNA in free RNA, in the L24 RNA complex and in 50S subunits. Implications for ribosomal assembly. *J. Mol. Biol.* **196**:125–136.

Garrett, R. A. 1983. Roles for ribosomal proteins. *Trends Biochem. Sci.* **8**:75–76.

Garrett, R. A., A. Christensen, and S. Douthwaite. 1984. Higher order structure in the 3′-terminal domain VI of the 23S rRNAs from *E. coli* and *Bacillus stearothermophilus*. *J. Mol. Biol.* **179**:689–712.

Gutell, R. R., and G. E. Fox. 1988. A compilation of large subunit RNA sequences presented in a structural format. *Nucleic Acids Res.* **16**(Suppl.):r175–r269.

Höpfl, P., W. Ludwig, K. H. Schleifer, and N. Larsen. 1989. Higher order structure 23S rRNA of *Pseudomonas cepacia* and other prokaryotes. *Eur. J. Biochem.* **185**:355–364.

Leffers, H., J. Egebjerg, A. Andersen, T. Christensen, and R. A. Garrett. 1988. Domain VI of *E. coli* 23S rRNA. Structure, assembly and function. *J. Mol. Biol.* **204**:507–522.

Leffers, H., J. Kjems, L. Østergaard, N. Larsen, and R. A. Garrett. 1987. Evolutionary relationships among archaebacteria. A comparative study of 23S rRNAs of a sulphur-dependent thermophile, an extreme halophile and a thermophilic methanogen. *J. Mol. Biol.* **195**:43–61.

Moazed, D., and H. F. Noller. 1987. Chloramphenicol, erythromycin, carbomycin and vernamycin B protect overlapping sites in the peptidyl transferase region of 23S ribosomal RNA. *Biochimie* **69**:879–884.

Moazed, D., and H. F. Noller. 1988. Interaction of tRNA with 23S RNA in the ribosomal A-site, P-site and E-site. *Cell* **57**:585–597.

Moazed, D., J. M. Robertson, and H. F. Noller. 1988. Interaction of elongation factors EF-G and EF-Tu with a conserved loop in 23S RNA. *Nature* (London) **334**:362–364.

Moazed, D., S. Stern, and H. F. Noller. 1986. Rapid chemical probing of conformation in 16S rRNA and 30S ribosomal subunits using primer extension. *J. Mol. Biol.* **187**:399–416.

Noller, H. F., J. Kop, V. Wheaton, J. Brosius, R. R. Gutell, A. M. Kopylov, F. Dohme, and W. Herr. 1981. The secondary structure of 23S ribosomal RNA. *Nucleic Acids Res.* **9**:6167–6189.

Schmidt, F. J., J. Thompson, K. Lee, J. Dijk, and E. Cundliffe. 1981. The binding site for ribosomal protein L11 within 23S RNA of *E. coli*. *J. Biol. Chem.* **256**:12301–12305.

Traub, P., and J. L. Sussmann. 1982. Adenine-guanine base pairing in ribosomal RNA. *Nucleic Acids Res.* **10**:2701–2708.

Vester, B., and R. A. Garrett. 1984. Structure of a protein L23-RNA complex located at the A-site domain of the ribosomal peptidyl transferase centre. *J. Mol. Biol.* **179**:431–452.

Vester, B., and R. A. Garrett. 1988. The importance of highly conserved nucleotides in the binding region of chloramphenicol at the peptidyl transferase centre of *E. coli* 23S rRNA. *EMBO J.* **7**:3577–3587.

Wower, J., S. S. Hixon, and R. A. Zimmermann. 1989. Labelling the peptidyltransferase centre of the *E. coli* ribosome, with photoreactive $tRNA^{Phe}$ derivatives containing azidoadenosine at the 3′-end of the acceptor arm: a model of the tRNA-ribosome complex. *Proc. Natl. Acad. Sci. USA* **86**:5232–5236.

Chapter 12

Ribosome Structure: Three-Dimensional Locations of rRNA and Proteins

MELANIE I. OAKES, ANDREW SCHEINMAN, THOMAS ATHA, GARY SHANKWEILER, and JAMES A. LAKE

Major advances have been made in our understanding of ribosome structure and function since the last ribosome meeting in Texas. Perhaps the major change since that meeting is that we now have highly detailed information about the functions of even individual nucleotides of rRNA.

In this chapter, we emphasize the detailed structural and biochemical information available for the *Escherichia coli* ribosome, relate it to the information now becoming available on the three-dimensional structure of rRNA, and present a model for approximately 40% of the small-subunit RNA based on our DNA hybridization electron microscopy mapping studies.

STRUCTURE OF THE RIBOSOME

There is now general agreement concerning the overall structure and morphology of the *E. coli* ribosome. The model shown in Fig. 1 (Lake, 1976) is now generally accepted. The history of development of three-dimensional ribosomal structures is detailed elsewhere (Lake, 1981).

The smaller (30S) subunit is divided into two unequal parts by an indentation that in micrographs consists of a region of accumulated negative stain (Vasiliev, 1974; Wabl, 1974; Lake et al., 1974). The two regions are the head, or the upper one-third, and the base, or lower two-thirds. A region of the subunit, called the platform, extends from the base of the small subunit and forms a cleft between it and the head (Lake and Kahan, 1975). All models (Vasiliev, 1974; Boublik and Hellman, 1978; Stöffler-Meilicke et al., 1983) are now in general agreement with our structure.

The large subunit, like the small subunit, is asymmetric (Lake, 1976). It consists of a central protuberance, or head, and dissimilar protrusions inclined approximately 50° to either side of the central protuberance. One of these, the L7/L12 stalk (at the right in the lower central panel of Fig. 1) contains the only multiple copy proteins present in the *E. coli* ribosome. In a projection perpendicular to this (upper center panel), the large subunit is characterized by a notch on the upper surface. All other large-subunit models (Boublik et al., 1976; Shatsky et al., 1979; Dabbs et al., 1981; Vasiliev et al., 1983a) are also asymmetric and in general agreement with the original model (Lake, 1976).

The small subunit is positioned asymmetrically on the large, in the monomeric ribosome (Lake, 1976, 1981; Fig. 1). The small-subunit platform contacts the large subunit, so that the partition between the head and body of the small subunit is approximately aligned with the notch of the large. All current models agree with this structure. Supporting experiments include double-labeling immunoelectron microscopy data (Kastner et al., 1981; Lake, 1982), three-dimensional analyses of ribosomes and ribosomal complexes (Vasiliev et al., 1983b; Bernabeu and Lake, 1982), and functional studies mapping bound ligands (Girshovich et al., 1981; Bernabeu and Lake, 1982; Evstafieva et al., 1983). This structure applies to ribosomes in several translational states (Vasiliev et al., 1983a).

Melanie I. Oakes, Andrew Scheinman, Thomas Atha, Gary Shankweiler, and James A. Lake ■ Molecular Biology Institute and Department of Biology, University of California at Los Angeles, Los Angeles, California 90024.

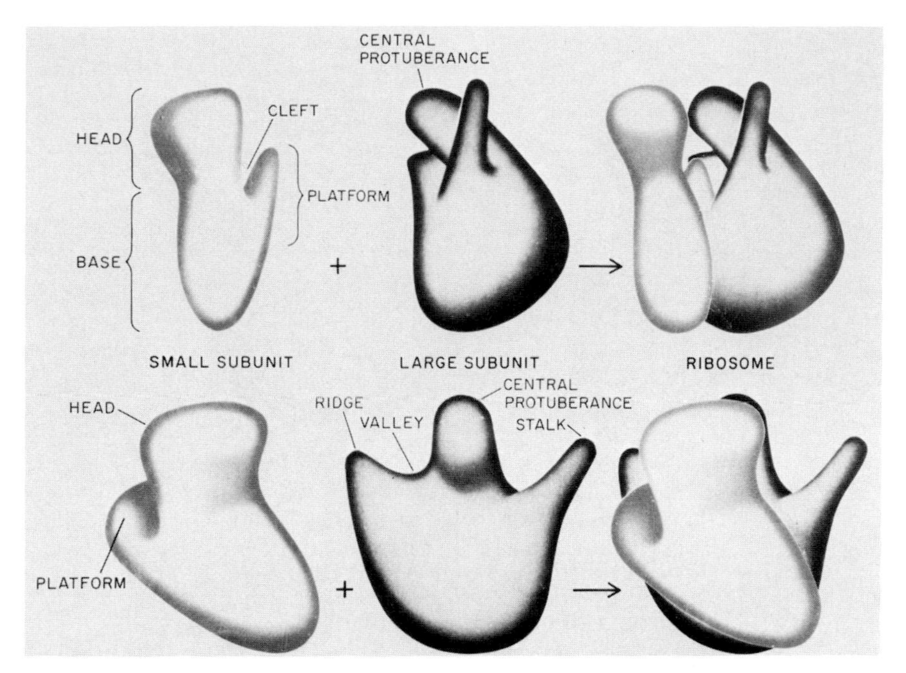

Figure 1. Three-dimensional model of the ribosome. This model gives an asymmetric shape to the two ribosomal subunits. The smaller subunit (left) includes a head, a base, and a platform. The larger subunit (second from left) includes a central protuberance, flanked by a ridge on one side and a stalk on the other. Two orientations of the model are shown. The length of a ribosome is about 250 Å.

RIBOSOMAL PROTEINS AND THE TRANSLATIONAL DOMAIN

Ribosomes from all organisms are divided into two general functional regions: the translational domain and the exit, or secretory, domain (Bernabeu and Lake, 1982). These domains are found at nearly opposite ends of the ribosome (Fig. 2). The translational domain encompasses the head and platform of the small subunit and the L7/L12 stalk, the central protuberance, and the L1 ridge of the large subunit. In general, all of the proteins of the *E. coli* small ribosomal subunit that have been mapped are located in this domain (Fig. 3). In the large subunit, many of the proteins are also found in the translational domain, although at least one, L17, is located in the exit domain (Fig. 3, right).

Ribosomal proteins have been localized by immunoelectron microscopy, neutron diffraction, chemical cross-linking, and other techniques (reviewed elsewhere in this volume). The locations mapped by neutron diffraction are particularly complete (Capel et al., 1987). In the limited space available, we will emphasize immunoelectron microscopy at the expense of the other techniques. This method (Wabl, 1974; Lake et al., 1974; Tischendorf et al., 1974) combines immunology and electron microscopy to allow one to map specific ribosomal protein and RNA components and to determine their locations in three dimensions. Using this technique, the authors, in collaboration with L. Kahan, W. Strycharz, and M. Nomura, have mapped 17 of the 21 small-subunit proteins and many of the large-subunit proteins (reviewed in Lake, 1981). Some of these localizations are shown in Fig. 3, and others are

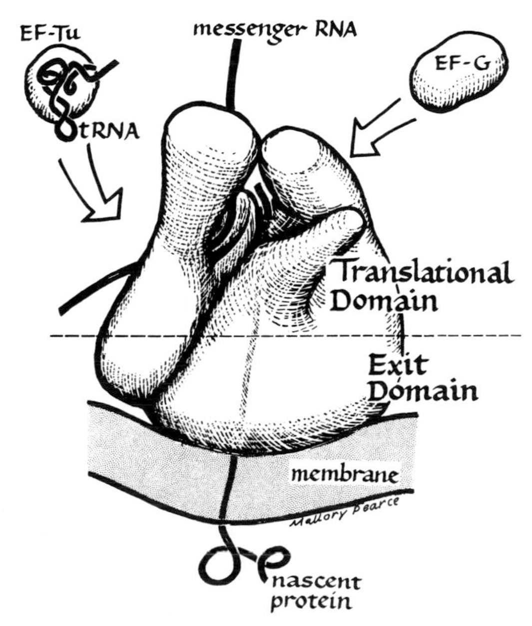

Figure 2. Diagrammatic representation of the exit and translational domains of the ribosome and their orientations with respect to the membrane-binding site. (Adapted from Bernabeu and Lake, 1982.)

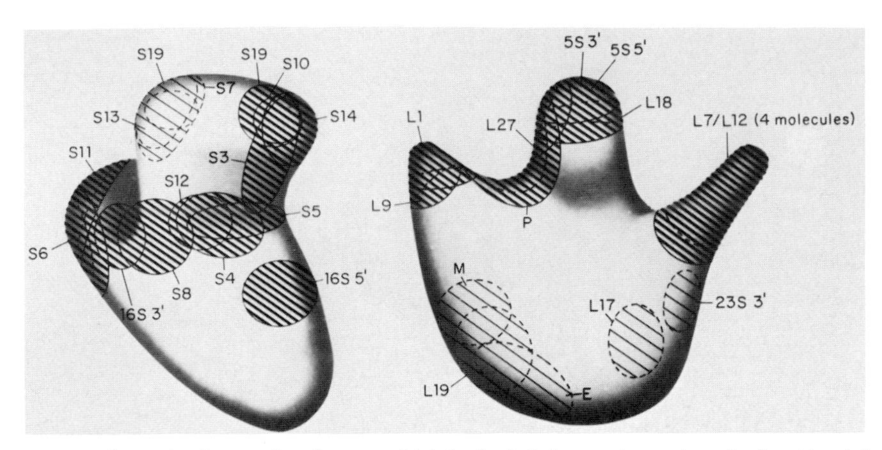

Figure 3. Summary map of protein sites on the ribosome. Lightly shaded sites are located on the far side of the subunits. P, M, and E, Peptidyltransferase, membrane-binding, and nascent protein exit sites, respectively; S and L, small- and large-subunit proteins, respectively; 16S 5′, 23S 3′, etc., 5′ and 3′ ends, respectively, of the rRNAs.

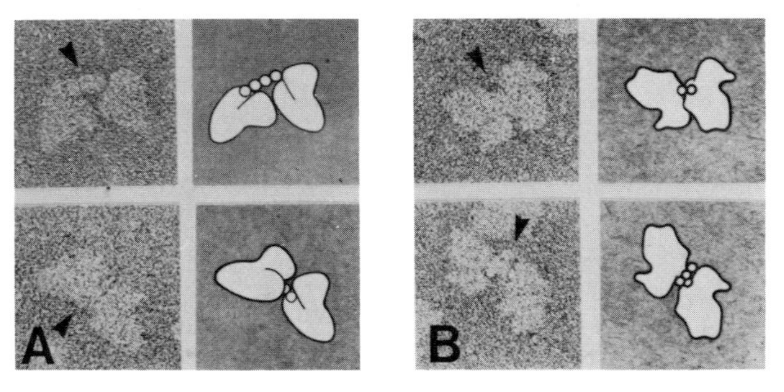

Figure 4. (A) Electron micrographs of eubacterial (*E. coli*) small subunits after hybridization with the 1400-region probe and labeling with avidin. (B) Electron micrographs of eucaryotic (*Saccharomyces cerevisiae*) small subunits and their reaction with the 1400-region probe.

discussed later in conjunction with locations of rRNA. Stöffler and co-workers have likewise mapped many proteins, and complete references to their work can be found in their chapters in this volume.

DNA HYBRIDIZATION ELECTRON MICROSCOPY

DNA hybridization electron microscopy is a useful technique to map single-stranded rRNA sites in three dimensions (Oakes et al., 1986). Thus far, seven 16S rRNA sequences have been localized on the surface of the 30S subunit. These sequences are: 16S rRNA 518–533, 686–703, 714–733, 787–803, 1392–1407, 1492–1505, and Shine-Dalgarno (Oakes et al., 1986; Oakes et al., 1987; Scheinman et al., 1988; Oakes and Lake, 1990). Another region that has been positioned by a variation of this technique is 16S rRNA 1531–1542 (Olson et al., 1988). In DNA hybridization electron microscopy, synthetic oligonucleotide probes, complementary to a specific rRNA sequence and carrying an attached biotin, are hybridized to ribosomal subunits (for functional applications, see the review by Hill et al. [1985]). The locations of biotins on subunits can then be mapped by electron microscopy to determine the three-dimensional site of attachment of the probe.

The *E. coli* sequence 1392–1407, the 1400 re-

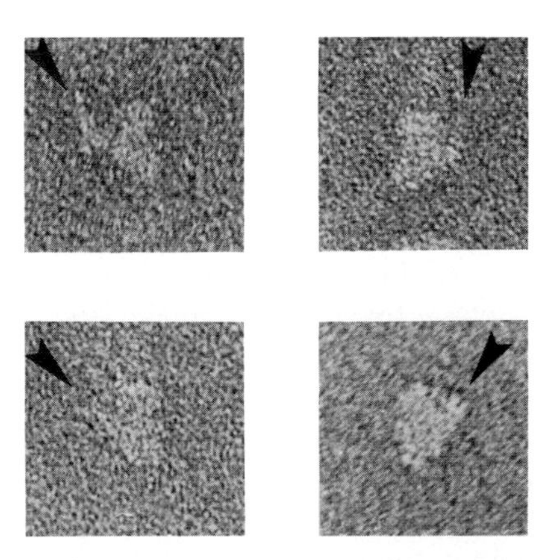

Figure 5. Electron micrographs of ribosomal subunits reconstituted in vitro from genetically engineered rRNA. The "16S" rRNA contains a 236-base-long yeast expansion sequence insert (indicated by arrows).

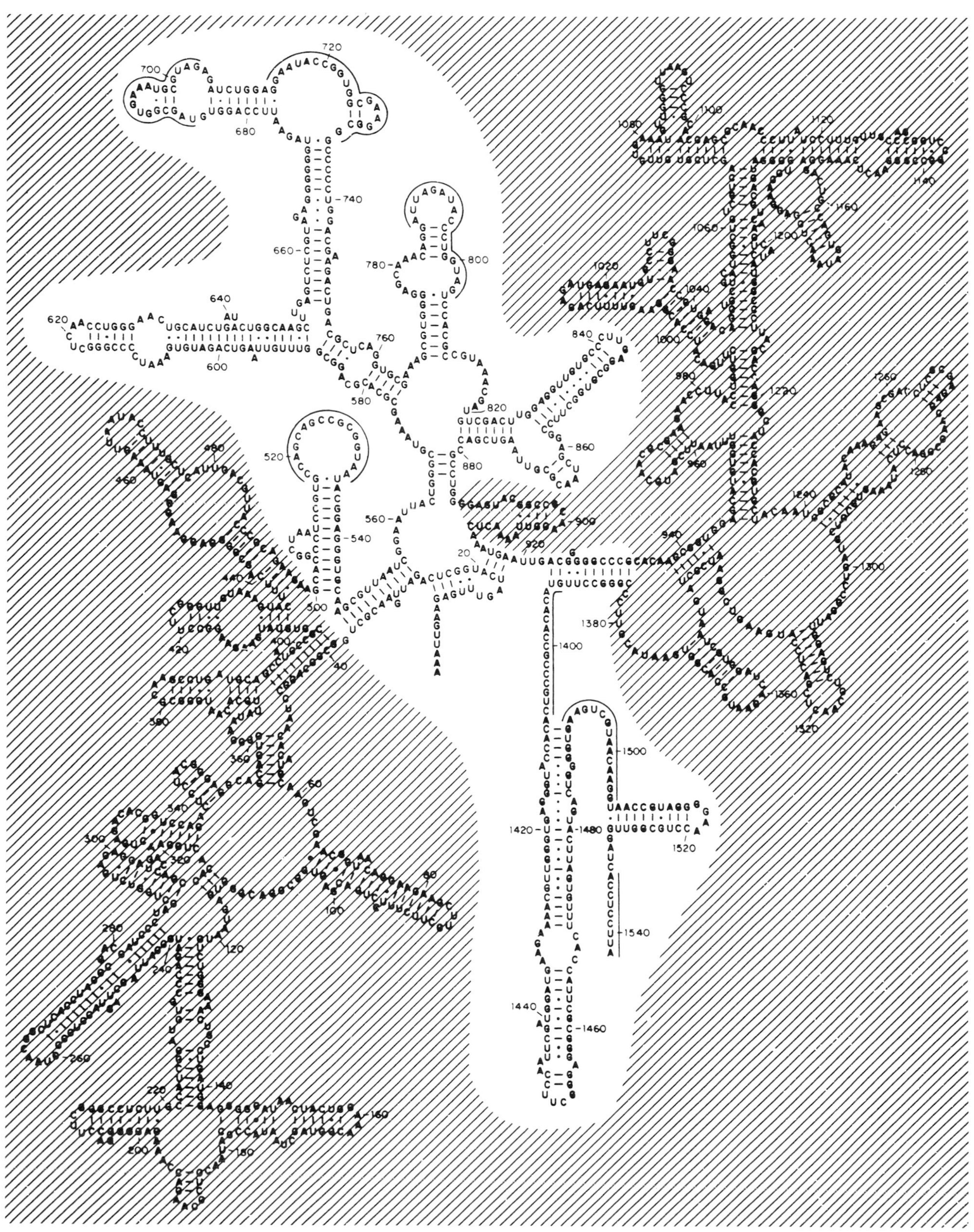

Figure 6. Secondary structure of 16S rRNA in *E. coli* 30S ribosomal subunits illustrating the regions mapped and included in the three-dimensional model. The underlined regions are the sequences that hybridized to complementary DNA probes. The hatched area shows regions excluded from the model. The secondary structure is from Stern et al., 1988a, Stern et al., 1988b, and Stern et al., 1988c.

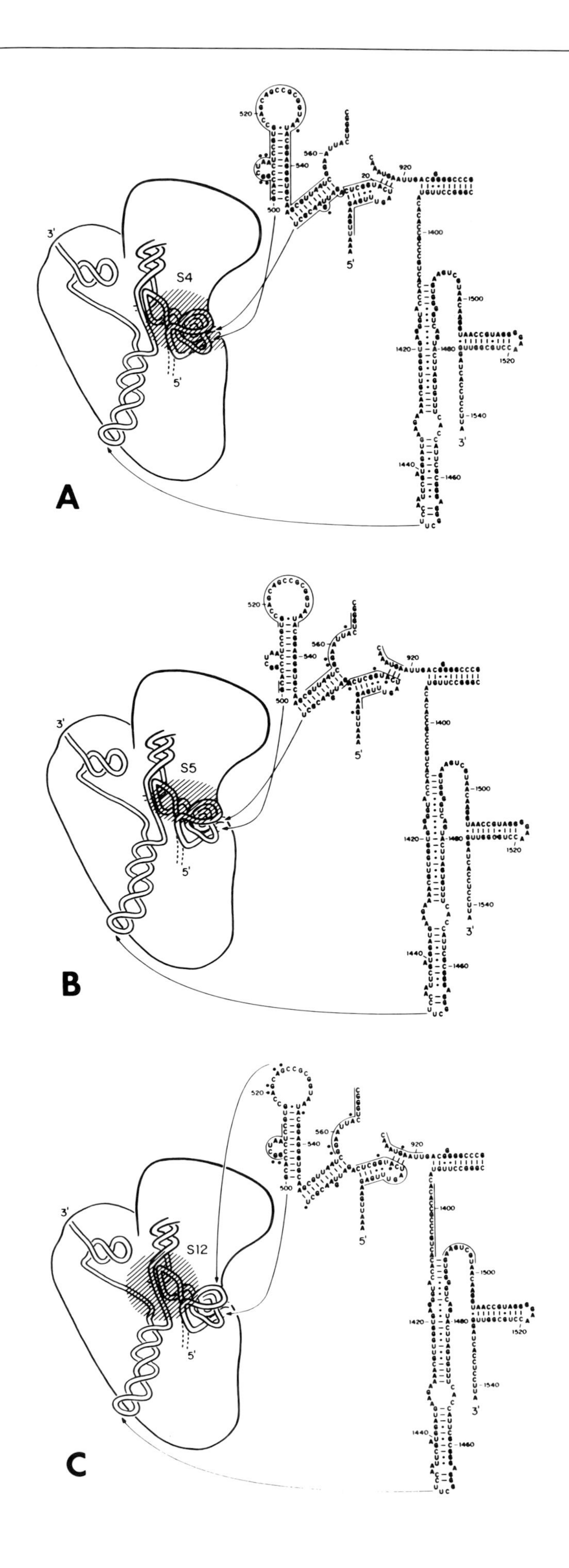

A
S4
3'
5'
B
S5
C
S12
520
540
560
500
920
1400
1420
1480
1500
1520
1540
1440
1460

722 and 723), we were able to reconstitute subunits. The size of this sequence allows it to be visualized easily and directly by electron microscopy. Those small subunits containing this insert have a platform with an additional structure attached (shown in Fig. 5). The ability to localize inserts rapidly by inspection of reconstituted subunits, 100% of which contain genetically engineered inserts, will be invaluable to our goal of mapping the conformation of all of the 16S and 23S rRNAs. Similar insert-mapping experiments with 23S rRNA also suggest that by the time of the next meeting on ribosomes, we may know the path of much of the 23S rRNA as well.

THREE-DIMENSIONAL LOCATIONS OF SEVEN REGIONS OF 16S rRNA

Thus far, we have used DNA hybridization electron microscopy to map a total of seven regions of 16S rRNA. Four of these regions are entirely located on the platform. This indicates both the relative accessibility of this region of the subunit and an experimental bias toward functionally significant regions of rRNA. The highly conserved regions of 16S rRNA that we have mapped may be constrained to be accessible to the surface because of their roles in protein synthesis. The regions that have been localized as well as the portion of 16S rRNA included in the model are shown in Fig. 6. The experimentally mapped regions are indicated by shading (Fig. 7), and parts of our proposed rRNA model are listed below.

Region 518–533 is located in a distinct part of the small subunit. These nucleotides appear at the level of the neck on the cytoplasmic side of the 30S subunit. This is consistent with the mapping of 7-methylguanosine G-526 (Trempe et al., 1982) and is also close to the elongation factor EF-Tu-binding site (Langer and Lake, 1986; Girshovich et al., 1986) and to ribosomal proteins S4 and S12 (Winkelmann et al., 1982), of which protein S4 protects these nucleotides (Stern et al., 1986). Since this region is proximal to the EF-Tu-binding site, known as the recognition (R) site, it may have a role in the initial tRNA recognition step of protein synthesis (Lake, 1977).

That sequence 1492–1505 is located in the cleft of the subunit is not altogether unexpected, since it is adjacent in the secondary structure model to sequence 1392–1407, which we had previously mapped to a similar region (used to illustrate DNA hybridization electron microscopy in this chapter). Within this region, two nucleotides, A-1491 and A-1493, have been shown to be protected by poly(U)-dependent tRNA binding (Moazed and Noller, 1986), characteristic of A-site binding. Nearby is also the Shine-Dalgarno region (probe 1534–1542), which has been extensively implicated in tRNA binding.

Two platform regions that mapped very close to each other are regions 686–703 and 714–733. Both are located near the top of the platform. Much experimental evidence regarding this region and its location within the 30S subunit is consistent with this result (these findings are detailed in Oakes and Lake, 1990). They are thought to be the location of the tRNA P site and are protected from modification by tRNA.

The related region, from nucleotides 787 to 803, also represents another region of 16S rRNA found to be on the platform. Within this region, A-794 and C-795 are protected by tRNA from chemical modification independent of the presence of mRNA (Moazed and Noller, 1986). As determined from these and other studies, and from the mapping of nucleotide 787 in the same region as nucleotides 686 and 714, part of this sequence may also have some function in the P site.

THREE-DIMENSIONAL MODEL FOR 16S rRNA

A variety of experimental approaches are contributing new information about the tertiary structure of rRNA. Recent models for the tertiary structure of 16S rRNA have been proposed (Expert-Bezançon and Wollenzien, 1985; Schuler and Brimacombe, 1988; Stern et al., 1988c) on the basis of data obtained from a variety of techniques, including RNA-RNA cross-linking, RNA-protein cross-linking, and ribosomal protein protection of rRNA. Two of these models also have incorporated neutron diffraction information on the placement of the ribosomal proteins (Capel et al., 1987). Our model has used DNA hybridization mappings as the primary source of data but has also incorporated data from all of these useful techniques. Two unusual features contained in it are (i) a recognition complex and (ii) a platform ring (Oakes and Lake, 1990; Oakes et al., 1990). These two structures are described below.

Figure 8. Diagram of regions of contact between ribosomal proteins involved in recognition (S4, S5, and S12) and the 5′ region of 16S rRNA. Solid dots indicate nucleotides protected from chemical probes by the respective proteins during assembly (Stern et al., 1988a; Stern et al., 1988b; Stern et al., 1988c); cross-hatched areas are sites mapped by us for the respective proteins. The outlined regions of nucleotides coincide with sites where proteins have been mapped by immunoelectron microscopy.

The Recognition Complex

The recognition complex is a folding of a region of the 16S molecule that contains the 520 loop and stem. This complex is positioned on the exterior or cytoplasmic surface of the small subunit. The base of the structure is formed by the coaxial stacking of helices 9–13/21–25, 17–19/916–918, and 27–37/547–556, similar to that predicted in earlier tertiary models (Noller and Lake, 1984). This pseudoknot structure (Pleij et al., 1985) forms a bridge from S4-binding domains to S7-binding domains. A primary constraint on the model, illustrated in Fig. 8, is that it places sequence 518–533 within the region on the 30S subunits where the same sequence was mapped by DNA hybridization electron microscopy. Since our mapping indicates that nucleotides within region 518–533 are located on the exterior or cytoplasmic side of the subunit, we therefore favor placing this region above the pseudoknot structure as indicated in Fig. 8. This placement differs from that of Stern et al. (1988c), who have placed these helices on the "interface" side. Our placement on the cytoplasmic surface is, however, consistent with immunoelectron microscopic localizations of ribosomal proteins S4, S5, and S12 and the modified nucleotide 7-methylguanosine G-527 (Winkelmann et al., 1982; Trempe et al., 1982).

The location on the 30S subunit of ribosomal protein S4, mapped by immunoelectron microscopy (Winkelmann et al., 1982), has been superimposed on our model in Fig. 8A in order to provide a test for it. Since S4 is known to protect several 16S rRNA regions from chemical modification during assembly, we have compared the overlap of the protein and rRNA mappings with the protection experiments of Stern et al. (1986). The outlined nucleotides that coincide with the protein mappings and hence could possibly interact with S4 are indicated by a solid line on the rRNA secondary structure. Nucleotides that are protected by S4 during assembly are indicated by closed circles. This same representation has been used for ribosomal proteins S5 and S12 as well (Fig. 8B and C, respectively). Some data suggest that S4, S5, and S12 form a functional complex. In particular, mutational alterations of both S4 and S5 are known to suppress the streptomycin dependence phenotype of some S12 mutations (Anderson et al., 1967; Gorini, 1971), and these three proteins appear to cooperate in ribosomal control of translational fidelity (Gorini, 1969; Kuwano et al., 1969). Furthermore, the recognition binding site for EF-Tu has been mapped on this exterior surface of the small subunit (Langer and Lake, 1986; Girshovich et al., 1986). Hence, our choice for the name of this region.

The Platform Ring

The name of this structure, i.e., platform ring, conveys its general shape. This structure is made from two stems and loops that literally ring the platform and define it. Central to the development of this part of the model was the realization that since the platform is thin (approximately 30 Å [3 nm]; Lake, 1976), it was unlikely to be more than a single RNA helix thick. This strongly constrained possible models. Since region 787 mapped to the top of the platform, we have placed the loop there. Region 787 is found at the top of the platform as shown in Fig. 7, and region 686 maps lower down on the platform on the convex surface. Likewise, the "elbow" at region 714–733 that links helices on either side of it has been placed on the convex lower surface of the platform. We have coaxially stacked helices 576–587/754–766 and 588–617/623–657 in the usual manner, since no unpaired nucleotides interrupt formation of the stacked helices. This allows more room for the lengthening of this region in small-subunit eucaryotic rRNA.

Many platform ribosomal proteins are associated with the central domain of rRNA. These include proteins S6, S11, S18, S21, and, to a lesser extent, S8. Detailed information regarding interactions with specific nucleotides has been gained largely through cross-linking (reviewed in Brimacombe et al., 1988) and protection from chemical probes (reviewed in Stern et al., 1988c). Figure 9 illustrates the locations of those proteins determined by immunoelectron microscopy (Oakes et al., 1990). Overlap of this region with the 16S rRNA is indicated on the side, as are the nucleotides that are protected by these respective proteins. Protection by the joint addition of proteins S6 and S18 is shown in Fig. 9A and B, since these studies were performed with both proteins present. Our protein mappings are low resolution, so the hatched area may include a larger area than the protein actually covers. Also, we have not attempted to determine whether a crossing strand of rRNA might protect a lower-lying strand. We simply have indicated all rRNA regions that lie within the cross-hatched areas and consequently might be protected by protein binding. In general, the model is in good agreement with the results of the protein protection (73 of 84 nucleotides are accounted for). Other reported interactions that are accommodated by our model are the cross-link between regions 690 and 790 (Atmadja et al., 1986), ribosomal protein cross-links between S11 and region 693–697 and 702–705 (Greuer et al., 1987; Osswald et al., 1987), and a cross-link between S21 and region 693–697 (Greuer et al., 1987).

The dimethyladenosine helix has been positioned within the platform ring by mappings. Three regions within the 3′-terminal end have been mapped on the surface of the 30S ribosomal subunit. The penultimate helix (1409–1492), the "cleft anchor" (Noller and Lake, 1984), is flanked at the 5′ and 3′ ends by two highly conserved regions. The 3′ flanking region enters into the platform, whereas the 5′ flanking end emerges from the subunit in the vicinity of the cleft (Oakes et al., 1986). Contained within these specific regions are the colicin E3 scission site (Bowman et al., 1971; Senior and Holland, 1974), a site protected by IF3 (Wickstrom, 1983) and the site of cross-linking of C-1400 to the anticodon of a P-site tRNA (Prince et al., 1982). At 1492, the rRNA leads back into the platform where we have mapped sequence 1506–1529. Additional evidence for placing the dimethyladenosine helix 1506–1529 on the platform comes from immunoelectron microscopy of the dimethyladenosines and of the 3′ end (Politz and Glitz, 1977; Evstafieva et al., 1983; Trempe and Glitz, 1981). The dimethyladenosine helix could be placed in the center of the platform ring in either of two orientations. We have positioned the dimethyladenosines toward the bottom of the platform, but this helix could also be placed in the opposite orientation with the dimethyladenosines pointing toward the top of the platform. In either orientation, the Shine-Dalgarno region is still accessible for binding to the mRNA upstream of the start codon.

In general, there is much similarity between other models (Stern et al., 1988c; Schuler and Brimacombe, 1988; Noller and Lake, 1984) and ours. One difference involves loop 518–533, which Stern et al. (1988c) placed on the 50S interface side of the 30S subunit, underneath the pseudoknot structure. Our model and the model of Schuler and Brimacombe (1988) place this region on the external or cytoplasmic side of the 30S subunit. Another region exhibiting some differences is the platform ring. Our model proposes a ring structure that is closed through interactions between adjacent loops. One can easily imagine specific base pairing through non-Watson-Crick interactions similar to those between the dimethyluridine and TUCG loops of tRNAs.

As a result of our experiments, we cannot help noting that the regions which have been most accessible to DNA probes are highly conserved and generally important for ribosome function. We hope to gather more information concerning the rRNA tertiary structure by finding more accessible regions and by genetically engineering rRNA for use with DNA hybridization electron microscopy.

We thank J. Washizaki for electron microscopy and M. Peris for photography. We thank W. Hill for organizing this meeting.

This work was supported by research grants from the National Science Foundation and the National Institutes of Health to J.A.L.

REFERENCES

Anderson, P., J. Davies, and B. D. Davis. 1967. Effect of spectinomycin on polypeptide synthesis in extracts of Escherichia coli. *J. Mol. Biol.* **29:**203–215.

Atmadja, J., W. Stiege, M. Zobawa, B. Greuer, M. Osswald, and R. Brimacombe. 1986. The tertiary folding of Escherichia coli 16S RNA, as studied by in situ intra-RNA cross-linking of 30S ribosomal subunits with bis-(2-chloroethyl)-methylamine. *Nucleic Acids Res.* **14:**659–673.

Bernabeu, C., and J. A. Lake. 1982. Nascent polypeptide chains emerge from the exit domain of the large ribosomal subunit: immune mapping of the nascent chain. *Proc. Natl. Acad. Sci. USA* **79:**3111–3115.

Boublik, M., and W. Hellman. 1978. Comparison of Artemia salina and Escherichia coli ribosome structure by electron microscopy. *Proc. Natl. Acad. Sci. USA* **75:**2829–2833.

Boublik, M., W. Hellman, and E. H. Roth. 1976. Localization of ribosomal proteins L7 L12 in the 50S subunit of Escherichia coli ribosomes by electron microscopy. *J. Mol. Biol.* **107:**479–490.

Bowman, C. M., J. E. Dahlberg, T. Ikemura, J. Konisky, and M. Nomura. 1971. Specific inactivation of 16S ribosomal RNA induced by colicin E3 in vivo. *Proc. Natl. Acad. Sci. USA* **68:**964–968.

Brimacombe, R., J. Atmadja, W. Stiege, and D. Schuler. 1988. A detailed model of the three-dimensional structure of Escherichia coli 16S ribosomal RNA in situ in the 30S subunit. *J. Mol. Biol.* **199:**115–136.

Capel, M. S., D. M. Engelman, B. R. Freeborn, M. Kjeldgaard, J. A. Langer, V. Ramakrishnan, D. G. Schindler, D. K. Schneider, B. P. Schoenborn, S. Yabuki, and P. B. Moore. 1987. A complete mapping of the proteins in the small ribosomal subunit of Escherichia coli. *Science* **238:**1403–1406.

Dabbs, E. R., R. Ehrlich, R. Hasenbank, B.-H. Schroeterm, M. Stöffler-Meilicke, and G. Stöffler. 1981. Mutants of Escherichia coli lacking ribosomal protein L1. *J. Mol. Biol.* **149:**553–578.

Evstafieva, A. G., I. N. Shatsky, A. A. Bogdanov, Y. P. Semenko, and V. D. Vasiliev. 1983. Localization of 5′ and 3′ ends of the ribosome bound segment of template polynucleotides by immune electron microscopy. *EMBO J.* **2:**799–804.

Expert-Bezançon, A., and P. Wollenzien. 1985. Three-dimensional arrangement of the Escherichia coli 16S ribosomal RNA. *J. Mol. Biol.* **184:**53–66.

Girshovich, A. S., E. S. Bochkareva, and V. D. Vasiliev. 1986. Localization of the elongation factor G on Escherichia coli ribosome. *FEBS Lett.* **197:**192–198.

Girshovich, A. S., T. V. Kurtschaliov, Y. A. Ovchinnikov, and V. D. Vasiliev. 1981. Localization of the elongation factor G on Escherichia coli ribosome. *FEBS Lett.* **130:**54–59.

Gorini, L. C. 1969. The contrasting role of strA and ram gene products in ribosomal functioning. *Cold Spring Harbor Symp. Quant. Biol.* **34:**101–111.

Gorini, L. 1971. Ribosomal discrimination of tRNAs. *Nature* (London) *New Biol.* **234:**261–264.

Greuer, B., M. Osswald, R. Brimacombe, and G. Stöffler. 1987. RNA-protein crosslinking in Escherichia coli 30S ribosomal subunits; determination of sites on the 16S RNA that are crosslinked to proteins S3, S4, S5, S7, S9, S11, S13, S19 and S21 by treatment with w-bis-(2-chloroethyl)-methylamine. *Nucleic Acids Res.* **15**:3241–3255.

Hill, W. E., W. E. Tapprich, and A. Tassanakajohn. 1985. Probing ribosomal structure and function, p. 233–252. *In* B. Hardesty and G. Kramer (ed.), *Structure, Function, and Genetics of Ribosomes.* Springer-Verlag, New York.

Kastner, B., M. Stöffler-Meilicke, and G. Stöffler. 1981. Arrangement of the subunits in the ribosome of Escherichia coli: demonstration by immunoelectron microscopy. *Proc. Natl. Acad. Sci. USA* **78**:6652–6656.

Kuwano, M., H. Endo, and Y. Ohrishi. 1969. Mutations to spectinomycin resistance which alleviate the restriction of an amber suppressor by streptomycin resistance. *J. Bacteriol.* **97**: 940–943.

Lake, J. A. 1976. Ribosomal structure determined by electron microscopy of Escherichia coli small subunits, large subunits and monomeric ribosomes. *J. Mol. Biol.* **105**:131–159.

Lake, J. A. 1977. Aminoacyl-tRNA binding at the recognition site is the first step of the elongation cycle of protein synthesis. *Proc. Natl. Acad. Sci. USA* **74**:1903–1907.

Lake, J. A. 1981. The ribosome. *Sci. Am.* **245**:84–97.

Lake, J. A. 1982. Ribosomal subunit orientations determined in the monomeric ribosome by single and by double-labeling immune electron microscopy. *J. Mol. Biol.* **161**:89–106.

Lake, J. A., and L. Kahan. 1975. Ribosomal proteins S5, S11, S13 and S19 localized by electron microscopy of antibody-labeled subunits. *J. Mol. Biol.* **99**:631–644.

Lake, J. A., M. Pendergast, L. Kahan, and M. Nomura. 1974. Localization of Escherichia coli ribosomal proteins S4 and S14 by electron microscopy of antibody-labeled subunits. *Proc. Natl. Acad. Sci. USA* **71**:4688–4692.

Langer, J., and J. A. Lake. 1986. Elongation factor TU localized on the exterior surface of the small ribosomal subunit. *J. Mol. Biol.* **187**:617–621.

Moazed, D., and H. F. Noller. 1986. Transfer RNA shields specific nucleotides in 16S ribosomal RNA from attack by chemical probes. *Cell* **47**:985–994.

Noller, H. F., and J. A. Lake. 1984. Ribosome structure and function: localization of rRNA. *Membr. Struct. Funct.* **6**:218–297.

Oakes, M. I., M. W. Clark, E. Henderson, and J. A. Lake. 1986. DNA hybridization electron microscopy: ribosomal RNA nucleotides 1392–1407 are exposed in the cleft of the small subunit. *Proc. Natl Acad. Sci. USA* **83**:275–279.

Oakes, M. I., L. Kahan, and J. A. Lake. 1990. DNA-hybridization electron microscopy: tertiary structure of 16S rRNA. *J. Mol. Biol.* **211**:907–918.

Oakes, M. I., and J. A. Lake. 1990. DNA-hybridization electron microscopy: localization of five regions of 16S rRNA on the surface of 30 S ribosomal subunits. *J. Mol. Biol.* **211**:897–906.

Oakes, M. L., A. Scheinman, M. Rivera, D. Soufer, G. Shankweiler, and J. A. Lake. 1987. Evolving ribosome structure and function: rRNA and the translation mechanism. *Cold Spring Harbor Symp. Quant. Biol.* **52**:675–685.

Olson, H. M., L. S. Lasater, P. A. Cann, and D. G. Glitz. 1988. Messenger RNA orientation on the ribosome. *J. Biol. Chem.* **263**:15196–15204.

Osswald, M., B. Greuer, R. Brimacombe, G. Stöffler, H. Baumert, and H. Fasold. 1987. RNA-protein crosslinking in Escherichia coli 30S ribosomal subunits: determination of sites on 16S that are crosslinked to proteins S3, S4, S5, S7, S8, S9, S11, S19 and S21 by treatment with methyl p-azidophenyl acetimidate. *Nucleic Acids Res.* **15**:3221–3240.

Pleij, C. W. A., K. Rietveld, and L. Bosch. 1985. A new principle of RNA folding based on pseudoknotting. *Nucleic Acids Res.* **13**:1717–1731.

Politz, S. M., and D. G. Glitz. 1977. Ribosome structure: localization of N(6),N(6)-dimethyladenosine by electron microscopy of a ribosome-antibody complex. *Proc. Natl. Acad. Sci. USA* **74**:1468–1472.

Prince, J. B., B. H. Taylor, D. L. Thrulow, J. Ofengand, and R. A. Zimmermann. 1982. Covalent crosslinking of tRNA val to 16S RNA at the ribosomal P site: identification of crosslinked residues. *Proc. Natl. Acad. Sci. USA* **79**:5450–5454.

Scheinman, A., G. W. Shankweiler, and J. A. Lake. 1988. Reconstitution of structurally intact small ribosomal subunits from in vitro transcribed rRNA containing an insert. *Abstr. Cold Spring Harbor Laboratory Symp. Ribosome Synthesis*, p. 157.

Schuler, D., and R. Brimacombe. 1988. The Escherichia coli 30S ribosomal subunit; an optimized three-dimensional fit between the ribosomal proteins and the 16S RNA. *EMBO J.* **7**:1509–1513.

Senior, B. W., and I. B. Holland. 1974. Effect of colicin E3 upon the 30S ribosomal subunit of Escherichia coli. *Proc. Natl. Acad. Sci. USA* **68**:959–963.

Shatsky, I. N., L. V. Mochalova, M. S. Kojouharova, A. A. Bogdanov, and V. D. Vasiliev. 1979. Localization of the 3′-end of Escherichia coli 16S rRNA by electron microscopy of antibody-labelled subunits. *J. Mol. Biol.* **133**:501–515.

Stern, S., L.-M. Changchien, G. R. Craven, and H. F. Noller. 1988a. Interaction of proteins S16, S17 and S20 with the 16S ribosomal RNA. *J. Mol. Biol.* **200**:291–299.

Stern, S., T. Powers, L.-M. Changchien, and H. F. Noller. 1988b. Interaction of ribosomal proteins S5, S6, S11, S12, S18 and S21 with the 16S rRNA. *J. Mol. Biol.* **201**:683–696.

Stern, S., B. Weiser, and H. F. Noller. 1988c. Model for the three-dimensional folding of 16S ribosomal RNA. *J. Mol. Biol.* **204**:447–481.

Stern, S., R. C. Wilson, and H. F. Noller. 1986. Localization of the binding site for protein S4 on 16S ribosomal RNA by chemical and enzymatic probing and primer extension. *J. Mol. Biol.* **192**:101–110.

Stöffler-Meilicke, M., M. Noah, and G. Stöffler. 1983. Location of eight ribosomal proteins on the surface of the 50S subunit from Escherichia coli. *Proc. Natl. Acad. Sci. USA* **80**:6780–6787.

Svensson, P., L.-M. Changchien, G. R. Craven, and H. F. Noller. 1988. Interaction of ribosomal proteins S6, S8, S15 and S18 with the central domain of 16S ribosomal RNA. *J. Mol. Biol.* **200**:301–308.

Tischendorf, G. W., M. Zeichardt, and G. Stöffler. 1974. Determination of the location of proteins L14, L17, L18, L19, L22 and L23 on the surface of the 50S ribosomal subunit of Escherichia coli by immune electron microscopy. *Mol. Gen. Genet.* **134**:187–208.

Trempe, M. R., and D. G. Glitz. 1981. Chloroplast ribosome structure: electron microscopy of ribosomal subunits and localization of N(6),N(6)-dimethyladenosine by immune electron microscopy. *J. Biol. Chem.* **256**:11873–11879.

Trempe, M. R., K. Ohgi, and D. G. Glitz. 1982. Ribosome structure: localization of 7-methylguanosine in the small subunits of Escherichia coli and chloroplast ribosome by immune electron microscopy. *J. Biol. Chem.* **257**:9822–9829.

Vasiliev, V. D. 1974. Morphology of the ribosomal 30S subparticle according to electron microscopic data. *Acta Biol. Med. Germ.* **33**:779–793.

Vasiliev, V. D., O. M. Selivanova, V. I. Baranov, and A. S. Spirin. 1983a. Structural study of translating 70S ribosomes for Escherichia coli. I. Electron microscopy. *FEBS Lett.* **155**:167–172.

Vasiliev, V. D., O. M. Selivanova, and S. N. Ryazantsev. 1983b. Structure of the Escherichia coli 50S ribosomal subunit. *J. Mol.*

Biol. **171**:561–569.

Wabl, M. R. 1974. Microscopic localization of two proteins on the surface of the 50S ribosomal subunit of Escherichia coli using specific antibody markers. *J. Mol. Biol.* **84**:241–247.

Wickstrom, E. 1983. Nuclease mapping of the secondary structure of the 49-nucleotide 3′ terminal cloacin fragment of Escherichia coli 16S RNA and its interactions with initiation factor 3. *Nucleic Acids Res.* **11**:2035–2052.

Winkelmann, D. A., L. Kahan, and J. A. Lake. 1982. Ribosomal protein S4 is an internal protein: localization by immune electron microscopy on the protein deficient subribosomal particles. *Proc. Natl. Acad. Sci. USA* **79**:5184–5188.

Chapter 13

New Possibilities for Neutron Scattering in the Study of RNA-Protein Interactions

I. N. SERDYUK, M. Y. PAVLOV, I. N. RUBLEVSKAYA, G. ZACCAI, R. LEBERMAN, and Y. M. OSTANEVICH

Neutron scattering was first applied to solve biological problems in 1969, when a group of physicists from the Federal Republic of Germany (Schelten et al., 1971) measured neutron scattering from hemoglobin solutions. Since then, several techniques have been developed and widely used to solve structural biological problems. Among these is the method of contrast variation, which is simple to use and implemented frequently (reviewed in Koch and Stuhrmann, 1979).

The method suggested by Engelman and Moore (1972) and Hoppe (1973) is more complicated. It was developed primarily for measuring distances between labeled components of a particle, using high particle concentrations. This technique has been applied to determine the distances between proteins of the 30S ribosome subunit of *Escherichia coli* (see Capel et al., 1988).

A third method used in small-angle neutron scattering (SANS) to obtain the form factor of a particle in concentrated solutions has been extensively used in studying polymer solutions (Williams et al., 1979; Akcasu et al., 1980).

There are three experimental schemes which correspond to the methods mentioned above. The first is based on subtracting the solvent scattering from that of the solution, which contains one type of particle, protonated or deuterated (Fig. 1A). This method of contrast variation and inverse contrast variation is based on a rather simple physical principle whereby the biological particle, or its component in solution, scatters neutrons only when its scattering density differs from that of the solvent. This technique is simple; moreover, the contrast can be varied by changing the scattering density of the entire solvent (different H_2O-D_2O mixtures) or the particle (growing in a deuterated medium).

Information on ribosome structure and its components has been obtained by this approach, including the distribution of RNA and protein, the structure of rRNA and proteins in solution and within ribosomal subunits, and conformational changes in ribosomal subunits as reconstitution and dynamics of the ribosome occur (reviewed in Hardesty and Kramer, 1986). The only disadvantage of this approach is that the resulting scattering curve usually contains contributions from particle association and interference and distortion of solvent structure near the particle surface. This, in turn, can lead to misinterpretation of the geometry and molecular mass of the particle.

The second scheme (Fig. 1B) is based on subtraction of the scattering of solution II from the scattering of solution I (Engelman and Moore, 1972; Hoppe, 1973). Solution I contains a mixture of particles of two types with deuterium-labeled and nonlabeled proteins. Solution II contains a mixture of particles of two other types in which only one of the proteins is labeled by deuterium. The main advantage of this approach is that distances can be measured between the labeled moieties in macromolecular complexes by using high macromolecule concentrations. Its main disadvantage is that it can only be used to measure interlabel distances.

In the third scheme (Fig. 1C), one also subtracts the scattering of the solvent from that of the solution, but in contrast to the first scheme, the solution contains a mixture of protonated and deuterated

I. N. Serdyuk, M. Y. Pavlov, and I. N. Rublevskaya ■ Institute of Protein Research, Academy of Sciences of the USSR, Pushchino, Moscow Region, USSR. **G. Zaccai** ■ Institut Laue-Langevin, Grenoble, France. **R. Leberman** ■ European Molecular Biology Laboratory Outstation, Grenoble, France. **Y. M. Ostanevich** ■ Joint Institute of Nuclear Physics, Dubna, USSR.

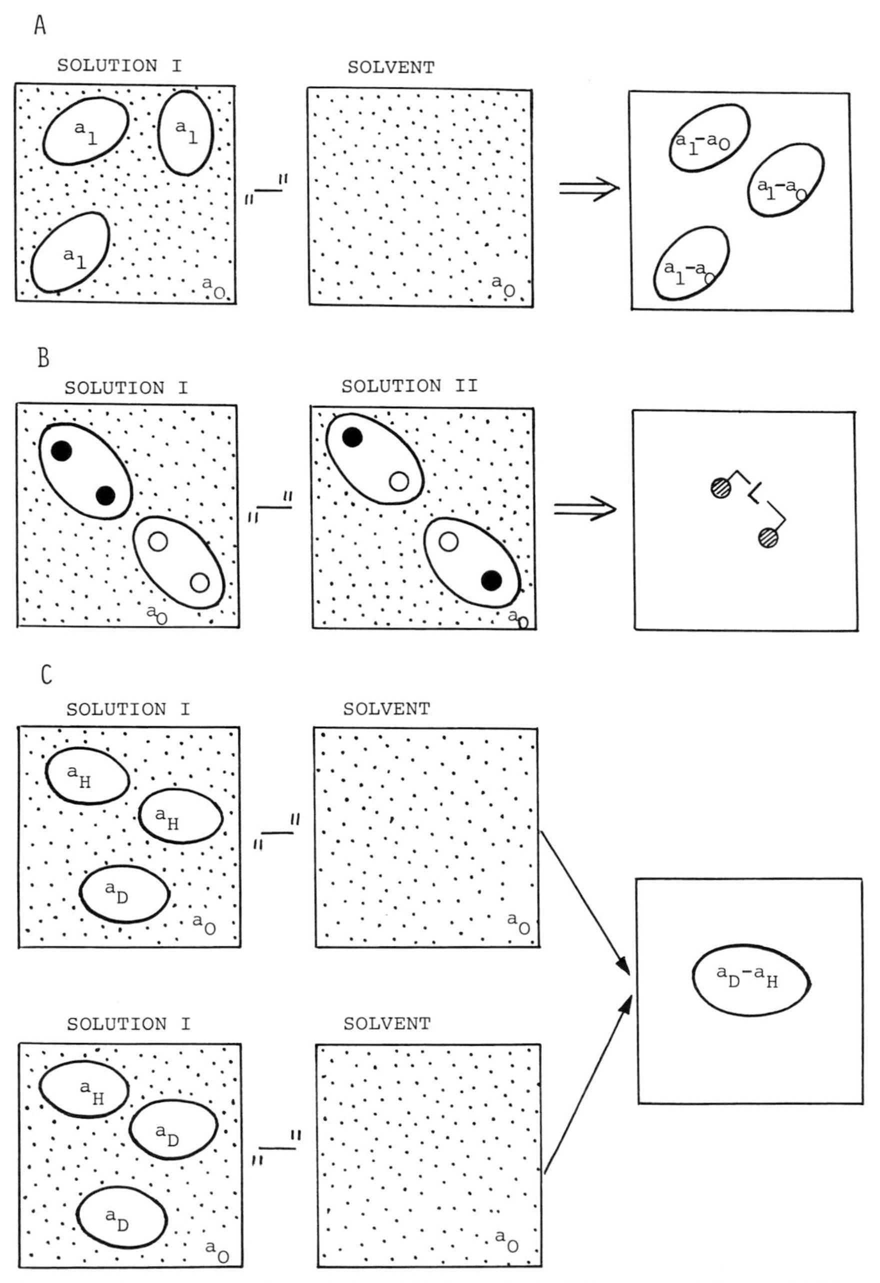

Figure 1. General scheme of contrast variation technique (A), the method of Engelman and Moore (1972) and Hoppe (1973) (B), and the method of double isotopic substitution (C). a_H or a_1, a_D, and a_0 are the scattering densities of protonated and deuterated particles and of the solvent, respectively.

macromolecules. We call this method double isotopic substitutions. A single-particle scattering function of macromolecules can be calculated at any finite concentration of macromolecules in solution if such a measurement is repeated using another ratio of deuterated to protonated macromolecules in the mixture. The main limitation of this approach is that it can be used only with homogeneous particles and therefore is of little use for biological samples.

We recently proposed a theoretical basis for a new approach in SANS for studying complex biological particles (Pavlov and Serdyuk, 1987; Serdyuk and Pavlov, 1988). This method is called the triple isotopic substitution method.

The experimental scheme of this approach (Fig. 2) is complex. Briefly, it is as follows. The scattering of solution I containing a mixture of protonated and deuterated particles and the scattering of solution II containing intermediately deuterated particles must be measured at the same concentration. At a defined

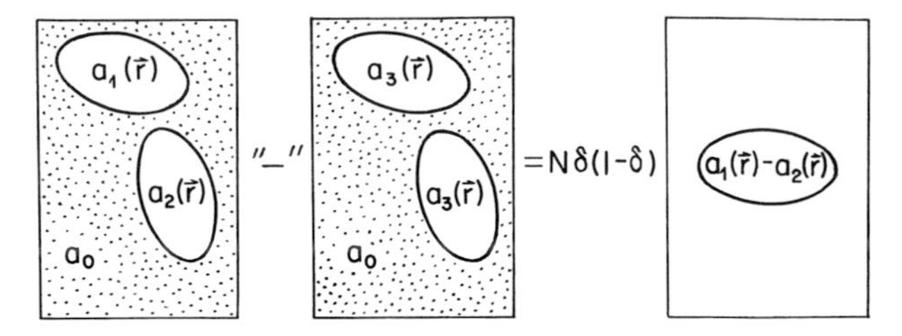

Figure 2. General scheme of the triple isotopic substitution method. $I_{1,2}(Q)$ and $I_3(Q)$ are the scattering intensities from solutions I and II, respectively; $a_1(\vec{r})$, $a_2(\vec{r})$, and $a_3(\vec{r})$ are the scattering densities of protonated, deuterated, and intermediately deuterated particles, respectively; a_0 is the scattering density of the solvent; $I_F(Q)$ is the scattering curve of a particle whose scattering density is equal to the difference between scattering densities of type 1 and type 2 particles.

fraction of deuterated particles in the mixture, the difference scattering curve coincides with the scattering curve of the particle whose scattering density is equal to the difference of scattering densities for deuterated and protonated particles. The proposed experimental scheme leads to interesting and novel predictions.

Below we discuss the main theoretical predictions of the new method, the protocol of the method, and experimental checking of the method, using elongation factor EF-Tu and the ternary complex between EF-Tu, tRNA, and GTP.

THEORETICAL PREDICTIONS

The theoretical principles underlying the method were derived by Pavlov and Serdyuk (1987) and can be written as follows:

$$I_{1,2}(Q) - I_3(Q) = N\delta(1 - \delta)I_F(Q) \quad (1)$$

$$a_3(\vec{r}) = (1 - \delta)a_1(\vec{r}) + \delta a_2(\vec{r}) \quad (2)$$

where $I_{1,2}(Q)$ is the scattering of solution I containing the mixture of protonated and deuterated particles whose scattering densities are $a_1(\vec{r})$ and $a_2(\vec{r})$, respectively, $I_3(Q)$ is the scattering of solution II containing intermediately deuterated particles of scattering density $a_3(\vec{r})$, N is the number of particles in solution, δ is the fraction of deuterated particles in solution I, and $I_F(Q)$ is the scattering curve of a particle whose scattering density $a_F(\vec{r})$ is equal to the difference between scattering densities $a_2(\vec{r})$ and $a_1(\vec{r})$:

$$a_F(\vec{r}) = a_2(\vec{r}) - a_1(\vec{r}) \quad (3)$$

For the given scattering densities $a_1(\vec{r})$, $a_2(\vec{r})$, and $a_3(\vec{r})$, equation 2 should be considered the equation for determining δ. For these equations to be valid, it is sufficient that the particle structure, interparticle interaction, and interaction of particles with the solvent do not depend on the extent of the particle deuteration.

These equations allow us to predict the following: the contribution of interparticle interference and particle association (dimerization) to the difference scattering curve, $I_{1,2}(Q) - I_3(Q)$, is eliminated because in equation 1, $I_3(Q)$ pertains only to one particle (prediction 1).

The exchange of labile hydrogen atoms between the particle and the solvent does not depend on particle deuteration, so the difference scattering curve will not depend on the solvent isotopic content (D_2O fraction in the H_2O-D_2O mixture) (prediction 2). This assumption results from the main equations, especially equation 3, since the scattering density $a_F(\vec{r})$ in this case does not depend on the D_2O fraction in the solution. From this, it also follows that the distortions of the solvent near the particle surface do not influence the difference scattering curve [because after subtracting $a_2(\vec{r}) - a_1(\vec{r})$, these distortions are canceled out in $a_F(\vec{r})$]. This prediction is important for studies of highly charged macromolecules such as tRNA, which are surrounded by a solvent layer whose properties differ from those of the solvent bulk (Zaccai and Xian, 1988).

Any small or large molecules added in equal concentrations to solutions I and II will be "invisible" in the scattering curves (prediction 3). This prediction also implies that a minor component of a binary complex can be studied.

Any component of a complex particle of constant scattering density in the particles of the three types will be invisible in our method at any D_2O fraction in the H_2O-D_2O mixture (prediction 4). This is also a simple consequence of equations 1 and 3, which state that the difference scattering curve pertains to a particle whose scattering density is the difference between the scattering densities of the particles of the first and the second types.

Thus, by using these four predictions, a protein or nucleic acid moiety of a nucleoprotein complex can be rendered invisible in ordinary light water. Hence, information about the protein conformation within the complex can be elicited.

The study of protein complexes with large ligands in light water offers methodological advantages if complex formation can be controlled quantitatively. However, any concern about detrimental effects of heavy water can be obviated.

It should be emphasized that the contrast in our method means the difference between the scattering densities of the particles of the two types of solution

I, i.e., $a_D(\vec{r}) - a_H(\vec{r})$. This is an important difference from the contrast variations techniques, where the contrast refers to the difference between the scattering densities of the particle and the solvent, i.e., $a_H(\vec{r})-a_0$ (for protonated particles). The difference between scattering densities of the protonated and completely deuterated proteins is twice as large as the contrast for protonated proteins in H_2O. This yields approximately the same difference intensity in both methods because of the term $\delta(1 - \delta)$ in equation 1.

PROTOCOL

To check the main prediction of the theory and to work out the protocol, we used EF-Tu (a protein with a molecular mass of 43 kilodaltons) from *E. coli* and the EF-Tu · GTP · Leu-tRNA^{Leu} ternary complex.

To obtain a completely deuterated EF-Tu, growth medium containing 99.8% D_2O and D-succinate as a carbon source was used. Medium containing 79% D_2O and H-glucose as a carbon source yielded intermediately deuterated EF-Tu. The measured match points γ_H, γ_{HD}, and γ_D are 40, 89, and 127% for undeuterated, intermediately deuterated, and deuterated EF-Tu (H-, HD-, and D-Tu), respectively.

The δ value was determined from the equation

$$(1 - \delta)\gamma_H + \delta\gamma_D = \gamma_{HD} \quad (4)$$

which for homogeneous particles is equivalent to the equation 2.

The key technical problem in this method is the precise mixing of the H- and D-Tu solutions to satisfy the predetermined δ. The demands for accuracy of δ in the mixture can be understood from the plot in Fig. 3. In 84% D_2O, when the excess scattering amplitudes of H-Tu and D-Tu molecules are equal in magnitude but are of opposite signs, the resulting

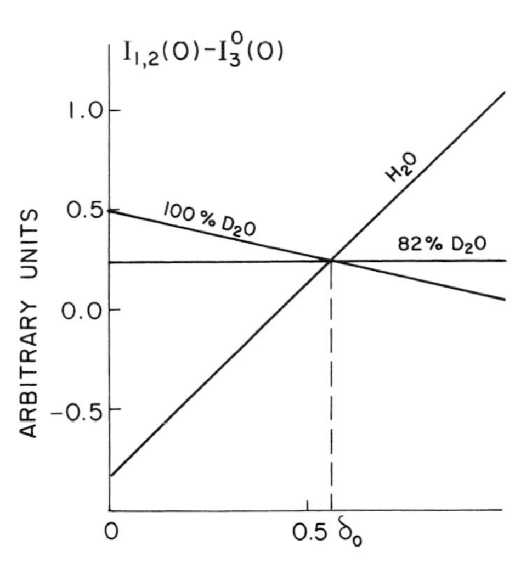

Figure 3. Dependence of the difference zero-angle scattering on fraction δ of deuterated particles in solution I.

scattering curve is fairly insensitive to the correct choice of δ. In contrast, in H_2O the demands for the accuracy of δ are most stringent, as small deviations in the scattering intensity of the mixture.

Thus, special attention was paid to developing the procedure for precise mixing of the H-Tu and D-Tu solutions. Initially, the optical densities of all solutions were adjusted to the same value by diluting the more concentrated ones. Then we performed SANS measurements in the same cuvettes in which UV measurements were carried out and determined δ by analyzing these measurements. Finally, solutions of H- and D-Tu were mixed by volume with a control by weight. The estimated precision of this mixing procedure was better than 0.5% for the δ value.

EXPERIMENTAL CHECKING

The results of the mixing experiments are represented in Table 1. The last two lines demonstrate that the difference scattering intensity $[I_{H,D}(0) - I_{HD}(0)]/$

Table 1. Experimental scattering cross sections[a]

Cross section at zero scattering angle in the absolute scale (barn/dalton)	H_2O, $\delta = 0.56$	69% D_2O, $\delta = 0.56$	79% D_2O, $\delta = 0.54$	79% D_2O, $\delta = 0.58$
$I_H(0)$	48.1 ± 0.8	21.1 ± 0.3	45.9 ± 0.7	45.9 ± 0.7
$I_D(0)$	387.5 ± 1.7	77.0 ± 0.8	55.7 ± 0.3	55.7 ± 0.3
$(1 - \delta)I_H(0) + \delta I_D(0)$	234.8	53.5	51.2	51.6
$I_{H,D}(0)$	233.2 ± 1.4	50.5 ± 0.4	44.9 ± 1.0	43.8 ± 1.2
$I_{HD}(0)$	191.7 ± 1.4	11.4 ± 0.3	3.4 ± 0.2	3.4 ± 0.2
$[I_{H,D}(0) - I_{HD}(0)]/[\delta(1 - \delta)]$	167.1 ± 4.3	160.5 ± 2.0	166.8 ± 0.9	165.2 ± 0.9
$I_F(0)_{calc}$	166.8	166.8	166.8	166.8

[a] $I_H(0)$, $I_D(0)$, $I_{H,D}(0)$, and $I_{HD}(0)$ at zero scattering angle for H-Tu, D-Tu, a mixture of H- and D-Tu, and HD-Tu at different contrasts in comparison with the results of mathematical mixing, $(1 - \delta)I_H(0) + \delta I_D(0)$. The last two lines represent comparison of the difference scattering, $[I_{H,D}(0) - I_{HD}(0)]/[\delta(1 - \delta)]$, obtained experimentally with the calculated difference scattering at zero angle, $I_F(0)_{calc}$.

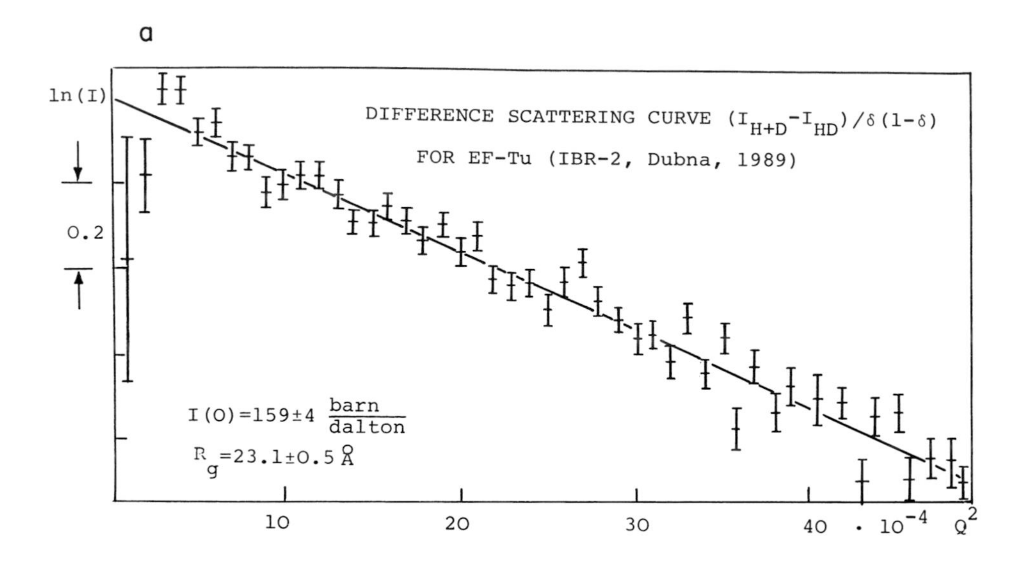

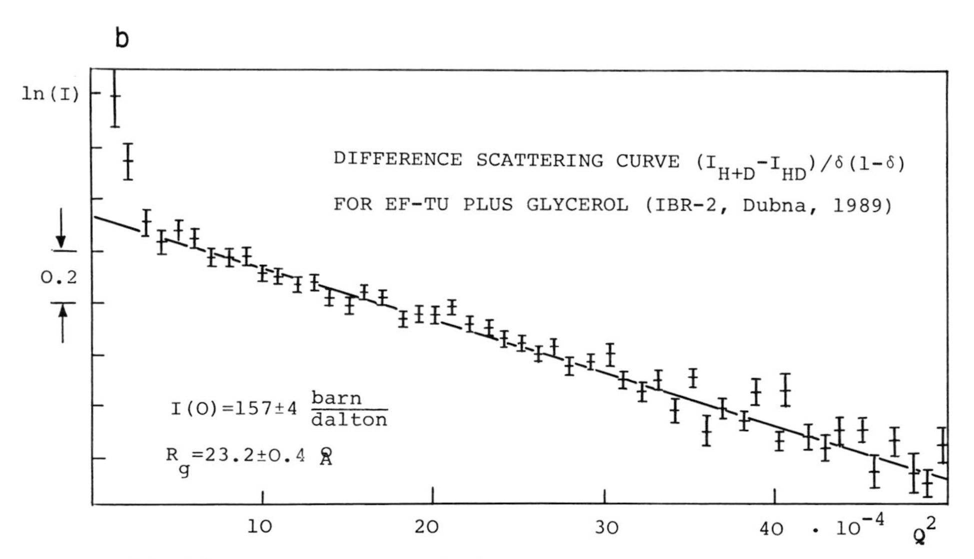

Figure 4. Comparison of the difference scattering curves for free EF-Tu (a) and for EF-Tu after addition of glycerol to both solutions (b).

$[\delta(1 - \delta)] = I_F(0)$ is, in fact, independent of the solvent isotopic content within a 5% error and corresponds to the calculated scattering intensity $I_F(0)$ of a particle whose scattering density is equal to the difference between the scattering densities of D- and H-Tu. This $I_F(0)$ can easily be calculated from the known match points for H-Tu and D-Tu and the known amino acid composition of the EF-Tu monomer, using the equation

$$I_F(0) = \{(\Sigma b_i - [\mathrm{MW} \cdot \bar{v}/\mathrm{NA}] \cdot a_{sol}) \cdot [(\gamma_{\dot{D}} - \gamma_{\dot{H}})/\gamma_{\dot{H}}]\}^2 \quad (5)$$

where $\bar{v} = 0.72$ cm^3/g is the partial specific volume of EF-Tu, NA is Avogadro's number, Σb_i is the sum of scattering lengths of all H-Tu atoms, MW is molecular weight (43,000), and a_{sol} is the scattering density of H_2O.

Thus, these data provide experimental confirmation of the predicted ability of the new method to extract the scattering curve pertaining to one particle. They also demonstrate that this curve is independent of the solvent isotopic content. However, the intensities for each particle depend strongly on the contrast as well as on the extent of its deuteration (Table 1); i.e., these data confirm predictions 1 and 2 of our method.

Let us compare the results of the so-called mathematical mixing present in line 3 of Table 1 with the results obtained experimentally for the real mixtures and shown in line 4.

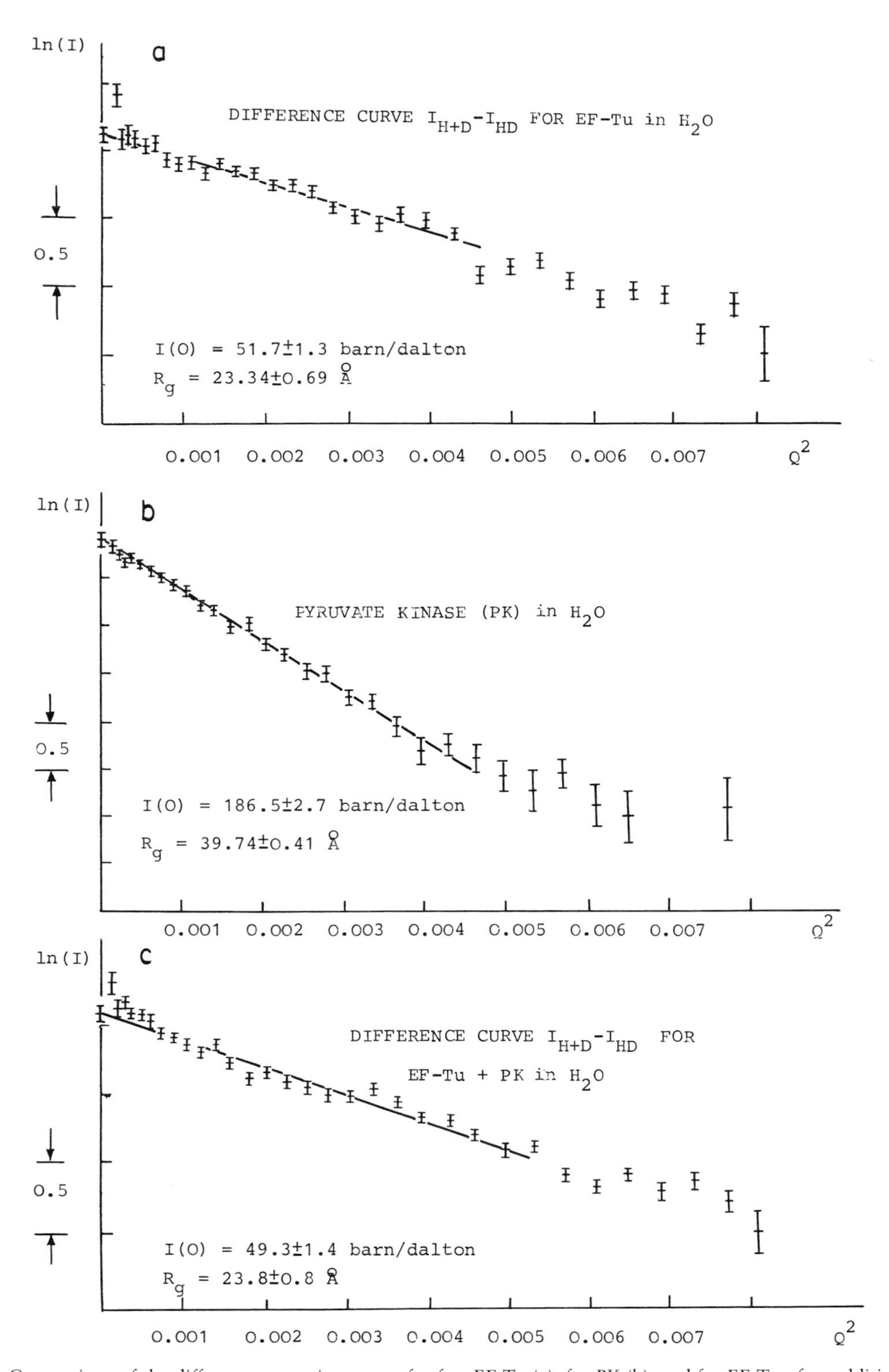

Figure 5. Comparison of the difference scattering curves for free EF-Tu (a), for PK (b), and for EF-Tu after addition of PK to both solutions I and II (c).

In mathematical mixing, we calculate the scattering of a mixture, suggesting that it can be obtained by a simple summation of intensities of H- and D-Tu solutions with $(1 - \delta)$ and δ weights. The comparison clearly demonstrates the trend for increasing the difference between the results of mathematical and real mixing with the increase of the D_2O fraction in solution.

Analysis shows that this trend toward large deviation at 69 and 84% D_2O, when excess scatter-

determination of biological quaternary structure by neutron scattering. *Proc. Natl. Acad. Sci. USA* **69**:1997–2002.

Hardesty, B., and G. Kramer (ed.). 1986. *Structure, Function, and Genetics of Ribosomes.* Springer-Verlag, New York.

Hoppe, W. 1973. Label triangulation method and the mixed isomorphous replacement principle. *J. Mol. Biol.* **78**:581–585.

Koch, M. H. J., and H. B. Stuhrmann. 1979. Neutron scattering studies of ribosomes. *Methods Enzymol.* **59**:670–706.

Pavlov, M. Y., and I. N. Serdyuk. 1987. Three-isotopic substitution method in small angle neutron scattering. *J. Appl. Crystallogr.* **20**:105–110.

Schelten, J., P. Schlecht, W. Schmatz, and Mayer. 1971. Neutron small-angle scattering by hemoglobin, p. 149–153. *In* E. Brodea (ed.), *Proceedings of the European Biophysics Congress*, vol. 1. Verlag Wiener Med. Akad., Vienna, Austria.

Serdyuk, I. N., and M. Y. Pavlov. 1988. A new approach in small angle neutron scattering: a method of triple isotopics substitutions. *Makromol. Chem. Macromol. Symp.* **15**:167–184.

Williams, C. E., M. Nierlich, J. P. Cotton, G. Jannink, F. Boue, M. Daoud, B. Farnoux, C. Pickot, P. G. de Gennes, M. Rinuado, M. Moan, and C. Wolff. 1979. Polyelectrolyte solutions: intrachain and interchain correlations observed by SANS. *J. Polym. Sci. Polym. Lett. Ed.* **17**:379–384.

Zaccai, G., and S. Xian. 1988. Structure of phenylalanine-accepting tRNA and of its environment in aqueous solvents with different salts. *Biochemistry* **27**:1316–1320.

Chapter 14

Structure, Function, and Evolution of Mammalian Ribosomes

IRA G. WOOL, YAETA ENDO, YUEN-LING CHAN, and ANTON GLÜCK

Importance attaches to obtaining a solution to the structure of ribosomes, since knowledge of the structure is believed, with cause, to be essential for a rational, molecular account of the function of the organelle in protein synthesis. It is by no means certain, perhaps not even likely, that knowledge of the structure will lead pari passu to an understanding of the function of ribosomes. However, it is difficult to imagine being able to describe function in precise molecular detail without knowledge of structure. For a solution of the structure, a requisite is knowledge of the sequences of nucleotides and of amino acids in the constituent nucleic acids and proteins. An attempt is under way to acquire the chemical data for mammalian (rat) ribosomes. The covalent structures of the four species of RNA have been established (Nazar et al., 1975; Aoyama et al., 1982; Chan et al., 1983b; Chan et al., 1984; Torczynski et al., 1983; Hadjiolov et al., 1984), and there are rational proposals for their secondary structures (Chan et al., 1984; Hadjiolov et al., 1984; Wool, 1986). As for the proteins, 82 have been isolated from the particles (Wool, 1979), and the amino acid sequences of 35 (Lin et al., 1982; Lin et al., 1983, 1984; Lin et al., 1987a; Lin et al., 1987b; Itoh et al., 1985; Kuwano et al., 1985; Nakanishi et al., 1985; Tanaka et al., 1985, 1988; Tanaka et al., 1986; Tanaka et al., 1987; Tanaka et al., 1989; Chan et al., 1987a; Chan et al., 1987b; Chan et al., 1987c; Chan and Wool, 1988; Devi et al., 1988, 1989a, 1989b; Gallagher et al., 1988; Rajchel et al., 1988; Aoyama et al., 1989a; Aoyama et al., 1989b; Glück et al., 1989; Nakamura et al., 1989; Paz et al., 1989) have been either determined directly from the proteins or deduced from the sequences of nucleotides in recombinant cDNAs. We shall review first the progress that has been made in determining the structure of the molecular components of mammalian ribosomes and then recent studies on the function of an important domain in 28S rRNA.

SEQUENCE OF AMINO ACIDS IN RAT RIBOSOMAL PROTEINS

A commitment has been made to determine the sequences of amino acids in all of the proteins in a single mammalian species, the rat. This is an arduous, tedious, time-consuming, expensive project. Moreover, now that there is a feasible plan for accomplishing the chore, it lacks excitement and intellectual challenge. How then do we justify this program? It is our abiding belief that the data are important—indeed, that all analytical and structural chemistry is laden with theory. We argue further that obtaining the data is an absolute prerequisite for determining the structure of mammalian ribosomes which is, in turn, essential for understanding their function. The fact that this information is already available for *Escherichia coli* ribosomes (Wittmann-Liebold, 1984) in no way diminishes the importance of obtaining the data for mammalian ribosomes. Indeed, each of the sets (*E. coli* and rat) is likely to enhance the value of the other, just as subsequent sequences of rRNAs (after determination of *E. coli* 16S and 23S rRNAs) were increasingly valuable because they made possible comparative analysis and hence led to the elucidation of the secondary structures of the nucleic acids. We are not so naive as to believe that determination of the sequences of amino acids in rat ribosomal proteins will lead inevitably to the structure of the particle or that the structure will tell us inevitably how the organelle functions in protein synthesis. However, we are convinced that without the basic chemistry, i.e., the sequences of the amino acids in the proteins, it will be impossible to fully

Ira G. Wool, Yuen-Ling Chan, and Anton Glück ■ Department of Biochemistry and Molecular Biology, The University of Chicago, Chicago, Illinois 60637. **Yaeta Endo** ■ Department of Biochemistry, Yamanashi Medical College, Yamanashi 409-38, Japan.

Table 3. Related ribosomal proteins: rat and yeast

Protein		Score[a] (SD units)		Identities/possible matches
Rat	Yeast[b]	RELATE	ALIGN	
S3	YS3[c]	6.6	6.7	9/20
S4	S7[c]	14.6	12.4	19/28
S6	S10	42.7	83.9	149/236
	S6[d]	41.1	72.2	164/237
S8	YS9[c]	20.3	24.9	30/49
S14	rp59	41.9	58.4	108/137
S17	rp51	32.2	52.5	76/132
S21	S25	17.8	32.6	45/83
	S28[d]	23.5	33.6	47/83
P0	A0	48.9	83.7	170/312
P1	A1	14.8	26.2	48/106
	L44′	14.6	31.6	42/106
P2	A2	20.3	26.4	52/106
	YPA1	17.7	29.8	58/109
	L44	20.0	26.4	52/106
	L45	13.7	28.2	57/109
	L40c[c,d]	13.5	21.9	18/40
L5	YL3 (CN1)[c]	13.3	12.9	19/30
	YL3 (CN2)[c]	9.5	11.5	25/82
L18	rp28	39.5	50.1	102/184
L18a	L17[c,d]	14.6	16.1	22/45
L19	L15[c,d]	13.9	9.6	14/29
	YL14[c]	4.3	6.8	15/30
L26	YL33[c]	10.6	11.2	20/40
L30	L32	28.9	38.5	61/105
L31	L34	26.0	38.0	65/112
L36a	rp44	26.3	45.8	74/101
L37	L27[c,d]	11.0	9.6	17/28
	YP55[c]	6.4	9.0	16/30
L39	L46	17.1	21.9	31/50
	L36[c,d]	15.0	21.1	22/42

[a] All are highly significant.
[b] The species is *Saccharomyces cerevisiae* unless otherwise indicated. Because of limitations of space, we do not provide references to the yeast sequences; they are available on request.
[c] Partial sequence.
[d] *Schizosaccharomyces pombe.*

can be correlated (Table 3); that is, 20 rat ribosomal proteins of the 35 whose primary structure has been determined have a homolog among the yeast sequences. All of the comparisons give highly significant scores; identities in the alignments range from 40 to 80%. The data are sufficient to provide confidence that most if not all of the ribosomal proteins from the two species are homologous and that it will be possible to establish a protein-to-protein correlation. This has at least one practical consequence: the correlation might be used to establish a uniform nomenclature for the two species, which would substitute a measure of order for the chaos that now confounds the designation of individual yeast ribosomal proteins.

Of the eight *Xenopus laevis* ribosomal protein amino acid sequences that have been determined in full or in part, six can be correlated with a rat ribosomal protein (Table 4).

It has been surmised that archaebacterial ribosomes are transition particles in the evolution of procaryotes and eucaryotes and that they have properties of both (Matheson, 1985). For example, if we consider the two 5S rRNA-binding proteins from *Halobacterium cutirubrum*, one, Hc L19, is related to *E. coli* L5, whereas the other, Hc L13, is related to rat L5 (Chan et al., 1987b). Nonetheless, the amino acid sequences of archaebacterial ribosomal proteins are in general closer to those of eucaryotes than to the procaryotic counterparts (Auer et al., 1989). We can correlate 12 of the rat ribosomal proteins with archaebacterial proteins (Table 5).

Until recently, common wisdom has had it that there are no strong sequence similarities between eucaryotic and procaryotic ribosomal proteins except for the A proteins (Wool and Stöffler, 1974; Lin et al., 1982; Auer et al., 1989). Thinking with regard to this issue had been strongly influenced by results of the comparison of the structures of rRNAs. The motif for these molecules is conservation of secondary structure rather than primary sequence (Noller, 1984), which is not to say there are no conserved sequences; most assuredly there are. The conserved sequences tend to be (but are not exclusively) in nonhelical regions, and there is evidence that they are important for function. There is a lesson here: it is possible that only the functionally important amino acids in ribosomal proteins, which need not be contiguous in the sequence, will be conserved. This may account for the difficulty in establishing homologies between eucaryotic and procaryotic ribosomal proteins.

Intuition leads one to surmise that ribosomes arose on a single occasion and hence that the proteins as well as the rRNAs of eucaryotic ribosomes are likely to be related to their procaryotic progenitors. The problem is how to trace the chemical spoor and

Table 4. Related ribosomal proteins: rat and *X. laevis*

Protein		Score[a] (SD units)		Identities/possible matches
Rat	*X. laevis*[b]	RELATE	ALIGN	
S3	S1[c]	31.4	39.5	68/116
S23/24	S19	52.1	57.3	129/132
URF S22	S22	43.9	55.1	113/119
L5	L5	75.2	118.4	267/295
L18	L14	69.0	54.4	130/156
L35a	L32[c]	36.7	31.1	65/70

[a] All are highly significant.
[b] Because of limitations of space, we do not provide references to the *X. laevis* sequences; they are available on request.
[c] Partial sequence.

Table 5. Related ribosomal proteins: rat and archaebacteria

Proteins		Score[a] (SD units)		Identities/possible matches
Rat	Archaebacteria[b]	RELATE	ALIGN	
S3	Hc S4[c]	9.0	9.1	11/40
S11	Hc S16[c]	15.5	10.0	17/31
	Hm S14	18.2	31.9	39/109
S14	Hm S19	21.1	32.3	60/126
S16	Hm S3	17.0	30.1	53/130
S23/24	Hm S15	10.7	15.3	25/102
P0	Hc L10e	14.2	30.1	80/312
	Hh ORFb	13.8	36.0	79/313
P1	Hc L20[c]	8.8	14.7	24/77
	Mt L7[c]	8.5	11.9	13/48
	Hh ORFc	7.3	15.2	31/105
P2	Hc L20[c]	8.3	11.3	16/76
	Mt L7[c]	9.3	7.7	13/48
	Mv L12	13.4	13.5	31/97
L5	Hc L13[c]	8.1	6.7	14/36
	Hm L12[c]	9.4	8.7	14/24
L7a	Hm S6	6.9	12.7	29/116
L26	Hm L16h	10.4	24.1	30/119
L30	Mv ORF1	14.6	31.4	39/105

[a] All are significant.
[b] Species: Hc, *Halobacterium cutirubrum*; Hh, *Halobacterium halobium*; Mt, *Methanobacterium thermoautotrophicum*; Mv, *Methanococcus vannielii*; Hm, *Halobacterium marismortui*. Because of limitations of space, we do not provide references to archaebacterial sequences; they are available on request. ORF, Open reading frame.
[c] Partial sequence.

unravel the mechanism by which the proteins evolved. For the ribosomal proteins, the analysis is complicated for three reasons. First, there are more of them. Second, there are still far fewer data than for the nucleic acids (the only complete set is for *E. coli*). Finally, the algorithms for the comparison of proteins are less satisfactory and less reliable, since one must deal with 20 different amino acids rather than four different nucleotides. Nonetheless, we have argued before for homology of eucaryotic and procaryotic ribosomal proteins (Wool and Stöffler, 1974; Lin et al., 1982; Wool, 1986). This seems certain for the acidic proteins (Matheson et al., 1980). The ribosomes of all species that have been analyzed have a protein or proteins homologous to *E. coli* L12; the rat equivalent is P2 (Table 6). There is convincing evidence that five other rat ribosomal proteins (S3, S11, S14, S16, and URF S22) are related to *E. coli* ribosomal proteins (Table 6). It is perhaps surprising that more proteins cannot be correlated, especially since the sequences of the entire *E. coli* set have been determined. One can only imagine that the divergence has been so great (the A proteins and a few others excepted) as to make it difficult to trace the relationships without improved methods. It is possible that the amino acid sequences of archaebacterial ribosomal proteins, since they seem intermediates between eucaryotic and procaryotic proteins, will provide a connecting link for the correlation and will in this way serve as a kind of molecular Rosetta stone.

RNA-PROTEIN INTERACTION: ANALYSIS OF THE RECOGNITION BY α-SARCIN OF A RIBOSOMAL DOMAIN CRITICAL FOR FUNCTION

α-Sarcin is a small, basic, cytotoxic protein produced by the mold *Aspergillus giganteus* (Olson and Goerner, 1965; Olson et al., 1965) that inhibits protein synthesis by inactivating ribosomes (Fernandez-Puentes and Vazquez, 1977; Conde et al., 1978; Hobden and Cundliffe, 1978). The inhibition is the result of the hydrolysis of a phosphodiester bond (Schindler and Davies, 1977) on the 3′ side of G-4325 (Endo and Wool, 1982; Chan et al., 1983a; Endo et al., 1983) which is in a single-stranded loop 459 residues from the 3′ end of 28S rRNA (Chan et al., 1983a; Wool, 1984). The cleavage site is embedded in a purine-rich, single-stranded segment of 14 nucleotides that is near universal (Wool, 1984; Raué et al., 1988). This is one of the most strongly conserved regions of rRNA; indeed, the ribosomes of all organisms that have been tested, including the producing fungus (Hobden, 1978), are sensitive to the toxin. α-Sarcin catalyzes the hydrolysis of only the one phosphodiester bond, and this single break accounts entirely for its cytotoxicity (Wool, 1984). This remarkable specificity is peculiar to α-sarcin; treatment of ribosomes with other RNases causes extensive digestion of rRNA. If, however, the substrate is naked RNA rather than 80S ribosomes or 60S subunits and the amount of the toxin is large, it cuts on the 3′ side of nearly every purine without regard to whether the nucleotides are in single- or double-stranded regions (Endo et al., 1983).

Table 6. Related ribosomal proteins: rat and *E. coli*

Protein		Score[a] (SD units)		Identities/possible matches
Rat	*E. coli*[b]	RELATE	ALIGN	
S3	S3	4.7	8.7	55/202; 73/202[c]
S11	S17	8.7	11.0	24/83; 36/83
S14	S11	11.1	20.5	53/127; 64/127
S16	S9	7.1	7.3	41/127; 52/127
URF S22	S10	6.9	12.5	26/102; 46/102
P2	L12[d]	4.3	2.8	28/113; 36/113

[a] All are, with one exception (P2 versus L12; ALIGN), significant.
[b] Amino acid sequences of the *E. coli* ribosomal proteins are given by Wittmann-Liebold (1984).
[c] The second set of figures is the number of similarities (S/T; D/E; K/R; I/L/V) in the alignment/number of possible matches.
[d] The NH_2 and carboxyl termini of *E. coli* L12 were transposed as given by Lin et al. (1982).

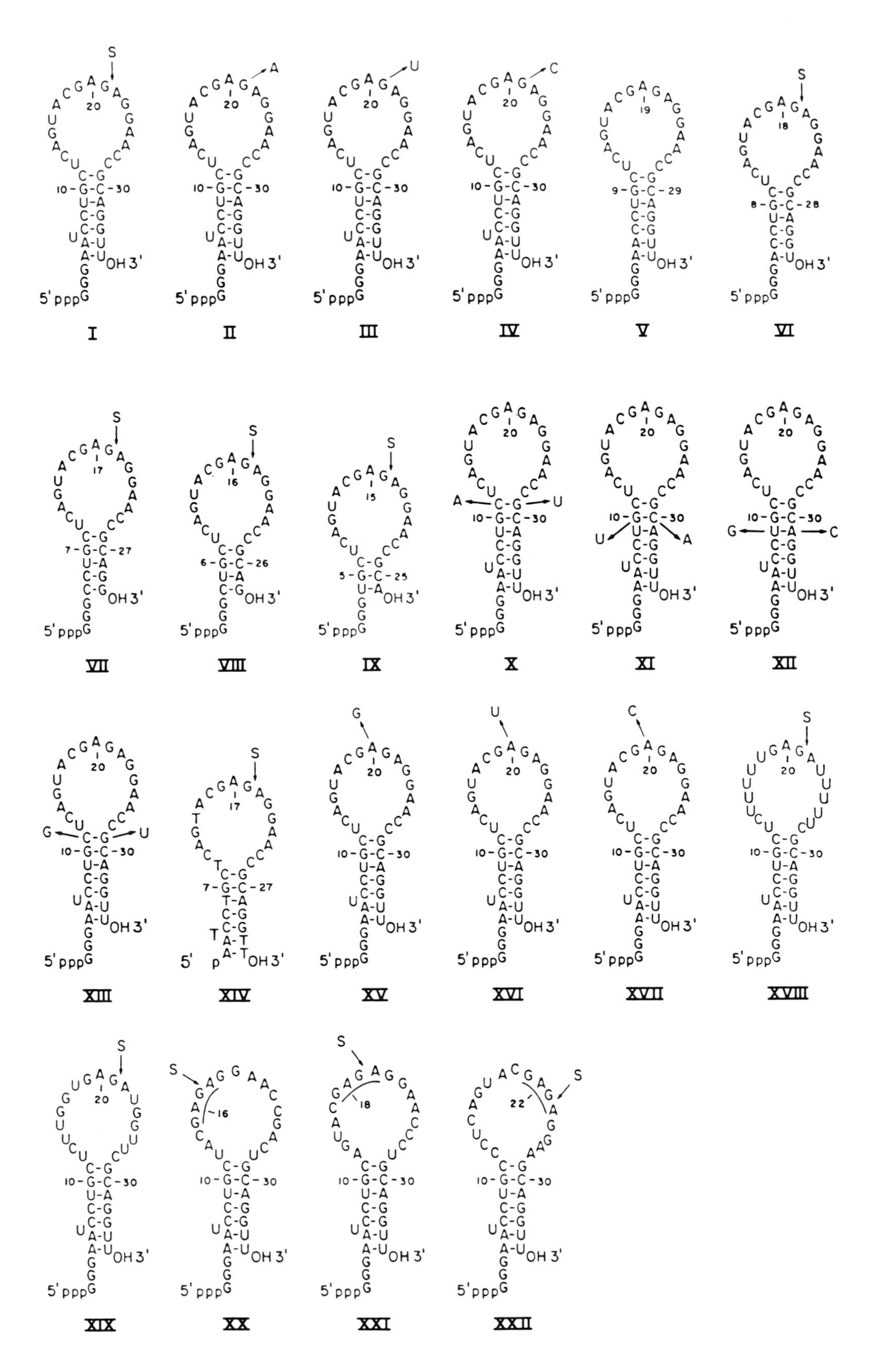

5' pppGGGCAGAGGUAUCAGUACGAGAGGAACCAUGGAGA OH 3'

XXIII

Figure 1. Structures of (I) an oligoribonucleotide that mimics the α-sarcin domain in 28S rRNA and of (II–XXIII) a series of variants.

constructed (Fig. 1, structures XV to XVII) with alterations of the ricin site A to G, U, or C. All of these variants were recognized by α-sarcin (Table 7).

The context was changed in another way: the tetranucleotide GAG(sarcin)A was left intact for reasons that will be apparent shortly, and the remainder of the universal portion of the loop sequence was engineered (Fig. 1) so that it was entirely uridines (structure XVIII) or, in a second mutant, uridines and guanosines (structure XIX). Neither oligonucleotide is a competent substrate for α-sarcin (Table 7). Thus, the context is an essential feature of recognition of the substrate by α-sarcin.

Contribution of Geometry of the Sequence GAGA in the Loop to Recognition by α-Sarcin of the Toxin Domain RNA

E. coli ribosomes are not sensitive to ricin (Gale et al., 1981); the ribosomes are not inactivated, and A-2660 in the α-sarcin/ricin domain is not depurinated (Endo et al., 1987). However, naked *E. coli* rRNA is a substrate for the *N*-glycosidase activity of the toxin (Endo and Tsurugi, 1988). The two sites of covalent modification, one each in 16S and 23S rRNAs, have stems with seven base pairs and loops that have the sequence GAGA. Not all of the sites in rRNA with this structure, however, are modified by ricin. A comparison of the two subsets, modified and unmodified by ricin, indicated that the tetranucleotide GAGA in the loop had to have a particular geometric relationship to, or orientation with, the stem, i.e., that the ricin-sensitive adenosine had to be centered over the axis of the helix at least as the structure is depicted in two dimensions.

In agreement with the prediction that comes from this observation, if we move the tetranucleotide GAGA either four (Fig. 1, structure XX) or two (structure XXI) bases closer to the 5′ end or two bases closer to the 3′ end (structure XXII), the specific response to α-sarcin is entirely lost (Table 7). Thus, recognition is not merely of a guanosine at the correct position in the sequence but of a particular conformation; in addition, there is a strong requirement for a particular geometry. This latter is an aspect of the recognition of RNA by a protein that, to our knowledge, has not been defined before.

The conclusion from these experiments is that recognition by α-sarcin of the toxin domain RNA requires, in the first instance, a stem and a single-stranded 17-member loop in which the sequence of at least 14 nucleotides (the universal sequence) affects binding and enzymatic activity; the stem needs only three base pairs, and the identity of the Watson-Crick interactions does not have a large influence. Perhaps most important is the observation that the tetranucleotide GAG(sarcin)A in the loop has to have the correct geometry with respect to the stem.

We propose that the structure of the α-sarcin domain in 28S rRNA is more complex than is depicted in the usual two-dimensional cartoons and, furthermore, that it is capable of undergoing reversible alterations.

That the structure of this domain is complex is suggested first by the enormous amount of ricin required to cleave the glycosidic bond in naked 28S rRNA or in the synthetic RNA. Depurination of A-4324 in 28S rRNA resident in the ribosome occurs at a ricin/substrate ratio of 0.001, whereas with naked 28S rRNA or with the wild-type oligoribonucleotide the ratio is 10, i.e., 10,000 times greater (Endo et al., 1987; Endo et al., 1988a). We do not know the reason for this incredible discrepancy, but the most likely explanation is that the ordered structure of the domain is different in ribosomes than it is in naked RNA. We note that the single-stranded loop is large (17 nucleotides) and hence unlikely to exist as such; it might well participate in a tertiary interaction with other regions of 28S rRNA.

That the domain might undergo transitions in its structure is suggested by an observation of the effect of α-sarcin on the association of 5.8S rRNA with 28S rRNA (Walker et al., 1983). In the large subunit of eucaryotic ribosomes, these two nucleic acids are noncovalently but stably associated (Wool, 1979). This interaction, which involves approximately 40 hydrogen bonds in two separate contact regions, is destabilized by the α-sarcin-catalyzed cleavage that is at a site more than 4,000 nucleotides away. Dissociation of 5.8S rRNA from 28S rRNA, which ordinarily requires treatment with 4 to 6 M urea, occurs spontaneously after α-sarcin action on ribosomes (Walker et al., 1983).

We cannot provide a coherent physical chemical explanation for this phenomenon. The cleavage at G-4325 appears to initiate a propagated change in the secondary or tertiary structure or both that in some way is transmitted through the molecule and leads to the collapse of 28S rRNA. This collapse could account for the destabilization of the 5.8S-28S rRNA complex and to the loss of function of the ribosomes. We also suggest that it interferes with the operation of an rRNA switch, i.e., with the ability to disrupt and reestablish a crucial tertiary interaction that involves the single-stranded region of the α-sarcin domain.

There are other observations that support this conjecture. Oligodeoxynucleotides complementary to the universal sequence in the loop of the α-sarcin domain will not bind to either *E. coli* (White et al.,

1988) or rat (Y. Endo and I. G. Wool, unpublished data) ribosomes suspended in buffer, which suggests but by no means proves that the structure is not simply single stranded. Occlusion by ribosomal proteins could account for the failure of the cDNA to bind to the site; however, this seems less likely, since α-sarcin and ricin have access to the domain. It is most important in this regard that if ribosomes are catalyzing protein synthesis, they will bind the complementary oligodeoxynucleotide (manifest as sensitivity to RNase H), suggesting a reversible change in the structure of the domain (Y. Endo, unpublished data). A similar observation has been made with *E. coli* ribosomes (C. Merryman and W. E. Hill, personal communication). Further support for our proposal comes from experiments with inhibitors of protein synthesis (cycloheximide, guanylyl-5′-imidodiphosphate [GMPPNP], and sparsomycin) in which α-sarcin is used as a probe of structure (Endo, unpublished data). The results of these experiments also indicate that there is a conformational change in the RNA during translocation, since ribosomes are sensitive to α-sarcin only when peptidyl-tRNA is in the A site before translocation. The conformational transition, if it occurs, could provide the motive force for changes in the structure of ribosomal 60S subunits and might underlie the movement required for the translocation of peptidyl-tRNA from the A to the P site and for the movement of mRNA one codon after each of the reiterative rounds in translation. In this paradigm, it is the elongation factors that initiate the reversible transition or switch in rRNA structure that propels translocation either directly or indirectly through the binding or the hydrolysis of GTP. Cleavage at G-4325 in 28S rRNA by α-sarcin or depurination at A-4324 by ricin might abolish the capacity to reversibly switch structures and in this way account for the catastrophic effect of the toxins on ribosome function.

This work was supported by Public Health Service grants GM 21769 and GM 33702 from the National Institutes of Health.

We are indebted to our colleagues V. Paz, J. Olvera, K. R. G. Devi, and Y. Aoyama, with whom we had the pleasure of working on some of the research described here. We are grateful to A. Timosciek for assistance with the preparation of the manuscript.

REFERENCES

Aoyama, K., S. Hidaka, T. Tanaka, and K. Ishikawa. 1982. The nucleotide sequence of 5S RNA from rat liver ribosomes. *J. Biochem.* **91**:363–367.

Aoyama, Y., Y. L., Chan, O. Meyuhas, and I. G. Wool. 1989a. The primary structure of rat ribosomal protein L18a. *FEBS Lett.* **247**:242–246.

Aoyama, Y., Y. L. Chan, and I. G. Wool. 1989b. The primary structure of rat ribosomal protein L34. *FEBS Lett.* **249**:119–122.

Auer, J., K. Lechner, and A. Böck. 1989. Gene organization and structure of two transcriptional units from *Methanococcus* coding for ribosomal proteins and elongation factors. *Can. J. Microbiol.* **35**:200–204.

Cahn, F., E. M. Schachter, and A. Rich. 1970. Polypeptide synthesis with ribonuclease-digested ribosomes. *Biochim. Biophys. Acta* **209**:512–520.

Chan, Y. L., Y. Endo, and I. G. Wool. 1983a. The sequence of the nucleotides at the α-sarcin cleavage site in rat 28S ribosomal ribonucleic acid. *J. Biol. Chem.* **258**:12768–12770.

Chan, Y. L., R. Gutell, H. F. Noller, and I. G. Wool. 1984. The nucleotide sequence of a rat 18S ribosomal ribonucleic acid gene and a proposal for the secondary structure of 18S ribosomal ribonucleic acid. *J. Biol. Chem.* **259**:224–230.

Chan, Y. L., A. Lin, J. McNally, D. Peleg, O. Meyuhas, and I. G. Wool. 1987a. The primary structure of rat ribosomal protein L19: a determination from the sequence of nucleotides in a cDNA and from the sequence of amino acids in the protein. *J. Biol. Chem.* **262**:1111–1115.

Chan, Y. L., A. Lin, J. McNally, and I. G. Wool. 1987b. The primary structure of rat ribosomal protein L5: a comparison of the sequence of amino acids in the proteins that interact with 5 S rRNA. *J. Biol. Chem.* **262**:12879–12886.

Chan, Y. L., A. Lin, V. Paz, and I. G. Wool. 1987c. The primary structure of rat ribosomal protein S8. *Nucleic Acids Res.* **15**: 9451–9459.

Chan, Y. L., J. Olvera, and I. G. Wool. 1983b. The structure of rat 28S ribosomal ribonucleic acid inferred from the sequence of nucleotides in a gene. *Nucleic Acids Res.* **11**:7819–7831.

Chan, Y. L., and I. G. Wool. 1988. The primary structure of rat ribosomal protein S6. *J. Biol. Chem.* **263**:2891–2896.

Conde, F. P., C. Fernandez-Puentes, M. T. V. Montero, and D. Vazquez. 1978. Protein toxins that catalytically inactivate ribosomes from eukaryotic microorganisms. Studies on the mode of action of alpha sarcin, mitogillin and restrictocin: response to alpha sarcin antibodies. *FEMS Microbiol. Lett.* **4**:349–355.

D'Eustachio, P., O. Meyuhas, F. Ruddle, and R. P. Perry. 1981. Chromosomal distribution of ribosomal protein genes in the mouse. *Cell* **24**:307–312.

Devi, K. R. G., Y. L. Chan, and I. G. Wool. 1988. The primary structure of rat ribosomal protein L18. *DNA* **7**:157–162.

Devi, K. R. G., Y. L. Chan, and I. G. Wool. 1989a. The primary structure of rat ribosomal protein L21. *Biochem. Biophys. Res. Commun.* **162**:364–370.

Devi, K. R. G., Y. L. Chan, and I. G. Wool. 1989b. The primary structure of rat ribosomal protein S4. *Biochim. Biophys. Acta* **1008**:258–262.

Dingwall, C., and R. A. Laskey. 1986. Protein import into the cell nucleus. *Annu. Rev. Cell Biol.* **2**:367–390.

Dudov, K. P., and R. P. Perry. 1984. The gene family encoding the mouse ribosomal protein L32 contains a uniquely expressed intron-containing gene and an unmutated processed gene. *Cell* **37**:457–468.

Endo, Y., Y. L. Chan, A. Lin, K. Tsurugi, and I. G. Wool. 1988a. The cytotoxins α-sarcin and ricin retain their specificity when tested on a synthetic oligoribonucleotide (35-mer) that mimics a region of 28S ribosomal ribonucleic acid. *J. Biol. Chem.* **263**: 7917–7920.

Endo, Y., P. W. Huber, and I. G. Wool. 1983. The ribonuclease activity of the cytotoxin α-sarcin. The characteristics of the enzymatic activity of α-sarcin with ribosomes and ribonucleic acids as substrates. *J. Biol. Chem.* **258**:2662–2667.

Endo, Y., K. Mitsui, M. Motizuki, and K. Tsurugi. 1987. The mechanism of action of ricin and related toxic lectins on eukaryotic ribosomes. The site and the characteristics of the modification in 28S ribosomal RNA caused by the toxins. *J. Biol. Chem.* **262**:5908–5912.

Endo, Y., and K. Tsurugi. 1987. RNA N-glycosidase activity of ricin A-chain. Mechanism of action of the toxin lectin ricin in eukaryotic ribosomes. *J. Biol. Chem.* **262:**8128–8130.

Endo, Y., and K. Tsurugi. 1988. The RNA N-glycosidase activity of ricin A-chain. The characteristics of the enzymatic activity of ricin A-chain with ribosomes and with rRNA. *J. Biol. Chem.* **263:**8735–8739.

Endo, Y., K. Tsurugi, T. Yutsudo, Y. Takeda, T. Ogasawara, and K. Igarashi. 1988b. Site of action of a Vero toxin (VT2) from *Escherichia coli* 0157:H7 and of Shiga toxin on eukaryotic ribosomes. *Eur. J. Biochem.* **171:**45–50.

Endo, Y., and I. G. Wool. 1982. The site of action of α-sarcin on eukaryotic ribosomes. The sequence at the α-sarcin cleavage site in 28S ribosomal ribonucleic acid. *J. Biol. Chem.* **257:**9054–9060.

Fernandez-Puentes, C., and D. Vazquez. 1977. Effects of some proteins that inactivate the eukaryotic ribosome. *FEBS Lett.* **78:**143–146.

Gale, E. F., E. Cundliffe, P. E. Reynolds, M. H. Richmond, and M. J. Waring. 1981. *The Molecular Basis of Antibiotic Action*, 2nd ed., p. 402–547. John Wiley & Sons, Inc., New York.

Gallagher, M. J., Y. L. Chan, A. Lin, and I. G. Wool. 1988. Primary structure of rat ribosomal protein L36a. *DNA* **7:** 269–273.

Glück, A., Y. L. Chan, A. Lin, and I. G. Wool. 1989. The primary structure of rat ribosomal protein S10. *Eur. J. Biochem.* **182:** 105–109.

Hadjiolov, A. A., O. I. Georgiev, V. V. Nosikov, and L. P. Yavachev. 1984. Primary and secondary structure of rat 28S ribosomal RNA. *Nucleic Acids Res.* **12:**3677–3693.

Hausner, T.-P., J. Atmadja, and K. H. Nierhaus. 1987. Evidence that the G^{2661} region of 23S rRNA is located at the ribosomal binding sites of both elongation factors. *Biochimie* **69:**911–923.

Hobden, A. N. 1978. Ph.D. dissertation. University of Leicester, Leicester, England.

Hobden, A. N., and E. Cundliffe. 1978. The mode of action of alpha sarcin and a novel assay of the puromycin reaction. *Biochem. J.* **170:**57–61.

Itoh, T., E. Otaka, and K. A. Matsui. 1985. Primary structures of ribosomal protein YS25 from *Saccharomyces cerevisiae* and its counterparts from *Schizosaccharomyces pombe* and rat liver. *Biochemistry* **24:**7418–7423.

Khan, A. S., and B. A. Roe. 1988. Aminoacylation of synthetic DNAs corresponding to *Escherichia coli* phenylalanine and lysine tRNAs. *Science* **241:**74–79.

Kuwano, Y., O. Nakanishi, Y. Nabeshima, T. Tanaka, and K. Ogata. 1985. Molecular cloning and nucleotide sequence of DNA complementary to rat ribosomal protein S26 messenger RNA. *J. Biochem.* **97:**983–992.

Lin, A., Y. L. Chan, R. Jones, and I. G. Wool. 1987a. The primary structure of rat ribosomal protein S12: the relationship of rat S12 to other ribosomal proteins and a correlation of the amino acid sequences of rat and yeast ribosomal proteins. *J. Biol. Chem.* **262:**14343–14351.

Lin, A., Y. L. Chan, J. McNally, D. Peleg, O. Meyuhas, and I. G. Wool. 1987b. The primary structure of rat ribosomal protein L7: the presence near the amino terminus of L7 of five tandem repeats of a sequence of 12 amino acids. *J. Biol. Chem.* **262:**12665–12671.

Lin, A., J. McNally, and I. G. Wool. 1983. The primary structure of rat liver ribosomal protein L37: homology with yeast and bacterial ribosomal proteins. *J. Biol. Chem.* **258:**10664–10671.

Lin, A., J. McNally, and I. G. Wool. 1984. The primary structure of rat liver ribosomal protein L39. *J. Biol. Chem.* **259:**487–490.

Lin, A., B. Wittmann-Liebold, J. McNally, and I. G. Wool. 1982. The primary structure of the acidic phosphoprotein P2 from rat liver 60S ribosomal subunits: comparison with ribosomal 'A' proteins from other species. *J. Biol. Chem.* **257:**9189–9197.

Matheson, A. T. 1985. Ribosomes of the archaebacteria, p. 345–377. *In* C. R. Woese and R. S. Wolfe (ed.), *The Bacteria: a Treatise on Structure and Function*, vol. 8. Academic Press, Orlando, Fla.

Matheson, A. T., W. Möller, R. Amons, and M. Yaguchi. 1980. Comparative studies on the structure of ribosomal proteins, with emphasis on the alanine-rich, acidic ribosomal 'A' protein, p. 297–332. *In* G. Chambliss, G. R. Craven, J. Davies, K. Davis, L. Kahan, and M. Nomura (ed.), *Ribosomes. Structure, Function, and Genetics.* University Park Press, Baltimore.

Milligan, J. F., D. R. Groebe, G. W. Witherell, and O. C. Uhlenbeck. 1987. Oligoribonucleotide synthesis using T7 RNA polymerase and synthetic DNA templates. *Nucleic Acids Res.* **15:**8783–8798.

Moazed, D., J. M. Robertson, and H. F. Noller. 1988. Interaction of elongation factors EF-G and EF-Tu with a conserved loop in 23S RNA. *Nature* (London) **334:**362–364.

Monk, R. J., O. Meyuhas, and R. P. Perry. 1981. Mammals have multiple genes for individual ribosomal proteins. *Cell* **24:**301–306.

Montanaro, L., S. Sperti, A. Mattioli, G. Testoni, and F. Stirpe. 1975. Inhibition by ricin of protein synthesis *in vitro*. Inhibition of the binding of elongation factor 2 and of adenosine diphosphate-ribosylated elongation factor 2 to ribosomes. *Biochem. J.* **146:**127–131.

Nakamura, H., T. Tanaka, and K. Ishikawa. 1989. Nucleotide sequence of cloned cDNA specific for rat ribosomal protein L7a. *Nucleic Acids Res.* **17:**4875.

Nakanishi, O., M. Oyanagi, Y. Kuwano, T. Tanaka, T. Nakayama, H. Mitsui, Y. Nabeshima, and K. Ogata. 1985. Molecular cloning and nucleotide sequences of cDNAs specific for rat liver ribosomal proteins S17 and L30. *Gene* **35:**289–296.

Nazar, R. N., T. O. Sitz, and H. Busch. 1975. Structural analyses of mammalian ribosomal ribonucleic acid and its precursors. Nucleotide sequence of 5.8S ribonucleic acid. *J. Biol. Chem.* **250:**8591–8597.

Noller, H. F. 1984. Structure of ribosomal RNA. *Annu. Rev. Biochem.* **53:**119–162.

Olson, B. H., and G. L. Goerner. 1965. Alpha sarcin, a new antitumor agent. I. Isolation, purification, chemical composition, and the identity of a new amino acid. *Appl. Microbiol.* **13:**314–321.

Olson, B. H., J. C. Jennings, V. Roga, A. J. Junek, and D. M. Schuurmans. 1965. Alpha sarcin, a new antitumor agent. II. Fermentation and antitumor spectrum. *Appl. Microbiol.* **13:** 322–326.

Paz, V., J. Olvera, Y. L. Chan, and I. G. Wool. 1989. The primary structure of rat ribosomal protein L26. *FEBS Lett.* **251:**89–93.

Peattie, D. A., S. Douthwaite, R. A. Garrett, and H. F. Noller. 1981. A "bulged" double helix in a RNA-protein contact site. *Proc. Natl. Acad. Sci. USA* **78:**7331–7335.

Proudfoot, N. J., and G. G. Brownlee. 1976. The 3′ non-coding region sequences in eukaryotic messenger RNA. *Nature* (London) **263:**211–214.

Rajchel, A., Y. L. Chan, and I. G. Wool. 1988. The primary structure of rat ribosomal protein L32. *Nucleic Acids Res.* **16:**2347.

Raué, H. A., J. Klootwijk, and W. Musters. 1988. Evolutionary conservation of structure and function of high molecular weight ribosomal RNA. *Prog. Biophys. Mol. Biol.* **51:**77–129.

Schindler, D. G., and J. E. Davies. 1977. Specific cleavage of ribosomal RNA caused by alpha sarcin. *Nucleic Acids Res.* **4:**1097–1110.

Seong, B. L., and U. L. RajBhandary. 1987. *Escherichia coli* formylmethionine tRNA: mutations in $^{GGG}_{CCC}$ sequence conserved in anticodon stem of initiator tRNAs affect initiation of protein synthesis and conformation of anticodon loop. *Proc. Natl. Acad. Sci. USA* **84**:334–338.

Stern, S., T. Powers, L. M. Changchien, and H. F. Noller. 1989. RNA-protein interactions in 30S ribosomal subunits: folding and function of 16S rRNA. *Science* **244**:783–790.

Tanaka, T., Y. Aoyama, Y. L. Chan, and I. G. Wool. 1989. The primary structure of rat ribosomal protein L37a. *Eur. J. Biochem.* **183**:15–18.

Tanaka, T., Y. Kuwano, K. Ishikawa, and K. Ogata. 1985. Nucleotide sequence of cloned cDNA specific for rat ribosomal protein S11. *J. Biol. Chem.* **260**:6329–6333.

Tanaka, T., Y. Kuwano, K. Ishikawa, and K. Ogata. 1988. Nucleotide sequence of cloned cDNA specific for rat ribosomal protein L27. *Eur. J. Biochem.* **173**:53–56.

Tanaka, T., Y. Kuwano, T. Kuzumaki, K. Ishikawa, and K. Ogata. 1987. Nucleotide sequence of cloned cDNA specific for rat ribosomal protein L31. *Eur. J. Biochem.* **162**:45–48.

Tanaka, T., K. Wakasugi, Y. Kuwano, K. Ishikawa, and K. Ogata. 1986. Nucleotide sequence of cloned cDNA specific for rat ribosomal protein L35a. *Eur. J. Biochem.* **154**:523–527.

Terao, K., T. Uchiumi, Y. Endo, and K. Ogata. 1988. Ricin and α-sarcin alter the conformation of 60S ribosomal subunits at neighboring but different sites. *Eur. J. Biochem.* **174**:459–463.

Torczynski, R., A. P. Bollon, and M. Fuke. 1983. The complete nucleotide sequence of the rat 18S ribosomal RNA gene and comparison with the respective yeast and frog genes. *Nucleic Acids Res.* **11**:4879–4890.

Wagner, M., and R. P. Perry. 1985. Characterization of the multigene family encoding the mouse S16 ribosomal protein: strategy for distinguishing an expressed gene from its processed pseudogene counterparts by an analysis of total genomic DNA. *Mol. Cell. Biol.* **5**:3560–3576.

Walker, T. A., Y. Endo, W. H. Wheat, I. G. Wool, and N. R. Pace. 1983. Location of 5.8S rRNA contact sites in 28S rRNA and the effect of α-sarcin on the association of 5.8S rRNA with 28S rRNA. *J. Biol. Chem.* **258**:333–338.

White, G. A., T. Wood, and W. E. Hill. 1988. Probing the α-sarcin region of *Escherichia coli* 23S rRNA with a cDNA oligomer. *Nucleic Acids Res.* **16**:10817–10831.

White, T. C., G. Rudenko, and P. Borst. 1986. Three small RNAs within the 10 kb trypanosome rRNA transcription unit are analogous to domain VII of other eukaryotic 28S rRNAs. *Nucleic Acids Res.* **14**:9471–9489.

Wiedemann, L. M., and R. P. Perry. 1984. Characterization of the expressed gene and several processed pseudogenes for the mouse ribosomal protein L30 gene family. *Mol. Cell. Biol.* **4**:2518–2528.

Wittmann-Liebold, B. 1984. Primary structure of *Escherichia coli* ribosomal proteins. *Adv. Protein Chem.* **36**:56–78.

Wool, I. G. 1979. The structure and function of eukaryotic ribosomes. *Annu. Rev. Biochem.* **48**:719–754.

Wool, I. G. 1984. The mechanism of action of the cytotoxic nuclease α-sarcin and its use to analyze ribosome structure. *Trends Biochem. Sci.* **9**:14–17.

Wool, I. G. 1986. Studies of the structure of eukaryotic (mammalian) ribosomes, p. 391–411. *In* B. Hardesty and G. Kramer (ed.), *Structure, Function, and Genetics of Ribosomes*. Springer-Verlag, New York.

Wool, I. G., and G. Stöffler. 1974. Structure and function of eukaryotic ribosomes, p. 417–460. *In* M. Nomura, A. Tissières, and P. Lengyel (ed.), *Ribosomes*. Cold Spring Harbor Laboratory, Cold Spring Harbor, N.Y.

III. PROBING rRNA FUNCTION

Raué et al., 1985; Raué et al., 1988). The regions on the various rRNA molecules recognized by these proteins have been identified in greater or lesser detail primarily by the classical "bind-and-chew" technique. This involves incubation of the rRNA–r-protein complex with an RNase, followed by sequence analysis of the fragments of the rRNA protected against digestion through their interaction with the protein. Although this type of analysis has accurately defined the limits of the binding site in several (though not all) cases, it provides little or no information on the molecular details of the rRNA–r-protein interaction. More detailed data, however, can be obtained by application of two other recent methods: chemical probing and the use of in vitro synthesized, structurally manipulated rRNA fragments in binding studies.

The first of these techniques, pioneered by Noller and co-workers (Stern et al., 1986), is based on the inability of reverse transcriptase to read certain modified nucleotides in an RNA template. This allows the rapid and accurate monitoring of even small changes in the sensitivity of specific nucleotides to a variety of structure-specific probes as a result of the binding of the r-protein. Thus, individual nucleotides contacting the r-protein can in principle be identified. The same technique can also be applied to study the interaction of rRNA with other types of molecule such as tRNA (Moazed and Noller, 1986, 1989) or antibiotics (Moazed and Noller, 1987a).

The second method follows from the fact that most r-protein binding sites encompass only a specific small region of the rRNA. By cloning a ribosomal DNA (rDNA) fragment encoding this region behind the bacteriophage T7 promoter and transcribing it in vitro, an rRNA fragment can be obtained that shows the same binding characteristics as the intact rRNA (Mougel et al., 1987; El-Baradi et al., 1987a; El-Baradi et al., 1987b). By manipulating the cloned rDNA fragment, one is then able to prepare various structural variants of the binding site whose interaction with the r-protein can be tested by in vitro methods such as nitrocellulose filter binding (Mougel et al., 1987; El-Baradi et al., 1987a).

Analysis of the r-protein binding sites on *E. coli* 16S and 23S rRNAs by one or more of the above techniques has shown that the majority of these binding sites are located in structurally conserved regions (Raué et al., 1985; Raué et al., 1988). General evidence for the functional conservation of these binding sites came first from heterologous reconstitution experiments showing that 16S rRNA and r-proteins from distantly related bacterial species could be assembled into biologically active 30S subunits (Nomura et al., 1968). Later, more detailed studies demonstrated specific binding to heterologous eubacterial rRNA for a number of individual *E. coli* r-proteins both from the small (ES7, ES8, ES15, ES17, and ES20) and large (EL1, EL11, and EL23) subunits (Zimmermann et al., 1980). (The two high-molecular-weight rRNA species from the small and large subunit are indicated as SSU and LSU rRNA, respectively. *E. coli* r-proteins are indicated by the prefix E. For yeast r-proteins, the nomenclature of Kruiswijk and Planta [1974] is used.) Similar results were obtained by using archaebacterial rRNAs, indicating functional conservation of at least some r-protein binding sites from one primary kingdom to another (Zimmermann et al., 1980; Thurlow and Zimmermann, 1982; Leffers et al., 1988). Recently, homologs of several of the r-proteins mentioned above, notably EL1, EL11, and EL23, have indeed been identified in a number of archaebacterial species on the basis of structural comparison (Hatakeyama and Kimura, 1988; Köpke and Wittmann-Liebold, 1988; Ramirez et al., 1989). The ability of these proteins to bind specifically to rRNA remains to be tested.

While phylogenetic comparison clearly reveals structural conservation of several of the known bacterial r-protein binding sites in eucaryotic rRNAs (Raué et al., 1988), originally no binding of any of the *E. coli* r-proteins (with the single exception of EL1; see below) to these rRNAs could be demonstrated (Zimmermann et al., 1980). Thus, it remained unclear whether the conservation of r-protein binding sites observed between eu- and archaebacteria extended to the eucaryotic kingdom as well.

As far as the LSU rRNA is concerned this situation has now changed. Studies on rRNA–r-protein interactions in *S. cerevisiae* 60S ribosomal subunits in our laboratory have so far uncovered two clear-cut examples of functional conservation of an r-protein binding site across the procaryote-eucaryote evolutionary barrier. The first of these concerns the binding site for *E. coli* r-protein EL23, whose equivalent on *S. cerevisiae* 26S rRNA is recognized by *S. cerevisiae* r-protein L25 (El-Baradi et al., 1985). The binding site for *E. coli* r-protein EL11 constitutes the second example. The *S. cerevisiae* equivalent of this site is recognized by *S. cerevisiae* r-protein L15 (El-Baradi et al., 1987a). Since, as discussed above, equivalents of these two binding sites are also present in archaebacterial rRNA, their conservation extends to all three primary kingdoms. The locations of the two sites, within domains II and III of LSU rRNA, respectively, are indicated in the secondary structure model for *S. cerevisiae* 26S rRNA shown in Fig. 1. The third site highlighted in Fig. 1 (domain IV) represents the structural equiva-

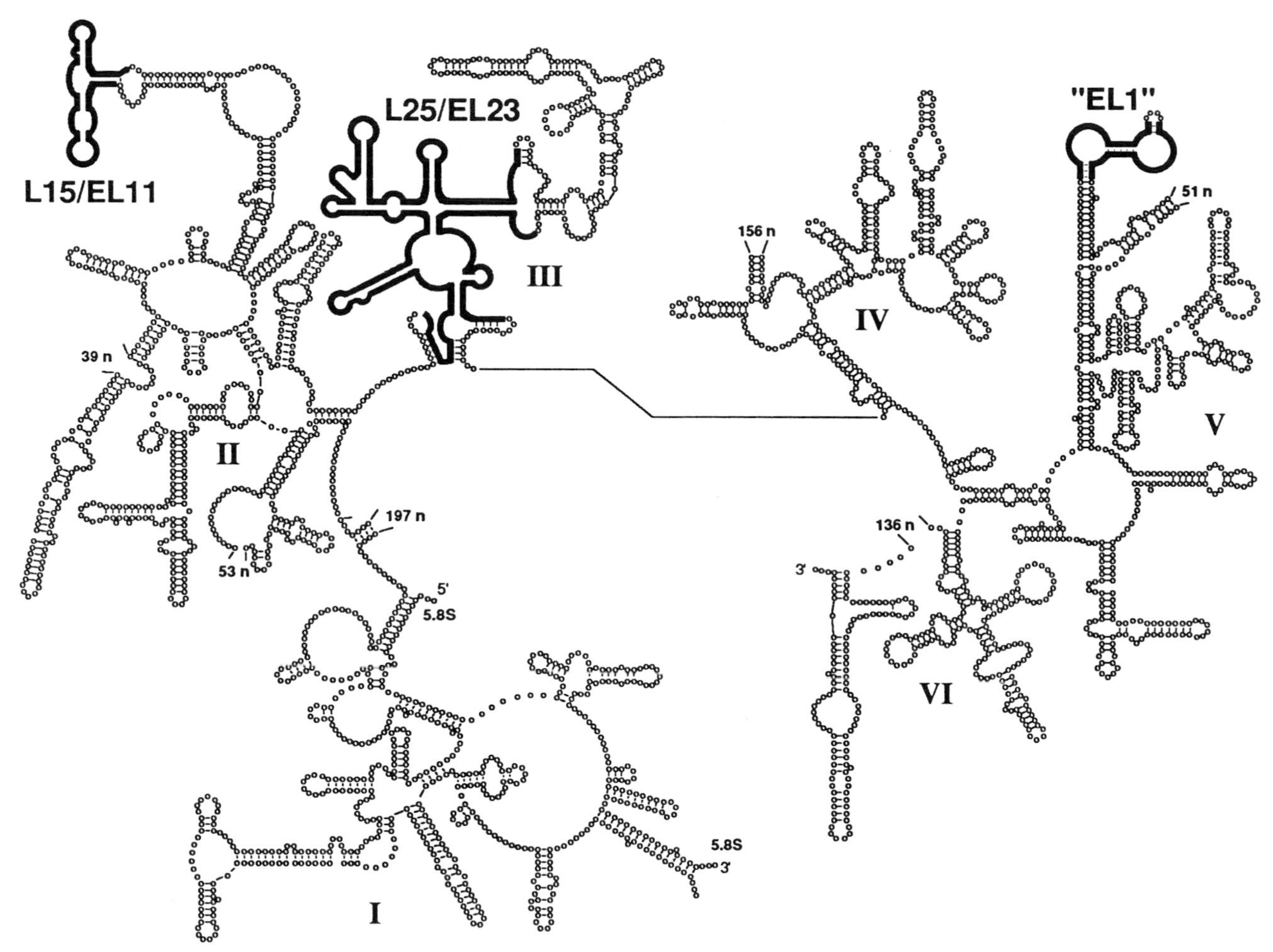

Figure 1. Locations of functionally conserved r-protein binding sites in LSU rRNA. The figure shows a secondary structure model for *S. cerevisiae* 26S/5.8S rRNA. Known conserved r-protein binding sites are highlighted (see text). Roman numerals indicate the various domains of the LSU rRNA structure.

lent of the binding site for *E. coli* EL1. Gourse et al. (1981) have shown that EL1 will bind to the corresponding region in *Dictyostelium discoideum* 26S rRNA. However, attempts to identify the *Dictyostelium* equivalent of EL1 have not been successful, leaving some doubt with respect to the function of this region as an r-protein binding site in eucaryotic rRNA.

The EL11/L15 site is one of the most strongly conserved features of the LSU rRNA, being clearly distinguishable even in the structurally highly divergent species from trypanosomal mitochondria (Raué et al., 1988). As discussed below, this region forms part of the so-called GTPase center of the large subunit and is involved in a number of different ribosomal functions. As far as r-protein binding is concerned, this strong conservation is manifest in the fact that the bacterial r-protein faithfully recognizes the equivalent site on the eucaryotic rRNA (El-Baradi et al., 1987a). Both EL11 and L15 bind also to the corresponding region in mouse 28S rRNA (El-Baradi et al., 1987a). The same region in *E. coli* 23S rRNA is known to serve as the binding site for r-protein EL10 as well (Beauclerk et al., 1984). Homologs for EL10 have recently been identified in both eucaryotes (*S. cerevisiae* and human) and archaebacteria (Rich and Steitz, 1987; Mitsui and Tsurugi, 1988; Ramirez et al., 1989) but have not been tested yet for rRNA binding.

Like EL11/L15, EL23 and L25 faithfully recognize the equivalents of their cognate binding sites on the respective heterologous rRNAs (El-Baradi et al., 1985). In contrast to the results obtained with EL11/L15, however, we were unable to demonstrate binding of either protein to the structural equivalents of their binding sites present in mouse 28S rRNA (El-Baradi et al., 1987b). Thus, the EL23/L25 site appears to be less strongly conserved than the EL11/L15 site.

Figure 2 shows a structural comparison of the

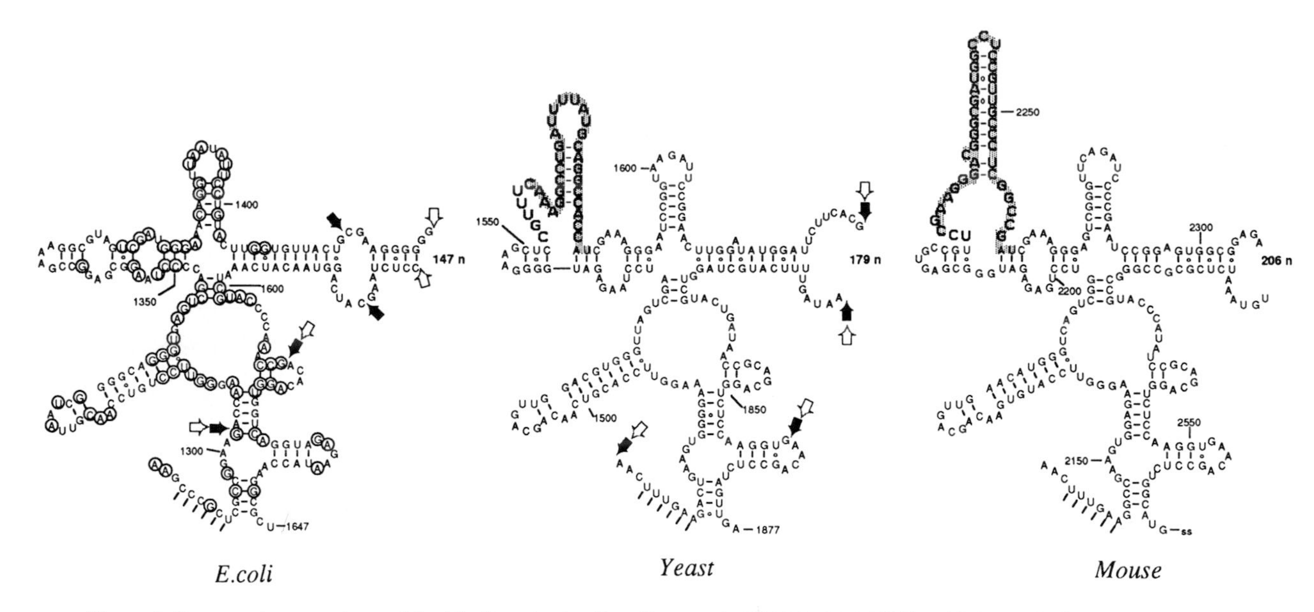

Figure 2. Structural comparison of the binding site for *E. coli* r-protein EL23 in 23S rRNA with its equivalents in *S. cerevisiae* 26S and mouse 28S rRNAs. Circled nucleotides are conserved in all three sequences. Numbering starts from the 5′ end of the LSU rRNA sequence in question. Arrows indicate the limits of protection by EL23 (filled) and L25 (open) against RNase T_1 (El-Baradi et al., 1985; El-Baradi et al., 1987b). The eucaryote-specific expansion segments are shown in boldface type. Shading outlines the portion of the *S. cerevisiae* expansion segment that was removed or replaced by the (shaded) mouse sequence in the constructs whose L25-binding capacity is shown in Fig. 3.

EL23-binding site on *E. coli* 23S rRNA with the corresponding regions in *S. cerevisiae* and mouse LSU rRNA, which are modeled to maximize structural resemblance (Raué et al., 1988). This comparison shows extensive conservation of secondary structure and even a considerable degree of similarity in sequence between the bacterial and the two eucaryotic sites. However, the latter contain an additional feature not found in any bacterial LSU rRNA, which belongs to a class of such structural elements found in eucaryotic SSU as well as LSU rRNAs. Whereas the position in the rRNA of each of these elements, dubbed expansion segments (Clark et al., 1984), is fixed (see Fig. 5 and 6), their structure can vary considerably from one eucaryotic rRNA species to another (Raué et al., 1988), a fact that is also evident from Fig. 2. The reason for the presence of one of these variable eucaryote-specific expansion segments (designated V9; see Fig. 6) in the middle of a strongly conserved r-protein binding site is unclear. Since L25 binds efficiently to *E. coli* 23S rRNA, the *S. cerevisiae*-specific expansion segment is unlikely to contain any element essential for interaction of this protein with its cognate binding site on 26S rRNA. On the other hand, it seems probable that the mouse-specific expansion segment is the primary cause of the failure of EL23 and L25 to bind to mouse 28S rRNA, even though the *S. cerevisiae*-specific expansion segment does not prevent binding of EL23 to *S. cerevisiae* 26S rRNA. Comparison of the *S. cerevisiae* and mouse "L25 sites" reveals little difference even at the level of primary structure except for the expansion segments (Fig. 2).

To analyze the effects of the V9 segment on r-protein binding further, we changed the structure of this segment by in vitro mutagenesis of a cloned *S. cerevisiae* rDNA fragment. The *S. cerevisiae* rRNA fragment containing the structurally altered expansion segment was then prepared by in vitro transcription, and its interaction with L25 was studied by nitrocellulose filter binding (El-Baradi et al., 1985).

Reduction of the expansion segment to only five nucleotides caused a slight increase in the efficiency of L25 binding (Fig. 3). This result on the one hand confirms the conclusion that the *S. cerevisiae* V9 expansion segment does not contain any elements essential for recognition of 26S rRNA by L25. On the other hand, the observed increase in binding efficiency is in agreement with our previous observation that L25 binds slightly more strongly to *E. coli* 23S (lacking a V9 segment) than to *S. cerevisiae* 26S rRNA. Therefore, rather than being required for L25 binding, the V9 segment appears to have a slight negative effect on interaction of this protein with *S. cerevisiae* 26S rRNA.

To our surprise, however, replacement of the *S. cerevisiae* expansion segment by its mouse counterpart did not significantly affect L25 binding (Fig. 3).

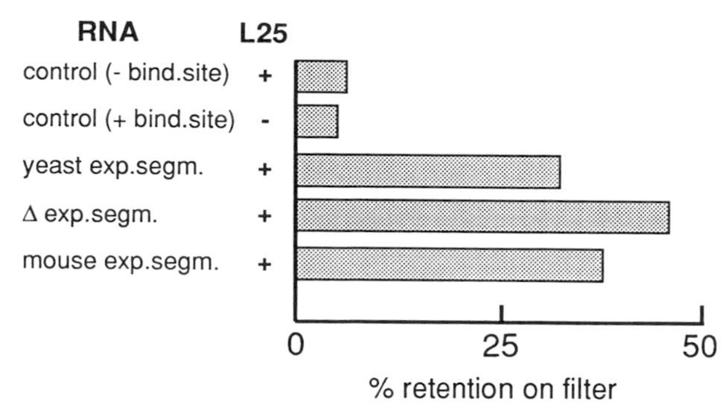

Figure 3. Effect of the structure of the expansion segment on the interaction of *S. cerevisiae* 26S rRNA with r-protein L25. The L25 expansion segment of *S. cerevisiae* 26S rRNA was either removed (except for five nucleotides; Δ exp.segm.) or replaced by its counterpart from mouse 28S rRNA (see Fig. 2). rDNA fragments containing the wild-type and modified L25 binding sites were cloned under control of the phage T7 promoter, and the ^{32}P-labeled rRNA fragments were prepared by in vitro transcription (El-Baradi et al., 1987a; El-Baradi et al., 1987b). The capacity of these fragments to be recognized by L25 was assayed by measuring their retention on a nitrocellulose filter, using L25 purified from 1.0 M LiCl core particles of *S. cerevisiae* 60S ribosomal subunits (El-Baradi et al., 1985). A fragment of *S. cerevisiae* 26S rRNA encompassing domain II was used as a negative control (-bind.site).

Thus, the presence of the mouse V9 expansion segment is in itself not sufficient to abolish interaction with the r-protein, a conclusion that leaves unresolved the precise reason for the failure of EL23 and L25 to bind to mouse 28S rRNA. An important further observation in this respect is that L25 also fails to bind to a synthetic *Tetrahymena thermophila* precursor-rRNA fragment containing the equivalent of its binding site but does recognize mature *T. thermophila* 26S rRNA in which three nucleotides have been removed from the tip of the V9 expansion segment by processing (V. C. Ware, personal communication). Replacement of the *S. cerevisiae* V9 segment by either its intact or processed *Tetrahymena* counterpart again did not have any discernible effect on L25 binding in vitro (W. Musters et al., unpublished data). This finding shows once more that blocking of L25 interaction is not due to the presence of the expansion segment per se. At the same time, however, the fact that L25 is able to recognize the mature (processed) but not the precursor form of the *Tetrahymena* equivalent of its binding site demonstrates a direct involvement of the V9 expansion segment in preventing binding of the r-protein. Further studies using both structurally altered "L25 binding sites" as well as the (yet to be identified) L25 homologs from mouse and *T. thermophila* are required to shed more light on the role of the V9 segment and the significance of the processing occurring within this segment in a number of organisms such as *Tetrahymena* spp., *Artemia salina*, and insects (Raué et al., 1988). The results described above, however, do indicate that the V9 expansion segment is likely to be more than just a neutral appendage whose presence in the rRNA is tolerated because it does not compromise ribosomal function.

Analysis Using In Vitro-Prepared r-Protein

Unambiguous identification of primary binding r-proteins and elucidation of their binding properties requires analysis of the interaction between rRNA and individual r-proteins. Although techniques for purification of individual r-protein species have been improved over the last few years, it still remains a laborious and time-consuming task. An additional severe impediment to such analysis is the fact that because of their low solubility, eucaryotic r-proteins are notoriously difficult to handle under physiological conditions. To circumvent these problems, we sought to analyze *S. cerevisiae* rRNA–r-protein interactions by using not only rRNA but also r-protein prepared in vitro.

cDNA encoding *S. cerevisiae* r-protein L25 was cloned under control of the phage T7 promoter and transcribed in vitro. The transcript was purified by polyacrylamide gel electrophoresis and used as template for in vitro synthesis of the protein in a wheat germ system in the presence of [^{35}S]methionine. Incorporation of the label into protein was checked by sodium dodecyl sulfate-polyacrylamide gel electrophoresis. Since this analysis showed L25 to be the only labeled product, samples of the in vitro translation system were incubated with various synthetic *S. cerevisiae* rRNA fragments without further purification. Binding of the protein to the rRNA fragment was assayed by sucrose gradient centrifugation. The results (Fig. 4) show that the in vitro-prepared r-protein specifically recognizes the L25 binding site, thus clearly demonstrating the feasibility of this approach for assaying the rRNA-binding capacity of a specific r-protein. We are currently using this method to identify further primary rRNA binding r-proteins from both the large and small subunits of *S. cerevisiae* ribosomes.

The same technique also offers a convenient way for detailed analysis of the structural elements of an r-protein required for its interaction with rRNA, a subject on which so far only scant information is available. Predetermined structural variants of the r-protein in question can easily be obtained by manipulating the cloned gene and can then be tested in vitro for their ability to recognize the binding site on the rRNA. We have analyzed the effect of various

Decoding and tRNA binding

Numerous studies addressing the functional role of the two universally conserved single-stranded sequences around positions 1400 and 1500 of SSU rRNA (Fig. 5) have convincingly demonstrated that these regions form an essential part of the decoding site on the ribosome (reviewed in Raué et al., 1988; Ofengand et al., 1988). The bulk of the evidence derives from characterization of mutations conferring resistance to a number of different aminoglycoside antibiotics, which are known to interfere with the decoding process (De Stasio et al., 1989); cross-linking of the anticodon of A- and P-site-bound tRNA to C-1400 or its equivalent in both pro- and eucaryotes (Ofengand et al., 1988); tRNA-stimulated cross-linking of mRNA to the 1394–1399 region of *E. coli* 16S rRNA (Stiege et al., 1988); and mutational analysis of most of the individual nucleotides in the 1400 and 1500 regions both in vivo and in vitro (Hui et al., 1988; Rottmann et al., 1988; Thomas et al., 1988; Denman et al., 1989b; De Stasio et al., 1989). Surprisingly, the latter studies show that the identity of several of the universally conserved nucleotides in the 1400 region of *E. coli* 16S rRNA can be changed without any undue effect on ribosomal function in vivo (Hui et al., 1987; Thomas et al., 1988; Denman et al., 1989b). This includes C-1400, which could be replaced by U. Other mutations (C-1395 → U; C-1407 → U), however, were found to be lethal even when the mutated rRNA was expressed at a moderate level. As mentioned above, these mutations could cause the formation of crippled subunits that prevent even the wild-type ribosomes from carrying out their appointed tasks (Thomas et al., 1988). G-1401 may be another such essential nucleotide, since its deletion blocks all ribosomal functions in vitro (Denman et al., 1989b). A C → U transition at position 1402 drastically affects elongation only (De Stasio et al., 1987). Finally, deletion studies show the precise length of the 1397–1404 region to be crucial for peptide chain initiation but not elongation in vitro (Denman et al., 1989b).

The highly conserved helix 45 at the 3′ end of SSU-rRNA also appears to play a role in decoding. This conclusion is based on the fact that bacterial ribosomes sensitive to kasugamycin because of a lack of dimethylation of the two A residues in the loop of this helix (Helser et al., 1971) are impaired in mRNA-dependent binding of fMet-tRNA and show increased translational fidelity (Van Knippenberg, 1986). The dimethylation of these two A residues is one of the few universally conserved modifications in SSU rRNA (Raué et al., 1988). Its functional significance in other classes of ribosomes has not been analyzed, however.

Point mutations causing resistance to streptomycin (which induces misreading) have been either created in or traced to the so-called 530 loop of helix 18 as well as the base of helix 27 in both *E. coli* (Melançon et al., 1988) and chloroplast (Montandon et al., 1985; Etzold et al., 1987) SSU rRNA, implicating these regions in the decoding process. As far as the 530 loop is concerned, further support for its involvement in decoding comes from the protection of several of its nucleotides by A-site-bound tRNA (Moazed and Noller, 1986). Moreover, an *S. cerevisiae* mitochondrial ochre suppressor mutation was found to be due to transition of the universally conserved G corresponding to *E. coli* G-517 in the mitochondrial SSU rRNA (Shen and Fox, 1989). On the other hand, Gravel et al. (1989) have recently shown that in vitro-synthesized *E. coli* 16S rRNA lacking part or all of helix 27 is assembled into 30S subunits that function normally in in vitro protein synthesis. Thus, helix 27 appears not to be essential for translation. Nevertheless, no transformed cells expressing this type of mutant 16S rRNA could be isolated, indicating that helix 27 does play an important, though hitherto undefined, role in vivo. Finally, translation is impaired by a G-791 → A transition in the conserved sequence at the tip of helix 24 which interferes with the initiation process (Tapprich et al., 1989). The tertiary structure models for SSU rRNA place this loop in close proximity to the 1400 region (Stern et al., 1988; Brimacombe, 1988).

Nucleotides protected against chemical modification by either A- or P-site-bound tRNA have been identified in all regions of *E. coli* 16S rRNA mentioned so far, providing additional evidence for their involvement in translation (Meier and Wagner, 1984; Moazed and Noller, 1986). Further protection was observed at several other positions, notably the tips of helices 23 and 31 (Moazed and Noller, 1986). The latter is of particular interest since the protected nucleotide (G-966 in *E. coli*) is modified in both pro- and eucaryotic SSU rRNAs, albeit not in the same manner (Raué et al., 1988). Although in vitro experiments have shown that modification of 16S rRNA is not a prerequisite for its assembly into active ribosomes (Denman et al., 1989a; Denman et al., 1989b), it may well play an important role in ensuring optimal ribosomal activity in vivo. In fact, the three-dimensional models for 16S rRNA place all of its modified nucleotides closely together in the proximity of the decoding region on the small ribosomal subunit (Stern et al., 1988; Brimacombe, 1988).

Subunit association

rRNA also appears to be directly involved in the association between the ribosomal subunits either by RNA-protein or by RNA-RNA interactions. Analysis of differences in sensitivity to chemical modification or enzymatic cleavage of the 16S rRNA in 30S subunits versus 70S ribosomes showed the protected sites to be clustered predominantly, though not exclusively, in the single-stranded loops of helices 23 and 24 already discussed above in another context (Meier and Wagner, 1984; Noller, 1984). Subsequent modification-selection experiments demonstrated that recruitment of 30S subunits into 70S ribosomes is severely impaired when specific nucleotides in these loops are chemically modified (Noller, 1984). As far as the helix 34 loop is concerned, additional evidence is provided by the observation that the G-791 → A transition discussed above also reduces the affinity of the mutant 30S subunits for their 50S counterparts (Tapprich et al., 1989). A similar effect is seen when the loop is occluded by hybridization to a complementary deoxyoligonucleotide (Tapprich and Hill, 1986).

Helix 44 also appears to be crucial for subunit association, since either removal of this helix (Zwieb et al., 1986) or perturbing its secondary structure (Meier et al., 1986) causes a severe reduction in the formation of 70S ribosomes.

Finally, chemical modification of G-1064 creates a substantial impediment for subunit association, implicating helix 34 in this process (Noller, 1984). Note that the same region also is involved in translation elongation and termination (see above).

Functional Analysis of LSU rRNA

Structural comparison of the more than 40 different LSU rRNAs sequenced to date produces a picture highly similar to that described above for its small-subunit counterpart (Fig. 6). The procaryotic/chloroplast and eucaryotic cytoplasmic species together define a conserved core decorated at specific sites with a total of 18 expansion segments that vary strongly in both sequence and secondary structure from one species to another (Raué et al., 1988). When the mitochondrial LSU rRNAs are also taken into account, the core is reduced, mainly by elimination of substantial parts of domains I to III and VI. The most extreme reduction is seen in trypanosomal mitochondrial LSU rRNAs, in which most of the elements characteristic of domains II and VI and all of those distinctive for domains I and III have vanished (Raué et al., 1988). Most of domains IV and V, however, is still clearly distinguishable even in these highly abbreviated species, indicating a pivotal role for these domains in ribosomal function. Three further structural elements appear to be universally conserved: helices 27 to 30 and 39 to 40 in domain II as well as helix 90 in domain VI (Fig. 6).

Domains I to III

No evidence directly implicating helices 27 to 30 in any ribosomal function has so far been reported. However, both RNA-RNA cross-linking data (Stiege et al., 1983) and chemical footprinting of the antibiotic vernamycin B (Moazed and Noller, 1987b) indicate that helix 30 is in close proximity to the central loop of domain V. Since a wealth of evidence exists for the involvement of this loop in peptidyl transfer and translocation (see below), by inference this proximity suggests a role for the helix 27-30 region in these processes. Given the mode of action of vernamycin B, participation in translocation is the most likely, though unproven, possibility.

Helices 39 and 40 have already been discussed in the context of rRNA–r-protein interaction. This region constitutes part of the so-called GTPase center of the ribosome, which is intimately involved in all phases of translation as well as the stringent response [i.e., production of (p)ppGpp] induced in bacterial cells by amino acid starvation. Important evidence for the direct participation of the rRNA in these processes has been furnished by studies on the interaction of thiostrepton with bacterial ribosomes. This antibiotic, which interferes with peptide chain initiation, elongation, and termination as well as (p)ppGpp synthesis (reviewed in Cundliffe, 1986), interacts directly with helices 39 and 40, as shown by binding and chemical footprinting studies (Cundliffe, 1986; Egebjerg et al., 1989), site-directed mutagenesis (Thompson et al., 1988), and analysis of mutations causing thiostrepton resistance (Thompson et al., 1982; Hummel and Böck, 1987b). Further support for a role of the rRNA in peptide chain formation is provided by the fact that A-site-bound tRNA protects several nucleotides in the helix 39-40 region against chemical modification (Moazed and Noller, 1989). Finally, cross-linking and chemical footprinting have demonstrated interaction of this region with elongation factor EF-G (Skjöld, 1983; Moazed et al., 1988). Since EF-G binding involves sequences in domains IV and VI as well (Bochkareva and Girshovich, 1984; Moazed et al., 1988; Leffers et al., 1988), the three domains appear to interact closely in the ribosome.

The exceptionally strong functional conservation of the helix 39-40 region already apparent from the r-protein binding studies is further underscored by experiments in our laboratory in which this part

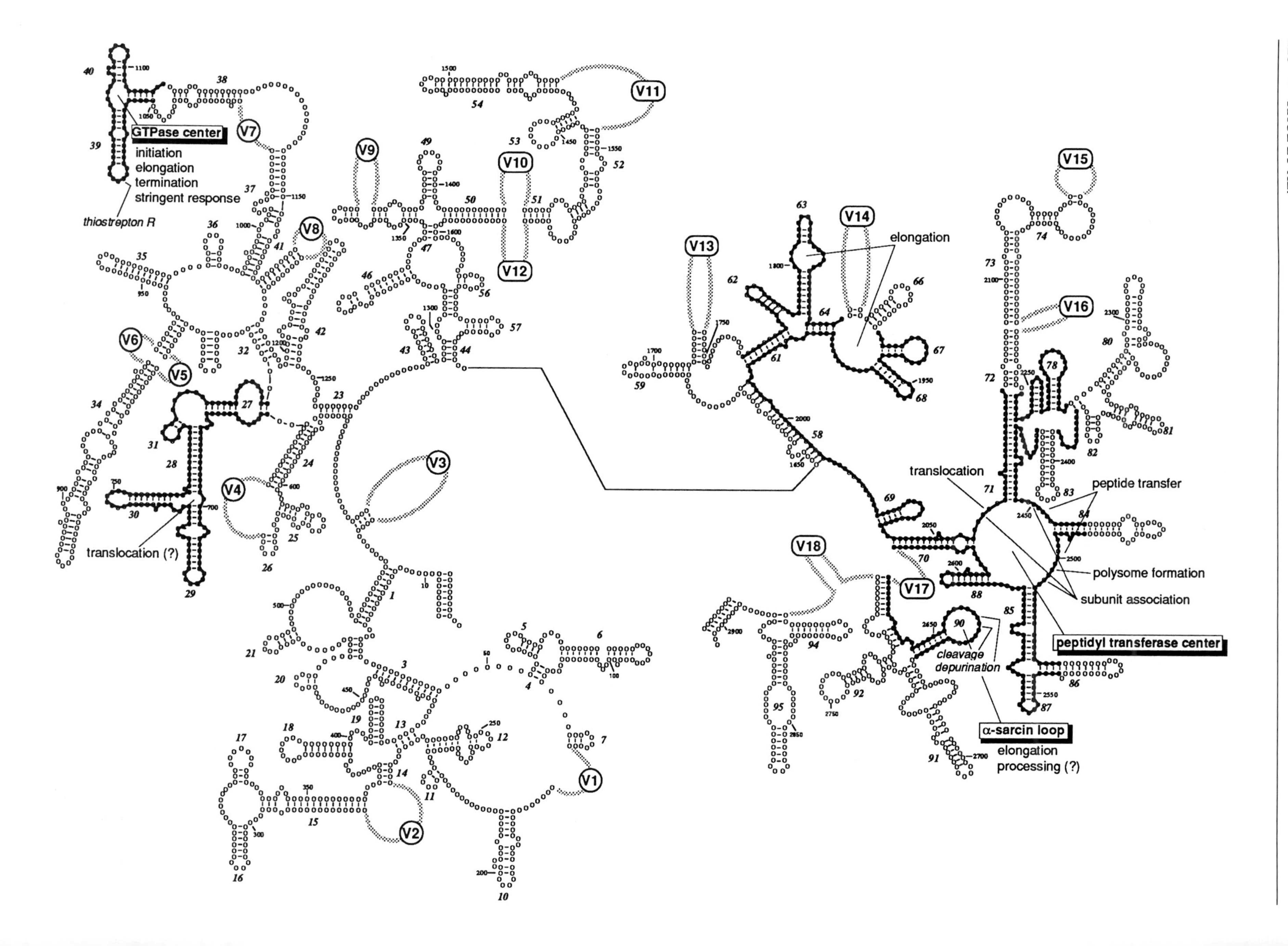
GTPase center
initiation
elongation
termination
stringent response
thiostrepton R
translocation (?)
elongation
translocation
peptide transfer
polysome formation
subunit association
peptidyl transferase center
cleavage
depurination
α-sarcin loop
elongation
processing (?)
V1
V2
V3
V4
V5
V6
V7
V8
V9
V10
V11
V12
V13
V14
V15
V16
V17
V18

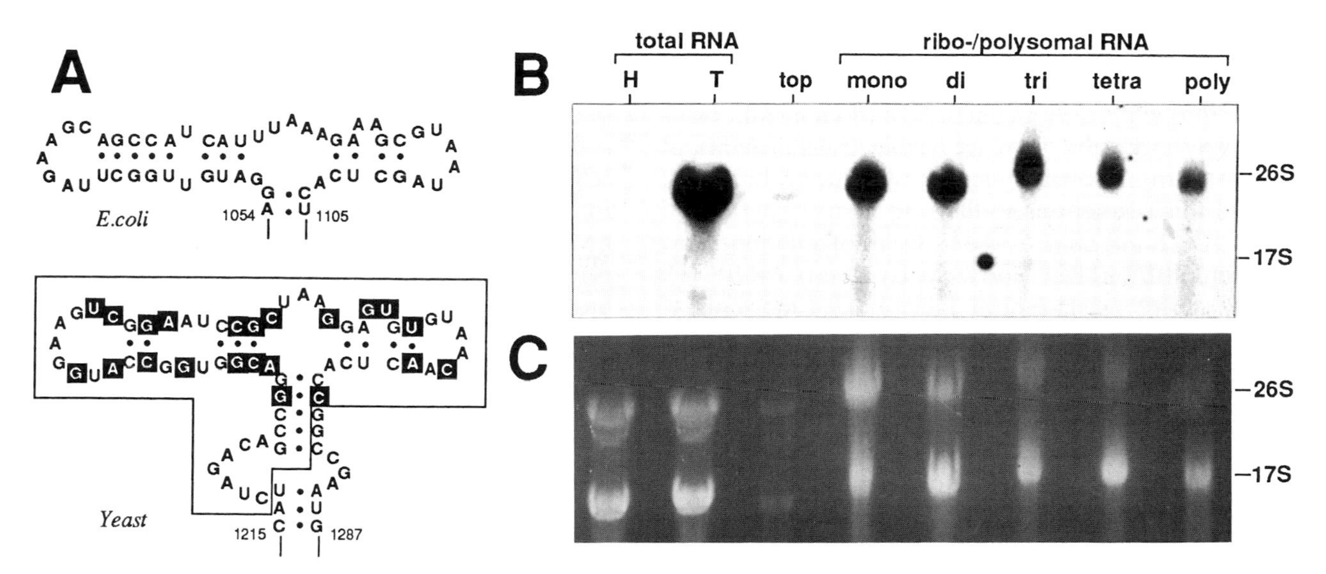

Figure 7. Functional exchange of the GTPase center of *S. cerevisiae* 26S rRNA against its *E. coli* equivalent. The region of *S. cerevisiae* 26S rRNA enclosed in the box was replaced by a synthetic oligonucleotide having the sequence of the equivalent region from *E. coli* 23S rRNA, using the *Xba*I and *Xme*III restriction sites flanking the helix 39-40 region (A). The mutated GTPase center was inserted into the tagged rDNA transcription unit. The distribution of the 26S rRNA transcript carrying the mutant GTPase center among polysomes from transformed cells was analyzed by Northern (RNA) hybridization with the oligonucleotide complementary to the 26S tag (B). As a reference, panel C shows the ethidium bromide staining of the RNA isolated from the various polysome fractions. Lanes H and T contain total RNA from untransformed host cells and transformed cells, respectively. Lane labeled "top" contains RNA from the top of the polysome gradient.

of *S. cerevisiae* 26S rRNA was replaced by its equivalent from *E. coli* by using the tagged gene system described above. Although this replacement results in the introduction of 22 point mutations, the polysomal distribution of the tagged 26S rRNA in vivo was indistinguishable from that of the wild-type species (Fig. 7). Thus, we judged the chimeric 26S rRNA to be fully functional. To our knowledge, this is the first example of a successful attempt at exchanging structural elements between pro- and eucaryotic rRNA.

Although domain III contains a conserved r-protein binding site, only a vestige of this domain is retained in mitochondrial LSU rRNA, in particular those from insect mitochondria (Raué et al., 1988). Clearly, this class of ribosomes can function perfectly well without domain III. Nevertheless, this domain is of functional interest because of the presence of expansion segment V9 in the middle of the conserved r-protein binding site. To establish whether this expansion segment might have some species-specific functional role in the ribosome, we have studied the fate in vivo of *S. cerevisiae* 26S rRNA carrying the mutations in V9 described earlier, using the tagged gene system. Analysis of the intracellular level of the various mutated 26S rRNAs failed to demonstrate any discernible effect either of removing the expansion segment or of replacing it by its counterpart from mouse 28S or *T. thermophila* 26S rRNA. Therefore, we conclude not only that the precise structure of the V9 expansion segment in *S. cerevisiae* 26S rRNA is immaterial for r-protein binding in vitro but that it is also likely to be of little or no consequence for ribosome assembly and function in vivo. A similar structural flexibility is shown by expansion segments V8 of 17S and V2 of 26S rRNA, as demonstrated by the successful tagging of these rRNAs by insertion of an oligonucleotide into either segment. This structural flexibility, however, cannot be generalized to include any change in any expansion segment. We have found that *S. cerevisiae* cells transformed with a 17S rRNA gene carrying a 19-base-pair (bp) insertion in expansion segment V3 (Fig. 5) did not accumulate significant amounts of the tagged transcript. Thus, the insertion interferes with either the biosynthesis of 17S rRNA or its assembly into 40S subunits. Sweeney and Yao (1989) have reported that attempts to transform *T. thermophila* cells with 26S genes containing a 119- or 2,300-bp insertion in expansion segment V13 (domain IV; Fig. 6) were unsuccessful, indicating a dominant defect in

Figure 6. Functional regions in LSU rRNA. The figure shows a schematic representation of the SSU rRNA structure based on the secondary structure model for *E. coli* 23S rRNA. Expansion segments are numbered consecutively from the 5′ end. Filled circles indicated universally conserved regions. Functional regions discussed in the text are labeled.

specific nucleotides in 16S ribosomal RNA from attack by chemical probes. *Cell* **47**:985–994.

Moazed, D., and H. F. Noller. 1987b. Interaction of antibiotics with functional sites in 16S ribosomal RNA. *Nature* (London) **327**:389–394.

Moazed, D., J. M. Robertson, and H. F. Noller. 1988. Interaction of elongation factors EF-G and EF-Tu with a conserved region in 23S RNA. *Nature* (London) **334**:362–364.

Moazed, D., B. J. Van Stolk, S. Douthwaite, and H. F. Noller. 1986. Interconversion of active and inactive 30S ribosomal subunits is accompanied by a conformational change in the decoding region of 16S rRNA. *J. Mol. Biol.* **191**:483–493.

Montandon, P. E., R. Wagner, and E. Stutz. 1986. *E. coli* ribosomes with a C912 to U change in the 16S rRNA are streptomycin resistant. *EMBO J.* **5**:3705–3708.

Muralikrishna, P., and E. Wickstrom. 1989. Escherichia coli initiation factor 3 protein binding to 30S ribosomal subunits alters the accessibility of nucleotides within the conserved central region of 16S rRNA. *Biochemistry* **28**:7505–7510.

Murgola, E. J., K. A. Hizaki, H. U. Göringer, and A. E. Dahlberg. 1988. Mutant 16S ribosomal RNA: a codon-specific translational suppressor. *Proc. Natl. Acad. Sci. USA* **85**:4162–4165.

O'Farrell, P. H. 1975. High resolution two-dimensional electrophoresis of proteins. *J. Biol. Chem.* **250**:4007–4021.

Petrullo, L. A., P. J. Gallagher, and D. Elsevier. 1983. The role of 2-methylthio-N^6-isopentenyladenosine in readthrough and suppression of nonsense codons in *Escherichia coli. Mol. Gen. Genet.* **190**:289–294.

Remaut, E., H. Tassao, and W. M. Fiers. 1983. Improved plasmid vectors with a thermoinducible expression and temperature-regulated runaway replication. *Gene* **22**:103–113.

Schüler, D., and R. Brimacombe. 1988. The *Escherichia coli* 30S subunit; an optimized three-dimensional fit between the ribosomal proteins and the 16S RNA. *EMBO J.* **7**:1509–1513.

Stark, M. J. R., R. L. Gourse, and A. E. Dahlberg. 1982. Site-directed mutagenesis of ribosomal RNA: analysis of ribosomal RNA deletion mutants using maxicells. *J. Mol. Biol.* **159**:417–439.

Steen, R., D. Jemiolo, R. Skinner, J. Dunn, and A. E. Dahlberg. 1986. Expression of plasmid-coded mutant ribosomal RNA in *E. coli*; choice of plasmid vectors and gene expression. *Prog. Nucleic Acid Res. Mol. Biol.* **33**:1–18.

Stern, S., T. Powers, L.-M. Changchien, and H. Noller. 1989. RNA-protein interactions in 30S ribosomal subunits: folding and function of 16S rRNA. *Science* **244**:783–790.

Tapio, S., and L. A. Isaksson. 1988. Antagonistic effects of mutant elongation factor Tu and ribosomal protein S12 on control of translational accuracy, suppression and cellular growth. *Biochimie* **70**:273–281.

Tapprich, W. E., D. J. Goss, and A. E. Dahlberg. 1989. Mutation at position 791 in *Escherichia coli* 16S ribosomal RNA affects processes involved in the initiation of protein synthesis. *Proc. Natl. Acad. Sci. USA* **86**:4927–4931.

Tapprich, W. E., and W. E. Hill. 1986. Involvement of bases 787–795 of *Escherichia coli* 16S ribosomal RNA in ribosomal subunit association. *Proc. Natl. Acad. Sci. USA* **83**:556–560.

Thompson, J., E. Cundliffe, and A. E. Dahlberg. 1988. Site-directed mutagenesis of *Escherichia coli* 23S ribosomal RNA at position 1067 within the GTP hydrolysis centre. *J. Mol. Biol.* **203**:457–465.

Trifonov, E. N. 1987. Translation framing code and frame-monitoring mechanism as suggested by the analysis of mRNA and 16S rRNA nucleotide sequences. *J. Mol. Biol.* **194**:643–652.

Valle, R. P. C., and M.-D. Morch. 1988. Stop making sense or regulation at the level of termination in eukaryotic protein synthesis. *FEBS Lett.* **235**:1–15.

Van Duin, J., and R. Wijnands. 1981. The function of ribosomal protein S21 in protein synthesis. *Eur. J. Biochem.* **118**:615–619.

Vester, B., and R. A. Garrett. 1988. The importance of highly conserved nucleotides in the binding region of chloramphenicol at the peptidyl transferase centre of *Escherichia coli* 23S ribosomal RNA. *EMBO J.* **7**:3577–3587.

Weiss, R. B., D. M. Dunn, J. F. Atkins, and R. F. Gesteland. 1987. Slippery runs, shifty stops, backwards steps and forward hops: −2, −1, +2, +5 and +6 ribosomal frameshifting. *Cold Spring Harbor Symp. Quant. Biol.* **52**:687–693.

Weiss, R. B., D. M. Dunn, A. E. Dahlberg, J. F. Atkins, and R. F. Gesteland. 1988. Reading frame switch caused by base-pair formation between the 3′ end of 16S rRNA and the mRNA during elongation of protein synthesis in Escherichia coli. *EMBO J.* **7**:1503–1507.

Wool, I. 1984. The mechanism of action of the cytotoxic nuclease alpha sarcin and its use to analyze ribosome structure. *Trends Biochem. Sci.* **9**:14–17.

Zoller, M. J., and M. Smith. 1983. Oligonucleotide-directed mutagenesis of DNA fragments cloned into M13 DNA. *Methods Enzymol.* **100B**:468–500.

Chapter 17

In Vitro Analysis of the Role of rRNA in Protein Synthesis: Site-Specific Mutation and Methylation

PHILIP R. CUNNINGHAM, CARL J. WEITZMANN, DIDIER NÈGRE, JOSEPH G. SINNING, VINCENT FRICK, KELVIN NURSE, and JAMES OFENGAND

The ribosome, a ribonucleoprotein complex of over 2×10^6 daltons containing 50 to 80 proteins and three or more RNAs, is one of the most complex catalytic species known. Its size and complexity reflect the central role of protein synthesis in all living systems and have long fascinated researchers concerned with the relationships between structure and function. For both historical and technical reasons, ribosomal proteins were considered to be the functional components of the ribosome until the beginning of this decade. Since then, advances in RNA technology have brought to light considerable evidence implicating regions of rRNA in the various functions of the ribosome (see other chapters in this volume).

In *Escherichia coli*, the 16S RNA contains three long highly conserved single-stranded sequences, one (positions 518 to 533) in the 5′ domain, and the remaining two (positions 1394 to 1408 and 1492 to 1505) in the 3′ minor domain. All three have been implicated in tRNA binding (Moazed and Noller, 1986; Krzyzosiak et al., 1987; Cunningham et al., 1988; Denman et al., 1989a) and other protein synthesis functions (Cunningham et al., 1988; Denman et al., 1989a). The 16S RNA also contains 10 methylated nucleotides (Fig. 1). Little is known about possible structural and functional roles for these methylated nucleotides except for the m^6_2A residues at positions 1518 and 1519 (Van Knippenberg, 1986). None of the methylated nucleotides appears to be essential for either ribosome assembly or function as carried out in vitro (Krzyzosiak et al., 1987; Mélançon et al., 1987; Ericson et al., 1989). Although methylation in vivo has been thought to occur during the later stages of assembly (Feunteun et al., 1974; Dahlberg et al., 1975), little is actually known about the temporal or sequential nature of the methylation reactions, and nothing is known about the substrate specificity of the methyltransferases other than that m^6_2A formation requires the 30S particle (Poldermans et al., 1979).

We have recently developed a totally in vitro system for facile site-specific base changes in *E. coli* 30S ribosomes (Krzyzosiak et al., 1987; Krzyzosiak et al., 1988). This system circumvents many of the problems inherent in the study of rRNA function in vivo (see chapters by Tapprich et al. and Zimmermann et al. in this volume). First, it allows the study of otherwise lethal mutations, since the 16S RNA gene is under the control of the T7 promoter and is not expressed in vivo. Second, all of the ribosomes are mutant. The approximately 50% background of wild-type ribosomes found in most in vivo studies is absent. Third, since ribosome assembly is separated from ribosome function, effects of mutations on each process can be monitored independently. Fourth, large quantities of purified mutant ribosomes are readily obtained so that the separate steps of the ribosomal protein synthesis cycle can be assayed.

This approach has been used to construct a number of mutations at and around C-1400, located in the center of one of the two highly conserved sequences in the 3′ minor domain (Ofengand et al., 1988; Cunningham et al., 1988; Denman et al., 1989a; Denman et al., 1989b). C-1400 is the site of specific cross-linking of the anticodon of tRNA (Prince et al., 1982; Ehresmann et al., 1984; Ehresmann and Ofengand, 1984) and is protected by P-site-bound tRNA (Moazed and Noller, 1986). The functional effects of these mutations have been described in detail (Denman et al., 1989a). A number of

Philip R. Cunningham, Carl J. Weitzmann, Didier Nègre, Joseph G. Sinning, Vincent Frick, Kelvin Nurse, and James Ofengand ■ Roche Institute of Molecular Biology, Roche Research Center, Nutley, New Jersey 07110.

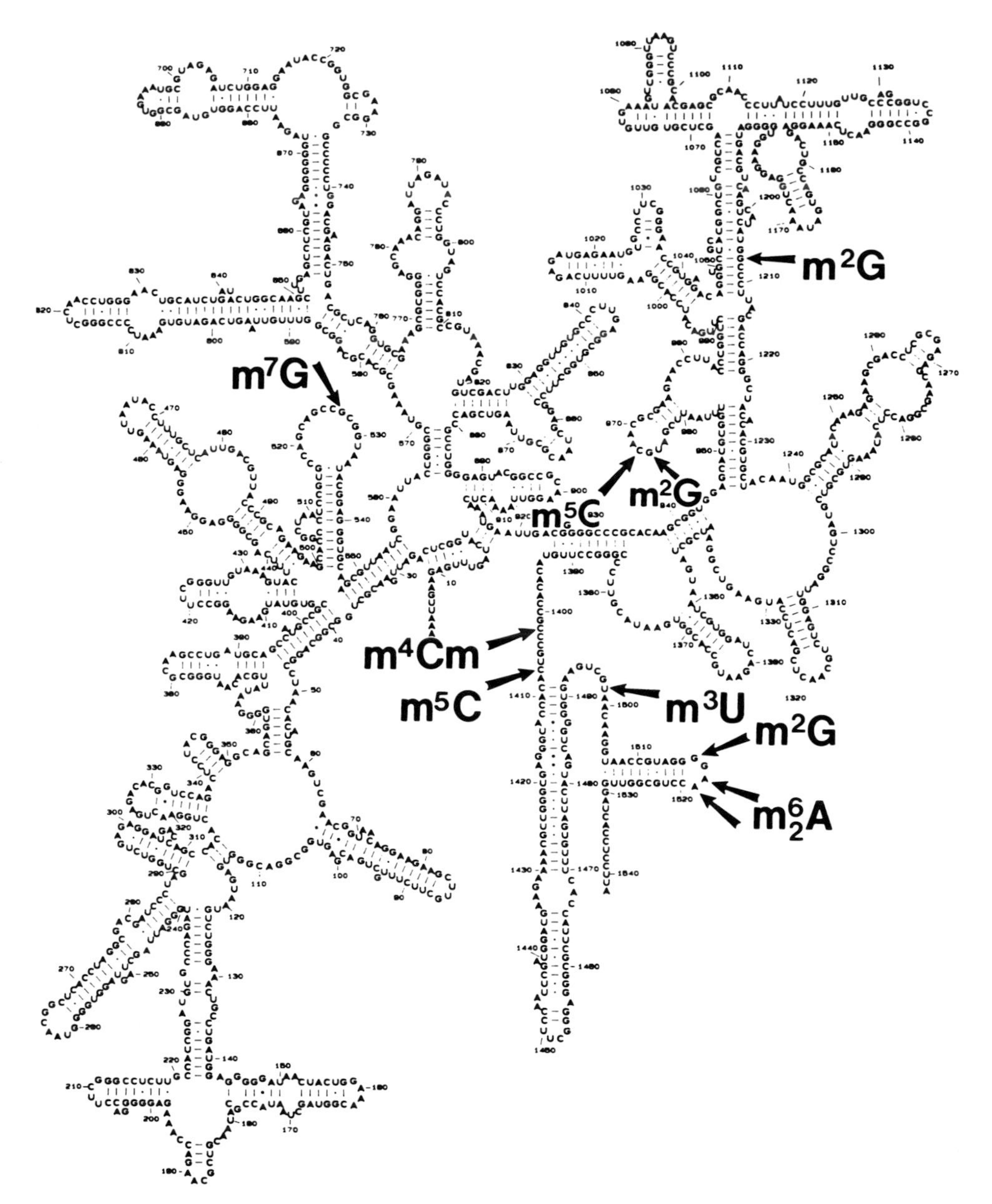

Figure 1. Known sites of methylation of *E. coli* 16S RNA. The secondary structure of 16S RNA is according to Moazed et al. (1986). The sites of methylation (Noller, 1984; Hsuchen and Dubin, 1980) are indicated by the standard modified base abbreviations.

additional mutations that involve both of the abovementioned conserved sequences of the 3′ minor domain as well as mutations in the m^6_2A-containing loop and stem have now been made and tested for function. These results will be summarized in this chapter. In addition, brief mention will be made of the application of these same techniques to the 50S subunit.

The system developed for site-specific mutagenesis proved equally useful for characterization of rRNA methyltransferases (Nègre et al., 1989, 1990) because 16S RNA synthesized in vitro and 30S ribosomes reconstituted from this RNA lack the methyl groups normally found in vivo. This allowed the use of homologous enzyme extracts and substrate and made it possible to obtain stoichiometric levels of methylation. Previous work in this field relied on the use of either heterologous substrates, which were then methylated at different sites, or partially methylated homologous substrates. Using this system, we have recently been able to copurify two 16S RNA methyltransferases, the m^5C-966 methyltransferase and the m^2G-967 methyltransferase (Nègre et al., 1990), and to separate them from each other (D. Nègre, K. Nurse, and J. Ofengand, unpublished results). The contrasting substrate specificities of these two methyltransferases have provided some new insights into the ribosome assembly process, which will be discussed. Some initial studies on the methylation of in vitro transcripts of 23S rDNA will also be summarized.

IN VITRO ASSEMBLY AND FUNCTIONAL ANALYSIS OF MUTANT 30S RIBOSOMES

Most of the mutants were made in the two long single-stranded conserved segments, 1394 to 1408 and 1492 to 1505, of the 3′ minor domain (Fig. 2) because of their involvement in various aspects of the protein synthesis cycle. These include sites protected from chemical modification by bound tRNA (Moazed and Noller, 1986), sites of mutation to antibiotic resistance and protection from chemical modification by bound antibiotic (Moazed and Noller, 1987, and references therein), and sites of involvement in the decoding process (Gornicki et al., 1984, and references therein). In addition, a series of mutations has been created at the site of m_2^6A formation and at the U-1512–G-1523 base pair which is universally conserved except in flagellate protozoans (Dams et al., 1988). All of the mutations constructed by us so far are listed in Fig. 2. Approximately three-fourths of these mutations have been transcribed into RNA, reconstituted into ribosomes, and tested for a defined set of protein synthesis functions (Table 1). None of the ribosomes were lacking any ribosomal proteins, as determined by high-performance liquid chromatography analysis (Denman et al., 1989b). Two of the mutants, A-1405 and U-1496, were not analyzed for protein content, but the double mutant A-1405 U-1496 was, and a normal complement of proteins was found.

The result of making all possible substitutions for C-1400, G-1401, and C-1402 showed that G-1401 was a critical nucleotide. Replacement of C-1400 with U, A, or G had little effect despite the close association of C-1400 or its equivalent with the decoding site in all species examined (Ofengand et al., 1988). Substitutions at position 1402 also were largely without effect. However, all of the substitutions for G-1401 severely depressed activity in all of the assays. Although all three positions are nearly universally conserved (Dams et al., 1988), the identity of the nucleotide at position 1401 appears uniquely important for ribosomal function. This is all the more remarkable because the essential G-1401 is sandwiched between two conserved but unessential

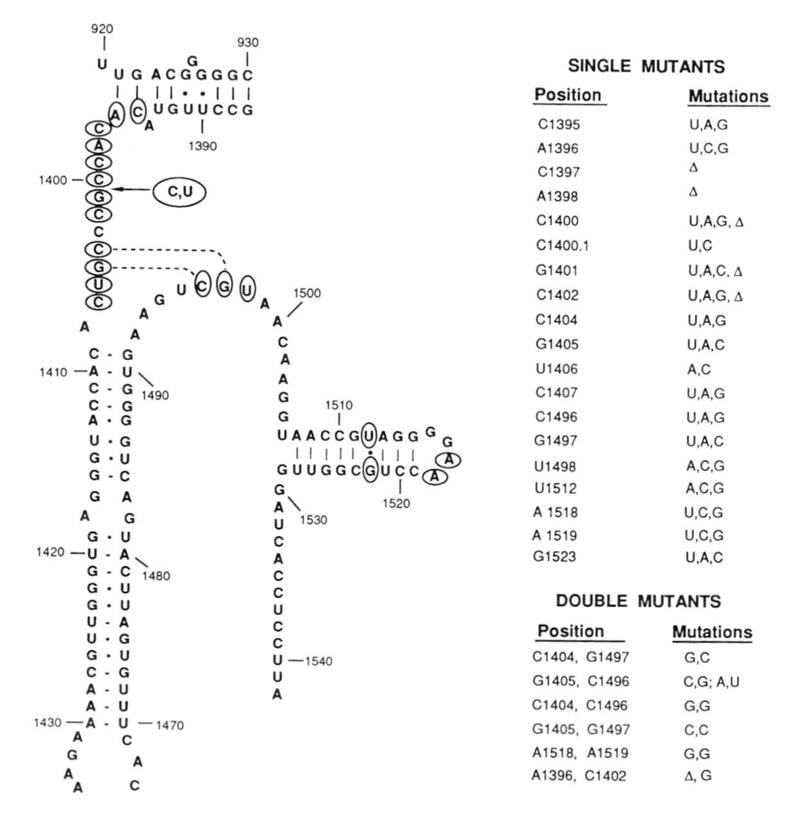

SINGLE MUTANTS

Position	Mutations
C1395	U,A,G
A1396	U,C,G
C1397	Δ
A1398	Δ
C1400	U,A,G, Δ
C1400.1	U,C
G1401	U,A,C, Δ
C1402	U,A,G, Δ
C1404	U,A,G
G1405	U,A,C
U1406	A,C
C1407	U,A,G
C1496	U,A,G
G1497	U,A,C
U1498	A,C,G
U1512	A,C,G
A 1518	U,C,G
A 1519	U,C,G
G1523	U,A,C

DOUBLE MUTANTS

Position	Mutations
C1404, G1497	G,C
G1405, C1496	C,G; A,U
C1404, C1496	G,G
G1405, G1497	C,C
A1518, A1519	G,G
A1396, C1402	Δ, G

Figure 2. Mutations constructed in the 3′ minor domain of 16S RNA. The secondary structure is from Stern et al. (1988). The mutated sites are circled. All of the single and double mutants constructed are listed. Position 1400.1 indicates the insertion between positions 1400 and 1401. The dashed line connects two putative base pairs (Noller et al., 1986). Δ, Deletion of the indicated residue.

Table 1. Functional effects of site-specific nucleotide substitutions[a]

Mutant	Effect (% of wild-type synthetic 30S activity)					
	tRNA binding				Peptide synthesis	
	Subunit association[b]	P site[c]	A site[c]	I site[d]	fMet-Val[e]	Phe-Val[c]
U-1395	40	50	50	40	110	30
C-1396	50	20	40	30	20	10
G-1396	50	60	110	50	10	60
U-1396	80	90	110	60	80	60
A-1400	110	140	80	ND[f]	140	120
G-1400	100	110	60	ND	60	140
U-1400	90	140	130	ND	110	130
A-1401	30	30	40	20	10	20
C-1401	20	20	10	5	10	20
U-1401	30	20	10	10	10	20
A-1402	70	70	70	60	90	80
G-1402	80	70	60	70	120	80
U-1402	100	100	80	110	100	100
G-1404	ND	40	40	ND	ND	130
C-1497	ND	30	20	ND	ND	170
G-1404 C-1497	ND	60	50	ND	ND	70
C-1405	ND	20	20	ND	10	80
G-1496	ND	10	20	ND	<5	110
C-1405 G-1496	ND	70	40	ND	10	90
A-1405	ND	60	60	ND	<5	110
U-1496	ND	60	40	ND	10	90
A-1405 U-1496	ND	100	ND	ND	ND	ND
G-1404 G-1496	ND	<5	10	ND	<5	110
C-1405 C-1497	ND	<5	10	ND	<5	180
A-1498	ND	80	90	ND	100	100
C-1498	ND	90	80	ND	80	100
G-1498	ND	90	90	ND	20	100
C-1518	ND	70	60	ND	40	120
G-1518	ND	60	50	ND	40	90
U-1518	ND	60	50	ND	30	110
C-1519	ND	70	60	ND	50	80
G-1519	ND	60	60	ND	50	80
U-1519	ND	70	70	ND	70	80

[a] Values for the 1400 series are from Denman et al. (1989a); all others are unpublished results.
[b] Assayed as described by Denman et al. (1989b) for series II, system A, except at 50 mM NH_4Cl.
[c] Assayed as described by Denman et al. (1989a).
[d] fMet-tRNA binding to 30S subunits dependent on initiation factors and mRNA, at 12 mM Mg^{2+}.
[e] Modified from Denman et al. (1989a) by substitution of a synthetic version of natural mRNA for the transcription system. A-1400, G-1400, and U-1400 were assayed exactly as described by Denman et al. (1989a).
[f] ND, Not determined.

residues. Consistent with these results, we have previously shown that G-1401 is a key reference point in the positioning of the anticodon loop of P-site-bound tRNA (Denman et al., 1989b). It is unlikely that the inactivating effect of replacing G-1401 by other nucleotides is a result of any gross structural perturbations, since the mutant particles sedimented at 30S and had not lost any ribosomal proteins. More subtle conformational effects are a possibility and are under investigation.

In agreement with our results, other studies have shown strong protection of G-1401 by P-site-bound tRNA (Meier and Wagner, 1984; Moazed and Noller, 1986) and that deletion of G-1401 caused complete inactivation in vitro (Denman et al., 1989a) and was lethal in vivo (Thomas et al., 1988). The double mutants A-1399 U-1401 and A-1399 C-1401 were also strongly inactivating in vivo (Rottmann et al., 1988). Also in agreement with the results of Table 1, mutation of C-1400 to U (Hui et al., 1988; Thomas et al., 1988) or m^4Cm-1402 to U (Jemiolo et al., 1985) had little or no effect in vivo. On the other hand, Hui et al. (1988) found A-1402 to be inactive in vivo, in contrast to our in vitro findings. Hui et al. also found that A-1400 and G-1400 mutants were much less active than the C-1400 wild type, whereas such large effects were not seen in vitro (Table 1). When comparing in vivo and in vitro data, however, one should keep in mind that assembly defects will also be detected as a functional loss in vivo. Just such an assembly defect was reported previously for the G-1400 mutant (Denman et al., 1989b).

U-1395 and the base changes at position 1396 were made to test the functional importance of the base-pairing scheme shown in Fig. 2, as well as the reported in vivo lethality of U-1395 (Thomas et al., 1988), an unexpected result since G-U should have been an acceptable replacement for G-C. U-1395 was approximately half as active as the wild type except for fMet-Val synthesis, which was unaffected, and Phe-Val copeptide synthesis, which was reduced by two-thirds. These results imply that the reported lethality conferred by this substitution did not result from a block in protein synthesis initiation. The high activity of U-1396 compared with G-1396 implies that the base pairing with U-921 shown in Fig. 2 is not a requirement for ribosomal function. The reduced activity of C-1396 compared with U-1396 suggests that the identity of the 1396 nucleotide, not its ability to base pair with U, is the important parameter. This position may play a role in one of the later steps of initiation, since G-1396, which was otherwise reasonably active, was specifically inhibited in fMet-Val synthesis but not in 30S initiation complex formation.

A series of single and double mutants at positions 1404, 1405, 1496, and 1497 was constructed to test the proposed occurrence of base pairing between these residues during translation (Noller et al., 1986). Substitutions at the normally methylated U-1498 were also made. These positions (Fig. 2) are near sites implicated in antibiotic action (De Stasio et al., 1989, and references therein). The functional effects of these mutations are shown in Table 1. The results support a functional role for the proposed 1404-1497 and 1405-1496 base pairing in P-site binding of tRNA, since base pair restoration by double mutation partially restored function. However, superimposed upon this effect there appears to be a strong inhibition of protein synthesis initiation by all of the substitutions tested. All of the mutations at position 1498 reduce tRNA binding and polypeptide synthesis by approximately 50%. In addition, there is a strong effect of C-1498 and G-1498 on initiation that is not shared by A-1498. We are currently examining this series of mutants more closely in an effort to determine which step of initiation is affected.

The two m_2^6A residues at positions 1518 and 1519 are among the most highly conserved residues known (Van Knippenberg et al., 1984; Raué et al., 1988), yet the absence of methylation of these bases only slightly affects ribosome function (Van Knippenberg, 1986). A similar result was observed when base substitutions at 1518 and 1519 were assayed in vitro. The 1518 substitutions reduced initiation significantly more than the 1519 substitutions, but in general all of the functional activities were only moderately decreased. Possible explanations for this lack of effect include a role for the two m_2^6A residues in some function(s) not tested in our assays or a role in optimizing ribosome function rather than in being required for it. We are currently transferring the mutations into pKK3535 in order to assess their effect in vivo.

Three large deletions of 16S RNA were also made. Each of the truncated RNAs could be reconstituted into a well-defined particle, as judged by sucrose gradient sedimentation, and contained a discrete set of proteins (Table 2). The proteins bound to mutant 1–526 are consistent with the protection studies of Stern et al. (1989), with the cross-linking results obtained by Brimacombe et al. (1988), and with the recognition site for S4 as described by Draper (1989). The two mutants deficient in part of the 3′ minor domain (1–1415 and 1–1509) bound approximately stoichiometric amounts of each of the proteins except for S21. The reduced amount of S21 is consistent with its putative interaction with the missing 3′ end of the RNA (Brimacombe et al., 1988) as well as interactions elsewhere (Stern et al., 1989). The truncation mutants were functionally inactive in each of the assays tested. In preliminary experiments, mutant 1–1415 also failed to associate with 50S, in agreement with the deletion studies of Zwieb et al. (1986) and the G-1416-to-U mutation of Rottmann et al. (1988). The inability of the mutant 30S to bind fMet-tRNA reaffirms the importance of the 1416–1542 segment in the initiation process, since the amount of missing S21 is insufficient to account for the total loss of binding activity. The inactivity of the 1–1509 30S subunits in all of the assays is not in agreement with the results of Mélançon et al. (1987)

and function of 16S rRNA. *Science* **244**:783–790.

Stern, S., B. Weiser, and H. F. Noller. 1988. Model for the three-dimensional folding of 16S ribosomal RNA. *J. Mol. Biol.* **204**:447–481.

Thomas, C. L., R. J. Gregory, G. Winslow, A. Muto, and R. A. Zimmermann. 1988. Mutations within the decoding site of *Escherichia coli* 16S rRNA: growth rate impairment, lethality and intragenic suppression. *Nucleic Acids Res.* **16**:8129–8146.

Van Knippenberg, P. H. 1986. Structural and functional aspects of the N^6,N^6 dimethyladenosines in 16S ribosomal RNA, p. 412–424. *In* B. Hardesty and G. Kramer (ed.), *Structure, Function, and Genetics of Ribosomes*. Springer-Verlag, New York.

Van Knippenberg, P. H., J. M. A. van Kimmenade, and H. A. Heus. 1984. Phylogeny of the conserved 3′ terminal structure of the RNA of small ribosomal subunits. *Nucleic Acids Res.* **12**:2595–2604.

Wagner, R., H. G. Gassen, C. Ehresmann, P. Stiegler, and J.-P. Ebel. 1976. Identification of a 16S RNA sequence located in the decoding site of 30S ribosomes. *FEBS Lett.* **67**:312–315.

Youvan, D. C., and J. E. Hearst. 1981. A sequence from *Drosophila melanogaster* 18S rRNA bearing the conserved hypermodified nucleoside amψ: analysis by reverse transcription and high-performance liquid chromatography. *Nucleic Acids Res.* **9**:1723–1741.

Zwieb, C., D. K. Jemiolo, W. F. Jacob, R. Wagner, and A. E. Dahlberg. 1986. Characterization of a collection of deletion mutants at the 3′-end of 16S ribosomal RNA of *Escherichia coli*. *Mol. Gen. Genet.* **203**:256–264.

Chapter 18

Probing Ribosome Structure and Function by Using Short Complementary DNA Oligomers

WALTER E. HILL, JENNIFER WELLER, THOMAS GLUICK, CHUCK MERRYMAN, RICHARD T. MARCONI, ANCHALEE TASSANAKAJOHN, and WILLIAM E. TAPPRICH

rRNA, probably in conjunction with ribosomal proteins, is primarily responsible for the function of the ribosome. The various ribosomal interactions between rRNA and tRNA, mRNA, ribosomal proteins, and translation factors are discrete and precise, frequently involving weak and transient interactions. It is now possible to probe the functions of various rRNA regions in situ by using several methods, including chemical modification (Noller et al., this volume), site-specific mutations (Tapprich et al., this volume) or cDNA probes (Hill et al., 1988). In addition, each of these techniques lends itself to a determination of rRNA structure within the ribosome. In this chapter, we will discuss the results emanating from probing rRNA function and structure by using cDNA oligomers.

GENERAL CONSIDERATIONS OF PROBE BINDING

Availability of Sites

Initially, we must address a fundamental dichotomy in ribosome function. In order for rRNA to be functional, the interactive regions of rRNA must be available to various ligands, implying that the rRNA regions should be exposed, single stranded, and conserved phylogenetically (Woese, 1980). Yet it may be argued that in a cell in which RNases abound, it would be very unwise to allow functional regions of rRNA to be exposed to the cellular milieu, since examination of the secondary structure of rRNA reveals most of the highly conserved regions to be single stranded.

How does the ribosome manage to maintain the functionality and availability of these regions yet protect them from roving RNases? One can imagine that this might be done by using a protecting device, such as a protein, or RNA-RNA interactions, or even exotic secondary structures. Alternatively, it is possible that the ribosome may limit the "window" of available rRNA to a size small enough to exclude RNases (most of which have diameters in excess of 35 Å [3.5 nm]) yet allow interactions of ligands such as tRNA and mRNA.

To approach this problem, it is necessary to determine what portion, if any, of a region is, in fact, accessible to cDNA oligomers. It is apparent from the studies of Noller's group (see Noller et al., this volume) that many regions are available to small chemical probes. We have concentrated initially on testing the highly conserved rRNA regions for availability to short DNA oligomers, generally hexamers, although longer probes have occasionally been used. Under our conditions, hexameric probes have a T_m high enough to allow binding to occur, whereas shorter probes lack sufficient coupling energy. Longer probes, although providing greater binding stability, do not provide the resolution necessary to precisely define available binding regions. From these studies, we have been able to determine various regions that are totally or partially available to DNA probes. These results are discussed in more detail below.

Specificity of Binding

Any assay of structure or function using cDNA probes requires that the complementary binding site be unambiguously identified. For any given target sequence, additional partially or totally identical sites

Walter E. Hill, Jennifer Weller, Thomas Gluick, Chuck Merryman, Richard T. Marconi, and Anchalee Tassanakajohn ■ Division of Biological Sciences, University of Montana, Missoula, Montana 59812. **William E. Tapprich** ■ Brown University, Providence, Rhode Island 02912.

may exist within the total rRNA molecule. To help ascertain the region or regions to which cDNA probes bind, RNase H has been used. RNase H cleaves the RNA involved in RNA-DNA base pairing, and the resulting rRNA fragments can be assayed to determine binding-site specificity. This technique is limited to those sites available to RNase H. We have found that site specificity for hexameric probes can best be demonstrated by using RNase H digestion of rRNA itself, in the absence of ribosomal proteins. Under these circumstances, all potential binding sites are available and the resulting fragments specify only those sites to which the probes readily bind.

Secondary Effects

It is difficult to determine whether secondary effects occur upon hybridization with a DNA oligomer. That is, when a DNA probe is bound to a complementary region, is it the presence of the DNA or an induced structural change in the rRNA that would impede binding of another ligand, such as tRNA? A secondary change of this nature may be expected, since in the process of forming a DNA-RNA hybrid, the single-stranded RNA would be linearized into an A-form double helix.

A long single-stranded region, such as bases 1393 to 1408 in 16S rRNA, may initially be in the A form and hybridization would not markedly perturb the structure. However, for a sequence found in a hairpin loop structure, the addition of a probe will linearize a portion of the loop, putting considerable strain on the stem, perhaps causing disruption of a portion of it. This would affect the structure of the rRNA markedly, perhaps causing long-range perturbations. Even with hexameric probes, such effects may take place.

Although various biophysical techniques may be used to characterize the ribosomal structures before and after probing (or chemical modification or mutagenesis) experiments, the resolution of such techniques is generally insufficient to provide definitive evidence of structural alterations occurring upon hybridization. However, careful experiments using multiple probes may shed some light on structural alterations in some cases. For instance, in the case of a hairpin loop, if a primary probe linearized a loop, forcing open the helical stem, a second probe to the stem region may bind in the presence of the primary probe.

Probes Longer than the Available Site

What happens with longer probes? If the target site is a hairpin loop region, the ends of the probe will probably dangle. But in putatively linear regions, what happens? If portions of the region are protected, do the ends of the probe melt through RNA-RNA base pairs, do they displace loosely bound proteins, or do they dangle?

In an effort to answer these questions, we have analyzed the region containing bases 1393 to 1408 in 16S rRNA by using probes that were longer than the exposed binding site. These results, presented below, show that in this case, the ends appear to dangle. However, this result may not be globally applicable, and additional experiments are necessary.

It should be noted that longer probes also allow increasing numbers of partially complementary sites to be bound. Therefore, it is even more essential that diligent care be taken to ensure the specificity of binding.

ANALYSIS OF rRNA AVAILABILITY

Site Availability

16S rRNA

Available binding sites on 16S rRNA were assayed by using cDNA probes of various lengths. Table 1 indicates the sites that we have analyzed to date and the extent to which each probe bound. A similar study has been made by Lasater et al. (1988), but longer probes were used and the site specificity was not determined.

From our studies, we noted that probe binding to a given site can be variable. To determine the cause of this, we have carefully studied the long, putatively single-stranded region surrounding base C-1400 in 16S rRNA by using overlapping hexameric probes. Recent results have shown that in this region, only bases 1397 to 1404 are fully available to hexameric probes, but the adjacent regions are protected in some manner from probe binding.

Comparison of these results with previous results obtained by using other probes in this region (Table 1) shows that the amount of probe binding is quite variable. Upon further analysis, we found that this generally depends on the ribosomes and how they were prepared. Preliminary evidence has shown that sites which are absolutely unavailable to hexameric probes may become partially available when the ribosomal subunits are salt washed or ethanol precipitated, suggesting the removal or displacement of a protecting ligand or a structural change. Sites that show partial availability may reflect partial removal of a transiently bound factor that is present on the ribosome to protect the labile rRNA. Thus, results obtained from probing fully protected, partially protected, and unprotected functional sites may

Table 1. Probe binding studies on active 30S subunits[a]

cDNA probe to 16S rRNA	% Binding to target[b]	cDNA probe to 16S rRNA	% Binding to target
518–523	15[c]	1397–1402	5[d]
519–524	11[c]	1398–1403	100[d]
520–525	12[c]	1399–1404	100[d]
521–525	28[c]	1399–1406	100[d]
522–527	10[c]	1394–1408	100[d]
523–528	13[c]	1400–1405	15[d]
524–529	6[c]	1401–1406	4[d]
525–530	6[c]	1402–1407	17[d]
526–531	4[c]	1403–1408	5[d]
527–532	13[c]	1404–1409	0[d]
528–533	13[c]	1410–1417	1[d]
517–526	11[c]	1490–1497	3[d]
518–526	30[c]	1495–1503	2[d]
526–533	12[c]	1498–1505	7[d]
787–795	38[c]	1530–1535	10[d]
		1530–1537	10[d]
815–823	20[c]	1532–1537	10[d]
		1533–1538	6[d]
919–924	20[d]	1534–1539	100[d]
923–928	2[d]	1534–1541	85[d]
1384–1389	13[d]	1535–1540	100[d]
1390–1395	11[d]	1536–1542	65[d]
1392–1397	0[d]	1530–1542	100[d]
1393–1398	2[d]		
1394–1399	2[d]	1393–1399	45[c]
1394–1402	0[d]	1394–1401	55[c]
1395–1400	10[d]	1394–1406	70[c]
1396–1401	7[d]	1395–1405	72[c]
		1396–1404	65[c]

[a] Subunit preparation is designated by footnotes *c* and *d*. See Tam and Hill (1981) for full preparative protocol.
[b] Expressed as picomoles of cDNA bound per picomole of subunit in the assay.
[c] Ethanol-precipitated subunits.
[d] Subunits pelleted out of sucrose by centrifugation.

possibly all be representative of various stages of rRNA activity. Indeed, the preparative technique may be chosen to allow studies of the functions of a particular site.

These results stand to reason, especially if one assumes that single-stranded rRNA regions have to be protected in some fashion. Are any additional sites, besides the region of 16S rRNA containing bases 1397 to 1404, fully exposed? To date, we have found only one additional site that gives stoichiometric (1:1 ratio of probe to subunit) binding consistently. This is found in the region containing bases 1534 to 1542 on 16S rRNA. We expect to find additional regions with similar availability as our studies progress.

The approximate size of the window of availability in the 1397–1404 region is slightly over 20 Å. These results may suggest that the simple method of providing a limited window of rRNA exposure is used as a method of protecting rRNA from RNase while permitting it to be functionally active.

It is also important to ask whether these probes can displace a ligand that may be protecting the bases adjacent to the exposed region. We analyzed this question by using probes to the 1393–1408 region of 16S rRNA that were longer than the exposed binding site (nucleotides 1397 to 1404). With a 15-mer probe, complementary to bases 1394 to 1408, we found binding to be 60%. A probe altered to create a single-base mismatch in the center of the binding region (at C-1400) caused binding to drop to 17%, whereas a single-base mismatch at position 1406, outside of the hexameric binding region, caused no decrease in the binding of the longer probe. This suggests that the ends do not react with complementary regions of rRNA, at least in this case. From this, it appears that these ends do not displace adjacent protecting ligands, and we can presume that the ends must dangle.

23S rRNA

Some probing of 23S rRNA in situ has also been done in our laboratory. Table 2 shows the cumulative results of probe studies on 23S rRNA. Once again, the specificity of each target region has been verified by using RNase H digestion analysis.

One of the most interesting regions is the so-called α-sarcin region (nucleotides 2654 to 2667). Under standard assay conditions, this region is inac-

Table 2. Probe binding studies on 50S subunits[a]

cDNA probe to 23S rRNA	% Binding to target[b]	cDNA probe to 23S rRNA	% Binding to target
801–806	49[c]	2455–2461	2[d]
802–807	56[c]	2461–2466	125[d]
803–808	58[c]	2467–2476	13[d]
803–811	70[c]	2468–2482	25[d]
804–809	37[c]	2469–2481	23[d]
805–810	48[c]	2472–2481	23[d]
806–811	62[c]	2489–2496	3[d]
		2492–2505	18[d]
2049–2057	16[d]	2497–2501	20[d]
2058–2062	11[d]	2497–2505	32[d]
		2500–2505	33[d]
2306–2313	50[c]		
		2589–2594	7[d]
2439–2447	26[d]	2607–2614	55[c]
2448–2454	21[d]	2654–2667	0[c]
2450–2455	16[d]	2653–2658	0[e]
		2656–2667	0[e]
		2659–2664	8[e]
		2662–2667	6[e]
		2750–2758	25[c]

[a–d] See Table 1, footnotes *a* to *d*.
[e] Ethanol-precipitated subunits subsequently salt washed in 0.5 M NH_4Cl.

Table 3. Functional assays using 30S subunits[a]

cDNA to 16S rRNA	% Binding of cDNA or tRNA[b]	Competitor added	% cDNA or tRNA subsequently bound
518–526	30[c]	$tRNA^{Phe}$	33
		Poly(U)	15
		$tRNA^{Phe}$-poly(U)	15
517–528	46[d]	Poly(AGU)	38
		MS2 mRNA	50
		Poly(AGU)-$tRNA^{fMet}$	37
		MS2-$tRNA^{fMet}$	44
	12[e]	Poly(AGU)	13
		MS2 mRNA	15
		Poly(AGU)-$tRNA^{fMet}$	13
		MS2-$tRNA^{fMet}$	13
526–533	12[c]	$tRNA^{Phe}$	13
		Poly(U)	3
		$tRNA^{Phe}$-poly(U)	4
787–795	38[c]	cDNA in subunit association reaction	Inhibited
		cDNA in protein biosynthesis reaction	Inhibited 23%
815–823	20[c]	Subunit association	20
		Protein biosynthesis	20
1393–1398	0[c,e]	Poly(U)	0
		$tRNA^{Phe}$	0
		Poly(U)-$tRNA^{Phe}$	0
1393–1399	45[c]	Displacements	
	(41)[e]	cDNA by $tRNA^{Phe}$[f]	41
	50[e]	$tRNA^{Phe}$ by cDNA[f]	50
	53[e]	cDNA[f] by $tRNA^{Phe}$	50
	(41)[e]	$tRNA^{Phe}$[f] by cDNA	40
1393–1401	45[c]	Displacements	
	(43)[e]	cDNA by $tRNA^{Phe}$[f]	41
		$tRNA^{Phe}$ by cDNA[f]	34
	38[e]	cDNA[f] by $tRNA^{Phe}$	25
		$tRNA^{Phe}$ by cDNA	43
1394–1406	74[c]	Displacements	
	(43)[e]	cDNA by $tRNA^{Phe}$	30
	74[c]	$tRNA^{Phe}$ by cDNA[f]	72
	(45)[e]	$tRNA^{Phe}$[f] by cDNA	30
	63[e]	cDNA[f] by $tRNA^{Phe}$	60
1394–1408	100[d]	Poly(U)	100
		MS2 mRNA	100
		$tRNA^{Phe}$	100
1395–1405	72[c]	Displacements	
		$tRNA^{Phe}$ by cDNA[f]	69
	(45)[e]	cDNA by $tRNA^{Phe}$[f]	28
1396–1401	0[d,e]	Poly(U)	0
		$tRNA^{Phe}$	0
		Poly(U)-$tRNA^{Phe}$	0
1398–1403	100[d]	Poly(U)	100
		MS2 mRNA	100
		$tRNA^{Phe}$	100
		$tRNA^{fMet}$	100
		Poly(U)-$tRNA^{Phe}$	20
		Displacements	
		cDNA[f] by $tRNA^{Phe}$	100
	(85)[d]	$tRNA^{Phe}$ by cDNA	85
1401–1406	47[d,e]	Displacements	
		cDNA by $tRNA^{Phe}$[f]	46
	60[e]	$tRNA^{Phe}$ by cDNA[f]	55
	54[e]	cDNA[f] by $tRNA^{Phe}$	49
	(42)[e]	$tRNA^{Phe}$[f] by cDNA	40

Continued on following page

Table 3.—*Continued.*

cDNA to 16S rRNA	% Binding of cDNA or tRNA[b]	Competitor added	% cDNA or tRNA subsequently bound
1402–1407	0[d,e]	Poly(U)	0
		$tRNA^{Phe}$	0
		Poly(U)-$tRNA^{Phe}$	0
1530–1542	100[d]	Poly(U)	100
		MS2 mRNA	100
1530–1535	0[d,e]	Poly(U)	0
		$tRNA^{Phe}$	0
		Poly(U)-$tRNA^{Phe}$	0
1532–1537	13[d]	Poly(U)	0
		$tRNA^{Phe}$	0
		Poly(U)-$tRNA^{Phe}$	0
1534–1541	85[d,e]	Poly(U)	85
		$tRNA^{Phe}$	85
		Poly(U)-$tRNA^{Phe}$	45
		MS2 mRNA	37
		$tRNA^{fMet}$	85
		MS2-$tRNA^{fMet}$	31
		Displacements	
	76[d]	cDNA[f] by $tRNA^{Phe}$	42
		$tRNA^{Phe}$[f] by cDNA	76

[a–d] See Table 1, footnotes *a* to *d*. A value in parentheses indicates a basal tRNA binding level to which the experimental value is compared. In the displacement reactions, poly(U) is always present in saturating amounts.
[e] Assay performed on tight-couple 70S ribosomes.
[f] Isotopically labeled species.

cessible to probes of variable size complementary to any portion of the region, yet it is readily cleaved with α-sarcin (Endo et al., 1983). However, if during an in vitro translation reaction probes complementary to portions of this region and RNase H are added, an rRNA fragment corresponding in size to that produced by α-sarcin is produced. These results suggest transient opening of this region.

One of the more exciting regions that we have studied is the so-called peptidyltransferase region in domain V (nucleotides 2448 to 2454 and 2497 to 2505). We have probed portions of this region, obtaining variable binding stoichiometries (Table 2). It appears that no portion of the region is as accessible to hexameric probes as were the two regions in 16S rRNA discussed above, but that portions of this region have at least partial availability to DNA oligomers.

Probes to the 23S rRNA region containing bases 801 to 811, although not binding stoichiometrically, also show significant binding. This region was selected for study since it was putatively a tRNA-binding site (see below). The results of overlapping hexamers in this region did not show any completely available regions under the salt wash conditions for the ribosomes.

FUNCTIONAL SITES OF rRNA

A straightforward method to determine the function of a specific site is to bind labeled probe to the site, then add a ligand such as tRNA or mRNA, and determine whether the probe is displaced. Alternatively, labeled ligands such as tRNA or mRNA can be bound and then the probe is added to see whether the tRNA or mRNA can be displaced.

16S rRNA Sites

Table 3 shows the cumulative results of studies on many 16S rRNA sites in the 30S ribosomal subunits. (Only those probes whose binding specificity has been verified by using RNase H are noted throughout.) From these results, certain conclusions can be drawn. Starting with the 3′ end of 16S rRNA, bases 1542 to 1537 are all open, both in the active and the inactive subunit conformation. This region contains the anti-Shine-Dalgarno site. Competition experiments with mRNA clearly show competition between probes to the anti-Shine-Dalgarno region and mRNA. Poly(U) does not compete, but poly(U)-directed $tRNA^{Phe}$ does compete with a probe complementary to bases 1534 to 1541.

We have found cDNA binding in the region containing bases 1492 to 1507, but the binding is strongly dependent on the method of ribosome preparation. We observed 20 to 30% binding when we used salt-washed or ethanol-precipitated 30S subunits but obtained no binding with unwashed subunits pelleted directly from sucrose. Preliminary results suggest that this region may be involved in IF3-mRNA interaction.

The region containing bases 1393 to 1407 has been extensively studied. This so-called decoding region is the longest single-stranded region in 16S rRNA and is also highly conserved. We have analyzed the complete region, using overlapping hexameric probes displaced by a single nucleotide. As noted above, the region containing nucleotides 1398 to 1404 is available to cDNA oligomers, both in the active and in the inactive form, and adjacent regions are totally unavailable. Poly(U)-directed $tRNA^{Phe}$ competes with those probes that cross C-1400. This finding was confirmed both with salt-washed and sucrose-pelleted subunits.

The region containing bases 787 to 795 was shown some time ago to be involved in subunit association (Tapprich and Hill, 1986), but the availability of this region varies according to the preparative method. The very highly conserved loop region containing bases 518 to 535 has been difficult to probe, indicating possible protecting groups or tertiary interactions. However, it is clear that the region containing bases 518 to 525 is more available for probe binding than is the rest of the region and that probes to this region may compete with poly(U), lending support to the proposal of Trifonov (1987) that this may be an mRNA interaction site.

We have also attempted to probe some regions that were inaccessible to probe binding under all conditions tried. Table 3 summarizes the data, showing relative binding of all regions. It should be emphasized that the fact that a region is not available to probe binding does not preclude it from being active in ribosome function. Transient binding of protecting groups is to be expected, but conditions for specific, gentle removal of these groups have yet to be determined.

23S rRNA Sites

Table 4 shows the cumulative results of probe studies on 23S rRNA. The region containing bases 807 to 809 has been suggested to be a possible binding site for the 3′ terminus of tRNA (Barta et al., 1984). We have shown that only probes having some complementarity with bases 807 to 809 show such competition. In addition, we have found that tRNA lacking the three or four 3′-terminal nucleotides can comfortably coexist with probes to the region containing bases 807 to 809, whereas tRNA that contains its full complement of nucleotides at the 3′ terminus cannot. These results suggest an interaction of the 3′ terminus of tRNA with this region of 23S rRNA. Unfortunately, this site is one of those for which we are unable to get good RNase H specificity results in situ.

One of the more promising regions of study is the so-called peptidyltransferase region mentioned above. Although binding does not occur stoichiometrically, tRNA competes favorably with probes to bases 2497 to 2505 (Marconi and Hill, 1989). In addition, some competition with chloramphenicol has been noted at the same region.

IN VITRO TRANSLATION STUDIES

We proposed that rRNA may be sterically protected from nonligands such as RNases, only exposing them during the period(s) in which they are used. This kind of switching mechanism has been postulated before, but hard data have been lacking.

To determine whether sites are transiently exposed during the course of translation, we have added DNA oligomers complementary to target regions of RNA that were otherwise unavailable. Adding RNase H to the translation mixture would, we hoped, allow cleavage of the target site to occur if the site were exposed during translation.

This experiment has been performed on the α-sarcin site on the 23S rRNA (bases 2654 to 2667). We have previously shown that this entire region is unavailable to probe binding using any size of probe, whether the ribosomes are salt washed, ethanol precipitated, or not (White et al., 1988). However, by adding (i) a 14-mer probe complementary to bases 2654 to 2667 and (ii) RNase H to the translation mixture, cleavage of 23S rRNA did occur at the α-sarcin site. It appears that the α-sarcin region is well protected when the ribosome is not translating, but at some stage(s) of translation, this site is at least partially exposed. At this stage only, the probe and RNase H can interact, causing cleavage of the rRNA. Similar results have emanated from the studies of Wool and his group in rabbit reticulocyte ribosomes (Wool et al., this volume). They also used antibiotics to show that such sensitivity to probes and RNase H occurs only when the peptidyl-tRNA is in the A sites before translocation.

Preliminary results on other nonavailable yet putatively functional sites have shown this to be true elsewhere as well. Such functional tests can be performed on all sites having probable function, even though the sites may not be available on the isolated subunits. By causing protein synthesis to pause at various stages, such as pre- or posttranslocation, this approach may reveal the exact time of exposure and function for these otherwise hidden regions of rRNA.

TOPOGRAPHY OF SITES

The detailed structure of the ribosome is still unknown. In recent years, the relative positions of the

Table 4. Functional assays using 50S subunits[a]

cDNA to 23S rRNA	% Binding of cDNA or tRNA[b]	Competitor added	% cDNA or tRNA subsequently bound
801–806	49[c]	Displacements	
	(45)	cDNA by tRNA$^{Phe\ d}$	45
		tRNA^{Phe} by cDNA[d]	49
802–807	56[c]	Displacements	
	(45)	cDNA by tRNA$^{Phe\ d}$	39
		tRNA^{Phe} by cDNA[d]	45
803–808	58[c]	Displacements	
	(45)	cDNA by tRNA$^{Phe\ d}$	37
		tRNA^{Phe} by cDNA[d]	44
803–811	90[c]	Phe-tRNA$^{Phe\ d}$ vs cDNA[d]	NC[e]
		Displacements	
	(45)[c]	cDNA by tRNA$^{Phe\ d}$	25
		tRNA^{Phe} by cDNA[d]	80
	(30)[c]	tRNA^{Phe}-CACCA[d] by cDNA	28
		tRNA^{Phe}-CACCA by cDNA[d]	90
804–809	37[c]	Displacements	
	(45)	cDNA by tRNA$^{Phe\ d}$	35
		tRNA^{Phe} by cDNA[d]	15
805–810	48[c]	Displacements	
	(45)	cDNA by tRNA$^{Phe\ d}$	36
		tRNA^{Phe} by cDNA[d]	24
806–811	61[c]	Displacements	
	(45)	cDNA by tRNA$^{Phe\ d}$	36
		tRNA^{Phe} by cDNA[d]	31
2306–2313	50[f]	cDNA in protein biosynthesis reaction	46
2448–2454	21[g]	tRNA^{Phe}	21
		Poly(U)	21
		tRNA^{Phe}-poly(U)	19
		Phe-tRNA^{Phe}	21
		Chloramphenicol	21
		Erythromycin	21
		Puromycin	21
2468–2482	25[g]	tRNA^{Phe}	25
		Poly(U)	2
		tRNA^{Phe}-poly(U)	5
		Phe-tRNA^{Phe}	25
		Chloramphenicol	28
		Erythromycin	25
		Puromycin	25
2497–2505	32[g]	tRNA^{Phe}	11
		Poly(U)	32
		tRNA^{Phe}-poly(U)	10
		Phe-tRNA^{Phe}	32
		Chloramphenicol	11
		Erythromycin	32
		Puromycin	21
2607–2614	55[f]	cDNA in protein biosynthesis reaction	NC
		tRNA^{Phe}	55
2654–2667	0[f]	tRNA^{Phe}	6
		Poly(U)	0
		tRNA^{Phe}-poly(U)	0
2750–2758	25[f]	cDNA in subunit association reaction	Inhibited
2606–2613	55[c]	Displacements	
	(40)[c]	cDNA by tRNA$^{Phe\ d}$	34
		tRNA^{Phe} by cDNA[d]	55

[a,b] See Table 1, footnotes *a* and *b*. A value in parentheses indicates a basal tRNA binding level to which the experimental value is compared. In the displacement reactions, poly(U) is always present in saturating amounts.
[c] Assay performed on tight-couple 70S ribosomes.
[d] Isotopically labeled species.
[e] NC, No change.
[f,g] See Table 1, footnotes *c* and *d*.

IV. INITIATION

IV. INITIATION

The translation of mRNA by ribosomes is customarily divided into three phases: initiation, elongation, and termination. Although the ribosome does not necessarily divide its work in this way, it is useful for those who study ribosome function. In this section, the process of initiation is discussed for both procaryotic and eucaryotic systems. A number of components in addition to ribosomes are involved in this process: mRNA and tRNA as well as initiation factors. The discrete roles of these factors are now being elucidated and the complexity of regulation at this step of translation is beginning to be understood.

Chapter 19

Aspects of Translation Initiation in *Escherichia coli*

P. H. VAN KNIPPENBERG

The initiation reaction of protein biosynthesis plays a major role in gene expression. It requires the assembly of a complex between a ribosome, an mRNA, and an aminoacylated initiator tRNA. The formation of this complex, which largely determines the rate of translation, differs in detail between eucaryotic cytoplasmic systems and procaryotes. I shall mainly discuss certain aspects of translation initiation in *Escherichia coli*. Several excellent reviews on initiation of protein biosynthesis have been written in recent years (Kozak, 1983; Maitra et al., 1982; Gualerzi et al., 1986; Gold and Stormo, 1987; Gold, 1988). Many of the basic facts of protein biosynthesis have been summarized in these reviews; where data and their interpretation are not controversial, I shall give them without specific reference. Those points, however, on which consensus has not been reached, or with respect to which my interpretation deviates from that of others, will be documented as much as possible with reference to the original literature.

BASIC FACTS OF TRANSLATION INITIATION IN *E. COLI*

Initiation of translation starts with the formation of a complex between the small ribosomal subunit (30S), mRNA, and fMet-$tRNA_f^{Met}$. These at least are the minimal macromolecular ingredients for initiation complex formation. In vitro, the mRNA can be mimicked by synthetic polynucleotides and the fMet-tRNA by an N-acylated aminoacyl-tRNA (and even by aminoacyl-tRNA at high Mg^{2+} concentrations).

At least three proteins, transiently associated with ribosomes (and called initiation factors IF1, IF2, and IF3), are required for efficient translation initiation on natural mRNA in vitro (reviewed in Maitra et al., 1982). Of these, only for IF2 has there been a consensus of opinion on the mode of action. This factor, complexed with GTP, binds to initiator tRNA (either fMet-$tRNA_f^{Met}$ or any N-acylated aminoacyl-tRNA) on the 30S ribosome. Binding to the natural initiator RNA, fMet-$tRNA_f^{Met}$, is preferred, however (Van der Laken et al., 1980). The free complex IF2 · GTP · fMet-tRNA can be shown to exist in the absence of ribosomes after fixation with glutaraldehyde (Van der Hofstad et al., 1977).

The function of factor IF3 has been, and probably still is, a matter of controversy. The protein is needed for the dissociation of 70S ribosomes into active subunits, and this may explain its necessity for the efficient translation of natural and synthetic mRNA in vitro (Sobura et al., 1977; Gualerzi et al., 1986; Gold, 1988). However, 70S dissociation is certainly not the only function of the protein (Gualerzi et al., 1986; Gold, 1988). Because of the various views on the mode of action of IF3, I shall discuss it in more detail below.

IF1 is a small protein that is transiently associated with the 70S ribosome and with the 30S subunit. It appears to accelerate IF3-dependent 70S dissociation and also increases the overall rate of the initiation reaction.

There is a general agreement that the three initiation factors have a clustered binding site on the 30S ribosome. They are located at the 50S site of the particle in the neighborhood of proteins S7, S11, S13, S18, S19, and S21 and the 3′ end of 16S rRNA (reviewed in Moore and Cashel, 1988).

The site where the initiation complex is formed on mRNA is protected by the ribosome against nuclease degradation and has been termed the ribosome-binding site (RBS). Hundreds of such sites for initiation of translation have now been sequenced,

P. H. Van Knippenberg ■ Department of Biochemistry, Gorlaeus Laboratories, Leiden University, Einsteinweg 5, 2333 CA Leiden, The Netherlands.

and more than 90% of them contain the classical AUG initiation triplet in phase with the reading frames of the respective genes (Gren, 1984). The remainder contain GUG or UUG, with a very few examples of AUA and AUU. In eucaryotes and in certain mRNA mutants in *E. coli*, ACG may also be used as initiation codon (summarized in Peabody, 1989).

It is probable that the list of in vivo start codons will be extended when more sequences become available. This would reconcile in vivo initiation events with much older data on the stimulation of fMet-tRNA binding to ribosomes by RNA triplets in vitro (Kellog et al., 1966; Ghosh et al., 1967). The role of the start codon in the initiation reaction will be discussed in more detail below (see also Van der Laken et al., 1980; Berkhout et al., 1986).

Most RBSs contain a polypurine sequence, the Shine-Dalgarno (SD) sequence or SD signal (Shine and Dalgarno, 1974), 5′ to the initiation codon that is complementary to nucleotides at the 3′ end of 16S rRNA. The uninterrupted complementarity varies from nine to only three nucleotides (including G · U pairs), and a few RBSs have been reported that do not seem to have an SD sequence at all (Waltz et al., 1976; Van Gemen et al., 1987). Nevertheless, selection of translation initiation sites in vivo clearly depends strongly on this base pair interaction (Hui and De Boer, 1987).

Statistical analysis of hundreds of RBSs has shown that nucleotide distribution in the RBS, apart from the SD signal and the initiation codon, is far from random (Scherer et al., 1980; Stormo et al., 1982; Gren, 1984; Schneider et al., 1986). This may have various reasons. The nonrandomness may be related to the avoidance of secondary structure involving the initiation signals (Looman et al., 1987; Gold, 1988; see below) or may have been selected because of additional interactions with 16S rRNA (Van Knippenberg, 1975; Petersen et al., 1988; Thanaray and Pandit, 1989). Possible interactions involving the 5′ end of 16S rRNA will be discussed below.

Numerous reports describe the effect of secondary structure affecting the RBS on the efficiency of translation (Iserentant and Fiers, 1980; Hall et al., 1982; Kastelein et al., 1983; Munson et al., 1984; Schottel et al., 1984; Tessier et al., 1984; Wood et al., 1984; Buell et al., 1985; Stanssens et al., 1985; Looman et al., 1986; McPheeters et al., 1986; Schoner et al., 1986; Shinedling et al., 1987; Preibish et al., 1988; De Smit and Van Duin, 1990; Spanjaard et al., 1989; Tomich et al., 1989). There is no doubt that shielding the important elements of the RBS (SD and initiation codon) in a secondary structure strongly reduces translation. One way of regulating gene expression exploits such secondary structures (Gryczan et al., 1980; Horinouchi and Weisblum, 1980; Schmidt et al., 1987). Secondary structure effects will also be discussed in more detail below.

Once a ternary complex between 30S, mRNA, and fMet-tRNA (a so-called 30S initiation complex, probably still containing the three initiation factors and GTP; see Maitra et al., 1982) has been formed, it is quickly joined by a 50S particle, thus giving rise to a 70S initiation complex. The initiation factors are released during the joining of the 50S, and GTP is hydrolyzed to GDP and P_i (Maitra et al., 1982). The 70S initiation complex is much more stable than the 30S complex and can readily be isolated.

IF3 AND THE INITIATION CODON

There has been a long-standing controversy between two lines of thought regarding the sequence of events in the formation of the 30S initiation complex. One school has advocated the IF3-dependent primary binding of mRNA to the 30S subunit (Vermeer et al., 1973a; Vermeer et al., 1973b), whereas other workers favored the view that fMet-tRNA is bound first to the particle, followed by the association of mRNA (Jay and Kaempfer, 1974). Gualerzi and co-workers (reviewed in Gualerzi et al., 1986) have presented evidence that would reconcile these different views: there would be no preference for the order of steps. The complex could be assembled by first binding either the mRNA or the fMet-tRNA. A weak ternary complex, fMet-tRNA · 30S · mRNA, the rapid formation of which requires the presence of IF3 according to Gualerzi et al. (1979), would be transformed into a stable complex upon interaction of the initiator tRNA with its cognate codon, AUG. The primary function of IF3, in this concept, is on the one hand to accelerate the formation of the stable initiation complex and on the other hand to discriminate against "artificial" complexes, such as between *N*-acetyl-Phe-tRNA and poly(U)-programmed 30S. Apart from the fact that the latter complexes play no role in vivo, factor IF2 is apparently also involved in discriminating in favor of fMet-tRNA binding (Van der Laken et al., 1980; see below).

The scheme of Gualerzi et al. does not account for the stable binding of fMet-tRNA to 30S ribosomes programmed with poly(U) (Van der Laken et al., 1979). The fMet moiety is incorporated into $fMet(Phe)_n$ when other ingredients for protein synthesis are added to this system (Van der Laken et al., 1979). However, IF3 strongly antagonizes the binding of fMet-tRNA in this system, suggesting that the major role of this factor is to safeguard against initiation with fMet-tRNA in response to a noncognate codon (Berkhout et al., 1986). It has indeed been shown that binding of natural mRNA (bacteriophage RNA) to 30S particles, and subsequent (initiation of) protein synthesis in vitro, can be achieved in the complete absence of IF3, provided that the 50S subunit is introduced after 30S-mRNA complex formation (Van Duin et al., 1980). In contrast, ribosomal protein S1 is an essential component for this binding.

Since IF3 acts also as a dissociation factor of 70S ribosomes (or as an antiassociation factor of ribosomal subunits) in some studies involving total ribosomes, it is not clear which activity of the factor has been measured (Sobura et al., 1977; Jay et al., 1980). Most evidence, in my view, points to a role of this factor in editing the initiation event. Lack of such proofreading may even account for the autoregulation of the synthesis of IF3 itself (Gold et al., 1984; Berkhout et al., 1986; Butler et al., 1986; Butler et al., 1987).

IS THE 5′ END OF 16S rRNA INVOLVED IN RBS RECOGNITION?

In 1975, I proposed a model suggesting base pairing between the 5′ end of 16S rRNA and initiation sites on phage RNA (Van Knippenberg, 1975). This interaction either would replace the SD interaction or would come in addition to it. Unfortunately, there was no experimental support for this model, and it soon became clear that it would not hold for other (bacterial) mRNAs. My curiosity was aroused again recently by a paper by other workers (Petersen et al., 1988) suggesting the same type of interaction (again in addition to the SD interaction; Fig. 1), although less extensive, but now applicable to all *E. coli* initiation sites. I do not know whether this proposal withstands statistical testing, but it is interesting to see whether one of the requirements of this model, i.e., the proximity of the 3′ and 5′ ends of 16S rRNA, is fulfilled. Many years ago I advocated the interaction between the 5′ and 3′ ends of 16S rRNA, as shown in Fig. 2A (unpublished data). This structure contains a so-called pseudoknot (Pleij et al., 1985). There are at least two reasons that make this configuration highly unlikely. First, in eubacteria other than *E. coli*, the pseudoknot base pairing should be extended to 6 base pairs because of an A→C change (Fig. 2A). However, bridging of a six-membered pseudoknot over the narrow groove of the helix by one nucleotide (U) is sterically impossible (Pleij et al., 1985). Second, an essential part of this interaction is involved in a different pseudoknot,

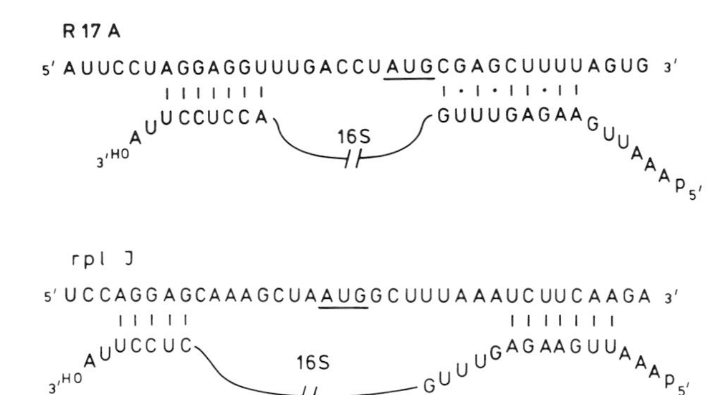

Figure 1. Models for interactions between initiation sites on *E. coli* mRNA and 16S rRNA. Other examples proposing an involvement of 5′ nucleotides of 16S rRNA have been described by Van Knippenberg (1975) and Petersen et al. (1988).

forming the heart of the 16S rRNA model (Stern et al., 1988). The latter interaction is shown for comparison in Fig. 2B.

An interaction between the 5′- and 3′-end regions that would probably not violate any of the existing data is shown in Fig. 2C. It uses the bridging nucleotides 5′-UGA of the 16S rRNA pseudoknot (Pleij et al., 1985; Stern et al., 1988) to base pair with the conserved sequence 5′-UCA just preceding the "SD nucleotides" at the 3′ end. It is interesting that in the inactive form of the 30S subunit, the UCA sequence near the 3′ end appears to be base paired to a UGA sequence that is only a few nucleotides away from the pseudoknot helix (Ericson and Wollenzien, 1989). A conformational switch releases this interaction and alters the inactive form of the 30S into an active form.

It remains interesting to see whether the 5′ end of 16S rRNA plays any role in initiation of translation.

EFFECTS OF SECONDARY STRUCTURE IN THE RBS

The reports on the (negative) effects of secondary structure sequestering the SD signal, initiation codon, or both are now so numerous (see above) that this matter can be considered settled. There is a strict correlation between the stability of a local secondary structure in the RBS and the level of gene expression (De Smit and Van Duin, 1990).

The overriding effect of secondary structure over all other features in an RBS is demonstrated by the data shown in Fig. 3 (Looman et al., 1986). A system, based on *lacZ* expression, was developed to test the efficiency of RBSs disconnected from the original mRNAs (Looman et al., 1985). The RBSs selected for their effect on expression contained two of such sites

Translational coupling

a. No coupling. Independent initiation

A — UAA ——//—— AUG — B

b. Coupling by removing secondary structure

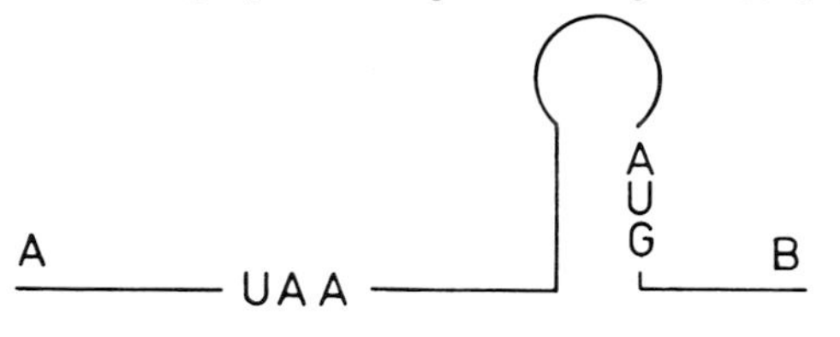

c. Coupling by forward scanning

A —— UAA—AUG —— B

d. Coupling by backward scanning

– – – – – A – – – – –|

—— AUG — UAA ——

|– – – – – B – – – – –

Figure 4. Various situations related to translational coupling. The two extreme cases of coupling discussed in the text are represented in panel b and in panels c and d.

distance between the stop codon of gene A and the start codon of gene B is large, yet translation of B is clearly dependent on translation of gene A (Fig. 4b). In this case, translational coupling probably is due to the prevention of secondary structure in the gene B initiation site by ribosomes that are translating cistron A. This situation is encountered in the L10 operon of *E. coli* (Yates et al., 1981; Nomura et al., 1984) and in the classical example of phage MS2 coat protein and replicase synthesis (Berkhout and Van Duin, 1985). (ii) The termination codon of gene A is very close to (or overlapping with) the initiation codon of gene B (Fig. 4c and d). It is very likely that in this situation the 30S ribosome does not disengage from the RNA but binds fMet-tRNA and starts a new elongation cycle from the AUG start codon. Clear examples of this mechanism have been described, such as for genes of the *trp* operon (Yanofsky and Crawford, 1987).

Intermediates between the extreme situations presented above may be encountered. It is likely that the 30S subunit, after release of the peptidyl-tRNA and the 50S (Martin and Webster, 1975), spends a while scanning the mRNA in the vicinity of the termination codon. The scanning may proceed randomly in both directions (M. R. Adhin and J. Van Duin, personal communication). The probability of a new initiation reaction may depend on a number of factors; e.g., the sticking time of the particle to the mRNA may be affected by the presence of an SD-like signal (or another sequence that can base pair with 16S rRNA; see above) and by the tendency of the mRNA to fold into a local secondary structure. The availability of IF2, fMet-tRNA, and IF3 may also be of importance for the chance that a new functional initiation complex will be formed. In agreement with the scenarios described before (Gold, 1988), I believe that the presence of IF3 is especially important to ensure that no false starts (i.e., those using non-AUG codons) are being made and that fMet-tRNA is guided to the correct initiation codon. However, if the necessary ingredients are not soon available or if the 30S has to wander along the mRNA for a prolonged period of time before the correct situation develops (i.e., the formation of an initiation complex), the chances will increase that the particle dissociates from the mRNA. Meanwhile, however, the terminating ribosome, or the scanning 30S particle, may have opened up a more distant site for independent initiation by free 30S subunits. There may hence be, depending on the exact conditions, an overlap between the two mechanisms of translational coupling.

IF2 AND IF3 JOINTLY GUARD AGAINST FALSE STARTS

It is obvious that certain reactions of polypeptide synthesis as they may occur in the test tube must be excluded in the cell. Translation of poly(U) at high Mg^{2+} concentrations can start with factor-independent entry of Phe-$tRNA^{Phe}$ in the ribosomal P site. If this type of initiation were to happen in the cell, it would create severe problems, as was pointed out several years ago by Gualerzi et al. (1971), who indeed suggested that IF3 might prevent such events from occurring. However, at physiological Mg^{2+} concentrations this reaction scarcely takes place, and it is very likely also that aminoacyl-tRNA in the cell is mostly present in a complex with EF-TU · GTP.

Translation of poly(U) at low Mg^{2+} concentrations takes place in the presence of *N*-acetyl-Phe-$tRNA^{Phe}$, provided that initiation factor IF2 is present. The binding of *N*-acetyl-Phe-$tRNA^{Phe}$ to poly(U)-programmed 30S ribosomes is strictly dependent on IF2 at low Mg^{2+} concentrations (Gualerzi et al., 1986). With a natural mRNA template (alfalfa mosaic virus RNA4), Castel et al. (1977) showed that an *E. coli* extract can start reading the mRNA at the second triplet (UUU) following the 5′ m^7G cap by using *N*-acetyl-Phe-$tRNA^{Phe}$ as an initiator. When fMet-tRNA was present, initiation took place at the AUG codon of the coat cistron, 12 triplets downstream from the cap. The leader is an extremely U+A-rich region, and in neither case is the initiation codon preceded by an SD signal. The initiation at both sites is stimulated by IF1, IF2, and IF3 in specific and similar ways (Castel et al., 1979). This clearly illustrates the potential of ribosomes to start with an N-blocked aminoacyl-tRNA at any place along an mRNA (as long as it is single stranded). It clarifies the mechanism of translational enhancement of the so-called Ω (5′ leader) sequences in alfalfa mosaic virus and other plant virus RNAs. They act in *cis* to enhance translation of foreign eucaryotic or procaryotic mRNAs both in vitro and in vivo and are functionally equivalent to an SD sequence (Sleat et al., 1988; Gallie and Kado, 1989).

The studies by Van der Laken et al. (1979, 1980) and by Berkhout et al. (1986) are especially relevant to the role of IF2 and IF3 in ensuring correct initiation. They showed that poly(U) can trigger binding of fMet-tRNA to 30S ribosomes, in addition to the well-established *N*-acetyl-Phe-tRNA binding. Both bindings are strictly dependent on IF2, although this factor has a preference for initiation with fMet-tRNA (Van der Laken et al., 1980). When IF3 is added to this system, it strongly inhibits the *N*-acetyl-Phe-tRNA binding (noncognate with respect to IF2) as well as the fMet-tRNA binding (noncognate with respect to the codon UUU). This suggests that 30S ribosomes fully equipped with IF2 and IF3 are only able to start translation with fMet-tRNA responding to AUG.

The results briefly reviewed above (Van der Laken et al., 1979; Van der Laken et al., 1980; Berkhout et al., 1986) suggest another pitfall for initiation of protein synthesis in the cell and a role for IF2 and IF3. It has been known for a long time that peptidyl-tRNA, probably arising from abortive elongations (Menninger, 1976), is lethal to the cell (Menninger, 1979). A peptidyl-tRNA hydrolase prevents the accumulation of peptidyl-tRNA (Atherly and Menninger, 1972). Why would peptidyl-tRNA be so toxic? I suggest that the peptidyl-tRNA behaves in the cell as a N-blocked aminoacyl-tRNA in the test tube and that it can potentially enter polypeptide synthesis again through the initiation reaction. At a high rate of protein synthesis, the amount of accumulating peptidyl-tRNA may become so high (Menninger, 1976), despite the presence of peptidyl-tRNA hydrolase, that faithful translation initiation is seriously threatened. The combined action of IF2 and IF3, however, would still guarantee proper initiation.

CONCLUSIONS

The ribosome is designed for the translation of mRNA. In vitro work has clearly shown that ribosomes can start translation in the absence of additional factors and using artificial initiator tRNA. The only requirement for the mRNA is that it be single stranded. Apparently, the initiating 30S can bind only to single-stranded RNA. Once it has done so, it will automatically bind (IF2-dependent) initiator tRNA. The only requirement for the latter is an N-blocked aminoacyl moiety, although fMet-tRNA is preferred.

The necessity for single strandedness of mRNA for initiation in vivo has been documented by numerous genetic studies. It is very likely that nucleotides on the RBS, other than the SD signal and the initiation codon, have been selected for structural reasons. The high preference for A residues in these regions is probably due to the fact that secondary structure in these sites must be avoided on the average.

The SD nucleotides may play a role in determining the "sticking" time of the 30S, important for the chance of a successful initiation. It is not excluded that interactions of the RBS with other parts of 16S rRNA (e.g., the 5′ end) also occur.

The role of initiation factors IF2 and IF3 might be to ensure correct initiation, i.e., using fMet-tRNA to start at the proper (mostly AUG) initiation codon.

The combined effects of the secondary structure of the RBS, the sticking time determined by mRNA-rRNA interactions, and the initiation codon determine the efficiency of initiation.

Although I am solely responsible for the views expressed in this chapter, many of the ideas were shaped by discussions with others in the Department of Biochemistry. I especially acknowledge the discussions with L. Bosch, J. Van Duin, B. Kraal, C. Pleij, and H. A. De Boer and their comments on the manuscript. Marianne Van der Ploeg and Tineke Van der Meer are gratefully acknowledged for typing the manuscript, and L. Welmers and J. J. Pot are thanked for the artwork.

REFERENCES

Atherly, A. G., and J. R. Menninger. 1972. Mutant *E. coli* strain with temperature sensitive peptidyl transfer RNA hydrolase. *Nature* (London) *New Biol.* **240:**245–246.

Berkhout, B., C. J. Van der Laken, and P. H. Van Knippenberg. 1986. Formylmethionyl-tRNA binding to 30S ribosomes programmed with homopolynucleotides and the effect of translational initiation factor 3. *Biochim. Biophys. Acta* **866**:144–153.

Berkhout, B., and J. Van Duin. 1985. Mechanism of translational coupling between coat protein and replicase genes of RNA bacteriophage MS_2. *Nucleic Acids Res.* **13**:6955–6967.

Buell, G., M.-F. Schulz, G. Selzer, A. Chollet, N. R. Movva, D. Semon, S. Escanez, and E. Kawashima. 1985. Optimizing the expression in E. coli of a synthetic gene encoding somatomedin-C (IGF-I). *Nucleic Acids Res.* **12**:1923–1938.

Butler, J. S., M. Springer, J. Dondon, M. Graffe, and M. Grunberg-Manago. 1986. Escherichia coli protein synthesis initiation factor IF3 controls its own gene expression at the translational level *in vivo*. *J. Mol. Biol.* **192**:767–780.

Butler, J. S., M. Springer, and M. Grunberg-Manago. 1987. AUU-to-AUG mutation in the initiator codon of the translation initiation factor IF-3 abolishes translational autocontrol of its own gene (inf C) in vivo. *Proc. Natl. Acad. Sci. USA* **84**: 4022–4025.

Castel, A., B. Kraal, P. R. M. Kerklaan, J. Klok, and L. Bosch. 1977. Initiation of polypeptide synthesis with various NH_2-blocked aminoacyl-tRNAs under the direction of alfalfa mosaic virus RNA4. *Proc. Natl. Acad. Sci. USA* **74**:5509–5513.

Castel, A., B. Kraal, A. Komieckny, and L. Bosch. 1979. Translation by E. coli ribosomes of alfalfa mosaic virus RNA4 can be initiated at two sites on the monocistronic message. *Eur. J. Biochem.* **101**:123–133.

De Boer, H. A., and H. M. Shepard. 1983. Strategies for optimizing foreign gene expression in Escherichia coli, p. 205–248. *In* A. M. Kroon (ed.), *Horizons in Biochemistry*, vol. 7. *Genes, Structure and Expression*. John Wiley & Sons, Inc., New York.

De Smit, M. H., and J. Van Duin. 1990. Control of prokaryotic translational initiation by mRNA secondary structure. *Prog. Nucleic Acid Res. Mol. Biol.* **38**:1–35.

Ericson, G., and P. Wollenzien. 1989. An RNA secondary structure switch between the inactive and active conformations of the Escherichia coli 30S ribosomal subunit. *J. Biol. Chem.* **264**: 540–545.

Gallie, D. E., and C. I. Kado. 1989. A translational enhancer derived from tobacco mosaic virus is functionally equivalent to a Shine-Dalgarno sequence. *Proc. Natl. Acad. Sci. USA* **86**: 129–132.

Ghosh, H. P., D. Söll, and H. G. Khorana. 1967. Studies on polynucleotides. LXVII. Initiation of protein synthesis *in vitro* by using ribopolynucleotides with repeating nucleotide sequences as messengers. *J. Mol. Biol.* **25**:275–298.

Gold, L. 1988. Posttranscriptional regulatory mechanisms in Escherichia coli. *Annu. Rev. Biochem.* **57**:199–233.

Gold, L., and G. Stormo. 1987. Translation initiation, p. 1302–1307. *In* F. C. Neidhardt, J. L. Ingraham, K. B. Low, B. Magasanik, M. Schaechter, and H. E. Umbarger (ed.), *Escherichia coli and Salmonella typhimurium: Cellular and Molecular Biology*. American Society for Microbiology, Washington, D.C.

Gold, L., G. Stormo, and R. Saunders. 1984. Escherichia coli translational initiation factor IF3: a unique case of translational regulation. *Proc. Natl. Acad. Sci. USA* **81**:7061–7065.

Gouy, M., and C. Gautier. 1982. Codon usage in bacteria: correlation with gene expressivity. *Nucleic Acids Res.* **10**:7055–7074.

Gren, E. J. 1984. Recognition of messenger RNA during translation initiation in Escherichia coli. *Biochimie* **66**:1–29.

Gryczan, T. J., G. Grandi, J. Hahn, R. Grandi, and D. Dubnau. 1980. Conformational alteration of mRNA structure and the posttranscriptional regulation of erythromycin-induced drug resistance. *Nucleic Acids Res.* **8**:6081–6097.

Gualerzi, C., C. L. Pon, and A. Kaji. 1971. Initiation factor dependent release of aminoacyl-tRNAs from complexes of 30S ribosomal subunits, synthetic polynucleotide and aminoacyl tRNA. *Biochem. Biophys. Res. Commun.* **45**:1312–1319.

Gualerzi, C. O., C. L. Pon, R. T. Pawlik, M. A. Canonaco, M. Paci, and W. Wintermeyer. 1986. Role of initiation factors in *Escherichia coli* translational initiation, p. 621–641. *In* B. Hardesty and G. Kramer (ed.), *Structure, Function, and Genetics of Ribosomes*. Springer-Verlag, New York.

Gualerzi, C., G. Risuleo, and C. Pon. 1979. Mechanism of the spontaneous and initiation factor 3-induced dissociation of 30S aminoacyl-tRNA. Polynucleotide ternary complexes. *J. Biol. Chem.* **254**:44–49.

Gutell, R. R., B. Weiser, C. R. Woese, and H. F. Noller. 1985. Comparative anatomy of 16S-like ribosomal RNA. *Prog. Nucleic Acid Res. Mol. Biol.* **32**:155–216.

Hall, M. N., J. Gabay, M. Débarbouillé, and M. Schwarz. 1982. A role for mRNA secondary structure in the control of translation initiation. *Nature* (London) **295**:616–618.

Horinouchi, S., and B. Weisblum. 1980. Posttranscriptional modification of mRNA conformation: mechanism that regulates erythromycin-induced resistance. *Proc. Natl. Acad. Sci. USA* **77**:7079–7083.

Hui, A., and H. A. De Boer. 1987. Specialized ribosome system: preferential translation of a single mRNA species by a subpopulation of mutated ribosomes in Escherichia coli. *Proc. Natl. Acad. Sci. USA* **84**:4762–4766.

Iserentant, D., and W. Fiers. 1980. Secondary structure of mRNA and efficiency of translation initiation. *Gene* **9**:1–12.

Jay, G., and R. Kaempfer. 1974. Sequence of events in initiation of translation: a role for initiator transfer RNA in the recognition of messenger RNA. *Proc. Natl. Acad. Sci. USA* **71**:3199–3203.

Jay, E., A. K. Seth, and G. Jay. 1980. Specific binding of a chemically synthesized prokaryotic ribosome recognition site. *J. Biol. Chem.* **255**:3809–3812.

Kastelein, R. A., B. Berkhout, and J. Van Duin. 1983. Opening the closed ribosomal binding site of the lysis cistron of bacteriophage MS2. *Nature* (London) **305**:741–743.

Kellog, D. A., B. P. Doctor, J. E. Loebel, and M. W. Nirenberg. 1966. RNA codons and protein synthesis. IX. Synonym codon recognition by multiple species of valine-, alanine-, and methionine-sRNA. *Proc. Natl. Acad. Sci. USA* **55**:912–919.

Kozak, M. 1983. Comparison of initiation of protein synthesis in procaryotes, eucaryotes, and organelles. *Microbiol. Rev.* **47**: 1–45.

Landick, R., and C. Yanofsky. 1987. Transcription attenuation, p. 1276–1301. *In* F. C. Neidhardt, J. L. Ingraham, K. B. Low, B. Magasanik, M. Schaechter, and H. E. Umbarger (ed.), *Escherichia coli and Salmonella typhimurium: Cellular and Molecular Biology*. American Society for Microbiology, Washington, D.C.

Looman, A. C., J. Bodlaender, L. J. Comstock, D. Eaton, P. Jhurani, H. A. De Boer, and P. H. Van Knippenberg. 1987. Influence of the codon following the AUG initiation codon on the expression of a modified *lac Z* gene in Escherichia coli. *EMBO J.* **6**:2489–2492.

Looman, A. C., J. Bodlaender, M. De Gruyter, A. Vogelaar, and P. H. Van Knippenberg. 1986. Secondary structure as primary determinant of the efficiency of ribosomal binding sites in Escherichia coli. *Nucleic Acids Res.* **14**:5481–5497.

Looman, A. C., M. De Gruyter, A. Vogelaar, and P. H. Van Knippenberg. 1985. Effects of heterologous ribosomal binding sites on the transcription and translation of the *lac Z* gene of Escherichia coli. *Gene* **37**:145–154.

Looman, A. C., and P. H. Van Knippenberg. 1986. Effects of GUG and AUG initiation codons on the expression of *lac Z* in Escherichia coli. *FEBS Lett.* **197**:315–320.

Maitra, M., E. A. Stringer, and A. Chaudhuri. 1982. Initiation factors in protein biosynthesis. *Annu. Rev. Biochem.* **51**:869–900.

Martin, J., and R. E. Webster. 1975. The *in vitro* translation of a terminating signal by a single Escherichia coli ribosome. *J. Biol. Chem.* **250**:8132–8139.

McPheeters, D. S., A. Christensen, E. T. Young, G. Stormo, and L. Gold. 1986. Translational regulation of expression of bacteriophage T4 lysozyme gene. *Nucleic Acids Res.* **14**:5813–5826.

Menninger, J. R. 1976. Peptidyl transfer RNA dissociates during protein synthesis from ribosomes of Escherichia coli. *J. Biol. Chem.* **251**:3392–3398.

Menninger, J. R. 1979. Accumulation of peptidyl-tRNA is lethal to *Escherichia coli. J. Bacteriol.* **137**:694–696.

Moore, P. B., and M. S. Cashel. 1988. Structure-function correlations in the small ribosomal subunit from Escherichia coli. *Annu. Rev. Biophys. Chem.* **17**:349–367.

Movva, N. R., K. Nakamura, and M. Inouye. 1980. Gene structure of the ompA protein, a major surface protein of Escherichia coli required for cell-cell interaction. *J. Mol. Biol.* **143**:317–328.

Munson, L. M., G. D. Stormo, R. L. Niece, and W. S. Reznikoff. 1984. *lac* Z translation initiation mutations. *J. Mol. Biol.* **177**:663–683.

Nomura, H., R. Gourse, and G. Baughman. 1984. Regulation of the synthesis of ribosomes and ribosomal components. *Annu. Rev. Biochem.* **53**:75–117.

Oppenheim, D. S., and C. Yanofsky. 1980. Translational coupling during expression of the tryptophan operon of Escherichia coli. *Genetics* **95**:785–795.

Peabody, D. S. 1989. Translation initiation at non-AUG triplets in mammalian cells. *J. Biol. Chem.* **264**:5031–5035.

Petersen, G. B., P. A. Stockwell, and D. F. Hill. 1988. Messenger RNA recognition in Escherichia coli: a possible second site of interaction with 16S ribosomal RNA. *EMBO J.* **7**:3957–3962.

Pleij, C. W. A., K. Rietveld, and L. Bosch. 1985. A new principle of RNA folding based on pseudoknotting. *Nucleic Acids Res.* **13**:1717–1731.

Preibish, G., H. Ishihara, D. Tipier, and M. Leineweber. 1988. Unexpected translation initiation within the coding region of eukaryotic genes expressed in Escherichia coli. *Gene* **73**:179–186.

Scherer, G. F. E., M. D. Walkinshaw, S. Arnott, and J. D. Morré. 1980. The ribosome binding sites recognized by E. coli ribosomes have regions with signal character in both the leader and protein coding segments. *Nucleic Acids Res.* **8**:3895–3907.

Schmidt, B. F., B. Berkhout, G. P. Overbeek, A. Van Strien, and J. Van Duin. 1987. Determination of the RNA secondary structure that regulates lysis gene expression in bacteriophage MS_2. *J. Mol. Biol.* **195**:505–516.

Schneider, T. D., G. D. Stormo, L. Gold, and A. Ehrenfeuchs. 1986. Information content of binding sites on nucleotide sequences. *J. Mol. Biol.* **188**:415–431.

Schoner, G., R. M. Belagaje, and R. G. Schoner. 1986. Translation of a synthetic two-cistron mRNA in Escherichia coli. *Proc. Natl. Acad. Sci. USA* **83**:8506–8510.

Schottel, J. L., J. J. Sninsky, and S. N. Cohen. 1984. Effects of alterations in the translation control region on bacterial gene expression: use of *cat* gene constructs transcribed from the lac promoter as a model system. *Gene* **28**:177–193.

Shine, J., and L. Dalgarno. 1974. The 3′ terminal sequence of E. coli 16S ribosomal RNA: complementarity to nonsense triplets and ribosome binding sites. *Proc. Natl. Acad. Sci. USA* **71**: 1342–1346.

Shinedling, S., M. Gayle, D. Pripnow, and L. Gold. 1987. Mutations affecting translation of the bacteriophage T4 rIIB gene cloned in Escherichia coli. *Mol. Gen. Genet.* **207**:224–232.

Sleat, D. E., R. Hull, P. C. Turner, and T. M. A. Wilson. 1988. Studies on the mechanism of translational enhancement by the 5′ leader sequence of tobacco mosaic virus RNA. *Eur. J. Biochem.* **175**:75–86.

Sobura, J. E., M. R. Chowdhurry, D. A. Hawley, and A. J. Wahba. 1977. Requirement of chain initiation factor 3 and ribosomal protein S1 in translation of synthetic and natural messenger RNA. *Nucleic Acids Res.* **4**:17–29.

Spanjaard, R. A., M. C. M. Van Dijk, A. J. Turion, and J. Van Duin. 1989. Expression of the rat interferon-α_1 gene in Escherichia coli controlled by the secondary structure of the translation initiation region. *Gene* **80**:345–351.

Stanssens, P., E. Remaut, and W. Fiers. 1985. Alterations upstream from the Shine-Dalgarno region and their effect on bacterial gene expression. *Gene* **36**:211–223.

Stanssens, P., E. Remaut, and W. Fiers. 1986. Inefficient translation initiation causes premature transcription termination in the *lacZ* gene. *Cell* **44**:711–718.

Stern, S., B. Weiser, and H. F. Noller. 1988. Model for the three-dimensional folding of 16S ribosomal RNA. *J. Mol. Biol.* **204**:447–481.

Stormo, G. D., T. D. Schneider, and L. M. Gold. 1982. Characterization of translation initiation sites in E. coli. *Nucleic Acids Res.* **10**:2971–2996.

Tessier, L. H., P. Sondermeyer, T. Faure, D. Dreyer, A. Benarente, D. Villeval, M. Courtney, and J.-P. Lecocq. 1984. The influence of mRNA primary and secondary structure on human IFN-γ gene expression in E. coli. *Nucleic Acids Res.* **12**:7663–7675.

Thanaray, T. A., and M. W. Pandit. 1989. An additional ribosome binding site on mRNA of highly expressed genes and a bifunctional site on the colicin fragment of 16S rRNA from Escherichia coli: important determinants of the efficiency of translation-initiation. *Nucleic Acids Res.* **17**:2973–2985.

Tomich, C.-S. C., E. R. Olson, M. K. Olsen, P. S. Kaytes, S. K. Rockenbach, and N. T. Hatzenbuhler. 1989. Effect of nucleotide sequences directly downstream from the AUG on the expression of bovine somatotropin in E. coli. *Nucleic Acids Res.* **17**:3179–3197.

Van der Hofstad, G. A. J. M., J. A. Foekens, L. Bosch, and H. O. Voorma. 1977. Cooperative effects of initiation factors and fMet-tRNA in the formation of the 40S initiation complex. *Eur. J. Biochem.* **77**:69–75.

Van der Laken, K., H. Bakker-Steeneveld, B. Berkhout, and P. H. Van Knippenberg. 1980. The role of the codon and initiation factor IF-2 in the selection of N-blocked aminoacyl-tRNA for initiation. *Eur. J. Biochem.* **104**:19–23.

Van der Laken, C., H. Bakker-Steeneveld, and P. Van Knippenberg. 1979. Polyuridylic acid-dependent binding of fMet-tRNA to Escherichia coli ribosomes and incorporation of formylmethionine into polyphenylalanine. *FEBS Lett.* **100**:230–234.

Van Duin, J., G. P. Overbeek, and C. Backendorf. 1980. Functional recognition of phage RNA by 30S ribosomal subunits in the absence of initiator tRNA. *Eur. J. Biochem.* **110**:593–597.

Van Gemen, B., H. J. Koets, C. A. M. Plooy, J. Bodlaender, and P. H. Van Knippenberg. 1987. Characterization of the *ksgA* gene of Escherichia coli determining kasugamycin sensitivity. *Biochimie* **69**:841–848.

Van Knippenberg, P. H. 1975. A possible role of the 5′ terminal sequence of 16S ribosomal RNA in the recognition of initiation sequences for protein synthesis. *Nucleic Acids Res.* **2**:79–85.

Vermeer, C., J. Boon, A. Talens, and L. Bosch. 1973a. Binding to the initiation factor IF-3 to Escherichia coli ribosomes and MS_2 RNA. *Eur. J. Biochem.* **40**:283–293.

Vermeer, C., R. J. De Kievit, W. J. Van Alphen, and L. Bosch. 1973b. Recycling of the initiation factor IF-3 on 30S ribosomal subunits of E. coli. *FEBS Lett.* **31**:273–276.

Waltz, A., V. Pirotta, and K. Ineichen. 1976. λ repressor regulates the switch between P_R and P_{rm} promotors. *Nature* (London) **262**:665–669.

Wood, C. R., M. A. Boss, T. P. Patel, and J. S. Emtage. 1984. The influence of messenger RNA secondary structure on expression of an immunoglobulin heavy chain in Escherichia coli. *Nucleic Acids Res.* **12**:3937–3950.

Yanofsky, C., and I. P. Crawford. 1987. The tryptophan operon, p. 1453–1472. *In* F. C. Neidhardt, J. L. Ingraham, K. B. Low, B. Magasanik, M. Schaechter, and H. E. Umbarger (ed.), *Escherichia coli and Salmonella typhimurium: Cellular and Molecular Biology*. American Society for Microbiology, Washington, D.C.

Yates, J. L., D. Dean, W. A. Strycharz, and M. Nomura. 1981. E. coli ribosomal protein L10 inhibits translation of L10 and L7/L12 mRNAs by acting at a single site. *Nature* (London) **294**:190–192.

Chapter 20

From Polynucleotide to Natural mRNA Translation Initiation: Function of *Escherichia coli* Initiation Factors

DIETER HARTZ, DAVID S. McPHEETERS, and LARRY GOLD

In vitro translation of natural mRNA by *Escherichia coli* ribosomes is dependent on the initiation factors IF1, IF2, and IF3, whereas polynucleotide-primed translation does not absolutely require these factors (Wahba et al., 1969a; Wahba et al., 1969b). Defining differences between polynucleotide-directed translation and natural mRNA-directed translation may lead to some insight into the functions of the initiation factors. Translation of polynucleotides has no defined initiation region, can start with elongator tRNAs (Wahba et al., 1969b), and does not require the dissociation of the 70S ribosome (Nomura and Lowry, 1967). In contrast, natural mRNA translation starts at a specific site on the mRNA, facilitated by the initiator tRNA which is base paired with the initiation codon. Furthermore, dissociation of the 70S ribosome may be required; the 30S subunit, the mRNA, the initiator tRNA, and the initiation factors first form an initiation complex before the 50S subunit is added and the initiation factors are released (Gualerzi and Pon, 1981). Thus, translation of polynucleotides in the absence of initiation factors uses a more primitive (or undefined) mechanism. Features on natural mRNA define the initiation regions, whereas the primary function of initiation factors is to limit initiation to the initiation codon by restricting the ribosomal P site to fMet-$tRNA_f^{Met}$.

THE RIBOSOME-BINDING SITE

Early nuclease digestion experiments of 70S initiation complexes on phage RNA showed that the mRNA was selectively protected against nucleolytic attack around translation initiation sites (Steitz, 1975). Measurement of initiator tRNA binding into 30S complexes on bacteriophage RNA showed a strong dependence on initiation factor IF3, whereas the trinucleotide AUG or polynucleotides did not require IF3 (Wahba et al., 1969b; Suttle and Ravel, 1974). It was concluded from these data that IF3 is responsible for the binding of the 30S subunit to natural mRNA. This conclusion was challenged by Lodish (1970), who showed that the distribution of initiation complexes on different initiation sites is an intrinsic property of the ribosome and not of the initiation factors. Furthermore, Van Duin et al. (1980) demonstrated that preformed 30S-mRNA binary complexes (formed without any tRNA or initiation factors) can be successfully chased into active translation elongation complexes. Calogero et al. (1988) showed recently that 30S binding to mRNA is practically unaffected by IF3. The idea that 30S subunits can find initiation sites in the absence of initiation factors is reasonable, since at least part of the recognition of mRNA involves simple base pairing between the Shine-Dalgarno region (Shine and Dalgarno, 1974) and the 3′ end of the 16S rRNA of the 30S subunit (Hui and DeBoer, 1987; for a review, see Gold, 1988). Other sequence elements within the ribosome-binding sites might provide additional interactions with the 16S rRNA, as proposed for the IF3 mRNA (Gold et al., 1984). Furthermore, with the toeprinting assay (Fig. 1), we detected ternary complexes between 30S subunits, mRNA, and initiator tRNA at true initiation sites in the absence of initiation factors (Winter et al., 1987; McPheeters et al., 1988; Hartz et al., 1988; Schaefer et al., 1989; Bläsi et al., 1989; Hartz et al., 1989). The toeprint that detects the location of the downstream edge of a ribosome on an mRNA usually appears 15 nucleo-

Dieter Hartz and Larry Gold ■ Department of Molecular, Cellular, and Developmental Biology, University of Colorado, Boulder, Colorado 80309. **David S. McPheeters** ■ Division of Biology, 147-75 California Institute of Technology, Pasadena, California 91125.

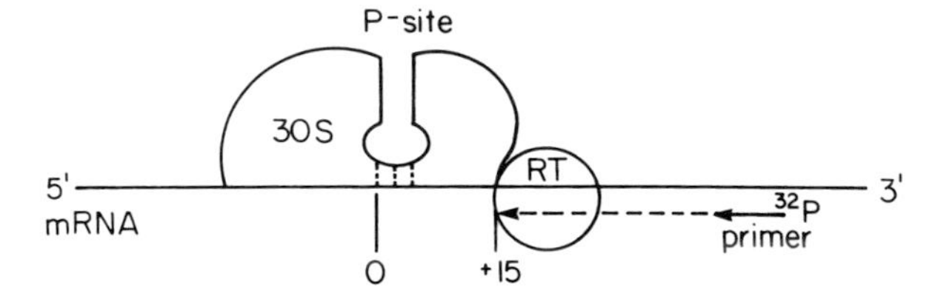

Figure 1. Model for the toeprinting method. Reverse transcriptase terminates cDNA synthesis when it encounters a 30S (or 70S) particle and tRNA bound on the template mRNA at the ribosome-binding site. The premature reverse transcriptase stop, called the toeprint, is usually located at +15 from the first base of the cognate codon recognized by the bound tRNA.

tides downstream from the first nucleotide of the initiation codon and thus agrees well with the mRNA nuclease digestion experiments and with statistical analyses of the information content in ribosome-binding sites (Schneider et al., 1986). The toeprinting studies show that the initiation site is recognized by an intrinsic property of the ribosome; toeprints with initiator tRNA occur only at true initiation sites. Sequence elements on the mRNA which specifically interact with the ribosome restrict the search for initiation codons to a certain distance from the Shine-Dalgarno sequence (Hartz et al., 1989). Shine-Dalgarno elements and other determinants are absent on most synthetic polynucleotides, which explains the lack of specific initiation domains on these templates.

tRNA-BINDING PROPERTIES OF 30S COMPLEXES

Translation initiation of natural mRNA always utilizes the initiator tRNA, fMet-tRNA$_f^{Met}$, which is able to initiate on AUG, GUG, UUG, CUG, AUA, and AUU (for a review, see Gold and Stormo, 1987) codons, whereas the elongator tRNA, Met-tRNA$_m^{Met}$, supplies methionine at internal AUG only (Clark and Marcker, 1966). Previous experiments with phage RNA showed that 30S subunits bind the initiator tRNA better than elongator tRNAs (Nomura and Lowry, 1967). It was concluded that the 30S subunit, when bound to a natural mRNA, has an intrinsic high affinity for the initiator tRNA at the ribosomal P site. Since synthetic polynucleotide-bound 30S particles were shown to readily accept elongator tRNAs into their P sites (Nomura and Lowry, 1967; Risuleo et al., 1976; Canonaco et al., 1986), synthetic polynucleotides were thought to be very different as templates. We have found that on several bacteriophage T4 mRNAs (Hartz et al., 1989), and on in vitro-prepared mRNA containing the T4 gene 32 initiation region (which we will refer to as gene 32 in vitro transcript), the initiator tRNA and also a variety of elongator tRNAs bind to the 30S P site. The only requirement for access of elongator tRNAs to the P site is the presence of a cognate codon near the Shine-Dalgarno sequence (Hartz et al., 1989). tRNA^{Cys} and tRNA^{Phe} compete well with tRNA$_f^{Met}$ for the ribosomal P site, as evidenced by the location of the 3′ edge of the 30S subunit on the mRNA at +15 from the cognate codons (Fig. 2A, lane 1). Even charged and formylated initiator tRNA is not selected over the uncharged elongator tRNAs (Fig. 2A, lane 3). Thus, the tRNA-binding properties of 30S subunits are similar with natural mRNA and polynucleotides as templates. That is, entry of elongator tRNAs to the 30S P site is a serious threat for the proper initiation process on natural mRNA. Use of an elongator tRNA would result either in a shorter protein product or in translation in the wrong reading frame. Below, we provide evidence that the selection of the initiator tRNA is provided by IF2 and IF3 (Hartz et al., 1989).

SELECTION OF THE INITIATOR tRNA BY IF3 ON 30S COMPLEXES

Some data indicating a role of IF3 in the discrimination of initiator from elongator tRNAs have been published previously. IF3 enhances the exchange rate of elongator tRNAs that are bound to polynucleotide-primed 30S complexes (Pon and Gualerzi, 1974). The corresponding initiator tRNA-containing complexes are more resistant to IF3-induced destabilization than are complexes with elongator tRNAs (Risuleo et al., 1976). We have evidence (Hartz et al., 1989) that IF3 selects the initiator tRNA on a natural mRNA template. Inclusion of IF3 in a toeprinting reaction that contained tRNA$_f^{Met}$, tRNA^{Cys}, and tRNA^{Phe} leads to selection of the initiator tRNA (Fig. 2A, lane 2); tRNA$_f^{Met}$ is as well selected as fMet-tRNA$_f^{Met}$ in the competition reaction (Fig. 2A, lane 4). Thus, IF3 action distinguishes the initiator tRNA from the elongator tRNAs without inspection of formylmethionine. At least part of the information for initiator tRNA discrimination is located in the anticodon stem-loop part of the initiator tRNA; a synthetic anticodon stem-loop fragment of the initiator tRNA is selected over tRNA^{Cys} and tRNA^{Phe} by IF3 (Hartz et al., 1989). The codon-anticodon interaction itself is also "measured" by IF3 (Berkhout et al., 1986; Hartz et al., unpublished data). Finally, IF3 selects tRNA$_f^{Met}$ in mixed tRNA incubations even after the complexes have formed (Hartz et al., 1989). The results from our tRNA competition experiments on the in vitro transcript of gene 32 mRNA are consistent with the data gained on polynucleotide

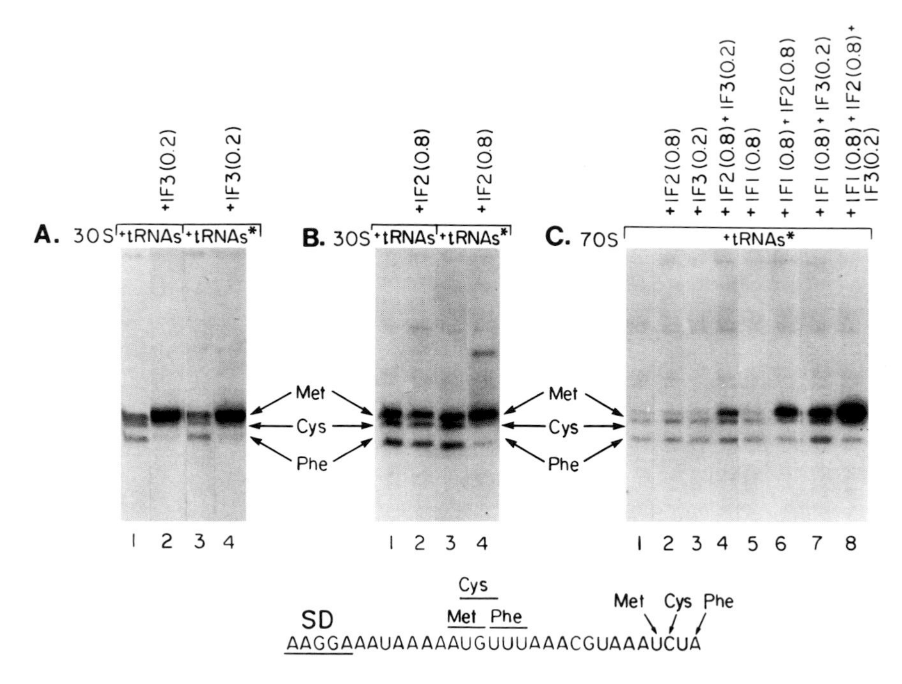

Figure 2. Selection of the initiator tRNA by *E. coli* initiation factors IF3 (A) and IF2 (B) on 30S complexes and by all three initiation factors (C) on 70S complexes. Toeprinting reaction mixtures (10 μl) were prepared in standard buffer (10 mM Tris-acetate [pH 7.4], 60 mM NH_4Cl, 6 mM β-mercaptoethanol, 10 mM magnesium acetate) and contained 6.7 nM gene 32 in vitro transcript annealed to 20 nM ^{32}P-labeled primer 32loopD, deoxynucleotide triphosphates at 0.375 μM each, plus 0.2 μM 30S (A and B) or 0.2 μM 70S (C; prepared by incubating 0.2 μM 30S with 0.24 μM 50S at 37°C for 15 min). In addition, a mixture of $tRNA_f^{Met}$, $tRNA^{Cys}$, and $tRNA^{Phe}$ (tRNAs) at 0.5 μM each or a mixture of fMet-$tRNA_f^{Met}$, $tRNA^{Cys}$, and $tRNA^{Phe}$ (tRNAs*) at 0.5 μM each was added. Furthermore, pure *E. coli* initiation factors were added at the concentrations (micromolar) indicated above the lanes. The reactions were preincubated at 37°C for 10 min before ca. 200 U of Moloney murine leukemia virus (M-MLV) reverse transcriptase was added, and the incubation was continued for 15 min. The reactions were stopped by adding 26 μl of formamide containing 0.12 M EDTA, 0.3× TBE buffer, 0.015% bromophenol blue, and 0.015% xylene cylanol and immediately heated at 95°C for 3 min. Samples of 5 μl were electrophoresed on an 8% polyacrylamide sequencing gel. Only the toeprint stops are shown, not the 5′ ends of the cDNA corresponding to the full-length transcript. The sequence of the gene 32 initiation domain is shown, with the Shine-Dalgarno sequence (SD) underlined. The cognate codons of $tRNA_f^{Met}$, $tRNA^{Cys}$, and $tRNA^{Phe}$ are marked, as are the positions of the corresponding toeprint stops. All data are from Hartz et al. (1989), which also contains further details on the procedures and the sources of the materials.

templates. IF3 action is not mandatory on polynucleotides, but it would favor translation initiation with initiator tRNA on polynucleotides that contain initiation codons (Dubnoff and Maitra, 1969). The role of IF3 in initiator tRNA selection has so far been underestimated because of reports that initiator tRNA binds most strongly to the 30S P site (Bretscher and Marcker, 1966; Nomura and Lowry, 1967), that IF2 is involved in the binding of the initiator tRNA (Jay and Kaempfer, 1974; see below), and that elongator tRNAs are complexed with EF-Tu and thus prevented from binding (but see Gnirke and Nierhaus, 1986).

SELECTION OF THE INITIATOR tRNA BY IF2 ON 30S COMPLEXES

IF2 has been suggested to be the carrier of initiator tRNA because IF2 is able to form binary complexes with initiator tRNA. Binary complex formation is dependent on aminoacylation and formylation of the tRNA (Sundari et al., 1976; Van der Hofstad et al., 1977; Petersen et al., 1983). However, IF2 probably acts at the 30S level (Gualerzi and Wintermeyer, 1986; Canonaco et al., 1986) and mainly influences the binding of the initiator tRNA to the 30S subunit (Vermeer et al., 1973). No direct demonstration of IF2 selecting the initiator tRNA on 30S subunits has been published. In fact, one report showed that IF2 stimulates the binding of several elongator tRNAs to polynucleotide-primed 30S subunits (Canonaco et al., 1986). In a toeprinting reaction, IF2 selects the initiator tRNA over $tRNA^{Cys}$ and $tRNA^{Phe}$ on the gene 32 in vitro transcript (Fig. 2B, lane 4). Selection is absolutely dependent on formylation and charging of the initiator tRNA (Fig. 2B, lane 2), in striking contrast to selection by IF3. The selective mechanism of IF2 is also different in that dissociation of the elongator tRNAs cannot be stimulated after the complexes have formed (Canonaco et al., 1986; Hartz et al., 1989). In essence, IF2 action not only discriminates the initiator tRNA from elongator tRNAs but also provides discrimination against uncharged and probably unformylated initiator tRNA. IF2 thus connects the fidelity of translation initiation to the intracellular redox state and the one-carbon metabolism via tetrahydrofolate (the formyl group donor); of course, methionine itself is the other important component of one-carbon metabolism (Danchin, 1973).

tRNA-BINDING PROPERTIES OF 70S COMPLEXES

Previously it was shown that neither initiator nor elongator tRNAs could form 70S ternary complexes directly on 70S ribosomes primed with natural mRNA (Nomura and Lowry, 1967; Nomura et al., 1967). However, 70S ribosomes primed with polynucleotides readily form complexes with elongator tRNAs, but not with fMet-$tRNA_f^{Met}$ if competing elongator tRNAs are present (Nomura and Lowry, 1967; Guthrie and Nomura, 1968). Using toeprinting, we were able to detect direct formation of 70S ternary complexes with $tRNA_f^{Met}$, $tRNA^{Cys}$, and $tRNA^{Phe}$ on the gene 32 in vitro transcript (Hartz et al., 1989). The 70S complexes give a weaker toeprinting signal than do 30S complexes (compare lanes 1 in Fig. 2A and C), which might have prevented their detection previously. In competition experiments with $tRNA^{Cys}$ and $tRNA^{Phe}$, fMet-$tRNA_f^{Met}$ (but not $tRNA_f^{Met}$) is bound less well than the elongator tRNAs. This result is reminiscent of the exclusion of fMet-$tRNA_f^{Met}$ on polynucleotides mentioned above and proves that the 70S complexes are formed directly (since 30S subunits would bind fMet-$tRNA_f^{Met}$ as well as elongator tRNAs [Fig. 2A, lane 3]). The 3′ edges of the 70S complexes are at the same position (+15 from the cognate codons) as in 30S complexes. The most likely explanation is that tRNAs bind to the same site (the P site) on 70S complexes as on 30S complexes and that the edge of the ribosome which reserve transcriptase encounters in toeprinting is composed of 30S material. We think that the 50S subunit is bound to the 30S subunit away from the mRNA (Gold, 1988). From our study, it is clear that (as in 30S complexes) a selection of initiator tRNA is needed on 70S complexes to define the proper translational initiation codon on natural mRNA.

SELECTION OF THE INITIATOR tRNA ON 70S COMPLEXES

Neither IF2 nor IF3 alone provides the selection of the initiator tRNA on 70S complexes on the gene 32 in vitro transcript (Fig. 2C, lanes 2 and 3). How then is initiator tRNA selection achieved, especially at the high Mg^{2+} concentrations used for in vitro translation of natural mRNA? (At high Mg^{2+} concentrations, most of the ribosomes are in the associated state; 70S formation would be favored.) The key component appears to be IF1, which by itself does not select the initiator tRNA (Fig. 2C, lane 5). If IF1 is added together with IF2, IF3, or both, selection is achieved (Fig. 2C, lanes 6 to 8). We favor the interpretation that IF1 facilitates the dissociation of 70S ribosomes so that IF2 and IF3 can resume their function on the 30S subunit. This idea is strengthened by the previous finding that IF1 enhances both forward and back rates of the 70S equilibrium with 50S and 30S (Godefroy-Colburn et al., 1975) and that 30S complex intermediates are formed if all three initiation factors are present (Nomura et al., 1967; Blumberg et al., 1979). In most of the literature, IF3 has been implicated as the major factor in 70S dissociation (Subramanian and Davis, 1970; Kaempfer, 1972). In toeprinting, we see an indication of that dissociation function, since IF2 added together with IF3 shows enhancement of the initiator tRNA signal (Fig. 2C, lane 4).

CONCLUDING REMARKS

30S and 70S complexes on natural mRNA show striking similarities in their tRNA-binding characteristics to corresponding complexes on polynucleotide templates. In both cases, the ribosomal P site is accessible to elongator tRNAs. Selection of the initiator tRNA (which also necessitates 70S dissociation) is a necessity on natural mRNA but not on polynucleotides (which often might not have an initiation codon). The initiation factors provide the selection of the initiator tRNA on natural mRNA (and on polynucleotides containing initiation codons [Dubnoff and Maitra, 1969]). Natural mRNA translation must have evolved from translation of simple polynucleotides. Changes in polynucleotide sequence to create ribosome-binding sites would restrict initiation to an initiation region, while further evolution of initiation factors would provide selection of an initiator tRNA and thus provide a completely defined start of translation.

We thank J. Binkley for preparing fMet-$tRNA_f^{Met}$ and D. Dix for providing IF3 and fMet-$tRNA_f^{Met}$. We also thank R. Traut for providing 30S and 50S ribosomal subunits and R. Saunders for constructing the plasmid containing the gene 32 initiation region. We are indebted to C. Gualerzi for his generous gift of large quantities of purified *E. coli* initiation factors. We are also grateful to E. Fish, T. Hollingsworth, D. Irvine, and C. Tuerk for their helpful discussions.

This work was supported by Public Health Service research grant GM28685 from the National Institutes of Health.

REFERENCES

Berkhout, B., C. J. Van der Laken, and P. H. Van Knippenberg. 1986. Formylmethionyl-tRNA binding to 30S ribosomes programmed with homopolynucleotides and the effect of translational initiation factor 3. *Biochim. Biophys. Acta* **866**:144–153.

Bläsi, U., K. Nam, D. Hartz, L. Gold, and R. Young. 1989. Dual translational initiation sites control function of the lambda-S gene. *EMBO J.* **8**:3501–3510.

Blumberg, B. M., T. Nakamoto, and F. J. Kezdy. 1979. Kinetics of initiation of bacterial protein synthesis. *Proc. Natl. Acad. Sci. USA* **76**:251–255.

Bretscher, M. S., and K. A. Marcker. 1966. Polypeptidyl-sribonucleic acid and amino-acyl-sribonucleic acid binding sites on ribosomes. *Nature* (London) **211**:380–384.

Calogero, R. A., C. L. Pon, M. A. Canonaco, and C. O. Gualerzi. 1988. Selection of the mRNA translational initiation region by Escherichia coli ribosomes. *Proc. Natl. Acad. Sci. USA* **85**: 6427–6431.

Canonaco, M. A., R. A. Calogero, and C. O. Gualerzi. 1986. Mechanism of translational initiation in procaryotes. Evidence for a direct effect of IF2 on the activity of the 30 S ribosomal subunit. *FEBS Lett.* **207**:198–204.

Clark, B. F. C., and K. A. Marcker. 1966. N-formyl-methionyl-sribonucleic acid and chain initiation in protein biosynthesis. Polypeptide synthesis directed by a bacteriophage ribonucleic acid in a cell-free system. *Nature* (London) **211**:378–380.

Danchin, A. 1973. Does formylation of initiator tRNA act as a regulatory signal in E. coli? *FEBS Lett.* **34**:327–332.

Dubnoff, J. S., and U. Maitra. 1969. Protein factors involved in polypeptide chain initiation in Escherichia coli. *Cold Spring Harbor Symp. Quant. Biol.* **34**:301–306.

Gnirke, A., and K. H. Nierhaus. 1986. tRNA binding sites on the subunits of Escherichia coli ribosomes. *J. Biol. Chem.* **261**: 14506–14514.

Godefroy-Colburn, T., A. D. Wolfe, J. Dondon, M. Grunberg-Manago, P. Dessen, and D. Pantaloni. 1975. Light-scattering studies showing the effect of initiation factors on the reversible dissociation of Escherichia coli ribosomes. *J. Mol. Biol.* **94**: 461–478.

Gold, L. 1988. Post-transcriptional regulatory mechanisms in E. coli. *Annu. Rev. Biochem.* **57**:199–233.

Gold, L., and G. Stormo. 1987. Translational initiation, p. 1302–1307. *In* F. C. Neidhardt, J. L. Ingraham, K. B. Low, B. Magasanik, M. Schaechter, and H. E. Umbarger (ed.), *Escherichia coli and Salmonella typhimurium: Cellular and Molecular Biology*, vol. 2. American Society for Microbiology, Washington, D.C.

Gold, L., G. Stormo, and R. Saunders. 1984. Escherichia coli translational initiation factor IF3: a unique case of translational regulation. *Proc. Natl. Acad. Sci. USA* **81**:7061–7065.

Gualerzi, C., and C. L. Pon. 1981. Protein biosynthesis in procaryotic cells: mechanism of 30S initiation complex formation in Escherichia coli, p. 805–826. *In* L. Balaban, J. L. Sussman, W. Traub, and A. Yonath (ed.), *Structural Aspects of Recognition and Assembly in Biological Macromolecules*. ISS, Rehovot, Israel.

Gualerzi, C. O., and W. Wintermeyer. 1986. Procaryotic initiation factor 2 stopped-flow study. *FEBS Lett.* **202**:1–6.

Guthrie, C., and M. Nomura. 1968. Initiation of protein synthesis: a critical test of the 30S subunit model. *Nature* (London) **219**:232–235.

Hartz, D., D. S. McPheeters, and L. Gold. 1989. Selection of the initiator tRNA by *Escherichia coli* initiation factors. *Genes Dev.* **3**:1899–1912.

Hartz, D., D. S. McPheeters, R. Traut, and L. Gold. 1988. Extension inhibition analysis of translation initiation complexes. *Methods Enzymol.* **164**:419–425.

Hui, A., and H. A. DeBoer. 1987. Specialized ribosome system: preferential translation of a single mRNA species by a subpopulation of mutated ribosomes in Escherichia coli. *Proc. Natl. Acad. Sci. USA* **84**:4762–4766.

Jay, G., and R. Kaempfer. 1974. Sequence of events in initiation of translation: a role for initiator transfer RNA in the recognition of messenger RNA. *Proc. Natl. Acad. Sci. USA* **71**:3199–3203.

Kaempfer, R. 1972. Initiation factor IF-3: a specific inhibitor of ribosomal subunit association. *J. Mol. Biol.* **71**:583–598.

Lodish, H. F. 1970. Specificity in bacterial protein synthesis: role of initiation factors and ribosomal subunits. *Nature* (London) **226**:705–707.

McPheeters, D. S., G. D. Stormo, and L. Gold. 1988. The autogenous regulatory site on the bacteriophage T4 gene 32 messenger RNA. *J. Mol. Biol.* **201**:517–535.

Nomura, M., and C. V. Lowry. 1967. Phage F2 RNA-directed binding of formylmethionyl-TRNA to ribosomes and the role of 30S ribosomal subunits in initiation of protein synthesis. *Proc. Natl. Acad. Sci. USA* **58**:946–953.

Nomura, M., C. V. Lowry, and C. Guthrie. 1967. The initiation of protein synthesis: joining of the 50S ribosomal subunit to the initiation complex. *Proc. Natl. Acad. Sci. USA* **58**:1487–1493.

Petersen, H. U., F. P. Wikman, G. E. Siboska, H. Worm-Leonhard, and B. F. C. Clark. 1983. Interaction between methionine-accepting tRNAs and proteins during initiation of procaryotic translation, p. 41–57. *In* B. F. C. Clark and H. U. Petersen (ed.), *Gene Expression*. Alfred Benzon Symposium 19. Munksgaard, Copenhagen.

Pon, C. L., and C. Gualerzi. 1974. Effect of initiation factor 3 binding on the 30S ribosomal subunits of Escherichia coli. *Proc. Natl. Acad. Sci. USA* **71**:4950–4954.

Risuleo, G., C. Gualerzi, and C. Pon. 1976. Specificity and properties of the destabilization, induced by initiation factor IF-3, of ternary complexes of the 30-S ribosomal subunit, aminoacyl-tRNA and polynucleotides. *Eur. J. Biochem.* **67**: 603–613.

Schaefer, E. M., D. Hartz, L. Gold, and R. D. Simoni. 1989. Ribosome-binding sites and RNA-processing sites in the transcript of the *Escherichia coli unc* operon. *J. Bacteriol.* **171**: 3901–3908.

Schneider, T. D., G. D. Stormo, L. Gold, and A. Ehrenfeucht. 1986. Information content of binding sites on nucleotide sequences. *J. Mol. Biol.* **188**:415–431.

Shine, J., and L. Dalgarno. 1974. The 3′-terminal sequence of 16S ribosomal RNA: complementarity to nonsense triplets and ribosome binding sites. *Proc. Natl. Acad. Sci. USA* **71**:1342–1346.

Steitz, J. A. 1975. Ribosome recognition of initiator tRNA regions in the RNA bacteriophage genome, p. 319–352. *In* N. Zinder (ed.), *RNA Phages*. Cold Spring Harbor Laboratory, Cold Spring Harbor, N.Y.

Subramanian, A. R., and B. D. Davis. 1970. Activity of initiation factor F3 in dissociating Escherichia coli ribosomes. *Nature* (London) **228**:1273–1275.

Sundari, R. M., E. A. Stringer, L. H. Schulman, and U. Maitra.

the calculated free energies for the presumed SD interactions have more negative values than in gram-negative bacteria; it has been suggested that this difference might be one of the reasons for the translational barrier that prevents the expression of gram-negative mRNAs in gram-positive systems (Band and Henner, 1984; Hager and Rabinowitz, 1985). However, the late transcripts of *E. coli* bacteriophage T7, which have strong SD regions, are poorly translated by *Bacillus subtilis* ribosomes, suggesting that the strength of the SD interaction alone is either not relevant or not sufficient to ensure efficient translation with gram-positive ribosomes. Accordingly, it has been reported that the mRNA for the *Streptococcus pneumoniae* highly expressed protein (PolA) begins just two nucleotides upstream from the initiation triplet (Lopez et al., 1989).

Concerning other procaryotic systems, in a survey of 196 protein-coding chloroplast DNA sequences, Bonham-Smith and Bourque (1989) found an SD sequence in ca. 92% of the genes. Its location was similar to that in *E. coli* (i.e., from −12 to −5 upstream from the AUG) in only 40% of the cases, however, whereas in the other cases it was found further upstream (up to 100 base pairs) from the initiation triplet. The presence of potential SD sequences far away from the initiation codons of chloroplast mRNAs has also been noticed by Ruf and Kössel (1988). The utilization of these far upstream SD sequences in base-pairing interactions with 30S subunits remains to be demonstrated, however. Finally, in ca. 6% of the cases, no SD sequence was found; thus, if these do not represent pseudogenes, one can conclude that the SD sequence is dispensable in many chloroplast mRNAs.

Unlike chloroplast 16S rRNA, none of the mitochondrial rRNAs sequenced so far has been found to have the anti-SD sequence (Dams et al., 1988). Accordingly, most of the mitochondrial mRNAs begin at or just before the initiation codon, and their binding to the small ribosomal subunit is thought to involve only the 30 or so nucleotides downstream from the initiation triplet. Interestingly, as in the case of bacterial mRNAs (see below), their interaction with ribosomes does not seem to require initiation factors (Liao and Spremulli, 1989); if it does, this may reflect a requirement for an mRNA helicase activity associated with the initiation factors which disrupts the mRNA secondary structure (Denslow et al., 1989).

SELECTION OF THE mRNA TIR BY RIBOSOMES

Several nonmutually exclusive mechanisms for mRNA selection by ribosomes during initiation can be envisaged. These include (i) base pairing between specific regions of the 16S rRNA and complementary sequences of the mRNA, (ii) formation of specific hydrogen or electrostatic bonds between the mRNA TIR and components of the ribosomal mRNA channel, (iii) restriction of possible interactions between ribosomes and internal regions of the mRNA by structural constraints allowing interaction only with the genuine mRNA TIR, and (iv) kinetic selection of productive 30S initiation complexes from among nonproductive pseudo-initiation complexes by the oncoming 50S subunits.

THE SD AND POTENTIAL SD-LIKE INTERACTIONS

The SD interaction was postulated over a decade ago (Shine and Dalgarno, 1974), but elegant and clear-cut evidence for its participation in initiation has been obtained only recently by use of genetically engineered ribosomes in which the 3′-end sequence of 16S rRNA was changed in parallel with the 5′ leader sequence of the mRNAs (Hui and de Boer, 1987; Jacob et al., 1987). Demonstration of the existence of the interaction is not sufficient, however, to define its role. This has been investigated by use of model mRNAs with and without the SD sequence (Calogero et al., 1988). It was found that the affinity of the ribosomes for the mRNAs containing the SD sequence is over an order of magnitude higher than that of the mRNA lacking this sequence and that its presence increases the amount of 30S and, to a lesser extent, 70S initiation complex formed at equilibrium. Nevertheless, the SD sequence confers only a modest advantage for in vitro translation and only at low mRNA input. If the mRNA concentration is properly selected, the mRNAs with and without SD sequences are translated at the same rate and with the same dependence on initiation factors (Fig. 1). Translation of these mRNAs also shows the same ionic optima and yields the same peptide product (not shown). Other experiments have also shown that the presence of the SD sequence is mechanistically irrelevant for the formation of 30S initiation complex and for the selection of the correct mRNA reading frame (Calogero et al., 1988). In conclusion, it was suggested that the function of the SD interaction is only to ensure millimolar concentrations of a potential initiation triplet near the ribosomal P site (Fig. 2).

Because of the competition of several potential mRNA TIRs for a limited number of native 30S subunits, however, the advantage offered by the SD sequence is probably greater in vivo than in vitro. In this connection, it should be recalled that mutations

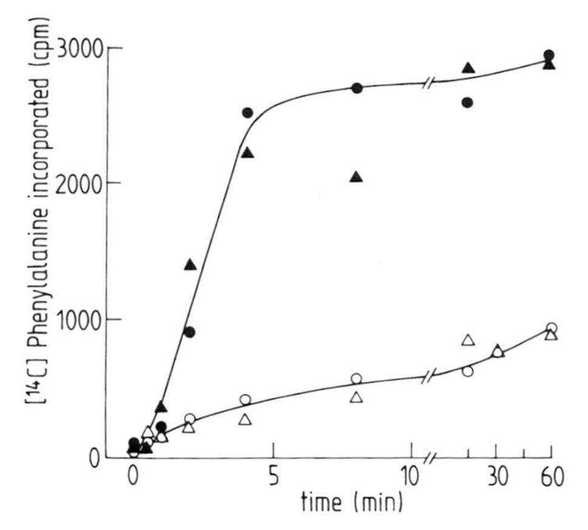

Figure 1. Time course of in vitro translation of model mRNAs (0.2 A_{260} unit each) with (△, ▲) and without (○, ●) SD sequence in the presence (▲, ●) or absence (△, ○) of initiation factors. Further experimental conditions are given by Calogero et al. (1988).

in the SD sequence of bacteriophage T7 0.3 mRNA reduced synthesis of the corresponding protein by ca. 90% in vivo (Dunn et al., 1978) but by no more than 20 to 40% in vitro (Ohsawa et al., 1984).

Other types of base pairing between mRNA and 16S rRNA have also been suggested. Examining 251 genes from *E. coli*, Petersen et al. (1988) found that 98.4 and 66.9% of the mRNAs contain, within the first 24 coding nucleotides, sequences of 3 and 4

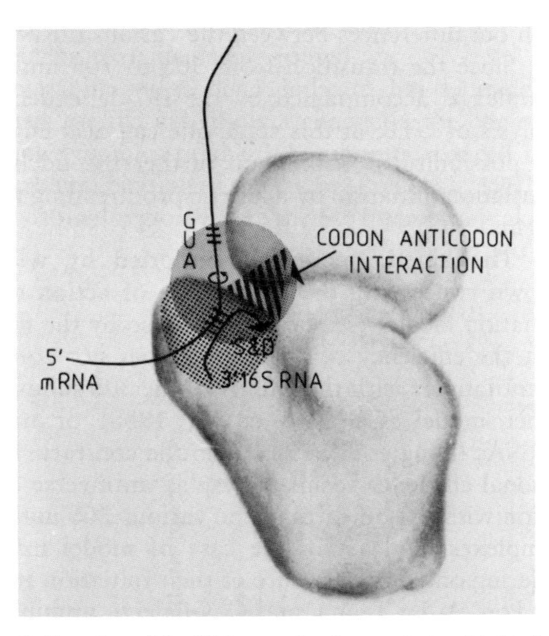

Figure 2. Function of the SD interaction in translational initiation. The stippled and striped areas indicate the presumed localization of the SD interaction and of the P-site decoding area, respectively. From Avogadro's number, it can be estimated that the concentration of the initiation triplet is ≧20 mM within the hemisphere described by the free rotation of the AUG triplet separated by a spacer of seven nucleotides from an SD sequence base paired to the 30S subunits.

nucleotides, respectively, complementary to some part of a 16-nucleotide-long sequence found at the 5′ terminus of 16S rRNA. On the other hand, Thanaraj and Pandit (1989) noticed the presence of the 5′-UGAUCC-3′ sequence (which is complementary to the 5′-GGAUCA-3′ sequence found at position 1529 of the 16S rRNA) upstream from the SD sequence in highly expressed genes. Finally, Gold et al. (1984) postulated that extensive annealing between *infC* mRNA and 16S rRNA could be a regulatory element in the expression of this gene.

These examples do not exhaust the list of potential base pairings between rRNA and mRNA proposed so far. The existence of any type of base pairing aside from the SD interaction, however, remains to be demonstrated.

OTHER TYPES OF mRNA-RIBOSOME INTERACTIONS

The mRNAs for several proteins (e.g., the lambda *c*I repressor transcribed from the p_{RM} promoter, *E. coli* DnaG, the gene V product of bacteriophage P2, *S. pneumoniae* PolA, and *Spirulina platensis* S12) begin directly with the initiation triplet AUG or have no detectable SD sequence (Ptashne et al., 1976; Smiley et al., 1982; Christie and Calendar, 1985; Lopez et al., 1989; Buttarelli et al., 1989), whereas many other mRNAs have very weak SD sequences. The trivial fact that these mRNAs and synthetic polynucleotides [e.g., poly(U)] bind to the ribosome and are translated shows that the SD sequence is not strictly necessary for mRNA-ribosome interaction and for translation initiation and that other means of mRNA recognition by ribosomes must exist.

Although the details of the model are still somewhat controversial, the available evidence suggests that the ribosome accommodates the mRNA in a U-shaped channel or trough (Evstafieva et al., 1983; Kang and Cantor, 1985; Olson et al., 1988). Very little is known, however, concerning the molecular components of this channel. At least two 30S proteins (S1 and S21) have been consistently implicated in mRNA binding (e.g., Backendorf et al., 1981; Subramanian, 1984) and it is likely that several stretches of the 16S rRNA, in addition to the 3′ terminus, line the mRNA-binding trough of the 30S subunit (Moazed and Noller, 1986). Thus, it is a rather obvious prediction that, depending on its particular structure, each mRNA binds to the 30S subunit by means of several alternative, more or less specific, interactions with both 16S rRNA and ribosomal proteins; these may involve specific RNA

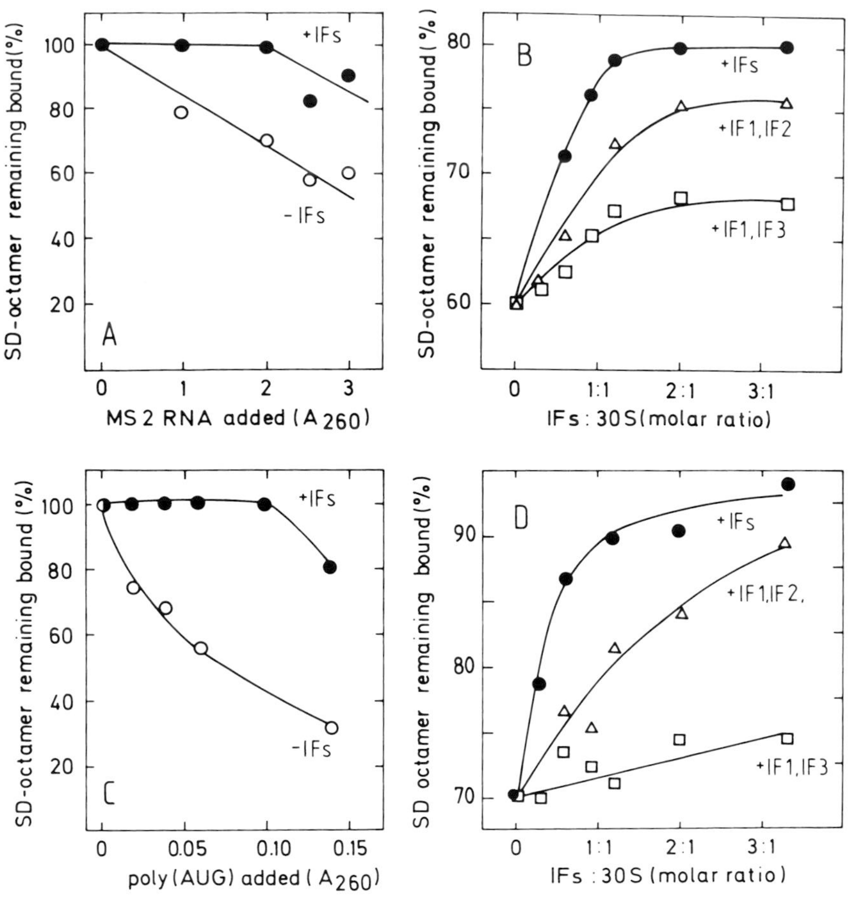

Figure 5. Competition between MS2 RNA or poly(AUG) and the SD octamer for ribosome binding. The amount of ^{32}P-labeled SD octamer bound to the 30S subunit was measured by sucrose gradient centrifugation. (A) After incubation with the indicated amounts of MS2 RNA in the presence (●) or absence (○) of initiation factors (IFs); (B) after incubation with 2.5 A_{260} units of MS2 RNA as a function of the indicated molar ratios of IF1 + IF2 + IF3/30S (●), IF1 + IF2/30S (△), and IF1 + IF3/30S (□); (C) after incubation with the indicated amounts of random poly(AUG) in the presence (●) or absence (○) of initiation factors; (D) after incubation with 0.04 A_{260} unit of poly(AUG) as a function of the indicated molar ratios of IF1 + IF2 + IF3/30S (●), IF1 + IF2/30S (△), and IF1 + IF3/30S (□). The amount of SD octamer bound in the controls (i.e., in the absence of competitors) corresponded to a stoichiometry of ~0.9 per 30S subunit and was virtually the same in the presence and absence of initiation factors. (Taken from Canonaco et al., 1989.)

ligands. Regardless of the mechanism, the results indicate that in the presence of initiation factors, the mRNA occupies a position different from that occupied in their absence and suggest the existence of a dual ribosome-binding site for mRNA. According to the scheme (Fig. 6), in the absence of factors, the mRNA preferentially occupies a ribosomal "stand-by site" roughly corresponding to the region where the SD interaction takes place. In the presence of factors, the mRNA is shifted away from this site toward another ribosomal site with only a slightly higher affinity for the template. No experimental characterization of this second site is available so far, but it is tempting to speculate that it is closer to the ribosomal P site, since it is there that the initiation factors are known to exert their kinetic influence on codon-anticodon interaction (see below).

FUNCTION OF THE INITIATION FACTORS

As mentioned above, the main role of the initiation factors is to affect kinetically codon-anticodon interaction at the P site of the 30S subunit and, ultimately, to influence which and how many 30S initiation complexes enter the elongation cycle after association with the 50S subunit. All three factors (but IF1 only in combination with the other two) stimulate the on rate of 30S initiation complex formation. The effect of IF2 is much greater for complexes containing aminoacyl-tRNAs having a blocked NH_2 group but is also observable with other aminoacyl-tRNAs (Gualerzi et al., 1986; Gualerzi et al., 1988; Pon and Gualerzi, 1988). As to the dissociation of the complexes, IF3 produces a large rate increase, especially when the bound aminoacyl-tRNA

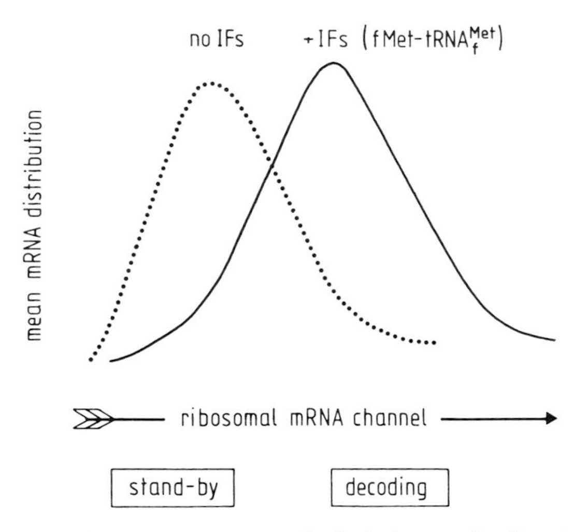

Figure 6. Alternative occupancy of a dual ribosome-binding site by mRNA. The figure shows a hypothetical mRNA mean occupancy of two partially overlapping binding sites in the mRNA channel. It is assumed that the statistical distribution of the mRNA on the ribosome can vary according to the presence or absence of the initiation factors (fMet-tRNA seems to have little or no effect) being shifted from a stand-by site corresponding to the SD interaction to the P-site decoding area. With different mRNAs, depending on the length of the spacer or on the secondary and tertiary structures of the mRNA, the two sites could overlap less or more or could even coincide. (Taken from Canonaco et al., 1989.)

is not the initiator fMet-tRNA, whereas IF2 slows down dissociation in all cases; hence, it appears as if the function of IF2 and IF3 were to lock in and unlock, respectively, the codon-anticodon interaction at the P site. This premise is illustrated by experiments in which we determined the rate of exchange between radioactive aminoacyl-tRNA prebound to 30S and subsequently added nonradioactive aminoacyl-tRNAs (Fig. 7). In some experiments, the exchanging aminoacyl-tRNAs were the same (Fig. 7A) or responded to the same codon (Fig. 7B), whereas in other cases the second aminoacyl-tRNA was not specified by the template present on the ribosome (Fig. 7C and D). Regardless of the experimental conditions, however, the exchange occurred much more rapidly in the absence of factors than in the presence of IF1 and IF2 but much more slowly than in the presence of all three factors. A notable exception is that in the presence of all three factors, virtually no exchange took place between 30S-bound fMet-tRNA and an excess of Phe-tRNA (Fig. 7D).

As a consequence of their activity, IF2 and IF3 function as discriminatory elements that allow the kinetic selection of the most fit initiation complexes over the less fit ones, with the spurious (noninitiation) complexes being an extreme example of the latter type.

TOWARD A BETTER UNDERSTANDING OF THE STRUCTURE-FUNCTION RELATIONSHIP IN INITIATION FACTORS IF2 AND IF1

Comparison of the primary structures of the IF2 proteins known so far shows that they are approximately the same size and share a high degree of sequence homology in the C-terminal two-thirds of the molecule but differ in the N-terminal one-third (Fig. 8). Large deletions from the 5′ terminus of *Bacillus stearothermophilus infB* (Fig. 8) resulted in IF2 fragments fully active in all basic translational activities of IF2. This fact and our finding (unpublished) that the NH_2-terminal part of *E. coli* IF2α can bind to DNA suggest that this region of the molecule might have some yet unidentified, species-specific function probably not related to translation.

The conserved portion of the IF2 molecule, on the other hand, is clearly divided into two domains, the central G domain containing the structural elements found in all GTP-binding sites of G proteins (Jurnak, 1985; March and Inouye, 1985) and a C-terminal domain. A proteolytic cleavage (arrow in Fig. 8) separates the two domains, producing a ~41-kilodalton G fragment and a ~24-kilodalton C fragment. The G fragment contains the intact GTP-binding site and is still capable of interacting with GTP, with 50S subunits, and very weakly with the 30S subunit and of carrying out ribosome-dependent GTPase activity. The C fragment, which is resistant to further proteolysis, contains the fMet-tRNA-binding site of IF2. The localization of this active site in the C fragment explains why an internal deletion close to the 3′ end of *infB* (Fig. 8) resulted in the complete loss of activity of the resulting protein. Despite its negligible affinity for 30S subunits and 70S ribosomes, the C fragment stimulates somewhat the ribosomal binding of fMet-tRNA in response to random poly(AUG), but it is inactive with MS2 RNA. Interestingly, the fMet-tRNA bound to ribosomes in the presence of the C fragment was found not to be puromycin reactive. The fragment is also inactive in GTPase activity and in promoting protein synthesis with either natural or synthetic templates.

The primary structures of IF1 from *E. coli*, *B. subtilis*, and three plant chloroplasts display extensive homology (Fig. 9). Starting from a synthetic modular gene (*infA**) expressing *E. coli* IF1 (Calogero et al., 1987), we have constructed several mutants by deleting amino acids from the carboxyl terminus or by making the following amino acid replacements: His-29→Tyr or Asp, His-34→Tyr or Asp, Phe-21→Lys, and Lys-38→Cys, Ser, or Arg (Fig. 9). The mutant proteins were tested in vitro in several partial reactions of the translation initiation

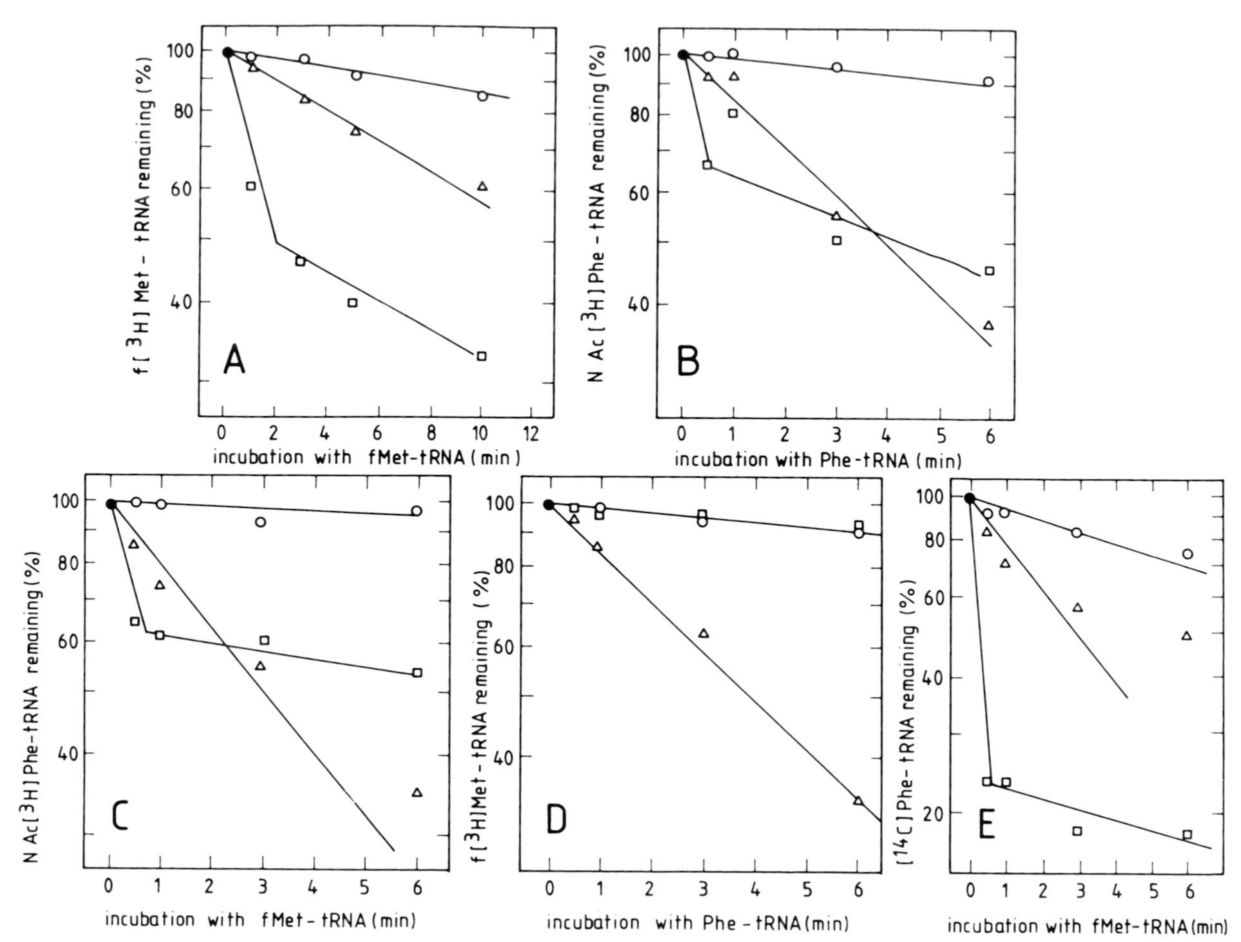

Figure 7. Effect of initiation factors on the rate of dissociation of 30S-bound aminoacyl-tRNAs in the presence of exogenously added aminoacyl-tRNAs. Ternary complexes were formed in the absence of initiation factors (△) or in the presence of IF1 and IF2 (○) or of IF1, IF2, and IF3 (□) by incubation of 30S subunits for 15 min with the radioactive aminoacyl-tRNAs indicated in the ordinates and poly(AUG) (A and D) and poly(U) (B, C, and E). The amount of ternary complex formed in each case was determined and, after subtracting the background, taken as 100% unexchanged. The nonradioactive aminoacyl-tRNAs indicated in the abscissas (a fivefold molar excess over the radioactive aminoacyl-tRNA) were then added to the complexes, and the amount of radioactive aminoacyl-tRNA remaining bound after incubation for the indicated times was determined. All experiments were carried out at 7 mM Mg^{2+} and 35°C.

pathway and for their capacity to stimulate MS2 RNA-dependent protein synthesis.

The results showed that (i) Arg-69 is part of the 30S ribosomal subunit-binding site of IF1, and its deletion results in the substantial loss of all IF1 functions; (ii) neither one of its two histidines is essential for binding of IF1 to the 30S ribosomal subunit, for stimulation of fMet-tRNA binding to

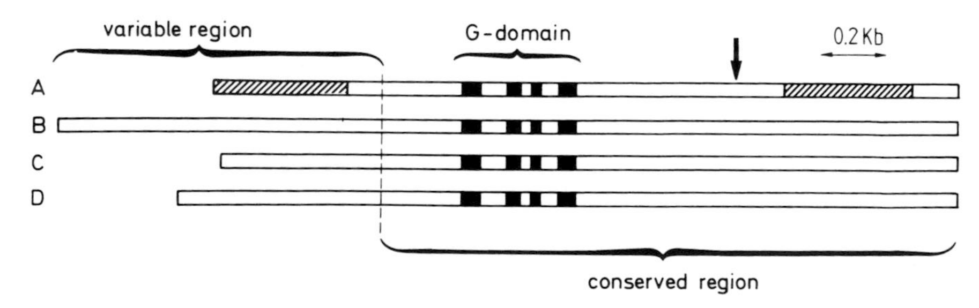

Figure 8. Schematic representation of *infB* from *B. stearothermophilus* (A), *E. coli* (IF2α [B] and IF2β [C]), and *Streptococcus faecium* (D). Black areas represent the conserved structural elements of the GTP-binding site; hatched areas represent the deletions introduced by genetic manipulations; the arrow marks the point of proteolytic cleavage that separates the G from the C domain.

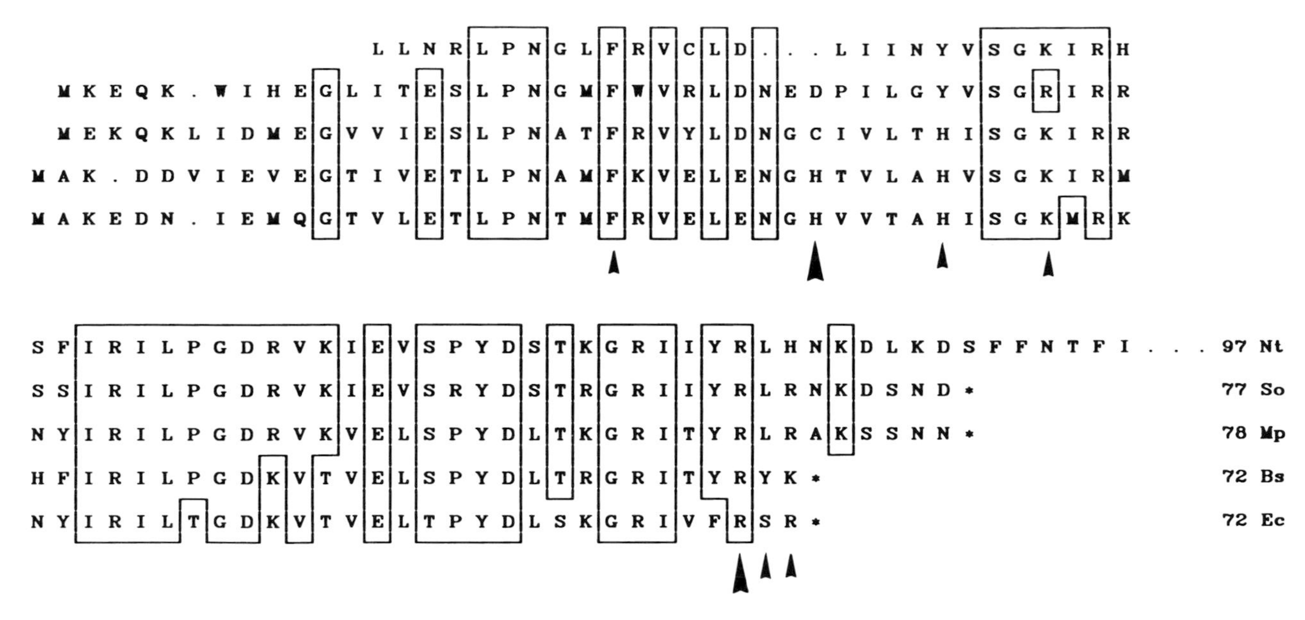

Figure 9. Comparison of the primary structures of proteins homologous to IF1 found in different procaryotic systems. Ec, *E. coli* (Pon et al., 1979); Bs, *B. subtilis* (Boylan et al., 1989); Mp, chloroplast of *Marchantia polymorpha* (Ohyama et al., 1986); So, chloroplast of *Spinacia oleracea* (Sijben-Müller et al., 1986); Nt, chloroplast of *Nicotiana tabacum* (EMBL data bank). Evidence for the in vivo existence of these proteins is available only in the case of *E. coli*. Identical residues are boxed. Arrowheads indicate the residues in the *E. coli* protein that have been changed or deleted by mutagenesis. The large arrowheads indicate amino acids found to be functionally important.

30S or 70S ribosomal particles, or for MS2 RNA-dependent protein synthesis; but (iii) His-29 is involved in the 50S subunit-induced ejection of IF1 from the 30S ribosomal subunit (Gualerzi et al., 1989).

REFERENCES

Backendorf, C., C. J. C. Ravensbergen, J. Van der Plas, J. H. van Boom, G. Veeneman, and J. Van Duin. 1981. Basepairing potential of the 3′ terminus of 16S RNA: dependence on the functional state of the 30S subunit and the presence of protein S21. *Nucleic Acids Res.* **9:**1425–1444.

Band, L., and D. J. Henner. 1984. *Bacillus subtilis* requires a "stringent" Shine-Dalgarno region for gene expression. *DNA* **3:**17–21.

Blumberg, B. M., T. Nakamoto, and I. S. Goldberg. 1975. Kinetic evidence for the obligatory formation of a 30S initiation complex in polyphenylalanine synthesis initiated with N-acetylphenylalanyl-tRNA. *Biochemistry* **14:**2889–2894.

Blumberg, B. M., T. Nakamoto, and F. J. Kézdy. 1979. Kinetics of initiation of bacterial protein synthesis. *Proc. Natl. Acad. Sci. USA* **76:**251–255.

Bonham-Smith, P., and D. P. Bourque. 1989. Translation of chloroplast-encoded mRNA: potential initiation and termination signals. *Nucleic Acids Res.* **17:**2057–2080.

Boylan, S. A., J. W. Suh, S. M. Thomas, and C. W. Price. 1989. Gene encoding the alpha core subunit of *Bacillus subtilis* RNA polymerase is cotranscribed with the genes for initiation factor 1 and ribosomal proteins B, S13, S11 and L17. *J. Bacteriol.* **171:** 2553–2562.

Buttarelli, F. R., R. A. Calogero, O. Tiboni, C. O. Gualerzi, and C. L. Pon. 1989. Characterization of the *str* operon genes from *Spirulina platensis* and their evolutionary relationship to those of other prokaryotes. *Mol. Gen. Genet.* **217:**97–104.

Calogero, R. A., C. L. Pon, M. A. Canonaco, and C. O. Gualerzi. 1988. Selection of the mRNA translation initiation region by *Escherichia coli* ribosomes. *Proc. Natl. Acad. Sci. USA* **85:** 6427–6431.

Calogero, R. A., C. L. Pon, and C. O. Gualerzi. 1987. Chemical synthesis and in vivo hyperexpression of a modular gene coding for *Escherichia coli* translational initiation factor IF1. *Mol. Gen. Genet.* **208:**63–69.

Canonaco, M. A., C. O. Gualerzi, and C. L. Pon. 1989. Alternative occupancy of a dual ribosomal binding site by mRNA affected by translation initiation factors. *Eur. J. Biochem.* **182:**501–506.

Canonaco, M. A., C. L. Pon, R. T. Pawlik, R. Calogero, and C. O. Gualerzi. 1987. Relationship between size of mRNA ribosomal binding site and initiation factor function. *Biochimie* **69:**957–963.

Christie, G. E., and R. Calendar. 1985. Bacteriophage P2 late promoters. II. Comparison of the four late promoter sequences. *J. Mol. Biol.* **181:**373–382.

Dams, E., L. Hendriks, Y. van de Peer, J.-M. Neefs, G. Smits, I. Vandenbempt, and R. DeWachter. 1988. Compilation of small ribosomal subunit RNA sequences. *Nucleic Acids Res.* **16**(Suppl.):r87–r173.

Denslow, N. D., G. S. Michaels, J. Montoya, G. Attardi, and T. W. O'Brien. 1989. Mechanism of mRNA binding to bovine mitochondrial ribosomes. *J. Biol. Chem.* **264:**8328–8338.

Dreyfus, M. 1988. What constitutes the signal for the initiation of protein synthesis on *Escherichia coli* mRNAs? *J. Mol. Biol.* **204:**79–94.

Dunn, J. J., E. Buzash-Pollert, and F. W. Studier. 1978. Mutations of bacteriophage T7 that affect initiation of synthesis of the gene 0.3 protein. *Proc. Natl. Acad. Sci. USA* **75:**2741–2745.

Ehresmann, C., H. Moine, M. Mougel, J. Dondon, M. Grunberg-Manago, J. P. Ebel, and B. Ehresmann. 1986. Crosslinking of initiation factor IF3 to *Escherichia coli* 30S ribosomal subunits by trans-diaminedichloroplatinum(II): characterization of two crosslinking sites in 16S rRNA: a possible way of functioning for IF3. *Nucleic Acids Res.* **14**:4803–4821.

Evstafieva, A. G., I. N. Shatsky, A. A. Bogdanov, Y. P. Semenkov, and V. D. Vasiliev. 1983. Localization of 5′ and 3′ ends of the ribosome-bound segment of template polynucleotides by immune electron microscopy. *EMBO J.* **2**:799–804.

Gold, L. 1988. Posttranscriptional regulatory mechanisms in *Escherichia coli*. *Annu. Rev. Biochem.* **57**:199–233.

Gold, L., G. Stormo, and R. Saunders. 1984. *Escherichia coli* translational initiation factor IF3: a unique case of translational regulation. *Proc. Natl. Acad. Sci. USA* **81**:7061–7065.

Gren, E. J. 1984. Recognition of messenger RNA during translational initiation in *Escherichia coli*. *Biochimie* **66**:1–29.

Gualerzi, C. O., R. A. Calogero, M. A. Canonaco, M. Brombach, and C. L. Pon. 1988. Selection of mRNA by ribosomes during prokaryotic translational initiation, p. 317–330. *In* M. F. Tuite, M. Picard, and M. Bolotin-Fukuhara (ed.), *Genetics of Translation*. Springer-Verlag, Berlin.

Gualerzi, C., and C. L. Pon. 1981. Protein biosynthesis in prokaryotic cells; mechanism of 30S initiation complex formation in *E. coli*, p. 805–826. *In* M. Balaban, J. L. Sussman, W. Traub, and A. Yonath (ed.), *Structural Aspects of Recognition and Assembly in Biological Macromolecules*. ISS, Rehovot, Israel.

Gualerzi, C., C. L. Pon, R. T. Pawlik, M. A. Canonaco, M. Pace, and W. Wintermeyer. 1986. Role of the initiation factors in *Escherichia coli* translational initiation, p. 621–641. *In* B. Hardesty and G. Kramer (ed.), *Structure, Function and Genetics of Ribosomes*. Springer Verlag, New York.

Gualerzi, C. O., R. Spurio, A. La Teana, R. Calogero, B. Celano, and C. L. Pon. 1989. Site-directed mutagenesis of *Escherichia coli* translation initiation factor IF1. Identification of the amino acids involved in its ribosomal binding and recycling. *Protein Eng.* **3**:133–138.

Hager, P. W., and J. C. Rabinowitz. 1985. Translational specificity in *Bacillus subtilis*, p. 1–32. *In* D. A. Dubnau (ed.), *The Molecular Biology of the Bacilli*, vol. 2. Academic Press, Inc., New York.

Hartz, D., D. S. McPheeters, R. Traut, and L. Gold. 1988. Extension inhibition analysis of translation initiation complexes. *Methods Enzymol.* **164**:419–425.

Hui, A., and H. A. de Boer. 1987. Specialized ribosome system: preferential translation of a single mRNA species by a subpopulation of mutated ribosomes in *Escherichia coli*. *Proc. Natl. Acad. Sci. USA* **84**:4762–4766.

Iserentant, D., and W. Fiers. 1980. Secondary structure of mRNA and efficiency of translation initiation. *Gene* **9**:1–12.

Jacob, W. F., M. Santer, and A. E. Dahlberg. 1987. A single base change in the Shine-Dalgarno region of 16S rRNA of *Escherichia coli* affects translation of many proteins. *Proc. Natl. Acad. Sci. USA* **84**:4757–4761.

Jurnak, F. 1985. Structure of the GDP domain of EF-Tu and location of the amino acids homolgous to *ras* oncogene proteins. *Science* **230**:32–36.

Kang, C., and C. R. Cantor. 1985. Structure of ribosome-bound messenger RNA as revealed by enzymatic accessibility studies. *J. Mol. Biol.* **181**:241–251.

Kastelein, R. A., B. Berkhout, G. P. Overbeek, and J. Van Duin. 1983. Effect of the sequences upstream from the ribosome-binding site on the yield of protein from the cloned gene for phage MS2 coat protein. *Gene* **23**:245–254.

Lang, V., C. Gualerzi, and J. E. G. McCarthy. 1989. Ribosomal affinity and translational initiation in *Escherichia coli*. *J. Mol. Biol.* **210**:659–663.

Liao, H.-X., and L. L. Spremulli. 1989. Interaction of bovine mitochrondrial ribosomes with messenger RNA. *J. Biol. Chem.* **264**:7518–7522.

Lopez, P., S. Martinez, A. Diaz, M. Espinosa, and S. A. Lacks. 1989. Characterization of the *polA* gene of *Streptococcus pneumoniae* and comparison of the DNA polymerase I it encodes to homologous enzymes from *Escherichia coli* and phage T7. *J. Biol. Chem.* **264**:4255–4263.

March, P. E., and M. Inouye. 1985. GTP binding membrane protein of *Escherichia coli* with sequence homology to initiation factor 2 and elongation factors Tu and G. *Proc. Natl. Acad. Sci. USA* **82**:7500–7504.

McPheeters, D. S., A. Christensen, E. T. Young, G. Stormo, and L. Gold. 1986. Translational regulation of expression of the bacteriophage T4 lysozyme gene. *Nucleic Acids Res.* **14**:5813–5826.

Moazed, D., and H. F. Noller. 1986. Transfer RNA shields specific nucleotides in 16S ribosomal RNA from attack by chemical probes. *Cell* **47**:985–994.

Moine, H., P. Romby, M. Springer, M. Grunberg-Manago, J. P. Ebel, C. Ehresmann, and B. Ehresmann. 1988. Messenger RNA structure and gene regulation at the translational level in *Escherichia coli*: the case of threonine-tRNA^{Thr} ligase. *Proc. Natl. Acad. Sci. USA* **85**:7892–7896.

Ohsawa, H., P. Herrlich, and C. Gualerzi. 1984. In vitro template activity of 0.3 mRNA from wild type and initiation mutants of bacteriophage T7. *Mol. Gen. Genet.* **196**:53–58.

Ohyama, K., H. Fukuzawa, T. Kohchi, H. Shirai, T. Sano, S. Sano, K. Umesono, Y. Shiki, M. Takeuchi, Z. Chang, S. I. Aota, H. Inokuchi, and H. Ozeki. 1986. Chloroplast gene organization deduced from complete sequence of liverwort *Marchantia polymorpha* chloroplast DNA. *Plant Mol. Biol. Rep.* **4**:148–175.

Olson, H. M., L. S. Lasater, P. A. Cann, and D. G. Glitz. 1988. Messenger RNA orientation on the ribosome. *J. Biol. Chem.* **263**:15196–15204.

Petersen, G. B., P. A. Stockwell, and D. F. Hill. 1988. Messenger RNA recognition in *Escherichia coli*: a possible second site of interaction with 16S ribosomal RNA. *EMBO J.* **7**:3957–3962.

Pon, C. L., and C. O. Gualerzi. 1988. Genetic and mechanistic aspects of translational initiation in bacteria, p. 137–150. *In* M. Bissell, G. Dehò, G. Sironi, and A. Torriani (ed.), *Gene Expression and Regulation. The Legacy of Luigi Gorini*. Elsevier Science Publishers, Amsterdam.

Pon, C. L., R. T. Pawlik, and C. Gualerzi. 1982. The topographical localization of IF3 on *Escherichia coli* 30S ribosomal subunits as a clue to its way of functioning. *FEBS Lett.* **137**:163–167.

Pon, C. L., B. Wittmann-Liebold, and C. Gualerzi. 1979. Structure-function relationships in *Escherichia coli* initiation factors. II. Elucidation of the primary structure of initiation factor IF1. *FEBS Lett.* **101**:157–160.

Ptashne, M., K. Backman, M. Z. Humayun, A. Jeffrey, R. Maurer, B. Meyer, and R. T. Saure. 1976. Autoregulation and function of a repressor in bacteriophage lambda. *Science* **194**:156–161.

Risuleo, G., C. Gualerzi, and C. Pon. 1976. Specificity and properties of the destabilization, induced by initiation factor IF3, of ternary complexes of the 30S ribosomal subunit, aminoacyl-tRNA and polynucleotides. *Eur. J. Biochem.* **67**:603–613.

Ruf, M., and H. Kössel. 1988. Occurrence and spacing of ribosome recognition sites in mRNAs of chloroplasts from higher plants. *FEBS Lett.* **240**:41–44.

Scherer, G. F. E., M. D. Walkinshaw, S. Arnott, and D. J. Morre. 1980. The ribosome binding sites recognized by *E. coli* ribosomes have regions with signal character in both the leader and protein coding segments. *Nucleic Acids Res.* **8**:3895–3907.

Schneider, T. D., G. D. Stormo, L. Gold, and A. Ehrenfeucht. 1986. The information content of binding sites on nucleotide sequences. *J. Mol. Biol.* **188**:415–431.

Shine, J., and L. Dalgarno. 1974. The 3′-terminal sequence of *Escherichia coli* 16S ribosomal RNA: complementarity to nonsense triplets and ribosome binding sites. *Proc. Natl. Acad. Sci. USA* **71**:1342–1346.

Sijben-Müller, G., R. B. Hallick, J. Alt, P. Westhoff, and R. G. Herrmann. 1986. Spinach plastid genes coding for initiation factor IF1, ribosomal protein S11 and RNA polymerase α-subunit. *Nucleic Acids Res.* **14**:1029–1044.

Smiley, B. L., J. R. Lupski, P. S. Svec, R. McMacken, and G. N. Godson. 1982. Sequences of the *Escherichia coli dnaG* primase gene and regulation of its expression. *Proc. Natl. Acad. Sci. USA* **79**:4550–4554.

Stanssens, P., E. Remaut, and W. Fiers. 1985. Alterations upstream from the Shine-Dalgarno region and their effect on bacterial gene expression. *Gene* **36**:211–223.

Steitz, J. A., A. J. Wahba, M. Laughrea, and P. B. Moore. 1977. Differential requirements for polypeptide chain initiation complex formation at the three bacteriophage R17 initiator regions. *Nucleic Acids Res.* **4**:1–15.

Stiege, W., K. Stade, D. Schüler, and R. Brimacombe. 1988. Covalent crosslinking of poly(A) to E. coli ribosomes and localization of the crosslinked site within the 16S RNA. *Nucleic Acids Res.* **16**:2369–2388.

Subramanian, A. R. 1984. Structure and functions of the largest *Escherichia coli* ribosomal protein. *Trends Biochem. Sci.* **9**: 491–494.

Thanaraj, T. A., and M. W. Pandit. 1989. An additional ribosome-binding site on mRNA of highly expressed genes and a bifunctional site on the colicin fragment of 16S rRNA from *Escherichia coli*: important determinants of the efficiency of translation initiation. *Nucleic Acids Res.* **17**:2973–2985.

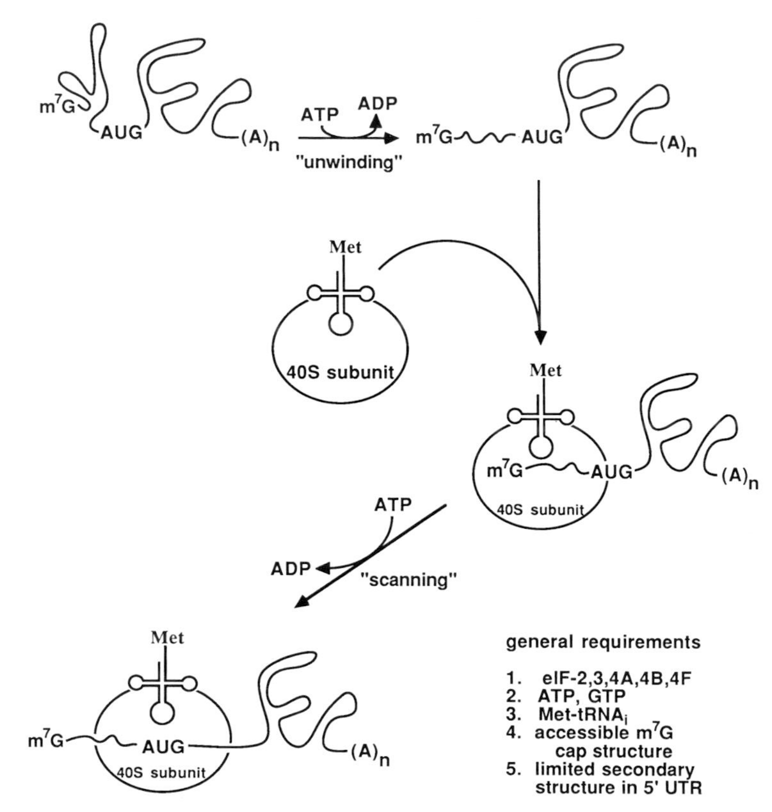

Figure 3. Standard initiation of protein synthesis in eucaryotes. Emphasized are the two unique characteristics of eucaryotic initiation. The first is the eIF-2, GTP-dependent binding of the initiator tRNA (Met-tRNA$_i$) to the 40S subunit; the second is the ATP-dependent movement of the 40S subunit along the mRNA to the AUG initiation codon. UTR, Untranslated region. See text for details.

elongation factors that are unique to particular factors (eIF-4D, EF-1α and EF-2) or are general modifications (methylation, phosphorylation, or glycosylation) that have not been reported in bacterial systems (Dever et al., 1989; Moldave, 1985; Pain, 1986; Sonenberg, 1988). This introduction of complexity may reflect fine-tune controls of translation, necessary in part because of the separation and compartmentalization of translation and transcription.

As a starting point for appreciating the difference in initiation between procaryotes and eucaryotes, a major element is that eucaryotic mRNAs are generally monocistronic, in contrast to bacterial systems, in which many, if not most, proteins are encoded by polycistronic mRNAs. The common elements of a eucaryotic mRNA are (i) a 5′ m^{7}G cap structure, (ii) a 5′ untranslated region of 50 to 150 bases, (iii) a single coding region, (iv) a 3′ untranslated region that may be 150 to 1,000 bases long, and (v) a poly(A) tail. In addition, it has been noted with globin mRNA that there is extensive secondary structure in the mRNA such that one can even perform thermal hyperchromicity studies, indicating that about 60% of the molecule is double stranded. Another complication is that although there tends to be sequence conservation at the 3′ end of the 18S rRNAs between species, there has been no identification of a complementary Shine-Dalgarno-like sequence in all eucaryotic mRNAs (the rarity of the few sequences that do show up would be expected statistically by chance).

Thus, the problem has emerged as to how the mRNA can be specifically attached to the 40S subunit. To this end, eucaryotes have evolved a number of protein factors and a more sophisticated ribosome. A generalized scheme is presented in Fig. 3. As a result of experiments with purified protein factors and in vitro experiments with mutated mRNAs, the following is thought to occur: (i) a protein factor (eIF-4F) recognizes and binds to the m^{7}G cap structure, and (ii) in the presence of ATP, eIF-4A, and eIF-4B, the mRNA is unwound.

These first two steps have been demonstrated to occur in the absence of ribosomes. However, it is possible that one or more of these factors is bound to the ribosome during the unwinding process (Merrick et al., 1987; Moldave, 1985; Pain, 1986; Ray et al., 1985; Rhoades et al., 1985; Sonenberg et al., 1983). In any event, it appears that the attachment of the mRNA to the 40S subunit occurs via protein-protein (or protein-ribosome) interactions, not by an interaction of the mRNA with the 18S rRNA.

The next phase involves movement of the mRNA on the ribosome to correctly position the AUG codon in the P site, a process referred to as scanning (Kozak, 1983, 1986, 1987). The characteristics of the scanning process are that it requires energy, from ATP hydrolysis, and that extensive secondary structure in the 5′ untranslated region inhibits this process (Baim et al., 1985; Kozak, 1986; Pelletier and Sonenberg, 1985). At present, it is not clear whether this process is driven by initiation factors (especially eIF-4A, -4B, and -4F) or whether this is an inherent capability of the 40S subunit. Recognition of the initiating AUG codon is via the ternary complex of eIF-2 · GTP · Met-tRNA$_f$, primarily via the codon-anticodon base pairing (Cigan et al., 1988; Cigan et al., 1989; Donahue et al., 1988). In most eucaryotic mRNAs, the initiating AUG is that which is nearest the 5′ end of the mRNA (Kozak, 1987). In sum, of the three essential steps, mRNA binding, scanning, and AUG codon recognition, in the first and third instances, the ribosome provides specificity by its interaction with initiation factors. Only the scanning process may represent a true ribosomal function, but in any event, this process has no equivalent in bacterial systems.

The foregoing discussion represents a rather simple statement of what occurs with the majority of initiation events in eucaryotes. This pathway is supported by both in vitro and in vivo experiments; readers wanting more detail are referred to the cited reviews (Kozak, 1983, 1987; Merrick et al., 1987; Moldave, 1985; Pain, 1986; Rhoads et al., 1985; Sonenberg, 1988). What is perhaps more intriguing are the exceptions to the above-described general pathway. One such exception is the translation of more than a single open reading frame in polycistronic mRNAs. The general interpretation of this process, called reinitiation (Kozak, 1987), is presented in Fig. 4. In this scheme, it is suggested that after termination, the 60S subunit dissociates but the 40S subunit remains bound and, in an ATP-dependent manner, migrates in a 3′ direction until the next AUG codon is found. At present, this mechanism poses two problems. First, how and when does the initiator tRNA become bound to the 40S subunit? Second, is the mechanism of scanning and Met-tRNA$_f$ binding really different in that the major initiation scheme requires Met-tRNA$_f$ binding before mRNA binding (Fig. 3), whereas reinitiation would necessitate mRNA binding before binding of the initiator tRNA? At present, there are insufficient in vitro or in vivo data to allow anything more than

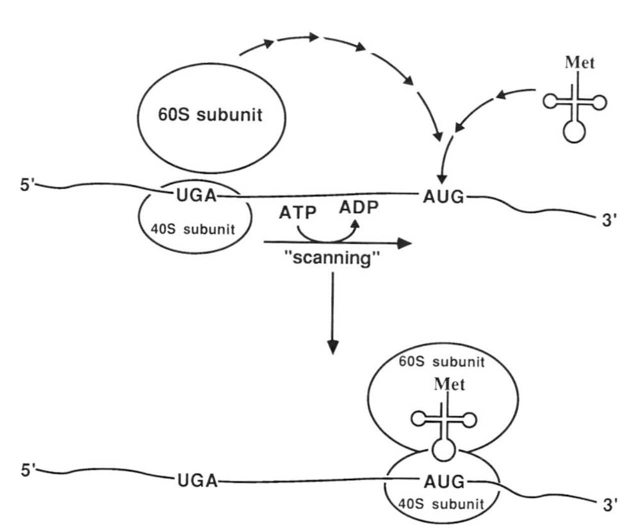

Figure 4. Reinitiation of protein synthesis in eucaryotes. Upon completion of translation of an upstream open reading frame (indicated by the stop codon UGA), the 60S subunit releases from the 40S subunit-mRNA complex. The 40S subunit moves in an ATP-dependent manner to the next AUG start codon. The temporal binding of the initiator tRNA is not known, nor is the protein responsible for its binding. Subsequent joining with a 60S subunit completes the reinitiation process.

speculations on the factors and requirements for this process, but the evidence from many viral systems and several mRNAs naturally under translational control (i.e., *GCN4* mRNA in yeast cells) makes it clear that this process occurs (Mueller and Hinnebusch, 1986).

An equally rare mechanism of initiation for which there are both in vivo and in vitro data, called internal initiation (Abramson et al., 1988; Pelletier and Sonenberg, 1988; Sonenberg, 1988), is depicted in Fig. 5. This scheme is formally very similar to

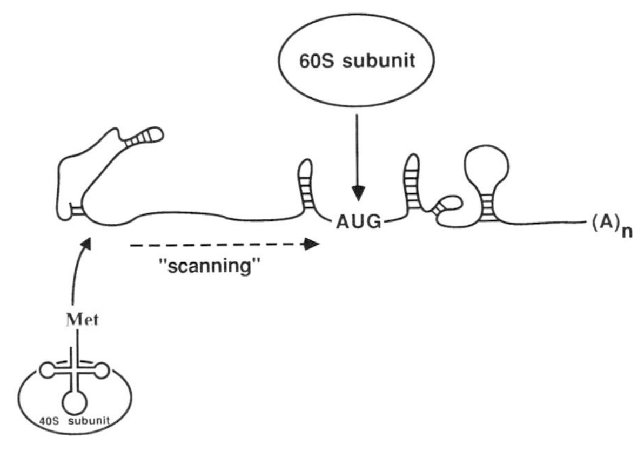

Figure 5. Internal initiation of eucaryotic protein synthesis. In this scheme, the initial binding of the 40S subunit is to an internal portion of the mRNA and not at the usual m^{7}G cap structure. This binding is thought to be favored by single-stranded regions in the mRNA. Subsequent ATP-dependent scanning allows selection of the starting AUG codon, and 60S subunit joining completes the initiation process. See text for details.

V. ELONGATION

Elongation is perhaps the most complex step in the process of translation. The tRNA-mRNA complex must work its way through the ribosome in a mechanical, reproducible fashion, but the details of how this happens are only now beginning to be understood. Instead of two tRNA-binding sites, there is now strong support for three sites. Determining the location of such sites, their relative positions, and their individual functions is an active area of research, but there is yet little evidence about the extent to which sites may map at discrete locations. The interaction of the elongation factors with tRNA and mRNA on the rRNA is beginning to be sorted out. Even ribosomal proteins seem to be showing their colors in this multistate functional step.

Chapter 24

Some Structural Aspects of Elongation

ANDERS LILJAS

The cycle of elongation is the central event in protein biosynthesis. Studies of elongation have extended over several decades, and current interest remains high. Textbook descriptions suggest that this process is well understood; however, contributions to this volume show that there are still several dilemmas and controversial findings. Some problems are: How many GTP molecules are hydrolyzed per peptide bond formed? What is the stoichiometry of EF-Tu and aminacyl-tRNA? How many binding sites for tRNA are there on the ribosome, and what are their role and location? Which components of the ribosome contribute to the activities during elongation?

A number of techniques have contributed to our understanding of elongation, and some are expected to improve our current knowledge. For investigation of the steps and mechanisms involved in elongation, an even more refined kinetic analysis will be required. However, it is also obvious that structural methods, in particular macromolecular crystallography, combined with sequence comparisons of components from different organisms will yield information about functional sites. The expectation that essential structures and moieties around the functional sites should be highly conserved has been rewarding. To test various ideas and hypotheses about functional groups and mechanisms, site-directed mutagenesis will remain invaluable.

This chapter presents a short summary of some structural aspects of elongation. The three main topics are the steps, the sites, and the components in elongation. Several reviews can be consulted for further details concerning translational elongation (Kaziro, 1978; Liljas, 1982; Kurland and Ehrenberg, 1984; Spirin, 1986). I will deal mainly with the elongation phase of procaryotic translation, but some comparisons with archaebacterial and eucaryotic systems will be made. One type of observation that I will repeatedly return to is that at least some region of the binding sites for tRNA is in the proximity of the thiostrepton-binding site on the 50S subunit.

THE STEPS

Figure 1 shows a simplified representation of the elongation cycle. This scheme contains the normally included pathway (full circle) plus the two inner cycles that are used to improve the fidelity of protein synthesis: the cycle of initial selection, which is used to test different ternary complexes against the codon in the A site (frequent event; heavy lines), and the cycle of proofreading, which can discard an aminoacyl-tRNA after the EF-Tu-carried GTP has been hydrolyzed (infrequent event; thin lines). The initiation and termination phases of protein synthesis are also included for completeness.

The Elongation Factors and GTP Hydrolysis

Before discussing the steps in elongation, I will briefly describe the function of the elongation factors. EF-Tu, EF-G, and IF2 are all able to bind GTP or GDP. In complex with GDP, they all have low affinity for the ribosome and dissociate from it (Kaziro, 1978). On the other hand, in complex with GTP, EF-Tu has high affinity for aminoacyl-tRNA, and this ternary complex has high affinity for the ribosomal A site. Likewise, EF-G in complex with GTP has high affinity for the ribosome in the translocation state of the elongation cycle, but it can also bind to 50S subunits in an uncoupled reaction (Kaziro, 1978). The factors are released from the ribosome after hydrolysis of the bound GTP molecule. The antibiotic kirromycin bound to EF-Tu makes the factor hydrolyze GTP even off the ribosome (Chinali et al., 1977). Since EF-G and IF2 have homologous regions

Anders Liljas ■ Molecular Biophysics, Chemical Center, Lund University, Box 124, S-221 00 Lund, Sweden.

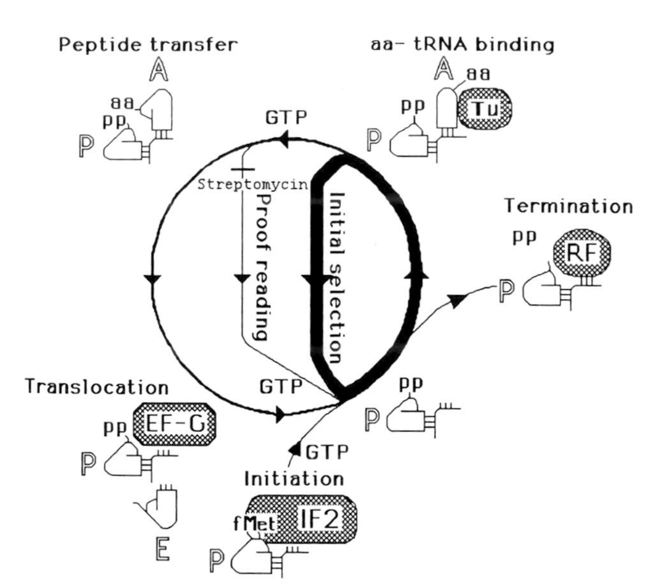

Figure 1. Simplified representation of the cycle of elongation. The tRNAs are symbolically represented and located in the A, P, and E sites. The factor proteins are hatched. Abbreviations: GTP, GTP hydrolysis; aa and pp, aminoacyl and polypeptidyl, respectively. The complete cycle resulting in the incorporation of one amino acid is represented by the outer circle. The initial phase is at the lower right, with a polypeptidyl tRNA in the P site. The codon to be used next is located in the empty A site. In the initial selection (inner loop, heavy line), different ternary complexes of EF-Tu, GTP, and aminoacyl-tRNA are tested against the A-site codon. When the GTP bound to EF-Tu is hydrolyzed, the elongating ribosome enters the proofreading cycle (thin inner loop), during which EF-Tu · GDP dissociates from the ribosome. The fate of the tRNA again depends on the codon-anticodon matching. If the matching is good, the tRNA will finally be properly bound into the fully functional A site. If the matching is less perfect, the dissociation rate of the tRNA is high enough for it to fall off before the amino acid is incorporated into the growing polypeptide in the peptidyl transfer step. Here the polypeptide bound to the tRNA in the P site is transferred to the amino acid bound to the tRNA in the A site. The final step in the elongation cycle is the translocation stimulated by EF-G. In this step, EF-G, by binding to the ribosome, causes the peptidyl-tRNA now located in the A site to move into the P site. At the same time, the mRNA is moved by one codon to expose the next one in the A site. The G factor also dissociates from the ribosome in conjunction with GTP hydrolysis. The relationship of the elongation cycle to the initiation and termination phases is also indicated.

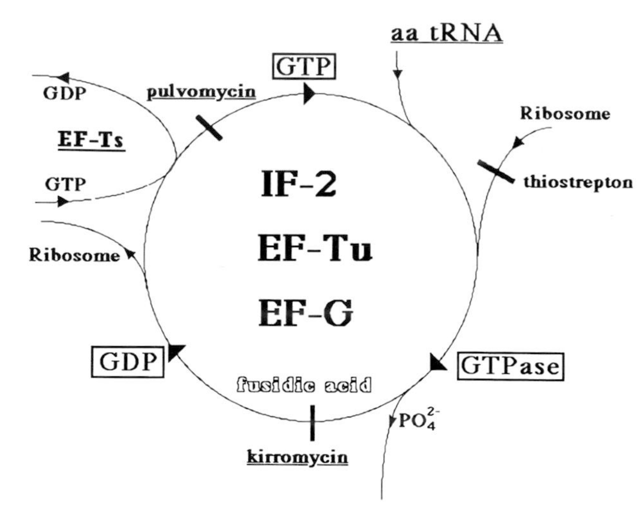

Figure 2. Simplified representation of the cycle of conformational changes that the GTP-binding factors go through. Antibiotics and tRNAs that interact exclusively with EF-Tu are underlined; fusidic acid that binds to EF-G is marked. The remainder are interactions common to all three factors. The names for the three main conformations are enclosed in boxes. A few antibiotics inhibiting these conformational changes or interactions are also indicated. aa, Aminoacyl.

corresponding to the GTP-binding domain of EF-Tu (reviewed in Liljas et al., 1986), one might expect that they also are able to hydrolyze their bound GTP molecules.

It is difficult to imagine the changes in affinities and the induction of GTP hydrolysis for the factors without associated conformational changes. Thus, EF-Tu, EF-G, and IF2 would have at least three main conformations: a GTP conformation, a GTPase conformation, and a GDP conformation. These three factors probably go through similar cycles of conformational changes (Fig. 2). The factors can also bind noncleavable analogs of GTP, GMPPCP (5′-guanylyl methylenediphosphate), and GMPPNP (5′-guanylyl imidodiphosphate). In these cases, the factor will remain bound to the ribosome, probably in the GTPase conformation, without the ability to proceed.

The factors can also be inhibited by antibiotics in different specific stages of their functional cycles. This gives opportunities to investigate the system in states where most molecules have the same conformation. Thus, when EF-G is inhibited by fusidic acid, the factor becomes locked on the ribosome even though the GTP molecule already is hydrolyzed and the translocation is performed (Bodley et al., 1970). EF-Tu can be inhibited in a similar manner by kirromycin, which inhibits the release of the factor from the ribosome after GTP hydrolysis (Wolf et al., 1977; Parmeggiani and Swart, 1985). Thiostrepton binds to the ribosome and prevents these three factors as well as release factors RF-1 and RF-2 from binding (Gale et al., 1981). Because there is only one thiostrepton bound per ribosome, and for a number of other reasons as reviewed by Liljas (1982), there is good reason to assume that their binding sites are partly overlapping.

Aminoacyl-tRNA Binding: Initial Selection

The cycle of elongation begins with a peptidyl-tRNA in the so-called P site and an empty A site, where the next codon to be translated is exposed. If this is a stop codon, one of the release factors will

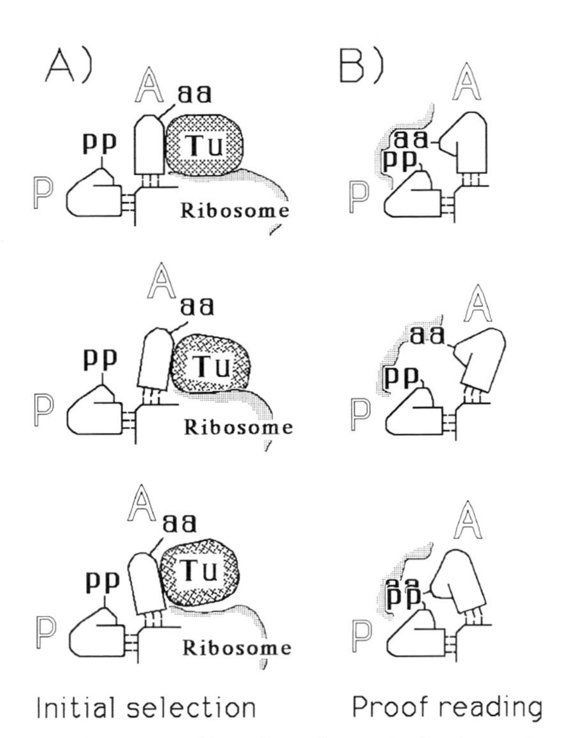

Figure 3. Illustration of how the codon-anticodon interaction can be used twice to achieve the fidelity needed. The top examples represent a cognate tRNA; the bottom examples represent two noncognate tRNAs. (A) The initial selection phase of aminoacyl-tRNA binding. The dissociation rate for the noncognate complexes will be so high that they will fall off before they can proceed to the subsequent steps in elongation. (B) The proofreading step in elongation. The codon-anticodon interaction determines the possibility for the tRNA in the A site to properly position its aminoacyl (aa) group in the peptidyl transfer site next to the nascent polypeptide (pp). The noncognate tRNAs have a high probability of falling off rather than becoming incorporated into the growing polypeptide.

bind and the elongation cycle is terminated by hydrolysis of the polypeptide from the tRNA (Fig. 1). If it is a codon corresponding to an amino acid, the A site has a high affinity for an aminoacyl-tRNA in complex with EF-Tu · GTP. A majority of the tRNA molecules are normally charged and found in the cytoplasm in the form of this ternary complex. These complexes are tested against the codon in the A site. If they are not recognized as properly matching, they will dissociate and return to the initial stage through the cycle of initial selection. After a number of such attempts, a complex will match the requirements and the elongation proceeds to the next branch point (Fig. 1). Figure 3A is a simple structural illustration of how the codon-anticodon could influence the affinity of the ribosome for the ternary complex, aminoacyl-tRNA · EF-Tu · GTP. The process can also be described in a reaction scheme (Fig. 4; Kurland and Ehrenberg, 1984). The reversible initial selection is the first step. The noncognate substrate, a ternary

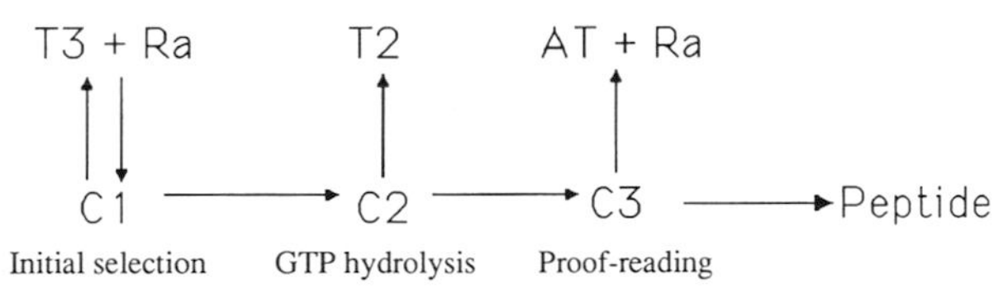

Figure 4. Scheme of the reactions in binding of the aminoacyl-tRNA (AT) to the ribosome (Kurland and Ehrenberg, 1984). T3 represents the ternary complex of EF-Tu, GTP, and aminoacyl-tRNA; T2 represents the binary complex of EF-Tu and GDP. Ra, Ribosomes with a peptidyl-tRNA in the P site and an empty A site; C1, ribosomes Ra in the initial selection phase with T3 bound at the A site; C2, ribosomes with EF-Tu hydrolyzing GTP; C3, ribosomes in the proofreading phase with aminoacyl-tRNA bound in the A site after dissociation of T2.

complex in which the anticodon does not properly match the codon in the A site (Fig. 3A), has a higher rate for dissociating from the ribosome than does the cognate substrate. Some further details of this scheme are discussed below.

Proofreading

It has long been recognized that the fidelity of protein synthesis cannot entirely depend on one step of codon-anticodon recognition. This would explain only part of the error frequency of about 10^{-3} to 10^{-4}. A second step, called the proofreading step, was first postulated on theoretical grounds (Hopfield, 1974; Ninio, 1975) and later experimentally observed (Thompson and Stone, 1977; Yates, 1979; Ruusala et al., 1982). The proofreading step or cycle will also return the ribosome to the state with an empty A site (Fig. 1 and 4). After the GTP molecule bound to EF-Tu has been hydrolyzed to GDP and phosphate, it will dissociate from the ribosome. The aminoacyl-tRNA has two possible paths to follow. In the case of a noncognate tRNA, it will probably dissociate from the ribosome, whereas a cognate tRNA will probably remain on the ribosome after EF-Tu dissociation.

Several mechanisms for this process have been discussed. Kurland and Ehrenberg (1984) present a model in which EF-Tu-GDP (T2) and noncognate aminoacyl-tRNA dissociate in the same step, whereas Spirin (1986) discusses a model in which aminoacyl-tRNA dissociates first, followed by EF-Tu. An early model by Thompson et al. (1981) in which EF-Tu · GDP dissociates first has some attraction (Fig. 4). This would make the T2 dissociation independent of whether the tRNA is cognate or noncognate. Simply, if EF-Tu has been induced to adopt the GTPase conformation (Fig. 2), the GTP will be hydrolyzed and the factor will be released from the aminoacyl-tRNA as well as from the ribosome. The aminoacyl-tRNA is now able to participate as the acceptor molecule in peptide transfer or fall off. One way to structurally visualize this model of proofreading is presented in Fig. 3B. The codon-anticodon interaction is now crucial for the -CCA end of the aminoacyl-tRNA to reach the peptidyl transfer site. Comparing this active site with those of enzymes, one realizes that differences in distance of less than 1 Å (0.1 nm) could determine whether a reaction will take place or not. Thus, incorrect base pairing in the anticodon end of the tRNA could easily induce errors in the position of opposite, aminoacyl end. Thus, proofreading may simply be achieved by the higher probability that an incorrectly matching tRNA will fall off rather than eventually find the peptidyl transfer site and get its amino acid incorporated.

According to the standard model, a cognate tRNA would need the hydrolysis of one or just over one GTP molecule, whereas a noncognate tRNA would need several (Ruusala et al., 1982). However, many measurements have given values significantly over 1 for cognate tRNA (Chinali and Parmeggiani, 1980; Thompson et al., 1981). In particular, very carefully controlled measurements have recently not been able to reproduce a figure close to 1 but have given values close to 2 for each incorporated cognate amino acid (Ehrenberg et al., 1990). This raises the question of what molecular mechanism is associated with the aminoacyl-tRNA binding and proofreading (Ehrenberg et al., this volume). Are two EF-Tu-bound GTP molecules hydrolyzed consecutively, or does EF-Tu form a pentameric complex with two factor molecules, two GTP molecules, and one aminoacyl-tRNA molecule in which both GTP molecules need to be hydrolyzed before the amino acid can be incorporated?

Peptidyl Transfer

The peptidyl transfer step, thoroughly reviewed by Spirin (1986), can occur only when a peptidyl-tRNA is located in the P site and an aminoacyl-tRNA has reached its proper orientation in the A site. It is also possible to achieve peptidyl transfer to short analogs of the aminoacyl end of tRNA such as puromycin. One aspect of the peptidyl transfer step is the stereochemistry of the tRNAs and the peptidyl and aminoacyl groups. It is obvious that the orientations must be identical regardless of which residues are involved. Thus, the growing polypeptide must initially adopt a repetitive conformation, probably a standard secondary structure for proteins. Lim and Spirin (1986) came to the conclusion that the growing polypeptide probably is synthesized as an α helix. This α helix can be excreted through the exit channel (Bernabeu and Lake, 1982), which probably corresponds to one of the channels observed in crystallized ribosomes (Yonath et al., 1987).

Translocation

The translocation concerns not only the peptidyl-tRNA in the A site and the deacylated tRNA in the P site but also the mRNA, which is moved by three nucleotides to expose the next codon in the A site. EF-G in complex with GTP catalyzes the translation reaction by binding to a site on the ribosome partly overlapping the site for EF-Tu. The factor dissociates from the ribosome after hydrolysis of its bound GTP molecule. The detailed mechanism is not fully understood, but a number of models have been suggested and are reviewed by Spirin (1985, 1986) and by Kurland et al. (this volume). It should be noted that even if the nonhydrolyzable GTP analogs are used, EF-G promotes translocation, which can be observed by using puromycin as the acceptor substrate. It is likely that the factor in this state has adopted the GTPase conformation (Fig. 2) without the ability to hydrolyze the GTP and be released from the ribosome. Fusidic acid, the antibiotic that binds to EF-G (Spirin, 1986), also permits translocation. In this case, the factor remains bound to the ribosome even though the GTP molecule has been hydrolyzed. Apparently, the factor cannot adopt the GDP conformation but remains in the GTPase or GTP conformation (Fig. 2).

THE SITES

In discussing the structural aspects of elongation, it is obvious that the physical sites on the ribosome for these activities are of great interest. The sites can be divided into the mRNA and tRNA sites and the factor sites. Both ribosomal subunits participate in the functional steps, and the sites are located on both subunits. It is clear that the tRNA molecules in the A and P sites must be bound at the ribosomal interface, since the codons of the mRNA are located on the 30S subunit and the peptidyl transfer site is located on the 50S subunit. Additional tRNA sites such as the E site are discussed by Rheinberger et al. and Noller et al. elsewhere in this volume.

The Decoding Site

The localization on the ribosome for the mRNA and in particular the two codons in the A and P sites has been improved considerably in recent years as a result of the advances in RNA sequencing and site-directed mutagenesis on rRNAs. It has long been known that binding of the mRNA is due primarily to the 30S particle. The first very precise information on codon location came from cross-linking studies (Prince et al., 1982) in which it was found that the 5′ base of the anticodon in the P site can be cross-linked to C-1400 of the 16S RNA. The same result has been obtained for other species (Ofengand et al., 1986). With the accumulation of rRNA sequences, it is clear that this region of the 16S RNA is highly conserved among species (Stern et al., 1989). Furthermore, it has been found that resistance against aminoglycosides, antibiotics that induce misreading, is due to modifications in vivo in this area (Beauclerk and Cundliffe, 1987). The multitude of rRNA sequences has also permitted the construction of a model of the secondary and even tertiary structures of the rRNAs (Stern et al., 1989; Brimacombe et al., 1988). The region from 1409 to 1491 essentially forms a long hairpin structure with internal base pairing. Mutations at the beginning and end of this hairpin can cause resistance to these antibiotics. The decoding site seems to involve mainly the cleft between the body and platform of the small subunit (Stern et al., 1989).

The proteins S12, S4, and S5 have for a long time been associated with the decoding of the message, since mutations in them have drastic effects on the accuracy of protein synthesis (reviewed in Tapio, 1989).

The A Site

The A site on the ribosome is fairly complicated. It is, first of all, able to bind an aminoacyl-tRNA. The normal process also includes a carrier for the tRNA, one or possibly two molecules of EF-Tu, which also interacts with parts of the ribosome. The factor interactions will be dealt with below. The aminoacyl-tRNA in the A site probably has different orientations or conformations during initial selection, proofreading, and peptide transfer. Thus, the site could be considered as several subsites. In the past, interest concerning the functional sites on the ribosome was focused on the proteins. Today, we have an increasing amount of information concerning the rRNA from footprinting and chemical labeling of the sites.

The A-site-bound tRNA shields a number of nucleotides in the 16S RNA. Positions 1408 and 1492 to 1494 at the base of the long hairpin loop in the cleft region of the 30S subunit (Stern et al., 1989) must be close to the codon-anticodon interaction site. Residues 529 to 531 are 70 to 100 Å (7 to 10 nm) away from this region in the current model but are nevertheless shielded. They can also be protected by the anticodon stem-loop fragment of the tRNA. The apparent discrepancy can be resolved by conformational changes in the tRNA or in the rRNA or possibly in both. It should be noted that the region around 530 is probably on the side of the 30S subunit away from the platform. In the 70S ribosome, this is close to the base of the L7/L12 stalk (Stern et al., 1989).

It was mentioned above that streptomycin induces misreading by inhibiting proofreading. This inhibition is the same as either preventing the aminoacyl-tRNA from falling off or inducing it to reach the peptidyl transfer site regardless of whether it is cognate or noncognate. Streptomycin resistance has been found not only for mutants of S12 but also in 16S RNA. Thus, a change from C to U in position 912 or a change at position 523 produces streptomycin resistance (reviewed in Stern et al., 1989). Clearly, streptomycin is located at the A site and is simultaneously not far from the thiostrepton at the base of the L7/L12 stalk (Stöffler and Stöffler-Meilicke, 1986).

Bhuta and Chladek (1982) have found that EF-Tu in complex with 3′ fragments of tRNA, such as CCA-Phe, hydrolyzes GTP when bound to the ribosome. This so-called uncoupled activity is not inhibited by thiostrepton as is the normal ternary complex, but GTP hydrolysis is stimulated. Thus, thiostrepton, which prevents the normal ternary complex from binding to the ribosome, does permit binding of a tRNA fragment. If no conformational changes are involved, this indicates that thiostrepton is bound at or near the ribosomal A site.

The P Site

The P-site tRNA has similarly been mapped by footprinting on the rRNAs (Stern et al., 1989). Positions close to the cleft between the platform and the head of the 30s subunit are protected by the anticodon loop and stem of the P-site tRNA.

The Peptidyl Transfer Site

In recent years, the main interest in peptide transfer has been shifted from ribosomal proteins as the essential constituents of the peptidyl transfer site. As a result, the rRNA and in particular the 23S RNA have been found to possess interesting properties (reviewed in Noller, 1984; Cundliffe, 1986; De Stasio et al., 1988). The first observations concern resistance against antibiotics that are inhibitors of peptidyl transfer. Thus, resistance against erythromycin is obtained by a modification of A-2058, a mutation of G-2057, or a deletion of residues 1219 to 1230. Similarly, resistance against chloramphenicol has been associated with residues 2447, 2451, 2452, 2503, and 2504 (Noller, 1984). Interestingly, all of these residues are located in the same central loop structure of domain V in the 23S RNA. It should be expected that some of the multitude of proteins that have been labeled from the polypeptidyl or aminoacyl moieties or from antibiotics will also be of relevance for this site.

The Factor-Binding Site

Elongation factors EF-Tu and EF-G, IF2, and RF-1 and -2 all seem all to compete for one unique site by having overlapping binding regions. This site has been located at the interface between the subunits, probably contacting both. The main contact area on the 50S subunit is located at the base of the L7/L12 stalk for both EF-Tu and EF-G (Spirin and Vasiliev, 1989). The cross-linking of EF-G to ribosomal proteins (S12, L6, L7/L12, L14, and L31) and RNA (around position 1068) gives further support to this view (Traut et al., 1986; Sköld, 1983).

It is also of interest to include some observations about RF-2, which competes for the same site. First, this protein somehow recognizes two termination codons located in the A site. Cross-linking with such termination codons has specifically labeled S6, S18, L2, L7/L12, L10, and L20 (Lang et al., 1989). Similar sets of proteins were also found to be related to RF-2 by other methods. Thus, not only is the factor located close to the stalk of the 50S subunit, but there are also indications that the decoding area of the mRNA is not far away.

As indicated by a number of observations in this chapter, it appears that the tRNA-related sites as well as the factor sites are located on the 70S particle in the valley between the 50S stalk and the nearest interface between the subunits. A similar conclusion, based partly on other arguments, has been reached by Spirin and Vasiliev (1989). It is premature to attempt a more detailed description of the tRNA- and factor-binding sites.

GTPase Induction

Some structure on the ribosome must contribute to induction of the active GTPase conformation in the factors. Removal of proteins such as L7/L12 or L11 from the ribosome strongly affects the GTPase activity. Furthermore, resistance against the antibiotics thiostrepton and micrococcin is obtained by the absence of protein L11 or by modifying A-1067 (Cundliffe and Thompson, 1981). As mentioned, thiostrepton prevents all factors from binding at the A site, whereas micrococcin, with a similar structure, stimulates the GTPase activity in EF-G (Cundliffe and Thompson, 1981). The binding of these antibiotics has been analyzed by shielding the rRNA against RNases or accessibility to chemical modifications. The modifications are found in the stem-loop structure in the region from 1040 to 1115 of the 23S RNA (Egebjerg et al., 1989).

ESSENTIAL COMPONENTS IN ELONGATION

Several components in the system have been implicated in various functions during elongation. I will focus on a few of the central components: the elongation factors, the rRNAs, and a few ribosomal proteins (S12, L10, L11, and L12).

The Factors

The tertiary structure is known for EF-Tu from *Escherichia coli* (LaCour et al., 1985; Jurnak, 1985). The structure is composed of three globular domains. One of them, the domain involved in GTP binding, shows clear homologies with regions in EF-G and IF2. However, RF-1 and -2 neither bind GTP nor display any homologous region (Liljas et al., 1986). The GTP-binding domain has been shown to be common to a large class of GTP-binding proteins, the so-called G proteins (Gilman, 1987). As discussed above, these factors must go through at least three conformational states in their functional cycle. We have no accurate information about more than one factor in one conformation. It remains an urgent task for the proper understanding of elongation to further explore the structures involved. If we take the example of EF-Tu, the situation is more complicated than just three conformational states. This factor also forms a complex with EF-Ts, which catalyzes the exchange between GDP and GTP (Kaziro, 1978). Furthermore, it is possible that the functional form of the factor is a dimer binding one aminoacyl-tRNA (Ehrenberg et al., 1990). On the other hand, there have been reports that in the presence of kirromycin, the factor can bind two tRNA molecules possibly related to the P- and A-site tRNAs on the ribosome (van Noort et al., 1986).

The details of factor binding on the ribosome are far from being understood. However, the similarity of one domain and the fact that the binding sites overlap on the ribosome might indicate a similarity in interaction and induction of conformational changes. One interesting difference between the tRNA-binding factors EF-Tu and IF2 on one hand and EF-G on the other is the insertion of 14 residues in EF-G in a loop near the tRNA-binding area of EF-Tu. The possibility that this extra feature is essential for the unique function of EF-G, translocation (Liljas, 1982), could be examined with site-directed mutagenesis.

The induction of GTPase activity in the factors remains poorly understood. The interactions with the proteins and antibiotics that influence this activity are very interesting, since GTP hydrolysis is the termination of both the initial selection of ternary complexes and the translocation reaction. Thus, GTP hydrolysis must be able to affect both the fidelity and the errors in the reading frame.

The rRNAs

The rRNAs are the focus of intense activity. Many of the functional sites have RNA components. Thus, the structures of the rRNAs and site-directed mutants will rapidly add to our understanding of the mechanisms involved in protein synthesis. I have summarized some of the available data indicating that the decoding site on the 30S subunit is not very far from the factor-binding site on the 30S and 50S subunits. The qualitative nature of the available data is a severe limitation in describing the events during elongation. The effects of the mutagenesis experiments cannot be fully exploited without a more accurate three-dimensional model of the ribosome, including the rRNAs.

The Ribosomal Proteins

A few proteins are of great interest for elongation. One of them is S12, which is involved in fidelity. Some streptomycin-resistant mutants are changed in S12. In addition, the protein is close to factors such as EF-G, as has been observed by cross-linking (Traut et al., 1986). It is not clear whether the effects on fidelity are mediated by the factor interaction or directly influence the codon-anticodon interaction. Homologous proteins to S12 have been found in bacteria, chloroplasts, and eucaryotes (Wittmann-Liebold et al., this volume). S12 binds to the region of the 16S RNA that previously was mentioned as important for fidelity and streptomycin resistance, around position 900 and around position 530 (Stern et al., 1989). The location of the protein, according to neutron-scattering data, is away from the platform and close to the surface of interaction for factors (Moore and Capel, 1988).

Proteins L11 and L10 are located at the base of the stalk of the 50S subunit (Stöffler and Stöffler-Meilicke, 1986), which is composed of the L7/L12 protein (Strycharz et al., 1978). L10 and the two dimers of L12 form a pentameric protein complex (Liljas, 1982). These proteins are bound to the same region of the 23S RNA as the antibiotics thiostrepton and micrococcin, the region from 1040 to 1115 (Egebjerg et al., in press). L10 is the protein in the pentameric complex that binds to the 23S RNA (Petterson, 1979). Apparently, this region of the RNA has a concentration of the elements that affect GTP hydrolysis. The detailed structure of these components is not known except for the C-terminal part of L12 (Leijonmarck and Liljas, 1988). Several electron microscopy studies have indicated that this C-terminal domain is located not only at the tip of the stalk but also at some location in the body of the 50S subunit (Möller et al., 1983; Traut et al., 1986).

L10, L11, and L12 seem to be universal in the ribosome (Matheson et al., this volume). L11 has been found in eucaryotes, eubacteria, and archaebacteria. A protein corresponding to L10 has also been found in all kingdoms, but the amino acid sequence is considerably longer in archaebacteria and eucaryotes than in *E. coli* (Shimmin et al., 1989; Rich and Steitz, 1987). L12 from eubacteria corresponds to proteins in archaebacteria and eucaryotes that share many properties, but the amino acid sequences are very difficult to align. Many attempts at alignment have been published, but the only substantial similarity is a short stretch that is rich in alanine and glycine, with no clear homology among the available sequences. This region of similarity has been identified as a hinge in the protein (Bushuev et al., 1989), but the surrounding domains are probably rearranged in archaebacteria and eucaryotes compared with the situation in eubacteria (Liljas et al., 1986).

A number of observations with regard to L12 give some further details of its function on the ribosome. First, cross-linking of EF-G has been done in both the presence and absence of fusidic acid (Traut et al., 1986). When EF-G is bound to the ribosome in complex with an uncleavable GTP analog, L7/L12 is cross-linked to the factor; however, it is not cross-linked when fusidic acid is used. The difference indicates that L7/L12 and EF-G may have different conformational states (Fig. 2). Possibly the factor has the GTPase conformation in complex with GMPPCP and the GTP conformation in the case with fusidic acid. In a study of trypsin accessibility of ribosomes in different states, it was found that in 70S ribosomes, L7/L12 is resistant to trypsin. However, when EF-G is added in complex with a noncleavable GTP analog, L7/L12 becomes accessible to trypsin. If EG-G is added together with fusidic acid, trypsin cannot cleave the protein (Gudkov and Gongadze, 1984). Apparently, L7/L12 undergoes a conformational change when EF-G binds with GMPPCP. In the case with fusidic acid, when the GTP is cleaved but with EF-G still bound to the ribosome, L7/L12 returns to the original state. Obviously, it appears that the factor and L7/L12 undergo coordinated conformational changes. ^{1}H nuclear magnetic resonance experiments give an idea of the conformational changes involved for the protein. L7/L12 is observed to have a flexible hinge region between residues 37 and 50 (Bushuev et al., 1989). Even in the whole ribosome, this flexibility can be observed as sharp resonances (Cowgill et al., 1984). Some or all of the L7/L12 molecules in the stalk are obviously able to express the hinge flexibility to a large extent. Spectra from 70S particles in complex with EF-G · GMPPCP indicate that this flexibility becomes strongly reduced (Gongadze et al., 1984). L7/L12 binds under these circumstances close enough to the factor to be cross-linked and loses its flexibility. This conformational change is then closely related to the change in factor conformation that leads to GTP hydrolysis.

Trypsin proteolysis experiments have also recently been performed on ribosomes in complex with EF-Tu. Here the findings are opposite to what has been found for EF-G (Gudkov and Bubunenko, 1989). Thus, with EF-Tu in complex with GMPPCP, L7/L12 is not cleaved; however, when 70S particles are incubated with EF-Tu · GTP · kirromycin, L7/L12 becomes susceptible to trypsin. Many more experiments clearly need to be done, but one current interpretation is that L7/L12 goes through a cyclic series of conformational states during the elongation cycle. Whether the two dimers of L7/L12 have different tasks or can alternate and whether the symmetry of the protein is related to a pentameric complex of EF-Tu_2 · GTP_2 · aminoacyl-tRNA (Ehrenberg et al., 1990) or to the internal symmetry of the tRNA molecule (Möller and Maassen, 1986) also remain to be explored.

I thank C. G. Kurland for support and a stimulating interaction over many years and look forward to a continued collaboration despite a new location. I also thank Måns Ehrenberg and Anatoly Gudkov for valuable discussions. I am grateful to John E. Johnson for linguistic corrections.

This work is supported by grants from the Swedish Natural Research Council.

REFERENCES

Beauclerk, A. A. D., and E. Cundliffe. 1987. Sites of action of two ribosomal RNA methylases responsible for resistance to aminoglycosides. *J. Mol. Biol.* **193:**661–671.

Bernabeu, C., and J. A. Lake. 1982. Nascent polypeptide chains emerge from the exit domain of the large ribosomal subunit: immune mapping of the nascent chain. *Proc. Natl. Acad. Sci. USA* **79:**3111–3115.

Bhuta, P., and S. Chladek. 1982. Effect of thiostrepton and 3′-terminal fragments of aminoacyl-tRNA on EF-Tu and ribosome dependent GTP hydrolysis. *Biochim. Biophys. Acta* **698:**167–172.

Bodley, J. W., F. J. Zieve, and L. Lin. 1970. Studies of translocation. IV. The hydrolysis of a single round of GTP in the presence of fusic acid. *J. Biol. Chem.* **245:**5662–5667.

Brimacombe, R., J. Atmadja, W. Stiege, and D. Schuler. 1988. A detailed model of the threedimensional structure of *Escherichia coli* 16S RNA *in situ* in the 30S subunit. *J. Mol. Biol.* **199:**115–136.

Bushuev, V. N., A. T. Gudkov, A. Liljas, and N. F. Sepetov. 1989. The flexible region of protein L12 from bacterial ribosomes studied by proton nuclear magnetic resonance. *J. Biol. Chem.* **264:**4498–4505.

Chinali, G., and A. Parmeggiani. 1980. The coupling with polypeptide synthesis of the GTPase activity dependent on elongation factor G. *J. Biol. Chem.* **255:**7455–7459.

Chinali, G., H. Wolf, and A. Parmeggiani. 1977. Effect of kirromycin on elongation factor Tu. Location of the catalytic center for ribosome elongation factor Tu GTPase activity on the elongation factor. *Eur. J. Biochem.* **75:**55–65.

Cowgill, C. A., R. G. Nichols, J. W. Kenny, P. Butler, E. M. Bradbury, and R. R. Traut. 1984. Mobile domains in ribosomes revealed by proton nuclear magnetic resonance. *J. Biol. Chem.* **259:**15257–15263.

Cundliffe, E. 1986. Involvement of specific portions of ribosomal RNA in defined ribosomal functions: a study utilizing antibiotics, p. 586–604. *In* B. Hardesty and G. Kramer (ed.), *Structure, Function, and Genetics of Ribosomes.* Springer-Verlag, New York.

Cundliffe, E., and J. Thompson. 1981. Concerning the mode of action of micrococcin upon bacterial protein synthesis. *Eur. J. Biochem.* **118:**47–52.

De Stasio, E. A., H. U. Goringer, W. E. Tapprich, and A. Dahlberg. 1988. Probing ribosome function through mutagenesis of ribosomal RNA, p. 17–41. *In* M. F. Tuite (ed.), *Genetics of Translation.* NATO ASI Series, vol. H14. Springer-Verlag KG, Berlin.

Egebjerg, J., S. Douthwaite, and R. A. Garrett. 1989. Antibiotic interactions of the GTPase-associated centre within *Escherichia coli* 23S RNA. *EMBO J.* **8:**607–611.

Egebjerg, J., S. Douthwaite, A. Liljas, and R. A. Garrett. Characterization of the binding sites of protein L11 and the $L10(L12)_4$ pentameric complex in the GTPase domain of the 23S ribosomal RNA from *Escherichia coli. J. Mol. Biol.*, in press.

Ehrenberg, M., A.-M. Rojas, J. Weiser, and C. G. Kurland. 1990. How many EF-Tu molecules participate in aminoacyl-tRNA binding and peptide bond formation in *Escherichia coli* translation? *J. Mol. Biol.* **211:**739–749.

Gale, E. F., E. Cundliffe, P. E. Reynolds, M. H. Richmond, and M. J. Waring. 1981. *The Molecular Basis of Antibiotic Action.* John Wiley & Sons, Inc., New York.

Gilman, A. G. 1987. G proteins: transducers of receptor-generated signals. *Annu. Rev. Biochem.* **56:**615–649.

Gongadze, G. M., A. T. Gudkov, V. N. Bushuev, and N. F. Sepetov. 1984. *Dokl. Akad. Nauk SSSR* **279:**230–232.

Gudkov, A. T., and M. G. Bubunenko. 1989. Conformational changes in ribosomes upon interaction with elongation factors. *Biochimie* **71:**779–785.

Gudkov, A. T., and G. M. Gongadze. 1984. The L7/L12 proteins change their conformation upon interaction of EF-G with ribosomes. *FEBS Lett.* **176:**32–36.

Hopfield, J. J. 1974. Kinetic proofreading: a new mechanism for reducing errors in biosynthetic processes requiring high specificity. *Proc. Natl. Acad. Sci. USA* **71:**4135–4139.

Jurnak, F. 1985. Structure of the GDP domain of EF-Tu and location of the amino acids homologous to *ras* oncogene protein. *Science* **230:**32–36.

Kaziro, Y. 1978. The role of guanosine 5′-triphosphate in polypeptide chain elongation. *Biochim. Biophys. Acta* **505:**95–127.

Kurland, C. G., and M. Ehrenberg. 1984. Optimization of translation accuracy. *Prog. Nucleic Acid Res. Mol. Biol.* **31:**191–219.

LaCour, T. F. M., J. Nyborg, S. Thirup, and B. F. C. Clark. 1985. Structural details of the binding of guanosine diphosphate to elongation factor Tu from *E. coli* as studied by X-ray crystallography. *EMBO J.* **4:**2385–2388.

Lang, A., C. Friemert, and H. G. Gassen. 1989. On the role of the termination factor RF-2 and the 16S RNA in protein synthesis. *Eur. J. Biochem.* **180**:547–554.

Leijonmarck, M., and A. Liljas. 1988. Structure of the C-terminal domain of the ribosomal protein L7/L12 from *Escherichia coli* at 1.7 Å. *J. Mol. Biol.* **195**:555–580.

Liljas, A. 1982. Structural studies of ribosomes. *Prog. Biophys. Mol. Biol.* **40**:161–228.

Liljas, A., S. Thirup, and A. T. Matheson. 1986. Evolutionary aspects of ribosome-factor interactions. *Chem. Scripta* **26B**: 109–119.

Lim, V. I., and A. S. Spirin. 1986. Stereochemical analysis of ribosomal transpeptidation. Conformation of nascent peptide. *J. Mol. Biol.* **188**:565–577.

Möller, W., and J. A. Maassen. 1986. On the structure, function and dynamics of L7/L12 from *Escherichia coli* ribosomes, p. 309–325. *In* B. Hardesty and G. Kramer (ed.), *Structure, Function, and Genetics of Ribosomes.* Springer-Verlag, New York.

Möller, W., P. I. Schrier, J. A. Maassen, A. Zantema, E. Schop, H. Reinalda, A. F. M. Cremers, and J. E. Mellema. 1983. Ribosomal proteins L7/L12 of *E. coli.* Localization and possible molecular mechanism in translation. *J. Mol. Biol.* **163**:553–573.

Moore, P. B., and M. S. Capel. 1988. Structure-function correlation in the small ribosomal subunit from *Escherichia coli. Annu. Rev. Biophys. Biophys. Chem.* **17**:349–367.

Ninio, J. 1975. Kinetic amplification of enzyme discrimination. *Biochmie* **57**:587–595.

Noller, H. 1984. Structure of ribosomal RNA. *Annu. Rev. Biochem.* **53**:119–162.

Ofengand, J., J. Ciesiolka, R. Denman, and K. Nurse. 1986. Structural and functional interactions of the tRNA-ribosome complex, p. 473–494. *In* B. Hardesty and G. Kramer (ed.), *Structure, Function, and Genetics of Ribosomes.* Springer-Verlag, New York.

Parmeggiani, A., and G. M. W. Swart. 1985. Mechanism of action of kirromycin-like antibiotics. *Annu. Rev. Microbiol.* **39**:557–577.

Petterson, I. 1979. Studies on the RNA and protein binding sites of the *E. coli* ribosomal protein L10. *Nucleic Acids Res.* **7**: 2637–2646.

Prince, J. B., B. H. Taylor, D. L. Thurlow, J. Ofengand, and R. A. Zimmermann. 1982. Covalent crosslinking of $tRNA_1^{Val}$ to 16S RNA at the ribosomal P-site: identification of crosslinked residues. *Proc. Natl. Acad. Sci. USA* **79**:5450–5454.

Rich, B. E., and J. A. Steitz. 1987. Human acidic ribosomal phosphosproteins P0, P1, and P2: analysis of cDNA clones, in vitro synthesis, and assembly. *Mol. Cell. Biol.* **7**:4065–4074.

Ruusala, T., M. Ehrenberg, and C. G. Kurland. 1982. Is there proofreading during polypeptide synthesis? *EMBO J.* **1**:741–745.

Shimmin, L. C., C. H. Newton, C. Ramirez, J. Yee, W. L. Downing, A. Louie, A. T. Matheson, and P. P. Dennis. 1989. Organization of genes encoding the L11, L1, L10, and L12 equivalent ribosomal proteins in eubacteria, archaebacteria, and eucaryotes. *Can. J. Microbiol.* **35**:164–170.

Sköld, S. E. 1983. Chemical crosslinking of elongation factor G to the 23S RNA in 70S ribosomes from *Escherichia coli. Nucleic Acids Res.* **11**:4923–4932.

Spirin, A. S. 1985. Ribosomal translocation: facts and models. *Prog. Nucleic Acid Res. Mol. Biol.* **32**:75–114.

Spirin, A. S. 1986. *Ribosome Structure and Protein Biosynthesis.* Benjamin Cummings Publishing Co., Inc., Menlo Park, Calif.

Spirin, A. S., and V. D. Vasiliev. 1989. Localization of functional centers on the procaryotic ribosome: immuno-electron microscopy approach. *Biol. Cell* **66**:215–223.

Stern, S., T. Powers, L.-M. Changchien, and H. Noller. 1989. RNA-protein interactions in 30S ribosomal subunits: folding and function of 16S rRNA. *Science* **244**:783–790.

Stöffler, G., and M. Stöffler-Meilicke. 1986. Immuno electron microscopy on *Escherichia coli* ribosomes, p. 28–46. *In* B. Hardesty and G. Kramer (ed.), *Structure, Function, and Genetics of Ribosomes.* Springer-Verlag, New York.

Strycharz, W. A., M. Nomura, and J. A. Lake. 1978. Ribosomal proteins L7/L12 localized at a single region of the large subunit by immune electron microscopy. *J. Mol. Biol.* **126**:123–140.

Tapio, S. 1989. The role of elongation factor Tu in control of translational accuracy. *Acta Univ Ups.* **178**:1–54.

Thompson, R. C., D. B. Dix, R. B. Gerson, and A. M. Karim. 1981. A GTPase reaction accompanying the rejection of Leu-$tRNA_2$ by UUU programmed ribosomes. *J. Biol. Chem.* **256**: 81–86.

Thompson, R. C., and P. J. Stone. 1977. Proofreading of the codon-anticodon interaction on ribosomes. *Proc. Natl. Acad. Sci. USA* **74**:198–202.

Traut, R. R., D. S. Tewari, A. Sommer, G. R. Gavino, H. M. Olson, and D. G. Glitz. 1986. Protein topography of ribosomal functional domains: effects of monoclonal antibodies to different epitopes in *Escherichia coli* protein L7/L12 on ribosome function and structure, p. 286–308. *In* B. Hardesty and G. Kramer (ed.), *Structure, Function, and Genetics of Ribosomes.* Springer-Verlag, New York.

van Noort, J. M., B. Kraal, and B. L. Bosch. 1986. GTPase center of elongation factor Tu is activated by occupation of the second tRNA binding site. *Proc. Natl. Acad. Sci. USA* **83**:4617–4621.

Wolf, H., G. Chinali, and A. Parmeggiani. 1977. Mechanism of the inhibition of protein synthesis by kirromycin: role of elongation factor Tu and ribosomes. *Eur. J. Biochem.* **75**:67–75.

Yates, J. L. 1979. Role of ribosomal protein S12 in discrimination of aminoacyl-tRNA. *J. Biol. Chem.* **254**:11550–11554.

Yonath, A., K. R. Leonard, and H. G. Wittmann. 1987. A tunnel in the large ribosomal subunit revealed by three dimensional image reconstruction. *Science* **236**:813–816.

Chapter 25

Allosteric Three-Site Model for the Ribosomal Elongation Cycle

HANS-JÖRG RHEINBERGER, UTE GEIGENMÜLLER, ANDREAS GNIRKE,
THOMAS-PETER HAUSNER, JAANUS REMME, HARUO SARUYAMA, and KNUD H. NIERHAUS

The detection (Rheinberger and Nierhaus, 1980) and characterization (Rheinberger et al., 1981; Grajevskaja et al., 1982; Rheinberger and Nierhaus, 1983; Kirillov et al., 1983; Lill et al., 1984; Rheinberger and Nierhaus, 1986a) of a third tRNA-binding site on *Escherichia coli* ribosomes has significantly changed our view of ribosomal function, in particular of the elongation cycle of protein biosynthesis. A detailed comparison between the classical model of Watson (1964) and the features of the three-site model has been given elsewhere (Nierhaus and Rheinberger, 1984; Nierhaus et al., 1986). Here we summarize our previous data and report on some recent experiments not yet published.

tRNA BINDING

During the elongation cycle, the ribosome has to deal with three kinds of tRNA: aminoacyl-tRNA, the substrate for protein biosynthesis; peptidyl-tRNA, which carries the growing peptide chain during elongation; and deacyl-tRNA, the tRNA product finally leaving the ribosome. The tRNA-binding capacity of poly(U)-programmed and nonprogrammed 70S tight-couple ribosomes (Rheinberger et al., 1981), as well as of 30S and 50S subunits from *E. coli* (Gnirke and Nierhaus, 1986), has been systematically analyzed by using all three kinds of tRNA mentioned above (for peptidyl-tRNA, a simple analog has been taken, i.e. *N*-acetylated Phe-tRNA [AcPhe-tRNA]). The results are summarized in Table 1.

30S Subunits

In the absence of mRNA, *E. coli* 30S subunits have no tRNA-binding capacity. Programmed 30S subunits expose one binding site, to which all three kinds of tRNA bind with nearly the same affinity. EF-Tu does not enhance the binding of Phe-tRNA to 30S particles. After 50S association, the tRNA prebound to 30S appears at the P site. Thus, in contrast to Kirillov et al. (1980), we find only one tRNA-binding site on the 30S subunit that can be defined as a prospective P site or as part of the ribosomal P site upon 50S association.

50S Subunits

The binding pattern of *E. coli* 50S differs from that of 30S subunits. Phe-tRNA and AcPhe-tRNA do not bind at all, either in the absence or in the presence of mRNA. However, deacylated tRNA binds to 50S in the presence of mRNA and even in its absence, although to a lesser extent. The exclusive binding of deacylated tRNA to 50S is likely to reflect the characteristics of the E site on 70S. Thus, the tRNA-binding site on the 50S subunit can be defined as a prospective E site or as part of the ribosomal E site upon 30S association (for additional evidence, see Kirillov and Semenkov, 1986). According to this view, the A site is generated only upon association of both subunits to form a 70S ribosome. In fact, the binding of an additional tRNA could be observed upon association (Gnirke and Nierhaus, 1986).

Hans-Jörg Rheinberger and Knud H. Nierhaus ■ Max-Planck-Institut für Molekulare Genetik, Abteilung Wittmann, Ihnestrasse 73, D-1000 Berlin-Dahlem, Federal Republic of Germany. **Ute Geigenmüller and Andreas Gnirke** ■ Washington University School of Medicine, 660 South Euclid Avenue, St. Louis, Missouri 63110-1093. **Thomas-Peter Hausner** ■ Department of Molecular Biophysics and Biochemistry, Yale University, 333 Cedar Street, New Haven, Connecticut 06510-8024. **Jaanus Remme** ■ Institute of Chemical Physics and Biophysics of the Academy of Sciences of Estonian SSR, Jakobistreet 2, Tartu, USSR. **Haruo Saruyama** ■ Hokkaido Green-Bio Institute, Naganuma-cho, Higashi 5 Kita 15, Hokkaido 069-13, Japan.

Table 1. tRNA-binding capacity of 30S and 50S subunits and of 70S ribosomes from *E. coli*[a]

Ribosome	mRNA	tRNA species	Binding sites		Remarks
			No.	Site(s)	
30S		AcPhe-tRNA	0		
		Phe-tRNA	0		
		$tRNA^{Phe}$	0		
	Poly(U)	AcPhe-tRNA	1	P*	Prospective P site (becomes part of P site upon 50S association)
	Poly(U)	Phe-tRNA	1	P*	Prospective P site (becomes part of P site upon 50S association)
	Poly(U)	$tRNA^{Phe}$	1	P*	Prospective P site (becomes part of P site upon 50S association)
50S		AcPhe-tRNA	0		
		Phe-tRNA	0		
		$tRNA^{Phe}$	1	E*	Prospective E site (saturates at 0.2 molecule per ribosome; becomes part of E site upon 30S association)
	Poly(U)	AcPhe-tRNA	0		
	Poly(U)	Phe-tRNA	0		
	Poly(U)	$tRNA^{Phe}$	1	E*	Prospective E site (saturates at 0.4 molecule per ribosome; becomes part of E site upon 30S association)
70S		AcPhe-tRNA	1	P	Saturates at 0.5 molecule per ribosome
		Phe-tRNA	0		
		$tRNA^{Phe}$	1	P	
	Poly(U)	AcPhe-tRNA	1	P or A	Exclusion principle
	Poly(U)	Phe-tRNA	2	P and A	
	Poly(U)	$tRNA^{Phe}$	3	P, E, and A	Binding sequence: P, E, A

[a] Data are taken from Rheinberger et al. (1981) and Gnirke and Nierhaus (1986).

70S Tight Couples

70S ribosomes have the remarkable property of being able to discriminate specifically between deacyl-tRNA, aminoacyl-tRNA, and peptidyl-tRNA (Rheinberger et al., 1981). Only one AcPhe-tRNA molecule is accepted by 70S ribosomes, and this can be located either at the P or at the A site, but not at both classical sites simultaneously. This exclusion principle for AcPhe-tRNA binding to 70S ribosomes probably reflects the fact that during elongation, the binding of two peptidyl-tRNAs to one and the same ribosome is not a physiological binding situation and would block protein biosynthesis (Geigenmüller et al., 1986). However, it seems to be possible to overcome this physiological barrier under special preparation and binding conditions (Kirillov and Semenkov, 1982). In the absence of mRNA, there is one binding site for AcPhe-tRNA, which can be identified as P site.

In contrast to AcPhe-tRNA, 70S ribosomes accept up to two Phe-tRNA molecules simultaneously at their P and A sites, respectively. The binding of the second Phe-tRNA molecule is followed by peptide bond formation. In the absence of mRNA, Phe-tRNA is not able to bind either to the P or to the A site.

With respect to deacyl-tRNA, it is possible to saturate 70S tight couples at a level of three tRNAs per ribosome. First, the P site is occupied. Second, a nonclassical site becomes saturated; we termed this the E site according to an old proposal by Wettstein and Noll (1965). Filling of the A site as the third site requires a large excess of tRNA over ribosomes. In the absence of mRNA, only the ribosomal P site is available for deacyl-$tRNA^{Phe}$.

To test whether this binding behavior with respect to the different kinds of tRNA is specific for eubacterial ribosomes (*E. coli*), ribosomes from the archaebacterium *Halobacterium halobium* were analyzed. Because of the high ionic strength required for the binding assays (6 M [!] monovalent cations), standard nitrocellulose filtration could not be applied, since the unbound tRNA was also collected quantitatively on the filters. Accordingly, centrifugation had to be used to separate the ribosomal complexes (Saruyama and Nierhaus, 1986). Despite the completely different binding and isolation conditions, essentially the same binding patterns were obtained for *H. halobium* ribosomes as for *E. coli* ribosomes (Fig. 1). The saturation values were one AcPhe-tRNA (at either the P or the A site), two Phe-tRNAs, and three deacyl-tRNAs per *H. halobium* 70S. In addition to earlier evidence (Wettstein and Noll, 1965), three tRNA-binding sites on eucaryotic ribosomes have recently been reported for a rabbit liver system (Rodnina et al., 1988). Therefore, the existence of a third ribosomal tRNA-binding site specific for deacy-

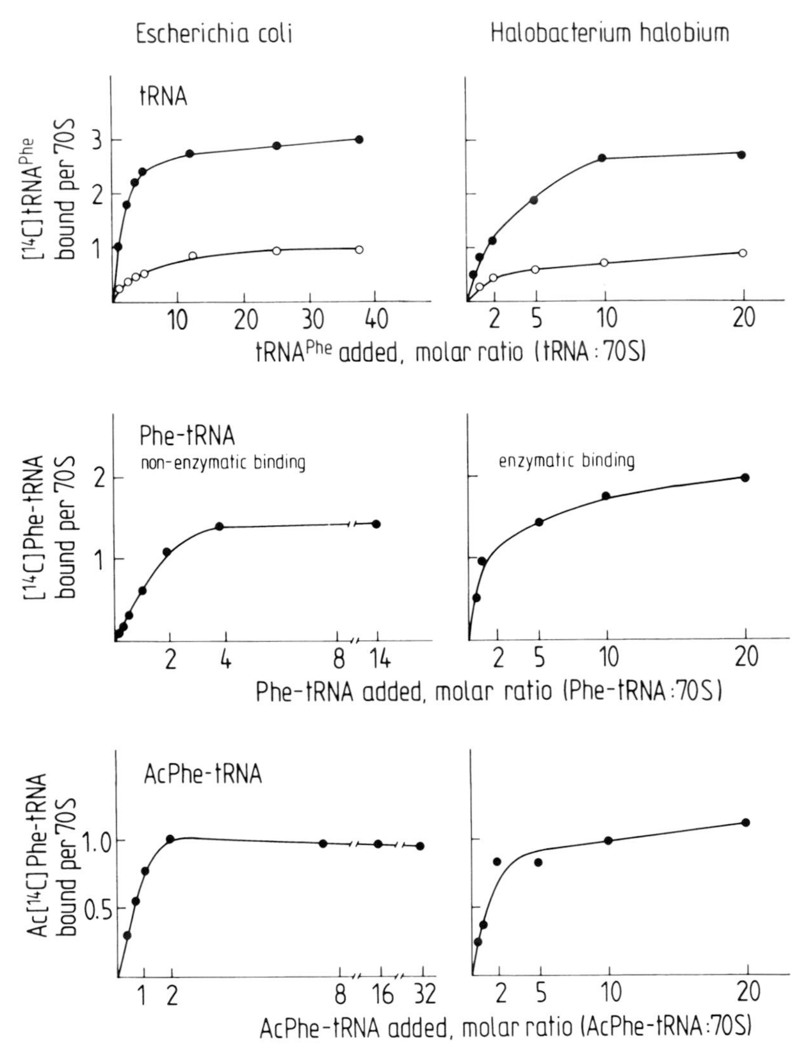

Figure 1. Comparison of tRNA saturation experiments using ribosomes from *E. coli* (left column; Rheinberger et al., 1981) or *H. halobium* (right column; Saruyama and Nierhaus, 1986).

lated tRNA appears to be a universal feature of protein biosynthesis.

FEATURES OF THE RIBOSOMAL E SITE

The E Site Is Specific for Deacylated tRNA

That the E site is specific for deacylated tRNA follows from the binding pattern of the various kinds of tRNA to the 70S ribosome, where only deacyl-tRNA is able to occupy a third site that differs from the A and P sites (Rheinberger et al., 1981). In contrast, the binding of aminoacyl- and peptidyl-tRNA can always be allocated to the classical A and P sites. If the P site is blocked by deacylated tRNA, both Phe-tRNA and AcPhe-tRNA are always directed to the ribosomal A site, whereas deacyl-tRNA, when given as the second tRNA, is directed to the E site (Rheinberger and Nierhaus, 1986a). The specificity of the E site for deacylated tRNA has also been repeatedly observed by others (Grajevskaja et al., 1982; Kirillov et al., 1983; Lill et al., 1984). Experiments using a heteropolymeric mRNA instead of the homopolymeric poly(U) have corroborated this observation. For instance, if a heteropolymeric mRNA $C_{17}AUGA_4C_{17}$, which contains the three codons AUG-AAA-ACC in its central region, is fixed with its ACC codon to the P site with the help of $tRNA^{Thr}$, then the E site can be quantitatively filled with the cognate $tRNA^{Lys}$ but not with AcLys-tRNA (Gnirke et al., 1989). Moreover, Lill et al. (1988) have shown that for binding a tRNA to the exit site, a free and intact 3′-CCA end is necessary.

Binding to the E Site Involves Codon-Anticodon Interaction

Early indirect evidence for the coded nature of deacyl-tRNA binding to the E site came from the observation that in the absence of poly(U), or in the presence of a noncognate messenger such as poly(A), ribosomes accept not more than one $tRNA^{Phe}$, the binding of which can be ascribed to the P site (Rheinberger et al., 1981). The E site becomes available for deacyl-tRNA only if a cognate messenger is present. A second line of evidence for codon-anticodon interaction at the E site comes from binding and chasing experiments using either cognate or noncognate tRNA as a substrate. If poly(U)- or poly(A)-programmed ribosomes are saturated at their P sites with the cognate deacyl-tRNA, then binding of the second tRNA, which can be shown to occur at the E site, strictly depends on the correct anticodon. Moreover, deacyl-tRNA bound to the E site either directly or via translocation can be chased from that site only by an excess of the cognate chasing substrate and not by the noncognate one (Rheinberger et al., 1986). Chasing experiments also showed that there are two codon-anticodon interactions simultaneously before (at the A and P sites) and after translocation (at the P and E sites; Rheinberger and Nierhaus, 1986b). Lill and Wintermeyer (1987) have reported that the association constant of cognate versus noncognate tRNAs at the E site differs by a factor of up to 45. Using the heteropolymer $C_{17}AUGA_4C_{17}$, we have recently demonstrated that quantitative filling of the E site by deacylated tRNA critically depends on the presence of a cognate codon at that site (Gnirke et al., 1989).

FUNCTIONAL ANALYSIS OF THE E SITE

During Translocation, Deacyl-tRNA Moves from the P to the E Site, Where It Remains Stably Bound

Using a homopolymer such as poly(U) or poly(A) as mRNA, we have previously reported that during translocation of a peptidyl-tRNA from the A site to the P site, deacylated tRNA is cotranslocated from the P to the E site. There it remains stably bound until the beginning of the next elongation round, where aminoacyl-tRNA binding to the A site triggers the release of deacylated tRNA from the E site (Rheinberger and Nierhaus, 1983, 1986a). As an alternative explanation, it has been suggested that during translocation, deacylated tRNA is in fact released from the P site but rebinds quantitatively to the A site, from which it is chased by aminoacyl-tRNA in the course of the next elongation round (Baranov and Ryabova, 1988).

To test the latter possibility, we repeated our translocation experiments in the presence of a heteropolymeric mRNA (Gnirke et al., 1989). If there are different codons at the E and A sites, respectively, any release event should be clearly visible, since rebinding to another site bearing another codon would be prevented. Figure 2 shows an example of a translocation reaction in the presence of the heteropolymer $C_{17}AUGA_4C_{17}$ as mRNA. The experiment was carried out at 6 mM Mg^{2+}. First, the reading frame is set by fixing the AUG codon at the P site with

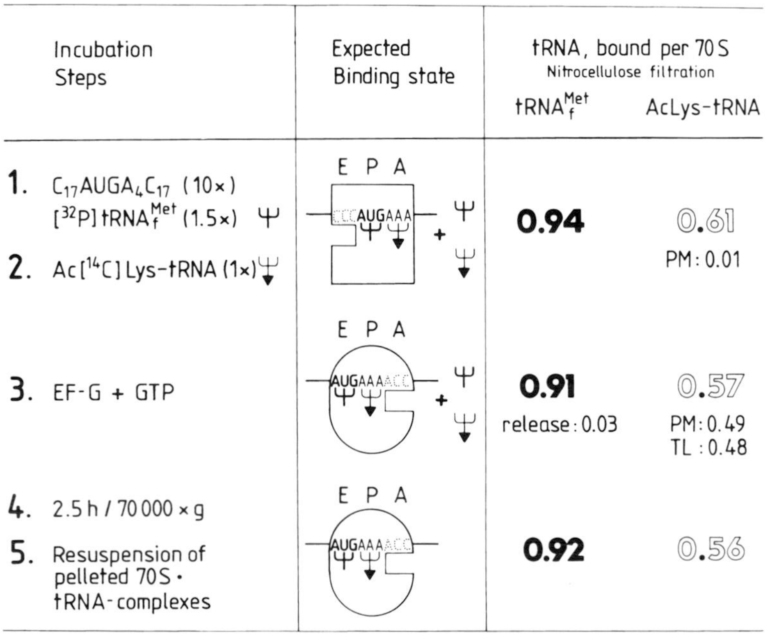

Incubation Steps	Expected Binding state	tRNA, bound per 70 S (Nitrocellulose filtration): $tRNA_f^{Met}$	AcLys-tRNA
1. $C_{17}AUGA_4C_{17}$ (10×) $[^{32}P]tRNA_f^{Met}$ (1.5×) 2. $Ac[^{14}C]Lys$-tRNA (1×)	E P A (CCCAUGAAA)	0.94	0.61 PM: 0.01
3. EF-G + GTP	E P A (AUGAAACCC)	0.91 release: 0.03	0.57 PM: 0.49 TL: 0.48
4. 2.5 h / 70 000 × g 5. Resuspension of pelleted 70S · tRNA-complexes	E P A (AUGAAACCC)	0.92	0.56

Figure 2. Translocation experiment using the heteropolymer mRNA $C_{17}AUGA_4C_{17}$. Before translocation, $[^{32}P]tRNA^{Met}$ was bound to the P site and $Ac[^{14}C]Lys$-tRNA was bound to the A site. PM, Puromycin reaction; TL, translocation. (From Gnirke et al., 1989.)

the help of a ^{32}P-labeled tRNA^{Met}. The P site becomes quantitatively occupied (tRNA bound per 70S = 0.94). Next, ^{14}C-labeled AcLys-tRNA is added in accordance with the codon exposed at the A site. The binding value obtained is 0.61. Note that Ac[^{14}C]Lys-tRNA is exclusively located at the A site, since the control puromycin reaction is almost zero. Upon addition of EF-G and GTP, the majority of the Ac[^{14}]Lys-tRNA initially bound becomes puromycin sensitive. The net amount of translocation as judged by the puromycin reaction before and after translocation is 0.48. In parallel, there is practically no release of [^{32}P]tRNA^{Met} (0.94 and 0.91 tRNA bound per 70S before and after, respectively, translocation, the net amount of tRNA release being 0.94 − 0.91 = 0.03). The translocation reaction thus exceeds the tRNA release by a factor of 16 (0.48 versus 0.03). This means that almost 95% of the cotranslocated [^{32}P]tRNA^{Met} remains bound at the E site. In addition, this binding can be shown to be perfectly stable, since posttranslocational ribosomal complexes can be pelleted from the incubation mixture and resuspended without any loss of either peptidyl-tRNA from the P site or deacyl-tRNA from the E site. We conclude that during translocation, deacylated tRNA is not released from the P site but rather is cotranslocated from the P site to the E site. The deacylated tRNA remains stably bound at the E site, from which it must be ejected by an active mechanism.

The E and A Sites Are Allosterically Linked in a Bidirectional Manner

It has been shown in both the homopolymeric and heteropolymeric mRNA-binding systems that after translocation, the binding of a new A-site ligand triggers the release of the deacyl-tRNA bound to the E site (Rheinberger and Nierhaus, 1983, 1986a; Gnirke et al., 1989).

Here we would like to focus on the bidirectionality of the interaction between the A and E sites in a somewhat more static binding assay (Fig. 3). When C_{17}-AUG-AAA-ACC-CCC-C_{12} is fixed with the ACC codon to the P site via tRNA^{Thr}, the A site can be readily filled with AcPro-tRNA (Fig. 3, experiment 1; AcPro-tRNA bound per 70S = 0.69). If, however, a cognate E-site ligand (tRNA^{Lys}) is prebound in addition to tRNA^{Thr}, subsequent binding of the A-site ligand AcPro-tRNA is considerably reduced (experiment 1; 0.69 and 0.23, respectively). Conversely, if in addition to tRNA^{Thr} at the P site, AcPro-tRNA is bound to the A site first (Fig. 3, experiment 2; AcPro-tRNA bound per 70S = 0.77), then subsequent binding of tRNA^{Lys} to the E site is impaired (experiment 2; 0.91 and 0.61, respectively).

The effect is somewhat less pronounced than in the reverse case, since binding of tRNA^{Lys} to the E site is so strong that it even triggers a partial release of AcPro-tRNA from the A site (0.77 and 0.56 for binding of AcPro-tRNA to the A site before and after, respectively, the addition of tRNA^{Lys}). Interestingly, the tRNA^{Thr} in the central position (P site) is not affected by any of the subsequent binding events. Since different codons are exposed at both lateral positions (A and E site), the observed effects must be due to an allosteric linkage in the sense of a negative cooperativity between these two sites.

The allosteric switch from low to high affinity at the A site, and the corresponding switch from high to low affinity at the E site, can be brought about by a temperature shift, and its time course can be followed kinetically under the conditions used (AcPhe-tRNA as A-site ligand, 15 mM Mg^{2+}) (Fig. 4). Figure 4A shows a schematic outline of the experiment. Poly(U)-programmed ribosomes are occupied with two [^{14}C]tRNA^{Phe} molecules at their P and E sites. Next, Ac[^{3}H]Phe-tRNA is added at either 0 or 37°C. At 0°C, the amount of [^{14}C]tRNA^{Phe} prebound remains essentially unchanged, and Ac[^{3}H]Phe-tRNA does not bind during the whole 60-min time course, indicating the low A-site affinity in the presence of an occupied E site. In contrast, at 37°C there is good binding of Ac[^{3}H]Phe-tRNA to the A site, with a stoichiometry of up to 0.6. In parallel, exactly the same amount of [^{14}C]tRNA^{Phe} is lost from the E site; i.e., the binding of one molecule of AcPhe-tRNA to the A site triggers the release of one molecule of deacylated tRNA from the E site.

INTEGRATION OF THE DESCRIBED FEATURES INTO AN ALLOSTERIC THREE-SITE MODEL

The main aspects of the three-site model of ribosomal elongation (presented schematically in Fig. 5B) can be summarized as follows. (i) In addition to the classical A and P sites, ribosomes contain a third tRNA-binding site, the E site, adjacent to the P site. (ii) When peptidyl-tRNA is translocated from the A site to the P site, deacyl-tRNA is not released from the P site but rather is cotranslocated from the P site to the E site, where it is tightly bound, undergoing codon-anticodon interaction. Therefore, in the course of elongation, a tRNA enters the A site as aminoacyl-tRNA, is translocated to the P site as peptidyl-tRNA, and is finally translocated as deacyl-tRNA to the E site, from which it leaves the ribosome. (iii) A and E site are allosterically linked via negative cooperativity; i.e., occupation of the A site reduces the affinity

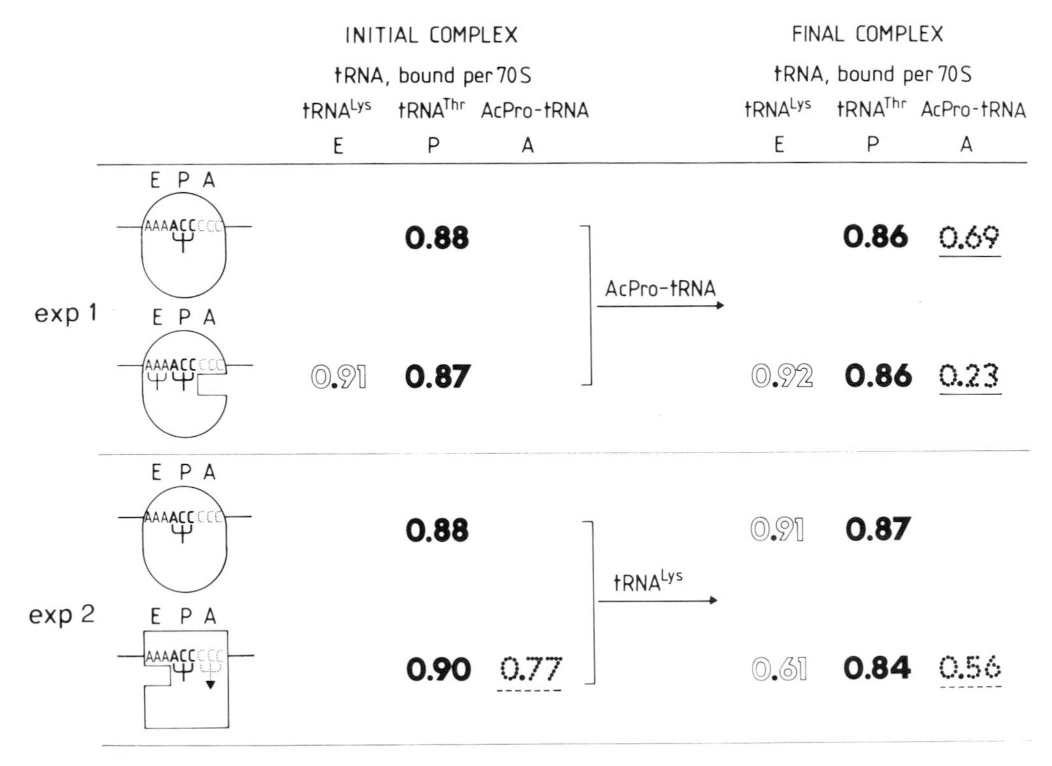

Figure 3. Allosteric interactions between A and E sites. When only the P site is occupied, the A site (first line in experiment 1) or the E site (first line in experiment 2) can readily be occupied as the second site. However, if the P and E sites are charged, the binding capacity of the A site is severely impaired (second line in experiment 1). Likewise, occupation of the P and A sites reduces E-site binding (second line in experiment 2). Note that in the course of the A site-E site interaction, binding at the P site remains unaffected. (From Gnirke et al., 1989.)

of the E site for deacyl-tRNA, and occupation of the E site reduces the affinity of the A site for aminoacyl-tRNA. It follows that two alternative functional states exist for the ribosome during the course of the elongation cycle: in the pretranslocational state, the A and P sites have a high affinity (the E site has low affinity); in the posttranslocational state, P and E are the high-affinity sites (the A site has low affinity). Both high-affinity sites (A and P or P and E) are occupied by tRNAs, and the adjacent tRNAs undergo codon-anticodon interaction simultaneously.

SIGNIFICANCE OF THE E SITE FOR THE ACCURACY OF mRNA MOVEMENT AND THE DECODING PROCESS

The complicated interplay between the A and E sites immediately raises the question of the importance of the E site. Two aspects of its physiological significance can be envisaged: proper translocation and proper A-site binding.

Belitsina et al. (1982) reported that some tRNAs can be translocated in the absence of mRNA. This might indicate that the tRNAs are moved actively in the course of translocation, whereas the mRNA follows passively. If so, it would make no sense to lose half of the binding energy between tRNA and mRNA by the release of tRNA from the ribosome during translocation, as suggested by the old two-site model; rather, the necessity for a tight coupling of tRNAs and mRNA would require two adjacent codon-anticodon interactions during the course of translocation. In fact, two adjacent codon-anticodon interactions in the pre- as well as posttranslocational state have been observed (Rheinberger and Nierhaus, 1986b).

The most surprising feature of the new model is the bidirectionality of the allosteric interplay between the A and E sites, i.e., not only that occupation of the A site reduces the E-site affinity, thus triggering the release of tRNA from the ribosome, but that the reverse is also true: occupation of the E site reduces the affinity of the A site.

We suggested that the latter effect, i.e., the E-site-induced reduction of the A-site affinity, is important for the accuracy of translation (U. Geigenmüller and K. H. Nierhaus, submitted for publication). The rationale for this hypothesis is depicted in Fig. 6.

The free energy (ΔG) for the A-site binding of an

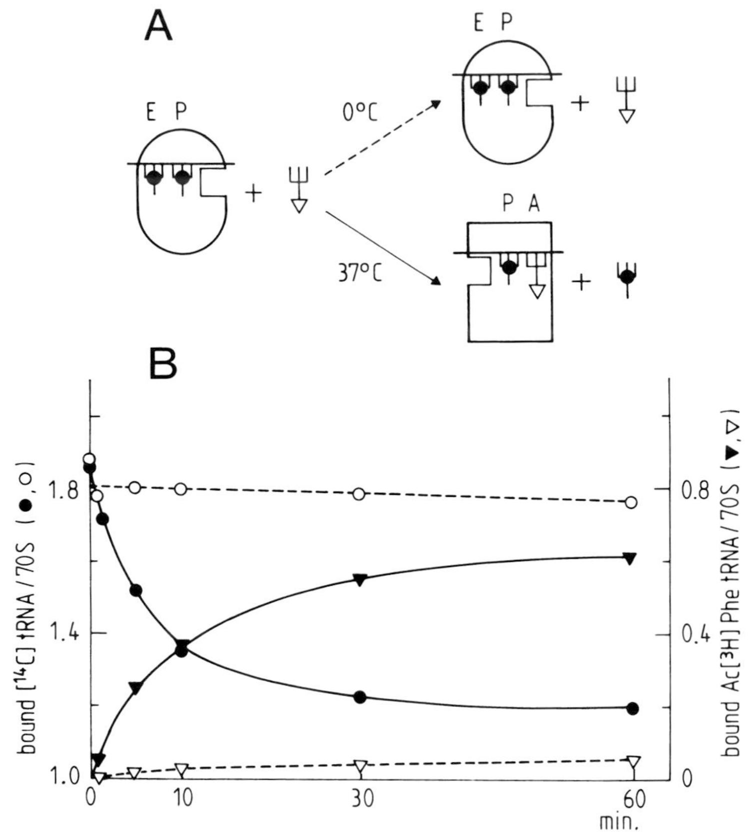

Figure 4. Allosteric interactions between the A and E sites: kinetics of Ac[^{3}H]Phe-tRNA binding to the A site. When the P and E sites are occupied with [^{14}C]tRNA^{Phe} in the presence of poly(U), Ac[^{3}H]Phe-tRNA does not bind at 0°C to the A site, indicating its low affinity. When activation energy is provided (incubation at 37°C), AcPhe-tRNA does bind; for each molecule of AcPhe-tRNA bound, one molecule of tRNA^{Phe} is released from the E site, demonstrating the negative cooperativity between the A and E sites. (From Rheinberger and Nierhaus, 1986a.)

aminoacyl-tRNA can be separated into two terms; the first term is caused by the interaction of tRNA with mRNA (codon-anticodon interaction), and the second is caused by the interaction of tRNA with the ribosome. It is mainly the second term that is responsible for a correct positioning of the tRNA on the ribosome and hence a successful peptide bond formation. However, this term does not discriminate between wrong and right tRNA. The discrimination is made by the first term, i.e., codon-anticodon interaction, where three classes of tRNAs can be distinguished. One class contains only a single tRNA species, namely, the cognate tRNA with the properly complementary anticodon. The second class contains up to six miscognate tRNAs with an anticodon similar to the cognate one, whereas the third class comprises the majority of the tRNAs (up to 50 in *E. coli*), representing the noncognate tRNAs, which have dissimilar anticodons. If both terms of the A-site affinity were active, one would expect significant A-site interaction of even the noncognate tRNAs, thus reducing the rate of translation and leading to an occasional incorporation of noncognate amino acids.

Now the E site comes in. We assume that occupation of the E site abolishes only the second term of the A-site affinity (tRNA-ribosome interaction) and does not impair codon-anticodon interaction. A successful codon-anticodon interaction would trigger the transition from low to high affinity of the A site and would thus generate the prerequisite for subsequent peptide bond formation. According to this mechanism, only cognate and (to a much lesser degree) miscognate tRNA can effectively trigger the transition, whereas the noncognate tRNA cannot. The A site practically does not exist for the noncognate tRNA, thus preventing any interference with rate and accuracy of translation.

We tested this hypothesis in the following way (Fig. 7). Poly(U)-programmed ribosomes carrying an AcPhe-tRNA at the P site were constructed. In one experiment, the E site was free (high-affinity A site; Rheinberger and Nierhaus, 1986a), whereas in the other experiment the E site was occupied with a deacylated tRNA (low-affinity A site). To both samples was added a mixture of ternary complexes containing cognate [^{14}C]Phe-tRNA and noncognate

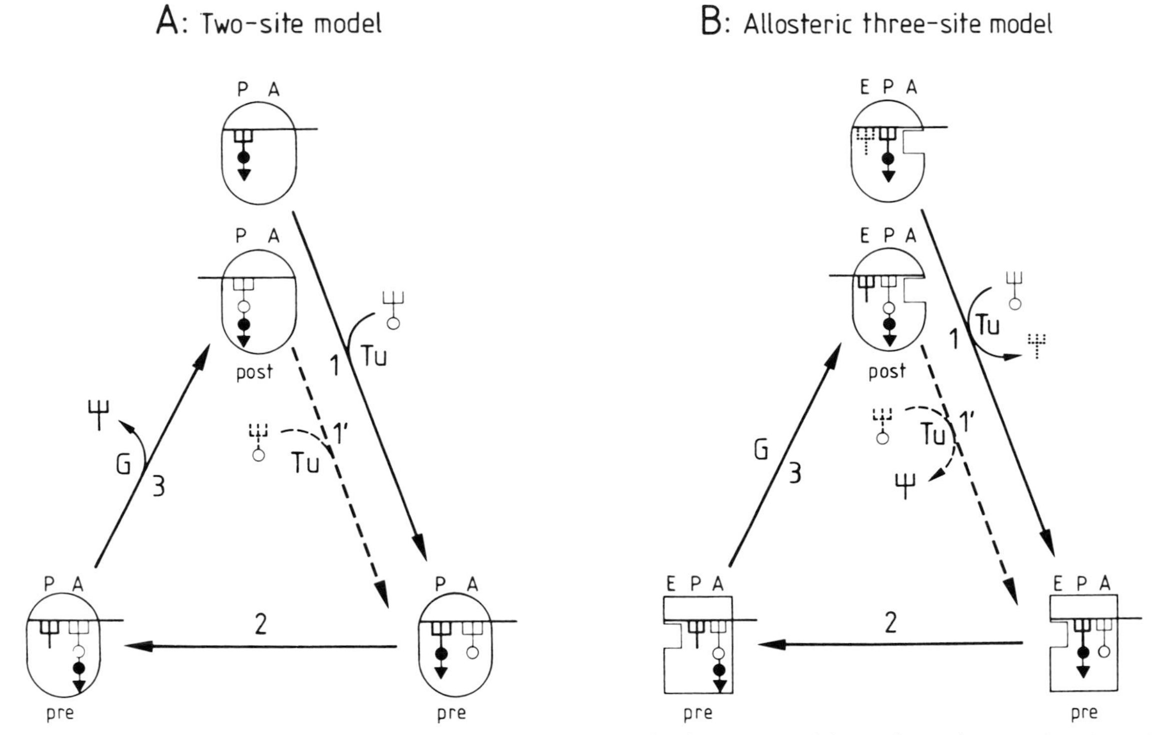

Figure 5. The three basic reactions of the elongation cycle: 1, A-site binding; 2, peptidyl transfer; and 3, translocation. (A) Two-site model; (B) allosteric three-site model.

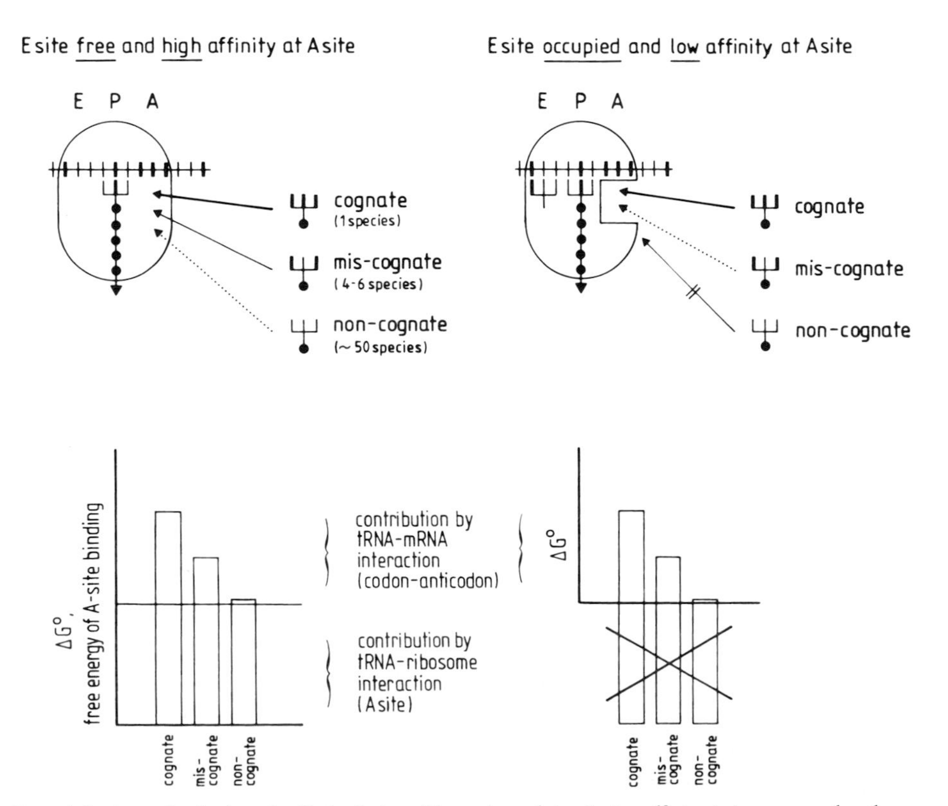

Figure 6. Outline of the hypothesis that the E-site-induced lowering of the A-site affinity is important for the accuracy of the selection of acyl-tRNAs at the A site (see text).

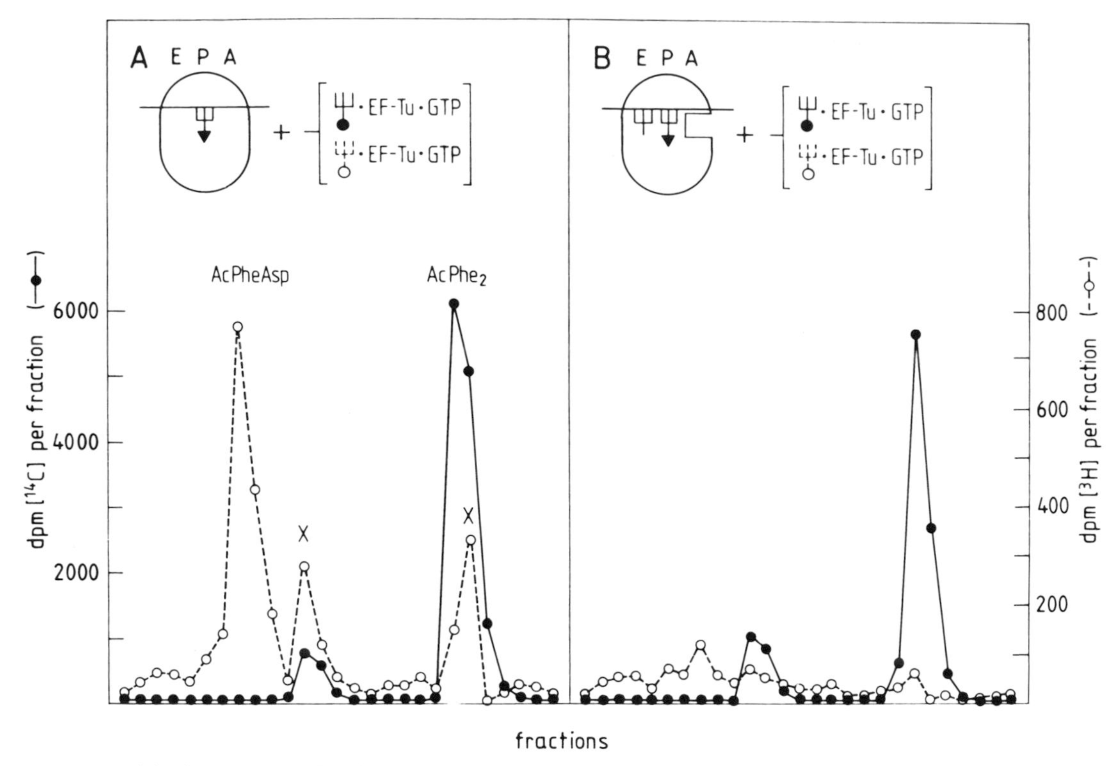

Figure 7. Test of the hypothesis outlined in Fig. 6. (A) When AcPhe-tRNA is present at the P site in the presence of poly(U) and the E site is free, the A site has a high affinity. Addition of a mixture of cognate and noncognate ternary complexes containing [^{14}C]Phe-tRNA and [^{3}H]Asp-tRNA, respectively, leads to a significant formation of wrong AcPhe-Asp dipeptides (detected by means of high-pressure liquid chromatography). (B) When P and E sites are occupied with AcPhe-tRNA and deacyl-tRNA (A site with low affinity), respectively, addition of the mixture of ternary complexes leads exclusively to cognate Ac-Phe$_2$ formation. (From Geigenmüller and Nierhaus, submitted.)

[^{3}H]Asp-tRNA (codon GA[C/U]). The amounts of acetylated dipeptides formed were assessed by means of high-pressure liquid chromatography.

When the E site was free (high affinity at the A site), significant AcPhe-Asp formation was observed (about 1% of the total dipeptide formation; Fig. 7). In sharp contrast, when the E site was occupied, no significant amounts of AcPhe-Asp were found, whereas the formation of cognate AcPhe-Phe was hardly affected. Clearly, an occupied E site prevents noncognate aminoacyl-tRNA from interacting with the A site, illustrating one important role of the E site during protein synthesis.

In a control experiment, miscognate tRNA^{Leu} instead of the cognate tRNA^{Phe} was added as the E-site ligand. Formation of the noncognate AcPhe-Asp was not reduced. We conclude that the trigger for the E-site-induced reduction of the A-site affinity is the presence of cognate tRNA at the E site, underlining the functional importance of codon-anticodon interaction at the E site.

If a mutation were to change a tRNA such that its affinity to the A site was increased without affecting the geometry of the anticodon loop, one would predict an increased misreading of this acyl-tRNA according to the allosteric three-site model. Precisely this has been found. Hirsh (1971) described a mutation that causes a G-24→A change in the D stem of tRNA^{Trp}, resulting in an active misreading. A careful analysis revealed that the geometry of the anticodon loop probably is not altered, whereas the affinity to the A site seems to be enhanced (Smith and Yarus, 1989a, 1989b). The allosteric three-site model offers a simple explanation: the alteration of tRNA^{Trp} (or its UAG suppressor derivative) in the D stem increases the affinity for the low-affinity A site, thus counteracting the codon specificity effect and increasing the frequency of misreading.

THE ALLOSTERIC THREE-SITE MODEL EXPLAINS THE INHIBITION MECHANISMS OF THE ANTIBIOTICS AMINOGLYCOSIDES, THIOSTREPTON, AND VIOMYCIN

Two different types of A-site occupation must be distinguished within the framework of the allosteric three-site model (Fig. 8). In one case, the P site (but not the E site) is occupied by a tRNA, and the A site adopts a high-affinity state that can already be easily charged with an aminoacyl-tRNA at 0°C (Rheinberger and Nierhaus, 1986a). In this state, the A site is occupied only once during protein synthesis, just after initiation; accordingly, this type of A-site binding is defined as i type (i for initiation; Hausner et al., 1988). The second and all subsequent A-site occupations during protein synthesis occur with tRNAs at both P and E sites, leaving the A site in a low-affinity state. This state cannot effectively be charged at 0°C but requires higher temperatures to provide activation energy for the allosteric transition from the post- to the pretranslocational state (Fig. 8); this type of A-site occupation is defined as e type (e for elongation).

In the past, the effects of various drugs on the A-site binding had been tested only under conditions

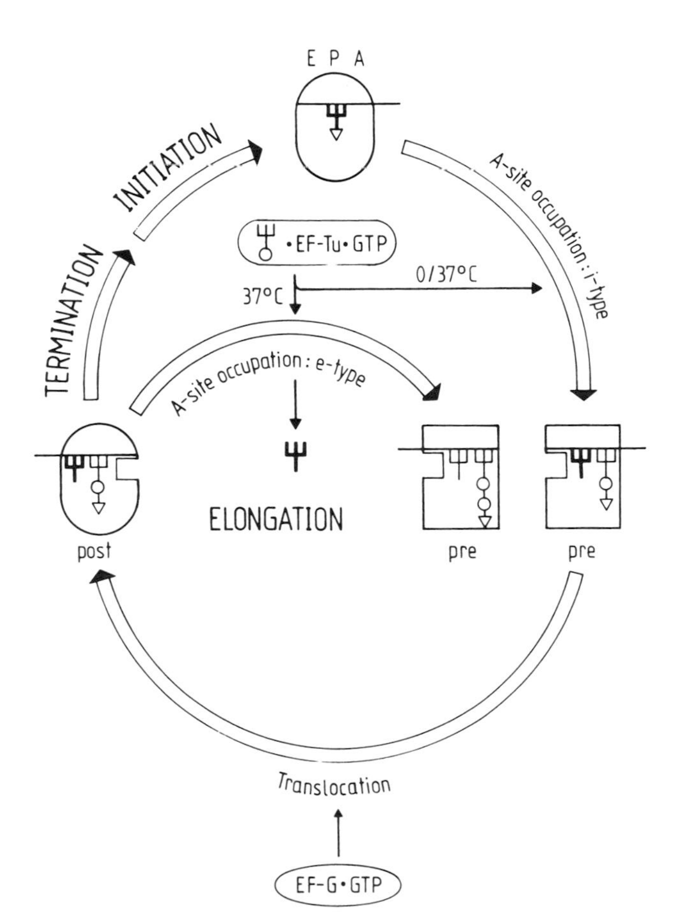

Figure 8. Two kinds of A-site occupation in the course of protein biosynthesis. In the i type, after initiation, the E site is free. Therefore, the A site has a high affinity and can be charged even at 0°C. In the e type, the second A-site occupation as well as all subsequent ones occur with occupied P and E sites (the A site has low affinity). Now A-site occupation requires a much higher activation energy, and in the course of the allosteric transition from the posttranslocational to the pretranslocational state, the deacyl-tRNA leaves the E site. (From Hausner et al., 1988.)

of the i-type occupation, which appears not to be an appropriate model for this reaction in the course of the elongation cycle. Therefore, we tested the reactions of the elongation cycle in the presence of various antibiotics, in particular using both types of A-site occupation (Hausner et al., 1988).

A-site binding of the i type was found to be affected very little by thiostrepton, viomycin, and a series of aminoglycosides, with at least one member of each of the known families (the neomycin family including hygromycins, the kanamycin and gentamicin families, and the streptomycin family; see Hausner et al., 1988, and references therein) (Fig. 9). Thiostrepton, viomycin, and the aminoglycosides hygromycin and neomycin blocked the subsequent translocation reaction. These observations are in agreement with published data. However, when effects on the e-type occupation of the A site were studied, severe inhibitions were found with thiostrepton, viomycin, and all aminoglycosides. It follows that the two nonrelated drugs thiostrepton and viomycin are inhibitors of the allosteric transitions in both directions; i.e., they block both the transition from the pre- to the posttranslocational state (translocation) and that from the post- to the pretranslocational state (e-type occupation of the A site).

The aminoglycosides are the classical inducers of misreading effects, but this feature is not the cause for their bactericidal activity (Fast et al., 1987). The simple picture that arises both from their common structure and from a defined ribosomal region, which has been determined to be at or near their binding sites (Moazed and Noller, 1987) and which harbors rRNA alterations conferring resistance (Cundliffe, 1987), is in contrast to the pleiotropic effects of the aminoglycosides that have been described (Davis, 1987). However, the data shown in Fig. 9 satisfyingly indicate that a common point of interference also exists and that the primary function inhibited by the aminoglycosides is the e-type occupation of the A site.

THE E SITE IS OCCUPIED IN NATIVE POLYSOMES

In the previous sections, we analyzed the tRNA-binding capabilities of the ribosome and the features of the binding sites in vitro. Now we discuss whether the E site is also occupied in vivo.

A distinctive corollary of the previous two-site model and the allosteric three-site model is the translocation reaction (Fig. 5), which according to the two-site model is coupled to the release of deacylated tRNA but according to the allosteric three-site model

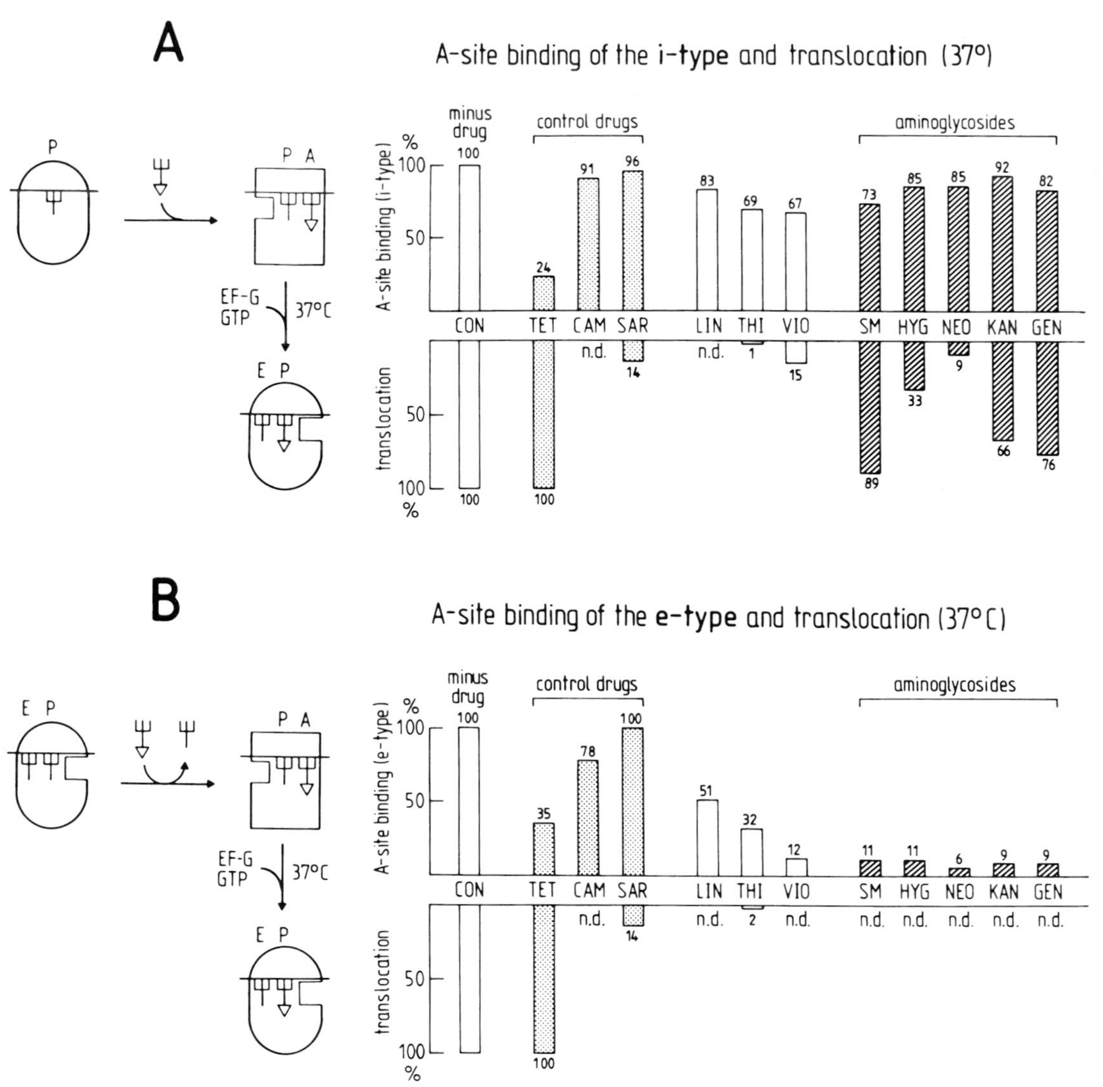

Figure 9. A-site occupation of the i type (A) and e type (B) in the presence of various antibiotics. CON, Control; TET, tetracycline; CAM, chloramphenicol; SAR, α-sarcin; LIN, lincomycin; THI, thiostrepton; VIO, viomycin; SM, streptomycin; HYG, hygromycin; NEO, neomycin; KAN, kanamycin; GEN, gentamicin. (From Hausner et al., 1988.)

is not. As a consequence, the previous model postulates that the posttranslocational ribosomes contain only one tRNA, whereas the allosteric three-site model predicts the presence of two tRNAs just as in the pretranslocational state. If a significant fraction of ribosomes were present in the posttranslocational state in native ribosomes and the number of tRNAs statistically bound per ribosome could be assessed, then this number should be able to discriminate between the two models.

Following this strategy, we grew *E. coli* cells in the presence of $^{32}P_i$ for about four generations, which uniformly labels all nucleic acids in the cells (Remme et al., 1989). The cells were harvested, suspended in buffer, and incubated for 3 min at 37°C in the presence of ^{3}H-amino acids but without any other carbon source. Under these restricted conditions, nascent peptidyl chains on the ribosome were preferentially labeled in vivo. The cells were carefully lysed, and the lysate was incubated with or without puromycin (10 min at 0°C) and subjected to a sucrose gradient centrifugation. The presence or absence of puromycin did not affect the ribosomal pattern (Fig. 10). However, the ^{3}H radioactivity was reduced to 40% in the polysomal region (22,083 versus 47,032 cpm), indicating that at least 60% of the ribosomes in the native polysomes were in the posttranslocational state. The two-site model predicts the statistical presence of 1.4 tRNAs per ribosome with respect to this fraction of posttranslocational ribosomes.

The RNAs were isolated from the polysomal ribosomes and the runoff 70S ribosomes and were subjected to a gel electrophoresis that easily separates 5S rRNA from the two tRNA families (tRNA1 with a short extra arm and an average length of 76 nucleotides; tRNA2 with a long extra arm and an average length of 88 nucleotides). Since the length of 5S rRNA is known (120 nucleotides) and each ribosome contains one molecule of 5S rRNA, the ^{32}P radioactivity derived from the tRNA and 5S bands can be used to calculate the molar ratio of tRNA to 5S rRNA, which is identical with the number of tRNAs bound per ribosome.

Figure 10 demonstrates that 1.98 and 2.2 tRNAs per ribosome were found in the polysomal region, whereas runoff ribosome carried only 0.61 tRNA. Runoff ribosomes do not contain mRNA, and these nonprogrammed ribosomes should not bind more than one tRNA per 70S (Rheinberger et al., 1981). The results demonstrate that the E site is occupied in posttranslocational ribosomes of native polysomes, clearly supporting the allosteric three-site model.

The physiological importance of the E site was also evident in experiments in which rRNA bases protected against chemical reagents were determined. When the E site was occupied with deacyl-tRNA, a distinct group of bases was protected (Moazed and Noller, 1989). Interestingly, analysis of native polysomes also showed protections within that group, indicating occupied E sites (Brow and Noller, 1983).

CONCLUSIONS

The allosteric three-site model appears to be the appropriate description of the ribosomal elongation cycle. It seems to be universally valid, since it has been shown for eubacterial (Rheinberger and Nierhaus, 1986; Gnirke et al., 1989) and archaebacterial (Saruyama and Nierhaus, 1986) systems. No detailed analysis is yet available for eucaryotic ribosomes, but a third binding site specific for deacylated tRNA has also been found for eucaryotic systems (Wettstein and Noll, 1965; Rodnina et al., 1988). One important feature of the allosteric three-site model is the reduction of the influence of the noncognate aminoacyl-tRNAs on rate and accuracy of translation. The inhibition mechanisms of some antibiotics (thiostrepton, viomycin, and aminoglycosides) can be better understood within the framework of the allosteric three-site model.

We thank H. G. Wittmann for continuous support, R. Brimacombe and H. Sternbach for help and advice, and Eva Philippi for the beautiful drawings.

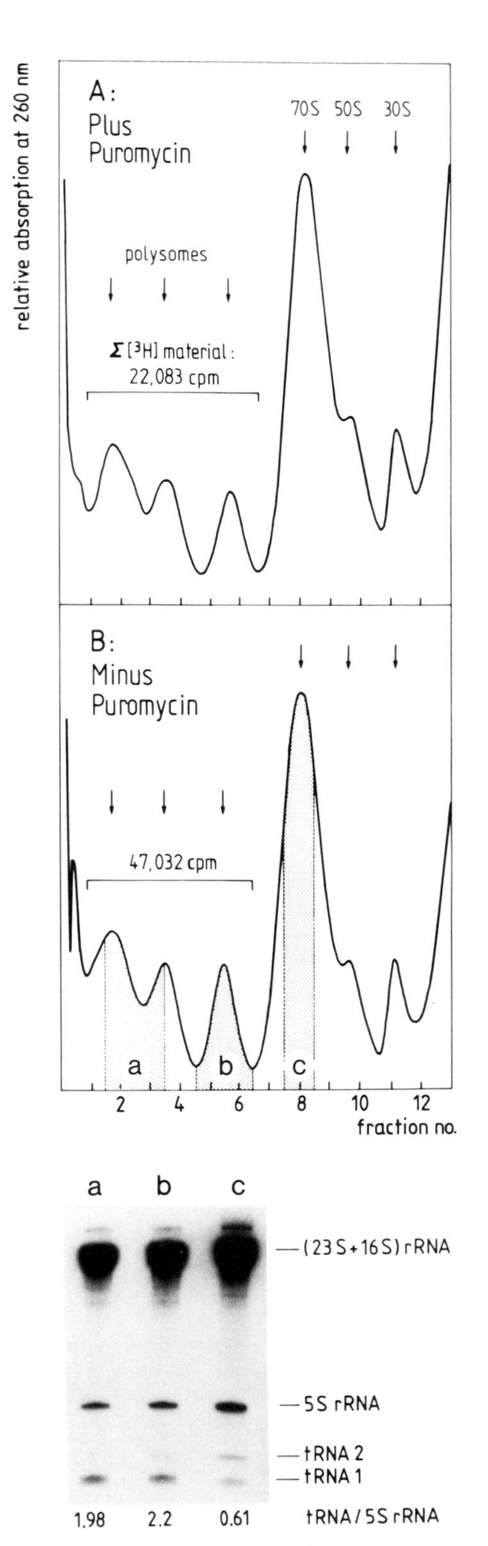

Figure 10. Sucrose gradient analysis of lysates from *E. coli* grown in the presence of $^{32}P_i$ and briefly incubated in the presence of ^{3}H-amino acids. The lysate was incubated in the presence (A) and absence (B) of puromycin. (Bottom) RNA analysis from polysomes (lanes a and b) and runoff 70S (lane c). (From Remme et al., 1989.)

REFERENCES

Baranov, V. I., and L. A. Ryabova. 1988. Is the three-site model for the ribosomal elongation cycle sound? *Biochimie* 70:259–265.

Belitsina, N. V., G. Z. Tnalina, and A. S. Spirin. 1982. Template-free ribosomal synthesis of polypeptides from aminoacyl-tRNAs. *BioSystems* **15**:233–241.

Brow, D. A., and H. Noller. 1983. Protection of ribosomal RNA from kethoxal in polyribosomes: implication of specific sites in ribosome function. *J. Mol. Biol.* **163**:27–46.

Cundliffe, E. 1987. On the nature of antibiotic binding sites in ribosomes. *Biochimie* **69**:863–869.

Davis, B. D. 1987. Mechanism of bactericidal action of aminoglycosides. *Microbiol. Rev.* **51**:341–350.

Fast, R., T. H. Eberhard, T. Ruusala, and C. G. Kurland. 1987. Does streptomycin cause an error catastrophe? *Biochimie* **69**: 131–136.

Geigenmüller, U., T.-P. Hausner, and K. H. Nierhaus. 1986. Analysis of the puromycin reaction: the ribosomal exclusion principle for AcPhe-tRNA binding re-examined. *Eur. J. Biochem.* **161**:715–721.

Gnirke, A., U. Geigenmüller, H.-J. Rheinberger, and K. H. Nierhaus. 1989. The allosteric 3-site model of the ribosomal elongation cycle: analysis with a heteropolymeric mRNA. *J. Biol. Chem.* **264**:7291–7301.

Gnirke, A., and K. H. Nierhaus. 1986. tRNA binding sites on the subunits of *Escherichia coli* ribosomes. *J. Biol. Chem.* **261**: 14506–14514.

Grajevskaja, R. A., Y. V. Ivanov, and E. M. Saminsky. 1982. 70S ribosomes of *Escherichia coli* have an additional site for deacylated tRNA binding. *Eur. J. Biochem.* **128**:47–52.

Hausner, T.-P., U. Geigenmüller, and K. H. Nierhaus. 1988. The allosteric three-site model for the ribosomal elongation cycle: new insight into the inhibition mechanisms of aminoglycosides, thiostrepton, and viomycin. *J. Biol. Chem.* **263**:13103–13111.

Hirsh, D. 1971. Tryptophan transfer RNA as the UGU suppressor. *J. Mol. Biol.* **58**:439–458.

Kirillov, S. V., E. M. Makarov, and Y. P. Semenkov. 1983. Quantitative study of interaction of deacylated tRNA with *Escherichia coli* ribosomes: role of 50S in formation of E-site. *FEBS Lett.* **157**:91–94.

Kirillov, S. V., V. I. Makhno, and Y. P. Semenkov. 1980. Mechanism of codon-anticodon-interaction in ribosomes: direct functional evidence that isolated 30S subunits contain two codon-dependent specific binding sites for transfer RNA. *Nucleic Acids Res.* **8**:183–196.

Kirillov, S. V., and Y. P. Semenkov. 1982. Non-exclusion principle of Ac-Phe-$tRNA^{Phe}$ interaction with the donor and acceptor sites of *Escherichia coli* ribosomes. *FEBS Lett.* **148**:235–238.

Kirillov, S. V., and Y. P. Semenkov. 1986. Extension of Watson's model for the elongation cycle of protein synthesis. *J. Biomol. Struct. Dynam.* **4**:263–269.

Lill, R., A. Lepier, F. Schwägele, M. Sprinzl, H. Vogt, and W. Wintermeyer. 1988. Specific recognition of the 3′-terminal adenosine of $tRNA^{Phe}$ in the exit site of *Escherichia coli* ribosomes. *J. Mol. Biol.* **203**:699–705.

Lill, R., J. M. Robertson, and W. Wintermeyer. 1984. tRNA binding sites of ribosomes from *Escherichia coli*. *Biochemistry* **23**:6710–6717.

Lill, R., and W. Wintermeyer. 1987. Destabilisation of codon-anticodon interaction in the ribosomal exit site. *J. Mol. Biol.* **196**:137–148.

Moazed, D., and H. F. Noller. 1987. Interaction of antibiotics with functional sites in 16S ribosomal RNA. *Nature* (London) **327**: 389–394.

Moazed, D., and H. F. Noller. 1989. Interaction of tRNA with 23S rRNA in the ribosomal A, P and E sites. *Cell* **57**:585–597.

Nierhaus, K. H., and H.-J. Rheinberger. 1984. An alternative model for the elongation cycle of protein biosynthesis. *Trends Biochem. Sci.* **9**:428–432.

Nierhaus, K. H., H.-J. Rheinberger, U. Geigenmüller, A. Gnirke, H. Saruyama, S. Schilling, and P. Wurmbach. 1986. Three tRNA binding sites involved in the ribosomal elongation cycle, p. 454–472. *In* B. Hardesty and G. Kramer (ed.), *Structure, Function, and Genetics of Ribosomes*. Springer-Verlag, New York.

Remme, J., T. Margus, R. Villems, and K. H. Nierhaus. 1989. The third ribosomal tRNA-binding site, the E site, is occupied in native polysomes. *Eur. J. Biochem.* **183**:281–284.

Rheinberger, H.-J., and K. H. Nierhaus. 1980. Simultaneous binding of three tRNA molecules by the ribosome of *Escherichia coli*. *Biochem. Int.* **1**:297–303.

Rheinberger, H.-J., and K. H. Nierhaus. 1983. Testing an alternative model of the ribosomal peptide elongation cycle. *Proc. Natl. Acad. Sci. USA* **80**:4213–4217.

Rheinberger, H.-J., and K. H. Nierhaus. 1986a. Allosteric interactions between the transfer-RNA binding sites A and E. *J. Biol. Chem.* **261**:9133–9139.

Rheinberger, H.-J., and K. H. Nierhaus. 1986b. Adjacent codon-anticodon interactions of both tRNAs present at the ribosomal A and P or P and E sites. *FEBS Lett.* **204**:97–99.

Rheinberger, H.-J., H. Sternbach, and K. H. Nierhaus. 1981. Three tRNA binding sites on *E. coli* ribosomes. *Proc. Natl. Acad. Sci. USA* **76**:5310–5314.

Rheinberger, H.-J., H. Sternbach, and K. H. Nierhaus. 1986. Codon-anticodon interaction at the ribosomal E site. *J. Biol. Chem.* **261**:9140–9143.

Rodnina, M. V., A. V. El'skaya, Y. P. Semenkov, and S. V. Kirillov. 1988. Number of tRNA binding sites on 80S ribosomes and their subunits. *FEBS Lett.* **231**:71–74.

Saruyama, H., and K. H. Nierhaus. 1986. Evidence that the three-site model for the ribosomal elongation cycle is also valid in the archaebacterium *Halobacterium halobium*. *Mol. Gen. Genet.* **204**:221–228.

Smith, D., and M. Yarus. 1989a. Transfer RNA structure and coding specificities. I. Evidence that a D-arm mutation reduces tRNA dissociation from the ribosome. *J. Mol. Biol.* **206**:489–501.

Smith, D., and M. Yarus. 1989b. Transfer RNA structure and coding specificity. II. A D-arm tertiary interaction that restricts coding range. *J. Mol. Biol.* **206**:503–511.

Watson, J. D. 1964. The synthesis of proteins upon ribosomes. *Bull. Soc. Chim. Biol.* **46**:1399–1425.

Wettstein, F. O., and H. Noll. 1965. Binding of transfer ribonucleic acid to ribosomes engaged in protein synthesis: number and properties of ribosomal binding sites. *J. Mol. Biol.* **11**:35–53.

Chapter 26

Structure and Function of rRNA in the Decoding Domain and at the Peptidyltransferase Center

ROBERT A. ZIMMERMANN, CHERYL L. THOMAS, and JACEK WOWER

Our understanding of the structure and function of procaryotic 16S and 23S rRNAs, and of their eucaryotic and organellar counterparts, has grown with astonishing speed within the past decade. Consider what we knew 10 years ago: the sequence of only one 16S rRNA had been completed, its predicted secondary structure was a matter of some controversy, and its three-dimensional organization was an almost total mystery (Noller, 1980). Moreover, only a handful of potential functional sites had been identified. There were nonetheless many tantalizing hints as to the importance of rRNA in the biological activity of the ribosome. Primary among these was the realization, as more and more sequences became available, that both the small- and large-subunit rRNAs were unusually well conserved in structure whether they came from procaryotes, eucaryotes, or eucaryotic organelles (Woese et al., 1983). The simplest explanation for this remarkable uniformity was that it reflected a conservation of function throughout evolution. And if this were so, then perhaps rRNA played the more important role in protein synthesis and the proteins were present to optimize its structure, to tune its activity, and to protect it from a hostile environment.

We now take the essential role of rRNA in translation almost for granted. The derivation of complete sequences for over 100 small- and large-subunit rRNAs, together with a vast array of other structural data obtained by the application of innovative techniques and much hard work, has confirmed the pattern of base pairing within these molecules, and we are well on the way to understanding their spatial configuration and the manner in which protein and rRNA are integrated within the ribosomal particles (Brimacombe et al., 1988; Stern et al., 1988). Perhaps most important, it is now possible to visualize in three dimensions how mRNA, tRNA, and various translation factors interact with both RNA and protein components of the ribosome as they carry out their appointed tasks in protein synthesis.

Structurally, the small- and large-subunit rRNAs are organized into domains, each of which contains several hundred bases (Noller, 1984). The structural domains are folded individually into rather compact units through base pairing between adjacent segments and are anchored to one another by tertiary interactions that include short helical stems whose strands are quite distant in the sequence, as well as more complex interactions about which we have little specific information. Although the 16S rRNA of *Escherichia coli* consists of three major domains and a small 3′ domain, most activities of the 30S subunit in translational initiation and decoding can be attributed to the 3′ minor domain, a conserved stretch of some 150 bases at the 3′ terminus. However, important contributions are thought to be made by the 5′ domain to the maintenance of translational accuracy, by the central domain to interaction with the 50S subunit, and by the 3′ major domain to elongation and termination (reviewed in Raué et al., 1988). Six domains have been distinguished in the *E. coli* 23S rRNA, and here various activities appear to be more equally shared. In particular, domain II contains the GTPase center, portions of domains IV and V comprise the peptidyltransferase center, and domain VI is involved in associations with elongation factors EF-Tu and EF-G (reviewed in Raué et al., 1988). Whether these categorical assignments are justified—or simply reflect our ignorance—remains uncertain.

This chapter will focus on the functional prop-

Robert A. Zimmermann, Cheryl L. Thomas, and Jacek Wower ■ Department of Biochemistry and Program in Molecular and Cellular Biology, University of Massachusetts, Amherst, Massachusetts 01003.

different tRNAs, reflecting the generality of this close structural relationship between anticodon and 16S rRNA (Ofengand et al., 1979; Prince et al., 1979); (ii) that *E. coli* $tRNA_1^{Val}$ can also be cross-linked to the base corresponding to C-1400 in eucaryotic small-subunit rRNAs (Ehresmann et al., 1984; Ciesiolka et al., 1985b); (iii) that covalent attachment of $tRNA_1^{Val}$ to C-1400 occurs equally well when the tRNA is complexed with 30S subunits instead of 70S ribosomes (Denman et al., 1988); and (iv) that A-site-bound $tRNA_1^{Val}$ bearing photoaffinity labels attached to the 5′ anticodon base by a 23- to 24-Å "leash" also cross-linked to C-1400, an observation consistent with both the expected distance between the A- and P-site anticodons and the exposed position of this rRNA residue in the translating ribosome (Ciesiolka et al., 1985a). As a consequence, the portion of the 30S subunit in the neighborhood of nucleotide C-1400 has often been referred to as the decoding site. Because of the presence of another remarkably well-conserved sequence around position 1500 (Dams et al., 1988), its probable proximity to C-1400 in the 30S subunit (Brimacombe et al., 1988; Stern et al., 1988), and the mounting evidence for its involvement in associations with a variety of ligands involved in the translational process (see below), the term "decoding domain" will be used to denote segments of the 16S rRNA occurring at positions 1390 to 1410 and 1490 to 1505, as well as nucleotides 920 to 930, with which a portion of the former sequence is base paired. The proximity of the 920–930 and 1390–1410 segments in active 30S particles was recently confirmed by the characterization of a psoralen-induced cross-link between nucleotides 924 and 1393 (Ericson and Wollenzien, 1989). The decoding domain has been mapped to the cleft between the platform and head of the 30S subunit by immunoelectron microscopy of ribosomal complexes containing either tRNA or oligonucleotides bound to residues 1392 to 1408 of the 16S rRNA (Gornicki et al., 1984; Oakes et al., 1986a), a region compatible with the sites to which mRNA has been localized by the same technique (Evstafieva et al., 1983; Olson et al., 1988). We can be quite confident that it is within the cleft that codon-anticodon recognition and interaction take place (Fig. 2).

Further indications of the functional significance of the highly conserved sequences surrounding positions 1400 and 1500 have accumulated at a rapid rate. To date, nearly every nucleotide in these segments has been implicated in the translational activity of the ribosome (Fig. 1). Among the ligands whose association with these regions has been either inferred or demonstrated are (in addition to tRNA) mRNA, initiation factor IF3, a number of antibiotics, and the 50S ribosomal subunit.

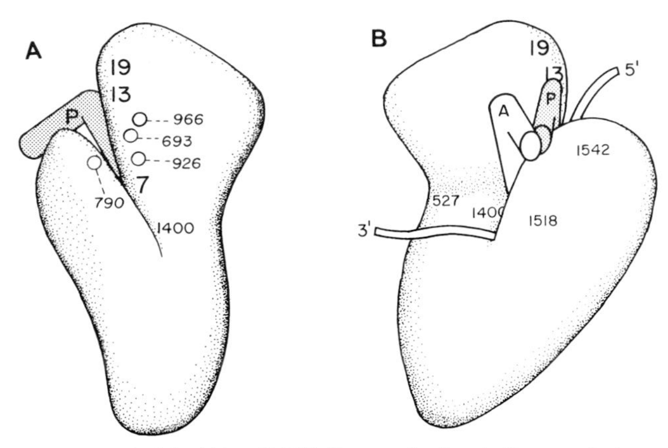

Figure 2. Model of *E. coli* 30S ribosomal subunit. Proposed placement of tRNA at the P site and the A site, as viewed through the cleft (A) and from the subunit interface (B). The locations of proteins S7, S13, and S19, as well as of nucleotides 527, 1400, 1518, and 1542 of the 16S rRNA, have been determined by immunoelectron microscopy (see Trempe et al., 1982; Gornicki et al., 1984; Stöffler and Stöffler-Meilicke, 1986; Oakes et al., 1986a; Oakes et al., 1986b). A P-site-bound $tRNA^{Phe}$ derivative containing 2-azidoadenosine at position 37 has been cross-linked to protein S7 (L. A. Sylvers, J. Wower, and R. A. Zimmermann, unpublished data); anticodon arm fragments with 2- or 8-azidoadenosine at position 43 labeled proteins S13 and S19 (see text). Positions of nucleotides 693, 780, 926, and 966, which are protected by tRNA at the P site (Moazed and Noller, 1986), are approximations based on the proposed 16S rRNA structures of Brimacombe et al. (1988) and Stern et al. (1988). Orientation of the mRNA is according to Olson et al. (1988). The subunit model depicted is that of Lake and co-workers (see Oakes et al. 1986a).

One very productive line of investigation has been the analysis of residues that become refractory to chemical modification in the presence of specific ligands. For example, a variety of bases are shielded when tRNA is bound to the P site (Meier and Wagner, 1984; Moazed and Noller, 1986). The strong protection of G-1401, along with weaker protection of C-1399 and C-1400 (Moazed and Noller, 1986), dovetails precisely with the tRNA-16S rRNA cross-link discussed above and most likely arises from the close approach of the anticodon to the rRNA. The inaccessibility of G-926 in this complex may also result from the anticodon loop, as G-926 is located quite near the 1400 region in the 16S rRNA secondary structure. We have found very recently that both 2- and 8-azidoadenosine incorporated at position 43 of an anticodon stem-loop fragment derived from yeast $tRNA^{Phe}$ can photochemically label protein S13 and, to a lesser extent, protein S19, when bound to the 30S subunit P site (J. Wower and R. A. Zimmermann, unpublished data). These data suggest that the shielding of G-693, A-794, C-795, and m^2G-966 by P-site-bound tRNA may be attributable to the coaxial anticodon-dihydrouridine helix, given the spatial relationship between these bases and the labeled proteins in the 30S subunit (Fig. 2) and the fact that the position of the photoaffinity probe in the tRNA fragment corresponds to the junction of the anticodon and dihydrouridine stems. Furthermore, Moazed and Noller (1986) have shown that A-site-bound tRNA protects mainly A-1408, A-1492, and A-1493 (as well as several nucleotides in the apical loop of the hairpin at positions 500 to 545; see below). Because the anticodons of tRNA at the A and P sites must be adjacent during polypeptide synthesis, the inaccessibility of A-1408 is consistent with the P-site protection at positions 1399 to 1401. According to this line of reasoning, A-1492 and A-1493 must be located in the same vicinity of the 30S subunit.

Valuable insights into the function of the 1400 and 1500 regions have been gained from the investigation of sites related to the action of a number of antibiotics (Fig. 1). The sensitivity of ribosomes to paromomycin and several other aminoglycosides, including kanamycin, requires maintenance of pairing between positions 1409 and 1491 at the beginning of the penultimate helix. Paromomycin-resistant mutants in which this conserved base pair is disrupted have now been described in yeast mitochondria, *Tetrahymena thermophila*, and *E. coli* (Li et al., 1982; Spangler and Blackburn, 1985; De Stasio et al., 1989). These results jibe neatly with the finding that the same antibiotic protects the adjacent resides, A-1408 and G-1494, from chemical modification when bound to 70S ribosomes (Moazed and Noller, 1987a). In *Streptomyces lividans*, resistance to kanamycin results from methylation of the bases corresponding to either G-1405 or A-1408 (Beauclerk and Cundliffe, 1987), and in *T. thermophila*, hygromycin resistance is associated with a U-to-C transition at position 1495 (Spangler and Blackburn, 1985). Once again, results of the chemical protection experiments are in close agreement, as kanamycin has been found to shield A-1408 and G-1494, whereas hygromycin protects G-1494 alone (Moazed and Noller, 1987a). Finally, a secondary cross-linking site for streptomycin was identified between residues 1394 and 1415 in addition to its main attachment site at positions 892 to 917 (Gravel et al., 1987). The fact that all of the drugs mentioned above are known to provoke misreading adds further impetus to the argument that the conserved segments at positions 1390 to 1410 and 1490 to 1505 lie close to the site of codon-anticodon interaction.

Several other findings also call our attention to the functional role of the 1400 and 1500 regions. First, the presence of mRNA in the decoding domain, never in serious doubt, has been confirmed by the covalent attachment of poly(A) message to nucleotides 1394 to 1399 (Stiege et al., 1988). This sequence is immediately adjacent to the site of anticodon cross-linking, reinforcing the notion that tRNA, rRNA, and mRNA are all in close contact at or near position 1400. Second, the specific involvement of the segment spanning nucleotides 1490 to 1505 in the initiation of translation is strongly implied by the presence of what is probably a major site of contact between the 30S subunit and IF3 (Fig. 1). Nuclease digestion of specific complexes between IF3 and a fragment encompassing the 49 3′-terminal residues of 16S rRNA indicates that the protein interacts with nucleotides 1494 to 1506 (Wickstrom, 1983), while physical measurements suggest that bases in the terminal stem-loop are involved as well (Wickstrom et al., 1986). Moreover, a cross-link between IF3 and the 16S rRNA has been localized to the sequence spanning residues 1506 to 1529 (Ehresmann et al., 1986). The interaction of IF3 within the segment bounded by nucleotides 1494 to 1529 is in good accord both with the placement of the terminal helix in recent 30S subunit models (Brimacombe et al., 1988; Stern et al., 1988) and with the mapping of IF3 to the cleft of the 30S subunit by immunoelectron microscopy (Emanuilov et al., 1978). Finally, as might be expected from its proximity to tRNA, the decoding domain comprises an important part of the subunit interface. In fact, bases throughout the 3′ minor domain have been implicated in 50S subunit association by chemical modification, damage selection, and the analysis of various mutations in the 16S rRNA (Herr et al., 1979; Poldermans et al., 1980; Meier and Wagner, 1985; Meier et al., 1986; Zwieb et al., 1986; Baudin et al., 1989).

GENETIC ANALYSIS OF THE DECODING DOMAIN CORRELATES WELL WITH STRUCTURAL AND FUNCTIONAL STUDIES

Notwithstanding the extensive body of information now available on the way in which the decoding domain interacts with other components of the translation apparatus, the task remains to piece together the individual facts into a unified model of 16S rRNA function within the ribosome. Can genetic analysis help us to sort out what is important from what is not? More specifically, can results derived from the systematic alteration of particular bases and base sequences be readily correlated with observations made by other techniques? Although both of these questions can probably be answered in the affirmative, at least two limitations must be kept in mind.

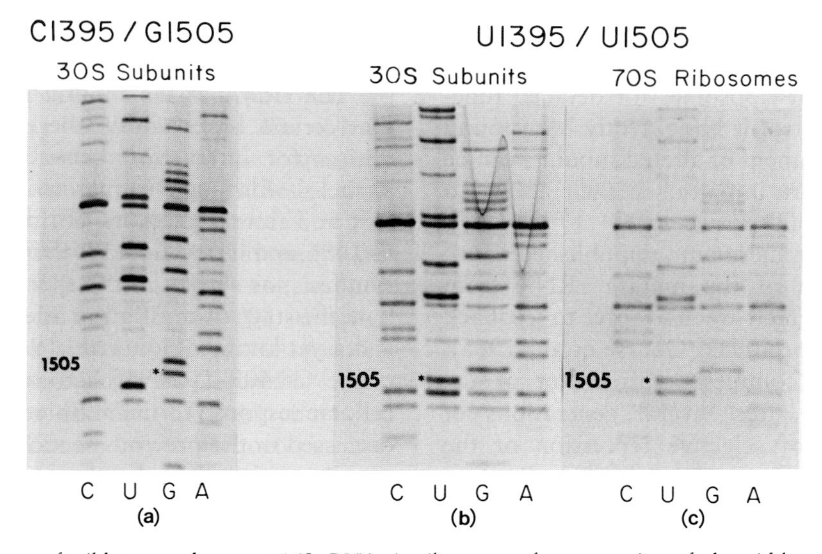

Figure 4. Occurrence of wild-type and mutant 16S rRNAs in ribosomes after expression of plasmid-borne *rrnB* operons for three generations. Autoradiograms show the sequences of the 1500 region of 16S rRNA from wild-type 30S subunits (control; a) and from 30S subunits and 70S ribosomes of the U-1395 U-1505 double mutant (b and c). RNA was sequenced by the primer-extension method, using avian myeloblastosis virus reverse transcriptase (Lane et al., 1985). The primer was complementary to bases 1520 to 1536 of the 16S rRNA. The gels indicate that the mutant 16S rRNA is the predominant species in the cell three generations after induction, implying that it is assembled into functional ribosomes.

activity in initiation-independent dipeptide synthesis. The only 30S subunits that were inactive in all of the assays were the ones that contained 16S rRNA with a deletion at position 1401, and even these particles were capable of associating with 50S subunits. Although all of the findings described above corroborate the importance of nucleotides 1395 to 1407 in 30S subunit function, perhaps by influencing the properties of the P site during polypeptide chain initiation, two discrepancies between the in vivo and in vitro results remain to be explained. First, the conclusion of Denman et al. (1989) that base substitutions at C-1400 had little influence on ribosome function in vitro conflicts with the experiments of Hui et al. (1988), who showed that replacement of C-1400 by either G or A was quite deleterious in vivo. Second, the deletion of C-1400, but not G-1401, was found to be lethal in vivo (Thomas et al., 1988), whereas in vitro, the deletion of G-1401 was far more detrimental to protein synthesis than was the absence of C-1400. One major difference between the two systems is, of course, the methylation of the 16S rRNA, which may tune the function of the 16S rRNA in a manner not yet understood. Perhaps more important, the concentrations of the translational substrates and ligands, and therefore the many equilibria among them, differ greatly in vivo and in vitro.

The lethal mutations, U-1395, U-1407, and ΔC-1400, have been further investigated in our laboratory. All three lead to cell death whether transcribed from the *rrnB* P1 and P2 promoters or the inducible λ p_L promoter (Thomas et al., 1988). When transiently expressed from p_L, the altered 16S rRNA is processed normally and incorporated into 30S subunits. These particles cannot, however, associate efficiently with 50S subunits in vivo or in vitro, suggesting that they are impaired either in association per se or in some preceding step such as 30S initiation complex formation. Moreover, as established by reverse transcriptase sequencing, at least 50% of the 16S rRNA present three to four generations after induction was from the wild-type, chromosomal rRNA operons. Interestingly, the lethal phenotype was suppressed by the replacement of G-1505 with U, C, or A. In the case of the double mutant U-1395 U-1505, over 90% of the 16S rRNA was mutant after four generations (Fig. 4). The virtual absence of chromosome-encoded 16S rRNA indicates that ribosomes containing the mutant 16S rRNA were active in protein synthesis, in other words, that the suppressor mutation corrects a functional defect arising from the first alteration. This conclusion is also consistent with the 16S rRNA composition of the mutant strains. Nomura and colleagues have provided evidence that repression of the strong *rrnB* P1 promoter is mediated by initiation-competent 30S subunits (Cole et al., 1987; Yamagishi et al., 1987). Nearly complete repression of chromosomal rRNA synthesis in cells expressing the double mutant therefore suggests that subunits assembled from the latter 16S rRNA are active in initiation; by contrast, 30S subunits containing only the lethal mutations appear to be poor repressors of the *rrn* promoters and hence are likely to be defective in initiation. How then can suppression of lethal mutations in the 1400 region by alterations in the 1500 region be explained? Given the likely proximity of the two sequences in the 30S particle, they may together comprise the binding site for an essential ligand such as IF3, whose binding site has been shown to include the sequence that encompasses the suppressor (Wickstrom, 1983; Ehresmann et al., 1986). In the lethal mutants, the altered 30S subunits might bind IF3 with above-normal affinity and thereby sequester it; the suppressor mutation might then readjust the equilibrium to physiological levels and permit the factor to cycle normally.

THE CONSERVED HAIRPIN AT POSITIONS 500 TO 545 OF THE 16S RNA IS INDIRECTLY LINKED TO THE DECODING DOMAIN

No discussion of the decoding domain would be complete without a consideration of the intriguing stem-loop structure encompassing nucleotides 500 to 545 of the 16S rRNA. Together with the segments surrounding positions 1400 and 1500, it comprises one of the three most highly conserved portions of the small-subunit rRNAs (Gutell et al., 1985). This feature consists of a stable helical stem, one of whose strands spills out to form a 6-base bulge loop, surmounted by an apical loop of 16 bases, most of which are invariant (Dams et al., 1988) (Fig. 5). In eubacteria, position 527 is occupied by m^{7}G. Although eucaryotes fail to methylate this residue, their small-subunit rRNAs invariably contain 2′-O-methyluridine at position 531, just 4 bases away. m^{7}G is also absent from most archaebacterial 16S rRNAs, and it is not known whether they are modified elsewhere in this region (Zueva et al., 1985). The inaccessibility of certain residues in the apical loop to base- and backbone-specific chemical reagents is indicative of higher-order structure (Moazed et al., 1986; Baudin et al., 1989). Indeed, a covariance involving nucleotides 505 to 507 and 524 to 526 suggests that these two sequences may be base paired to form a pseudoknot, or other tertiary interaction, during at least some stage of the ribosome cycle (Woese and Gutell, 1989). Other bases in the apical loop, accessible in free 16S rRNA, become unreactive in the 30S subunit (Moazed et al., 1986). Although much of the protection can be attributed to the influence of protein S12 (Stern et al., 1989), this region is also part of the binding site of protein S4 (Stern et al., 1986).

Despite the fact that the 500–545 hairpin is situated on the opposite side of the 30S subunit from

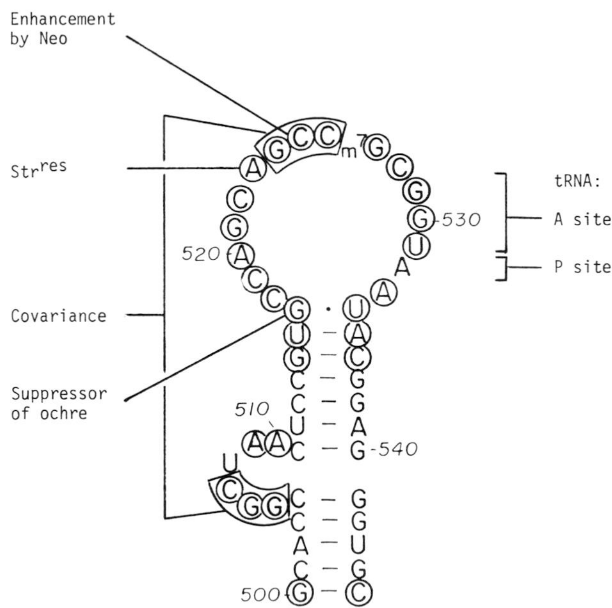

Figure 5. Sites of functional interest in the stem-loop structure at positions 500 to 545 of *E. coli* 16S rRNA. tRNA, Protection from chemical modification by tRNA^{Phe} bound to the A or P site (Moazed and Noller, 1986); Enhancement by Neo, reactivity to dimethyl sulfate increased when neomycin and related aminoglycoside antibiotics are bound to ribosome (Moazed and Noller, 1987a); Str^{res}, substitution of C at this position leads to streptomycin-resistant ribosomes (Gauthier et al., 1988; Melançon et al., 1988); Suppressor of ochre, transition at the corresponding site in yeast mitochondrial small-subunit rRNA results in suppression of ochre mutations (Shen and Fox, 1989); Covariance, coordinated base changes in the two trinucleotide sequences have maintained the potential for Watson-Crick pairing between them in 16S-like rRNAs from a wide variety of organisms (Woese and Gutell, 1989). Encircled bases are highly conserved in evolution (Gutell et al., 1985).

the decoding domain, as judged from the location of m^{7}G-527 by immunoelectron microscopy (Trempe et al., 1982), and separated from it by 60 to 70 Å (Moazed and Noller, 1986), there is evidently some connection or communication between the two regions (see Fig. 2). The interaction of the tRNA with the P site, for instance, renders A-532 moderately inaccessible to chemical modification, and A-site-bound tRNA shields G-529, G-530, and U-531 in a poly(U)-dependent fashion (Moazed and Noller, 1986). In addition, the error-inducing antibiotics neomycin, kanamycin, gentamicin, and paromomycin, whose sites of interaction with the 30S subunit presumably lie within the decoding domain, enhance the chemical reactivity of C-525 (Moazed and Noller, 1987a). Even more surprising, ribosomes containing 16S rRNA in which A-523 is replaced by C become resistant to streptomycin (Gauthier et al., 1988; Melançon et al., 1988). The binding site for this antibiotic, which also induces miscoding, is located close to the decoding domain, most likely between

Melançon, P., C. Lemieux, and L. Brakier-Gingras. 1988. A mutation in the 530 loop of *Escherichia coli* 16S ribosomal RNA causes resistance to streptomycin. *Nucleic Acids Res.* **16**:9631–9639.

Moazed, D., and H. F. Noller. 1986. Transfer RNA shields specific nucleotides in 16S ribosomal RNA from attack by chemical probes. *Cell* **47**:985–994.

Moazed, D., and H. F. Noller. 1987a. Interaction of antibiotics with functional sites in 16S ribosomal RNA. *Nature* (London) **327**:389–394.

Moazed, D., and H. F. Noller. 1987b. Chloramphenicol, erythromycin, carbomycin and vernamycin B protect overlapping sites in the peptidyl transferase region of 23S ribosomal RNA. *Biochimie* **69**:879–884.

Moazed, D., and H. F. Noller. 1989. Interaction of tRNA with 23S rRNA in the ribosomal A, P and E sites. *Cell* **57**:585–597.

Moazed, D., J. M. Robertson, and H. F. Noller. 1988. Interaction of elongation factors EF-G and EF-Tu with a conserved loop in 23S RNA. *Nature* (London) **334**:362–364.

Moazed, D., S. Stern, and H. F. Noller. 1986. Rapid chemical probing of conformation in 16S ribosomal RNA and 30S ribosomal subunits using primer extension. *J. Mol. Biol.* **187**: 399–416.

Noller, H. F. 1980. Structure and topography of ribosomal RNA, p. 3–22. *In* G. Chambliss, G. R. Craven, J. Davies, K. Davis, L. Kahan, and M. Nomura (ed.), *Ribosomes. Structure, Function, and Genetics.* University Park Press, Baltimore.

Noller, H. F. 1984. Structure of ribosomal RNA. *Annu. Rev. Biochem.* **53**:119–162.

Nomura, M., J. Sidikaro, K. Jakes, and N. Zinder. 1974. Effects of colicin E3 on bacterial ribosomes, p. 805–814. *In* M. Nomura, A. Tissières, and P. Lengyel (ed.), *Ribosomes.* Cold Spring Harbor Laboratory, Cold Spring Harbor, N.Y.

Oakes, M., M. W. Clark, E. Henderson, and J. A. Lake. 1986a. DNA hybridization electron microscopy: ribosomal RNA nucleotides 1392–1407 are exposed in the cleft of the small subunit. *Proc. Natl. Acad. Sci. USA* **83**:275–279.

Oakes, M., E. Henderson, A. Scheinman, M. Clark, and J. A. Lake. 1986b. Ribosome structure, function and evolution: mapping ribosomal RNA, proteins, and functional sites in three dimensions, p. 47–67. *In* B. Hardesty and G. Kramer (ed.), *Structure, Function, and Genetics of Ribosomes.* Springer-Verlag, New York.

Ofengand, J., and R. Liou. 1980. Evidence for pyrimidine-pyrimidine cyclobatane dimer formation in the covalent cross-linking between transfer ribonucleic acid and 16S RNA at the ribosomal P site. *Biochemistry* **19**:4814–4822.

Ofengand, J., R. Liou, J. Kohut III, I. Schwartz, and R. A. Zimmermann. 1979. Covalent cross-linking of tRNA to the ribosomal P site: mechanism and site of reaction in tRNA. *Biochemistry* **18**:4322–4332.

Olson, H. M., L. S. Lasater, P. A. Cann, and D. G. Glitz. 1988. Messenger RNA orientation on the ribosome. Placement by electron microscopy of antibody-complementary oligodeoxynucleotide complexes. *J. Biol. Chem.* **263**:15196–15204.

Poldermans, B., H. Bakker, and P. H. van Knippenberg. 1980. Studies on the function of two adjacent N^6,N^6-dimethyladenosines near the 3′ end of 16S ribosomal RNA of *Escherichia coli.* IV. The effect of the methyl groups on ribosomal subunit interaction. *Nucleic Acids Res.* **8**:143–151.

Prince, J. B., S. S. Hixson, and R. A. Zimmermann. 1979. Photochemical crosslinking of $tRNA^{Lys}$ and $tRNA_2^{Glu}$ to 16S RNA at the P site of *Escherichia coli* ribosomes. *J. Biol. Chem.* **254**:4745–4749.

Prince, J. B., B. H. Taylor, D. L. Thurlow, J. Ofengand, and R. A. Zimmermann. 1982. Covalent crosslinking of $tRNA_1^{Val}$ at the ribosomal P site: identification of crosslinked residues. *Proc. Natl. Acad. Sci. USA* **79**:5450–5454.

Raué, H. A., J. Klootwijk, and W. Musters. 1988. Evolutionary conservation of structure and function of high molecular weight ribosomal RNA. *Prog. Biophys. Mol. Biol.* **51**:77–129.

Rottman, N., B. Kleuvers, J. Atmadja, and R. Wagner. 1988. Mutants with base changes at the 3′-end of the 16S RNA from *Escherichia coli.* Construction, expression and functional analysis. *Eur. J. Biochem.* **177**:81–90.

Schmidt, F. J., J. Thompson, K. Lee, J. Dijk, and E. Cundliffe. 1981. The binding site for ribosomal protein L11 within 23S ribosomal RNA of *Escherichia coli. J. Biol. Chem.* **256**:12301–12305.

Shen, Z., and T. D. Fox. 1989. Substitution of an invariant nucleotide at the base of the highly conserved '1530-loop' of 15S rRNA causes suppression of yeast mitochondrial ochre mutations. *Nucleic Acids Res.* **17**:4535–4539.

Shine, J., and L. Dalgarno. 1974. The 3′-terminal sequence of *Escherichia coli* 16S ribosomal RNA: complementarity to nonsense triplets and ribosome binding sites. *Proc. Natl. Acad. Sci. USA* **71**:1342–1346.

Sköld, S. E. 1983. Chemical crosslinking of elongation factor G to the 23S RNA in 70S ribosomes from *Escherichia coli. Nucleic Acids Res.* **11**:4923–4932.

Sogin, M. L., and J. C. Edman. 1989. A self-splicing intron in the small subunit rRNA gene of *Pneumocystis carinii. Nucleic Acids Res.* **17**:5349–5359.

Spangler, E. A., and E. H. Blackburn. 1985. The nucleotide sequence of the 17S ribosomal RNA gene of *Tetrahymena thermophila* and the identification of point mutations resulting in resistance to the antibiotics paromomycin and hygromycin. *J. Biol. Chem.* **260**:6334–6340.

Steiner, G., E. Kuechler, and A. Barta. 1988. Photo-affinity labelling at the peptidyl transferase centre reveals two different positions for the A- and P-sites in domain V of 23S rRNA. *EMBO J.* **7**:3949–3955.

Steitz, J. A. 1980. RNA · RNA interactions during polypeptide chain initiation, p. 479–495. *In* G. Chambliss, G. R. Craven, J. Davies, K. Davis, L. Kahan, and M. Nomura (ed.), *Ribosomes. Structure, Function, and Genetics.* University Park Press, Baltimore.

Steitz, J. A., and K. Jakes. 1975. How ribosomes select initiator regions in mRNA: base pair formation between the 3′ terminus of 16S rRNA and the mRNA during the initiation of protein synthesis in *Escherichia coli. Proc. Natl. Acad. Sci. USA* **72**: 4734–4738.

Stern, S., T. Powers, L.-M. Changchien, and H. F. Noller. 1989. RNA-protein interactions in 30S ribosomal subunits: folding and function of 16S rRNA. *Science* **244**:783–790.

Stern, S., B. Weiser, and H. F. Noller. 1988. Model of the three-dimensional folding of 16S ribosomal RNA. *J. Mol. Biol.* **204**:447–481.

Stern, S., R. C. Wilson, and H. F. Noller. 1986. Localization of the binding site for protein S4 on 16S ribosomal RNA by chemical and enzymatic probing and primer extension. *J. Mol. Biol.* **192**:101–110.

Stiege, W., J. Atmadja, M. Zobawa, and R. Brimacombe. 1986. Investigation of the tertiary folding of *Escherichia coli* ribosomal RNA by intra-RNA cross-linking *in vivo. J. Mol. Biol.* **191**: 135–138.

Steige, W., C. Glotz, and R. Brimacombe. 1983. Localization of a series of intra-RNA cross-links in the secondary and tertiary structure of 23S RNA, induced by ultraviolet irradiation of *Escherichia coli* 50S ribosomal subunits. *Nucleic Acids Res.* **11**:1687–1706.

Stiege, W., L. Stade, D. Schüler, and R. Brimacombe. 1988.

Covalent cross-linking of poly(A) to *Escherichia coli* ribosomes and localization of the cross-link site within the 16S RNA. *Nucleic Acids Res.* **16**:2369–2388.

Stöffler, G., and M. Stöffler-Meilicke. 1986. Immuno electron microscopy on *Escherichia coli* ribosomes, p. 28–46. *In* B. Hardesty and G. Kramer (ed.), *Structure, Function, and Genetics of Ribosomes.* Springer-Verlag, New York.

Tapprich, W. E., D. J. Goss, and A. E. Dahlberg. 1989. Mutation at position 791 in *Escherichia coli* 16S ribosomal RNA affects processes involved in the initiation of protein synthesis. *Proc. Natl. Acad. Sci. USA* **86**:4927–4931.

Thanaraj, T. A., and M. W. Pandit. 1989. An additional ribosome-binding site on mRNA of highly expressed genes and a bifunctional site on the colicin fragment. *Nucleic Acids Res.* **17**: 2973–2985.

Thomas, C. L., R. J. Gregory, G. Winslow, A. Muto, and R. A. Zimmermann. 1988. Mutations within the decoding site of *Escherichia coli* 16S rRNA: growth rate impairment, lethality and intragenic suppression. *Nucleic Acids Res.* **16**:8129–8146.

Thompson, J., E. Cundliffe, and A. E. Dahlberg. 1988. Site-directed mutagenesis of *Escherichia coli* 23S ribosomal RNA at position 1067 within the GTP hydrolysis centre. *J. Mol. Biol.* **203**:457–465.

Thompson, R. C. 1988. EFTu provides an internal kinetic standard for translational accuracy. *Trends Biochem. Sci.* **13**:91–93.

Trempe, M. R., K. Ohgi, and D. G. Glitz. 1982. Ribosome structure: localization of 7-methyl guanosine in the small subunits of *Escherichia coli* and chloroplast ribosomes by immunoelectron microscopy. *J. Biol. Chem.* **257**:9822–9829.

van Knippenberg, P. H. 1986. Structural and functional aspects of the N^6,N^6 dimethyladenosines in 16S ribosomal RNA, p. 412–424. *In* B. Hardesty and G. Kramer (ed.), *Structure, Function, and Genetics of Ribosomes.* Springer-Verlag, New York.

Vester, B., and R. A. Garrett. 1988. The importance of highly conserved nucleotides in the binding region of chloramphenicol at the peptidyl transfer centre of *Escherichia coli* 23S ribosomal RNA. *EMBO* J. **7**:3577–3587.

Walleczek, J., D. Schüler, M. Stöffler-Meilicke, R. Brimacombe, and G. Stöffler. 1988. A model for the spatial arrangement of the proteins in the large subunit of the *Escherichia coli* ribosome. *EMBO J.* **7**:3571–3576.

Weiss, R. B., D. M. Dunn, A. E. Dahlberg, J. F. Atkins, and R. F. Gesteland. 1988. Reading frame switch caused by base-pair formation between the 3′ end of 16S rRNA and the mRNA during elongation of protein synthesis in *Escherichia coli*. *EMBO J.* **7**:1503–1507.

Wickstrom, E. 1983. Nuclease mapping of the secondary structure of the 49-nucleotide 3′ terminal cloacin fragment of *Escherichia coli* 16S RNA and its interactions with initiation factor 3. *Nucleic Acids Res.* **11**:2035–2052.

Wickstrom, E., H. A. Heus, C. A. G. Haasnoot, and P. H. van Knippenberg. 1986. Circular dichroism and 500-MHz proton magnetic resonance studies of the interaction of *Escherichia coli* translational initiation factor 3 protein with the 16S ribosomal RNA 3′ cloacin fragment. *Biochemistry* **25**:2770–2777.

Woese, C. R., and R. R. Gutell. 1989. Evidence for several higher order structural elements in ribosomal RNA. *Proc. Natl. Acad. Sci. USA* **36**:3119–3122.

Woese, C. R., R. Gutell, R. Gupta, and H. F. Noller. 1983. Detailed analysis of the higher-order structure of 16S-like ribosomal ribonucleic acids. *Microbiol. Rev.* **47**:621–669.

Wower, I. 1984. The study of RNA-protein interactions in the *Escherichia coli* ribosome. Ph.D. dissertation. University of Leeds, Leeds, United Kingdom.

Wower, I., J. Wower, M. Meinke, and R. Brimacombe. 1981. The use of 2-iminothiolane as an RNA-protein cross-linking reagent in *Escherichia coli* ribosomes, and the localisation on 23S RNA of sites cross-linked to proteins L4, L6, L21, L23, L27 and L29. *Nucleic Acids Res.* **9**:4285–4302.

Wower, J., S. S. Hixson, and R. A. Zimmermann. 1989. Labeling the peptidyl transferase center of the *Escherichia coli* ribosome with photoreactive $tRNA^{Phe}$ derivatives containing azidoadenosine at the 3′ end of the acceptor arm: a model of the tRNA-ribosome complex. *Proc. Natl. Acad. Sci. USA* **36**:5232–5236.

Yamagishi, M., H. A. de Boer, and M. Nomura. 1987. Feedback regulation of rRNA synthesis. A mutational alteration in the anti-Shine-Dalgarno region of the 16S rRNA gene abolishes regulation. *J. Mol. Biol.* **198**:547–550.

Zueva, V. S., A. S. Mankin, A. A. Bogdanov, D. L. Thurlow, and R. A. Zimmermann. 1985. Occurrence and location of 7-methylguanine residues in small-subunit ribosomal RNAs from eubacteria, archaebacteria and eukaryotes. *FEBS Lett.* **188**: 233–238.

Zwieb, C., and A. E. Dahlberg. 1984. Structural and functional analysis of *Escherichia coli* ribosomes containing small deletions around position 1760 in the 23S ribosomal RNA. *Nucleic Acids Res.* **12**:7135–7152.

Zwieb, C., D. K. Jemiolo, W. F. Jacob, R. Wagner, and A. E. Dahlberg. 1986. Characterization of a collection of deletion mutants at the 3′-end of 16S ribosomal RNA of *Escherichia coli*. *Mol. Gen. Genet.* **203**:256–264.

Chapter 27

Role of the tRNA Exit Site in Ribosomal Translocation

WOLFGANG WINTERMEYER, ROLAND LILL, and JAMES M. ROBERTSON

Ribosomes from *Escherichia coli* possess, in addition to the P and A sites, a third tRNA-binding site that specifically binds deacylated tRNA (Rheinberger et al., 1981; Grajevskaja et al., 1982; Kirillov et al., 1983; Lill et al., 1984). The site has been termed the exit site (E site), since the deacylated tRNA leaving the P site during translocation appears to be bound to this site before it is released from the ribosome. On the basis of binding studies with isolated ribosomal subunits, Kirillov and Semenkov (1986) proposed that the tRNA interaction with the E site takes place predominantly on the 50S subunit.

Questioning the classic two-site model of the elongation cycle, Rheinberger and Nierhaus (1983) proposed that after translocation, the E-site-bound tRNA is required to prevent slippage of the mRNA, this fixation constituting a decisive feature of a three-site model of elongation proposed by these authors. In contrast, Lill et al. (1986) suggested that intermediate E-site binding facilitates the release of the tRNA from the P site, where it is bound in an extremely stable fashion. In the latter model, the E-site-bound state of the leaving tRNA is considered a transient intermediate rather than a stable product of translocation; the function of the E site is seen as catalyzing the tRNA release from the P site, i.e., as a feature of the mechanism of translocation.

The two models of E-site function differ in the weight they put on both the stability of the E-site complexes and the contribution of codon-anticodon interaction to tRNA binding in this site. In this chapter, we summarize our pertinent data, describe the important contribution of the 3′ terminus of the tRNA to E-site binding, and address the functional role of this interaction. Finally, we present a model describing the molecular mechanism of tRNA release from the P site during translocation.

CHARACTERISTICS OF tRNA BINDING IN THE E SITE

The tRNA-Binding Properties of the E Site Differ from Those of Both the A and P Sites

The E site is distinguished from the other two sites by a number of properties suggesting that the tRNA molecule in the E site is bound in a qualitatively different way from the binding to both A and P sites: (i) it binds deacylated tRNA with a very high preference over aminoacylated tRNA; (ii) the tRNA is bound in a labile fashion, since it dissociates relatively rapidly and is readily exchanged in chasing experiments (Robertson et al., 1986a; Robertson and Wintermeyer, 1987); (iii) the affinity is not ionic strength dependent (Lill et al., 1986); and (iv) the contribution of codon-anticodon interaction to the free energy of tRNA binding is small (Lill and Wintermeyer, 1987). Some pertinent experimental results are summarized below.

In our hands, the E-site complex is quite unstable; therefore, we have used fluorescence measurements rather than nitrocellulose filtration to study the E-site interaction of tRNA. The results discussed were obtained with two fluorescent derivatives of $tRNA^{Phe}$, one ($tRNA^{Phe}_{Prf\text{-}37}$) from yeast cells carrying proflavine covalently bound at position 37 next to the anticodon, and the second ($tRNA^{Phe}_{Prf\text{-}16/20}$) from *Escherichia coli*.

The basic experiment that introduces the E site as an intermediate binding site for the deacylated tRNA released from the P site during translocation is depicted in Fig. 1. To a pretranslocation complex carrying fluorescent $tRNA^{Phe}_{Prf\text{-}37}$ in the P site and *N*-acetylated Phe-tRNA (AcPhe-$tRNA^{Phe}$) in the A site, elongation factor EF-G is added, and the fate of the deacylated tRNA is monitored by the polariza-

Wolfgang Wintermeyer ■ Institut für Molekularbiologie, Universität Witten/Herdecke, D-5810 Witten, Federal Republic of Germany. **Roland Lill** ■ Institut für Physiologische Chemie, Universität München, D-8000 Munich 2, Federal Republic of Germany. **James M. Robertson** ■ Applied Biosystems, Inc., Foster City, California 94403.

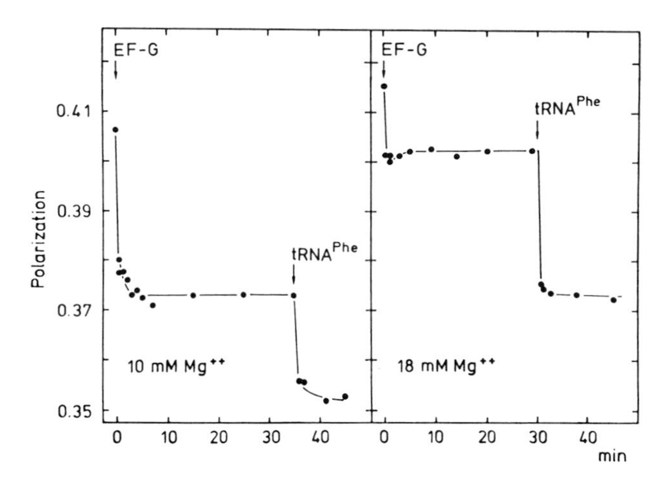

Figure 1. Release of $tRNA^{Phe}_{Prf\text{-}37}$ from the P site during EF-G-dependent translocation. The pretranslocation complex was prepared by incubating 70S ribosomes from *E. coli* (Robertson and Wintermeyer, 1981) at 25°C in buffer A (see Table 1) containing either 10 or 18 mM Mg^{2+} and poly(U) (1 A_{260} unit per ml) with an equimolar amount of $tRNA^{Phe}_{Prf\text{-}37}$ (30 min) and then with a twofold molar excess of AcPhe-$tRNA^{Phe}$ (*E. coli*) (60 min); the concentration of pretranslocation complex was 0.15 μM. Translocation was initiated by the addition of EF-G (0.8 μM) and GTP (0.5 mM), followed by chasing of the fluorescent tRNA from the E site by addition of nonfluorescent $tRNA^{Phe}$ in eightfold excess. Polarization of proflavine fluorescence was monitored as described previously (Robertson and Wintermeyer, 1981). At the higher Mg^{2+} concentration, the higher final level is due to binding to the A site of the $tRNA^{Phe}_{Prf}$ released from the P site.

tion of fluorescence. The initial high value of the fluorescence polarization reflects the firm binding of the tRNA in the P site and the fixation of the anticodon by binding to the mRNA (Robertson and Wintermeyer, 1981); the lowering of the polarization

Table 1. Affinities of the E site of *E. coli* 70S ribosomes programmed with poly(U) for $tRNA^{Phe}$ from *E. coli* and yeast cells (25°C)

Binding constant (10^6 M^{-1})		Buffer[a]
E. coli	Yeast	
28	3.5	A (20 mM Mg^{2+})
43	5.0	C
3.5	0.9	A (10 mM Mg^{2+})
5.0	1.3	B

[a] Buffer A, 50 mM Tris hydrochloride (pH 7.5)–50 mM KCl–90 mM NH_4Cl–1 mM dithioerythritol–10 or 20 mM magnesium acetate. Buffer B, 50 mM Tris hydrochloride (pH 7.5)–30 mM KCl–30 mM NH_4Cl, 1 mM dithioerythritol–10 mM magnesium acetate. Buffer C, 10 mM Tris hydrochloride (pH 7.5)–35 mM NH_4Cl–1 mM dithioerythritol–20 mM magnesium acetate. Data are taken from Lill et al. (1986).

reflects the release of the tRNA from the P site. From the further decrease of the signal observed upon addition of an excess of unlabeled $tRNA^{Phe}$, it is clear that part of the tRNA released from the P site remains bound on the ribosome. Since the dissociation due to the chase is rapid, it must be at a site other than the P and A sites, i.e., the E site.

The position of the equilibrium, i.e., the relative concentrations of E-site-bound and unbound tRNA, is strongly influenced by the experimental conditions, in particular by the Mg^{2+} concentration. When this concentration is lowered, then the amount of tRNA found in the E site after translocation is also decreased (Fig. 1 and 2), reflecting the Mg^{2+} dependence of the affinity of the E site (Table 1). The binding constants also show that the binding strength of the E site is not appreciably influenced by the ionic strength of the buffer. Furthermore, the homologous

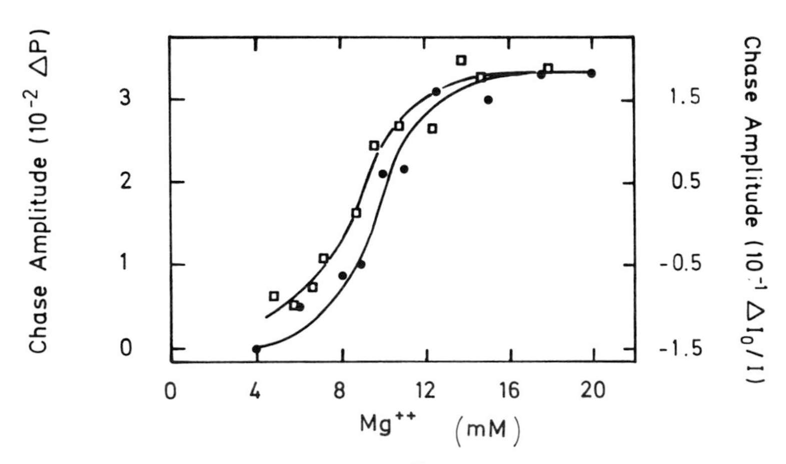

Figure 2. Mg^{2+} dependence of the E-site occupancy with $tRNA^{Phe}$ after translocation. The translocation experiment of Fig. 1 was performed at various concentrations of Mg^{2+}, and the E-site-bound fluorescent tRNA was chased by adding an excess of nonfluorescent $tRNA^{Phe}$. Two fluorescent $tRNA^{Phe}$ derivatives were used, $tRNA^{Phe}_{Prf\text{-}37}$ (yeast) (●; the fluorescence polarization change, ΔP, is plotted; left ordinate) and $tRNA^{Phe}_{Prf\text{-}16/20}$ (*E. coli*) (□; the relative fluorescence change is plotted; right ordinate). The values are corrected for the actual amount of pretranslocation complex present, which varied with the Mg^{2+} concentration. (From Robertson and Wintermeyer, 1987.)

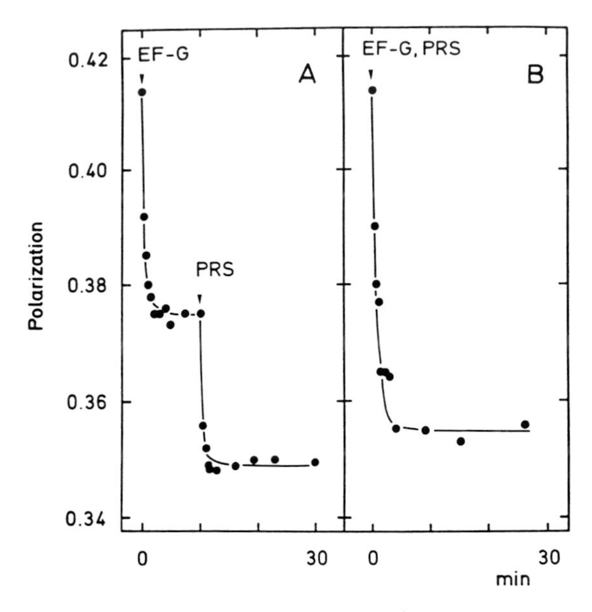

Figure 3. EF-G-induced release of $tRNA^{Phe}_{Prf\text{-}37}$ from the P site in the presence of phenylalanyl-tRNA synthetase (PRS). The experiment was performed as for Fig. 1 (10 mM Mg^{2+}) except that purified phenylalanyl-tRNA synthetase from yeast cells (0.14 μM) was added either after (A) or together with (B) EF-G·GTP. ATP (1 mM) and L-phenylalanine (0.2 mM) were present in the buffer. (From Robertson and Wintermeyer, 1987.)

$tRNA^{Phe}$ is bound more strongly than is the heterologous one.

Labile binding of tRNA to the E site has also been reported by Kirillov et al. (1983) and is probably the reason for failure to detect the E-site-bound tRNA (Spirin, 1985; Baranov and Ryabova, 1988). On the other hand, Rheinberger and Nierhaus (1983, 1986) have reported E-site complexes with an apparently higher stability against dissociation. These differences may be attributed to different methods of preparing the ribosomes. Furthermore, the buffer composition seems to influence the stability of the complexes, since in a buffer containing spermine and spermidine in addition to 5 mM Mg^{2+}, rather stable complexes were obtained (Gnirke et al., 1989).

A readily displaceable equilibrium between E-site-bound and unbound $tRNA^{Phe}$ is also demonstrated by the observation (Fig. 3) that the provision of charging conditions by adding phenylalanyl-tRNA synthetase, phenylalanine, and ATP very efficiently removes the deacylated $tRNA^{Phe}$ from the ribosome after translocation. This result also shows, as did previous results (Robertson and Wintermeyer, 1987), that occupancy of the A site is not required for complete dissociation of the tRNA from the E site, at least not in our ribosome system, in contrast to reports from Nierhaus and co-workers (Rheinberger and Nierhaus, 1983, 1986; Gnirke et al., 1989).

The Contribution of Codon-Anticodon Interaction to E-Site Binding of tRNA Is Smaller than That of Both P and A Sites

In the P site, the presence of the cognate codon increases the affinity of $tRNA^{Phe}$ (or $AcPhe\text{-}tRNA^{Phe}$) by 2 and 4 orders of magnitude at 20 and 10 mM Mg^{2+}, respectively (Lill et al., 1986), bringing about a very substantial stabilization of the cognate complex (Table 2). The affinity increase due to codon-anticodon interaction in the A site is difficult to measure, since coded binding of $Phe\text{-}tRNA^{Phe}$ to the A site is somewhat weaker than to the P site, and non- or miscoded binding is below the detection limit of about 10^5 M^{-1} (Lill et al., 1986). Thus, only a lower limit can be estimated for the energetic contribution of codon-anticodon interaction in the A site (Table 2); the actual value probably is larger.

To assess the energetic contribution of codon-anticodon interaction in the E site, we measured the binding constants of a number of tRNAs in the E sites of ribosomes programmed with either poly(U) or poly(A) by determining the competition of a given tRNA for E-site binding of the fluorescent $tRNA^{Phe}_{Prf\text{-}16/20}$. Not only $tRNA^{Phe}$ but also a number of noncognate tRNAs are able to compete, although with different efficiencies (Fig. 4). This is shown in some detail by the equilibrium titrations depicted in Fig. 5. $AcPhe\text{-}tRNA^{Phe}$, which does not significantly bind to the E site, has no effect, thus proving that a competition for the E site is being studied in these experiments. A summary and comparison of the binding constants derived from the titration curves are given in Tables 3 and 4.

Comparison of the E-site binding affinities of cognate and noncognate tRNAs suggests that there is a weak interaction with the mRNA in the E site. In addition, the data show an up to 10-fold variation in the intrinsic affinities of the noncognate tRNAs. For instance, in the poly(A) system, the cognate $tRNA^{Lys}$ and the noncognate $tRNA^{Phe}$, both from *E. coli*, exhibit exactly the same affinity (Fig. 5), in keeping

Table 2. Contribution of codon-anticodon interaction to the binding of $tRNA^{Phe}$ in the ribosomal A, P, and E sites[a]

Site	Approx stabilization at Mg^{2+} concn of:		ΔG^0 (kJ/mol) at Mg^{2+} concn of:	
	10 mM	20 mM	10 mM	20 mM
A	$>10^2$		>14	
P	10^4	10^2	23	11
E	20	4	7	3

[a] Data for both the A and P sites are from Lill et al. (1986) and were measured in buffer A with 10 or 20 mM Mg^{2+}. Values for the E site (Lill and Wintermeyer, 1987) are averaged from the data of Tables 3 and 4 (buffer containing 20 mM Mg^{2+}) and similar data obtained in buffer B.

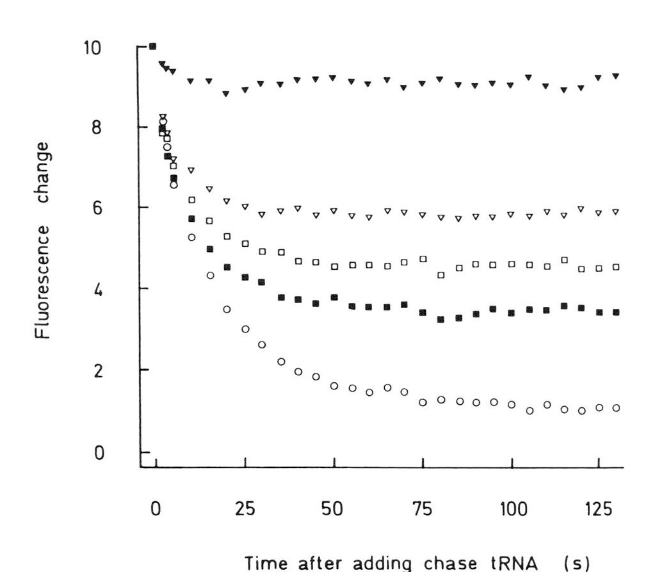

Figure 4. Time course of chasing $tRNA^{Phe}_{Prf-16/20}$ from the E site of poly(U)-programmed ribosomes by competing nonlabeled tRNAs. For E-site binding, $tRNA^{Phe}_{Prf-16/20}$ (0.11 μM) was incubated (buffer A [20 mM Mg^{2+}]) with ribosomes (0.16 μM) carrying AcPhe-$tRNA^{Phe}$ in both the P and A sites for 5 min, and the fluorescence signal was measured. Then the indicated nonlabeled tRNAs were added in a small volume (3% of the total mixture) to a final concentration of 2.1 μM (mixing time, 2 to 3 s) and measurements were taken every second; one fifth of the datum points is displayed. Symbols: ▼, Gly_1; ▽, Ser_y; □, Tyr; ■, Leu_5; ○, Phe. (From Lill and Wintermeyer, 1987.)

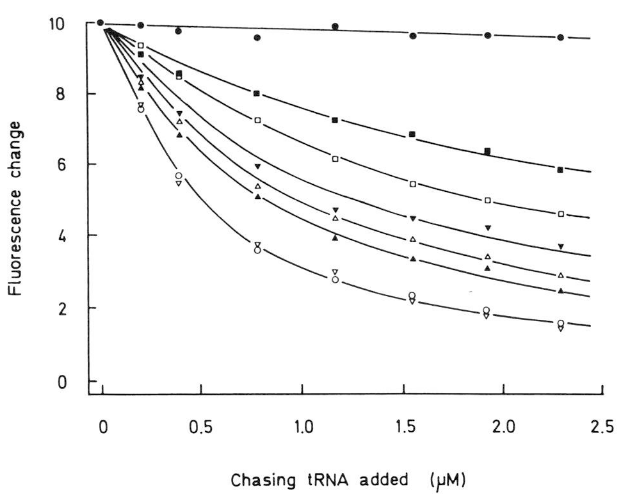

Figure 5. Chasing of $tRNA^{Phe}_{Prf-16/20}$ from the E site of poly(A)-programmed ribosomes by nonlabeled tRNAs competing for E-site binding. The P site of the ribosomes (0.14 μM) was blocked by incubation with 1.2 μM $tRNA^{Lys}$ per ribosome (buffer A [20 mM Mg^{2+}]). Then $tRNA^{Phe}_{Prf-16/20}$ (0.2 μM) was allowed to equilibrate with the E site for 5 min. Finally, the indicated amounts of nonfluorescent tRNA were added, and the fluorescence change was measured. Ordinate values are given as fractions of the maximal fluorescence change, set to 10. Data were evaluated by computer fitting (continuous lines) as described by Lill and Wintermeyer (1987); the respective binding constants obtained are summarized in Table 3. Control titrations were performed with either buffer or AcPhe-$tRNA^{Phe}$ (*E. coli*), which, because of the specificity of the E site for deacylated tRNA, does not bind. The data for AcPhe-$tRNA^{Phe}$ are corrected for a 10% content of deacylated $tRNA^{Phe}$ (100% charging = 1,750 pmol per A_{260} unit). Symbols: ●, control; ■, Gly_1; □, Tyr; ▼, Cys; △, Val_2; ▲, Thr; ○, Phe; ▽, Lys. (From Lill and Wintermeyer, 1987.)

with the finding that in the poly(U) system, $tRNA^{Lys}$ is one of the weakest E-site binders. Nevertheless, we note that in both poly(U) and poly(A) systems, the respective cognate tRNA exhibits the strongest binding (*E. coli* $tRNA^{Phe}$) or belongs to the tRNAs binding most strongly ($tRNA^{Lys}$), indicating some stabilizing effect of codon-anticodon interaction. On the average, an affinity increase due to codon-anticodon interaction of about 4-fold (20 mM Mg^{2+}) or 20-fold (10 mM Mg^{2+}) is estimated (Table 2).

These values are rather small in comparison with the free energies of base pairing between codon and anticodon in the other two ribosomal sites, thus highlighting the relative unimportance of tRNA-mRNA interaction in the E site and reinforcing the previous statement that tRNA binding in this site differs fundamentally from the binding to both the A and P sites.

The 3′-Terminal Adenine of $tRNA^{Phe}$ Forms Hydrogen Bonds in the E Site, Probably with 23S rRNA

The specificity for deacylated tRNA is the most obvious distinction of the E site (and one all groups studying the E site agree on); the binding of charged tRNA could not be detected. This, and the reported

Table 3. Binding constants of various tRNAs for the E site of poly(A)-programmed ribosomes (buffer A [20 mM Mg^{2+}]; 25°C)[a]

tRNA	Anticodon	K (10^6 M^{-1})	K_{Lys}/K_{non}
Lys	UUU	6	1
Leu_5	CAA	3	2
Thr	GGU	3	2
Tyr	GUA	1	6
Val_1	UAC	1	6
Phe	GAA	6	1
Phe (yeast)	GAA	0.5	12
Cys	GCA	2	3
Leu_2	GAG	2.5	2.4
$Ser_{1/2}$ (yeast)	IGA	4	1.5
Val_2	GAC	2.2	2.7
Gly_1	CCC	0.5	12

[a] Binding constants were estimated from the titration curves of Fig. 5 by least-squares fitting. K_{Lys}/K_{non} denotes how much more weakly than the cognate $tRNA^{Lys}$ a particular tRNA binds. Unless otherwise indicated, tRNAs were from *E. coli*. In the anticodon sequences, modifications are omitted. Data are taken from Lill and Wintermeyer (1987).

intractable properties of the hypermodified nucleotides. However, the situation changed dramatically when recombinant technology became available and the primary sequence of the rRNA of *Escherichia coli* was completely determined (Brosius et al., 1980). In addition, our group developed two techniques that allowed identification of modified nucleotides within an RNA without the necessity of isolating each modified nucleotide per se (Barta et al., 1984).

Hybridization techniques

We first established a simple and general method for mapping the affinity-labeled nucleotide to a restricted region on a large RNA (Barta et al., 1984). This approach uses hybridization of affinity-labeled 23S RNA to isolated ribosomal DNA fragments. Specific restriction fragments are obtained from plasmid pKK123, which contains the 3′ two-thirds of the 23S RNA gene from the *rrnB* operon (Brosius et al., 1980). These fragments are separated electrophoretically on an agarose gel, blotted onto nitrocellulose, and hybridized to 23S RNA photoaffinity labeled with BP-[^{3}H]Phe-tRNA. After digestion of the unhybridized parts of the RNA with RNase, only hybrids with affinity-labeled radioactive RNA are visible upon autoradiography. Determination of the sizes of these hybrids usually allows the localization of this fragment on 23S RNA. By using four restriction enzymes, it was possible with this method to limit the site of reaction to a 183-base fragment spanning the nucleotides from positions 2442 to 2625 in 23S RNA (Barta et al., 1984). A similar method was used to localize the region of cross-link of a 5′-azido-2-nitrobenzoyl-[^{3}H]Phe-tRNA to nucleotides 2445 to 2668 on 23S RNA (Hall et al., 1985).

Primer extension analysis using reverse transcriptase

The primer extension method was based on observations made by Youvan and Hearst (1979, 1981) that reverse transcriptase slows down or halts chain elongation one base before encountering a modified nucleotide. We therefore expected that the photoaffinity probe cross-linked to a nucleotide of the 23S RNA acts as a barrier for the enzyme. Thus, the newly created cDNA product can be distinguished from background bands. From the size of the cDNA product, the site of cross-link can be precisely identified. Since the cross-link site had already been assigned to a 183-base fragment, a 124-base-pair restriction fragment (corresponding to nucleotides 2715 to 2839) 3′ to nucleotide 2625 was used as a primer for reverse transcription. Three additional bands, designated I, II, and III, appeared when photo-

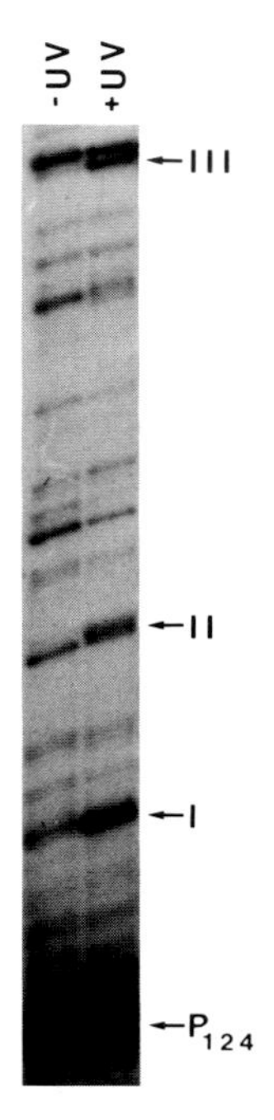

Figure 2. Autoradiograph of a primer extension analysis performed on 23S RNA. A reverse transcriptase reaction was performed on 23S RNA photo-cross-linked with BP-Phe-tRNA (lane +UV) and on 23S RNA from nonirradiated ribosomes (lane −UV), using a primer corresponding to the region from nucleotides 2715 to 2839 (P_{124}). cDNA bands resulting from the irradiation process are designated I, II, and III.

cross-linked RNA was used as template (Fig. 2). Band I appeared also in the control experiment when ribosomes were irradiated alone and was due to an internal cross-link involving U-2613 from domain V and T-747 from domain II (unpublished data). Evidence of close proximity of these two regions has also been provided by others (Stiege et al., 1983; Douthwaite et al., 1985). Bands II and III are caused by the photoreaction with BP-Phe-tRNA and reflect labeling from the A and P sites, respectively (see also below).

These results demonstrate that the chain elongation method can be used to scan any RNA with a suitable set of primers for identifying modified nucleotides. This method has subsequently been used by Noller's group in their elegant modification-protection experiments for elucidating the secondary structures of rRNAs (reviewed in Noller, 1984). It is noteworthy that reverse transcription on unmodified rRNA also produces a reproducible pattern of bands that do not result from stops at modified nucleotides. Some of these bands could be correlated to stops at strong helices according to the secondary structure model (Noller, 1984), but some of them fall in single-stranded regions (unpublished results). One can speculate that in these cases tertiary interactions might hinder the reverse transcriptase, but this has yet to be proven.

Table 1. Binding of BP-[^{3}H]Phe-tRNA to ribosomes and photo-cross-linking to 23S RNA[a]

Assay conditions	Binding (cpm) or sensitivity	Photo-cross-linking (cpm) or sensitivity
P site		
−Poly(U), −UV	21,000	ND
+Poly(U), −UV	183,000	13,000
+Poly(U), +UV	196,000	152,000
+Poly(U), +puromycin, +UV	35,000	5,000
% Puromycin sensitivity	80	95
A-site		
−Poly(U), −UV	52,000	ND
+Poly(U), −UV	149,000	9,000
+Poly(U), +UV	168,000	150,000
+Poly(U), +puromycin, +UV	135,000	42,000
% Puromycin sensitivity	20	72

[a] Values are normalized to 10 pmol of ribosomes or 23S RNA, respectively. Specific activity of BP-[^{3}H]Phe-tRNA was 110 Ci/mmol (about 90,000 cpm/pmol). For the puromycin controls, the antibiotic was added after binding of BP-Phe-tRNA but before irradiation. ND, Not done.

Specific Labeling of the A and P Sites on 23S RNA

Binding of BP-Phe-tRNA to either the A or P site had to be optimized in order to obtain specific photoaffinity labeling for both sites. P-site binding was found to be optimal at 6 mM Mg^{2+} and was checked by the puromycin assay. Specific A-site binding required a threefold molar excess of uncharged tRNA and 10 mM Mg^{2+}. Under these conditions, uncharged tRNA binds to the P site, thereby blocking this site. A-site binding was checked by the puromycin assay before and after incubation with EF-G and GTP. Determinations of binding and photo-cross-linking yields were described in detail earlier (Steiner et al., 1988; Kuechler et al., 1988).

Data for the binding and photo-cross-linking of BP-Phe-tRNA to either the A or P site are presented in Table 1. In both sites, more than 70% of the BP-Phe-tRNA bound became cross-linked to 23S RNA. The specificity of cross-linking was demonstrated by its dependence on poly(U) and on irradiation. Binding of BP-Phe-tRNA under P-site conditions was suppressed >80% by preincubation with puromycin, whereas photo-cross-linking was completely inhibited. In contrast, A-site binding was only slightly inhibited by puromycin, indicating that most of the BP-Phe-tRNA was bound to this site. Furthermore, upon addition of EF-G and GTP, puromycin sensitivity increased up to 80% as a result of the translocation of BP-Phe-tRNA from the A to the P site (data not shown). Unexpectedly, photo-cross-linking at the A site was inhibited to 72%, although binding was reduced only by 20%. This result can be best explained by assuming that puromycin (an analog of the 3′ end of Tyr-tRNA) was competing with the 3′ end of BP-Phe-tRNA for a common binding site. This interpretation was supported by experiments showing that the inhibition of photo-cross-linking could be reversed by removing puromycin from isolated A-site complexes before irradiation (unpublished data). These data show that the puromycin sensitivity of photo-cross-linking alone is not a valid criterion for assignment of the BP-Phe-tRNA to either the A or the P site.

RNA was isolated from ribosomes specifically labeled with BP-[^{3}H]Phe-tRNA from either the A or the P site and was used as a template for the primer extension analysis with reverse transcriptase (Fig. 3). As a primer, we used a 5′-^{32}P-labeled 56-base-pair *Ava*II fragment corresponding to nucleotides 2607 to 2663 of 23S RNA, which is closer to the region of cross-link and allows a better resolution of the affinity-labeled nucleotides than does the 124-base restriction fragment used previously. RNA from nonirradiated ribosomes served as a control and was also used to perform dideoxy sequencing reactions (Fig. 3, lanes C, U, A, and G). The cDNA transcripts obtained were separated on a sequencing gel, and their lengths were determined by comparison with the sequencing lanes. It can be seen that under A-site conditions, two additional bands appear above background, corresponding to stops at nucleotides U-2586 and U-2585, thus identifying U-2584 and U-2585 as affinity-labeled bases. In addition to these two strong bands, two minor bands appear as a result of reactions at A-2503 and U-2504. When the experiment was carried out in the presence of puromycin, the intensity of the bands decreased, as expected from the cross-linking data in Table 1.

With RNA affinity labeled at the P site used as a template, fragments stopping at nucleotides C-2452 and A-2453 appear in this lane, indicating that nucleotides A-2451 and C-2452 are being affinity labeled. In addition, minor bands from photoreactions at U-2504, G-2505, and U-2506 can be detected. As expected, these bands disappear when the experiment is carried out in the presence of puromycin (Fig. 3, lane Puro).

Inhibition of the Photo-Cross-Link by Antibiotics

The observation that puromycin had an effect on photo-cross-linking but not on the binding of BP-Phe-tRNA to the A site led us to infer that other antibiotics might also have an influence on photo-cross-linking. If photo-cross-linking does occur near the peptidyltransferase site, antibiotics known to inhibit peptide bond formation should affect affinity

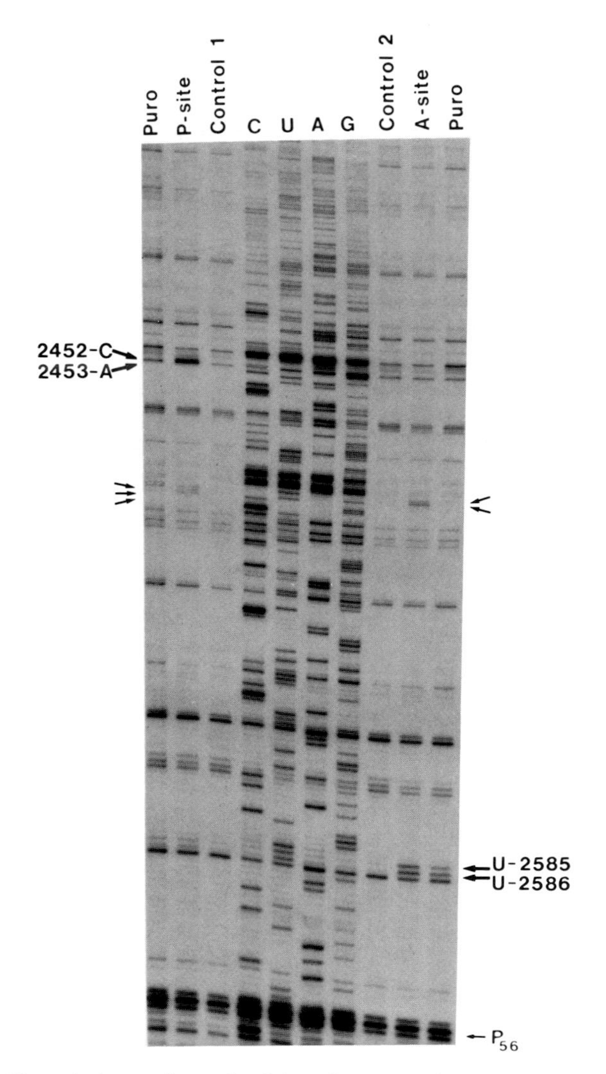

Figure 3. Autoradiograph of the gel pattern of reverse transcripts obtained in the primer extension experiment, using 23S RNA photoaffinity labeled at the A site and at the P site as a template. Specific binding and photo-cross-linking were performed as described in the text. Reverse transcriptase reactions were primed with a 56-base-pair *Ava*II DNA fragment corresponding to positions 2607 to 2663 on 23S RNA. Lanes C, U, A, and G are dideoxy sequencing lanes and refer to the nucleotide sequence of 23S rRNA. rRNA from nonirradiated ribosomes (lane Control 1) and from ribosomes irradiated in the absence of BP-[^{3}H]Phe-tRNA (lane Control 2) served as controls. In lane Puro, 1 mM puromycin was added after binding but before irradiation. Stops at the designated nucleotides are due to photo-cross-linking to the ensuing nucleotide. Arrows mark minor cross-linking sites.

labeling. In addition to these inhibitors, antibiotics known to affect elongation and tRNA binding were also tested. The antibiotics were added before irradiation, and the extent of photo-cross-linking was determined (Steiner et al., 1988). Known inhibitors of the peptidyltransferase reaction such as chloramphenicol (Vasquez, 1979), tiamulin (Hoegenauer, 1974; Hodgin and Hoegenauer, 1975), sparsomycin,

Table 2. Inhibition of photo-cross-linking of BP-[^{3}H]Phe-tRNA to 23S RNA by antibiotics

Antibiotic	Inhibition[a]	
	A site	P site
Inhibitors of peptidyltransferase and/or translocation		
Chloramphenicol	+++	++++
Lincomycin	−	+++
Tiamulin	++++	++++
Sparsomycin	++	+++
Vernamycin B	++++	++++
Erythromycin	++	++
Inhibitors of tRNA binding		
Fusidic acid	−	−
Gentamicin	−	+
Kanamycin	+	−
Streptomycin	−	−
Thiostrepton	−	−
Tetracycline	++++	++++
Inhibitor of cell wall synthesis		
Ampicillin	−	−

[a] ++++, Complete; +++, strong; ++, medium; +, weak; −, absent.

and lincomycin (Cundliffe, 1980, 1986) had a pronounced effect on cross-linking under A- and P-site conditions, the only exception being lincomycin at the A site (Table 2). Erythromycin and vernamycin B, which are generally considered to inhibit translocation (Andersson and Kurland, 1987; Vester and Garrett, 1987), showed inhibition at both the A and P sites. In contrast, the inhibitors of EF-Tu-dependent aminoacyl-tRNA binding generally showed no or only weak effects. Only tetracycline inhibited photo-cross-linking, most probably because of its own intrinsic photoreactivity at light of 320 nm (Steiner et al., 1988).

SUMMARY

The data from these photoaffinity labeling studies with BP-Phe-tRNA revealed two different sites on 23S RNA labeled either from the P site (A-2451 and C-2452) or from the A site (U-2584 and U-2585). In addition, a third region with minor reactivity was labeled from both sites. The specificity of the labeling was established by its dependence on poly(U), irradiation, and precise positioning in either the A or the P site. The sites labeled are in close proximity to the peptidyltransferase center, as demonstrated by the fact that cross-linked BP-Phe-tRNA still can form peptide bonds and by the ability of peptidyltransferase inhibitors to influence greatly the photoaffinity reaction.

The affinity-labeled sites are located at different

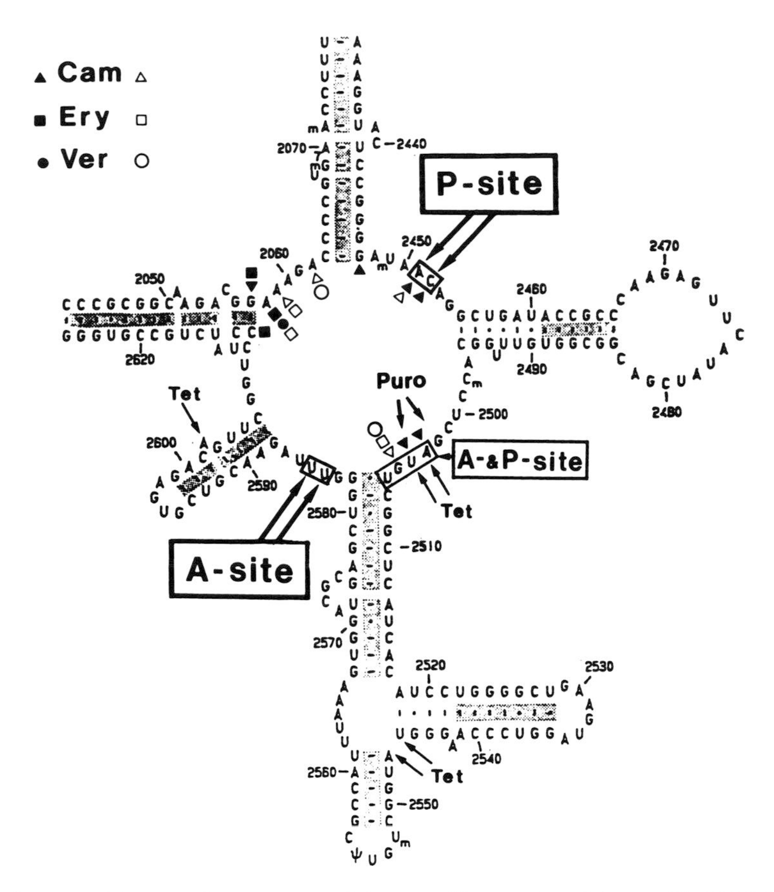

Figure 4. Schematic diagram of the secondary structure of the central loop region of domain V of 23S rRNA. The nucleotides specifically affinity labeled by A- and P-site-bound BP-[^{3}H]Phe-tRNA are boxed and designated (Steiner et al., 1988). Arrows point to nucleotides modified by irradiating ribosomes with tetracycline alone (Steiner et al., 1988). Cam, Chloramphenicol; Ery, erythromycin; Ver, vernamycin. Filled symbols indicate nucleotides whose mutation confers resistance to the respective antibiotic (reviewed in Noller, 1984; Douthwaite et al., 1985; Sor and Fukuhara, 1984; Ettayebi et al., 1985); open symbols designate antibiotics whose binding to the ribosome causes an alteration of reactivity of the respective nucleotide toward chemical modification (Moazed and Noller, 1987). Arrows labeled Puro designate nucleotides photoaffinity labeled by *p*-azidopuromycin (Hall et al., 1988).

positions according to the primary sequence but are brought together by the secondary structure folding of 23S RNA (Noller, 1984). All nucleotides involved are unpaired and highly conserved in a variety of species and reside in the central loop of domain V (Noller, 1984). Dissection of this region into a bipartite structure is of great importance for any molecular model of the peptidyltransferase center.

The high conservation of this region is strongly indicative of its involvement in an essential ribosomal function. Genetic experiments with mutants conferring antibiotic resistance placed this region in the vicinity of peptidyltransferase (closed symbols in Fig. 4). Mutants resistant to chloramphenicol, an inhibitor of peptidyltransferase, and erythromycin and vernamycin B, which affect translocation, have been shown to possess nucleotide changes in this region (Sigmund and Morgan, 1984; Sor and Fukuhara, 1984; Ettayebi et al., 1985). These antibiotics also inhibit chemical modification of 23S RNA in ribosomes at sites indicated in Fig. 4 by open symbols (Moazed and Noller, 1987). Furthermore, nucleotides involved in binding puromycin (Hall et al., 1988) and tetracycline (Steiner et al., 1988) are also located in this region. These data correlate well with our observations that the antibiotics discussed above are also good inhibitors of the BP-Phe-tRNA photocross-link (Table 2).

The final proof that the central loop of domain V of 23S RNA is indeed a main component of the peptidyltransferase region came recently from a detailed analysis of the tRNA-binding sites on the 50S subunit (Moazed and Noller, 1989; see also Noller et al., this volume). Most of the nucleotides identified as being protected from chemical modifications by binding of tRNA derivatives to the A, P, and E sites were located in domain V. Specifically, the protection of A-2451 was dependent on the aminoacyl moiety of

Chapter 29

Movement of tRNA through Ribosomes during Peptide Elongation

BOYD HARDESTY, O. W. ODOM, and JOHN CZWORKOWSKI

BACKGROUND

The key to ribosome function is to understand when, where, and how the tRNAs move during the discrete reaction steps by which peptides are elongated. The results of early studies of peptide elongation, and the logical imperatives of the tRNA-mediated process by which the genetic message encoded into mRNA is translated into a peptide, prompted Watson (1964) to propose the two-site model that became the principal dogma in the field. The central features of this model were that the peptide linked through its carboxyl group to a tRNA bound into the peptidyl (P or donor) site was transferred to the free amino group of an amino acid attached to the next tRNA bound into the ribosomal acceptor (A) site. The basic two-site model evolved to include the two GTP-dependent reactions carried out by EF-Tu and EF-G that were thought to mediate movement of the tRNAs into, between, and out of the A and P ribosomal sites. The location of acyl-tRNA in the A or P site was defined functionally by its reactivity in the peptidyltransferase reaction; P-site acyl-tRNA would function as a donor, whereas A-site aminoacyl-tRNA would act as the acceptor. However, problems with this simple model began to emerge. Aminoacyl-tRNA could be bound to ribosomes with EF-Tu and a nonhydrolyzable analog of GTP such that it functioned as neither the donor nor the acceptor in the peptidyltransferase reaction. The ribosomal site to which it was bound, therefore, was called the entry or E site by Hardesty and co-workers (Hardesty et al., 1969). Later, it was found that deacylated tRNA could bind to ribosomes to which *N*-acetylated Phe-tRNA (AcPhe-tRNA) had been bound previously into the P site (Rheinberger and Nierhaus, 1980). This led to the formal definition of the exit site, as originally proposed by Wettstein and Noll (Wettstein and Noll, 1965).

Many techniques and empirical approaches have been used to correlate the functionally defined tRNA-binding sites with physical locations on the ribosome. These have resulted in a very extensive literature, one central feature of which has been to focus attention on the functional importance of the rRNAs. Dahlberg (1989) has recently considered results indicating that rRNA, rather than the ribosomal proteins, constitutes the primary sites of interaction between tRNA, mRNA, and ribosomes. In an elegant series of experiments, Moazed and Noller have delineated the effect of tRNA binding on the chemical reactivity of specific bases of the 16S RNA (Moazed and Noller, 1986) and 23S RNA (Moazed and Noller, 1989) in the subunits of 70S ribosomes.

We have used fluorescence techniques, particularly nonradiative energy transfer between probes, to study the binding and movements of the tRNAs during peptide elongation. These techniques have a particular advantage because of their excellent sensitivity and applicability to the reaction conditions in which the individual steps of peptide elongation can be monitored as they occur in solution. They have the special advantage of allowing measurements to be made on only those ribosomes that bind tRNA and are functionally active. The basic experimental approach is to bind to ribosomes acylated or deacylated $tRNA^{Phe}$ to which a fluorescent probe is covalently attached at a specific site and then to measure differences in fluorescence quantum yield, anisotropy, or energy transfer that are caused by formation of the complex. Generally, the latter is to an acceptor probe covalently attached at a specific site on a ribosomal protein or RNA.

The results from an early series of experiments were unanticipated from the classical two-site model (Hardesty et al., 1986). One of the most striking

Boyd Hardesty, O. W. Odom, and John Czworkowski ■ Clayton Foundation Biochemical Institute, Department of Chemistry and Biochemistry, University of Texas at Austin, Austin, Texas 78712.

Table 1. Distances between three points on $tRNA^{Phe}$ and S21 or L11

Protein	Distance (Å)		
	dhU labeled	s^4U_8-C_{13} cross-link	acp^3U_{47} labeled
Fluorescein-S21			
Deacylated	54	58	64
AcPhe	>57	65	75
AcPhe + puromycin	52	60	76[a]
L11			
Deacylated	64	69	73
AcPhe	>59	>69	>88
AcPhe + puromycin	>60	>69	>88[a]

[a] Not reactive with puromycin.

results was the difference in the quantum yield of fluorescence from a coumarin probe attached to the dihydrouridine (dhU) positions of deacylated $tRNA^{Phe}$ or AcPhe-tRNA bound to ribosomes. These tRNA species are predicted to be bound to the same site (P) by the classical two-site model; however, the fluorescence quantum yield of the coumarin-labeled deacylated species was about 1.6-fold greater than the corresponding value for the AcPhe-tRNA species. Coumarin fluorescence is particularly sensitive to the local environment (solvent hydrophobicity, charge density and distribution, hydration, etc.). Thus, the differences in quantum yield clearly demonstrated that deacylated $tRNA^{Phe}$ and AcPhe-tRNA bound to ribosomes are in different sites or conformations or both. The transition between the two states could be monitored easily as AcPhe-tRNA was deacylated during transfer of the AcPhe to puromycin or to incoming Phe-tRNA during the peptidyltransferase reaction. Distance measurements were carried out by energy transfer from probes attached to each of three points located near the center of $tRNA^{Phe}$ (dhU loop of yeast $tRNA^{Phe}$, acp^3U_{47} of *Escherichia coli* $tRNA^{Phe}$, s^4U_8-C_{13} cross-link of *E. coli* $tRNA^{Phe}$) to acceptor probes linked to the single cysteine of protein L11 or S21. The results, summarized in Table 1, showed that deacylated $tRNA^{Phe}$ and AcPhe-tRNA were physically in different positions on the ribosome. Furthermore, with the exception of the acp^3U-labeled $tRNA^{Phe}$ species, which is inexplicably unreactive with puromycin, the $tRNA^{Phe}$ generated in the peptidyltransferase reaction from AcPhe-tRNA moves into a conformation, site, or position that is similar or identical to that of deacylated $tRNA^{Phe}$ bound directly to the ribosomes. Considered together, these data strongly indicate that the tRNA moves during the peptidyltransferase reaction, rather than the alternative explanation that the changes in energy transfer are all due only to changes in the relative orientation of the different sets of probes. The observations led us to propose the displacement model for peptide elongation (Hardesty et al., 1986), the central feature of which is that the tRNA rather than the peptide is moved relative to the ribosome during the peptidyltransferase reaction. Our purpose has been to further define the movement of the tRNAs during the reactions of peptide elongation and specifically to test the displacement model critically.

FINDINGS

Our first objective has been to document further the apparent movement of the $tRNA^{Phe}$ during the peptidyltransferase reaction. Different pairs of probes were used, with the energy donor generally attached to sites on $tRNA^{Phe}$ and the acceptor probe on a ribosomal component. Coumarin probes were covalently attached to the 5′-terminal phosphate and to s^4U_8 of *E. coli* $tRNA^{Phe}$. Proflavine was substituted for wybutine at position 37 in the anticodon loop of yeast $tRNA^{Phe}$. The positions of the attachment sites for these three probes are along the long axis of $tRNA^{Phe}$ crystal structure near a plane through the anticodon loop, whereas the probes used for Table 1 were distributed close to a plane near the center of the molecule and perpendicular to this axis. Considered with the data in Table 1, the results should provide an indication of how the tRNA moves during the peptidyltransferase reaction.

Probes to function as the acceptor in energy transfer experiments were covalently attached to ribosomal proteins S21 and L1. Labeled S21 has proven to be one of the most useful proteins in energy transfer experiments from probes on tRNA. It is centrally located in the cluster of 30S proteins at or near the interfacing side of the small subunit (Capel et al., 1988) and appears to have a relatively large radius of gyration, suggesting that it may be elongated. Under the conditions used, S21 binds tightly to 70S ribosomes and to both 30S and 50S subunits (Odom et al., 1984). Binding appears to be in the order of 70S > 30S > 50S, with at least 20-fold tighter binding to 70S ribosomes than to either subunit. Dissociation constants are below 80 nM for either subunit but are too low to be quantitatively determined by the fluorescence techniques used. L1 is located near the lateral corner of the 50S subunit opposite the L7/L12 stalk but near the central protuberance (Stöffler and Stöffler-Meilicke, 1986; Oakes et al., 1986). Protein S21 was specifically labeled on the thiol group of its single cysteine residue. However, it should be noted that L1 contains no cysteine; thus, amino group-reactive reagents were used to

loop and S21 during the peptidyltransferase reaction. This result prompts the question: When in the reaction cycle of peptide elongation do the anticodon stem and presumably the mRNA move? Generally, this movement has been attributed to the reaction catalyzed by EF-G. After peptide transfer, this reaction clearly activates the ribosome complex for the synthesis of the next peptide bond, and the results of early experiments in which the mRNA was protected from RNase by the ribosome indicated that the EF-G reaction causes movement of the mRNA by about three nucleotides in the 5′ direction (Thach and Thach, 1971). If the EF-G reaction does cause movement of the mRNA, how is it accomplished? The effect of frameshift suppressor mutation with a fourth nucleotide in the tRNA anticodon (Riddle and Carbon, 1973) strongly suggests that the codon-anticodon interaction is involved with and ultimately responsible for mRNA movement. Very little is known in physical terms about what changes take place in the ribosome-tRNA complex during the EF-G reaction. EF-G binds to the 50S subunit in what Wittmann (1986) has called the GTPase center near the base of L7/L12 stalk and L11. Footprinting experiments place its binding site near positions 1067 and 2660 in domains II and IV of the 23S RNA (Moazed et al., 1988), and it appears to be shielded by the dhU loop of tRNA in ribosomes (Robertson and Wintermeyer, 1987). These observations implicate the large subunit in EF-G function. However, there is evidence that EF-G also interacts on the interfacing surface of the 30S subunit on the body near the deep cleft (Girshovich et al., 1981). The one clear effect the EF-G reaction appears to have on ribosome conformation is to cause an increase of approximately 3 Å in the radius of gyration of the protein component of the small ribosomal subunit (Serdyuk and Spirin, 1986).

We have used *E. coli* $tRNA^{Phe}$ labeled with fluorescein on acp^3U_{47} to study the effects of EF-G · GTP on ribosome-bound tRNA. Typical results are depicted in Fig. 1. There is a marked increase in fluorescence anisotropy upon binding of the tRNA to ribosomes, from about 0.12 for the free tRNA to about 0.19 in the presence of poly(U) or 0.15 in its absence. As judged by anisotropy, the tRNA bound in the presence of poly(U) is stable for an extended period in the presence or absence of a sevenfold molar excess of unlabeled $tRNA^{Phe}$ or after incubation with EF-G · GTP in the absence of excess unlabeled $tRNA^{Phe}$. However, a drop in anisotropy to about 0.155 is observed if the ribosomes are incubated with both EF-G · GTP and excess unlabeled $tRNA^{Phe}$.

This anisotropy is very close to the value that

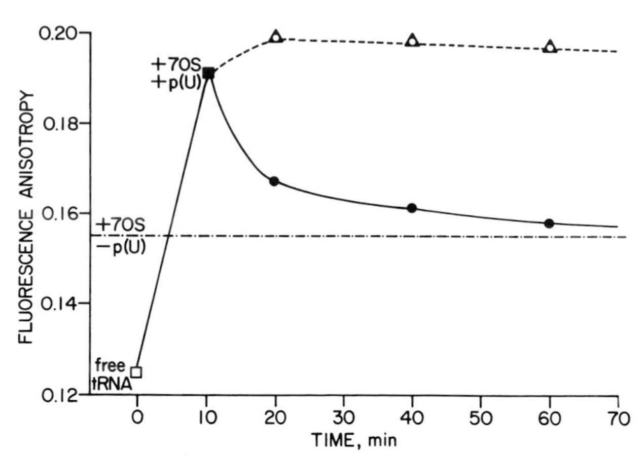

Figure 1. Displacement of OH-tRNA. $tRNA^{Phe}$ from *E. coli* was labeled on its acp^3U_{47} base by reaction with fluorescein-5-isothiocyanate (FITC) as described by Plumbridge et al. (1980), and then the product was purified by HPLC (Odom et al., 1988). Ribosomal subunits were obtained as previously described (Odom et al., 1980). Typically, 40 pmol of labeled $tRNA^{Phe}$ was dissolved in 0.50 ml of buffer (50 mM Tris hydrochloride [pH 7.6], 15 mM $MgCl_2$, 160 mM NH_4Cl, 5 mM 2-mercaptoethanol) containing 0.24 mg of poly(U) per ml. A fourfold molar excess of each of the ribosomal subunits (i.e., 160 pmol) was added, followed by either 160 pmol of EF-G (with 160 μM GTP), 1,100 pmol of unlabeled yeast deacylated OH-$tRNA^{Phe}$, or both, as indicated. All incubations were carried out at 37°C. Fluorescence measurements were made at 20°C on an SLM model 8000 spectrofluorometer. In other experiments, excess *E. coli* $tRNA^{Phe}$ in the deacylated form or as AcPhe-$tRNA^{Phe}$, prepared by enzymatic aminoacylation (Robbins et al., 1981) and chemical acetylation (Rappoport and Lapidot, 1974), produced very similar results. The fluorescence anisotropy of the acp^3U_{47}-FITC-labeled $tRNA^{Phe}$ remained above the lower limit indicated after 70S ribosomes were added to the mixture. Anisotropy was about 0.16 in mixtures containing 70S ribosomes even in the presence of excess unlabeled tRNA and absence of poly(U). Symbols: □, free acp^3U_{47}-FITC-labeled OH-tRNA; ■, labeled OH-tRNA bound to 70S-poly(U); △, after addition of EF-G · GTP, no excess unlabeled OH-tRNA; ○, after addition of excess unlabeled OH-tRNA, no EF-G; ●, after addition of both EF-G · GTP and excess unlabeled OH-tRNA.

was obtained if poly(U) was omitted from the incubation mixture. Qualitatively similar results are observed with yeast $tRNA^{Phe}$ in which 1-aminoanthracene is substituted for wybutine at position 37 in the anticodon loop. Shown in Table 4 are the results from experiments in which ribosomal subunits as well as 70S ribosomes were used. Little or no effect is seen with 30S subunits, but 50S subunits cause an increase in anisotropy to 0.155, the value observed in the absence of poly(U). At lower Mg^{2+} concentrations, such as 6 mM, much of the labeled $tRNA^{Phe}$ appears to completely dissociate from the ribosomes (data not shown). Considered together, these results appear to indicate that the codon-anticodon interaction on the 30S subunit affects the mobility of the fluorescent probe on the acp^3U_{47} to cause the increase in anisotropy from 0.155 to 0.19 or slightly

Table 4. Fluorescence anisotropy of (acp^3U_{47}-FITC)-OH-$tRNA^{Phe}$ in various mixtures[a]

Mixture	Component(s)	Anisotropy
a	Buffer only	0.127
b	+30S subunits	0.132
c	+50S subunits	0.155
d	+70S	0.163
e	+70S, +poly(U)	0.190
f	Isolated 70S complex	0.219

[a] In each mixture, the buffer was 50 mM Tris hydrochloride (pH 7.6)–15 mM $MgCl_2$–160 mM NH_4Cl–5 mM 2-mercaptoethanol containing in 0.50 ml 15 pmol of labeled tRNA. In mixtures b to e, 120 pmol of 30S, 50S, or 70S was added. Mixture *e* also contained 0.24 mg of poly(U) per ml. The isolated 70S-poly(U)-$tRNA^{Phe}$ complex was obtained by gel filtration over Sephacryl S-300 in the buffer described above.

higher. In the presence of excess $tRNA^{Phe}$, EF-GTP causes a relaxation of this probe.

The effect of the EF-G reaction on deacylated tRNA ultimately leading to its release from the ribosomes is anticipated from consideration of most models of ribosome function. However, Fig. 2 shows results that we did not anticipate and find difficult to interpret. It shows the results of experiments very similar to those shown in Fig. 1 but carried out with AcPhe-tRNA labeled on acp^3U_{47}. The fluorescence anisotropy of this species free in solution, 0.13, is near that observed for the deacylated form; however, on binding to ribosome in the presence of poly(U), this increases to 0.25, a value that is experimentally distinct from 0.19, the corresponding value observed with the deacylated tRNA species. In the absence of poly(U), the anisotropy is 0.155 for both the acylated and deacylated species. In the presence of excess unlabeled $tRNA^{Phe}$ and either deacylated or AcPhe-$tRNA^{Phe}$, the EF-G reaction causes a drop in the anisotropy of labeled AcPhe-tRNA to about 0.19, the value seen for deacylated tRNA with poly(U). This value is clearly distinct from the value of 0.155 observed for both the deacylated and AcPhe-tRNA species in the absence of poly(U). The data seem to show that either deacylated tRNA or AcPhe-tRNA can promote an EF-G · GTP-dependent change in the mobility of the acp^3U_{47} probe on AcPhe-tRNA that would be in the ribosomal P site by the classical model. The data suggest that two AcPhe-tRNAs might bind simultaneously to the same ribosome. The change in fluorescence anisotropy is to the level seen with codon-directed binding of deacylated $tRNA^{Phe}$, apparently indicating that the codon-anticodon interaction is maintained. This may reflect movement of the tRNA and presumably the mRNA that is promoted by the EF-G reaction, as reported previously (Thach and Thach, 1971).

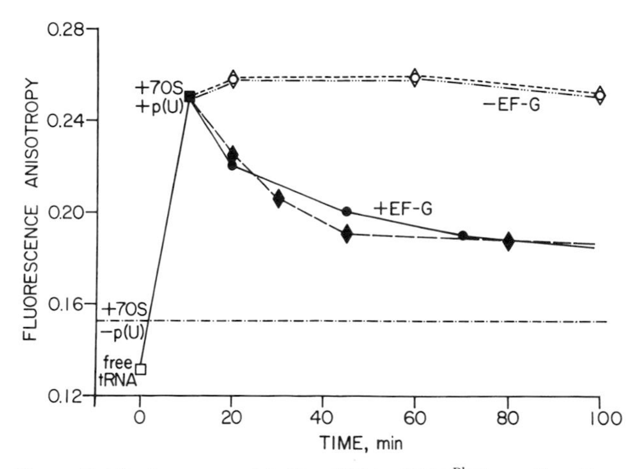

Figure 2. Displacement of AcPhe-tRNA. $tRNA^{Phe}$ from *E. coli* was labeled, aminoacylated, acetylated, and purified as described in the legend to Fig. 1. Binding experiments were also carried out analogously to those depicted in Fig. 1. Symbols: □, free acp^3U_{47}-FITC-labeled AcPhe-$tRNA^{Phe}$; ■, labeled AcPhe-$tRNA^{Phe}$ bound to 70S-poly(U); ○, after addition of excess unlabeled yeast $tRNA^{Phe}$, no EF-G; ◇, after addition of excess unlabeled AcPhe-$tRNA^{Phe}$, no EF-G; ●, after addition of EF G · GTP and excess unlabeled yeast OH-$tRNA^{Phe}$; ◆, after addition of EF-G · GTP and excess unlabeled AcPhe-$tRNA^{Phe}$.

SUMMARY

The data summarized above lead to the following conclusions. (i) The 5′ end, dhU loop, s^4U_8, and TΨC loop of the tRNA move toward S21 and L1 during the peptidyltransferase reaction, whereas the anticodon loop undergoes much less movement at this point in the reaction cycle. (ii) The position of the peptide relative to S21 does not change appreciably during the peptidyltransferase reaction. (iii) S21 is relatively close to the tRNA and is probably positioned between the ribosomal subunits.

The data also indicate but do not firmly establish the following. (iv) The tRNA itself may undergo a conformational change associated with the movement indicated under (i). (v) The EF-G reaction causes a poly(U)-dependent change in the interaction of deacylated tRNA in the 70S ribosomal complex that is dependent on binding of a second tRNA to the complex. The codon-anticodon interaction involving 30S subunits appears to be disrupted. (vi) The EF-G reaction also causes relaxation of the ribosome · poly(U) · AcPhe-tRNA complex that may be associated with movement of the anticodon and mRNA. (vii) The nascent peptide may be positioned near the bottom of the 50S subunit distal to the central protuberance.

We are indebted to G. Kramer and B. Davison for discussion and assistance in preparing the manuscript and to R. Keegan for the artwork.

This work was supported by grant DMB 88-18579 from the National Science Foundation.

with the idea that the two $(L7/12)_2$ dimers carry out different functions in factor interaction.

Finally, it is quite possible that one of the two EF-Tu molecules in the extended ternary complex is involved in a first step of the movement of peptidyl-tRNA from the A to the P site (Kurland et al., this volume; Moazed and Noller, 1989a, 1989b).

Accordingly, the hydrolysis of two molecules of GTP in EF-Tu function per elongation cycle may reflect the need to remove quickly from the ribosome first one molecule of EF-Tu participating in A-site binding and then another one catalyzing a partial movement of tRNA, perhaps to a hybrid A/P site (Moazed and Noller, 1989a, 1989b; Kurland et al., this volume).

Recent observations indicate that tRNA has three binding sites on the ribosome. During the elongation cycle, tRNA passes from the A to the P to the E site before it leaves the 70S particle (Gnirke et al., 1989; Lill and Wintermeyer, 1987; Moazed and Noller, 1989a, 1989b). There is a strong, negative cooperativity between binding to the A and E sites (Gnirke et al., 1989) so that there are never more than two tRNAs bound simultaneously to the ribosome. These data indicate that EF-Tu catalyzes not only the binding of an aminoacyl-tRNA to the A site but also the release of a deacylated tRNA from the E site. This complex reaction sequence may occur in two distinct steps, each of which requires one cycle of EF-Tu and the second of which may involve a structural rearrangement of peptidyl-tRNA in the A site. It is furthermore not improbable that the $(L7/12)_4$ complex connects the A and E sites and plays a key role in the negative cooperativity observed between them.

This work was supported by grants from the Swedish Cancer Society and the Swedish Natural Science Research Council.

REFERENCES

Andersson, D. I., S. G. E. Andersson, and C. G. Kurland. 1986a. Functional interactions between mutated forms of ribosomal protein S4, S5 and S12. *Biochimie* **68**:705–713.

Andersson, D. I., and C. G. Kurland. 1983. Ram ribosomes are defective proofreaders. *Mol. Gen. Genet.* **191**:378–381.

Andersson, D. I., H. W. van Verseveld, A. H. Stouthammer, and C. G. Kurland. 1986b. Suboptimal growth with hyperaccurate ribosomes. *Arch. Microbiol.* **144**:96–101.

Beres, L., and J. Lucas-Lenard. 1973. Studies on the fluorescence of the Y-base of yeast phenyl-alanine transfer ribonucleic acid. Effect of pH, aminoacylation and interaction with elongation factor Tu. *Biochemistry* **12**:3998–4002.

Björk, G. R., J. U. Ericson, C. E. D. Gustafsson, T. G. Hagervall, Y. H. Jönsson, and P. M. Wikström. 1987. Transfer RNA modification. *Annu. Rev. Biochem.* **56**:263–287.

Bohman, K. T., T. Ruusala, P. C. Jelenc, and C. G. Kurland. 1984. Kinetic impairment of restrictive streptomycin resistant ribosomes. *Mol. Gen. Genet.* **198**:90–99.

Buckingham, R. H., and H. Grosjean. 1986. The accuracy of mrRNA-trRNA recognition, p. 83–126. *In* T. B. L. Kirkwood, R. F. Rosenberger, and D. J. Galas (ed.), *Accuracy in Molecular Processes.* Chapman & Hall, Ltd., London.

Chinali, G., H. Wolf, and A. Parmeggiani. 1977. Effect of kirromycin on elongation factor Tu. Location of the catalytic center for ribosome elongation-factor-Tu GTPase activity on the elongation factor. *Eur. J. Biochem.* **75**:55–65.

Dix, D. B., and R. C. Thompson. 1986. Elongation factor Tu guanosine 3′-diphosphate 5′-diphosphate complex increases the fidelity of proofreading in protein biosynthesis: mechanism for reducing translational errors introduced by amino acid starvation. *Proc. Natl. Acad. Sci. USA* **83**:2027–2031.

Duisterwinkel, F. J., J. M. De Graaf, P. J. M. Schratleu, B. Kraal, and L. Bosch. 1981. A mutant of elongation factor Tu which does not immobilize the ribosome upon binding of kirromycin. *Eur. J. Biochem.* **117**:7–12.

Edelman, P., and J. Gallant. 1977. Mistranslation in *Escherichia coli. Cell* **10**:131–137.

Ehrenberg, M., D. Andersson, K. Bohman, P. C. Jelenc, T. Ruusala, and C. G. Kurland. 1986. Ribosomal proteins tune rate and accuracy in translation, p. 573–585. *In* B. Hardesty and G. Kramer (ed.), *Structure, Function, and Genetics of Ribosomes.* Springer-Verlag, New York.

Ehrenberg, M., and C. G. Kurland. 1984. Cost of accuracy determined by a maximal growth rate constraint. *Q. Rev. Biophys.* **17**:45—82.

Ehrenberg, M., A.-M. Rojas, J. Weiser, and C. G. Kurland. 1990. How many EF-Tu molecules participate in aminoacyl-tRNA binding and peptide bond formation in *E. coli* translation. *J. Mol. Biol.* **211**:739–749.

Gavrilova, L. P., O. E. Kostiashkina, V. E. Koteliansky, N. M. Rutkevitch, and A. S. Spirin. 1976. Factor-free (non-enzymic) and factor-dependent systems of translation of polyuridylic acid by *Escherichia coli* ribosomes. *J. Mol. Biol.* **101**:537–552.

Gavrilova, L. P., I. N. Perminova, and A. S. Spirin. 1981. Elongation factor Tu can reduce translation errors in poly(U)-directed cell-free systems. *J. Mol. Biol.* **149**:69–78.

Gnirke, A., U. Geigenmüller, H.-J. Rheinberger, and K. H. Nierhaus. 1989. The allosteric three-site model for the ribosomal elongation cycle. *J. Biol. Chem.* **264**:7291–7300.

Gordon, J. 1969. Hydrolysis of guanosine 5′-triphosphate associated with binding of aminoacyl transfer ribonucleic acid to ribosomes. *J. Biol. Chem.* **244**:5680–5686.

Hopfield, J. J. 1974. Kinetic proofreading: a new mechanism for reduced errors in biosynthetic processes requiring high specificity. *Proc. Natl. Acad. Sci. USA* **71**:261–264.

Hughes, D. 1984. Eternal suppressors of −1 and +1 frameshift mutations. A genetic analysis in bacteria. Ph.D. thesis. Dublin University, Dublin, Ireland.

Hughes, D. 1987. Mutant forms of tuf A and tuf B independently suppress nonsense mutations. *J. Mol. Biol.* **197**:611–615.

Hughes, D., J. F. Atkins, and S. Thompson. 1987. Mutants of elongation factor Tu promote ribosomal frameshifting and nonsense readthrough. *EMBO J.* **6**:4235–4239.

Hughes, D., and C. G. Kurland. 1989. Novel mutants of EF-Tu, p. 51–56. *In* L. Bosch, B. Kraal, and A. Parmeggiani (ed.), *The Guanine Nucleotide Binding Proteins.* Plenum Publishing Corp., New York.

Jelenc, P. C., and C. G. Kurland. 1979. Nucleoside triphosphate regeneration decreases the frequency of translation errors. *Proc. Natl. Acad. Sci. USA* **76**:3174–3178.

Jukes, T. H. 1973. Possibilities for the evolution of the genetic code from a preceding form. *Nature* (London) **246**:22–26.

Kakniashvili, D. G., S. K. Smailov, and L. P. Gavrilova. 1986. The excess GTP hydrolyzed during mistranslation is expended at the stage of EF-Tu-promoted binding of non-cognate aminoacyl-tRNA. *FEBS Lett.* **196**:103–107.

Kaziro, Y. 1978. The role of guanosine 5-triphosphate in polypeptide chain elongation. *Biochim. Biophys. Acta* **505**:95–127.

Kim, S. H. 1975. Symmetry recognition hypothesis model for tRNA binding to aminoacyl-tRNA synthetase. *Nature* (London) **256**:679–681.

Kurland, C. G. 1979. Reading frame errors on ribosomes, p. 97–108. *In* J. E. Celis and J. D. Smith (ed.), *Nonsense Mutations and tRNA Suppressors.* Academic Press, Inc., New York.

Kurland, C. G., and M. Ehrenberg. 1984. Optimization of translation accuracy. *Prog. Nucleic Acids Res. Mol. Biol.* **31**: 191–219.

Kurland, C. G., and M. Ehrenberg. 1987. Growth-optimizing accuracy of gene expression. *Annu. Rev. Biophys. Biophys. Chem.* **16**:291–317.

Liljas, A. 1982. Structural studies of ribosomes. *Prog. Biophys. Mol. Biol.* **40**:161–228.

Liljas, A., L. A. Kirsebom, and M. Leijonmarck. 1986. Structural studies of the factor binding domain, p. 379–390. *In* B. Hardesty and G. Kramer (ed.), *Structure, Function, and Genetics of Ribosomes.* Springer-Verlag, New York.

Lill, R., and W. Wintermeyer. 1987. Destabilization of codon-anticodon interaction in the ribosomal exit site. *J. Mol. Biol.* **196**:137–148.

Miller, D. L., and H. Weissbach. 1977. Factors involved in the transfer of aminoacyl-tRNA to the ribosome, p. 323–373. *In* H. Weissbach and S. Pestka (ed.), *Molecular Mechanisms of Protein Synthesis.* Academic Press, Inc., New York.

Moazed, D., and H. F. Noller. 1989a. Interaction of tRNA with 23S rRNA in the ribosomal A, P and E sites. *Cell* **57**:585–597.

Moazed, D., and H. F. Noller. 1989b. Intermediate states in the movement of tRNA in the ribosome. *Nature* (London) **342**: 142–148.

Möller, W., and J. A. Maassen. 1986. On the structure, function and dynamics of L7/12 from *Escherichia coli* ribosomes, p. 309–325. *In* B. Hardesty and G. Kramer (ed.), *Structure, Function, and Genetics of Ribosomes.* Springer-Verlag, New York.

Murgola, E. J. Mutant glycine tRNAs and other wonders of translational suppression. *In* J. D. Cherayil (ed.), *Transfer RNAs and Other Soluble RNAs.* CRC Press, Inc., Boca Raton, Fla., in press.

Nierhaus, K. H., H.-J. Rheinberger, U. Geigenmüller, A. Gnirke, H. Saruyama, S. Schilling, and P. Wurmrback. 1986. Three tRNA binding sites involved in the ribosomal elongation cycle, p. 454–472. *In* B. Hardesty and G. Kramer (ed.), *Structure, Function, and Genetics of Ribosomes.* Springer-Verlag, New York.

Ninio, J. 1975. Kinetic amplification of enzyme discrimination. *Biochimie* **57**:587–595.

Pingoud, A., C. Urbanke, G. Krauss, F. Peters, and G. Maass. 1977. Ternary complex formation between elongation factor Tu, GTP and aminoacyl-tRNA: an equilibrium study. *Eur. J. Biochem.* **78**:403–409.

Rojas, A.-M. 1988. Kinetic modulation of accuracy in protein synthesis by ppGpp. Ph.D. thesis. Uppsala University, Uppsala, Sweden.

Ruusala, T., D. I. Andersson, M. Ehrenberg, and C. G. Kurland. 1984. Hyperaccurate ribosomes inhibit growth. *EMBO J.* **3**: 2575–2580.

Ruusala, T., M. Ehrenberg, and C. G. Kurland. 1982. Is there proofreading during polypeptide synthesis? *EMBO J.* **1**: 741–745.

Ruusala, T., and C. G. Kurland. 1984. Streptomycin preferentially perturbs ribosomal proofreading. *Mol. Gen. Genet.* **198**: 100–104.

Swart, G. W. M. 1987. The polypeptide chain elongation factor Tu from *Escherichia coli*. Characterization of mutants and protein engineering. Ph.D. thesis. Leiden University, Leiden, The Netherlands.

Tapio, S., N. Bilgin, and M. Ehrenberg. 1990. Impaired *in vitro* kinetics of EF-Tu mutant Aa. *Eur. J. Biochem.* **188**:347–354.

Tapio, S., and L. A. Isaksson. 1987. Antagonistic effects of mutant elongation factor Tu and ribosomal protein S12 on control of translational accuracy, suppression and cellular growth. *Biochimie* **70**:273–281.

Tapio, S., and C. G. Kurland. 1986. Mutant EF-Tu increases missense error *in vitro*. *Mol. Gen. Genet.* **205**:186–188.

Thompson, R., and P. J. Stone. 1977. Proofreading of the codon-anticodon interaction on ribosomes. *Proc. Natl. Acad. Sci. USA* **74**:198–202.

Thompson, R. C. 1988. EF-Tu provides an internal kinetic standard for translational accuracy. *Trends Biochem. Sci.* **13**:91–93.

Vacher, J., H. Grosjean, R. H. Buckingham, and C. Houssier. 1984. The effect of point mutations affecting *Escherichia coli* tryptophan tRNA anticodon-anticodon interactions and on UGA suppression. *J. Mol. Biol.* **177**:329–342.

van de Klundert, J. A. M., P H. van der Meide, P. van de Putte, and L. Bosch. 1978. Mutants of *Escherichia coli* altered in both genes coding for the elongation factor Tu. *Proc. Natl. Acad. Sci. USA* **75**:4470–4473.

van der Meide, P. H., F. J. Duisterwinkel, J. M. De Graaf, B. Kraal, L. Bosch, J. Douglass, and T. Blumenthal. 1981. Molecular properties of two mutant species of the elongation factor Tu. *Eur. J. Biochem.* **117**:1–6.

Vijgenboom, E., and L. Bosch. 1989. Translational frameshifts induced by mutant species of the polypeptide chain elongation factor Tu of *Escherichia coli*. *J. Biol. Chem.* **264**:13012–13017.

Vijgenboom, E., T. Vink, B. Kraal, and L. Bosch. 1985. Mutants of the elongation factor EF-Tu, a new class of nonsense suppressors. *EMBO J.* **4**:1049–1052.

Wagner, E. G. H., P. C. Jelenc, M. Ehrenberg, and C. G. Kurland. 1982. Rate of elongation of polyphenylalanine *in vitro*. *Eur. J. Biochem.* **122**:193–197.

Weissbach, H., B. Redfield, and N. Brot. 1971. Aminoacyl-tRNA-Tu-GTP interaction with ribosomes. *Arch. Biochem. Biophys.* **145**:676–684.

Woese, C. 1970. Molecular mechanics of translation: a reciprocating ratchet mechanism. *Nature* (London) **226**:817–820.

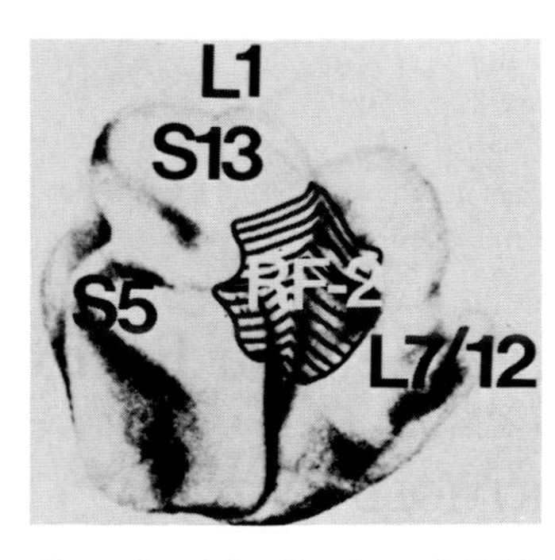

Figure 1. Three-dimensional localization of RF-2 on the *E. coli* ribosome. The binding site of RF-2 as determined by immunoelectron microscopy is shown by the hatching. The protein positions are given to mark the orientation of the subunits in this view.

Stop codons can be misread at low efficiency by translational readthrough with a normal charged tRNA, for example, $tRNA^{Trp}$ (Engelberg-Kulka and Schoulaker-Schwarz, 1988).

The picture that emerges is that external factors determine whether a termination codon is actually recognized as a stop signal. These factors may act either in *cis* to the codon in its ribosomal environment (the context of the codon in the mRNA) (Brown et al., in press) or in *trans* (tRNAs or polypeptide release factors, even the physiological condition of the cell). How does the release factor manage to ensure that in most cases the termination codons are correctly read as stop? As a starting point to answer that question, we have determined where the factor interacts with the ribosome in bacteria and how that relates to the position of the codon.

WHERE IS THE RELEASE FACTOR ON THE RIBOSOME?

70S-Binding Domain

An idea of where RF-2 binds to the 70S particle has been obtained by immunoelectron microscopy. First, the release factor was chemically cross-linked to the ribosome in a functional complex with UGA (Kastner et al., 1987), and immunocomplexes were isolated with a specific antibody to RF-2. Examination of these resulting complexes by electron microscopy revealed antibody-labeled 70S ribosomes in different projection forms (Kastner et al., 1990). Since these forms had been characterized (Stöffler and Stöffler-Meilicke, 1984), the three-dimensional localization of RF-2 on the 70S model could be established (Fig. 1). The RF-2-binding site was at the ribosomal subunit interface comprising the base of the stalk region of the large subunit and the head-neck region of the concave side of the small subunit. This study had several important implications; first, the factor seemed to penetrate further into the cavern of the large subunit than the biochemical studies had indicated, and second, the domain of the 30S subunit in contact with the release factor was on the side opposite the decoding site. What was not possible to deduce from such a study was how far the factor extended toward the decoding site.

50S-Binding Domain

Early biochemical studies identified L7/L12 of the bacterial ribosome as an important determinant for ribosomal binding of the release factor (Brot et al., 1974). Subsequently L11, situated at the base of the stalk, was shown to modulate the binding of RF-1 (UAG and UAA) and RF-2 (UGA and UAA) differentially (Tate et al., 1983b; Tate et al., 1984). If the highly exposed tyrosine 7 of L11 was specifically modified, the binding of RF-1 was drastically reduced (Tate et al., 1986). The results of cross-linking studies were consistent with a binding domain for the factors encompassing the stalk and that part of the body of the subunit nearby (Stöffler et al., 1982). Recently, we have shown with polyclonal and monoclonal antibodies that it is the body dimer of L7/L12 rather than the stalk dimer that is important for release factor interaction and that the C-terminal domains of the proteins in this dimer are more important than the N-terminal domains (Tate et al., 1990).

These studies suggested that the 50S subunit provided the important determinants for release factor binding. However, binding to this subunit alone had not been achieved, explained by a requirement for a small-subunit-bound termination codon.

30S-Binding Domain

A number of studies implicated proteins of the 30S subunit falling into two general domains, one on each side of the subunit, as important for release factor interaction. Reexamination of this question with antibodies specific for the 30S proteins implicated some epitopes of S5, S10, S3, and S4 as being in the vicinity of the release factor (Tate et al., 1988). These are consistent with the 70S-binding site as deduced from the electron micrographs.

Covalent cross-linking had stabilized the release factor on the 70S ribosome. Could it trap the factor in a functional complex on the individual subunits, despite our inability to isolate complexes in the absence of cross-linking? The cross-linking to the large subunit was very weak, as determined by an immunodot blot of samples from fractions of a

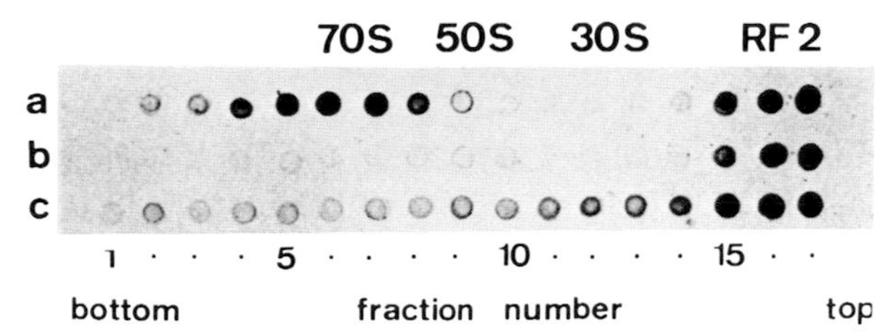

Figure 2. Immunodot blot of covalently cross-linked RF-2–ribosome complexes. After covalent cross-linking of RF-2 with 70S (row a), 50S (row b), and 30S (row c) ribosomes or subunits, the complexes were separated on a sucrose gradient and the factor was detected in samples of the fractions with a polyclonal immunoglobulin G against RF-2.

sucrose gradient (Fig. 2, row b), compared with that to 70S ribosomes (row a), whereas significant cross-linking occurred to 30S subunits (row c). This cross-linking was dependent on codon, but there were also intersubunit cross-links; multimers of the subunits spread down the gradient, and these also contained release factor provided that codon was present. The implication of this result is that there is a binding domain on the small subunit for the release factor in addition to that characterized biochemically on the 50S subunit.

Release Factor-Binding Domains and Other Functional Sites

Although the low efficiency of cross-linking made an analysis of these cross-linked particles by electron microscopy difficult, an alternative strategy was informative. The 70S immunocomplexes were dissociated at a low magnesium ion concentration so that those which contained RF cross-linked to an individual subunit could be analyzed at the subunit level rather than as a 70S ribosome. A high proportion of the ribosomes did not dissociate, but sufficient lacked intersubunit cross-linking for examination of images with antibodies linked to one of the individual subunits. This strategy provided the RF-binding domains on the individual subunits and thereby better information on how far the factor extended across the small subunit and into the cavern of the large subunit. A summary of the binding domains is shown in Fig. 3.

From images of the small subunit in several orientations, the factor site was deduced to be at the head-neck region of the small lobe but extending across to the middle of the subunit. This interpretation was consistent with the site assigned from the images of the 70S immunocomplexes and the more recent biochemical studies (Tate et al., 1988), but it was still not clear whether the factor could reach the

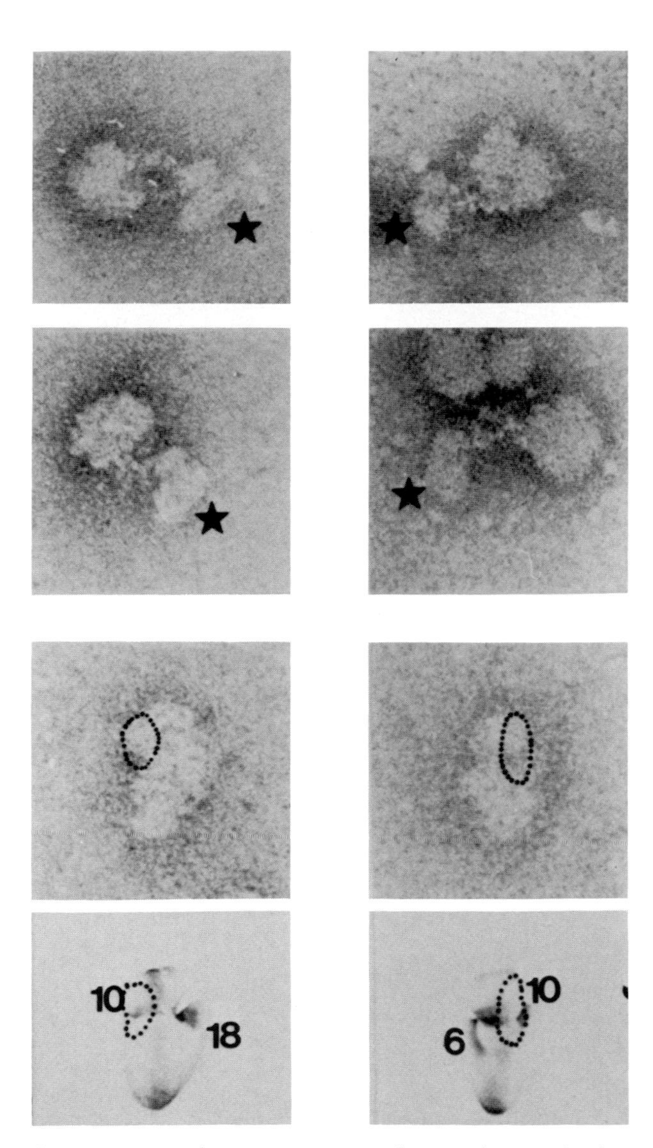

Figure 3. Immunoelectron microscopy showing the RF-2-binding domain on the 30S subunit after dissociation of 70S · RF-2 immunocomplexes. Selected images in which a 70S ribosome not held by intersubunit cross-links had dissociated and the antibody was bound to the 30S subunit (star) are shown in two orientations. An image and a model of the subunit in these orientations are shown in the bottom two rows, with protein positions added to show the particular views. The RF-2-binding site is circled.

decoding site in the cleft of the large lobe of the subunit (Gornicki et al., 1984).

The images of the large subunit confirmed that the factor extended from the base of the stalk well across the subunit and would be able to make contact with the peptidyltransferase center, together with which it mediates the hydrolysis of the final ester bond between tRNA and polypeptide. A summary diagram illustrating these conclusions is shown in Fig. 4.

Recently, we have examined by immunoelectron microscopy the cross-linked complexes of RF-2 and

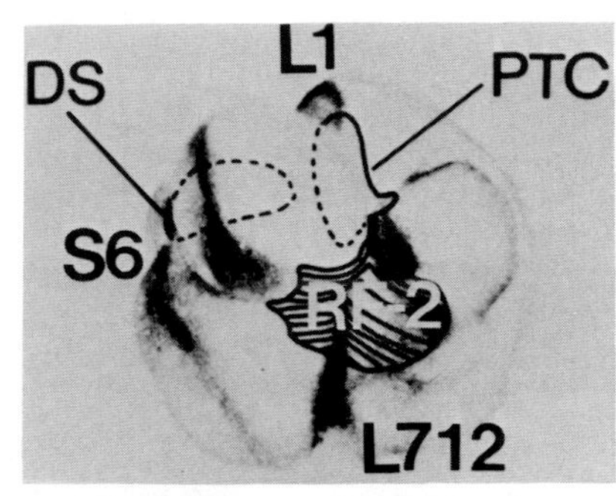

Figure 4. Localization of RF-2 on the *E. coli* ribosome compared with the decoding site and peptidyltransferase center. The binding site of RF-2 as determined by immunoelectron microscopy is shown by the hatching. The protein positions are given to mark the orientation of the subunits in this view. The positions of the peptidyltransferase and decoding sites at the 70S ribosome interface are indicated by the dotted lines.

30S subunits isolated from the sucrose gradients (Fig. 2) after reaction with anti-RF-2. In addition to the particles with the antibody bound to the head and upper body of the small lobe, consistent with the previous observations, significant numbers of particles have the antibody bound to the large lobe of the subunit near the decoding site. This finding indicates that the factor could extend from the small lobe over to the decoding site and the termination codon. Presumably, these epitopes on the release factor molecule are less accessible to antibody when it is bound in a functional 70S complex, and hence they are only seen on the other micrographs at a very low frequency.

Antibiotics and the Release Factor 30S Subunit Domain

A footprinting technique has been used to determine the binding sites of a number of antibiotics that affect protein synthesis through an interaction with the 30S subunit (Moazad and Noller, 1987). For example, spectinomycin binds within the nucleotides of helix 34 of the model of Brimacombe for the three-dimensional structure of the rRNA (Brimacombe, 1988; Stern et al., 1988). This helix has been placed at the head-neck region of the small-subunit lobe, that is, within the binding domain deduced for RF-2. Significantly, we have found that RF-2 activity is stimulated twofold by spectinomycin exclusively in UGA-dependent reactions (Brown et al., 1989). RF-2 activities dependent on UAA were not affected, nor were RF-1 activities, supporting our previous results which suggested that its domain may be somewhat lower on the small subunit (Tate et al., 1988). Spectinomycin-resistant mutants have been constructed by alteration in one nucleotide, 1192 in helix 34, and this rRNA carried on a plasmid is found in about half of the ribosomes of the cell (Makosky and Dahlberg, 1987). Two of the spectinomycin-resistant mutants in which a C has been replaced by a U or A exhibit lower release factor activities than does the wild type.

The region of the small subunit encompassing helix 34 may contribute to the decoding of specifically UGA by RF-2. Compelling support for this conclusion comes from Murgola et al. (1988), who characterized a defective ribosomal mutant that suppressed UGA stop codons. The ribosome had a deletion of a single nucleotide (1054) in helix 34 of the small-subunit rRNA. Disruption of the rRNA in this region caused a significant perturbation in the decoding of the UGA codon as a stop signal, a function carried out exclusively in the cell by RF-2.

Two other antibiotics, hygromycin and neomycin, bind at the decoding site of the small subunit, around the critical area of the rRNA encompassing C-1400. This is quite distant from helix 34 containing the spectinomycin site. These antibiotics are also strongly inhibitory of termination reactions, although unlike spectinomycin they do not discriminate between the two factors RF-1 and RF-2 or any of the three codons UAA, UAG, and UGA. A function common to the two factors is being affected, and this supports a proposal that the factors may extend over to the decoding site although their primary binding domain is at the small lobe side of the subunit. Long-range effects resulting from the antibiotic interactions extending through to the other side of the subunit may occur, however, and hence affect release factor functions.

A summary of the antibiotic-binding sites, important functional sites, and the release factor-binding domain is shown in Fig. 5.

SITE OF THE TERMINATION CODON ON THE RIBOSOME

Stop codons can have several fates as a signal in the cell. Competition between external agents, release factor, and tRNA (Beaudet and Caskey, 1970) suggests that the nonsense codon initially occupies a site in common with sense codons, likely to be the decoding region of the A site. This does not exclude the possibility that the position of the codon on the ribosome may be influenced through a conformational change when it is decoded by release factor. Indeed, this may enable the factor to compete successfully with other agents such as natural suppressing tRNAs.

Radiolabeled termination codons have been

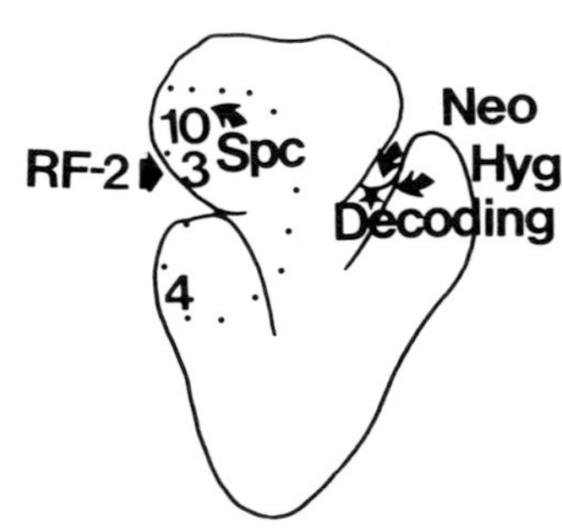

Figure 5. Antibiotic sensitivity of release factor-dependent termination. On the interface side of the 30S subunit near the decoding site, the sites of neomycin (Neo) and hygromycin (Hyg), which inhibit both release factor and all codon-mediated events, are shown. On the opposite side of the subunit, the binding site of spectinomycin (Spc), which stimulates only RF-2- and UGA-dependent termination and falls within the major binding domain of the factor, is shown. The binding sites of antibodies to S3, S4, and S10, which affect the release factor-mediated events, are also shown.

cross-linked to the 30S ribosomal components S6 and S18 when in a functional complex with RF-2 (Lang et al., 1989). These proteins are near the decoding site, consistent with the codon initially occupying this site. However, the codon also cross-linked to proteins L7/L12, L10, and L2 of the large subunit. The first two of these proteins are in contact with the other side of the 30S subunit distant from the decoding site. Indeed, the results of our antibody studies were consistent with proteins from this side of the subunit, S3 and S10, being in contact with RF-2. Earlier ribosomal proteins L7/L12, L2, and L11 were also found cross-linked to RF-2 in a 70S functional complex, and weak complexes were also found with S6, S17, and S18 (Stöffler et al., 1982).

Identification of the two groups of ribosomal proteins suggests that the codon may occupy two sites. S6 and S18 are placed quite distant from the L7/L12 stalk of the large subunit on current models of the 70S ribosome, and it would not be possible for a codon in a single site to be part of cross-linked complexes with ribosomal components from each area, although our immunoelectron microscopy studies suggest that it is possible for RF-2 to span the distance. Is this evidence for a significant conformational change occurring during the termination event? This could either bring the decoding site into much closer contact with the release factor on the other side of the ribosome or shift the termination codon out of the decoding site during the event.

Further evidence that the stop codon may be recognized outside the decoding site by release factor has been provided by studies in which both the A and P sites are occupied by codons and cognate tRNAs (Tate et al., 1983a). A number of surprising results emerged from this study. First, termination was prevented neither by occupation of the A site by a sense codon nor by displacement of the termination codon from the A-site codon position by spacer nucleotides between the P-site-bound AUG and UAA. Moreover, a tRNA occupying the A site did not interfere with UAA-directed release factor binding to these particles. In contrast, a ternary complex containing elongation factor EF-Tu abolished the binding of the release factor. These data suggest that the release factor and elongation factor have functional overlapping binding sites but that the A-site-binding domains of the release factor and the tRNA are exclusive. The ability of the release factor to recognize the displaced codon, that is, a termination codon out of phase, suggests it may have flexibility in its codon recognition domain and can adjust to this enforced change in codon position without loss of specificity (Tate et al., 1983b).

More recently, short artificial mRNAs of the form $AUGU_mUAA_n$ rather than oligonucleotide fragments and triplet codons have been used in similar studies (Buckingham et al., 1987). A continuous fragment of AUGUAA supported termination poorly in contrast to AUG and UAA, in two fragments, features of the in vitro assay as originally developed (Caskey et al., 1968). The other members of this series $AUGU_{1-5}UAA_n$, all supported the release of formylmethionine by RF-1, further attesting to the ability of the release factor to recognize termination codons out of phase. Furthermore, addition of exogenous UAA stimulated the termination reaction when these oligonucleotides were bound to the ribosome, indicating that the release factor can circumvent the sense or nonsense codon in the A site and apparently preferentially recognize a codon bound in an alternative position (Buckingham et al., 1987). Since this occurs on the same ribosome, it supports the hypothesis that there is conformational flexibility in the release factor to find the codon.

Although the long-standing dogma is that in vivo EF-G-mediated translocation is a prerequisite for the termination event, freeing the A site of polypeptidyl-tRNA complex and bringing the termination codon into the site, a dipeptide can in fact be released from the ribosome programmed with $AUGU_3UAA_n$ or with $AUGU_4UAA_n$ without prior translocation. With the first oligonucleotide, the release factor must recognize the displaced codon in phase, but out of phase with the other oligonucleotide (Buckingham et al., 1987).

These studies are carried out on isolated reactions of protein synthesis, and although the influences of some of the components of the natural system are missing, they indicate considerable flexibility in the spatial relationship between termination codon and release factor in the recognition event. Despite this, the specificity of recognition, RF-1 with UAG or UAA and RF-2 with UGA and UAA, is faithfully maintained.

ARE THE BINDING SITES OF THE RELEASE FACTOR AND THE ELONGATION FACTORS SIMILAR?

The ribosome-binding sites of elongation factors EF-Tu and EF-G and that of the release factors are overlapping, since binding of one type of factor to the ribosome excludes binding of the other (Tate et al., 1973; Tate et al., 1983a). Electron microscopy of the factor-ribosome complexes, however, suggests there are clear differences in their ribosome-binding domains. EF-G can bind to the 50S subunit and has been mapped on this subunit to bind at the base of the stalk in the region of L11 (Girshovich et al., 1981), whereas the EF-Tu-binding domain extends to the tip of the stalk (Girshovich et al., 1986). The RF-2-binding site, in contrast, is directed more toward the central protuberance of the 50S subunit and the head of the small subunit (Kastner et al., 1990). The two monoclonal antibodies against epitopes in the N- and C-terminal domains of L7/L12 (Olson et al., 1986) have been very informative by revealing subtle differences in the ribosome-binding domains of the three factors. The anti-N-terminal antibody inhibits EF-Tu functions more strongly, EF-G functions similarly, and release factor functions less strongly at lower ratios than does the anti-C-terminal species (Sommer et al., 1985; Nag et al., 1985; Tate et al., 1990). Also, the antibiotic hygromycin inhibits release factor binding but not binding of either EF-Tu or EF-G (Cabanas et al., 1978).

It is perhaps not surprising, since the release factors and the two elongation factors have little homology in primary sequences and different functional domains, that there should be differences in the binding domains of the three factors. Nevertheless, they each bind in the same region of the ribosome quite distant from the decoding site, where each has its own special function. This seems to support strongly the suggestion (Spirin, 1983) that the two tRNA-binding sites extend back toward the side of the ribosome where the factors are situated; Stern et al. (1988), in their detailed three-dimensional model of the 16S rRNA, believe that this arrangement and the more convential placement of the two tRNAs are both possible.

DIRECT INTERACTION OR INDIRECT RECOGNITION OF THE TERMINATION CODON BY THE RELEASE FACTOR

From the time when Scolnick and Caskey (1969) demonstrated a specific codon incorporation into a ribosomal complex with each release factor, it was assumed that there would be a direct interaction between the factor and codon. Surprisingly, the evidence for this even today remains equivocal. Equilibrium dialysis studies of Capecchi and Klein (1969) suggested a weak specificity for the factors to bind the codons in the absence of the ribosome.

A direct interaction between single-stranded codon and the protein release factor was questioned by Shine and Dalgarno (1974), who observed that consecutive nucleotides at the end of the 3′ terminus of the 16S rRNA of *Escherichia coli* and 18S rRNA of a number of eucaryotes were complementary to UAA (16S), to UAA and UGA (18S), and, with wobble base pairing, to UAG. They suggested that termination codon recognition may involve interaction of the codon with the rRNA rather than the release factor. To test this idea, Tate et al. (1983c) used antibodies raised against a hapten on the terminal adenosine of the 16S rRNA, and also an oligonucleotide complementary to the terminal sequence, and observed no effects on the release factor-mediated termination event in vitro. The radiolabeled codon was cross-linked to RF-2 when in a functional ribosomal complex, suggesting that there was close contact. In this experiment there were also many other products cross-linked to the codon, and it is difficult to conclude unequivocally that the cross-linking was specific (Lang et al., 1989).

Although RF-2 will not form a stable complex with the 50S subunit, we have detected such a complex at low efficiency by using the cross-linking reagent dimethylsuberimidate in a sensitive enzyme-linked immunosorbent assay. Significantly, the reaction is stimulated by the codon (Fig. 6). The fact that here the codon is out of its natural 30S environment yet can influence the binding is suggestive of a direct interaction.

The best evidence for direct interaction of codon and release factor was a study in the absence of ribosomes with reticulocyte release factor indicating that UGA_n had a stronger affinity for the factor than for suppressor $tRNA^{Ser}$ (Mizutani and Hitaka, 1988). No demonstration of codon specificity was reported in this study, nor were controls with other proteins shown. In our experience such controls are essential, since a protein like bovine serum albumin will bind codon in a filter binding assay considerably

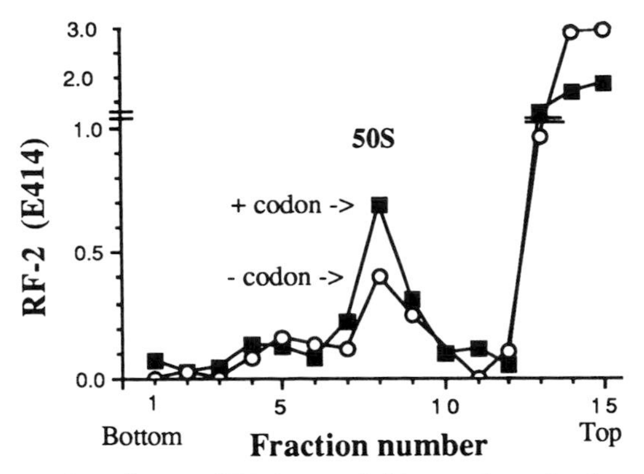

Figure 6. Stimulation of RF-2 cross-linking to the 50S subunit by termination codon. RF-2 was covalently cross-linked to 50S subunits with dimethylsuberimidate in the presence or absence of UAA, and the complexes were separated by sucrose gradient centrifugation. RF-2 was detected in the ribosomal complex with an enzyme-linked immunosorbent assay, using polyclonal antibody against RF-2.

above background, and the results must still be interpreted with caution.

The definitive evidence to resolve whether there is a codon-binding domain on the release factor would be the codon labeling of a specific peptide within the release factor molecule. We are currently attempting to affinity label the G of UGA and UAG with azido phenyl glyoxal to demonstrate such a specific interaction between the codons and the bacterial release factors within a functional ribosomal complex, but after stabilizing the factor in the complex with a protein cross-linking reagent. The modified codons function, albeit at reduced efficiency, in the termination reactions in vitro and stimulate the incorporation of the release factor in the cross-linked complex.

FUNCTIONAL DOMAINS ON THE RELEASE FACTOR

Although the primary sequences of the release factors have been deduced from the DNA sequences of the cloned genes (Craigen et al., 1985), no information is available on their tertiary structures. Little is known of the functional domains for codon recognition, ribosome binding, interaction with the peptidyltransferase center of the ribosome, or interaction with the third release factor, RF-3. We are using gene fusions of *trpE*-ΔRF to produce antigens for raising antibodies to specific regions of the molecule (Moffat et al., 1988). One such antibody inhibited release factor functions (Moffat, 1988). Short peptides within the sequence of the protein have been used to identify the epitopes recognized and thereby narrow down the specific region of the protein implicated in the function. Antibodies have been raised to these peptides to create more specific antibody probes. Such an approach to date has identified two functionally important regions of the release factors, corresponding to amino acids 215 to 236 and 239 to 268 (RF-1) and 234 to 255 and 263 to 286 (RF-2). Significantly, these are in the region of very high homology between the two proteins.

DOES THE TERMINATION CODON FORM A DOUBLE-STRANDED COMPLEX WITH rRNA?

If the first ribosome were indeed made entirely of RNA, and there was a distinct mechanism for stopping synthesis of a protein, then a double-stranded interaction between the program or mRNA and the synthetic apparatus, the rRNA, was likely. Indeed, modified trinucleotides have been used to determine which changes affect the ability of the termination codon to function as a stop signal. The specificity closely resembled that of Watson-Crick and wobble base pairing (Smrt et al., 1970). This was strongly suggestive of a nucleic acid-nucleic acid interaction occurring during the termination event.

Shine and Dalgarno (1974) proposed such an interaction on the basis of their sequencing of the 3′ terminus of the 16S rRNA, although as described above, an oligonucleotide complementary to the sequence was unable to block the triplet codon-directed termination event in vitro (Tate et al., 1983c). Cleavage of the 16S rRNA with cloacin DF13, which nicks the rRNA 49 nucleotides from the 3′ terminus, rendered ribosomes inactive for the binding of the release factors dependent on termination codons specifically (Caskey et al., 1977). However, this treatment also affects a number of other ribosomal functions, such as binding of aminoacyl-tRNA, and the effect on the termination reaction may not be specific.

The termination codon in a functional release factor complex could cross-link with the 16SrRNA, indicating that the two RNAs must be in close contact (Lang et al., 1989).

A specific UGA suppressor activity in a ribosome had a base deletion at nucleotide 1054 of the 16S rRNA, and Murgola et al. (1988) proposed that an interaction might occur between UGA and complementary tandem UCA triplets in the 16S rRNA between nucleotides 1199 and 1204, which are in the same base-paired stem as the deleted nucleotide. If such an interaction occurs in this region where the release factor has been shown to bind, then there must be further regions for the codons UAA and UAG. These rRNA sequences are quite distant from

colum does not have two tandem 5′-UCA-3′ triplets, or even one, in the 1200 region. Instead it has the hexamer 5′-CUACUA-3′ (Gutell et al., 1985). This seems significant, since in that organism UGA is not a termination signal; it codes instead for Trp (Yamao et al., 1985). (iii) Structural studies indicate that no ribosomal proteins bind to the 1054 helix, only "around" it (Noller et al., 1987; Brimacombe et al., 1988; Powers et al., 1988). That could allow for entry or at least proximity of mRNA and RF-2. Furthermore, the presence in that area of conserved triplets complementary to UAA and UAG (e.g., at positions 956 to 958 and 1090 to 1092) led us to suggest that the helical region encompassing residue 1054 is a "domain for termination" for peptide chain termination signaled not only by UGA but also by UAA or UAG. This could mean that the action of both RF-1 and RF-2 involves that same region. This suggestion is reasonable from an evolutionary point of view when one considers that for eucaryotic terminations, there is only one codon-dependent release factor (Caskey et al., 1987). Furthermore, it is consistent with the positioning of both release factors on the 30S subunit (see Tate et al., this volume). (iv) Finally, it is comforting that a recently published three-dimensional model of 16S rRNA (Brimacombe et al., 1988) shows the 1054 helix located reasonably close to the 1400 region, to which A- and P-site tRNAs have been cross-linked (Ofengand et al., 1986). In any case, the functional possibilities of this domain are expanded by the involvement of helix 34, and possibly also immediately adjacent regions, in what appear to be protein-induced conformational rearrangements (Powers et al., 1988) and in spectinomycin stimulation of RF-2 activity in UGA-dependent termination reactions (Tate et al., this volume).

SITE-DIRECTED MUTAGENESIS OF TRIPLETS AT 1199 TO 1201 AND 1202 TO 1204

A major prediction of our model for peptide chain termination is that mutations in the appropriate 16S rRNA triplet (the one complementary to a particular termination codon) away from strict complementarity to the codon should lead to a decrease in or the elimination of termination at that codon. Such an effect might be manifested by suppression of nonsense mutations or by the production of new and larger proteins as a result of readthrough of some normal termination signals.

For the first set of mutants, we concentrated on UGA termination (Murgola et al., 1989). Simple base changes were made at positions 1199 and 1202 in the 16S rRNA by oligonucleotide-directed mutagenesis. The U residues at 1199 and 1202 were independently changed to C residues, resulting in the two mutations C-1199 and C-1202. In each case, a 5′-CCA-3′ triplet was produced. When these mutations were present in the *rrnB*-containing, high-copy-number plasmid pKK3535, the strains containing them grew more slowly than the cells containing the wild-type plasmid. In the case of ΔC-1054, high-copy expression was detrimental to otherwise wild-type cells, one manifestation being heterogeneity of colony morphology on solid growth media. These negative growth effects were reduced or eliminated when the plasmids were introduced into a strain containing a chromosomal *pcnB* mutation, which lowers the copy number to a few per cell (Lopilato et al., 1986; Liu and Parkinson, 1989). The double mutation C-1199 C-1202 was constructed and cloned into a lambda repressor-controlled conditional expression vector. It is lethal in high copy number, but the lethality is prevented in a *pcnB* background. The finding that the double mutation is lethal while each single-base substitution mutation is viable indicates that at least one operative triplet is essential but that either one can serve reasonably well.

TRANSLATION OF UGA CODONS

The suppression abilities of the three site-directed mutations were examined in the presence of *lacZ* UGA, UAA, and UAG mutations. All three suppressed UGA, but not UAA or UAG. To detect possible readthrough of natural UGA termination signals, we chose to examine first the effect of ΔC-1054, which presumably affects the accessibility of the 1199–1204 triplets. Cells were pulse-labeled with [^{35}S]methionine, and the radioactive proteins were then separated by two-dimensional gel electrophoresis. The mutant profile exhibited new high-molecular-weight proteins as well as a variety of novel smaller peptides (Tapprich et al., this volume). (The small peptides could be stable breakdown products of larger proteins or the result of aberrant translational events, such as frameshifting or premature chain termination [see below].) Although the altered products are yet to be identified, the preliminary result indicates the rRNA mutant has a general effect on UGA readthrough of several cellular mRNAs.

CAN THE 1199–1201 AND 1202–1204 TRIPLETS "MISREAD" SENSE CODONS?

If the base-pairing model for peptide termination is correct, one can imagine that even in wild-type cells, 16S rRNA sometimes misreads sense codons that are related to UGA by one base difference and thereby leads to premature chain termination. Such action may be difficult to observe in wild-type cells. But if the new triplets, produced by site-directed changes of the 1199–1201 and 1202–1204 triplets, are complementary to sense codons, premature termination may be detectable at the appropriate sense codon. That may seem tantamount to "fooling the release factor." But if base pairing is the predominant factor in recognition of the termination codon, it should be possible. An in vivo test that we have begun to use is the introduction of the mutant (5′-CCA-3′) rRNA plasmids (plus control plasmids) into strains containing *trpA* UGG missense mutations and each of several glycine tRNA UGG-specific suppressors. Premature termination at UGG by, for example, C-1199 would be manifested as decreased suppression of the UGG mutation. In our first tests, involving examination of growth on solid media and also enzymatic assays of the *trpA* gene product, we have observed a decrease of UGG suppression in the presence of C-1199 but not of ΔC-1054 as compared with wild-type 16S RNA (pKK3535). Under the conditions we used, the decrease is not very large (approximately threefold) but, since these first tests were done with *pcnB* strains, the next determinations will be modified to increase plasmid copy number without introducing detrimental effects. Furthermore, the double mutation C-1199 C-1202 has yet to be tested in this way (in a *pcnB* background).

Finally, the site-directed mutant constructs that yield the triplets 5′-CUA-3′ and 5′-UUA-3′ at 1199 to 1201 and 1202 to 1204 may lead to increased termination at UAG and UAA codons, respectively. This possibility will be examined in vivo with similar *trpA* mutations and cognate suppressor tRNAs.

The preliminary observations concerning premature chain termination and the demonstration of UGA readthrough and the suppression of nonsense mutations support the proposed codon-specific involvement of 16S rRNA residues 1199 to 1204 in UGA termination.

A DOMAIN FOR TERMINATION

Since we have suggested that the region encompassing residues 1054 and 1199 to 1204 is a domain for termination at all three nonsense codons (Murgola et al., 1989), we are searching now for rRNA suppressors of UAA or UAG mutations. To this end, and to avoid suppressors derivable from the $tRNA^{Glu}$ gene present in the *rrnB* operon cloned in pKK3535, we have inactivated the tRNA gene by inserting foreign DNA sequences into a unique restriction site (*Xba*I) present in the D stem of the tRNA gene. An especially useful construct is the one in which the $tRNA^{Glu}$-inactivating insert is the gene coding for chloramphenicol acetyltransferase (CAT). This CAT construct allows enough expression of the CAT gene to result in resistance to chloramphenicol, thereby allowing the continuous monitoring of the presence of the insert. Two important questions then needed to be answered: Did the CAT insertion inactivate the $tRNA^{Glu}$ gene? Does the CAT insertion leave (reasonably) intact the expression and processing of the rRNAs? To answer the first question, we asked whether the CAT construct can compete in vivo with a glycine tRNA suppressor of GAA and GAG (glutamate codons) missense mutations in *trpA*. With wild-type pKK3535 at high copy number, the decrease in suppression is dramatic; with the CAT construct, no decrease is observed. To answer the second question, we used two different in vivo tests. The first test was to determine whether the CAT construct at high copy number decreases the UGA suppression exhibited by the chromosomally located ΔC-1054 suppressor mutation. It does. The second test required recombining the chromosomal ΔC-1054 mutation into the 16S rRNA gene of the CAT construct and asking whether that CAT construct acts as the original UGA suppressor. It does.

Some of the UAA or UAG suppressors that we will obtain should reveal the triplet(s) complementary to UAA and/or UAG. Others should indicate other regions involved in termination, either in the immediate area (analogous to ΔC-1054) or in other domains that may interact with the 1054 helix. Two interesting rRNA UAA suppressors have been characterized recently in yeast mitochondria (Shen and Fox, 1989; A. Gargouri and P. Slonimski, personal communication). It is not certain yet that they are UAA specific, but both are base substitutions at the same residue in the 530 (*E. coli* numbering) stem-loop region. Evidence has been presented that the 530 loop may be involved in allosterically induced conformational relationships with the decoding region (Moazed and Noller, 1986; Allen and Noller, 1989).

SUMMARY AND CONCLUSION

Our characterizations of the ΔC-1054 mutant isolated in vivo and of several mutants made by site-directed mutagenesis support our rRNA-mRNA base-pairing model of peptide chain termination. The question of whether a termination codon is recognized by a protein-RNA interaction (release factor-mRNA) or by an RNA-RNA interaction (rRNA-mRNA) has never been resolved (Steege and Söll, 1979; Craigen and Caskey, 1987; Tate et al., this volume). The suggestion of Shine and Dalgarno (1974) that the 3′-terminal triplet of *E. coli* 16S rRNA might be involved in base pairing with each of the three termination codons was rendered unlikely, if not totally untenable, by subsequent comparative sequence analyses of 16S-like RNAs (reviewed in Steege and Söll, 1979) and also by experiments performed by Tate and colleagues, who examined termination by using 16S RNA whose 3′ end was either modified or base paired with a complementary oligonucleotide (Tate et al., this volume).

So the assumption continued to remain strong that release factors directly recognize the termination codons. Our results suggest otherwise. Our model is consistent with what is known about release factors, even as of this writing (see Tate et al., this volume). The model does not contradict, for example, the notion that release factors act in a codon-specific way. In fact, if the model is correct, the release factors could act in a codon-specific way without directly recognizing the termination codon. How they work and whether they contribute to the base pairing remains to be worked out. It could be that release factors participate by recognizing the base-paired triplet duplex formed between the codon and the appropriate rRNA triplet. Alternatively, it is possible that RF-2, for example, recognizes or binds to the highly conserved 1053–1054 bulge to bring about the accessibility of the complementary triplets. In any case, it is at least reasonable to think that it is rRNA that recognizes the codon, i.e., that defines (in the terminology of Noller and Woese [1981]) codon recognition, and to think that release factors are necessary to assist in the recognition, that is, to facilitate codon recognition.

In conclusion, our results strongly indicate that a major determinant in the recognition of a termination codon is antiparallel base pairing between the codon and a complementary triplet in 16S rRNA. They also suggest the possibility that a codon-dependent involvement of rRNA in elongation (other than for premature termination) will eventually be uncovered.

We are grateful to H. U. Göringer for his contributions, including the communication of unpublished results, to George Pennabble for occasional technical assistance and helpful discussions, to Walter J. Pagel for editorial consultation, and to Martha S. Trinkle for assistance in preparation of the manuscript.

This work was supported by American Cancer Society grant NP167 to E.J.M. and by Public Health Service grants from the National Institute of General Medical Sciences to E.J.M. (GM21499) and to A.E.D. (GM19756).

REFERENCES

Allen, P. N., and H. F. Noller. 1989. Mutations in ribosomal proteins S4 and S12 influence higher order structure of 16S ribosomal RNA. *J. Mol. Biol.* **208:**457–468.

Barta, A., G. Steiner, J. Brosius, H. F. Noller, and E. Kuechler. 1984. Identification of a site on 23S ribosomal RNA located at the peptidyl transferase center. *Proc. Natl. Acad. Sci. USA* **81:**3607–3611.

Brimacombe, R., J. Atmadja, W. Stiege, and D. Schüler. 1988. A detailed model of the three-dimensional structure of *Escherichia coli* 16S ribosomal RNA *in situ* in the 30S subunit. *J. Mol. Biol.* **199:**115–136.

Caskey, C. T., W. C. Forrester, and W. Tate. 1987. Peptide chain termination, p. 149–158. *In* B. F. C. Clark and H. U. Petersen (ed.), *Gene Expression: the Translational Step and Its Control.* Munksgaard, Copenhagen.

Cech, T. R. 1987. The chemistry of self-splicing RNA and RNA enzymes. *Science* **236:**1532–1539.

Craigen, W. J., and C. T. Caskey. 1987. The function, structure and regulation of *E. coli* peptide chain release factors. *Biochimie* **69:**1031–1041.

Dahlberg, A. E. 1989. The functional role of ribosomal RNA in protein synthesis. *Cell* **57:**525–529.

Dams, E., L. Hendriks, Y. Van de Peer, J.-M. Neefs, G. Smits, I. Vandenbempt, and R. De Wachter. 1988. Compilation of small ribosomal subunit RNA sequences. *Nucleic Acids Res.* **16:** r87–r173.

De Stasio, E. A., H. U. Göringer, W. E. Tapprich, and A. E. Dahlberg. 1988. Probing ribosome function through mutagenesis of ribosomal RNA, p. 17–41. *In* M. F. Tuite, M. Bolotin-Fukuhara, and M. Picard (ed.), *Genetics of Translation.* NATO ASI Series H, vol. 14. Springer-Verlag KG, Berlin.

Gutell, R. R., B. Weiser, C. R. Woese, and H. F. Noller. 1985. Comparative anatomy of 16S-like ribosomal RNA. *Prog. Nucleic Acid Res. Mol. Biol.* **32:**155–166.

Hui, A., and H. A. de Boer. 1987. Specialized ribosome system: preferential translation of a single mRNA species by a subpopulation of mutated ribosomes in *Escherichia coli. Proc. Natl. Acad. Sci. USA* **84:**4762–4766.

Jacob, W. F., M. Santer, and A. E. Dahlberg. 1987. A single base change in the Shine-Dalgarno region of 16S rRNA of *Escherichia coli* affects translation of many proteins. *Proc. Natl. Acad. Sci. USA* **84:**4757–4761.

Kurland, C. G., and M. Ehrenberg. 1984. Optimization of translation accuracy. *Prog. Nucleic Acid Res. Mol. Biol.* **31:**191–219.

Liu, J., and J. S. Parkinson. 1989. Genetics and sequence analysis of the *pcnB* locus, an *Escherichia coli* gene involved in plasmid copy number control. *J. Bacteriol.* **171:**1254–1261.

Lopilato, J., S. Bortner, and J. Beckwith. 1986. Mutations in a new chromosomal gene of *Escherichia coli* K-12, *pcnB*, reduce plasmid copy number of pBR322 and its derivatives. *Mol. Gen. Genet.* **205:**285–290.

Moazed, D., and H. F. Noller. 1986. Transfer RNA shields specific nucleotides in 16S ribosomal RNA from attack by chemical probes. *Cell* **47:**985–994.

Moazed, D., and H. F. Noller. 1989. Interaction of tRNA with 23S rRNA in the ribosomal A, P, and E sites. *Cell* **57:**585–597.

Moazed, D., S. Stern, and H. F. Noller. 1986. Rapid chemical probing of conformation in 16S ribosomal RNA and 30S ribosomal subunits using primer extension. *J. Mol. Biol.* **187:** 399–416.

Moore, P. B. 1985. Polypeptide polymerase: the structure and function of the ribosome in 1985. *Proc. Robert A. Welch Found.* **29:**185–214.

Murgola, E. J. 1985. tRNA, suppression, and the code. *Annu. Rev.*

Genet. **19**:57–80.

Murgola, E. J., H. U. Göringer, A. E. Dahlberg, and K. A. Hijazi. 1989. Ribosomal RNA and UGA-dependent peptide chain termination. *UCLA Symp. Mol. Cell. Biol. N. Ser.* **94**:221–229.

Murgola, E. J., K. A. Hijazi, H. U. Göringer, and A. E. Dahlberg. 1988. Mutant 16S ribosomal RNA: a codon-specific translational suppressor. *Proc. Natl. Acad. Sci. USA* **85**:4162–4165.

Noller, H. F., S. Stern, D. Moazed, T. Powers, P. Svensson, and L.-M. Changchien. 1987. Studies on the architecture and functions of 16S rRNA. *Cold Spring Harbor Symp. Quant. Biol.* **52**:695–708.

Noller, H. F., and C. R. Woese. 1981. Secondary structure of 16S ribosomal RNA. *Science* **212**:403–411.

Ofengand, J., J. Ciesiolka, R. Denman, and K. Nurse. 1986. Structural and functional interactions of the tRNA-ribosome complex, p. 473–494. *In* B. Hardesty and G. Kramer (ed.), *Structure, Function, and Genetics of Ribosomes.* Springer-Verlag, New York.

Powers, T., S. Stern, L.-M. Changchien, and H. F. Noller. 1988. Probing the assembly of the 3′ major domain of 16S rRNA *J. Mol. Biol.* **201**:697–716.

Shen, Z., and T. D. Fox. 1989. Substitution of an invariant nucleotide at the base of the highly conserved '530-loop' of 15S rRNA causes suppression of yeast mitochondrial ochre mutations. *Nucleic Acids Res.* **17**:4535–4539.

Shine, J., and L. Dalgarno. 1974. The 3′-terminal sequence of *Escherichia coli* 16S ribosomal RNA: complementarity to nonsense triplets and ribosome binding sites. *Proc. Natl. Acad. Sci. USA* **71**:1342–1346.

Steege, D. A., and D. G. Söll. 1979. Suppression, p. 433–485. *In* R. F. Goldberger (ed.), *Biological Regulation and Development*, vol. 1. Plenum Publishing Corp., New York.

Weiss, R. B., D. M. Dunn, A. E. Dahlberg, J. F. Atkins, and R. F. Gesteland. 1988. Reading frame switch caused by base pair formation between the 3′ end of 16S rRNA and the mRNA during elongation of protein synthesis in *Escherichia coli. EMBO J.* **7**:1503–1507.

Yamao, F., A. Muto, Y. Kawauchi, M. Iwami, S. Iwagami, Y. Azumi, and S. Osawa. 1985. UGA is read as tryptophan in *Mycoplasma capricolum. Proc. Natl. Acad. Sci. USA* **82**:2306–2309.

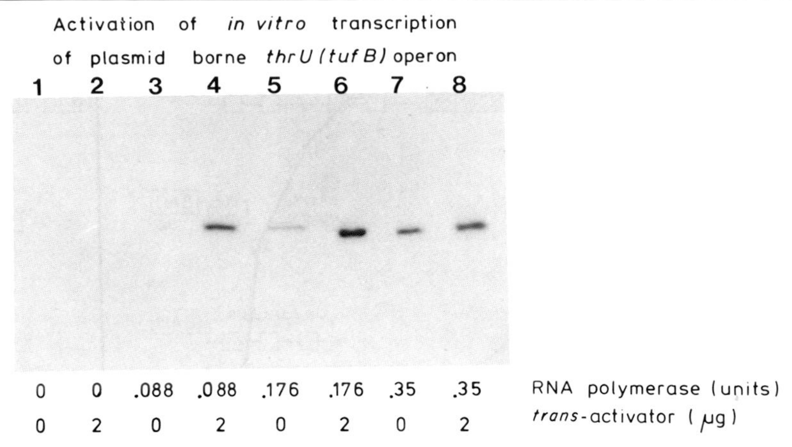

Figure 3. Stimulation by the protein of in vitro transcription of the plasmid-borne *thrU(tufB)* operon. In vitro transcription was performed with *E. coli* RNA polymerase and a supercoiled DNA template derived from a plasmid harboring the operon. After transcription, the reaction mixture was extracted with phenol-chloroform and the RNA product was hybridized with an end-labeled DNA probe (+26 to +109). The hybrid was treated with S1 nuclease and analyzed by electrophoresis on a 7% (wt/vol) polyacrylamide-urea gel (compare Lamond, 1985).

UAS compete for binding to the protein (Fig. 5). Increasing the amount of unlabeled DNA results in the disappearance of first complex III, followed by complexes II and I successively. A similar competition was observed between DNA fragments of the *tyrT* UAS and DNA fragments of the *thrU(tufB)* UAS. The specific binding of one and the same protein to the three *cis*-acting regions strongly suggests that transcription of the three operons is controlled by a similar if not identical *trans*-activation mechanism.

How general is this novel regulation mechanism? An attractive possibility is that most and perhaps all rRNA and tRNA operons share a common *trans* activator, enabling coordination of operon expression. In agreement with such a possibility is the striking sequence conservation of the promoter upstream regions of the seven *rrn* operons. The UBP-binding sites on a single UAS, although similar, are not identical. In the case of the *thrU(tufB)* operon, this finding suggests that the three sites differ in affinity for UBP. A common feature of the UBP target sequences is that they are A+T rich and display bending or kinking of the DNA helix (Gourse et al., 1986; Bossi and Smith, 1984). Footprinting with DNase I indicates that this bending becomes more pronounced upon UBP binding (Vijgenboom, 1989). Possibly, DNA bending facilitates the binding of the RNA polymerase to its binding site. Furthermore, the extent of DNA bending may be a factor in determining the extent of *trans* activation, which may differ for the various stable RNA operons. However, at present we do not know the number or the nature of the proteins involved. Their role in the regulation of transcription initiation remains to be established.

trans ACTIVATION OF TRANSCRIPTION SHOWS LARGE VARIATIONS DURING THE BACTERIAL GROWTH CYCLE

By fusing the *thrU(tufB)* operon to the promoterless *galK* gene, promoter activity was measured

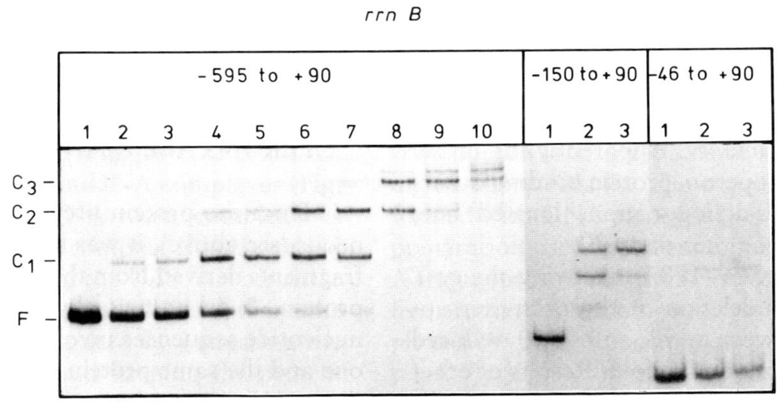

Figure 4. Protein binding to DNA fragments derived from the UAS of the *rrnB* operon. Three end-labeled DNA fragments, comprising positions −595 to +90, −150 to +90, and −46 to +90, were incubated with increasing amounts of the EF-Tu · GDP preparation. Reaction mixtures were analyzed as described by Vijgenboom et al. (1988).

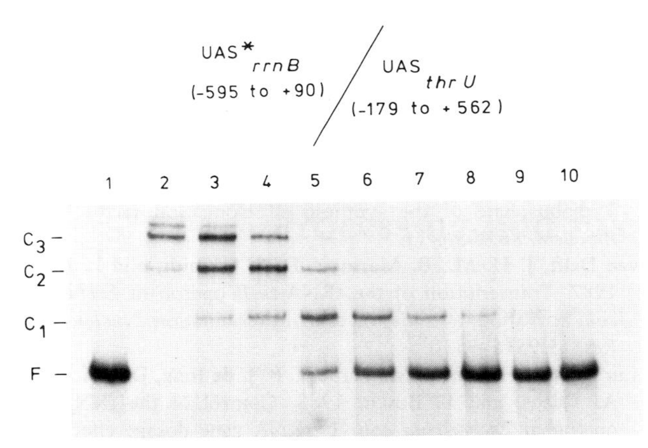

Figure 5. Competition for protein binding between DNA fragments derived from the *rrnB* UAS and DNA fragments derived from the *thrU(tufB)* UAS. A constant amount of end-labeled DNA fragment from the *rrnB* UAS (−98 to +50) was incubated with increasing amounts of nonlabeled DNA fragment from the *thrU(tufB)* UAS. For further experimental details, see the legend to Fig. 4.

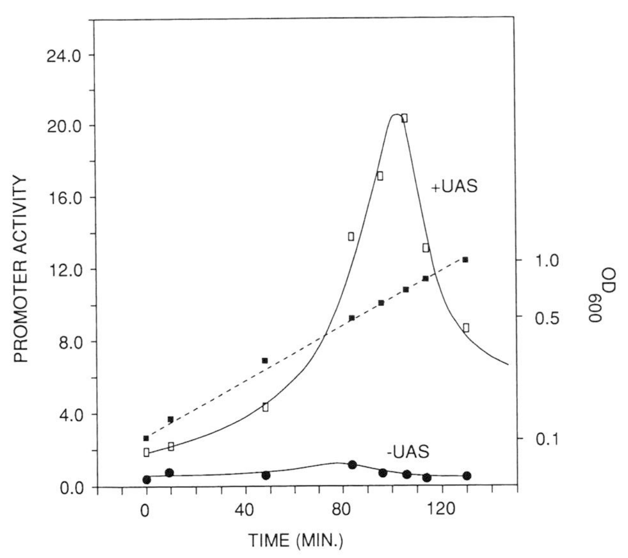

Figure 6. Variation of promoter activity of the plasmid-borne *thrU(tufB')-galK* operon fusion during the growth cycle. Promoter activities were determined by measuring galactokinase activities of cells transformed with pDS10 or a UAS deletion derivative thereof (see legend to Fig. 2). Transformants were grown in LC medium plus glucose. At zero time, the stationary culture was resuspended in fresh medium. Symbols: -------, optical density at 600 nm (OD_{600}); ——, promoter activity of transformants as indicated.

under different physiological conditions. It was found that this activity can vary considerably (Fig. 6). When growth of stationary cultures is reinitiated upon dilution in fresh medium, a burst of transcription of the *thrU(tufB)* operon is observed, which is followed by a rapid decline. This burst in promoter activity is not seen upon deletion of the UAS, showing that it is *trans* activation of transcription that displays this rather dramatic change (Nilsson et al., 1990). A similar burst is observed after a nutritional shift-up (our unpublished results).

Great variations in the activities of the P1 and P2 promoters of the *rrnB* operon have been described by Lukacsovich et al. (1987). Although they did not report on the mechanism involved and did not describe the DNA determinants in detail, they were aware of the fact that they were dealing with "a hitherto poorly characterized type of regulation that is associated with outgrowth from the stationary phase or metabolic shift-up." The great similarity in the behavior of the *rrnB* operon and that of the *thrU(tufB)* operon confirms that the two operons are under the control of the same mechanism.

We believe that the *trans*-activation mechanism meets the cellular demands in rRNA and tRNA under changing environmental conditions with great efficiency. Changes can be very rapid. Modulation of promoter activity may be affected by changes in the level of *trans*-activator protein. Preliminary observations in this laboratory suggest that the UBP concentration drops to low levels when the cells enter the stationary phase (our unpublished experiments). The intrinsic strength of the stable RNA promoters being rather low (compare our introductory comments), modulation of the UAS-dependent *trans* activation may modulate transcription within relatively wide margins.

CONCLUDING REMARKS

The novel regulatory mechanism controlling transcription of the *rrnB*, *tyrT*, and *thrU(tufB)* operons clearly differs from ribosome feedback and growth rate-dependent regulation or stringent control. The latter regulations repress transcription, in contrast to the activation by the mechanism described here. Conceivably, the operation of two opposing mechanisms creates optimal conditions for fine tuning the synthesis of stable RNA under various physiological conditions.

These investigations have been supported in part by the Commission of the European Communities, Biotechnology Action Programme, Directorate-General "Science, Research and Development," Brussels. L.N. is the recipient of a long-term EMBO fellowship.

Our thanks are due to A. Travers, J. H. M. van Delft, and R. L. Gourse for providing the plasmids harboring, respectively, the *tyrT*, *thrU(tufB)*, and *rrnB* operons and their deletion derivatives. We also thank B. Kraal for many helpful discussions and suggestions.

ADDENDUM IN PROOF

UBP has been identified as FIS (factor for inversion stimulation) (Nilsson et al., 1990).

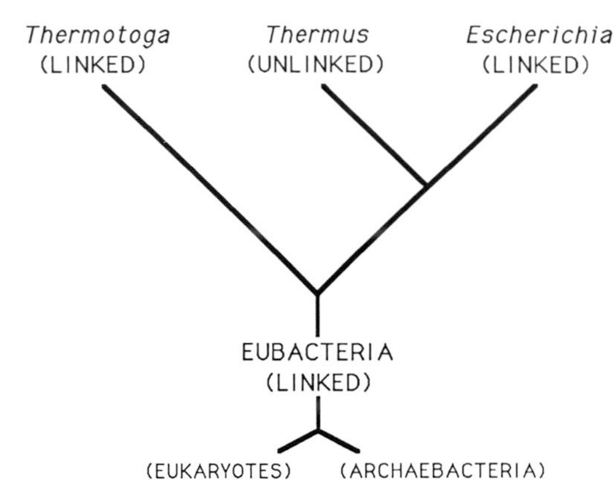

Figure 3. A phylogeny of rRNA gene organization. The network corresponds to the phylogenetic branching order of the indicated organisms, according to rRNA sequence comparisons. In *Thermotoga maritima* (Achenbach-Richter, 1988) and in *E. coli* (King et al., 1986), the rRNA genes are physically closely associated (text) and so are referred to as linked; in *Thermus thermophilus*, the 16S gene is uncoupled physically and probably transcriptionally from the 23S and 5S rRNA (Hartmann et al., 1987).

other rRNA genes has not been detected in eucaryotes, but only a few early-diverging representatives of this, the third primary lineage, have been inspected for that trait. The diversity of rRNA gene organization has been studied mostly with representatives of the eubacteria. Consequently, the ancestral character of the linked state is clearest there. Figure 3 shows the evolutionary relationships based on rRNA sequences of some eubacterial genera (Woese, 1987) and the types of rRNA gene organization in those organisms. As indicated in the figure, the rRNA operon of *Thermus thermophilus* is interrupted between the 16S and 23S rRNA genes, and there is evidence that the 16S and still-linked 23S-5S unit are independently transcribed (Hartmann et al., 1987). Yet in *E. coli*, a more peripherally branching organism in the rRNA relatedness tree, and in *Thermotoga maritima*, a more deeply branching organism, the rRNA genes are closely associated (King et al., 1986; Achenbach-Richter, 1988). The close physical association of the *Thermotoga* rRNA genes suggests transcriptional linkage, as occurs in *E. coli*. The most parsimonious conclusion from these considerations is that the linked state is ancestral in the eubacteria. The occurrence of the linked state in the archaebacteria (Fig. 2) as well suggests that the rRNA genes in the common ancestor of both archaebacteria and eubacteria also were linked.

EXCISION OF INTERVENING SEQUENCES FROM BACTERIAL 23S rRNA

The nucleolytic processing of rRNAs sometimes occurs not only at their termini but also within the mature sequences. In some cases, introns are excised and the integrity of the rRNA phosphodiester chain is restored by splicing (Cech, 1987). In other cases, extra sequences within the rRNAs are excised during maturation, but the break in the rRNA chain is not repaired. The result is a fragmented rRNA. The break in the rRNA phosphodiester chain evidently does not affect the function of the ribosome. The global integrity of such a fragmented rRNA presumably is maintained by secondary and tertiary structure.

An extreme example of a fragmented rRNA is the LSU rRNA of trypanosomatids, which consists of six separate rRNA species that correspond in their aggregate molecular length to 23S-like rRNAs in other organisms (Spencer et al., 1987; Campbell et al., 1987). Fragmented LSU rRNAs have long been recognized in insects (Lanversin and Jacq, 1989, and references therein) and in the crustacean *Artemia salina* (Nelles et al., 1984). The occurrence of the eucaryotic 5.8S rRNA is another example of such fragmentation (reviewed in Walker and Pace, 1983). This RNA is homologous to the 5′ end of the bacterial 23S rRNA, yet the 5.8S rRNA gene is commonly (but not always; see Vossbrinck et al., 1987) separated from the rest of the 23S-like sequence (e.g., the mammalian 28S rRNA; Fig. 1) by hundreds of base pairs of an evolutionary conserved internal transcribed spacer. Removal of the spacer during maturation of the rRNA results in the 5.8S and 28S (in mammals) rRNAs. This 23S-like rRNA is thereby fragmented relative to its bacterial counterpart.

Fragmented 23S rRNAs also occur in some bacteria (reviewed in Pace, 1973). The cause of the fragmentation has been unknown; it was not even clear whether the fragmentation is specific or rather is due to nonspecific degradation during isolation of the RNA. We investigated this phenomenon and discovered that the long-reported nicked 23S rRNAs in bacteria can result from the excision of intervening sequences that seem analogous to internal transcribed spacers (Burgin et al., 1990).

It was previously reported by Winkler (1979) that intact 23S rRNA is not recovered from *Salmonella typhimurium*. A screen of the sizes of rRNAs from several other *Salmonella* spp. by denaturing gel electrophoresis revealed that some of those, too, lack 23S rRNA (Burgin et al., 1990; Smith et al., 1988). Instead, smaller RNAs with the abundance and cumulative molecular length of 23S rRNA (ca. 2,900 nucleotides) are observed (Fig. 4A). The RNAs from *E. coli* indicate the positions in the gel of the canonical 16S (1.6-kilobase [kb]) and 23S (2.9-kb) rRNAs. *S. typhimurium* (e23566) entirely lacks the 23S rRNA band but contains discrete fragments (2.4, 1.7,

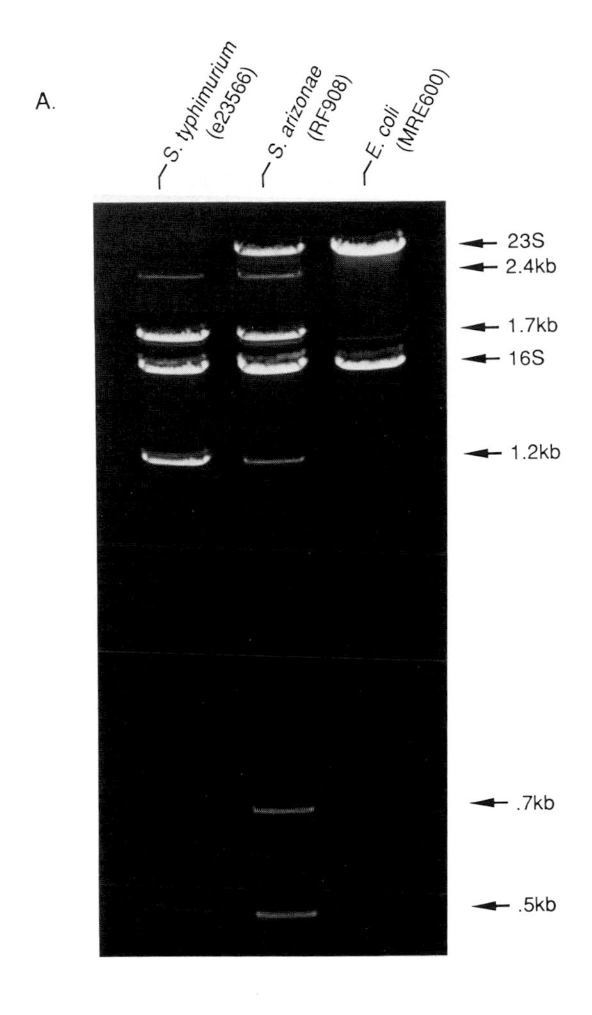

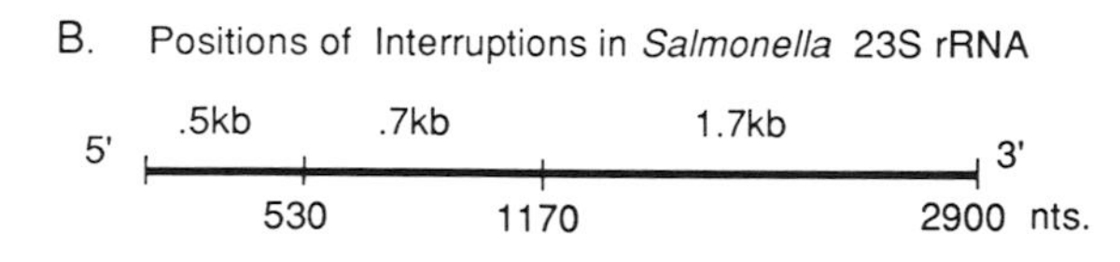

Figure 4. Demonstration that *Salmonella* spp. have fragmented 23S rRNAs. (A) Bacterial cultures were grown, and RNA was extracted and resolved by polyacrylamide gel electrophoresis as described by Burgin et al. (1990). The gel was stained with ethidium bromide and photographed. *E. coli* 16S and 23S rRNA and *Salmonella* 23S rRNA fragments are indicated by arrows. (B) The potential breakage points in the *Salmonella* 23S rRNA that account for the observed fragment sizes are indicated. nts., Nucleotides.

1.2, 0.7, and 0.5 kb in size) that are not present in *E. coli. Salmonella arizonae* contains some intact 23S rRNA but in reduced amounts relative to the level in *E. coli*, as well as the same array of fragments that are obtained from *S. typhimurium.* All of the fragments seen in Fig. 4A are present in the same stoichiometries after treatment of bacterial cultures with rifampin (an inhibitor of RNA synthesis) before harvest (data not shown). Thus, the abundant fragments are all metabolically stable species, not intermediates of transcription or processing. Fragmented 23S rRNAs were encountered in other *Salmonella* spp., including *S. strasbourg* (RF761), *S. freetown* (RF838), *S. bern* (RF844), *S. djakarta* (RF868), *S. flint* (RF870), and *S. wassenaar* (RF873 and RF874) (Burgin et al., 1990). The sizes of the putative 23S rRNA fragments in these organisms are the same as seen in *S. typhimurium* and *S. arizonae*; however, the relative stoichiometries of the fragments vary from organism to organism.

A Northern (RNA) blot analysis (not shown), using hybridization probes specific for the appropriate segments of the *E. coli* 23S rRNA gene, showed that the RNA fragments seen in Fig. 4A indeed correspond to discretely fragmented 23S rRNA. Figure 4B summarizes the distribution of breakage points in the *Salmonella* 23S rRNA that can account for the observed fragment sizes and relative abundances. The fragmentation patterns can be understood as the result of two breakage points in the 23S rRNA and heterogeneity with regard to the number of breaks present in particular 23S rRNA molecules. Thus, one type of fragmented 23S rRNA (2.9 kb in total) would consist of 0.5- and 2.4-kb fragments, another would consist of 1.2- and 1.7-kb fragments, and a third would consist of 0.5-, 0.7-, and 1.7-kb fragments. The heterogeneity in the fragment patterns would derive from the occurrence of breaks in only some of the transcripts of the multiple rRNA operons in the genomes of the *Salmonella* spp. (The genomes of *Salmonella* spp. are nearly identical to that of *E. coli* and so are expected to contain seven rRNA operons [Pace and Pace, 1971].)

To elucidate the basis for the fragmentation of the 23S rRNAs, single rRNA operons from *S. typhimurium* and *S. arizonae* were cloned and the relevant regions of the 23S rRNA genes were sequenced (Burgin et al., 1990). The mature rRNA sequences from the two *Salmonella* spp. are nearly identical to one another and to the sequence of the 23S rRNA gene of *E. coli* (97% similarity to each other and to the *E. coli* sequence). However, both of the *Salmonella* rRNA genes contain sequence blocks that are present neither in the *E. coli* 23S rRNA genes nor in the 23S rRNA genes of some of the other *Salmonella* spp. The cloned *S. typhimurium* 23S rRNA gene contains an extra sequence, approximately 90 base pairs in length, at nucleotide position approximately 1170 (*E. coli* numbering). The cloned *S. arizonae* gene contains two novel elements, one at position approximately 530 and another, as in *S. typhimurium*, at position approximately 1170. The sequences and possible secondary structures of the transcripts of the novel *Salmonella* elements are shown in Fig. 5,

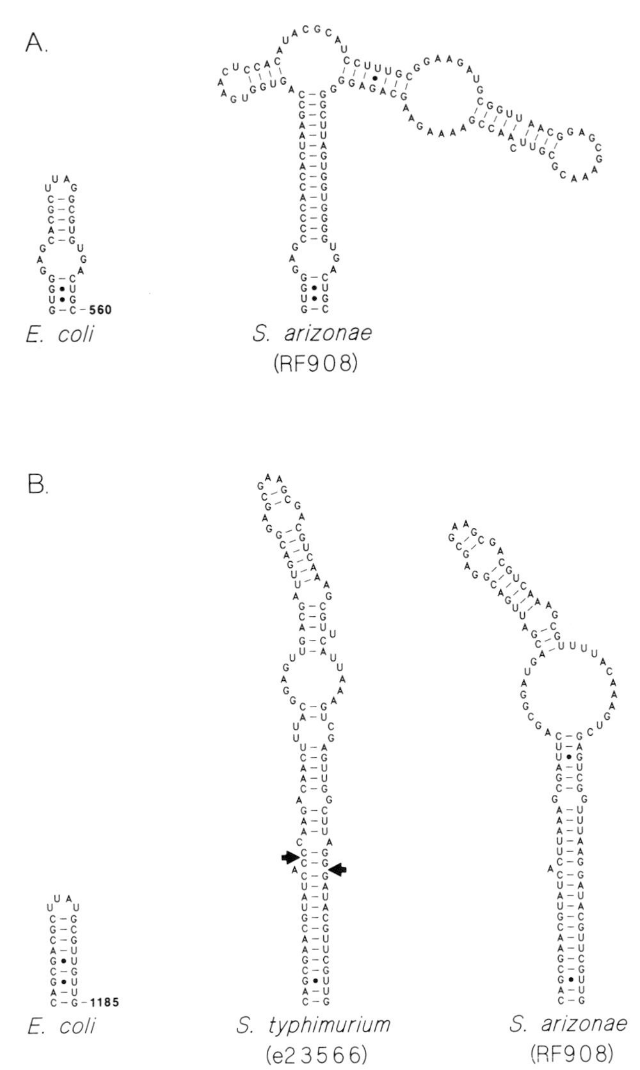

Figure 5. Nucleotide sequences and possible secondary structures of the novel *Salmonella* spp. elements. Numbers indicate nucleotide positions in *E. coli* 23S rRNA. (A) Element in *S. arizonae* at position ca. 530 and the corresponding region in *E. coli* 23S rRNA. (B) Elements at position ca. 1170 and the corresponding region in *E. coli*. Arrows indicate the positions of in vivo cleavage sites that generate the 1.2-kb and 1.7-kb fragments in *S. typhimurium*.

along with the corresponding 23S rRNA regions of *S. typhimurium* or *E. coli* that lack the elements. None of the extra sequence elements contains an open reading frame for protein synthesis.

The excision of the extra sequence at nucleotide position approximately 1170 in the *S. typhimurium* 23S rRNA has been characterized in detail (Burgin et al., 1990). The terminal sequences of the rRNA fragments that result from the excision were determined by primer extension and by direct sequencing. These termini, the sites of action of the enzyme that releases the extra sequence, are indicated by arrows in Fig. 5. According to the secondary structure model in Fig. 5, the cleavages occur in the opposite strands of a duplex stem that is an extension of a conserved stem in intact 23S rRNA. Although there are approximately 84 extra nucleotides in this region of the *S. typhimurium* sequence compared with that of *E. coli*, only 78 nucleotides are removed. Six nucleotides originating from the extra sequence in the *S. typhimurium* pre-rRNA remain in the processed rRNA fragments.

The novel, intervening sequence elements in the *Salmonella* rRNA genes and in the pre-rRNAs are intronlike in that they interrupt the linear continuity of the 23S rRNA gene, and they are excised during posttranscriptional processing of the rRNA. The excision of "conventional" introns would be followed by ligation of the resulting termini. However, ligation of these rRNA fragments is not necessary for the integrity of the rRNA superstructure because the fragments remain associated through secondary and tertiary contacts between them. The excised sequences do not fall into the category of internal transcribed spacers because they interrupt a gene transcript that is normally intact, and they occur sporadically. For brevity, we hereafter refer to these intervening sequences in the DNA and in the RNA as IVSs.

The two cleavages indicated by arrows in Fig. 5 resemble those carried out by RNase III (see above), which indeed proved to be the agent responsible for removal of that IVS from the 23S rRNA, thereby fragmenting the rRNA (Burgin et al., 1990). The involvement of RNase III was determined in vivo by inspection of the fragmentation pattern of the *S. typhimurium* rRNA expressed in *E. coli* strains that contain or lack RNase III activity. The *S. typhimurium* 23S rRNA is fragmented in normal *E. coli*, but it is intact in a mutant with a defective RNase III. Moreover, the IVS at nucleotide position approximately 1170 was shown to be specifically excised from 23S RNA in vitro by purified RNase III.

IVSs AND THE EVOLUTION OF rRNA OPERONS

If the occurrence of fragmented 23S rRNA is diagnostic for the presence of IVSs such as described here, then such elements seem to be widespread in bacteria. 23S rRNAs that are fragmented (or absent) have been reported for *Agrobacterium* and *Rhodobacter* spp., which, like *Salmonella* spp., are representatives of the purple bacteria phylogenetic group, one of the dozen or so major phylogenetic groups of eubacteria so far outlined by rRNA sequence comparisons (Woese, 1987). Fragmented 23S

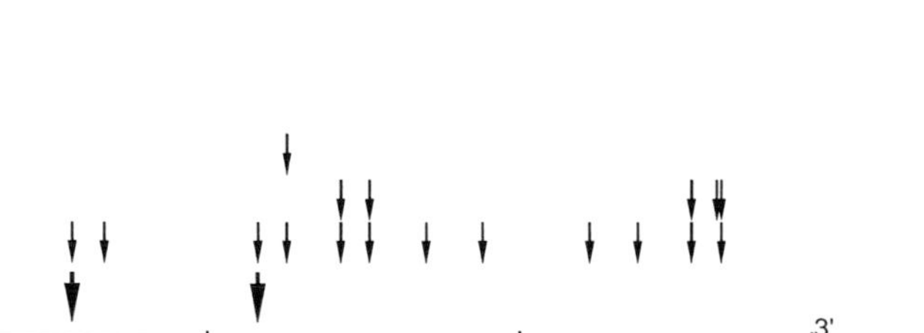

Figure 6. Variable regions in LSU rRNAs. The horizontal line represents the eubacterial LSU rRNA. The small arrows indicate the homologous positions at which breaks occur in the LSU rRNAs of eucaryotes as a consequence of the excision of an internal transcribed spacer (ITS II; Gerbi, 1985) or other extra sequences in insects (Lanversin and Jacq, 1989) or trypanosomatids (*Crithidia fasciculata* [Spencer et al., 1987] and *Trypanosoma brucei* [Campbell et al., 1987]) or at which expansion segments commonly occur (Gerbi, 1985). The bold arrows indicate the positions at which occur the *Salmonella* IVSs described.

rRNAs also have been reported to occur in *Anacystis nidulans* (Doolittle, 1973) and *Micrococcus luteus* (D.-C. Yang and C. R. Woese, personal communication). These latter organisms are representatives of still others of the major eubacterial phyla, the cyanobacteria and the gram-positive bacteria, respectively (Woese, 1987). The sources of these fragmentations have not been traced at the sequence level, so it is not certain that they result from the removal of IVSs as described above for *Salmonella* spp.

Although IVSs may occur widely in eubacteria, their occurrence is sporadic. Within a particular *Salmonella* strain, some rRNA operons have one IVS, some operons have two such elements, and others have none. Close relatives of *Salmonella* spp., such as *E. coli* MRE600, do not possess extra sequences that result in fragmentation of their 23S rRNAs. This sporadic occurrence of the IVSs in very closely related bacteria (and within the multiple rRNA operons of the same isolate) not only suggests that the elements have no utility to the cell but also is strong evidence that the IVSs were acquired relatively recently in evolution. It has been argued that fragmented LSU rRNAs in some eucaryotes may reflect the state of the primordial rRNAs (Spencer et al., 1987; Gray et al., 1988). According to that notion, the fragmented rRNAs observed in some modern organisms would be a remnant of an ancestral rRNA composed of the products of multiple genes. However, fragmented rRNAs occur only sporadically in eucaryotes or bacteria, so we believe it is more likely that the common-ancestral LSU RNA was intact.

The origins of the described IVSs are unknown. It is possible that they are, in essence, the footprints of transposable elements that inserted into and then excised from the 23S rRNA genes. The insertion and subsequent excision of transposable elements commonly results in the creation of inverted sequence repeats at the site of the excision (Kleckner, 1981). Such inverted repeats in the DNA would result in the presence of base-paired hairpins in RNA transcribed from that DNA, similar to the manifestation of the IVSs in the *Salmonella* 23S rRNAs. Such extra sequences in the cellular RNAs need not be deleterious to the cell if they reside in nonfunctional sequences or if they are excised.

Whatever their origin, elements such as these IVSs are likely to have been involved in the evolution of rRNA. rRNAs from different organisms differ substantially in length. The eucaryotic 28S rRNA, for instance, is substantially larger than the homologous bacterial 23S rRNA (ca. 4,000 versus ca. 3,000 nucleotides). The larger size of the eucaryotic rRNA is largely due to the occurrence of eucaryote-specific sequence blocks at discrete sites in rRNA genes that otherwise are homologous to the bacterial genes. These eucaryote-specific (sometimes organism-specific) sequences have been called expansion segments (Gerbi, 1985; Hancock and Dover, 1988; Raué et al., 1988). The two characterized *Salmonella* IVSs occur at positions in the 23S rRNA gene that are homologous to two of the positions at which expansion segments occur in eucaryotes (Fig. 6). It seems possible that during the course of evolution, elements analogous to the *Salmonella* IVSs may have been inserted into the rRNA genes and, in the absence of rRNA processing, may have given rise to expansion segments. Such newly acquired sequences would be useless to the cell, at least initially. If not deleterious, the newly acquired sequences could be propagated in a lineage and, in the absence of function, could evolve rapidly. Most expansion segments in eucaryotic rRNAs are highly variable in sequence and occurrence, and so even in extant organisms they may be functionless. Other expansion segments may have acquired function. Some of the expansion segments in eucaryotic large-subunit rRNAs are removed during RNA processing and have been dubbed transcribed spacers or fragmentation spacers (Lanversin and Jacq, 1989). Such spacers would be directly analogous to the *Salmonella* IVSs.

We have termed the extra sequences in the *Salmonella* 23S rRNA genes intervening sequences because they interrupt the normally linear continuity of genes. Clearly the IVSs are not transcribed spacers; they do not separate functionally distinct molecules. In their small sizes, informationless character, and evolutionary volatility, the *Salmonella* IVSs resemble tRNA introns in eucaryotes and archaebacteria and many mRNA introns in eucaryotes. Introns in tRNAs and mRNAs differ from these IVSs (and from one another) in their mechanisms of excision from precursor RNAs and in the fact that the sequences flanking the introns are spliced after the excision. However, the mechanisms of excision and the necessity for splicing are not properties of the introns; rather, they are dictated by the host cells and the genes in which the introns reside. Introns that undergo excision coupled with splicing are a subset of a broader group of similar elements that would include the *Salmonella* IVSs, some internal transcribed spacers, untranslated intragenic sequences (Huang et al., 1988), and insertion elements such as transposons. All of these elements are similar in that they can interrupt the normally linear continuity of gene sequences. Collectively, such elements must have had substantial impact on the course of the evolution of genomes.

Our research is supported by Public Health Service grant GM34527 from the National Institutes of Health and by grant N14-87-K-0813 from the Office of Naval Research.

We thank Jim Brown for assistance with artwork.

REFERENCES

Achenbach-Richter, L. 1988. Phylogenetic relationships of thermophilic bacteria. Ph.D. thesis. University of Illinois, Urbana.

Björk, G. R. 1984. Modified nucleosides in RNA—their formation and function, p. 291–320. *In* D. Apirion (ed.), *Processing of RNA*. CRC Press, Inc., Boca Raton, Fla.

Burgin, A. B., K. Parodos, D. J. Lane, and N. R. Pace. 1990. The excision of intervening sequences from *Salmonella* 23S ribosomal RNA. *Cell* **60:**405–414.

Campbell, D. A., K. Kubo, C. G. Clark, and J. C. Boothroyd. 1987. Precise identification of cleavage sites involved in the unusual processing of trypanosome ribosomal RNA. *J. Mol. Biol.* **196:**113–124.

Cech, T. 1987. The chemistry of self-splicing RNA and RNA enzymes. *Science* **236:**1532–1539.

Doolittle, W. F. 1973. Postmaturational cleavage of 23S ribosomal ribonucleic acid and its metabolic control in the blue-green alga *Anacystis nidulans*. *J. Bacteriol.* **113:**1256–1263.

Dunn, J. J. 1976. RNase III cleavage of single-stranded RNA: effect of ionic strength on the fidelity of cleavage. *J. Biol. Chem.* **251:**3807–3814.

Gegenheimer, P., and D. Apirion. 1981. Processing of procaryotic ribonucleic acid. *Microbiol. Rev.* **45:**502–541.

Gegenheimer, P., N. Watson, and D. Apirion. 1977. Multiple pathways for primary processing of ribosomal RNA in *Escherichia coli*. *J. Biol. Chem.* **252:**3064–3073.

Gerbi, S. A. 1985. Evolution of ribosomal DNA, p. 419–517. *In* R. J. MacIntyre (ed.), *Evolution of Ribosomal RNA*. Plenum Publishing Corp., New York.

Gray, N. W., B. H. Poppo, J. C. Collings, T. K. Heinenen, D. F. Spencer, and M. N. Schnare. 1988. Ribosomal RNA genes in pieces, p. 521–530. *In* A. Kotyk, J. Skoda, U. Poces, and V. Kostka (ed.), *Highlights of Modern Biochemistry*. VSP International Publishers, Zeist, The Netherlands.

Hadjiolov, A. A. 1985. *The Nucleolus and Ribosome Biogenesis*. Springer-Verlag, New York.

Hancock, J. M., and G. A. Dover. 1988. Molecular coevolution among cryptically simple expansion segments of eukaryotic 26S/28S rRNAs. *Mol. Biol. Evol.* **5:**377–391.

Hartmann, R. K., N. Ulbrich, and V. A. Erdmann. 1987. An unusual rRNA operon constellation: in *Thermus thermophilus* HB8 the 23S/5S rRNA operon is a separate entity from the 16S rRNA operon. *Biochimie* **69:**1097–1104.

Huang, W. H., S. Ao, S. Casjens, R. Orlandi, R. Zeikus, R. Weiss, D. Winge, and M. Fang. 1988. A persistent untranslated sequence within bacteriophage T4 DNA topoisomerase gene 60. *Science* **239:**1005–1012.

King, T. C., R. Sirdeskmukh, and D. Schlessinger. 1986. Nucleolytic processing of ribonucleic acid transcripts in procaryotes. *Microbiol. Rev.* **50:**428–451.

Kleckner, N. 1981. Transposable elements in prokaryotes. *Annu. Rev. Genet.* **15:**341–404.

Lanversin, G., and B. Jacq. 1989. Sequence and secondary structure of the central domain of *Drosophila* 26S rRNA: a universal model for the central domain of the large rRNA containing the region in which the central break may happen. *J. Mol. Evol.* **28:**403–417.

Nelles, L., C. Van Broeckhoven, R. de Wachter, and A. Vandenberghe. 1984. Location of the hidden break in the large subunit ribosomal RNA of *Artemia salina*. *Naturwissenschaften* **71:** 634–635.

Neumann, H., A. Gierl, J. Tu, J. Liebrock, D. Staiger, and W. Zillig. 1983. Organization of the genes for ribosomal RNA in archaebacteria. *Mol. Gen. Genet.* **192:**66–72.

Pace, B., and N. R. Pace. 1971. Gene dosage for 5S ribosomal ribonucleic acid in *Escherichia coli* and *Bacillus megaterium*. *J. Bacteriol.* **105:**142–149.

Pace, N. R. 1973. Structure and synthesis of the ribosomal ribonucleic acid of procaryotes. *Bacteriol. Rev.* **37:**562–603.

Pace, N. R. 1984. Protein-polynucleotide recognition and the RNA processing nucleases in prokaryotes, p. 1–30. *In* D. Apirion (ed.), *Processing of RNA*. CRC Press, Inc., Boca Raton, Fla.

Pace, N. R., and B. Pace. 1990. Ribosomal RNA terminal maturase: ribonuclease M5 from Bacillus subtilis. *Methods Enzymol.* **181:**366–374.

Pace, N. R., and D. Smith. 1990. Ribonuclease P: function and variation. *J. Biol. Chem.* **265:**3587–3590.

Raué, H. A., J. Klootwijk, and W. Musters. 1988. Evolutionary conservation of structure and function of high molecular weight ribosomal RNA. *Prog. Biophys. Mol. Biol.* **51:**77–129.

Smith, N. H., D. B. Crichton, D. C. Old, and C. F. Higgins. 1988. Ribosomal RNA patterns of *Escherichia coli*, *Salmonella typhi-*

murium and related entrobacteriaceae. *J. Med. Microbiol.* **26:** 223–228.

Spencer, D. F., J. C. Collings, M. N. Schnare, and N. W. Gray. 1987. Multiple spacer sequences in the nuclear large subunit ribosomal RNA gene of *Crithidia fasciculata*. *EMBO J.* **6:** 1063–1071.

Tu, J., and J. Zillig. 1982. Organization of rRNA structural genes in the archaebacterium *Thermoplasma acidophilum*. *Nucleic Acids Res.* **10:**7231–7245.

Vossbrinck, C. R., J. V. Maddox, S. Friedman, B. A. Debrunner-Vossbrinck, and C. R. Woese. 1987. Ribosomal RNA sequence suggests microsporidia are extremely ancient eukaryotes. *Nature* (London) **326:**411–414.

Walker, T. A., and N. R. Pace. 1983. 5.8S ribosomal RNA. *Cell* **33:**320–322.

Winkler, M. E. 1979. Ribosomal ribonucleic acid isolated from *Salmonella typhimurium*: absence of the intact 23S species. *J. Bacteriol.* **139:**842–849.

Woese, C. R. 1987. Bacterial evolution. *Microbiol. Rev.* **51:** 221–271.

Chapter 36

rRNA Processing in *Escherichia coli*

ANAND K. SRIVASTAVA and DAVID SCHLESSINGER

During the assembly of functional particles, rRNA undergoes a variety of metabolic steps that are included under the general heading of processing. Studies of processing are now venerable parts of molecular biology, but the significance of rRNA processing, and therefore of the studies of processing, has remained questionable.

By 1971, both indirect measurements of the transcription time of rRNA and direct observation of species of rRNA found in growing and in chloramphenicol-treated cells had suggested that like other RNA molecules, whether stable structural entities or unstable mRNAs, rRNA is transcribed in large precursor molecules. Extra sequences must be removed to produce mature rRNA. At a 1971 meeting in Aarhus, there were already relevant discussions of the possible significance of the RNA metabolism associated with processing. The dominant work in the field at that time (and in many respects since then) was the preeminent discovery by Nomura and his colleagues that rRNA and ribosomal proteins could be mixed together to reconstitute active ribosomes. (The disparity in studies at the time can be appreciated if one considers that another major group at the meeting reported that the wrong ratios of sodium and potassium ions irreversibly inactivated their preparations of ribosomes. Sol Spiegelman commented at the time, "He [Nomura] takes the whole thing apart and puts it back together, and these guys can't even manage the concentration of sodium chloride.") Since the results of Traub and Nomura (1969) clearly demonstrated that mature rRNA and proteins were enough to form a ribosome, it seemed likely that pre-rRNAs were an epiphenomenon of only peripheral interest. At that meeting, our group suggested that precursor sequences might nevertheless be important, perhaps in providing a favorable route to the formation of the mature rRNA and ribosomes; otherwise, why would the cell make long precursors if it wanted mature 16S and 23S rRNAs? But such arguments seemed rather ad hoc. In fact, Wally Gilbert made the countersuggestion at the time that extra sequences in pre-rRNA might be just leftovers of the transcription process that necessarily had to be discarded.

In the intervening years, a number of lines of work have suggested that precursor sequences and processing are important in various ways, although the evidence remains fragmentary. The origins of various suggestions have been given elsewhere; here we will try to give a straightforward overview. Work on the rRNA of *Escherichia coli* is discussed, with only a few comments on other species.

rRNA OPERONS AND THEIR TRANSCRIPTS

The *E. coli* genome contains seven rRNA transcriptional units (*rrnA*, *-B*, *-C*, *-D*, *-E*, *-G*, and *-H*), each at a different chromosomal location (for details, see King et al., 1986). All of the operons have been isolated and characterized, and most have been sequenced. Each has two promoters, P1 and P2, about 120 base pairs apart. The 16S rRNA gene is located about 190 nucleotides downstream of promoter P2 and is followed by a spacer region containing (i) tRNA, (ii) the 23S rRNA, and (iii) another spacer region containing a 5S rRNA sequence. In each operon, 16S and 23S rRNA genes are each present in one copy, but the numbers of tRNA and 5S RNA genes vary. Several different tRNAs (spacer tRNA) are found between 16S and 23S rRNA, and still others can be present distal to the 5S rRNA gene (Lund and Dahlberg, 1977; Morgan et al., 1980). The pathway of processing of the tRNAs is similar to

Anand K. Srivastava and David Schlessinger ■ Department of Molecular Microbiology, Washington University School of Medicine, St. Louis, Missouri 63110.

that of other tRNAs, using RNases P, D, etc. (Altman, 1981; King et al., 1986), and will not be discussed further here.

In all bacteria studied to date, each *rrn* operon is transcribed into long precursor molecules that include all of the spacer elements. In *E. coli*, the 30S pre-rRNA contains 22% such extra sequences. Precursor sequences at the 5′ and 3′ ends of 16S and 23S rRNAs contain complementary sequence tracts that form strong base-paired stems enclosing the sequence of the mature species (Bram et al., 1980; Young and Steitz, 1978). In detail, for *rrnB* the 146 nucleotides upstream of 16S rRNA include 131 involved in stem formation. Those and the 43 nucleotides immediately following mature 16S rRNA are identical in the four operons studied thus far. The stem bracketing 23S rRNA involves 114 nucleotides on the 5′ side and 71 nucleotides 3′ to the 23S nucleotides, and actually it includes eight base pairs involving the 5′- and 3′-terminal nucleotides of mature 23S rRNA. The sequences are conserved in the various operons (Woese, 1987).

The conservation of features of the operon and its transcripts extends to all eubacteria studied thus far; in particular, cotranscription of 16S and 23S rRNAs is the rule, and the long transcripts invariably contain spacer tRNAs and sequences that enclose the mature 16S and 23S rRNA species in double-stranded stems. Unsurprisingly, these features tend to be involved in the regulation of transcription and processing, and thus of ribosome formation.

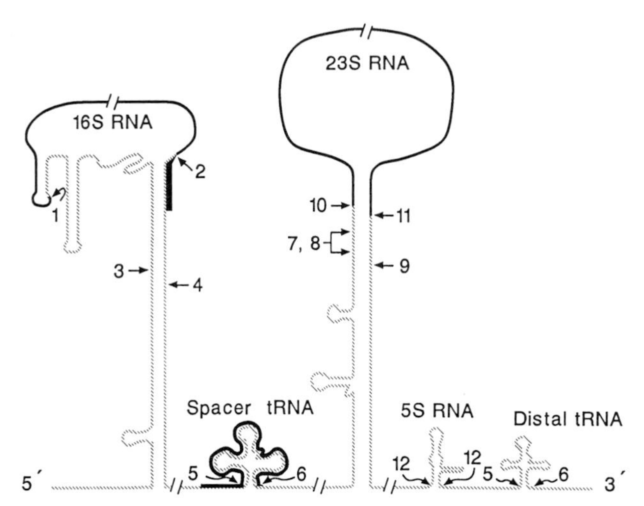

Figure 1. Schematic of the structure of an *rrn* operon and major processing steps for 16S and 23S RNAs in relation to spacer tRNA. The drawing is not to scale so as to indicate by number the positions of primary processing cleavages by RNase III (3, 4, 7, 8, and 9) and secondary processing to produce the mature termini of 16S rRNA (1 [5′ end] and 2 [3′ end]), 23S rRNA (10 [5′ end] and 11 [3′ end]), and 5S rRNA (12). The RNase P cleavage site (5) is shown at the 5′ ends of the tRNAs; one additional point of nuclease cleavage (6) at the 3′ ends is also indicated. Mature 16S and 23S rRNA sequences are indicated by filled lines; other precursor sequences are marked by hatched lines; two regions required for 16S rRNA formation (distal to the 3′ end and from a point upstream through the spacer tRNA; see text) are represented by thick filled lines superimposed on the hatched lines. Additional details of the secondary structures are given by King et al. (1984) and Srivastava and Schlessinger (1989a).

STUDY OF PROCESSING PATHWAYS

In *E. coli* cells, processing is rapid and most rRNA is mature. Only 1 to 2% of rRNA is in the form of very long precursors (King and Schlessinger, 1983). The processing is multistep: (i) initial cleavages lead to formation of intermediate precursors of rRNA from the long transcripts and (ii) maturation steps produce mature termini of rRNA from precursors, either by direct action of single enzymes or with the preliminary action of additional enzymes (Fig. 1).

Initial studies of processing used sucrose gradient and gel electrophoretic fractionation of products of processing. Newer technology now allows the study of processing steps at the nucleotide level, using Northern (RNA) hybridization, nuclease protection assays, primer extension analysis, fingerprinting, etc. Two approaches are especially useful: (i) detection of processing intermediates in steady-state cellular RNA (i.e., following maturation reactions that had occurred in vivo) and (ii) detection of maturation in vitro, using isolated precursor particles as a substrate to define the minimum requirements for the release of pre-rRNA fragments. The studies are facilitated by using plasmids containing cloned rRNA operons and permitting selective and conditional expression of the plasmid-borne rRNA under the control of *E. coli*, lambda, or T7 promoters (Steen et al., 1986). Such plasmids can be reintroduced into the cell in intact or mutated form, and the expression of the rRNA can be studied without interference with or effect on endogenous rRNA metabolism.

PRODUCTION OF p16S AND p23S rRNAs BY RNase III

The isolation of an *E. coli* mutant (AB 301/105) defective in RNase III (Kindler et al., 1973) allowed identification of an endoribonuclease RNase III as an early participant in in vivo processing of rRNA and also permitted the detection of the full 30S pre-rRNA (Dunn and Studier, 1973; Nikolaev et al., 1973), which reaches appreciable steady-state levels in the absence of RNase III. In addition, 30S pre-rRNA was cleaved with purified RNase III in vitro to produce species slightly larger than mature 16S and 23S rRNAs (Dunn and Studier, 1973; Nikolaev et al., 1974). These later proved to be the same as major processing intermediates found in wild-type cells.

In wild-type cells, processing begins even before transcription of the *rrn* operon is complete (Apirion and Gegenheimer, 1984; King and Schlessinger, 1983). RNase III separates precursor 16S (p16S), p23S, ptRNA, and p5S species. Secondary processing events then produce the mature rRNA. A critical difference between the initial RNase III action and final maturation is that accurate RNase III cleavage occurs in vitro with naked pre-rRNA or with preribosomes as the substrate, but final maturation depends on the preformation of the complex of rRNA and ribosomal proteins (Schlessinger, 1980). Thus, while all steps of rRNA processing occur at the level of ribonucleoprotein particles, ribosomal proteins are requisite for only the final steps.

RNase III cuts in the stems that bracket the mature 16S and 23S rRNA sequences (Bram et al., 1980; Young and Steitz, 1978). Comparison of the RNase III-processing sites in different operons shows close sequence homology. Sequence variation between different operons tends to be at positions thought to be unpaired, but much of the unpaired sequences outside of the stems also show strong conservation.

p16S has much longer extra sequences than does p23S at both its 5′ and 3′ termini. RNase III cleavage leaves 115 nucleotides at its 5′ end and 33 at its 3′ end, with a base-paired stem of 26 nucleotides still remaining. In contrast, p23S rRNA is left with only three or seven nucleotides at its 5′ end and seven to nine at its 3′ end (Fig. 1; Sirdeshmukh and Schlessinger, 1985a). This species retains a base-paired stem of only 17 nucleotides, including eight base pairs formed by complementary terminal nucleotides of mature sequence. The difference in the amounts of precursor sequence in the two RNAs may be related to the locations of their termini in ribosomal particles; the 5′ and 3′ ends of mature 16S rRNA are well separated, whereas the termini of 23S rRNA are in close proximity (Brimacombe et al., 1988; Sirdeshmukh and Schlessinger, 1985b).

E. coli cells lacking RNase III do not excise the normal RNA precursors from nascent transcripts as in wild-type cells. Nevertheless, these cells also have p16S and p23S RNAs slightly larger than their wild-type counterparts (Gegenheimer et al., 1977). These species contain additional sequences at both termini and extend beyond the normal sites of RNase III cleavage. These species are probably produced from larger precursors by nucleases acting nonspecifically at single-stranded regions in the absence of RNase III.

PATHWAY OF 23S rRNA FORMATION

The initial cleavage by RNase III is indispensable for the maturation of 23S rRNA; maturation fails completely in the absence of the enzyme (King et al., 1984). (Fortunately for the cell, the unprocessed 23S RNA in RNase III-deficient cells is functional enough to maintain strain viability [see below].) Studies of 23S rRNA processing have exploited the RNase III-deficient strain for both in vivo and in vitro experiments. Its heterogeneous population of 23S precursors have 20 to over 97 additional nucleotides at their 5′ ends (King et al., 1984), produced in vivo at single-stranded segments in the RNA secondary structure.

Mature 5′ termini are not formed in vitro when isolated pre-23S rRNA or 50S or 70S ribosomes are treated with purified RNase III or a salt-extracted preparation of ribosomal proteins from wild-type cells (Sirdeshmukh and Schlessinger, 1985a). Such incubations produce only the species characteristic of RNase III action, three or seven nucleotides longer than the mature 5′ end. Instead, mature 5′ termini form when incubations are carried out in protein synthetic conditions or, more simply, with polysomes as a substrate (Srivastava and Schlessinger, 1988; Sirdeshmukh and Schlessinger, 1985a). In the latter case, the reaction proceeds in ordinary buffered salts, without the need for protein synthetic conditions. These findings imply that (i) a soluble enzyme or factor that converts the RNase III-cleaved product to mature form seems to be required (most likely, the reaction is endonucleolytic, since no intermediate species was detected) and (ii) ribosomes in polysomes adopt a conformation that facilitates the maturation cleavage and lose that conformation when they are in the form of free unmatured 50S ribosomes (see below).

At the 3′ end of p23S rRNA, exonucleolytic action produces the mature terminus after RNase III has cleaved the stem. Intermediates are observed that lack successive nucleotides up to the stem remaining in mature 23S rRNA. The mechanism is confirmed by the observation of the same intermediate species at low levels in RNA from wild-type cells. The exonucleolytic action occurs in buffered salt solutions but seems to be more efficient in protein synthetic conditions, perhaps in coordination with maturation of the 5′ terminus (Sirdeshmukh and Schlessinger, 1985a, 1985b).

PATHWAY OF 16S rRNA FORMATION

Unlike 23S rRNA, 16S rRNA matures with no required order of processing reactions. At first this seemed improbable, since cleavage of nascent transcripts by RNase III in wild-type cells is certainly much faster than the formation of mature termini (King and Schlessinger, 1983). Kinetic order, however, was not reflected in any ordered processing mechanism in this case. This became clear when the pathway was detailed in the RNase III-deficient strain. In the absence of RNase III, no cleavages were seen in precursor-specific sequences in the vicinity of its cleavage sites, but 16S molecules formed with normal mature termini and at the same rate as in the wild-type strain. These results implied that enzymes involved in final maturation act independently of RNase III.

The mechanism of 16S rRNA processing is not known in detail, but intact precursor fragments that extend from the mature 5′ and 3′ termini to sequences far beyond the RNase III cleavage sites are released during processing reactions in vitro (Srivastava and Schlessinger, 1989a). The same fragments are seen in the RNase III-deficient strain in vivo. Thus, mature termini are apparently formed by single endonucleolytic cleavages, one at the 5′ end and another at the 3′ end, with the concomitant release of precursor fragments and without formation of any substantially longer intermediate species. These results extend the finding that mature 16S rRNA forms at the same rate independent of prior RNase III cleavage.

The enzymes involved in the formation of 5′ and 3′ termini of 16S rRNA have been partially isolated and characterized. At the 5′ end of 16S rRNA, an endonuclease (RNase M16) has been studied (Dahlberg et al., 1978). It seems to be deficient in a particular mutant that accumulates a "16.3S" p16S rRNA precursor with 66 extra 5′ nucleotides. The enzyme preparation does not cleave purified 16.3S rRNA or longer precursors but cleaves the precursor in 30S or 70S ribosomes, in keeping with the notion that maturation is dependent on ribosome assembly.

The 16.3S rRNA has a mature 3′ end, suggesting that different enzymes form the two termini of 16S rRNA. An enzymatic activity that forms 3′ mature termini in vitro has been partially purified from crude extracts (Hayes and Vasseur, 1976). In vitro maturation of the 3′ end of 16S rRNA, like that of 23S rRNA, occurs efficiently in protein synthetic conditions (Hayes and Vasseur, 1976). The significance of this effect on processing is discussed below.

PATHWAY OF 5S rRNA FORMATION

5S rRNA matures through a multistep process. The product of RNase III cleavage in wild-type cells is a 9S precursor molecule (p5S) with 85 nucleotides at its 5′ end and extra 3′ nucleotides extending to the terminator site. In operons with distal tRNA at the ends of transcripts, it is cleaved from the 5S moiety, probably by RNase P action at the 5′ terminus of the tRNA. The 9S precursor is not normally seen in wild-type cells, but it accumulates in a mutant temperature sensitive in RNase E activity when the mutant is grown at a nonpermissive temperature (Ghora and Apirion, 1979). RNase E rapidly cleaves p5S rRNA in wild-type cells to produce a species with three extra nucleotides adjacent to each end of the 5S RNA sequence (Roy et al., 1983). Additional species found in cells in which protein synthesis is inhibited have one, two, or three extra nucleotides, suggesting that final maturation of 5S RNA probably involves 5′ and 3′ exonucleases (Feunteun et al., 1972). Similar to 16S and 23S rRNA, p5S rRNA has been reported in polysomes in cells, and maturation most likely occurs in polysomes in this case as well.

FEATURES OF PRECURSORS REQUIRED FOR RECOGNITION DURING PROCESSING

The precise requirements for RNase III cleavage are still not clear. It has been shown to make cleavages at staggered sites in the double-stranded stems, probably involving a separate cut in each strand (Fig. 1). It remains most likely that the specificity is primarily determined by the double-stranded stem itself, not by the sequence, so that deletion in one strand prevents the RNase III action on the other strand as well (Srivastava and Schlessinger, 1989b). This would be consistent with results for 23S rRNA processing, where even a small deletion in the stem eliminates RNase III cleavage (Stark et al., 1984). However, the primacy of double-stranded structure does not preclude alteration in RNase III action when the conformation of the rest of the rRNA is altered. For example, cleavage by RNase III is restricted when the distal spacer tRNA in transcript is deleted in whole or in part (Szymkowiak et al., 1988). Processing can proceed at both the 5′ and 3′ termini, however, when the base-paired stem is totally absent; and 5′ termini are formed independently of the sequence or processing events at the 3′ end (Srivastava and Schlessinger, 1989b). Any requirement for the stem structure for rRNA maturation must therefore be more subtle.

Morgan has presented some provocative speculations about the juxtaposition, in both ribosomal DNA and the early leftward transcription unit of lambda, of box A sequences (Friedman and Olson, 1983; Li et al., 1984) and sequences involved in double-stranded RNase III sites (E. Morgan, personal communication). He proposes that at a box A sequence the RNA polymerase grabs hold of the distal nascent RNA sequence and keeps hold of the 5′ portion until the 3′ portion of the double-stranded region has been synthesized. With no free ends available during the synthetic process, susceptibility of the transcript to RNases would be lessened, and the formation of the intramolecular stem at which RNase III acts would be facilitated by the continued proximity of the 5′ stem-forming sequence when the polymerase reached the complementary 3′ sequence tract. In this way, the RNA polymerase would help to direct the formation of the secondary structure of pre-rRNA molecules and would thus help to stabilize nascent molecules until the more protective double-stranded stem was formed. This model is somewhat analogous to that of Horwitz et al. (1987) for *boxA* function in bacteriophage lambda phage development. It is appealing and is consistent with data for the lambda system and with studies of Morgan on effects of box A mutations on ribosomal DNA expression. As he points out, however (Morgan, personal communication), these effects on ribosome formation must be second order, and in this regard it is significant that stable constructs can be formed and properly processed in the absence of the double-stranded stem (see above).

In contrast to the apparent dispensability of the double-stranded stems, we have found, in analyses of the expression of deletion constructs, that the distal tRNA must be present, and at a minimum spacer distance, to permit formation of the 3′ terminus of 16S pre-rRNA (Fig. 1; Srivastava and Schlessinger, 1989b). Proper maturation also requires part but not all of the 16S rRNA sequence itself (Gourse et al., 1982; Stark et al., 1982).

The maturation of 16S pre-rRNA, which requires mature rRNA, spacer RNA, and distal tRNA sequence at a proper distance, is thus very different from the model case of 5S RNA of *Bacillus subtilis*. The recognition signals for cleavage of that 5S pre-rRNA lie completely in the mature sequence, and precursor-specific segments adjacent to either terminus can be completely replaced by oligonucleotides without blocking processing (Meyhack et al., 1978; Altman et al., 1982).

In the absence of the stem structure, the tRNA must be more than 23 nucleotides from the 3′ end of the 16S rRNA sequence, but 39 nucleotides is at least sufficient to support accurate processing. Presumably, the tRNA is folded into a standard cloverleaf configuration even in pre-rRNA, and the spacer distance would permit the tRNA to interact properly with other parts of the pre-rRNA (Fig. 2). We have

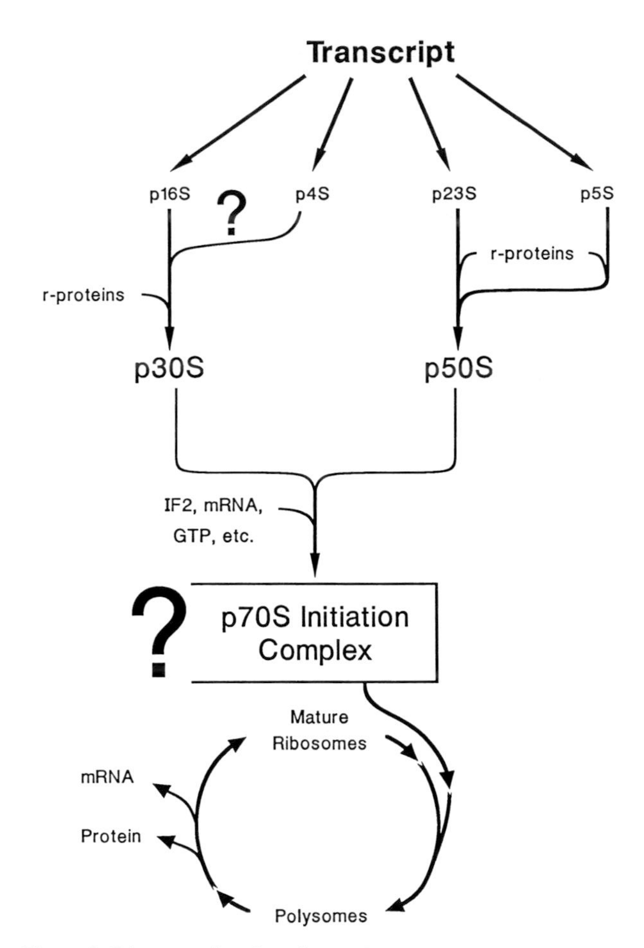

Figure 2. Diagram showing that primary processing can occur with cues for enzymatic cleavage in the pre-rRNA itself, producing the p30S and p50S preribosomes with the addition of ribosomal proteins. A small question mark indicates that the p4S spacer sequence may remain bound to the nascent p30S particle and contribute to its maturation (see text). Secondary processing (maturation) occurs later, possibly in polysomes, with the speculative suggestion of an initiation complex containing the preribosomes (larger question mark; see text).

also suggested that pre-tRNA could interact with or even help to form the nascent P site in a pre-30S ribosome.

The evolutionarily conserved double-stranded stem might also participate in this process. Although it is dispensable for 16S rRNA processing, the stem might increase the efficiency of the process. For example, the stem might have evolved to angle the nascent tRNA in pre-rRNA toward its site of interaction with the ribosome. According to this speculative model, tRNA would tend to remain attached to pre-16S rRNA through an intact intervening spacer until cleavage occurred at the 3′ end of 16S rRNA (or at RNase III or RNase P cleavage sites). Thus, the processing steps would provide a route for the maturation of ribosomes and a checkpoint to ensure that only properly assembled ribosomes are added to the cellular pool (King et al., 1986).

MATURATION IN POLYSOMES

How Functional Are pre-rRNAs?

All 50S ribosomes of the viable RNase III-deficient mutant contain unmatured 23S rRNA. Therefore, p23S rRNA must be functional in vivo (King et al., 1984). In contrast, ribosomes containing pre-16S rRNA are not biologically active: processing to mature 16S rRNA molecules is obligate for competence in protein synthesis (Wireman and Sypherd, 1974; Nomura and Held, 1974). This may be because the 16S rRNA and 30S ribosome are more intimately involved in the initiation phase of protein synthesis, with demanding conformational requirements for the Shine-Dalgarno interaction, attachment of initiation factors, etc. It may be relevant that the 10 nucleotides of precursor sequence immediately preceding the 5′ mature terminus are base paired with neighboring mature 16S rRNA sequence. This could help to ensure nonfunctionality until processing is complete.

Ribosomes with p23S rRNA only function enough for cells to survive. They are probably not as efficient as mature ribosomes, but it is not surprising that p23S rRNAs are found in polysomes extracted from cells (Sirdeshmukh and Schlessinger, 1985b). It is quite surprising, however, that the inactive p16S rRNA is also found in polysomes (Mangiarotti et al., 1974). These results raise the general question of the extent to which precursors of structural rRNAs can function in cells before processing is completed.

The Polysomal Substrate

The situation has been somewhat clarified by the findings that mature termini of both 16S and 23S rRNAs can be formed from preribosomes incubated with enzyme preparations in vitro, but maturation of both termini of 16S rRNA and the 3′ end of 23S rRNA is certainly more efficient under conditions of in vitro protein synthesis (Hayes and Vasseur, 1976; Sirdeshmukh and Schlessinger, 1985a). Most extreme is the case of maturation of the 5′ end of 23S rRNA, which occurs appreciably on isolated pre-50S ribosomes only in the presence of 30S ribosomes and other components required for protein synthesis (Sirdeshmukh and Schlessinger, 1985a).

The relationship between maturation events and protein synthesis could hold if one of the components of protein synthetic mixture, such as GTP, were required for maturation. Alternatively, preribosomes might be found in polysomes because they can participate in some partial reaction of protein synthesis and thereby achieve an RNA conformation required for maturation reaction. The latter alternative is favored by our finding that polysomes rather than free ribosomes are the preferred substrate for maturation. With polysomes instead of 50S particles as the substrate, maturation can be carried out in the absence of protein synthetic conditions (Srivastava and Schlessinger, 1988). This implies that the final maturation steps occur only after preribosomes join in polysomes (Fig. 2). The formation of ribosomes is thus directly linked to their incorporation into the protein synthetic machinery. The in vitro requirement of protein synthesis for maturation with free ribosomes presumably serves to form the polysome substrate.

The exact nature of the substrate for maturation is still unclear. However, it seems possible that even 30S preribosomes, which cannot translate mRNA, can join in 70S initiation complexes and mature in that form. The complex normally forms by the successive accretion of mRNA, initiation factors, initiator tRNA, and 50S ribosomes combined with 30S ribosomes; an analogous complex containing preribosomes may be an adequate substrate for the processing reaction.

COREGULATION OF rRNA PROCESSING AND PROTEIN SYNTHESIS

Because cleavage and trimming of the large precursor to mature rRNA naturally precedes rRNA function, it has been thought that the production and activity of ribosomes are distinct and noninteractive processes. However, recent findings in *E. coli* show a connection between the synthesis of proteins and maturation of rRNA, suggesting that the processing and function of ribosomes must be reciprocally dependent (Srivastava and Schlessinger, 1988).

All procaryotic 16S, 23S, and 5S rRNAs mature after ribosomes are formed (Srivastava and Schlessinger, 1988; Ceccarelli et al., 1978; Mangiarotti et al., 1974; Feunteun et al., 1972). Even in eucaryotic yeast cells, in which ribosome assembly occurs in nucleoli, the final maturation of 18S rRNA has been reported to occur in the cytoplasm (Udem and Warner, 1973). It is quite possible in all of these cases that preribosomes join to mRNA before maturation is complete. Direct evidence is lacking, but if this were so, the continued movement of active ribosomes to free up initiation sites on mRNA could be obligate for continued binding and rapid maturation of preribosomes. This implies that the rates of protein synthesis and rRNA processing are interrelated or mutually controlled. In this way, bacteria might regulate quantity and rate of production of ribosomes at the level of maturation. The rate of protein synthesis may directly limit the rate of processing; immature ribosomes, in turn, may even limit the movement of mature component ribosomes on mRNA. Significance can be assigned to such a link between processing and ribosome function in the regulation of cell physiology only if polysomes containing pre-rRNA translate less efficiently than matured polysomes. Thus far, the evidence is indirect but suggestive; the RNase III-deficient strain, which contains polysomes with only pre-23S rRNA, grows more slowly and shows defects in translation of β-galactosidase and other mRNAs (Gitelman and Apirion, 1980; Talkad et al., 1978; Silengo et al., 1974). (The results are inconclusive because defects are corrected when RNase III is restored to the strains, but the effects can be indirect rather than based on altered polysome function in the mutant.)

POSSIBLE LINK OF PROCESSING TO AUTOREGULATION OF RIBOSOME CONTENT

In exponentially growing *E. coli*, the number of ribosomes per amount of cellular protein increases with growth rate (see reviews by Nomura et al. [1984] and Lindahl and Zengel [1986]). At any time, the number of active ribosomes in the cells is proportional to the rate of synthesis of total rRNA. Only small quantities of free ribosomal proteins and rRNA exist in growing cells, and the rate of net synthesis is therefore just sufficient to meet the needs for new ribosomes. It therefore seems that the functional activity of ribosomes is somehow connected with their own production.

Several mechanisms have been suggested to regulate the formation of new rRNA and ribosomes. Elsewhere in this volume, Nomura discusses the way in which he and his colleagues have deciphered the mechanism of coregulation of rRNA and ribosomal proteins. The regulation of rRNA, in turn, has recently been discussed in terms of possible inhibition of RNA transcription by the action of guanosine tetraphosphate on RNA polymerase or, more generally, by an autoregulatory process through a negative-feedback loop responding to the level of functional ribosomes.

In their formulation, Nomura and co-workers proposed a principal feedback control of rRNA synthesis based on the cellular pool of free, nontranslating ribosomes. Consistent with the feedback model, the rate of rRNA synthesis is gene dosage independent when multiple copies of rDNA are introduced into cells on plasmids (Jinks-Robertson et al., 1983).

The molecular mechanism involved in the feedback regulation remains in doubt, but recent work has focused on translational capacity rather than the number of ribosomes as the important variable. For example, ribosomes unable to participate in initiation should be unable to effect feedback. In support of this notion, (i) ribosomes with a mutational alteration in the anti-Shine-Dalgarno region at the 3′ end of 16S rRNA are unable to participate in feedback regulation in vivo (Yamagishi et al., 1987) and (ii) lowering the IF2 concentration in a cell to a level sufficient to support growth at only about one-third of the normal growth rate yields to an increase in rRNA content and a large accumulation of nontranslating ribosomes (Cole et al., 1987). In addition, induction of IF2 synthesis in IF2-deficient cells causes a repression of rRNA synthesis as ribosomes become more active. It has been suggested that not just excess free ribosomes and IF2 but probably all of the components involved in translation initiation are needed for rRNA feedback regulation.

New emerging evidence suggests that processing steps in bacteria may also act to regulate steps in nascent ribosome formation, perhaps again at the initiation step in protein synthesis (Srivastava and Schlessinger, 1988). We have considered that the rates of ribosome formation and protein synthesis are coregulated at the step where preribosomes join the translational machinery to become mature. Effects of IF2 level or altered Shine-Dalgarno sequences might interfere simultaneously with the formation of initiation complexes on preribosomes. This would affect the rate of maturation and thus would affect the function of ribosomes and subsequent rRNA synthesis by feedback regulation.

PROSPECTS

Recent work emphasizes the power of the available technology, which seems adequate to resolve most or all remaining questions about the mechanism of processing. For example, with respect to the portions of transcripts required for processing, the extension of the use of deletion constructs should permit one to find out (i) whether intact spacer tRNA is required for 16S rRNA maturation (and if not, which portions are required); (ii) whether the maturation process will accept any tRNA, only the one in the corresponding operon, or any of the group of tRNAs present in the seven operons; and (iii) whether spacer tRNA, the 5S RNA, or the 3′ distal tRNA has any corresponding role in the maturation of 23S rRNA.

In a similar way, the availability of substrates and rapid assays for maturation in vitro provides a direct route to the purification of enzymes or other factors involved in the formation of mature 16S and 23S rRNA termini. This could aid in the subsequent analysis of the integration of the processes involved in gene expression.

As the entire field begins to move toward the challenge provided by eucaryotic systems, others (Jacobs, 1989; Warner, 1989) have reviewed the way in which compartmentalization into nucleus, cytoplasm, and mitochondria can lead to independent evolution of the pathways that lead to ribosomes in eucaryotic cells. It is premature to try to extend the incomplete results with *E. coli*, but it is already certain that both ordered and stochastic mechanisms will be encountered, that the machinery for processing will be found to include ribonucleoprotein complexes (small nuclear ribonucleoproteins instead of RNase P or polysomal complexes), and that as a result, precursor-specific sequences will be found to be involved in the mechanism.

Work in our laboratory has been sustained by National Science Foundation grant PMS PCM 8406949.

We thank T. C. King, R. Sirdeshmukh, and E. Morgan for their efforts and discussions that have much benefited us.

REFERENCES

Altman, S. 1981. Transfer RNA processing enzymes. *Cell* **23**:3–4.

Altman, S., C. Guerrier-Takada, H. M. Frankfort, and H. D. Robertson. 1982. RNA processing nuclease, p. 243–274. *In* S. M. Linn and R. J. Roberts (ed.), *Nuclease.* Cold Spring Harbor Laboratory, Cold Spring Harbor, N.Y.

Apirion, D., and P. Gegenheimer. 1984. Molecular biology of RNA processing in prokaryotic cells, p. 36–62. *In* D. Apirion (ed.) *Processing of RNA.* CRC Press, Inc., Boca Raton, Fla.

Bram, R. J., R. A. Young, and J. A. Steitz. 1980. The ribonuclease III site flanking 23S sequence in the 30S ribosomal precursor RNA of E. coli. *Cell* **19**:393–401.

Brimacombe, R., J. Atmadja, W. Stiege, and D. Schuler. 1988. A detailed model of the three-dimensional structure of *Escherichia coli* 16S ribosomal RNA *in situ* in the 30S subunit. *J. Mol. Biol.* **199**:115–136.

Ceccarelli, A., G. P. Dotto, F. Altruda, C. Perlo, L. Silengo, E. Turco, and G. Mangiarotti. 1978. Immature 50S subunits in *Escherichia coli* polysomes. *FEBS Lett.* **93**:348–350.

Cole, J. R., C. L. Olsson, J. W. B. Hershey, M. Grunberg-Manago, and M. Nomura. 1987. Feedback regulation of rRNA synthesis in *Escherichia coli*: requirement for initiation factor IF2. *J. Mol. Biol.* **198**:383–392.

Dahlberg, A. E., J. E. Dahlberg, E. Lund, H. Tokimatsu, A. B. Rabson, P. C. Calvert, F. Reynolds, and M. Zahalak. 1978. Processing of the 5′ end of *Escherichia coli* 16S ribosomal RNA. *Proc. Natl. Acad. Sci. USA* **75**:3598–3602.

Dunn, J. J., and F. W. Studier. 1973. T7 early RNAs and *Escherichia coli* ribosomal RNAs are cut from large precursor RNAs *in vitro* by ribonuclease III. *Proc. Natl. Acad. Sci. USA* **70**:3296–3300.

Feunteun, J., B. R. Jordan, and R. Monier. 1972. Study of maturation of 5S precursors in *Escherichia coli. J. Mol. Biol.* **70**:465–474.

Friedman, D. I., and E. R. Olson. 1983. Evidence that a nucleotide sequence, "box A," is involved in the action of the nusA protein. *Cell* **34**:143–149.

Gegenheimer, P., N. Watson, and D. Apirion. 1977. Multiple pathways for primary processing of ribosomal RNA in *Escherichia coli. J. Biol. Chem.* **252**:3064–3073.

Ghora, B. K., and D. Apirion. 1979. Identification of a novel RNA molecule in a new RNA processing mutant of *Escherichia coli* which contains 5S rRNA sequence. *J. Biol. Chem.* **254**:1951–1956.

Gitelman, D. R., and D. Apirion. 1980. The synthesis of some proteins is affected in RNA processing mutant of *Escherichia coli. Biochem. Biophys. Res. Commun.* **96**:1063–1070.

Gourse, R. L., M. J. R. Stark, and A. E. Dahlberg. 1982. Site-directed mutagenesis of ribosomal RNA: construction and characterization of deletion mutants. *J. Mol. Biol.* **159**:397–416.

Hayes, F., and M. Vasseur. 1976. Processing of 17S *Escherichia coli* precursor RNA in the 27-S pre-ribosomal particle. *Eur. J. Biochem.* **61**:433–442.

Horwitz, R. J., J. Li, and J. Greenblatt. 1987. An elongation control particle containing the N gene transcriptional antitermination protein of bacteriophage lambda. *Cell* **51**:631–641.

Jacobs, H. D. 1989. Do ribosomes regulate mitochondrial RNA synthesis? *BioEssays* **11**:27–34.

Jinks-Robertson, S., R. L. Gourse, and M. Nomura. 1983. Expression of rRNA and tRNA genes in *E. coli*: evidence for feedback regulation by products of rRNA operons. *Cell* **33**:865–876.

Kindler, P., T. U. Kiel, and P. H. Hofschneider. 1973. Isolation and characterization of a ribonuclease III deficient mutant of *Escherichia coli. J. Biol. Chem.* **251**:53–69.

King, T. C., and D. Schlessinger. 1983. S1 nuclease mapping analysis of ribosomal RNA processing in wild-type and processing deficient *Escherichia coli. J. Biol. Chem.* **258**:12034–12042.

King, T. C., R. Sirdeshmukh, and D. Schlessinger. 1984. RNase III cleavage is obligate for maturation but not for function of *Escherichia coli* pre-23S rRNA. *Proc. Natl. Acad. Sci. USA* **81**:185–188.

King, T. C., R. Sirdeshmukh, and D. Schlessinger. 1986. Nucleolytic processing of ribonucleic acid transcripts in procaryotes. *Microbiol. Rev.* **50**:428–451.

Li, S. C., C. L. Squires, and C. Squires. 1984. Antitermination of *E. coli* rRNA transcription is caused by a control region segment containing lambda nut-like sequences. *Cell* **38**:851–860.

Lindahl, L., and M. Zengel. 1986. Ribosomal RNA gene in Escherichia coli. *Annu. Rev. Genet.* **20**:297–326.

Lund, E., and J. E. Dahlberg. 1977. Spacer transfer RNAs in ribosomal RNA transcripts of *E. coli*: processing of 30S ribosomal RNA *in vitro. Cell* **11**:247–262.

Mangiarotti, G., E. Turco, A. Ponzetto, and F. Altruda. 1974. Precursor 16S RNA in active 30S ribosomes. *Nature* (London) **247**:147–148.

Meyhack, B., B. Pace, O. C. Ullenbeck, and N. R. Pace. 1978. Use of T4 RNA ligase to construct model substrates for a ribosomal RNA maturation endonuclease. *Proc. Natl. Acad. Sci. USA* **75**:3045–3049.

Morgan, E. A., T. Ikemura, L. E. Post, and M. Nomura. 1980. tRNA genes in rRNA operons of *Escherichia coli*, p. 259–266. *In* P. R. Schimmel, D. Soll, and J. N. Abelson (ed.), *Transfer RNA: Biological Aspects.* Cold Spring Harbor Laboratory, Cold Spring Harbor, N.Y.

Nikolaev, N., D. Schlessinger, and P. K. Wellauer. 1974. 30S pre-ribosomal RNA of *Escherichia coli* and products of cleavage by RNase III: length and molecular weight. *J. Mol. Biol.* **86**:741–747.

Nikolaev, N., L. Silengo, and D. Schlessinger. 1973. Synthesis of large precursor to ribosomal RNA in a mutant of *Escherichia* *coli. Proc. Natl. Acad. Sci. USA* 70:3361–3365.

Nomura, M., R. Gourse, and G. Baughman. 1984. Regulation of the synthesis of ribosomes and ribosomal components. *Annu. Rev. Biochem.* 53:75–117.

Nomura, M., and W. A. Held. 1974. Reconstitution of ribosomes: studies of ribosomes structure, function and assembly, p. 193–223. *In* M. Nomura, A. Tissières, and P. Lengyel (ed.), *Ribosomes.* Cold Spring Harbor Laboratory, Cold Spring Harbor, N.Y.

Roy, M. K., B. Singh, B. K. Ray, and D. Apirion. 1983. Maturation of 5S rRNA: ribonuclease E cleavages and their dependence on precursor sequences. *Eur. J. Biochem.* **131**:119–127.

Schlessinger, D. 1980. Processing of ribosomal RNA transcripts in bacteria, p. 767–780. *In* G. Chambliss, G. R. Craven, K. Davies, L. Kahan, and M. Nomura (ed.), *Ribosomes. Structure, Function, and Genetics.* University Park Press, Baltimore.

Silengo, L., N. Nikolaev, D. Schlessinger, and F. Imamato. 1974. Stabilization of mRNA with polar effects in an *Escherichia coli* mutant. *Mol. Gen. Genet.* **134**:7–19.

Sirdeshmukh, R., and D. Schlessinger. 1985a. Ordered processing of *Escherichia coli* 23S rRNA *in vitro. Nucleic Acids Res.* **13**:5041–5054.

Sirdeshmukh, R., and D. Schlessinger. 1985b. Why is processing of 23S ribosomal RNA in *Escherichia coli* not obligate for its function? *J. Mol. Biol.* **186**:669–672.

Srivastava, A. K., and D. Schlessinger. 1988. Coregulation of processing and translation: mature 5′ termini of *Escherichia coli* 23S rRNA form in polysomes. *Proc. Natl. Acad. Sci. USA* 74:7144–7148.

Srivastava, A. K., and D. Schlessinger. 1989a. Processing pathways of *Escherichia coli* pre-16S rRNA. *Nucleic Acids Res.* **17**: 1649–1663.

Srivastava, A. K., and D. Schlessinger. 1989b. *Escherichia coli* 16S rRNA 3′-end formation requires a distal transfer RNA sequence at a proper distance. *EMBO J.* **8**:3159–3166.

Stark, M. J. R., R. L. Gourse, and A. E. Dahlberg. 1982. Site-directed mutagenesis of ribosomal rRNA: analysis of ribosomal RNA deletion mutant using maxicells. *J. Mol. Biol.* **159**:417–439.

Stark, M. J. R., R. J. Gregory, R. L. Gourse, D. L. Thurlow, T. C. Zwieb, R. A. Zimmermann, and A. E. Dahlberg. 1984. Effects of site-directed mutation in the central domain of 16S ribosomal RNA upon ribosomal protein binding, RNA processing, and 30S subunit assembly. *J. Mol. Biol.* **178**:303–322.

Steen, R., D. K. Jemiolo, R. H. Skinner, J. J. Dunn, and A. E. Dahlberg. 1986. Expression of plasmid-coded mutant ribosomal RNA in *E. coli*: choice of plasmid vectors and gene expression systems. *Prog. Nucleic Acids Res. Mol. Biol.* **33**:1–18.

Szymkowiak, C., R. L. Reynolds, M. J. Chamberlin, and R. Wagner. 1988. The $tRNA^{Glu}{}_2$ gene in the *rrnB* operon of *E. coli* is a prerequisite for correct RNase III processing *in vitro. Nucleic Acids Res.* **16**:7885–7899.

Talkad, V., D. Achord, and D. Kennell. 1978. Altered mRNA metabolism in ribonuclease III-deficient strains of *Escherichia coli. J. Bacteriol.* **135**:528–541.

Traub, P., and M. Nomura. 1969. Structure and function of *E. coli* ribosomes studies *in vitro. J. Mol. Biol.* **40**:391–413.

Udem, S. A., and J. R. Warner. 1973. The cytoplasmic maturation of a ribosomal precursor ribonucleic acid in yeast. *J. Biol. Chem.* **248**:1412–1416.

Warner, J. R. 1989. Synthesis of ribosomes in *Saccharomyces cerevisiae. Microbiol. Rev.* **53**:256–271.

Wireman, J. W., and P. S. Sypherd. 1974. In vitro assembly of 30S ribosomal particles from precursor 16S RNA of *Escherichia coli. Nature* (London) **247**:552–554.

Woese, C. R. 1987. Bacterial evolution. *Microbiol. Rev.* **51**: 221–271.

Yamagishi, M., H. A. de Boer, and M. Nomura. 1987. Feedback regulation of RNA synthesis: a mutational alteration in the anti-Shine Dalgarno region of the 16S rRNA gene abolishes regulation. *J. Mol. Biol.* **198**:547–550.

Young, R. A., and J. A. Steitz. 1978. Complementary sequences 1700 nucleotides apart from a ribonuclease III cleavage site in *Escherichia coli* ribosomal precursor RNA. *Proc. Natl. Acad. Sci. USA* 75:3593–3597.

Chapter 37

Functional Analysis of the Transcribed Spacers of *Saccharomyces cerevisiae* Ribosomal DNA: It Takes a Precursor To Form a Ribosome

W. MUSTERS, R. J. PLANTA, H. van HEERIKHUIZEN, and H. A. RAUÉ

The cytoplasmatic ribosomes of eucaryotic organisms are highly structured particles consisting of about 80 different components. Assembly of these supramolecular complexes takes place largely within the nucleolus, where the rRNA genes (except for the 5S rRNA gene) are transcribed by RNA polymerase I into a single precursor rRNA that associates with both ribosomal and nonribosomal proteins to form a preribosomal particle (Hadjiolov, 1985). Conversion of this particle into mature, functional ribosomal 40S and 60S subunits is an intricate process consisting of a number of chronologically ordered steps. Although most of the details of this process remain to be elucidated, the processing and modification of the rRNA molecules have been studied extensively (Hadjiolov, 1985; Raué et al., 1988; Klootwijk and Planta, 1989).

Eucaryotic cells may contain up to 1,000 copies of each of the rRNA genes, which (except for the 5S rRNA gene) are arranged in a cluster of tandemly repeated units (Long and Dawid, 1980). Each unit constitutes a single operon encompassing one gene each for 17S-18S, 5.8S, and 26S-28S rRNAs, preceded by an external transcribed spacer (ETS) and separated by two internal transcribed spacers (ITS1 and ITS2), which are posttranscriptionally removed. Figure 1 shows the organization of the rRNA genes in *Saccharomyces cerevisiae*, as well as the various steps required for the removal of the ETS and ITS sequences. As can be seen, formation of the mature rRNA molecules proceeds by an ordered series of nucleolytic cleavages, first separating the individual species and then trimming them to their final size. Although not shown in Fig. 1, it is important to remember that this whole process involves ribonucleoprotein particles rather than naked RNA. Thus, rRNA maturation is accompanied by the association with ribosomal proteins in an orderly, though so far largely unknown, sequence. Moreover, association and dissociation of nonribosomal proteins may also play an important role in ensuring the successful completion of ribosomal subunit formation.

Comparison of the sequences of complete ribosomal transcription units from a variety of organisms has demonstrated that the mature sequences are under considerable evolutionary constraint (Gerbi, 1985). However, much less evolutionary conservation is apparent in the ETS and ITS regions (Furlong and Maden, 1983; Verbeet et al., 1984). This observation raises the question of whether the eucaryotic transcribed spacer regions are dispensable for ribosome formation (as appears to be the case for at least the in vitro assembly of procaryotic ribosomes) or whether these regions are active participants in one or more ribosomal maturation steps and hence are essential to eucaryotic ribosome biogenesis. The following sections describe experiments that provide at least a partial answer to this question.

A SYSTEM FOR FUNCTIONAL ANALYSIS OF *S. CEREVISIAE* (PRE-)rRNA SEQUENCES

Over the last few years, a number of elegant in vivo and in vitro approaches to unraveling the structure-function relationships of procaryotic rRNA have been worked out (see chapters by Tapprich et al., Hill et al., and Cunningham et al. in this volume). However, none of these systems as such is applicable to eucaryotes, either because in vitro assembly of eu-

W. Musters, R. J. Planta, H. van Heerikhuizen, and H. A. Raué ■ Biochemisch Laboratorium, Vrije Universiteit, de Boelelaan D83, 1081 HV Amsterdam, The Netherlands.

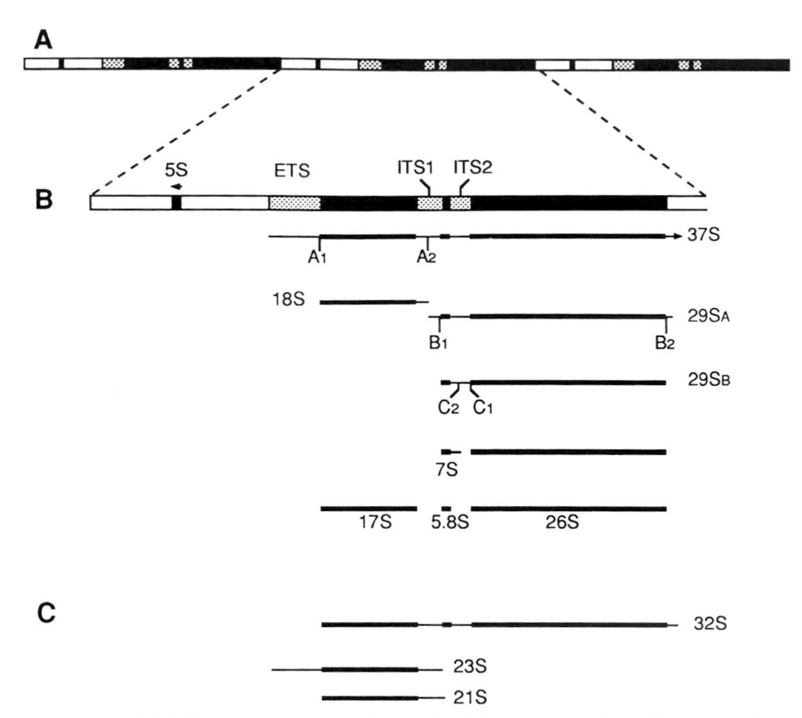

Figure 1. Genetic organization of rDNA and processing of pre-rRNA in *S. cerevisiae*. (A) Schematic representation of three tandemly repeated rDNA units (of the 50 to 200 repeats present on chromosome XII). White, shaded, and black bars represent nontranscribed spacers, transcribed spacers, and rRNA genes, respectively. (B) A single rDNA unit aligned with the products of the various steps of the major processing pathway in wild-type cells. Nomenclature is according to Veldman et al. (1981). (C) Unusual processing intermediates that are found either in cells harboring rDNA units from which a part of ITS1 has been deleted (32S) or in *snr10.3* strains (21S and 23S [Tollervey, 1987]). See text for details. Nomenclature of these unusual intermediates is according to Tollervey (1987).

caryotic ribosomal subunits so far has proved to be an intractable problem or because the large number of rRNA genes produces a level of wild-type transcripts that totally obscures any aberrant behavior of the products of the few mutant genes that can be introduced into a eucaryotic host cell by transformation. A complete replacement of the many ribosomal DNA (rDNA) units in eucaryotes by mutated rDNA units is virtually impossible except in the case of *Tetrahymena* species. In this organism, all of the rRNA genes that are expressed in the macronucleus are derived from a few copies of the rDNA unit that are present in the micronucleus (Yao, 1986). The advantages offered by this system have been elegantly exploited by Sweeney and Yao (1989) to study the effect of insertions in the *Tetrahymena* rDNA. A major disadvantage of this approach, however, is that rDNA mutations causing severe defects in the formation of functional ribosomes can only be scored indirectly, since they will not produce viable transformants.

Recently, we have succeeded in developing a system for *S. cerevisiae* that makes it possible to monitor the fate of mutant (pre-)rRNA transcripts in vivo, even in the presence of as much as a 500-fold excess of wild-type molecules (Musters et al., 1989). To achieve this goal, we have constructed a vector carrying a complete *S. cerevisiae* rDNA unit, including the nontranscribed spacer, which separates two consecutive units in the chromosome (Fig. 2). The crucial difference between the extrachromosomal unit and those present in the host cell rRNA genome is the presence of a tag in the form of an oligonucleotide insertion in either the 17S or 26S rRNA gene or both. The effect of rDNA mutations can thus be determined by introducing them into the tagged rDNA unit and monitoring the fate of the tagged rRNA by hybridization with an oligonucleotide complementary to the tag. For obvious reasons, the presence of the tag by itself should not affect functioning of the rRNA. Therefore, we chose a variable region (or expansion segment [Clark et al., 1984]) as the most logical target for the insertion. Figure 2 shows the locations of the target sites in domain III of 17S and domain I of 26S rRNA. In both cases, the inserted oligonucleotide was designed to differ as much as possible from the rest of the *S. cerevisiae* rDNA sequence while at the same time satisfying two requirements. First, it should contain a unique restriction site (*Kpn*I and *Xho*I, respectively, were used)

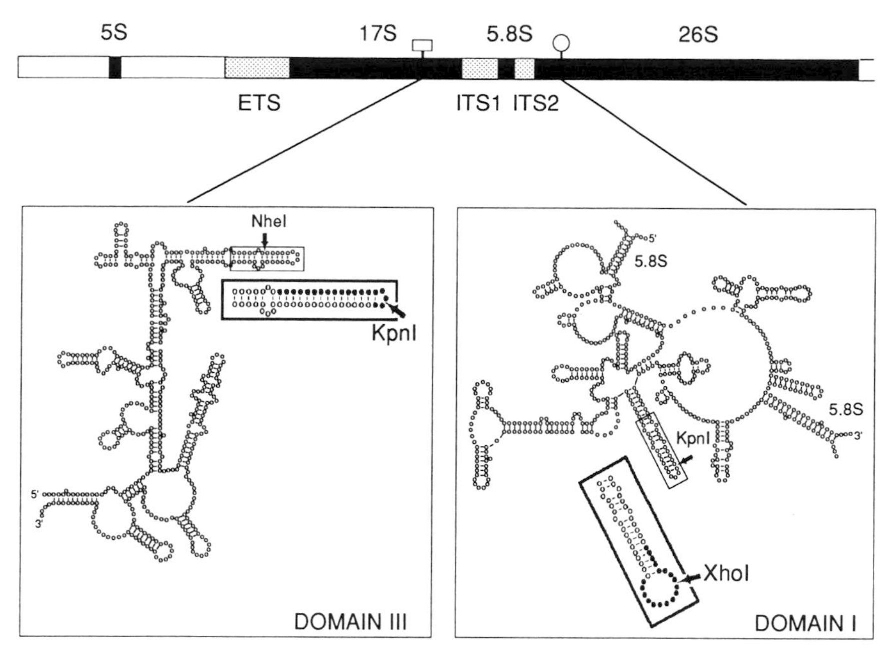

Figure 2. Tagging of the *S. cerevisiae* rRNA genes. The 17S and 26S rRNA genes were tagged by insertion of a short oligonucleotide (having a sequence unique in the rDNA) into an expansion segment of domain III of the 17S and domain I of the 26S rRNA gene, respectively. The position of each of the tags is marked in the schematic representation of the rDNA unit. The lower panels show the secondary structure models for the two domains, the restriction site at which the tag was inserted, and the putative secondary structure of the region containing the tag (insets). The tag itself is indicated by the filled circles.

for convenient detection of its presence during the cloning procedure and to allow reduction of the inserted sequence to a single copy of the tag. Second, it should disturb the secondary structure of the rRNA at the site of the insertion as little as possible. To ascertain the absence of any discernible effect of the tag on ribosome function, the polysomal distribution of the ribosomes carrying the tagged rRNAs was compared with that of wild-type ribosomes. The two patterns were found to be identical for both the tagged 26S rRNA (Musters et al., 1989) and the tagged 17S rRNA (W. Musters et al., unpublished data), showing that the system is indeed suitable for studying the effect of rDNA mutations on the formation and functioning of ribosomes.

In addition to the analysis of mutations in the ETS and ITS regions described in the following sections, we have used this system to study the effects of structural changes in the mature rRNAs on ribosome formation and function. Some of these studies, in particular those involving mutagenesis of the highly conserved GTPase center in domain II and an expansion segment in domain III of 26S rRNA, are described by Raué et al. elsewhere in this volume.

DELETIONS IN THE TRANSCRIBED SPACERS OF *S. CEREVISIAE* rDNA

ETS

In all rRNA operons so far analyzed, the small-subunit (SSU) rRNA gene is preceded by an ETS. The length of this element varies widely among different species, from several hundred nucleotides (nt) in procaryotes to about 3,600 nt in mammals. In *S. cerevisiae*, the ETS has a length of 699 nt (Klootwijk and Planta, 1989).

Several observations suggest that the ETS is not merely a dispensable element of the primary ribosomal transcript but serves additional functions in the processing and assembly of the precursor rRNA. For instance, the earliest processing steps in mammals consist of cleavages within the ETS rather than at its 3′ end (Kass et al., 1987), which suggests that the ETS in these organisms plays a role in the early stages of (pre)ribosome assembly. Moreover, the discovery of an evolutionarily conserved block of at least 50 nt amid otherwise divergent sequences in the ETS of different yeast species (Verbeet et al., 1984), as well as the fact that the ETS in some organisms was implicated to be associated with proteins (Herrera and Olson, 1986; Jordan, 1987; our unpublished results), also indicates that this element, despite its rapid removal from the primary transcript in *S. cerevisiae* (Veinot-Drebot et al., 1988), might serve an important role in the maturation of the preribosome.

To investigate the possible functions of the *S. cerevisiae* ETS, we constructed four deletion mutants. Three of these lack various parts of the ETS; in the fourth mutant, the complete ETS was deleted except for 20 nt at its 5′ end (required for promoter activity)

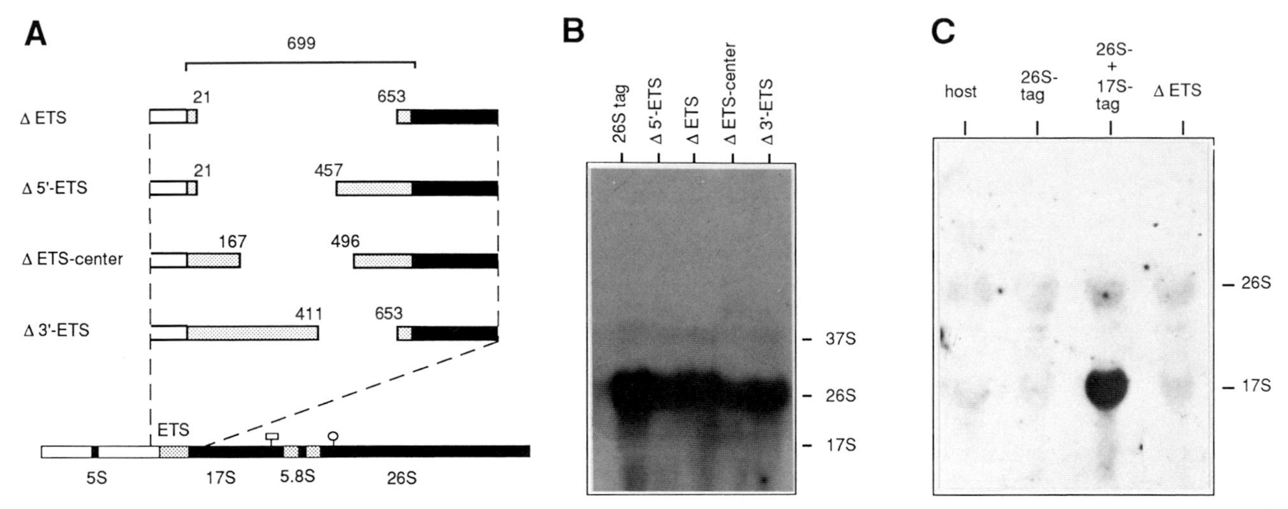

Figure 3. Functional analysis of the ETS. (A) Diagrams of tagged rDNA units carrying four different deletions in the ETS. Numbers indicate the endpoints of the deletions, the 5′ end of the ETS being position 1. (B) Northern blot of RNA isolated from cells transformed with the various mutant rDNA units and from cells containing the intact, tagged rDNA unit (26S tag). The oligonucleotide complementary to the 26S tag was used as a probe. (C) Northern blot of RNA isolated from untransformed host cells, cells transformed with the intact rDNA unit carrying either the 26S tag or both the 17S and 26S tags, and cells transformed with the rDNA unit lacking nearly the entire ETS (ΔETS). The oligonucleotide complementary to the 17S tag was used as a probe.

and 46 nt flanking the 17S rRNA gene (Fig. 3). When these deletions were tested in the rDNA unit carrying a tag in the 26S rRNA gene, neither of them affected the production of tagged 26S rRNA (Fig. 3). Moreover, Northern (RNA) analysis of polysomal RNA fractions from cells that harbor the tagged operon lacking the complete ETS demonstrated that 60S subunits carrying the tagged 26S rRNA were fully functional.

Although these results show the ETS to be dispensable as far as formation of 60S subunits is concerned, 40S subunit formation does require the presence of the ETS in *cis*. This can be concluded from the fact that an rRNA operon with a tag in the 17S rRNA gene and carrying the complete ETS (Fig. 3C) or any of the three partial ETS deletions did not produce detectable levels of tagged 17S rRNA. Further experiments, using additional ETS deletion mutants, are necessary to identify the region of the ETS involved in 40S subunit formation and to elucidate its precise role in this process.

ITS1

In all of the nuclear rDNA units known to date, the genes coding for the large-subunit (LSU) rRNA are cotranscribed with the genes coding for the SSU rRNA, and in each case the SSU rRNA precedes the LSU rRNA in the primary transcript. Whereas in procaryotes and mitochondria several other genes may be found in the region that separates the rRNA genes (mostly tRNA genes), and the rRNA genes in the mitochondrial genome of *Chlamydomonas reinhardtii* are fragmented and scrambled among protein-coding and tRNA genes (Boer and Gray, 1988), in nuclear rDNA the intergenic region encompasses only a 5.8S rRNA gene flanked by two spacers, ITS1 and ITS2. ITS1, which separates the 3′ end of the 17S-18S rRNA gene from the 5′ end of the 5.8S rRNA gene, in *S. cerevisiae* has a length of 362 nt (Veldman et al., 1980) and contains an internal processing site that is cleaved at an early stage of rRNA maturation (Fig. 1, site A2). Subsequent processing events occurring in the nucleus generate the mature 5′ and 3′ ends of 5.8S and 26S rRNAs. Formation of the mature 3′ end of 17S rRNA, however, is delayed until the preribosomal particles have reached the cytoplasm (Udem and Warner, 1973; Trapman and Planta, 1976).

To investigate the role of ITS1 in ensuring the proper processing of *S. cerevisiae* pre-rRNA, we constructed a doubly tagged rDNA unit from which 160 nt spanning processing site A2 were deleted (Fig. 4A; cf. Fig. 1). Normal levels of tagged 26S rRNA accumulated in cells containing this construct, whereas no significant amounts of tagged 17S rRNA could be detected (Fig. 4). Reverse transcriptase mapping of transcripts of the tagged 26S rRNA gene (data not shown, but obtained in a similar way as the results described in Fig. 5) revealed correct processing at the 5′ end of 26S rRNA to have occurred. However, the removal of ITS2 from the transcripts of this construct seemed to take place in a somewhat dif-

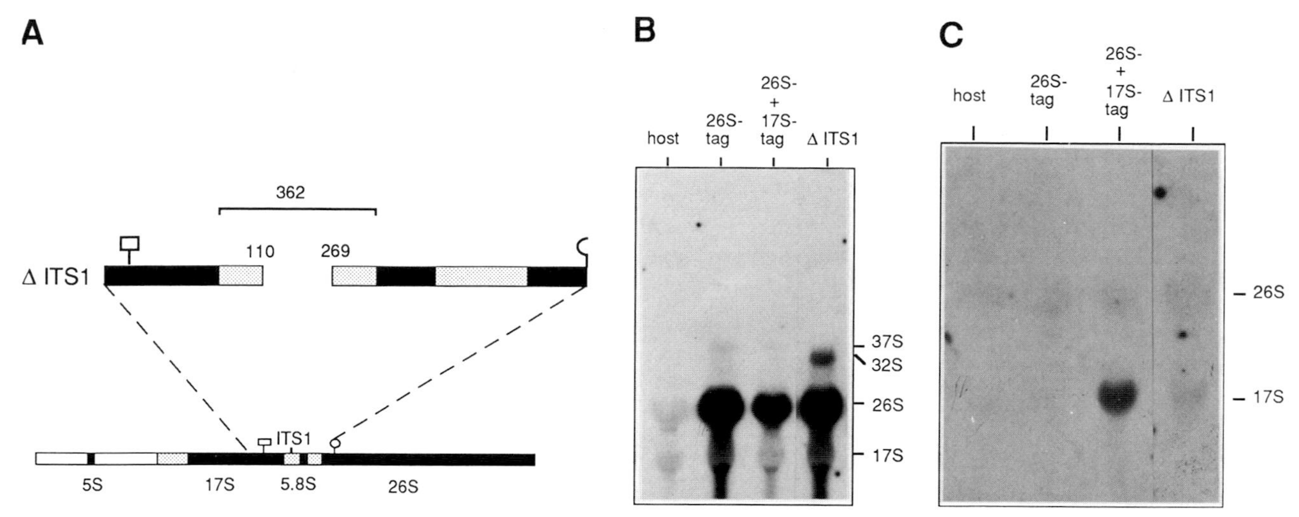

Figure 4. Functional analysis of ITS1. (A) Diagram of a tagged rDNA unit carrying a 160-base-pair deletion of the central part of ITS1. Numbers indicate the endpoints of the deletion, the 5′ end of ITS1 being position 1. (B) Northern blot of RNA isolated from untransformed host cells, cells transformed with the intact rDNA unit carrying either the 26S tag or both the 17S and 26S tags, and cells transformed with the mutant (ΔITS1) rDNA unit. The oligonucleotide complementary to the 26S tag was used as a probe. The autoradiograph has deliberately been overexposed to visualize the precursor bands. (C) Twin of the blot shown in panel B but probed with the oligonucleotide complementary to the 17S tag.

ferent way, since the same experiment demonstrated an elevated level of 29SB pre-rRNA. Finally, a tagged 32S precursor (Fig. 1C), not detected in the cells containing the unmutated tagged operon (Fig. 4), was found to accumulate in these cells.

These observations correlate well with the results described by Tollervey (1987), who has found that processing at sites A1 and A2 is assisted by small nuclear R10 RNA and that in *snr10⁻* mutants an alternative pathway using site B1 at an earlier stage is employed with greater efficiency. Our results suggest that the 29SB precursor generated by this pathway is efficiently used for the formation of 60S subunits, which is hardly surprising since it also is an intermediate of the normally used processing scheme. However, the SSU rRNA precursors generated by the alternative pathway (23S and 21S; Fig. 1C) do not give rise to the formation of 40S subunits, and we predict from our results that the cold sensitivity of *snr10⁻* mutants is due to a defect in the assembly of the small ribosomal subunits.

Furthermore, since processing at A1 was affected in *snr10⁻* mutants but not in our experiments (we detect the 32S RNA but not the 23S rRNA precursor), it can be inferred that the cleavages at sites A1 and A2 can proceed independently, as opposed to what has previously been suggested (Veldman et al., 1981).

ITS2

Whereas the LSU RNA in procaryotes is a single, continuous molecule, in eucaryotes it contains one or several discontinuities, resulting from the excision of transcribed spacers from the primary transcript. The most common example is represented by ITS2, the removal of which separates a 5.8S rRNA from the remainder of the LSU RNA in the cytoplasmatic ribosomes of all but one eucaryotic organism (Vossbrinck and Woese, 1986). Association between 5.8S rRNA (which is homologous to the 5′-terminal region of procaryotic LSU RNA) and 26S-28S rRNA is maintained by base pairing. Thus, ITS2 shares some characteristics with mRNA introns and is therefore sometimes called a pseudo-intron (Veldman et al., 1981).

To investigate whether ITS2 plays a distinct role in the process of ribosome formation in *S. cerevisiae*, we have constructed rDNA units with a tagged 26S rRNA gene, in which various parts of ITS2 have been deleted (Fig. 5). A precise removal of ITS2, fusing the genes for 5.8S and 26S rRNAs (analogous to the procaryotic arrangement of genes), prevented the accumulation of tagged rRNA (data not shown). This finding demonstrates that it is not possible to assemble fused LSU RNA molecules into 60S subunits in *S. cerevisiae*. Essentially the same results were obtained by using several partial deletions of ITS2, comprising either the middle part of ITS2 (encompassing processing site C2; Fig. 1), 50 nt at the 3′ end of ITS2, or all of ITS2 except for the 50 nt at its 3′ end. Since all of these deletions caused the rapid degradation of tagged (pre-)rRNA, we infer that the presence of an intact ITS2 in the early stages of 60S subunit formation is essential and that its removal should

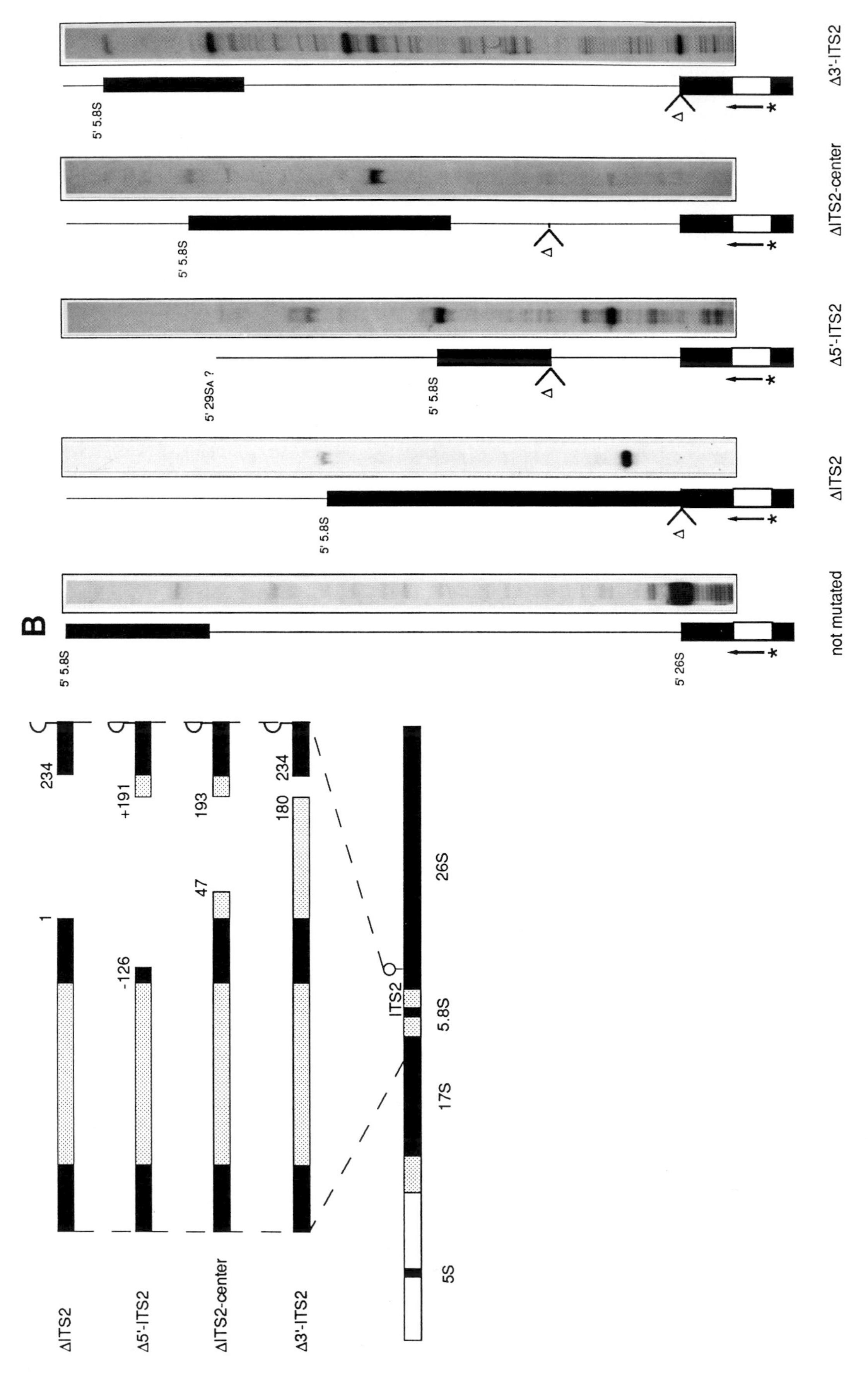
A
ΔITS2
Δ5'-ITS2
ΔITS2-center
Δ3'-ITS2
1
-126
47
234
+191
193
180
234
5S
17S
5.8S
ITS2
26S
B
5' 5.8S
5' 26S
not mutated
5' 5.8S
ΔITS2
5' 29SA ?
5' 5.8S
Δ5'-ITS2
5' 5.8S
ΔITS2-center
5' 5.8S
Δ3'-ITS2

Figure 5. Functional analysis of ITS2. (A) Diagram of four different deletions in ITS2 constructed in an rDNA unit carrying the 26S tag. Note that in Δ5′-ITS2 a large part of the 5.8S rRNA gene has also been deleted. Numbers indicate the endpoints of the deletions, the 5′ end of ITS2 being position 1. (B) 5′-End mapping of tagged transcripts with reverse transcriptase. A radiolabeled oligonucleotide, complementary to the 26S tag (arrow), was hybridized to total cellular RNA isolated from *S. cerevisiae* cells containing either the intact tagged operon (not mutated) or the deleted constructs and extended by reverse transcriptase. Next to each autoradiograph, a schematic representation of the pre-rRNA is given in such a way that the endpoints of the mature sequences (black bars) coincide with the expected positions of the corresponding reverse transcription products. The position of the deletion in the schematic pre-rRNA is indicated by (Δ).

take place only in later stages of eucaryotic ribosome formation.

To investigate in which way the deletions affect rRNA processing, we have mapped the 5′ ends of the tagged transcripts by using a reverse transcriptase assay (Fig. 5). For each mutant studied, a signal corresponding to the 5′ end of 5.8S rRNA is detected. This signal is derived from the 29SB precursor (Fig. 1). None of the mutants shows correct formation of the 5′ end of mature, tagged 26S rRNA except for the Δ3′-ITS2 mutant, in which a low amount of this 5′ end is formed. Preliminary results from control experiments using in vitro-synthesized RNA that contains the deletions have shown that all other signals observed are caused by artificial stops of the reverse transcriptase as a result of the high degree of secondary structure of the template RNA. Thus, the (deleted) 29SB precursor is formed in all mutants, but subsequent processing steps are blocked (ΔITS2, ΔITS2-center, and Δ5′-ITS2) or proceed with very low efficiency (Δ3′-ITS2). We therefore conclude that ITS2 must contain some structural element(s) required for its proper removal from the primary transcript and that the relevant information is located in an internal region of ITS2 rather than at its boundaries.

As mentioned above, Yao and co-workers (Yao and Yao, 1989; Sweeney and Yao, 1989) have recently described a system allowing functional analysis of (pre-)rRNA sequences in *Tetrahymena* cells (see also Raué et al. [this volume] for a description of the details of this system). Interestingly, these authors found a 119-nt insertion in ITS2 to cause a growth defect as well as the accumulation of processing intermediates. Moreover, it proved impossible to obtain dominant transformants by using mutant genes that contain larger insertions in ITS2. Both their and our results demonstrate that ITS2 makes an important contribution to the formation of eucaryotic ribosomes, and with these two in vivo systems at hand, we should soon be able to reach a better understanding of how this contribution is brought about.

CONCLUDING REMARKS

We have studied the transcribed spacers of the *S. cerevisiae* nuclear rDNA and found that large deletions in each of the spacer regions interfere with the formation of either the small ribosomal subunit (ETS and ITS1) or the large ribosomal subunit (ITS2). Despite the large extent of variation in the primary structures of these spacers, even among closely related species, this must mean that these elements serve a specific function in the process of ribosome formation, which explains their presence in virtually all eucaryotic nuclear rDNA units. Our next task will be to identify the relevant structural elements within the spacers by studying more subtle mutations, for which a better knowledge of the (evolutionary conservation of) the secondary structure of the spacers would be very helpful.

Finally, an intriguing possibility of the tools that are now available is to test the evolutionary conservation of spacer function by checking whether transcribed spacers derived from other organisms can functionally substitute for the transcribed spacers of *T. thermophila* or *S. cerevisiae*. The next years are likely to bring us more detailed information on these umbilical cords of the ribosome.

We express our gratitude to Co Klootwijk for his original ideas that initiated this project as well as for his supervision in the initial stage of the project. We also thank Kathy Boon for her excellent assistance in performing most of the experiments.

This study was supported in part by the Netherlands Foundation for Chemical Research, with financial aid from the Netherlands Organization for Scientific Research.

REFERENCES

Boer, P. H., and M. W. Gray. 1988. Scrambled ribosomal gene pieces in *Chlamydomonas reinhardtii* mitochondrial DNA. *Cell* 55:399–411.

Clark, C. G., B. W. Tague, V. C. Ware, and S. A. Gerbi. 1984. *Xenopus laevis* 28S ribosomal RNA: a secondary structure model and its evolutionary and functional implications. *Nucleic Acids Res.* 12:6197–6220.

Furlong, J. C., and B. E. H. Maden. 1983. Patterns of major divergence between the internal transcribed spacers of ribosomal DNA in *Xenopus borealis* and *Xenopus laevis*, and of minimal divergence within ribosomal coding regions. *EMBO J.* 2:443–448.

Gerbi, S. A. 1985. Evolution of ribosomal DNA, p. 419–517. *In* R. J. MacIntyre (ed.), *Molecular Evolutionary Genetics*. Plenum Publishing Corp., New York.

Hadjiolov, A. A. 1985. *The Nucleolus and Ribosome Biogenesis. Cell Biology Monographs*, vol. 12. Springer-Verlag, New York.

Herrera, A. H., and M. O. J. Olson. 1986. Association of protein C23 with rapidly labelled nucleolar RNA. *Biochemistry* 25:

6258–6264.

Jordan, G. 1987. At the heart of the nucleolus. *Nature* (London) 329:489–490.

Kass, S., N. Craig, and B. Sollner-Webb. 1987. Primary processing of mammalian rRNA involves two adjacent cleavages and is not species specific. *Mol. Cell. Biol.* 7:2891–2898.

Klootwijk, J., and R. J. Planta. 1989. Isolation and characterization of yeast ribosomal RNA precursors and preribosomes. *Methods Enzymol.* **180**:96–109.

Long, E. O., and I. B. Dawid. 1980. Repeated genes in eukaryotes. *Annu. Rev. Biochem.* **49**:727–764.

Musters, W., J. Venema, G. van der Linden, H. van Heerikhuizen, J. Klootwijk, and R. J. Planta. 1989. A system for the analysis of yeast ribosomal DNA mutations. *Mol. Cell. Biol.* **9**:551–559.

Raué, H. A., J. Klootwijk, and W. Musters. 1988. Evolutionary conservation of structure and function of high molecular weight ribosomal RNA. *Prog. Biophys. Mol. Biol.* **51**:77–129.

Sweeney, R., and M.-C. Yao. 1989. Identifying functional regions of rRNA by insertion mutagenesis and complete gene replacement in *Tetrahymena thermophila. EMBO J.* **8**:933–938.

Tollervey, D. 1987. A yeast small nuclear RNA is required for normal processing of pre-ribosomal RNA. *EMBO J.* **6**:4169–4175.

Trapman, J., and R. J. Planta. 1976. Maturation of ribosomes in yeast. I. Kinetic analysis by labelling of high molecular weight rRNA species. *Biochim. Biophys. Acta* **442**:265–274.

Udem, S. A., and J. R. Warner. 1973. The cytoplasmic maturation of a ribosomal precursor ribonucleic acid in yeast. *J. Biol. Chem.* **248**:1412–1416.

Veinot-Drebot, L. M., R. A. Singer, and G. C. Johnston. 1988. Rapid initial cleavage of nascent pre-rRNA transcripts in yeast. *J. Mol. Biol.* **199**:107–113.

Veldman, G. M., R. C. Brand, J. Klootwijk, and R. J. Planta. 1980. Some characteristics of processing sites in ribosomal precursor RNA of yeast. *Nucleic Acids Res.* **8**:2907–2920.

Veldman, G. M., J. Klootwijk, H. van Heerikhuizen, and R. J. Planta. 1981. The nucleotide sequence of the intergenic region between the 5.8S and 26S rRNA genes of the yeast ribosomal RNA operon. Possible implications for the interaction between 5.8S and 26S rRNA and the processing of the primary transcript. *Nucleic Acids Res.* **9**:4847–4862.

Verbeet, M. P., H. van Heerikhuizen, J. Klootwijk, R. D. Fontijn, and R. J. Planta. 1984. Evolution of yeast ribosomal DNA: molecular cloning of the rDNA units of Kluyveromyces lactis and Hansenula wingei and their comparison with the rDNA units of other Saccharomycetoideae. *Mol. Gen. Genet.* **195**: 116–125.

Vossbrinck, C. R., and C. R. Woese. 1986. Eukaryotic ribosomes that lack a 5.8S RNA. *Nature* (London) **320**:287–288.

Yao, M.-C. 1986. Amplification of ribosomal RNA genes, p. 179–201. *In* J. G. Gall (ed.), *The Molecular Biology of Ciliated Protozoa.* Academic Press, Inc., Orlando, Fla.

Yao, M.-C., and C.-H. Yao. 1989. Accurate processing and amplification of cloned germ line copies of ribosomal DNA injected into developing nuclei of *Tetrahymena thermophila. Mol. Cell. Biol.* **9**:1092–1099.

Chapter 38

Genetic Approaches to Ribosome Biosynthesis in the Yeast *Saccharomyces cerevisiae*

JONATHAN R. WARNER, DIANE M. BARONAS-LOWELL, FRANCIS J. ENG, STEWART P. JOHNSON, QIDA JU, and BERNICE E. MORROW

Perhaps the clearest lesson of this entire meeting is the fundamental similarity of ribosomes derived from all organisms and the implication that all ribosomes have evolved from a primeval ur-ribosome. The similarity is unmistakable for the RNA components of the ribosome, as shown so elegantly in the chapters in this volume by Noller et al., Pace and Burgin, Lake, and Gouy and Li. The relationships between the ribosomal proteins of organisms from different kingdoms are less clear because we must depend only on sequence rather than structure. There are a few examples of proteins homologous between eucaryotes and eubacteria, more between eucaryotes and archaebacteria. The relationships are treated extensively in this volume by Matheson et al. and Wittman-Liebold et al. Nevertheless, within the eucaryotes it is clear from a large number of examples (reviewed by Warner, 1989, and by Wool et al., this volume) that there is likely to be a one-to-one correlation between the ribosomal proteins, from yeasts through plants and invertebrates to mammals. Indeed, a mouse ribosomal protein will substitute for its yeast homolog (Fleming et al., 1989).

That being the case, conclusions derived from the study of one eucaryotic organism are likely to be applicable to many others. The use of the potent genetic techniques available for the simple budding yeast *Saccharomyces cerevisiae* may lead us to fundamental insights about the function and the regulation of the ribosomal proteins of all eucaryotes.

S. cerevisiae has several advantages for genetic analysis, many of which derive from the ease with which site-specific recombination can be accomplished (Botstein and Fink, 1988). Thus, it is possible to alter a gene, either substantially for a gene disruption or subtly by site-directed mutagenesis, and then to replace the wild-type gene with the altered one. Consequently, one can follow the effects of an altered gene not only in the few moments after injection or transfection of DNA but indefinitely, as the cell continues to grow. The availability of both haploid and diploid states facilitates the isolation of mutants, revertants, and suppressors and their sorting into complementation groups. The ability to recover all four products of meiosis permits rapid and powerful genetic analysis of both dominant and recessive mutations.

The structure of yeast ribosomes has recently been reviewed by Lee (1990), and their synthesis has been reviewed by Warner (1989). This paper will review selected aspects of the function and the synthesis of the components of the ribosome that have been carried out in this laboratory. Additional aspects are reviewed in this volume by Culbertson et al., Finley et al., Raué et al., and Musters et al.

STRUCTURE AND FUNCTION OF RIBOSOMAL PROTEINS

Phosphorylation of S10

Aside from the acidic proteins, only two of the ribosomal proteins in eucaryotic cells are phosphorylated (Warner, 1989; Zinker and Warner, 1976). One of these, termed S6 in most organisms but S10 in yeasts (yS10), is multiply phosphorylated, and its phosphorylation responds to a number of physiological effectors such as hormones (e.g., insulin [Smith et

Jonathan R. Warner, Diane M. Baronas-Lowell, Francis J. Eng, Qida Ju, and Bernice E. Morrow ■ Department of Cell Biology, Albert Einstein College of Medicine, 1300 Morris Park Avenue, Bronx, New York 10461. **Stewart P. Johnson** ■ Department of Pathology, Duke University Medical Center, Durham, North Carolina 27710.

```
                   230   **    *    *   *
RAT S6:     ---------KRRRLSSLRASTSKSESSQK COOH
                   227   **
YEAST S10:  ---------RKRRASSLKA COOH
                   227
YEAST S10*: ---------RKRRAAALKA COOH
```

Figure 1. C-terminal sequences of rat S6 and yS10. Phosphorylation sites are indicated by asterisks. The first phosphorylation sites of S6 were determined by Wettenhall and Morgan (1984), and the locations of all of them have recently been published (Krieg et al., 1988). The location of the phosphorylation sites in yS10 was surmised by comparison with those of S6 and determined as described in the text.

al., 1979]), tumor promoters (e.g., phorbol esters [Blenis and Erikson, 1985]), and oncogenes (Blenis and Erikson, 1986). In general, cells stimulated to proliferate, and sometimes to differentiate, have more extensively phosphorylated S6 (Thomas et al., 1982). An analogous situation exists in *S. cerevisiae*. yS10 is not phosphorylated in cells in stationary phase (Johnson and Warner, 1987). Dilution of such cells into fresh medium, leading to renewed growth, brings about rapid phosphorylation of yS10, which is maintained through log phase. However, despite a great deal of work, no convincing function for the phosphorylation of S6 has been demonstrated. Using the genetic techniques available in *S. cerevisiae*, we have attempted to determine a function for the phosphorylation of yS10 (Johnson and Warner, 1987).

The sequence of yS10 is 62% identical to that of S6 (Chan and Wool, 1988; Leer et al., 1985). The phosphorylation sites of S6 are shown in Fig. 1 (Krieg et al., 1988). It is clear that yS10 is shorter than S6 by 10 amino acids and is missing three of the phosphorylation sites. This is consistent with the fact that yS10 is phosphorylated only twice, whereas S6 can be phosphorylated five times.

To assess the function of the phosphorylation of yS10, the two phosphorylated serines were exchanged for alanines, using site-directed mutagenesis to yield yS10* (Fig. 1) (Johnson and Warner, 1987; Kruse et al., 1985). This was done for each of the two copies of the yS10 gene in the cell, and the wild-type genes were replaced by the mutant genes (Rothstein, 1983). The yS10 of the resultant cells was entirely devoid of phosphate. Nevertheless, the cells were indistinguishable from wild-type cells in their growth rate, response to heat shock, and a variety of other characteristics. Since to the best of our knowledge the role of the small subunit is primarily to establish a translation initiation complex, it has been suggested that the phosphorylation of S6 could alter the affinity of the subunit for different mRNAs, leading to an altered spectrum of proteins synthesized (Palen and

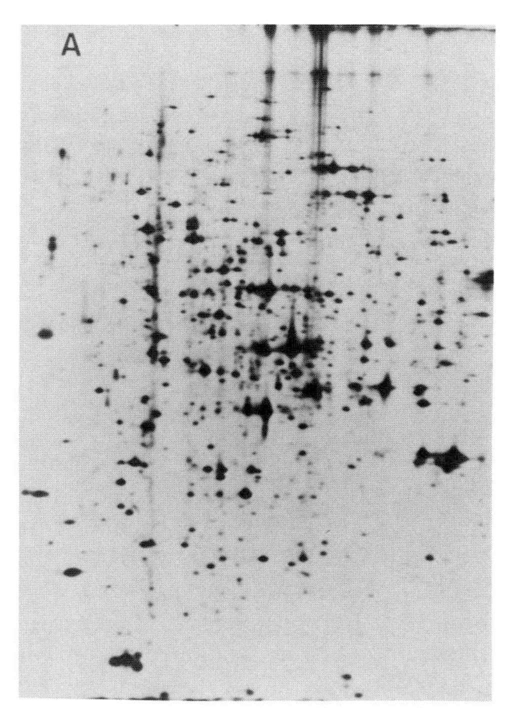

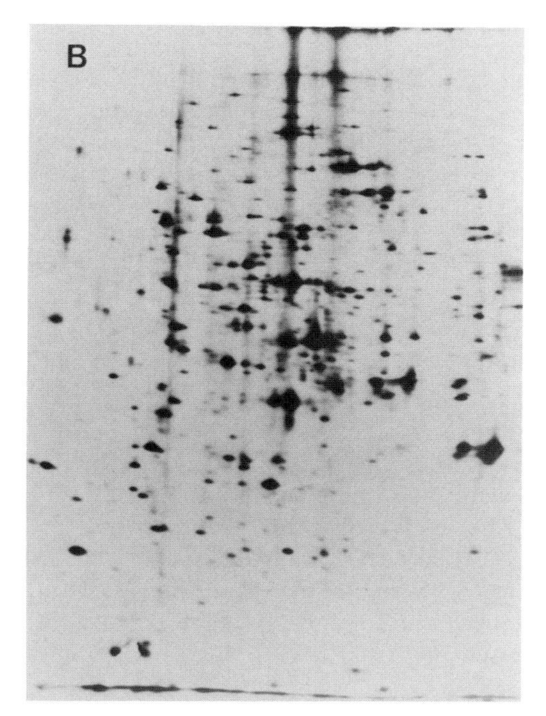

Figure 2. Proteins synthesized by ribosomes with and without phosphorylation of their yS10. Wild-type cells (A) or cells both of whose genes for yS10 had been converted to yS10* (B) were pulsed with [^{35}S]methionine for 15 min. Samples were prepared and analyzed by two-dimensional electrophoresis as described previously (Garrels, 1989; Johnson et al., 1988) at the Quest Protein Database facility at Cold Spring Harbor Laboratory.

```
L30A:   MKVEIDSFSGAKIYPGRGTLFVRGDSKIFRFQNSKSASLFKQRKNPRRIAWTVLFRKHHK   60
L30B:   ----V-------------------------------------------------------   60

L30A:   KGITEEVAKKRSRKTVKAQRPITGASLDLIKERRSLKPEVRKANREEKLKANKEKKKAEK  120
L30B:   --------------------------------------------------------R---  120

L30A:   AARKAEKAKSAGTQSSKFSKQQAKGAFQKVAATSR  155
L30B:   ------------V-G--V-----------------  155
```

Figure 3. Sequences of the two L30 proteins as predicted from the nucleotide sequences of the genes *RPL30A* and *RPL30B*.

Traugh, 1987). To determine whether such was the case, we examined the spectrum of proteins synthesized by cells with and without phosphorylation of yS10 by using two-dimensional polyacrylamide gel electrophoresis (Fig. 2). The patterns were the same (Johnson et al., 1988). Careful examination of nearly 1,000 spots did not reveal any whose synthesis was reproducibly altered by the presence of the phosphates on yS10.

We conclude that the phosphorylation of yS10 has no apparent effect on the yeast cell. If there is any effect, it is subtle indeed. What are the implications for the role of phosphorylation of S6? One possibility that now has some experimental support (Palen and Traugh, 1987; Thomas and Thomas, 1986) is that the additional three phosphates play some role in translation (Krieg et al., 1988). Alternatively, the phosphorylation of S6 may be simply a by-product of the activation of kinases by the agents listed above. Protein kinase C, which is activated by phorbol esters among other things, can phosphorylate a prodigious number of substrates, including S6 (Nishizuka, 1986), not all of which may be involved in the biological function of the kinase.

Dispensability of L30

In *S. cerevisiae*, most ribosomal proteins are coded for by two genes, both of which are functional (reviewed in Warner, 1989). This is in contrast to the situation in mammals, where there is usually only a single functional gene for each ribosomal protein, although there are many pseudogenes (Dudov and Perry, 1984). We have cloned (Fried et al., 1981), sequenced, and analyzed *RPL30A* and *RPL30B*, the two copies of the gene for ribosomal protein L30 (formerly known as RP29) (Mitra and Warner, 1984; D. Baronas-Lowell and J. R. Warner, submitted for publication). Sequences of the two proteins are compared in Fig. 3. As is usually the case, the two genes are slightly divergent; 35 of 467 nucleotides differ within the coding region, leading to conservative replacement of 5 of 155 amino acids. Immediately outside the coding region as well as within the intron, the genes diverge completely, suggesting that the duplication is ancient. Northern (RNA) analysis indicates that *RPL30A* and *RPL30B* are transcribed in a ratio of at least 3:1.

Each gene has been disrupted, *RPL30A* with *URA3* and *RPL30B* with *HIS3* (Rothstein, 1983). When strains carrying each disrupted gene were mated and the resultant diploids were allowed to sporulate, we were surprised to find four viable spores. Southern analysis indicated that a cell carrying both disrupted genes is viable. To demonstrate directly that cells could survive without L30, we analyzed the proteins of ribosomes isolated from the spores of a tetrad (Fig. 4). It is clear that in the double disruption there is no detectable L30. These cells grow with a doubling time roughly 40% longer than that of the wild-type cells. Experiments are under way to determine whether the lengthened doubling time is due to a slowing of protein synthesis, to an increase in the error frequency, or to a defect in the assembly of ribosomes.

Extensive genetic analysis of *Escherichia coli* has demonstrated that cells can survive the loss of any one of several ribosomal proteins, with variable effects on the growth rate (Dabbs, 1986). These experiments are more difficult to do in eucaryotic cells. Only in *S. cerevisiae* has it been measured directly. In this case, of at least 11 other ribosomal proteins that have been examined, only S27 is dispensable, and the cells without S27 are very sick (Finley et al., 1989; Finley et al., this volume). Thus, the case of L30 is unusual and may lead to insight into the structure of the eucaryotic ribosome.

N-Acetylation of Yeast Ribosomal Proteins

As purified from cell extracts, many eucaryotic proteins carry an acetyl group on the amino terminus. In some cases this acetylation has been carried out by the cell; in others it has occurred during purification of the protein. Little is known about the N-terminal acetylation of ribosomal proteins in eucaryotic cells.

In *S. cerevisiae*, two genes, *NAT1* and *ARD1*, have recently been shown to be necessary for the N-terminal acetylation of proteins (Mullen et al., 1989). Deletion of either gene leads to a lack of N-terminal acetylation. We have made use of such

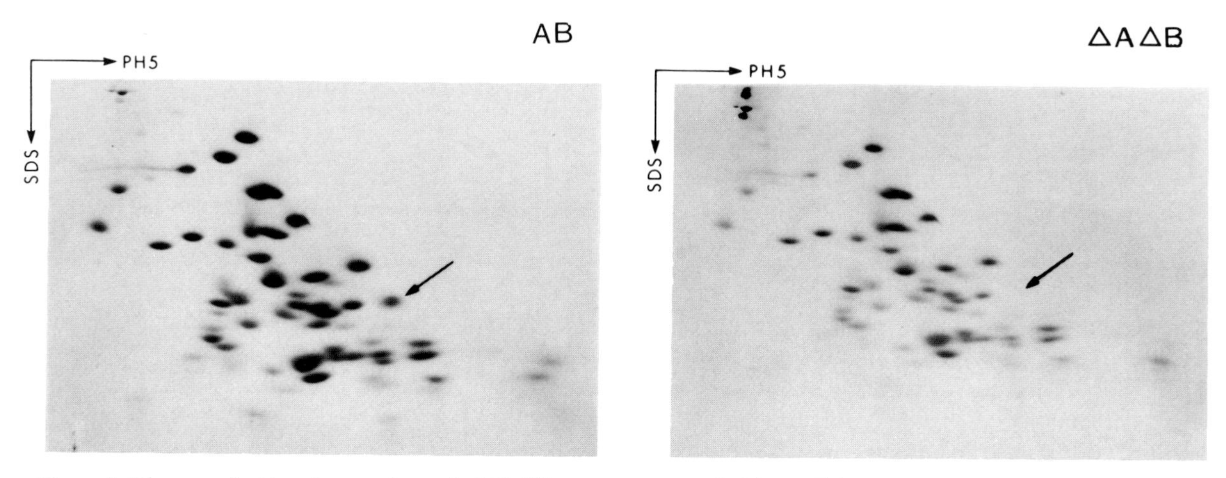

Figure 4. Ribosomes lacking ribosomal protein L30. Ribosomes were purified from all four spores of a tetrad derived from the sporulation of a diploid in which one copy of each *RPL30* gene had been disrupted. Proteins were purified and analyzed by two-dimensional electrophoresis (Gorenstein and Warner, 1976). Only those from the wild-type (AB) and the doubly disrupted (ΔAΔB) cells are shown. Ribosomal protein L30 is indicated by the arrow. (The former name of this protein was RP29 [Gorenstein and Warner, 1976]. We are grateful to J. Lee for the positive identification.) SDS, Sodium dodecyl sulfate.

deletions to determine which ribosomal proteins of *S. cerevisiae* have N-terminal acetyl groups. For cells carrying a deletion of *NAT1*, at least 11 of the ribosomal proteins migrate more rapidly toward the cathode, as expected if the amino terminus is unblocked (Fig. 5). Deletion of *NAT1* or *ARD1* has little effect on the growth of the cell. Thus, it appears that the acetylation of ribosomal proteins is not essential for the function of the ribosome.

REGULATION OF RIBOSOMAL PROTEINS

Autogenous Regulation of Splicing by L32

Introns are rare in *S. cerevisiae*. Of the several hundred nonribosomal genes that have been studied, only eight have introns. Yet 24 of the 31 ribosomal protein genes examined have an intron (reviewed in Warner, 1989; Woolford, 1989). Indeed, a number of mutations that were originally isolated as having a defect in ribosome biosynthesis (Warner and Udem, 1972) were actually mutations in various components of the splicing apparatus (Rosbash et al., 1981), i.e., the genes *PRP2-11* (formerly known as *RNA2-11*). It has been suggested that the paucity of introns in *S. cerevisiae* is due to recombination into the genome of reverse transcripts of spliced mRNAs (Fink, 1987). If that is the case, some selective effect must be responsible for the maintenance of introns in the ribosomal protein genes. We have demonstrated the possible basis of such selection for the gene

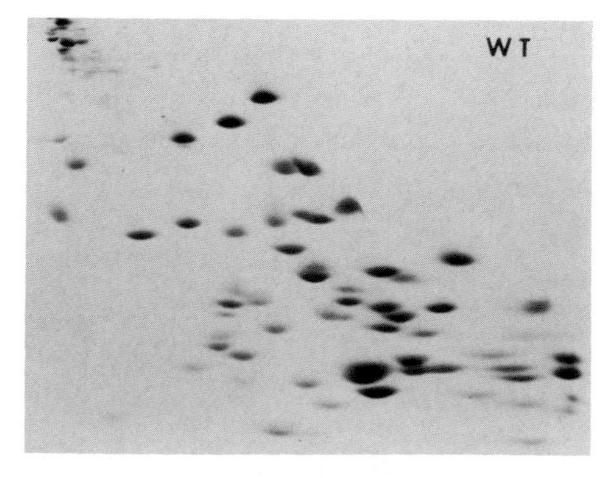

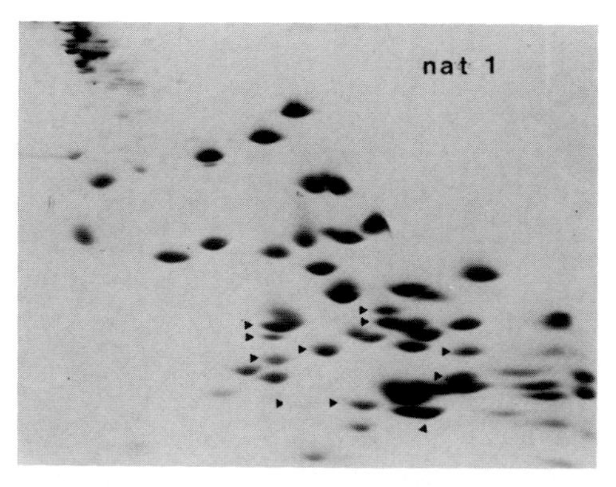

Figure 5. N-acetylation of yeast ribosomal proteins. Proteins were purified from the ribosomes of wild-type (WT) and mutant (*nat1*) cells and analyzed on two-dimensional gels (Gorenstein and Warner, 1976). Migration in the first dimension (horizontal) is from left (anode) to right (cathode) at pH 5.0 in the presence of 8 M urea. Eleven proteins from the mutant migrate more rapidly in the first dimension (arrowheads). A similar pattern was obtained by using proteins purified from cells carrying a deletion of *ARD1*, the other gene necessary for N-terminal acetylation.

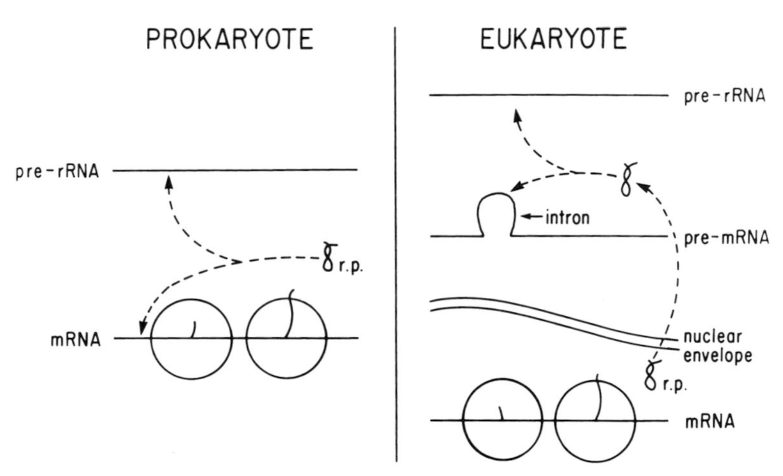

Figure 6. Model of autogenous regulation of splicing as compared with autogenous regulation of translation. See text for details. r.p., Ribosomal protein.

RPL32. Excess copies of *RPL32* lead to a substantial excess of unspliced transcripts but only a marginal excess of mRNA. This is because the product of the gene, ribosomal protein L32, inhibits the splicing of the transcript (Dabeva et al., 1986). This phenomenon could be considered a eucaryotic analog of the autogenous regulation of translation in procaryotes (Nomura et al., 1984) (Fig. 6). The rationale is that in eucaryotic cells, ribosome assembly and translation occur in different compartments. Therefore, competition between the assembling ribosome and the mRNA for a newly synthesized ribosomal protein is unlikely. An attractive alternative is a competition between the assembling ribosome and the unspliced transcript that can take place entirely within the nucleus. This may be a special case of regulation by alternative splicing (reviewed in Bingham et al., 1988). In *Xenopus laevis*, there is another case in which the synthesis of a ribosomal protein is regulated at splicing. Injection into an oocyte of many copies of the gene for ribosomal protein L1 leads to the accumulation of transcripts in which the third of nine introns is retained (Amaldi et al., 1989; Bozzoni et al., 1984).

Our recent work has attempted to understand the structural basis of the regulation of splicing; i.e., what sequences in the transcript of *RPL32* are necessary for the regulation of splicing, and what secondary structure might they take up? As one approach, we have constructed reciprocal gene fusions between *RPL32* and *RPS10A* (Fig. 7). Analysis of the RNA from cells cotransformed by the two plasmids revealed that the splicing of the transcript of the *LS* gene was subject to regulation and that it was responsive to the presence of the *SL* gene, which overproduces ribosomal protein L32. Thus, the 5′ exon and the 5′ half of the intron are sufficient to determine regulation. Deletion analysis has shown that the first 75 nucleotides of the *RPL32* transcript are sufficient to support the regulation of splicing (Fig. 8A). A number of gene fusions and site-directed mutations have demonstrated the importance of the nucleotides indicated in Fig. 8A for the regulation of splicing. By careful, but not necessarily justified, choice of parameters for the RNA-folding program of Zuker (1989), it is possible to obtain the structure shown in Fig. 8B, which shows the interaction of all of the sites thus far known to be involved in the regulation of splicing.

Splicing appears to initiate by the interaction of the small nuclear RNA U1 with sequences just downstream from the 5′ splice site (Siliciano and Guthrie, 1988; Woolford, 1989). The important aspect of the model shown in Fig. 8B is that these nucleotides are hidden by interaction with those near the beginning

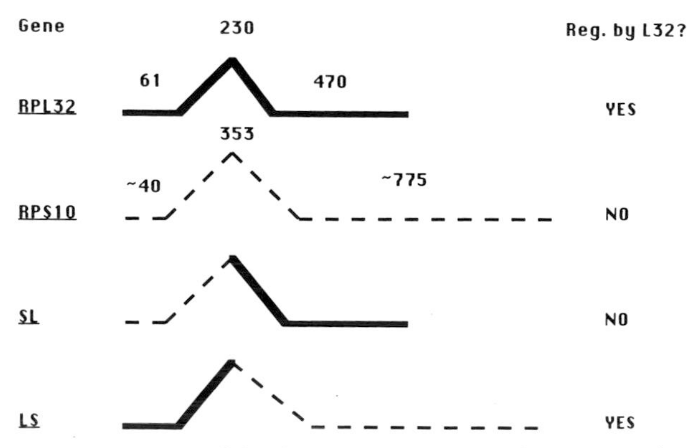

Figure 7. Diagram of the chimeric genes used to demonstrate that the sequences involved in the regulation of splicing are in the 5′ exon, the 5′ half of the intron, or both. The introns are in frame, that of *RPL32* being between codons 1 and 2 and that of *RPS10A* being between codons 2 and 3. When these genes are introduced into the cell, *SL* gives rise to functional L32 and *LS* gives rise to functional yS10.

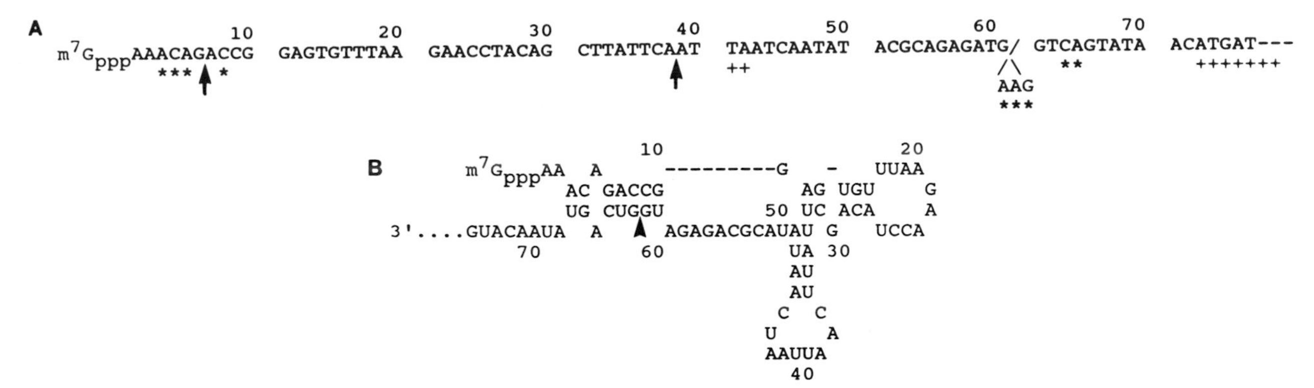

Figure 8. Portions of the transcript of *RPL32* involved in the regulation of its splicing. (A) The sequence of the first 75 nucleotides of the transcript, showing the splice site after nucleotide 61, as well as the nucleotides involved in (*) and dispensable for (+) the regulation of splicing. The AAG in the lower line blocks regulation of splicing if inserted just upstream of the splice site. Arrows indicate sites at which upstream sequences of *CYC1* were fused. In both cases, regulation of splicing was blocked. (B) Possible structure of the 5′ portion of the transcript that could occlude the 5′ splice site (arrowhead) to prevent interaction with U1 small nuclear RNA.

of the transcript. The structure shown in Fig. 8B is not very stable, as expected, since in the absence of excess L32 the transcript is spliced very effectively ($t_{1/2}$ of about 15 s [Dabeva et al., 1986]). We postulate that L32 binds to the structure shown in Fig. 8B to stabilize it and prevent its interaction with U1. The model predicts that there should be a homologous structure in rRNA. Indeed, there are sequences in 25S rRNA that resemble strongly the sequences that interact in Fig. 8B. Unfortunately, one is in an expansion segment whose structure remains unknown (Hogan et al., 1984).

rRNA TRANSCRIPTION

An Enhancer of rRNA Transcription

In nearly all eucaryotes, the genes for rRNA are arranged in a tandem array of several to several hundred copies. In *S. cerevisiae*, the tandem array consists of 100 to 200 copies of a transcription unit of about 6,600 nucleotides separated by a spacer of about 2,500 nucleotides (Fig. 9) (reviewed in Warner, 1989). The spacer includes the gene for 5S RNA, which is transcribed in the direction opposite that of the major rRNA. Unlike the spacer regions described for many other eucaryotes (reviewed in Sollner-Webb and Tower, 1986), the spacer in *S. cerevisiae* has no repeated sequences and no spacer promoter elements that can act in vivo.

In an effort to determine the sequences involved in initiating and regulating the transcription of rRNA, we constructed an RNA polymerase I transcription unit that yielded an identifiable product (containing bacteriophage T7 sequences) (Elion and Warner, 1984). Analysis of its transcription revealed that a specific enhancer of RNA polymerase I transcription lies 2 kilobase pairs (kbp) upstream of the transcription initiation site (Fig. 9). This enhancer has many of the properties of a polymerase II enhancer in that it is active in either orientation and either upstream or downstream of the gene. It is the only enhancer in *S. cerevisiae* to work from a downstream position (Elion and Warner, 1986). It differs from the 60- to 81-bp repeats described as an enhancer of polymerase I transcription in *X. laevis*, which for the most part do not stimulate transcrip-

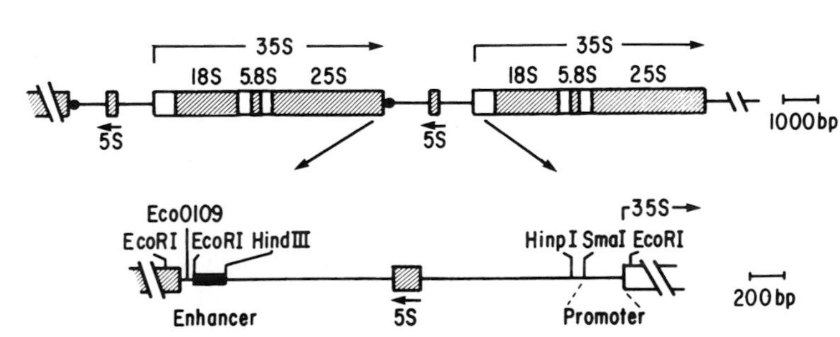

Figure 9. The ribosomal DNA repeat of *S. cerevisiae* showing the transcription units. Below is a blowup of the spacer region showing the location of the *Eco*RI-*Hin*dIII fragment that acts as an enhancer (see text).

	[T7A rRNA]	[T7B rRNA]
RR93	1.0	1.0
RR89	28.0±4.0	16.0±1.0
RR95	21.0±2.0	10.0±1.0
RR96	11.0±3.0	21.0±3.0
RR99	85.0±7.0	31.0±3.0
RR101	42.0±5.0	51.0±6.0
RR102	76.0±5.0	82.0±12.0
RR103	21.0±2.0	28.0±2.0

Figure 10. Diagram of the tandem constructs inserted into the *URA3* locus to test the effects of the enhancer. Bacteriophage T7 sequences are indicated by hatched and black regions; the RNA polymerase I transcription units are marked by arrows. The regions marked T contain the sequence leading to 3′ processing of the RNA polymerase I transcript. The arrowheads represent 190-bp *Eco*RI-*Hin*dIII fragments carrying the enhancer (Fig. 9). RR93 serves as the control. The relative concentrations of transcripts of the T7 A and T7 B genes are shown to the right (Johnson and Warner, 1989).

tion in *cis* but rather inhibit transcription in *trans* (Labhart and Reeder, 1984).

The 190-bp fragment that has been identified as an enhancer lies very near the end of the transcription unit (Fig. 9). It is itself transcribed at least to some degree (Johnson and Warner, 1989; Kempers-Veenstra et al., 1986; Mestel et al., 1989), and it appears to play a role in the termination process (Kempers-Veenstra et al., 1986). In addition, it contains sequences active as an RNA polymerase I promoter in vitro (Swanson and Holland, 1983), although there is little evidence for their use in vivo (Skryabin et al., 1984).

The association of terminator and enhancer function suggested that there might be some sort of coupling between termination and initiation (Elion and Warner, 1987; Kempers-Veenstra et al., 1986), which seems plausible for the transcription of a tandem repeat of active genes. To inquire into the nature of such coupling, we constructed an artificial tandem repeat containing two marked genes that were then integrated into the genome at the *URA3* locus (Johnson and Warner, 1989). A single enhancer (RR89) can stimulate transcription of two genes, even when it is upstream (RR95) or downstream (RR96) of both (Johnson and Warner, 1989) (Fig. 10). Termination is not essential for enhancer function, as is evident from RR95, in which the enhancer is not transcribed by RNA polymerase I. As a working model, we propose that there is an association between many (all?) terminator-enhancer elements and the promoter elements, looping out both the rRNA transcription unit and the spacer with its 5S RNA transcription unit, as in a lampbrush chromosome (Fig. 11). In this model, the enhancer acts by bringing the sites of termination and initiation together either to facilitate a direct passage of RNA polymerase I (presumably with its associated transcription factors) from one transcription unit to another (not necessarily the next) or simply to generate a localized high concentration of RNA polymerase I.

Proteins Associated with the Enhancer

Most models of enhancer function invoke protein factors that interact in some way both with DNA and with RNA polymerase. We have identified two proteins that bind the yeast rRNA enhancer, termed REB1 and REB2 (Morrow et al., 1989). REB1 has been purified to homogeneity. It binds not only to the enhancer but also to a similar sequence about 200 nucleotides upstream of the site of initiation of 35S RNA. In both cases, the protein protects a sequence CCGGGTA, as well as flanking sequences, in a footprint assay. Surprisingly, binding of REB1 to the promoter site protects from either nuclease or chemical cleavage a pair of otherwise hypersensitive nucleotides immediately adjacent to the transcription initiation site 200 nucleotides downstream (Morrow et al., 1989).

By screening a lambda gt11 library with a double-stranded oligonucleotide to which *REB1* binds (Vinson et al., 1988), we have recently cloned the

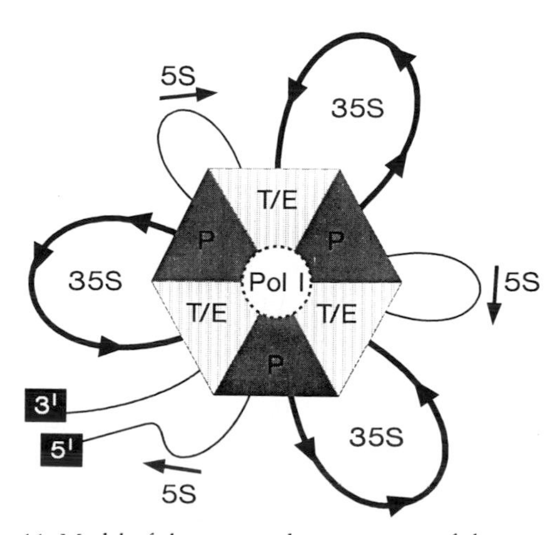

Figure 11. Model of the proposed arrangement of three copies of the ribosomal DNA repeat in growing cells of *S. cerevisiae*. P, Promoter element; T/E, terminator-enhancer element; Pol I, RNA polymerase I. The large loops are the 35S RNA transcription units; the small ones indicate the spacer containing the 5S RNA transcription unit. See text for details.

REB1 gene. *REB1* is essential for growth. Partial sequence analysis of *REB1* has revealed substantial similarity to the oncogene *myb*. Although *myb* is widespread and has been widely studied, no clear function has been attributed to it, except that it is generally active in proliferating cells (Gewirtz and Calabretta, 1988). Further analysis of *REB1* is proceeding.

The enhancer of rRNA transcription is complex. No one region of the 190-bp sequence is essential for full function (Mestel et al., 1989; S. P. Johnson, B. E. Morrow, and J. R. Warner, unpublished data). Furthermore, we have no direct evidence of a role for *REB1* in rRNA transcription. Nevertheless, its close association with both enhancer and promoter, as well as its indispensability, suggest that it may be important for maintaining a structure such as that conjectured in Fig. 11.

CONCLUSIONS

Many aspects of the structure and function of ribosomes are similar in procaryotes and eucaryotes. Yet many aspects of the synthesis of ribosomes differ between procaryotes and eucaryotes, reflecting the fundamental consequences of the evolution of the nuclear envelope that separates transcription from translation. The availability of the powerful genetic techniques for study of *S. cerevisiae* may help us not only to learn much about the structure and synthesis of ribosomes in eucaryotic cells but also, in the very analysis of ribosome synthesis, to learn much about the fundamental nature of eucaryotic cells.

The research in this laboratory was supported by Public Health Service grant GM25532 from the National Institutes of Health, by grant MV-323S from the American Cancer Society, and by Public Health Service grant 2P30CA13330 to the Chanin Cancer Center. D.M.B.-L. and F.J.E. are supported by Public Health Service training grant 5T32 GM07128 from the National Institutes of Health.

We are grateful to many colleagues for their contributions to this work over the years: Lefa Alksne, Mariana Dabeva, Cornelia Kruse, Lenore Neigeborn, Ke Shuai, and Mary Studeny. We also thank Julius Marmur for discussion and insight, Cal McLaughlin for help with the Quest Protein Database, Rolf Sternglanz for the *nat1* and *ard1* strains, and Ann Gorgoglione for secretarial assistance.

REFERENCES

Amaldi, F., I. Bozzoni, E. Beccari, and P. Pierandrei-Amaldi. 1989. Expression of ribosomal protein genes and regulation of ribosome biosynthesis in *Xenopus* development. *Trends Biochem. Sci.* **14:**175–178.

Bingham, P. M., T.-B. Chou, I. Mims, and Z. Zachar. 1988. On/off regulation of gene expression at the level of splicing. *Trends Genet.* **4:**134–138.

Blenis, J., and R. L. Erikson. 1985. Regulation of a ribosomal protein S6 kinase activity by the Rous sarcoma virus transforming protein, serum, or phorbol ester. *Proc. Natl. Acad. Sci. USA* **82:**7621–7625.

Blenis, J., and R. L. Erikson. 1986. Stimulation of ribosomal protein S6 kinase activity by pp60$^{v\text{-}src}$ or by serum: dissociation from phorbol ester-stimulated activity. *Proc. Natl. Acad. Sci. USA* **83:**1733–1737.

Botstein, D., and G. R. Fink. 1988. Yeast: an experimental organism for modern biology. *Science* **240:**1439–1443.

Bozzoni, I., P. Fragapane, F. Annesi, P. Pierandrei-Amaldi, F. Amaldi, and E. Beccari. 1984. Expression of two *Xenopus laevis* ribosomal protein genes in injected frog oocytes. A specific splicing block interferes with the L1 RNA maturation. *J. Mol. Biol.* **180:**987–1005.

Chan, Y.-L., and I. G. Wool. 1988. The primary structure of rat ribosomal protein S6. *J. Biol. Chem.* **263:**2891–2896.

Dabbs, E. 1986. Mutant studies on the prokaryotic ribosome, p. 733–748. *In* B. Hardesty and G. Kramer (ed.), *Structure, Function, and Genetics of Ribosomes*. Springer-Verlag, New York.

Dabeva, M. D., M. A. Post-Beittenmiller, and J. R. Warner. 1986. Autogenous regulation of splicing of the transcript of a yeast ribosomal protein gene. *Proc. Natl. Acad. Sci. USA* **83:**5854–5857.

Dudov, K. P., and R. P. Perry. 1984. The gene family encoding the mouse ribosomal protein L32 contains a uniquely expressed intron-containing gene and an unmutated processed gene. *Cell* **37:**457–468.

Elion, E. A., and J. R. Warner. 1984. The major promoter element of rRNA transcription in yeast lies 2 kb upstream. *Cell* **39:** 663–673.

Elion, E. A., and J. R. Warner. 1986. An RNA polymerase I enhancer in *Saccharomyces cerevisiae*. *Mol. Cell. Biol.* **6:**2089–2097.

Elion, E. A., and J. R. Warner. 1987. Characterization of a yeast RNA polymerase I enhancer. *UCLA Symp. Mol. Cell. Biol. N. Ser.* **52:**21–29.

Fink, G. R. 1987. Pseudogenes in yeast? *Cell* **49:**5–6.

Finley, D., B. Bartell, and A. Varshavsky. 1989. The tails of ubiquitin precursors are ribosomal proteins whose fusions to ubiquitin facilitate ribosome biogenesis. *Nature* (London) **338:** 394–401.

Fleming, G., P. Belhumeur, D. Skup, and H. M. Fried. 1989. Functional substitution of mouse ribosomal protein L27′ for yeast ribosomal protein L29 in yeast ribosomes. *Proc. Natl. Acad. Sci. USA* **86:**217–221.

Fried, H. M., N. J. Pearson, C. H. Kim, and J. R. Warner. 1981. The genes for fifteen ribosomal proteins of *Saccharomyces cerevisiae*. *J. Biol. Chem.* **256:**10176–10183.

Garrels, J. I. 1989. The QUEST system for quantitative analysis of two-dimensional gels. *J. Biol. Chem.* **264:**5269–5282.

Gewirtz, A. M., and B. Calabretta. 1988. A c-myb antisense oligodeoxynucleotide inhibits normal human hematopoiesis *in vitro*. *Science* **242:**1303–1306.

Gorenstein, C., and J. R. Warner. 1976. Coordinate regulation of the synthesis of eukaryotic ribosomal proteins. *Proc. Natl. Acad. Sci. USA* **73:**1547–1551.

Hogan, J. J., R. R. Gutell, and H. F. Noller. 1984. Probing the conformation of 26S rRNA in yeast 60S ribosomal subunits with kethoxal. *Biochemistry* **23:**3330–3335.

Johnson, S. P., C. McLaughlin, and J. R. Warner. 1988. Thoughts on the phosphorylation of a ribosomal protein, p. 145–157. *In* M. F. Tuite, M. Picard, and M. Bolotin-Fukuhara (ed.), *Genetics of Translation: New Approaches*. Springer-Verlag, New York.

Johnson, S. P., and J. R. Warner. 1987. Phosphorylation of the *Saccharomyces cerevisiae* equivalent of ribosomal protein S6 has no detectable effect on growth. *Mol. Cell. Biol.* **7:**1338–1345.

Johnson, S. P., and J. R. Warner. 1989. Unusual enhancer function in yeast ribosomal RNA transcription. *Mol. Cell. Biol.* **9:** 4986–4993.

Kempers-Veenstra, A. E., J. Oliemans, H. Offenberg, A. F. Dekker, P. W. Piper, R. J. Planta, and J. Klootwijk. 1986. 3′-End formation of transcripts from the yeast rRNA operon. *EMBO J.* **5:**2703–2710.

Krieg, J., J. Hofsteenge, and G. Thomas. 1988. Identification of the 40 S ribosomal protein S6 phosphorylation sites induced by cycloheximide. *J. Biol. Chem.* **263:**11473–11477.

Kruse, C., S. P. Johnson, and J. R. Warner. 1985. Phosphorylation of the yeast equivalent of ribosomal protein S6 is not essential for growth. *Proc. Natl. Acad. Sci. USA* **82:**7515–7519.

Labhart, P., and R. H. Reeder. 1984. Enhancer-like properties of the 60/81 bp elements in the ribosomal gene spacer of *Xenopus laevis*. *Cell* **37:**285–289.

Lee, J. C. 1990. Ribosomes from *Saccharomyces cerevisiae*, p. 489–539. *In* A. H. Rose and R. Harrison (ed.), *The Yeasts*, 2nd ed., vol. 4. Academic Press, Inc., New York.

Leer, R. J., M. M. C. van Raamsdonk-Duin, C. M. T. Molenaar, H. M. A. Witsenboer, W. H. Mager, and R. J. Planta. 1985. Yeast contains two functional genes coding for ribosomal protein S10. *Nucleic Acids Res.* **13:**5027–5039.

Mestel, R., M. Yip, J. P. Holland, E. Wang, J. Kang, and M. J. Holland. 1989. Sequences within the spacer region of yeast rRNA cistrons that stimulate 35S rRNA synthesis in vivo mediate RNA polymerase I-dependent promoter and terminator activities. *Mol. Cell. Biol.* **9:**1243–1254.

Mitra, G., and J. R. Warner. 1984. A yeast ribosomal protein gene whose intron is in the 5′ leader. *J. Biol. Chem.* **259:**9218–9224.

Morrow, B. E., S. P. Johnson, and J. R. Warner. 1989. Proteins that bind to the yeast rDNA enhancer. *J. Biol. Chem.* **264:** 9061–9064.

Mullen, J. R., P. S. Kayne, R. P. Moerschell, S. Tsunasawa, M. Gribskov, M. Colavito-Shepanski, M. Grunstein, F. Sherman, and R. Sternglanz. 1989. Identification and characterization of genes and mutants for an N-terminal acetyltransferase from yeast. *EMBO J.* **8:**2067–2075.

Nishizuka, Y. 1986. Studies and perspectives of protein kinase C. *Science* **233:**305–312.

Nomura, M., R. Gourse, and G. Baughman. 1984. Regulation of the synthesis of ribosomes and ribosomal components. *Annu. Rev. Biochem.* **53:**75–117.

Palen, E., and J. A. Traugh. 1987. Phosphorylation of ribosomal proteins S6 by cAMP-dependent protein kinase and mitogen-stimulated S6 kinase differentially alters translation of globin mRNA. *J. Biol. Chem.* **262:**3518–3523.

Rosbash, M., P. K. W. Harris, J. L. Woolford, Jr., and J. L. Teem. 1981. The effect of temperature sensitive RNA mutants on the transcription products from cloned ribosomal protein genes of yeast. *Cell* **24:**679–686.

Rothstein, R. J. 1983. One-step gene disruption in yeast. *Methods Enzymol.* **101:**202–211.

Siliciano, P. G., and C. Guthrie. 1988. 5′ splice site selection in yeast: genetic alterations in base pairing with U1 reveal additional requirements. *Genes Dev.* **2:**1258–1267.

Skryabin, K. G., M. A. Eldarov, V. L. Larionov, A. A. Bayev, J. Klootwijk, V. C. H. F. de Regt, G. M. Veldman, R. J. Planta, O. I. Georgiev, and A. A. Hadjiolov. 1984. Structure and function of the nontranscribed spacer regions of yeast rDNA. *Nucleic Acids Res.* **12:**2955–2968.

Smith, C. J., P. J. Wejksnora, J. R. Warner, C. S. Rubin, and O. M. Rosen. 1979. Insulin-stimulated protein phosphorylation in 3T3-L1 preadipocytes. *Proc. Natl. Acad. Sci. USA* **76:**2725–2729.

Sollner-Webb, B., and J. Tower. 1986. Transcription of cloned eukaryotic ribosomal RNA genes. *Annu. Rev. Biochem.* **55:** 801–830.

Swanson, M. E., and M. J. Holland. 1983. RNA polymerase I-dependent selective transcription of yeast ribosomal DNA. Identification of a new cellular ribosomal RNA precursor. *J. Biol. Chem.* **258:**3242–3250.

Thomas, G., J. Martin-Perez, M. Siegmann, and A. M. Otto. 1982. The effect of serum, EGF, PGF_{2a} and insulin on S6 phosphorylation and the initiation of protein and DNA synthesis. *Cell* **30:**235–242.

Thomas, G., and G. Thomas. 1986. Translational control of mRNA expression during the early mitogenic response in Swiss mouse 3T3 cells: identification of specific proteins. *J. Cell Biol.* **103:**2137–2144.

Vinson, C. R., K. L. LaMarco, P. F. Johnson, W. H. Landschulz, and S. L. McKnight. 1988. *In situ* detection of sequence-specific DNA binding activity specified by a recombinant bacteriophage. *Genes Dev.* **2:**801–806.

Warner, J. R. 1989. Synthesis of ribosomes in *Saccharomyces cerevisiae*. *Microbiol. Rev.* **53:**256–271.

Warner, J. R., and S. A. Udem. 1972. Temperature sensitive mutations affecting ribosome synthesis in *Saccharomyces cerevisiae*. *J. Mol. Biol.* **65:**243–257.

Wettenhall, R. E. H., and F. J. Morgan. 1984. Phosphorylation of hepatic ribosomal protein S6 on 80 and 40S ribosomes: primary structure of S6 in the region of the major phosphorylation sites for cAMP-dependent protein kinases. *J. Biol. Chem.* **259:**2084–2091.

Woolford, J. L. 1989. Nuclear pre-mRNA splicing in yeast. *Yeast* **5:**439–457.

Zinker, S., and J. R. Warner. 1976. The ribosomal proteins of *Saccharomyces cerevisiae*: phosphorylation and exchangeable proteins. *J. Biol. Chem.* **251:**1799–1807.

Zuker, M. 1989. On finding all optimal foldings of an RNA molecule. *Science* **244:**48–52.

Chapter 39

A Role for U3 Small Nuclear Ribonucleoprotein in the Nucleolus?

SUSAN A. GERBI, ROCCO SAVINO, BARBARA STEBBINS-BOAZ, CLAUS JEPPESEN, and RAFAEL RIVERA-LEÓN

snRNPs are small nuclear ribonucleoproteins found in the nuclei of eucaryotic cells (reviewed in Busch et al., 1982; Birnstiel, 1988). The snRNA component of these particles usually is rich in uridine; therefore, the various snRNA species have been named U1, U2, U3, etc. The U snRNAs are metabolically stable and are synthesized by RNA polymerase II. However, unlike other RNA polymerase II transcripts, the snRNAs lack a poly(A) tail at their 3′ ends and possess a unique 2,2,7-trimethylguanosine cap structure (m^3G) that is added in the cytoplasm to their 5′ ends (Mattaj, 1986). An exception to this is U6 snRNA, which is synthesized by RNA polymerase III and has a nonnucleotide cap structure.

Anti-Sm/RNP autoantibodies from patient sera immunoprecipitate U1, U2, U4/U6, and U5 snRNPs, all of which share a single-stranded consensus Sm protein binding site $PuA(U)_nGPu$ in their snRNA moieties. U1, U2, U4/U6, and U5 snRNPs assemble in a 1:1:1:1 stoichiometry into spliceosomes, which mediate intron removal and splicing together of exons during mRNA maturation. Other snRNPs play additional roles in mRNA maturation: U7 snRNP helps form the 3′ end of histone pre-mRNA (reviewed in Birnstiel, 1988; Mowry and Steitz, 1988), whereas U11 snRNP is implicated in polyadenylation (Christofori and Keller, 1988).

Unlike the snRNPs listed above, which are present in the nucleoplasm, U3 snRNP is localized within the nucleolus (Weinberg and Penman, 1968; Prestayko et al., 1970; Zieve and Penman, 1976). Since the nucleoplasmic snRNPs participate in mRNA maturation, it seems likely that U3 snRNP may play a role in rRNA maturation, which occurs in the nucleolus. In this chapter, we review the evidence that the nucleolus is the site of ribosome biogenesis, the biochemical steps for rRNA maturation, the cytological components of the nucleolus where these events occur, and the structure and possible function(s) of U3 snRNA.

HISTORICAL OVERVIEW: RIBOSOME BIOGENESIS OCCURS IN THE NUCLEOLUS

The nucleolus is a prominent, spherical, non-membrane-bounded object located near the nuclear envelope within the nuclei of eucaryotic cells. It has fascinated biologists over the past 200 years, since its description in 1781 by Fontana in his "Traité sur le venim de la vipère." The first 100 years saw the production of about 700 papers on this subject, which were reviewed in a classic work by Montgomery (1898); several excellent reviews have been published in more modern times (Vincent, 1955; Perry, 1967, 1981; Busch and Smetana, 1970; Miller, 1981; Bielka, 1982; Jordan and Cullis, 1982; Goessens, 1984; Hadjiolov, 1985; Sommerville, 1986; Reimer et al., 1987b).

Over 50 years ago, Heitz (1931, 1933) and McClintock (1934) coined the term "nucleolus organizer region" (NOR) for the specific chromosomal site(s) where the nucleolus forms at telophase. During mitosis the nucleolus disappears, and the NOR can generally be recognized as a secondary constriction (Heitz, 1931, 1933; McClintock, 1934) usually located near the telomere of a condensed chromosome (Lima-de-Faria, 1976). At telophase, there is a good correspondence between the number of nucleoli that reform and the number of NORs per cell, but later in interphase the number of nucleoli may decrease as they coalesce with one another (Anastassova-Kristeva, 1977; Sigmund et al., 1979; Wachtler et al., 1982). In contrast, the number of nucleoli may

Susan A. Gerbi, Rocco Savino, Barbara Stebbins-Boaz, Claus Jeppesen, and Rafael Rivera-León ■ Division of Biology and Medicine, Brown University, Providence, Rhode Island 02912.

increase substantially during oogenesis in many organisms, reflecting extrachromosomal rDNA amplification (reviewed in Jordan and Cullis, 1982). The nucleolus is the cytological manifestation of NOR activity during interphase and fails to appear after UV or laser microbeam inactivation of NORs (Burns et al., 1970; Sakharov et al., 1972). In the latter cases as well as in the anucleolate mutant of *Xenopus laevis* (Elsdale et al., 1958), which lacks almost all of its rDNA (Wallace and Birnstiel, 1966; Birnstiel et al., 1966; Steele et al., 1984; Tashiro et al., 1986), fibrous pseudonucleoli form instead (Jones, 1965; Hay and Gurdon, 1967). These observations showed that there is a framework or matrix of the nucleolus whose components are specified other than by the NOR; the nucleolar matrix was shown to contain not just protein but also RNA (Hay and Gurdon, 1967). Shortly in this chapter we will describe the biochemical identity of some of these nucleolar proteins and snRNAs.

Clues about the function of the nucleolus first came from the cytochemical studies of Caspersson (1950) and Brachet (1957) showing that the nucleolus contains RNA; they hypothesized that this RNA might be related to cytoplasmic RNA. Indeed, other cytological studies noted that the granular component of the nucleolus (Porter, 1954; Gall, 1956; Swift, 1959) appeared similar to the "Palade granules" (equivalent to ribosomes) of the cytoplasm (Palade, 1955; Palade and Siekevitz, 1956). Subsequent pulse-chase experiments (Woods and Taylor, 1959; Perry et al., 1961; Perry, 1962) and base composition comparisons (Edström et al., 1961) proved that nucleolar RNA became stable, cytoplasmic RNA. Thus, the role for the nucleolus in ribosome biogenesis became established. Shortly thereafter, with the advent of the technique of RNA-DNA hybridization, it was shown that the NOR contains the rRNA genes (ribosomal DNA [rDNA]); these experiments used *Drosophila melanogaster* (Ritossa and Spiegelman, 1965; Ritossa et al., 1966) and *X. laevis* (Wallace and Birnstiel, 1966; Birnstiel et al., 1966) carrying deletions of rDNA. Development of the methodology of in situ hybridization allowed the direct visualization by light microscopy of rDNA within the nucleoli of amphibia (Gall and Pardue, 1969; John et al., 1969), flies (Pardue et al., 1970), and other organisms. In the same year, by electron microscopy of spread chromatin, one could see active transcription of rDNA (Miller and Beatty, 1969). The active transcription units looked like Christmas trees on their sides and contained nascent rRNA lateral fibrils (Angelier et al., 1979) bound to the rDNA axis by granules of RNA polymerase I (Miller and Hamkalo, 1972a, 1972b; Franke et al., 1976; Franke et al., 1979). Each active transcription unit was separated from the next by an apparently nontranscribed spacer. At the chromosomal NOR, such a tandem array comprises the several hundred rDNA units previously quantitated by molecular hybridization. In the following section, we will examine the formation and processing of the rDNA transcripts.

BIOCHEMICAL EVENTS IN rRNA PRODUCTION

Initiation

With the developmental activation of rRNA synthesis, it can be seen by electron microscopy that the promoter of each transcription unit is activated independently of adjacent transcription units: some units are fully loaded with RNA polymerase I, whereas other units nearby may still be inactive (Meyer and Hennig, 1974; McKnight and Miller, 1976; Scheer et al., 1976). At least in *Xenopus* oocytes, RNA polymerase I itself does not seem to be limiting: the rDNA content of a *Xenopus* oocyte nucleus can be increased 2 to 3 orders of magnitude by injection of cloned rDNA, and half of the added rDNA becomes densely packed with RNA polymerase I transcripts (Trendelenburg and Gurdon, 1978). Species-specific transcription factors assist RNA polymerase I to initiate at the start site, and this is a field of experimental study for many (reviewed in Jacob, 1986; Sollner-Webb and Tower, 1986). Identification of 5′-terminal triphosphate as a capping substrate (Reeder et al., 1977) and direct demonstration of 5′-terminal triphosphate (Batts-Young and Lodish, 1978; Niles, 1978; Klootwijk et al., 1979) show that the start site for initiation of transcription is identical to the 5′ end of the initial rRNA precursor which can be isolated.

Termination

After initiation, RNA polymerase I transcribes the rDNA to produce a large pre-rRNA precursor containing, from its 5′ to 3′ ends, the external transcribed spacer (ETS), 18S, 5.8S embedded within the internal transcribed spacers (ITS1 and ITS2), and 28S (Fig. 1). After much debate, this polarity of the pre-rRNA precursor was finally established (Trapman and Planta, 1975; Dawid and Wellauer, 1976; Reeder et al., 1976).

The 3′ end of 28S rRNA from many organisms has the consensus sequence GAUUUAU$_{OH}$, and transcription of *Xenopus* rDNA does not terminate properly if the stretch of U's is shortened to only two U's (Bakken et al., 1982). Moreover, since the 3′ end of the *Xenopus* pre-rRNA precursor is identical to the

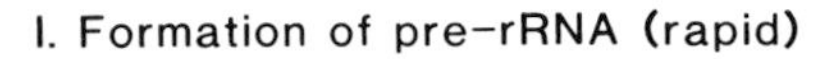

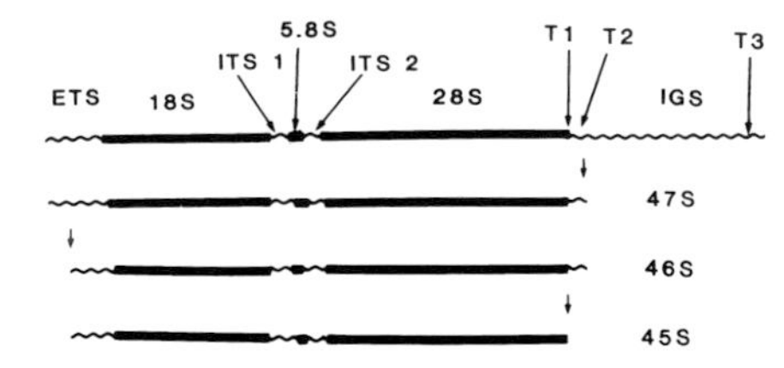

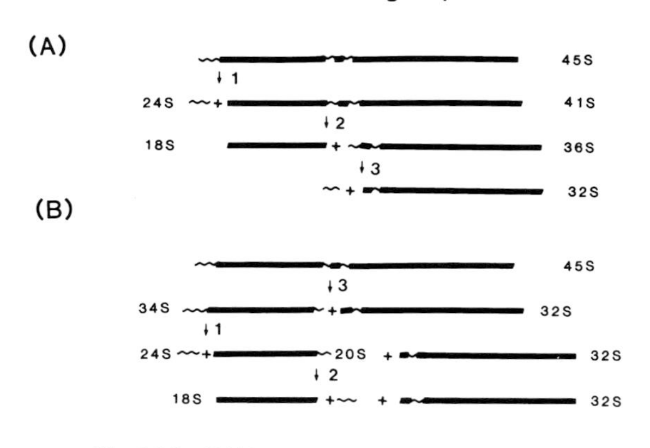

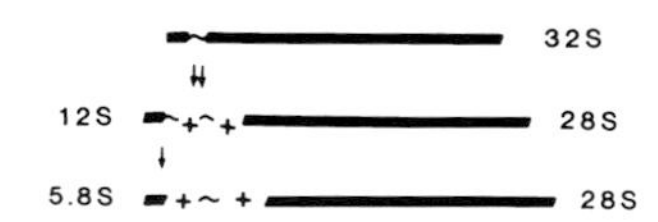

Figure 1. Eucaryotic rRNA processing. Shown is a composite of data from many different organisms, and the sizes given are generally for mouse RNA (Bowman et al., 1983). rRNA processing can be divided into three major steps. Step I, formation of pre-rRNA, is rapid. In *Xenopus*, RNA polymerase I continues until termination site T3, but there is rapid processing at T2 (Labhart and Reeder, 1986). There is no evidence that RNA polymerase I continues past the counterpart of T2 in mouse or yeast rDNA (see text). Next, there is cleavage within the ETS of many organisms (see text), but it is unknown whether this also happens in *Xenopus* and yeast. Finally, the 3′ end of the pre-rRNA is clipped at the end of the 28S region. Step II is also rapid, and the order of cleavages is variable between and even within organisms. Mouse L cells and *Xenopus* and *Drosophila* cells follow step II.(A), but HeLa cells follow II.(B). In either case, the net result is formation of mature 18S rRNA and 32S pre-rRNA. Step III is slower and involves cleavages in ITS2 to form mature 28S rRNA and then cleavage at the 5′ end of ITS2, resulting in mature 5.8S RNA.

3′ end of mature 28S rRNA (Sollner-Webb and Reeder, 1979), it was commonly believed that termination of RNA polymerase I transcription occurred at this point. However, investigators have more recently come to the conclusion that although this sequence is used as a recognition signal for pre-rRNA 3′-end formation, it is not the site for termination itself. Instead, RNA polymerase I continues transcription further downstream into the nontranscribed spacer (which therefore has been renamed the intergenic spacer, or IGS). Rapid processing events ensue such that the 3′ end of the predominant pre-rRNA precursor does indeed match the 3′ end of 28S rRNA. Transcription beyond the 3′ end of 28S has now been reported for a large number of organisms: yeasts (Klootwijk et al., 1979; Kempers-Veenstra et al., 1986), *D. melanogaster* (Tautz and Dover, 1986), mice (Grummt et al., 1985), humans (Bartsch et al., 1987), and *X. laevis* (Labhart and Reeder, 1986). In *X. laevis*, runoff experiments showed that RNA polymerase I spans almost all of the intergenic spacer and is not released until termination site T3, which is 215 base pairs (bp) upstream of the promoter for the adjacent rDNA unit (Labhart and Reeder, 1986).

Although RNA polymerase I transcribes *Xenopus* rDNA to T3, rapid processing occurs such that the nascent pre-rRNA precursor is released from the polymerase at site T2. In *X. laevis*, site T2 is only 235 bp downstream of the 3′ end of 28S (Labhart and Reeder, 1986), but T3 must be present to allow this processing step at T2 (Labhart and Reeder, 1987a). Both the T2 and T3 areas of *X. laevis* contain the nucleotide box GACTTGC, which is necessary for function of T2 (Labhart and Reeder, 1987a) and T3 (Labhart and Reeder, 1987b). Similar details for termination of rDNA transcription are also being worked out in mouse (Henderson and Sollner-Webb, 1986; Grummt et al., 1986a; Grummt et al., 1986b; Kuhn et al., 1988) and yeast (Planta and Raué, 1988) cells, but in these cases RNA polymerase I does not seem to proceed past the counterpart of the T2 region.

The pre-rRNA precursor that ends at the T2 site in *X. laevis* is rapidly processed by cleavage at T1 to yield the predominant 40S rRNA precursor (Labhart and Reeder, 1986). In mouse cells, the first step in processing the rRNA precursor that ends at the T2 counterpart involves removal of only 10 nucleotides (nt) (Kuhn and Grummt, 1989). The rapidity of the T1 processing event in *X. laevis* may be visualized by electron microscopy of spread chromatin, where three RNA polymerases lacking nascent RNA fibrils are seen at the end of the rDNA "Christmas tree" (Trendelenburg, 1982), presumably between T1 and T2. However, the same method shows only a few polymerases between T2 and T3 (McKnight et al., 1980), in contrast to the predictions of nuclear runoff experiments (Labhart and Reeder, 1986).

Maturation of the rRNA Precursor

The processing of pre-rRNA has been reviewed previously by others (Perry, 1976, 1981; Hadjiolov and Nikolaev, 1976; Hadjiolov, 1980; Crouch, 1984). Generally, the steps of rRNA maturation occur after termination of transcription and release of the rRNA precursor; an exception to this is *Dictyostelium* rRNA, which begins its processing during transcription, as seen in Miller spreads (Grainger and Maizels, 1980). Also, processing of the intron in *Tetrahymena* rDNA occurs during or shortly after transcription (Cech and Rio, 1979; Din et al., 1979) by autocatalytic self-splicing (Kruger et al., 1982; Cech et al., 1983) before further pre-rRNA processing.

The predominant rRNA precursor seen after a brief radioactive pulse varies from 37S (yeasts) to 40S (*X. laevis*) to 45S (mouse). Early pulse-chase experiments demonstrated that the rRNA precursor is rapidly processed into 18S and 32S RNAs; the subsequent processing of 32S into 5.8S and 28S RNAs occurs more slowly (for examples: Perry, 1962; Scherrer et al., 1963; Weinberg et al., 1967; Maden, 1968; Loening et al., 1969; Loening, 1975; Trapman and Planta, 1976). Figure 1 summarizes the rRNA-processing steps, which will be described below.

In mouse cells, immediately after the 47S pre-rRNA is released at termination site T2, there is cleavage at about +650 which removes the 5′ end of the ca. 4 kilobase-long ETS (Miller and Sollner-Webb, 1981). Cleavage within the ETS has also been described in *Tetrahymena* (Sutiphong et al., 1984), *Neurospora* (Tyler and Giles, 1985), *Physarum* (Blum et al., 1983; Blum et al., 1986), human (Kass et al., 1987), and rat (Stroke and Weiner, 1989) cells, but it is not yet clear whether it occurs also in *Xenopus* cells (B. Sollner-Webb, personal communication). The +650 cleavage happens at either of two alternative sites that are 6 bp apart (Kass et al., 1987). Although the sequence of the ETS processing region has not been evolutionarily conserved, the S-100 supernatant factor that catalyzes this cleavage event is not species specific, and the ETS of human rDNA can be cleaved by mouse S-100 (Kass et al., 1987). In mouse cells, after the +650 cut has occurred, the 46S pre-rRNA then loses 600 nt at its 3′ end to become the predominant 45S pre-rRNA (Gurney, 1985).

The next set of rapid cleavages occurs at the 5′ and 3′ ends of 18S rRNA and at the 5′ boundary of 5.8S RNA (Fig. 1). The order in which these three cuts happen is variable. The pathway shown in II.(A) of Fig. 1 occurs in mouse L cells (Wellauer et al., 1974) and in *Xenopus* (Loening et al., 1969; Wellauer and Dawid, 1974) and *Drosophila* (Levis and Penman, 1978), but the pathway shown in II.(B) occurs in HeLa cells (Wellauer and Dawid, 1973). The pathway can vary even in the same cell type (Wellauer et al., 1974; Winicov, 1976; Dudov et al., 1978; Bowman et al., 1981); in one example, a temperature-sensitive mutant of BHK cells follows pathway II.(A) at 38.5°C and II.(B) at 33.5°C (Winicov, 1976). In mouse L cells, after the ETS is severed by cut 1 from the rest of the rRNA precursor, it is degraded from its 3′ end (Bowman et al., 1983).

The subsequent maturation of 32S pre-rRNA into 5.8S and 28S occurs more slowly (Fig. 1). It might be that 32S must first be liberated from the larger rRNA precursor to allow hydrogen bonding of the 5′ and 3′ ends of 5.8S RNA to 28S rRNA (Pene et al., 1968; Pace et al., 1977; Sitz et al., 1981; Peters et al., 1982; Walker et al., 1982). Also, it is at the 32S step that 5S RNA joins the maturing ribosome (Knight and Darnell, 1967).

The 5′ end of 32S and of mature 5.8S RNA of mouse L cells has a 6- to 7-nt heterogeneity, suggesting variability in the cleavage site at the 3′ end of ITS1 (Bowman et al., 1983); similar heterogeneity at the 5′ end of 5.8S RNA has also been reported for HeLa cells (Khan and Maden, 1977), *Xenopus* (Boseley et al., 1978; Ford and Mathieson, 1978), yeast (Veldman et al., 1980), and rat (Smith et al., 1984). The first step in processing mouse L-cell 32S pre-rRNA involves cuts at 300 nt upstream of the 3′ end of ITS2 and at the 5′ end of 28S rRNA (Bowman et al., 1983); it is unknown which of these cuts occurs first. The cut site at the 3′ end of ITS2 shows a 5- to 6-nt heterogeneity (Bowman et al., 1983). A cleavage site occurring within ITS2 has also been reported for yeast (Trapman et al., 1975) and rat (Reddy et al., 1983a) cells. After these first two cleavages, a cut occurs at the 3′ end of 5.8S RNA to remove the remaining portion of ITS2 (Bowman et al., 1983).

Most eucaryotes share the general processing schemes described above, resulting in the production of 5.8S, 18S, and 28S rRNAs. Processing steps that occur after this point can be found in some organisms but are not universally shared by all eucaryotes (Clark and Gerbi, 1982). For example, several lower eucaryotes remove a few nucleotides in the center of 28S rRNA to create 28Sα and 28Sβ (Ishikawa, 1977; deLanversin and Jacq, 1983; Nelles et al., 1984; Ware et al., 1985; Fujiwara and Ishikawa, 1986; deLanversin and Jacq, 1989), which remain hydrogen bonded together (Shine and Dalgarno, 1973). The material removed by this processing step coincides with an expansion segment (nucleotide stretches present in eucaryotic rRNA but missing in *Escherichia coli* rRNA [Ware et al., 1983; Clark et al., 1984; Clark, 1987; Michot and Bachellerie, 1987]). Insects also have a second, unique, late processing event that subdivides 5.8S RNA (Pavlakis et al., 1979; Jordan et al., 1980).

In addition to the cleavage events described above, modifications of rRNA occur during its maturation. During or shortly after transcription, methylations are found in the *Xenopus* 40S pre-rRNA; 105 of these are 2′-*O* methyls of ribose, and 10 are base methylations (Maden et al., 1977). Extensive work has now mapped these methylation sites within the *Xenopus* rRNA sequence (Atmadja et al., 1984; Maden, 1988). There are also about 100 sites in the *Xenopus* 40S pre-rRNA where uridine is converted to pseudouridine (Khan et al., 1978). None of these modifications occur in the ETS and ITS; they have been found only in evolutionarily conserved regions of the rRNAs (Khan et al., 1978). Yeast 37S pre-rRNA has somewhat fewer modifications: 67 methyls and 47 pseudouridines (Klootwijk and Planta, 1973a, 1973b; Brand et al., 1979). The function of these modifications is still unknown. Early studies in which tissue culture cells were deprived of methionine are difficult to interpret because the observed block in rRNA processing could have been an indirect effect of blocking protein synthesis. However, when cycloleucine, which is an inhibitor of *S*-adenosyl-L-methionine, was added to cultured cells, it abolished 95% of 45S pre-rRNA methylation, had no effect on rRNA synthesis, but did reduce the rate of rRNA processing (Caboche and Bachellerie, 1977). Of course, the role of rRNA modifications might be not just at the rRNA-processing level alone, but for the ultimate functions of rRNA within the mature ribosome. *E. coli* ribosomes have far fewer methylations than do eucaryotes, and surprisingly, reconstitution of *E. coli* 30S ribosomal subunits with unmodified 16S rRNA synthesized in vitro indicates that ribosome function can be maintained even when it lacks the 13 methylations usually present on the 16S rRNA (Krzyzosiak et al., 1987; Melançon et al., 1987).

Most of the ribosomal proteins are transported from the cytoplasm to the nucleolus and added to the preribosome during and shortly after synthesis of the rRNA precursor. Miller spreads show that nascent rRNA fibrils are somewhat shorter than expected if they were naked RNA and suggest that they may be already coated with some ribosomal proteins (Miller and Hamkalo, 1972a, 1972b). Electron microscopy with antibodies against specific ribosomal proteins directly proved this idea (Chooi and Leiby, 1981). Rapidly labeled 30S RNP isolated from nucleoli contains 45S pre-rRNA and seems to be a precursor to the 80S RNP particle (Bachellerie et al., 1971; Bachellerie et al., 1975; Rodrigues-Pousada et al., 1979). Additional ribosomal proteins presumably are added to this 30S RNP to create 80S RNP, which has 45S pre-rRNA, over two-thirds of the 60S ribosomal proteins, and about half of the 40S ribosomal proteins already in place (Kuter and Rodgers, 1976; Auger-Buendia and Longuet, 1978; Auger-Buendia et al., 1979; Fujisawa et al., 1979; Lastick, 1980; Todorov et al., 1983). Most of the remaining ribosomal proteins are not added until later, when rRNA processing is completed. The 80S RNP particle is the precursor to 55S RNP (Warner, 1974; Craig, 1974), which contains 32S rRNA (Warner and Soeiro, 1967) and 5S RNA (Knight and Darnell, 1967). 55S RNP constitutes 70 to 80% of the nucleolar population of preribosomes, whereas 80S RNP constitutes only 10 to 20% (Hadjiolov, 1985). Processing of 32S pre-rRNA and addition of remaining ribosomal proteins results in conversion of 55S RNP into the 60S large ribosomal subunit found subsequently in the cytoplasm. Although 80S RNP is also the initial precursor for the 40S small ribosomal subunit, it is harder to find a direct precursor RNP of the 40S subunit in the nucleolus; probably there is rapid transport to the cytoplasm. Ribosomal proteins might not be absolutely essential for rRNA processing, because a relaxed yeast mutant has been found that can slowly process pre-rRNA into naked 18S and 25S rRNA when protein synthesis is inhibited (Waltschewa et al., 1983).

CYTOCHEMISTRY OF THE NUCLEOLUS

Cytochemical experiments (reviewed by Goessens, 1984; Hadjiolov, 1985; Reimer et al., 1987b) are being carried out by several investigators to study the macromolecules that mediate ribosome biogenesis and maturation and to define their localization within subcompartments of the nucleolus. These subcompartments are the fibrillar center (FC), composed of 5-nm diameter fibrils, a dense fibrillar component (DFC) that surrounds the FC, and a granular component (GC) that comprises about 70% of the nucleolar volume (Jordan and McGovern, 1981; Lepoint and Goessens, 1982) and contains the preribosome particles. In growing cells, the nucleolar subcompartments can be distinguished only by electron microscopy, but they are visible by light microscopy in stationary cells or cells treated with the rRNA synthesis inhibitor actinomycin D, which segregates the fibrillar and granular components (reviewed by Simard et al., 1974; Daskal, 1979). Autoradiography after pulse-labeling shows nascent RNA labeled at the periphery of the DFC and in the FC (Fakan, 1978; Mirre and Stahl, 1978; Lepoint and Goessens, 1978; Hernandez-Verdun and Bouteille, 1979; Fakan and Puvion, 1980; Mirre and Knibiehler, 1981; Thiry et al., 1985) before synthesis of 45S pre-rRNA has been completed (Royal and Simard, 1975). Subsequently, the pulse-label can be chased into 32S

Table 1. Some nucleolar macromolecules

Name, molecular mass	Location in nucleolus (interphase)	Mitotic location
rDNA	FC	NOR
RNA polymerase I	FC	NOR
Topoisomerase I	FC	NOR
Nucleolin (C23), 110 kDa	DFC	NOR, prenucleolar bodies
Fibrillarin, 34 kDa	DFC	Prenucleolar bodies
180-kDa protein	DFC	Cytosol
145-kDa protein	GC	Prenucleolar bodies
Ribocharin, 40 kDa	GC	Chromosomal surfaces
B23 (rat), 37 kDa } Similar	GC	Chromosomal surfaces
NO38 (*X. laevis*), 38 kDa } Similar	GC	Chromosomal surfaces
7-2 RNP	GC	

pre-rRNA in the GC (Royal and Simard, 1975). Therefore, pre-rRNA formation occurs in the FC and later events of rRNA processing occur in the GC. Some of the nucleolar proteins that may mediate these events are described below and summarized in Table 1.

Fibrillar Center

DNA is only in the FC of the nucleolus, as shown by immunoelectron microscopy using a monoclonal antibody against single- and double-stranded DNA (Scheer et al., 1987; Thiry et al., 1988). This confirmed earlier observations after [^{3}H]thymidine labeling, DNA staining, or in situ hybridization for rDNA (reviewed in Hernandez-Verdun, 1983, and Goessens, 1984). Topoisomerase I is utilized for rDNA transcription (Rose et al., 1988; Zhang et al., 1988); during interphase it is found in the FC of the nucleolus as well as in the nucleoplasm (Rose et al., 1988), whereas at the end of mitosis it is also found at the NOR (Guldner et al., 1986). Of course, the main enzyme essential for rDNA transcription is RNA polymerase I, and it is not surprising that it localizes with rDNA in the nucleolar FC or at the NOR during mitosis (Scheer and Rose, 1984; Scheer and Raška, 1987; Reimer et al., 1987b). As antibodies to rDNA transcription factors become available, they will doubtless also be found in the FC of the nucleolus.

rDNA is not transcribed during mitosis, but much of the transcriptional machinery remains associated with it at the NOR and may act as a nucleation site for reformation of the nucleolus at the end of telophase. Injection of antibody against RNA polymerase I into dividing cells caused release of this enzyme from the NOR and prevented coalescence of prenucleolar bodies (with DFC components) at the NOR (Benavente et al., 1987). RNA polymerase I and/or its attached nascent rRNA seems to be needed not only to organize the nucleolus but also to maintain it, since injection of anti-RNA polymerase I into the interphase nucleus mimics nucleolar disintegration of prophase (Benavente et al., 1988). Although the NOR of most eucaryotes usually contains several hundred tandem copies of rDNA, recently it was shown in *Drosophila* transformants that a single copy of rDNA is sufficient to organize a mininucleolus (Karpen et al., 1988). Thus, no sequences other than rDNA seem necessary for this function.

Dense Fibrillar Component

Nucleolin, or C23, is a 110-kilodalton (kDa) (713-amino-acid) protein that is responsible for much of the silver staining of the nucleolus (Lischwe et al., 1979). It is found within the DFC (Lischwe et al., 1981; Spector et al., 1984; Escande et al., 1985) and might act as a bridge between rDNA and its nascent transcript (Egyhazi et al., 1988). The nucleolin gene has been cloned and sequenced (Lapeyre et al., 1987; Bourbon et al., 1988; Caizergues-Ferrer et al., 1989); it is thought that the N-terminal region of nucleolin may bind to the intergenic spacer of rDNA chromatin (Olson and Thompson, 1983; Olson et al., 1983; Erard et al., 1988), whereas the C-terminal region binds to preribosomes containing 45S pre-rRNA (Prestayko et al., 1974; Bugler et al., 1982; Bourbon et al., 1983; Herrera and Olson, 1986; Bugler et al., 1987). Proteolysis of nucleolin (Bouche et al., 1984; Suzuki et al., 1985) or injection of an antibody against it (Egyhazi et al., 1988) stimulates rDNA transcription; perhaps nucleolin acts as a feedback regulator to couple rDNA transcription with rRNA processing (Egyhazi et al., 1988). Nucleolin can induce chromatin decondensation by displacement of histone H1 (Erard et al., 1988). During mitosis, some nucleolin remains with the rDNA at the NOR (Ochs et al., 1983; Gas et al., 1985); at the end of mitosis, nucleolin is found in prenucleolar bodies that participate in nucleolar reformation (Gas et al., 1985; Ochs et al., 1985a).

Like nucleolin, fibrillarin is found in prenucleolar bodies (Gas et al., 1985; Benavente et al., 1987; Reimer et al., 1987a) and may play a role in nucleolar reformation at interphase. Fibrillarin is a 34-kDa protein (Lischwe et al., 1985; Ochs et al., 1985b) which carries this name because it is located in the fibrillar region of the nucleolus (Ochs et al., 1985b; Reimer et al., 1987a) within the DFC (Reimer et al., 1987b). Fibrillarin has been implicated in binding to U3 snRNA (Lischwe et al., 1985; Parker and Steitz, 1987) (see below, Structure and Possible Functions of U3 snRNA). Likely counterparts to fibrillarin in plants (Gultinan et al., 1988) and in yeasts (Aris and Blobel, 1988) have recently been described.

Still another constituent of the nucleolar DFC is a 180-kDa protein that during mitosis is diffusely distributed in the cytoplasm (Schmidt-Zachmann et al., 1984). Its function is not yet known.

Granular Component

One protein in the nucleolar GC is a 145-kDa protein which seems to form filamentous structures that could be a nucleolar karyoskeleton upon which preribosomal particles could attach during their maturation (Franke et al., 1981; Krohne et al., 1982; Benavente et al., 1984). Even though the 145-kDa protein is found in the nucleolar GC, at the end of mitosis it is present in prenucleolar bodies (Benavente et al., 1984) along with some of the constituents of the nucleolar DFC (nucleolin and fibrillarin).

Ribocharin is a 40-kDa protein also found in the nucleolar GC (Hügle et al., 1985). It may play a role in maturation of the 60S ribosomal subunit, because it is found associated with the precursor of this subunit and seems to dissociate from the subunit, when the latter leaves the nucleolus, in order to be reutilized on another 60S subunit precursor (Hügle et al., 1985). During mitosis, ribocharin is distributed over the chromosomal surfaces (Hügle et al., 1985).

Protein B23 is a 37-kDa molecule found in the nucleolar GC of rat cells (Prestayko et al., 1974; Spector et al., 1984; Morris et al., 1985; Fields et al., 1986), where it is associated with ribosomal precursor particles (Ochs et al., 1983). During mitosis, it associates with the chromosomal surfaces (Ochs et al., 1983; Spector et al., 1984).

Xenopus protein NO38 shows striking similarity to protein B23 of rat cells (Schmidt-Zachmann et al., 1987). It is an acidic 38-kDa protein also located in the nucleolar GC, where it associates with 40S and 60S ribosomal subunit precursors (Schmidt-Zachmann et al., 1987). A cDNA clone encoding NO38 has been sequenced and shows that the N-terminal region resembles the histone-binding protein nucleoplasmin; indeed, a monoclonal antibody against NO38 also precipitates nucleoplasmin (Schmidt-Zachmann et al., 1987). Just as nucleoplasmin plays a role in binding the basic histone proteins for their transfer to DNA, so too may NO38 promote the assembly of the basic ribosomal proteins on pre-rRNA (Schmidt-Zachmann et al., 1987). During mitosis, *Xenopus* NO38, like its rat counterpart B23, is found on the chromosomal surfaces.

The 7-2 RNP has also been immunolocalized to the nucleolar GC (Reimer et al., 1988). Autoimmune serum against this complex precipitated a 40-kDa protein that may or may not be related to the other proteins of the nucleolar GC that are mentioned above and have a similar molecular weight (Reimer et al., 1988). Also present in the precipitate from this autoimmune serum was 7-2 RNA (Hashimoto and Steitz, 1983; Reddy et al., 1983b; Reimer et al., 1988). 7-2 RNA is an RNA polymerase III transcript (Hashimoto and Steitz, 1983) previously known to be located in the nucleolus (Reddy et al., 1981). The possible function of 7-2 RNP in ribosome maturation remains to be discovered.

Nucleolar Reformation

At the end of mitosis, the nucleolus must be reformed. The constituents of the nucleolar FC (rDNA, RNA polymerase I, topoisomerase I) are already present at the chromosomal NOR and attract the prenucleolar bodies that contain many of the structural components of the nucleolar framework (nucleolin, fibrillarin, and 145-kDa protein). When ribosome biogenesis begins again in interphase, the preribosomal subunits and their associated proteins (ribocharin, rat B23, and *Xenopus* NO38) appear as granules composing the nucleolar GC. A leukemia protein for processing viral RNA has a nucleolar targeting sequence of two amino acids embedded in the middle of the nuclear targeting sequence of 20 highly basic and proline-rich amino acids at the N terminus (Siomi et al., 1988). Whether a similar nucleolar targeting sequence exists in some or all of the nucleolar proteins enumerated above remains unknown. Although the topographical location in nucleolar subcompartments is known for the proteins mentioned above, it is not clear whether they act independently or instead may associate into complexes that also may contain small RNA components. Information about one of these small RNAs, U3, is discussed below.

STRUCTURE AND POSSIBLE FUNCTIONS OF U3 snRNA

Since U3 snRNA is localized in the nucleoli of eucaryotic cells (Weinberg and Penman, 1968; Prestayko et al., 1970; Zieve and Penman, 1976), it has been hypothesized that it may play a role in ribosome biogenesis. U3 snRNA is quite abundant (about 2×10^5 copies per mammalian cell [Weinberg and Penman, 1968]) and is complexed with proteins. Preliminary work on the proteins in the U3 snRNP has begun (Parker and Steitz, 1987); this particle does not contain the anti-Sm antigen present in many nucleoplasmic snRNPs. It has been proposed that the 34-kDa protein fibrillarin is part of the U3 snRNP (Lischwe et al., 1985; Parker and Steitz, 1987). However, only 10 to 20% of cellular U3 snRNA could be precipitated by antibody against this protein (Parker and Steitz, 1987), creating the worry that fibrillarin might only be a contaminant loosely associated with U3 snRNA. Moreover, a paradox exists in that fibrillarin is localized in the FC of the nucleolus (Reimer et al., 1987a), but antibody against the trimethyl G cap of U snRNAs (Lührmann et al., 1982) is localized in the GC of the nucleolus (U. Scheer, personal communication, cited in Reimer et al., 1987b). To resolve this paradox, we are currently doing in situ hybridizations with a probe against U3 snRNA to determine in which nucleolar compartment it resides. It is important to address the question of U3 localization to clarify whether U3 might be involved in the early events of ribosome biogenesis that occur in the FC or in the later events in the GC (see above). Of course, it might turn out that U3 is in both nucleolar compartments and plays roles in both early and late steps of ribosome biogenesis. In accord with this possibility, U3 snRNA is found associated with <10S to >80S RNPs (Tyc and Steitz, 1989).

The sequence of U3 snRNA has been determined for organisms of a wide taxonomic range (Table 2). The U3 snRNAs are generally a little over 200 nt (as shown in Fig. 2 for *X. laevis*) but are 255 nt in *Schizosaccharomyces pombe* and 328 nt in *Saccharomyces cerevisiae* due primarily to a block of inserted sequence in the middle of the yeast U3 snRNA. Sequence alignments have identified four regions that are highly conserved in all of the U3 snRNAs: boxes A, B, C, and D (Wise and Weiner, 1980; Hughes et al., 1987; Jeppesen et al., 1988; Tyc and Steitz, 1989) (Fig. 2). Box A could be extended 3 nt more if a one-base mismatch in *Dictyostelium discoideum* is tolerated. Boxes B and C each contain a two-base mismatch in yeasts (see sequence alignment in Jeppesen et al., 1988). We did not include box D in our original sequence alignment (Jeppesen et al., 1988)

Table 2. U3 snRNA sequences

Organism	Reference
Plant	
Broad bean (partial sequence)	Kiss et al., 1985
Unicellular animals	
Yeast (*Saccharomyces cerevisiae*)	Hughes et al., 1987
Yeast (*Schizosaccharomyces pombe*)	Porter et al., 1988
Slime mold (*Dictyostelium discoideum*)	Wise and Weiner, 1980
Ciliated protozoan (*Tetrahymena thermophila*)	J. Engberg, unpublished data (cited in Raué et al., this volume)
Insect	
Silk moth (*Bombyx mori*; partial sequence)	Adams et al., 1985
Amphibia	
Toad (*Xenopus laevis*)	Jeppesen et al., 1988
Toad (*Xenopus borealis*)	Jeppesen et al., 1988
Mammals	
Mouse	Mazan and Bachellerie, 1988
Rat	Reddy et al., 1979; Stroke and Weiner, 1985
Human	Suh et al., 1986

because there are one- to two-base mismatches in *Xenopus*, *Dictyostelium*, and *S. pombe* within the short six-base consensus sequence (Tyc and Steitz, 1989).

Sequences of functional importance are often conserved because a mutation within such regions may be lethal and hence not perpetuated during evolutionary time. Therefore, boxes A to D are candidates for functionally important sites within U3 snRNA. To determine whether these conserved boxes are exposed and available for interactions with other RNA molecules (such as pre-rRNA), we needed to determine the secondary structure of U3 snRNP. We chose not to rely on computer-generated models, which are often inaccurate for RNA structure prediction (Turner et al., 1987). A better theoretical approach utilizes compensatory base changes that occurred during evolution to preserve base-paired stems. When we compared taxonomically diverse U3 snRNA sequences, compensatory base changes supported the stems in the 3′ half of U3 snRNA but not at the 5′ end, even though a potential stem could be proposed (Fig. 2; see also Fig. 7 in Jeppesen et al., 1988). Porter et al. (1988) also noted that the 5′ end of U3 snRNAs from unicellular eucaryotes cannot readily be folded into a single hairpin stem. Experimental data independently obtained from *Xenopus* and HeLa cells support the U3 snRNA structure drawn in Fig. 2. Nucleotides in U3 snRNP that can be modified by the chemicals dimethyl sulfoxide, CMCT

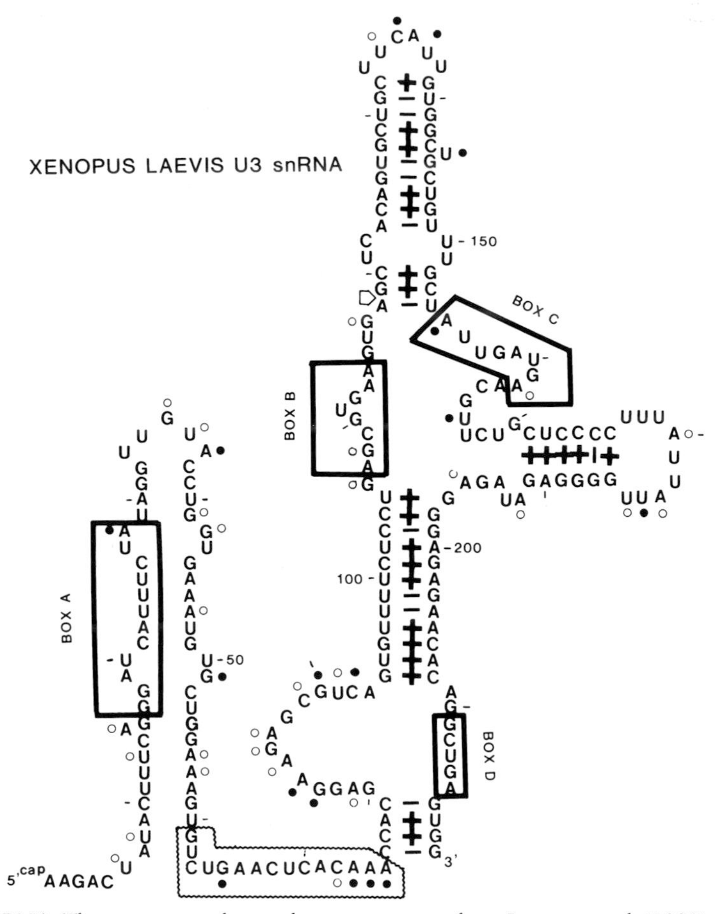

Figure 2. *X. laevis* U3 snRNA. The sequence and secondary structure are from Jeppesen et al. (1988) except that residue 210, originally published as A, is generally G on the basis of our recent sequencing of genomic clones. The sequences in boxes A to D are evolutionarily conserved; the open arrow indicates the area of additional sequence found in yeast U3 snRNAs. Our secondary structure model of U3 snRNA is supported by compensatory base changes (+ indicates base pair) and chemical modification of U3 snRNP (●, strong; ○, medium [Jeppesen et al., 1988]). The 15-nt synthetic oligonucleotide used for oocyte injection (see text) is complementary to nt 61 to 75, indicated by a wavy-line enclosure.

[hexyl-3-(2-morpholinoethyl) carbodiimide metho-*p*-toluene sulfonate], or kethoxal are neither base paired nor tightly protected by proteins (Parker and Steitz, 1987; Jeppesen et al., 1988); Fig. 2 shows our results on these modified sites. Cobra venom nuclease cuts base-paired stems and generally confirms (Parker and Steitz, 1987) our secondary structure model shown in Fig. 2 for U3 snRNA. Boxes A to D all occur in single-stranded regions in this model and are available for potential interactions with pre-rRNA. However, the most highly exposed region of U3 snRNP is found at residues 65 to 75 (Fig. 2) (Parker and Steitz, 1987; Jeppesen et al., 1988), which is a sequence that is not evolutionarily conserved.

Functional roles have been postulated for each of the conserved boxes. Tyc and Steitz (1989) noticed that box D and part of box C (UGAUGA) are present not only in U3 snRNAs but also in several other RNA species that seem to reside in the nucleolus: HeLa cell U8 and U13 snRNAs (Tyc and Steitz, 1989); HeLa cell RNA X and RNA Y (RNA Y has box D but it is unknown whether it also has box C [Tyc and Steitz, 1989]); yeast snR128 and snR190 (Zagorski et al., 1988); and mouse 4.5S hybRNA (Maxwell and Martin, 1986; Trinh-Rohlik and Maxwell, 1988). It is known that the HeLa cell RNAs listed above can all be immunoprecipitated with antifibrillarin (Tyc and Steitz, 1989), so boxes C and D might represent binding sites for this protein. In addition, boxes C and D could be important as transport signals to target these RNAs to the nucleolus.

Earlier on, another function was posited for box C of U3 snRNA: just as U1 snRNA base pairs at the 5′ end of introns for splicing, perhaps, by analogy, box C of U3 snRNA base pairs to the 5′ end of ITS2 to aid in removal of this spacer during processing (Bachellerie et al., 1983; Tague and Gerbi, 1984). This proposal was based on the observations that 5.8S RNA seems to have become separated from the main body of 28S rRNA during evolution, by the insertion of ITS2 (Nazar, 1980; Jacq, 1981; Clark and Gerbi, 1982), and also that the complementarity between box C and the 5′ end of ITS2 could account for the hydrogen bonding found experimentally between U3 snRNA and 28S-35S rRNA (Prestayko et al., 1970; Zieve and Penman, 1976). However, the U3 box C-ITS2 complementarity cannot be drawn for unicellular eucaryotes, thereby calling this model into question (Tague and Gerbi, 1984); moreover, yeast U3 snRNA seems to be hydrogen bonded to the initial precursor, 37S pre-rRNA (Tollervey, 1987). Other models for U3-ITS2 base pairing have also been proposed (Crouch et al., 1983; Kupriyanova and Timofeeva, 1988) but do not hold up well to taxonomic comparisons.

There has also been speculation about the functions of U3 box B. It was suggested that box B might base pair near the termination processing region (T2) downstream of the 3′ end of 28S rRNA (Parker and Steitz, 1987), but the 7-bp complementarity is weakened by a mismatch in *Xenopus borealis* (Jeppesen et al., 1988). Moreover, this complementarity resides between T1 and T2 (e.g., at +83 to +89 in *X. laevis*, where T2 is at +235) rather than at T2 itself. Finally, treatment of nuclear extract to destroy U3 snRNP (i.e., micrococcal nuclease digestion or addition of sense or antisense U3 RNA, U3-specific oligonucleotides, or antifibrillarin) did not affect in vitro termination or 3′ processing (Kuhn and Grummt, 1989).

Recently another putative function for U3 box B has been deduced, namely, that it may bind to the α-sarcin region near the 3′ end of 28S rRNA (Parker et al., 1988). In this experiment in vitro transcripts of regions of rRNA were incubated with a HeLa cell extract (presumably containing U3 snRNP) and then immunoprecipitated with antifibrillarin after RNase T_1 digestion; the only rRNA fragments precipitated were from a 20-nt stretch that begins 4 nt downstream of the α-sarcin site (Parker et al., 1988). 5.8S RNA may interact with this region of rRNA (Choi, 1985), and cleavage of 28S rRNA with α-sarcin dissociates 5.8S from 28S rRNA (Walker et al., 1983). Thus, the exciting possibility is raised that U3 snRNP may bind to the α-sarcin region of 28S rRNA to mediate conformational changes needed for 5.8S processing. However, as pointed out by Parker et al. (1988), this result should be treated cautiously. Artificial in vitro conditions were used in this experiment, and the rRNA was neither full length nor complexed with ribosomal proteins. It is clear that the immunoprecipitation was mediated by some RNA in the nuclear extract because it was micrococcal nuclease sensitive, but the identity of the mediator RNA is unproven. Not only is there some doubt that fibrillarin is a bona fide component of the U3 snRNP (see beginning of this section), but it is also plausible that antifibrillarin may have immunoprecipitated 5.8S RNA, which is known to have some sequence and secondary structure similarity to U3 (Kupriyanova and Timofeeva, 1988).

A different experimental approach, using psoralen cross-linking, suggested a functional role for box A in U3 snRNP. Although U3 snRNA can base pair with rRNA (Prestayko et al., 1970; Zieve and Penman, 1976), most of this interaction is dependent on proteins: 65% of total U3 snRNP is bound to preribosomes, but 85% of this bound U3 snRNA can be removed by phenol extraction (Epstein et al., 1984). The advantage of psoralen cross-linking is that it can be done in vivo, and it can stabilize otherwise weak interactions. Recent experiments indicated that 5 to 7% of the total amount of U3 snRNA could be psoralen cross-linked to nucleolar RNA (R. L. Maser and J. P. Calvet, personal communication) which was <18S to >28S in size (Maser and Calvet, personal communication; Stroke and Weiner, 1989). Hybridization demonstrated that the cross-link was close to the ETS-processing site: between +438 and +695 in human rRNA (Maser and Calvet, 1989; +415 to +422 is the ETS-processing site in human rRNA [Kass et al., 1987]) and between +767 and +1149 in rat rRNA (Stroke and Weiner, 1989; +790 to +795 is the ETS-processing site in rat rRNA [Stroke and Weiner, 1989]). These ETS-processing regions in human, rat, and mouse (+651 to +657 [Kass et al., 1987]) cells all share the 13-bp sequence GAUCGAUGUGGUG (Stroke and Weiner, 1989). However, this sequence is not conserved at the ETS-processing region of nonmammalian rRNA (see above, Maturation of the rRNA Precursor). Although the psoralen cross-link has not been defined to better than a few hundred nucleotides within the ETS, sequencing has pinpointed the cross-links in rat U3 snRNA to residues U-13, C-14, and U-23, which are all in or very close to box A (Stroke and Weiner, 1989). The sequence of U3 box A is not complementary to the ETS-processing site itself, but in rat cells it is complementary to 75 bases further downstream in the ETS (nt 18 to 29, spanning U3 box A, complements nt 864 to 876 of the ETS [Stroke and Weiner, 1989]). This complementarity is not evolutionarily conserved when the counterpart region in the ETS of mouse and human rDNAs (Kass et al., 1987) is examined. On the other hand, it is possible that U3 box A is simply very close to the ETS-processing site but not base paired there, since a psoralen cross-link could also be formed in this situation (Garrett-Wheeler et al., 1984). There is precedence for enzymatic activity of RNA in processing without apparent base pairing, as M1 RNA of *E. coli* RNase P can catalyze the cleavage of tRNA precursors to generate mature 5′ termini (Guerrier-Takada et al., 1983), though M1 RNA and pre-tRNA lack complementarity at this site of interaction.

The psoralen cross-linking data described above strongly suggest that box A of U3 snRNA is in close proximity, even if not base paired, to the ETS-processing site. This does not preclude box B or box C of U3 snRNA also being close to other regions of pre-rRNA. Such other contact sites could be missed because psoralen cross-linking has strict stereochemical requirements (Cimino et al., 1985). To test the functional roles of the conserved boxes in U3 snRNA, it is necessary to use an in vivo approach because an in vitro system for the complete maturation of preribosomes is not yet available. One in vivo approach could be the use of U3 snRNA conditional mutants of yeast cells. We have chosen a different in vivo approach that is applicable to *Xenopus*. We wish to inject synthetic oligonucleotides complementary to each of the conserved boxes into *Xenopus* oocytes and rely on the endogenous RNase H to cleave the in vivo heteroduplex, thereby destroying the integrity of U3 snRNA at this site. Then we will assay which, if any, of the rRNA processing steps are blocked because of this cleavage of U3 snRNA. To work out the details of this method, we have first injected an oligonucleotide complementary to nt 61 to 75, which is the most highly exposed region in *Xenopus* U3 snRNP (Fig. 2). Subsequent primer extension shows that we were indeed able to cleave all U3 snRNA molecules in the oocyte by this approach. Even though the 5′ one-third of the U3 snRNA is now disjoined from the rest of the molecule, preliminary experiments show that mature 18S, 5.8S, and 28S rRNAs still seem to be made. However, a 20S intermediate that includes ETS and 18S rRNA (Moss, 1983; Labhart and Reeder, 1986) is decreased in amount. It is plausible that separation of the box A domain from the rest of U3 snRNA has interfered with ETS processing. Nonetheless, this step might not be obligatory for processing, because mature 18S rRNA still can be formed. If this proves to be the case, it is tempting to speculate that U3 snRNA might play additional roles for other steps that are obligatory in the pre-rRNA-processing pathway, especially since yeast U3 snRNA is essential for cell viability (Hughes et al., 1987).

It is important to remember that proteins are present in the putative interaction between U3 snRNP and preribosomes; they can help to stabilize RNA-RNA interactions (Epstein et al., 1984) and perhaps also play functional roles for ribosome maturation. Moreover, other nucleolar snRNPs (e.g., U8 and U13 in humans [Tyc and Steitz, 1989] and yeast nucleolar snRNAs [Tollervey, 1987; Zagorski et al., 1988]) may also be a part of the rRNA processing complex, by analogy to the spliceosome used for mRNA maturation. snRNPs may not only play a role in cleavages of pre-rRNA but also perhaps help in conformational changes of the preribosome necessary for its maturation. Future experiments should define the probable roles of nucleolar snRNPS in the mechanism of ribosome biogenesis. All molecules involved in this maturation complex should be examined, not just the snRNPs but also the preribosomes themselves. With the advent of techniques for transformation of eucaryotes with mutagenized rDNA (Sweeney and Yao, 1989; Musters et al., 1989; Musters et al., this volume), it should prove possible to identify the pre-rRNA sequences necessary for proper processing.

We thank Jim Calvet, Ulrich Scheer, Barbara Sollner-Webb, Joan Steitz, and Alan Weiner for sharing their unpublished data with us; Rudi Planta for helpful discussion; Judy Furlong for her efforts at the onset of our work on *Xenopus* U3 snRNA; and Johnie Sanders for typing this manuscript.

Our work is supported by Public Health Service grant GM20261 from the National Institutes of Health.

ADDENDUM IN PROOF

A role for U3 snRNP in ETS processing as previously suggested by psoralen cross-linking has now been confirmed (S. Kass, K. Tyc, J. A. Steitz, and B. Sollner-Webb, *Cell*, in press). Another role for U3 snRNP has also been demonstrated by our antisense oligonucleotide injections into *Xenopus* oocytes: correct cleavage at the ITS1-5.8S boundary depends on intact U3 snRNA; disruption of U3 snRNA increases a 36S RNA precursor and decreases the 32S and 20S RNA products of this cleavage (R. Savino and S. A. Gerbi, *EMBO J.*, in press).

REFERENCES

Adams, D. S., R. J. Herrera, R. Lührmann, and P. M. Lizardi. 1985. Isolation and partial characterization of U1-U6 small RNAs from *Bombyx mori*. *Biochemistry* **24:**117–125.

Anastassova-Kristeva, M. 1977. The nucleolar cycle in man. *J. Cell Sci.* **25:**103–110.

Angelier, N., D. Hemon, and M. Bouteille. 1979. Mechanisms of transcription in nucleoli of amphibian oocytes as visualized by high resolution autoradiography. *J. Cell Biol.* **80:**277–290.

Aris, J. P., and G. Blobel. 1988. Identification and characterization of a yeast nucleolar protein that is similar to a rat liver nucleolar protein. *J. Cell Biol.* **107:**17–31.

Atmadja, J., R. Brimacombe, and B. E. H. Maden. 1984. *Xenopus laevis* 18S ribosomal RNA: experimental determination of secondary structure elements and locations of methyl groups in the secondary structure model. *Nucleic Acids Res.* **12:**2649–2667.

Auger-Buendia, M.-A., and M. Longuet. 1978. Characterization of proteins from nucleolar preribosomes of mouse leukaemia cells by two-dimensional polyacrylamide gel electrophoresis. *Eur. J. Biochem.* **85:**105–114.

Auger-Buendia, M.-A., M. Longuet, and A. Tavitian. 1979. Kinetic studies on ribosomal protein assembly in preribosomal particles and ribosomal subunits of mammalian cells. *Biochim.*

Biophys. Acta **563**:113–128.

Bachellerie, J. P., C. Martin-Prevel, and J. Zalta. 1971. Cinétique de l'incorporation d'uridine (^{3}H) dans les fractions subnucleolaires de cellules d'hepatome ascitique du rat. *Biochimie* **53**: 383–389.

Bachellerie, J.P., B. Michot, and F. Raynal. 1983. Recognition signals for mouse pre-rRNA processing: a potential role for U3 nucleolar RNA. *Mol. Biol. Rep.* **9**:79–86.

Bachellerie, J. P., M. Nicoloso, and J. P. Zalta. 1975. Early nucleolar preribosomal RNA-protein in mammalian cells. *Eur. J. Biochem.* **55**:119–129.

Bakken, A., G. Morgan, B. Sollner-Webb, J. Roan, S. Busby, and R. H. Reeder. 1982. Mapping of transcription initiation and termination signals on *Xenopus laevis* ribosomal DNA. *Proc. Natl. Acad. Sci. USA* **79**:56–60.

Bartsch, I., C. Schoneberg, and I. Grummt. 1987. Evolutionary changes of sequences and factors that direct transcription termination of human and mouse ribosomal genes. *Mol. Cell. Biol.* **7**:2521–2552.

Batts-Young, B., and H. F. Lodish. 1978. Triphosphate residues at the 5′ ends of rRNA precursor and 5S RNA from *Dictyostelium discoideum. Proc. Natl. Acad. Sci. USA* **75**:740–744.

Benavente, R., G. Krohne, R. Stick, and W. W. Franke. 1984. Electron microscopic immunolocalization of a karyoskeletal protein of molecular weight 145,000 in nucleoli and prenucleolar bodies of Xenopus laevis. *Exp. Cell Res.* **151**:224–235.

Benavente, R., G. Reimer, K. M. Rose, B. Hügle-Dörr, and U. Scheer. 1988. Nucleolar changes after microinjection of antibodies to RNA polymerase I into the nucleus of mammalian cells. *Chromosoma* **97**:115–123.

Benavente, R., K. M. Rose, G. Reimer, B. Hügle-Dörr, and U. Scheer. 1987. Inhibition of nucleolar reformation after microinjection of antibodies to RNA polymerase I into mitotic cells. *J. Cell Biol.* **105**:1483–1491.

Bielka, H. 1982. *The Eukaryotic Ribosome.* Springer-Verlag KG, Berlin.

Birnstiel, M. L. (ed.). 1988. *Small Nuclear Ribonucleoprotein Particles.* Springer-Verlag KG, Berlin.

Birnstiel, M. L., H. Wallace, J. L. Sirlin, and M. Fischberg. 1966. Localization of the ribosomal DNA complements in the nucleolar organizer region of *Xenopus laevis. Natl. Cancer Inst. Monogr.* **23**:431–448.

Blum, B., G. Pierron, T. Seebeck, and R. Braun. 1986. Processing in the external transcribed spacer of ribosomal RNA from *Physarum polycephalum. Nucleic Acids Res.* **14**:3153–3166.

Blum, B., T. Seebeck, R. Braun, P. Ferris, and V. Vogt. 1983. Localization and DNA sequence around the initiation site of ribosomal RNA transcription in *Physarum polycephalum. Nucleic Acids Res.* **11**:8519–8533.

Boseley, P. G., A. Tuyns, and M. L. Birnstiel. 1978. Mapping of the *Xenopus laevis* 5.8S rDNA by restriction and DNA sequencing. *Nucleic Acids Res.* **5**:1121–1137.

Bouche, G., M. Caizergues-Ferrer, B. Bugler, and F. Amalric. 1984. Interrelations between the maturation of a 100 kD nucleolar protein and pre rRNA synthesis in CHO cells. *Nucleic Acids Res.* **12**:3025–3035.

Bourbon, H. M., B. Bugler, M. Caizergues-Ferrer, F. Amalric, and J. P. Zalta. 1983. Maturation of a 100 kDa protein associated with preribosomes in CHO cells. *Mol. Biol. Rep.* **9**:39–47.

Bourbon, H. M., B. Lapeyre, and F. Amalric. 1988. Structure of the mouse nucleolin gene: the complete sequence reveals that each RNA binding domain is encoded by two independent exons. *J. Mol. Biol.* **200**:627–638.

Bowman, L. H., W. E. Goldman, G. I. Goldberg, M. B. Herbert, and D. Schlessinger. 1983. Location of the initial cleavage sites in mouse pre-rRNA. *Mol. Cell. Biol.* **3**:1501–1510.

Bowman, L. H., B. Rabin, and D. Schlessinger. 1981. Multiple ribosomal RNA cleavage pathways in mammalian cells. *Nucleic Acids Res.* **9**:4951–4966.

Brachet, J. 1957. *Biochemical Cytology,* p. 535. Academic Press, Inc., New York.

Brand, R. C., J. Klootwijk, C. O. Sibom, and R. J. Planta. 1979. Pseudouridylation of yeast ribosomal precursor RNA. *Nucleic Acids Res.* **7**:121–134.

Bugler, B., H. M. Bourbon, B. Lapeyre, M. O. Wallace, J. H. Chang, F. Amalric, and M. O. J. Olson. 1987. RNA binding fragments from nucleolin contain the ribonucleoprotein consensus sequence. *J. Biol. Chem.* **262**:10922–10925.

Bugler, B., M. Caizergues-Ferrer, G. Bouche, H. Bourbon, and F. Amalric. 1982. Detection and localization of a class of proteins immunologically related to a 100 kD nucleolar protein. *Eur. J. Biochem.* **128**:475–480.

Burns, M. W., Y. Ohnuki, D. E. Rounds, and R. S. Olson. 1970. Modification of nucleolar expression following laser microirradiation of chromosomes. *Exp. Cell. Res.* **60**:133–142.

Busch, H., R. Reddy, L. Rothblum, and Y. C. Choi. 1982. SnRNAs, snRNPs and RNA processing. *Annu. Rev. Biochem.* **51**:617–654.

Busch, H., and K. Smetana. 1970. *The Nucleolus.* Academic Press, Inc., New York.

Caboche, M., and J. P. Bachellerie. 1977. RNA methylation and control of eucaryotic RNA biosynthesis. *Eur. J. Biochem.* **74**: 19–29.

Caizergues-Ferrer, P. Mariottini, C. Curie, B. Lapeyre, N. Gas, F. Amalric, and F. Amaldi. 1989. Nucleolin from *Xenopus laevis*: cDNA cloning and expression during development. *Genes Dev.* **3**:324–333.

Caspersson, T. O. 1950. *Cell Growth and Cell Function,* p. 185. W. W. Norton and Co., New York.

Cech, T. R., and D. C. Rio. 1979. Localization of transcribed regions on extrachromosomal rRNA genes of *Tetrahymena thermophila* by R-loop mapping. *Proc. Natl. Acad. Sci. USA* **76**:5051–5055.

Cech, T. R., N. K. Tanner, I. Tinoco, B. R. Weir, M. Zuker, and P. S. Perlman. 1983. Secondary structure of Tetrahymena ribosomal RNA intervening sequence: structural homology with fungal mitochondrial intervening sequences. *Proc. Natl. Acad. Sci. USA* **80**:3903–3907.

Choi, Y. C. 1985. Structural organization of ribosomal RNAs from Novikoff hepatoma. II. Characterization of possible binding sites of 5S rRNA and 5.8S rRNA to 28S rRNA. *J. Biol. Chem.* **260**:12773–12779.

Chooi, W. Y., and K. R. Leiby. 1981. An electron microscopic method for localization of ribosomal proteins during transcription of ribosomal DNA: a method for studying protein assembly. *Proc. Natl. Acad. Sci. USA* **78**:4823–4827.

Christofori, G., and W. Keller. 1988. 3′ cleavage and polyadenylation of mRNA precursors *in vitro* requires a poly(A) polymerase, a cleavage factor, and a snRNP. *Cell* **54**:875–889.

Cimino, G. D., H. B. Gamper, S. T. Isaacs, and J. E. Hearst. 1985. Psoralens as photoreactive probes of nucleic acid structure and function: organic chemistry, photochemistry, and biochemistry. *Annu. Rev. Biochem.* **54**:1151–1193.

Clark, C. G. 1987. On the evolution of ribosomal RNA. *J. Mol. Evol.* **25**:343–350.

Clark, C. G., and S. A. Gerbi. 1982. Ribosomal RNA evolution by fragmentation of the 23S progenitor: maturation pathway parallels evolutionary emergence. *J. Mol. Evol.* **18**:329–336.

Clark, C. G., B. W. Tague, V. C. Ware, and S. A. Gerbi. 1984. *Xenopus laevis* 28S ribosomal RNA: a secondary structure model and its evolutionary and functional implications. *Nucleic Acids Res.* **12**:6197–6220.

Craig, N. C. 1974. Ribosomal RNA synthesis in eucaryotes and its regulation. *MTP Int. Rev. Sci. Ser. 1 Biochem.* 6:255–288.

Crouch, R. 1984. Ribosomal RNA processing in eukaryotes, p. 214–226. *In* D. Apirion (ed.), *Processing of RNA*. CRC Press, Inc., Boca Raton, Fla.

Crouch, R. J., S. Kanaya, and P. L. Earl. 1983. A model for the involvement of the small nucleolar RNA (U3) in processing eukaryotic ribosomal RNA. *Mol. Biol. Rep.* 9:75–78.

Daskal, Y. 1979. Drug effects on nucleolar and extranucleolar chromatin, p. 107–125. *In* H. Busch, S. T. Crooke, and Y. Daskal (ed.), *Effects of Drugs on the Cell Nucleus*. Academic Press, Inc., New York.

Dawid, I. B., and P. K. Wellauer. 1976. A reinvestigation of 5′-3′ polarity in 40S ribosomal RNA precursor of *Xenopus laevis*. *Cell* 8:443–448.

deLanversin, G., and B. Jacq. 1983. Séquence de la région de la coupure centrale du précurseur de l'ARN ribosomique 26S de Drosophile. *C.R. Acad. Sci. Ser. III* 296:1041–1044.

deLanversin, G., and B. Jacq. 1989. Sequence and secondary structure of the central domain of *Drosophila* 26S rRNA: a universal model for the central domain of the large rRNA containing the region in which the central break may happen. *J. Mol. Evol.* 28:403–417.

Din, N., W. Kaffenberger, and W. Eckert. 1979. The intervening sequence in the 26S rRNA coding region of *T. thermophila* is transcribed within the largest stable precursor for rRNA. *Cell* 18:525–532.

Dudov, K. P., M. D. Dabeva, A. A. Hadjiolov, and B. N. Todorov. 1978. Processing and migration of ribosomal RNA in the nucleolus and nucleoplasm of rat liver nuclei. *Biochem. J.* 171:375–383.

Edström, J. E., W. Grampp, and N. Schor. 1961. The intracellular distribution and heterogeneity of ribonucleic acid in starfish oocytes. *J. Biophys. Biochem. Cytol.* 11:549–557.

Egyhazi, E., A. Pigon, J.-H. Chang, S. Ghaffari, T. D. Dreesen, S. E. Wellman, S. T. Case, and M. O. J. Olson. 1988. Effects of anti-C23 (nucleolin) antibody on transcription of ribosomal DNA in *Chironomus* salivary gland cells. *Exp. Cell Res.* 178: 264–272.

Elsdale, T. R., M. Fischberg, and S. Smith. 1958. A mutation that reduces nucleolar number in *Xenopus laevis*. *Exp. Cell Res.* 14:642–643.

Epstein, P., R. Reddy, and H. Busch. 1984. Multiple states of U3 RNA in Novikoff hepatoma nucleoli. *Biochemistry* 23:5421–5425.

Erard, M., P. Belenguer, M. Caizergues-Ferrer, A. Pantaloni, and F. Amalric. 1988. A major nucleolar protein, nucleolin, induces chromatin decondensation by binding of histone H1. *Eur. J. Biochem.* 175:525–530.

Escande, M. L., N. Gas, and B. J. Stevens. 1985. Immunolocalization of the 100K nucleolar protein in CHO cells. *Biol. Cell* 53:99–110.

Fakan, S. 1978. High resolution autoradiography studies on chromatin functions, p. 3–53. *In* H. Busch (ed.), *The Cell Nucleus*, vol. 5. Academic Press, Inc., New York.

Fakan, S., and E. Puvion. 1980. The ultrastructural visualization of nucleolar and extranucleolar RNA synthesis and distribution. *Int. Rev. Cytol.* 65:255–299.

Fields, A. P., S. H. Kaufmann, and J. H. Shaper. 1986. Analysis of the internal nuclear matrix. Oligomers of a 38kD nucleolar polypeptide stabilized by disulfide bonds. *Exp. Cell Res.* 164: 139–153.

Ford, P. Y., and T. Mathieson. 1978. The nucleotide sequence of 5.8S ribosomal RNA from *Xenopus laevis* and *Xenopus borealis*. *Eur. J. Biochem.* 87:199–214.

Franke, W. W., J. A. Kleinschmidt, H. Spring, G. Krohne, C. Grund, M. F. Trendelenburg, M. Stoehr, and U. Scheer. 1981. A nucleolar skeleton of protein filaments demonstrated in amplified nucleoli of *Xenopus laevis*. *J. Cell Biol.* 90:289–299.

Franke, W. W., U. Scheer, H. Spring, M. F. Trendelenburg, and G. Krohne. 1976. Morphology of transcriptional units of rDNA. *Exp. Cell Res.* 100:233–244.

Franke, W. W., U. Scheer, H. Spring, M. F. Trendelenburg, and H. Zentgraf. 1979. Organization of nucleolar chromatin, p. 49–95. *In* H. Busch (ed.), *The Cell Nucleus*, vol. 7. Academic Press, Inc., New York.

Fujisawa, T., K. Imai, V. Tanaka, and K. Ogata. 1979. Studies on the protein components of 110S and total ribonucleoprotein particles of rat liver. *J. Biochem.* (Tokyo) 85:277–286.

Fujiwara, H., and H. Ishikawa. 1986. Molecular mechanism of introduction of the hidden break into the 28S rRNA of insects: implication based on structural studies. *Nucleic Acids Res.* 14:6393–6401.

Gall, J. G. 1956. Small granules in the amphibian oocyte nucleus and their relationship to RNA. *J. Biophys. Biochem. Cytol.* 2(Suppl):393–395.

Gall, J. G., and M. L. Pardue. 1969. Formation and detection of RNA-DNA hybrid molecules in cytological preparations. *Proc. Natl. Acad. Sci. USA* 63:378–383.

Garrett-Wheeler, R., R. E. Lockard, and A. Kumar. 1984. Mapping of psoralen cross-linked nucleotides in RNA. *Nucleic Acids Res.* 12:3405–3423.

Gas, N., M. L. Escande, and B. J. Stevens. 1985. Immunolocalization of the 100 kDA nucleolar protein during the mitotic cycle in CHO cells. *Biol. Cell* 53:209–218.

Goessens, G. 1984. Nucleolar structure. *Int. Rev. Cytol.* 87: 107–108.

Grainger, R. M., and N. Maizels. 1980. *Dictyostelium* ribosomal RNA is processed during transcription. *Cell* 20:619–623.

Grummt, I., A. Kuhn, I. Bartsch, and H. Rosenbauer. 1986a. A transcription terminator located upstream of the mouse rDNA initiation site affects rRNA synthesis. *Cell* 47:901–911.

Grummt, I., U. Maier, A. Öhrlein, N. Hassouna, and J.-P. Bachellerie. 1985. Transcription of mouse rDNA terminates downstream of the 3′ end of 28S RNA and involves interaction of factors with repeated sequences in the 3′ spacer. *Cell* 43: 801–810.

Grummt, I., H. Rosenbauer, I. Niedermayer, U. Maier, and A. Öhrlein. 1986b. A repeated 18 bp sequence motif in the mouse rDNA spacer mediates binding of a nuclear factor and transcription termination. *Cell* 45:837–846.

Guerrier-Takada, C., K. Gardiner, T. Marsh, N. Pace, and S. Altman. 1983. The RNA moiety of ribonuclease P is the catalytic subunit of the enzyme. *Cell* 35:849–857.

Guldner, H. H., C. Szostecki, H. P. Vosberg, H. J. Lakomek, E. Penner, and F. A. Bautz. 1986. Scl 70 autoantibodies from scleroderma patients recognize a 95 kDa protein identified as DNA topoisomerase I. *Chromosoma* 94:132–138.

Gultinan, M. J., M. E. Schelling, N. Z. Ehtesham, J. C. Thomas, and M. E. Christensen. 1988. The nucleolar RNA-binding protein B-36 is highly conserved among plants. *Eur. J. Cell Biol.* 46:547–553.

Gurney, T. 1985. Characterization of mouse 45S ribosomal RNA subspecies suggests that the first processing cleavage occurs 600±100 nucleotides from the 5′ end and the second 500±100 nucleotides from the 3′ end of a 13.9 kb precursor. *Nucleic Acids Res.* 13:4905–4919.

Hadjiolov, A. A. 1980. Biogenesis of ribosomes in eukaryotes, p. 1–80. *In* D. B. Roodyn (ed.), *Subcellular Biochemistry*, vol. 7. Plenum Publishing Corp., New York.

Hadjiolov, A. A. 1985. *The Nucleolus and Ribosome Biogenesis*. Springer-Verlag KG, Vienna.

Hadjiolov, A. A., and N. Nikolaev. 1976. Maturation of ribosomal ribonucleic acids and the biogenesis of ribosomes. *Prog. Biophys. Mol. Biol.* **31**:95–144.

Hashimoto, C., and J. A. Steitz. 1983. Sequential association of nucleolar 7-2 RNA with two different antoantigens. *J. Biol. Chem.* **258**:1379–1382.

Hay, E. D., and J. B. Gurdon. 1967. Fine structure of the nucleolus in normal and mutant *Xenopus* embryos. *J. Cell Sci.* **2**:151–162.

Heitz, E. 1931. Nukleolen und Chromosomen in der Gattung *Vicia. Planta* **15**:495–505.

Heitz, E. 1933. Über totale und partielle somatische Heteropyknose, sowie strukturelle Geschlechtschromosomen bei *Drosophila funebris. Z. Zellforsch. Mikrosk. Anat.* **19**:720–742.

Henderson, S., and B. Sollner-Webb. 1986. A transcriptional terminator is a novel element of the promoter of the mouse ribosomal RNA gene. *Cell* **47**:891–900.

Hernandez-Verdun, D. 1983. The nucleolar organizer region. *Biol. Cell* **49**:191–202.

Hernandez-Verdun, D., and M. Bouteille. 1979. Nucleologenesis in chick erythrocyte nuclei reactivated by cell fusion. *J. Ultrastruct. Res.* **69**:164–179.

Herrera, A. H., and M. O. J. Olson. 1986. Association of protein C23 with rapidly labeled nucleolar RNA. *Biochemistry* **25**: 6258–6263.

Hughes, J. M. X., D. A. M. Konings, and G. Cesareni. 1987. The yeast homologue of U3 snRNA. *EMBO J.* **6**:2145–2155.

Hügle, B., U. Scheer, and W. W. Franke. 1985. Ribocharin: a nuclear M_r 40,000 protein specific to precursor particles of the large ribosomal subunit. *Cell* **41**:615–627.

Ishikawa, H. 1977. Evolution of ribosomal RNA. *Comp. Biochem. Physiol.* **58B**:1–7.

Jacob, S. T. 1986. Transcription of eukaryotic ribosomal RNA gene. *Mol. Cell Biochem.* **70**:11–20.

Jacq, B. 1981. Sequence homologies between eukaryotic 5.8S rRNA and the 5′ end of prokaryotic 23S rRNA: evidences for a common evolutionary origin. *Nucleic Acids Res.* **9**:2913–2932.

Jeppesen, C., B. Stebbins-Boaz, and S. A. Gerbi. 1988. Nucleotide sequence determination and secondary structure of *Xenopus* U3 snRNA. *Nucleic Acids Res.* **16**:2127–2148.

John, H. A., M. L. Birnstiel, and K. W. Jones. 1969. RNA-DNA hybrids at the cytological level. *Nature* (London) **223**:582–587.

Jones, K. W. 1965. The role of the nucleolus in the formation of ribosomes. *J. Ultrastruct. Res.* **13**:257–262.

Jordan, B. R., M. Latil-Damotte, and R. Jourdan. 1980. Coding and spacer sequences in the 5.8S-2S region of *Sciara coprophila* ribosomal DNA. *Nucleic Acids Res.* **8**:3565–3573.

Jordan, E. G., and C. A. Cullis. 1982. *The Nucleolus.* Cambridge University Press, Cambridge.

Jordan, E. G., and J. McGovern. 1981. The quantitative relationship of the fibrillar centres and other nucleolar components to changes in growth conditions, serum deprivation and low doses of actinomycin D in cultured diploid human fibroblasts (strain MRC-5). *J. Cell Sci.* **52**:373–389.

Karpen, G. H., J. E. Schaefer, and C. D. Laird. 1988. A *Drosophila* rRNA gene located in euchromatin is active in transcription and nucleolus formation. *Genes Dev.* **2**:1745–1763.

Kass, S., N. Craig, and B. Sollner-Webb. 1987. Primary processing of mammalian rRNA involves two adjacent cleavages and is not species specific. *Mol. Cell. Biol.* **7**:2891–2898.

Kempers-Veenstra, A. E., J. Oliemans, H. Offenberg, A. F. Dekker, P. W. Piper, R. J. Planta, and J. Klootwijk. 1986. 3′-End formation of transcripts from the yeast rRNA operon. *EMBO J.* **5**:2703–2710.

Khan, M. S. N., and B. E. H. Maden. 1977. Nucleotide sequence relationships between vertebrate 5.8S ribosomal RNAs. *Nucleic Acids Res.* **4**:2495–2505.

Khan, M. S. N., M. Salim, and B. E. H. Maden. 1978. Extensive homologies between the methylated nucleotide sequences in several vertebrate rRNAs. *Biochem. J.* **169**:531–542.

Kiss, T., M. Toth, and F. Solymosy. 1985. Plant small nuclear RNAs: nucleolar U3 snRNA is present in plants: partial characterization. *Eur. J. Biochem.* **152**:259–266.

Klootwijk, J., P. deJonge, and R. J. Planta. 1979. The primary transcript of the ribosomal repeating unit in yeast. *Nucleic Acids Res.* **6**:27–39.

Klootwijk, J., and R. J. Planta. 1973a. Analysis of the methylation sites in yeast ribosomal RNA. *Eur. J. Biochem.* **39**:325–330.

Klootwijk, J., and R. J. Planta. 1973b. Modified sequences in yeast ribosomal RNA. *Mol. Biol. Rep.* **1**:187–191.

Knight, E., and J. E. Darnell. 1967. Distribution of 5S RNA in HeLa cells. *J. Mol. Biol.* **28**:491–500.

Krohne, G., R. Stick, J. A. Kleinschmidt, R. Moll, W. W. Franke, and P. Hausen. 1982. Immunological localization of a major karyoskeletal protein in nucleoli of oocytes and somatic cells of *Xenopus laevis. J. Cell Biol.* **94**:749–754.

Kruger, K., P. J. Grabowski, A. J. Zaug, J. Sands, D. E. Gottschling, and T. R. Cech. 1982. Self-splicing RNA: autoexcision and autocyclization of the rRNA intervening sequence of *Tetrahymena. Cell* **31**:147–157.

Krzyzosiak, W., R. Denman, K. Nurse, W. Hellmann, M. Boublik, C. W. Gehrke, P. F. Agris, and J. Ofengand. 1987. *In vitro* synthesis of 16S ribosomal RNA containing single base changes and assembly into a functional 30S ribosome. *Biochemistry* **26**:2353–2364.

Kuhn, A., and I. Grummt. 1989. 3′-End formation of mouse pre-rRNA involves both transcription termination and a specific processing reaction. *Genes Dev.* **3**:224–231.

Kuhn, A., A. Normann, I. Bartsch, and I. Grummt. 1988. The mouse ribosomal gene terminator consists of three functionally separable sequence elements. *EMBO J.* **7**:1497–1502.

Kupriyanova, N. S., and M. Y. Timofeeva. 1988. 32S pre-rRNA processing: a dynamic model for interaction with U3RNA and structural rearrangements of spacer regions. *Mol. Biol. Rep.* **13**:91–96.

Kuter, D. J., and A. Rodgers. 1976. The protein composition of HeLa ribosomal subunits and nucleolar precursor particles. *Exp. Cell Res.* **102**:205–212.

Labhart, P., and R. H. Reeder. 1986. Characterization of three sites of RNA 3′ end formation in the Xenopus ribosomal gene spacer. *Cell* **45**:431–443.

Labhart, P., and R. H. Reeder. 1987a. Ribosomal precursor 3′ end formation requires a conserved element upstream of the promoter. *Cell* **50**:51–57.

Labhart, P., and R. H. Reeder. 1987b. A 12-base-pair sequence is an essential element of the ribosomal gene terminator in *Xenopus laevis. Mol. Cell. Biol.* **7**:1900–1905.

Lapeyre, B., H. Bourbon, and F. Amalric. 1987. Nucleolin, the major nucleolar protein of growing eukaryotic cells: an unusual protein structure revealed by the nucleotide sequence. *Proc. Natl. Acad. Sci. USA* **84**:1472–1476.

Lastick, S. M. 1980. The assembly of ribosomes in HeLa cell nucleoli. *Eur. J. Biochem.* **113**:175–182.

Lepoint, A., and G. Goessens. 1978. Nucleologenesis in Ehrlich tumor cells. *Exp. Cell Res.* **117**:89–94.

Lepoint, A., and G. Goessens. 1982. Quantitative analysis of Ehrlich tumor cell nucleoli during interphase. *Exp. Cell Res.* **137**:456–459.

Levis, R., and S. Penman. 1978. Processing steps and methylation in the formation of the rRNA in cultured *Drosophila* cells. *J. Mol. Biol.* **121**:219–238.

Lima-de-Faria, A. 1976. The chromosome field. I. Prediction of the location of ribosomal cistrons. *Hereditas* **83**:1–22.

Lischwe, M. A., R. L. Ochs, R. Reddy, R. G. Cook, L. C. Yeoman, E. M. Tan, M. Reichlin, and H. Busch. 1985. Purification and partial characterization of a nucleolar scleroderma antigen (M_r=34,000;pI 8.5) rich in N^G, N^G-dimethylarginine. *J. Biol. Chem.* **260**:14304–14310.

Lischwe, M. A., R. L. Richards, R. K. Busch, and H. Busch. 1981. Localization of phosphoprotein C23 to nucleolar structures and to the nucleolus organizer regions. *Exp. Cell Res.* **136**:101–109.

Lischwe, M. A., K. Smetana, M. O. J. Olson, and H. Busch. 1979. Protein C23 and B23 are the major nucleolar silver staining proteins. *Life Sci.* **25**:701–708.

Loening, U. E. 1975. The mechanism of synthesis of ribosomal RNA. *FEBS Symp.* **33**:151–157.

Loening, U. E., K. W. Jones, and M. L. Birnstiel. 1969. Properties of the rRNA precursor in *Xenopus laevis*: comparison to the precursor in mammals and in plants. *J. Mol. Biol.* **45**:353–366.

Lührmann, R., B. Appel, P. Bringmann, J. Rinke, R. Reuter, and S. Rothe. 1982. Isolation and characterization of rabbit anti-m2'2'7G antibodies. *Nucleic Acids Res.* **10**:7103–7113.

Maden, B. E. H. 1968. Ribosome formation in animal cells. *Nature* (London) **219**:685–689.

Maden, B. E. H. 1988. Locations of methyl groups in 28S rRNA of *Xenopus laevis* and man: clustering in the conserved core of the molecule. *J. Mol. Biol.* **201**:289–314.

Maden, B. E. H., M. S. N. Khan, D. G. Hughes, and J. P. Goddard. 1977. Inside 45S ribonucleic acid. *Biochem. Soc. Symp.* **42**: 165–179.

Maser, R. L., and J. P. Calvet. 1989. U3 small nuclear RNA can be psoralen cross-linked *in vivo* to the 5'- external transcribed spacer of pre-ribosomal RNA. *Proc. Natl. Acad. Sci. USA* **86**:6523–6527.

Mattaj, I. W. 1986. Cap trimethylation of U snRNA is cytoplasmic and dependent on U snRNP protein binding. *Cell* **46**:905–911.

Maxwell, E. S., and T. E. Martin. 1986. A low-molecular-weight RNA from mouse ascites cells that hybridizes to both 18S mRNA and mRNA sequences. *Proc. Natl. Acad. Sci. USA* **83**:7261–7265.

Mazan, S., and J. P. Bachellerie. 1988. Structure and organization of mouse U3B RNA functional genes. *J. Biol. Chem.* **263**: 19461–19467.

McClintock, B. 1934. The relation of a particular chromosomal element to the development of the nucleoli in *Zea mays*. *Z. Zellforsch Mikrosk. Anat.* **21**:294–328.

McKnight, S. L., R. A. Hipskind, and R. Reeder. 1980. Ultrastructural analysis of ribosomal gene transcription in vitro. *J. Biol. Chem.* **255**:7907–7911.

McKnight, S. L., and O. L. Miller, Jr. 1976. Ultrastructural patterns of RNA synthesis during early embryogenesis of *Drosophila melanogaster*. *Cell* **8**:305–319.

Melançon, P., M. Gravel, G. Boileau, and L. Brakier-Gingras. 1987. Reassembly of active 30S ribosomal subunits with an unmethylated *in vitro* transcribed 16S rRNA. *Biochem. Cell Biol.* **65**:1022–1030.

Meyer, G. F., and W. Hennig. 1974. The nucleolus in primary spermatocytes of *Drosophila hydei*. *Chromosoma* **46**:121–144.

Michot, B., and J. P. Bachellerie. 1987. Comparisons of large subunit rRNAs reveal some eukaryote-specific elements of secondary structure. *Biochimie* **69**:11–23.

Miller, K. G., and B. Sollner-Webb. 1981. Transcription of mouse rRNA genes by RNA polymerase I: in vitro and in vivo initiation and processing sites. *Cell* **27**:165–174.

Miller, O. L., Jr. 1981. The nucleolus, chromosomes and visualization of genetic activity. *J. Cell Biol.* **91**:15S–27S.

Miller, O. L., Jr., and B. R. Beatty. 1969. Visualization of nucleolar genes. *Science* **164**:955–957.

Miller, O. L., Jr., and B. A. Hamkalo. 1972a. Electron microscopy of active genes. *FEBS Symp.* **23**:367–378.

Miller, O. L., Jr., and B. A. Hamkalo. 1972b. Visualization of RNA synthesis on chromosomes. *Int. Rev. Cytol.* **33**:1–25.

Mirre, C., and B. Knibiehler. 1981. Ultrastructural antoradiographic localization of the rRNA transcription sites in the quail nucleolar components using two RNA antimetabolites. *Biol. Cell* **42**:73–78.

Mirre, C., and A. Stahl. 1978. Peripheral RNA synthesis of fibrillar center in nucleoli of Japanese quail oocytes and somatic cells. *J. Ultrastruct. Res.* **64**:377–387.

Montgomery, T. H. 1898. Comparative cytological studies with special regard to the morphology of the nucleolus. *J. Morphol.* **15**:265–560.

Morris, G. E., N. T. Man, and L. P. Head. 1985. Monoclonal antibodies against a nucleolar protein from differentiating chick muscle cells. *J. Cell Sci.* **76**:105–113.

Moss, T. 1983. A transcriptional function for the repetitive ribosomal spacer in Xenopus laevis. *Nature* (London) **302**: 223–228.

Mowry, K. L., and J. A. Steitz. 1988. snRNP mediators of 3' end processing: functional fossils? *Trends Biochem. Sci.* **13**:447–451.

Musters, W., J. Venema, G. Van der Linden, H. Van Heerikhuizen, J. Klootwijk, and R. J. Planta. 1989. A system for the analysis of yeast ribosomal DNA mutations. *Mol. Cell. Biol.* **9**:551–559.

Nazar, R. N. 1980. A 5.8S rRNA-like sequence in prokaryotic 23S rRNA. *FEBS Lett.* **119:212–214.**

Nelles, L., C. Van Broeckhoven, R. deWachter, and A. Vandenberghe. 1984. Location of the hidden break in large subunit ribosomal RNA of *Artemia salina*. *Naturwissenschaften* **71**: 634–635.

Niles, E. G. 1978. Isolation of a high specific activity 35S rRNA precursor from *Tetrahymena pyriformis* and identification of its 5'-terminus pppAp. *Biochemistry* **16**:3215–3219.

Ochs, R. L., M. A. Lischwe, P. O'Leary, and H. Busch. 1983. Localization of nucleolar phosphoproteins B23 and C23 during mitosis. *Exp. Cell Res.* **146**:139–149.

Ochs, R. L., M. A. Lischwe, E. Shen, R. E. Carroll, and H. Busch. 1985a. Nucleologenesis: composition and fate of prenucleolar bodies. *Chromosoma* **92**:330–336.

Ochs, R. L., M. A. Lischwe, W. H. Spohn, and H. Busch. 1985b. Fibrillarin: a new protein of the nucleolus identified by autoimmune sera. *Biol. Cell* **54**:123–134.

Olson, M. O., J. Z. M. Rivers, B. A. Thompson, N. Y. Kao, and S. T. Case. 1983. Interaction of nucleolar phosphoprotein C23 with cloned segments of rat ribosomal deoxyribonucleic acid. *Biochemistry* **22**:3345–3351.

Olson, M. O. J., and B. A. Thompson. 1983. Distribution of proteins among chromatin components of nucleoli. *Biochemistry* **22**:3187–3193.

Pace, N. R., T. A. Walker, and E. Schroeder. 1977. Structure of the 5.8S RNA component of the 5.8S-28S ribosomal RNA junction complex. *Biochemistry* **16**:5321–5328.

Palade, G. E. 1955. A small particulate component of the cytoplasm. *J. Biophys. Biochem. Cytol.* **1**:59–68.

Palade, G. E., and P. Siekevitz. 1956. Liver microsomes. *J. Biophys. Biochem. Cytol.* **2**:171–200.

Pardue, M. L., S. A. Gerbi, R. A. Eckhardt, and J. G. Gall. 1970. Cytological localization of DNA complementary to rRNA in polytene chromosomes of Diptera. *Chromosoma* **29**:269–290.

Parker, K. A., J. P. Bruzik, and J. A. Steitz. 1988. An *in vitro* interaction between the human U3 snRNP and 28S rRNA sequences near the α-sarcin site. *Nucleic Acids Res.* **16**:10493–10509.

Parker, K. A., and J. A. Steitz. 1987. Structural analyses of the human U3 ribonucleoprotein particle reveal a conserved sequence available for base pairing with pre-rRNA. *Mol. Cell. Biol.* 7:2899–2913.

Pavlakis, G. N., B. R. Jordan, R. M. Wurst, and J. N. Vournakis. 1979. Sequence and secondary structure of *Drosophila melanogaster* 5.8S and 2S rRNAs and the processing site between them. *Nucleic Acids Res.* 7:2213–2238.

Pene, J. J., E. Knight, and J. E. Darnell. 1968. Characterization of a new low molecular weight RNA in HeLa cell ribosomes. *J. Mol. Biol.* 33:609–623.

Perry, R. P. 1962. The cellular sites of ribosomal and 4S RNA. *Proc. Natl. Acad. Sci. USA* **48**:2179–2186.

Perry, R. P. 1967. The nucleolus and the synthesis of ribosomes. *Prog. Nucleic Acid Res. Mol. Biol.* 6:219–257.

Perry, R. P. 1976. Processing of RNA. *Annu. Rev. Biochem.* 45:605–629.

Perry, R. P. 1981. RNA processing comes of age. *J. Cell Biol.* 91:28s–38s.

Perry, R. P., A. Hell, and M. Errera. 1961. The role of the nucleolus in ribonucleic acid and protein synthesis. I. Incorporation of cytidine into normal and nucleolar inactivated HeLa cells. *Biochim. Biophys. Acta* 49:47–57.

Peters, M. A., T. A. Walker, and N. R. Pace. 1982. Independent binding sites in mouse 5.8S ribosomal ribonucleic acid for 28S ribosomal ribonucleic acid. *Biochemistry* **21**:2329–2355.

Planta, R. J., and H. A. Raué. 1988. Control of ribosome biogenesis in yeast. *Trends Genet.* 4:64–68.

Porter, G. L., P. J. Brennwald, K. A Holm, and J. A. Wise. 1988. The sequence of U3 from *Schizosaccharomyces pombe* suggests structural divergence of this snRNA between metazoans and unicellular eukaryotes. *Nucleic Acids Res.* **16**:10131–10151.

Porter, K. R. 1954. Electron microscopy of basophilic components of cytoplasm. *J. Histochem. Cytochem.* 2:346–371.

Prestayko, A. W., G. R. Klomp, D. J. Schmoll, and H. Busch. 1974. Comparison of proteins of ribosomal subunits and nucleolar preribosomal particles from Novikoff hepatoma ascites cells by two-dimensional polyacrylamide gel electrophoresis. *Biochemistry* **13**:1945–1952.

Prestayko, A. W., M. Tonato, and H. Busch. 1970. Low molecular weight RNA associated with 28S nucleolar RNA. *J. Mol. Biol.* 47:505–515.

Reddy, R., D. Henning, and H. Busch. 1979. Nucleotide sequence of nucleolar U3B RNA. *J. Biol. Chem.* **254**:11097–11105.

Reddy, R., W.-Y. Li, D. Henning, Y. C. Choi, K. Nogha, and H. Busch. 1981. Characterization and subcellular localization of 7-8S RNAs of Novikoff hepatoma. *J. Biol. Chem.* **256**:8452–8457.

Reddy, R., L. I. Rothblum, C. S. Subrahmanyam, M. H. Liu, D. Henning, B. Cassidy, and H. Busch. 1983a. The nucleotide sequence of 8S RNA bound to preribosomal RNA of Novikoff hepatoma. *J. Biol. Chem.* **258**:584–589.

Reddy, R., E. M. Tan, D. Henning, K. Nogha, and H. Busch. 1983b. Detection of nucleolar 7-2 ribonucleoprotein and a cytoplasmic 8-2 ribonucleoprotein with autoantibodies from patients with scleroderma. *J. Biol. Chem.* **258**:1383–1386.

Reeder, R. H., T. Higashinakagawa, and O. L. Miller, Jr. 1976. The 5′-3′ polarity of the *Xenopus* rRNA precursor molecule. *Cell* 8:449–454.

Reeder, R. H., B. Sollner-Webb, and H. Wahn. 1977. Sites of transcription initiation *in vivo* on *Xenopus laevis* rDNA. *Proc. Natl. Acad. Sci. USA* 74:5402–5406.

Reimer, G., K. M. Pollard, C. A. Penning, R. L. Ochs, M. A. Lischwe, H. Busch, and E. M. Tan. 1987a. Monoclonal autoantibody from NB/NZW F1 mouse and some human scleroderma sera target a M_r 34,000 nucleolar protein of the U3-ribonucleoprotein particle. *Arthritis Rheum.* 30:793–800.

Reimer, G., I. Raška, U. Scheer, and E. M. Tan. 1988. Immunolocalization of 7-2 ribonucleoprotein in the granular component of the nucleolus. *Exp. Cell Res.* **176**:117–128.

Reimer, G., I. Raška, E. M. Tan, and U. Scheer. 1987b. Human autoantibodies: probes for nucleolus structure and function. *Virchows Archiv B* 54:131–143.

Ritossa, F., K. Atwood, D. Lindsley, and S. Spiegelman. 1966. On the chromosomal distribution of DNA complementary to ribosomal and soluble RNA. *Natl. Cancer Inst. Monogr.* **23**:449–472.

Ritossa, F., and S. Spiegelman. 1965. Localization of DNA complementary to rRNA in the nucleolus organizer region of *Drosophila melanogaster*. *Proc. Natl. Acad. Sci. USA* **53**:737–745.

Rodrigues-Pousada, C., M. L. Cyrne, and D. Hayes. 1979. Characterization of preribosomal ribonucleoprotein particles from *Tetrahymena pyriformis*. *Eur. J. Biochem.* **102**:389–397.

Rose, K. M., J. Szopa, F.-S. Han, Y.-C. Cheng, A. Richter, and U. Scheer. 1988. Association of DNA topoisomerase I in ribosomal gene transcription. *Chromosoma* **96**:411–416.

Royal, A., and R. Simard. 1975. RNA synthesis in the ultrastructural and biochemical components of the nucleolus of chinese hamster ovary cells. *J. Cell Biol.* **66**:577–585.

Sakharov, V. N., L. N. Voronkova, and U. S. Chentsov. 1972. Ultrastructure of intranuclear bodies formed during cell division of cells irradiated with an ultraviolet microbeam. *Rep. Moscow Univ. Biol. Sci.* N5:56–59.

Scheer, U., K. Messner, R. Hazan, I. Raška, P. Hausmann, H. Falk, E. Spiess, and W. Franke. 1987. High sensitivity immunolocalization of double and single-stranded DNA by a monoclonal antibody. *Eur. J. Cell Biol.* **43**:358–371.

Scheer, U., and I. Raška. 1987. Immunocytochemical localization of RNA polymerase I in the fibrillar centers of nucleoli. *Chromosomes Today* 9:284–294.

Scheer, U., and K. Rose. 1984. Localization of RNA polymerase I in interphase cells and mitotic chromosomes by light and electron microscopic immunocytochemistry. *Proc. Natl. Acad. Sci. USA* **81**:1431–1435.

Scheer, U., M. F. Trendelenburg, and W. W. Franke. 1976. Regulation of transcription of genes of ribosomal RNA during amphibian oogenesis. *J. Cell Biol.* **69**:465–489.

Scherrer, K., H. Latham, and J. E. Darnell. 1963. Demonstration of an unstable RNA and of a precursor to ribosomal RNA in HeLa cells. *Proc. Natl. Acad. Sci. USA* **49**:240–248.

Schmidt-Zachmann, M. S. B. Hügle, U. Scheer, and W. W. Franke. 1984. Identification and localization of a novel nucleolar protein of high molecular weight by a monoclonal antibody. *Exp. Cell Res.* **153**:327–346.

Schmidt-Zachmann, M. S., B. Hügle-Dörr, and W. W. Franke. 1987. A constitutive nucleolar protein identified as a member of the nucleoplasmin family. *EMBO J.* 6:1881–1890.

Shine, J., and L. Dalgarno. 1973. Occurrence of heat-dissociable ribosomal RNA in insects: the presence of three polynucleotide chains in 26S RNA from cultured *Aedes aegypti* cells. *J. Mol. Biol.* 75:57–72.

Sigmund, J., H. G. Schwarzacher, and A. V. Mikelsaar. 1979. Satellite association frequency and number of nucleoli depend on cell cycle duration and NOR activity. *Hum. Genet.* **50**: 81–91.

Simard, R., Y. Langelier, M. Rosemonde, N. Maestracci, and A. Royal. 1974. Inhibitors as tools in elucidating the structure and function of the nucleus, p. 447–487. *In* H. Busch (ed.), *The Cell Nucleus*, vol. 3. Academic Press, Inc., New York.

Siomi, H., H. Shida, S. H. Nam, T. Nosaka, M. Maki, and M. Hatanaka. 1988. Sequence requirements for nucleolar localization of human T cell leukemia virus type 1 pX protein, which regulates viral RNA processing. *Cell* **55**:197–209.

Sitz, T. O., N. Banjeree, and R. N. Nazar. 1981. Effect of point mutations on 5.8S ribosomal ribonucleic acid secondary structure and the 5.8S-28S ribosomal ribonucleic acid junction. *Biochemistry* **20:**4029–4033.

Smith, S. D., N. Banerjee, and T. O. Sitz. 1984. Gene heterogeneity: a basis for alternative 5.8S rRNA processing. *Biochemistry* **23:**3648–3652.

Sollner-Webb, B., and R. Reeder. 1979. The nucleotide sequence of the initiation and termination sites for ribosomal RNA transcription in *Xenopus laevis. Cell* **18:**485–499.

Sollner-Webb, B., and J. Tower. 1986. Transcription of cloned eukaryotic ribosomal RNA genes. *Annu. Rev. Biochem.* **55:** 801–830.

Sommerville, J. 1986. Nucleolar structure and ribosome biogenesis. *Trends Biochem. Sci.* **11:**438–442.

Spector, D. L., R. L. Ochs, and H. Busch. 1984. Silver staining, immunofluorescence, and immunelectron microsopic localization of nucleolar phosphoproteins B23 and C23. *Chromosoma* **90:**139–148.

Steele, R. E., P. S. Thomas, and R. H. Reeder. 1984. Anucleolate frog embryos contain ribosomal DNA sequences and a nucleolar antigen. *Dev. Biol.* **102:**409–416.

Stroke, I., and A. M. Weiner. 1985. Genes and pseudogenes for rat U3A and U3B small nuclear RNA. *J. Mol. Biol.* **184:**183–193.

Stroke, I. L., and A. M. Weiner. 1989. The 5′ end of U3 snRNA can be crosslinked *in vivo* to the external transcribed spacer of rat ribosomal RNA precursors. *J. Mol. Biol.* **210:**497–512.

Suh, D., H. Busch, and R. Reddy. 1986. Isolation and characterization of a human U3 small nucleolar RNA gene. *Biophys. Biochem. Res. Commun.* **137:**1133–1140.

Sutiphong, J., C. Matzura, and E. G. Niles. 1984. Characterization of a crude selective Pol I transcription system from *Tetrahymena pyriformis. Biochemistry* **23:**6319–6327.

Suzuki, N., H. Matsui, and T. Hosoya. 1985. Effects of androgen and polyamines on the phosphorylation of nucleolar proteins from rat ventral prostates with particular reference to a 110-kDa phosphoprotein. *J. Biol. Chem.* **260:**8050–8055.

Sweeney, R., and M.-C. Yao. 1989. Identifying functional regions of rRNA by insertion mutagenesis and complete gene replacement in *Tetrahymena thermophila. EMBO J.* **8:**933–938.

Swift, H. 1959. Studies on nuclear fine structure. *Brookhaven Symp. Biol.* **12:**134–152.

Tague, B. W., and S. A. Gerbi. 1984. Processing of the large rRNA precursor: two proposed categories of RNA-RNA interactions in eukaryotes. *J. Mol. Evol.* **20:**362–367.

Tashiro, K., K. Shiokawa, K. Yamana, and Y. Sakai. 1986. Structural analysis of ribosomal DNA homologues in nucleolusless mutant of *Xenopus laevis. Gene* **44:**299–306.

Tautz, D., and G. A. Dover. 1986. Transcription of the tandem array of ribosomal DNA in *Drosophila melanogaster* does not terminate at any fixed point. *EMBO J.* **5:**1267–1273.

Thiry, M., A. Lepoint, and G. Goessens. 1985. Re-evaluation of the site of transcription in Ehrlich tumor cell nucleoli. *Biol. Cell* **54:**57–64.

Thiry, M., U. Scheer, and G. Goessens. 1988. Localization of DNA within Ehrlich tumour cell nucleoli by immunoelectron microscopy. *Biol. Cell* **63:**27–34.

Todorov, I., F. Noll, and A. A. Hadjiolov. 1983. The sequential addition of ribosomal proteins during the formation of the small ribosomal subunit in Friend erythroleukemia cells. *Eur. J. Biochem.* **131:**271–275.

Tollervey, D. 1987. A yeast small nuclear RNA is required for normal processing of pre-ribosomal RNA. *EMBO J.* **6:**4169–4175.

Trapman, J., P. DeJonge, and R. J. Planta. 1975. On the biosynthesis of 5.8S ribosomal RNA in yeast. *FEBS Lett.* **57:**26–30.

Trapman, J., and R. J. Planta. 1975. Detailed analysis of the ribosomal RNA synthesis in yeast. *Biochim. Biophys. Acta* **414:**115–125.

Trapman, J., and R. J. Planta. 1976. Maturation of ribosomes in yeast. I. Kinetic analysis by labelling of high molecular weight rRNA species. *Biochim. Biophys. Acta* **442:**265–274.

Trendelenburg, M. F. 1982. Chromatin structure of *Xenopus* rDNA transcription termination sites. Evidence for a two-step process of transcription termination. *Chromosoma* **86:**703–715.

Trendelenburg, M. F., and J. B. Gurdon. 1978. Transcription of cloned *Xenopus* ribosomal genes visualized after injection into oocyte nuclei. *Nature* (London) **276:**292–294.

Trinh-Rohlik, Q., and E. S. Maxwell. 1988. Homologous genes for mouse 4.5S hybRNA are found in all eukaryotes and their low molecular weight RNA transcripts intermolecularly hybridize with eukaryotic 18S ribosomal RNAs. *Nucleic Acids Res.* **16:**6041–6056.

Turner, D. H., N. Sugimoto, J. A. Jaeger, C. E. Longfellow, S. M. Freier, and R. Kierzek. 1987. Improved parameters for prediction of RNA structure. *Cold Spring Harbor Symp. Quant. Biol.* **52:**123–133.

Tyc, K., and J. A. Steitz. 1989. U3, U8 and U13 comprise a new class of mammalian snRNPs localized in the cell nucleolus. *EMBO J.* **8:**3113–3119.

Tyler, B. M., and N. H. Giles. 1985. Structure of a *Neurospora* RNA polymerase I promoter defined by transcription *in vitro* with homologous extracts. *Nucleic Acids Res.* **13:**4311–4331.

Veldman, G. M., R. C. Brand, J. Klootwijk, and R. J. Planta. 1980. Some characteristics of processing sites in ribosomal precursor RNA of yeast. *Nucleic Acids Res.* **8:**2907–2920.

Vincent, W. S. 1955. Structure and chemistry of nucleoli. *Int. Rev. Cytol.* **4:**269–298.

Wachtler, F., H. G. Schwarzacher, and A. Ellinger. 1982. The influence of the cell cycle on structure and number of nucleoli in cultured human lymphocytes. *Cell Tissue Res.* **225:**155–163.

Walker, T. A., Y. Endo, W. H. Wheat, I. G. Wool, and N. R. Pace. 1983. Location of 5.8S rRNA contact sites in 28S rRNA and the effect of α-sarcin on the association of 5.8S rRNA with 28S rRNA. *J. Biol. Chem.* **258:**333–338.

Walker, T. A., K. D. Johnson, G. J. Olsen, M. A. Peters, and N. R. Pace. 1982. Enzymatic and chemical structure mapping of mouse 28S ribosomal ribonucleic acid contacts in 5.8S ribosomal ribonucleic acid. *Biochemistry* **21:**2320–2329.

Wallace, H. R., and M. L. Birnstiel. 1966. Ribosomal cistrons and the nucleolar organizer. *Biochim. Biophys. Acta* **114:**296–310.

Waltschewa, L., O. Georgiev, and P. Venkov. 1983. Relaxed mutant of Saccharomyces cerevisiae: proper maturation of ribosomal RNA in absence of protein synthesis. *Cell* **33:**221–230.

Ware, V. C., R. Renkawaitz, and S. A. Gerbi. 1985. rRNA processing: removal of only nineteen bases at the gap between 28Sα and 28Sβ rRNAs in *Sciara coprophila. Nucleic Acids Res.* **13:**3581–3597.

Ware, V. C., B. W. Tague, C. G. Clark, R. L. Gourse, and S. A. Gerbi. 1983. Sequence analysis of 28S ribosomal DNA from the amphibian *Xenopus laevis. Nucleic Acids Res.* **11:**7795–7817.

Warner, J. R. 1974. The assembly of ribosomes in eucaryotes, p. 461–488. *In* M. Nomura, A. Tissières, and P. Lengyel (ed.), *Ribosomes.* Cold Spring Harbor Laboratory, Cold Spring Harbor, N.Y.

Warner, J. R., and R. Soeiro. 1967. Nascent ribosomes from HeLa cell. *Proc. Natl. Acad. Sci. USA* **58:**1981–1990.

Weinberg, R. A., U. Loening, M. Willems, and S. Penman. 1967. Acrylamide gel electrophoresis of HeLa cell nucleolar RNA. *Proc. Natl. Acad. Sci. USA* **58:**1088–1095.

Weinberg, R. A., and S. Penman. 1968. Small molecular weight monodisperse nuclear RNA. *J. Mol. Biol.* **38:**289–304.

Wellauer, P. K., and I. B. Dawid. 1973. Secondary structure maps of RNA. Processing of HeLa rRNA. *Proc. Natl. Acad. Sci. USA* **70:**2827–2831.

Wellauer, P. K., and I. B. Dawid. 1974. Secondary structure maps of rRNA and rDNA. I. Processing of *Xenopus laevis* rRNA and structure of single-stranded rDNA. *J. Mol. Biol.* **89:**379–395.

Wellauer, P. K., I. B. Dawid, D. E. Kelley, and R. P. Perry. 1974. Secondary structure maps of rRNA. II. Processing of mouse L-cell rRNA and variations in the processing pathway. *J. Mol. Biol.* **89:**397–407.

Winicov, I. 1976. Alternate temporal order in rRNA maturation. *J. Mol. Biol.* **100:**141–155.

Wise, J. A., and A. M. Weiner. 1980. Dictyostelium small nuclear RNA D2 is homologous to rat nucleolar RNA U3 and is encoded by a dispersed multigene family. *Cell* **22:**109–118.

Woods, P. S., and J. H. Taylor. 1959. Studies of ribonucleic acid metabolism with tritium-labeled cytidine. *Lab. Invest.* **8:**309–318.

Zagorski, J., D. Tollervey, and M. Fournier. 1988. Characterization of an *SNR* locus in *Saccharomyces cerevisiae* that specifies both dispensable and essential small nuclear RNAs. *Mol. Cell. Biol.* **8:**3282–3290.

Zhang, H., J. C. Wang, and L. F. Liu. 1988. Involvement of DNA topoisomerase I in transcription of human ribosomal RNA genes. *Proc. Natl. Acad. Sci. USA* **85:**1060–1064.

Zieve, G., and S. Penman. 1976. Small RNA species of the HeLa cell: metabolism and subcellular localization. *Cell* **8:**19–31.

Chapter 40

Nucleocytoplasmic Transport of Ribosomal Subunits

VASSIE C. WARE and ARATI KHANNA-GUPTA

Our knowledge of the nucleolar events involved in the biogenesis of eucaryotic ribosomes is most complete for the earlier stages of this process, particularly for the rRNA component of the ribosome, yet major gaps remain in our understanding of the factors that regulate the synthesis of rRNA, the processing and posttranscriptional modifications of rRNA, and the rRNA-protein interactions that govern the processing and assembly of ribosomal ribonucleoprotein (rRNP) particles. The formation of rRNPs depends on an extensive cast of nonribosomal and ribosomal components, mostly synthesized in the cytoplasm, that must be imported into the nucleolus for participation in rRNP assembly. Assembled ribosomal subunits exit the nucleolus and must then be transported across the nuclear envelope into the cytoplasm, where they interact to form mature, functionally active ribosomes. Details of the later steps in ribosomal subunit maturation just before their translocation into the cytoplasm as well as the mechanics of the transport process itself are rather limited. Compelling questions related to what nuclear factors, if any, effect nuclear rRNP export and what components of rRNPs (proteins, RNA, or both) interact with the transport machinery remain unanswered.

Many of the nucleolar events involved in ribosome biogenesis have been reviewed by Gerbi et al. (this volume) and will not be reiterated here. Nucleocytoplasmic rRNP transport is an intriguing area in ribosome biogenesis that has received little attention. Therefore, we have confined our discussions to the later events in ribosome maturation in the nucleus and to the export of rRNPs into the cytoplasm, omitting aspects of ribosome biogenesis that include the targeting of ribosomal proteins to the nucleolus. Since the mechanics of rRNP translocation remain elusive, the nuclear export of other RNAs and RNPs will be considered where such data may provide clues to understanding the transport of rRNPs. Finally, we address several questions being considered in our laboratory related to the regulation of the kinetics of ribosomal subunit transport and to the putative involvement of rRNA in the transport process.

NUCLEOCYTOPLASMIC TRANSPORT OF NUCLEAR RNPs

The transport of nuclear RNPs through the nuclear envelope has been the subject of several reviews (Wunderlich et al., 1976; Clawson and Smuckler, 1982; Maul, 1982; Newport and Forbes, 1987). Several investigators have hypothesized that nuclear RNPs are closely associated with the nuclear matrix. There is evidence that RNA is not freely diffusible within the nucleus or cytoplasm but is associated with structural elements of the nuclear matrix and cytoskeleton (Fey et al., 1986). The matrix may play a role in RNP processing and in the intranuclear movement of RNPs toward the nuclear pore complexes (Wunderlich et al., 1976; Herlan et al., 1979). Although several hypotheses that address the problem of RNP nucleocytoplasmic transport have been advanced, relatively few are based on direct experimental data. According to a general gating mechanism proposed by Wunderlich and Speth (1972), the nuclear RNPs first bind to specific pore complex constituents until a threshold of bound RNP particles is reached. Above a critical concentration, the RNPs are translocated through the nuclear pore complex and released on the cytoplasmic side of the nuclear envelope. Agutter (1985) likewise has proposed, on the basis of the nondiffusible nature of RNA, that the efflux of RNA occurs in three stages: first, transport to the nuclear envelope along the nuclear matrix; second, translocation through the pores; and third, transport along the cytoskeleton.

Vassie C. Ware and Arati Khanna-Gupta ■ Department of Biology and Center for Molecular Bioscience and Biotechnology, Lehigh University, Bethlehem, Pennsylvania 18015.

Regardless of the type of RNP to be translocated (messenger RNP [mRNP] or rRNP), the transport machinery at the level of the nuclear pore appears to have a number of standard features. Translocation is generally considered as an energy-requiring process, although the nucleocytoplasmic transport of rRNP from isolated nuclei can also be induced by varying the ratio of calcium and magnesium ions in the medium (Wunderlich and Herlan, 1977). These RNPs are too large in diameter to pass through the 9-nm-diameter diffusion channel presented by the nuclear pore complex. Therefore, large RNPs must conceivably undergo massive deformations in shape during their translocation through the nuclear pore complexes or the pore complex itself must undergo extensive expansion (Clawson and Smuckler, 1982). The transport of these RNPs appears strictly vectorial, unlike the movements of several small nuclear RNPs (snRNPs), which include nuclear snRNA synthesis, transport of the snRNA into the cytoplasm for assembly with snRNP proteins, and reentry of the snRNP into the nucleus for function (DeRobertis et al., 1982).

The principles governing the transport of RNPs appear to share some common features with the mechanism proposed for importation of proteins into the nucleus (Newmeyer and Forbes, 1988). Using an in vitro system, Newmeyer and Forbes have demonstrated that protein import to the nucleus occurs in two distinct steps: binding and translocation. Binding of the nuclear-bound protein in question to the pore complex requires the presence of a nuclear import signal sequence and occurs in the absence of ATP. Translocation of the bound protein, however, has the strict requirement for ATP.

Dworetzky and Feldherr (1988) have recently demonstrated that RNA- and protein-coated gold particles are capable of traversing the same nuclear pores, implying that the same nuclear pore is competent to recognize RNA as well as protein, although the translocation mechanism for each species might be quite distinct within that pore. Several nuclear proteins, such as nucleoplasmin, have been shown to contain nuclear import signals that interact with signal receptors presumably in the nuclear pore complex (Dingwall et al., 1982). It is therefore reasonable to assume that nuclear pore receptors for RNAs or RNPs exist as well.

Several specific interactions involving both the RNP and pore complex constituents must occur. It remains to be demonstrated which components of the pore complex interact with the RNPs, although specific protease inhibitors have been shown to block the release of RNPs from isolated nuclei, suggesting that some protein interactions are essential (Giese et al., 1979). The involvement of protein components in the nuclear pore complex in RNA and protein transport is further suggested by the observations of Featherstone et al. (1988). These investigators showed that a monoclonal antibody against a nuclear pore complex inhibits the nucleocytoplasmic transport of both RNA (5S and tRNA) and the karyophilic protein nucleoplasmin while having no measurable impact on the diffusion capacity of the nuclear pore complex. Other features of nuclear pore complexes have recently been reviewed by Newport and Forbes (1987).

The complexities involved in the formation of RNPs in general have obscured efforts to elucidate the precise intracellular movements of these particles to date. Only limited progress has been made in identifying some of the components involved in RNP transport, particularly for the transport of mature mRNA from the nucleus (reviewed in Schröder et al., 1987). Most of the conclusions about rRNP transport are based primarily on the permeability properties of the nuclear envelope and the sizes of the rRNP transit forms (Paine and Horowitz, 1980). Only a limited number of in vivo RNA and RNP transport studies have been reported to date.

RNP Structure and Transport

Which features of RNPs are recognized by the transport machinery? Are there specific primary structural features on the RNP (housed in protein, RNA, or both) that are required, or is the general shape of RNPs conducive for transport? These questions remain unanswered for the majority of RNPs, but some hints are emerging from studies of mRNP, tRNA, and snRNP transport.

mRNP transport

The efflux of mRNP from isolated nuclei has been studied in a number of laboratories (see Schröder et al., 1987, for an extensive review of this subject). Work from several laboratories has suggested that the poly(A) tail of mRNA is involved in the transport of this species across the nuclear envelope. Evidence is mounting that mRNP transport can occur only after the release of intranuclear poly(A)-associated proteins and that different poly(A)-binding proteins are compartmentalized in the nucleus or cytoplasm (reviewed in Schröder et al., 1987). A significant finding is that the poly(A) tail of mRNA or synthetic poly(A) stimulates the nucleoside triphosphatase activity found in intact nuclear envelopes (Bernd et al., 1982). In addition, poly(A)-coated gold particles, microinjected into the germinal vesicles of *Xenopus* oocytes, traversed the nuclear envelope unidirectionally, whereas bovine serum albumin- or ovalbumin-coated gold particles remained intranuclear (Dworetzky and Feldherr, 1988). Together, these results support the hypothesis that the poly(A) tail in mRNP is a vital structural feature involved in transporting most mRNPs across the nuclear envelope, although other features must be considered for mRNPs that lack poly(A) tails (e.g., histone mRNAs).

tRNA transport

Zasloff (1983) investigated the movement of $tRNA^{Met}$ from the nucleus to the cytoplasm in microinjected *Xenopus* oocytes. He obtained evidence for a saturable, carrier-mediated, temperature-dependent transport system for translocating tRNAs into the cytoplasm. Furthermore, Zasloff's group (Tobian et al., 1985) has constructed mutations within the highly conserved D stem-loop and T stem-loop of tRNA that impair the kinetics of tRNA transport into the cytoplasm. The microinjection studies of Dworetzky and Feldherr (1988) also used tRNA-coated gold particles, observing rapid transport of these particles into the cytoplasm via nuclear pores. Control particles (coated with bovine serum albumin or ovalbumin) were not transferred to the cytoplasm, indicating that the nucleoplasmic surfaces of nuclear pores have the ability to discriminate between macromolecules that have the signals required for nuclear export and those that do not. The nature of those signals, however, remains a mystery.

snRNP transport

The biogenesis of snRNPs has been studied extensively in *Xenopus* oocytes. In oocytes and early embryos, the overwhelming abundance of snRNP proteins are located in the cytoplasm (Fritz et al., 1984). Early in embryogenesis, snRNA synthesis commences in the nucleus; later, assembled snRNPs migrate into the nucleus. The acquired ability of snRNP proteins residing in the cytoplasm to enter the nucleus at this stage of development is related to snRNA-protein interactions. A specific sequence of U2 snRNA is required for protein binding and for nuclear targeting of the assembled U2 snRNP (Mattaj and DeRobertis, 1985); however, the RNA sequence alone is insufficient for targeting. Clearly, protein binding must modulate the structure of the RNA sequence somehow to account for the inability of the RNA sequence to confer nuclear localization on its own. These observations reinforce the point that transport information on RNPs may not reside in a single macromolecular component but may represent a composite of sites on several components.

5S RNA transport

The synthesis of 5S RNA and its assembly into ribosomes have been studied extensively (reviewed in Hadjiolov, 1985, and Sommerville, 1986). In *Xenopus* oocytes, 5S synthesis begins on extranucleolar chromatin at an earlier developmental stage in oogenesis than the onset of pre-rRNA synthesis. As such, the 5S RNA is transported from the nucleus to the cytoplasm, where it is stored in nonribosomal particles awaiting the onset of rRNA synthesis and other ribosomal components. Stored cytoplasmic RNA is recruited back to the nucleus, where it participates in ribosome biogenesis along with the other RNAs (Allison et al., in press). Protein-RNA interactions appear vital to the targeting of 5S RNA to the nucleus. Using truncated 5S RNA species, these investigators have shown that an internal stretch in 5S RNA is required for nuclear influx and retention. What additional factors may be required is unknown. The reemergence of 5S RNA into the cytoplasm is as a component of the large ribosomal subunit. At this time, it is unknown whether inclusion of 5S RNA into the large subunit is a necessary step for large-subunit transport.

rRNP transport

As previously stated, the rRNA transport aspect of ribosome biogenesis is perhaps the least understood step in the pathway. rRNPs are generally thought to traverse the nuclear envelope as ribosomal subunits and not as intact ribosomes, judging from the appearance of newly synthesized 40S subunits in the cytoplasm before the appearance of newly synthesized 60S subunits in pulse-chase experiments (Leick and Andersen, 1970; Wunderlich, 1972; Eckert et al., 1975). These data, however, do not address the mechanism of transport at the pore or any putative interactions that may take place at that site. Clawson and Smuckler (1982) proposed that some of the RNA of rRNPs may have a surface location and may be involved in interacting directly with the pore complex to effect nuclear rRNP export. The possibility that RNA plays a role in rRNP transport has recently been revisited in our laboratory and will be discussed below.

Nucleolar Proteins Implicated in rRNP Transport

Two general classes of proteins contribute to the structural integrity of the nucleolus. One class comprises proteins that are resident in distinct nucleolar compartments, thereby functioning with rDNA or with precursor ribosomal particles or as structural nucleolar proteins forming the nucleolar skeletal framework. Members of the other class of proteins are not generally confined to the nucleolus but may be shuttled between other cellular compartments. Recycled processing and transport factors as well as ribosomal proteins are members of this group.

Nonribosomal nucleolar proteins that are involved in the processing or packaging of rRNPs are generally believed to dissociate from the preribosomal subunits before rRNP nucleocytoplasmic export. An example of one such protein is ribocharin, a 40-kilodalton acidic protein that associates with precursors of the 60S subunit (Hugle et al., 1985). Since ribocharin has been localized to the granular component of the nucleolus and to the nucleoplasm, but not to the cytoplasm, it has been proposed that this protein may play a role in the nucleocytoplasmic transport or later stages of processing of the 60S subunit.

Not all nonribosomal "nucleolar" proteins with apparent processing-packaging or transport functions remain intranuclear. Recently, Borer et al. (1989) demonstrated that two major nucleolar proteins may function in the nucleus as well as in the cytoplasm. Nucleolin or C23, a phosphorylated multifunctional protein, has been implicated in rRNA transcription (Bouche et al., 1984) and in the binding of nascent 45S pre-rRNA (Bugler et al., 1987); NO38 (B23), a 38-kilodalton protein, is reputed to be involved in the assembly or intranuclear transport of pre-RNPs. It was determined that both proteins apparently shuttle back and forth between the nucleus and cytoplasm, implicating them in the transport of ribosomal components across the nuclear envelope. At what level these proteins interact with rRNPs or the transport machinery is unresolved.

NEW APPROACHES TO UNDERSTANDING rRNP TRANSPORT

The assembly of ribosomal subunits and their translocation across the nuclear envelope is a complex process. There is a dearth of information about many of the steps in this pathway to the cytoplasm. Our laboratory has focused on several aspects of the transport problem, particularly on the putative role that rRNA may play in effecting rRNP transport. We have described a transport assay in microinjected mature *Xenopus* oocytes (Khanna-Gupta and Ware, 1989) in which radiolabeled rRNPs from various sources are microinjected into the nucleus and their redistribution into the cytoplasmic compartment is monitored over time. Using this system, we can ask several questions about the kinetics of rRNP transport. Microinjection of in vitro-manipulated rRNPs will provide an opportunity to examine the importance of rRNP structure in rRNP transport and will offer the possibility of identifying putative nuclear transport factors. Several of our findings are discussed below.

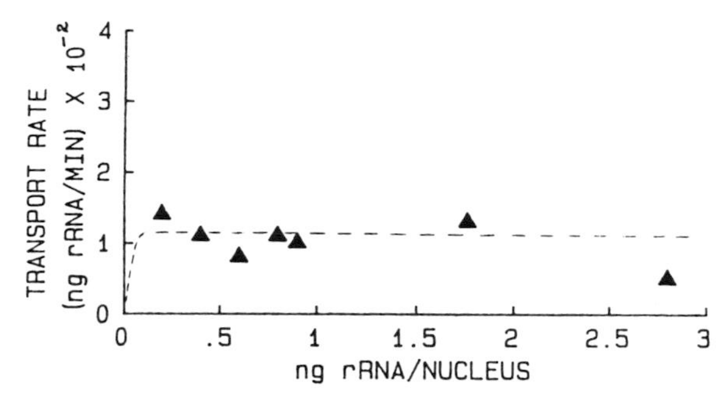

Figure 1. Rate of *Xenopus* ribosome transport as a function of the concentration of ribosomes injected in *X. laevis* oocytes. *Xenopus* ribosomes (0.2 to 2.8 ng of RNA; 0.4×10^3 to 6×10^3 cpm) were injected in a volume of 20 nl into the nuclei of oocytes. After incubation of injected oocytes for up to 60 min in Holtfreter medium at 25°C, the oocytes were fixed in cold 1% trichloroacetic acid and manually dissected; radioactivity present in either pooled or individual nuclear and cytoplasmic fractions was determined by liquid scintillation counting. $t_{1/2}$ values calculated from transport data were used to calculate the transport rate.

Kinetics of rRNP Transport

We examined the kinetics of the process by which ribosomes are exported from the nucleus to the cytoplasm, using the *Xenopus* oocyte system as described (Khanna-Gupta and Ware, 1989). Radiolabeled ribosomes from *Xenopus laevis*, *Tetrahymena thermophila*, and *Escherichia coli* were microinjected into the germinal vesicles of oocytes. Microinjected eucaryotic mature 80S ribosomes were redistributed into the oocyte cytoplasm by an apparent carrier-mediated translocation process that exhibits saturation kinetics as increasing amounts of ribosomes are injected (Fig. 1). It is assumed that in the cases where 80S ribosomes were injected, the particles dissociated in the nucleus into constituent subunits, on the basis of the widely held view that ribosomal subunits are the transportable species. This assumption, however, was not tested directly in these experiments. *Tetrahymena* ribosomes were competent to traverse the nuclear envelope, suggesting that the basic rRNP transport mechanism has been evolutionarily conserved. In fact, the transport kinetics for microinjected *Tetrahymena* ribosomes were nearly identical to those of injected *Xenopus* ribosomes, indicating that the biochemical nature of the interaction of the presumed carrier with the ribosomal subunit has been conserved throughout the evolution of the eucaryotes.

Our results with microinjected 70S ribosomes

from *E. coli* were consistent with the hypothesis that procaryotic ribosomes lack the signals required for nuclear export, since these ribosomes remained intranuclear in our experiments. In fact, in coinjection experiments using differentially radiolabeled *E. coli* 70S ribosomes and *Tetrahymena* 80S ribosomes, only the latter particles were transported. We have interpreted these results to suggest that the transport machinery may not efficiently recognize procaryotic 70S ribosomes as transportable species because of the absence of the necessary transport information in those particles, although alternative interpretations are possible.

A surprising result was obtained when we coinjected small and large subunits from *T. thermophila* into *Xenopus* oocytes. We monitored the transport kinetics of one radiolabeled subunit when coinjected with its unlabeled partner subunit and discovered that its transport rate was remarkably faster in the presence of the unlabeled partner subunit than when the unlabeled partner subunit was absent. For example, in the presence of unlabeled 60S subunits from *T. thermophila*, radiolabeled 40S subunits from *T. thermophila* were transported from the nucleus much more rapidly; the 40S subunit $t_{1/2}$ decreased from 162 min to 19 min. The change in kinetics was similar when the reciprocal experiment was done using radiolabeled 60S subunits. The transport rate of heterologous ribosomal subunits was clearly facilitated or accelerated by the presence of both subunits. We had not anticipated any synergistic rate effects, yet this conclusion was further supported in experiments in which heterologous partner subunits were introduced into the nucleus 1 h after the injection of the opposing partner subunits. A significant change in the rate of transport was noted as well.

It appears that although the basic transport mechanism has been conserved, there may be species-specific interactions that play a role in the process. The nature of these putative interactions is unclear; however, we have proposed that subunit interactions at the level of the nuclear envelope may be involved. The possibility that subunits interact at some late stage in rRNP transport is supported by the observation that in experiments in which only one heterologous subunit was injected, the rate of export was much slower than when the heterologous partner subunit was present in the nucleus as well. It would appear that *Xenopus* 40S subunits were not as effective as *Tetrahymena* 40S subunits in facilitating the transport of *Tetrahymena* 60S subunits.

Although unexpected, our findings do not contradict any previous claims based on pulse-chase data about the nature of the nucleocytoplasmic transport of rRNPs as independent units across the nuclear envelope. In examining the earlier pulse-chase experiments, however, it is impossible to evaluate the contribution of unlabeled or partially labeled ribosomal subunits to the rate of transport of the newly synthesized 40S or 60S subunits. It is possible that our heterologous studies have identified a mechanism that serves to regulate the rate of transport of subunits as ribosome biogenesis progresses. Elucidation of the exact nature of the facilitation effect must await further experimentation.

Role of rRNA in rRNP Transport

Our interest in the RNA component of the ribosome stems from a general interest in the structure and function of eucaryotic rRNA. There is scarcely any doubt that rRNA is intimately involved in ribosome function, as numerous regions within 16S rRNA and 23S rRNA of *E. coli* have been implicated in protein synthesis (reviewed in Dahlberg, 1989, and elsewhere in this volume). Since several of the implicated regions in procaryotic rRNAs are conserved in structure in the homologous eucaryotic rRNAs, it is likely that similar functions may be ascribed to the eucaryotic rRNAs as well (Clark et al., 1984). Despite considerable advances in understanding the role of rRNA in translation, virtually nothing is known about the possible contribution of other rRNA sequences to processes outside of the translation realm. Especially for eucaryotes are the extra sequences (called expansion segments by Clark et al. [1984]) in rRNA good candidates for areas involved in eucaryote-specific functions (e.g., membrane binding, interaction with components of the secretory apparatus, nucleocytoplasmic transport, and interaction with species-specific proteins). Expansion segments as well as other segments of rRNA may be associated with other aspects of ribosomal organization.

As originally proposed by Clawson and Smuckler (1982), surface RNA may be instrumental in the rRNP transport process, interacting with components in the nuclear pore complex. In preliminary studies to examine the role of rRNA in the rRNP transport process, we have microinjected radiolabeled rRNPs, which were subjected to limited micrococcal nuclease digestion in vitro, into *Xenopus* oocytes (Fig. 2). Although the extent of damage to these particles was not ascertained except to note that for the *Tetrahymena* rRNPs, only 10 to 15% of the radioactivity was removed as judged by trichloroacetic acid precipitation of parallel samples, the results give a preliminary indication that surface RNA may be important for the transport process. The nuclease treatment may alter the shape or disrupt a potential

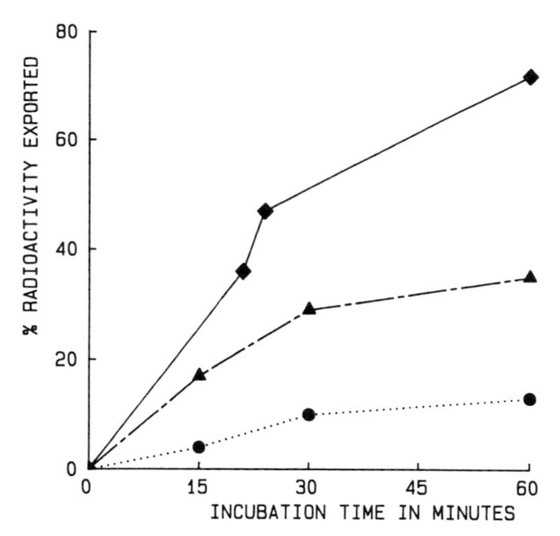

Figure 2. Kinetics of nucleocytoplasmic transport of *Tetrahymena* ribosomes treated with micrococcal nuclease. Radiolabeled rRNPs were treated with various concentrations of micrococcal nuclease for 1 min (♦, no enzyme; ▲, 1 U; ●, 5 U). Nuclease-treated rRNPs were then microinjected into oocytes and analyzed as described in the legend to Fig. 1.

interaction such that the digested rRNP is less efficiently recognized by the transport machinery. If this is the case, it would be important to know what sequences are present on the ribosomal subunit surface.

In the absence of a detailed sketch of the three-dimensional structure of eucaryotic rRNA within the ribosome, it is an arduous task to identify sequences that are exposed on the surface and are candidates for areas that may be involved in ribosomal subunit transport. We have attempted to target specific single-stranded areas on the ribosome surface by using a synthetic cDNA oligomer-binding approach to mask the availability of exposed sequences on radiolabeled rRNPs. The probe hybridization technique has been used successfully to probe the availability of single-stranded RNA sequences on the surface of *E. coli* ribosomes and to disrupt several ribosomal functions (e.g., subunit association [Tapprich and Hill, 1986]). We have also used oligomer-directed RNase H digestion of rRNPs to disrupt the availability of exposed RNA sequences on the ribosome or subunit surface. We have garnered information from a variety of sources in order to maximize our chances of synthesizing probes that will bind to ribosomes or ribosomal subunits. Once rRNPs are bound by probe, as determined by nitrocellulose filtration or sucrose gradients, and the specificity of binding to rRNPs has been determined in RNase H and RNA sequencing experiments, one can ask whether the probe-rRNP complex or oligomer-directed RNase H-treated rRNP is competent for transport across the *Xenopus* nuclear envelope. Through the analysis of cDNA probe effects on rRNP transport, we hope to identify regions within rRNA that effect the transport of rRNPs.

Our analysis of probe effects on 60S subunit transport is just beginning. So far, we have found several probes that bind specifically to ribosomes or to 60S subunits. Although not all probes have as yet been tested in our transport assay system, we have identified one region in 28S rRNA within the 5′ part of the molecule that, when disrupted, has a dramatic impact on the kinetics of ribosome transport in *Xenopus* oocytes (A. Khanna-Gupta and V. C. Ware, unpublished data) while having no measurable impact on in vitro protein synthesis in rabbit reticulocytes (C. DeHoratius and V. C. Ware, unpublished data). Other probes that bind have no apparent effect on nucleocytoplasmic transport kinetics in our system. Further analysis of these putative exposed regions is in progress. It is our hope that the approaches described here coupled with genetic strategies to mutate specific regions in rRNA will yield fruitful results, enabling us to advance our understanding of the role of rRNA in the transport of rRNPs.

We acknowledge other members of the laboratory, Jeanne Garvey and Caryn DeHoratius, for their contributions to this work.

This work was supported in part by Public Health Service grant GM38574 from the National Institutes of Health to V.C.W.

REFERENCES

Agutter, P. S. 1985. RNA processing, RNA transport, and nuclear structure, p. 539–589. *In* E. A. Smuckler and G. A. Clawson (ed.), *Nuclear Envelope Structure and RNA Maturation.* Alan R. Liss, Inc., New York.

Allison, L. A., P. J. Romaniuk, and A. H. Bakken. Fate of stored 5S RNA during ribosome biogenesis in oocytes of Xenopus laevis. I. The localization and movement of 5S RNA. *Dev. Biol.*, in press.

Bernd, A., H. C. Schroder, R. K. Zahn, and W. E. G. Muller. 1982. Modulation of the nuclear envelope nucleoside triphosphatase by poly(A)-rich mRNA and by microtubule protein. *Eur. J. Biochem.* **129:**43–49.

Borer, R. A., C. F. Lehner, H. M. Eppenberger, and E. A. Nigg. 1989. Major nucleolar proteins shuttle between nucleus and cytoplasm. *Cell* **56:**379–390.

Bouche, G., M. Caizergues-Ferrer, B. Bugler, and F. Amalric. 1984. Interrelations between the maturation of a 100 kDa nucleolar protein and pre rRNA synthesis in CHO cells. *Nucleic Acids Res.* **12:**3025–3035.

Bugler, B., H. Bourbon, B. Lapeyre, M. O. Wallace, J.-H. Chang, F. Amalric, and M. O. J. Olson. 1987. RNA binding fragments from nucleolin contain the ribonucleoprotein consensus sequence. *J. Biol. Chem.* **262:**10922–10925.

Clark, C. G., B. W. Tague, V. C. Ware, and S. A. Gerbi. 1984. Xenopus laevis 28S ribosomal RNA: a secondary structure model and its evolutionary and functional implications. *Nucleic Acids Res.* **12:**6197–6220.

Clawson, G. A., and E. A. Smuckler. 1982. A model for nucleocytoplasmic transport of ribonucleoprotein particles, p. 271–

278. *In* G. G. Maul (ed.), *The Nuclear Envelope and the Nuclear Matrix*. Alan R. Liss, Inc., New York.

Dahlberg, A. 1989. The functional role of ribosomal RNA in protein sysnthesis. *Cell* **57**:525–529.

DeRobertis, E. M., S. Lienhard, and R. F. Parisot. 1982. Intracellular transport of microinjected 5S and small nuclear RNAs. *Nature* (London) **295**:572–577.

Dingwall, C. S., S. Sharnick, and R. Laskey. 1982. A polypeptide domain that specifies migration of nucleoplasmin in the nucleus. *Cell* **30**:449–458.

Dworetzky, S. I., and C. M. Feldherr. 1988. Translocation of RNA-coated gold particles through the nuclear pores of oocytes. *J. Cell Biol.* **106**:575–584.

Eckert, W. A., W. W. Franke, and U. Scheer. 1975. Nucleocytoplasmic translocation of RNA in Tetrahymena pyriformis and its inhibition by actinomycin D and cycloheximide. *Exp. Cell Res.* **94**:31–46.

Featherstone, C., M. K. Darby, and L. Gerace. 1988. A monoclonal antibody against the nuclear pore complex inhibits nucleocytoplasmic transport of protein and RNA in vivo. *J. Cell Biol.* **107**:1289–1297.

Fey, E. G., D. A. Ornelles, and S. Penman. 1986. Association of RNA with the cytoskeleton and the nuclear matrix. *J. Cell Sci. Suppl.* **5**:99–119.

Fritz, A., R. Parisot, D. Newmeyer, and E. M. DeRobertis. 1984. Small nuclear U-ribonucleoproteins in Xenopus laevis development: uncoupled accumulation of the protein and RNA components. *J. Mol. Biol.* **178**:273–285.

Giese, G., G. Herlan, and F. Wunderlich. 1979. Nuclear RNA release: inactivation by temperature and protease inhibitors. *Hoppe Seyler's Z. Physiol. Chem.* **360**:266–267.

Hadjiolov, A. A. 1985. *The Nucleolus and Ribosome Biogenesis.* Springer-Verlag, New York.

Herlan, G., W. A. Eckert, W. Kaffenberger, and F. Wunderlich. 1979. Isolation and characterization of an RNA-containing nuclear matrix from Tetrahymena macronuclei. *Biochemistry* **18**:1782–1788.

Hugle, B., U. Scheer, and W. W. Franke. 1985. Ribocharin: a nuclear M_r 40,000 protein specific to precursor particles of the large ribosomal subunit. *Cell* **41**:615–627.

Khanna-Gupta, A., and V. C. Ware. 1989. Nucleocytoplasmic transport of ribosomes in a eukaryotic system: is there a facilitated transport process? *Proc. Natl. Acad. Sci. USA* **86**: 1791–1795.

Leick, V., and S. B. Andersen. 1970. Pools and turnover rates of nuclear ribosomal RNA in Tetrahymena pyriformis. *Eur. J. Biochem.* **14**:460–464.

Mattaj, I. W., and E. M. DeRobertis. 1985. Nuclear segregation of U2 snRNA requires binding to specific snRNP proteins. *Cell* **40**:111–118.

Maul, G. G. 1982. Aspects of a hypothetical nucleocytoplasmic transport mechanism, p. 1–13. *In* G. G. Maul (ed.), *The Nuclear Envelope and Nuclear Matrix.* Alan R. Liss, Inc., New York.

Newmeyer, D. D., and D. J. Forbes. 1988. Nuclear import can be separated into distinct steps in vitro: nuclear pore binding and translocation. *Cell* **52**:641–653.

Newport, J., and D. J. Forbes. 1987. The nucleus: structure, function, and dynamics. *Annu. Rev. Biochem.* **56**:535–565.

Paine, P. L., and S. B. Horowitz. 1980. The movement of material between nucleus and cytoplasm, p. 299–338. *In* D. M. Prescott and L. Goldstein (ed.), *Cell Biology: a Comprehensive Treatise.* Academic Press, Inc., New York.

Schröder, H. C., M. Bachmann, B. Diehl-Seifert, and W. E. G. Muller. 1987. Transport of mRNA from nucleus to cytoplasm. *Prog. Nucleic Acid Res. Mol. Biol.* **34**:89–142.

Sommerville, J. 1986. Nucleolar structure and ribosome biogenesis. *Trends Biochem. Sci.* **11**:438–442.

Tapprich, W. E., and W. E. Hill. 1986. Involvement of bases 787–795 of Escherichia coli 16S ribosomal RNA in ribosomal subunit association. *Proc. Natl. Acad. Sci. USA* **83**:556–560.

Tobian, J. A., L. Drinkard, and M. Zasloff. 1985. tRNA nuclear transport: defining the critical regions of human $tRNA_i^{met}$ by point mutagenesis. *Cell* **43**:414–422.

Wunderlich, F. 1972. The macromolecular envelope of Tetrahymena pyriformis GL in different physiological states. V. Nuclear pore complexes—a controlling system in protein biosynthesis? *J. Membr. Biol.* **7**:220–230.

Wunderlich, F., R. Berezney, and H. Kleinig. 1976. The nuclear envelope: an interdisciplinary analysis of its structure, composition, and functions, p. 241–333. *In* D. Chapmena and D. F. H. Wallach (ed.), *Biological Membranes.* Academic Press, Inc., New York.

Wunderlich, F., and G. Herlan. 1977. A reversibly contractile nuclear matrix. *J. Cell Biol.* **73**:271–278.

Wunderlich, F., and V. Speth. 1972. The macronuclear envelope of Tetrahymena pyriformis GL in different physiological states. IV. Structural and functional aspect of nuclear pore complexes. *J. Microsc.* **13**:361–382.

Zasloff, M. 1983. tRNA transport from the nucleus in a eukaryotic cell: carrier-mediated translocation process. *Proc. Natl. Acad. Sci. USA* **80**:6436–6440.

VIII. ANTIBIOTIC MECHANISMS AND PROBES

VIII. ANTIBIOTIC MECHANISMS AND PROBES

The early studies on the interaction between antibiotics and ribosomes were designed to focus on how antibiotics functioned. More recently, antibiotics have been used as probes to help elucidate the mechanisms of action of the ribosome itself. By determining where the antibiotics are binding to the ribosome, and knowing the particular stage of translation which is interrupted, the functions of certain sections of rRNA can be described. In addition, some detailed interactions between ribosomes and specific ligands have become very well defined.

Chapter 41

Recognition Sites for Antibiotics within rRNA

ERIC CUNDLIFFE

"Antibiotics are selective agents provided by Nature for our enlightenment and we can hope, from a study of their actions, not only to improve our knowledge of the application of known principles, but also to find new principles." E. F. Gale (1966).

Much of what is known about the modes of action of ribosome inhibitors (and other antibiotics) has been reviewed at length elsewhere (Gale et al., 1981); only the more recent literature is specifically cited here. In order to utilize such knowledge in addressing ribosomal structure-function relationships, it becomes immediately important to find out what it is that specific antibiotics bind to within the particle. The methods available for doing so and the nature of the evidence that they provide were reviewed in a previous ribosome book (Cundliffe, 1986), and the problems of identifying the binding sites for small molecules (M_r usually less than 1,000) within the ribosome were emphasized. At that time, it was concluded that there is almost no evidence that antibiotics make their primary interactions with ribosomal proteins, although given the multiplicity and chemical diversity of such drugs, it would perhaps be surprising if none of them were to make meaningful contacts with proteins. Nevertheless, against a background of intense interest in RNA catalysis, it was concluded that a growing number of ribosome inhibitors appear to bind primarily to rRNA. This chapter starts from that premise. Before proceeding, however, the binding of erythromycin to ribosomal protein L15 ($K_a = 5 \times 10^4$ M^{-1}) should be noted (Teraoka and Nierhaus, 1978). This remains a unique observation. No other ribosomal protein has ever been shown to bind any native antibiotic molecule.

Perhaps the earliest indication that changes affecting rRNA could influence antibiotic activity and ribosomal function came from studies of the action of colicin E3, which cleaves 16S RNA between residues A-1493 and G-1494 (Bowman et al., 1971; Senior and Holland, 1971), and also from analysis of the ribosomes of kasugamycin-resistant (*ksgA*) strains of *Escherichia coli*. The *ksgA* product is an enzyme that dimethylates each of a geminal pair of adenosines (residues 1518 and 1519) close to the 3′ end of 16S rRNA, and ribosomes containing undermethylated RNA are resistant to kasugamycin (Helser et al., 1972). Also at about that same time, resistance to MLS antibiotics (macrolides, lincosamides, and streptogramin type B compounds) in *Staphylococcus aureus* was shown to involve overmethylation of 23S rRNA (Lai et al., 1973). Since then, methylation of 23S or 16S rRNA at specific sites leading to antibiotic resistance has been observed in various antibiotic-producing organisms (i.e., *Streptomyces* spp. and other actinomycetes) that need some way of defending themselves against their toxic products (reviewed in Cundliffe, 1989); in some cases, a direct effect of such methylation on the binding of antibiotics to ribosomes has been demonstrated (see below). Other observations consistent with the binding of antibiotics to rRNA have come from footprinting studies, in which various drugs (bound to ribosomes or their subunits) have been shown to protect specific rRNA residues from chemical or nuclease attack or even to enhance such attack. The results of that work, pioneered by Noller and colleagues, marry well with the methylation data (see below and Table 2). There are also impressive correlations with other data derived from genetic studies.

When the *rrnH* ribosomal operon of *E. coli* was isolated, subjected to mutagenesis, and then reintroduced into *E. coli* on a multicopy plasmid, transformants resistant to various antibiotics could be selected. The mutations were then mapped within the plasmid-located rRNA genes and characterized by

Eric Cundliffe ■ Department of Biochemistry and Leicester Biocentre, University of Leicester, Leicester LE1 7RH, United Kingdom.

Returning briefly to the matter of how thiostrepton affects ribosomal function, it is interesting to note that 23S RNA chemical footprints obtained with the antibiotic micrococcin differed in only one significant respect from those obtained with thiostrepton. Micrococcin promoted attack by dimethyl sulfate at A-1067, whereas thiostrepton was protective (Egebjerg et al., 1989). These results correlate in a curious way with effects of the two drugs on the EF-G-dependent ribosomal GTPase activity, which is stimulated in vitro by micrococcin (Cundliffe and Thompson, 1981) and inhibited by thiostrepton, which apparently prevents the binding of EF-G to the ribosome. Thus, the drug blocks the formation of ribosome · EF-G · GDP complexes that are otherwise stabilized by the presence of fusidic acid (Bodley et al., 1970) or by the inclusion of nonhydrolyzable analogs of GTP.

Somewhat similar effects are produced by α-sarcin, a cytotoxic peptide produced by *Aspergillus giganteus* that inactivates ribosomes by cleaving 23S-like rRNA at a specific site near the 3′ end. Cleavage occurs after residue G-2661 (Chan et al., 1983) within a highly conserved sequence that forms the so-called α-sarcin loop (residues 2653 to 2667). Immediately adjacent to G-2661 is the site of action of ricin, a toxic glycoprotein found in the seeds of castor beans. Ricin is a specific *N*-glycosidase that inactivates eucaryotic ribosomes by removing the adenine base from residue A-2660 of 23S-like RNA (Endo et al., 1987). α-Sarcin and ricin both affect eucaryotic ribosomal functions that involve either of the elongation factors EF-1 and EF-2 (reviewed in Gale et al., 1981), and functions of *E. coli* ribosomes dependent on EF-Tu or EF-G were also blocked specifically by α-sarcin (Hausner et al., 1987). The latter authors concluded that cleavage of 23S RNA by α-sarcin blocks the binding of EF-Tu and EF-G to mutually exclusive ribosomal sites, an effect that had earlier been attributed to thiostrepton. Indeed, there is more than a passing similarity between the performance of ribosomes containing RNA cleaved by α-sarcin and those inhibited by thiostrepton.

Evidently, there are functional interactions between the thiostrepton region and the α-sarcin loop within 23S rRNA. For example, when fusidic acid was used to stabilize ribosome · EF-G · GDP complexes, residues G-2655, A-2660, and G-2661 of the α-sarcin loop were protected from chemical attack in addition to residues A-1067 and A-1069 of the thiostrepton site (Moazed et al., 1988). However, when aminoacyl-tRNA · EF-Tu · GTP complexes were stabilized on the ribosome in the presence of kirromycin, footprints were observed only in the α-sarcin loop, at residues G-2655, A-2660, G-2661, and A-2665 (Moazed et al., 1988). These results are consistent with the idea that EF-Tu and EF-G bind alternately to the α-sarcin loop of 23S RNA during the peptide chain elongation cycle, although earlier, factor EF-G had been cross-linked to 23S RNA at or near A-1067 (Sköld, 1983). Moreover, proteins L11 and L7/L12, which are crucially involved in EF-G-dependent GTPase activity, had also been shown to bind to the 1067 region of 23S rRNA (Beauclerk et al., 1984) as does thiostrepton. One model (Moazed and Noller, 1989) has the α-sarcin loop interacting with EF-Tu when the ternary complex aminoacyl-tRNA · EF-Tu · GTP attempts to enter the A site, thereby delaying occupancy of the A′ site (and peptide bond formation) to allow time for translational proofreading, whereas thiostrepton has been proposed to block ribosomal EF-G-dependent GTPase activity by preventing the binding of EF-G to 23S RNA (Moazed et al., 1988).

Functional interactions between domains II and VI of 23S RNA, involving the thiostrepton and α-sarcin sites, may well be made and broken in a repetitive manner during normal ribosomal function, perhaps facilitated by conformational rearrangements occurring within the individual domains. It is interesting to speculate that thiostrepton might interrupt that sequence of events by stabilizing one of the intermediates and preventing subsequent transitions (see Cundliffe, 1986). Such a model might be entirely consistent with the suggestion (Hausner et al., 1988) that thiostrepton prevents allosteric transitions of the ribosome that are normally associated with translocation and A-site occupancy. It is also of more than semantic interest to reconsider whether the sobriquet "GTPase center" is better applied to the 1067 region of 23S RNA (as in numerous publications from several laboratories) or to the α-sarcin loop (as hinted by Moazed et al. [1988]) or to a tertiary interaction involving both domains (Egebjerg et al., 1989).

THE PEPTIDYLTRANSFERASE CENTER

Chloramphenicol competes with lincomycin and with macrolide antibiotics (such as erythromycin, spiramycin, and carbomycin) for binding to procaryotic ribosomes. Evidently, these drugs bind to closely related sites on the 50S ribosomal subunit, and the K_a values range from about 2×10^5 to 5×10^5 M^{-1} for lincomycin or chloramphenicol to 10^7 to 10^8 M^{-1} for the macrolides. There is also a much weaker site ($K_a = 5 \times 10^3$ M^{-1}) of unknown significance for chloramphenicol on the 30S subunit (reviewed in Cundliffe, 1980). As we shall see, evidence is accumulating that these drugs all interact

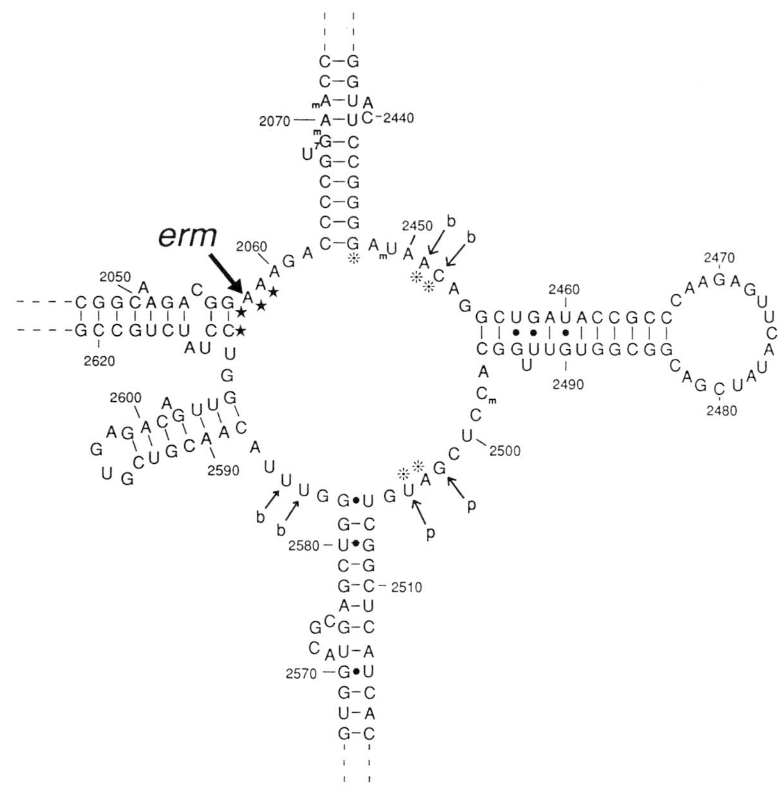

Figure 2. Peptidyltransferase region of domain V of 23S rRNA. The site of action (residue A-2058) of the erythromycin resistance methylase (erm) from *Streptomyces erythraeus* is indicated. Residues altered by mutation to give antibiotic resistance: ★, resistance to macrolides or lincomycin; *, resistance to chloramphenicol. Sites of attachment of affinity probes are also shown for a benzophenone derivative of aminoacyl-tRNA (b) and for *p*-azidopuromycin (p). See text for details.

with 23S RNA within the ribosome, a suggestion that was probably first made in the context of erythromycin (Mao and Putterman, 1969). However, the evidence so far is indirect; such binding has never been detected with free RNA (see, for example, Teraoka and Nierhaus, 1978).

Chloramphenicol, lincomycin, and the macrolides all inhibit peptidyltransferase assays in vitro (commonly "puromycin reactions"), although erythromycin does so only under special conditions, usually involving oligopeptidyl donor substrates. Other differences in the actions of these drugs were noted in studies involving the ribosomal binding of aminoacyl oligonucleotides, ostensibly into the P′ and A′ sites where petidyltransferase recognizes its substrates. The macrolides also displace peptidyl-tRNA from the ribosome in vivo (Menninger and Otto, 1982), an effect not seen with chloramphenicol. In summary, the macrolides, lincomycin, and chloramphenicol all affect the peptidyltransferase activity of ribosomes but not necessarily in identical fashion, and the details of how they do so are not completely resolved. However, chloramphenicol appears to block specifically the recognition of A-site substrates by the peptidyltransferase (reviewed in Gale et al., 1981).

For these various reasons, the ribosomal neighborhood into which chloramphenicol, lincomycin, and the macrolides all bind is generally referred to as the peptidyltransferase domain. Fundamentally, it involves the central portion of domain V of 23S rRNA (Fig. 2), the earliest indications of which came from studies of antibiotic resistance.

In *Streptomyces erythraeus*, the bacterium that produces erythromycin, residue A-2058 of 23S rRNA is dimethylated on N6, and the ribosomes are resistant to MLS antibiotics but, interestingly, not to chloramphenicol (Skinner et al., 1983). Dimethylated ribosomes are apparently insensitive to lincomycin, highly resistant to erythromycin, and less so to the more potent macrolides such as tylosin and spiramycin. Presumably, such graded responses reflect the extent to which dimethylation of A-2058 affects the ribosomal affinity for the various antibiotics and implies a critical role for that residue in the binding and action of MLS compounds. Other studies indicating the importance in this context of A-2058, and adjacent residues within 23S RNA, involved antibiotic-resistant strains of *E. coli* (containing mutagenized copies of the *rrnH* operon) and also the analysis of resistance mutations occurring in mitochondria or

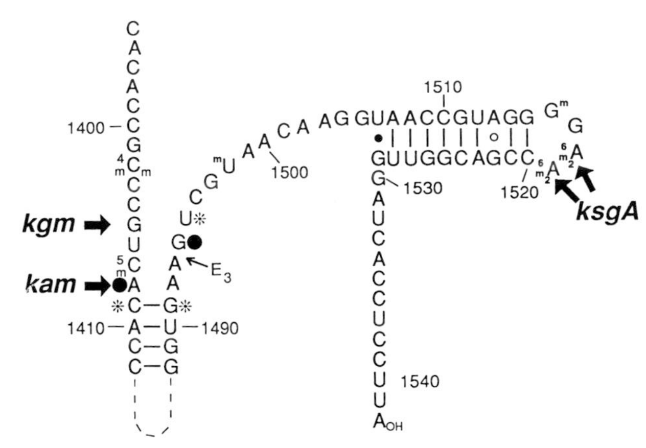

Figure 3. Aminoglycoside (or decoding) region of 16S rRNA. Sites at which methylation confers resistance to antibiotics: kgm, the kanamycin-gentamicin resistance methylase from *Micromonospora purpurea*; kam, the kanamycin-apramycin resistance methylase from *Streptomyces tenjimariensis* (see text for details). Also shown are the sites of action of the *ksgA* gene product (see text; failure to methylate these two sites gives resistance to kasugamycin). Symbols: ●, residues protected by aminoglycosides; *, residues changed by mutation to give resistance to aminoglycosides or hygromycin. E_3, Site of cleavage of 16S RNA by colicin E3.

vation that ribosome-bound aminoacyl-tRNA also does so is readily reconciled with powerful evidence that aminoglycosides affect the decoding process. Moreover, formation of a UV-induced cross-link between C-1400 of 16S rRNA and the anticodon wobble base of *N*-acetylvalyl-tRNA bound in the P site (Prince et al., 1982) implies that the highly conserved sequence of 16S RNA around residue G-1400 is within about 4 Å (0.4 nm) of the ribosomal site(s) where codon-anticodon recognition occurs. Binding of aminoglycosides to 16S rRNA at the sites summarized in Fig. 3, which also lie within conserved sequences, could readily disrupt that decoding process.

The action of streptomycin on bacterial ribosomes has been studied in extraordinary detail (reviewed in Gale et al., 1981). Among the many effects attributed to the drug, the misreading of mRNA codons both in vivo and in vitro is probably the best known; other effects can probably also be accounted for by disturbance of the aminoacyl acceptor and accuracy assessment activities of the ribosomal A site during the decoding process. A detailed analysis revealed that streptomycin specifically reduces the translational proofreading ability of ribosomes (Ruusala and Kurland, 1984), in which the initial choice of incoming aminoacyl-tRNA (as a ternary complex with EF-Tu and GTP) is reviewed, whereas other "misreading" drugs such as kanamycin affect the initial choice of ternary complex in addition to impairment of proofreading (Jelenc and Kurland, 1984).

Streptomycin binds to a single, primary site on the bacterial ribosome, although weaker secondary interactions may also occur. The binding affinity, estimated by using dihydrostreptomycin, was about 10^6 M^{-1} for 70S ribosomes and slightly less for 30S subunits (Schreiner and Nierhaus, 1973). The consequences of such binding are apparently determined by ribosomal protein S12, which is altered in streptomycin-resistant (*strA*) mutants. That change is definitely the cause of resistance in such strains (Ozaki et al., 1969), although the drug does not appear to bind to S12. Rather, streptomycin binds to S12-deficient ribosomal core particles, and other proteins such as S3 and S5 are required (Schreiner and Nierhaus, 1973). Mutations affecting ribosomal proteins S4 and S5 confer *ram* phenotypes (ribosomal ambiguity) that to some extent mimic the effects of streptomycin and can also be reversed by *strA* mutations. Moreover, ribosomes from *ram* strains bind streptomycin with enhanced affinity (Böck et al., 1979). For these various reasons, proteins S4, S5, and S12 have featured prominently over the years in discussions concerning the site of streptomycin binding.

More recently, two streptomycin sites have been identified within 16S-like rRNA, prompting recollection of Luigi Gorini's repeated assertion that two molecules of the drug bind to 16S rRNA of *E. coli* (Garvin et al., 1974). In chloroplasts, mutational changes affecting residue 912 of 16S RNA have been found associated with streptomycin resistance, and it has been subsequently confirmed by site-directed mutagenesis that one of those base changes gives streptomycin-resistant ribosomes in *E. coli* (Table 3). Significantly, when bound to 70S ribosomes or to 30S subunits, streptomycin protects residues 911 to 915 of 16S RNA from chemical attack (Moazed and Noller, 1987b). In other studies, ribosomal proteins S5 and S12 were bound to 16S rRNA, and chemical footprints were obtained around the pseudoknot in the 16S RNA secondary structure that is immediately adjacent to the 911–915 region (reviewed in Noller et al., 1987). The other streptomycin site is at position 523 not far from where protein S4 also gave footprints. This residue had undergone substitution in streptomycin-resistant *Chlamydomonas* chloroplasts, and again, the consequences of introducing a similar change into *E. coli* 16S RNA were confirmed (Table 3). Residue 523 lies within one of the most highly conserved rRNA sequences that is itself alongside sites that are protected by aminoacyl-tRNA, ostensibly bound in the A site (Moazed and Noller, 1986). Although streptomycin itself did not affect the chemical reactivity of the 523 loop of 16S RNA, other misreading antibiotics such as neomycin did so, the reactivity of C-525 with dimethyl sulfate being weakly stimulated (Moazed and Noller, 1987b). By themselves, these data do not prove that streptomycin binds primarily (or even at all) to 16S rRNA, but that is certainly a plausible speculation.

THE 30S P SITE

During peptide chain elongation, the ribosomal P site can be regarded as a relatively passive site into which peptide tRNA is parked pending peptidyl transfer. There is no reason why specific antibiotics should not prevent the recognition of donor substrates by the peptidyltransferase; others might act in the P site and prevent displacement of deacylated tRNA from the P to the E site or otherwise inhibit translocation; the problem is to devise assay systems to recognize possible inhibitory effects in such detail. However, the polypeptide chain initiation process seems to involve the insertion of fMet-tRNA directly into the P site, and some inhibitors of initiation might therefore be P site specific. Unfortunately, there are remarkably few antibiotics that are known to affect this process at all, let alone specifically. One such is edeine, a basic peptide that clearly does other things inside the cell besides inhibiting protein synthesis (reviewed in Gale et al., 1981). Edeine binds extremely tightly to ribosomes ($K_a \sim 10^{10}$ M^{-1}) and apparently blocks the P site on the smaller subunit during polypeptide chain initiation. Another candidate could be pactamycin, which appears to be a specific inhibitor of polypeptide chain initiation in eucaryotic systems when used at carefully defined concentrations (Macdonald and Goldberg, 1970). However, at elevated inputs, and in procaryotic systems under any conditions, it is not clear that the drug specifically affects initiation. Pactamycin binds to a single site on the 30S subunit (whether or not mRNA is present) and to free 70S ribosomes, but does not bind to 50S subunits or to 70S particles complexed with mRNA (Stewart and Goldberg, 1973). These two drugs give interesting, overlapping chemical footprints. Edeine strongly protects residues G-693, A-794, C-795, and G-926 of 16S RNA within 70S ribosomes, and additional sites are weakly protected within 30S particles (Moazed and Noller, 1987b). Pactamycin protects G-693 and C-795 (J. Woodcock and D. Moazed, personal communication; J. Egebjerg, personal communication). All four of the edeine sites (two of which are also protected by pactamycin) were independently designated as P site related from the manner in which they were also protected by tRNA (Moazed and Noller, 1986). Given what is known of the modes of action of the two drugs, that fits.

MISCELLANEOUS SITES

Resistance to spectinomycin arises spontaneously in *E. coli* as a result of mutations affecting *rpsE*, the gene encoding ribosomal protein S5. Other mutations affecting protein S5, including *ram*, give different phenotypes (reviewed in Hummel and Böck, 1989). The first RNA mutations giving rise to spectinomycin resistance were obtained by random mutagenesis of the *rrnH* operon of *E. coli* and involved base substitution at position 1192 within 16S rRNA (Table 3). Subsequently, by site-directed mutagenesis, the level of resistance to spectinomycin due to changes at position 1192 was shown to decrease in the order G U A. In the secondary structure of 16S RNA, residues 1191 to 1193 are paired with residues 1065 to 1063, and mutational changes affecting either side of this helical stem have been found associated with spectinomycin resistance in chloroplasts. For example, in *Chlamydomonas* spp., substitutions at position 1191 gave high-level resistance, whereas changes at residue 1193 conferred lower resistance levels. In tobacco, high-level resistance was associated with a base change at position 1064, and again, a change at position 1193 was less effective (Table 3). To complete the picture, spectinomycin protected residues C-1063 and G-1064 in chemical footprinting studies (Moazed and Noller, 1987b). The mode of action of spectinomycin is not completely understood, although effects on translocation have been reported, and there remains room for discussion concerning its effects. Therefore, the function of the spectinomycin helix within 16S rRNA cannot, at present, be defined with any real confidence.

Kasugamycin evidently interacts with the 3′-terminal portion of 16S RNA within the 30S ribosomal subunit. In keeping with other evidence that the 3′ end of 16S RNA is involved in decoding mRNA sequences, effects of kasugamycin on the fidelity of translation have been reported; interestingly, the net effect is the opposite of that seen with aminoglycosides. Thus, kasugamycin enhances the accuracy of protein synthesis (van Buul et al., 1984), although how this effect is related to the destabilization of 30S initiation complexes (see Gale et al., 1981) is not clear.

Tetracycline-resistant ribosomes apparently do not exist, which makes the site identified by chemical footprinting of 16S RNA in the presence of the drug all the more interesting (Moazed and Noller, 1987b). There is also a viomycin site within 23S RNA (Moazed and Noller, 1987a) that is presumably somehow involved in translocation. Even with some of the antibiotics discussed in detail above, weak effects on chemical attack (protection or enhancement) were observed within rRNA at additional sites not mentioned in this article. Those sites may well be significant in the action of the drugs and may be important in aiding the construction of tertiary folded models for 16S and 23S rRNAs.

CONCLUDING REMARKS

There is direct, unequivocal evidence that thiostrepton binds to a specific site in 23S rRNA and that such binding is the cause of its inhibitory effects. That interaction probably was the first example of noncovalent, single-site binding of any small molecule to any nucleic acid. A wealth of evidence, less direct but quite convincing, suggests that other antibiotics also interact with RNA within the ribosome and, among them, affect most of the things that ribosomes do. In other words, RNA has the capacity to interact with small molecules of diverse chemical types and is present in most (possibly all) of the active sites of the ribosomal enzyme.

To conclude where this article began: Having commented on the potential usefulness of antibiotics as probes in biochemical systems, Gale (1966) pointed out that studies of the interactions of drugs with DNA "show us a new principle whereby we can inactivate a functional macromolecule by distortion of its structure. There may be further examples of the same general principle, to emerge from studies of the interactions of ribosomes *or other forms of active RNA* [italics added] with streptomycin, chloramphenicol, tetracycline etc."

Right on!

Work from my laboratory cited in this article was supported by the Medical Research Council, the Science and Engineering Research Council, and The Wellcome Trust.

I am indebted to David Holmes for producing the figures, to Sylvia Dexter for typing the manuscript, and to past and present members of my group for the pleasure of their company.

REFERENCES

Anderson, B., D. Crowfoot-Hodgkin, and M. A. Viswamitra. 1970. The structure of thiostrepton. *Nature* (London) **225:** 233–235.

Beauclerk, A. A. D., and E. Cundliffe. 1987. Sites of action of two ribosomal RNA methylases responsible for resistance to aminoglycosides. *J. Mol. Biol.* **193:**661–671.

Beauclerk, A. A. D., and E. Cundliffe. 1988. The binding site for ribosomal protein L2 within 23S ribosomal RNA of *Escherichia coli*. *EMBO J.* 7:3589–3594.

Beauclerk, A. A. D., E. Cundliffe, and J. Dijk. 1984. The binding site for ribosomal protein complex L8 within 23S ribosomal RNA of *Escherichia coli*. *J. Biol. Chem.* **259:**6559–6563.

Blanc, H., C. T. Wright, M. J. Bibb, D. C. Wallace, and D. A. Clayton. 1981. Mitochondrial DNA of chloramphenicol-resistant mouse cells contains a single nucleotide change in the region encoding the 3′ end of the large ribosomal RNA. *Proc. Natl. Acad. Sci. USA* **78:**3789–3793.

Böck, A., A. Petzet, and W. Piepersberg. 1979. Ribosomal ambiguity (*ram*) mutations facilitate dihydrostreptomycin binding to ribosomes. *FEBS Lett.* **104:**317–321.

Bodley, J. W., F. J. Zieve, L. Lin, and S. T. Zieve. 1970. Studies on translocation. III. Conditions necessary for the formation and detection of a stable ribosome-G factor-guanosine diphosphate complex in the presence of fusidic acid. *J. Biol. Chem.* **245:** 5656–5661.

Bowman, C. M., J. E. Dahlberg, T. Ikemura, J. Konisky, and M. Nomura. 1971. Specific inactivation of 16S ribosomal RNA induced by colicin E3 *in vivo*. *Proc. Natl. Acad. Sci. USA* **68:**964–968.

Chan, Y.-L., Y. Endo, and I. G. Wool. 1983. The sequence of the nucleotides at the α-sarcin cleavage site in rat 28S ribosomal ribonucleic acid. *J. Biol. Chem.* **258:**12768–12770.

Cseplö, A., T. Etzold, J. Schell, and P. H. Schreier. 1988. Point mutations in the 23S rRNA genes of four lincomycin resistant *Nicotiana plumbaginifola* mutants could provide new selectable markers for chloroplast transformation. *Mol. Gen. Genet.* **214:** 295–299.

Cundliffe, E. 1980. Antibiotics and prokaryotic ribosomes: action, interaction and resistance, p. 555–581. *In* G. Chambliss, G. R. Craven, J. Davies, K. Davis, L. Kahan, and M. Nomura (ed.), *Ribosomes. Structure, Function, and Genetics*. University Park Press, Baltimore.

Cundliffe, E. 1986. Involvement of specific portions of ribosomal RNA in defined ribosomal functions: a study utilizing antibiotics, p. 586–604. *In* B. Hardesty and G. Kramer (ed.), *Structure, Function, and Genetics of Ribosomes*. Springer-Verlag, New York.

Cundliffe, E. 1989. How antibiotic-producing organisms avoid suicide. *Annu. Rev. Microbiol.* **43:**207–233.

Cundliffe, E., and J. Thompson. 1981. Concerning the mode of action of micrococcin upon bacterial protein synthesis. *Eur. J. Biochem.* **118:**47–52.

De Stasio, E. A., D. Moazed, H. F. Noller, and A. E. Dahlberg. 1989. Mutations in 16S ribosomal RNA disrupt antibiotic-RNA interactions. *EMBO J.* **8:**1213–1216.

Douthwaite, S., J. B. Prince, and H. F. Noller. 1985. Evidence for functional interaction between domains II and V of 23S ribosomal RNA from an erythromycin-resistant mutant. *Proc. Natl. Acad. Sci. USA* **82:**8330–8334.

Dujon, B. 1980. Sequence of the intron and flanking exons of the mitochondrial 21S rRNA gene of yeast strains having different alleles at the omega and *rib-1* loci. *Cell* **20:**185–197.

Egebjerg, J., S. Douthwaite, and R. A. Garrett. 1989. Antibiotic interactions at the GTPase-associated centre within *Escherichia coli* 23S rRNA. *EMBO J.* **8:**607–611.

Endo, Y., K. Mitsui, M. Motizuki, and K. Tsurugi. 1987. The mechanism of action of ricin and related toxic lectins on eukaryotic ribosomes. *J. Biol. Chem.* **262:**5908–5912.

Ettayebi, M., S. M. Prasad, and E. A. Morgan. 1985. Chloramphenicol-erythromycin resistance mutations in a 23S rRNA gene of *Escherichia coli*. *J. Bacteriol.* **162:**551–557.

Etzold, T., C. C. Fritz, J. Schell, and P. H. Schreier. 1987. A point mutation in the chloroplast 16S rRNA gene of a streptomycin resistant *Nicotiana tabacum*. *FEBS Lett.* **219:**343–346.

Fromm, H., M. Edelman, D. Aviv, and E. Galun. 1987. The molecular basis for rRNA-dependent spectinomycin resistance in *Nicotiana* chloroplasts. *EMBO J.* **6:**3233–3237.

Gale, E. F. 1966. The object of the exercise, p. 1–21. *In* B. A. Newton and P. E. Reynolds (ed.), *Biochemical Studies of Antimicrobial Drugs*. Cambridge University Press, Cambridge.

Gale, E. F., E. Cundliffe, P. E. Reynolds, M. H. Richmond, and

M. J. Waring. 1981. *The Molecular Basis of Antibiotic Action.* John Wiley & Sons, London.

Garvin, R. T., D. K. Biswas, and L. Gorini. 1974. The effects of spectinomycin or dihydrostreptomycin binding to 16S RNA or to 30S ribosomal subunits. *Proc. Natl. Acad. Sci. USA* **71:** 3814–3818.

Gauthier, A., M. Turmel, and C. Lemieux. 1988. Mapping of chloroplast mutations conferring resistance to antibiotics in *Chlamydomonas*: evidence for a novel site of streptomycin resistance in the small subunit RNA. *Mol. Gen. Genet.* **214:** 192–197.

Hall, C. C., D. Johnson, and B. S. Cooperman. 1988. [^{3}H]-p-azidopuromycin photoaffinity labeling of *Escherichia coli* ribosomes: evidence for site-specific interaction at U-2504 and G-2502 in domain V of 23S ribosomal RNA. *Biochemistry* **27:**3983–3990.

Hausner, T.-P., J. Atmadja, and K. H. Nierhaus. 1987. Evidence that the G-2661 region of 23S rRNA is located at the ribosomal binding sites of both elongation factors. *Biochimie* **69:**911–923.

Hausner, T.-P., U. Geigenmüller, and K. H. Nierhaus. 1988. The allosteric three-site model for the ribosomal elongation cycle. New insights into the inhibition mechanisms of aminoglycosides, thiostrepton and viomycin. *J. Biol. Chem.* **263:**13103–13111.

Helser, T. L., J. E. Davies, and J. E. Dahlberg. 1972. Mechanism of kasugamycin resistance in *Escherichia coli. Nature* (London) *New Biol.* **235:**6–9.

Hensens, O. D., G. Albers-Schönberg, and B. F. Anderson. 1983. The solution conformation of the peptide antibiotic thiostrepton: a ^{1}H NMR study. *J. Antibiot.* **36:**799–813.

Hummel, H., and A. Böck. 1987a. Thiostrepton resistance mutations in the gene for 23S ribosomal RNA of halobacteria. *Biochimie* **69:**857–861.

Hummel, H., and A. Böck. 1987b. 23S ribosomal RNA mutations in halobacteria conferring resistance to the anti-80S ribosome targeted antibiotic anisomycin. *Nucleic Acids Res.* **15:**2431–2443.

Hummel, H., and A. Böck. 1989. Ribosomal changes resulting in antimicrobial resistance, p. 193–225. *In* L. E. Bryan (ed.), *Microbial Resistance to Drugs.* Springer-Verlag KG, Berlin.

Jelenc, P. C., and C. G. Kurland. 1984. Multiple effects of kanamycin on translational accuracy. *Mol. Gen. Genet.* **194:** 195–199.

Kearsey, S. E., and I. W. Craig. 1981. Altered ribosomal RNA genes in mitochondria from mammalian cells with chloramphenicol resistance. *Nature* (London) **290:**607–608.

Lai, C.-J., B. Weisblum, S. R. Fahnestock, and M. Nomura. 1973. Alteration of 23S ribosomal RNA and erythromycin-induced resistance to lincomycin and spiramycin in *Staphylococcus aureus. J. Mol. Biol.* **74:**67–72.

Li, M., A. Tzagoloff, K. Underbrink-Lyon, and N. C. Martin. 1982. Identification of the paromomycin-resistance mutation in the 15S rRNA gene of yeast mitochondria. *J. Biol. Chem.* **257:**5921–5928.

Macdonald, J. S., and I. H. Goldberg. 1970. An effect of pactamycin on the initiation of protein synthesis in reticulocytes. *Biochem. Biophys. Res. Commun.* **41:**1–8.

Makosky, P. C., and A. E. Dahlberg. 1987. Spectinomycin resistance at site 1192 in 16S ribosomal RNA of *E. coli*: an analysis of three mutants. *Biochimie* **69:**885–889.

Mao, J. C.-H., and M. Putterman. 1969. The intermolecular complex of erythromycin and ribosome. *J. Mol. Biol.* **44:** 347–361.

Melançon, P., C. Lemieux, and L. Brakier-Gingras. 1988. A mutation in the 530 loop of *Escherichia coli* 16S ribosomal RNA causes resistance to streptomycin. *Nucleic Acids Res.* **16:**9631–9639.

Menninger, J. R., and D. P. Otto. 1982. Erythromycin, carbomycin, and spiramycin inhibit protein synthesis by stimulating dissociation of peptidyl-tRNA from ribosomes. *Antimicrob. Agents Chemother.* **21:**811–818.

Moazed, D., and H. F. Noller. 1986. Transfer RNA shields specific nucleotides in 16S ribosomal RNA from attack by chemical probes. *Cell* **47:**985–994.

Moazed, D., and H. F. Noller. 1987a. Chloramphenicol, erythromycin, carbomycin and vernamycin B protect overlapping sites in the peptidyl transferase region of 23S ribosomal RNA. *Biochimie* **69:**879–884.

Moazed, D., and H. F. Noller. 1987b. Interaction of antibiotics with functional sites in 16S ribosomal RNA. *Nature* (London) **327:**389–394.

Moazed, D., and H. F. Noller. 1989. Interaction of tRNA with 23S rRNA in the ribosomal A, P and E sites. *Cell* **57:**585–597.

Moazed, D., J. M. Robertson, and H. F. Noller. 1988. Interaction of elongation factors EF-G and EF-Tu with a conserved loop in 23S RNA. *Nature* (London) **334:**362–364.

Montandon, P.-E., P. Nicolas, P. Schürmann, and E. Stutz. 1985. Streptomycin resistance of *Euglena gracilis* chloroplasts: identification of a point mutation in the 16S rRNA gene in an invariant position. *Nucleic Acids Res.* **13:**4299–4310.

Montandon, P. E., R. Wagner, and E. Stutz. 1986. *E. coli* ribosomes with a C912 to U base change in the 16S rRNA are streptomycin resistant. *EMBO J.* **5:**3705–3708.

Noller, H. F. 1984. Structure of ribosomal RNA. *Annu. Rev. Biochem.* **53:**119–162.

Noller, H. F., S. Stern, D. Moazed, T. Powers, P. Svensson, and L.-M. Changchien. 1987. Studies on the architecture and function of 16S rRNA. *Cold Spring Harbor Symp. Quant. Biol.* **52:**695–708.

Ozaki, M., S. Mizushima, and M. Nomura. 1969. Identification and functional characterization of the protein controlled by the streptomycin-resistant locus in *E. coli. Nature* (London) **222:** 333–339.

Pestka, S., D. Weiss, and R. Vince. 1976. Partition of ribosomes in two-polymer aqueous phase systems. *Anal. Biochem.* **71:**137–142.

Prince, J. B., B. H. Taylor, D. L. Thurlow, J. Ofengand, and R. A. Zimmermann. 1982. Covalent crosslinking of $tRNA_1^{val}$ to 16S RNA at the ribosomal P site: identification of crosslinked residues. *Proc. Natl. Acad. Sci. USA* **79:**5450–5454.

Ruusala, T., and C. G. Kurland. 1984. Streptomycin preferentially perturbs ribosomal proofreading. *Mol. Gen. Genet.* **198:**100–104.

Schreiner, G., and K. H. Nierhaus. 1973. Protein involved in the binding of dihydrostreptomycin to ribosomes of *Escherichia coli. J. Mol. Biol.* **81:**71–82.

Senior, B. W., and I. B. Holland. 1971. Effect of colicin E3 on the 30S ribosomal subunit of *Escherichia coli. Proc. Natl. Acad. Sci. USA* **68:**959–963.

Sigmund, C. D., M. Ettayebi, and E. A. Morgan. 1984. Antibiotic resistance mutations in 16S and 23S ribosomal RNA genes of *Escherichia coli. Nucleic Acids Res.* **12:**4653–4663.

Skinner, R., E. Cundliffe, and F. J. Schmidt. 1983. Site of action of a ribosomal RNA methylase responsible for resistance to erythromycin and other antibiotics. *J. Biol. Chem.* **258:**12702–12706.

Sköld, S. E. 1983. Chemical crosslinking of elongation factor G to the 23S RNA in 70S ribosomes from *Escherichia coli. Nucleic Acids Res.* **11:**4923–4932.

Slott, E. F., Jr., R. O. Shade, and R. A. Lansman. 1983. Sequence analysis of mitochondrial DNA in a mouse cell line resistant to chloramphenicol and oligomycin. *Mol. Cell. Biol.* **3:**1694–1702.

Sor, F., and H. Fukuhara. 1982. Identification of two erythromycin resistance mutations in the mitochondrial gene coding for the large ribosomal RNA in yeast. *Nucleic Acids Res.* **10:**6571–6577.

Sor, F., and H. Fukuhara. 1984. Erythromycin and spiramycin resistance mutations of yeast mitochondria: nature of the *rib2* locus in the large ribosomal RNA gene. *Nucleic Acids Res.* **12:**8313–8318.

Spangler, E. A., and E. H. Blackburn. 1985. The nucleotide sequence of the 17S ribosomal RNA gene of *Tetrahymena thermophila* and the identification of point mutations resulting in resistance to the antibiotics paromomycin and hygromycin. *J. Biol. Chem.* **260:**6334–6340.

Stark, M. J. R. 1979. Properties of the ribosomes of bacterial mutants resistant to thiostrepton. Ph.D. thesis. University of Leicester, Leicester, England.

Steiner, G., E. Kuechler, and A. Barta. 1988. Photo-affinity labeling at the peptidyl transferase centre reveals two different positions for the A- and P- sites in domain V of 23S rRNA. *EMBO J.* **7:**3949–3955.

Stewart, M. L., and I. H. Goldberg. 1973. Pactamycin binding to *Escherichia coli* ribosomes: interference by formation of the protein synthesizing complex with f2 viral RNA. *Biochim. Biophys. Acta* **294:**123–137.

Stiege, W., C. Glotz, and R. Brimacombe. 1983. Localisation of a series of intra-RNA cross-links in the secondary and tertiary structure of 23S RNA, induced by ultraviolet irradiation of *Escherichia coli* 50S ribosomal subunits. *Nucleic Acids Res.* **11:**1687–1706.

Teraoka, H., and K. H. Nierhaus. 1978. Proteins from *Escherichia coli* ribosomes involved in the binding of erythromycin. *J. Mol. Biol.* **126:**185–193.

Thompson, J., E. Cundliffe, and A. E. Dahlberg. 1988. Site-directed mutagenesis of *Escherichia coli* 23S ribosomal RNA at position 1067 within the GTP hydrolysis centre. *J. Mol. Biol.* **203:**457–465.

Thompson, J., E. Cundliffe, and M. Stark. 1979. Binding of thiostrepton to a complex of 23S rRNA with ribosomal protein L11. *Eur. J. Biochem.* **98:**261–265.

Thompson, J., F. Schmidt, and E. Cundliffe. 1982. Site of action of a ribosomal RNA methylase conferring resistance to thiostrepton. *J. Biol. Chem.* **257:**7915–7917.

Tori, K., K. Tokura, Y. Yoshimura, Y. Terui, K. Okabe, H. Otsuka, K. Matsushita, F. Inagaki, and T. Miyazawa. 1981. Structures of siomycin-B and -C and thiostrepton-B determined by NMR spectroscopy and carbon-13 signal assignments of siomycins, thiostreptons, and thiopeptin-B_a. *J. Antibiot.* **34:**124–129.

van Buul, C. P. J. J., W. Vissert, and P. H. van Knippenberg. 1984. Increased translational fidelity caused by the antibiotic kasugamycin and ribosomal ambiguity in mutants harbouring the *ksgA* gene. *FEBS Lett.* **177:**119–124.

Vester, B., and R. A. Garrett. 1987. A plasmid-coded and site-directed mutation in *Escherichia coli* 23S RNA that confers resistance to erythromycin: implications for the mechanism of action of erythromycin. *Biochimie* **69:**891–900.

Vester, B., and R. A. Garrett. 1988. The importance of highly conserved nucleotides in the binding region of chloramphenicol at the peptidyl transferase centre of *Escherichia coli* 23S ribosomal RNA. *EMBO J.* **7:**3577–3587.

Chapter 42

Antibiotic Probes of *Escherichia coli* Ribosomal Peptidyltransferase

BARRY S. COOPERMAN, CARL J. WEITZMANN, and CARMEN L. FERNÁNDEZ

Antibiotics and their chemically reactive derivatives have great potential as affinity label probes of ribosomal structure and function, since identifying the site of interaction with the ribosome of an antibiotic of known inhibitory function provides strong evidence for the identification of the functional site. Elsewhere we have published comprehensive reviews of affinity labeling of ribosomes (Cooperman, 1978, 1980, 1987, 1988; Cooperman et al., 1986; Cooperman et al., 1989). Here we focus on affinity labeling by antibiotics directed toward the peptidyltransferase center of *Escherichia coli* ribosomes, placing particular stress on some recent photoaffinity labeling results obtained in our laboratory for four different antibiotics, puromycin, *p*-azidopuromycin, chloramphenicol, and tetracycline (Fig. 1), and on related work of others with other peptidyltransferase antibiotics and with tRNAs derivatized with reactive groups at their 3′ ends.

AFFINITY LABELING STUDIES WITH PEPTIDYLTRANSFERASE ANTIBIOTICS

Photoaffinity Labeling by Puromycin and *p*-Azidopuromycin

Puromycin is an inhibitor of peptidyltransferase that is both a structural and a functional analog of the 3′ end of Tyr-tRNA^{Tyr}. It has two inhibitory modes of action, acting as a substrate in accepting, at its α-amino position, a peptidyl group from P-site-bound peptidyl-tRNA to form an aborted peptide chain (Allen and Zamecnik, 1962; Nathans, 1964) and competing with aminoacyl-tRNA for binding to the A′ site (defined as the site of binding of the 3′ end of aminoacyl-tRNA within the peptidyltransferase center [Steiner et al., 1988]). We have used both native puromycin and *p*-azidopuromycin (Fig. 1), a functionally competent derivative in which a *p*-azidophenyl group replaces the *p*-methoxy group in puromycin, as photoaffinity labels. Although puromycin photoincorporates into both protein and rRNA, only protein labeling has been examined in detail. Using relatively high light doses because of the limited photoreactivity of puromycin, two proteins have been shown to be labeled from relatively high affinity sites, L23 and S14, with L23 labeling proceeding in considerably higher yield (Jaynes et al., 1978; Weitzmann and Cooperman, 1985). These proteins are labeled from two different sites on the ribosome, with only L23 being labeled from a site on the 50S subunit.

In related work using *p*-azidopuromycin as a photoaffinity label for ribosomes, we found that the protein labeled to the greatest extent is again L23 (Nicholson et al., 1982a, 1982b). Such labeling proceeds via an azide-dependent photochemistry not available to puromycin and is additional evidence for the site specificity of L23 labeling. The other major labeled proteins are L18 (and possibly L22) (Weitzmann, 1989) and L15.

There is a good correlation between the ability of nonradioactive ligands to bind to the A′ site and to inhibit photoincorporation of [^{3}H]puromycin into L23. Thus the 3′-O-phenylalanyl ester of the dinucleoside phosphate CpA, which is a very good ligand for the A′ site, acting as an acceptor toward peptidyl-tRNA, is also a strong inhibitor of puromycin photoincorporation into L23 (Jaynes et al., 1978). Similarly, puromycin aminonucleoside (which lacks the O-methyltyrosine moiety of puromycin; Fig. 1) inhibits photoincorporation of [^{3}H]puromycin into L23 and is also a competitive inhibitor for puromycin as a substrate in peptidyltransferase (Nicholson et al., 1982b; Weitzmann and Cooperman, 1985), whereas

Barry S. Cooperman and Carmen L. Fernández ■ Department of Chemistry, University of Pennsylvania, Philadelphia, Pennsylvania 19104. Carl J. Weitzmann ■ Roche Institute of Molecular Biology, Roche Research Center, Nutley, New Jersey 07110.

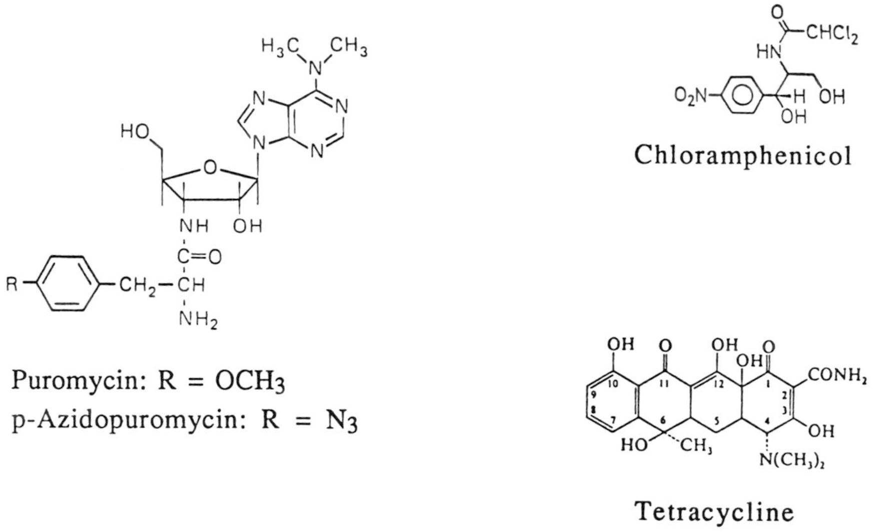

Puromycin: R = OCH_3
p-Azidopuromycin: R = N_3

Chloramphenicol

Tetracycline

Figure 1. Antibiotic structures.

N^6,N^6-dimethyladenosine (in which the 3′-amino group of puromycin aminonucleoside is replaced by a 3′-hydroxyl group) inhibits neither photoincorporation of [^{3}H]puromycin nor peptidyltransferase. Finally, [^{3}H]puromycin aminonucleoside shows strong photoincorporation into L23, whereas N^6,N^6-[^{3}H]-dimethyladenosine shows no such photoincorporation (although it does photoincorporate into S14).

Recently, we have used a reconstitution approach to investigate the functional consequences of photoaffinity labeling of L23 (Weitzmann and Cooperman, 1990). Ribosomes were photolyzed with puromycin, and reverse-phase high-performance liquid chromatography (HPLC) was used to prepare puromycin-L23 in a form not contaminated with unmodified L23. This work exploited a previous result that tetracycline specifically stimulates puromycin photoincorporation into L23 in an intact ribosome, a point we return to below. Puromycin-L23 was then used in place of unmodified L23 in reconstituting 50S subunits, and the activities of the reconstituted 50S subunits were tested in two assays, the stimulation of poly(U)-dependent binding of Phe-tRNA^{Phe} to 30S subunits and mRNA-independent peptidyltransferase (Table 1). Although we were unable to achieve a stoichiometry of more than 0.5 puromycin-L23 per 50S particle, largely because of contamination of protein pools with unmodified L23, the results obtained support the conclusions that puromycin-L23 has little effect on peptidyltransferase but does prevent 50S stimulation of Phe-tRNA^{Phe} binding. Omitting L23 altogether has little effect on peptidyltransferase, in agreement with previous results of Hampl et al. (1981), but does lead to a small loss of tRNA-binding stimulation (Table 1). Earlier (Jaynes et al., 1978; Nicholson et al., 1982b), we had shown that neither puromycin nor *p*-azidopuromycin, when photoincorporated into ribosomes (including, of course, into L23), functions as an acceptor substrate in peptidyltransferase. Thus, our overall conclusion is that although L23 is not directly involved in peptidyltransferase, it is close enough to the peptidyltransferase center that puromycin-L23 can interfere with tRNA binding.

p-Azidopuromycin, like puromycin, photoincorporates into both rRNA and protein. We have used the two-step procedure recently introduced by ourselves and others (Barta et al., 1984; Hall et al., 1985; Hall et al., 1988; Gravel et al., 1987; Steiner et al., 1988) to identify major sites of *p*-azidopuromycin photoincorporation into 23S rRNA. In the first step, ribosomal DNA restriction fragments were used as hybridization probes to localize the major site of labeling as falling between 23S rRNA bases 2445 and

Table 1. Relative activities of reconstituted 50S subunits[a]

Protein pool	Stimulation of tRNA binding	Peptidyltransferase activity
TP50-L23 + L23	1.27 ± 0.04 (3)	1.10 ± 0.03 (2)
TP50-L23 + hv-L23	1.24 ± 0.09 (3)	1.07 ± 0.13 (3)
TP50-L23 + L23*[b]	0.64 ± 0.07 (3)	0.97 ± 0.23 (4)
Native 50S subunits	2.9 ± 1.1 (6)[c]	14.3 ± 5.0 (5)[d]

[a] Data are from Weitzmann and Cooperman (1990). Activities are relative to that of TP50-L23, taken as 1. Ranges given are average deviations. Numbers in parentheses indicate the number of independent determinations.
[b] Average L23* uptake, 0.46 per 50S particle.
[c] Typically, 10 pmol of native 50S subunits added to 13 pmol of 30S subunits increased tRNA binding from 0.3 to 1.9 pmol out of a total of 7 pmol of tRNA in the reaction mixture.
[d] In the typical assay using a 1-h incubation, 20 pmol of native 50S subunits catalyzed transfer of 1.2 to 1.4 pmol of *N*-acetylphenylalanine to puromycin out of a total of 2.5 pmol of *N*-acetyl-Phe-tRNA added to the reaction mixture.

2668. Since added puromycin inhibits photoincorporation of *p*-azidopuromycin into this fragment to the same extent as it inhibits photoincorporation into each of proteins L23, L18/22, and L15, labeling of each of these proteins appears to proceed from a single site (Hall et al., 1988). In the second step, oligonucleotide primers complementary to 23S RNA residues 2741 to 2725 or 2613 to 2597 were hybridized to photoaffinity-labeled rRNA, and the resulting heteroduplexes were used as substrates for reverse transcriptase. Bases G-2502 and U-2504 were identified as sites of photoincorporation (or, more properly, as sites of interaction [Hall et al., 1988]) by using a classical sequencing gel to detect a halt or pause in DNA polymerization (Youvan and Hearst, 1979, 1981).

Affinity Labeling by Chloramphenicol

Chloramphenicol inhibits peptidyltransferase. Inhibition is thought to result from its ability to interfere with the correct binding of the 3′ end of aminoacyl-tRNA to the A′ site, but the exact nature of such inhibition is unclear. A recent detailed study of the inhibitory effect of chloramphenicol on the peptidyltransferase reaction, using puromycin as an acceptor, led to the conclusion that chloramphenicol inhibits principally by forming a complex with a much reduced V_{max} and is not, strictly speaking, a competitive inhibitor of puromycin (Drainas et al., 1987). Binding studies have shown that chloramphenicol has a single high-affinity site (K_d, ~2 μM) located on the 50S subunit, presumably at the peptidyltransferase center (K_i values for peptidyltransferase are also 1 to 2 μM), as well as one or more low-affinity sites, at least one of which is on the 30S subunit (Lessard and Pestka, 1972; Grant et al., 1979a).

A number of affinity labeling experiments have been reported that (i) used either electrophilic analogs in which iodoacetyl or bromoacetyl groups replace the dichloroacetyl group of the native antibiotic or the photolabile analog in which a 4-azido group replaces the 4-nitro group (Fig. 1) or (ii) exploited the intrinsic photolability of native chloramphenicol. Although a number of 50S proteins were reported to have been labeled in these studies, including L1, L2, L11, L16, L19, and L27 (Pongs et al., 1973; Sonenberg et al., 1973; Sonenberg et al., 1974; Nielsen et al., 1978; Wilchek and Bayer, 1978; LeGoffic et al., 1980; Bouthier de la Tour et al., 1985), no results were obtained that unambiguously identified 50S proteins at the high-affinity site. Accordingly, we have decided to reinvestigate the photoaffinity labeling of the chloramphenicol site, taking full advantage of the superior analytical capabilities offered by the use of HPLC to separate ribosomal proteins (Kerlavage et al., 1983; Capel et al., 1988; Cooperman et al., 1988).

Our initial studies have concentrated on the photoincorporation of native chloramphenicol. For this purpose, we used a modification of the method of LeGoffic et al. (1980) to synthesize [^{3}H]chloramphenicol at high specific radioactivity (1.4 Ci/mmol), permitting us to detect the photoincorporation at very low levels of covalent incorporation (0.01 to 0.1%) into any individual protein. As a result, we are able to study photoincorporation by using low concentrations of [^{3}H]chloramphenicol (<10 μM) and low light doses. These conditions are important in maximizing labeling taking place from the high-affinity rather than the low-affinity sites as well as in maximizing the fraction of labeling occurring by incorporation of the native chloramphenicol molecule into the native ribosome structure. It is important to note that chloramphenicol undergoes photodecomposition via formation of *p*-nitrobenzaldehyde (Mubarak et al., 1982) and that irradiation of ribosomes in the presence of chloramphenicol results in the loss of peptidyltransferase activity (Wilchek and Bayer, 1978).

Chloramphenicol has two optically active centers (Fig. 1). Of the four stereoisomers, only one, the D-*threo* form, has high antibiotic activity. This same stereochemical specificity is displayed by the high-affinity site for chloramphenicol binding mentioned above. Evidence for site specificity in affinity labeling studies is typically sought by determining whether a nonradioactive ligand will compete for the binding of the corresponding radioactive ligand, as manifested by a reduction in the photoincorporation of radioactivity in the presence of the nonradioactive ligand. In the case of chloramphenicol, this approach is complicated by the photodynamic action of the nitrophenyl moiety within the chloramphenicol molecule, which gives rise to nonspecific increases in photoincorporation of [^{3}H]chloramphenicol. To address this problem, we took advantage of the stereochemical specificity of chloramphenicol binding to the ribosome, using two different approaches (C. L. Fernández and B. S. Cooperman, unpublished data).

In the first approach, we subtracted the labeling pattern (using reverse-phase HPLC analysis) obtained for photoincorporation of D-*threo*-[^{3}H]-chloramphenicol in the presence of excess nonradioactive D-*threo*-chloramphenicol from that obtained in the presence of excess nonradioactive L-*erythro*-chloramphenicol. Here the reasoning was that the nonspecific effects of the *p*-nitrophenyl group should be the same in both experiments and give rise to no differential

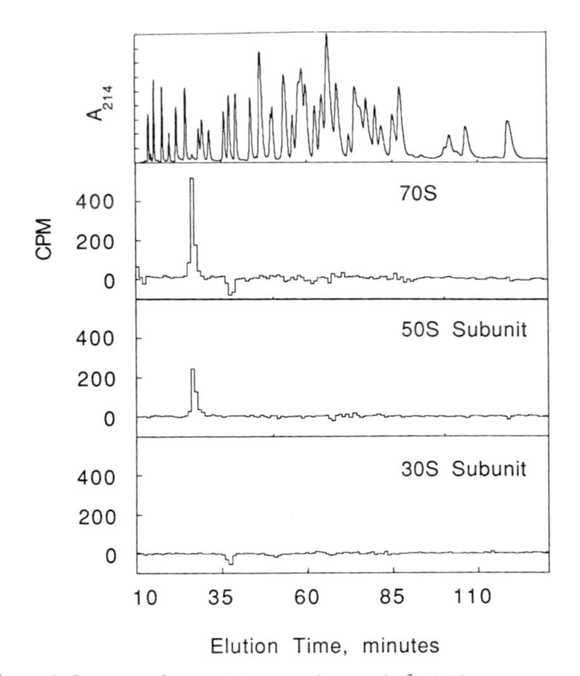

Figure 2. Reverse-phase HPLC analysis of [^{3}H]chloramphenicol photoaffinity labeling of 70S ribosomes. The top panel shows a reverse-phase HPLC pattern of 70S proteins. The bottom three panels show the radioactivity due to covalently attached [^{3}H]chloramphenicol, expressed as the difference in labeling occurring in the presence of excess nonradioactive L-*erythro*-chloramphenicol minus that occurring in the presence of nonradioactive D-*threo*-chloramphenicol, in total protein from 70S ribosomes and 50S and 30S subunits. (From Cooperman et al., 1989.)

labeling. By contrast, a positive peak of labeling would result only from specific site labeling, since the D-*threo* isomer but not the L-*erythro* isomer should inhibit such labeling. These expectations were well borne out by the experimental results (Fig. 2). The positive peak of radioactivity in the difference chromatogram shown elutes slightly behind L27 in chromatograms of either 70S or 50S protein, both derived from labeled 70S ribosomes (further evidence identifying this peak as labeled L27 was provided by additional analysis by ion-exchange HPLC and by one-dimensional polyacrylamide gel electrophoresis [PAGE]). From a technical point of view, the closeness of the base line to zero over most of the difference chromatogram is a measure of the quantification achievable by using reverse-phase HPLC to analyze photoaffinity-labeled ribosomal proteins. The small negative peak seen in the difference chromatogram of 30S protein is real and corresponds to an enhanced labeling of S14 on photolysis in the presence of nonradioactive D-*threo*-chloramphenicol. We have previously observed a similar stereospecific effect of the enhancement of [^{3}H]puromycin photoincorporation into S14 by D-*threo*-chloramphenicol but not L-*erythro*-chloramphenicol (Grant et al., 1979b). The second approach involved subtracting labeling obtained by using D-*erythro*-[^{3}H]chloramphenicol from labeling obtained by using D-*threo*-[^{3}H]chloramphenicol. Although the results of this analysis were less straightforward because the D-*erythro* isomer gives generally higher levels of photoincorporation, in agreement with previous observations (LeGoffic et al., 1980), the specificity of labeling of the L27 peak by the D-*threo* isomer was quite apparent. Our results thus support the notion that L27 is present at or near the peptidyltransferase center.

Our current efforts are directed toward synthesizing photolabile derivatives of chloramphenicol that retain strong binding to the high-affinity site and afford high levels of photoincorporation. Such derivatives would permit us to carry out reconstitution studies paralleling those described above for puromycin (as well as immunoelectron microscopy studies; see below). Using inhibition of high-affinity binding of [^{3}H]chloramphenicol as a criterion, we have demonstrated that the 4-azido analog of chloramphenicol mentioned above does not bind strongly to the high-affinity site, accounting for the inconclusive results reported earlier with this compound (Nielsen et al., 1978; Wilchek and Bayer, 1978).

Results with Other Antibiotics

Tiamulin is an antibiotic that blocks the assembly of functional initiation complexes but also binds to a site that overlaps the binding sites of puromycin and chloramphenicol. It also appears to interfere with binding of the 3′ end of tRNA to both the A′ and P′ sites (Steiner et al., 1988). The *N*-bromoacetyl derivative of tiamulin reacts covalently with 50S proteins L2 and L27, as identified by two-dimensional PAGE analysis (Högenauer et al., 1981). The functional significance of such labeling was demonstrated in two ways. First, the presence of excess tiamulin inhibited covalent reaction with the bromoacetyl derivative, thus demonstrating that labeling occurred from a saturable site. Second, and quite significant, no affinity labeling could be detected by using ribosomes isolated from tiamulin-resistant strains. In a third approach, ribosomes incubated with high concentrations of the bromoacetyl derivative were found to lack peptidyltransferase activity, but under the conditions used it was not clear that 50S labeling was restricted to proteins L2 and L27.

A series of recent papers has appeared describing photoaffinity labeling of *E. coli* ribosomes with macrolides (Siegrist et al., 1985; Tejedor and Ballesta, 1985a, 1985b, 1986; Arévalo et al., 1988). It has been proposed that macrolides can be divided into two groups with respect to their inhibitory effects on protein synthesis. The erythromycin group, which is characterized by the attachment of monosaccharides to the aglycone ring, inhibits peptide chain elongation only after formation of several peptide bonds and is thought to be bound close to, but not at, the peptidyltransferase center. Fluorescence energy transfer experiments suggest a distance of 23 Å (2.3 nm) (Wells and Cantor, 1980). Members of the spiramycin group, on the other hand, which is characterized by the attachment of disaccharides, are thought to be typical peptidyltransferase inhibitors. Erythromycin-resistant mutants that are spiramycin sensitive provide evidence that the ribosome-binding sites of the two groups of macrolides are not identical.

In recent work, Arévalo et al. (1988) have demonstrated that two different photolabile derivatives of erythromycin, one containing a 4-nitroguaiacol group (compound I) and the other containing an aryl azide (compound II), specifically photoincorporate into protein L22, with the affinity for compound I being somewhat higher. A combination of PAGE and reverse-phase HPLC approaches was used to identify labeled proteins (RNA was also labeled, but the positions of labeling were not determined). In both cases, the labeling of L22 was inhibited by added erythromycin. Ribosomes from erythromycin-resistant mutants are known to contain L22 with altered electrophoretic mobility and to bind erythromycin without being inhibited. Photoaffinity labeling of such ribosomes with compound I led to specific incorporation into a protein peak with the expected altered mobility, providing very strong evidence for the identification of L22 as the site of labeling. Less clear evidence, with respect to both identity and specificity, was also obtained for the labeling of protein L15 by compound I. Here it is worth recalling the experiments of Teroaka and Nierhaus (1978) showing that protein L15 in solution binds to erythromycin.

Earlier work by Tejedor and Ballesta (1985a, 1985b, 1986) focused on the spiramycin-type macrolides. By reduction of the aldehyde function, they prepared ^{3}H-labeled dihydro derivatives of carbomycin A, niddamycin, and tylosin. All three of these derivatives undergo light-independent incorporation into ribosomal protein, and the first two also show photoinduced incorporation. All five labeling reactions are consistent in showing L27 as the major labeled protein (as identified by PAGE and reverse-phase HPLC), and all show site specificity (as evidenced by the ability of unlabeled macrolides to inhibit covalent labeling). A similar study by Siegrist et al. (1985) investigating the photoincorporation of ^{3}H-labeled dihydrorosaramicin gave disappointingly nonspecific results.

SUMMARY OF AFFINITY LABELING AND RELATED STUDIES

Proteins at the Peptidyltransferase Center

The results of affinity labeling experiments leading to the identification of 50S proteins at the peptidyltransferase center for the antibiotics discussed above, as well as for affinity labels placed at the 3′ ends of tRNAs bound in the A and P sites, which are also directed toward the peptidyltransferase center, are presented in Table 2. Here we include only those for which the identity of the covalently labeled ribosomal component is well established (except as otherwise indicated) and for which there is at least some evidence that incorporation occurs from a functionally important site of the ribosome. The labeled proteins meeting these criteria are L2, L11, L15, L16, L18, L22, L23, and L27. We next consider the relationship of these proteins to other studies of 50S structure and function, including protein-protein cross-linking, immunoelectron microscopy, neutron diffraction using selectively deuterated ribosomes containing pairs of protonated proteins, the protein assembly map of 50S subunits, and reconstitution studies of peptidyltransferase.

By far the largest number of protein cross-linking results have been reported by Traut et al. (1986). A summary of their results linking proteins within the peptidyltransferase domain is shown in Fig. 3A, where the placement of the proteins within the 50S particle is also based on immunoelectron microscopy results. As can be seen, five of the proteins implicated by affinity labeling studies, L2, L16, L18, L23, and L27, are depicted as being quite close to one another. On the basis of these results, it appears that L15 should also be included in this group, since Traut et al. (1986) have identified a fairly prominent L15-L16 cross-link and at least weak cross-links between L15 and L2, L18, and L23. Only proteins L11 and L22 do not show multiple cross-links within the group of eight affinity-labeled proteins. By contrast, Walleczek et al. (1989), summarizing an extensive set of cross-linking experiments, have identified only two relevant cross-links among the eight proteins of interest, L18-L22 and L16-L27. On the basis of their cross-linking data and the results of Stöffler and Stöffler-Meilicke (1986) mapping protein epitopes within the 50S subunit by immunoelectron microscopy, Walleczek et al. (1988) have constructed a formal model for the placement of proteins within the 50S subunit (Fig. 3B). In this

Table 2. 50S proteins identified at or near the peptidyltransferase center by affinity labeling

L protein	Presence as identified with given ribosomal ligand and reagent type[a]								
	Puromycin, photolabile[b]	Chloramphenicol		Tiamulin, electrophilic[c]	Spiromycin-type macrolides		Erythromycin, photolabile[d]	3′ end of tRNA	
		Photo-labile[e]	Electro-philic[f]		Photo-labile[g]	Electro-philic[h]		Photo-labile[i]	Electro-philic[j]
2			+	+					+
11[k]								+	
15	+						(+)		+
16			(+)						+
18	+							+	
22	(+)						+		
23	+								
27		(+)	+	+	+	+		+	+

[a] Parentheses indicate some uncertainty in protein identification.
[b] Jaynes et al., 1978; Grant et al., 1979b; Nicholson et al., 1982a, 1982b; Weitzmann and Cooperman, 1985.
[c] Högenauer et al., 1981.
[d] Arévalo et al., 1988.
[e] Sonenberg et al., 1974; Wilchek and Bayer, 1978; Nielsen et al., 1978; LeGoffic et al., 1980; Fernández and Cooperman, unpublished data.
[f] Pongs et al., 1973; Sonenberg et al., 1973.
[g] Siegrist et al., 1985; Tejedor and Ballesta, 1985a.
[h] Tejedor and Ballesta, 1986.
[i] Hsiung and Cantor, 1974; Hsiung et al., 1974; Wower et al., 1988.
[j] Cooperman, 1980.
[k] See text.

model, there is a clear clustering of L2, L15, L16, and L27 and possibly L22, whereas L18 is somewhat further away and L11 and L23 are quite distant.

Nowotny et al. (1989) have recently summarized their ongoing neutron diffraction studies with a list of 90 pairwise distances between 50S proteins ranging from 34 to 162 Å, with a median value of 76 Å. Only 11 distances were ≤46 Å; these included four, L2-L16 (42 Å), L2-L27 (42 Å), L15-L27 (46 Å), and L18-L22 (45 Å), linking some of the eight proteins listed above. With respect to the assembly map of the 50S subunit, there is a strong dependence on protein L2 of the uptake of L15, a strong dependence on L15 of the uptake of L16, L18, and L27, and a weaker dependence of L16 uptake on L11 and of L11 uptake on L15 (Herold and Nierhaus, 1987). Finally, reconstitution experiments directed toward defining the essential protein components of the peptidyltransferase center were initially interpreted as implicating L2, L3, L4, L15, and L16 as primary candidates (Schulze and Nierhaus, 1982). More recent work (Lotti et al., 1983; Tate et al., 1987) suggests that

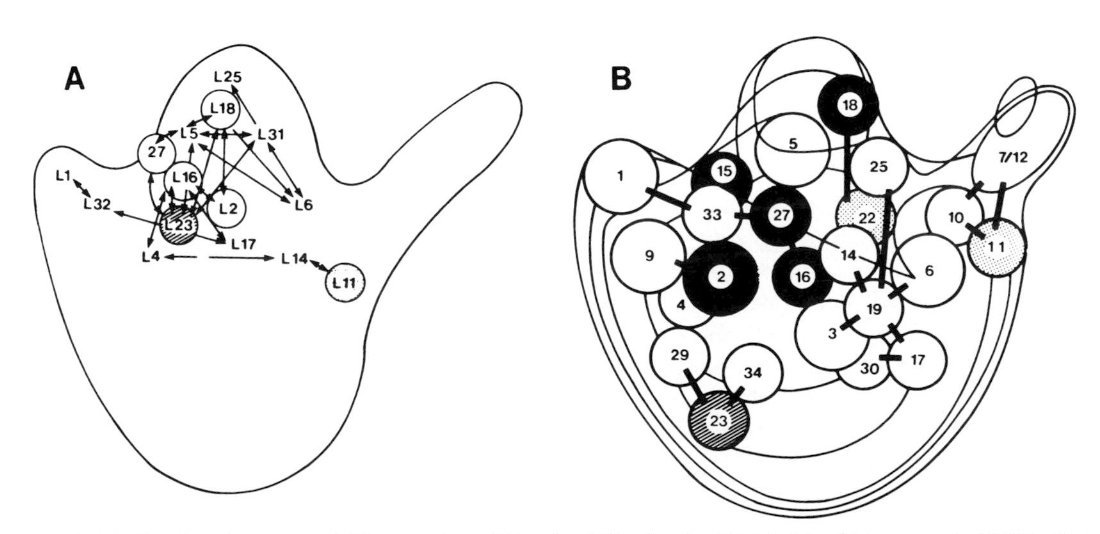

Figure 3. Models for the placement of 50S proteins within the 50S subunit. (A) Model of Traut et al. (1986). Proteins implicated at the peptidyltransferase center are circled. (B) Model of Walleczek et al. (1988). Proteins implicated at the peptidyltransferase center are darkened. Protein L11 is probably not within the peptidyltransferase center. The location of protein L23 is controversial.

although L15 and L16 may be necessary for optimal peptidyltransferase activity, only L2, L3, or L4 can be essential for this activity.

In summarizing these results, it seems clear that the proteins that can be placed at or near the peptidyltransferase center with the least ambiguity are L2 (which may be essential for activity), L15, L16, and L27. The accumulated evidence is also strong for L18 and L22. Aside from the early photoaffinity results of Hsiung and Cantor (1974) and Hsiung et al. (1974), there is little evidence supporting the inclusion of L11 within this center. Since the identification of L11 was based solely on two-dimensional PAGE, with no allowance made for possible changes in protein migration as a result of chemical modification, our inclination is to drop L11 from the list of proteins at or near the peptidyltransferase center.

This leaves L23, the placement of which is clearly controversial. The evidence for placing it within the peptidyltransferase center comes from the photoaffinity labeling results with puromycin and *p*-azidopuromycin discussed above, the cross-linking results of Traut et al. (1986), and immunoelectron microscopy results of Olson and co-workers (Olson et al., 1982; Olson et al., 1985) showing that antibody to N^6,N^6-dimethyladenosine (which recognizes puromycin) binds to 50S subunits that have been photoaffinity labeled with either puromycin or *p*-azidopuromycin in the region of the 50S subunit defined by the L2, L15, L16, and L27 cluster (Fig. 3). In addition, tetracycline, which, as we have already noted, specifically stimulates puromycin photoincorporation into L23 (Grant et al., 1979b; Weitzmann and Cooperman, 1990), has recently been shown to specifically photomodify 23S rRNA bases (Steiner et al., 1988) falling within the putative peptidyltransferase center (see below).

On the other hand, the most recent immunoelectron microscopy studies of 50S subunits, using antibody to L23 as a probe (Stöffler-Meilicke et al., 1983; Hackl and Stöffler-Meilicke, 1988), place L23 near the base of the 50S subunit and on the side facing away from the 30S subunit, quite far from the peptidyltransferase center, and Walleczek et al. (1989) report no cross-links placing L23 close to the center. We are left with three possibilities. The first is that in the immunoelectron microscopy studies of Olson et al. (1982, 1985), the bound antibody binds to puromycin photoincorporated into other sites but not into L23. For reasons summarized in these latter publications we think this unlikely, particularly since in the puromycin-labeled 50S subunit, 70% of photoincorporation was into L23. The second possibility is that the placement of L23 by using anti-L23 is in error, either because the anti-L23 preparation is contaminated with antibodies to other ribosomal proteins or because an incorrect site has been identified. Thus, Nag et al. (1987) have shown that the region in which L23 has been placed by anti-L23 binding has considerable nonspecific binding capacity for antibody Fab or Fc arms, and that this poses particular problems for immunoelectron microscopy studies using antibodies of low affinity. The third possibility is that both sets of results are correct and that L23 is an elongated protein. Thus, Nagano et al. (1988) have proposed that L23 could have a dumbbell-like structure of overall length 90 Å and running in parallel to the excretion channel (Carazo et al., 1988; Yonath and Wittmann, 1988), with the domain labeled by puromycin in the peptidyltransferase center and the domain to which anti-L23 binds at the peptide exit site. Further work will clearly be needed to resolve this question.

RNA at the Peptidyltransferase Center

p-Azidopuromycin is the only peptidyltransferase antibiotic affinity label having a known site of covalent incorporation into rRNA, at positions U-2504 and G-2502. These bases fall within the central loop of domain V of 23S rRNA, a region that is linked to the peptidyltransferase center by a considerable body of other results (Fig. 4), including (i) the photoincorporation of 3-(4′-benzoylphenyl)propionyl-Phe-tRNA into bases U-2584 and U-2585 when bound at the A site, bases A-2451 and C-2452 when bound at the P site, and bases A-2503 through U-2506 when bound at the either the A or the P site (Steiner et al., 1988), (ii) the location of site-specific mutants conferring resistance to direct inhibitors of peptidyltransferase, chloramphenicol and anisomycin (bases C-2452 and A-2453 [Hummel and Böck, 1987]), and an indirect inhibitor of peptidyltransferase (erythromycin), and (iii) sites protected from chemical modification in the presence of chloramphenicol, vernamycin, carbomycin, and erythromycin (Moazed and Noller, 1987) and by tRNA derivatives bound in either the A or the P site (Moazed and Noller, 1989). In addition, a second photolabile Phe-$tRNA^{Phe}$ derivative has been shown to photoincorporate in the region from 2445 to 2668 (Hall et al., 1985), although in this case the precise sites of incorporation were not determined.

RNA and Protein at the Peptidyltransferase Center

Two of the photoaffinity labeling experiments discussed above provide direct links between several of the proteins identified at the peptidyltransferase center and the central loop of domain V. Thus, *p*-azidopuromycin appears to label several ribosomal

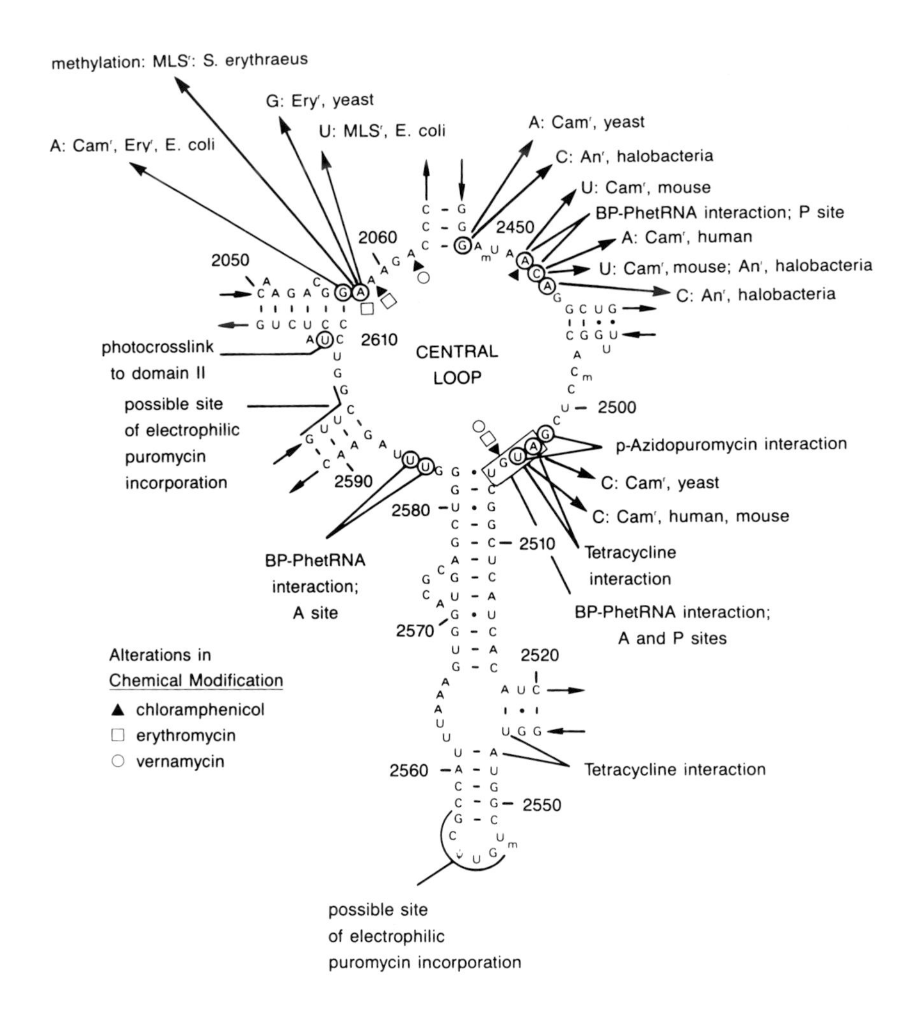

Figure 4. Schematic diagram of the secondary structure of the central loop of domain V of 23S rRNA adapted from Hall et al. (1988) and Steiner et al. (1988). Sites of affinity labeling by *p*-azidopuromycin and by 3-(4′-benzoylphenyl)propionyl-Phe-tRNA bound in either the A or the P site are indicated, as are the sites of photomodification induced by tetracycline, two possible sites of covalent incorporation by an electrophilic derivative of puromycin, and a site of cross-linking to domain II. Nucleotides which when mutated confer resistance to specific antibiotics are indicated (Ettayebi et al., 1985; Hummel and Böck, 1987; Vester and Garrett, 1988). Symbols identify nucleotides having altered reactivity toward chemical reaction in the presence of the indicated antibiotic (Moazed and Noller, 1987).

components (proteins L23, L18 [and possibly L22], and L15 and bases U-2504 and G-2502) from a single site (Nicholson et al., 1982b; Hall et al., 1988). Furthermore, tetracycline, which has been shown to photomodify bases A-2503 and U-2504 (Steiner et al., 1988), also specifically stimulates the photoincorporation of puromycin into protein L23 (Grant et al., 1979a; Weitzmann and Cooperman, 1990). This last result is surprising, given the weight of evidence that tetracycline inhibits A-site (and to lesser extent P-site) binding of tRNA chiefly through its binding to a unique high-affinity site located on the 30S subunit (Goldman et al., 1983; Epe et al., 1987; Buck and Cooperman, 1990). On the other hand, tetracycline is known to bind to several other sites on the ribosome, and it now appears that a site at the central loop of domain V is of particularly high affinity. It is an interesting question whether this site is important for the action of tetracycline as a protein synthesis inhibitor.

CONCLUSIONS

The results summarized above illustrate the success of affinity labeling studies in identifying specific ribosomal components at or near the peptidyltransferase center of the *E. coli* ribosome. These results, taken together with related results of other studies, provide direct evidence for the presence of proteins L2, L15, L16, L18, L22, L23, and L27, as well as the central loop of domain V of 23S rRNA, at or near the peptidyltransferase center. While some problems remain unresolved, it is clear that increased attention by workers in the field both to the definitive identification of labeled ribosomal components and to control experiments demonstrating that functional site labeling has been achieved has provided results having greater reliability than was true in the past. Information gained from properly conducted affinity label experiments is of major importance in the construction of detailed functional models of the ribosome.

This work was supported by Public Health Service grant AI-16806 from the National Institutes of Health.

REFERENCES

Allen, D. W., and P. C. Zamecnik. 1962. The effect of puromycin on rabbit reticulocyte ribosomes. *Biochim. Biophys. Acta* **55:** 865–874.

Arévalo, M. A., F. Tejedor, F. Polo, and J. P. G. Ballesta. 1988. Protein components of the erythromycin binding site in bacterial ribosomes. *J. Biol. Chem.* **263:**58–63.

Barta, A., G. Steiner, J. Brosius, H. F. Noller, and E. Kuechler. 1984. Identification of a site on 23S ribosomal RNA located at the peptidyl transferase center. *Proc. Natl. Acad. Sci. USA* **81:**3607–3611.

Bouthier de la Tour, C., M.-L. Capmau, and F. LeGoffic. 1985. Affinity labeling of bacterial ribosome with bromamphenicol. *Eur. J. Med. Chem.* **20:**213–218.

Buck, M. A., and B. S. Cooperman. 1990. Single protein omission reconstitution studies of tetracycline binding to the 30S subunit of *Escherichia coli* ribosomes. *Biochemistry* **29:**5374–5379.

Capel, M. S., D. B. Datta, C. R. Nierras, and G. R. Craven. 1988. Ion-exchange high-performance liquid chromatographic separation of ribosomal proteins. *Methods Enzymol.* **164:**532–541.

Carazo, J. M., T. Wagenknecht, M. Radermacher, V. Mandiyan, M. Boublik, and J. Frank. 1988. Three-dimensional structure of 50S *Escherichia coli* ribosomal subunits depleted of proteins L7/L12. *J. Mol. Biol.* **201:**393–404.

Cooperman, B. S. 1978. Affinity labeling studies on *Escherichia coli* ribosomes, p. 81–115. *In* E. van Tamelen (ed.), *Bioorganic Chemistry. A Treatise to Supplement Bioorganic Chemistry, An International Journal*, vol. 4. Academic Press, Inc., New York.

Cooperman, B. S. 1980. Functional sites on the *E. coli* ribosome as defined by affinity labeling, p. 531–554. *In* G. Chambliss, G. R. Craven, J. Davies, K. Davis, L. Kahan, and M. Nomura (ed.), *Ribosomes. Structure, Function, and Genetics*. University Park Press, Baltimore.

Cooperman, B. S. 1987. Photoaffinity labeling of *Escherichia coli* ribosomes. *Pharmacol. Ther.* **34:**271–302.

Cooperman, B. S. 1988. Affinity labeling of ribosomes. *Methods Enzymol.* **164:**341–361.

Cooperman, B. S., M. A. Buck, C. L. Fernandez, C. J. Weitzmann, and B. F. D. Ghrist. 1989. Antibiotic photoaffinity labeling probes of *Escherichia coli* ribosomal structure and function, p. 123–139. *In* P. E. Nielsen (ed.), *Photochemical Probes in Biochemistry*. Kluwer, Dordrecht, The Netherlands.

Cooperman, B. S., C. C. Hall, A. R. Kerlavage, C. J. Weitzmann, J. Smith, T. Hasan, and J. D. Friedlander. 1986. Photoaffinity labeling of *Escherichia coli* ribosomes: new approaches and results, p. 362–378. *In* B. Hardesty and G. Kramer (ed.), *Structure, Function, and Genetics of Ribosomes*. Springer-Verlag, New York.

Cooperman, B. S., C. J. Weitzmann, and M. A. Buck. 1988. Reversed-phase high-performance liquid chromatography of ribosomal proteins. *Methods Enzymol.* **164:**523–532.

Drainas, D., D. L. Kalpaxis, and C. Coutsogeorgopoulos. 1987. Inhibition of ribosomal peptidyltransferase by chloramphenicol: kinetic studies. *Eur. J. Biochem.* **164:**53–58.

Epe, B., P. Woolley, and H. Hornig. 1987. Competition between tetracycline and tRNA at both P and A sites of the ribosome of *Escherichia coli. FEBS Lett.* **213:**443–447.

Ettayebi, M. S., M. Prasad, and E. A. Morgan. 1985. Chloramphenicol-erythromycin resistance mutation in a 23S rRNA gene of *Escherichia coli. J. Bacteriol.* **162:**551–557.

Goldman, R. A., T. Hasan, C. C. Hall, W. A. Strycharz, and B. S. Cooperman. 1983. Photoincorporation of tetracycline into *Escherichia coli* ribosomes. Identification of the major proteins photolabeled by native tetracycline and tetracycline photoproducts and implications for the inhibitory action of tetracycline on protein synthesis. *Biochemistry* **22:**359–368.

Grant, P. G., B. S. Cooperman, and W. A. Strycharz. 1979a. On the mechanism of chloramphenicol-induced changes in the photoinduced affinity labeling of *Escherichia coli* ribosomes by puromycin. *Biochemistry* **18:**2154–2159.

Grant, P. G., W. A. Strycharz, E. N. Jaynes, Jr., and B. S. Cooperman. 1979b. Antibiotic effects on the photoinduced affinity labeling of *Escherichia coli* ribosomes by puromycin. *Biochemistry* **18:**2149–2154.

Gravel, M., P. Melançon, and L. Brakier-Gingras. 1987. Cross-linking of streptomycin to the 16S ribosomal RNA of *Escherichia coli. Biochemistry* **26:**6227–6232.

Hackl, W., and M. Stöffler-Meilicke. 1988. Immunoelectron microscopic localisation of ribosomal proteins from *Bacillus stearothermophilus* that are homologous to *Escherichia coli* L1, L6, L23, and L29. *Eur. J. Biochem.* **174:**431–435.

Hall, C. C., D. Johnson, and B. S. Cooperman. 1988. [^{3}H]-p-azidopuromycin photoaffinity labeling of *Escherichia coli* ribosomes: evidence for site-specific interaction at U-2504 and G-2502 in Domain V of 23S ribosomal RNA. *Biochemistry* **27:**3983–3990.

Hall, C. C., J. E. Smith, and B. S. Cooperman. 1985. Mapping labeled sites in *Escherichia coli* ribosomal RNA: distribution of methyl groups and identification of a photoaffinity-labeled RNA region putatively at the peptidyltransferase center. *Biochemistry* **24:**5702–5711.

Hampl, H., H. Schulze, and K. H. Nierhaus. 1981. Ribosomal components from *Escherichia coli* 50S subunits involved in the reconstitution of peptidyl transferase activity. *J. Biol. Chem.* **256:**2284–2288.

Herold, M., and K. H. Nierhaus. 1987. Incorporation of six additional proteins to complete the assembly map of the 50S subunit from *Escherichia coli* ribosomes. *J. Biol. Chem.* **262:** 8826–8833.

Högenauer, G., H. Egger, C. Ruf, and B. Stumper. 1981. Affinity labeling of *Escherichia coli* ribosomes with a covalently binding derivative of the antibiotic pleuromutilin. *Biochemistry* **20:** 546–552.

Hsiung, N., and C. R. Cantor. 1974. A new simpler photoaffinity analogue of peptidyl tRNA. *Nucleic Acids Res.* **1:**1753–1762.

Hsiung, N., S. A. Reines, and C. R. Cantor. 1974. Investigation of ribosomal peptidyl transferase centre using a photoaffinity label. *J. Mol. Biol.* **88:**841–855.

Hummel, H., and A. Böck. 1987. Ribosomal RNA mutations in halobacterium conferring resistance to the anti-80S ribosome targeted antibiotic anisomycin. *Nucleic Acids Res.* **15:**2431–2443.

Jaynes, E. N., Jr., P. G. Grant, G. Giangrande, R. Wieder, and B. S. Cooperman. 1978. Photoinduced affinity labeling of the *Esche-*

richia coli ribosome puromycin site. *Biochemistry* **17**:561–569.

Kerlavage, A. R., C. J. Weitzmann, and B. S. Cooperman. 1983. Reverse phase high performance liquid chromatography of *Escherichia coli* ribosomal proteins: standardization of 70S, 50S and 30S protein chromatograms. *J. Biol. Chem.* **258**:6313–6318.

LeGoffic, F., M.-L. Capmau, L. Chausson, and D. Bonnet. 1980. Photo-induced affinity labeling of *Escherichia coli* ribosomes by chloramphenicol. *Eur. J. Biochem.* **106**:667–674.

Lessard, J. L., and S. Pestka. 1972. Studies on the formation of transfer ribonucleic acid-ribosome complexes. XXIII. Chloramphenicol, aminoacyl-oligonucleotides, and *Escherichia coli* ribosomes. *J. Biol. Chem.* **247**:6909–6912.

Lotti, M., E. R. Dabbs, R. Hasenbank, M. Stöffler-Meilicke, and G. Stöffler. 1983. Characterization of a mutant from *Escherichia coli* lacking protein L15 and localization of protein L15 by immuno electron microscopy. *Mol. Gen. Genet.* **192**:295–300.

Moazed, D., and H. F. Noller. 1987. Chloramphenicol, erythromycin, carbomycin and vernamycin B protect overlapping sites in the peptidyl transferase region of 23S ribosomal RNA. *Biochimie* **69**:879–884.

Moazed, D., and H. F. Noller. 1989. Interaction of tRNA with 23S rRNA in the ribosomal A, P, and E sites. *Cell* **57**:585–597.

Mubarak, S. I. M., J. B. Standford, and J. K. Sundgen. 1982. Some aspects of the photochemical degradation of chloramphenicol. *Pharm. Helv. Acta* **57**:226–230.

Nag, B., D. S. Tewari, A. Sommer, H. M. Olson, D. G. Glitz, and R. R. Traut. 1987. Probing ribosome function and the location of *Escherichia coli* ribosomal protein L5 with a monoclonal antibody. *J. Biol. Chem.* **262**:9681–9687.

Nagano, K., M. Harel, and M. Takezawa. 1988. Prediction of three-dimensional structure of *Escherichia coli* ribosomal RNA. *J. Theor. Biol.* **134**:199–256.

Nathans, D. 1964. Puromycin inhibition of protein synthesis: incorporation of puromycin into peptide chains. *Proc. Natl. Acad. Sci. USA* **51**:585–592.

Nicholson, A. W., C. C. Hall, W. A. Strycharz, and B. S. Cooperman. 1982a. Photoaffinity labeling of *Escherichia coli* ribosomes by an aryl azide analogue of puromycin. On the identification of the major covalently labeled ribosomal proteins and mechanism of photoincorporation. *Biochemistry* **21**:3797–3808.

Nicholson, A. W., C. C. Hall, W. A. Strycharz, and B. S. Cooperman. 1982b. Photoaffinity labeling of Escherichia coli ribosomes by an aryl azide analogue of puromycin. Evidence for the functional site specificity of labeling. *Biochemistry* **21**: 3809–3817.

Nielsen, P. E., V. Leick, and O. Buchardt. 1978. On photoaffinity labeling of *Escherichia coli* ribosomes using an azidochloramphenicol analogue. *FEBS Lett.* **94**:287–290.

Nowotny, V., P. Nowotny, H. Bov, K. H. Nierhaus, and R. P. May. 1989. The quaternary structure of the ribosome from E. coli: a neutron small-angle scattering study. *Physica B* **156-157**:499–501.

Olson, H. M., P. G. Grant, B. S. Cooperman, and D. G. Glitz. 1982. Immunoelectron microscopic localization of puromycin binding on the large subunit of the *Escherichia coli* ribosome. *J. Biol. Chem.* **257**:2649–2656.

Olson, H. M., A. W. Nicholson, G. S. Cooperman, and D. G. Glitz. 1985. Localization of sites of photoaffinity labeling of the large subunit of *Escherichia coli* ribosomes by arylazide derivative of puromycin. *J. Biol. Chem.* **260**:10326–10331.

Pongs, O., R. Bald, and V. A. Erdmann. 1973. Identification of chloramphenicol-binding protein in *Escherichia coli* ribosomes by affinity labeling. *Proc. Natl. Acad. Sci. USA* **70**:2229–2233.

Schulze, H., and K. H. Nierhaus. 1982. Minimal set of ribosomal components for reconstitution of the peptidyltransferase activity. *EMBO J.* **1**:609–613.

Siegrist, S., N. Moreau, and F. LeGoffic. 1985. Photo-induced affinity labeling of *Escherichia coli* ribosomes by dihydrorosaramicin, a macrolide related to erythromycin. *Eur. J. Biochem.* **153**:131–135.

Sonenberg, N., M. Wilchek, and A. Zamir. 1973. Mapping of *Escherichia coli* ribosomal components involved in peptidyl transferase activity. *Proc. Natl. Acad. Sci. USA* **70**:1423–1426.

Sonenberg, N., A. Zamir, and M. Wilchek. 1974. A photo-induced reaction of chloramphenicol with E. coli ribosomes: covalent binding of the antibiotic and inactivation of peptidyl transferase. *Biochem. Biophys. Res. Commun.* **59**:693–696.

Steiner, G., E. Kuechler, and A. Barta. 1988. Photo-affinity labelling at the peptidyl transferase centre reveals two different positions of the A- and P-sites in domain V of 23S rRNA. *EMBO J.* **7**:3949–3955.

Stöffler, G., and M. Stöffler-Meilicke. 1986. Immuno electron microscopy on *Escherichia coli* ribosomes, p. 28–46. *In* B. Hardesty and G. Kramer (ed.), *Structure, Function, and Genetics of Ribosomes.* Springer-Verlag, New York.

Stöffler-Meilicke, M., M. Noah, and G. Stöffler. 1983. Location of eight ribosomal proteins on the surface of the 50S subunit from *Escherichia coli. Proc. Natl. Acad. Sci. USA* **80**:6780–6784.

Tate, W. P., V. G. Sumpter, C. N. A. Trotman, M. Herold, and K. H. Nierhaus. 1987. The peptidyltransferase centre of the *Escherichia coli* ribosome. *Eur. J. Biochem.* **165**:403–408.

Tejedor, F., and J. P. G. Ballesta. 1985a. Ribosome structure: binding site of macrolides studied by photoaffinity labeling. *Biochemistry* **24**:467–472.

Tejedor, F., and J. P. G. Ballesta. 1985b. Components of the macrolide binding site on the ribosome. *J. Antimicrob. Chemother.* **16**(Suppl. A):53–62.

Tejedor, F., and J. P. G. Ballesta. 1986. Reaction of some macrolide antibiotics with the ribosome. Labeling of the binding site components. *Biochemistry* **25**:7725–7731.

Teroaka, H., and K. H. Nierhaus. 1978. Proteins from *Escherichia coli* ribosomes involved in the binding of erythromycin. *J. Mol. Biol.* **126**:185–193.

Traut, R. R., D. S. Terari, A. Sommer, G. R. Gavino, H. M. Olson, and D. G. Glitz. 1986. Protein topography of ribosomal functional domains: effects of monoclonal antibodies to different epitopes in *Escherichia coli* protein L7/L12 on ribosome function and structure, p. 286–308. *In* B. Hardesty and G. Kramer (ed.), *Structure, Function, and Genetics of Ribosomes.* Springer-Verlag, New York.

Vester, B., and R. A. Garrett. 1988. The importance of highly conserved nucleotides in the binding region of chloramphenicol at the peptidyl transfer centre of *Escherichia coli* 23S ribosomal RNA. *EMBO J.* **7**:3577–3587.

Walleczek, J., T. Martin, B. Redl, M. Stöffler-Meilicke, and G. Stöffler. 1989. Comparative cross-linking study on the 50S ribosomal subunit from *Escherichia coli. Biochemistry* **28**: 4099–4105.

Walleczek, J., D. Schüler, M. Stöffler-Meilicke, R. Brimacombe, and G. Stöffler. 1988. A model for the spatial arrangement of the proteins in the large subunit of the *Escherichia coli* ribosome. *EMBO J.* **7**:3571–3576.

Weitzmann, C. J. 1989. Reconstitution of ribosomal subunits containing puromycin-labeled protein L-23. Ph.D. thesis. University of Pennsylvania, Philadelphia.

Weitzmann, C. J., and B. S. Cooperman. 1985. On the structural specificity of puromycin binding to *Escherichia coli* ribosomes. *Biochemistry* **24**:2268–2274.

Weitzmann, C. J., and B. S. Cooperman. 1990. Reconstitution of *Escherichia coli* 50S ribosomal subunits containing puromycin-modified L23: functional consequences. *Biochemistry* **29**:3458–3465.

Wells, B. D., and C. R. Cantor. 1980. Ribosome binding by tRNAs with fluorescent labeled 3′-termini. *Nucleic Acids Res.* **8**:3229–3246.

Wilchek, M., and E. A. Bayer. 1978. *Affinity Labeling from the Isolated Protein to the Cell*, p. 201–223. Academic Press, Inc., New York.

Wower, J., S. S. Hixson, and R. A. Zimmermann. 1988. Photochemical cross-linking of yeast $tRNA^{Phe}$ containing 8-azidoadenosine at positions 73 and 76 to the *Escherichia coli* ribosome. *Biochemistry* **27**:8114–8121.

Yonath, A., and H. G. Wittman. 1988. Crystallographic and image reconstruction studies of ribosomal particles from bacterial sources. *Methods Enzymol.* **164**:95–117.

Youvan, D. C., and J. E. Hearst. 1979. Reverse transcriptase pauses at N2-methylguanine during *in vitro* transcription of *E. coli* 16S ribosomal RNA. *Proc. Natl. Acad. Sci. USA* **76**:3751–3754.

Youvan, D. C., and J. E. Hearst. 1981. A sequence from *Drosophila melanogaster* 18S rRNA bearing the conserved hypermodified nucleoside amψ: analysis by reverse transcription and high-performance liquid chromatography. *Nucleic Acids Res.* **9**:1723–1741.

Chapter 43

Peptidyltransferase Inhibitors: Structure-Activity Relationship Analysis by Chemical Modification

J. P. G. BALLESTA and E. LAZARO

Chemical modification has been extensively used as a way to improve the therapeutic activity of antibiotics, in some instances, as with the β-lactams, with a great deal of success. In most cases, however, a trial-and-error strategy has been used rather than a rational program based on a detailed understanding of the antibiotic mode of action at the molecular level. Nevertheless, the efforts dedicated to this endeavor have yielded a wealth of data establishing structure-activity relationships in many antibiotics that have been extremely useful for a better understanding not only of the antibiotic mode of action but also of the underlying process that they inhibit.

In the field of protein synthesis inhibitors, chemical modification has been frequently used to study structure-activity relationships, and there are extensive reviews of the work carried out on a number of different antibiotics (Magerlein, 1977; Cox et al., 1977; Omura, 1984). This chapter will discuss recent work performed in different laboratories related to the chemical modification of antibiotics that inhibit peptide bond formation by ribosomes. Emphasis will be placed on studies in which antibiotic modification has been used as a tool to study different aspects of the drug-ribosome interaction rather than on those focused on improvement of therapeutic activity. The more recent work has concentrated on sparsomycin, the macrolide group of antibiotics, and puromycin.

SPARSOMYCIN

Sparsomycin is a broad-spectrum antibiotic active against eucaryotic, bacterial, and archaebacterial cells, including *Sulfolobus solfataricus*, a species that is resistant to most antibiotics (Cammarano et al., 1985). Sparsomycin is a highly specific ribosomal peptidyltransferase inhibitor that blocks the binding of substrates to the A site while stimulating strongly the interaction at the P site (reviewed in Ottenheijm et al., 1986). This peculiar mode of action made sparsomycin an interesting and extensively used tool in ribosome structure-function studies. Unfortunately, the toxicity of the drug reduced its clinical interest, which led to a halt in its production and thus limited its use as a research tool in basic studies.

The interesting biological properties of sparsomycin come together in a peculiar chemical structure containing two sulfur atoms, one of them acting as a sulfoxide; this feature is rarely found in biological molecules and therefore attracted the attention of researchers in synthetic organic chemistry. Recently, two groups have independently achieved complete synthesis of the drug (Helquist and Shekhani, 1979; Ottenheijm et al., 1981; Liskamp et al., 1981; Hwang et al., 1985); consequently, the drug is available once more. In addition, numerous sparsomycin analogs carrying single structural alterations have been synthesized, enabling detailed analysis of structure-function relationships. These studies have provided interesting data related not only to the molecular requirements for activity but also to some structural characteristics of the ribosomel-binding site of the drug.

Any alteration at either the oxygen or nitrogen atoms in the modified uracil ring totally inactivates the drug by abolishing its affinity for ribosomes (van den Broek et al., 1987) (Fig. 1). On the other hand, it is possible to remove the C-6 methyl group in the ring with only partial loss of activity (van den Broek et al., 1987). These data suggest that the drug binds to the ribosome by hydrogen bonding through the O and N atoms in the uracil ring, probably interacting with the rRNA.

The *trans* geometry of the double bond in the

J. P. G. Ballesta and E. Lazaro ■ Centro de Biología Molecular, Canto Blanco, Madrid 28049, Spain.

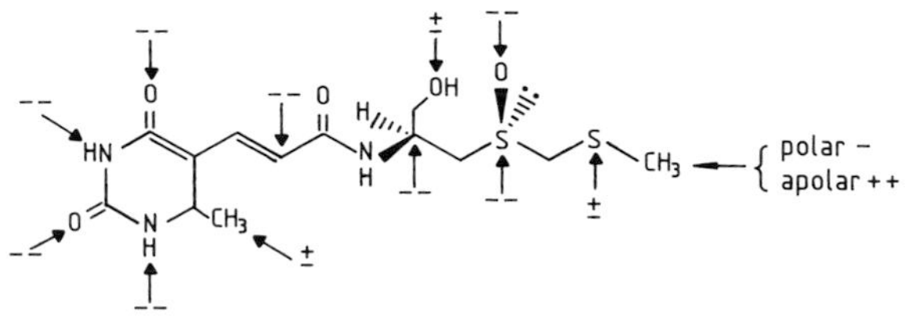

Figure 1. Summary of structure-activity results for sparsomycin. +, Stimulation; −, inhibition.

lateral chain is critical for activity, and as a result *cis*-sparsomycin is nearly inactive (van den Broek et al., 1987). It is interesting, however, that *cis*-sparsomycin competes with sparsomycin for binding to ribosomes to a greater extent than expected considering its low biological activity (Table 1), suggesting that this sparsomycin analog is able to interact with the ribosome, probably through the unmodified uracyl ring, but requires the right conformation in the rest of the molecule to block ribosome function.

By using homochiral preparations, it has been possible to determine unambiguously that of the four stereoisomers, resulting from the presence of two chiral centers in the molecule, only the ScRs configuration is active (Ash et al., 1984b; van den Broek et al., 1987). In addition, it has been shown, disproving previous results (Lee and Vince, 1978), that the presence of a sulfoxide function on the chiral sulfur atom is required for biological activity, since its reduction to sulfide diminishes drastically the activity of the drug (van den Broek et al., 1987). Similarly, the correct position of the sulfoxide is important; when the relative position of this function on the two sulfur atoms is reversed, activity is markedly affected (Flynn and Ash, 1983; van den Broek et al., 1987). As in the case of *cis*-sparsomycin, some of the sparsomycin derivatives that have altered structural configurations but still carry an unmodified uracyl ring conserve a high ribosome-binding capacity disproportionate to their low biological activity, which indicates the role of that part of the molecule in the interaction with ribosomes (Table 1).

Table 1. Binding and inhibitory activities of some sparsomycin analogs

Analog	Chirality	Relative activity[a]	
		Binding[b]	Activity[c]
cis-Sparsomycin	ScRs	5.3	70
Deoxosparsomycin	ScRs	6.8	104
Benzylsparsomycin	ScSs	18	100

[a] Relative activity is expressed as the ratio of the ED_{50} (concentration that produces 50% inhibition of the test) of the analog to the ED_{50} of sparsomycin.
[b] Competition tests for binding to bacterial ribosomes with labeled sparsomycin were performed.
[c] Inhibition of polyuridylic acid-dependent polymerization systems.

Modification of the sparsomycin hydroxymethylene function by introducing groups of different size and chemical character has little effect on the activity of the drug when tested on bacterial ribosomes (Lazaro et al., 1987; van den Broek et al., 1989a). Interestingly, the same sparsomycin derivatives have a drastically reduced activity when tested on ribosomes from yeast cells and from *S. solfataricus* (Lazaro et al., 1987). Contrary to the findings discussed above, in which modifications of sparsomycin did not differentiate between ribosomes, the sparsomycin analogs carrying alterations at this position are able to distinguish structural differences in the peptidyltransferase centers of ribosomes from different kingdoms. By using appropriate photoreactive sparsomycin analogs, it might be possible to identify ribosomal domains more affected by evolution from other, more conserved regions. Another consequence of these results is that in this respect *S. solfataricus* ribosomes resemble those of yeasts rather than those of eubacteria.

The most interesting results have been obtained by introducing modifications affecting the *S*-methyl end of the sparsomycin molecule, and several drug derivatives altered in this part of the structure have been synthesized (Lin and Dubois, 1977; Vince et al., 1977; Ash et al., 1984a; van den Broek et al., 1987; van den Broek et al., 1989b). The use of homochiral preparations has provided conclusive evidence that an increase in the hydrophobic properties of the drug by introducing aromatic and aliphatic substituents at the sulfoxide end of the molecule causes a strong stimulation of biological activity (Ash et al., 1984a; van den Broek et al., 1987; van den Broek et al., 1989b). Conversely, the introduction of a polar function markedly decreases and eventually completely abolishes the biological activity of sparsomycin (van den Broek et al., 1987). The stimulatory effect of the hydrophobic substituents is especially evident in the eucaryotic systems. In *Saccharomyces cerevisiae* and L1210 lymphoid leukemia cells, some of the lipophilic sparsomycin derivatives show up to a 10-fold increase in in vitro and in vivo activity (Zylicz, 1988; van den Broek et al., 1989b). This dramatic increase in activity has renewed interest in the therapeutic properties of these compounds, and some of them are now in preclinical tests as antitumor agents (Zylicz, 1988).

The stimulatory effect of the hydrophobic substitutents on the activity of sparsomycin seems to be due to an increase in the affinity of the drug for ribosomes, as indicated by binding studies performed

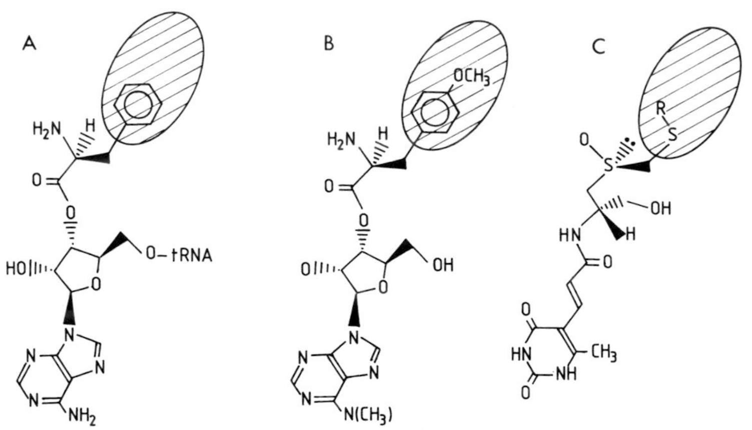

Figure 2. Chemical structures of the aminoacyl-tRNA 3′-terminal aminoacyladenosine (A), puromycin (B), and sparsomycin (C), indicating the relative positions of the different moieties and of the proposed hydrophobic pocket (hatched area).

with radioactive derivatives (E. Lazaro, L. A. G. M. van den Broek, H. C. J. Ottenheijm, and J. P. G. Ballesta, unpublished results). Moreover, it has been shown that the introduction of a hydrophobic substituent in the terminal position of an otherwise inactive sparsomycin analog improves the ability of the compound to compete for binding to ribosomes, although in some cases its activity does not increase proportionally (Table 1). These results suggest that the hydrophobic substitutent provides an additional site of interaction between the molecule and the ribosome, and therefore there must be a suitable region at the peptidyltransferase center capable of binding the new lipophilic group in the drug.

This region might correspond to the postulated recognition site for the hydrophobic side chains of amino acids in the aminoacyl-RNA that has been proposed on the basis of the relative affinity of the 3′-end fragments from different aminoacyl-tRNAs (Rychlik et al., 1970; Chladek, 1980). Sparsomycin can be considered an analog of the aminoacyladenine moiety of acceptor tRNA, having overlapping binding sites (Flynn and Ash, 1983; Ash et al., 1984b); comparison of the steric configurations of the molecules (Fig. 2) shows that the hydrophobic substituents in the sparsomycin derivatives are in a position equivalent to that of the lateral amino acid chain in the aminoacyl-tRNA, whereas the uracil ring corresponds to the adenine. Similar considerations apply to puromycin, which has a steric configuration analogous to that of sparsomycin (Fig. 2). In support of the role of the hydrophobic region in the interaction of these antibiotics, the L-tyrosyl analog of puromycin (O-demethylpuromycin), which carries an unblocked hydroxyl group that decreases its lipophilicity, is considerably less active than the original drug (Vara et al., 1985).

It has been proposed that chloramphenicol as well as other peptidyltransferase inhibitors, including anisomycin and puromycin, mimic the peptide bond and inhibit the ribosome function by binding to the active center, where the bond-forming reaction takes place (Vester and Garrett, 1988). Sparsomycin can also be fit into this model (Fig. 3); in each case, a lipophilic ring is found in the position corresponding to the postulated hydrophobic region of the peptidyltransferase center. If this model is correct, it can be proposed that as in the cases of puromycin and sparsomycin, the activities of other drugs could be manipulated by changing the hydrophobicity in the aromatic ring. Removal of the O-methyl group in anisomycin would inactivate the drug, as in the case of puromycin; on the other hand, the introduction of more hydrophobic groups would increase activity, as in the case of sparsomycin.

An additional useful application of some semisynthetic sparsomycin analogs results from the rela-

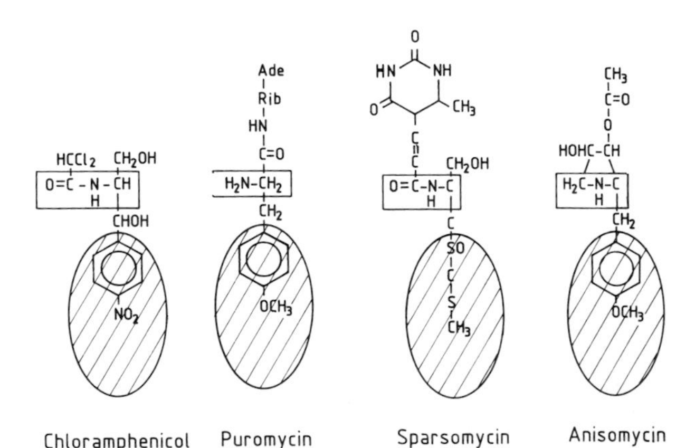

Figure 3. Isostructural regions of chloramphenicol, puromycin, sparsomycin, and anisomycin according to the model of inhibition of peptide bond formation proposed by Vester and Garrett (1988). The relative position of the peptidyltransferase hydrophobic pocket is marked (hatched area).

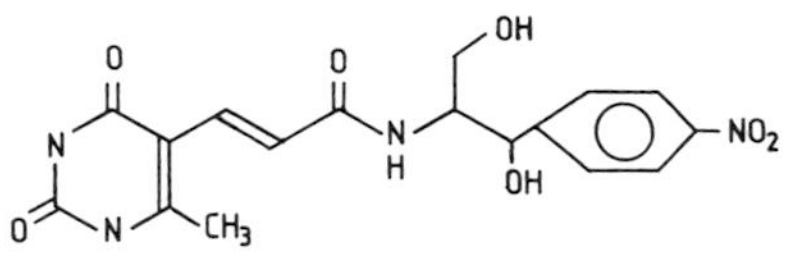

Figure 4. Chemical structure of sparsophenicol.

tive ease with which they can be radiolabeled. The sparsomycin derivative carrying a phenol group at the terminal position is readily labeled by standard radioiodination methods. By using this labeled derivative, the kinetics of binding to 70S and 80S ribosomes has been studied; in both cases, a K_d value of around 2×10^{-7} to 3×10^{-7} M was found (Lazaro et al., unpublished results). Similarly, it has been possible to show that contrary to one model proposed for the mode of action of sparsomycin, which implicated a chemical reaction between the drug and a protein component in the peptidyltransferase center (Flynn and Ash, 1983), no such covalent binding of the radioactive derivative to ribosomes takes place (Lazaro et al., unpublished results).

Related to the chemical modification of sparsomycin is the reported synthesis of sparsophenicol (Fig. 4), a hybrid antibiotic carrying the uracil moiety of sparsomycin and the nitrophenyl function of chloramphenicol (Zemlicka and Bhuta, 1982). This compound fits in the model of Vester and Garrett (1988) of inhibition for chloramphenicol and puromycin (Fig. 3) and is indeed a good inhibitor of the puromycin reaction, although it has virtually no antibacterial and antitumor activity, probably because of low permeability (Zemlicka and Bhuta, 1982).

MACROLIDES

The macrolides form a large group of protein synthesis inhibitors that have similar chemical characteristics but display dissimilar inhibitory properties when tested in different in vivo and in vitro assay systems (reviewed by Vazquez [1979] and Gale et al. [1981]).

They can be broadly divided structurally and functionally into two relatively homogeneous groups, typically represented by erythromycin and spiramycin, respectively. The erythromycin group has a 14-atom lactone ring that is glycosylated in two positions by different monosaccharides. The second group carries a 16-member lactone ring, usually glycosylated at only one position by a disaccharide. A few macrolides that have a 12-atom lactone ring, such as methymycin, have been functionally assigned to the first group.

The 16-atom lactone macrolides are considered typical inhibitors of peptide bond formation and are good inhibitors of most protein synthesis model systems, such as puromycin reaction and the polyuridylic acid-dependent polymerization of phenylalanine. On the other hand, erythromycin is a poor inhibitor of these assays unless donor substrates of specific characteristics are used, and it has been proposed that this drug is an inhibitor of peptidyl-tRNA translocation.

Functional differences can be also detected by using in vivo tests. Thus, it is interesting that the ability to induce MLS resistance (cross-resistance to macrolides, lincosamines, and streptogramins) in bacteria seems to be limited to erythromycin among the macrolides. According to a proposed model for the induction of MLS resistance, in order to allow the expression of a 23S RNA *N*-methyltransferase, the inducing antibiotic must block the translating ribosome in the middle of a short mRNA that leads the methylase messenger (Kamimiya and Weisblum, 1988). The antibiotic must therefore allow the synthesis of the first few peptide bonds and stall the ribosome at the appropriate position. In agreement with these requirements, erythromycin, but not the spiramycin-type macrolides, seems to require the presence of an oligopeptide in the donor substrate to be able to inhibit the peptide bond formation model assays (Vazquez, 1974).

Despite these differences, all macrolides seem to compete for binding to the ribosome, indicating the existence of similar or overlapping sites of interaction. The similarity between the binding sites is confirmed by the fact that cross-resistance to different macrolides can be induced by modification of similar nucleotides in the 23S RNA (Cundliffe, this volume), although it is also true that dissociated resistance to different macrolides has been reported (Garrod, 1957) and specific erythromycin-resistant mutants have been described (Apirion, 1967; Otaka et al., 1969; Pardo and Rosset, 1974).

It has been proposed that macrolides bind by the aglycone ring to common or closely related binding sites in the ribosome and that differences in their modes of action are due to the different sizes of the sugar residues linked to the lactone ring. As a way of identifying common as well as specific components in the macrolide-binding sites, rRNA protection and affinity labeling experiments have been performed.

The first type of approach has shown that erythromycin and carbomycin, a 16-atom ring macrolide, protect overlapping but not identical regions in the highly conserved central loop of domain V in 23S RNA. Both antibiotics protect from chemical modification nucleotides A-2058, A-2059, and G-2505; in

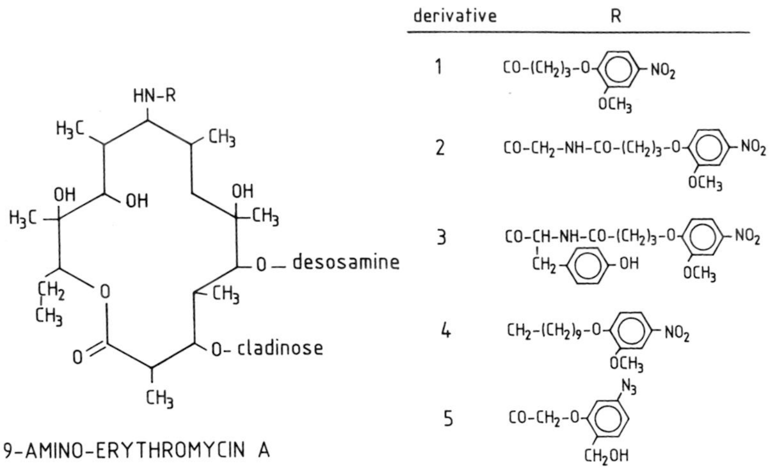

Figure 5. Chemical structures of photoreactive derivatives of erythromycin.

addition, carbomycin protects A-2062 and A-2451, suggesting that the 16-atom macrolide-binding site covers a larger region in the rRNA (Moazed and Noller, 1987).

By using affinity labeling, it has been found that whereas erythromycin derivatives specifically label protein L22 (Arévalo et al., 1988), carbomycin derivatives mostly label protein L27 (Arévalo et al., 1989). Although these results can be at least partially attributable to the different relative positions of the photoreactive groups in the antibiotic molecules, they suggest differences in the protein environment of the binding site for the two macrolide types.

On the whole, these results seem to indicate that, as expected, the 14- and 16-atom ring macrolides probably have overlapping binding sites on the ribosome with a common region, probably composed of RNA, and a specific region, perhaps composed mostly of protein. However, we are still far from understanding the differences in the modes of action of the macrolides and how they are related to the structures of their ribosome-binding sites.

Chemical modification has been used extensively to study the structure-activity relationships of macrolides, and basic facts on the relative importance of the different functions have been established and reviewed (Omura, 1984). Here we will summarize recent reports that in some cases open up interesting perspectives for an understanding of the mode of action of the macrolides.

A number of erythromycin derivatives have been synthesized that carry photoreactive groups linked to the lactone ring through aliphatic chains of different lengths (Fig. 5). Although the use of these compounds as affinity labels will not be discussed here, the results obtained in the assays performed to test their activity can be helpful in understanding the role of the lateral residues on the activity of macrolides (Arévalo et al., 1989).

Even though the introduction of these groups affects the overall activity of erythromycin as a protein synthesis inhibitor, the presence of some substituents increases the inhibition of standard peptide bond formation model reactions, such as the fragment reaction and the puromycin reaction, which are insensitive to the unmodified drug but very sensitive to the spiramycin-type macrolides (Table 2). In addition, the extent of inhibition is roughly proportional to the relative size of the substituents, determined mostly by the length of the chain linking the photoreactive group to the antibiotic molecule.

These results support the concept that the differences in the mode of action of the two macrolide groups can be attributed to the different glycosidic substituents, the size of the substituent being important in determining the activity of the drug.

Related to the differences in mode of action of the macrolides, as related to their saccharide substit-

Table 2. Inhibition of peptide bond formation model systems by erythromycin derivatives

Antibiotic	% of inhibition		
	Puromycin reaction at:		Fragment reaction at 10^{-5} M
	10^{-6} M	10^{-5} M	
Erythromycin	0	3	0
Erythromycinamine	1	4	0
Derivative 1	13	25	11
Derivative 2	18	30	24
Derivative 3	18	27	15
Derivative 4	7	14	0
Derivative 5	15	23	0
Spiramycin	53	85	48

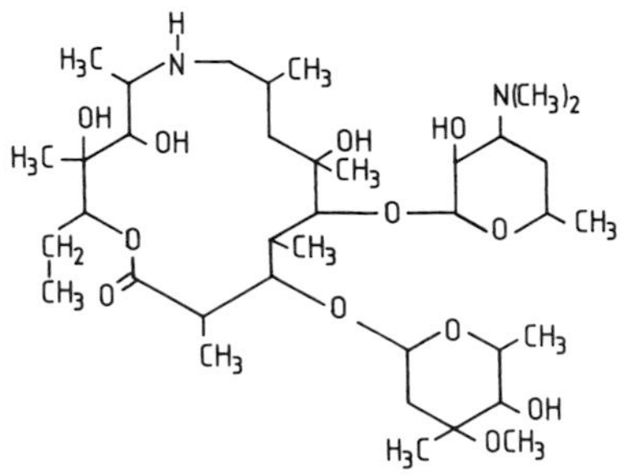

Figure 6. Chemical structure of azithromycin (10-dihydro-10-deoxo-11-azaerythromycin A).

uents, is the report on the preparation of a series of semisynthetic erythromycin derivatives. These are generically called azythromycins and were obtained by Beckmann rearrangement of the erythromycin oxime, resulting in new macrolides that carry an expanded 15-atom lactone ring (Djokic et al., 1986, 1988). The rest of the erythromycin structural components are practically unmodified in the new compounds; therefore, an intermediate group of macrolides having a larger lactone ring but conserving the saccharide substituents of the small-size aglycone drugs is produced (Fig. 6). Unfortunately, most of the biological assays performed with azythromycins have been designed for testing their in vivo activity as antibacterial agents (Djokic et al., 1987). The results, apart from indicating some advantageous properties of the new drugs, are of little value in the analysis of their interactions with ribosomes. Very few data on the activity of azythromycins in in vitro systems are available, and experimental results using the peptide bond formation model reactions are totally missing. It is interesting, however, that azythromycins are inactive when tested in polyuridylic acid-dependent systems but inhibit polyadenylic and polycytidylic acid-dependent polymerizations (S. Djokić, personal communication). This is a characteristic of the 14-member lactone macrolides and indicates that despite having a larger lactone ring, azythromycins are functionally equivalent to the erythromycin-type macrolides. This supports the notion that it is the lateral substituents and not the size of the lactone ring that are responsible for the functional peculiarities of the different macrolide groups. Additional work using azythromycins in other model systems will undoubtedly provide data relevant to the relative roles of the lactone ring and the sugar substituents on the inhibitory activity of macrolides.

Extensive chemical manipulation of tylosin, a 16-atom lactone ring macrolide, and other related antibiotics has also been performed (Kirst et al., 1986; Kirst et al., 1987; Kirst et al., 1988; Omura et al., 1983). Some of the derivatives obtained carry chemical modifications, such as altered sugar substituents, that can be very useful for studying their roles in antibiotic activity at the ribosomal level. Unfortunately, as in the case of azythromycins, most of the biological tests performed have been focused on assaying their in vivo activity, although some studies on the ribosome-binding capacity of different tylosin derivatives have been performed (Omura et al., 1983). Interestingly, these results indicate that modifications in the sugar residues do not drastically affect the affinity of the drug for the ribosome, confirming that they have a secondary role in this interaction, although they substantially affect the inhibitory activity of the antibiotic. The possibilities for use of many of these derivatives to analyze the role of the sugar residues in defining the characteristic mode of action of a macrolide drug are very promising but have not been fully explored.

It has recently been shown that certain 6-O-methyl-11,12 cyclic carbamate derivatives of erythromycin are able to bind and inhibit 70S ribosomes from MLS-resistant cells (Goldman and Kadam, 1989), confirming that the substituents in the lactone ring play a critical role in determining the mode of action of macrolides.

The data therefore seem to confirm the existence of a common ribosome-binding site for all macrolides that is modulated by the different saccharide molecules glycosylating the lactone ring of these drugs. This interaction site cannot be located very close to the P site of the peptidyltransferase center (i.e., the site of interaction of the 3′ end of the peptidyl-tRNA molecule) because peptide bond formation would then be inhibited by both types of macrolides. The drug-binding site is probably displaced some distance along the exit route of the nascent peptide chain, and only the 16-atom ring macrolides, carrying larger disaccharide substituents, are able to interfere with peptide bond formation. The results indicating that 14-atom ring macrolides are able to inhibit peptide bond formation when donor substrates carrying peptides of a certain length are used (Contreras and Vazquez, 1977) support this view.

PUROMYCIN

Structure-function relationships in puromycin, as far as the nucleotide and the amino acid moieties of the molecule are concerned, show the important role of these two regions in the activity of the drug (Nathans and Neidle, 1963).

Recently, there has been an effort to study the role of the sugar moiety by chemical manipulation of the puromycin molecule. Thus, carbocyclic analogs

tic int
riboso
Cox, D.
coside
Djokic,
series.
10-de
tivity
1015.
Djokic,
Erythr
oxime
Trans.
Djokic,
Mrvos
elucid
mycin
Flynn, G
moiety
cin. Bi
Gale, E.
M. J.
p. 402
Garrod,
Med. J
Goldman
macrol
B-resis
growth
Helquist,
sparso
Hwang,
synthes
serine
Kamimiy
control
rRNA
teriol.
Kirst, H.
J. R. T
Dauper
1986.
16-mer
biot. 3
Kirst, H.
J. L. C
Pekarel
rolides
effectiv
Kirst, H.
R. M.
and F.
and str
tives of
Lazaro, E
Lielevel
ships o
Biochin
Lee, C. K.
the pu
J. Med.
Lin, C.-C.
antitum
Med. C
Liskamp,
1981. S

Vince, R., J. Brownell, and C. L. Lee. 1977. Inhibition of protein synthesis. The first active sparsomycin analog. *Biochem. Biophys. Res. Commun.* 75:563–566.

Vince, R., S. Daluge, and J. Brownell. 1986. Carbocyclic puromycin: synthesis and inhibition of protein biosynthesis. *J. Med. Chem.* **29**:2400–2403.

Zemlicka, J., and A. Bhuta. 1982. Sparsophenicol: a new synthetic hybrid antibiotic inhibiting ribosomal peptide synthesis. *J. Med. Chem.* **25**:123–125.

Zylicz, Z. 1988. Sparsomycin and its analogues. A preclinical study on novel anticancer drugs. Ph.D. dissertation. University of Nijmegen, Nijmegen, The Netherlands.

IX. TRANSLATIONAL FIDELITY

be indirect, mediated through the rRNA (Kurland, 1974). This interpretation fell on deaf ears because proteins were "in" and because in the 1970s there was no technology to systematically study the functions of separate domains of the rRNA. On the other hand, the beginnings of such a technology were emerging (Noller et al., 1971; Noller and Chaires, 1972).

A decade later the pendulum has swung, and now it is the rRNA that is "in" (see Moore, 1988). An impressive body of sequence information has been accumulated. Among other things, this information has revealed an astonishing and highly suggestive conservation through the different phyla of rRNA structures, both primary and higher-order structures (Woese et al., 1983; Brimacombe et al., 1983; Noller, 1984). More to the point, a chemical methodology has been developed to identify sites in the RNA structures that are near neighbors to tRNA, as well as the other effector molecules that interact with the ribosome during translation.

The relevant studies have involved chemical modification of rRNA in situ to establish a catalog of accessible sites in the unperturbed ribosome. Then the masking or enhancement of chemically accessible sites by ribosome-bound effector molecules provides information about the location in the RNA structure of domains to which the effector molecules may be associated. Such studies have recently culminated in a tour de force by Moazed and Noller (1989a, 1989b). This Dynamic Duo has identified five distinct RNA sites, three on the 50S subunit and two on the 30S subunit, to which tRNA molecules are associated during the different intermediate stages of translation. These sites will be discussed in some detail below. For the moment, they are introduced as direct evidence for an extensive interaction between the tRNA and the ribosome during translation.

That such interactions are relevant to the phenotypes of the accuracy mutants has been suggested by a number of observations. First, the classic restrictive mutants in S12 were originally selected as streptomycin-resistant mutants (Gorini, 1971). However, it was demonstrated more recently that mutation in 16S RNA also confers resistance in *E. coli* to the error-enhancing antibiotic streptomycin (Montandon et al., 1986). In addition, the suggestion that this antibiotic interacts directly with 16S RNA rather than with proteins (Garvin et al., 1974) is supported by the effect of streptomycin on the chemical reactivity of a short string of nucleotides in 16S RNA that is also protected by bound tRNA (Moazed and Noller, 1987). Finally, Allen and Noller (1989) have shown that mutant forms of S4 and S12 with altered accuracy phenotypes affect the chemical reactivity of sites in 16S RNA in the unperturbed ribosome, and among these is the region associated with bound streptomycin as well as tRNA.

Accordingly, the most recent results are consistent with the view that tRNA is bound to the ribosome at RNA sites that supplement the codon-anticodon interaction. In particular, some of these same rRNA sites are responsive to structural changes in proteins that are altered in the classical accuracy mutants. They are also protected from chemical modification by the error-enhancing antibiotic streptomycin. Such observations lend support to the view that the translation apparatus is an RNA machine. Nevertheless, it would be well to bear in mind that there is a striking similarity between our current enthusiasm for RNA and our attitude toward protein in the 1970s.

The Substrate

One of the nagging little mysteries of protein synthesis has been why tRNAs are so big and not the simple oligonucleotides that Crick (1958) had anticipated. Gorini's view of the interaction between tRNA and ribosome (1971) provides the beginnings of a solution to this mystery. Thus, the suppression frequencies for a given nonsense codon obtained in either wild-type or mutant ribosome backgrounds depend on characteristics of the suppressor tRNA besides its anticodon. For example, different suppressor tRNA species with the same anticodon can be selected from different wild-type isoacceptor species, and it is observed that ribosomes will process them with different efficiencies (Gorini, 1971). From this, Gorini inferred that the ribosome is sensitive to the molecular context into which an anticodon has been inserted. He attributed this discrimination to a ribosomal "screening" function which depends on interactions distant from the anticodon.

A quite respectable literature concerning the effect of structural alterations of tRNA on translational efficiency and fidelity has been elaborated in the meantime (Smith, 1979; Yarus, 1982; Murgola, 1985; Buckingham and Grosjean, 1986). It is quite evident from this data set that the codon-programmed ribosome is sensitive to the whole tRNA molecule and is not simply testing the anticodon. In this sense, it is not out of place to think of the aminoacyl-tRNAs as substrates that are fitted into stereospecific sites on the working surfaces of the ribosome. Of course, the fascinating thing about the ribosomal decoding site is its capacity to match each of 40 different substrate molecules to its cognate codon at the same site.

What Gorini and subsequent workers discovered

is that the precision of this match depends on the degree to which the anticodon and the "body" of the tRNA molecule are tuned to each other. As will be argued below, the evolution of a relatively large tRNA structure in which the sites for interaction with ribosome and mRNA are functionally coupled to each other is an essential feature of the translocation process that supports the processivity of translation.

TIME AND ENERGY

The discriminating functions of Gorini's ribosomal screen were difficult to understand in simple physical terms, but through hindsight it is now possible to understand his model as an attempt to formulate a geometric solution to a kinetic problem (see Kurland, 1988). Indeed, it is striking how static our views of the ribosome were before Ninio made his debut in the accuracy arena.

Branch Points

Ninio (1974) introduced a very simple kinetic scheme consisting of two variables to describe the selection of tRNA species at a ribosomal discrimination site. With the aid of this model, he could account for the relative magnitudes of the suppression frequencies resulting from the interaction of the different ribosome and tRNA mutants in Gorini's experiments.

The basic idea is that when an aminoacyl-tRNA is bound at a ribosomal discriminating site, at least two different events can follow: either the tRNA proceeds along the path of peptide bond formation or it leaves the binding site. In other words, the intermediate binding state is a kinetic branch point. According to this formulation, the accuracy of the selection depends on the preference of the cognate tRNA to proceed forward as well as the preference of the noncognate species to dissociate from the site. The physical basis of these preferential flows can be described as follows.

For a codon-dependent selection, one of the discriminating factors will be the relative stability of cognate versus noncognate codon-anticodon interactions. This parameter can be thought of as determining the relative tendency of competing tRNA species to dissociate from the codon. The second relevant parameter describes the tendency of the tRNA at sites distinct from the anticodon to dissociate from the ribosomal site. Here, we may imagine that the ribosomal interaction is the same for all tRNA species. Nevertheless, the stability of this interaction will influence the degree to which the specificity of codon-anticodon interactions can be expressed. For example, if the ribosomal affinity is too large, it will stabilize tRNA binding to a degree that will mask the specificity of the codon-anticodon interaction. At the other extreme, if the ribosome affinity is very weak, the codon-dependent discrimination can be maximized, but even correctly matched tRNA species will be retained with low efficiency. As a consequence of this interplay, the efficient and accurate codon-dependent acquisition of tRNA species at a ribosomal discrimination site will depend on an optimal balance between the codon-specific and the codon-independent interactions.

As we hope is evident from this interpretation of Ninio's ideas, a straightforward account can now be made of the ribosomal mutant phenotypes originally described by Gorini. According to this interpretation, a restrictive mutant would be one that supports a relatively loose interaction of tRNA at the ribosomal binding site, which encourages the tRNAs to dissociate more readily. Accordingly, the contribution of the codon-anticodon interaction to the stability at the discrimination site is more dominant and therefore, a relatively greater accuracy is achieved. In contrast, the *ram* mutants, according to this interpretation, would have relatively stable tRNA-ribosome interactions, which would tend to lower the dissociation rate of tRNA species from the discriminating site. This would tend to lower the relative weight of the codon-anticodon interaction, and this in turn will lower the accuracy of the discriminating site.

In summary, the branch point provides a kinetic option for each tRNA molecule as it passes through the discrimination site: either it can dissociate or it can be forwarded. Structural changes of the tRNA or of the ribosome can influence the accuracy of the selection by altering the kinetic constants along either of the branches. In this view, the ratio of rate constants of the one branch over the other is what distinguishes the flow of a cognate from a noncognate tRNA over the branch point. We return below to an experimental verification of this kinetic view, but first we must take up another problem.

Proofreading

For some time now, the evidence has suggested that the accuracy of the codon-anticodon interaction is much less than the accuracy of protein synthesis. This created a problem which may have amused the remaining vitalists in the field but which exercised the rest of us. The solution to this enigma was so clever that it now seems obvious. What Hopfield (1974) and Ninio (1975) suggest is that it does not matter if the accuracy of the codon-anticodon interaction per se is not as great as that of protein synthesis. Instead,

virtually indistinguishable from one another (Andersson and Kurland, 1983; Ruusala et al., 1984; Bohman et al., 1984). In contrast, correlations were obtained for the restrictive mutants between (i) the ease with which they are kinetically saturated by ternary complex (Michaelis-Menten constant) and (ii) their error rates. The more accurate the mutant, the greater the concentration of aminoacyl-tRNA ternary complex required to support the maximum rate of translation with cognate species. In effect, the rate factor (Fersht, 1977) for the ternary complex-ribosome interaction is reduced in hyperaccurate mutants. This initially created a serious problem of interpretation.

Until recently, it has been considered axiomatic that ribosomes must function at their maximum rates in vivo (Maaløe, 1979). One argument for this is the relatively large size of the ribosome pool, which represents a biosynthetic mass investment that should be used at maximum efficiency. To this could be added the relatively small size of ternary complex compared with ribosomes. Such disproportion in size was taken to mean that the kinetic efficiency of the smaller component is less important than that of the larger. Therefore, ternary complex could be accumulated in vivo at concentrations high enough to saturate ribosome kinetics without lowering significantly the mass efficiency of the translation system as a whole. Thus, from the perspective of the Copenhagen school, the lower rate factors of restrictive ribosome mutants could not make up a significant cost of accuracy for the cell.

The situation was saved by a more rigorous theory (Ehrenberg and Kurland, 1984). To begin, it was shown there exist in principle optimal arrangements of both the ternary complex and ribosome concentrations such that a maximum rate of translation per mass of total biosynthetic machinery is obtained. In addition, it was shown that the optimal arrangements are not in general obtained when the ribosomes are saturated by ternary complex. Instead, the optimal arrangement depends on the relative metabolic commitment of the cell to protein synthesis. This in turn means that the arrangement of the translation system depends on the quality of the growth medium and that it is therefore a growth rate-dependent parameter.

The variation of the ribosomal saturation level with growth rate is straightforward in the model of Ehrenberg and Kurland (1984): at the lowest growth rates, the system is expected to operate with an equal mass of ternary complex and ribosomes, far from kinetic saturation. As the growth rate increases, the commitment to protein synthesis increases. This is reflected in progressively higher concentrations of ribosomes and in a lowering of the mass ratio of ternary complex to ribosomes but a progressively higher degree of kinetic saturation for the ribosomes. At the very highest conceivable growth rates, at which the cell is doing little else than synthesizing proteins, it may approach kinetic saturation of ribosomes by ternary complex.

This model of growth optimization finds support in the data of Bremer and colleagues, who argued for a long time that the rate of protein synthesis is growth rate dependent (Churchward et al., 1981; Churchward et al., 1982). This conclusion has been confirmed most recently by Pedersen (1984). In addition, the variations of the ternary complex and ribosome concentrations predicted by this model are in remarkable agreement with the data of Neidhardt et al. (1977).

The relevance of the growth optimization model to accuracy is straightforward; it defines the kinetic conditions under which the costs of accuracy will be expressed. Since ribosomes in general are starved for ternary complexes to match the codons being translated in vivo, the fact that the maximum rates of mutant ribosome translation are unaffected by mutation to the restrictive phenotype is irrelevant to their phenotype. In contrast, the observation that enhanced proofreading leads to lower saturation of ribosomes by ternary complexes in vitro is highly relevant. This correlation should be reflected in correspondingly lower elongation rates for the restrictive ribosomes in vivo.

Again, the data are uncomplicated. A series of restrictive mutants was analyzed in vitro and in vivo. A linear correspondence was obtained between four relevant parameters: the intensity of the proofreading flow for cognate species in vitro, an increase in the Michaelis-Menten constant for the ternary complex-ribosome interaction in vitro, the decrease in the rate at which polypeptide is elongated in vivo, and the growth rate depression of the mutants (Andersson et al., 1986b). Therefore, for these mutants there seems to be little question that the cost of enhanced accuracy in vivo is a decreased growth rate attending a lower efficiency of translation.

Optimal Accuracy

Evidently, the optimal strategies for the translation system include the selection of appropriate concentrations as well as preferred structures. The connections between the concentrations and the structures are kinetics. As we have seen, there are signs of a kinetic optimization for accuracy that is nested in the growth optimization of the translation system as a whole. In fact, the essential premise of growth optimization, which is to maximize a given rate of protein synthesis per total mass of translation apparatus, can be used to construct an optimization for the accuracy of translation.

One can imagine all sorts of destructive effects of proteins containing missense substitutions. However, we assume that on the whole, the effects of missense substitutions are simply to lower the average activity of proteins. It follows from this assumption that the global cost of missense errors will be to lower the efficiency of protein biosynthesis. However, we have also seen that the cost of enhanced accuracy is a lower kinetic efficiency of translation. Accordingly, we make the conjecture that these two costs are equivalent and that they can be traded for each other in an optimization of translational accuracy (Ehrenberg and Kurland, 1984). In particular, we suggest that under any particular growth condition, the accuracy is optimized when the cost of further lowering the missense error rate by a small increment is equal to the cost of that error increment on the functional efficiency of the proteins.

The notion of an optimal accuracy phenotype for ribosomes disposes of the superficial contradiction that the restrictive mutant phenotype initially presented. It also is essential to modern views of error feedback in cells, to which we turn in the next section. However, there is one detail of the accuracy optimization that should be discussed before moving on. This concerns the growth rate dependence of accuracy.

We have seen that the optimal accuracy of translation is in principle growth rate dependent because the metabolic commitment to protein synthesis is growth rate dependent. In general, we expect the accuracy optimum to increase as the quality of the growth medium decreases (Ehrenberg and Kurland, 1984). This expectation follows from the fact that at the slower growth rates, protein synthesis is a smaller fraction of the metabolic effort of the cell. Hence, an incremental decrease in the efficiency of translation associated with higher accuracy would have a comparably small impact on the overall metabolic efficiency of the cell. Of course, this expectation depends on whether or not the cell has the capacity to regulate its translational error rate.

The stringent response that produces the guanine nucleotide analogs called the "magic spots" has for a long time been thought to improve the accuracy of translation during starvation for amino acids (Gallant, 1979). Furthermore, these guanine analogs are accumulated in bacteria at steady-state concentrations that become progressively higher as the growth rates of bacteria are reduced (Ryals et al., 1982). Therefore, the magic spots were of interest to us as potential physiological regulators of ribosomal accuracy. Indeed, it was reported recently that the magic spots could modulate the intensity of proofreading in vitro (Dix and Thompson, 1986). However, these small effects on proofreading have been attributed to a design error in the protocol used by Dix and Thompson (1986), and it was observed in a steady-state system that the presence of magic spots has no influence on the proofreading flows under physiologically relevant conditions (Rojas, 1988; A.-M. Rojas, M. Ehrenberg, and C. M. Kurland, unpublished data).

Accordingly, we were not surprised when attempts to observe a systematic variation in the accuracy of translation as a function of the growth rate failed (Mikkola and Kurland, 1988). In these studies, the nonsense suppression frequencies supported by wild-type, *ram*, and restrictive mutants were found to be virtually invariable over a threefold variation of the growth rate. If the conclusions of this solitary study are general, we might infer that bacteria do not regulate the accuracy of their ribosomes as a function of the growth state. The alternative is that the ribosomal performance characteristics are selected to support an accuracy that represents some sort of weighted average of the optima corresponding to the states most often encountered by the bacteria.

One Catastrophe Less

A venerable shibboleth of the accuracy world is the error catastrophe. It was first revealed by Orgel (1963) in a conjecture about the consequences of those errors of gene expression that are expressed in the devices responsible for gene expression. Orgel saw that when the errors in the construction of a device such as a ribosome raise the functional error level of that device, a positive error loop is generated that eventually bursts into a cascade of errors. The force of this conjecture derived from the then current conviction that error-containing devices are more error prone than are canonical devices. Accordingly, the gloomy perspective opened by this conjecture is that all organisms are sitting on a time bomb that eventually will set off a cascade of errors ending in the heat death of the cells, cancer, and the like.

Orgel himself was the first to point out that such a scenario is not inevitable, and in a review of the relevant experimental literature, he made it clear that no unambiguous example of such an error catastrophe had been identified (Orgel, 1970, 1973). Nevertheless, a small band of theorists soldiered on to explore error-damping schemes with which cells might avert the dread catastrophe (Hoffman, 1974; Kirkwood and Holliday, 1975). The introduction of the kinetic proofreading scheme in a curious way lent impetus to the work of the catastrophists because they inferred that an energy-driven system would be more vulnerable to error feedback in times of need (Kirkwood et al., 1984).

The phenomenology of translation errors does not fit well with the catastrophist view. Thus, a small number of measurements of the missense substitution rates have been made at two specific sites in two ribosomal proteins, and they have been found to be close to 1 per 1,000 (Bouadloun et al., 1983). If these rates are representative, they imply that every ribosome is assembled with an average of 10 different errors in its protein complement. In other words, it would be very difficult to find two ribosomes in a cell that are the same. Nevertheless, these cells show no signs of error catastrophe, nor do those of the error-prone *ram* phenotype that have demonstrably higher error rates than do wild-type bacteria (Kurland, 1987). More to the point, wild-type bacteria have been cultured in lethal concentrations of streptomycin, and ribosomes were prepared from them as the cells died (Fast et al., 1987). It was observed under these conditions that even though the ribosomes accumulated missense substitutions in their proteins at a higher rate than normal, their performance characteristics were not significantly different from those of the untreated control ribosomes. Finally, when the death rates of bacteria with different accuracy phenotypes are compared, no influence of mutant phenotypes is observed (Jörgensen and Kurland, 1987). In summary, all of the available data suggest that the translation system is buffered in such a way that positive error feedback is not expressed.

An explanation for the observed stability of translational accuracy is the optimal character of the ribosome performance characteristics. If the accuracy of ribosomes is selected at some intermediate level rather than at some theoretical maximum, an arbitrary missense substitution in one of its proteins could in principle raise the accuracy of the new ribosome with the same likelihood as it could lower it. According to this view, the statistical consequence of random missense substitutions is that the accuracy of the ribosomes will tend to remain close to that of the initial population of ribosomes.

In summary, during the 26 years of its academic history, the error catastrophe has not been observed in a single organism. Furthermore, we would not expect to see this event in nature because accuracy-optimized biosynthetic devices are not in general vulnerable to positive error loops. Perhaps enough has been said by now about error catastrophes.

PROCESSIVITY

So far, we have had very little to say about the absolute error rates in cells. One reason for this reticence is that there is precious little information available. Earlier, estimates of the missense rate had a tendency to cluster around 3×10^{-4} (Loftfield, 1963; Loftfield and Vanderjagt, 1972; Edelman and Gallant, 1977; Ellis and Gallant, 1982). However, the most recent estimates have tended to be more variable and often approach values closer to 10^{-3} (Bouadloun et al., 1983; Parker et al., 1983; Parker and Holtz, 1984). The space of missense errors with all of its undiscovered structure still waits to be explored. It is likely to remain in this virgin state until the analytical methods for the measurement of missense substitutions are stripped of their heroic character.

As we have seen, an error rate of the order of 10^{-3} implies that an average bacterial ribosome has 10 missense substitutions. On the other hand, a single polypeptide with a chain length of 500 would accumulate errors less readily; roughly two-thirds of these sorts of molecules would be error free at the above-stated missense rate. Given this variability for the penetrance of errors into structures of different size, we might wonder what determines the average missense rate. The answer depends very much on what the dominant functional consequences are for missense substitutions in proteins.

The fact is that protein structures seem to be more robust than we initially might be inclined to believe. For example, when nonsense mutations dispersed throughout the structures of proteins such as β-galactosidase, *lac* repressor, or a bacterial RNA polymerase are suppressed by mutant tRNA suppressors, collections of missense-substituted proteins can be generated for functional analyses (Langridge and Campbell, 1969; Miller et al., 1979; Nene and Glass, 1983). The telling observation is that most such substituted proteins are functional to a degree that approaches the normal activity of the canonical structure. More modern crystallographic studies of proteins confirm this tendency. Thus, in general the native fold is not lost in domains harboring individual amino acid substitutions (Matthews, 1987).

Perhaps the most relevant example of the functional neutrality of random missense substitutions is found in a study carried out by Langridge and Campbell (1969). They measured the distribution of missense and nonsense mutants that are recovered from A-174 *lac* point mutants. Had these been randomly distributed according to the relative frequency of codons, the missense-to-nonsense ratio would be close to 20 to 1. What they observed is the reciprocal ratio; only 5% missense mutations were observed. This implies that roughly 1 in 400 missense mutations is sufficiently defective to be scored as defective as a nonsense mutant in this collection. In other words, most missense substitutions are, if not silent, pretty quiet.

Such observations provide the impetus to a conjecture that is central to our view of ribosome function. Thus, we have the impression that there is a disproportion between (i) the elaborate mechanism for the selection of the correct aminoacyl-tRNA by codon-programmed ribosomes and (ii) the relative indifference of protein structures to the occasional amino acid missense substitution. Thus, the selection mechanism may ensure more than that the correct amino acid is placed in the polypeptide sequence. Our conjecture is that it is at least equally important that the codon be properly matched with a tRNA species that can support the orderly movements that make up the most vulnerable aspect of the translation process. In particular, we suggest that the destructive consequences of choosing the wrong tRNA species may be greater than those following the insertion of the wrong amino acid. To explore this suggestion, we must first discuss other kinds of translation errors.

Dud Synthesis

Any accident that interrupts the orderly addition of the successive amino acids in a polypeptide chain before its completion is with overwhelming probability a lethal event for that chain. Thus, almost all incomplete fragments will be functionally inactive duds. It is the production of such duds that makes nonsense mutants so easily detectable. There are at least four distinguishable translation errors that can generate duds: (i) abortive termination, as for example when a sense codon is mistranslated as a termination codon; (ii) reading frame errors, which with very high probability activate an out-of-frame termination codon close to the shift; (iii) drop-off events in which the nascent polypeptide still bound to a tRNA species is lost from the working surface of the ribosome; and (iv) premature transcriptional termination or cleavage, leading to incomplete mRNA species.

We have no idea how often abortive termination occurs, nor do we have an estimate of the global frequency with which incomplete mRNA fragments are generated. Menninger (1976) has suggested that the drop-off frequencies are in the range of 10^{-4} to 10^{-3} per codon. Kurland (1979) has estimated a global reading frame error frequency of close to 3×10^{-5} per codon. Neither of these figures is very reliable, since both are based on exceedingly limited data and rest on questionable assumptions. Manley (1978) measured the accumulation of incomplete β-galactosidase fragments and estimated that 31% of the initiated chains ended prematurely. This corresponds to a dud frequency of 3.5×10^{-4} per codon. However, Manley cautioned that his estimate might be too low because fragments smaller than half of full length might have escaped detection and because it was difficult to estimate the influence of peptide turnover on the estimates of recovery of truncated products.

We have developed a new assay that permits us to estimate the total number of events that interrupt the processivity of the elongation process (Jörgensen and Kurland, in press). A virtue of this assay is that it does not depend on the recovery of duds; instead, it measures the attenuation of polypeptide synthesis as a function of the length of the mRNA. For example, it is possible to compare the amount of protein produced from a monomeric *lacZ* gene and that from a dimeric *lacZ* gene. The difference between the two represents the sum of the duds that were produced in going from the end of the monomer length to the end of the dimer. For wild-type bacteria, we find that of the order of one-fourth of the chains are never completed, and this corresponds to a dud frequency of close to 2.5×10^{-4} per codon. Approximately half of this attrition is due to premature termination of mRNA.

Such a loss of processivity is somewhat lower than might be expected from the data of Manley (1978). It is possible that the difference can be attributed to strain idiosyncracies. For example, we find that a strain carrying a mutation in the *rpsL* gene, a partially streptomycin-dependent mutant, produces duds at three times the rate of wild type. This means that only one-fourth of the β-galactosidase chains are completed by this strain in the absence of antibiotic; in the presence of antibiotic, the processivity is improved by a factor of two. Likewise, Tsung et al. (1989) report a loss of processivity of roughly 50% of all β-galactosidase starts in their strain, which contains a mutation in *rpsE*; unfortunately, they did not report any data for their wild-type strain.

In summary, we can measure the totality of accidents that interrupt the processing of a polypeptide chain. The frequency of all such events for a 500-amino-acid-long polypeptide corresponds to a loss of roughly 12% of all initiated chains. This is an impressive number because if the duds are produced randomly along the chain, it means that 6% of the total biochemical investment in translation is wasted in the production of duds. Translation by a rapidly growing bacterium represents more than half of the total mass and energy investment of the cell (Maaløe,

1979; Tempest and Neijssel, 1987). In other words, more than 3% of the cell activity is wasted in the production of duds.

Compare this with the consequences of a 10^{-3} missense rate. At such an error rate, 40% of the polypeptides with a length of 500 amino acids will contain at least one substitution. If we may generalize the conclusion of the study by Langridge and Campbell (1969) (see above), then we assume that only 1 in 400 missense events will generate an inactive product. This means that for a missense rate of 10^{-3}, roughly 0.1% of the polypeptides with a length of 500 amino acids are wasted, which corresponds to a 0.05% loss of the cellular resources. In other words, this comparison suggests that the consequences of missense substitutions are negligible compared with those following the errors of processivity.

The weighting factor of 400 upon which this comparison rests is obtained from the discrepancy between the expected and observed frequencies with which missense and nonsense mutants were recovered in a collection of *lac* mutants (Langridge and Campbell, 1969). Even if this weighting factor turns out to be highly biased, it is not likely to approach values smaller than 10. Therefore, it seems safe to suggest that the major physiological drain caused by ribosome malfunction in translation is that represented in the production of dud polypeptides. The next step of our argument is to associate malfunctions of tRNA in the production of dud polypeptides.

We note in passing that there is a fascinating class of frameshift events that we shall not discuss here. They represent a group of high-frequency events with distinctive sequence requirements that set them apart from the more mundane movement errors that occur during elongation (Jacks and Varmus, 1985; Craigen and Caskey, 1986; Weiss et al., 1987; Curran and Yarus, 1988). These are discussed in detail elsewhere in this volume.

The tRNA Link

The paradigmatic error of processivity is a reading frame error. Likewise, the first evidence that tRNA structure is important to processivity was the involvement of tRNA mutants in the suppression of frameshift mutants (Yourno and Heath, 1969; Riddle and Roth, 1970, 1972a, 1972b; Riddle and Carbon, 1973). Here, for example, it was observed that a +1 frameshift suppressor is a mutant tRNA with an extra C in its anticodon loop. The suppressor tRNA seems to use, at a significant frequency, the four C's in the anticodon in place of the normal three C's to read the extra base in the mRNA and return the system to the correct reading frame. Such observations led naturally to the idea that the orderly movement of mRNA relative to the ribosomal decoding site is guided in part by the codon-anticodon interaction. Indeed, it now became useful to view mRNA advance and tRNA translocation as coupled movements.

At about the same time, it also had been noticed that frameshift mutants in the *lacZ* gene are somewhat leaky; that is, that they are spontaneously suppressed at low levels (Gorini, 1971; Atkins et al., 1972). Furthermore, the spontaneous suppression rates of the frameshift mutants were reduced when restrictive alleles of *rpsL* were present and enhanced when error-prone alleles of *rpsD* were present (Atkins et al., 1972). These and other observations were used to support the interpretation that the accuracy of tRNA selection and mRNA movement are coupled to each other (Kurland, 1979). Here, the interpretation was that "when either the codon-anticodon interaction or the conformation of the tRNA does not correspond to some standard configuration, we expect the movement of the mRNA to be perturbed" (Kurland, 1979).

Convincing evidence has since appeared to confirm the two principal predictions of this view of mRNA movement: first, that codon-anticodon mismatching stimulates reading frame shifts has been verified in the case of amino acid starvation-induced missense errors by Weiss and Gallant (1983); and second, a large number of mutant tRNA species with alterations in the anticodon loop or with alterations well separated from the anticodon loop have been shown to influence the determination of reading frame (Murgola, 1985, in press; Buckingham and Grosjean, 1986; Curran and Yarus, 1987; Tucker et al., 1989).

It was suggested on the basis of mutant tRNA behavior that there is a built-in tendency of the translation system to move mRNA by three nucleotides even if three correctly base-paired nucleotide interactions are not formed (Kurland, 1979). The idea here is that the codon-tRNA complex forms a structure on the ribosome that provides the "measure" of the standard movement. According to this view, the size of the tRNA molecule is related to this function of guiding the movements of mRNA.

It is also relevant that there have been a number of experiments carried out to characterize the drop-off reaction and to analyze the effects of ribosome mutants as well as antibiotics on the frequency of this event (Menninger, 1976, 1978; Menninger et al., 1983). These data have been analyzed in some detail elsewhere (Kurland and Ehrenberg, 1985). At present, we are a little wary of these results because the drop-off estimates are based on indirect measurements that are very difficult to carry out. Furthermore, we have found in an analysis of the mode of action of erythromycin very clear evidence that this drug does not stimulate random drop-off of peptidyl-tRNA from ribosomes (Andersson and Kurland, 1987) as had been claimed earlier on the basis of the indirect drop-off assay (Menninger and Otto, 1982). It would be very useful if a direct assay of this event could be developed.

Moving Branch Points

It is difficult to believe that an accurate movement of the tRNA-mRNA complex occurs in a single leap from one ribosomal site to the next. It is easier to imagine that it occurs in a series of steps during which intermediate states are characterized by the retention of one contact with the first site while ligands are formed in the second site (Kurland and Ehrenberg, 1985, 1987). Furthermore, according to this view it is important that the partial movements of the tRNA-mRNA complex into the different binding state be reversible. In that case, the ease or difficulty of maintaining the intermediate states can be coupled to the quality of the movement. For example, an intermediate state corresponding to a four-nucleotide movement would be presumed to be much less stable than that for a three-nucleotide movement. The physical basis for this difference in stability would be found in the structure of the tRNA, which we assume adopts more stable conformations in response to a canonical movement than it does in reading frame shifts.

We assume that the stability of the intermediate states is accordingly expressed in the corresponding rate constants (Kurland and Ehrenberg, 1985, 1987). Furthermore, we recognize the configuration of a kinetic branch point in this flow scheme for the mRNA movement. In effect, the accuracy of the movement can be described in a manner that is completely analogous to the tRNA selections that we described above. The difference is that the correct and incorrect intermediate complexes now correspond to intermediate states in which the mRNA has moved three or some other number of nucleotides with respect to the ribosome.

An essential prediction of this view of the RNA movements on the ribosome is that there are intermediate states with hybrid binding sites for the mRNA-tRNA complexes (Kurland and Ehrenberg, 1985, 1987). The hybrid sites are expected to be made up, for example, of a 30S A-site contribution and a 50S P-site contribution. This prediction has been verified in detail by Moazed and Noller (1989b). However, there are important differences between the theoretical expectations and the experimental observations.

First, our kinetic model for the mRNA movements was formulated within the framework of a model of a two-tRNA-site ribosome. It followed from this that we would expect there to be only one hybrid site. In contrast, Moazed and Noller (1989a, 1989b) have observed two such sites: 30S A plus 50S P sites in one configuration and 30S P plus 50S E sites in another configuration. Second, we had anticipated that these hybrid configurations for tRNA binding might be unstable transient intermediates. However, the hybrid binding sites observed by Moazed and Noller (1989a, 1989b) are quite stable.

Their flow scheme is as follows: after the tRNA is accepted at the A site, formation of the peptide bond after the exit of EF-Tu from the ribosome is accompanied by a shift of tRNA to a hybrid site made up of 50S P-site and 30S A-site contributions. In this model the movement of the mRNA-tRNA complex bound at the 30S A site to a 30S P site is mediated by EF-G, and this leaves the complex in the P site. There is much more in the relevant data, but the essential point is that the movement of the mRNA is incremental. Therefore, the kinetic schemes that we have introduced earlier to represent mRNA movement seem to be worth pursuing.

One of the surprising details in the scheme described by Moazed and Noller (1989a, 1989b) is the movement of the tRNA into a hybrid state consisting of a 50S P site and a 30S A site after the release of EF-Tu · GDP from the ribosome and associated with the formation of the peptide bond. It is possible that this movement is mediated by an EF-Tu · GTP cycle that is distinct from the EF-Tu · GTP cycle associated with the proofreading of the tRNA. This suggestion follows from the observations indicating that two EF-Tu · GTP complexes may accompany a cognate aminoacyl-tRNA into the ribosome (Ehrenberg et al., 1990). Thus, the unexpected complexity of the EF-Tu stoichiometry in translation may have a functional counterpart in the newly discovered richness of ribosomal binding states for tRNA.

EF-G Emerges

As we have seen, according to Moazed and Noller (1989a), the movement of the mRNA-tRNA complex from the 30S A site to the 30S P site is mediated by EF-G. We assume that this rearrangement corresponds to the movement of the mRNA relative to the decoding site. We naturally wish to identify the role, if any, that EF-G and its GTP reaction might play in determining the accuracy of this step.

We have shown that during polypeptide elongation the maximal turnover rate of EF-G is comparable to that of EF-Tu (Bilgin et al., 1988), with the consequence that both the tRNA selection steps associated with EF-Tu and the translocation steps associated with EF-G appear to take roughly the same minimum times on the ribosome. Furthermore, we have identified domains in the 30S ribosomal subunit that are likely to be associated with its functions in translation. Thus, spectinomycin inhibits EF-G function during translation, and mutations in S5 as well as at position 1192 in 16S RNA provide varying degrees of relief for this inhibition (Bilgin et al., 1990). Since S5 is also known to exist in mutant forms that influence the proofreading flows of EF-Tu (Andersson et al., 1986a), this protein is associated with both of the factor-related selection steps on the 30S subunit.

As discussed above, the movements of mRNA are conceivably governed by the differential flows over virtual branch points in a way that is reminiscent of tRNA selections. Likewise, mRNA movement could be mediated by a protein factor that is associated with a GTP reaction precisely as in the case for the tRNA selection. Accordingly, it was suggested that the EF-G-dependent GTP reaction is not merely associated with the translocation of tRNA movement, which is the conventional view. Rather, the conjecture was that this factor cycle supports the accuracy of mRNA movement, conceivably by a proofreading mechanism (Kurland and Ehrenberg, 1985, 1987).

That a proofreading function would be advantageous is motivated by the idea that the accuracy of the mRNA movement depends on the accuracy and stability of the codon-anticodon interaction. If proofreading is required to support the accuracy of tRNA selection, it seems reasonable to suppose that it would also be required to edit transient misalignments of the codon-anticodon interaction during transitions of the RNAs from one site to another on the ribosome. Indeed, detailed kinetic schemes for such reactions can be described, and they have provided the impetus to search for mutants of EF-G with which to test these ideas.

Recently, a mutant of EF-G has been identified that has a restrictive influence on frameshift leakiness: it lowers by a factor of 3 the spontaneous suppression rate for several different reading frame mutants in *lacZ* (A. A. Richter and C. G. Kurland, submitted for publication). This mutant provides the first direct evidence that the accuracy of mRNA movement is influenced by EF-G function. Furthermore, the mutant factor shows characteristic kinetic changes in vitro. We therefore are encouraged by these results to pursue our ideas about mRNA movement.

REFERENCES

Allen, P. A., and H. F. Noller. 1989. Mutations in ribosomal proteins S4 and S12 influence the higher order structure of 16S ribosomal RNA. *J. Mol. Biol.* **208:**457–468.

Andersson, D. I., S. G. E. Andersson, and C. G. Kurland. 1986a. Functional interactions between mutated forms of ribosomal protein S4, S5 and S12. *Biochimie* **68:**705–713.

Andersson, D. I., and C. G. Kurland. 1983. Ram ribosomes are defective proofreaders. *Mol. Gen. Genet.* **191:**378–381.

Andersson, D. I., H. W. van Verseveld, A. H. Stouthammer, and C. G. Kurland. 1986b. Suboptimal growth with hyperaccurate ribosomes. *Arch. Microbiol.* **144:**96–101.

Andersson, S. G. E., and C. G. Kurland. 1987. Elongating ribosomes *in vivo* are refractory to erythromycin. *Biochimie* **69:** 901–904.

Atkins, J. F., D. Elseviers, and L. Gorini. 1972. Low activity of β-galactosidase in frameshift mutants of *Escherichia coli. Proc. Natl. Acad. Sci. USA* **69:**1192–1195.

Bilgin, N., L. A. Kirsebom, M. Ehrenberg, and C. G. Kurland. 1988. Mutations in ribosomal proteins L7/L12 perturb EF-G and EF-Tu functions. *Biochimie* **70:**611–618.

Bilgin, N., A. A. Richter, M. Ehrenberg, A. E. Dahlberg, and C. G. Kurland. 1990. Ribosomal RNA and protein mutants resistant to spectinomycin. *EMBO J.* **9:**735–739.

Birge, E. A., and C. G. Kurland. 1969. Altered ribosomal protein in streptomycin-dependent *Escherichia coli. Science* **166:**1282–1284.

Birge, E. A., and C. G. Kurland. 1970. Reversion of a streptomycin-dependent strain of *Escherichia coli. Mol. Gen. Genet.* **109:**356–369.

Bohman, K. T., T. Ruusala, P. C. Jelenc, and C. G. Kurland. 1984. Kinetic impairment of restrictive streptomycin resistant ribosomes. *Mol. Gen. Genet.* **198:**90–99.

Bollen, A., T. Cabezón, M. de Wilde, R. Villarroel, and A. Herzog. 1975. Alteration of ribosomal protein S17 by mutation linked to neamine resistance in Escherichia coli. *J. Mol. Biol.* **99:**795–806.

Bouadloun, F., D. Donner, and C. G. Kurland. 1983. Codon-specific missense errors *in vitro. EMBO J.* **2:**1351–1356.

Brimacombe, R., P. Maly, and C. Zweib. 1983. The structure of ribosomal RNA and its organization relative to ribosomal protein. *Prog. Nucleic Acid Res. Mol. Biol.* **28:**1–48.

Buckingham, R. H., and H. Grosjean. 1986. The accuracy of messenger RNA-transfer RNA recognition, p. 83–126. *In* T. B. L. Kirkwood, R. F. Rosenberger, and D. J. Galas (ed.), *Accuracy in Molecular Processes.* Chapman & Hall, Ltd., London.

Churchward, G., H. Bremer, and R. Young. 1982. Macromolecular composition of bacteria. *J. Theor. Biol.* **94:**651–670.

Churchward, G., E. Estiva, and H. Bremer. 1981. Growth rate-dependent control of chromosome replication initiation in *Escherichia coli. J. Bacteriol.* **145:**1232–1238.

Craigen, W. J., and C. T. Caskey. 1986. Expression of peptide chain release factor 2 requires high-efficiency frameshift. *Nature* (London) **322:**273–275.

Crick, F. H. C. 1958. On protein synthesis. *Symp. Soc. Exp. Biol.* **12:**138–163.

Curran, J. F., and M. Yarus. 1987. Reading frame selection and transfer RNA anticodon loop stacking. *Science* **238:**1545–1550.

Curran, J. F., and M. Yarus. 1988. Use of tRNA suppressors to probe regulation of *Escherichia coli* release factor 2. *J. Mol. Biol.* **203:**75–83.

Deuser, E., G. Stöffler, H. G. Wittmann, and G. Apirion. 1970.

Altered S4 proteins in *Escherichia coli* mutants from streptomycin dependence to independence. *Mol. Gen. Genet.* **109**:298–302.

Dix, B. D., and R. C. Thompson. 1986. Elongation factor Tu guanosine 3′-diphosphate 5′-diphosphate complex increases the fidelity of proofreading in protein biosynthesis: mechanism for reducing translational errors introduced by amino acid starvation. *Proc. Natl. Acad. Sci. USA* **83**:2027–2031.

Edelman, P., and J. Gallant. 1977. Mistranslation in *Escherichia coli*. *Cell* **10**:131–137.

Ehrenberg, M., and C. Blomberg. 1980. Thermodynamic constraints on kinetic proofreading in biosynthetic pathways. *Biophys. J.* **31**:333–358.

Ehrenberg, M., and C. G. Kurland. 1984. Cost of accuracy determined by a maximal growth rate constant. *Q. Rev. Biophys.* **17**:45–82.

Ehrenberg, M., A.-M. Rojas, J. Weiser, and C. G. Kurland. 1990. How many EF-Tu molecules participate in aminoacyl-tRNA binding and peptide bond formation in *Escherichia coli* translation. *J. Mol. Biol.* **211**:739–749.

Ellis, N., and J. Gallant. 1982. An estimate of the global error frequency in translation. *Mol. Gen. Genet.* **188**:169–172.

Fast, R., T. H. Eberhard, T. Ruusala, and C. G. Kurland. 1987. Does streptomycin cause an error catastrophe? *Biochimie* **69**: 131–136.

Fersht, A. 1977. *Enzyme Structure and Mechanism*, p. 91. W. H. Freeman & Co., San Francisco.

Gallant, J. A. 1979. Stringent control in *E. coli*. *Annu. Rev. Genet.* **13**:393–415.

Garvin, R. T., D. K. Biswas, and L. Gorini. 1974. The effects of streptomycin or dihydrostreptomycin to 16S RNA or to 30S ribosomal subunits. *Proc. Natl. Acad. Sci. USA* **71**:3814–3818.

Gorini, L. 1971. Ribosomal discrimination of tRNAs. *Nature* (London) *New Biol.* **234**:261–264.

Gorini, L., and E. Kataja. 1964a. Phenotypic repair by streptomycin of defective genotypes in *Escherichia coli*. *Proc. Natl. Acad. Sci. USA* **51**:487–493.

Gorini, L., and E. Kataja. 1964b. Streptomycin-induced oversuppression in *Escherichia coli*. *Proc. Natl. Acad. Sci. USA* **51**: 995–1001.

Hoffman, G. W. 1974. On the origin of the genetic code and the stability of transporation apparatus. *J. Mol. Biol.* **80**:349–362.

Hopfield, J. J. 1974. Kinetic proofreading: a new mechanism for reduced errors in biosynthetic processes requiring high specificity. *Proc. Natl. Acad. Sci. USA* **71**:4135–4139.

Jacks, T., and H. E. Varmus. 1985. Expression of Rous sarcoma virus pol gene by ribosomal frameshifting. *Science* **230**:1237.

Jelenc, P. C., and C. G. Kurland. 1979. Nucleoside triphosphate regeneration decreases the frequency of translation errors. *Proc. Natl. Acad. Sci. USA* **76**:3174–3178.

Jörgensen, F., and C. G. Kurland. 1987. Death rates of bacterial mutants. *FEMS Microbiol. Lett.* **40**:43–46.

Jörgensen, F., and C. G. Kurland. Processivity errors of gene expression in *E. coli*. *J. Mol. Biol.*, in press.

Kirkwood, T. B. L., and R. Holliday. 1975. The stability of the translation process. *J. Mol. Biol.* **97**:257–265.

Kirkwood, T. B. L., R. Holliday, and R. F. Rosenberger. 1984. Stability of the cellular translation process. *Int. Rev. Cytol.* **92**:93–132.

Kirsebom, L. A., and L. A. Isaksson. 1985. Involvement of ribosomal protein L7/L12 in control of translation accuracy. *Proc. Natl. Acad. Sci. USA* **82**:717–721.

Kühberger, R., W. Piepersberg, A. Petzet, P. Buckel, and A. Böck. 1979. Alteration of ribosomal protein L6 in gentamicin-resistant strains of *Escherichia coli*. Effects on fidelity of protein synthesis. *Biochemistry* **18**:187–193.

Kurland, C. G. 1974. Functional organization of the 30S ribosomal subunit, p. 309–321. *In* M. Nomura, A. Tessières, and P. Lengyel (ed.), *Ribosomes*. Cold Spring Harbor Laboratory, Cold Spring Harbor, N.Y.

Kurland, C. G. 1979. Reading frame errors on ribosomes, p. 97–108. *In* J. E. Celis and J. D. Smith (ed.), *Nonsense Mutations and tRNA Suppressors*. Academic Press, Inc., New York.

Kurland, C. G. 1987. Strategies for efficiency and accuracy in gene expression. *Trends Biochem. Sci.* **12**:126–128.

Kurland, C. G. 1988. The screenless screen, p. 89–96. *In* M. Bissell, G. Dehò, G. Sironi, and A. Torriani (ed.), *Gene Expression and Regulation: The Legacy of Luigi Gorini*. Elsevier Science Publishers, Amsterdam.

Kurland, C. G., and M. Ehrenberg. 1985. Constraints on the accuracy of messenger RNA movement. *Q. Rev. Biophys.* **18**:423–450.

Kurland, C. G., and M. Ehrenberg. 1987. Growth-optimizing accuracy of gene expression. *Annu. Rev. Biophys.* **16**:291–317.

Kurland, C. G., R. Rigler, M. Ehrenberg, and C. Blomberg. 1976. Allosteric mechanism for codon-dependent tRNA selection on ribosomes. *Proc. Natl. Acad. Sci. USA* **72**:4248–4251.

Langridge, J., and J. H. Campbell. 1969. Classification and intragenic position of mutations in the β-galactosidase gene of Escherichia coli. *Mol. Gen. Genet.* **103**:339–347.

Loftfield, R. 1963. The frequency of errors in protein synthesis. *Biochem. J.* **89**:82–92.

Loftfield, R., and D. Vanderjagt. 1972. The frequency of errors in protein synthesis. *Biochem. J.* **128**:1353–1356.

Maaløe, O. 1979. Regulation of the protein-synthesizing machinery—ribosomes, tRNA, factors, and so on, p. 487–542. *In* R. F. Goldberger (ed.), *Biological Regulation and Development*. Plenum Publishing Corp., New York.

Manley, J. L. 1978. Synthesis and degradation of termination and premature termination fragments of β-galactosidase in vitro. *J. Mol. Biol.* **125**:407–432.

Matthews, B. W. 1987. Structural basis of protein stability and DNA-protein interaction. *Harvey Lect.* **81**:33–51.

Menninger, J. R. 1976. Peptidyl transfer RNA dissociates during protein synthesis from ribosomes of *Escherichia coli*. *J. Biol. Chem.* **251**:3392–3398.

Menninger, J. R. 1978. The accumulation as peptidyl-transfer RNA of isoaccepting transfer RNA families in *Escherichia coli* with temperature-sensitive peptidyl-transfer RNA hydrolase. *J. Biol. Chem.* **253**:6808–6813.

Menninger, J. R., A. B. Caplan, P. K. E. Gingrich, and A. G. Atherly. 1983. Tests of the ribosome editor hypothesis. II. Relaxed (relA) and stringent (relA+) *E. coli* differ in rates of dissociation of peptidyl-tRNA from ribosomes. *Mol. Gen. Genet.* **190**:215–221.

Menninger, J. R., and D. P. Otto. 1982. Erthromycin, carbomycin, and spiramycin inhibit protein synthesis by stimulating the dissociation of peptidyl-tRNA from ribosomes. *Antimicrob. Agents Chemother.* **21**:811–818.

Mikkola, R., and C. G. Kurland. 1988. Media dependence of translational mutant phenotype. *FEMS Microbiol. Lett.* **56**: 265–270.

Miller, J. H., C. Coulondre, M. Hofer, U. Schmeissner, H. Sommer, A. Schmitz, and P. Lu. 1979. Genetic studies of the lac repressor. *J. Mol. Biol.* **131**:191–222.

Moazed, D., and H. F. Noller. 1987. Interaction of antibiotics with functional sites in 16S ribosomal RNA. *Nature* (London) **327**: 389–394.

Moazed, D., and H. F. Noller. 1989a. Interaction of tRNA with 23S rRNA in the ribosomal A, P and E sites. *Cell* **57**:585–597.

Moazed, D., and H. F. Noller. 1989b. Intermediate states in the movement of tRNA in the ribosome. *Nature* (London) **342**:

142–148.

Montandon, P. E., R. Wagner, and E. Stutz. 1986. *E. coli* ribosomes with a C912 to U base change in the 16S rRNA are streptomycin resistant. *EMBO J.* **5:**3705–3708.

Moore, P. B. 1988. The ribosome returns. *Nature* (London) **331:**223–227.

Murgola, E. J. 1985. tRNA, suppression and the code. *Annu. Rev. Genet.* **19:**57–80.

Murgola, E. J. Mutant glycine tRNAs and other wonders of translational suppression. *In* J. D. Cherayil (ed.), *Transfer RNAs and Other Soluble RNAs*. CRC Press, Inc., Boca Raton, Fla., in press.

Neidhardt, F. C., P. L. Bloch, S. Pedersen, and S. Reeh. 1977. Chemical measurement of steady-state levels of ten aminoacyl-transfer ribonucleic acid synthetases in *Escherichia coli. J. Bacteriol.* **129:**378–387.

Nene, V., and R. E. Glass. 1983. Relaxed mutants of *Escherichia coli* RNA polymerase. *FEBS Lett.* **153:**307–310.

Ninio, J. 1974. A semiquantitative treatment of missense and nonsense suppression in the strA and ram ribosomal mutants of *Escherichia coli*. Evaluation of some molecular parameters of translation *in vitro*. *J. Mol. Biol.* **84:**297–313.

Ninio, J. 1975. Kinetic amplification of enzyme discrimination. *Biochimie* **57:**587–595.

Noller, H. F. 1984. Structure of ribosomal RNA. *Annu. Rev. Biochem.* **53:**119–162.

Noller, H. F., and J. B. Chaires. 1972. Functional modification of 16S RNA by kethoxal. *Proc. Natl. Acad. Sci. USA* **69:**3115–3118.

Noller, H. F., C. Chang, G. Thomas, and J. Aldridge. 1971. Chemical modification of the transfer RNA and polyuridylic acid binding site of *Escherichia coli* 30S ribosomal subunits. *J. Mol. Biol.* **61:**669–679.

Orgel, L. E. 1963. The maintenance of the accuracy of protein synthesis and its relevance to ageing. *Proc. Natl. Acad. Sci. USA* **49:**517–521.

Orgel, L. E. 1970. The maintenance of the accuracy of protein synthesis and its relevance to ageing: a correction. *Proc. Natl. Acad. Sci. USA* **67:**1476.

Orgel, L. E. 1973. Ageing of clones of mammalian cells. *Nature* (London) **243:**441–445.

Ozaki, M., S. Mizushima, and M. Nomura. 1969. Identification and functional characterization of the protein controlled by the streptomycin resistant locus in *Escherichia coli*. *Nature* (London) **222:**333–339.

Parker, J., and G. Holtz. 1984. Controle of basal-level codon misreading in *Escherichia coli*. *Biochem. Biophys. Res. Commun.* **121:**487–492.

Parker, J., T. C. Johnston, P. T. Borgia, G. Holtz, E. Remaut, and W. Fiers. 1983. Codon usage and mistranslation. *J. Biol. Chem.* **258:**10007–10012.

Pedersen, S. 1984. *Escherichia coli* ribosomes translate *in vivo* with variable rates. *EMBO J.* **3:**2895–2898.

Riddle, D. L., and J. Carbon. 1973. Frameshift suppression: a nucleotide addition in the anticodon of a glycine transfer RNA. *Nature* (London) *New Biol.* **242:**230–234.

Riddle, D. L., and J. R. Roth. 1970. Suppression of frameshift mutations in *Salmonella typhimurium*. *J. Mol. Biol.* **54:**131–144.

Riddle, D. L., and J. R. Roth. 1972a. Frameshift suppressors. *J. Mol. Biol.* **66:**483–493.

Riddle, D. L., and J. R. Roth. 1972b. Frameshift suppressors. *J. Mol. Biol.* **66:**495–506.

Rojas, A.-M. 1988. Kinetic modulation of accuracy in protein synthesis by ppGpp. Ph.D. thesis. Uppsala University, Uppsala, Sweden.

Ruusala, T., D. I. Andersson, M. Ehrenberg, and C. G. Kurland. 1984. Hyperaccurate ribosomes inhibit growth. *EMBO J.* **3:** 2575–2580.

Ruusala, T., M. Ehrenberg, and C. G. Kurland. 1982. Is there proofreading during polypeptide synthesis? *EMBO J.* **1:**741–745.

Ruusala, T., and C. G. Kurland. 1984. Streptomycin preferentially perturbs ribosomal proofreading. *Mol. Gen. Genet.* **198:**100–104.

Ryals, J., R. Little, and H. Bremer. 1982. Control of rRNA and tRNA synthesis in *Escherichia coli* by guanosine tetraphosphate. *J. Bacteriol.* **151:**1261–1268.

Smith, D., and M. Yarus. 1989. A D-arm mutation changes tRNA coding specificity through alteration of kinetic proofreading by the ribosome. *J. Mol. Biol.* **206:**489–501.

Smith, J. D. 1979. Suppressor tRNAs in prokaryotes, p. 109–126. *In* J. E. Celis and J. D. Smith (ed.), *Nonsense Mutations and tRNA Suppressors*. Academic Press, Inc., New York.

Tempest, D. W., and O. M. Neijssel. 1987. Growth yield and energy distribution, p. 797–806. *In* F. C. Neidhardt, J. L. Ingraham, K. B. Low, B. Magasanik, M. Schaechter, and H. E. Umbarger (ed.), *Escherichia coli and Salmonella typhimurium: Cellular and Molecular Biology*. American Society for Microbiology, Washington, D.C.

Thompson, R. C., and P. J. Stone. 1977. Proofreading of the codon-anticodon interaction on ribosomes. *Proc. Natl. Acad. Sci. USA* **74:**198–202.

Tsung, K., S. Inouye, and M. Inouye. 1989. Factors affecting the efficiency of protein synthesis in *Escherichia coli*. *J. Biol. Chem.* **264:**4428–4433.

Tucker, S. D., E. J. Murgola, and F. T. Pagel. 1989. Missense and nonsense suppressors can correct frameshift mutations. *Biochimie* **71:**729–739.

Wagner, E. G. H., P. C. Jelenc, M. Ehrenberg, and C. G. Kurland. 1982. Rates of elongation of polyphenylalanine *in vitro*. *Eur. J. Biochem.* **122:**193–197.

Weiss, R., and J. Gallant. 1983. Mechanism of ribosome frameshifting during translation of the genetic code. *Nature* (London) **302:**389–393.

Weiss, R. B., D. M. Dunn, J. F. Atkins, and R. F. Gesteland. 1987. Slippery runs, shifty stops, backward steps, and forward hops: −2, −1, +1, +2, +5, and +6 ribosomal frameshifting. *Cold Spring Harbor Symp. Quant. Biol.* **52:**687–693.

Woese, C. R., R. R. Gutell, R. Gupta, and H. F. Noller. 1983. Detailed analysis of the higher-order structure of 16S-like ribosomal ribonucleic acids. *Microbiol. Rev.* **47:**621–669.

Yarus, M. 1982. Translational efficiency of transfer RNA's. *Science* **218:**646–652.

Yarus, M., and R. C. Thompson. 1983. Precision of protein biosynthesis, p. 23–63. *In* J. Beckwith, J. Davies, and J. A. Gallant (ed.), *Gene Function in Procaryotes*. Cold Spring Harbor Laboratory, Cold Spring Harbor, N.Y.

Yourno, J., and S. Heath. 1969. Nature of the *hisD3015* frameshift mutation in *Salmonella typhimurium*. *J. Bacteriol.* **100:** 460–468.

Zimmermann, R. A., T. Garvin, and L. Gorini. 1971. Alterations of a 30S ribosomal protein accompanying the *ram* mutation in *Escherichia coli*. *Proc. Natl. Acad. Sci. USA* **68:**2263–2267.

Chapter 45

Codon Choice and Gene Expression: Synonymous Codons Differ in Translational Efficiency and Translational Accuracy

DANIEL B. DIX, LINDA K. THOMAS, and ROBERT C. THOMPSON

The binding of aminoacyl-tRNA (aa-tRNA) to mRNA-programmed ribosomes is determined largely by the interaction between the codon on the mRNA and the anticodon on the tRNA. This binding reaction is believed to have great significance in determining the rate and the fidelity of aa-tRNA incorporation into protein. Both are important issues for the *Escherichia coli* cell, where approximately one-half of the macromolecular mass of the cell is devoted to the protein biosynthesis machinery.

The genetic code is highly degenerate; in the extreme case, the incorporation of serine, arginine, or leucine may be directed by six different codons. As genes from procaryotes and eucaryotes have been sequenced, it has become apparent that synonymous codons for different genes are not used with equal frequencies. In all organisms for which there are sufficient data, the translational apparatus shows a significant bias in the codons used in highly expressed genes (de Boer and Kastelein, 1986; Grosjean and Fiers, 1982; Post et al., 1979). Indeed, the extent of codon bias increases with the level of gene expression (Bennetzen and Hall, 1982; Gouy and Gautier, 1982; Ikemura, 1981). Attempts to explain the existence of codon bias have focused on its potential for modulating gene expression (Holm, 1986). For this potential to be realized, codons should vary in their speed or accuracy of translation.

CODON CHOICE AND TRANSLATION EFFICIENCY IN VIVO

Proof that the choice of codons can affect the rate of translation was provided by Pedersen (1984), who measured the translation times of mRNAs rich in rare codons (*lacI* and *bla*) and mRNAs that contain few rare codons (*fus*, *tsf*, *tuf*, and *rpsA*). When both sets of genes were expressed at high levels, the mRNAs that contain the rare codons were found to be translated about 50% the rate of those containing few such codons. Further evidence for an effect of codon choice of translation has been provided by Varenne et al. (1982), who found that pauses in the synthesis of colicin E1 in vivo occur in regions of the mRNA that contain rare codons.

Codon choice not only affects the rate of translation but also changes the level of gene expression. Robinson et al. (1984) inserted two oligonucleotides into the chloramphenicol acetyltransferase (CAT) structural gene, one encoding four arginine codons (CGU) favored in highly expressed genes and the other with four rarely used arginine codons (AGG). When maximally induced, the CAT-derived gene containing the CGU codons produced three times more CAT protein than did the AGG-containing gene. Perhaps the most convincing evidence for whether biased codon usage affects gene expression levels has been provided by Hoekema et al. (1987). They studied the effects of codon usage in the highly expressed phosphoglycerate kinase gene in *S. cerevisiae*. When 39% of the codons were replaced with synonymous codons that are never found in highly expressed genes, a 10-fold reduction in phosphoglycerate kinase protein levels was observed.

Codon choice can also influence the regulation of gene expression through transcription attenuation mechanisms. Expression of these operons is regulated by the rate of translation of specific codons in the leader regions. When rarely used codons in the leader regions of the leucine operon from *Salmonella typhimurium* (Carter et al., 1986) or *ilv* operon of *Serratia marcescens* (Harms and Umbarger, 1987) were re-

Daniel B. Dix, Linda K. Thomas, and Robert C. Thompson ■ Molecular, Cellular, and Developmental Biology, University of Colorado, Boulder, Colorado 80309.

placed by synonymous frequently used codons, changes were observed in the regulation of the leucine operon. When three of four control leucine codons in the leader mRNA of the former operon were changed from CUA to CUG, the basal level of operon expression was reduced and a higher degree of starvation was required to elicit a starvation response (Carter et al., 1986). The loss of regulation was attributed to ribosomes traversing the leucine leader mRNA faster when it contains CUG codons, since the concentration of Leu-tRNAs that can read the CUG codon is believed to be greater than the concentration of tRNAs able to read CUA codons. A similar result was obtained when the rare CUA codon was changed to the common CUG leucine codon in the *ilv* operon (Harms and Umbarger, 1987). Although this change should not affect the secondary structure proposed for the leader mRNA, the presence of the CUG codon reduced repression of the *ilv* operon by approximately 70%. The relative translational efficiency of individual codons has also been investigated using the *pyrE* attenuator (Bonekamp et al., 1985). When three rare (AGG) arginine codons were replaced with frequently used CGT arginine codons, transcriptional readthrough increased eightfold after all three arginine substitutions were made.

All of the data discussed above are consistent with the hypothesis that codons commonly used in highly expressed genes are translated more rapidly. Sorensen et al. (1989) have recently proved this by measuring transit times after inserting a small DNA fragment from the *rpsA* gene into the *lacZ* gene. One 72-base-pair fragment is translated in a frame whose sequences are primarily composed of commonly used codons, whereas a second 57-base-pair fragment is frameshifted by +1 base, which produces a sequence primarily containing infrequently used codons. The mRNA containing the uncommon codons slowed β-galactosidase translation by 3 s (approximately 4%). Since the insert containing the uncommon codons has a large number of codons that call for the *mia*-modified tRNA (position 37 of the tRNA), the experiments were also done in a *miaA* background to amplify the differences between the two mRNAs. In the *miaA* background, the mRNA containing the uncommon codons was observed to retard translation of β-galactosidase by 17 s (approximately 14%) relative to the mRNA containing common codons.

CODON CHOICE AND TRANSLATIONAL ACCURACY IN VIVO

Codon choice has also been suggested to affect translational accuracy. The best in vivo evidence that some inherent specificity exists in the translational apparatus has been provided by Parker and colleagues, who studied mistranslation of AAU and AAC codons during asparagine starvation by using *E. coli* that had been infected with the RNA virus MS2 (Parker et al., 1980; Parker et al., 1983). This synonymous codon pair is of particular interest since in highly expressed genes, the AAC codon is used 13 times more frequently than the AAU codon to code for asparagine. By sequencing the mistranslated protein, it was determined that Lys-tRNA (anticodon UUU) was incorporated at AAU codons eight times more frequently than at AAC codons. However, Parker and Precup (1986) have also presented a possible counterexample for the importance of accuracy in determining codon choice. They examined the misincorporation of Leu-tRNA at the synonymous phenylalanine codons UUU and UUC in the *argI* gene of *E. coli* under conditions of phenylalanine starvation. After sequencing the N terminus of ornithine transcarbamylase, they observed that the UUC codon incorporated more leucine than the UUU codon, which is used only one-third as often in highly expressed genes. Thus, although in both synonymous codon pairs one of the codons is more accurate in aa-tRNA selection, in the first example the highly used codon is more accurate while in the second the highly used codon is less accurate.

CODON CHOICE AFFECTS TRANSLATIONAL EFFICIENCY AND TRANSLATIONAL ACCURACY IN VITRO

Use of Synthetic mRNAs

Interpretation of the in vivo results obtained by using natural mRNAs is generally quite complex, since the codons studied are in different contexts and the products may be selectively subject to degradation (Menninger, 1977). To overcome these limitations, we have recently modified an in vitro experimental system, which was originally designed to measure the rate of reaction of poly(U)-programmed ribosomes with aa-tRNA ternary complexes, to determine the rate of reaction of ribosomes programmed with defined, synonymous codons with aa-tRNA ternary complexes. This has allowed us to determine the rate and accuracy of translation of synonymous codons in a system not subject to the complications of different contexts and product degradation and capable of assigning any differences observed to individual steps of translation. This modified in vitro translation system also enables one to examine the rate and accuracy of translation at a

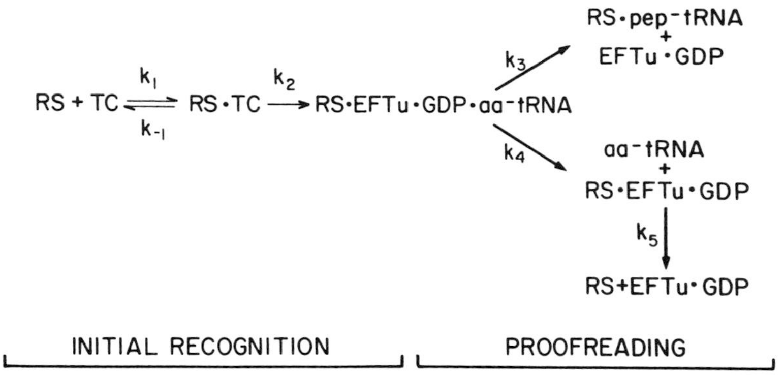

$$RS + TC \underset{k_{-1}}{\overset{k_1}{\rightleftharpoons}} RS\cdot TC \xrightarrow{k_2} RS\cdot EFTu\cdot GDP\cdot aa\text{-}tRNA$$

$$\xrightarrow{k_3} RS\cdot pep\text{-}tRNA + EFTu\cdot GDP$$

$$\xrightarrow{k_4} aa\text{-}tRNA + RS\cdot EFTu\cdot GDP \xrightarrow{k_5} RS + EFTu\cdot GDP$$

INITIAL RECOGNITION | PROOFREADING

Figure 1. Mechanism for aa-tRNA binding to ribosomes. RS, Ribosomes programmed with mRNA and with fMet-tRNA in the P site; TC, ternary complex of EF-Tu · GTP · aa-tRNA.

greater number of codons than presently available using in vivo mRNAs. The accuracy of synonymous codon pairs can also be examined with a greater number of near-cognate aa-tRNAs, providing assurance that one of the codons is inherently more accurate.

A complex mechanism exists for the selection of an aa-tRNA and involves two independent discrimination steps (Hopfield, 1974; Thompson and Stone, 1977; Thompson and Dix, 1982; Thompson et al., 1981; Ruusala et al., 1982). A simplified version of the mechanism is shown in Fig. 1, where RS denotes the ribosome programmed with mRNA and with peptidyl-tRNA in the P site and TC denotes a ternary complex composed of elongation factor EF-Tu, GTP, and aa-tRNA. The first discrimination step, called initial recognition, follows the binding of an EF-Tu · GTP · aa-tRNA ternary complex to the programmed ribosome (k_1). Either this ternary complex may be released from the ribosomes (k_{-1}) or it may be accepted and its GTP hydrolyzed (k_2). Cognate ternary complexes would be accepted, while near-cognate ternary complexes would preferentially be released. For those ternary complexes that are accepted, there is a second discrimination step, called proofreading. In this step, the aa-tRNA may be either released from the ribosome (k_4) or incorporated into peptide (k_3). As in initial discrimination, cognate aa-tRNAs would be accepted and near-cognate aa-tRNAs would preferentially be released.

The synthetic mRNAs used to study codon choice are 52 nucleotides long and are identical except for the three bases that immediately follow the translational initiation codon. The mRNAs were prepared as described by Thomas et al. (1988) and were designed to have the general sequence 5′-GG GAGACCGGAAGCUUGGGCUGCAGGAGGAU UUAAUCAUGXYZAAGAUCUCG-3′, where XYZ is the codon to be located in the A site by interactions between the ribosome and the Shine-Dalgarno sequence GGAGGA and between the initiator tRNA and the AUG codon. In the future, these mRNAs will be abbreviated to indicate the codon to be translated. The activity of these synthetic mRNAs was tested first by checking their ability to properly initiate protein synthesis and second by showing that they direct the specific binding of ternary complex to the second codon. These mRNAs were able to direct initiation, as indicated by fMet-tRNA binding to ribosomes, and this binding was dependent on the presence of both initiation factors and mRNA (Table 1). These mRNAs were also able to direct binding of a cognate ternary complex (Table 2). When initiated ribosomes containing the UUU or UUC codons were mixed with EF-Tu · GTP · Phe-tRNA, Phe-tRNA was observed to bind to the ribosomes with an approximate molar equivalent of GTP hydrolyzed.

Translational Efficiency: Kinetic Studies Involving Synthetic mRNAs and Cognate aa-tRNAs

To measure differences in the translational efficiency between synonymous codons, Phe-tRNA(GAA) ternary complexes were reacted with AUGUUU- and AUGUUC-programmed ribosomes initiated with fMet-tRNA, and Leu-tRNA$_2^{Leu}$ (GAG) ternary complexes were reacted with AUGCUC- and AUGCUU-

Table 1. Dependence of initiation complex formation on initiation factors and mRNA[a]

Reactants	mRNA addition	[^{3}H]fMet-tRNA bound (pmol)
RS, fMet-tRNA, GTP		0.2
RS, fMet-tRNA, GTP + IFs		0.06
RS, fMet-tRNA, GTP + IFs	AUGUUU	1.4
RS, fMet-tRNA, GTP + IFs	AUGUUC	1.1

[a] Data are taken from Thomas et al. (1988). IFs, Initiation factors.

Table 2. Ability of mRNAs to direct binding of cognate aa-tRNA ternary complexes to the ribosomal A site[a]

mRNA	aa-tRNA	GTP hydrolyzed (pmol)	aa-tRNA bound (pmol)
Poly(U)	Phe	1.2	0.9
AUGUUU	Phe	0.66	0.6
AUGUUC	Phe	1.66	1.4

[a] Ribosomes were programmed either with poly(U) and acetyl-Phe-tRNA or with fMet-tRNA, initiation factors, and synthetic mRNA containing the sequence indicated. Data are taken from Thomas et al. (1988).

programmed ribosomes initiated with fMet-tRNA. To measure the rate of this reaction, excess ribosome · mRNA · fMet-tRNA complexes were reacted with ternary complex in a rapid mixing apparatus (Eccleston et al., 1980), and the reaction was stopped between 0.2 and 30 s later with excess EDTA. Analysis of the reaction for $^{32}P_i$ and ^{3}H-peptide allowed us to reconstruct the progress of the reactions leading to GTP hydrolysis and peptide formation, respectively. Simulation of the rate of product formation according to the above-described mechanism enabled us to determine the apparent rate constant for GTP hydrolysis, k_{GTP}, and peptide formation, k_{PEP}, for each aa-tRNA–codon pair. Ribosomes programmed with the UUC codon hydrolyzed GTP faster than did ribosomes programmed with the UUU codon. The difference between UUC and UUU codons was almost independent of the experimental conditions used (Table 3). In contrast, the rates of peptide formation, k_{PEP}, for ribosomes programmed with the UUU and UUC codons were very similar (Table 3).

A similar result was observed for the CUU and CUC synonymous codon pair. When Leu-tRNA$_2^{Leu}$ ternary complex inserts were mixed with CUC-programmed ribosomes, a faster rate of GTP hydrolysis (k_{GTP}) was observed than when Leu$_2$ ternary complex was mixed with CUU-programmed ribosomes. Again similar rates of peptide formation (k_{PEP}) were measured (Table 3).

The difference between the two classes of ribosomes, C-terminating versus U-terminating codon-programmed ribosomes, can be attributed to different rate constants for ternary complex binding to ribosomes (k_1). In cognate reactions the elementary rate constant for GTP cleavage (k_2) is much faster than that for ternary complex dissociation (k_{-1}) from the ribosome, so the observed rate constant k_{GTP} [$= k_1k_2/(k_{-1} + k_2)$] must therefore equal k_1 (Eccleston et al., 1985; Thompson and Karim, 1982). To determine whether any of these steps other than k_1 is influenced by the codon translated, we substituted GTPγS for GTP in the ternary complex to measure the rate constants k_1, k_{-1}, and k_2 (Thompson and Karim, 1982). The use of GTPγS slows the cleavage of the pyrophosphate bond and allows k_2 to be measured directly. Stabilization of the ribosome · ternary complex complex to hydrolysis also allows k_{-1} to be calculated, since if excess nonradioactive GTPγS ternary complex is added as a chase after the RS · TC complex forms, the rate of GTPγS hydrolysis equals $k_{-1} + k_2$. Finally, if excess

Table 3. Rate constants for the reaction of programmed ribosomes with cognate aa-tRNA ternary complexes[a]

Codon[b]	aa-tRNA	Ionic condition[c]	Temp (°C)	k_{GTP} (10^{-6} M^{-1} s^{-1})	k_{PEP}, (s^{-1})
UUU	Phe	M5	5	2.8 ± 0.2	1.1 ± 0.1
UUC	Phe	M5	5	4.6 ± 0.2	1.1 ± 0.2
UUU	Phe	M5	25	14 ± 1	ND[d]
UUC	Phe	M5	25	27 ± 1	ND
UUU	Phe	M3/S2/P10	5	11 ± 1	6 ± 1
UUC	Phe	M3/S2/P10	5	17.5 ± 1	5.5 ± 1
UUU	Phe	M3/S2/P10	25	95 ± 5	ND
UUC	Phe	M3/S2/P10	25	150 ± 5	ND
CUU	Leu$_2$	M5	5	2.5 ± 0.3	1.0 ± 0.1
CUC	Leu$_2$	M5	5	4.5 ± 0.2	1.0 ± 0.1
CUU	Leu$_2$	M5	25	17 ± 4	ND
CUC	Leu$_2$	M5	25	27 ± 3	6 ± 2
CUU	Leu$_2$	M3/S2/P10	5	14 ± 1	ND
CUC	Leu$_2$	M3/S2/P10	5	22 ± 1	ND

[a] Data are taken from Thomas et al. (1988).
[b] Ribosomes were programmed with fMet-tRNA, initiation factors, and synthetic mRNA containing the variable codon indicated.
[c] M, Millimolar $MgCl_2$; S, millimolar spermidine; P, millimolar putrescine.
[d] ND, Not done.

Table 4. Rate constants and proofreading ratios for the reaction of ribosomes with near-cognate aa-tRNA ternary complexes[a]

Codon	aa-tRNA	Anticodon	Mg^{2+} (mM)	Temp (°C)	k_{GTP} (10^{-4} M^{-1} s^{-1})	Ratio, GTP hydrolysis/peptide formation
UUU	Leu_2	GAG	5	5	0.7	271
UUC	Leu_2	GAG	5	5	5.1	94
UUU	Leu_2	GAG	5	25	17	ND
UUC	Leu_2	GAG	5	25	100	ND
UUU	Leu_2	GAG	10	5	1.1	ND
UUC	Leu_2	GAG	10	5	7.5	ND
UUU	Leu_4	NAA	5	5	2.5	97
UUC	Leu_4	NAA	5	5	5.5	159
UUU	Leu_5	NAA	5	5	1.0	55
UUC	Leu_5	NAA	5	5	20	28
CUU	Phe	GAA	5	5	0.7	95
CUC	Phe	GAA	5	5	5.5	60
CUU	Leu_1	CAG	5	5	4.0	65
CUC	Leu_1	CAG	5	5	15.0	130

[a] Ribosomes were programmed with fMet-tRNA, initiation factors, and synthetic mRNA containing the variable codon indicated. ND, Not done. Data are taken from Dix and Thompson (1989).

nonradioactive GTPγS ternary complex is added as a quench in a time course as described above, k_1 can be measured directly by nitrocellulose filtration. When ribosomes programmed with UUU and UUC are mixed with EF-Tu · GTP · Phe-tRNA, we observe that the Phe ternary complex binds (k_1) 1.7-fold faster to UUC-programmed ribosomes, that Phe-tRNA dissociates (k_{-1}) 1.8-fold slower from UUC-programmed ribosomes, and that no difference exists in the GTP cleavage step (k_2).

We conclude that significant differences do exist between the members of synonymous codon pairs with respect to the ability to direct incorporation of their respective cognate amino acids into nascent protein. The differences are observed in the rate at which a ternary complex is bound (k_1), as indicated by the faster rate of GTP hydrolysis when Phe ternary complex is mixed with UUC-programmed ribosomes. The rates at which bound aa-tRNAs are incorporated into peptide do not differ significantly.

Translational Accuracy: Kinetic Studies Involving Synthetic mRNAs and Near-Cognate aa-tRNAs

To determine whether synonymous codons also differ in the ability to discriminate against near-cognate aa-tRNAs, we examined the accuracy of translation at phenylalanine codons UUU and UUC and the leucine codons CUU and CUC. As before, ribosomes were bound to the synthetic mRNAs in the presence of initiation factors and fMet-tRNA, and these programmed ribosomes were reacted with EF-Tu · GTP · aa-tRNA ternary complexes. The progress of the GTPase and peptidyltransferase reactions was again monitored by stopping the reaction with excess EDTA at times between 2 s and 80 min. The reaction was then analyzed for the presence of $^{32}P_i$ and [^{3}H]peptidyl-tRNA. The rate of GTP hydrolysis (k_{GTP}), relative to k_{GTP} for the cognate ternary complex, is a measure of the accuracy in initial discrimination, and the proofreading ratio, GTP hydrolyzed per peptide formed, is an indication of the accuracy in proofreading.

To compare the accuracy of recognition of the 5′ base of the codons UUU and UUC, we reacted ribosomes programmed with these codons with a Leu-$tRNA_2^{Leu}$ ternary complex. Both reactions require a U-G mispair at the 5′ position of the codon-anticodon complex. UUC-programmed ribosomes appeared to be less accurate because the apparent rate of GTP hydrolysis by these ribosomes was sevenfold higher than that of UUU-programmed ribosomes (Table 4). This was true regardless of the experimental conditions. It also appeared that UUC-programmed ribosomes were less accurate in proofreading, since the proofreading ratio of these ribosomes was one-third that of the synonymous UUU-programmed ribosomes.

To compare the accuracy of recognition of the wobble base, we challenged UUU- and UUC-programmed ribosomes with ternary complexes containing Lelu-$tRNA_5^{Leu}$ and Leu-$tRNA_4^{Leu}$, which normally read UUA and UUG codons (Blank and Soll, 1971; Natale and Elilat, 1976). Again we observed that UUC-programmed ribosomes were the most prone to make errors in initial discrimination. The k_{GTP}s for UUC-programmed ribosomes reacting with Leu-$tRNA_4^{Leu}$ and Leu-$tRNA_5^{Leu}$ are 2- and 17-fold higher, respectively, than the analogous k_{GTP}s for UUU-programmed ribosomes (Table 4). UUC-programmed ribosomes are also more prone to errors in the proofreading reaction with Leu-tRNA$_5^{Leu}$, since their proofreading ratio is one-half that of UUU-programmed ribosomes. However, the situation is reversed in proofreading Leu-tRNA$_2^{Leu}$, since UUU-programmed ribosomes are more likely to incorporate Leu-tRNA$_4^{Leu}$ than are UUC-programmed ribosomes.

To determine whether this pattern of error proneness is a general phenomenon, we extended our studies to the reaction of CUC- and CUU-programmed ribosomes with near-cognate ternary complexes. When these ribosomes react with Phe-tRNA(GAA) ternary complex, creating a C-A mispair in the 5′ position of the codon-anticodon complex, ribosomes programmed with the CUC codon react faster. The apparent second-order rate constant for GTP hydrolysis (k_{GTP}) for CUC-programmed ribosomes is eightfold higher than the k_{GTP} for CUU-programmed ribosomes (Table 4). Thus, CUC-programmed ribosomes are more likely to bind Phe-tRNA during initial discrimination. These ribosomes are also less accurate than CUU-programmed ribosomes in proofreading, as shown by their lower proofreading ratio (Table 4).

The accuracy of translation of the wobble base with CUC- and CUU-programmed ribosomes was determined by reacting them with Leu$_1$ ternary complex (anticodon CAG). Although the correct amino acid would be incorporated into peptide, this reaction still shows all of the characteristics of a translational error, since the reaction of CUC- and CUU-programmed ribosomes with Leu-tRNA$_1^{Leu}$ ternary complex requires C-C and U-C mispairs, respectively, in the wobble position of the codon-anticodon interaction. Like UUC-programmed ribosomes, CUC-programmed ribosomes are less accurate in initial discrimination, since the k_{GTP} for CUC-programmed ribosomes is fourfold greater than that for CUU-programmed ribosomes. However, CUU-programmed ribosomes have a proofreading ratio that is one-half that of CUC-programmed ribosomes, indicating that CUU-programmed ribosomes are less accurate in the proofreading reaction.

Codon-induced differences in the rate of GTP hydrolysis are more marked in near-cognate reactions than in cognate reactions. This results from the fact that the apparent rate of GTP hydrolysis in cognate and near-cognate reactions is dependent on different sets of the elementary rate constants. In the cognate reaction, $k_2 >> k_{-1}$, so $k_{GTP} = k_1$ (Eccleston et al., 1985; Thompson and Karim, 1982); in the near-cognate reaction, $k_{GTP} = k_1k_2/(k_{-1} + k_2)$. As discussed above, when we substituted GTPγS for GTP in the reaction of UUC- and UUU-programmed ribosomes with Phe ternary complex, we observed that the k_2s for synonymous codons were similar but that the C-terminating codon bound a Phe ternary complex 1.7-fold faster (k_1) and that the C-terminating codon released Phe ternary complex 1.8-fold slower (k_{-1}). The preference of the cognate ternary complex for the C-terminating codon is based solely on its ability to bind faster, whereas the preference of the near-cognate ternary complex for that codon depends on both its ability to bind faster and its ability to be released more slowly. The codon that is the most efficient is therefore the least accurate.

No such clear pattern emerges in the proofreading reaction. C-terminating codons were observed to be more inaccurate than U-terminating codons in proofreading of 5′ mispairs in the codon-anticodon interaction, as indicated by their lower ratio of GTP hydrolysis to peptide formation. However, in proofreading 3′ mispairs, we found two examples in which the ribosome programmed with a U-terminating codon was less accurate. When Leu$_1$ ternary complex was mixed with CUC- and CUU-programmed ribosomes, a C-U mispair appeared to be more stable after GTP hydrolysis than a C-C mispair. When Leu$_4$ ternary complex was mixed with UUC- and UUU-programmed ribosomes, we observed that Leu-tRNA$_4^{Leu}$ was more likely to be misincorporated with UUU- than with UUC-programmed ribosomes. The reversal of specificity that sometimes occurs between U-terminating and C-terminating codons in initial recognition and proofreading probably indicates the existence of a subtle difference between the codon-anticodon interactions in the RS · ternary complex and RS · aa-tRNA complexes.

CONCLUSIONS

Ribosomes programmed by pairs of synonymous codons differ in the ability to bind cognate aa-tRNAs and in the ability to discriminate against near-cognate aa-tRNAs. For the two synonymous codon pairs examined, the C-terminating codon is not only more efficient in cognate reactions but less accurate in near-cognate reactions. Since the C-terminating codons are preferentially used in highly expressed mRNAs in *E. coli*, the results suggest not only that synonymous codons play a role in modulating gene expression but that maximization of translational accuracy has not been a universal determinant of codon choice.

REFERENCES

Bennetzen, J. L., and B. D. Hall. 1982. Codon selection in yeast. *J. Biol. Chem.* **257**:3026–3031.

Blank, H. U., and D. Soll. 1971. Purification of five leucine transfer ribonucleic acid species from *Escherichia coli* and their acylation by heterologous leucyl-transfer ribonucleic acid synthetase. *J. Biol. Chem.* **246:**4947–4950.

Bonekamp, F., H. D. Andersen, T. Christensen, and K. F. Jensen. 1985. Codon-defined ribosomal pausing in *Escherichia coli* detected by using the pyrE attenuator to probe the coupling between transcription and translation. *Nucleic Acids Res.* **13:** 4113–4123.

Carter, P. W., J. M. Bartkus, and J. M. Calvo. 1986. Transcription attenuation in *Salmonella typhimurium*: the significance of rare leucine codons in the Leu region. *Proc. Natl. Acad. Sci. USA* **83:**8127–8131.

de Boer, H. A., and R. A. Kastelein. 1986. Biased codon usage: an exploration of its role in optimization of translation, p. 225–285. *In* W. S. Reznikoff and L. Gold (ed.), *Maximising Gene Expression.* Butterworth Publishing, Stoneham, Mass.

Dix, D. B., and R. C. Thompson. 1989. Codon choice and gene expression: synonymous codons differ in translational accuracy. *Proc. Natl. Acad. Sci. USA* **86:**6888–6892.

Eccleston, J. F., D. B. Dix, and R. C. Thompson. 1985. The rate of cleavage of GTP on the binding of Phe-tRNA · elongation factor Tu · GTP to poly(UUU)-programmed ribosomes of *Escherichia coli. J. Biol. Chem.* **260:**16237–16241.

Eccleston, J. F., R. G. Messerschmidt, and D. W. Yates. 1980. A simple rapid mixing device. *Anal. Biochem.* **106:**73–77.

Gouy, M., and G. C. Gautier. 1982. Codon usage in bacteria: correlation with gene expressivity. *Nucleic Acids Res.* **10:**7055–7075.

Grosjean, H., and W. Fiers. 1982. Preferential codon usage in prokaryotic genes; the optimal codon-anticodon interaction energy and the selective codon usage in efficiently expressed genes. *Gene* **18:**199–209.

Harms, E., and H. E. Umbarger. 1987. Role of codon choice in the leader region of the *ilvGMEDA* operon of *Serratia marcescens. J. Bacteriol.* **169:**5668–5677.

Hoekema, A., R. A. Kastelein, M. Vasser, and H. A. de Boer. 1987. Codon replacement in the *PGK1* gene of *Saccharomyces cerevisiae*: experimental approach to study the role of biased codon usage in gene expression. *Mol. Cell. Biol.* **7:**2914–2924.

Holm, L. 1986. Codon usage and gene expression. *Nucleic Acids Res.* **14:**3075–3087.

Hopfield, J. J. 1974. Kinetic proofreading: a new mechanism for reducing errors in biosynthetic processes requiring high specificity. *Proc. Natl. Acad. Sci. USA* **71:**4135–4139.

Ikemura, T. 1981. Correlation between the abundance of *Escherichia coli* transfer RNAs and the occurrence of the respective codons in its protein genes: a proposal for a synonymous codon choice that is optimal for the *E. coli* translational system. *J. Mol. Biol.* **158:**573–597.

Menninger, J. R. 1977. Ribosomes editing and the error catastrophe hypothesis of cellular aging. *Mech. Ageing Dev.* **6:**131–142.

Natale, P. O., and D. Elilat. 1976. Patterns of *E. coli* leucine tRNA isoacceptors following bacteriophage MS2 infection. *Nucleic Acids Res.* **3:**917–931.

Parker, J., T. C. Johnston, and P. T. Borgia. 1980. Mistranslation in cells infected with the bacteriophage MS2: direct evidence of Lys for Asn substitution. *Mol. Gen. Genet.* **180:**275–281.

Parker, J., T. C. Johnston, P. T. Borgia, G. Holtz, E. Remaut, and W. Fiers. 1983. Codon usage and mistranslation. *J. Biol. Chem.* **258:**10007–10012.

Parker, J., and J. Precup. 1986. Mistranslation during phenylalanine starvation. *Mol. Gen. Genet.* **204:**70–74.

Pedersen, S. 1984. *Escherichia coli* ribosomes translate *in vivo* with variable rate. *EMBO J.* **3:**2895–2898.

Post, L. E., G. D. Strycharz, M. Nomura, H. Lewis, and P. P. Dennis. 1979. Nucleotide sequence of the ribosomal protein gene cluster adjacent to the gene for RNA polymerase subunit B in *Escherichia coli. Proc. Natl. Acad. Sci. USA* **76:**1697–1701.

Robinson, M., R. Lilley, S. Little, J. S. Emtage, B. Yarranton, P. Stephens, A. Millican, M. Eaton, and G. Humphreys. 1984. Codon usage can effect efficiency of translation of genes in *Escherichia coli. Nucleic Acids Res.* **12:**6663–6671.

Ruusala, T., M. Ehrenberg, and C. G. Kurland. 1982. Is there proofreading during polypeptide synthesis? *EMBO J.* **1:**741–745.

Sorensen, M. A., C. G. Kurland, and S. Pedersen. 1989. Codon usage determines translation rate in *Escherichia coli. J. Mol. Biol.* **207:**365–377.

Thomas, L. K., D. B. Dix, and R. C. Thompson. 1988. Codon choice and gene expression: synonymous codons differ in their ability to direct aminoacylated-transfer RNA binding to ribosomes *in vitro. Proc. Natl. Acad. Sci. USA* **85:**4242–4246.

Thompson, R. C., and D. B. Dix. 1982. Accuracy of protein biosynthesis. A kinetic study of the reactions of poly(U)-programmed ribosomes with a leucyl-$tRNA_2$ elongation factor Tu · GTP complex. *J. Biol. Chem.* **257:**6677–6682.

Thompson, R. C., D. B. Dix, R. B. Gerson, and A. M. Karim. 1981. A GTPase reaction accompanying the rejection of Leu-$tRNA_2$ by UUU-programmed ribosomes. *J. Biol. Chem.* **256:** 81–86.

Thompson, R. C., and A. M. Karim. 1982. The accuracy of protein biosynthesis is limited by its speed: high fidelity selection by ribosomes of aminoacyl-tRNA ternary complexes containing GTPγS. *Proc. Natl. Acad. Sci. USA* **79:**4922–4926.

Thompson, R. C., and P. J. Stone. 1977. Proofreading of the codon-anticodon interaction on ribosomes. *Proc. Natl. Acad. Sci. USA* **74:**198–202.

Varenne, S., M. Knibiehler, D. Cavard, J. Morton, and C. Lazdunski. 1982. Variable rate of polypeptide chain elongation for colicins A, E2, & E3. *J. Mol. Biol.* **159:**57–70.

Chapter 46

The Ribosome's Rubbish

ROBERT WEISS, DIANE DUNN, JOHN ATKINS, and RAY GESTELAND

Working on ribosomes is like exploring an ancient metropolis. There is a landscape that has definition in both past and present, and we can now peer into this world trying to assign function to form. The often-made analogy to a machine sets the spectacle in motion with a cacophony of whirring helices, snapping high-energy bonds, and bustling factors. The architecture is as enigmatic as it is austere, and we now must shade our eyes from the dazzling glint reflecting off the edifices of the subunits. Yet we are aliens in the ribosome's world, not fully cognizant of what lies before us. Speaking for ourselves, we prefer to inquire into the design of this apparatus not by looking at its triumphs but rather by sifting through the ribosome's waste. This endeavor is similar to learning about a civilization by digging up its garbage; it's informative, and it's fun.

One particularly fascinating piece of junk found on the ribosome's scrap heap consists of botched translocations (a.k.a. reading frame errors). These aborted polypeptides carry the residue of their missteps and abound in vivo and in vitro, although the cell is usually fairly efficient at cleaning up the ribosome's mess with rapid recycling pathways. As collectors of this curious type of ribosomal flotsam, we have spent years amassing examples of this debris and have honed our skills in hope of finding even more. It's relatively easy to find reading frame errors when you know what you're doing: just build a frameshift mutation in a reporter gene, express it in the cell of one's choice, recover the frameshifted polypeptides, and sequence them to find the signature of a ribosomal frameshift. The hard part is getting a discrete signal, since there is so much garbage when one gets deep into the problem.

As is often the case, one person's garbage is another person's gold; so it is with ribosomal frameshifts. Evolution has evidently taken some of these no-account cast-offs from the translation process and put them to good use in numerous genes (Dunn and Studier, 1983; Jacks and Varmus, 1985; Craigen et al., 1985; Clare and Farabaugh, 1985; Mellor et al., 1985; Moore et al., 1987; Brierly et al., 1987; Huang et al., 1988; Clare et al., 1988; Sekine and Ohtsubo, 1989). Not only are we privileged to help decipher the inner workings of the ribosome from its detritus, but also the findings may have relevance in the larger world of molecular biology; what luck!

READING FRAME ERRORS

One view of the ribosome's framing mechanism postulates that the zero reading frame is set and maintained because codon-anticodon pairing occurs only in strictly defined ribosomal sites. Reading frame is thus passively maintained through each cycle of decoding, transpeptidation, and translocation as a result of this strict definition of sites and continual codon-anticodon pairing of tRNA and mRNA within them. The P-A-site ribosome would keep at least one codon-anticodon pair at all times, whereas the E-P-A-site ribosome would keep at least two (Nierhaus et al., 1987). The conventional engine of ribosome movement, the passage of tRNA from A to P sites (or P-A to E-P sites), relies on continual codon-anticodon pairing both to move the mRNA and to maintain the frame. Some would say that these are the only things required for both an engine and a guide, a passive mRNA that moves only because the tRNA moves (Maizels and Weiner, 1987); others would disagree, postulating the necessity for an rRNA-based framing device outside the A and P sites (Trifonov, 1987), although push me/pull you scenarios become proble-

Robert Weiss, Diane Dunn, John Atkins, and Ray Gesteland ■ Howard Hughes Medical Institute and Department of Human Genetics, University of Utah Medical Center, Salt Lake City, Utah 84132.

matic with ribosome-mRNA contacts flanking the decoding sites. Unconventional engines (Woese, 1970) also rely on continual codon-anticodon pairing and site definition for framing, and engines outside the decoding sites (i.e., ones that would move mRNA directly) appear, at the moment, unnecessary. What do reading frame errors tell us about how the ribosome actually moves and maintains frame?

Forming codon-anticodon pairs displaced from the zero frame of the mRNA appears to be the common cause underlying most reading frame errors caused by normal tRNAs. Thus, tRNA-mRNA interplay has dominated the analysis of reading frame errors, but it is now clear that forces outside the A and P sites are coming into play (Jacks et al., 1987; Weiss et al., 1987; Weiss et al., 1988a; Jacks et al., 1988; Brierly et al., 1989). Classification based on when tRNA-mRNA pairing occurs splits reading frame errors into two categories: ones that appear to happen during and ones that happen after decoding. The ones that happen during decoding were the first to be examined in detail (Atkins et al., 1979; Weiss and Gallant, 1983). They seem to result from incoming tRNAs base pairing correctly with the mRNA but with a codon offset one nucleotide from the zero frame (Dayhuff et al., 1986; Weiss et al., 1988b). For instance, AGC-decoding Ser-tRNA can cause −1 shifts at the sequence A-GCA (Dayhuff et al., 1986) or +1 shifts at the sequence AAG-C, and the rate of both events can be manipulated by changing the ratio of the AGC-decoding Ser-tRNA to zero frame-decoding tRNA (Ala-tRNA or Lys-tRNA, respectively). Evidently, some tRNAs can override the ribosome's definition of the A site and change the frame through an improperly located codon-anticodon pairing. The ability of different tRNAs to accomplish this type of read varies widely (Atkins et al., 1979; Weiss and Gallant, 1986) for reasons that are poorly understood.

The second class of reading frame errors appears to happen after a tRNA is properly decoded, and tRNAs that decode strings of like nucleotides are obvious culprits in this type of misbehavior (Fox and Weiss-Brummer, 1980; Atkins et al., 1983; Jacks and Varmus, 1985; Weiss et al., 1987). For example, the ribosome may limit the first class of error by a strict spatial definition of the A site, perhaps through conserved tRNA-rRNA contacts and stabilization of a spatially localized A-site codon-anticodon helix (Noller et al., 1987). Thus, decoding out of frame, as observed on the sequence AAG-C by an AGC-reading Ser-tRNA, as well as +1 decoding on the sequence AAA-A, would be curtailed by incoming AAA-reading Lys-tRNA. However, once an AAA-decoding Lys-tRNA has passed the A-site selection process, it is still in danger of misaligning its anticodon on the mRNA by slipping or sliding if codon-anticodon pairing should break down post-A site.

Such post-A-site slippage underlies most of the examples of high-level ribosomal frameshifting so far observed, and these types of slips have been shown to be particularly sensitive indicators of ribosome-mRNA contacts outside the decoding sites. In *Escherichia coli*, stop codons adjacent to repetitive strings of bases aggravate slipping on these strings (Weiss et al., 1987). This is one component of the ribosomal frameshift found in the RF2 gene. The other component is base pairing between a region of the mRNA 3 nucleotides 5′ of the shift site and the anti-Shine-Dalgarno region of 16S rRNA (Weiss et al., 1987; Weiss et al., 1988a), and this serves as one example of a push me/pull you problem. The other example comes from the −1 frameshifts observed in many retroviral gene overlaps (Jacks and Varmus, 1985; Moore et al., 1987; Brierly et al., 1989). Members of this broad class of ribosomal frameshifts have in common a requirement for tandem slippery codons, implying that the frameshift occurs at a point in the translation cycle where there is a strict requirement for simultaneous pairing of adjacent tRNAs (Jacks et al., 1988). *E. coli* ribosomes can also be triggered to slip −1 by tandem slippery codons (Weiss et al., 1989), suggesting functional conservation of this requirement. The other element observed in the retroviral class is flanking 3′ mRNA structures, either stems (Jacks et al., 1988) or pseudoknots (Brierly et al., 1989), which elevate the rate of shifting at the tandem slippery codons, perhaps with a push me/pull you mechanism similar to that seen with the RF2 shift. Procaryotic ribosomes have been inferred to melt mRNA secondary structures pre-A site from the work of many groups on bacterial attenuators (Landick and Yanofsky, 1988) and on translational control regions of inducible antibiotic resistance genes (Alexieva et al., 1988; Gryczan et al., 1980; Horinouchi and Weisblum, 1980), and the positioning of this melting activity from the A-site codon (6 to 10 nucleotides) seems remarkably conserved between procaryotic and eucaryotic ribosomes. From the analysis of the slippage-based ribosomal frameshifts emerges a picture of the ribosome in which codon-anticodon pairing is maintained (the tandem slippery codon rule almost cries for the E site!) throughout the cycle and in which the ribosome contacts the mRNA both 3′ of the tRNA sites, where mRNA structure is melted, and 5′ of the tRNA sites, for reasons yet unfathomed.

p163UAG stop hop

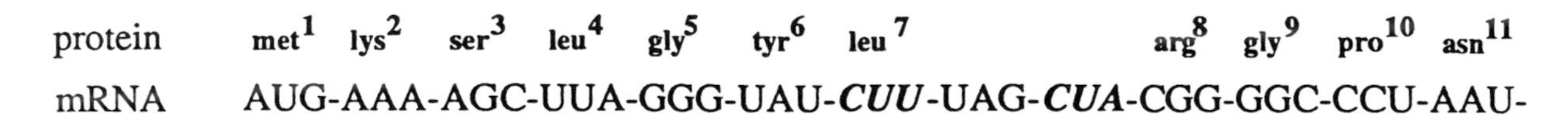

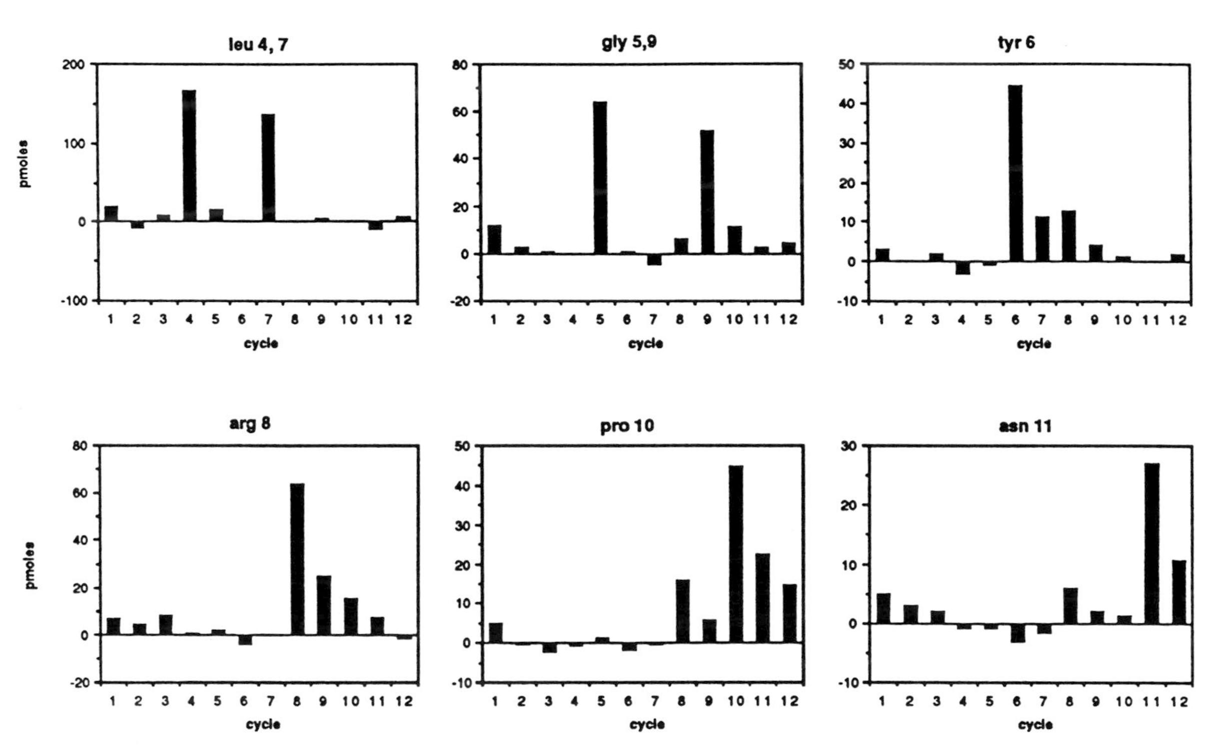

Figure 1. An unusual piece of translational junk: the stop hop. The protein sequence resulting from bypass of a *lacZ* amber mutation is shown. Fabrication of this amber-*lacZ* reporter gene and verification by plasmid DNA sequencing were done as described by Weiss et al. (1987). Frameshifted β-galactosidase was purified from saturated broth cultures grown at 37°C with an anti-β-galactosidase column (0.5-ml bed volume; Protosorb; Promega Biotec), eluted with 3 volumes of 0.1 M $NaCO_2$, and spun dialyzed in Centricon 30 ultrafiltration cartridges (Amicon Corp.). The protein sequence was determined on an ABI 470a/900/120 liquid-pulse protein sequencer equipped with on-line high-performance liquid chromatography analyzer and data analysis modules; these data were baseline corrected, and phenylthiohydantoin yields were calculated from peak height. Histograms of relevant phenylthiohydantoin-derived amino acids through 14 cycles of Edman degradation are depicted; the major amino acid present in each cycle is indicated above the mRNA sequence.

tRNA HOPPING

Whereas slipping on repetitive strings is a conspicuous type of error, less obvious is the type of event shown in Fig. 1. Here, an amber stop is placed in the path of the ribosome, eight codons from the start. This nonsense mutant displays a medium level of residual β-galactosidase activity (approximately 1% [Weiss et al., 1987]), which is a clear sign that some ribosomes are getting through: that's our signal to come and examine the trash. The major protein sequence translated from this message is shown above the mRNA (Fig. 1), and it is apparent that the two are not completely colinear. The amino acids found at positions 8 to 11 correspond to codons 10 to 13, as if the ribosome has skipped over codons 8 and 9. A fanciful version of the possible mechanics of such an event is shown in Fig. 2. The interpretation implies that the peptidyl-tRNA base paired at the CUU-7 codon detaches and re-forms base pairs with the first two positions of the CUA-9 codon; i.e., the tRNA hops. This event is formally similar to the class II slips discussed above in that the tRNA enters correctly through the A site and then forms an alternate codon-anticodon pairing, but it is the distance traveled (here +6 nucleotides, as compared with +1 or −1 slips) that is striking. With normal tRNAs, +6 hopping over stop codons as well as +5 hopping within a coding region have been observed (Weiss et al., 1987), and with mutant *E. coli* Val-tRNA, *hopR1* (Falahee et al., 1988; O'Connor et al.,

Figure 2. One interpretation of stop hopping. This fanciful view of hopping tRNAs depicts a CUU-decoding Leu-tRNA hopping six nucleotides down the mRNA onto a CUA leucine codon.

1989), +2, +3, +6, and +9 hops have been observed. All of these examples have in common a requirement for at least a two-of-three position match of the codons at the "takeoff" and "landing" sites. The observation of mutant Val-tRNA increasing the level of observed hopping strengthens the conclusion that these observed hops are translational events. The hopping rate of the *hopR1* Val-tRNA can be as high as 20% on some sequences (O'Connor et al., 1989). Such high rates of hopping could obviously have a severely disturbing influence on maintaining frame, especially in light of the relaxed requirements for triplet codon-anticodon pairing at the landing site (Fig. 1; Weiss et al., 1987; O'Connor et al., 1989).

The unanswered question regarding tRNA hopping is how the ribosome manages to squelch it. Two observations may help constrain future models. One, hopping is directional: only hops 3′ have been observed (leaving gaps in the polypeptide sequence instead of duplications), even though 5′ hops have been searched for directly (Weiss et al., 1987; O'Connor et al., 1989). This implies that the message, if it is diffusing past the decoding sites, does so only in one direction. Two, when *hopR1* Val-tRNA is given the choice of multiple landing sites (for instance, +3 and +6 targets), it jumps to each with comparably high frequency (O'Connor et al., 1989). This may suggest that the hopping tRNA just scans in the local vicinity for pairing opportunities just 3′ of the decoding site, rather than that the ribosome becomes a loose cannon upon the deck of mRNA. Clearly, more evidence is needed to determine what is curtailing tRNA hopping.

GENE 60: THE MOTHERSHIP ARRIVES

In the final scenes of the Steven Spielberg film *Close Encounters of the Third Kind*, scientists gather to await the arrival of extraterrestrial visitors atop Devil's Tower, Wyoming. Spritely spacecraft, the size of Winnebagos, arrive and perform fly-bys before ascending back into the clouds. The scientists rejoice. Their jubilation makes them momentarily unaware of the booming arrival of the main craft, which is approximately the size of Rhode Island. A crescendo of awe and horror similar to that displayed by the film's fictional scientists may be felt by some ribosomologists as they try to come to grips with the nearly 100% efficient, ribosome long jump of 50 internal nucleotides within the coding region of gene 60 of bacteriophage T4 (Huang et al., 1988).

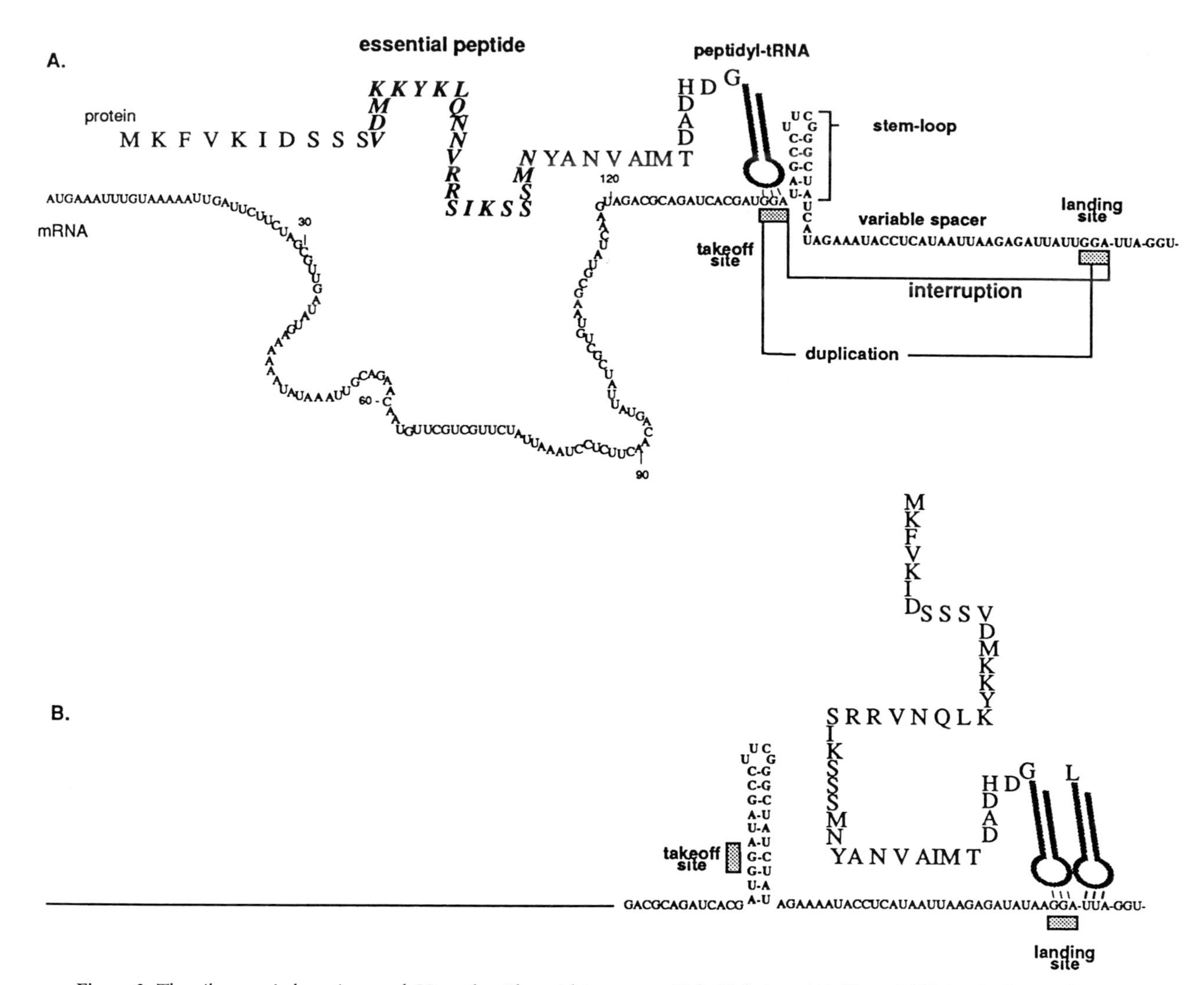

Figure 3. The ribosome's long jump of 50 nucleotides within a gene 60-*lacZ* fusion. (A) The mRNA beginning at the start codon of a gene 60-*lacZ* fusion through the 3′ junction of the interruption is shown with some essential components of the bypass mechanism. Peptidyl-tRNA is shown base paired to the GGA codon at the 5′ junction. Also shown are the GGA duplicated in the takeoff and landing sites, the stem-loop at the 5′ junction, and the leader peptide. Mutational analysis suggests that all of these components play a role in bypassing the interruption (Weiss et al., in press). (B) Completion of the bypass is shown, with the peptidyl-tRNA now base paired with the GGA codon at the 3′ junction.

Within gene 60, the ribosome passes from a glycine codon that specifies amino acid 46 of the mature protein to a leucine codon specifying amino acid 47 but located 50 nucleotides 3′ of codon 46. When a segment of gene 60 is cloned into *lacZ*, full bypass is observed (Huang et al., 1988; Weiss et al., in press), implying that *E. coli* ribosomes in uninfected cells can negotiate the interruption. A mutational analysis of the gene 60-*lacZ* fusion (Weiss et al., in press) has delineated some of the mRNA components necessary for bypass (summarized in Fig. 3). Although no information 3′ of the interruption is required for bypass, almost the entire 5′ coding region (the 46-amino-acid leader) is necessary. Preliminary findings indicate that a peptide sequence translated from within this sequence may be one key component (Weiss et al., in press) in this segment. The sequence of the interruption is remarkably flexible, tolerating small deletions as well as insertions of at least 16 nucleotides with relatively little effect on bypass. The crucial features of the interruption seem to be a stem-loop structure inside the 5′ border and a duplication of a GGA glycine codon flanking the interruption. Such a duplication is suggestive of hopping tRNAs, and it is possible that gene 60 is some form of very fancy hopping. In analogy with the other types of high-level ribosomal frameshifts, it is possible that the 5′ leader of gene 60 is playing a push me/pull you role. Just as the flanking elements with the RF2 and retroviral shifts impinge on a basic ribosome-mRNA contact to effectively stimulate slipping in the decoding sites, so too might the 5′ region of gene 60 be stimulating hopping. But clearing 50 nucleotides at nearly 100% efficiency is a lot of stimulation. Further analysis of the gene 60 ribosome jump should prove valuable.

PROSPECTS

The facetious analogy of reading frame errors to the ribosome's rubbish has perhaps served to illustrate some of the oddities that ribosomes are capable of producing. The study of reading frame errors over the past several years has revealed an unexpected detail of ribosome-mRNA-tRNA interplay both inside and outside the decoding sites. One must be amazed at the cleverness of certain genes in thwarting the normally stately course of the ribosome with high-level slips and hops. And how fortunate we are to have examples of such grossly unusual translational events to dissect and examine. One wonders what else awaits us out there in the ribosome's trash pile.

REFERENCES

Alexieva, Z., E. J. Duvall, N. P. Ambulos, U. J. Kim, and P. S. Lovett. 1988. Chloramphenicol induction of cat-86 requires ribosome stalling at a specific site in the leader. *Proc. Natl. Acad. Sci. USA* **85:**3057–3060.

Atkins, J. F., R. F. Gesteland, B. R. Ried, and C. W. Anderson. 1979. Normal tRNAs promote ribosomal frameshifting. *Cell* **18:**1119–1131.

Atkins, J. E., B. P. Nichols, and S. Thompson. 1983. The nucleotide sequence of the first externally suppressible −1 frameshift mutant, and of some nearby leaky frameshift mutants. *EMBO J.* **2:**1345–1350.

Brierly, I., M. E. G. Boursnell, M. M. Binns, B. Bilimoria, V. C. Blok, T. D. K. Brown, and S. C. Inglis. 1987. An efficient ribosomal frameshifting signal in the polymerase-encoding region of the coronavirus IBV. *EMBO J.* **6:**3779–3785.

Brierly, I., P. Digard, and S. C. Inglis. 1989. Ribosomal frameshifting signal: requirement for an RNA pseudoknot. *Cell* **57:**537–547.

Clare, J. J., M. Belcourt, and P. J. Farabaugh. 1988. Efficient translation frameshifting occurs within a conserved sequence of the overlap between the two genes of yeast Ty1 transposon. *Proc. Natl. Acad. Sci. USA* **85:**6816–6820.

Clare, J. J., and P. Farabaugh. 1985. Nucleotide sequence of a yeast Ty element: evidence for an unusual mechanism of gene expression. *Proc. Natl. Acad. Sci. USA* **82:**2829–2833.

Craigen, W. J., R. G. Cook, W. P. Tate, and C. T. Caskey. 1985. Bacterial peptide chain release factors: conserved primary structure and possible frameshift regulation of release factor 2. *Proc. Natl. Acad. Sci. USA* **82:**3616–3620.

Dayhuff, T. J., J. F. Atkins, and R. F. Gesteland. 1986. Characterization of ribosomal frameshift events by protein sequence analysis. *J. Biol. Chem.* **261:**7491–7500.

Dubnau, D. 1985. The *ermC* leader. *EMBO J.* **4:**533–537.

Dunn, J. J., and F. W. Studier. 1983. Complete nucleotide sequence of bacteriophage T7 DNA and the locations of T7 genetic elements. *J. Mol. Biol.* **166:**477–535.

Falahee, M. B., R. B. Weiss, M. O'Connor, S. Doonan, R. F. Gesteland, and J. F. Atkins. 1988. Mutants of translational components that alter reading frame by two steps forward or one step back. *J. Biol. Chem.* **263:**18099–18103.

Fox, T. D., and B. Weiss-Brummer. 1980. Leaky +1 and −1 frameshift mutations at the same site in a yeast mitochondrial gene. *Nature* (London) **288:**60–63.

Gryczan, T. J., G. Grandi, J. Hahn, and D. Dubnau. 1980. Conformational alteration of mRNA structure and the posttranscriptional regulation of erythromycin-induced drug resistance. *Nucleic Acids Res.* **8:**6081–6097.

Horinouchi, S., and B. Weisblum. 1980. Posttranscriptional modification of mRNA conformation: mechanism that regulates erythromycin-induced resistance. *Proc. Natl. Acad. Sci. USA* **77:**7079–7083.

Huang, W. M., S. Z. Ao, S. Casjens, R. Orlandi, R. Zeikus, R. Weiss, D. Winge, and M. Fang. 1988. A persistent untranslated sequence within bacteriophage T4 DNA topoisomerase gene 60. *Science* **239:**1005–1012.

Jacks, T., H. D. Madhani, F. R. Masiarz, and H. E. Varmus. 1988. Signals for ribosomal frameshifting in the Rous sarcoma virus gag-pol region. *Cell* **55:**447–458.

Jacks, T., K. Townley, H. E. Varmus, and J. Majors. 1987. Two efficient ribosomal frameshifts are required for synthesis of mouse mammary tumor virus gag-related polypeptides. *Proc. Natl. Acad. Sci. USA* **84:**4298–4302.

Jacks, T., and H. E. Varmus. 1985. Expression of Rous sarcoma virus pol gene by ribosomal frameshifting. *Science* **230:**1237–1242.

Landick, R., and C. Yanofsky. 1988. Transcription attenuation, p. 1276–1301. *In* F. C. Neidhardt, J. L. Ingraham, B. Magasanik, K. B. Low, M. Schaechter, and H. E. Umbarger (ed.), *Escherichia coli and Salmonella typhimurium: cellular and molecular biology*. American Society for Microbiology, Washington, D.C.

Maizels, N., and A. Weiner. 1987. Peptide-specific ribosomes, genomic tags, and the origin of the genetic code. *Cold Spring Harbor Symp. Quant. Biol.* **52:**743–752.

Mellor, J., S. M. Fulton, M. J. Dobson, W. Wilson, S. M. Kingsman, and A. J. Kingsman. 1985. A retrovirus-like strategy for expression of a fusion protein encoded by yeast transposon Ty1. *Nature* (London) **313:**243–246.

Moore, R., M. Dixon, R. Smith, G. Peters, and C. Dickson. 1987. Complete nucleotide sequence of a milk-transmitted mouse mammary tumor virus: two frameshift suppression events are required for translation of *gag* and *pol*. *J. Virol.* **61:**480–490.

Nierhaus, K. H., R. Brimacombe, V. Nowotny, C. L. Pon, H. J. Rheinberger, B. Wittmann-Liebold, and H. G. Wittmann. 1987. New aspects of structure, assembly, evolution and function of ribosomes. *Cold Spring Harbor Symp. Quant. Biol.* **52:**665–674.

Noller, H. F., S. Stern, D. Moazed, T. Powers, P. Svensson, and L. M. Changchien. 1987. Studies on the architecture and function of 16S rRNA. *Cold Spring Harbor Symp. Quant. Biol.* **52:** 695–708.

O'Connor, M., R. F. Gesteland, and J. F. Atkins. 1989. tRNA hopping: enhancement by an expanded anticodon. *EMBO J.* **8:**4315–4323.

Sekine, Y., and E. Ohtsubo. 1989. Frameshifting is required for production of the transposase encoded by insertion sequence 1. *Proc. Natl. Acad. Sci. USA* **86:**4609–4613.

Trifonov, E. N. 1987. Translation framing code and frame-monitoring mechanism as suggested by the analysis of mRNA and 16S rRNA nucleotide sequences. *J. Mol. Biol.* **194:**643–652.

Weiss, R. B., D. M. Dunn, J. F. Atkins, and R. F. Gesteland. 1987.

Slippery runs, shifty stops, backward steps and forward hops: −2, −1, +5 and +6 ribosomal frameshifting. *Cold Spring Harbor Symp. Quant. Biol.* **52:**687–693.

Weiss, R. B., D. M. Dunn, A. E. Dahlberg, J. F. Atkins, and R. F. Gesteland. 1988a. Reading frame switch caused by base-pair formation between the 3′ end of 16S rRNA and the mRNA during elongation of protein synthesis. *EMBO J.* **7:**1503–1507.

Weiss, R. B., D. M. Dunn, M. Shuh, J. F. Atkins, and R. F. Gesteland. 1989. *E. coli* ribosomes re-phase on retroviral frameshift signals at rates ranging from 2 to 50 percent. *New Biol.* **1:**159–170.

Weiss, R. B., and J. A. Gallant. 1983. Mechanism of ribosome frameshift during translation of the genetic code. *Nature* (London) **302:**389–393.

Weiss, R. B., and J. A. Gallant. 1986. Frameshift suppression in aminoacyl-tRNA limited cells. *Genetics* **112:**727–739.

Weiss, R. B., W. M. Huang, and D. M. Dunn. A nascent peptide is required for ribosomal bypass of the coding gap in bacteriophage T4 gene *60*. *Cell,* in press.

Weiss, R., D. Lindsley, B. Falahee, and J. Gallant. 1988b. On the mechanism of ribosomal frameshifting at hungry codons. *J. Mol. Biol.* **203:**403–410.

Woese, C. 1970. Molecular mechanics of translation: a reciprocating ratchet model. *Nature* (London) **226:**817–820.

Chapter 47

Effects of Codon Context on the Suppression of Nonsense and Missense Mutations in the *trpA* Gene of *Escherichia coli*

R. H. BUCKINGHAM, E. J. MURGOLA, P. SORENSEN, F. T. PAGEL, K. A. HIJAZI, B. H. MIMS, N. FIGUEROA, D. BRECHEMIER-BAEY, and E. COPPIN-RAYNAL

Observations that the nucleotide sequence around nonsense codons can produce significant effects on the efficiency of nonsense codon suppression provided the first clues that surrounding sequence, or codon context, might be important in protein synthesis, at least in polypeptide chain termination, and possibly also during the process of chain elongation. The early findings of Salser (1969), Salser et al. (1969), Yahata et al. (1970), and others (see Bossi, 1983) led to the isolation of context mutants that permitted greatly increased suppression by *supE* at a particular site (Bossi and Roth, 1980). This work was greatly extended by the systematic study of suppression at 42 different *lacI* amber sites in *Salmonella typhimurium* (Bossi, 1983) and the same sites together with 14 *lacI* opal sites in *Escherichia coli* (Miller and Albertini, 1983). These data, later reexamined by Stormo et al. (1986), showed clearly that the nature of the two bases following an amber codon strongly influences suppression. On the other hand, a study by Edelmann et al. (1987) of amber suppression by *supE44* at different sites in T4 genes 22 and 23, although consistent with the previously observed effect of the 3′ neighboring base, suggested an important influence of unidentified aspects of context outside the immediate downstream codon.

Compared with the effects of context on nonsense suppression, little is known concerning the extent to which context may influence missense suppression. This is unfortunate, because a clear demonstration of effects on missense suppression would help resolve a major ambiguity as to the mechanism of context effects. It is uncertain whether codon context affects primarily the termination process (for example, the functioning of release factors) or the translation of nonsense codons by suppressor tRNAs. Apart from clarifying the nature of context effects, this distinction is particularly important because if present during tRNA selection, context effects would potentially act at every step in the elongation of the polypeptide chain rather than exclusively at termination. The different mechanisms that have been suggested for context effects will be discussed in detail below; however, it seems from currently available evidence that both processes, tRNA selection and release factor function, are sensitive to context effects.

By comparing suppression at two missense sites (211 and 234) in the *trpA* gene of *E. coli* by a variety of tRNA suppressors, Murgola et al. (1984) found evidence for different efficiencies of suppression at the two sites. Some examples could not easily be explained by the suppressor inserting (partially or completely) an amino acid different from glycine, the wild-type amino acid, at both sites 211 and 234. We present in this chapter further evidence for context effects on missense suppression, free from possible objections related to the nature of the amino acid inserted. These data add weight to the suggestion that context can directly affect tRNA selection.

Nevertheless, recent ingenious experiments by Martin et al. (1988) provide evidence that release factor activity is also sensitive to codon context, in this case, the context around UAA codons. These authors show that the relative activity of release factors RF-1 and RF-2 at different UAA codons

R. H. Buckingham, P. Sorensen, and D. Brechemier-Baey ■ Institut de Biologie Physico-Chimique, Paris, France. **E. J. Murgola, F. T. Pagel, K. A. Hijazi, and B. H. Mims** ■ Department of Molecular Genetics, University of Texas M. D. Anderson Cancer Center, Houston, Texas 77030. **N. Figueroa** ■ Centre de Génétique Moleculaire du Centre National de la Recherche Scientifique, 91190 Gif-sur-Yvette, France. **E. Coppin-Raynal** ■ Laboratoire de Génétique de l'Université Paris-Sud, 91405 Orsay, France.

varies significantly according to the site of the codon. Furthermore, a correlation can be shown between the relative activity of the factors and the efficiency of suppression at different sites, suggesting that release factor sensitivity to context is at least partly responsible for the variation in efficiency of suppression observed.

CONTEXT MUTATIONS ADJACENT TO CODON 234 IN THE *trpA* GENE

We undertook a study of context effects at site 234 in the tryptophan synthetase (TS) α subunit, the *trpA* gene product, for several reasons. First, most previous studies (Bossi, 1983; Miller and Albertini, 1983; Murgola et al., 1984), by comparing many different sites in an mRNA, dealt with an unknown number of sequence variables and hence begged the question of precisely how far upstream or downstream the nucleotide sequence influences the translation of a given codon. Consequently, it seemed to us necessary to undertake a study in which defined changes are introduced in the neighborhood of a single site, a site, furthermore, at which both missense and nonsense codons exist. This clearly simplifies the task of understanding what changes are associated with given effects and resolves the potential problem of the differential activity of amino acids at different sites. The *trpA* gene has been the subject of detailed investigation for many years, and therefore a multitude of nonsense and missense mutations are available at a number of amino acid positions, including 234. The two sites, 233 and 235, adjacent to this position are both occupied by Ser residues, which may be coded for by six different codons. It is thus possible to introduce, by site-directed mutagenesis, a considerable number of changes in the immediate vicinity of codon 234 without changing the amino acid sequence of the protein synthesized.

Furthermore, the TS α subunit lends itself particularly well to comparison of the efficiencies of suppression of different constructions. The physiological reaction of TS, conversion of indole-3-glycerol phosphate to Trp, requires both subunits, α and β, of the tetrameric $\alpha_2\beta_2$ complex. However, each subunit exhibits two measurable activities: a catalytic activity that is "one half" of the physiological reaction, and a stimulatory activity, i.e., stimulation of the other subunit in its own half reaction (see Murgola, 1985). These properties, together with the coexpression of *trpA* and *trpB* (which encodes the β subunit), allow valuable internal controls in the measurement of the efficiency of suppression of both nonsense and missense mutations in the *trpA* gene.

So far, we have constructed complete defined sets of context mutations adjacent to four codons at position 234: the nonsense codons UGA and UAA and the missense codons AGA (Arg) and AAA (Lys). Within a certain range of TS activities, bacterial growth rate under defined conditions can reflect the cellular TS activity and hence the efficiency of suppression of TS mutants. We have therefore used growth rate to provide an initial indication of different suppression efficiencies between context mutants.

The UGA-234 context mutants were derived from the AGA-234 series, obtained by site-directed mutagenesis (see below). Specifically, the UGA-234 series was obtained by selecting for reversion to Trp^+ of each AGA context mutation in the presence of *glyT*(SuUGA/G). The Trp^+ colonies were then screened for suppressed mutants (versus *trpA* revertants) and for failure to grow on medium containing 5-methyltryptophan and limiting indole (Murgola and Yanofsky, 1974; Murgola and Jones, 1978). Strains were constructed that contained the context mutations and UGA/G reading tRNA suppressors derived from each of the $tRNA^{Gly}$ genes *glyT*, *glyU*, and *glyV*. The growth of these strains was studied on solid glucose minimal medium after replicating of patches so as to obtain different strengths of initial inoculum. This allowed a ranking order to be established for growth rates within each series and also permitted comparisons to be made between different series.

The observations can be summarized as follows. Considerable differences in growth rate are seen within each context series. The ranking order within each series is the same for each of the suppressors, with the proviso that some combinations of context mutant and suppressor lead to such poor growth that observations become difficult. Within the series, we observe the order of suppression shown in Table 1.

The change from U to A in the adjacent downstream position greatly favors suppression (Table 1), in agreement with the results of others (Bossi and Roth, 1980; Bossi, 1983; Miller and Albertini, 1983). Indeed, the two best contexts allowed some growth on medium containing 5-methyltryptophan and low indole, not seen with the rest of the series. However, the observations are remarkable for the facts that (i) all UGA-234 contexts are readily distinguishable by growth rate and (ii) some changes quite apart from those just mentioned are dramatic. Particularly notable is the effect of introducing the upstream codon UCA, which leads to a large increase in growth rate compared with the other codons UCN-233. There are also clear differences between all four codons UCN-235; thus, the third codon base in the downstream codon makes a significant contribution to the context effect in addition to the first two bases.

Comparisons of growth rate show *glyT*(SuUG A/G) to be the most efficient of the three suppressors, sufficiently that differences in growth between the two fastest strains (Table 1) are less easy to observe than in the case of the other two suppressors, probably because growth of the suppressed mutants is approaching that of the wild-type strain. Growth on glucose minimal medium indicates that *glyV*(SuUGA/G) is more efficient than *glyU*(SuUGA/G); however, when growth rates are compared on medium containing 5-methyltryptophan and low indole, the order is reversed. This result suggests that *glyU* (SuUGA/G) is more efficient than *glyV*(SuUGA/G) in suppressing *trpA*(UGA234) but that the suppressor tRNA is partially mischarged with an amino acid other than glycine. Although growth rate comparisons such as those we report here do have certain advantages over other methods of comparing suppression efficiencies, it is clearly necessary to support these data with measurements of TS enzymatic activity in suppressed mutant *trpA* strains. Such experiments are in progress. Analogous context mutants adjacent to UAA-234 and UAG-234 are also under study to determine whether our observations with UGA-234 can be extended to these other nonsense codons.

In addition to our observations on the suppression of nonsense mutants, we have obtained clear evidence that the suppression of missense mutants can also be subject to pronounced context effects dependent on the codons adjacent to the missense site. Two missense series have so far been constructed

Table 1. Suppression by *glyT*(SuUGA/G), *glyU*(SuUGA/G), and *glyV*(SuUGA/G) of the *trpA* mutation UGA-234 in different contexts

Codon position			Order of suppression efficiencies[a]
233	234	235	
	UGA	AGU	12
	UGA	AGC	11
UCA	UGA		9
	UGA	UCA	8
	UGA	UCC	7
AGC	UGA		6
	UGA	UCU	5
UCG	UGA		4
UCC	UGA		3
AGU	UGA		2
UCU	UGA	UCG[b]	1

[a] The increment between successive strains is an indication of the difference in growth rate. An increment of 1 indicates a significant and reproducible difference; an increment of 2 indicates a very pronounced difference.
[b] Wild-type context.

Table 2. Suppression by *glyT*(SuAAA/G) of the *trpA* mutation AAA-234 in different contexts

Codon position			Order of suppression efficiencies[a]
233	234	235	
UCC	AAA		7
	AAA	UCC	7
	AAA	UCA	6
AGC	AAA		6
	AAA	AGC	5
	AAA	AGU	5
AGU	AAA		4
	AAA	UCU	4
UCU	AAA	UCG[b]	3
UCA	AAA		2
UCG	AAA		1

[a] The increment between successive strains is an indication of the difference in growth rate. No increment indicates differences that were small or poorly reproducible, an increment of 1 indicates a significant and reproducible difference, and an increment of 2 represents a very pronounced difference.
[b] Wild-type context.

and studied. The first series, the AGA-234 context mutants, was obtained by oligonucleotide-directed site-specific mutagenesis. A second series of the same context mutants but with AAA-234 was then derived by selecting for Trp$^+$ revertants in the presence of *glyT*(SuAAA/G) and screening for *trpA* mutants that were Trp$^-$ in a *glyT*(SuAGA/G) or Su$^-$ background but suppressed by *glyT*(SuAAA/G). By DNA sequence analysis, it was verified that all missense mutants had the expected sequence in the vicinity of position 234. Growth rate was then used to look for context effects on suppression of AAA-234 by *glyT* (SuAAA/G) and AGA-234 by *glyT*(SuAGA/G) or *glyU*(SuAGA/G). Significant and reproducible effects were observed in the case of the AAA-234 series. The order of suppression efficiencies (Table 2) is clearly very different from that for the UGA-234 series, but once again a wide range of growth rates is apparent. On the other hand, no growth differences were observed in the case of the AGA-234 series with either suppressor.

SIGNIFICANCE AND MECHANISM OF CONTEXT EFFECTS ON SUPPRESSION

The effect of codon context on translation has been most clearly demonstrated in studies of suppression. Although this approach has clear advantages experimentally, in the sense that suppression may be readily linked to a measurable phenotype, it is natural to question whether it has much relevance to the normal process of translation. We will discuss first nonsense suppression and then the significance of context effects on missense suppression.

The importance of nonsense codon suppression

as a biological strategy in both procaryotes and eucaryotes is beyond doubt, and the role that nonsense codon context plays has been clearly demonstrated (reviewed in Valle and Morch, 1988). It is evident that the efficiency of suppression reflects the outcome of two competing processes. In nonsense suppression, the competition is between the factor-dependent termination of elongation and the tRNA-dependent translation of the nonsense codon. At which of these levels context acts is a subject of debate. One proposal is that adjacent bases stack on the minihelix composed of the codon-anticodon bases and contribute a stabilizing factor whose magnitude depends on the nature of the bases. The importance of dangling bases has been known for many years from the studies by Grosjean et al. (1978) of the interaction in solution between tRNA molecules with complementary anticodons. In an even simpler system, the interaction between self-complementary oligonucleotides, it is seen that dangling bases do indeed stabilize the interaction and that the greatest effect requires an adjacent purine followed by a pyrimidine (Freier et al., 1983). The clear parallel with the effect on nonsense suppression of the two downstream bases led Stormo et al. (1986) to propose that this was the principal mechanism behind context effects. However, as mentioned above, the recent work of Martin et al. (1988) suggests that a significant part of the context effect on nonsense suppression is concerned with release factor function.

Where context effects due to adjacent upstream bases are concerned, one hypothesis that has been widely discussed evokes interaction between P-site- and A-site-bound tRNAs (Feinstein and Altman, 1978; Bossi, 1983; Murgola et al., 1984; Smith and Yarus, 1989). Such a mechanism could account for our observations on the effects of changing codon 233, due to alterations in the Ser isoacceptor involved. We have little to add at present to the extensive discussions in the literature concerning the possible mechanisms for effects of context (Bossi, 1983; Miller and Albertini, 1983; Stormo et al., 1986; Martin et al., 1988; Murgola et al., 1988; Murgola et al., 1989) except to point out the possibility of a codon-specific involvement of rRNA. Context effects may, however, be less direct than is commonly supposed. It is conceivable that interactions between the messenger and the ribosome, or the P-site-bound tRNA and the ribosome, may alter the conformation of the ribosomal A site and affect ternary complex binding. We know little as yet of how the conformation of the codon in the A site may depend on nearby sequence. RNA sequences do display different degrees of flexibility and may adopt alternative conformations according to surrounding sequence (Hartmann and Lavery, 1989).

Our results on the effect of the adjacent codons on the suppression of UGA-234 are consistent with the observations of others to the extent that the two codons AGU/C-235 favor suppression more than does any of the codons UCN-235. However, we observe in addition pronounced effects of changing the third (wobble) base of both the upstream and downstream codons. One may ask why such effects did not emerge from comparisons of numerous sites (Bossi, 1983; Miller and Albertini, 1983). A detailed discussion of this point should await more quantitative data on the efficiency of suppression; however, one possibility is that the contribution to the UGA context of these positions depends on other elements of the context and is perhaps a function of the particular codon of which they are part. In other words, it is possible that changing the wobble U to an A in the codon UCU, for example, is quite different in effect from making the same change in another codon.

An essential difference between nonsense suppression and missense suppression is that in the case of missense suppression, the measurable phenotype is the result of competition between two rather similar processes, involving either the normal cognate tRNA species or the mutant suppressor species. Thus, it is perhaps surprising that missense suppression should be sensitive to context effects. Presumably, such effects arise when translation by one of the competing species is more affected by context than is translation by the other. Conversely, situations in which missense suppression appears insensitive or little sensitive to context, such as the AGA-234 context series that we have examined, may simply reflect similar effects on the translation efficiency of the two species. These considerations also complicate any attempt to compare the effects seen on missense suppression and nonsense suppression, even at the same site in a messenger, such as those we report here. Thus, the fact that the strong purine-pyrimidine effect of the adjacent downstream base seen in nonsense suppression is not reproduced in the missense observations we report here does not necessarily mean that the elongation process is much less affected by changes in this position than is the termination process. On the contrary, the observation of pronounced effects of nearby sequence in some instances on missense suppression indicates that the context effects are not restricted to termination. This is the most important conclusion to be drawn from our data concerning the AAA-234 context series. Otherwise, the considerations outlined above concerning the possible mechanisms of context effects apply equally to missense suppression.

The areas in which context effects on elongation are most likely to be of biological significance concern the efficiency and accuracy of gene expression. Missense suppression creates an abnormal situation by introducing a new competing species, the missense suppressor. In the normal situation, the events competing with normal translation include mistranslation by noncognate species of tRNA and other accidents of translation such as frameshifting and peptidyl-tRNA drop-off (Buckingham and Grosjean, 1986). The probability of such accidents is expected to depend on context. Two analyses of coding sequences provide support for the idea that constraints exist on codon context in *E. coli* genes (Shpaer, 1986; Yarus and Folley, 1985) as well as on the choice of synonymous codons (Staden, 1984). This has led to the suggestion that in highly expressed genes, *E. coli* tends to avoid poorly translated codons if more efficiently translated synonymous codons are available or, in the absence of choice, to increase the efficiency of translation by the optimal choice of codon context (Shpaer, 1986). It seems likely that the effects of codon context contribute to the evolutionary pressures exerted on the organism, which lead to the statistically significant constraints on codon context displayed by coding sequences.

We gratefully acknowledge the Commission of the European Communities for contract 85100037 under the Stimulation program, the Centre National pour la Recherche Scientifique (UA 1139), NATO Scientific Affairs Division (grant 84/712), the Fondation pour la Recherche Médicale, the Danish Natural Science Research Council (grants 11-70382 and 11-7697), E. I. du Pont de Nemours and Co., Inc., the American Cancer Society (grant NP167), and the National Institute for General Medical Sciences (Public Health Service grant GM 21499).

REFERENCES

Bossi, L. 1983. Context effects: translation of UAG codon by suppressor tRNA is affected by the sequence following UAG in the message. *J. Mol. Biol.* **164:**73–87.

Bossi, L., and J. R. Roth. 1980. The influence of codon context on genetic code translation. *Nature* (London) **286:**123–127.

Buckingham, R. H., and H. Grosjean. 1986. The accuracy of mRNA-tRNA recognition, p. 83–126. *In* T. B. L. Kirkwood, R. F. Rosenberger, and D. J. Galas (ed.), *Accuracy in Molecular Processes: Its Control and Relevance to Living Organisms.* Chapman & Hall, Ltd., London.

Edelmann, P., R. Martin, and J. Gallant. 1987. Nonsense suppression context effects in Escherichia coli bacteriophage T4. *Mol. Gen. Genet.* 207:517–518.

Feinstein, S. I., and S. Altman. 1978. Context effects on nonsense codon suppression in *E. coli. Genetics* **88:**201–209.

Freier, S. M., B. J. Burger, D. Alkema, T. Neilson, and D. H. Turner. 1983. Effects of 3′ dangling end stacking on the stability of GGCC and CCGG double helices. *Biochemistry* **22:**6198–6206.

Grosjean, H., S. de Henau, and D. M. Crothers. 1978. On the physical basis for ambiguity in genetic coding interactions. *Proc. Natl. Acad. Sci. USA* **75:**610–614.

Hartmann, B., and R. Lavery. 1989. The conformation and stability of ribonucleic acids: modeling base sequence effects in double stranded helices. *J. Biomol. Struct. Dynam.* 7:363–380.

Martin, R., M. Weiner, and J. Gallant. 1988. Effects of release factor context at UAA codons in *Escherichia coli. J. Bacteriol.* 170:4714–4717.

Miller, J. H., and A. M. Albertini. 1983. Effects of surrounding sequence on the suppression of nonsense codons. *J. Mol. Biol.* **164:**59–71.

Murgola, E. J. 1985. tRNA, suppression, and the code. *Annu. Rev. Genet.* **19:**57–80.

Murgola, E. J., H. U. Göringer, A. E. Dahlberg, and H. A. Hijazi. 1989. Ribosomal RNA and UGA-dependent peptide chain termination, p. 221–229. *In* T. Cech (ed.), *Molecular Biology of RNA.* Alan R. Liss, Inc., New York.

Murgola, E. J., H. A. Hijazi, H. U. Göringer, and A. E. Dahlberg. 1988. Mutant 16S ribosomal RNA: a codon specific translational suppressor. *Proc. Natl. Acad. Sci. USA* **85:**4162–4165.

Murgola, E. J., and C. I. Jones. 1978. A novel method for detection and characterization of ochre suppressors in *Escherichia coli. Mol. Gen. Genet.* **159:**179–184.

Murgola, E. J., F. T. Pagel, and K. A. Hijazi. 1984. Codon context effects in missense suppression. *J. Mol. Biol.* **175:**19–27.

Murgola, E. J., and C. Yanofsky. 1974. Selection for new amino acids at position 211 of the tryptophan synthetase chain of *Escherichia coli. J. Mol. Biol.* **86:**775–784.

Salser, W. 1969. The influence of reading context upon the suppression of nonsense codons. *Mol. Gen. Genet.* **105:**125–130.

Salser, W., M. Fluck, and R. Epstein. 1969. The influence of reading context upon the suppression of nonsense codons. III. *Cold Spring Harbor Symp. Quant. Biol.* **34:**513–520.

Shpaer, E. G. 1986. Constraints on codon context in *Escherichia coli* genes: their possible role in modulating the efficiency of translation. *J. Mol. Biol.* **188:**555–564.

Smith, D., and M. Yarus. 1989. tRNA-tRNA interactions within cellular ribosomes. *Proc. Natl. Acad. Sci. USA* **86:**4397–4401.

Staden, R. 1984. Measurements of the effects that coding for a protein has on a DNA sequence and their use for finding genes. *Nucleic Acids Res.* **12:**551–567.

Stormo, G. D., T. D. Schneider, and L. Gold. 1986. Quantitative analysis of the relationship between nucleotide sequence and functional activity. *Nucleic Acids Res.* **14:**6661–6679.

Valle, R. P. C., and M.-D. Morch. 1988. Stop making sense, or regulation at the level of termination in eukaryotic protein synthesis. *FEBS Lett.* **235:**1–15.

Yahata, H., Y. Ocada, and A. Tsugita. 1970. Adjacent effect on suppression efficiency. II. Study on ochre and amber mutants of T4 phage lysozyme. *Mol. Gen. Genet.* **106:**208–212.

Yarus, M., and M. S. Folley. 1985. Sense codons are found in specific contexts. *J. Mol. Biol.* **182:**529–540.

Chapter 48

Effect of Protein Overexpression on Mistranslation in *Escherichia coli*

GREGG BOGOSIAN, BERNARD N. VIOLAND, PATRICIA E. JUNG, and JAMES F. KANE

Since the advent of modern recombinant DNA techniques, the use of *Escherichia coli* for the construction of microbial strains designed to overproduce desired proteins is becoming increasingly common and widespread. A recent review of such production systems, restricted to those using the *trp* promoter, listed nearly 80 different proteins overexpressed in *E. coli* (Somerville, 1988). The number of literature and patent reports of such systems is increasing at a dizzying pace, with *E. coli* protein production systems now being exploited on a commercial as well as a research scale.

E. coli is popular with microbial genetic engineers because it is so well characterized and so amenable to genetic manipulation. Surprisingly, despite the impressive data base that exists on *E. coli*, there are two major areas where relevant information is relatively scant. The first involves the use of *E. coli* in large-scale fermentations, an application of fairly recent origin which is in need of much further research. The second neglected area concerns how the forced overproduction of a protein affects *E. coli*. It is the purpose of this chapter to review the small inroads that have been made into this area, especially with respect to the effects of protein overproduction on translational fidelity in *E. coli*.

Mistranslation can be defined as the incorporation into a growing polypeptide chain of an amino acid residue which is different from the amino acid normally corresponding to the codon being translated. Mistranslation can occur by two general mechanisms. In the first mechanism, the codon is read incorrectly and a different amino acid is incorporated at that point in the polypeptide chain. This type of mistranslation is most commonly exemplified by ribosomal ambiguity and certain types of suppression. In the second mechanism, the codon is read correctly, but the tRNA molecule is charged with the "wrong" amino acid. This can involve either a tRNA with a mutated anticodon (illustrated by tRNA suppressors) or a tRNA that has been incorrectly charged. *E. coli* strains engineered to overproduce proteins have been observed to exhibit increased mistranslation by both of these general mechanisms.

MISCHARGING ERRORS

Mischarging of tRNA by aminoacyl-tRNA synthetases is restricted to charging of the cognate tRNA with a noncognate amino acid; the charging of the cognate amino acid onto a noncognate tRNA is not thought to occur (Cramer et al., 1979; Fersht, 1981; Yarus, 1979). There are, as is so often the case in science, some fascinating exceptions to this rule. Glutamic acid is inserted more often than tryptophan at amber (TAG) codons in *supU(trpT)* strains by misacylation of this mutant tryptophan tRNA by glutamyl-tRNA synthetase (Soll and Berg, 1969; Yaniv et al., 1974; Celis et al., 1976; Knowlton et al., 1980); the charging ratio is 9:1 Glu to Trp. A mutation has been isolated in *glnS*, the structural gene for glutaminyl-tRNA synthetase, which misacylates the mutant tyrosine tRNA in *supF(tyrT)* strains, inserting glutamine at amber codons (Hoben et al., 1984); the wild-type glutaminyl-tRNA synthetase is known to misacylate other tRNAs (but not tyrosine tRNA) at a very low level (Inokuchi et al., 1984).

Mischarging of a mutant tRNA with one of the 20 canonical amino acids is the underlying basis for tRNA nonsense, missense, and frameshift suppressors. In addition, certain aminoacyl-tRNA syn-

Gregg Bogosian, Bernard N. Violand, Patricia E. Jung, and James F. Kane ■ Animal Sciences Division, Monsanto Corporation, Mail Zone BB3M, 700 Chesterfield Village Parkway, Chesterfield, Missouri 63198.

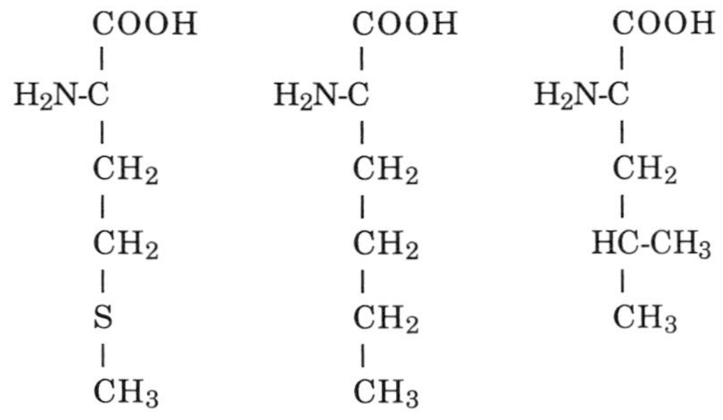

methionine	norleucine	leucine
COOH	COOH	COOH
I	I	I
H_2N-C	H_2N-C	H_2N-C
I	I	I
CH_2	CH_2	CH_2
I	I	I
CH_2	CH_2	HC-CH_3
I	I	I
S	CH_2	CH_3
I	I	
CH_3	CH_3	

Figure 1. Structures of methionine, norleucine, and leucine.

thetases have been observed to charge their cognate, wild-type tRNAs with noncognate amino acids. Normally, this type of error is prevented by the proofreading function of aminoacyl-tRNA synthetases. This proofreading is not perfect, however, as evidenced by the mischarging in *E. coli* of isoleucine tRNA with valine (Flossdorf and Kula, 1973), of alanine tRNA with serine and glycine (Tsui and Fersht, 1981), and of valine tRNA with alanine, cysteine, isoleucine, serine, and threonine (Owens and Bell, 1970).

There are many analogs of certain amino acids which can be mischarged onto the corresponding tRNA and incorporated into protein in place of the normal amino acid residue (reviewed in Richmond, 1962, and Pine, 1978). For the most part, these analogs must be supplied exogenously in order to observe their misincorporation. Recently, however, two exceptions have been described.

Incorporation of Selenocysteine into Protein

The cysteine analog selenocysteine was found in formate dehydrogenase from *E. coli* (Zinoni et al., 1986; Zinoni et al., 1987). The structural gene for formate dehydrogenase, *fdhF*, was sequenced and found to contain an in-frame opal (TGA) stop codon at position 140; formate dehydrogenase contains 715 amino acids. Selenocysteine was found to be the amino acid occupying position 140 in the polypeptide chain. It has been determined that selenocysteine is synthesized from a serine residue covalently bound to a novel opal-suppressing serine tRNA molecule (Leinfelder et al., 1988; Leinfelder et al., 1989; Heider et al., 1989). Thus, although the amino acid selenocysteine is synthesized by *E. coli*, it is incorporated into protein not by mischarging but rather via the postcharging modification of serine.

Misincorporation of Norleucine into Protein

Incorporation of the methionine analog norleucine (Fig. 1) into bacterial proteins in place of methionine residues was originally observed over 30 years ago, using exogenously supplied norleucine (Munier and Cohen, 1956, 1959; Cohen and Munier, 1959; Cowie et al., 1959). It was subsequently established that norleucine can substitute for methionine in protein both at internal residues and at the amino terminus (Trupin et al., 1966; Kerwar and Weissbach, 1970). Norleucine was shown to substitute for methionine in the acylation of methionine tRNA and was able to be formylated after acylation; methionine was found to compete with norleucine in the acylation reaction (Trupin et al., 1966). Formylnorleucyl-tRNA was shown to initiate protein synthesis, and the formyl group could be removed from the nascent peptide (Kerwar and Weissbach, 1970). Others confirmed that norleucine is incorporated at internal and amino-terminal methionine residues of proteins and that norleucine can be charged onto both $tRNA^{Met}$ and $tRNA^{fMet}$ (Brown, 1973; Barker and Bruton, 1979). Whether the amino-terminal norleucyl residue can be processed off the nascent peptide by methionine aminopeptidase seems to depend on the nature of the second amino acid residue in the protein (Brown, 1973; Barker and Bruton, 1979).

Using exogenously supplied norleucine, several proteins containing norleucine at all of their methionine residues have been made: staphylococcal nuclease from *Staphylococcus aureus* (Anfinsen and Corley, 1969), β-galactosidase (Naider et al., 1972) and adenylate kinase (Gilles et al., 1988) from *E. coli*, and human epidermal growth factor from a recombinant strain of *E. coli* (Koide et al., 1988). In all of these cases, the biological activities and other properties of the norleucine-containing proteins were identical to those of the unsubstituted proteins. Bovine somatotropin (BST) isolated from an overproducing recombinant strain of *E. coli* grown in minimal medium was found to have norleucine comprising up to 14% of the methionine content, indicating that *E. coli* was capable of biosynthesizing norleucine and incorporating it into protein (Bogosian et al., 1989). BST contains five methionine residues (including the amino-terminal residue); the norleucine misincorporation was found to be evenly distributed at these sites. In addition, norleucine was found to be present in the total cell protein from this strain.

Earlier studies with *Serratia marcescens* had established that norleucine was biosynthesized in mutants that overproduced leucine and that mutants unable to make leucine were also unable to make norleucine (Kisumi et al., 1976a, 1977; Kisumi et al., 1976b). In *E. coli*, the leucine biosynthetic enzymes were shown to be responsible for the formation of norleucine (Bogosian et al., 1989). It was also shown that the levels of these enzymes were very high in the

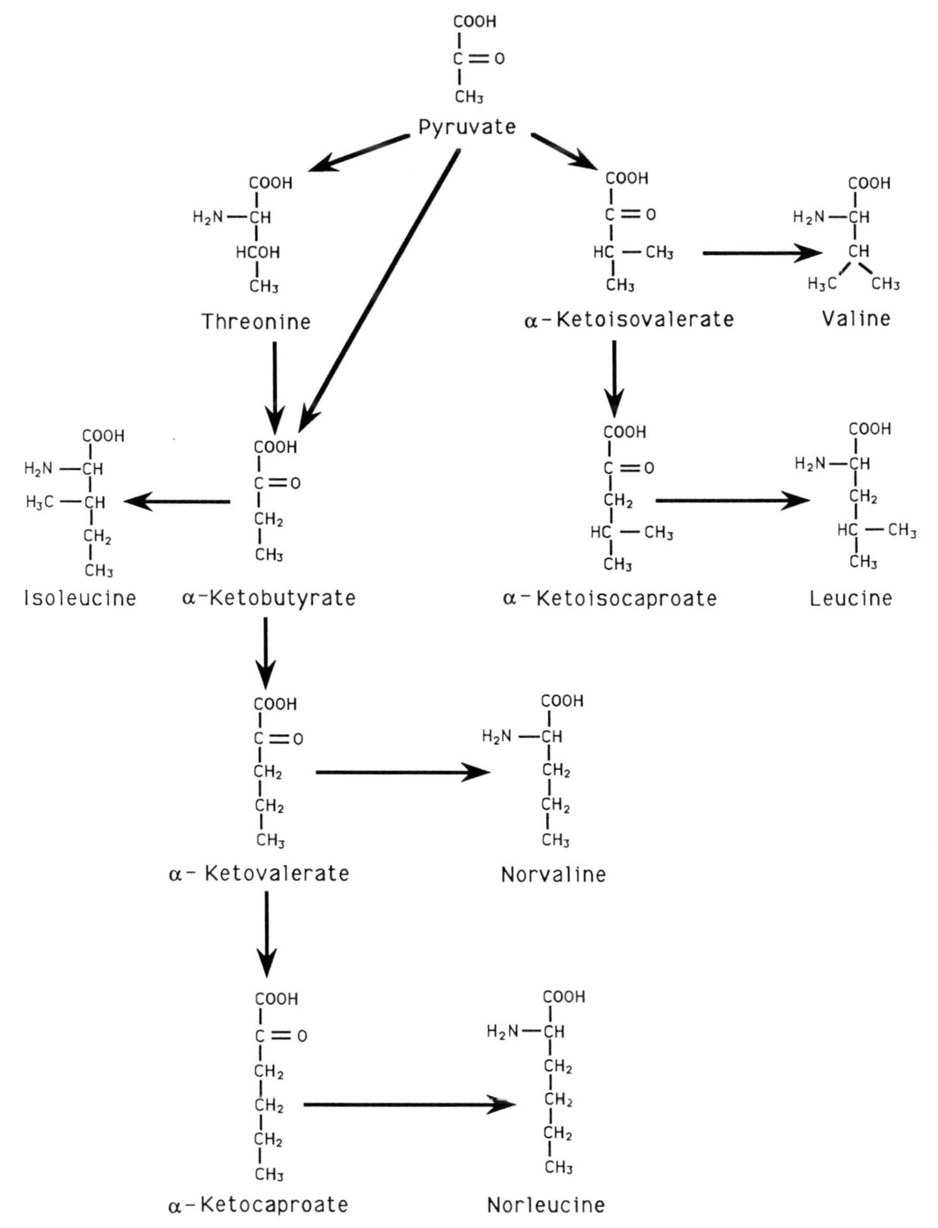

Figure 2. Proposed pathway of norleucine and norvaline biosynthesis in *E. coli*. Starting with pyruvate, the established routes of synthesis of threonine, isoleucine, valine, and leucine are shown. Also depicted is the proposed route of conversion of α-ketobutyrate to α-ketovalerate and α-ketocaproate by one or two passes, respectively, through the keto acid chain elongation process catalyzed by the enzymes of leucine biosynthesis. Upon transamination, these latter two keto acids would yield, respectively, norvaline and norleucine. Also depicted is the proposed conversion of pyruvate to α-ketobutyrate by the keto acid chain elongation process. (Taken with permission from Bogosian et al., 1989.)

strain overproducing BST; furthermore, the leucine biosynthetic enzymes were found to have broad substrate specificities, permitting their participation in the formation of amino acids other than leucine. A norleucine biosynthetic pathway was proposed wherein the leucine biosynthetic enzymes convert pyruvate and α-ketobutyrate to α-ketocaproate, which was subsequently transaminated to give norleucine (Fig. 2). Deletion of the leucine operon was found to prevent the formation of norleucine. BST is a leucine-rich protein, with leucine comprising 27 of the 191 amino acid residues of the protein. The leucine content of native *E. coli* proteins from the strain overproducing BST was found to be just over half that of BST. Apparently, induction of high-level synthesis of the leucine-rich BST led to the derepression of the leucine biosynthetic enzymes (Fig. 3), with the subsequent formation of norleucine at levels high enough to be incorporated into protein. Supplementation of the culture medium with excess methionine

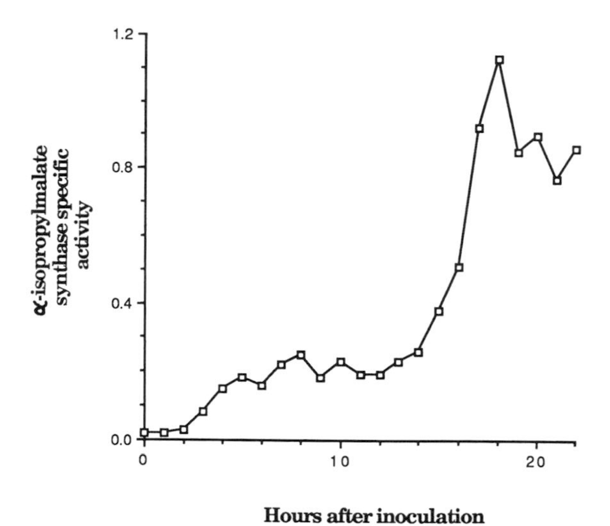

Figure 3. α-Isopropylmalate synthase specific activity in W3110G (pBGH1) grown in a fermentor. Samples were taken at hourly intervals starting with the time point designated, h 0 being the point of inoculation of the fermentor. High-level synthesis of BST was induced by the addition of indoleacrylate at h 12. Growth and assay conditions are described elsewhere (Bogosian et al., 1989). The specific activity of α-isopropylmalate synthase is expressed as units of enzyme per milligram of protein. One unit is the amount of enzyme that generates 1 μM of coenzyme A per min. (Taken with permission from Bogosian et al., 1989.)

was found to be capable of preventing norleucine misincorporation, presumably by swamping out the norleucine in competition for charging onto $tRNA^{Met}$ and $tRNA^{fMet}$.

Recombinant strains of *E. coli* overproducing other proteins have been found to exhibit norleucine misincorporation. These include porcine somatotropin (Violand et al., 1989), which has a leucine content similar to that of BST, and interleukin-2 (Lu et al., 1988; Tsai et al., 1988), which has 22 leucine residues in a total of 134 amino acid residues. Strains overproducing interferon, which has a relatively low leucine content, exhibit a low level of norleucine misincorporation (Tsai et al., 1988).

Norleucine misincorporation arose as a result of imbalances in the amino acid pools in the cell. Overproduction of a leucine-rich protein led to a situation in which the level of norleucine was sufficiently higher than the level of methionine to bring about significant misincorporation of norleucine for methionine in proteins.

MISSENSE ERRORS

Substitution of Lysine for Arginine

Missense errors are difficult to detect in large proteins because of technological limitations in characterizing such proteins at the amino acid sequence level. Small proteins are much more amenable to such examination, and recently two small proteins overexpressed in *E. coli* have been found to exhibit an unprecedented type of missense error.

Atrial peptide III (APIII) consists of 24 amino acid residues; it has arginine residues at positions 7, 10, and 23 and has no lysine residues. When purified from an overproducing strain of *E. coli* and examined by reverse-phase high-performance liquid chromatography, the APIII was observed to consist of one major protein peak and two minor peaks. Protein sequencing of the material in these peaks revealed that the major peak consisted of APIII itself, whereas the minor peaks consisted of APIII with one or two of the arginine residues substituted by lysine. One of the minor peaks consisted of equimolar amounts of APIII with lysine at residue 7 and APIII with lysine at residue 10; the other minor peak had lysine at both residues 7 and 10 (A. Hershman, N. Horn, B. Larsen, and N. Siegel, unpublished observations). These three minor species comprised as much as 11% of the total APIII obtained from the overproducing strain.

Lysine misincorporation was never observed at position 23 in APIII. There were two arginine codons utilized in the APIII structural gene, AGA at positions 7 and 10 and CGT at position 23. In *E. coli*, the arginine CGT codon is used with high frequency, whereas the arginine AGA codon is used very rarely (McPherson, 1988). Insulinlike growth factor 1 (IGF-1) consists of 70 amino acid residues, with arginine residues at positions 21, 36, 37, 50, 55, and 56. When overproduced in *E. coli* by use of a structural gene with the rare AGA codon at all six positions, lysine-for-arginine substitution was observed at a level of up to 12% of the total IGF-1 protein (Seetharam et al., 1988). In this case, protein sequencing was used to confirm lysine misincorporation at positions 21, 36, and 37; the automated protein sequencer used could not reach residues 50, 55, and 56 with reliability.

The lysine misincorporation observed with APIII and IGF-1 could be due to either of two distinct mechanisms. Ribosomal ambiguity could be allowing a lysine tRNA that normally reads the lysine codon AAA to read the rare arginine codon AGA. This would involve a second-position wobble of the middle U in the anticodon with the G in the AGA codon, as proposed by McPherson (1988). Alternatively, the arginyl-tRNA synthetase could be mischarging the arginine AGA tRNA with lysine.

All six of the rare arginine AGA codons in IGF-1 were changed to the preferred arginine CGT codon. IGF-1 obtained from overproducing strains of *E. coli* by using this altered structural gene was free of lysine misincorporation. When one AGA codon (at position

Table 1. Amino acid composition of a 1% Casamino Acids solution[a]

Amino acid[b]	Level			
	Unhydrolyzed[c]		Hydrolyzed[d]	
	mM	μg/ml	mM	μg/ml
Alanine	1.76	160	2.55	220
Arginine	0.77	130	1.36	240
Aspartic acid[e]	2.69	360	3.41	450
Cysteine	0	0	0	0
Glutamic acid[e]	5.18	760	9.94	1,500
Glycine	1.13	90	1.56	120
Histidine	0.42	70	0.96	150
Isoleucine	0.88	120	2.27	300
Leucine	2.31	300	3.96	520
Lysine	2.09	310	3.96	580
Methionine	0.71	110	1.25	190
Phenylalanine	0.82	140	1.35	220
Proline	2.97	340	5.91	680
Serine	1.78	190	2.76	290
Threonine	1.08	130	1.96	230
Tyrosine	0.37	70	0.92	170
Valine	1.50	180	3.34	390

[a] The Casamino Acids used was the technical grade from Difco Laboratories, Detroit, Mich. Approximately 44% of the sample was composed of higher-molecular-weight material, presumably small peptides, which was converted by acid hydrolysis to individual amino acids. The total weight of the amino acids was 6.2 mg/ml out of the total 10 mg/ml used to prepare the 1% solution of Casamino Acids. According to the supplier, this material is approximately 37% sodium chloride by weight, which accounts for the remainder.
[b] The methods of amino acid analysis used do not yield tryptophan.
[c] Averages of determinations by three liquid chromatographic methods: pre-column and post-column derivatization with o-phthaldialdehyde and pre-column derivatization with phenylthiohydantoin. The three methods yielded nearly identical results.
[d] The sample was dried and then hydrolyzed in vacuo with 6 N HCl at 115°C for 24 h.
[e] The conditions used for the amino acid analyses convert asparagine to aspartic acid and glutamine to glutamic acid.

36) was retained, the IGF-1 obtained was also free of lysine misincorporation (Seetharam et al., 1988). When the two AGA codons in APIII were replaced with CGT codons, lysine misincorporation was eliminated from that peptide as well (D. McPherson, A. Hershman, N. Horn, B. Larsen, and N. Siegel, unpublished observations).

As with the norleucine misincorporation discussed above, amino acid pool imbalances seemed to play a role in lysine misincorporation. The IGF-1-overproducing strain was grown in a culture medium supplemented with Casamino Acids (Difco Laboratories) at a level of 1% (Wong et al., 1988). This supplementation imparted to the *E. coli* cells distinct amino acid pool imbalances, as the amino acid composition of Casamino Acids revealed a lysine-to-arginine ratio of nearly 3:1 (Table 1). In contrast, the free amino acid pool in a wild-type strain of *E. coli* had essentially equal levels of arginine and lysine (Table 2). In addition, the high growth rate obtained in such a medium contributed to the level of lysine misincorporation. With the APIII-overproducing strain, when the growth rate was slowed without making any medium changes, the level of lysine misincorporation was reduced from 15 to 7%. In addition, when the culture medium was supplemented with arginine to a level of 1.8 g/liter, the lysine misincorporation into APIII was reduced to 1% (M. Kuo, P. Wang, G. Tong, C. Nolan, and M. Gustafson, unpublished observations).

Insertion of Glutamine at an Amber Stop Codon

The recombinant plasmid used in the BST-overproducing strain discussed above (see Misincorporation of Norleucine into Protein) was designated pBGH1. This plasmid carries the BST structural gene under the control of the *E. coli trp* promoter (Seeburg et al., 1983). To achieve high-level synthesis of BST, a wild-type strain of *E. coli* harboring pBGH1 was grown in a minimal medium, and BST synthesis was turned on by addition of the gratuitous inducer indoleacrylate, a tryptophan analog (Kane and Bogosian, 1987; Calcott et al., 1988). With this system, nearly 30% of the total cellular protein consisted of BST.

Table 2. Amino acid levels in *E. coli* and Casamino Acids

Amino acid[a]	Level (%)[b]	
	E. coli[c]	Casamino Acids[d]
Alanine	1.9	5.4
Arginine	0.9	2.9
Aspartic acid[e]	19	7.2
Glutamic acid[e]	58	21
Glycine	3.2	3.3
Histidine	0.3	2.0
Isoleucine	0.4	4.8
Leucine	2.0	8.3
Lysine	1.4	8.3
Methionine	1.1	2.6
Phenylalanine	0.9	2.8
Proline	1.0	12
Serine	0.9	5.8
Threonine	1.0	4.1
Tyrosine	0.7	1.9
Valine	6.8	7.0

[a] The methods of analysis are as described in Table 1, footnote *a*. Cysteine and tryptophan could not be determined from the *E. coli* sample and are omitted from both columns.
[b] Percentage of the total molar concentration of the 16 amino acids quantitated.
[c] The analysis was performed on strain W3110 grown in a 10-liter fermentor in a chemically defined minimal medium with glucose as the carbon source; exact growth conditions and preparation of the amino acid sample are given elsewhere (Bogosian et al., 1989). The sample was taken when the culture was in the logarithmic phase of growth at an optical density of 50 (determined spectrophotometrically at 550 nm); this corresponds to a dry cell weight of 11 to 12 g/liter. About 0.7% of the dry cell weight was composed of free amino acids.
[d] Data converted from the analysis of hydrolyzed Casamino Acids given in Table 1.
[e] The conditions used for the amino acid analyses convert asparagine to aspartic acid and glutamine to glutamic acid.

```
                                       Gln
Phe  Gly  Glu  Ala  Ser  Cys  Ala  Phe  |   Lys  Leu  Asn  Ser

TTC  GGG  GAG  GCC  AGC  TGC  GCA  TTC  TAG  AAG  CTT  AAT  TCT
```

Figure 4. Nucleotide sequence in the vicinity of the amber stop codon of the BST structural gene on plasmid pBGH1. The nucleotide sequence is divided into codons representing the reading frame of the BST structural gene. The amino acid residues corresponding to those codons are given above the nucleotide sequence. Normally, the carboxy terminus of the BST protein is the phenylalanine residue encoded immediately before the amber stop codon. Also depicted is the readthrough of the stop codon by misincorporation of a glutamine residue. The first four codons and their corresponding amino acid residues are shown in frame downstream of the stop codon. The nucleotide sequence AAGCTT, located immediately past the stop codon, is a recognition site for the restriction endonuclease *Hin*dIII, at coordinate 874 on pBGH1 (Calcott et al., 1988; also see Fig. 5). Following the *Hin*dIII site is the nucleotide sequence AATTC, which is the remnant of a filled-in *Eco*RI site from the transcription terminator fragment used in the construction of pBGH1 (Bogosian and Kane, 1987; Calcott et al., 1988).

Downstream of the BST structural gene on pBGH1 is a transcriptional terminator consisting of two tandem *lacUV5* promoters (Calcott et al., 1988). The translational stop codon used for the BST structural gene on pBGH1 is an amber (TAG) codon. The nucleotide sequence of the tandem *lacUV5* promoter region has been determined (Bogosian and Kane, 1987) and reveals that there is an in-frame open reading frame of 81 amino acids located beyond the amber stop codon of the BST structural gene; this reading frame ends within the two *lacUV5* promoters at an ochre (TAA) codon.

BST consists of 191 amino acids and has a molecular weight of 22,000. When isolated from an overproducing strain of *E. coli* harboring pBGH1, the material was found to contain a protein of molecular weight 30,000 as well as the expected protein of molecular weight 22,000. By Western (immunoblot) analysis using antibodies directed against BST, both molecular weight species were found to be immunoreactive. The 30,000-molecular-weight material constituted approximately 3% of the total isolated BST.

By protein sequencing, both BST species were found to have identical amino-terminal amino acid sequences, corresponding to that expected for normal BST. By tryptic mapping, the larger BST species had 10 more tryptic peptides than the normal BST. These additional tryptic peptides were isolated and sequenced. They were found to consist of the amino acids encoded by the open reading frame located downstream of the amber stop codon on pBGH1. One of the tryptic peptides contained the eight carboxy-terminal amino acids of normal BST, followed by a glutamine and a lysine residue (trypsin cleaves after lysine or arginine residues in proteins). On pBGH1, following the amber stop codon for the BST structural gene is a lysine AAG codon (Fig. 4). Apparently, the 30,000-molecular-weight BST species arose as a result of a misincorporation event in which a glutamine residue was inserted at an amber stop codon. The calculated molecular weight of the additional amino acids encoded by this downstream open reading frame is 8,000, which accounts for the size of the larger BST species observed.

As mentioned above, this downstream open reading frame ends with an ochre codon. No readthrough of this ochre codon could be detected. When a variant of pBGH1 was made with an ochre codon inserted immediately upstream of the amber codon at the end of the BST structural gene, no readthrough was observed.

The *E. coli* strain used for overproducing BST was W3110 (Bogosian et al., 1989), a wild-type strain that does not contain any amber suppressors. The suppressor mutation *supE* (in the gene *glnV*) inserts glutamine at amber codons. Mutants of bacteriophage lambda with amber mutations that were suppressed by *supE* strains were not suppressed by W3110. Amber mutations in the *lacZ* gene that were suppressed by *supE* strains were not suppressed by W3110. Apparently, when BST is overproduced, wild-type glutaminyl-tRNA^{Gln} misreads amber stop codons with a significant frequency.

A PLASMID TO MONITOR MISTRANSLATION: pXT31

To investigate the observed misincorporation of glutamine at the amber stop codon of the BST structural gene on pBGH1, a special derivative of this plasmid was constructed. Using an in vivo selection technique (Berman and Jackson, 1984), *lacZ* (the structural gene for β-galactosidase) was placed on pBGH1 within, and in frame with, the open reading frame downstream of the BST structural gene; this variant of pBGH1 was designated pXT31 (Fig. 5). The particular *lacZ* gene used in this construction lacked an ATG initiation codon, ribosome-binding site, and promoter and so was phenotypically Lac$^-$; this defective *lacZ* gene can be activated by putting it

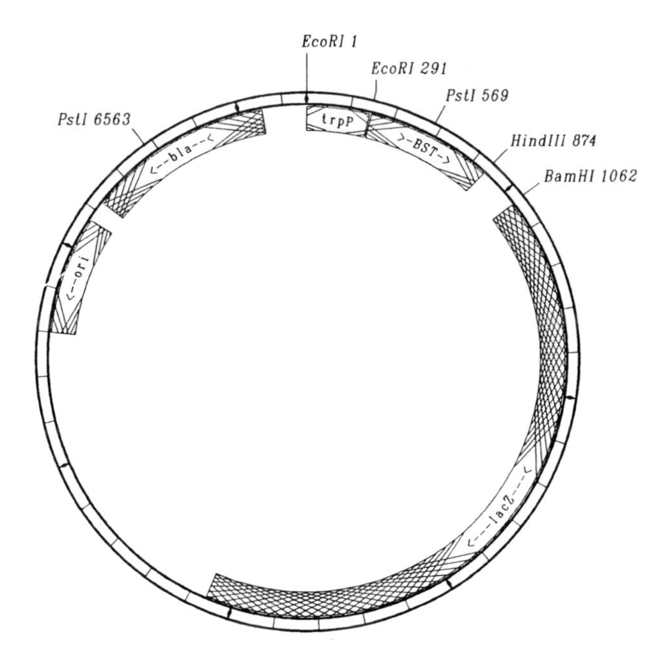

Figure 5. Map of plasmid pXT31. The plasmid is 7,310 base pairs long. Key restriction endonuclease cleavage sites and their map coordinates are shown, as well as blocks corresponding to genes or regulatory sequences. Arrows within the blocks indicate the orientation of these sequences, which include the origin of replication (*ori*), the *trp* promoter (*trpP*), and the structural genes for β-lactamase (*bla*) (determinant of ampicillin resistance), bovine somatotropin (BST), and β-galactosidase (*lacZ*). As discussed in the text, the *lacZ* gene is located in frame with and 65 codons downstream of the amber stop codon of the BST structural gene.

in frame with an expressed gene (Berman and Jackson, 1984). The exact location of the *lacZ* gene on pXT31 was determined by DNA sequencing of the relevant portion of the plasmid. On pXT31, there are 65 codons of the open reading frame downstream of the BST structural gene, followed by the *lacZ* gene; the first *lacZ* codon present encodes the valine residue at position 9 of β-galactosidase.

A derivative of *E. coli* W3110 that has a deletion of the entire *lac* operon was constructed; this strain was designated LBB84. Thus, when harboring pXT31, the only source of β-galactosidase in LBB84 was from the plasmid. LBB84 harboring pXT31 was found to express β-galactosidase. By sodium dodecyl sulfate-polyacrylamide gel electrophoresis examination of the whole-cell proteins, only one species of β-galactosidase was observed. By Western analysis, this protein, as well as BST itself, was found to be immunoreactive with antibodies against BST, demonstrating that pXT31 expresses BST as well as a fusion protein consisting of BST at the amino terminus and β-galactosidase at the carboxy terminus. This fusion protein is formed as a result of misincorporation of glutamine at the amber stop codon of the BST structural gene on pXT31. This misincorporation event was determined to be dependent on expression of the BST protein. When a portion of the *E. coli trpE* gene, also with an amber stop codon, was substituted for the BST structural gene on pXT31, no readthrough was observed and no fusion protein was produced. pXT31 thus enables one to use the relatively simple assay of β-galactosidase activity to monitor mistranslation elicited by expression of BST and as manifested by readthrough of an amber codon.

RESPONSE OF pXT31 TO ENHANCERS OF MISTRANSLATION

When grown in a chemically defined minimal medium with glucose as the carbon source (Bogosian et al., 1989), LBB84 harboring pXT31 expressed 300 to 350 U of β-galactosidase. The mutation *supE42* or *supE44*, both of which encode glutamine-inserting amber suppressors, were transduced into LBB84; when the resulting strains were transformed with pXT31, the level of β-galactosidase produced was increased to 1,300 to 1,500 U. Thus, readthrough of the amber codon on pXT31 responded to classical suppressors.

Agents that are known to increase mistranslation were also tested on LBB84 harboring pXT31. Streptomycin is known to suppress mutations in sensitive strains at sublethal levels (Gorini and Kataja, 1964; Davies et al., 1964; Gorini and Kataja, 1965; Davies et al., 1965; Gorini, 1966). When LBB84 harboring pXT31 was grown in minimal medium with increasing concentrations of streptomycin, the β-galactosidase levels increased from about 300 to over 1,800 U just as the streptomycin was beginning to affect growth (Fig. 6). At levels of streptomycin that severely inhibited growth, the levels of β-galactosidase fell sharply. This latter effect probably reflected general inhibition of translation at these levels of streptomycin. Ethanol is known to affect the accuracy of translation in vitro (So and Davie, 1964). Growth of LBB84 harboring pXT31 in minimal medium with increasing levels of ethanol resulted in an increase in β-galactosidase activity from about 300 to over 4,000 U in a fashion inversely proportional to the effect on growth (Fig. 7).

EFFECT OF MEDIUM COMPOSITION ON pXT31

All of the assays described above were performed in a chemically defined minimal medium (Bogosian et al., 1989). As mentioned above, when LBB84 harboring pXT31 was grown in this type of medium, the level of β-galactosidase typically produced was 300 to 350 U. When this medium was supplemented with Casamino Acids to a concentration of 1%, the level of β-galactosidase increased nearly threefold, to 1,100 to 1,200 U. An independent plasmid-borne *trp-lac* operon fusion did not exhibit such a response, indicating that the Casamino Acids were not affecting transcription from the *trp* promoter. The amino acid composition of Casamino Acids is very different from the free amino acid pools in an *E. coli* cell (Table 2). The effect of Casamino Acids on readthrough of the amber stop codon was not due to the increased growth rate in such a medium. When LBB84 harboring pXT31 was grown in L broth, a complex medium containing yeast extract and tryptone (Lennox, 1955), the level of β-galactosidase produced was 300 to 350 U, the same as in minimal medium. It appears the amino acid pool bias imparted by Casamino Acids caused the increased readthrough observed.

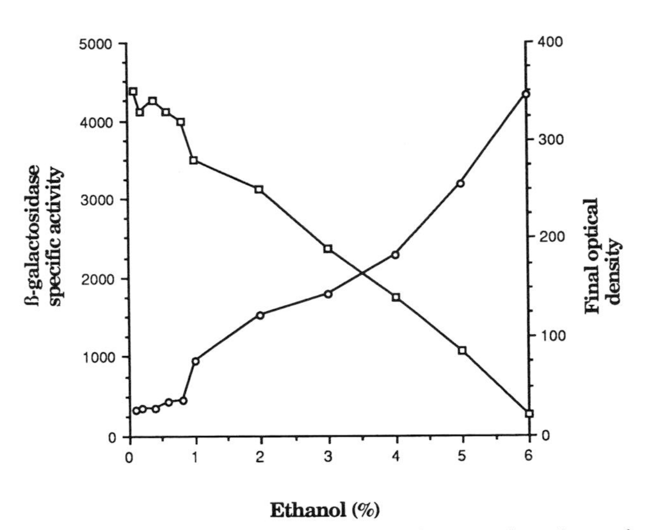

Figure 7. Growth and β-galactosidase production of a culture of LBB84(pXT84) in response to increasing concentrations of ethanol. See the legend to Fig. 6 for growth and β-galactosidase assay conditions. Symbols: ○, β-galactosidase specific activity; □, final optical density.

READTHROUGH OF THE AMBER CODON IN RESTRICTIVE *rpsL* MUTANTS

Streptomycin resistance can arise in *E. coli* by mutation of the *rpsL* gene, encoding ribosomal protein S12 (Traub and Nomura, 1968; Birge and Kurland, 1969; Ozaki et al., 1969). Certain types of such streptomycin-resistant mutants are known to restrict the leakiness of nonsense mutations (Gorini and Kataja, 1964; Gorini, 1969; Breckenridge and Gorini, 1970); such mutant strains are said to possess hyperaccurate ribosomes. Two streptomycin-resistant *rpsL* mutants were selected, with moderate (*rpsL2203*) or strong (*rpsL2204*) restriction of nonsense suppressors. Both mutant *rpsL* alleles were transduced into LBB84 to yield LBB147 (with *rpsL2203*) and LBB150 (with *rpsL2204*). When LBB147 and LBB150 were transformed with pXT31 and assayed for β-galactosidase production, activity had dropped from the normal 300 to 350 U to approximately 260 and 40 U, respectively. When LBB150 was transformed with pBGH1, production of the 30,000-molecular-weight readthrough species of BST was reduced to undetectable levels, but the overall yield of BST was drastically reduced, from over 30% to less than 6% of total cellular protein. This latter effect of the strongly restrictive *rpsL2204* mutation is probably due to the inhibition of growth and translation exhibited by strains with hyperaccurate ribosomes (Galas and Branscomb, 1976; Ruusala et al., 1984; Bohman et al., 1984). At low levels of expression, BST is rapidly turned over in *E. coli*.

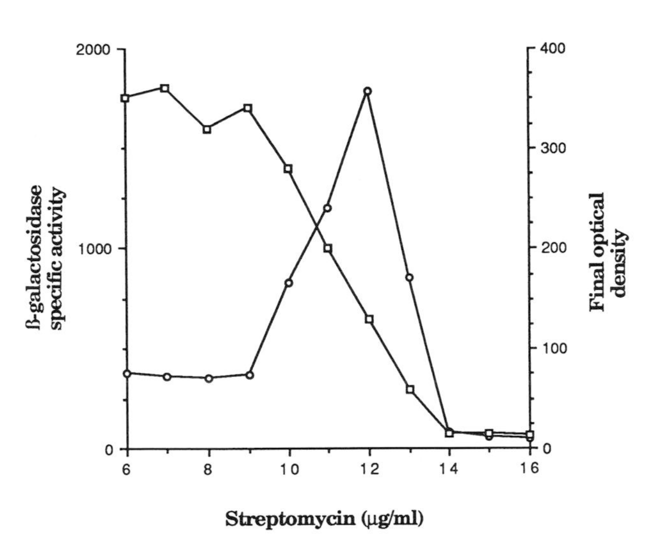

Figure 6. Growth and β-galactosidase production of a culture of LBB84(pXT31) in response to increasing concentrations of streptomycin. The cultures were grown at 37°C on a rotary shaker at 300 rpm in 250-ml baffled sidearm flasks containing 50 ml of a chemically defined minimal medium with glucose as the carbon source. After 10 h of growth, the optical density of the culture was determined with a Klett-Summerson photoelectric colorimeter equipped with a red no. 66 filter, and a sample was taken for determination of β-galactosidase activity. β-Galactosidase assays were performed as described by Epstein (1967). The specific activity units are 200 nmol of *o*-nitrophenol liberated per ml of culture per A_{660} unit per min. Symbols: ○, β-galactosidase specific activity; □, final optical density.

SUPPRESSION OF A FRAMESHIFT MUTATION BY BST OVERPRODUCTION

It has been proposed that imbalances in amino acid or aminoacyl-tRNA pools force increases in

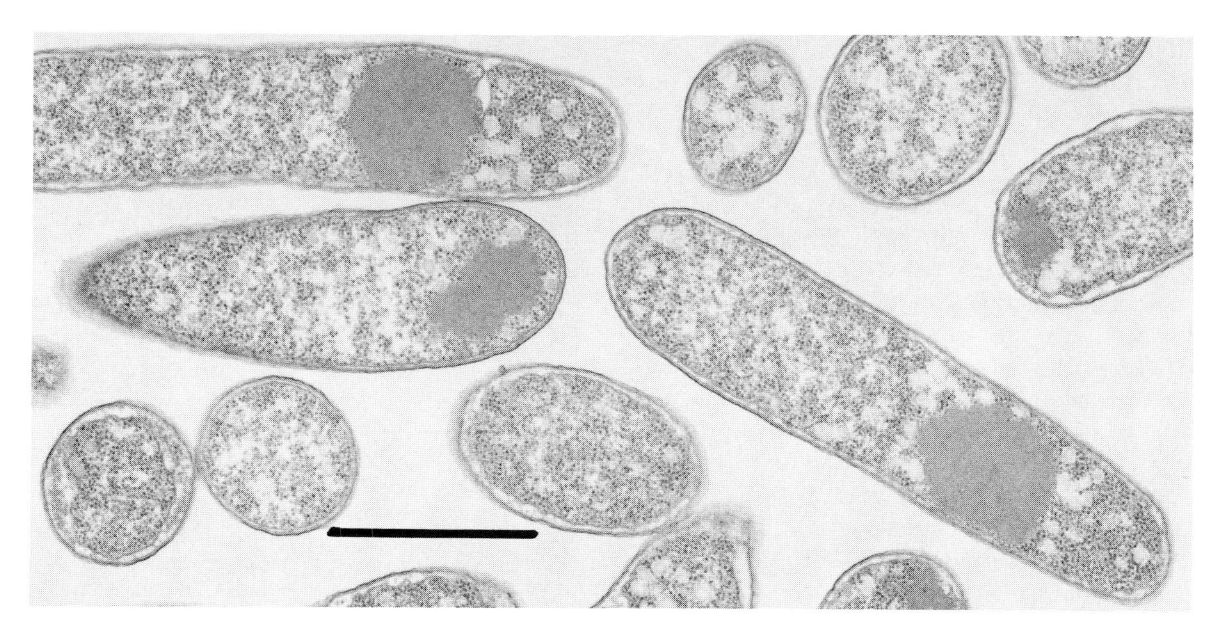

Figure 8. Electron micrograph of *E. coli* cells containing inclusion bodies of bovine somatotropin. Strain W3110(pBGH1) was grown in a fermentor and induced with indoleacrylate as described by Bogosian et al. (1989). The cells shown are from a sample taken 6 h after induction. Bar, 1 μm.

missense errors, with the resulting ribosomal frameshifting at positions of noncognate aminoacyl-tRNA binding detectable by increased leakiness of frameshift mutants (Kurland and Gallant, 1986). If BST-induced imbalances in the amino acid pools were to affect translational fidelity, this effect could have global manifestations within the cell. This possibility was tested by using a *lacZ* frameshift mutation. The rapid petri plate assay of Kurland and Gallant (1986) was used to screen a set of *lacZ* frameshift mutations to identify those which were most suppressed by increases in mistranslation; streptomycin was used to effect the increase in mistranslation for these tests. Three *lacZ* frameshift mutations were identified that responded strongly to streptomycin: *lacZ43*, *lacZ84*, and *lacZ95* (obtained from B. Bachmann of the *E. coli* Genetic Stock Center). Of these, *lacZ84* gave the best response and was used for further analysis. The *lacZ84* mutant allele was moved into strain W3110 to yield strain LBB95. When this strain was grown in a minimal medium, the level of β-galactosidase was about 20 U. When LBB95 was transformed with pBGH1, there was no effect on the level of β-galactosidase. When high-level synthesis of BST was induced by the addition of indoleacrylate, the level of β-galactosidase increased to 230 to 260 U. Indoleacrylate had no effect on the β-galactosidase levels in LBB95 itself. These findings support the contention that overproduction of BST results in amino acid pool imbalances, which in turn either directly or indirectly increase mistranslation in a general fashion.

INCLUSION BODIES AND MISTRANSLATION

Many proteins, when overexpressed in *E. coli*, accumulate as inclusion bodies (Marston, 1986; Kane and Hartley, 1988; Krueger et al., 1989). This is true of many foreign proteins, two native *E. coli* proteins, the sigma subunit of RNA polymerase and the osmoreceptor encoded by the *envZ* gene, and the bacteriophage lambda *c*II protein. When a culture of W3110 harboring pBGH1 is induced for high-level synthesis of BST with indoleacrylate, inclusion bodies consisting largely of BST are formed (Fig. 8). It has been proposed that inclusion bodies are random aggregates of abnormal proteins (Kane and Hartley, 1988; Krueger et al., 1989); however, such a model does not take into account the fact that inclusion bodies form at the poles of the cell (Fig. 8).

When the level of mistranslation was increased by the incorporation of certain amino acid analogs, inclusion bodies composed of proteins containing the analogs were formed. This occurred when an arginine auxotroph was exposed to the arginine analog canavanine in an arginine-free medium (Schachtele et al., 1968; Prouty et al., 1975) and when a proline auxotroph was exposed to the proline analog azetidine-2-carboxylic acid in a proline-free medium (Rabinovitz et al., 1969). We have observed that these two analogs will also cause the formation of inclusion bodies in a wild-type strain such as W3110 when grown in a minimal medium.

Certain mutant proteins from *E. coli* have been found to accumulate in inclusion bodies when overproduced: the X90 nonsense mutant of β-galactosidase (Cheng et al., 1981; Cheng, 1983), mutants of chemotaxis proteins CheB and CheY (Krueger et al., 1989), rifampin-resistant β subunit of RNA polymerase (Kashlev et al., 1989), and the *trpB8* mutant of the β subunit of tryptophan synthase (a glycine-to-arginine substitution at position 281) (G.-P. Zhao and R. L. Somerville, personal communication). We have observed that temperature-sensitive *lexA* mutants of *E. coli* form inclusion bodies when grown at 42°C but not at 37°C.

Overexpression of foreign proteins, abnormal proteins such as missense mutants, and mistranslated proteins induces the heat shock response and a concomitant increase in proteolysis in *E. coli* (Goldberg, 1972; Gottesman and Zipser, 1978; Baker et al., 1984; Goff and Goldberg, 1985; Straus et al., 1988). Certain heat shock proteins of *E. coli* are involved in the assembly of macromolecular structures (Pelham, 1986; Chandrasekhar et al., 1986; Goloubinoff et al., 1989); the term "chaperonin" has been applied to such proteins (Hemmingsen et al., 1988). It may be that chaperonin heat shock proteins of *E. coli* are involved in the assembly and breakdown of inclusion bodies. An *E. coli rpoH* mutant, unable to mount a normal heat shock response, exhibited decreased breakdown of inclusion bodies composed of mutant RNA polymerase β subunit (Kashlev et al., 1989).

INCLUSION BODIES AND CELL DIVISION

The formation of inclusion bodies in *E. coli* occurs at the poles of the cell (Fig. 8). Cells containing inclusion bodies at both poles do not divide, although they may elongate to many times their normal length. Under the light microscope, cells containing a single inclusion body at one pole have been seen to continue to divide at the other pole. Inclusion bodies have also been observed beginning to form in the center of a dividing cell, in the area of the invaginating peptidoglycan layer.

As described above, the formation of inclusion bodies in *E. coli* is an inducible phenomenon. For example, the addition of indoleacrylate to a culture of W3110 harboring plasmid pBGH1, or the addition of canavanine or azetidine-2-carboxylic acid to a culture of W3110, induced the formation of polar inclusion bodies and the cessation of cell division. Removal of the inducing agent led to the breakdown of the inclusion bodies and a resumption of cell division and growth.

The observed inhibition of cell division in cells containing inclusion bodies may be due to a stress response elicited by the production of abnormal or foreign proteins. Induction of the SOS response in *lon* mutants resulted in extensive cell filamentation (Howard-Flanders et al., 1964). Mutations in *sulA* blocked this type of filamentation (Donch et al., 1971; Johnson and Greenberg, 1975; Johnson, 1977; Gottesman et al., 1981). SulA protein levels were increased by SOS and heat shock responses; the SulA protein has been hypothesized to be an inhibitor of cell division (Schoemaker et al., 1984). Because of the involvement of the SulA protein in stress responses, expression of the *sulA* gene was monitored in a strain overproducing a foreign protein. An *E. coli* strain harboring a *sulA-lacZ* fusion (a gift of M. Marinus) was transformed with plasmid pBGH1. When high-level synthesis of BST was induced with indoleacrylate, the level of β-galactosidase produced increased almost fivefold, from 60 U to 280 to 300 U. No response to indoleacrylate was observed in the absence of pBGH1. The increased expression of the *sulA* gene in response to the overexpression of BST may inhibit cell division in such strains.

SUMMARY

High-level expression of abnormal or foreign proteins in *E. coli* has been associated with a variety of changes in cellular physiology and morphology. These include the eliciting of stress responses such as the SOS system and the heat shock response, with corresponding increases in proteolysis, inhibition of cell division, and cell filamentation. Depending on the amino acid composition of the protein overexpressed, the codons utilized in the structural gene for that protein, and the composition of the culture medium used, imbalances in the pools of amino acids and other key metabolites may arise and contribute to increases in mistranslation.

The identification and quantitation of such mistranslation events have only recently been achieved with overproducing strains of *E. coli*. The small number of known cases to date, discussed in this review, indicate that a great deal more research is needed in this area. Other responses of *E. coli* to protein overexpression, such as increases in proteolysis and the formation of inclusion bodies, are also in need of more attention. The successful exploitation of *E. coli* and other organisms for the efficient production of high-quality proteins of research and commercial interest is dependent on an understanding of the processes used by these organisms to produce these proteins.

The unpublished observations on APIII were obtained in the Department of Biological Sciences at Monsanto Corp. We thank

Ned R. Siegel for protein sequencing results, Cheryl Harris-Gordon, Donald E. Willis, and James F. Zobel for amino acid analyses, Kamila S. Kavka for synthesizing the primers used for DNA sequencing, Gabrielle R. Neises for electron microscopy, and Joseph F. Wiese, Michael R. Schlittler, John H. Moran, Rodney D. Shively, Sheila D. Kohlman, and Michael P. Macke for technical assistance.

REFERENCES

Anfinsen, C. B., and L. G. Corley. 1969. An active variant of staphylococcal nuclease containing norleucine in place of methionine. *J. Biol. Chem.* **244:**5149–5153.

Baker, T. A., A. D. Grossman, and C. A. Gross. 1984. A gene regulating the heat shock response in *Escherichia coli* also affects proteolysis. *Proc. Natl. Acad. Sci. USA* **81:**6779–6783.

Barker, D. G., and C. J. Bruton. 1979. The fate of norleucine as a replacement for methionine in protein synthesis. *J. Mol. Biol.* **133:**217–231.

Berman, M. L., and D. E. Jackson. 1984. Selection of *lac* gene fusions in vivo: *ompR-lacZ* fusions that define a functional domain of the *ompR* gene product. *J. Bacteriol.* **159:**750–756.

Birge, E. A., and C. G. Kurland. 1969. Altered ribosomal protein in streptomycin dependent *E. coli. Science* **166:**1282–1284.

Bogosian, G., and J. F. Kane. 1987. Nucleotide sequence of the *Eco*RI fragment from pLJ3 bearing two tandem *lacUV5* promoters. *Nucleic Acids Res.* **15:**7185.

Bogosian, G., B. N. Violand, E. J. Dorward-King, W. E. Workman, P. E. Jung, and J. F. Kane. 1989. Biosynthesis and incorporation into protein of norleucine by *Escherichia coli. J. Biol. Chem.* **264:**531–539.

Bohman, K., T. Ruusala, P. C. Jelenc, and C. G. Kurland. 1984. Kinetic impairment of restrictive streptomycin-resistant ribosomes. *Mol. Gen. Genet.* **198:**90–99.

Breckenridge, L., and L. Gorini. 1970. Genetic analysis of streptomycin resistance in *Escherichia coli. Genetics* **65:**9–25.

Brown, J. L. 1973. The modification of the amino terminal region of *Escherichia coli* proteins after initiation with methionine analogues. *Biochim. Biophys. Acta* **294:**527–529.

Calcott, P. H., J. F. Kane, G. G. Krivi, and G. Bogosian. 1988. Parameters affecting production of bovine somatotropin in *Escherichia coli* fermentations. *Dev. Ind. Microbiol.* **29:**257–266.

Celis, J. E., C. Colondre, and J. H. Miller. 1976. Suppressor su+7 inserts tryptophan in addition to glutamine. *J. Mol. Biol.* **104:**729–734.

Chandrasekhar, G. N., K. Tilly, C. Woolford, R. Hendrix, and C. Georgopoulos. 1986. Purification and properties of the groES morphogenetic protein of *Escherichia coli. J. Biol. Chem.* **261:** 12414–12419.

Cheng, Y.-S. E. 1983. Increased cell buoyant densities of protein overproducing *Escherichia coli* cells. *Biochem. Biophys. Res. Commun.* **111:**104–111.

Cheng, Y.-S. E., D. Y. Kwoh, T. J. Kwoh, B. C. Soltvedt, and D. Zipser. 1981. Stabilization of a degradable protein by its overexpression in *Escherichia coli. Gene* **14:**121–130.

Cohen, G. N., and R. Munier. 1959. Effets des analogues structuraux d'aminoacides sur la croissance, la synthèse de protéines et la synthèse d'enzymes chez *Escherichia coli. Biochim. Biophys. Acta* **31:**347–356.

Cowie, D. B., G. N. Cohen, E. T. Bolton, and H. De Robichon-Szulmajster. 1959. Amino acid analog incorporation into bacterial proteins. *Biochim. Biophys. Acta* **34:**39–46.

Cramer, F., F. von der Haar, and G. L. Igloi. 1979. Mechanism of aminoacyl-tRNA synthetases: recognition and proofreading processes, p. 267–279. *In* P. R. Schimmel, D. Soll, and J. N. Abelson (ed.), *Transfer RNA: Structure, Properties, and Recognition.* Cold Spring Harbor Laboratory, Cold Spring Harbor, N.Y.

Davies, J., W. Gilbert, and L. Gorini. 1964. Streptomycin, suppression, and the code. *Proc. Natl. Acad. Sci. USA* **51:**883–890.

Davies, J., L. Gorini, and B. N. Davis. 1965. Misreading of RNA codewords induced by aminoglycoside antibiotics. *Mol. Pharmacol.* **1:**93–106.

Donch, J. J., Y. S. Chung, M. H. L. Green, J. Greenberg, and G. Warren. 1971. Genetic analysis of *sul* mutants of *Escherichia coli* B. *Genet. Res.* **17:**185–193.

Epstein, W. 1967. Transposition of the *lac* region of *Escherichia coli.* IV. Escape from repression in bacteriophage-carried *lac* genes. *J. Mol. Biol.* **30:**529–543.

Fersht, A. R. 1981. Editing mechanisms in the aminoacylation of tRNA, p. 247–254. *In* P. R. Schimmel, D. Soll, and J. N. Abelson (ed.), *Transfer RNA: Structure, Properties, and Recognition.* Cold Spring Harbor Laboratory, Cold Spring Harbor, N.Y.

Flossdorf, J., and M.-R. Kula. 1973. Ultracentrifuge studies on binding of aliphatic amino acids to isoleucyl-tRNA synthetase from *Escherichia coli* MRE600. *Eur. J. Biochem.* **36:**534–540.

Galas, D. J., and E. W. Branscomb. 1976. Ribosome slowing by mutation to streptomycin resistance. *Nature* (London) **262:** 617–619.

Gilles, A.-M., P. Marliere, T. Rose, R. Sarfati, R. Longin, A. Meier, S. Fermandjian, M. Monnot, G. N. Cohen, and O. Barzu. 1988. Conservative replacement of methionine by norleucine in *Escherichia coli* adenylate kinase. *J. Biol. Chem.* **263:**8204–8209.

Goff, S. A., and A. L. Goldberg. 1985. Production of abnormal proteins in *E. coli* stimulates transcription of *lon* and other heat shock genes. *Cell* **41:**587–595.

Goldberg, A. L. 1972. Degradation of abnormal proteins in *Escherichia coli. Proc. Natl. Acad. Sci. USA* **69:**422–426.

Goloubinoff, P., A. A. Gatenby, and G. H. Lorimer. 1989. GroE heat-shock proteins promote assembly of foreign prokaryotic ribulose bisphosphate carboxylase oligomers in *Escherichia coli. Nature* (London) **337:**44–47.

Gorini, L. 1966. The action of streptomycin on protein synthesis in vivo. *Bull. N.Y. Acad. Med.* **42:**633–637.

Gorini, L. 1969. The contrasting role of *strA* and *ram* gene products in ribosomal functioning. *Cold Spring Harbor Symp. Quant. Biol.* **34:**101–111.

Gorini, L., and E. Kataja. 1964. Phenotype repair by streptomycin of defective genotypes in *E. coli. Proc. Natl. Acad. Sci. USA* **51:**487–493.

Gorini, L., and E. Kataja. 1965. Suppression activated by streptomycin and related antibiotics in drug sensitive strains. *Biochem. Biophys. Res. Commun.* **18:**656–663.

Gottesman, S., E. Halpern, and P. Trisler. 1981. Role of *sulA* and *sulB* in filamentation by *lon* mutants of *Escherichia coli* K-12. *J. Bacteriol.* **148:**265–273.

Gottesman, S., and D. Zipser. 1978. Deg phenotype of *Escherichia coli lon* mutants. *J. Bacteriol.* **133:**844–851.

Heider, J., W. Leinfelder, and A. Bock. 1989. Occurrence and functional compatibility within Enterobacteriaceae of a tRNA species which inserts selenocysteine into protein. *Nucleic Acids Res.* **17:**2529–2540.

Hemmingsen, S. M., C. Woolford, S. M. van der Vies, K. Tilly, D. T. Dennis, C. P. Georgopoulos, R. W. Hendrix, and R. J. Ellis. 1988. Homologous plant and bacterial proteins chaperone oligomeric protein assembly. *Nature* (London) **333:**330–334.

Hoben, P., H. Uemura, F. Yamao, A. Cheung, and R. Swanson. 1984. Misaminoacylation by glutaminyl-tRNA synthetase: relaxed specificity in wild-type and mutant enzymes. *Fed. Proc.*

43:2972–2976.

Howard-Flanders, P., E. Simson, and K. Theriot. 1964. A locus that controls filament formation and sensitivity to radiation in *Escherichia coli* K12. *Genetics* **49:**237–246.

Inokuchi, H., P. Hoben, F. Yamao, H. Ozeki, and D. Soll. 1984. Transfer RNA mischarging mediated by a mutant *Escherichia coli* glutaminyl-tRNA synthetase. *Proc. Natl. Acad. Sci. USA* **81:**409–413.

Johnson, B. F. 1977. Fine structure mapping and properties of mutations suppressing the *lon* mutation in *Escherichia coli* K-12 and B strains. *Genet. Res.* **30:**273–286.

Johnson, B. F., and J. Greenberg. 1975. Mapping of *sul*, the suppressor of *lon* in *Escherichia coli. J. Bacteriol.* **122:**570–574.

Kane, J. F., and G. Bogosian. 1987. Bovine somatotropin production: selecting the best host-vector system. *Biopharm. Manufact.* **1**(Nov.):26–51.

Kane, J. F., and D. L. Hartley. 1988. Formation of recombinant protein inclusion bodies in *Escherichia coli. Trends Biotechnol.* **6:**95–101.

Kashlev, M. V., A. I. Gragerov, and V. G. Nikiforov. 1989. Heat shock response in *Escherichia coli* promotes assembly of plasmid encoded RNA polymerase β-subunit into RNA polymerase. *Mol. Gen. Genet.* **216:**469–474.

Kerwar, S. S., and H. Weissbach. 1970. Studies on the ability of norleucine to replace methionine in the initiation of protein synthesis in *E. coli. Arch. Biochem. Biophys.* **141:**525–532.

Kisumi, M., M. Sugiura, and I. Chibata. 1976a. Biosynthesis of norvaline, norleucine, and homoisoleucine in *Serratia marcescens. J. Biochem.* **80:**333–339.

Kisumi, M., M. Sugiura, and I. Chibata. 1977. Norleucine accumulation by a norleucine-resistant mutant of *Serratia marcescens. Appl. Environ. Microbiol.* **34:**135–138.

Kisumi, M., M. Sugiura, J. Kato, and I. Chibata. 1976b. L-Norvaline and L-homoisoleucine formation by *Serratia marcescens. J. Biochem.* **79:**1021–1028.

Knowlton, R. G., L. Soll, and M. Yarus. 1980. Dual specificity of su+7 tRNA. Evidence for translational discrimination. *J. Mol. Biol.* **139:**705–720.

Koide, H., S. Yokoyama, G. Kawai, J.-M. Ha, T. Oki, S. Kawai, T. Miyake, T. Fuwa, and T. Miyazawa. 1988. Biosynthesis of a protein containing a nonprotein amino acid by *Escherichia coli*: L-2-aminohexanoic acid at position 21 in human epidermal growth factor. *Proc. Natl. Acad. Sci. USA* **85:**6237–6241.

Krueger, J. K., M. H. Kulke, C. Schutt, and J. Stock. 1989. Protein inclusion body formation and purification. *Biopharm. Manufact.* **2**(Mar.):40–45.

Kurland, C. G., and J. A. Gallant. 1986. The secret life of the ribosome, p. 127–157. *In* T. B. L. Kirkwood, R. F. Rosenberger, and D. J. Galas (ed.), *Accuracy in Molecular Processes.* Chapman & Hall, Ltd., London.

Leinfelder, W., T. C. Stadtman, and A. Bock. 1989. Occurrence *in vivo* of selenocysteyl-tRNA^{Ser} in *Escherichia coli. J. Biol. Chem.* **264:**9720–9723.

Leinfelder, W., E. Zehelein, M.-A. Mandrand-Berthelot, and A. Bock. 1988. Gene for a novel tRNA species that accepts L-serine and cotranslationally inserts selenocysteine. *Nature* (London) **331:**723–725.

Lennox, E. S. 1955. Transduction of linked genetic characters of the host by bacteriophage P1. *Virology* **1:**190–206.

Lu, H. S., L. B. Tsai, W. C. Kenney, and P.-H. Lai. 1988. Identification of unusual replacement of methionine by norleucine in recombinant interleukin-2 produced by *E. coli. Biochem. Biophys. Res. Commun.* **156:**807–813.

Marston, F. A. O. 1986. The purification of eukaryotic polypeptides synthesized in *Escherichia coli. Biochem. J.* **240:**1–12.

McPherson, D. T. 1988. Codon preference reflects mistranslational constraints: a proposal. *Nucleic Acids Res.* **16:**4111–4120.

Munier, R., and G. N. Cohen. 1956. Incorporation d'analogues structuraux d'aminoacides dans les protéines bactériennes. *Biochim. Biophys. Acta* **21:**592–593.

Munier, R., and G. N. Cohen. 1959. Incorporation d'analogues structureaux d'aminoacides dans les protéines bactériennes au cours de leur synthès *in vivo. Biochim. Biophys. Acta* **31:** 378–391.

Naider, F., Z. Bohak, and J. Yariv. 1972. Reversible alkylation of a methionine residue near the active site of β-galactosidase. *Biochemistry* **11:**3202–3208.

Owens, S. L., and F. E. Bell. 1970. Specificity of the valyl ribonucleic acid synthetase from *Escherichia coli* in the binding of valine analogues. *J. Biol. Chem.* **245:**5515–5523.

Ozaki, M., S. Mizushima, and M. Nomura. 1969. Identification and functional characterization of the protein controlled by the streptomycin-resistant locus in *E. coli. Nature* (London) **222:** 333–339.

Pelham, H. R. B. 1986. Speculations on the functions of the major heat shock and glucose-regulated proteins. *Cell* **46:**959–961.

Pine, M. J. 1978. Comparative physiological effects of incorporated amino acid analogs in *Escherichia coli. Antimicrob. Agents Chemother.* **13:**676–685.

Prouty, W. F., M. J. Karnovsky, and A. L. Goldberg. 1975. Degradation of abnormal proteins in *Escherichia coli.* Formation of protein inclusions in cells exposed to amino acid analogs. *J. Biol. Chem.* **250:**1112–1122.

Rabinovitz, M., A. Finkleman, R. L. Reagan, and T. R. Breitman. 1969. Amino acid antagonist death in *Escherichia coli. J. Bacteriol.* **99:**336-338.

Richmond, M. H. 1962. The effect of amino acid analogues on growth and protein synthesis in microorganisms. *Bacteriol. Rev.* **26:**398–420.

Ruusala, T., D. Anderson, M. Ehrenberg, and C. G. Kurland. 1984. Hyperaccurate ribosomes inhibit growth. *EMBO J.* **3:** 2575–2580.

Schachtele, C. F., D. L. Anderson, and P. Rogers. 1968. Mechanism of canavanine death in *Escherichia coli.* II. Membrane-bound canavanyl-protein and nuclear disruption. *J. Mol. Biol.* **33:**861–872.

Schoemaker, J. M., R. C. Gayda, and A. Markovitz. 1984. Regulation of cell division in *Escherichia coli*: SOS induction and cellular location of the SulA protein, a key to *lon*-associated filamentation and death. *J. Bacteriol.* **158:**551–561.

Seeburg, P. H., S. Sias, J. Adelman, H. A. de Boer, J. Hayflick, P. Jhurani, D. V. Goeddel, and H. L. Heyneker. 1983. Efficient bacterial expression of bovine and porcine growth hormones. *DNA* **2:**37–45.

Seetharam, R., R. A. Heeren, E. Y. Wong, S. R. Braford, B. K. Klein, S. Aykent, C. E. Kotts, K. J. Mathis, B. F. Bishop, M. J. Jennings, C. E. Smith, and N. R. Siegel. 1988. Mistranslation in IGF-1 during over-expression of the protein in *Escherichia coli* using a synthetic gene containing low frequency codons. *Biochem. Biophys. Res. Commun.* **155:**518–523.

So, A. G., and E. W. Davie. 1964. The effect of organic solvents on protein biosynthesis and their influence on the amino acid code. *Biochemistry* **3:**1165–1169.

Soll, L., and P. Berg. 1969. Recessive lethal nonsense suppressor in *Escherichia coli* which inserts glutamine. *Nature* (London) **223:**1340–1342.

Somerville, R. L. 1988. The *trp* promoter of *Escherichia coli* and its exploitation in the design of efficient protein production systems. *Biotechnol. Genet. Eng. Rev.* **6:**1–41.

Straus, D. B., W. A. Walter, and C. A. Gross. 1988. *Escherichia coli* heat shock gene mutants are defective in proteolysis. *Gene*

Dev. **2**:1851–1858.

Traub, P., and M. Nomura. 1968. Streptomycin resistance mutation in *Escherichia coli*: altered ribosomal protein. *Science* **160**:198–199.

Trupin, J., H. Dickerman, M. Nirenberg, and H. Weissbach. 1966. Formylation of amino acid analogues of methionine sRNA. *Biochem. Biophys. Res. Commun.* **24**:50–55.

Tsai, L. B., H. S. Lu, W. C. Kenney, C. C. Curless, M. L. Klein, P.-H. Lai, D. M. Fenton, B. W. Altrock, and M. B. Mann. 1988. Control of misincorporation of *de novo* synthesized norleucine into recombinant interleukin-2 in *E. coli*. *Biochem. Biophys. Res. Commun.* **156**:733–739.

Tsui, W. C., and A. R. Fersht. 1981. Probing the principles of amino acid selection using the alanyl-tRNA synthetase from *Escherichia coli*. *Nucleic Acids Res.* **9**:4627–4637.

Violand, B. N., N. R. Siegel, G. Bogosian, W. E. Workman, and J. F. Kane. 1989. Detection of norleucine incorporation into recombinant proteins synthesized in *E. coli*, p. 315–326. *In* T. E. Hugli (ed.), *Techniques in Protein Chemistry*. Academic Press, Inc., San Diego, Calif.

Wong, E. Y., R. Seetharam, C. E. Kotts, R. A. Heeren, B. K. Klein, S. R. Braford, K. J. Mathis, B. F. Bishop, N. R. Siegel, C. E. Smith, and W. C. Tacon. 1988. Expression of secreted insulin-like growth factor-1 in *Escherichia coli*. *Gene* **68**:193–203.

Yaniv, M., W. R. Folk, P. Berg, and L. Soll. 1974. A single modification of a tryptophan-specific transfer RNA permits aminoacylation by glutamine and translation of the codon UAG. *J. Mol. Biol.* **86**:245–260.

Yarus, M. 1979. The relationship of the accuracy of aminoacyl-tRNA synthesis to that of translation, p. 501–509. *In* P. R. Schimmel, D. Soll, and J. N. Abelson (ed.), *Transfer RNA: Structure, Properties, and Recognition*. Cold Spring Harbor Laboratory, Cold Spring Harbor, N.Y.

Zinoni, F., A. Birkmann, W. Leinfelder, and A. Bock. 1987. Cotranslational insertion of selenocysteine into formate dehydrogenase from *Escherichia coli* directed by a UGA codon. *Proc. Natl. Acad. Sci. USA* **84**:3156–3160.

Zinoni, F., A. Birkmann, T. C. Stadtman, and A. Bock. 1986. Nucleotide sequence and expression of the selenocysteine-containing polypeptide of formate dehydrogenase (formate-hydrogen-lyase-linked) from *Escherichia coli*. *Proc. Natl. Acad. Sci. USA* **83**:4650–4654.

Chapter 49

Frameshift Suppression†

MICHAEL R. CULBERTSON, PETER LEEDS, MARK G. SANDBAKEN, and PATRICIA G. WILSON

In theory, errors in translation can occur in any of the three phases of protein synthesis. Some translational errors resemble the types of mistakes commonly found in written language. For example, one is constantly reminded by less erudite students of the incomplete sentence (incorrect initiation or termination), the dangling participle (incorrect termination and reinitiation), the incorrect choice of word (misincorporation), or the run-on sentence (readthrough). There is, however, one type of translational error, the frameshift, that has no counterpart in written language. This is due to the punctuationless nature of the genetic code. With rare exception, once the reading frame is preset, bases are slavishly read in groups of three with no compensatory mechanism for resetting the reading frame to accommodate an extra or missing letter in the genetic alphabet.

The triplet reading frame is nonetheless maintained by specific translational components. One way in which these components have been identified and studied is through the isolation of mutations called frameshift suppressors. In general, frameshift suppressors compensate for an elongation error caused by the presence of a frameshift mutation, usually a +1 nucleotide insertion or −1 deletion, in the coding region of mRNA. In general, a frameshift mutation results in the production of a nonfunctional, prematurely terminated protein product. In cases where a frameshift mutation causes a detectable phenotype, revertants (suppressors) can be isolated that reverse or partially reverse the negative effects of the frameshift mutation, leading to production of a functional or partially functional product.

When the mutation causing reversion is in a gene distinct from the one containing the frameshift mutation itself, it is called an extragenic frameshift suppressor. Through the study of different types of extragenic frameshift suppressors, it has been shown that tRNA, rRNA, ribosomal proteins, and soluble protein factors are all important in the maintenance of the reading frame during elongation. In this chapter, we will survey the literature on extragenic frameshift suppressors in an attempt to highlight what these studies tell us about the mechanism of reading frame maintenance.

+1 AND −1 FRAMESHIFT MUTATIONS

The existence of well-characterized frameshift mutations was a necessary prerequisite to the isolation of extragenic frameshift suppressors. +1 frameshift mutations for which sequence information is now available were induced with acridine half-mustards (ICR-191 or ICR-170) in *Escherichia coli* (Calos and Miller, 1981), *Salmonella typhimurium* (Roth, 1974), and *Saccharomyces cerevisiae* (Mathison and Culbertson, 1985). These mutagens induce +1 frameshifts by insertion of a G · C base pair in runs of contiguous G · C base pairs. Some examples of suppressible ICR-induced +1 frameshifts are shown in Table 1. −1 frameshifts can also be induced with ICR-191 in bacteria (Calos and Miller, 1981). Thus far, only one ICR-induced −1 mutation, *trpE91* in *S. typhimurium* (Atkins et al., 1983), is suppressible by extragenic frameshift suppressors (Falahee et al., 1988; O'Mahony et al., 1989a) (Table 1).

An important point to keep in mind regarding suppressible frameshift mutations is that the position in the mRNA at which correction of the reading frame occurs can differ from the position of the frameshift mutation itself. The potential sites of action of an extragenic suppressor are, however,

† Publication no. 3063 of the Laboratory of Genetics, University of Wisconsin, Madison.

Michael R. Culbertson, Peter Leeds, Mark G. Sandbaken, and Patricia G. Wilson ■ Laboratory of Molecular Biology, University of Wisconsin, 1525 Linden Drive, Madison, Wisconsin 53706.

Table 1. Suppression at alternative sites in the frameshift windows of representative +1 and −1 mutations

Suppression	Mutation	Sites[a]	References
sufJ-mediated suppression of *hisD* mutations	*hisD*[+]	...Trp Asn Ser Cys Ser Pro Glu... ...UGG AAC AGC UGU AGU CCU GAA...	Bossi and Roth, 1981; Bossi and Smith, 1984
	hisD3749 (+1C)	...Trp Asn Ser Cys Ser Pro end ...UGG AAC AGC UGU AGC CCC UGA	
	hisD3749-S6 sufJ (+1C; A to C)	...Trp Thr Ala Val Ala Pro Glu... ...UGG ACCA GCU GUA GCC CCU GAA...	
	hisD3749-S7 sufJ (+1C; G to C)	...Trp Asn Thr Val Ala Pro Glu... ...UGG AAC ACCU GUA GCC CCU GAA...	
	hisD3749-S15 sufJ (+1C; G to C)	...Trp Asn Ser Cys Thr Pro Glu... ...UGG ACC AGC UGU ACCC CCU GAA...	
SUF8-mediated suppression of *his4-713*[b]	*HIS4*[+]	...Pro Thr Tyr..........Thr Pro Glu... ...CCA ACC UAU (23 triplets) ACC CCU GAA...	Donahue et al., 1981; Cummins et al., 1985; Winey et al., 1986; Mathison et al., 1989
	his4-713 (+1C)	...Pro Thr Tyr..........Thr Pro end ...CCA AAC UAU (23 triplets) ACC CCC UGA...	
	his4-713 SUF8	...Pro Pro Met ....Pro Pro Glu ...CCAA CCU AUG (23 triplets) CCC CCU GAA...	
sufS- and *hopR*-mediated suppression of *trpE91*	*trpE*[+]	...Leu Gln Gly Val Val Asn... ...UUA CAG GGA GUG GUG AAC...	Falahee et al., 1988; O'Mahony et al., 1989a; Atkins et al., 1983
	trpE91 (−1G)	...Leu Gln Gly Val end ...UUA CAG GGA GUG UGA ACA...	
	trpE91 sufS	...Leu Gln Gln Ser Val Asn... ...UUA CAG GG AGU GUG AAC...	
	trpE91 hopR1	...Leu Gln Gly Val Asn... ...UUA CAG GGA GUGUG AAC...	

[a] Underlined single bases represent the sites of single-base insertion or deletion. Underlined two-, four-, and five-base codons represent the sites of action of the suppressors.
[b] Suppression at the CCAA codon was proposed by Curran and Yarus (1987).

limited by what is called the frameshift window, which has two boundaries. The 3′ boundary is defined by the first stop codon brought in frame as the result of the frameshift mutation. The 5′ boundary is defined by the first codon at which correction of the reading frame would allow translation to proceed unhindered by subsequent stop codons. Correction of the reading frame can in theory occur anywhere in the frameshift window. Of course, correction at a site other than the frameshift mutation itself will result in amino acid substitutions between the site of correction and the frameshift mutation. In order for suppression to occur by this mechanism, the substitutions must be at least partially compatible with the function of the gene product.

The nomenclature used to designate potential reading frames has proven to be a common source of confusion. The "zero" reading frame refers to the normal open reading frame that codes for the wild-type gene product. A +1 frameshift (base insertion) shifts translation to the −1 reading frame. This is because codon reading, which still occurs in groups of three, is shifted back 1 nucleotide (nt) toward the 5′ end of the mRNA. Similarly, a −1 frameshift (base deletion) shifts translation to the +1 reading frame.

ALTERED tRNAs THAT SUPPRESS +1 FRAMESHIFTS

tRNAs Containing 8-nt Anticodon Loops

Almost all wild-type tRNAs have a conserved structure for the anticodon stem and loop consisting of 5 base pairs in the stem and 7 nt in the loop. The only known exceptions to this structural motif derive from some unusual mitochondrial tRNAs that have 8 or 9 nt in the anticodon loop (Li and Tzagoloff, 1979; Sumner-Smith et al., 1984). A large number of frameshift suppressor mutations that correct +1 frameshifts correspond to tRNAs containing an extra nucleotide in the anticodon loop. Suppressor tRNAs of this type differ from the majority of wild-type tRNAs by having an 8-nt rather than a 7-nt loop. The in vivo behavior of suppressor tRNA containing an 8-nt anticodon loop is illustrated below by four suppressors: *sufD* and *sufJ* in *S. typhimurium*, *SUF16* in *S. cerevisiae*, and *glyT* missense suppressors in *E. coli*.

sufD, the first frameshift suppressor to be described, was an altered glycine tRNA encoded by an allele of the *glyU* gene (Riddle and Carbon, 1973). The suppressor tRNA contains the anticodon sequence 3′-CCCC-5′ as a result of insertion of an extra C residue in the anticodon loop. *sufD* suppresses +1 frameshifts in glycine codons of the *hisD* gene, presumably at the site of the frameshift, by reading the 4-nt codon 5′-GGGG-3′. These results demonstrated that 4-nt translocation could result from changes in tRNA structure, but the details of the frameshift mechanism were unclear.

Another bacterial frameshift suppressor, *sufJ*, encodes an altered threonine tRNA that contains an eighth nucleotide in the anticodon loop by virtue of an insertion in the anticodon stem (Bossi and Smith, 1984). The insertion is such that the normal number of 5 base pairs is retained in the stem, but one distal stem nucleotide is displaced into the 5′ side of the anticodon loop, creating an 8-nt anticodon loop (Fig. 1). *sufJ* suppresses a variety of +1 frameshifts in *hisD* by decoding a 4-nt threonine codon located in the frameshift window of each suppressible allele at a site distinct from the +1 frameshift mutation (Table 1). Genetic selection for an alternate set of suppressible sites has revealed that *sufJ* can decode 5′-ACCA-3′, 5′-ACCU-3′, and 5′-ACCC-3′ codons, using a 3′-UGGU-5′ anticodon (Bossi and Roth, 1981). This result suggests that *sufJ* can correct a +1 frameshift in the absence of pairing at the fourth nucleotide position.

The *SUF16-1* frameshift suppressor allele encodes an altered glycine tRNA that contains an 8-nt anticodon loop as a result of a +1 insertion in the loop (Fig. 1) (Gaber and Culbertson, 1982). This tRNA, which contains the anticodon sequence 3′-CCCG-5′, suppresses *his4-211* (5′-GGGU-3′) at the site of the frameshift but fails to suppress *his4-519* (5′-GGGG-3′) (Donahue et al., 1981; Gaber and Culbertson, 1984; Mathison and Culbertson, 1985). The suppressor therefore initially appeared to exhibit a requirement for fourth nucleotide base pairing. This turned out not to be the case.

Additional alleles of *SUF16* were isolated that contained all possible base substitutions at the fourth position (N) in the 8-nt anticodon loop (3′-CCCN-5′) (Gaber and Culbertson, 1984). These alleles were tested for suppression of +1 frameshifts in the *his4* gene, including all possible 5′-GGGN-3′ glycine codons (Table 2). The resulting matrix of suppression data showed that fourth-nucleotide base pairing enhances the efficiency of 4-nt codon reading but is not required for the basal level of suppression. G · G at the fourth position was the only combination that did not result in suppression, apparently for reasons other than the lack of potential for standard base pairing.

The phenotypic behavior of the *sufJ* and *SUF16* suppressors suggested a model in which tRNA plays a central role in specifying translational step size. This is accomplished primarily through the conserved primary and secondary structures of the anticodon stem and loop (Gaber and Culbertson, 1984). It was proposed that the anticodon loop adopts a conformation that serves as a yardstick to measure codon length. Thus, an extra nucleotide in the anticodon loop causes the tRNA to read a 4-nt codon through displacement of 4 nt on the message rather than through the formation of 4 base pairs between the anticodon and codon.

Further insight regarding frameshift suppression came from the analysis of some unusual mutations in the *glyT* tRNA gene that suppress missense mutations (Prather et al., 1981; Murgola et al., 1983). When the sequences of the mutant tRNAs were determined, it was found that some contained a base insertion on the 3′ side of the anticodon loop (Fig. 1). Insertion on this side of the loop changed the coding specificity of the tRNA and conferred the ability to read a 5′-UGG-3′ tryptophan rather than a 5′-GGA/G-3′ glycine codon. Thus, it was shown for the first time that an 8-nt tRNA can read a 3-nt codon. This is consistent with the finding that a naturally occurring yeast mitochondrial tRNA containing an 8-nt anticodon loop selectively decodes a 3-nt codon (Li and Tzagoloff, 1979).

In addition to suppressing missense mutations, some of the *glyT* suppressors described above also suppress at least one +1 frameshift mutation in the *trpA* gene (Tucker et al., 1989). This finding indicates that *glyT* suppressor tRNAs that contain an 8-nt anticodon loop can cause either 3- or 4-nt translocation. One possible explanation for this dual decoding ability is that 8-nt tRNAs can adopt at least two alternate anticodon loop conformations that direct either 3- or 4-nt translocation. This point will be further elaborated in a subsequent section.

Another interesting and somewhat puzzling feature of the *glyT* suppressors is that in some cases they correct the reading frame within a frameshift window that is devoid of cognate mRNA sequences (Tucker et al., 1989). The mechanism of frameshift suppression in these cases is therefore complicated by two requirements for suppression that are superimposed on each other: noncognate reading and frameshifting. Further information on noncognate reading and its relationship to frameshifting must await identification of the noncognate sequences within the frame-

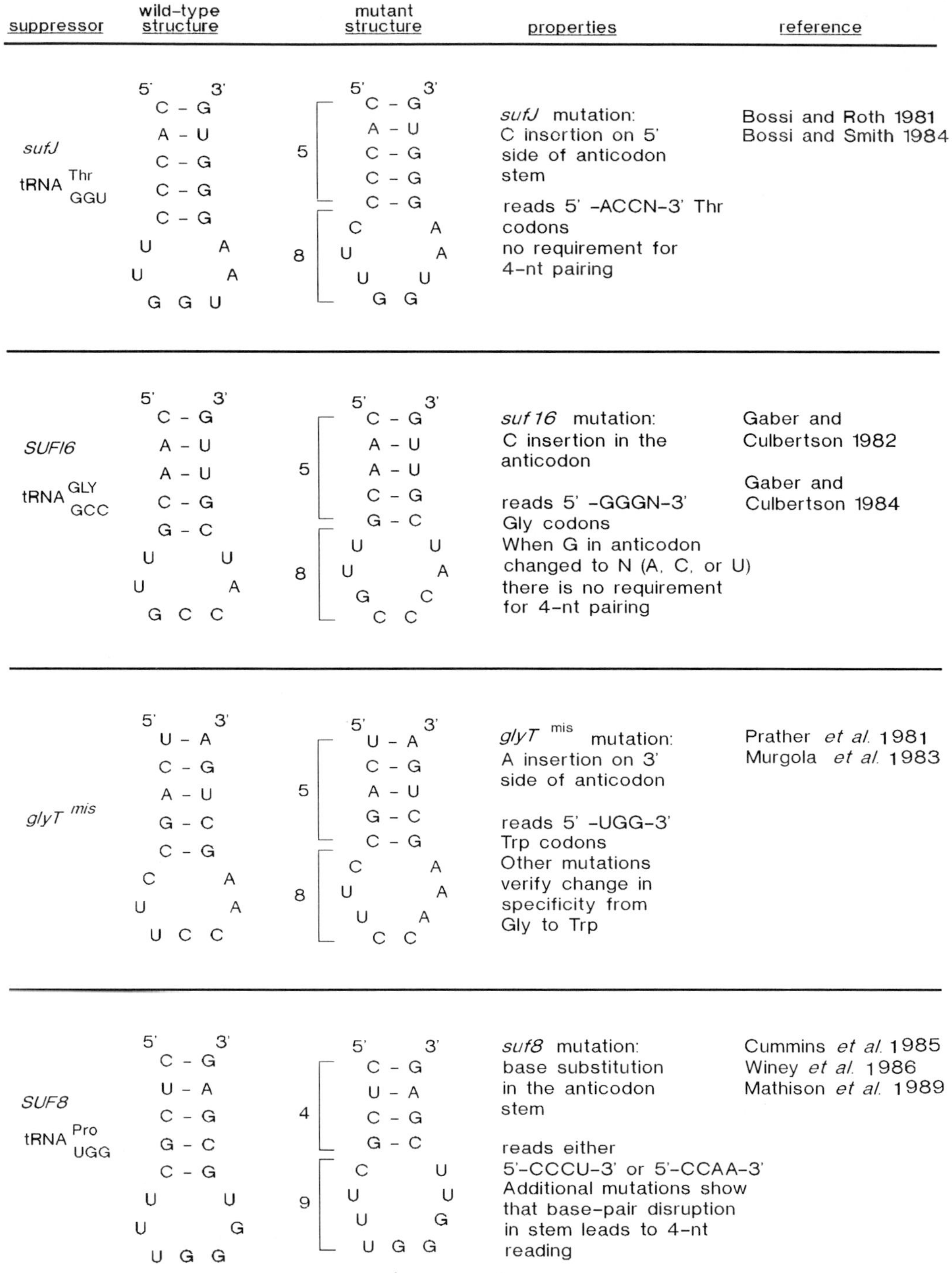

suppressor	wild-type structure	mutant structure	properties	reference
sufJ $tRNA^{Thr}_{GGU}$	5′ 3′ C - G A - U C - G C - G C - G U A U A G G U	5′ 3′ [5] C - G A - U C - G C - G C - G [8] C A U A U U G G	*sufJ* mutation: C insertion on 5′ side of anticodon stem reads 5′ -ACCN-3′ Thr codons no requirement for 4-nt pairing	Bossi and Roth 1981 Bossi and Smith 1984
SUF16 $tRNA^{GLY}_{GCC}$	5′ 3′ C - G A - U A - U C - G G - C U U U A G C C	5′ 3′ [5] C - G A - U A - U C - G G - C [8] U U U A G C C C	*suf16* mutation: C insertion in the anticodon reads 5′ -GGGN-3′ Gly codons When G in anticodon changed to N (A, C, or U) there is no requirement for 4-nt pairing	Gaber and Culbertson 1982 Gaber and Culbertson 1984
glyT ^{mis}	5′ 3′ U - A C - G A - U G - C C - G C A U A U C C	5′ 3′ [5] U - A C - G A - U G - C C - G [8] C A U A U A C C	*glyT* ^{mis} mutation: A insertion on 3′ side of anticodon reads 5′ -UGG-3′ Trp codons Other mutations verify change in specificity from Gly to Trp	Prather *et al.* 1981 Murgola *et al.* 1983
SUF8 $tRNA^{Pro}_{UGG}$	5′ 3′ C - G U - A C - G G - C C - G U U U G U G G	5′ 3′ [4] C - G U - A C - G G - C [9] C U U U U G U G G	*suf8* mutation: base substitution in the anticodon stem reads either 5′-CCCU-3′ or 5′-CCAA-3′ Additional mutations show that base-pair disruption in stem leads to 4-nt reading	Cummins *et al.* 1985 Winey *et al.* 1986 Mathison *et al.* 1989

Figure 1. Representative suppressor tRNAs containing 8- or 9-nt anticodon loops.

shift window that interact with the suppressor tRNAs.

Finally, it was also shown by Tucker et al. (1989) that a variety of nonsense and missense suppressors that contain nothing more than simple base substitutions in the anticodon loop can suppress one or more *trpA* frameshift mutations. Noncognate reading is required in some of these examples. These results indicate that suppression of +1 frameshifts can be mediated by altered tRNAs that contain the normal number of 7 nt in the anticodon loop.

tRNAs Containing 9-nt Loops

Mutations in a redundant set of yeast genes encoding proline tRNA(UGG) were shown to confer

Table 2. Codon recognition by *SUF16* suppressor tRNAs

his4 codons (5′-3′)	Suppression by given *SUF16* anticodon bases (3′-5′)[a]			
	CCC G	CCC C	CCC A	CCC U
GGG G	−	++	+	+
GGG C	++	+	+	+
GGG A	+	+	+	++
GGG U	+	++	++	+

[a] −, No detectable suppression; +, suppression; ++, more efficient suppression. The results show that pairing at the fourth nucleotide position is not required for codon recognition and 4-nt translocation. In general, the efficiency of suppression is enhanced by fourth-nucleotide pairing (for further details and a discussion of the exceptions to this statement, see Gaber and Culbertson, 1984).

suppression of a +1 frameshift by a novel mechanism (Cummins et al., 1985; Winey et al., 1986; Winey et al., 1989). All of the mutations isolated by in vivo selection resulted in suppressor tRNA containing a 9-nt anticodon loop without the introduction of any extra nucleotides (Fig. 1). The canonical mutation, *SUF8-1*, was a substitution of U for G at a position corresponding to the last nucleotide on the 3′ side of the anticodon stem. The net effect of the mutation was to cause disruption of base pairing in the last base pair of the stem (C-G versus CU) adjacent to the anticodon loop. Thus, this tRNA contains a 4-base-pair anticodon stem with a net displacement of 2 nt into the anticodon loop, resulting in a 9-nt loop (Winey et al., 1986).

To test whether the unusual anticodon stem-loop structure in *SUF8-1* tRNA represents a general structural motif that consistently results in frameshift suppression, additional mutations were obtained by in vitro mutagenesis of the *SUF8* gene (Mathison et al., 1989). Sixteen alleles were constructed that encode mutant tRNAs containing all possible base combinations in the last base pair of the anticodon stem adjacent to the anticodon loop (Table 3 and Fig. 2). Six of the alleles failed to confer frameshift

Table 3. Phenotypes of nt 31-nt 39 combinations (suppression of *his4-713*)[a]

His⁺ phenotype	His⁻ phenotype
G G	C-G
A A	G-C
U U	A-U
C C	U-A
A C	G · U
C A	U · G
G A	
A G	
U C	
C U	

[a] See Fig. 2; see Table 1 for further information on *SUF8*-mediated suppression of *his4-713*. For further details, see Mathison et al. (1989).

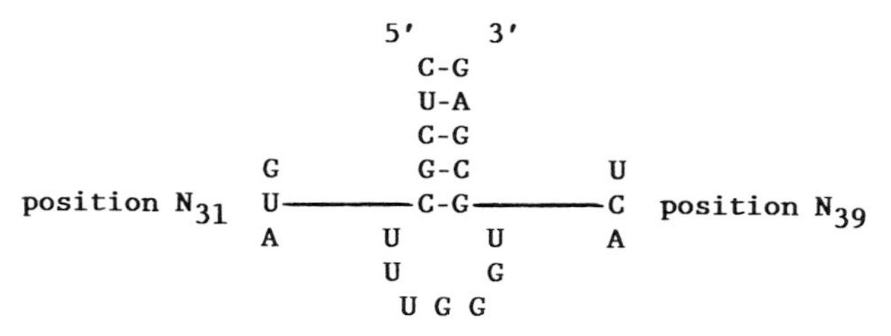

Figure 2. *SUF8* anticodon stem and loop (see Table 3). For further details, see Mathison et al. (1989).

suppression, including those containing the base combinations C-G, G-C, A-U, U-A, G · U, and U · G. Lack of suppression most likely results from maintenance of normal tRNA secondary structure, particularly the 5-base-pair anticodon stem and 7-nt anticodon loop. These tRNAs resemble the wild type and probably read a 3-nt codon. The remaining alleles, all of which contain base pair disruptions in the anticodon stem, result in frameshift suppression and therefore read a 4-nt codon.

There is still uncertainty regarding which bases serve as the anticodon nucleotides in the *SUF8* alleles. There are two possibilities. The first was suggested by Winey et al. (1986), and the second was suggested by Curran and Yarus (1987). The 9-nt loop contains the sequence 3′-GGGU-5′, which could allow the tRNA to act at the site of the suppressible *his4-713* frameshift mutation by reading 5′-CCCU-3′. Alternatively, the anticodon sequence 3′-GGUU-5′ could act at a single 4-nt sequence 5′-CCAA-3′ that is located 26 codons upstream of the frameshift mutation in a frameshift window 69 codons in length (Table 1). It is not known whether the resulting amino acid substitutions would be compatible with *HIS4* function.

Another +1 frameshift mutation, *his4-712*, is not suppressed by *SUF8* alleles even though it results in a 5′-CCCU-3′ sequence at the site of the mutation (Cummins et al., 1980; Donahue et al., 1981; Mathison and Culbertson, 1985). This mutation differs from *his4-713* only by its position in the *his4* gene. The *his4-712* mutation lies within a frameshift window 38 codons in length and, like *his4-713*, contains a 5′-CCAA-3′ sequence within the window 9 codons upstream of the frameshift. Thus, if *SUF8* alleles read the 4-nt codon 5′-CCAA-3′, it must mean that substitution of 9 amino acids upstream of *his4-712* is incompatible with *HIS4* function, whereas substitution of 26 amino acids upstream of *his4-713* is compatible with *HIS4* function.

There is one naturally occurring mitochondrial tRNA in *Schizosaccharomyces pombe* that contains a 9-nt anticodon loop (Sumner-Smith et al., 1984). The fact that this tRNA reads a 3-nt codon suggests that, like the *glyT* missense suppressors, *SUF8* alleles might read both 3- and 4-nt codons. However, this possibility has not been directly tested.

ALTERED tRNAs THAT SUPPRESS −1 FRAMESHIFTS

The *S. typhimurium* mutation *trpE91*, a −1 frameshift resulting from deletion of one G residue, has been used to isolate extragenic suppressors, some of which map in tRNA genes (Table 1) (Atkins et al., 1983; Falahee et al., 1988; Hughes et al., 1989). The manner in which these suppressors restore the reading frame was assessed by determining both the nature of the suppressor mutations and the sequences of amino acids within the frameshift window that result from the action of a given suppressor. This was accomplished by constructing cassettes of the *trpE91* frameshift window fused to *lacZ* such that a frameshift would be required to produce β-galactosidase. The fusion proteins resulting from suppression were purified and subjected to N-terminal sequence analysis into the region of the frameshift window.

One class of −1 extragenic suppressors, designated *sufS*, mapped in the *glyT* gene (O'Mahony et al., 1989a; O'Mahony et al., 1989b). Several different mutations were identified (J. Atkins, personal communication). One type, represented by *sufS601*, results in a U*-to-C base substitution in the first anticodon nucleotide. Other mutations were located outside the anticodon loop. For example, *sufS605* was a single-base insertion of U in the TUC loop, with a concomitant U-to-pseudouridine modification in the D loop. *sufS617* and *sufS625* were C-to-U and C-to-A base substitutions, respectively, in the TUC arm. Finally, *sufS627* was a G-to-A base substitution in the first 5′ nucleotide of the acceptor stem (O'Mahony et al., 1989b). Since none of these mutations changes the number of nucleotides in the anticodon loop, it is clear that suppression of −1 frameshifts can occur without directly altering anticodon loop size.

sufS601 suppresses a *trpE91-lacZ* fusion by producing the amino acid sequence Gln-Gly-Ser instead of Gln-Gly-Val in response to the mRNA sequence 5′-CAG <u>GG</u>A GUG-3′ located upstream of the −1 frameshift mutation (Table 1) (Falahee et al., 1988). Thus, the net consequence of *sufS*-mediated suppression is a compensatory change that corrects the reading frame and produces a protein with a Ser-for-Val amino acid substitution. This could occur by one of two possible mechanisms. The suppressor glycine tRNA either reads the doublet 5′-GG-3′ (underlined above) or reads the triplet 5′-GGG-3′ through a backspacing mechanism involving realignment on the message. The latter model would indicate that the first G in the sequence 5′-CAG GGA-3′ is read twice, once as the last letter of the CAG Gln codon and once as the first letter of a GGG Gly codon in the −1 reading frame. Genetic evidence favors the former case of doublet decoding, since suppression still occurs when the G in CAG is changed to other nucleotides (O'Mahony et al., 1989a).

Additional suppressors that correct the *trpE91* frameshift are designated *hopR* and *hopE*, which correspond to alleles of two loci encoding valine tRNA in either *E. coli* or *S. typhimurium*. Two classes of suppressor mutations were uncovered (M. O'Connor, personal communication). One consists of a nucleotide insertion of U or A in the anticodon. In addition to *trpE91* suppression, these mutations also behave as classical 8-nt tRNAs and can suppress appropriate +1 frameshifts. The other class consists of nucleotide substitutions in the aminoacyl stem, resulting in a 5′-ACA-3′ or 5′-UCA-3′ terminus in place of the normal 5′-CCA-3′ terminus.

hopR1, which contains an 8-nt anticodon loop (O'Connor, personal communication), suppresses a *trpE91-lacZ* fusion by inserting valine in response to the 5-nt sequence 5′-GUGUG-3′ (Table 1) (Falahee et al., 1988). This is formally equivalent to a shift into the +2 reading frame which, combined with the shift to the +1 reading due to the *trpE91* mutation, restores the proper frame and results in a protein that is deleted for one amino acid residue, valine. This could happen in either of two ways. First, the tRNA could read a 5-nt codon. The second and perhaps more likely possibility suggested by Falahee et al. (1988) is that an error in tRNA alignment results from the presence of two overlapping GUG codons sandwiched within the 5′-GUGUG-3′ sequence. Correction of the *trpE91* reading frame by this type of suppressor may be a consequence of two concerted effects, one related to the altered structure of the tRNA and the other related to the particular error-prone nature of the codon and the context in which it resides at the site of suppression in the mRNA.

ROLE OF ANTICODON STEM-LOOP IN STEP SIZE DETERMINATION

Curran and Yarus (1987) have proposed a model that explains the behavior of tRNAs containing 7-, 8-, and 9-nt anticodon loops in terms of the stereochemical configuration of the anticodon hairpin loop (Fig. 3). The model is based on the premise that all normal tRNAs contain an anticodon arm with two parts, an A-type anticodon stem helix followed by a stack of 5 nt on the 3′ side of the

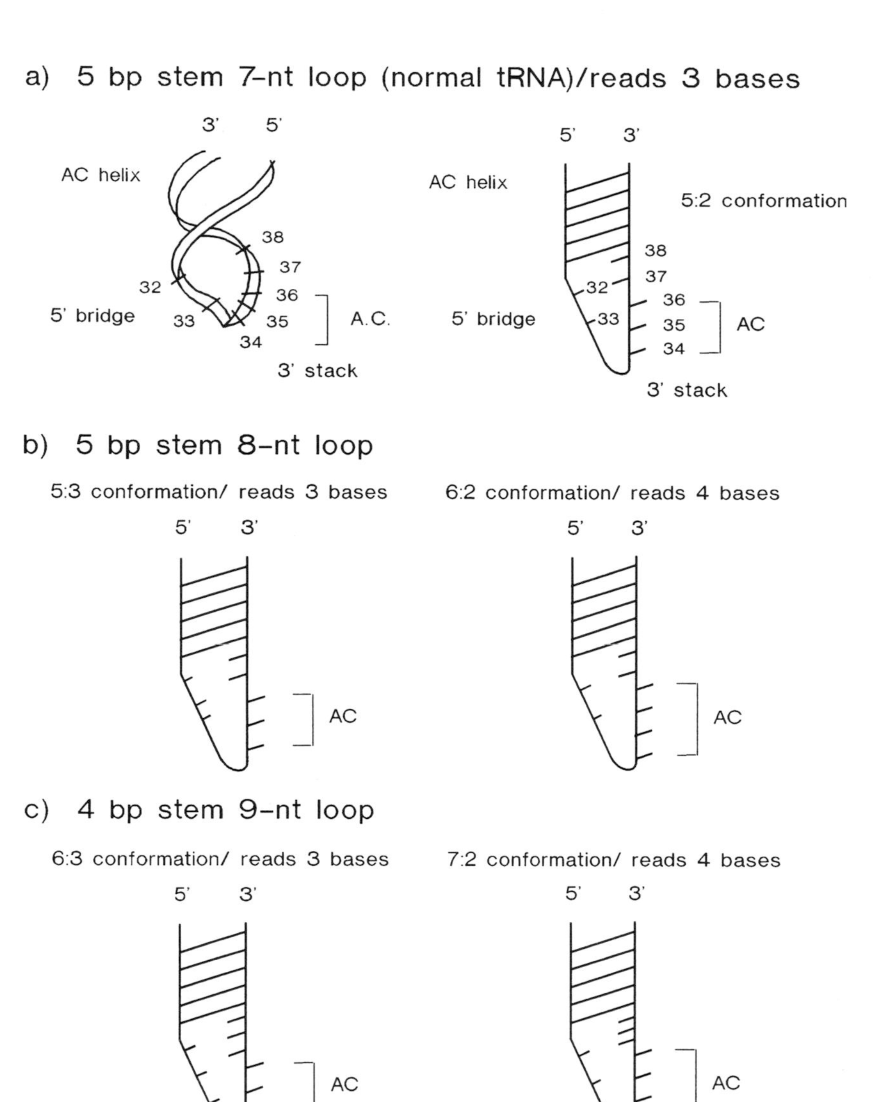

Figure 3. Role of the anticodon loop stack in frameshift suppression. The figure summarizes the structural features of normal tRNAs (a), 8-nt tRNAs (b), and 9-nt tRNAs (c) as proposed by Curran and Yarus (1987). Conformation is designated by the number of nucleotides in the 5′ bridge and the number of nucleotides in the 3′ stack (e.g., 5:2). AC, Anticodon.

anticodon loop. The remaining 2 nt on the 5′ side of the loop serve as a bridge to complete the loop. Overall, the structure is referred to as the 5:2 conformation. As a rule, the anticodon nucleotides are the three distal-most nucleotides in the 5-nt 3′ stack. Given this structure, the prediction is that all tRNAs with 7-nt loops will confer 3-nt translocation on the mRNA. Although this is true for the most part, there are now several exceptions, discussed in previous sections, of tRNAs containing a 7-nt anticodon loop that cause either 2- or 4-nt translocation.

Additional structural possibilities exist for tRNAs containing expanded anticodon loops. Loops of 8 nt are predicted to assume two active conformations, one consisting of a 5-nt 3′ stack with a 3-nt 5′ bridge (the 5:3 conformation) and the other consisting of a 6-nt 3′ stack with a 2-nt 5′ bridge (the 6:2 conformation). The two alternate conformations provide a physical basis for the apparent capacity of 8-nt frameshift suppressor tRNAs to read either 3- or 4-nt codons. The model predicts that the 5:3 conformation leads exclusively to 3-nt translocation, whereas the 6:2 conformation leads exclusively to 4-nt translocation. The choice of conformation may be dictated in part by the codon being translated. For example, the 6:2 conformation may be stabilized by the potential for fourth-nucleotide pairing, which would explain the observed increase in suppressor efficiency that often results from pairing at all four anticodon-codon positions (Gaber and Culbertson, 1984).

To test the model, 8-nt tRNAs were constructed

by inserting single nucleotides on the 5′ side of the anticodon loop of *su7*, an *E. coli* glutamine-inserting amber suppressor (Curran and Yarus, 1987). Using appropriate mutations in *lacZ* to assay for translational step size, they showed that 8-nt tRNAs can indeed read both 3- and 4-nt codons.

Anticodon loop stacking also predicts the coding specificity of suppressor tRNAs containing an 8-nt loop. Nucleotide insertions on the 5′ side of loop expand the 5′ bridge but do not affect the 3′ stack and therefore should not alter coding specificity. Frameshift suppressors containing such insertions in fact retain their coding specificity, as is the case, for example, for *SUF16* alleles that retain the ability to decode glycine codons. However, insertions on the 3′ side of the stack should alter specificity, as is the case for the *glyT* missense suppressors that confer a change in specificity from glycine to tryptophan codons.

For tRNAs containing a 9-nt anticodon loop, at least two conformational isomers are possible, including a 6:3 and a 7:2 conformation (Fig. 3). Alleles of *SUF8* may adopt the 7:2 conformation, which would position the nucleotides 3′-GGUU-5′ at the same place as a 4-nt anticodon in the 6:2 conformation of 8-nt tRNAs. The prediction, therefore, is that *SUF8* alleles may read a 5′-CCAA-3′ codon in the *his4-713* frameshift window, as discussed in a previous section, with nucleotide pairing at all four positions. Such pairing may help stabilize the 7:2 conformation. The 6:3 conformation may be used by the 9-nt *S. pombe* mitochondrial $tRNA^{Leu}$, which reads a 3-nt codon. Similarly, *SUF8* might also adopt a 6:3 conformation under certain circumstances, allowing it to read a 3-nt codon.

Although the anticodon loop stacking model provides a rational explanation for suppressor tRNAs containing extra anticodon loop nucleotides, it provides no simple explanation for some of the more recently identified classes of suppressors. One of these classes is represented by suppressor tRNAs containing simple base substitutions in the anticodon loop that result in suppression of either +1 or −1 frameshift mutations. Since these tRNAs contain seven anticodon loop nucleotides, they should not act as frameshift suppressors according to the model of Curran and Yarus (1987). Furthermore, for some of these suppressors there is nothing obvious in the mRNA sequence or context in the frameshift windows that offer a mechanistic explanation for how these suppressors act.

The model of Curran and Yarus (1987) also fails to account for the behavior of a number of suppressors that act on the −1 frameshift mutation *trpE91*. Many of these suppressor mutations are located outside the anticodon loop, and therefore their behavior cannot be adequately explained solely by reference to anticodon loop stacking. These types of suppressor mutations might act in two different ways. They may perturb anticodon loop conformation by acting at a distance, in which case the nature of the structural perturbations is not easily predicted. Second, they might affect the interaction of tRNA with the elongation factor EF-Tu (EF-1) or contact points outside the anticodon loop between tRNA and ribosomal components. According to this view, the interaction between anticodon and codon may be indirectly affected. Finally, the results so far on some suppressors (e.g., *hopR* and *hopE*) of *trpE91* indicate that codon sequence and context may play a role. It would therefore appear that additional genetic tests will be necessary before we can sort out the tRNA-mRNA interactions that lead to *trpE91* suppression.

PROTEIN-MEDIATED FRAMESHIFT SUPPRESSION

Some frameshift suppressors influence reading frame selection by mechanisms other than a direct change in the primary structure of tRNA. In bacteria, these suppressors include the S4 *ram* mutants (Atkins et al., 1972), the *supK* tRNA methylase mutant, which now appears to be an allele of the *prfB* gene coding for peptide chain release factor RF-2 (Atkins and Ryce, 1974; O'Mahony et al., 1989b; Kawakami et al., 1988), the *ksgA* rRNA methylation mutant (van Buul et al., 1984), and the paramomycin resistance mutation *par*ʳ*-454* in yeast mitochondrial rRNA (Weiss-Brummer and Huttenhofer, 1989).

Frameshift suppressor studies also suggest a role for soluble translation factors in reading frame selection, including EF-Tu in bacteria, EF-1 α (designated throughout as EF-1) in yeasts, and some previously unknown "factorlike" proteins in yeasts, including SUF12P and possibly UPF1P. We have selected these factors for a more detailed discussion because future discovery of their role in reading frame selection promises to be instructive.

Elongation Factor EF-Tu (EF-1)

EF-Tu, which binds aminoacyl-tRNA and GTP to form a ternary complex, plays a central role in the recognition of cognate tRNA and the rejection of noncognate tRNA during elongation. Initial recognition and proofreading can be rationalized according to the rate constants that characterize ribosome-ternary complex interactions (reviewed in Thompson, 1988). Although the kinetics of ternary complex function provide a rational basis for ensuring incor-

Table 4. Suppressor mutations in elongation factor EF-1[a]

Mutation	Amino acid substitution	Suppression of:	
		Frameshift mutations	Nonsense mutations
TEF2-1	Glu-286 to Lys	+	+
TEF2-2	Glu-317 to Lys	+	+
TEF2-3	Glu-40 to Lys	+	−
TEF2-4	Glu-122 to Lys	+	+
TEF2-10	Glu-122 to Gln	+	+
TEF2-7	Thr-142 to Ile	+	+
TEF2-9	Glu-295 to Lys	+	−
TEF2-13	Asp-130 to Asn	+	−
TEF2-16	Glu-291 to Lys	+	−

[a] For further details, see Sandbaken and Culbertson (1988).

poration of the correct amino acid, we also need to account for the genetic observation that translational suppressors often suppress both nonsense and frameshift mutations. This leads to consideration of the possibility that a common mechanism ensures both accurate selection of cognate tRNA and accurate maintenance of the reading frame.

If EF-Tu-mediated proofreading does play a role in reading frame selection, one would expect mutated forms of EF-Tu to confer frameshift suppression. This prediction was tested by screening for suppression in *S. typhimurium* strains carrying kirromycin resistance mutations in *tufA* and *tufB*, the duplicate structural genes for EF-Tu (Hughes et al., 1987). Many of the mutations conferred suppression of the −1 frameshift *trpE91*. Some of the mutations were tested further and were shown to suppress a limited number of +1 frameshifts and nonsense mutations.

Additional information on the role of EF-Tu in reading frame selection was obtained by mutational analysis of *S. cerevisiae* EF-1, the eucaryotic functional homolog of bacterial EF-Tu (Sandbaken and Culbertson, 1988). The *TEF2* gene, one of two duplicate genes encoding yeast EF-1, was mutagenized in vitro with hydroxylamine and introduced into yeast cells on a low-copy-number plasmid to test for variants that confer frameshift suppression. *TEF2* mutations were located by DNA sequence analysis at eight different sites within the coding region of the gene (Table 4). About half of the mutations were represented more than once among independent isolates, indicating near saturation of the sites that can mutate to confer frameshift suppression in response to hydroxylamine. When the mutations were tested further, it was found that about half of them suppressed nonsense mutations as well as frameshift mutations.

The most informative result of this study was revealed in the nonrandom distribution of mutations within three localized regions of the protein (Fig. 4). These regions were assessed with respect to the predicted secondary structure of EF-1 and the location of consensus elements for GTP and tRNA interaction and, where possible, with respect to the known crystallographic structure of EF-Tu (Jurnak, 1985). One region, represented by *TEF2-3*, lies near the GTP-binding consensus element Gly-X-X-X-X-Gly-Lys. This region is important in the interaction of the protein with the phosphate in GTP. A second region, represented by *TEF2-4*, *-10*, *-13*, *-7*, and *-15*, lies between two GTP consensus elements, Asp-X-Pro-Gly and Asn-Lys-X-Asp. This region is important in the interaction of the protein with the guanine ring of GTP. A third region, represented by *TEF2-1*, *-16*, *-9*, and *-2*, resides near the tRNA-binding site.

On the basis of these results, it was predicted that the mutations may affect the interaction of EF-1 with either GTP or aminoacyl-tRNA and thereby influence the coordination of error-limiting functions of the protein. However, the mutations must act in subtle ways. With only one exception (*TEF2-16*), strains containing any of the *TEF2* alleles are viable in the absence of wild-type EF-1, indicating that essential functions of the protein are not significantly disrupted. Furthermore, the amino acid substitutions are conservative and should result in minor but clearly important alterations in protein structure.

In two separate studies in *S. cerevisiae*, it has also been shown that increased wild-type *TEF2* gene dosage enhances mistranslation of frameshift and nonsense mutations (S. Liebman, personal communication; M. Sandbaken, unpublished data). However, decreased *TEF2* gene dosage had no detectable effect on mistranslation. Dosage-dependent suppression required expression of the *TEF2* gene, but 5′ flanking sequences including the promoter were not required. These results indicate that the intracellular level of *TEF2* mRNA or EF-1 protein affects translational accuracy. The underlying mechanism for dosage-dependent suppression has not yet been investigated.

In summary, we have not yet proven that EF-Tu-mediated proofreading ensures accurate maintenance of the reading frame, but current data are consistent with that view. We have not been able to identify an independent mechanism that would limit frameshift errors. This issue will undoubtedly be addressed in greater detail in the future by determining the biochemical effects of mutations in EF-Tu and EF-1 on the functions of the protein. Furthermore, amino acid microsequence technology is now available to assess in detail the nature of frameshift events caused by mutations in EF-Tu (EF-1).

A.

Gly-X-X-X-X-Gly-Lys 18 24 Asp-X-Pro-Gly 80 84 Asn-Lys-X-Asp 134 144

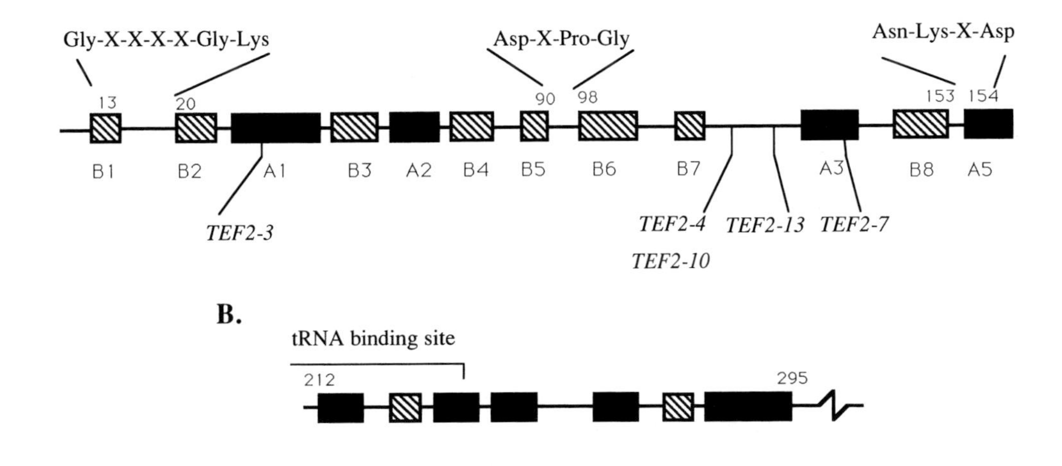

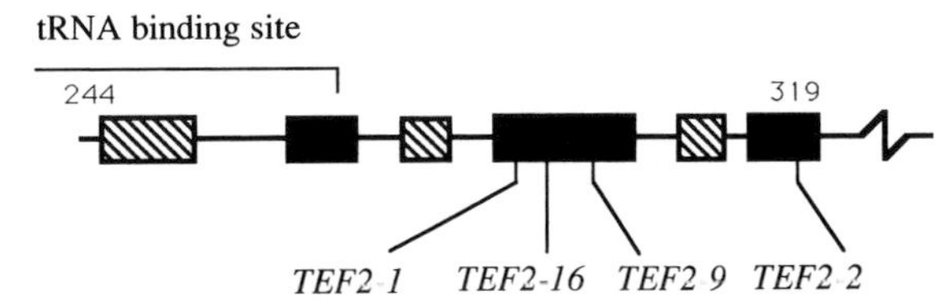

Figure 4. Frameshift suppressor mutations in elongation factor EF-1. The positions of EF-1 suppressor mutations are shown relative to regions of secondary structure predicted by Chou and Fasman (1978). The numbers above each line refer to the positions of the amino acids relative to the N terminus of the protein. (A) Crystallographic structure of the GTP-binding domain of EF-Tu · GDP (Jurnak, 1985) (top) compared with the predicted secondary structure of the corresponding region of EF-1 (bottom). Positions of the GTP-binding consensus sequences are shown. (B) Predicted secondary structure of a portion of EF-Tu (top) compared with that of EF-1 (bottom) (Chou and Fasman, 1978). The tRNA-binding domain of EF-Tu and the corresponding region of EF-1 are shown. Solid boxes, α Helix; hatched boxes, β sheet.

Role of "Factorlike" Proteins in Frameshift Suppression

Studies of frameshift suppressors in *S. cerevisiae* have revealed the existence of two previously unknown proteins, SUF12P and UPF1P, that may play a role in reading frame selection (Cummins et al., 1980; Wilson and Culbertson, 1988a, 1988b; Culbertson et al., 1980; P. Leeds, unpublished data). The *SUF12* and *UPF1* genes encode products whose structures and properties resemble those of known soluble translation factors, yet they appear to perform unique functions that do not correspond to the functions of known factors.

The *SUF12* gene was identified through the isolation of mutations that suppress +1 frameshift mutations in glycine (5′-GGN-3′) and proline (5′-CCN-3′) codons, as well as UGA and UAG nonsense mutations (Cummins et al., 1980). The *SUF12* gene was also identified by an independent approach that yielded a *suf12* allele conferring temperature-sensitive growth (Kikuchi et al., 1988). $SUF12^+$ is an essential, single-copy gene that codes for a 77-kilodalton protein. Both the size and relatively low abundance of SUF12P indicate that it is not a previously identified ribosomal protein. The most striking feature of SUF12P is its apparent relationship to EF-1. The protein consists of a unique, acidic N-terminal domain fused to other domains that are colinear with and similar in sequence to EF-1, including the domains for GTP binding and hydrolysis and tRNA binding. However, the phenotypes of *suf12* and *tef1-tef2* null alleles indicate that *suf12* cannot substitute for the function of EF-1 (Wilson and Culbertson, 1988a; Sandbaken and Culbertson, 1988). Recent studies demonstrate that SUF12P is associated with actively translating ribosomes (polysomes) (P. Wilson, unpublished data). Furthermore, a strain containing a temperature-sensitive allele of *suf12* exhibits stage-specific arrest during the cell cycle at a point following "start 1" in late G1 but before the onset of S phase, indicating a potentially important role for *suf12* in coordinating translation with global cellular events such as progression through the cell cycle (Kikuchi et al., 1988). The phenotypes of the suppressor mutations, the structural similarity between SUF12P and EF-1, and the association of SUF12P with ribosomes all suggest a direct role for SUF12P in translational fidelity.

The *UPF1* gene is one of four genes originally defined by mutations that enhance the efficiency of frameshift suppression in strains carrying *SUF1*, a suppressor glycine tRNA containing a 4-nt anticodon (Culbertson et al., 1980; Mendenhall et al., 1987; Leeds, unpublished data). The *UPF1* gene has now been cloned (Leeds, unpublished data). The DNA sequence indicates that UPF1P encodes a 110-kilodalton protein, too large to correspond to any known ribosomal protein. UPF1P contains several interesting structural motifs, including consensus elements for nucleotide binding and multiple Zn^{2+} fingers that could play a role in RNA binding similar to that found for yeast initiation factor eIF-2β (Donahue et al., 1988). The codon bias index for UPF1P indicates that it is present in relatively low intracellular abundance.

When a strain containing a *upf1* null allele was analyzed, it was found that loss of UPF1P function did not inhibit cell growth. However, loss of function did result in suppression of at least one frameshift mutation as well as all three classes of nonsense mutations (Leeds, unpublished data). Thus, the original *upf1* mutations identified by their ability to enhance *SUF1*-mediated frameshift suppression were most likely loss-of-function alleles that confer suppression by themselves in the absence of *SUF1*.

A large number of additional suppressor genes conferring phenotypes similar to those of *suf12* or *upf1* have been identified and await analysis (e.g., *suf13*, *suf14*, *upf2*, *upf3*, and *upf4* [Culbertson et al., 1980; Cummins et al., 1980, unpublished data; Leeds, unpublished data]). It will be important to provide evidence that these genes play a direct role in translation per se. The original selection scheme that uncovered *UPF1*, which was based on enhancement of tRNA-mediated frameshift suppression, might just as easily uncover mutations that affect the synthesis or breakdown of mRNA as opposed to translation. In such a case, suppression might result from the overexpression of mRNA containing a nonsense or frameshift mutation that is inherently leaky at a low level. Experiments are in progress with *upf1* to distinguish between these possibilities.

There is a lesson in the findings presented above that is still incompletely revealed. One clue regarding the nature of this lesson comes from the observation that all of the non-tRNA frameshift suppressors in procaryotes affect known translational components such as ribosomal proteins, rRNA, tRNA base modification enzymes, or known soluble factors. By contrast, studies of eucaryotic frameshift suppressors are turning up new proteins with as yet undefined functions. Does this mean that eucaryotes require additional factors to control ribosome assembly or function in a manner that influences translational fidelity? A related question has to do with our definition of "essential" factor. *SUF12*, an essential gene in vivo, was never uncovered as an essential factor in vitro, perhaps because accuracy is not as critical in an in vitro system, and loss of accuracy is not easily monitored. The hope is that further genetic analysis will reveal how many factors have been missed and will provide a more complete view of translation.

The writing of this article was supported by the College of Agricultural and Life Sciences, University of Wisconsin, Madison, and Public Health Service grant GM26217 to M.R.C. from the National Institutes of Health.

REFERENCES

Atkins, J., D. Elseviers, and L. Gorini. 1972. Low activity of β-galactosidase in frameshift mutants of *Escherichia coli*. *Proc. Natl. Acad. Sci. USA* **69:**1192–1195.

Atkins, J., B. Nichols, and S. Thompson. 1983. The nucleotide sequence of the first externally suppressible −1 frameshift mutant, and of some nearby leaky frameshift mutants. *EMBO J.* **2:**1345–1350.

Atkins, J., and S. Ryce. 1974. UGA and non-triplet suppressor reading of the genetic code. *Nature* (London) **249:**527–530.

Bossi, L., and J. Roth. 1981. Four-base codons ACCA, ACCU, and ACCC are recognized by frameshift suppressor *sufJ*. *Cell* **25:** 489–496.

Bossi, L., and D. Smith. 1984. Suppressor *sufJ*: a novel type of tRNA mutant that induces translational frameshifting. *Proc. Natl. Acad. Sci. USA* **81:**6105–6109.

Calos, M., and J. Miller. 1981. Genetic and sequence analysis of frameshift mutations induced by ICR-191. *J. Mol. Biol.* **153:** 39–66.

Chou, P., and G. Fasman. 1978. Prediction of the secondary structure of proteins from their amino acid sequence. *Adv. Enzymol.* **47**:45–149.

Culbertson, M., K. Underbrink, and G. Fink. 1980. Frameshift suppression in *Saccharomyces cerevisiae*. II. Genetic properties of Group II suppressors. *Genetics* **95**:833–853.

Cummins, C., M. Culbertson, and G. Knapp. 1985. Frameshift suppressor mutations outside the anticodon in yeast proline tRNAs containing an intervening sequence. *Mol. Cell. Biol.* **5**:1760–1771.

Cummins, C., R. Gaber, M. Culbertson, R. Mann, and G. Fink. 1980. Frameshift suppression in *Saccharomyces cerevisiae*. III. Isolation and genetic properties of Group III suppressors. *Genetics* **95**:855–875.

Curran, J., and M. Yarus. 1987. Reading frame selection and transfer RNA anticodon loop stacking. *Science* **238**:1545–1550.

Donahue, T., A. Cigan, E. Pabich, and B. Valavicius. 1988. Mutations at a Zn(II) finger motif in the yeast eIF-2B gene alter ribosomal start-site selection during the scanning process. *Cell* **54**:621–632.

Donahue, T., P. Farabaugh, and G. Fink. 1981. Suppressible glycine and proline four base codons. *Science* **212**:455–457.

Falahee, M., R. Weiss, M. O'Connor, S. Doonan, R. Gesteland, and J. Atkins. 1988. Mutants of translational components that alter reading frame by two steps forward or one step back. *J. Biol. Chem.* **263**:18099–18103.

Gaber, R., and M. Culbertson. 1982. The yeast frameshift suppressor gene *SUF16-1* encodes an altered tRNA containing the four-base anticodon 3′-CCCG-5′. *Gene* **19**:163–172.

Gaber, R., and M. Culbertson. 1984. Codon recognition during frameshift suppression in *Saccharomyces cerevisiae*. *Mol. Cell. Biol.* **4**:2052–2061.

Hughes, D., J. Atkins, and S. Thompson. 1987. Mutants of elongation factor Tu promote ribosomal frameshifting and nonsense readthrough. *EMBO J.* **6**:4235–4239.

Hughes, D., S. Thompson, M. O'Connor, T. Tuohy, B. Nichols, and J. Atkins. 1989. Genetic characterization of frameshift suppressors with new decoding properties. *J. Bacteriol.* **171**:1028–1034.

Jurnak, F. 1985. Structure of the GDP domain of EF-Tu and location of the amino acids homologous to ras oncogene proteins. *Science* **230**:32–36.

Kawakami, K., Y. Jonsson, G. Bjork, H. Ikeda, and Y. Nakamura. 1988. Chromosomal location and structure of the operon encoding peptide-chain-release factor 2 of *Escherichia coli*. *Proc. Natl. Acad. Sci. USA* **71**:5620–5624.

Kikuchi, Y., H. Shimatake, and A. Kikuchi. 1988. A yeast gene required for the G_1-to-S transition encodes a protein containing an A-kinase target site and GTPase domain. *EMBO J.* **7**:1175–1182.

Li, M., and A. Tzagoloff. 1979. Assembly of the mitochondrial membrane system: sequences of yeast mitochondrial valine and an unusual threonine tRNA gene. *Cell* **18**:47–53.

Mathison, L., and M. Culbertson. 1985. Suppressible and nonsuppressible +1 G · C base pair insertions induced by ICR-170 at the *his4* locus in *Saccharomyces cerevisiae*. *Mol. Cell. Biol.* **5**:2247–2256.

Mathison, L., M. Winey, C. Soref, M. Culbertson, and G. Knapp. 1989. Mutations in the anticodon stem affect removal of introns from pre-tRNA in *Saccharomyces cerevisiae*. *Mol. Cell. Biol.* **10**:4220–4228.

Mendenhall, M., P. Leeds, H. Fen, L. Mathison, M. Zwick, C. Sleiziz, and M. Culbertson. 1987. Frameshift suppressor mutations affecting the major glycine transfer RNAs of *Saccharomyces cerevisiae*. *J. Mol. Biol.* **194**:41–58.

Murgola, E., N. Prather, B. Mims, F. Pagel, and K. Hijazi. 1983. Anticodon shift in tRNA: a novel mechanism in missense and nonsense suppression. *Proc. Natl. Acad. Sci. USA* **80**:4936–4939.

O'Mahony, D., D. Hughes, S. Thompson, and J. Atkins. 1989a. Suppression of a −1 frameshift mutation by a recessive tRNA suppressor which causes doublet decoding. *J. Bacteriol.* **171**:3824–3830.

O'Mahony, D., B. Mims, S. Thompson, E. Murgola, and J. Atkins. 1989b. Glycine tRNA mutants with normal anticodon loop size cause −1 frameshifting. *Proc. Natl. Acad. Sci. USA* **86**:7979–7983.

Prather, N., E. Murgola, and B. Mims. 1981. Nucleotide insertion in the anticodon loop of a glycine transfer RNA causes missense suppression. *Proc. Natl. Acad. Sci. USA* **78**:7408–7411.

Riddle, D., and J. Carbon. 1973. A nucleotide addition to the anticodon of glycine tRNA. *Nature* (London) *New Biol.* **242**:230–237.

Roth, J. 1974. Frameshift mutations. *Annu. Rev. Genet.* **8**:319–346.

Sandbaken, M., and M. Culbertson. 1988. Mutations in elongation factor EF1 affect the frequency of frameshifting and amino acid misincorporation in *Saccharomyces cerevisiae*. *Genetics* **120**:923–934.

Sumner-Smith, M., H. Hottinger, I. Willis, T. Koch, R. Arentzen, and D. Soll. 1984. The *sup8* $tRNA^{Leu}$ gene of *Schizosaccharomyces pombe* has an unusual intervening sequence and reduced pairing in the anticodon stem. *Mol. Gen. Genet.* **197**:447–452.

Thompson, R. 1988. EFTu provides an internal kinetic standard for translational accuracy. *Trends Biochem. Sci.* **13**:91–93.

Tucker, S., E. Murgola, and F. Pagel. 1989. Missense and nonsense suppressors can correct frameshift mutations. *Biochimie* **71**:729–739.

van Buul, C., W. Visser, and P. van Knippenberg. 1984. Increased translational fidelity caused by the antibiotic kasugamycin and ribosomal ambiguity mutants harboring the *ksgA* gene. *FEBS Lett.* **177**:119–124.

Weiss-Brummer, B., and A. Huttenhofer. 1989. The paramomycin resistance mutation (*par*^{r}*-454*) in the 15S rRNA gene of the yeast *Saccharomyces cerevisiae* is involved in ribosomal frameshifting. *Mol. Gen. Genet.* **217**:362–369.

Wilson, P., and M. Culbertson. 1988a. SUF12 suppressor protein of yeast: a fusion protein related to the EF-1 family of elongation factors. *J. Mol. Biol.* **199**:559–573.

Wilson, P., and M. Culbertson. 1988b. Translational frameshifting and codon recognition, p. 415–430. *In* M. Tuite, M. Picard, and M. Bolotin-Fukuhara (ed.), *Genetics of Translation: New Approaches*. NATO ASI Series H, vol. 14. Springer-Verlag KG, Berlin.

Winey, M., L. Mathison, C. Soref, and M. Culbertson. 1989. Distribution of introns in frameshift suppressor proline tRNA genes in *Saccharomyces cerevisiae*. *Gene* **76**:89–97.

Winey, M., M. Mendenhall, C. Cummins, M. Culbertson, and G. Knapp. 1986. Splicing of a yeast proline tRNA containing a novel suppressor mutation in the anticodon stem. *J. Mol. Biol.* **192**:49–63.

X. EVOLUTION OF RIBOSOMES

X. EVOLUTION OF RIBOSOMES

In the past decade, rRNA, specifically the small subunit rRNA, has taken a preeminent role in the elucidation of phylogenetic trees and evolutionary comparisons. It is useful because it is ubiquitous, very ancient, and slow to evolve. As noted in the chapters that follow, there is some question as to exactly how evolutionary information inherent in sequences is best extracted from them. The problem clearly demands further study.

The origin and evolution of ribosomal protein genes is another fascinating area of study, in particular because it provides insight into the relationships of widely divergent microorganisms as well as into the origin of cellular organelles.

Chapter 50

Evolutionary Relationships among Primary Lineages of Life Inferred from rRNA Sequences

MANOLO GOUY and WEN-HSIUNG LI

Determining the number of extant primary lineages of life and their relationships is a fundamental step toward understanding the early diversification of life on earth. Phylogenetic analysis of the nucleotide sequences of the highly conserved genes encoding rRNAs has greatly increased our understanding of this issue, but it remains highly controversial. There are two opposite views. From analysis of the small-subunit (SSU) rRNAs, Woese and co-workers (see Woese, 1987) propose that eubacteria, archaebacteria, and eucaryotes are the three primary lines of descent (Fig. 1a). In this view, i.e., the archaebacterial tree, the three main types of archaebacteria (extremely thermophilic sulfur-dependent organisms, extreme halophiles, and methanogens) form a monophyletic group, distinct from both eucaryotes and eubacteria. In contrast, Lake (1988) claims that archaebacteria are paraphyletic and that the relationship of the eocytes (Lake's terminology for sulfur-dependent bacteria), halophiles, and methanogens to other life forms can be represented by Fig. 1b. In this view, i.e., the eocyte tree, all life forms are separated into two taxonomic divisions of the highest rank: a proto-eucaryotic group comprising eucaryotes and eocytes, and an essentially procaryotic group containing eubacteria, halophiles, and methanogens. This hypothesis was initially based on differences between sulfur-dependent and halophilic organisms in three-dimensional ribosomal structure as seen in electron microscopy (Lake et al., 1984, 1985). It has received considerable attention (Penny, 1988) after publication of an analysis of SSU rRNAs (Lake, 1988) by a new tree-making method, termed evolutionary parsimony (EP) (Lake, 1987a).

Here we review recent studies on this controversy. First, we discuss methods of tree reconstruction.

METHODS OF PHYLOGENETIC TREE RECONSTRUCTION

Three different approaches have been used to infer the universal evolutionary tree. One, known as the distance matrix approach, is based on pairwise evolutionary distances. In the method of Fitch and Margoliash (1967) and in the least-squares method (Olsen, 1987), one tries to find which unrooted tree and which set of branch lengths on that tree most closely match observed distances between sequence pairs. In the neighbor-joining (NJ) method (Saitou and Nei, 1987), one sequentially clusters sequence pairs that minimize the total branch length of the tree. In these three methods, pairwise evolutionary distances are obtained by correcting raw sequence differences for multiple hits, either by the Jukes-Cantor formula (Jukes and Cantor, 1969) or by the two-parameter model of Kimura (1980), which allows for unequal rates of transitions and transversions. One may also use a more sophisticated method such as the six-parameter method (Gojobori et al., 1982).

The second approach, maximum parsimony (MP), is to find a tree that requires the minimum number of substitutions to explain the nucleotide pattern found in the sequences under study (Fitch, 1977).

The third approach, EP, was proposed by Lake (1987a). This method applies only to four lines of descent, although each line may contain multiple species. For each of the three possible unrooted trees, a χ^2 with 1 df is computed from nucleotide configurations among the four lineages, and a tree is taken to be the true tree if the associated χ^2 is significant whereas the other two χ^2 values are not.

Lake (1987a, 1988) argues that his EP method is

Manolo Gouy ■ Laboratoire de Biométrie, Université Lyon I, 69622 Villeurbanne Cedex, France. **Wen-Hsiung Li** ■ Center for Demographic and Population Genetics, University of Texas, P.O. Box 20334, Houston, Texas 77225.

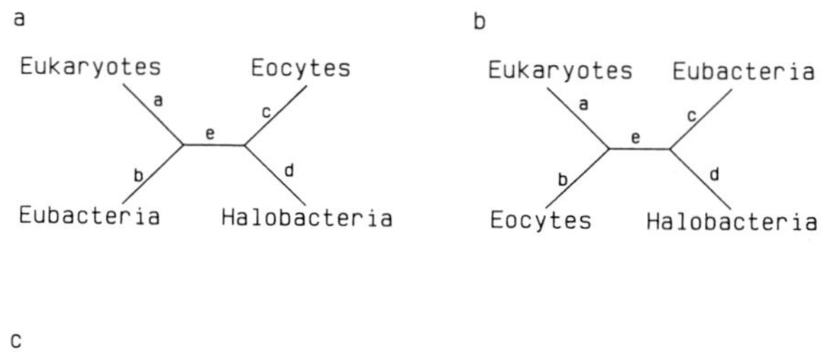

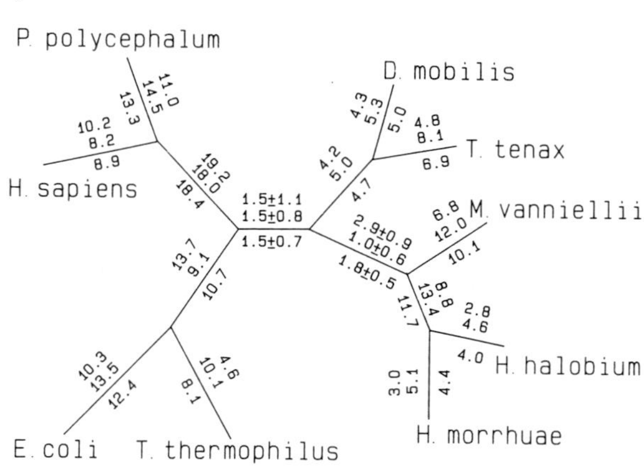

Figure 1. (a) The archaebacterial tree. For simplicity, methanogens are not included. (b) The eocyte tree. The third formally possible alternative, the halobacterial tree, clusters eucaryotes with halophiles and methanogens. (c) Unrooted tree inferred by the NJ and MP methods, reproduced from Gouy and Li (1989). The first, second, and third (below the branch) values on each branch represent the estimated numbers of substitutions per 100 sites for the SSU data, the LSU data, and the combined data, respectively. The standard errors indicate that a branch length is significantly greater than 0 at the 5% level if the mean is two times the standard error or greater. The species used are *Homo sapiens* and *Physarum polycephalum* (eucaryotes), *Escherichia coli* and *Thermus thermophilus* (eubacteria), *Desulfurococcus mobilis* and *Thermoproteus tenax* (sulfur-dependent bacteria), *Methanococcus vannielii* (methanogen), and *Halobacterium halobium* and *Halococcus morrhuae* (halophiles).

rate invariant, whereas distance matrix and MP methods are subject to the effect of unequal rates among lineages (Felsenstein, 1978). Jin and Nei (1990) conducted a simulation study of the efficiencies of the NJ, MP, and EP methods under a model tree similar to that of Fig. 1a. Their study shows that if a proper distance measure is used, the NJ method performs well, and better than the EP method, even under unequal rates of evolution among lineages. In another study (Li et al., 1987), it was concluded that the NJ method fails under extremely unequal rates of evolution. It appears now that this effect occurred because the authors used observed differences between sequences uncorrected for multiple hits. The study of Jin and Nei (1990) shows that use of a distance corrected for multiple hits improves greatly the accuracy of tree reconstruction by the NJ method. In a later section, we shall review simulation results that were obtained under conditions similar to those employed to obtain the data used to infer the universal tree.

SEQUENCE ALIGNMENT

Published phylogenetic analyses of rRNAs differ in procedures for multiple sequence alignments. Most authors (e.g., Olsen, 1987; Cedergren et al., 1988; Leffers et al., 1987; Gouy and Li, 1989) use the consensus secondary structure folding models of SSU (Dams et al., 1988) and large-subunit (LSU) (Leffers et al., 1987) rRNA molecules to verify whether aligned nucleotides are phylogenetically homologous. Only sites belonging to the same secondary structure motif (i.e., hairpin or single-stranded loop) across all species are retained for further analysis. Therefore, the fast-evolving parts of the molecule, where it is often difficult to obtain a multiple alignment, are discarded. The criterion of evolutionary conservation of secondary structure elements constitutes an objective means for selection of sites for analysis. A minority of sites, however, are more subjectively chosen because the boundaries of retained regions are not exactly defined by this criterion. Gouy and Li (1989) have checked that their conclusion was not significantly affected by the selection of sites or the alignment procedure.

The SSU rRNA analysis of Lake (1988) used two alignment procedures. One is based on published interkingdom alignments, and the other is a star alignment using the eocyte *Thermoproteus tenax* sequence as a reference. The latter approach implicitly assumes that the *T. tenax* sequence is the ancestral molecule and makes no attempt to achieve a global optimum because in aligning a sequence against *T. tenax*, no reference is made to any other sequences. This approach is obviously inferior to the alignment based on the consensus secondary structure, which has been thoroughly studied by many authors. Moreover, in Lake's analysis, the whole SSU rRNA molecule, except two divergent regions, was kept in the alignment. As also noted by Olsen (1987), discarding highly divergent regions before attempting to build the tree is preferable because these regions have lost most of their evolutionary information as a result of numerous multiple substitutions and so cannot be reliably aligned between very distantly related organisms.

PHYLOGENIES OBTAINED FROM PAIRWISE EVOLUTIONARY DISTANCES

Both the SSU (Woese, 1987; Olsen, 1987; Gouy and Li, 1989) and LSU (Leffers et al., 1987; Gouy and Li, 1989) rRNAs have been used to determine the evolutionary relationships among the primary lineages of life by application of distance matrix methods. All of these analyses consistently support the archaebacterial tree (Fig. 1c); that is, archaebacteria form a monophyletic taxon. Olsen (1987) studied in great detail, with SSU rRNAs only, whether this result could have been biased by the assumption of equal rate of substitution at all sites along the molecule. He developed an algorithm for computing pairwise evolutionary distances from observed sequence differences (i.e., to correct them for multiple hits) under a model in which substitution rates are log-normally distributed over the positions in the sequence. He showed that the archaebacterial tree is still predicted when this more elaborate procedure is used to build the tree. He also tested the efficiency of this procedure by applying it to recover the evolutionary relationship between the mitochondrion of the ciliate *Paramecium tetraurelia* and eubacteria. This procedure clusters correctly the mitochondrial sequence to its closest relatives, α-group purple bacteria, even when the tree contains very distant organisms such as archaebacteria or eucaryotes. In contrast, the Jukes-Cantor correction results in underestimating the largest pairwise distances, which in turn produces an incorrect placement of the mitochondrial lineage in the evolutionary tree.

Gouy and Li (1989) evaluated the sampling error attached to the central branch of the tree in Fig. 1c by using the method of Li (1989). This showed that when SSU and LSU rRNAs are pooled for analysis, the length of the central branch is statistically greater than zero. The statistical significance of the archaebacterial tree is also supported by bootstrapping (Felsenstein, 1985) either the SSU or LSU data. When more species are included in the analysis, a larger and probably more accurate estimate of that branch length is obtained (Fig. 2).

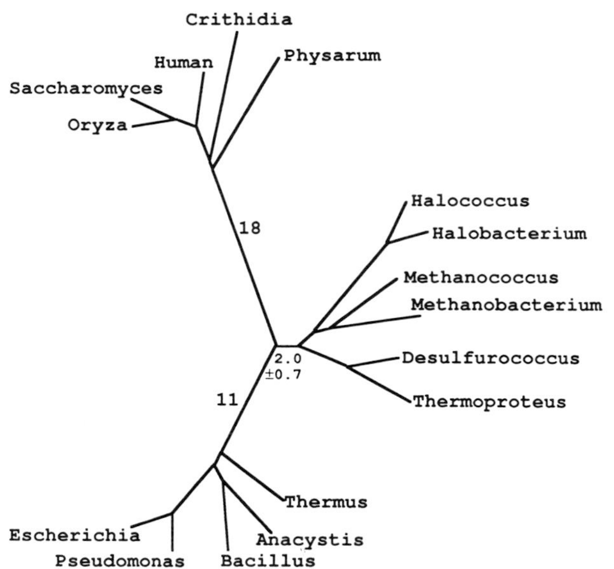

Figure 2. Unrooted universal evolutionary tree inferred from SSU and LSU rRNAs analyzed by the NJ method at 2,517 homologous sites (873 from SSU and 1,644 from LSU rRNAs). The tree-making procedure is that of Gouy and Li (1989). Branch lengths are proportional to evolutionary distances and are in terms of substitutions per 100 sites. The species used are those in Fig. 1c plus *Saccharomyces cerevisiae* (a fungus), *Oryza sativa* (a plant), and *Crithidia fasciculata* (a flagellate) for eucaryotes and *Pseudomonas aeruginosa* (α-group purple bacterium), *Bacillus subtilis* (gram-positive bacterium), and *Anacystis nidulans* (cyanobacterium) for eubacteria. This data set contains all procaryotes for which both the SSU and LSU rRNA sequences have been published. *C. fasciculata* and *P. polycephalum* are the protists with known SSU and LSU rRNA sequences that diverged most from higher eucaryotes. Many of these sequences can be found in recent compilations of LSU (Gutell and Fox, 1988) and SSU (Dams et al., 1988) rRNAs. Other sources are Johansen et al. (1988) for *P. polycephalum*, Toschka et al. (1988) for *P. aeruginosa*, and Höpfl et al. (1988) and Murzina et al. (1988) for *T. thermophilus*. Sequences were extracted from GenBank (release 59) and the EMBL nucleotide sequence library where possible.

PHYLOGENIES OBTAINED FROM THE PARSIMONY PRINCIPLE

The parsimony principle has been applied to SSU rRNAs by Wolters and Erdmann (1986) and by Olsen (1987). These authors reached divergent conclusions. Wolters and Erdmann reported three positions in which eucaryotes share a nucleotide with the sulfur-dependent bacteria, whereas eubacteria share another nucleotide with halophiles-methanogens. This analysis supported the eocyte tree. Olsen used only the most slowly changing positions of SSU rRNA, those having more than 93% conservation within eucaryotes and also within eubacteria. He found nine positions supporting the archaebacterial tree, three for the eocyte tree, and two for the halobacterial tree. Therefore, as noted by Olsen, more data than the SSU rRNA sequences are needed to obtain a clear discrimination between alternative phylogenies.

The MP method was applied to SSU and LSU rRNAs by Cedergren et al. (1988) and by Gouy and Li (1989). The analysis of Cedergren et al. included mitochondrial rRNAs which resulted in a substantial reduction of the number of sites retained in LSU rRNA because this molecule has been extensively transformed during its evolution in the mitochondrial ribosome. Therefore, Cedergren et al. reported only a moderate statistical support for the archaebacterial tree from LSU rRNAs, though a significant support from SSU rRNAs. Our results based on 2,517 homol-

Table 1. MP analysis of combined SSU and LSU rRNAs[a]

Tree topology	Total no. of changes	No. of supporting informative sites
Archaebacterial[b]	4,367	56
Eocyte	4,396	27
Halobacterial	4,398	25

[a] rRNA data set as in Fig. 2.
[b] The most parsimonious tree, which differs from that of Fig. 2 only in having *Methanobacterium* spp. clustered with halobacteria.

ogous sites from the two molecules are presented in Table 1. The archaebacterial tree is very strongly supported, since it is compatible with twice as many informative sites as the other two alternatives are. Analysis of SSU and LSU rRNAs separately also consistently supports the archaebacterial tree.

PHYLOGENIES OBTAINED FROM THE EP METHOD

Lake (1988) applied the EP method to SSU rRNA sequences and obtained strong support for the eocyte tree. We applied the EP method to both the SSU and LSU rRNAs either separately or jointly (Table 2; Gouy and Li, 1989). In agreement with Lake's results, the EP method supports the eocyte tree when applied to SSU rRNAs. However, when applied to LSU rRNAs, the archaebacterial tree is strongly favored, with one χ^2 highly significant, the other two being nonsignificant. This strong discrepancy arises probably because the EP method is sensitive to the assumption of equal transversional rates. This assumption is essential for EP to be applicable (Lake, 1987a; Jin and Nei, 1990). Lake (this volume) argues that the LSU data are not suitable for the present purpose because LSU sequences evolve faster than SSU sequences. However, we note from Table 2 that for the LSU data, the χ^2 value is extremely large for the archaebacterial tree but small for the eocyte tree, and so a most difficult point is why this set of data provides not even slight signals for the eocyte tree. On the other hand, for the SSU data, the χ^2 value is significant not only for the eocyte tree but also for the halobacterial tree. Thus, this set of data does not provide more consistent results than the LSU data. At any rate, when the two sets of data are combined, the EP method also favors the archaebacterial tree ($P = 4 \times 10^{-4}$).

RELIABILITY OF TREE-BUILDING METHODS ASSESSED BY COMPUTER SIMULATIONS

Lake (1988) has argued that distance-based and MP methods are more sensitive than the EP method to effects of unequal rates of substitution among lineages. This is true only when the degree of sequence divergence is large (Olsen, 1987; Li et al., 1987). When highly divergent regions are trimmed off from SSU and LSU rRNAs, as described above, the degree of divergence becomes moderate (about 50% for the largest distance, that between eucaryotes and eubacteria, after correction for multiple hits).

Our computer simulation shows that in this situation, NJ and MP perform better than EP. We use either the archaebacterial or the eocyte tree as a model tree (Fig. 1a and b). In the former case, we compute branch lengths by the NJ method, whereas in the latter case we compute branch lengths by the operator metrics method of Lake (1987b). We then simulate nucleotide changes along each branch according to the two-parameter model of Kimura (1980). Consistent with Olsen (1987), we assume that substitution frequencies in each lineage vary among sites according to a log-normal distribution. We then apply each of the three tree-building methods to the simulated sequence sets. The probabilities of recovering the model tree are shown in Table 3. When the model tree is as inferred by the NJ method from the combined SSU and LSU rRNAs, both the NJ and MP methods have excellent recovery scores, whereas the EP method misses the tree in more than 40% of the replicates (Table 3, line 1). When the model tree is as inferred by the NJ method from the SSU rRNA (line 3), the NJ method still performs well,

Table 2. Application of the EP method to SSU and LSU rRNAs[a]

rRNA molecule	Archaebacterial: χ^2 with 1 df	Archaebacterial: P^b	Eocyte: χ^2 with 1 df	Eocyte: P	Halobacterial: χ^2 with 1 df	Halobacterial: P
SSU	1.6	0.20	6.8	0.009	3.9	0.048
LSU	23.1	$<5 \times 10^{-6}$	0.5	0.46	3.0	0.08
SSU + LSU	12.3	0.0004	3.6	0.059	6.4	0.012

[a] rRNA data set as in Fig. 2.
[b] Probability of exceeding the χ^2 value.

Table 3. Reliability of NJ, MP, and EP tree-building methods assessed by simulation[a]

Model tree	Sequence length	Branch length (%)					Probability (%)		
		a	b	c	d	e	NJ	MP	EP
Archaebacterial	2,500	29.5	21.5	10.7	17.5	2.0	97	98	58
Eocyte	2,500	27.7	7.3	18.7	13.1	1.9	87	83	63
Archaebacterial	900	30.0	21.5	8.8	14.4	1.5	81	82	39
Eocyte	900	26.1	3.9	24.8	12.7	4.1	94	90	69

[a] Branches a to e are identified in Fig. 1a and b. The fraction of transitions was 55%; 500 replicates were simulated in each case. The rate of substitution was made to vary along sequence sites according to a log-normal distribution. The width of the distribution was set such that 95% of the sites have a rate between one-eighth and eight times the median rate for any given lineage. Unlike previous data of Gouy and Li (1989), observed sequence differences were corrected for multiple hits according to the two-parameter method of Kimura (1980) before application of the NJ method.

whereas the EP method is barely better than the 33% score obtained by random choice. The performance of the EP method improves when the model tree is that predicted by EP for SSU rRNAs (Table 3, line 4). However, the other two methods still perform better than EP.

The reliability of the three tree-making methods has also been tested empirically (Gouy and Li, 1989) by applying them to the SSU rRNAs from human, *Drosophila melanogaster*, rice, and *Physarum polycephalum*. Both the NJ and MP methods gave the correct tree, which clusters humans with *D. melanogaster*, whereas the EP method puts humans and rice in a sister group. Lake (this volume) argues that if the EP method fails, the NJ and MP methods necessarily must fail because the EP method applies to more general conditions. The example given above contradicts that claim. Furthermore, the simulation of Jin and Nei (1990) shows that NJ is superior to EP under various conditions.

THE ARCHAEBACTERIAL TREE IS STRONGLY SUPPORTED BY rRNA SEQUENCE ANALYSES

In sum, the eocyte tree does not seem tenable. It is favored only when the EP method is applied to SSU rRNA. On the other hand, all three methods, NJ, MP, and EP, cluster LSU rRNAs and combined SSU plus LSU molecules according to the archaebacterial tree. Both distance-based methods and the parsimony principle support the archaebacterial tree at a statistically significant level when the two rRNA molecules are pooled. The strong incongruency in the tree topologies inferred by EP from SSU and LSU rRNAs, the computer simulation results, and the empirical test all indicate that the EP method is rather inefficient for phylogenetic reconstruction, at least in this particular situation.

TOWARD ROOTING THE UNIVERSAL EVOLUTIONARY TREE

Currently, there is much debate about the root of the tree. One view is that archaebacteria represent the "ur-kingdom" from which eubacteria and eucaryotes have arisen, the eucaryotic line from within sulfur-dependent organisms and the eubacterial line from the methanogen-halophile group (Woese and Wolfe, 1985). The view of Lake (1988) is similar except that he does not assume that archaebacteria represent the ur-kingdom. This view is obviously not compatible with the branching order shown in Fig. 2. Another possibility is that the root lies in the central branch of Fig. 2 (Woese, 1987). Both this view and the view that eucaryotes, eubacteria, and archaebacteria were all derived from a progenote (Woese, 1987) imply that the eucaryotic lineage is as old as the eubacterial one, but eucaryotes probably did not appear before 2 billion years ago, whereas eubacteria go back at least 3.5 billion years (Schopf et al., 1983). A more plausible hypothesis is that archaebacteria and eucaryotes were derived from eubacteria, because they share marked molecular and cellular resemblances (Cavalier-Smith, 1987). For example, the archaebacterial genes coding for RNA polymerase core subunits are much more similar to the eucaryotic than to the eubacterial counterparts (Pühler et al., 1989). This hypothesis is also supported by a recent analysis of the evolution of the H^+-ATPase in all three primary lineages (Gogarten et al., 1989; Inatomi et al., 1989). Under this hypothesis, sulfur-dependent organisms branched off shortly after the divergence between archaebacteria and eucaryotes, and methanogens branched off shortly after that. This supports the observation that archaebacteria form a monophyletic but highly diversified group (Woese, 1987). The greater lengths of the halophile and methanogen branches than of the sulfur-dependent branch may explain why the resemblance between archaebacteria and eucaryotes is most pronounced in the case of the sulfur-dependent archaebacteria (Woese and Wolfe, 1985; Lake, 1988).

REFERENCES

Cavalier-Smith, T. 1987. The origin of eukaryote and archaebacterial cells. *Ann. N.Y. Acad. Sci.* **503:**17–54.

Cedergren, R., M. W. Gray, Y. Abel, and D. Sankoff. 1988. The evolutionary relationships among known life forms. *J. Mol. Evol.* **28:**98–112.

Dams, E., L. Hendriks, Y. Van der Peer, J.-M. Neefs, G. Smits, I. Vandenbempt, and R. De Wachter. 1988. Compilation of small ribosomal subunit RNA sequences. *Nucleic Acids Res.* **16:** r87–r173.

Felsenstein, J. 1978. Cases in which parsimony or compatibility methods will be positively misleading. *Syst. Zool.* **27:**401–410.

Felsenstein, J. 1985. Confidence limits on phylogenies: an approach using the bootstrap. *Evolution* **39:**783–791.

Fitch, W. M. 1977. On the problem of discovering the most parsimonious tree. *Am. Nat.* **3:**223–257.

Fitch, W. M., and E. Margoliash. 1967. Construction of phylogenetic trees: a method based on mutational distances as estimated from cytochrome c sequences is of general applicability. *Science* **155:**279–284.

Gogarten, J. P., H. Kibak, P. Dittrich, L. Taiz, E. J. Bowman, B. J. Bowman, M. F. Manolson, R. J. Poole, T. Date, T. Oshima, J. Konishi, K. Denda, and M. Yoshida. 1989. Evolution of the vacuolar H^+-ATPase: implications for the origin of eukaryotes. *Proc. Natl. Acad. Sci. USA* **86:**6661–6665.

Gojobori, T., K. Ishii, and M. Nei. 1982. Estimation of average number of nucleotide substitutions when the rate of substitution varies with nucleotide. *J. Mol. Evol.* **18:**414–423.

Gouy, M., and W.-H. Li. 1989. Phylogenetic analysis based on rRNA sequences supports the archaebacterial rather than the eocyte tree. *Nature* (London) **339:**145–147.

Gutell, R. R., and G. E. Fox. 1988. A compilation of large subunit RNA sequences presented in a structural format. *Nucleic Acids Res.* **16:**r175–r269.

Höpfl, P., N. Ulrich, R. K. Hartmann, W. Ludwig, and K. H. Schleifer. 1988. Complete nucleotide sequence of a 23S ribosomal RNA gene from *Thermus thermophilus* HB8. *Nucleic Acids Res.* **16:**9043.

Inatomi, K.-I., S. Eya, M. Maeda, and M. Futai. 1989. Amino acid sequence of the α and β subunits of *Methanosarcina barkeri* ATPase deduced from cloned genes. *J. Biol. Chem.* **264:**10954–10959.

Jin, L., and M. Nei. 1990. Limitations of the evolutionary parsimony method of phylogenetic analysis. *Mol. Biol. Evol.* **7:** 82–102.

Johansen, T., S. Johansen, and F. B. Haugli. 1988. Nucleotide sequence of the *Physarum polycephalum* small subunit ribosomal RNA as inferred from the gene sequence: secondary structure and evolutionary implications. *Curr. Genet.* **14:**265–273.

Jukes, T. H., and C. R. Cantor. 1969. Evolution of protein molecules, p. 21–132. *In* H. N. Munro (ed.), *Mammalian Protein Metabolism*. Academic Press, Inc., New York.

Kimura, M. 1980. A simple method for estimating evolutionary rates of base substitutions through comparative studies of nucleotide sequences. *J. Mol. Evol.* **16:**111–120.

Lake, J. A. 1987a. A rate-independent technique for analysis of nucleic acid sequences: evolutionary parsimony. *Mol. Biol. Evol.* **4:**167–191.

Lake, J. A. 1987b. Determining evolutionary distances from highly diverged nucleic acid sequences: operator metrics. *J. Mol. Evol.* **26:**59–73.

Lake, J. A. 1988. Origin of the eukaryotic nucleus determined by rate-invariant analysis of rRNA sequences. *Nature* (London) **331:**184–186.

Lake, J. A., M. W. Clark, E. Henderson, S. P. Fay, M. Oakes, A. Scheinman, J. P. Thornber, and R. A. Mah. 1985. Eubacteria, halobacteria, and the origin of photosynthesis: the photocytes. *Proc. Natl. Acad. Sci. USA* **82:**3716–3720.

Lake, J. A., E. Henderson, M. Oakes, and M. W. Clark. 1984. Eocytes: a new ribosome structure indicates a kingdom with a close relationship to eukaryotes. *Proc. Natl. Acad. Sci. USA* **81:**3786–3790.

Leffers, H., J. Kjems, L. Østergaard, N. Larsen, and R. A. Garrett. 1987. Evolutionary relationships amongst archaebacteria: a comparative study of 23S ribosomal RNAs of a sulphur-dependent extreme thermophile, an extreme halophile and a thermophilic methanogen. *J. Mol. Biol.* **195:**43–61.

Li, W.-H. 1989. A statistical test of phylogenies estimated from sequence data. *Mol. Biol. Evol.* **6:**424–435.

Li, W.-H., K. H. Wolfe, J. Sourdis, and P. M. Sharp. 1987. Reconstruction of phylogenetic trees and estimation of divergence times under nonconstant rates of evolution. *Cold Spring Harbor Symp. Quant. Biol.* **52:**847–856.

Murzina, N. V., D. P. Vorozheykina, and N. I. Matvienko. 1988. Nucleotide sequence of *Thermus thermophilus* HB8 gene coding 16S rRNA. *Nucleic Acids Res.* **16:**8172.

Olsen, G. J. 1987. The earliest phylogenetic branchings: comparing rRNA-based evolutionary trees inferred with various techniques. *Cold Spring Harbor Symp. Quant. Biol.* **52:**825–837.

Penny, D. 1988. What was the first living cell? *Nature* (London) **331:**111–112.

Pühler, G., H. Leffers, F. Gropp, P. Palm, H.-P. Klenk, F. Lottspeich, R. A. Garrett, and W. Zillig. 1989. Archaebacterial DNA-dependent RNA polymerases testify to the evolution of the eukaryotic nuclear genome. *Proc. Natl. Acad. Sci. USA* **86:**4569–4573.

Saitou, N., and M. Nei. 1987. The neighbor-joining method: a new method for reconstructing phylogenetic trees. *Mol. Biol. Evol.* **4:**406–425.

Schopf, J. W., J. M. Hayes, and M. R. Walter. 1983. Evolution of earth's earliest ecosystems: recent progress and unsolved problems, p. 361–384. *In* J. W. Schopf (ed.), *Earth's Earliest Biosphere: Its Origin and Evolution*. Princeton University Press, Princeton, N.J.

Toschka, H. Y., P. Höpfl, W. Ludwig, K. H. Schleifer, N. Ulbrich, and V. A. Erdmann. 1988. Complete nucleotide sequence of a 16S ribosomal RNA gene from *Pseudomonas aeruginosa*. *Nucleic Acids Res.* **16:**2348.

Woese, C. R. 1987. Bacterial evolution. *Microbiol. Rev.* **51:** 221–271.

Woese, C. R., and R. S. Wolfe. 1985. Archaebacteria, p. 561–564. *In* C. R. Woese and R. S. Wolfe (ed.), *The Bacteria*, vol. 8. Academic Press, Inc., New York.

Wolters, J., and V. A. Erdmann. 1986. Cladistic analysis of 5S rRNA and 16S rRNA secondary and primary structure—the evolution of eukaryotes and their relation to archaebacteria. *J. Mol. Evol.* **24:**152–166.

Chapter 51

Origin of the Eucaryotic Nucleus: rRNA Sequences Genotypically Relate Eocytes and Eucaryotes

JAMES A. LAKE

Great interest has been focused on the origin of the nucleus and its relationship to its closest procaryotic relatives. In particular, recent studies and analyses (Lake, 1988) of the early evolution of cells suggest that the traditional procaryote-eucaryote classification of organisms may have to be revised. The origins of modern organisms must be known if we are to understand the beginnings of ribosomes and protein synthesis. Furthermore, the origin of the nuclear rRNA genes must necessarily revolve about details of tree reconstruction and be resistant to artifacts of unequal rates (Penny, 1988; Felsenstein, 1988). This consideration has been an important catalyst for formulating new methods of tree reconstruction.

For some time, it has been generally appreciated that the traditional grouping of organisms into procaryotes and eucaryotes was based on a negative definition (Eldridge and Cracraft, 1986; Alsopp, 1969) and might be incorrect. Eucaryotes are defined by a positive character, i.e., the presence of a nucleus, but procaryotes are not; procaryotes are organisms that lack a nucleus. This view is, of course, not novel, and similar claims have been made about the invertebrates (animals lacking backbones) as well as about many other groups. If all known organisms were sorted into one of two boxes according to whether they had a nucleus, one would find that the box of eucaryotes was a coherent group, since they were defined by a positive character, the nucleus. The procaryotic box, on the other hand, would be filled with all cells that lacked a nucleus, and these might or might not be related. This possibility could be tested when sequences and treeing algorithms that were appropriate for sufficiently deep trees became available. Indeed, evolutionary parsimony analysis of rRNA sequences indicated that one group of cells that had been placed in the procaryotic box, the eocytes, appeared instead to belong in the same box as the eucaryotes (Lake, 1988, 1990a). This finding was unexpected since the eocytes are extreme thermophiles, living at temperatures greater than 85°C, while typically metabolizing sulfur by chemically reducing it with hydrogen to form hydrogen sulfide. In this tree, eocytes are the sister group of eucaryotes.

TREE RECONSTRUCTION METHODS

Determining the tree that relates all organisms is a lofty aspiration. Obtaining robust tree reconstruction algorithms and understanding the artifacts that can result when rates of evolution differ among the various taxa are central to pursuing this quest.

Parsimony and augmented-distance tree reconstruction algorithms are widely used methods. Hence, evolutionary systematists were astonished when Felsenstein (1978) first showed that parsimony can predict the incorrect tree when rates are different in juxtaposed branches of a tree. He showed that when reconstruction algorithms fail, long branches will be placed with long ones and short branches with short ones no matter which tree is correct.

To understand these effects, it is useful to describe the technical details of reconstructing trees from sequences by parsimony analysis. Parsimony functions by looking for patterns. First consider a particular nucleotide position in a set of aligned sequences. For four organisms (Fig. 1), the pattern AAGG is interpreted as supporting the placement of

James A. Lake ■ Molecular Biology Institute and Department of Biology, University of California at Los Angeles, Los Angeles, California 90024.

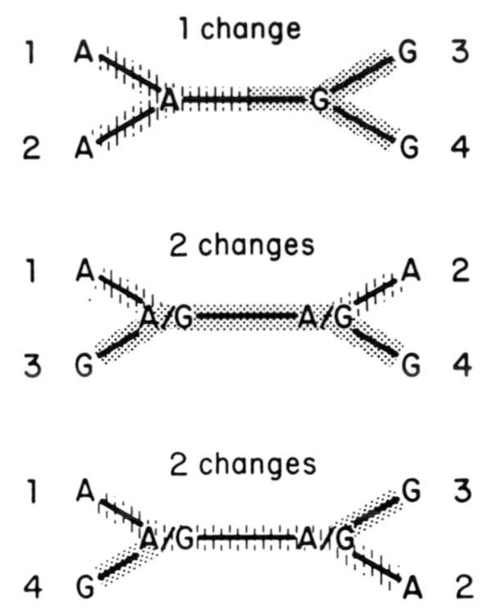

Figure 1. Parsimony requires the minimum number of nucleotide changes: the three possible tree topologies that relate four taxa. Taxa are shown as numbers 1 to 4. Nucleotides represent a nucleotide at a single homologous sequence position. The top tree requires only one nucleotide change, whereas the bottom two require at least two changes. Hence, the top is most parsimonious.

taxon 1 with taxon 2 and of taxon 3 with taxon 4. This is seen in the unrooted (meaning that a starting time point is not indicated) tree that joins taxa 1 and 2 on the left by noting that only a single change is sufficient to produce the pattern. In comparison, the other two possible trees both require at least two changes. Hence, the most parsimonious (or minimum-change) tree requires only a single change.

To understand how parsimony can fail, consider the four aligned sequences in Fig. 2a that are calculated from the (1,2)(3,4) tree discussed above. When the central branch is long, as in Fig. 2a, then patterns like UUAA (at position 4) support the (1,2)(3,4) tree as being most parsimonious. On the other hand, if the central branch is short and juxtaposed branches are of unequal lengths, then parsimony can fail. In Fig. 2b, for example, taxa 2 and 4 are rapidly evolving (long branches) and taxa 1 and 3 are slowly evolving (short branches). Since 1 and 3 are slow, a nucleotide at the tip of 1 will be the same as one at the tip of 3 most of the time. In contrast, since 2 is fast almost to the point of being random, one calculates that three times out of four it will differ from those at 1 and 3. Also, 4 will often differ from the nodal nucleotide and will, by random chance, match 2 one time out of four. Thus, we expect to have a pattern (like GCGC in Fig. 2b) that will support the incorrect tree at 3/4 × 1/4 (or 3/16) of the positions. This artifact occurs with the algorithms of the parsimony, of the distance matrix, and of the augmented-distance matrix methods (widely used to construct multikingdom trees [Woese, 1987; Pace et al., 1986]).

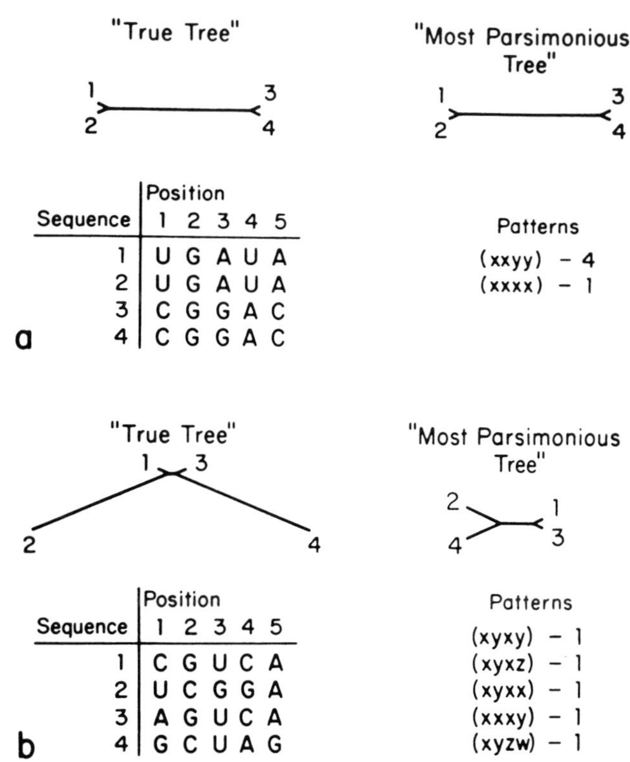

Figure 2. Examples of how parsimony correctly selects a tree and how it can fail. Branch lengths represent the relative probabilities of a nucleotide difference at any one position. The patterns observed in the aligned sequences and the number of their occurrences are shown adjacent to the sequences. In panel a, parsimony predicts the correct tree. In panel b, the tree predicted by parsimony places long-branch taxa together and short-branch taxa together in a topology different from that of the starting tree.

EXAMPLES OF UNEQUAL RATE EFFECTS

It is now widely appreciated that unequal rate effects occur frequently, and they are probably even more common than is realized. Two recently published trees illustrate this. In both cases, the trees violate fundamental evolutionary relationships long known from traditional analyses, and both were derived using augmented-distance matrix methods.

The first example is taken from the work of Field et al. (1988) on the origin of the metazoans. Their tree, calculated by using the augmented-distance technique of Olsen (1987) and Woese (1987), is reproduced in Fig. 3 in unrooted form. As a result of this tree, they proposed that the metazoans are polyphyletic. This contradicts the traditional view that metazoans are monophyletic and that all multicellular organisms are derived from a common ancestor. Their tree places long branches with long branches (the faster-evolving coelomates were placed with the faster-evolving slime mold) and short ones with short ones (the slower-evolving coelenterates were placed with a set of slower-evolving organisms

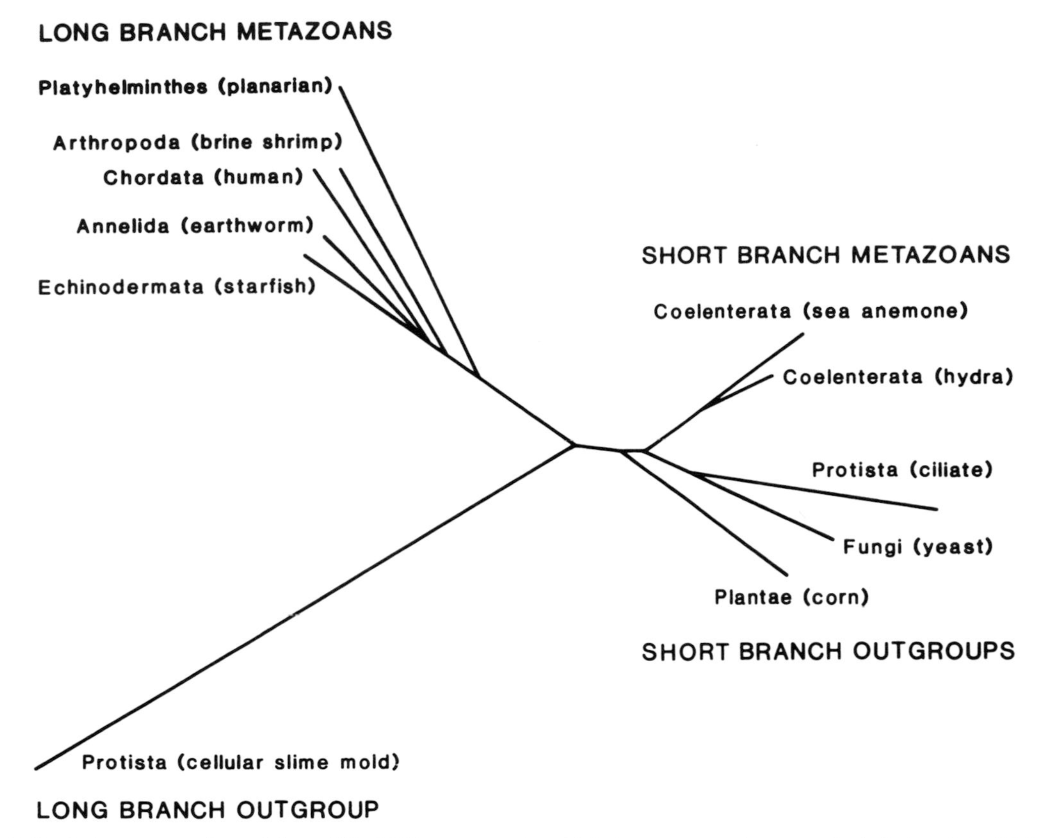

Figure 3. Evolutionary tree derived from 18S rRNA sequences in which the metazoans have a polyphyletic origin (Field et al., 1988). This tree contradicts the well-established, traditional monophyly of the metazoans. Shown is an unrooted version of the tree, with the branch lengths drawn to scale. In this tree, the long-branch metazoans are next to the longest-branch outgroup (the slime mold) and the shortest-branch metazoans are next to the shortest-branch outgroups. This grouping suggests that unequal rate effects may have caused artifactual selection of an incorrect tree.

including a yeast and a plant). This grouping indicates that unequal rate effects could have been responsible for this outcome. Furthermore, it violates one of the fundamental findings of traditional evolutionary biology, suggesting that unequal rates may be at fault. Many systematists have expressed skepticism about the polyphyletic tree (Nielsen, 1989; Walker, 1989; Bode and Steele, 1989). Recent reinterpretations of the data using methods other than augmented-distance matrix support the traditional monophyletic view (Patterson, 1989; Ghiselin, 1989; Lake, 1989, 1990b; reviewed in Patterson, 1990) and suggest that the augmented-distance algorithm is unreliable. This same distance algorithm was used to reconstruct the archaebacterial tree.

A second example of possible unequal rate effects is taken from the analysis of DNA-dependent RNA polymerase sequences (Pühler et al., 1989). In this tree (shown in unrooted form in Fig. 4), the eubacteria are a cluster of organisms that emerge from within the eucaryotes (or vice versa, depending on the root). This violates the established principle of the monophyly of the eucaryotes and suggests that the eubacteria are a subgroup of the eucaryotes (or vice versa). If this is true, then a radical change in our view of the eucaryotic cell is required. Again, the more reasonable interpretation is that since the two longest branches of the tree (the eubacterial cluster and the eucaryotic polymerase I branch) are clustered, unequal rate effects artifactually created this tree.

EVOLUTIONARY PARSIMONY IS RELATIVELY UNAFFECTED BY UNEQUAL RATE EFFECTS

Evolutionary parsimony was developed as a tree reconstruction method that was less affected by unequal rates (Lake, 1987). It is much more robust than other methods (Cavender, 1989) and "relatively free of the major criticisms" that have been levied against the other two methods (parsimony and distance matrix) (Holmquist et al., 1988). It functions in quite a different way from previous approaches and does so by analyzing patterns of nucleotide substitution. Since it is based on the symmetry properties of the

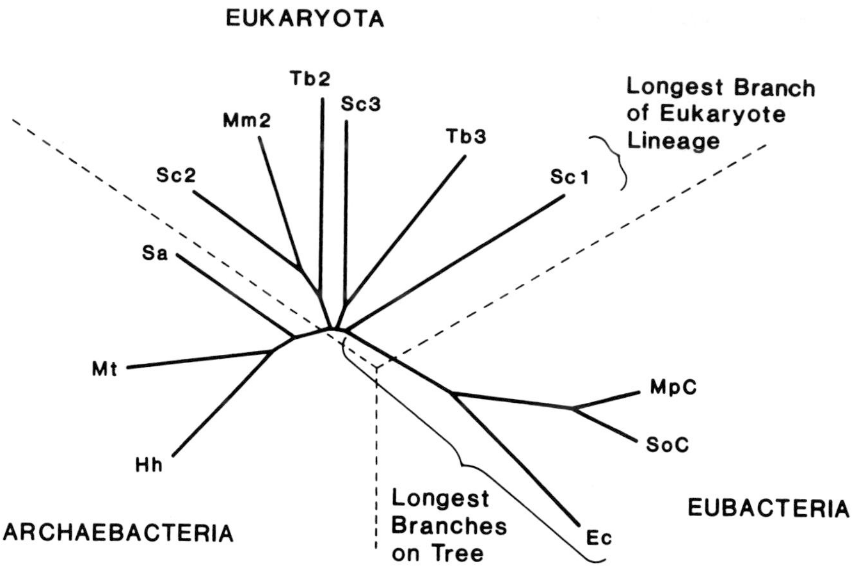

Figure 4. Unrooted phylogenetic tree derived from the DNA-dependent RNA polymerase components corresponding to the *Escherichia coli* β and β' components. This tree violates the traditional view that the eucaryotes are a monophyletic group. The longest branch on the tree, the eubacteria, emerges from within the eucaryotes (or vice versa, depending on the rooting). Eubacteria merge from the second-longest branch on the tree (polymerase I of yeasts), which suggests that unequal rate effects are responsible for this tree topology. This seems a simpler explanation, given the many characters that unite the eucaryotes.

trees, it is nearly independent of lengths of peripheral branches of a tree. Evolutionary parsimony counts transversions (e.g., U to G) rather than transitions (e.g., U to C), since transversions are normally more conserved than are transitions. Like parsimony, it counts patterns like UUAA; unlike traditional parsimony, it also considers three additional patterns: patterns like UCAG (which are added) and patterns like UCAA and UUAG (which are subtracted).

Table 1 (reconstructed from data of Li et al., 1987) illustrates the evolutionary parsimony method and compares it with the parsimony and distance matrix methods. In this example, only transversions are scored. When peripheral branch lengths are nearly equal (lengths of 40 and 80% are adjacent), standard methods equal or outperform evolutionary parsimony. As sequence lengths are increased from 1,000 to 4,000 nucleotides, all methods get a higher percentage of solutions correct. In contrast, in the Felsenstein zone where parsimony fails (shown in the last eight lines of Table 1), both parsimony and distance matrix typically yield less than 10% correct topologies when the sequence is 100 nucleotides long. When the length is increased to 4,000 nucleo-

Table 1. Probability of obtaining the correct (unrooted) tree with four taxa[a]

Branch length (%)[b]					Length[c]	Probability (%) of obtaining correct tree[d]		
a	b	c	d	Central		MPV	DMV	EP
40	80	40	80	10	1,000	50	48	41
					4,000	68	51	52
20	80	20	80	10	1,000	9	4	52
					4,000	0	0	73
8	80	8	80	10	1,000	0	0	59
					4,000	0	0	85
8	80	8	80	20	1,000	23	4	79
					4,000	3	0	97
6	60	6	60	6	1,000	1	0	68
					4,000	0	0	93

[a] Data are taken from Li et al. (1987).
[b] In terms of the number of nucleotide substitutions per site.
[c] Number of nucleotides in each sequence. The number of replicates conducted is 400 for a length of 1,000 nucleotides and 100 for a length of 4,000 nucleotides.
[d] MPV, Maximum parsimony method using transversional differences only; DMV, distance matrix method using transversional differences only; EP, evolutionary parsimony method. Probability of transitions is 33%.

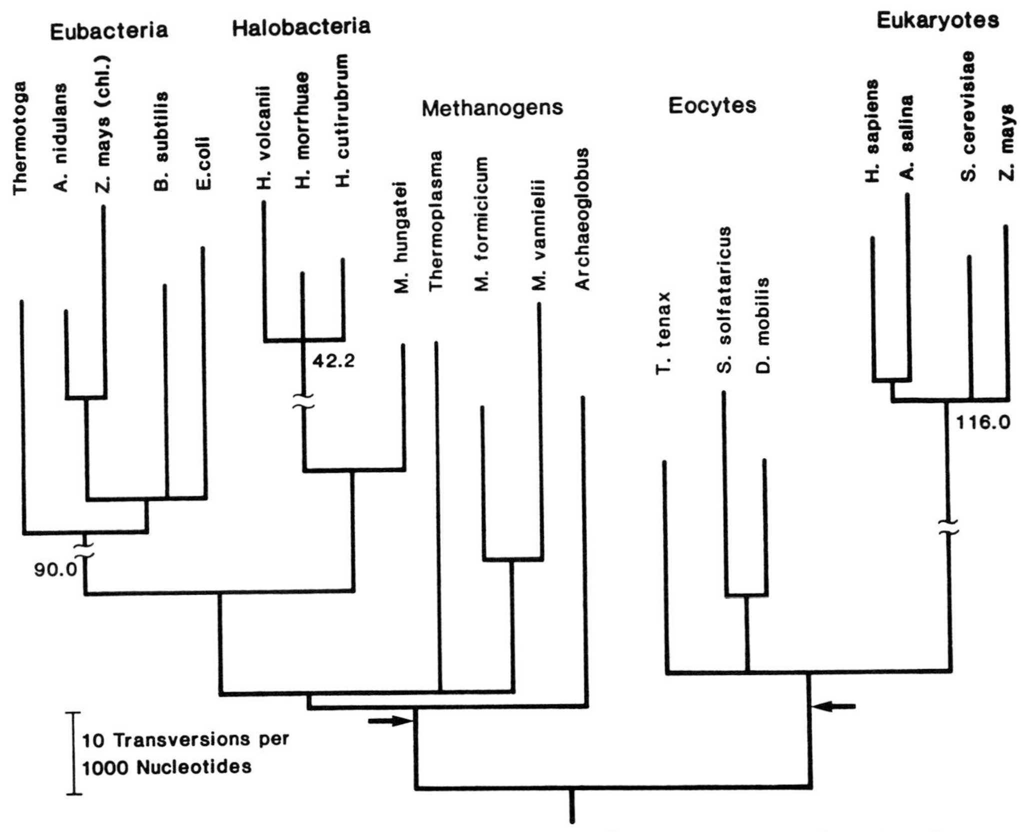

Figure 5. The rooted evolutionary tree that relates all known groups of extant organisms. Eubacteria and eucaryotes (and halobacteria to a lesser extent) are "fast-clock" organisms; hence, the lengths of the three long branches have been shortened to fit into the figure. The rate-independent technique of evolutionary parsimony was used in conjunction with the neighborliness procedure (Fitch, 1981) to reconstruct the unrooted tree. Rooting was based on the simultaneous application of two criteria, parsimony (Fitch, 1977) and the relative rate test (Wilson et al., 1977).

tides, parsimony and distance matrix perform even worse and yield no correct solutions. This is the diabolical feature of unequal rate effects: more data give the incorrect result more strongly. In contrast, evolutionary parsimony gets increasingly larger percentages of the solutions as the sequences are lengthened. For sequences of approximately 17,000 nucleotides (corresponding to the effective length of the 32 rRNA sequences used to derive our eocyte tree), evolutionary parsimony determined nearly 100% of the trees correctly.

THE EOCYTE TREE

With this theoretical background, relationships among the known groups of organisms were tested (Lake, 1988). The resulting tree (Fig. 5) indeed has the features that make standard methods consistently fail. The eubacteria and the eucaryotes ("fast-clock" organisms) both have branches (substitution rates) up to 10 times longer than their "slow-clock" neighbors (or sister groups). The eucaryotes are juxtaposed with the eocytes on one side of the tree, and the eubacteria, halobacteria, and methanogens are on the other.

The tree relating eubacteria, halobacteria, eocytes, and eucaryotes could be joined in one of three possible topologies corresponding to three competing theories. These are the eocyte tree (eocytes-eucaryotes; eubacteria-halobacteria), the archaebacterial tree (eubacteria-eucaryotes, etc.), and the halobacterial tree (halobacteria-eucaryotes, etc.). Since conflicting results had been reported for this branch, it was analyzed in detail.

Tree topologies can be affected by sequence alignments. Hence, two different alignments of small-subunit sequences were tested. In the star alignment, rRNA sequences were aligned with respect to a standard (*Thermoproteus tenax*) sequence, using a global computer algorithm. The literature alignment used the published eubacterial-eucaryotic alignment of Woese and collaborators. Both gave results similar to those obtained when we exchanged alignments with Woese and Olsen (Olsen, 1987). (Gouy and Li [1989] reported similar results for their alignment.) The significance plots in Fig. 4 indicate strong support for the eocyte tree with both alignments (between 0.002 and 0.02%) but not for either the archaebacterial or halobacterial tree.

COMPARISON WITH ANALYSES OF OTHERS

Both Olsen and Woese (Woese, 1987) and Gouy and Li (1989) have used secondary structure alignments. These have the major disadvantage that sequences are selected by eye. This makes the alignment highly subjective and dependent on which positions were paired. For the more conserved small-subunit rRNA sequences this may not matter, since Gouy and Li (1989) also obtain the eocyte tree by evolutionary parsimony with their alignment. For the less conserved large-subunit sequences, it is unwarranted. Hence, Bachellerie and Michot (1989) obtained the eocyte tree by using the secondary structure alignments of the most conservative regions of large-subunit rRNAs, whereas Gouy and Li (1989) obtained the archaebacterial one. Furthermore, the process involves subjectively deciding which nucleotide positions are good and which are not. The regions that Gouy and Li used, some as short as six nucleotides, are peppered across the structure. Their choice is highly subjective, and small changes affect the resulting tree topology. For these reasons, our alignments were done entirely by computer in an order-independent manner. Gouy and Li (1989) also used a computer alignment, but it was order dependent, further favoring the artifactual selection of the archaebacterial tree. Furthermore, contrary to their claims, the alignment of expansion sequences among the various eucaryotes has no bearing on the tree topologies, since they occur in only one of the four groups and a sequence must be present in all four to influence the selection of a tree.

The significance plot in Fig. 6A illustrates another property of the analysis. In our analysis, we examined some 1,400,000 nucleotide quartets, corresponding to an effective sequence length of 17,000 nucleotides. Although Gouy and Li (1989) reported that "the eocyte tree is strongly favored" by evolutionary parsimony analyses of their complete set of small-subunit rRNAs, they used simulations with conspicuously unrealistic parameters to discount this result. For example, in our original analysis of rRNA data (Fig. 6A), more than 1,400,000 nucleotide quartets were analyzed. Gouy and Li used just 900 quartets for their simulations. Of course, if one simulates only a sequence 900 nucleotides long (Fig. 6), evolutionary parsimony will support no tree. If one uses a realistic number, it functions robustly, and its strong support for the eocyte tree cannot be dismissed.

Likewise, simulations supporting the methods favored by Gouy and Li (1989) for reconstruction are erroneous. For example, they calculate branch lengths by using augmented distances to estimate the number of substitutions. Shoemaker and Fitch (1989) have shown that because of invariable sites, these lengths underestimate the actual distances by factors of 3 to 10. Since unequal rate effects increase rapidly with sequence divergence, and since they have underestimated branch lengths, the simulations of Gouy and Li (1989) are unable to detect the failure of algorithms and can provide no support for the archaebacterial tree.

Gouy and Li (1989) argued that they compensated for the faster evolution of large-subunit sequences by visually selecting conservative positions to be analyzed. However, an independent augmented-distance analysis (Bachellerie and Michot, 1989) of the "most conserved domains" at the 3′ end of the large-subunit sequence (48 sequences were used; Gouy and Li [1989] report 9) supports the eocyte tree. Furthermore, Gouy and Li themselves found (see Gouy and Li, this volume) that when the most slowly evolving eubacterial large-subunit sequences are used (*Anacystis nidulans*, *Bacillus subtilis*, and *Micrococcus luteus*), evolutionary parsimony supports the halobacterial tree (at the 5% level) and not the archaebacterial tree. Clearly, their analysis of the rapidly evolving large-subunit sequences is subject to uncertainties of alignment and sequence selection.

It seems clear that the large-subunit analyses of Gouy and Li (1989), confounded by high rates of evolution, are doubtful. On the other hand, Gouy and Li (1989) have confirmed my (Lake, 1988) earlier analysis of small subunit sequences (Fig. 6B); i.e., evolutionary parsimony supports the eocyte tree, and parsimony and distance matrix analyses support the archaebacterial tree. Given the large body of data supporting evolutionary parsimony as the preferred method of analysis, I favor the eocyte tree.

THE EOCYTE TREE IS CONSISTENT WITH BIOLOGICAL PROPERTIES

The most distinctive feature of the tree is the greatly differing branch lengths leading to the major groups (Fig. 5). Both the eubacteria and the eucaryotes are at the end of extremely long branches. (Note the gaps inserted into them in order to draw them on the tree conveniently.) In fact, the eucaryotic branch is about 10 times longer than that of its sister group, the eocytes. Similarly, the eubacterial branch is considerably longer than either the methanogen or the halobacterial branch. Juxtaposed unequal branches

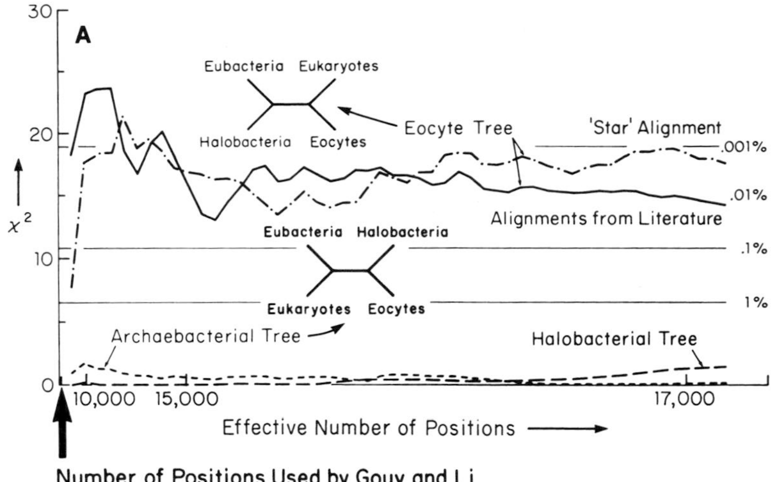

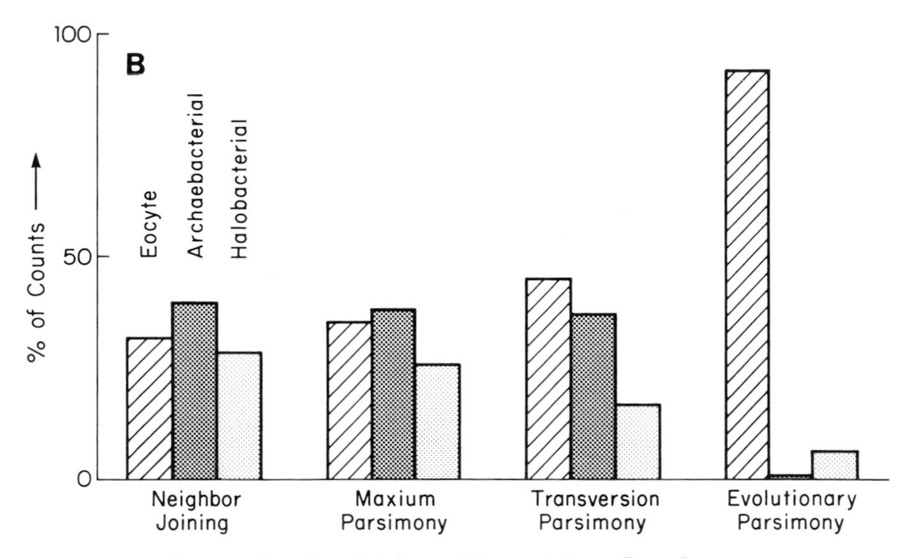

Figure 6. Analysis of the central branch relating eubacteria, halobacteria, eocytes, and eucaryotes. A total of 1,401,504 nucleotide quartets were analyzed. Alignments were constructed for 32 representative small-subunit rRNA sequences. These are described elsewhere (Lake, 1988). (A) Significance plot illustrating the effect of alternative alignments on tree selection. The significance plot shows our alignment and the alignment of others that might have been expected to support the archaebacterial tree. This analysis supports the eocyte tree strongly at a level of between 0.01 and 0.001%. For reference, I have noted the region of the plot where Gouy and Li (1989) selected their simulations (arrow). (B) Comparison of results from various methods, since tree reconstruction algorithms vary in sensitivity to unequal rate effects. Distance matrix (equivalent to neighbor joining for four taxon trees) and maximum parsimony (Fitch, 1977), most sensitive to unequal rate artifacts (Li et al., 1987), support the archaebacterial tree but distribute their counts nearly equally among all three trees, which indicates that unequal rates are biasing them. Transversion parsimony (Fitch, 1977; Brown et al., 1982), less sensitive to the unequal rate effects, favors the eocyte tree. Evolutionary parsimony, least sensitive, exclusively supports the eocyte tree. Taken together, these results indicate that when the biases are properly accounted for, the eocyte tree is strongly favored. Total counts were as follows: neighbor joining (distance matrix, 482,044; maximum parsimony, 116,110; transversion parsimony, 52,915; and evolutionary parsimony, 13,411.

of the tree are precisely the features that in the past have led others to select an incorrect tree. The great differences in rates seen in this tree have made this complication very difficult to untangle.

A number of properties, including sequence data from other sources, support the eocyte tree. When rRNA operons from eocytes were sequenced (Leinfelder et al., 1985; Larsen et al., 1986), it was surprising to find that they resembled those of eucaryotes rather than those of other bacteria. Eocytic rRNA operons contain a 16S RNA, lack a tRNA spacer, and are followed by 23S RNA (Fig. 7). The 5S RNA is elsewhere on the chromosome (Gerbi, 1985). The pattern is generally similar to that of eucaryotic

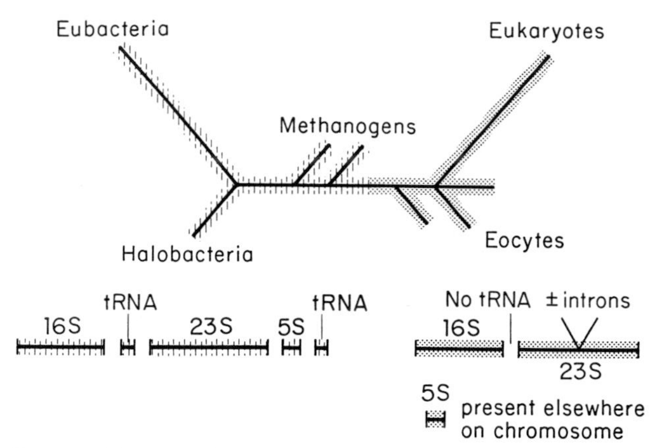

Figure 7. Schematic diagram showing the most parsimonious tree relating the molecular organizations of ribosomal operons. Eocytes and eucaryotes share a typically "eucaryotic" arrangement, whereas halobacteria, methanogens, and eubacteria are characteristically of "eubacterial" arrangement. Deviations from these patterns are for the most part minor, with *Thermoplasma* spp. and a cryptomonad being notable exceptions.

rRNA operons. The sole eucaryotic exception is the genus *Cryptomonas* (see Gray and Schnare, this volume), which has a linked 5S. Whether this 5S has a separate polymerase III promoter is not known, however. This pattern contrasts strongly with that of eubacteria, halobacteria, and methanogens, in which the typical pattern is 16S, tRNA, 23S, 5S, and tRNA (with some minor variations on this theme). An exception to this pattern is the genus *Thermoplasma*.

Other molecular sequences also support the eocyte tree. For example, gas vesicles have been found in some eubacteria (cyanobacteria), in halobacteria, and in a methanogen. Protein sequences determined for the major vesicle protein (Walker et al., 1984; Das Sarma et al., 1987) indicate that this component is homologous in halobacteria and eubacteria. Furthermore, the tree derived from these sequences places eubacteria with eubacteria and halobacteria with their kind. This argues against the protein having been laterally transferred (except possibly early in the divergence of eubacteria and halobacteria). In addition, eubacteria and halobacteria share

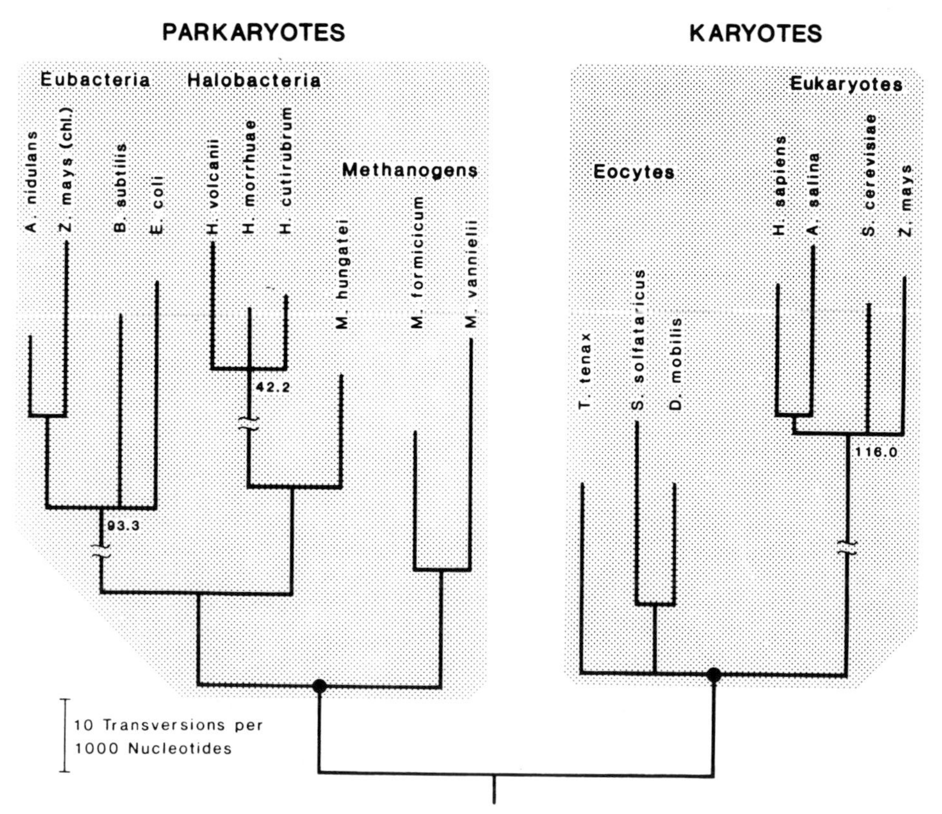

Figure 8. Diagram showing that the procaryotic classification is not supported by the tree relating all extant organisms but that an appropriate subdivision can be made by introducing two monophyletic groups. One, the parkaryotes, contains the eubacteria, halobacteria, and methanogens (including the genera *Thermoplasma* and *Archaeoglobus*). The other, the karyotes, consists of the eocytes and the eucaryotes. Group membership is, of course, defined phylogenetically and not metabolically; e.g., the sulfur-metabolizing *Thermoplasma* spp. (for 16S sequence, see Zimmermann et al., this volume) are with the eubacteria, halobacteria, and methanogens.

many biochemical pathways. Among these are pathways for arginine deiminase, for fatty acid esters, and for carotenoid synthesis (for original references, see Lake, 1986).

While many properties support the eocyte tree, two properties contradict it. First, the eucaryotic rhodopsins have structural similarities with the bacteriorhodopsin of halobacteria (Findlay and Pappin, 1986). This argues for the halobacterial tree and against both the eocyte and archaebacterial trees. Second, ether lipids found in halobacteria, methanogens, and eocytes argue for the archaebacterial tree and against the halobacterial and eocyte trees. If both of these exceptions were valid, no tree could be correct. This suggests a need for caution in dealing with eucaryotic properties and a consideration of the chimeric, organellar origins of some nuclear DNAs. Except for these two counterexamples, numerous shared properties (synapomorphies) representing fundamental molecular properties support this tree (Lake, 1986).

THE PROCARYOTES ARE NOT A PROPER PHYLOGENTIC GROUP

One of the most interesting aspects of the eocyte tree is that it has revised the traditional procaryote-eucaryote view of life. The tree does not consist of two branches, one leading to all organisms that lack a nucleus and the other to those that have a nucleus. Instead, the anucleate eocytes are the sister group of the nucleated eucaryotes. The left side of the tree shown in Fig. 8 consists of the anucleate eubacteria, halobacteria, and methanogens. The tree suggests that two supergroups are defined by the deep central split.

The necessity of using the correct tree topology to interpret the origin of mechanisms during the evolution of modern organisms cannot be underestimated. For example, if one is to decide whether introns are an early development or whether they came late, one must first know the topology of the tree. If one uses different trees, then one will reach different conclusions, since the tree forms the backbone for evolutionary testing and inquiry. As an example, the eocyte tree predicts that sulfur metabolism is primitive for more than 98% of the possible trees (Lake, 1988), whereas the archaebacterial tree predicts a sulfur-metabolizing ancestor for only 33% of the most parsimonious character assignments.

In the past, determining the evolutionary tree that relates all organisms has been hindered by a lack of sequence analysis techniques able to discriminate among deep branches. Such analysis is now possible, and one can hope that with this knowledge will come a new and deeper comprehension of the origins of life.

This work was supported by research grants from the National Institutes of Health and the National Science Foundation.

REFERENCES

Alsopp, A. 1969. Phylogenetic relationships of the procaryota and the origin of the eucaryotic cell. *New Phytol.* **68**:591–612.

Bachellerie, J.-P., and B. Michot. 1989. Evolution of large subunit ribosomal RNA structure: the 3′ domain contains elements of secondary structure specific to the major phylogenetic groups. *Biochimie* **71**:701–709.

Bode, H. R., and R. E. Steele. 1989. Phylogeny and molecular data. *Science* **243**:549.

Brown, W. M., E. M. Prager, A. Wang, and A. C. Wilson. 1982. Mitochondrial DNA sequences of primates: tempo and the mode of evolution. *J. Mol. Evol.* **18**:225–239.

Cavender, J. A. 1989. Mechanized derivation of linear invariants. *Mol. Biol. Evol.* **6**:301–316.

Das Sarma, S., T. Damerval, J. G. Jones, and N. T. Demarsac. 1987. A plasmid-encoded gas vesicle protein gene in a halophilic archaebacterium. *Mol. Microbiol.* **1**:365–370.

Eldridge, N., and J. Cracraft. 1986. *Phylogenetic Patterns and the Evolutionary Process.* Columbia University Press, New York.

Felsenstein, J. 1978. Cases in which parsimony or compatibility methods will be positively misleading. *Syst. Zool.* **27**:401–410.

Felsenstein, J. 1988. Perils of molecular introspection. *Nature* (London) **335**:118.

Field, K. G., G. J. Olsen, D. J. Lane, S. J. Giovannoni, M. T. Ghiselin, E. C. Raff, N. R. Pace, and R. A. Roff. 1988. Molecular phylogeny of the animal kingdom. *Science* **239**: 748–753.

Findlay, J. B. C., and D. J. C. Pappin. 1986. The opsin family of proteins. *Biochem. J.* **238**:625–642.

Fitch, W. 1977. On the problem of generating the most parsimonious tree. *Am. Nat.* **111**:223–257.

Fitch, W. M. 1981. A nonsequential method for constructing trees and hierarchial classifications. *J. Mol. Evol.* **18**:30–37.

Gerbi, S. A. 1985. Evolution of ribosomal RNA, p. 419–517. *In* R. J. MacIntyre (ed.), *Molecular Evolutionary Genetics.* Plenum Publishing Corp., New York.

Ghiselin, M. T. 1989. Summary of our present knowledge of metazoan phylogeny, p. 261–272. *In* B. Fernholm, K. Bremer, and H. Jornvall (ed.), *The Hierarchy of Life. Proceedings from Nobel Symposium 70.* Excerpta Medica, Amsterdam.

Gouy, M., and W.-H. Li. 1989. Phylogenetic analysis based on rRNA sequences supports the archaebacterial rather than the eocyte tree. *Nature* (London) **339**:145–147.

Holmquist, R., M. Miyamoto, and M. Goodman. 1988. Analysis of higher-primate phylogeny from transversion differences in nuclear and mitochondrial DNA by Lake's methods of evolutionary parsimony and operator metrics. *Mol. Biol. Evol.* **5**: 217–236.

Lake, J. A. 1986. In defence of bacterial phylogeny. *Nature* (London) **321**:658.

Lake, J. A. 1987. A rate-independent technique for analysis of nucleic acid sequences: evolutionary parsimony. *Mol. Biol. Evol.* **4**:167–178.

Lake, J. A. 1988. Origin of the eukaryotic nucleus determined by rate-invariant analysis of rRNA sequences. *Nature* (London) **331**:184–186.

Lake, J. A. 1989. Origin of the multicellular animals, p. 273–278.

In B. Fernholm, K. Bremer, and H. Jornvall (ed.), *The Hierarchy of Life. Proceedings from Nobel Symposium 70.* Excerpta Medica, Amsterdam.

Lake, J. A. 1990a. Eocyte or archaebacterial tree? *Nature* (London) **343:**418–419.

Lake, J. A. 1990b. Origin of the Metazoa. *Proc. Natl. Acad. Sci. USA* **87:**763–766.

Larsen, N., H. Leffers, J. Kjems, and R. A. Garrett. 1986. Evolutionary divergence between the rRNA operons of *Halococcus morrhuae* and *Desulfurococcus mobilis. Syst. Appl. Microbiol.* **7:**49–57.

Leinfelder, W., M. Jarsch, and A. Bock. 1985. The phylogenetic position of the sulfur-dependent archaebacterium *Thermoproteus tenax*: sequence of the 16S rRNA gene. *Syst. Appl. Microbiol.* **6:**164–170.

Li, W. H., K. H. Wolfe, J. Sourdis, and P. M. Sharp. 1987. Reconstruction of phylogenetic trees and estimation of divergence times under non-constant rates of evolution. *Cold Spring Harbor Symp. Quant. Biol.* **52:**847–856.

Nielsen, C. 1989. Phylogeny and molecular data. *Science* **243:**548.

Olsen, G. J. 1987. Earliest phylogenetic branchings: comparing rRNA-based evolutionary trees inferred with various techniques. *Cold Spring Harbor Symp. Quant. Biol.* **52:**805–824.

Pace, N. R., D. A. Stahl, D. J. Lane, and G. J. Olsen. 1986. The analysis of natural microbial populations by rRNA sequence. *Adv. Microbiol. Ecol.* **9:**1–55.

Patterson, C. 1989. Phylogenetic relationships of major groups: conclusions and prospects, p. 471–488. *In* B. Fernholm, K. Bremer, and H. Jornvall (ed.), *The Hierarchy of Life. Proceedings from Nobel Symposium 70.* Excerpta Medica, Amsterdam.

Patterson, C. 1990. Reassessing relationships. *Nature* (London) **344:**199–200.

Penny, D. 1988. What was the first living cell? *Nature* (London) **331:**111–112.

Pühler, G., F. Lottspei, R. A. Garrett, and W. Zillig. 1989. Archaebacterial DNA-dependent RNA polymerases testify to the evolution of the eukaryotic nuclear genome. *Proc. Natl. Acad. Sci. USA* **86:**4569–4573.

Shoemaker, J. S., and W. M. Fitch. 1989. Evidence from nuclear sequences that invariable sites should be considered when sequence divergence is calculated. *Mol. Biol. Evol.* **6:**270–289.

Walker, J. E., P. K. Hayse, and A. E. Walsby. 1984. Homology of gas vesicle proteins in cyanobacteria and halobacteria. *J. Gen. Microbiol.* **130:**2709–2715.

Walker, W. F. 1989. Phylogeny and molecular data. *Science* **243:**548–549.

Wilson, A., S. Carlson, and T. White. 1977. Biochemical evolution. *Annu. Rev. Biochem.* **46:**573–639.

Woese, C. R. 1987. Bacterial evolution. *Microbiol. Rev.* **51:** 221–271.

Chapter 52

Evolution of the Modular Structure of rRNA

MICHAEL W. GRAY and MURRAY N. SCHNARE

In considering how rRNA has evolved to its present form and function, a number of unique structural features have to be rationalized. In the first place, one has to account for the length of rRNA molecules, which in most organisms are continuous polyribonucleotide chains several kilobases in size. Prototypical small-subunit (SSU) and large-subunit (LSU) rRNAs are exemplified by the 16S and 23S rRNAs of *Escherichia coli*, which are 1,542 and 2,904 nucleotides in length, respectively (Noller, 1984). A key question in ribosome evolution is how such long molecules originated and how they acquired the complex secondary and tertiary structures that are required for their many interactions with each other, with partner ribosomal proteins, and with other components of the translational apparatus (mRNA, tRNA, and protein synthesis factors). Our current notions of chemical and biochemical evolution suggest that rRNA molecules must have started small and grown large, i.e., that relatively short polynucleotide chains (perhaps only tens of nucleotides long) were somehow incorporated into progressively longer molecules, eventually to become hundreds and then thousands of nucleotides in length. How such a process could have occurred is mostly a matter of conjecture at the present time. Although it appears that functional ribosomes or ribosomelike entities must have existed at the very earliest stages of evolution, it is not at all clear how closely this original (primordial) ribosome resembled contemporary ones in structure and activity. It has, in fact, been suggested that the primordial ribosome may have consisted largely or solely of RNA (Crick, 1968; Woese, 1972).

INTERSPERSED CONSERVED AND VARIABLE REGIONS IN rRNA

When one compares the primary and potential secondary structures of homologous rRNA species, both within and between major evolutionary lineages, large portions of the structure are seen to be conserved. For example, in a comparison that includes all known eucaryotic nuclear-encoded LSU rRNAs and *E. coli* 23S rRNA, almost four-fifths (79%) of the *E. coli* secondary structure is precisely represented among the collection of nucleocytoplasmic LSU rRNAs, with a larger proportion of the 3′ half of the molecule conserved than of the 5′ half (84% versus 76%) (Fig. 1). As such comparisons are broadened to include other lineages, and particularly when mitochondrial sequences are considered, there is a decrease in the proportion of overall structure that is conserved. Nevertheless, in comparisons that encompass all known lineages (archaebacterial, eubacterial, nucleocytoplasmic, plastid, and mitochondrial) and that include the truncated and highly divergent animal mitochondrial rRNAs (Gutell and Fox, 1988), the globally conserved regions are still equivalent to roughly one-third of the total *E. coli* secondary structure (Gray et al., 1984; Cedergren et al., 1988).

These conserved regions of precisely corresponding secondary structure alternate with variable regions that differ markedly in length, base composition, and potential secondary structure (Fig. 2). The comparison shown in Fig. 1 defines 35 discrete variable regions, certain of which correspond to the divergent regions previously identified in eucaryotic nucleocytoplasmic LSU rRNAs (Michot et al., 1984). Within the secondary structure, the variable regions tend to be located peripherally and often involve (all or most of) discrete hairpin loops. Long-range hydrogen-bonding interactions tie together conserved segments that are remote from one another in the primary sequence, thus defining a conserved core of secondary structure. Within this core are located almost all of the functional sites that have been

Michael W. Gray and Murray N. Schnare ■ Department of Biochemistry, Dalhousie University, Halifax, Nova Scotia B3H 4H7, Canada.

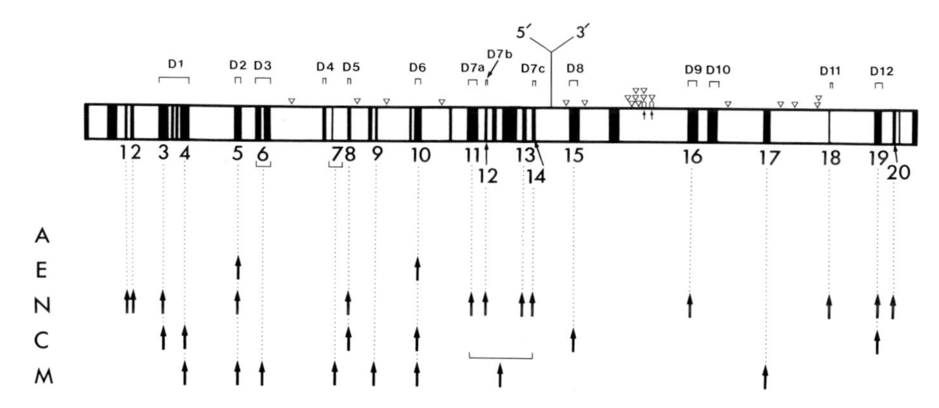

Figure 1. Distribution of conserved (open) and variable (solid) regions in a comparison of known eucaryotic nucleocytoplasmic and *E. coli* LSU rRNAs, mapped to the linear sequence of the latter (2,904 nucleotides). Conserved regions are defined as those that correspond precisely in primary or potential secondary structure among all compared sequences (listed in Table 1). Numbers refer to variable regions that are the sites of discontinuity resulting from rRNA processing, with large vertical arrows indicating the known distribution of these breaks in different evolutionary lineages: A, archaebacterial; E, eubacterial; M, mitochondrial; C, chloroplast; N, nucleocytoplasmic. A detailed compilation of these discontinuities will be presented elsewhere (M. W. Gray and M. N. Schnare, unpublished data). Open triangles indicate the known sites of introns and insertion elements in LSU rRNA genes (Table 2; C. Lemieux, personal communication). Small vertical arrows denote elements whose insertion results in a short (13- to 14-nucleotide) duplication of the target site. D1, D2, etc., refer to the divergent regions previously identified in eucaryotic 28S rRNA (Michot et al., 1984). The vertical line marked 5′,3′ divides the LSU rRNA into 5′ and 3′ sections (see Gutell and Fox, 1988).

implicated in translation (Dahlberg, 1989) as well as almost all of the modified nucleoside constituents of rRNA (Raué et al., 1988). These observations imply that the conserved core is the basic functional unit of rRNA molecules and that it represents that portion of contemporary rRNA molecules that was first established in the course of ribosome evolution. A corollary is that the variable regions are largely dispens-

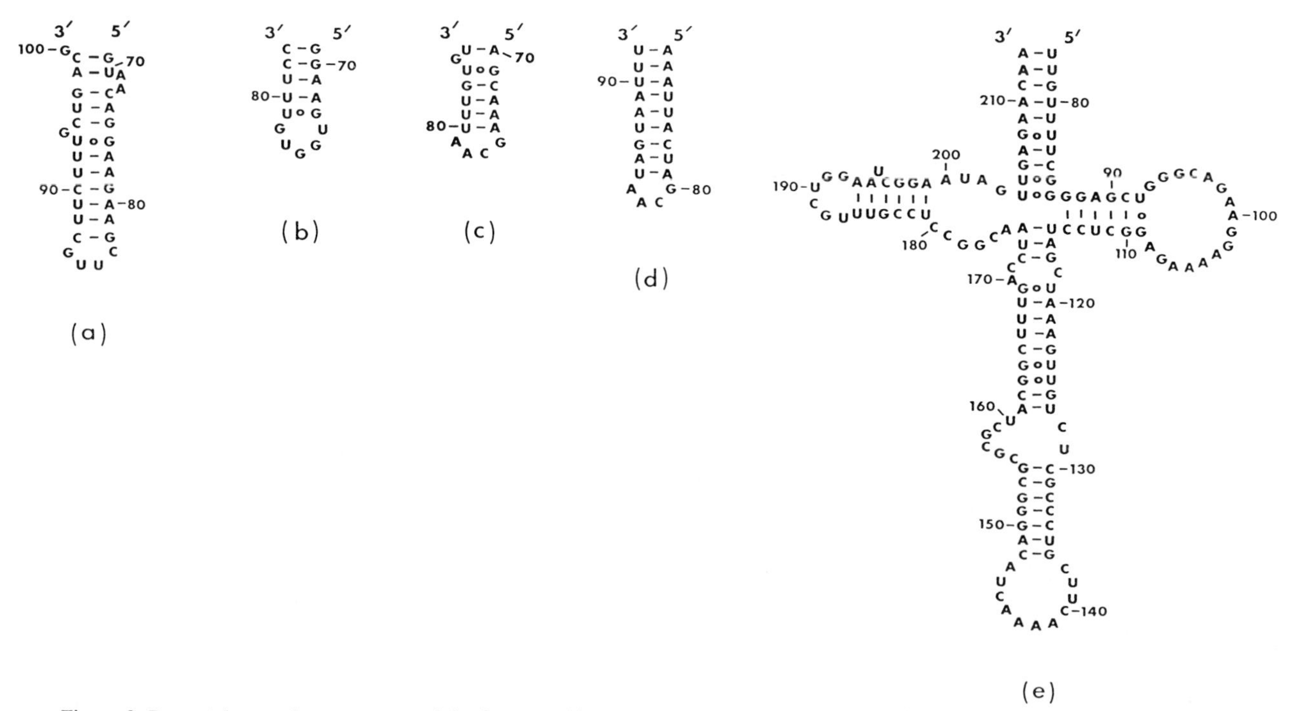

Figure 2. Potential secondary structure of the first variable region (V1; see Spencer et al., 1984) in eubacterial-like SSU rRNAs. (a) *E. coli*; (b) maize chloroplast; (c) *C. reinhardtii* chloroplast; (d) *Euglena gracilis* chloroplast; (e) wheat mitochondria. The secondary structures of diagrams a to d are those of Gutell et al. (1985). This variable region is recognized by the fact that the flanking structure is highly conserved; among these five SSU rRNAs, the primary sequence is 90 to 100% identical for 20 nucleotides upstream and downstream of V1.

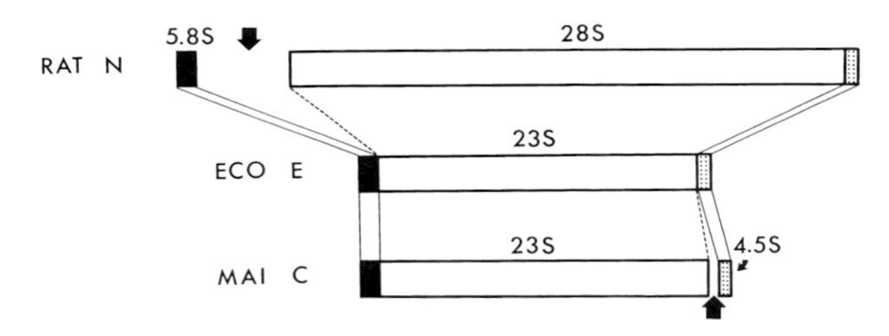

Figure 3. Structural relationship between the LSU rRNAs of rat cytoplasm (RAT N; 5.8S-28S), *E. coli* (ECO E; 23S), and maize chloroplast (MAI C; 23S-4.5S). The diagram emphasizes that eucaryotic 5.8S rRNA and plant chloroplast 4.5S rRNA are the structural equivalents of, respectively, the 5′-terminal and 3′-terminal domains of *E. coli* 23S rRNA. Vertical arrows denote the positions of ITSs whose removal during posttranscriptional processing generates the separate small 5.8S and 4.5S rRNAs.

able, at least insofar as basic ribosome function is concerned, and that functional constraints on structural divergence are minimal within variable regions.

CONTINUOUS AND DISCONTINUOUS rRNAs

In *E. coli* and most other procaryotes, the LSU (23S) and SSU (16S) rRNAs are single, continuous polynucleotide chains (Noller, 1984). However, in almost all eucaryotes (but see Vossbrinck and Woese, 1986), the corresponding LSU rRNA is split into at least two pieces, termed 5.8S (~160 nucleotides long) and 25S-28S. The discovery of 5.8S rRNA (Pene et al., 1968; Sy and McCarty, 1970), a seemingly unique constituent of eucaryotic 80S ribosomes (Oakden and Lane, 1973, and references therein), predated by a number of years the recognition that this small RNA is the structural equivalent of the 5′-terminal region of *E. coli* 23S rRNA (Fig. 3). This correspondence was first explicitly noted by Nazar (1980) and was subsequently confirmed by other observations (Cox and Kelly, 1981; Jacq, 1981; Walker, 1981; Nazar, 1982; Clark and Gerbi, 1982). A split LSU rRNA also occurs in plant chloroplasts. In this case, as noted by MacKay (1981) and others (Machatt et al., 1981; Edwards et al., 1981; Clark and Gerbi, 1982), the 3′-terminal ~100 nucleotides are detached from the rest of the LSU rRNA and appear as a unique species, 4.5S rRNA (Fig. 3). The 5.8S and 4.5S coding regions effectively represent distinct modules of structure, separated from the remainder of the LSU rRNA coding sequence by internal transcribed spacer (ITS) sequences (Fig. 3). These ITSs are cotranscribed with the flanking rRNA coding regions, and their transcripts are then excised during posttranscriptional processing of the precursor rRNA, generating split rRNAs that must continue to interact in the final tertiary structure. In the case of the 5.8S-28S complex, it is evident that extensive secondary structure interactions (base pairing) ensure that the two rRNA components remain tightly bound to one another after spacer excision (see Gutell and Fox, 1988). However, in the 23S-4.5S example, there are no postulated secondary structure interactions between the two partners (Gutell and Fox, 1988). In this case, tertiary RNA-RNA or RNA-protein interactions are presumably responsible for integration of 4.5S rRNA into the large ribosomal subunit.

As additional systems have been investigated, examples of more extensively fragmented LSU rRNAs have been discovered. The most extreme case documented to date is seen in *Euglena gracilis*, a protist whose cytoplasmic LSU rRNA is a collection of 14 separate RNA species (Schnare and Gray, in press). As with the seven-component cytoplasmic LSU rRNA of *Crithidia fasciculata* (Spencer et al., 1987), it is possible to model a secondary structure in which the various pieces of *Euglena gracilis* LSU rRNA pair with one another in a predictable fashion, through specific hydrogen-bonding interactions, to form a noncovalent network that reconstitutes the conserved functional core discussed above.

In contrast to LSU rRNA, there are as yet no reports of discontinuous SSU rRNAs in the ribosomes of bacteria, chloroplasts, or the eucaryotic cytoplasm. So far, split SSU rRNAs are confined to the mitochondrial lineage: a two-component mitochondrial SSU rRNA in *Tetrahymena pyriformis* (a ciliate protozoan) (Schnare et al., 1986) and a four-component one in *Chlamydomonas reinhardtii* (a unicellular green alga) (Boer and Gray, 1988).

The finding of highly fragmented rRNA molecules that are discontinuous within variable regions yet are obviously functional is consistent with the idea that there are reduced functional constraints within variable regions as compared with the conserved core. Not only are variable regions allowed to diverge in structure, but they may be effectively removed without appreciably altering ribosome function.

VARIABLE REGIONS AND SPACERS IN rRNA GENES

When the positions of discontinuities in LSU rRNA are mapped, without exception the breaks are found to be localized to the variable regions defined by sequence comparisons (e.g., Spencer et al., 1987). This is also true for SSU rRNA (Schnare et al., 1986; Boer and Gray, 1988). Figure 1 shows the positions of all known discontinuities in LSU rRNA; it can be seen that these are rather uniformly distributed

Table 1. Comparison of length and base composition of various regions within eucaryotic LSU rRNA genes[a]

Organism	LSU rRNA gene region									
	25S-28S rRNA[b]		ITS2		D8[c]		D10[d]		Conserved[e]	
	Length (nucleotides)	G+C %	Length (nucleotides)	G+C %	Length (nucleotides)	G+C %	Length (nucleotides)	G+C %	Length (nucleotides)	G+C %
Rat	4,785	66	765	80	598	84	97	80	239	52
Mouse	4,712	67	1,089	75	617	82	102	79	240	52
Xenopus laevis	4,110	66	262	88	341	83	97	75	240	51
Drosophila melanogaster	3,900	39	385	20	222	32	223	29	240	47
Caenorhabditis elegans	3,509	49	383	48	170	52	92	45	240	47
Rice	3,377	59	233	77	145	81	87	63	239	51
Tomato	3,381	57	217	71	157	75	81	60	240	50
Saccharomyces carlsbergensis	3,393	48	234	39	159	57	84	46	240	49
Physarum polycephalum	3,788	53	492	50	168	61	242	57	239	49
Prorocentrum micans	3,408	48	195	56	147	53	87	41	240	49
Crithidia fasciculata	3,906	51	416	45	223	59	265	55	240	51
Euglena gracilis	3,889	57	337	61	189	65	156	56	240	51

[a] References to the sequences are given by Gutell and Fox (1988), with the following exceptions: *P. micans* (Maroteaux et al., 1985; Lenaers et al., 1989); *D. melanogaster* (Tautz et al., 1988); tomato (Kiss et al., 1988, 1989); rice (Takaiwa et al., 1985); *S. carlsbergensis* (Veldman et al., 1981); *X. laevis* (Hall and Maden, 1980); rat (Subrahmanyam et al., 1982); and *E. gracilis* (Schnare et al., in press).
[b] Only the mature rRNAs were used in calculations of length for *Crithidia*, *Euglena*, and *Drosophila* spp.
[c] Corresponds to *E. coli* positions 1713 to 1745.
[d] Corresponds to *E. coli* positions 2197 to 2229.
[e] Corresponds to *E. coli* positions 1888 to 2118.

throughout the length of the LSU rRNA sequence. Almost two-thirds (20 of 35) of the variable regions are sites of discontinuity in one or more organisms. Interestingly, there are several cases in which discontinuities are present at the same location in different evolutionary lineages. For example, discontinuities at the position of eucaryotic divergent region D2 (Fig. 1, no. 5) have been found in eubacterial, nucleocytoplasmic, and mitochondrial LSU rRNAs, whereas some eubacterial, chloroplast, and mitochondrial LSU rRNAs are split within region D6 (Fig. 1, no. 10).

It is important to note that except in the special case where rRNA coding modules are rearranged (Boer and Gray, 1988; see below), discontinuity results from excision of nucleotides (ITSs) during processing rather than from a single phosphodiester bond cleavage (incision). Thus, there is a strict correlation between the locations of variable regions in mature rRNA and spacers in rRNA genes. We argue (see below) that this correlation exists because there is an evolutionary relationship between variable regions and spacers.

Other characteristics suggest a structural correspondence between ITS sequences and variable regions in rRNA. In some ribosomal DNA (rDNA) units there is a rather striking correlation between the G+C content of spacers and variable regions (Table 1; see also Cox and Kelly, 1981; Tautz et al., 1988). Thus, a high ITS2 G+C content (as, for example, in vertebrate animals) is accompanied by a correspondingly high G+C content in variable regions D8 and D10, whereas a low G+C content in ITS2 (as, for example, in *Drosophila melanogaster*) is reflected in a low G+C content in D8 and D10. This is suggestive of parallel evolution of ITSs and variable regions in these particular cases.

The divergent regions in eucaryotic cytoplasmic LSU rRNAs have also been called expansion segments (Clark et al., 1984) because they are often very much larger than their eubacterial counterparts and thus effectively expand the size of eucaryotic 25S-28S rRNA over that of bacterial 23S rRNA (Gerbi, 1986). In certain eucaryotes (particularly vertebrate animals), these expansion segments display a high level of cryptic simplicity (Tautz et al., 1986), a condition characterized by the presence of interspersed short repetitive sequences that are apparently generated by a slippage mechanism during replication. It has been shown that the pattern of cryptic simplicity is very similar within the expansion segments of a particular species, suggesting that these segments coevolve (Hancock and Dover, 1988). Moreover, a similar pattern of intraspecies cryptic simplicity has been found in noncoding (spacer) regions and within LSU rRNA expansion segments (Tautz et al., 1988). This is another indication that at least some coding and spacer sequences are evolving in parallel.

Table 2. Localization of introns and insertion elements relative to the 23S rRNA sequence of *E. coli*[a]

Organism	Nucleotide position(s)	Reference(s)
Introns		
Saccharomyces cerevisiae (M)	2449	Dujon, 1980
Kluyveromyces thermotolerans (M)	2449	Jacquier and Dujon, 1983
Aspergillus nidulans (M)	2449	Netzker et al., 1982
Neurospora crassa (M)	2449	Burke and RajBhandary, 1982
Podospora anserina (M)	2449, 1699	Cummings et al., 1989
Chlamydomonas reinhardtii (C)	2593	Allet and Rochaix, 1979
Desulfurococcus mobilis (A)	1952	Kjems and Garrett, 1988
Physarum polycephalum (N)	1925, 1949, 2449	Otsuka et al., 1983; Muscarella and Vogt, 1989
Tetrahymena sp. (N)	1925	Nielsen and Engberg, 1985
Insertion elements		
Ascaris lumbricoides (N)	1957–1969	Neuhaus et al., 1987
Bombyx mori (N)	1991–2004, 1930	Fujiwara et al., 1984; Eickbush and Robins, 1985
Calliphora erythrocephala (N)	1991–2004	Smith and Beckingham, 1984
Drosophila virilis (N)	1991–2004	Rae et al., 1980
Drosophila melanogaster (N)	1991–2004, 1930	Rae, 1981; Dawid and Rebbert, 1981; Roiha et al., 1981

[a] Introns are inserted after the listed nucleotide position (*E. coli* coordinates). A, Archaebacterial; C, chloroplast; M, mitochondrial; N, nuclear. Insertion elements are retrotransposons that interrupt a fraction of the LSU rRNA genes in some invertebrate (mainly insect) species (see Xiong and Eickbush, 1988). Their transposition often results in duplication of the target site within the 28S rRNA gene, as indicated. Most such interrupted genes appear to be transcriptionally inactive (see Jamrich and Miller, 1984).

MODULAR STRUCTURE OF rRNA GENES

The interspersed pattern of conserved and variable domains seems to suggest that rRNA molecules are constructed in a modular fashion. This impression draws further support from the existence of ITSs, whose excision generates individual, interacting structural domains. In almost all cases, these domains are arrayed in a strictly conserved transcriptional order at the genome level. However, in two systems that we have characterized, rRNA structural modules are actually rearranged at the genome level relative to their usual order of transcription. In *T. pyriformis* mitochondria, the 5′-terminal 280 nucleotides of the LSU rRNA gene are located downstream of the rest of the LSU coding sequence and separated from it by a tRNA gene (Heinonen et al., 1987). In *C. reinhardtii*, the mitochondrial SSU and LSU rRNA genes are both split into a number of modules (four and at least eight, respectively) that are scrambled among one another and among protein-coding and tRNA genes throughout a 6-kilobase-pair region of the mitochondrial DNA (Boer and Gray, 1988). In these two cases, constraints on the order of module transcription must be relaxed. The genomic rearrangement of individual rRNA domains further reinforces the concept of a modular construction of rRNA genes.

INTRONS VERSUS TRANSCRIBED SPACERS IN rRNA GENES

ITSs may be considered a type of intervening sequence in rRNA genes in the sense that they interrupt the genic regions that are destined to become the mature rRNAs and so are eliminated at the RNA level during posttranscriptional processing of the rRNA precursor. Introns (Table 2) represent another type of intervening sequence, and these have been found in the rDNAs of eucaryotes (nuclear, chloroplast, and mitochondrial rRNA genes) and an archaebacterium. There are, however, a number of distinct differences between ITSs and introns in rDNA, differences substantial enough that it seems unlikely that the two originated in the same way in evolution.

(i) The sites of intron insertion are localized exclusively to the conserved core of rDNA, in marked contrast to the distribution of ITSs (Fig. 1). Intron localization may reflect a requirement for a highly conserved target sequence. For example, in [*omega*$^+$] strains of *Saccharomyces cerevisiae*, an intron-encoded mitochondrial endonuclease (transposase) recognizes and cleaves an 18-base-pair target site within the peptidyltransferase center of an intronless ([*omega*$^-$]) mitochondrial LSU rRNA gene (Colleaux et al., 1988), precipitating a unidirectional gene conversion event that results in the insertion of the [*omega*$^+$] intron between positions 2449 and 2450 (*E. coli* coordinates). Other rDNA introns (Muscarella and Vogt, 1989) and insertion elements (Xiong and Eickbush, 1988) (Table 2) have recently been shown to be mobile and to encode site-specific endonucleases that function in intron-insertion element transposition.

(ii) The mechanisms of spacer and intron processing are probably different. Most introns in rDNA are of the group I self-splicing type; they display elements of conserved secondary structure that play a role in their catalytic activity (Cech and Bass, 1986). There is little evidence of conserved secondary structure among positionally equivalent spacer sequences in different lineages and little support for an involvement of secondary structure in spacer excision, although admittedly not a lot is known about the biochemistry of this process. Some differences have been observed in the nature of the end groups (e.g., 5′-P or 5′-OH) generated by spacer excision (Spencer et al., 1987), suggesting that different ITSs in the same molecule may be removed by different pathways.

(iii) The conserved target site of intron insertion may also be required during intron excision at the RNA level. In the course of the excision-splicing reactions that remove group I introns, exon sequences flanking the insertion site are thought to base pair with an internal guide sequence in the intron (Cech and Bass, 1986). In contrast, processing sites for spacer excision appear to be located wholly within variable regions and not to involve adjacent exons.

(iv) Operationally, introns differ from spacers in that intron excision at the RNA level is accompanied by ligation of the flanking coding sequences, whereas spacer excision is not. Because ITSs and introns have distinctive localizations within the secondary structure of rRNA, quite different consequences are likely to ensue from disruption of the phosphodiester backbone during their removal. Covalent continuity may be essential within the conserved core, as evidenced by the loss of ribosome function that can result from phosphodiester bond breakage within the core. A particularly good example is the ribosome inactivation induced by the toxin α-sarcin, which cleaves LSU rRNA at a single position within a highly conserved region (Chan et al., 1983) that interacts with elongation factors (Moazed et al., 1988). The need for covalent integrity within the conserved core would explain why intron removal carries with it an absolute requirement for exon ligation. In contrast, because functional constraints are much lower in variable regions, discontinuity can be tolerated, eliminating the need for a mechanism whereby adjacent exons are ligated together after spacer excision.

SPACERS IN rDNA: PRIMITIVE OR DERIVED?

The discovery of ITSs in rDNA has prompted considerable speculation about their evolutionary origin. One view is that ITSs have been acquired by rRNA genes that were originally continuous. In this scenario, sequences introduced into variable regions are tolerated because of the reduced functional constraints that exist in these regions. Functions required to excise these spacers must then have been introduced at the same time as the spacers themselves; alternatively, sites for processing by indigenous enzymes must have been fortuitously present in the introduced spacers or else were created as a result of mutational changes in the course of evolution.

An alternative view is that ITSs are the evolutionary relics of a modular and split organization of the primordial rRNA genes (Gray et al., 1989). According to this scheme, contemporary rRNA genes that are highly fragmented (e.g., Spencer et al., 1987) are representative of a more primitive state. In other cases, where rRNA genes contain few or no ITSs, the presumption is that spacers or the ability to excise spacers, or both, have been lost in the course of evolution.

Although it is not possible at this point to rigorously prove whether spacers in general (or any given spacer in particular) are primitive or derived, the view that ITS sequences are ancient is attractive because it provides a ready explanation for two of the unique features of SSU and LSU rRNA molecules noted earlier: (i) their large size and (ii) the interspersion of variable and conserved regions. Starting with an originally highly fragmented rRNA gene in which individual modules specifying the conserved core were separated by ITSs in the genome, loss of the capacity for ITS excision would lead to the incorporation of spacer sequences into the final mature rRNA. These unexcised spacer sequences would then correspond to the variable regions in contemporary high-molecular-weight rRNA species. Such a spacer → variable region succession is consistent with the fact that discontinuities in present-day rRNAs map exclusively to the positions of variable regions. The result, over evolutionary time, would be a transition from a large number of relatively low-molecular weight rRNA "pieces" to a smaller number of higher-molecular-weight rRNA molecules, i.e., a progressive increase in the size of rRNA molecules, with concomitant replacement of intermolecular interactions by intramolecular pairings in the establishment of the core secondary and tertiary structure. Proposals outlining an evolutionary transition of this sort have been presented elsewhere (Clark, 1987; Gray et al., 1989).

The study of split rRNAs and their genes has provided a number of possible examples of different phases in rRNA evolution. Thus, the scrambled rRNA genes in *C. reinhardtii* mitochondria could represent a partial reversion to an ancient stage in which modules of rRNA structure were independently transcribed and the transcripts separately processed, subsequently associating to form the conserved rRNA core. The multiply fragmented nuclear LSU rRNA genes of *Crithidia fasciculata* and *Euglena gracilis* may reflect a transitional stage, in which separate modules of structure had been brought under the control of a single promoter and in which transcription, posttranscriptional processing, and ribosome assembly were more tightly coupled. Nevertheless, the end result of excision of multiple ITSs would be much the same as in the case of the *C. reinhardtii* mitochondria: a network of small rRNA pieces forming the conserved functional core by way of noncovalent intermolecular interactions. Finally, the single SSU and LSU rRNAs of *E. coli* may be viewed as the ultimate result of a progressive loss of spacer excision ability.

Although we may never be able to distinguish definitively whether ITS sequences in rDNA were always present or have been more recently acquired, phylogenetic considerations are providing some insights into this question. Phylogenetic trees based on rRNA sequences have recently emphasized the deep divergence and antiquity of certain of the unicellular eucaryotes (protists) (Sogin et al., 1986; Sogin et al., 1989). Among these are the trypanosoid protozoans (such as *Crithidia fasciculata*) and the euglenoid protozoans (such as *Euglena gracilis*), which appear to have separated from one another and from the main branch of eucaryotes at a very early stage in the evolution of this lineage (Sogin et al., 1986). The fact that these two organisms possess a separate 5.8S rRNA supports the idea that ITS2 was present in the LSU rRNA gene of the common ancestor of them and other eucaryotes. Also, because other spacers are shared by the LSU rRNA genes of *Crithidia fasciculata* and *Euglena gracilis*, we would argue that these particular spacers were already present in the common ancestor of the two organisms.

How far back we will be able to trace the existence of spacers is uncertain at this point. An even more divergent eucaryote, the microsporidian *Vairimorpha necatrix* (Vossbrinck et al., 1987), possesses a 23S-size LSU rRNA that lacks a separate 5.8S rRNA species (Vossbrinck and Woese, 1986). We cannot tell whether spacers were completely lost from the LSU rRNA gene of *V. necatrix* after its divergence from the remainder of the eucaryotes or were selectively gained in the latter. The recent finding of spacer sequences in eubacterial rRNA genes (Burgin et al., 1990) is particularly intriguing because it at least allows the possibility that some rDNA spacers predated the divergence of procaryotes and eucaryotes.

We thank Lisa Laskey for assistance in preparation of the manuscript.

This work was supported by grant A8387 from the Natural Sciences and Engineering Research Council of Canada to M.W.G., who is a Fellow in the Evolutionary Biology Program of the Canadian Institute for Advanced Research.

REFERENCES

Allet, B., and J.-D. Rochaix. 1979. Structure analysis at the ends of the intervening DNA sequences in the chloroplast 23S ribosomal genes of C. reinhardii. *Cell* **18:**55–60.

Boer, P. H., and M. W. Gray. 1988. Scrambled ribosomal RNA gene pieces in Chlamydomonas reinhardtii mitochondrial DNA. *Cell* 55:399–411.

Burgin, A. B., K. Parodos, D. J. Lane, and N. R. Pace. 1990. The excision of intervening sequences from Salmonella 23S ribosomal RNA. *Cell* **60:**405–414.

Burke, J. M., and U. L. RajBhandary. 1982. Intron within the large rRNA gene of N. crassa mitochondria. A long open reading frame and a consensus sequence possibly important in splicing. *Cell* **31:**509–520.

Cech, T. R., and B. L. Bass. 1986. Biological catalysis by RNA. *Annu. Rev. Biochem.* **55:**599–629.

Cedergren, R., M. W. Gray, Y. Abel, and D. Sankoff. 1988. The evolutionary relationships among known life forms. *J. Mol. Evol.* **28:**98–112.

Chan, Y.-L., Y. Endo, and I. G. Wool. 1983. The sequence of the nucleotides at the α-sarcin cleavage site in rat 28 S ribosomal ribonucleic acid. *J. Biol. Chem.* **258:**12768–12770.

Clark, C. G. 1987. On the evolution of ribosomal RNA. *J. Mol. Evol.* **25:**343–350.

Clark, C. G., and S. A. Gerbi. 1982. Ribosomal RNA evolution by fragmentation of the 23S progenitor: maturation pathway parallels evolutionary emergence. *J. Mol. Evol.* **18:**329–336.

Clark, C. G., B. W. Tague, V. C. Ware, and S. A. Gerbi. 1984. *Xenopus laevis* 28S ribosomal RNA: a secondary structure model and its evolutionary and functional implications. *Nucleic Acids Res.* **12:**6197–6220.

Colleaux, L., L. d'Auriol, F. Galibert, and B. Dujon. 1988. Recognition and cleavage site of the intron-encoded *omega* transposase. *Proc. Natl. Acad. Sci. USA* **85:**6022–6026.

Cox, R. A., and J. M. Kelly. 1981. Mature 23 S rRNA of prokaryotes appears homologous with the precursor of 25-28 rRNA of eukaryotes. Comments on the evolution of 23-28 rRNA. *FEBS Lett.* **130:**1–6.

Crick, F. H. C. 1968. The origin of the genetic code. *J. Mol. Biol.* **38:**367–379.

Cummings, D. J., J. M. Domenico, and J. Nelson. 1989. DNA sequence and secondary structures of the large subunit rRNA coding regions and its two class I introns of mitochondrial DNA from *Podospora anserina*. *J. Mol. Evol.* **28:**242–255.

Dahlberg, A. E. 1989. The functional role of ribosomal RNA in protein synthesis. *Cell* **57:**525–529.

Dawid, I. B., and M. L. Rebbert. 1981. Nucleotide sequences at the boundaries between gene and insertion regions in the rDNA of Drosophila melanogaster. *Nucleic Acids Res.* **9:**5011–5020.

Dujon, B. 1980. Sequence of the intron and flanking exons of the mitochondrial 21S rRNA gene of yeast strains having different alleles at the ω and *rib*-1 loci. *Cell* **20:**185–197.

Edwards, K., J. Bedbrook, T. Dyer, and H. Kössel. 1981. 4.5S rRNA from *Zea mays* chloroplasts shows structural homology with the 3′-end of prokaryotic 23S rRNA. *Biochem. Int.* **2:** 533–538.

Eickbush, T. H., and B. Robins. 1985. *Bombyx mori* 28S ribosomal genes contain insertion elements similar to the Type I and

Type II elements of *Drosophila melanogaster*. *EMBO J.* **4:** 2281–2285.

Fujiwara, H., T. Ogura, N. Takada, N. Miyajima, H. Ishikawa, and H. Maekawa. 1984. Introns and their flanking sequences of *Bombyx mori* rDNA. *Nucleic Acids Res.* **12:**6861–6869.

Gerbi, S. A. 1986. The evolution of eukaryotic ribosomal DNA. *BioSystems* **19:**247–258.

Gray, M. W., P. H. Boer, J. C. Collings, T. Y. K. Heinonen, D. F. Spencer, and M. N. Schnare. 1989. Ribosomal RNA genes in pieces, p. 521–530. *In* A. Kotyk, J. Škoda, V. Pačes, and V. Kostka (ed.), *Highlights of Modern Biochemistry.* VSP International Science Publishers, Zeist, The Netherlands.

Gray, M. W., D. Sankoff, and R. J. Cedergren. 1984. On the evolutionary descent of organisms and organelles: a global phylogeny based on a highly conserved structural core in small subunit ribosomal RNA. *Nucleic Acids Res.* **12:**5837–5852.

Gutell, R. R., and G. E. Fox. 1988. A compilation of large subunit RNA sequences presented in a structural format. *Nucleic Acids Res.* **16**(Suppl.):r175–r269.

Gutell, R. R., B. Weiser, C. R. Woese, and H. F. Noller. 1985. Comparative anatomy of 16-S-like ribosomal RNA. *Prog. Nucleic Acid Res. Mol. Biol.* **32:**155–216.

Hall, L. M. C., and B. E. H. Maden. 1980. Nucleotide sequence through the 18S-28S intergene region of a vertebrate ribosomal transcription unit. *Nucleic Acids Res.* **8:**5993–6005.

Hancock, J. M., and G. A. Dover. 1988. Molecular coevolution among cryptically simple expansion segments of eukaryotic 26S/28S rRNAs. *Mol. Biol. Evol.* **5:**377–391.

Heinonen, T. Y. K., M. N. Schnare, P. G. Young, and M. W. Gray. 1987. Rearranged coding segments, separated by a transfer RNA gene, specify the two parts of a discontinuous large subunit ribosomal RNA in *Tetrahymena pyriformis* mitochondria. *J. Biol. Chem.* **262:**2879–2887.

Jacq, B. 1981. Sequence homologies between eukaryotic 5.8S rRNA and the 5′ end of prokaryotic 23S rRNA: evidence for a common evolutionary origin. *Nucleic Acids Res.* **9:**2913–2932.

Jacquier, A., and B. Dujon. 1983. The intron of the mitochondrial 21S rRNA gene: distribution in different yeast species and sequence comparison between *Kluyveromyces thermotolerans* and *Saccharomyces cerevisiae*. *Mol. Gen. Genet.* **192:**487–499.

Jamrich, M., and O. L. Miller, Jr. 1984. The rare transcripts of interrupted rRNA genes in *Drosophila melanogaster* are processed or degraded during synthesis. *EMBO J.* **3:**1541–1545.

Kiss, T., M. Kis, and F. Solymosy. 1988. Nucleotide sequence of the 17S-25S spacer region from tomato rDNA. *Nucleic Acids Res.* **16:**7179.

Kiss, T., M. Kis, and F. Solymosy. 1989. Nucleotide sequence of a 25S rRNA gene from tomato. *Nucleic Acids Res.* **17:**796.

Kjems, J., and R. A. Garrett. 1988. Novel splicing mechanism for the ribosomal RNA intron in the archaebacterium Desulfurococcus mobilis. *Cell* **54:**693–703.

Lenaers, G., L. Maroteaux, B. Michot, and M. Herzog. 1989. Dinoflagellates in evolution. A molecular phylogenetic analysis of large subunit ribosomal RNA. *J. Mol. Evol.* **29:**40–51.

Machatt, M., J.-P. Ebel, and C. Branlant. 1981. The 3′-terminal region of bacterial 23S ribosomal RNA: structure and homology with the 3′-terminal region of eukaryotic 28S rRNA and with chloroplast 4.5S rRNA. *Nucleic Acids Res.* **9:**1533–1549.

MacKay, R. M. 1981. The origin of plant chloroplast 4.5 S ribosomal RNA. *FEBS Lett.* **123:**17–18.

Maroteaux, L., M. Herzog, and M.-O. Soyer-Gobillard. 1985. Molecular organization of dinoflagellate ribosomal DNA: evolutionary implications of the deduced 5.8S rRNA secondary structure. *BioSystems* **18:**307–319.

Michot, B., N. Hassouna, and J.-P. Bachellerie. 1984. Secondary structure of mouse 28S rRNA and general model for the folding of the large rRNA in eukaryotes. *Nucleic Acids Res.* **12:** 4259–4279.

Moazed, D., J. M. Robertson, and H. F. Noller. 1988. Interaction of elongation factors EF-G and EF-Tu with a conserved loop in 23S RNA. *Nature* (London) **334:**362–364.

Muscarella, D. E., and V. M. Vogt. 1989. A mobile group I intron in the nuclear rDNA of Physarum polycephalum. *Cell* **56:** 443–454.

Nazar, R. N. 1980. A 5.8 S rRNA-like sequence in prokaryotic 23 S rRNA. *FEBS Lett.* **119:**212–214.

Nazar, R. N. 1982. Evolutionary relationship between eukaryotic 29-32 S nucleolar rRNA precursors and the prokaryotic 23 S rRNA. *FEBS Lett.* **143:**161–162.

Netzker, R., H. Köchel, N. Basak, and H. Küntzel. 1982. Nucleotide sequence of *Aspergillus nidulans* mitochondrial genes coding for ATPase subunit 6, cytochrome oxidase subunit 3, seven unidentified proteins, four tRNAs and L-rRNA. *Nucleic Acids Res.* **10:**4783–4794.

Neuhaus, H., F. Müller, A. Etter, and H. Tobler. 1987. Type I-like intervening sequences are found in the rDNA of the nematode *Ascaris lumbricoides*. *Nucleic Acids Res.* **15:**7689–7707.

Nielsen, H., and J. Engberg. 1985. Sequence comparison of the rDNA introns from six different species of *Tetrahymena*. *Nucleic Acids Res.* **13:**7445–7455.

Noller, H. F. 1984. Structure of ribosomal RNA. *Annu. Rev. Biochem.* **53:**119–162.

Oakden, K. M., and B. G. Lane. 1973. Chain termini of the satellite RNA from yeast ribosomes. *Can. J. Biochem.* **51:** 520–528.

Otsuka, T., H. Nomiyama, H. Yoshida, T. Kukita, S. Kuhara, and Y. Sakaki. 1983. Complete nucleotide sequence of the 26S rRNA gene of *Physarum polycephalum*: its significance in gene evolution. *Proc. Natl. Acad. Sci. USA* **80:**3163–3167.

Pene, J. J., E. Knight, Jr., and J. E. Darnell, Jr. 1968. Characterization of a new low molecular weight RNA in HeLa cell ribosomes. *J. Mol. Biol.* **33:**609–623.

Rae, P. M. M. 1981. Coding region deletions associated with the major form of rDNA interruption in Drosophila melanogaster. *Nucleic Acids Res.* **9:**4997–5010.

Rae, P. M. M., B. D. Kohorn, and R. P. Wade. 1980. The 10 kb Drosophila virilis 28S rDNA intervening sequence is flanked by a direct repeat of 14 base pairs of coding sequence. *Nucleic Acids Res.* **8:**3491–3504.

Raué, H. A., J. Klootwijk, and W. Musters. 1988. Evolutionary conservation of structure and function of high molecular weight ribosomal RNA. *Prog. Biophys. Mol. Biol.* **51:**77–129.

Roiha, H., J. R. Miller, L. C. Woods, and D. M. Glover. 1981. Arrangements and rearrangements of sequences flanking the two types of rDNA insertion in *D. melanogaster*. *Nature* (London) **290:**749–753.

Schnare, M. N., J. R. Cook, and M. W. Gray. Fourteen internal transcribed spacers in the circular ribosomal DNA of *Euglena gracilis*. *J. Mol. Biol.*, in press.

Schnare, M. N., and M. W. Gray. Sixteen discrete RNA components in the cytoplasmic ribosome of *Euglena gracilis*. *J. Mol. Biol.*, in press.

Schnare, M. N., T. Y. K. Heinonen, P. G. Young, and M. W. Gray. 1986. A discontinuous small subunit ribosomal RNA in *Tetrahymena pyriformis* mitochondria. *J. Biol. Chem.* **261:**5187–5193.

Smith, V. L., and K. Beckingham. 1984. The intron boundaries and flanking rRNA coding sequences of *Calliphora erythrocephala* rDNA. *Nucleic Acids Res.* **12:**1707–1724.

Sogin, M. L., H. J. Elwood, and J. H. Gunderson. 1986. Evolutionary diversity of eukaryotic small-subunit rRNA genes. *Proc. Natl. Acad. Sci. USA* **83:**1383–1387.

Sogin, M. L., J. H. Gunderson, H. J. Elwood, R. A. Alonso, and D. A. Peattie. 1989. Phylogenetic meaning of the kingdom concept: an unusual ribosomal RNA from *Giardia lamblia*. *Science* **243**:75–77.

Spencer, D. F., J. C. Collings, M. N. Schnare, and M. W. Gray. 1987. Multiple spacer sequences in the nuclear large subunit ribosomal RNA gene of *Crithidia fasciculata*. *EMBO J.* **6**: 1063–1071.

Spencer, D. F., M. N. Schnare, and M. W. Gray. 1984. Pronounced structural similarities between the small subunit ribosomal RNA genes of wheat mitochondria and *Escherichia coli*. *Proc. Natl. Acad. Sci. USA* **81**:493–497.

Subrahmanyam, C. S., B. Cassidy, H. Busch, and L. I. Rothblum. 1982. Nucleotide sequence of the region between the 18S rRNA sequence and the 28S rRNA sequence of rat ribosomal DNA. *Nucleic Acids Res.* **10**:3667–3680.

Sy, J., and K. S. McCarty. 1970. Characterization of 5.8-S RNA from a complex with 26-S ribosomal RNA from *Arbacia punctulata*. *Biochim. Biophys. Acta* **199**:86–94.

Takaiwa, F., K. Oono, and M. Sugiura. 1985. Nucleotide sequence of the 17S-25S spacer region from rice rDNA. *Plant Mol. Biol.* **4**:355–364.

Tautz, D., J. M. Hancock, D. A. Webb, C. Tautz, and G. A. Dover. 1988. Complete sequences of the rRNA genes of *Drosophila melanogaster*. *Mol. Biol. Evol.* **5**:366–376.

Tautz, D., M. Trick, and G. A. Dover. 1986. Cryptic simplicity in DNA is a major source of genetic variation. *Nature* (London) **322**:652–656.

Veldman, G. M., J. Klootwijk, H. van Heerikhuizen, and R. J. Planta. 1981. The nucleotide sequence of the intergenic region between the 5.8S and 26S rRNA genes of the yeast ribosomal RNA operon. Possible implications for the interaction between 5.8S and 26S rRNA and the processing of the primary transcript. *Nucleic Acids Res.* **9**:4847–4862.

Vossbrinck, C. R., J. M. Maddox, S. Friedman, B. A. Debrunner-Vossbrinck, and C. R. Woese. 1987. Ribosomal RNA sequence suggests microsporidia are extremely ancient eukaryotes. *Nature* (London) **326**:411–414.

Vossbrinck, C. R., and C. R. Woese. 1986. Eukaryotic ribosomes that lack a 5.8S RNA. *Nature* (London) **320**:287–288.

Walker, W. F. 1981. Proposed sequence homology between the 5′-end regions of prokaryotic 23S rRNA and eukaryotic 28S rRNA. *FEBS Lett.* **126**:150–151.

Woese, C. R. 1972. The emergence of genetic organization, p. 301–341. *In* C. Ponnamperuma (ed.), *Exobiology*. North-Holland Publishing Co., Amsterdam.

Xiong, Y., and T. H. Eickbush. 1988. Functional expression of a sequence-specific endonuclease encoded by the retrotransposon R2Bm. *Cell* **55**:235–246.

Chapter 53

Sequence Comparison and Evolution of Ribosomal Proteins and Their Genes

BRIGITTE WITTMANN-LIEBOLD, ANDREAS K. E. KÖPKE, EVELYN ARNDT, WOLFGANG KRÖMER, TOMOMITSU HATAKEYAMA, and HEINZ-GÜNTER WITTMANN

Ribosomes and their components provide an excellent means of studying the evolution of the organisms, since they are essential components for all cells. The ribosomes from the three kingdoms, eubacteria, eucaryotes, and archaebacteria (Woese and Fox, 1977; Woese and Olsen, 1986), vary considerably in number of constituents and sedimentation coefficients (Wittmann, 1986). Ribosomes contain approximately 55 to 85 different macromolecules (three to four RNA strands and numerous proteins) of various sizes and compositions (Table 1). Depending on the organism, the proteins vary in net charge from very basic to highly acidic (Giri et al., 1984; Wittmann-Liebold, 1986). Sequence comparison of the rRNA and proteins from different species gives a valuable tool for evolutionary studies of the organisms. Since proteins are composed of 20 different amino acids whereas RNAs consist of four bases only, the evolutionary relationships can be traced more easily by sequence comparison of the proteins. Moreover, homologous proteins which perform the same functions are usually more conserved than homologous RNA molecules, and this is reflected by predominantly conserved amino acid residues or highly conserved regions in the primary structures.

This chapter summarizes the comparison data obtained from the ribosomal protein sequences of the archaebacteria *Halobacterium marismortui* and *Methanococcus vannielii* and their eubacterial and eucaryotic counterparts. The comparisons are based on more than 300 complete and many partial sequences of ribosomal proteins from the three kingdoms. Furthermore, the results on ribosomal protein gene organization in the archaebacteria are compared with those of the *Escherichia coli* genome and the gene arrangement in eucaryotes.

THE RIBOSOMAL CONSTITUENTS

The best-studied ribosome is that of *E. coli*. Its mass consists of two-thirds RNA and one-third protein. It contains three different RNAs (5S, 16S, and 23S) and 55 proteins (Wittmann, 1986). In addition, a few short polypeptides might be attached (Wada and Sako, 1987). All primary structures of these components are known (Noller et al., 1986; Wittmann-Liebold, 1986). The proteins of *E. coli* have unique primary structures which show no statistically significant sequence relationships among each other (Wittmann-Liebold et al., 1984). Their shapes and secondary structures have been investigated (Giri et al., 1984), and their locations on the ribosomal subunits have been studied by immunoelectron microscopy (Stöffler and Stöffler-Meilicke, 1986) and neutron scattering (Moore et al., 1986; Nowotny et al., 1986). Recently, detailed knowledge on the topography of the ribosomal constituents was gained by the analysis of RNA-RNA, RNA-protein, and protein-protein cross-links generated in situ (Walleczek et al., 1988; Brimacombe et al., 1986; Noller et al., 1986; Brockmöller and Kamp, 1988; Pohl and Wittmann-Liebold, 1988). A computer graphic model for the spatial structure of the 30S subunit of *E. coli* was developed (Schüler and Brimacombe, 1988).

For some proteins, mainly from *Bacillus stearothermophilus*, crystals are available for X-ray structure analysis (Wittmann, 1986). Intensive efforts are now being made to determine the spatial structure of the ribosomal particles by X-ray structure analysis of three-dimensional crystals and by image reconstruction of two-dimensional crystalline sheets (Yonath et

Brigitte Wittmann-Liebold, Andreas K. E. Köpke, Evelyn Arndt, Wolfgang Krömer, Tomomitsu Hatakeyama, and Heinz-Günter Wittmann ■ Max-Planck-Institut für Molekulare Genetik, Ihnestrasse 73, D-1000 Berlin-Dahlem, Federal Republic of Germany.

Table 1. Size and composition of ribosomes

Organism	Organelle	Ribosome	No. of proteins	RNAs
Bacteria		70S	50–60	5S, 16S, 23S
Plants	Chloroplasts	70S	50–60	5S, 16S, 23S, (4.5S)
	Mitochondria	75S	Unknown	5S, 18S, 26S
Protozoans and fungi	Mitochondria	75S	65–75	15S–17S, 21S–24S
Mammals	Mitochondria	55S	80–90	12S, 16S
Eucaryotes (cytoplasm)		80S	75–90	5S, 5.8S, 17–18S, 26S–28S

al., 1986; Yonath and Wittmann, 1989; Yonath et al., this volume).

Since only limited amounts of ribosomes can be isolated from different sources and large-scale purification of all ribosomal proteins is difficult, it did not seem feasible to obtain full sets of protein sequence information from various organisms by applying amino acid sequencing only. The combined approach of (i) partial amino acid sequencing in order to design oligonucleotide probes for cloning of the corresponding gene and (ii) the fast methods of nucleotide sequencing made investigations of ribosomal proteins from different sources possible. Furthermore, genes for ribosomal proteins in procaryotes are often organized in operons, which allowed several sequences to be obtained by gene walking. Therefore, many proteins have been sequenced recently from ribosomes of *B. stearothermophilus* (Wittmann-Liebold, 1986; Krömer et al., in press); from archaebacteria such as *H. marismortui* (Kimura et al., 1989), *Halobacterium halobium* (Itoh, 1988; Mankin, 1989; Spiridonova et al., 1989), *Sulfolobus solfataricus* (Ramirez et al., 1989), and *M. vannielii* (Auer et al., 1989; Köpke and Wittmann-Liebold, 1989; Köpke et al., 1989; Baier et al., in press; Matheson et al., this volume); and from various eucaryotes (Wool et al., this volume; Planta et al., 1986; Warner, 1989). In addition to the already completed *E. coli* set, almost complete sets of proteins can be expected in the near future for *B. stearothermophilus*, *H. marismortui*, *M. vannielii*, and rat liver ribosomes. Comparison of these entire sequence sets of ribosomal proteins will yield detailed and reliable information on the evolution of these ribosomal components. Already the available protein sequences have given interesting insights into evolutionary aspects of ribosomes and their proteins.

NOMENCLATURE OF RIBOSOMAL PROTEINS

Systematic names have been given to those ribosomal proteins deriving from archaebacteria or eucaryotes that show significant amino acid sequence similarities to eubacterial protein in order to express this homology and to group the equivalent proteins from the various organisms in one ribosomal protein family. The systematic name is composed of (i) three characters to describe the organism from which the protein originates, e.g., Eco (*E. coli*), Hma (*H. marismortui*), and Rno (*Rattus norvegicus*); (ii) the prefix L or S to indicate a ribosomal protein from the large or small subunit, respectively; and (iii) the same number as for the equivalent *E. coli* protein. Accordingly, protein Mva L23 is the ribosomal protein of *M. vannielii* that is homologous to protein L23 of *E. coli*.

In cases where no clear homology can be assigned for a protein derived from the archaebacterial or eucaryotic kingdom, the vulgar name (rat, mou[se], ham[ster], etc.) has been kept to allow for easy differentiation. Proteins with no obvious relationship to the eubacterial ribosomal proteins that were originally named according to the migration in a two-dimensional gel or high-performance liquid chromatography (HPLC) chromatogram got only two characters, e.g., HS or HL, indicating the organism (one character) and the subunit (S or L), followed by the protein number of the migration in the gel or HPLC trace. By this means, it is possible to differentiate unrelated from related proteins (e.g., in the data base) and to call all proteins belonging to one protein family, such as protein family S11, by one command in appropriate computer programs that only select for systematic names. This is important if multiple alignments or predictions are to be made by computer.

EUBACTERIAL RIBOSOMAL PROTEINS

The amino acid sequences of 39 proteins from the *B. stearothermophilus* ribosome have been completed (Table 2). Examination of the sequences revealed that they are similar but not always identical in length and net charge to the *E. coli* counterparts. Although *B. stearothermophilus* is a gram-positive organism, the sequences can be aligned easily with the corresponding proteins from the gram-negative

Table 2. Comparison of ribosomal proteins derived from *E. coli* and *B. stearothermophilus*

Protein	Residue		Identity (%)	Reference for Bst protein[a]
	Eco	Bst		
S2	240	234	53	Kimura, personal communication
S3	232	218	55	Krömer et al., in press
S4	203	198	52	Kimura, personal communication
S5	166	166	55	Kimura, 1984
S7	177	155	50	Kimura, personal communication
S8	129	130	47	Kimura, personal communication
S9	129	129	54	Kimura and Chow, 1984
S11	128	128	68	Kimura et al., 1988
S12	123	138	76	Kimura and Kimura, 1987c
S13	117	119	58	Brockmöller and Kamp, 1988
S15	88	88	57	Kimura, personal communication
S16	82	88	45	Kimura, personal communication
S17	83	85	75	Hirano, personal communication
S18	74	77	55	McDougall et al., 1989
S19	91	91	71	Hirano et al., 1987; Krömer et al., in press
S20	86	88	51	Kimura, personal communication
S21	70	56	43	Hirano, personal communication
L1	233	232	50	Kimura et al., 1985a
L2	272	275	60	Kimura et al., 1985b
L5	176	179	59	Kimura and Kimura, 1987b
L6	176	177	49	Kimura et al., 1981
L9	148	147	33	Kimura, et al., 1980
L11	141	130	62	Kimura, personal communication
L12	120	122	63	Garland et al., 1987
L14	123	122	69	Kimura et al., 1985a
L15	144	146	44	Kimura et al., 1985a
L17	127	119	49	Kimura and Chow, 1984
L18	117	120	53	Kimura and Kimura, 1987b
L20	117	119	62	Pon et al., 1989
L22	110	113	50	Krömer et al., in press
L23	100	95	28	Kimura et al., 1985a
L24	103	103	45	Kimura et al., 1985a
L27	84	87	55	Kimura and Chow, 1984
L29	63	66	46	Kimura et al., 1985a
L30	58	62	53	Kimrua, 1984
L32	56	56	25	Tanaka et al., 1984
L34	63	54	15	Kimura, personal communication
L35[b]	64	62	48	Pon et al., 1989
L36[b]	37	37	27	Tanaka et al., 1984

[a] For the *E. coli* ribosomal protein sequences, see Wittmann-Liebold (1986) and Giri et al. (1984).
[b] Wada and Sako, 1987.

E. coli. However, the similarity between the equivalent proteins ranges, depending on the protein pair, from only 20% identity (that is, 20 identical amino acids are found at identical positions in 100 amino acids compared) to as high as 76% (Table 2).

Interestingly, high sequence similarities were found for some ribosomal proteins of functional importance, such as S12 and L2, proving that their functions are maintained through conservation of the essential amino acids in the respective sequence regions (active sites). Other proteins important for ribosomal structure or function, such as S13 and S19, were also found to be highly conserved (Brockmöller and Kamp, 1988; Hirano et al., 1987). By cross-linking studies on the 30S subunits of *E. coli* and *B. stearothermophilus* with the bifunctional reagent diepoxybutane, it was shown that the S13-S19 protein pairs of both 30S subunits were cross-linked at identical amino acid residues. These were a histidine and a cysteine at corresponding positions in *E. coli* and *B. stearothermophilus*, respectively (Brockmöller and Kamp, 1988; Pohl and Wittmann-Liebold, 1988). Analysis of the cross-linked proteins at the amino acid level proved not only that the primary sequences of these two proteins in both organisms are highly homologous but also that they form identical or very similar three-dimensional structural domains.

In contrast, no counterpart to the largest ribosomal protein in *E. coli*, S1, could be detected in the *B. stearothermophilus* ribosome (Isono and Isono, 1976), although this protein is involved in the binding of mRNA to the 30S subunit in *E. coli* (Suryanarayana and Subramanian, 1983). Other ribosomal proteins, such as L32 and L34, showed weak but significant sequence similarities to their *E. coli* and other eubacterial counterparts.

CHLOROPLAST RIBOSOMAL PROTEINS

All ribosomal proteins of known primary structure derived from various chloroplasts (more than 50 sequences; see Table 3) showed significant similarities to their eubacterial counterparts. Again, the degree of sequence conservation depends on the protein examined (Wittmann-Liebold, 1986; Subramanian et al., this volume). Furthermore, chloroplast ribosomes have a set of proteins similar to those of *E. coli* (Table 1). Therefore, no doubts remain about the phylogenetic relationship of these organelles and their ribosomal proteins to the eubacterial kingdom. As found for the *E. coli* and *B. stearothermophilus* ribosomal proteins, S12 is one of the most conserved ribosomal proteins of chloroplasts and exhibits a similar degree of conservation.

EUCARYOTIC RIBOSOMAL PROTEINS

Several eucaryotic ribosomal proteins from mammals have been sequenced completely (Table 3; for more details, see Wool et al., this volume). In addition, a number of complete sequences are known for ribosomal proteins from yeasts (Planta et al., 1986; Warner, 1989), the African clawed frog *Xenopus laevis*, the brine shrimp *Artemia salina*, and the fruit fly *Drosophila melanogaster* (Table 3). Many more partial sequences from these and other species have been determined. Unfortunately, no complete set of ribosomal proteins for any of these eucaryotic organisms is yet available. Therefore, conclusions about the relationship of the eucaryotic ribosome to that of the other kingdoms are still preliminary. In cases where a eucaryotic protein has not been found to correspond to eubacterial proteins, this obviously could be due to the lack of sequence information. Alternatively, the sequences of these proteins might have undergone considerable changes during evolution, with no or only very distantly related sequence relationships, but they might still serve similar functions and might even have similar three-dimensional structures. On the other hand, some of the eucaryotic proteins do not have counterparts in the eubacterial kingdom, since the number of proteins is considerably higher in eucaryotic than in eubacterial ribosomes (Wittmann, 1986) (Table 1).

Table 3. Completely sequenced ribosomal proteins from different organisms[a]

Group	Organism	No. of sequenced proteins
Eubacteria	*Escherichia coli*	55
	Bacillus stearothermophilus	39
	Bacillus subtilis	3
	Mycoplasma capricolum	3
	Micrococcus lysodeikticus	2
	National Research Council of Canada isolate	1
	Rhodopseudomonas sphaeroides	1
	Salmonella typhimurium	1
	Streptomyces griseus	1
	Thermus aquaticus	1
	Desulfovibrio vulgaris	1
	Providencia sp.	1
Chloroplasts	*Marchantia polymorpha*	20
	Nicotiana tabacum	20
	Zea mays	14
	Spinacia oleracea	5
	Spirodella oligorhiza	2
	Euglena gracilis	2
	Chlamydomonas reinhardtii	1
Mitochondria	*Saccharomyces cerevisiae*	13
	Plants	6
	Neurospora crassa	6
	Paramecium sp.	3
	Drosophila melanogaster	2
	Tetrahymena thermophila	1
Eucaryotes	Rat liver	35
	Saccharomyces cerevisiae	15
	Human	7
	Mouse	3
	Drosophila melanogaster	5
	Artemia salina	2
	Chinese hamster	2
	Xenopus laevis	2
	Candida utilis	1
	Tetrahymena thermophila	1
	Schizosaccharomyces pombe	1
Archaebacteria	*Halobacterium marisomortui*	30
	Methanococcus vannielii	21
	Halobacterium halobium	7
	Halobacterium cutirubrum	4
	Sulfolobus acidocaldarius and *S. solfataricus*	5

[a] From the RIB data base, Max Planck Institute of Molecular Genetics.

MITOCHONDRIAL RIBOSOMAL PROTEINS

Mitochondrial ribosomes vary considerably in their sedimentation coefficients as well as in the number and size of their RNAs and proteins (Table 1). Only a few proteins have been investigated in detail, e.g., by HPLC and N-terminal amino acid sequencing (Graack et al., 1988). Different approaches have been used to clone the corresponding genes; complementation assays of mitochondrion-lacking mutants, immunological screening of gene expression libraries, and direct cloning of nuclear DNA restriction fragments by using oligonucleotide probes deduced from N-terminal amino acid sequences have been successful (Grohmann et al., 1989; Myers et al., 1987; Partaledis and Mason, 1988). Other genes have been identified by their similarities to eucaryotic ribosomal protein genes, but the actual occurrence of their translational products as constituents of the mitochondrial ribosome has not been proven in most cases (Table 4). A significant relatedness to bacterial ribosomal proteins is not a general feature of mitochondrial ribosomal proteins, since almost half of the known proteins lack any relationship to proteins of other organisms. Whereas in plants a significant number of the genes for mitochondrial ribosomal proteins are located in the mitochondrial DNA, both in animals and in fungi all except one of the genes are located in the nuclear DNA.

ARCHAEBACTERIAL RIBOSOMAL PROTEINS

The best-studied archaebacterial ribosomal proteins derive from *H. marismortui* (Table 5) (Kimura et al., 1989) and *M. vannielii* (Table 6) (Lechner et al., 1989; Auer et al., 1989; Köpke and Wittmann-Liebold, 1989; Köpke et al., 1989; Baier et al., in press; Matheson et al., this volume; J. Auer and A. K. E. Köpke, unpublished results). In addition, several complete sequences were obtained from *H. halobium* (Itoh, 1988; Mankin, 1989; Spiridonova et al., 1989), *H. cutirubrum* (Ramirez et al., 1989), and *S. solfataricus* ribosomes (Ramirez et al., 1989).

Ribosomal Proteins from *H. marismortui*

Thirty proteins deriving from *H. marismortui* ribosomes have already been fully sequenced in our laboratory. Table 5 lists these together with the related proteins from other archaebacterial, eubacterial, and eucaryotic ribosomes as revealed by statistical comparison analyses (for details, see Köpke and Wittmann-Liebold, 1989; Wittmann-Liebold, 1988); also shown is the percentage of identical residues in comparable sequence regions for each of the proteins. The proteins are named in Table 5 according to their migration in a two-dimensional gel electrophoresis system as described by Hatakeyama and Kimura (1988) and are shown in Fig. 1. To those proteins that could be correlated to the *E. coli* ribosomal proteins, a name equivalent to the *E. coli* nomenclature was assigned (Hma proteins).

Ribosomal Proteins from *M. vannielii*

About 30 *M. vannielii* proteins were isolated from total 30S and 50S protein mixtures by HPLC procedures and in part rechromatographed to sequencer purity (Kamp and Wittmann-Liebold, 1988). Their N-terminal sequences were established (R. M. Kamp and B. Wittmann-Liebold, unpublished data), by employing a liquid-phase sequencer equipped with in-machine on-line detection of the phenylthiohydantoin-derived amino acids (Wittmann-Liebold and Ashman, 1985) to determine partial amino acid sequences for the synthesis of oligonucleotide probes and localization of the genes. Deduced sequences were obtained for six of these proteins (Mva L2, L3, L4, L10, L12, and L23; Table 6) from nucleotide sequencing of the genes in our group. In addition, other proteins from *M. vannielii* recently became available (Auer et al., 1989; Lechner et al., 1989; Matheson et al., this volume; J. Auer, unpublished results). The amino acid sequences translated from the gene sequences were found to agree well with the N-terminal sequences determined by protein-chemical methods for the purified proteins, e.g., for Mva S8, L10, L12, L14, L22, L23, L24, L29, and L30 (Table 7).

SEQUENCE COMPARISON OF RIBOSOMAL PROTEINS FROM ARCHAEBACTERIA, EUBACTERIA, AND EUCARYOTES

The comparison values of the archaebacterial sequences and those of the other corresponding ribosomal proteins (Tables 5, 8, and 9) prove that the sequence similarities (35 to 60% identity) are highest within the archaebacterial kingdom. Some ribosomal proteins of the archaebacterial kingdom show similar sequence relationships to their eubacterial and eucaryotic equivalents (e.g., Hma S11, S17, L2, L15, and L23); however, more show a closer sequence relationship to proteins of the eucaryotic ribosome than to that of eubacteria (e.g., Hma S8, L3, L4, and L14). The Mva proteins in general are much more similar to the eucaryotic than to the eubacterial equivalents, whereas this relationship is not as pronounced for the Hma proteins. In addition, several archaebacterial proteins have no counterparts in eubacterial proteins (e.g., HS6, HS12, HS13, HS15, HL21/22, HL24, HL29, HL30, HL31, and HL32). Consequently, a systematic name according to the eubacterial nomenclature of ribosomal proteins cannot be assigned to these proteins. Sequence comparisons with these proteins indicate that some of them are significantly related to eucaryotic proteins (e.g., HS6 to Rat S12, HS12 to yeast [Yea] S16A, HS15 to

Table 4. Cloned mitochondrial ribosomal protein genes[a]

Species	Gene	Gene location[b]	Identity of amino acid residues in		Reference(s)
			%	To:	
Plants					
Oenothera sp.	S4[c]	mt.	?	Plastid rpS4	Schuster and Brennicke, 1987
Wheat, maize	S12[c]	mt.	57	Eco S12	Gualberto et al., 1988
Maize	S13[c]	mt.	38	Eco S13	Bland et al., 1986
Tobacco	S13[c]	mt.	39	Eco S13	Bland et al., 1986
Wheat	S13[c]	mt.	38	Eco S13	Bonen, 1987
Broad bean	S14[c]	mt.	41	Eco S14	Wahleithner and Wolstenholme, 1988
Fungi					
Neurospora crassa	S5[c,d]	mt.		No similarity found	Burke and RajBhandary, 1982
	cyt-21	nuc., LG IL	35	Eco S16	Kuiper et al., 1988
	cyt-22	nuc.	?	Eco L?	Kuiper et al., 1988; data not shown
	MRP3	nuc.	20	Eco S1[e]	Kreader et al., 1989
	MRP15	nuc.	?	Eco S5 and Eco S6	Kreader et al., 1989; data not shown
Saccharomyces cerevisiae	*var1*	mt.		No similarity found	Hudspeth et al., 1982
	MRP1	nuc., chr. IV		No similarity found	Myers et al., 1987
	MRP2	nuc., chr. XVI	34	Eco S14[f]	Myers et al., 1987
	MRP7	nuc.	49	Eco L27[g]	Fearon and Mason, 1988
			57	Bst L27	
	MRP-L4	nuc., chr. XII		No similarity found	H.-R. Graack, L. Grohmann, and M. Kitakawa, unpublished data
	MRP-L8	nuc., chr. X	42	Eco L17	Kitakawa et al., 1990
			46	Bst L17	
	MRP-L9	nuc., chr. VII or XV	38	Eco L3[h]	Graack et al., unpublished data
	MRP13	nuc.		No similarity found	Partaledis and Mason, 1988
	MRP-L20	nuc., chr. XI		No similarity found	Kitakawa et al., 1990
	MRP-L27	nuc., chr. X		No similarity found	Graack et al., unpublished data
	MRP-L31	nuc., chr. XI		No similarity found	Grohmann et al., 1989
	YMR31	nuc., chr. VI		No similarity found	Matsushita et al., 1989
	YMR44	nuc., chr. XIII or XVI		No similarity found	Matsushita et al., 1989
	MRP B[c]	nuc., chr. XI	20	Eco S2	Abraham et al., 1989
	MRP C[i]	nuc., chr. XI		No similarity found	Abraham et al., 1989
Protozoa					
Paramecium sp.	*rps12*[c]	mt.	42	Eco S12	Pritchard et al., 1989
	rps14[c]	mt.	20	Eco S14	Pritchard et al., 1989
	rps12[c]	mt.	21	Eco L2	Pritchard et al., 1989
Tetrahymena thermophila	*mt-L14*[c]	mt.	35	Eco L14	Suyama and Jenney, 1989
Invertebrate					
Drosophila melanogaster	*tko*[c]	nuc.	41	Eco S12	Royden et al., 1987
			46	*Euglena gracilis* chloroplast S12	

[a] According to Grohmann et al. (in press).
[b] mt., Mitochondrial; nuc., nuclear; LG, linkage group; chr., chromosome.
[c] Identified by search for similarity to known protein sequences only.
[d] For the sequenced gene, coding of the S5 protein has been postulated by comparison of the deduced size, hydrophobicity, basicity, and amino acid composition of its translational product with the features of the S5 protein. Evidence for that assumption by amino acid sequencing of the S5 protein has not been given yet.
[e] Similarity found between the *E. coli* protein and 96 amino acids of the N terminus.
[f] Similarity found between the *E. coli* protein and 78 amino acids of the N terminus.
[g] Similarity found between the *E. coli* protein and 85 amino acids of the N terminus.
[h] Similarity found between the *E. coli* protein and 209 amino acids of the C terminus.
[i] Identified on the basis of *MRP*-like properties.

Table 5. Comparison of ribosomal proteins from *H. marismortui* with those from other sources[a]

H. marismortui[b]	Sequence similarity[c] to (identities in percent): Archaebacteria	Eubacteria	Eucaryotes[d]	Reference
HS1 (Hma S3)	Mva S3 (42) Hha S3 (71)	Eco S3 (25) Bst S3 (25)	?	Arndt et al., 1990
HS6	?		Rat S12 (20)	Kimura et al., 1987
HS11 (Hma S15)	?	Eco S15 (26) Bst S15 (28)	Yea S15 (N terminus)	Arndt et al., 1986
HS12	?		Yea S16A (35)	Kimura et al., 1987
HS13	?		Yea S10 (27)	Kimura et al., 1989
HS14 (Hma S17)	Mva S17 (42)	Eco S17 (37) Bst S17 (32)	Rno S17 (36) [S11]	Kimura and Kimura, 1987a
HS15	?		Xla S19 (24)	Kimura et al., 1987
HS16 (Hma S8)	Mva S8 (52) Sac S8	Eco S8 (25) Bst S8 (33)	Sce S8 (40) [S24]	Kimura and Kimura, 1987a
HS17 (Hma S9)	?	Eco S9 (29) Bst S9 (37)	?	Kimura and Langner, 1984
HS18 (Hma S19)	Hha S19 (81)	Eco S19 (37) Bst S19 (33)	?	Arndt et al., 1990
HS19 (Hma S11)	?	Eco S11 (44) Bst S11 (42) Bsu S11 (43)	Sce S11 (49) [rp59] Cha S11 (46) [S14]	Kimura et al., 1988
HS20 (Hma S10) (N terminus)	Mva S10 (44)	Eco S10 (36)	?	Kimura, personal communication
HS? (Hma S5)	Mva S5 (49)	Eco S5 (28) Bst S5 (35)	?	Scholzen, personal communication
HL1 (Hma L3)	Mva L3	Eco L3 (28)	Sce L3 (34) [rp1]	Arndt et al., 1990
HL4 (Hma L2)	Mva L2 (59)	Eco L2 (36) Bst L2 (37)	Spo L2 (41) [K5]	Arndt et al., 1990
HL6 (Hma L4)	Mva L4 (48)	Eco L4 (21)	Sce L4 (34) [L2] Xla L4 (33) [L1] Dme L4 (32) [L1]	Arndt et al., 1990
HL9 (Hma L15)	Mva L15 (43)	Eco L15 (28) Bst L15 (27)	Sce L15 (28) [L29]	Hatakeyama et al., 1989
HL16 (Hma L24)	Mva L24 (47)	Eco L24 (21) Bst L24 (24)	Sce L24 (39) [L33] (N terminus)	Hatakeyama et al., 1988; Arndt, unpublished data
H17 (Hma L5)	Mva L5 (44)	Eco L5 (38) Bst L5 (33)	Sce L5 (38) [L16]	Hatakeyama, unpublished data
HL20 (Hma L30)	Mva L30 (40)	Eco L30 (25) Bst L30 (23)	?	Hatakeyama et al., 1989
HL21/22	?		Yea L9 (27)	Hatakeyama et al., 1989
HL23 (Hma L22)	Hha L22 (65) Mva L22 (42)	Eco L22 (21) Bst L22 (29)	?	Hatakeyama et al., 1988
HL24	Mva ORF (42)		Rat L19 (35)	Hatakeyama et al., 1989
HL25 (Hma L23)	Mva L23 (36)	Eco L23 (35) Bst L23 (43)	Sce L23 (42) [L25] Cut L23 (40) [L25]	Hatakeyama and Kimura, 1988

Continued on following page

Xla S19, HL21/22 to Yea L9, HL24 to Rat L19, and HL30 to Yea L34 and Rat L31), whereas others exhibit no sequence similarities to any of the proteins of the other kingdoms (e.g., HL31 and HL32). Therefore, the archaebacterial proteins have been divided into four groups as given for the Hma proteins in Fig. 2. Those sequences for which no equivalent proteins were found at all, neither in the eucaryotic nor in the

Table 5.—Continued.

H. marismortui[b]	Sequence similarity[c] to:			Reference
	Archaebacteria	Eubacteria	Eucaryotes[d]	
HL27 (Hma L14)	Mva L14 (52)	Eco L14 (37) Bst L14 (37)	Sce L14 (44) [L17]	Arndt, unpublished data
HL29	?		Yea L28 (28)	Hatakeyama and Kimura, 1988
HL30	?		Yea L34 (25) Rat L31 (27)	Hatakeyama, unpublished data
HL31	?		?	Hatakeyama and Kimura, 1988
HL32	?		?	Hatakeyama et al., 1989
HL33 (Hma L29)	Mva L29 (41) Hha L29 (72)	Eco L29 (32) Bst L29 (44)	?	Hatakeyama et al., 1988

[a] To archaebacterial and eucaryotic proteins that are significantly related to eubacterial equivalents, systematic names according to the *E. coli* nomenclature were assigned as follows: Hma, *H. marismortui*; Sce, *Saccharomyces cerevisiae* (Yea); Spo, *Schizosaccharomyces pombe*; Rno, *Rattus norvegicus* (Rat); Cha, Chinese hamster; Xla, *Xenopus laevis*; Dme, *Drosophila melanogaster*; Cut, *Candida utilis*; Hha, *Halobacterium halobium*; Mva, *Methanococcus vannielii*; Sac, *Solfolobus acidocaldarius*. Other designations: Eco, *Escherichia coli*; Bst, *Bacillus stearothermophilus*; Bsu, *Bacillus subtilis*. References for proteins: *E. coli* (Wittmann-Liebold, 1986); *B. stearothermophilus* (Table 2); *H. marismortui* (last column, this table); *M. vannielii* (Table 6; Matheson et al., this volume); *H. halobium* (Spiridonova et al., 1989; Mankin, 1989); *S. cerevisiae* (Planta et al., 1986; Warner, 1989); *S. pombe* (Gatermann et al., 1989); mammals (Wool et al., this volume).
[b] HS and HL refer to migration of the small and large ribosomal subunit proteins, respectively, in two-dimensional polyacrylamide gel electrophoresis as described by Hatakeyama and Kimura (1988); see Fig. 1.
[c] Given in parentheses as percentage of identical residues in corresponding regions of the sequences.
[d] Original designations of the renamed eucaryotic ribosomal proteins are given in brackets.

eubacterial ribosome, and those that have corresponding proteins only in eubacteria might be related to a yet unknown eucaryotic protein sequence. On the other hand, clearly some archaebacterial proteins have no significantly related counterpart in the eubacterial ribosome, since for *E. coli* the primary structures of all ribosomal proteins have been determined (Wittmann-Liebold, 1986). On the other hand, the sequences might have so diverged during evolution that their phylogenetic relationship cannot be traced by direct amino acid comparison.

Ribosomal Protein Family L2

Protein L2 is highly conserved throughout evolution, as shown by immunological methods (Stöffler and Stöffler-Meilicke, 1986). Proteins Mva L2 and

Table 6. Examples of *M. vannielii* ribosomal proteins that show sequence similarities to proteins in other archaebacteria and other kingdoms[a]

Mva protein	Homologous ribosomal protein			Reference for Mva protein
	Eubacteria	Eucaryotes	Archaebacteria	
L1	Eco L1 Bst L1			Baier et al., in press
L2	Eco L2 Bst L2	Spo L2	Hma L2	Köpke and Wittmann-Liebold, 1989
L3	Eco L3	Spo L3	Hma L3	Heckinbach and Köpke, unpublished data
L4	Eco L4 Bst L4	Sce L4 Xla L4 Dme L4	Hma L4	Köpke, unpublished data
L10	Eco L10	Man L10	Hha L10 Sac L10	Köpke et al., 1989
L12	Eco L12 Bst L12	Man L12	Hha L12 Sac L12	Strobel et al., 1988
L23	Eco L23 Bst L23	Yea L23	Hma L23	Köpke and Wittmann-Liebold, 1988

[a] See also Matheson et al., this volume.

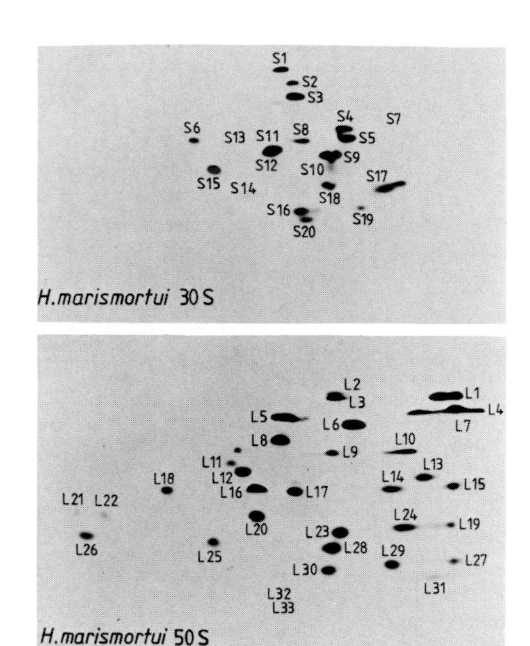

Figure 1. Two-dimensional polyacrylamide gel electrophoresis of total protein mixture derived from *H. marismortui* 50S ribosomal subunits according to Hatakeyama and Kimura (1988). The gel is adapted for resolution of the mainly acidic ribosomal proteins of halophilic organisms. Top right, origin; right to left, first dimension; top to bottom, second dimension.

Hma L2 are very similar to the L2 proteins in eubacteria and chloroplasts, with sequence similarities of >30%. Recently an L2 equivalent has been detected in *Schizosaccharomyces pombe* (Gatermann et al., 1989). All members of the L2 family gave excellent alignments of sequences (Köpke and Wittmann-Liebold, 1989) (Fig. 3). The pairwise similarities of this family deduced from a multiple alignment of all known L2 sequences are given in Table 8.

Ribosomal Protein Family L23

Similarly, the L23 ribosomal proteins have counterparts in all kingdoms, but the sequences are less conserved than that of the L2 protein family. The eucaryotic L23 proteins have an extended N terminus of about 50 residues and share a common gap with the archaebacterial proteins in their C-terminal parts in comparison with the eubacterial counterparts (Fig. 4). On the other hand, the archaebacterial sequences have the same lengths as the eubacterial and chloroplast L23 proteins (Fig. 4). The similarity values for each protein pair are given in Table 9. In this case, the methanococcal protein is more similar to the yeast protein than to the Hma protein if only identical amino acids are counted. Although the archaebacterial L23 sequences share more identical amino acids with the eucaryotic than with the eubacterial L23 sequences, the differences are small in this case. In contrast, rather low similarities are obtained between the L23 proteins from chloroplasts on one hand and those of the archaebacteria and eucaryotes on the other (Köpke and Wittmann-Liebold, 1988).

Ribosomal Protein Families L3, L4, and L24

The L3 proteins from *M. vannielii* and halobacteria show low but statistically significant similarities to that of *E. coli*, whereas the L4 and Hma L24 proteins do not. However, Mva L4 and Eco L4 are only significantly similar in a region located about 50 amino acids from the N terminus, and the archaebacterial L24 equivalents can be significantly aligned only in the regions of positions 40 to 100 with the N-terminal sequences of the eubacterial L24 proteins. However, the L3, L4, and L24 proteins from both archaebacteria are well conserved within this kingdom and are also strongly sequence related to their eucaryotic counterparts (Table 5).

Ribosomal Protein Families L10 and L12

Whereas identification of the corresponding *E. coli* protein for the archaebacterial L4 proteins and for Mva L10 was difficult but possible, the L12 protein derived from *M. vannielii* has undergone such drastic changes during evolution that no significant sequence similarity has been found between the known members of the eubacterial ribosome on one hand and this archaebacterial protein and the eucaryotic equivalents on the other. However, from the amino acid composition of the Mva L12 protein (high content of alanine and glutamic acid; very few arginines) and the localization by immunoelectron microscopy (M. Stöffler-Meilicke, unpublished results), it is obvious that this protein is the L12 equivalent in *M. vannielii* constituting the stalk of the 50S subunit. In addition, the archaebacterial and eucaryotic L10 and L12 proteins have a common feature that is not observed in eubacteria: a virtually identical C-terminal region in these two proteins from the same organism (for details, see Rich and Steitz, 1987; Itoh, 1988; Köpke and Wittmann-Liebold, 1989; Matheson et al., this volume). The C-terminal amino acids, however, differ for different organisms. This points to a coevolution of these sequence regions (Köpke et al., 1989).

Overexpression of Mva L12 in *E. coli* and Incorporation into *H. marismortui* Ribosomes

Recently, the gene coding for protein L12 derived from *M. vannielii* was overexpressed in *E. coli* (Köpke et al., 1990). Interestingly, the expressed protein was N-terminally blocked by acetylation of

Table 7. N-terminal amino acid sequences of *M. vannielii* proteins[a]

Mva protein	N-terminal amino acid sequence
S8	Ser-Leu-Met-Asp-Pro-Leu-Ala-Asn Initiator methionine of the gene sequence is not present in the mature protein
L10	Met-Ile-Asp-Ala-Lys-Ser-Glu-His-Lys-Ile-Ala
L12	Met-Glu-Tyr-Ile-Tyr-Ala-Ala-Leu-Leu-Leu-Asn-Ser-Ala-Asn
L14	Met-Lys-Gly-Leu-Gly-Ser-Thr-Ile-Val-Arg-Ser-Leu-Pro-Asn-Gly-Ala-Arg-Leu-Val-Cys-Ala-Asp-Asn-Thr-Gly-Ala-Lys-Glu-Leu-Glu-Val-Ile
L22	Ala-Lys-Leu-Gly-Tyr-Lys-Val-Glu-Ala-Asp-Pro-Ser-Lys-Thr-Ala-Lys-Ala-Met-Gly-Arg-Thr-Leu- Initiator methionine of the gene sequence is not present in the mature protein
L23	Met-Asp-Ala-Phe-Asp-Val-Ile-Lys-Thr-Pro-Ile-Val-Ser-Glu-Lys-Thr-Met-Lys-Leu-
L24	Val-Leu-Thr-Asn-Ser-Lys-Gln-Pro-Arg-Lys-Gln-Arg-Lys-Ala-Leu-Tyr-Asn-Ala-Pro-Leu-His-Leu-Arg-Asn-Ser-Val-Met-Ser- Initiator methionine of the gene sequence is not present in the mature protein
L29	Ala-Ile-Leu-Lys-Ala-Ser-Glu-Ile-Arg-Glu-Phe-Ser-Ile-Asp-Glu-Met-Asn-Glu-Lys-Ile-Ala-Glu-Leu Initiator methionine of the gene sequence is not present in the mature protein
L30	Ala-Tyr-Ala-Val-Val-Arg-Val-Arg-Gly-Ser-Val-Gly-Val-Arg-Gly-Asp-Ile-Ala-Asp-Thr-Met-Lys Initiator methionine of the gene sequence is not present in the mature protein

[a] Total protein mixtures (100 μg each) of the 30S and 50S ribosomal subunits were separated by reverse-phase HPLC on Vydac C_4. The N-terminal sequences of the protein fractions were determined by liquid-phase sequencing in a cup sequencer (constructed in the Max Planck Institute workshop) and equipped with automatic conversion and on-line isocratic HPLC detection, using an acetonitrile or 2-propanol solvent system. Half of the released phenylthiohydantoin-derived amino acids were injected and positively identified at 254 and 313 nm (underlined).

the N-terminal methionine, although the genuine Mva protein carries a free N-terminal methionine, whereas the equivalent protein in *E. coli* (Eco L7) is the acetylated form of Eco L12 (beginning with acetylserine). Therefore, it was interesting to see whether the overexpressed protein can be used for replacement of the L12 protein in other ribosomes. It could be shown that the genuine L12 protein in *H. marismortui* 50S subunits can be substituted by the L12 protein of *M. vannielii*, whereas no incorporation of *E. coli* or *B. stearothermophilus* L12 proteins into *H. marismortui* ribosomes was achieved (Köpke et al., 1990).

Table 8. Sequence similarities of L2 proteins from eucaryotes, archaebacteria, and eubacteria from the L2 multiple protein alignment[a]

	Spo	Mva	Hma	Bst	Eco	Mpo	Nta
Spo	100	46	40	24	25	22	25
Mva		100	58	32	32	31	32
Hma			100	33	34	31	33
Bst				100	58	54	51
Eco					100	48	47
Mpo						100	73
Nta							100

[a] Similarities were calculated by the program DISTANCES. Only identical amino acids were counted. The percentage of matched amino acids to compared amino acids is shown. Designations: Spo, *Schizosaccharomyces pombe*; Mva, *Methanococcus vannielii*; Hma, *Halobacterium marismortui*; Bst, *Bacillus stearothermophilus*; Eco, *Escherichia coli*; Mpo, *Marchantia polymorpha* chloroplasts; Nta, *Nicotiana tabacum* chloroplasts.

Results of Sequence Comparisons

The sequence analysis of the archaebacterial proteins and their comparison with the other known ribosomal sequences lead to the following interesting findings.

(i) All halophilic and methanogenic proteins

Table 9. Sequence similarities of ribosomal proteins L23 from eucaryotes, archaebacteria, and eubacteria from the L23 multiple protein alignment[a]

	Sce	Cut	Mva	Hma	Eco	Bst	Mpo	Nta
Sce	100	85	55	45	29	31	23	13
Cut		100	53	43	32	33	23	14
Mva			100	51	44	40	27	22
Hma				100	40	46	26	23
Eco					100	37	27	24
Bst						100	43	35
Mpo							100	62
Nta								100

[a] Similarities were calculated as for Table 8. Identical amino acids and conservative exchanges were counted. The percentage of matched amino acids to compared amino acids is shown. Designations: Sce, *Saccharomyces cerevisiae*; Cut, *Candida utilis*; Mva, *Methanococcus vannielii*; Hma, *Halobacterium marismortui*; Eco, *Escherichia coli*; Bst, *Bacillus stearothermophilus*; Mpo, *Marchantia polymorpha* chloroplasts; Nta, *Nicotiana tabacum*.

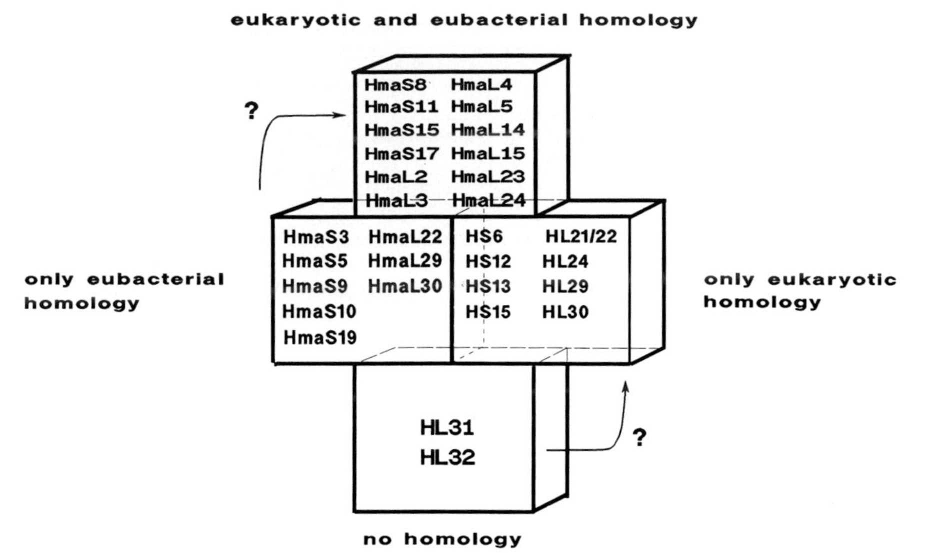

Figure 2. Homology of 30 ribosomal proteins from *H. marismortui* to corresponding eubacterial and eucaryotic proteins. For nomenclature, see Table 5, footnote *a*.

studied so far are much more closely related to each other (with approximately 50% identical amino acids) compared with the equivalent proteins from other archaebacterial species such as *S. solfataricus*. The halophilic ribosomal proteins contain a surplus of glutamic and aspartic acids which are often in positions usually occupied by lysine or arginines in other organisms. As a consequence, these ribosomal proteins are mainly acidic. On the other hand, the Mva proteins are mainly basic and richer in hydrophobic residues. Despite these changes in net charge, the proteins in general are strongly related to each other.

(ii) A considerable number of proteins are well sequence related to their equivalents in the eubacterial and eucaryotic kingdoms. This usually was obvious from visual inspection of the sequences and was additionally proven by statistical analyses. These proteins can be well aligned; examples of this group of proteins are given in Fig. 3 to 6, in which the sequences of the L2, L23, S8, and S11 protein families are aligned. In general, these archaebacterial proteins share about 25 to 40% identical amino acids in comparable positions with eubacterial equivalents and 25 to 50% identities with the eucaryotic counterparts. In addition, conservative replacements of similar amino acids (e.g., glutamic acid for aspartic acid, serine for threonine, valine for isoleucine, and lysine for arginine) are frequently noticed. Whereas some of these proteins have almost the same lengths as their eubacterial or eucaryotic counterparts, other families contain proteins of different lengths. In these cases, the changes in the protein sizes are considerable (see protein family S15 [Wittmann-Liebold, 1988]). In some cases, the eucaryotic proteins have sequence extensions of more than 50 residues compared with archaebacterial proteins and extensions of more than 150 amino acids compared with eubacterial proteins (see the protein families L18 and L30 [Wittmann-Liebold, 1988]).

(iii) Furthermore, the net charges and the content of hydrophobic residues of a protein vary to a large extent among the ribosomes of the various organisms. The halophilic ribosomes have mainly acidic proteins, whereas the methanogenic or thermophilic ribosomal proteins contain a greater number of hydrophobic residues. The high content of acidic amino acids in halophilic ribosomal proteins (which often replace basic residues of their basic *E. coli* or *B. stearothermophilus* equivalents) helps to adapt the ribosome to the extreme salt conditions in the cell by various mechanisms, such as by chelate formation between metal ions and the highly charged acidic residues (Rainer Jaenicke, personal communication).

(iv) Several archaebacterial proteins were found to be only homologous to eubacterial sequences. It is likely that a corresponding eucaryotic protein has not yet been sequenced, and hence no equivalent protein could be found among the eucaryotic sequences.

(v) Some archaebacterial proteins were found which are similar to eucaryotic but not to eubacterial proteins. Since all ribosomal proteins from *E. coli* have been completely sequenced, these archaebacterial proteins probably have no equivalents in the eubacterial ribosome. It is not clear whether these sequences underwent such drastic changes during evolution that no significant similarities to the eubac-

```
   Spo                                      MG.RVIRAQR.KSGGIFQAHTRLRKGAAQLRTLDFA
   Mva                                      MGKRLISQNRGRGTPKYRSPSHKRKGEVKYRSYDEM
   Hma                                       GRRIQGQRRGRGTSTFRAPSHR...YKADLEHRKV
   Bst  AIKKYKPTSNGRRGMTVLDFSEITTDQPEKSLLAPLKKRAGRNNQGKITV.RHQGGGHKR....QYRIIDFK
   Eco  AVVKCKPTSPGRRHVVKVVNPELHKGKPFAPLLEKNSKSGGRNNNGRITT.RHIGGGHKQ....AYRIVDFK
   Mpo MAIRLYRAYTPGTRNRSVPKFDEIVKCQPQKKLTY.NKHIKKGRNNRGIITSQHRGGGHKR....LYRKIDFC
   Nta MAIHLYKTSTPSTRNGTVDS...QVKSNPRNNLIYGQHHCGKGRNARGIITARHRGGGHKR....LYRKIDFR
Con L2 .........................................................................

   Spo FRNPYHYRTDVETFVATEGMYTGQFVYCGKNAALTVGNVLPVGEMPEGTIISNVEEKAGDRGALGRSSGNYVI
   Mva FAN.....GEERLVLIPEGISVGEQIECGISAEIKPGNVLPLGEIPEGIPVYNIETIPGDGGKLVRAGGCYAH
   Hma FED.....GDRRLIILAPEGVGVGDELQVGVDAEIAPGNTLPLAEIPEGVPVCNVESSPGDGGKFARASGVNAC
   Bst YAD.....GEKRYIIAPKNLKVGMEIMSGPDADIKIGNALPLENIPVGTLVHNIELKPGRGGQLVRAAGTSAC
   Eco YKD.....GERRYILAPKGLKAGDQIQSGVDAAIKPGNTLPMRNIPVGSTVHNVEMKPGKGGQLARSAGTYVC
   Mpo YED.....GEKRYILYPRGIKLDDTIISSEEAPILIGNTLPLTNMPLGTAIHNIEITPGKGGQLVRAAGTVAK
   Nta YGD.....GEKRYILHPRGAIIGDTIVSGTEVPIKMGNALPLTDMPLGTAIHNIEITLGKGGQLARAAGAVAK
Con L2 ................................GN.LP....P.G....N.E...G..G...R..G....

   Spo RGVVGIVAGGGRIDKPLLKAGRAFHKYRVKRNCWPRTRGVAMNPVDHPHGGGN.HQHVGHSTTVPRQSAPGQK
   Mva RATIGVVAGGGRKEKPFVKAGKKHHSLSAKAVAWPKVRGVAMNAVDHPYGGGR.HQHLGKPSSVSRNTSPGRK
   Hma RATIGVVGGGGRTDKPFVKAGNKHHKMKARGTKWPNVRGVAMNAVDHPFGGGG.RQHPGKPKSISRNAPPGRK
   Bst RATVGEVGNEQHELVNIGKAG.....RARWLGIRPTVRGSVMNPVDHPHGGGEGKAPIGRKSPMTPWKPTLGY
   Eco RATLGEVGNAEHMLRVLGKAG.....AARWRGVRPTVRGTAMNPVDHPGGGHEGR..NFGKHPVTPWGVQTKG
   Mpo LATIGQIGNVDVNNLRIGKAG.....SKRWLGKRPKVRGVVMNPIDHPHGGGEGRAPIGRKKPLTPWGHPALG
   Nta SATVGQVGNVGVNQKSLGRAG.....SKRWLGKRPVVRGVVMNPVDHPHGGGEGRAPIGRKKPTTPWGYPALG
Con L2 ....G..............AG............P..RG..MN..DHP.GG.....................
```

Figure 3. Sequence alignment of the ribosomal protein family L2 showing comparison of the eucaryotic sequences of *Schizosaccharomyces pombe* (Spo) with the archaebacterial sequences of *Methanococcus vannielii* (Mva) and *Halobacterium marismortui* (Hma), with the eubacterial sequences of *Bacillus stearothermophilus* (Bst) and *Escherichia coli* (Eco), and with the chloroplast sequences of *Marchantia polymorphus* (Mpo) and *Nicotiana tabacum* (Nta). The line Con L2 shows amino acid residues that are conserved in all sequences.

terial sequences remained, or whether these proteins have been added to the archaebacterial or eucaryotic ribosome. It is often difficult to find a significant relationship of a new protein sequence from the archaebacterial and eucaryotic kingdoms with the eubacterial protein sets. If the alignments were not convincing and did not reach significant scores, it was generally found that multiple alignments (using also protein sequences derived from archaebacteria) helped to clarify the relationships among the three kingdoms.

(vi) Several ribosomal proteins from archaebacteria did not show any sequence similarity with eubacterial or eucaryotic proteins. A relationship of these proteins to that of eucaryotes cannot be excluded, since it is likely that for these proteins no corresponding eucaryotic ribosomal protein has been sequenced so far. Another possibility is that some of these proteins are unique in archaebacteria and may form a separate domain on this ribosome.

(vii) The sequence data for archaebacterial ribosomal proteins led to the conclusion that individual ribosomal proteins have evolved at different rates during evolution and may have been incorporated at different evolutionary stages of ribosome biogenesis. This complicates the evaluation of an evolutionary tree summarizing the data of all ribosomal proteins, since different proteins show different degrees of similarity to their counterparts in the other kingdoms.

Phylogenetic Trees for Ribosomal Proteins

Phylogenetic trees were constructed by using the distance data matrix obtained by the program DISTANCES (University of Wisconsin Genetics Computer Group program package). DISTANCES calculates the ratio of matching amino acids for every protein pair of a multiple protein alignment (program LINEUP). The proteins used for this study had equivalents in all kingdoms which could be unambiguously aligned (protein families S8, S11, S17, L2, L3, L5, and L23). Although all ribosomal protein families showed different degrees of conservation, the branching orders of the resulting trees were always identical. In addition, no difference in the branching order was

```
    1                                                                                                        100
Sce MAPSAKATAA KKAVVKGTNG KKALKVRTSA TFRLPKTLKL ARAPKYASKA VPHYNRLDSY KVIEQPITSE TAMKKVEDGN ILVFQVSMKA NKYQIKKAVK
Cut MAPSTKAASA KKAVVKGSNG SKALKVRTST TFRLPKTLKL TRAPKYARKA VPHYQRLDNY KVIVAPIASE TAMKKVEDGN TLVFQVDIKA NKHQIKQAVK
Mva                                                            MDAF DVIKTPIVSE KTMKLIEEEN RLVFYVERKA TKEDIKEAIK
Hma                                                              SW DVIKHPHVTE KAMNDMDFQN KLQFAVDDRA SKGEVADAVE
Eco                                                        MIREERLL KVLRAPHVSE KASTAMEKSN TIVLKVAKDA TKAEIKAAVQ
Bst                                                           MKDPR DIIKRPIITE NTMN.LIGQK KYTFEVDVKA NKTEVKDAVE
Mpo                                                               M NQVKYPVLTE KTIR.LLEKN QYSFDVNIDS NKTQIKKWIE
Nta                                                               M DGIKYAVFTD KSIR.LLGKN QYTSNVESGS TRTEIKHWVE
Con .......... .......... .......... .......... .......... .......... .......... .......... .....V.... ..........

    101                                          150
Sce ELY.....EV NI........ ....LVRPNG TKKAYVRLTA DYDALDIANR IGYI
Cut DLY.....EV DVLAV..... ..NTLIRPNG TKKAYVRLTA DHDALDIANK IGYI
Mva QLFNAEVAEV NT........ ....NITPKG QKKAYIKLKD EYNAGEVAAS LPIY
Hma EQYDVTVEQV NT........ ....QNTMDG EKKAVVRLSE DDDAQEVASR IGVF
Eco KLFEVEVEVV NTLVVKGKVK RHGQRIGRRS DKKAYVTLKE GQNLDFVGGA E
Bst KIFGVKVEKV NIMNYKGKFK RVGRYSGYTN RKKAIVTLTP DSKEIELFEV
Mpo LFFNVKVISV NSHRLPKKKK KIGTTTGYTV RYKRMIIKLQ SGYSIPLFSN K
Nta LFFGVKVIAM NSHRLPGKSR RMGPIMGHTM HYRRMIITLQ PGYSIPPLRK KRT
Con .......... .......... .......... .......... .......... ....
```

Figure 4. Sequence alignment of the ribosomal protein family L23 showing comparison of the eucaryotic sequences of *Saccharomyces cerevisiae* (Sce) and *Candida utilis* (Cut) with those of the archaebacteria *Methanococcus vannielii* (Mva) and *Halobacterium marismortui* (Hma), with those of the eubacteria *Escherichia coli* (Eco) and *Bacillus stearothermophilus* (Bst), and with those of chloroplasts of *Marchantia polymorphus* (Mpo) and *Nicotiana tabacum* (Nta). The consensus sequence (bottom line) shows amino acids that are conserved in all sequences.

observed when either only identical residues or also conserved residues were taken into account as a match. Typical phylogenetic trees for the ribosomal protein families S11, L12, and L2 (Fig. 7) show the branching order observed in common for all ribosomal protein families examined so far. However, the common tree representing the data from all ribosomal protein families gives only an average in the

```
    1                                                                                                        100
Sce MTRSSVLADA LNAINNAEKT GKRQVLIRPS SKVIIKFLQV MQKHGYIGEF EYIDDHRSGK IVVQLNGR.. .......... .........L NKCGVISPRF
Mva MSLMDPLANA LNHVSNCEGV GKNVAYLKPA SKLIGRVLKV MQDQGYIGNF EYIEDGKAGV YKVDLIGQ.. .......... .........I NKCGAVKPRY
Hma .TGNDPFANA LSALNNAESV GHLEQTVSPA SNEIGSVLEV FYDRGYIDGF SFVDDGKAGE FEVELKGA.. .......... .........I NECGPVKPRY
Bst .VMTDPIADM LTAIRNANMV RHEKLEV.PA SKIKREIAEI LKREGFIRDY EYIEDNKQGI LRIFLKYG.P NERV...... ......ITGL KRISKPGLRV
Eco .SMQDPIADM LTRIRNGQAA NKAAVTM.PS SKLKVAIANV LKEEGFIEDF K.VEGDTKPE LELTLKYF.Q GKAV...... ......VESI QRVSRPGLRI
Zma .MGKDTIADL LTSIRNADMN KKGTVRV.VS TNITENIVKI LLREGFIESV RKHQESNRYF LVSTLRHQRR KTRKGIYRTR .......TFL KRISRPGLRI
Mca .MTTDVIADM LTRIRNANQR YLKTVSV.PS SKVKLEIARI LKEEGFISDF TVEGDVKKTI NI.ELKYQ.G KTRV...... ......IQGL KKISKPGLRV
Mpo .MGNDTIANM ITSIRNANLG KIKTVQV.PA TNITRNIAKI LFQEGFIDNF IDNKQNTKDI LILNLKYQ.G KKKK...... ....SYITTL RRISKPGLRI
Nta .MGRDTIAEI ITSIRNADMD RKRVVRI.AS TNITENIVQI LLREGFIENV RKHREKN... .....KYF.L VLTLRHRRNR KRPYRNILNL KRISRPGLRI
Con .......A.. .....N.... .......... .......... ....G.I... .......... .......... .......... .......... ........R.

    101                                          150
Sce NVKIGDIEKW TANLLPARQF GYVILTTSAG IMDHEEARRK HVSGKILGFV Y
Mva AVKYQEFEKF EKRYLPAKGF GLLIVSTPKG LMTHDEARTA GVGGRLISYV Y
Hma SAGADEFEK. ..RFLPARDY GTLVVTTSVG IMSHYEAREQ GVGGQVIAYV Y
Bst YVKAHEVPRV L......NGL GIAILSTSQG VLTDKEARQK GTGGEIIAYV I
Eco YKRKDELPKV M......AGL GIAVVSTSKG VMTDRAARQA GLGGEIICYV A
Zma YANYQGIPKV L......GGM GIAILSTSRG IMTDREARLN RIGGEVLCYI W
Mca YAQANEIPQV L......NGL GISIVSTSQG IMTGKKARLA NAGGEVLAFI .
Mpo YSNHKEIPKV L......GGM GIVILSTSRG IMTDREARQK KIGGELLCYV W
Nta YSNYQRIPRI L......GGM GIVILSTSRG IMTDREARLE GIGGEILCYI W
Con .......... .......... G.....T..G ......AR.. ...G...... .
```

Figure 5. Sequence alignment of the ribosomal protein family S8 showing comparison of the eucaryotic sequence in *Saccharomyces cerevisiae* (Sce) with the archaebacterial sequences of *Methanococcus vannielii* (Mva) and *Halobacterium marismortui* (Hma), with the eubacterial sequences of *Mycoplasma capricolum* (Mca), *B. stearothermophilus* (Bst), and *Escherichia coli* (Eco), and with the chloroplast proteins of *Zea mays* (Zma), *Marchantia polymorphus* (Mpo), and *Nicotiana tabacum* (Nta). The consensus sequence Con (bottom line) shows amino acids that are conserved in all sequences.

```
     1                                                                                                        100
Hsa  MAPRKGKEKK EEQVISLGPQ VAEGENVFGV CHIFASFNDT FVHVTDLSG. KETICRVTGG MKVKADRDES SPYAAMLAAQ DVAQRCKELG ITALHIKLRA
Cha  MAPRKGKEKK EEQVISLGPQ VAEGENVFGV CHIFASFNDT FVHVTDLSG. KETICRVTGG MKVKADRDES SPYAAMLAAQ DVAQRCKELG ITALHIKLRA
Sce  .......... ....MSNVVQ ARDNSQVFGV ARIYASFNDT FVHVTDLSG. KETIARVTGG MKVKADRDES SPYAAMLAAQ DVAAKCREVG ITAVHVKIRA
Hma  .......... .......... SEETEDINGI AHVHASFNNT IITITDQTGA QETLAKSSGG TVVKQNRDEA SPYAAMQMAE VVAEKALDRG VEGVDVRVRG
Bsu  ........MA AARKSNTRKR RVKKNIESGI AHIRSTFNNT IVTITDTHG. .NAISWSSAG ALGFRGSRKS TPFAAQMAAE TAAKGSIEHG LKTLEVTVKG
Bst  .......... .ARRTNTRKR RVRKNIDTGI AHIRSTFNNT IVTITDVHG. .NALGWATAG VSGFKGSRKS TPFAAQMAAE AAAKASMEHG MKTVEVNVKG
Eco  .......... .AKAPIRARK RVRKQVSDGV AHIHASFNNT IVTITDRQG. .NALGWATAG GSGFRGSRKS TPFAAQVAAE RCADAVKEYG IKNLEVMVKG
Sol  .MAKPIPKIG SRRNGRISSR KSARKIPKGV IHVQASFNNT IVTVTDVRG. .RVVSWASAG TCGFRGTKRG TPFAAQTAAG NAIRTVVEQG MQRAEVMIKG
Nta  .MAKAIPKIS SRRNGRIGSR KGARRIPKGV IHVQASFNNT IVTVTDVRG. .RVVSWSSAG TSGFKGTRRG TPFAAQTAAA NAIRTVVDQG MQRAEVMIKG
Mpo  .........M PKSVKKINLR KGKRRLPKGV IHIQASFNNT IVTVTDIRG. .QVVSWSSAG ACGFKGTKKS TPFAAQTAAE NAIRILIDQG MKQAEVMISG
Con  .......... .......... ........G. ......FN.T ....TD..G. .........G .......... .P.AA...A. .........C ..........

     101                                          150
Hsa  TGGNRTKTPG PGAQSALRAL ARSGMKIGRI EDVTPIPSDS TRRKGGRRGR RL
Cha  TGGNRTKTPG PGAQSALRAL ARSGMKIGRI EDVTPIPSDS TRRKGGRRGR RL
Sce  TGGTRTKTPG PGGQAALRAL ARSGLRIGRI EDVTPVPCDS TRKKGGRRGR RL
Hma  PGGNLQTSPG PGAQATIRAL ARAGLEIGRI EDVTPTPHNG TRAPKNSGF. ..
Bsu  PG........ SGREAAIRAL QAAGLEVTAI RDVTPVPHNG CRPPKRRRV. ..
Bst  PG........ AGREAAIRAL QAAGIEITAI KDVTPIPHDG CRPPKRRRV. ..
Eco  PG........ PGRESTIRAL NAAGFRITNI TDVTPIPHNG CRPPKKRRV. ..
Sol  PS........ LGRDAALRAI RRSGILLSFV RNVTPMPHNG CRPPKKRRV. ..
Nta  PG........ LGRDAALRAI RRSGILLTFV RDVTPMPHNG CRPPKKRRV. ..
Mpo  PG........ PGRDTALRAI RRSGIILSFV RDVTPMPHNG CRPPRKRRV. ..
Con  .......... .G.....RA. ...G...... .DVTP.P... .R........ ..
```

Figure 6. Sequence alignment of the ribosomal protein family S11. The sequences of the eubacterial proteins (Eco, *Escherichia coli*; Bst, *Bacillus stearothermophilus*; Bsu, *Bacillus subtilis*) and those from chloroplast ribosomes (Sol, *Spinacia oleracea*; Mpo, *Marchantia polymorpha*; Nta, *Nicotiana tabacum*) are compared with sequences of *Halobacterium marismortui* (Hma) and eucaryotic proteins (Sce, *Saccharomyces cerevisiae*; Cha, Chinese hamster; Hsa, *Homo sapiens*). All sequences can be aligned well and show a considerable degree of homology. The archaebacterial and eucaryotic proteins have an insertion at positions 103 to 110. The N-terminal regions of the proteins are quite different from each other. The consensus sequence Con (bottom line) shows amino acids that are conserved in all sequences.

branch points, since all proteins show different degrees of conservation (Fig. 7).

GENE ORGANIZATION OF RIBOSOMAL PROTEINS IN ARCHAEBACTERIA

Recently, the organization of various ribosomal protein genes in archaebacterial genomes was resolved. The analyses were performed on *H. marismortui* (Arndt et al., 1990), *H. halobium* (Itoh, 1988; Mankin, 1989), *H. cutirubrum* (Shimmin et al., 1989), *Sulfolobus acidocaldarius* (Shimmin et al., 1989), and *M. vannielii* (Auer et al., 1989; Lechner et al., 1989; Köpke and Wittmann-Liebold, 1989; Köpke et al., 1989; Auer and Köpke, unpublished results). In our laboratory, the genes of the *rif* operon and of the S10 operon in *M. vannielii* were analyzed. Furthermore, genes of the S10 operon, the adjacent *spc* operon, and the *rif* operon (Arndt and Weigel, 1990) were localized in *H. marismortui* mainly by using oligonucleotide probes based on partial amino acid sequence information of the corresponding proteins for hybridization, whereas other groups used hybridization with heterologous probes or immunological screening of an expression library. The data obtained allow the following conclusions.

(i) The overall ribosomal gene organization is similar for all archaebacterial genomes studied so far, although they derive from different groups of the archaebacteria: the halobacteria, methanogenes, and thermoacidophiles.

(ii) Most ribosomal genes in these archaebacteria are clustered in transcription units similar to that of the *E. coli* genome. An operon was identified which carries the genes for the ribosomal proteins L11, L1, L10, and L12, similar to the *rif* operon of *E. coli* (Fig. 8). The promoter regions are located at different positions compared with the *E. coli* genome or even the various other archaebacterial gene clusters. The results are similar for *H. halobium* (Itoh, 1988), *H. cutirubrum* (Shimmin et al., 1989), *M. vannielii* (Köpke et al., 1989; Baier et al., in press), and *S. acidocaldarius* (Shimmin et al., 1989). Gene arrangements similar to those of the *E. coli str*, S10, and *spc* operons are found for both the *H. marismortui* and *M. vannielii* genomes. The gene orders within the S10 operon and the adjacent *spc* region of the archaebacterial clusters are given in Fig. 9.

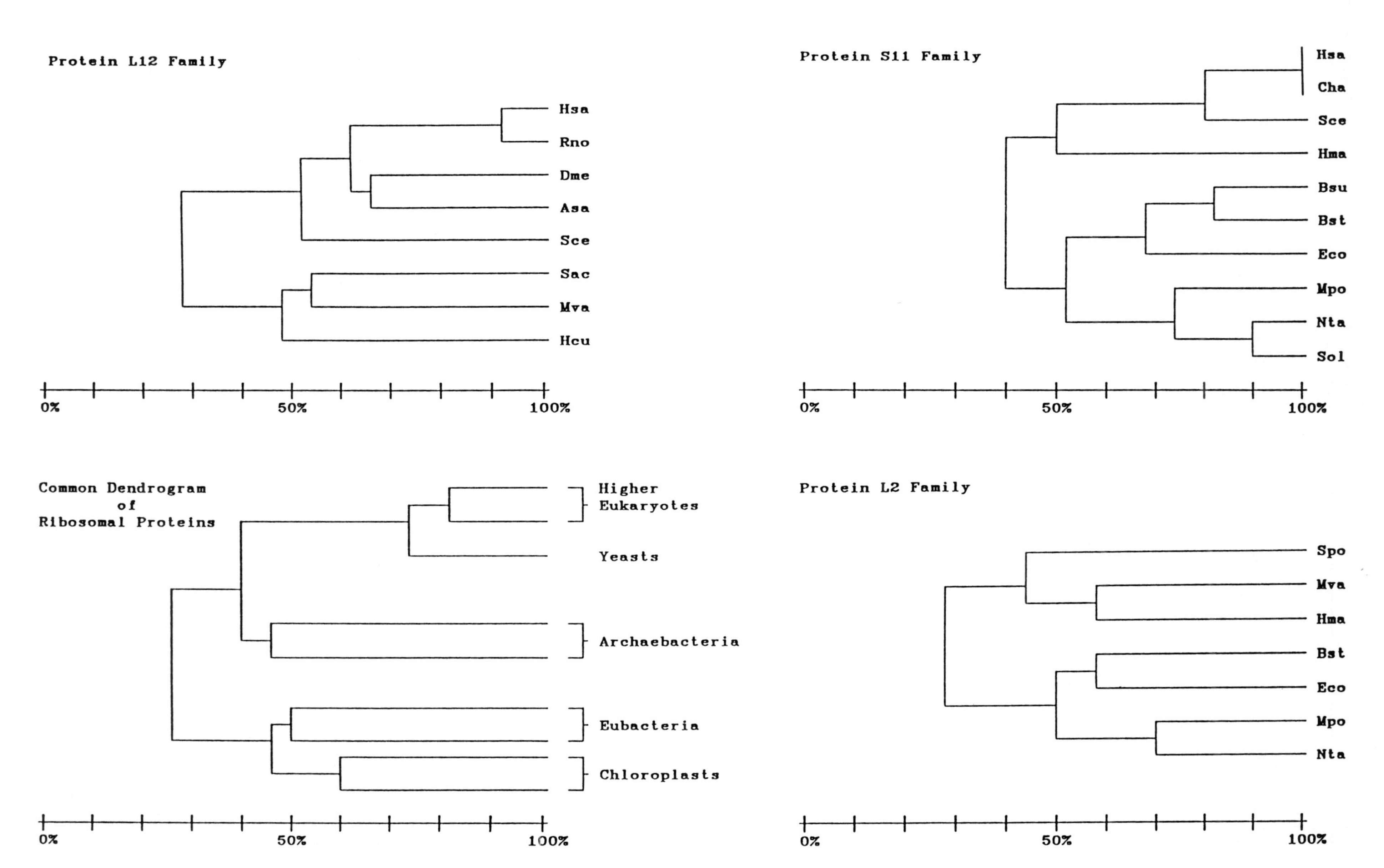

Protein L12 Family
Hsa
Rno
Dme
Asa
Sce
Sac
Mva
Hcu
0%
50%
100%
Protein S11 Family
Hsa
Cha
Sce
Hma
Bsu
Bst
Eco
Mpo
Nta
Sol
0%
50%
100%
Common Dendrogram
of
Ribosomal Proteins
Higher
Eukaryotes
Yeasts
Archaebacteria
Eubacteria
Chloroplasts
0%
50%
100%
Protein L2 Family
Spo
Mva
Hma
Bst
Eco
Mpo
Nta
0%
50%
100%

(iii) Some genes that are located within a given cluster of *E. coli*, such as the genes coding for S10 and L16, are missing in the equivalent archaebacterial operon. The gene for S10 is transferred into the adjacent gene cluster with the gene order ORF1 ORF2 S12 S7 EF-2 EF-1α S10 (Lechner et al., 1989). Furthermore, there are some additional open reading frames which code for proteins sequence-related to eucaryotic ribosomal proteins. In *H. marismortui*, the S10 gene is also missing and the S10-like operon starts with a promoter sequence prior to two new open reading frames that have no similarities to any other known ribosomal protein and are located 5′ to the L3 gene (Arndt et al., 1990). Although the gene located at the 3′ end of the L3 gene codes for protein L4 in *E. coli*, only weak amino acid sequence similarities to the *E. coli* counterpart were found for the deduced sequence in *H. marismortui* and in *M. vannielii* (Arndt et al., 1990; A. K. E. Köpke, unpublished results). In the *M. vannielii spc*-like operon, the order of the genes is similar to that of the *E. coli* cluster, but three new open reading frames are included (ORFa and -b between the genes for proteins L29 and S17) (see Matheson et al., this volume). In the case of the *H. marismortui* cluster, only one open reading frame ORF3 flanks the genes for L29 and S17. In addition, the S10 and *str* operons lie closer together in both archaebacterial genomes (Auer et al., 1989; E. Arndt, unpublished results). These results demonstrate clearly that a quite similar genome organization for ribosomal protein genes is present in eubacteria and in archaebacteria, i.e., that the overall feature of these genomes is conserved.

(iv) The various archaebacterial genomes differ considerably in the G+C content of their DNA (the halophilic DNA is G+C rich, and the methanogenic DNA is A+T rich). Accordingly, *H. marismortui* ribosomal protein genes predominantly use codons with G and C, whereas those from *M. vannielii* contain preferably A+T-rich codons. Interestingly, these differences at the nucleotide level are not reflected at the amino acid level, since the primary structures of the ribosomal proteins are strongly conserved within the archaebacterial kingdom. These results show that for structural and evolutionary studies of ribosomal proteins, a comparison at the amino acid level is more suitable than a comparison of nucleotide sequences.

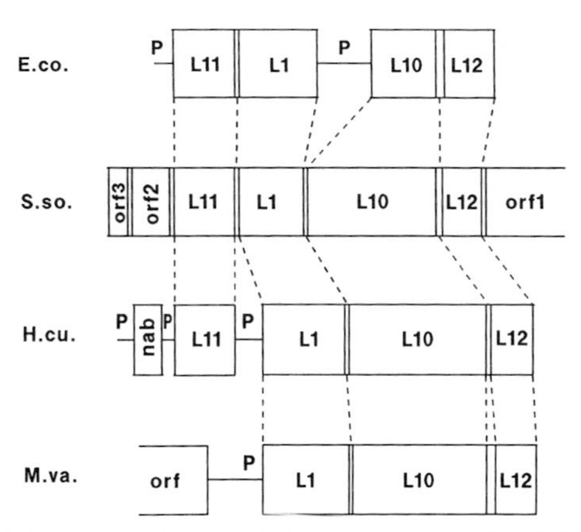

Figure 8. Gene organization of the *rif* operons in *E. coli* (E.co.), *S. solfataricus* (S.so.), *H. cutirubrum* (H.cu.), and *M. vannielii* (M.va.). P, Promoter; nab, nucleic acid binding protein.

(v) Three promoterlike structures were found upstream of the two open reading frames in the S10 operon of *H. marismortui* (Arndt et al., 1990). The start of the mRNA was determined by S1 nuclease mapping. The putative promoter regions are similar to other promoter sequences in archaebacteria, but no conserved −10 and −35 regions, present in most eubacterial promoters, were found. Instead, an A+T-rich region spaced by 21 to 32 base pairs from the transcriptional start was obtained which resembles the TATA box in eucaryotes (Thomm et al., 1989).

GENERAL CONCLUSION

Ribosomal protein sequences have evolved at different rates, depending on the individual protein and its function. Evolutionary trees were constructed for the different ribosomal protein families by pairwise sequence comparisons of the homologous ribosomal proteins from organisms of the different kingdoms. They enabled us to construct a preliminary common phylogenetic tree. Data for the different ribosomal protein families can be summarized as follows.

(i) The ribosomal proteins from gram-positive and gram-negative eubacteria are closely sequence

Figure 7. Phylogenetic dendrograms for ribosomal protein families. Dendrograms for the ribosomal protein families S11, L2, and L12 are shown, and a common tree comprising data of several investigated protein families is presented. The degree of conservation between two proteins of various organisms is demonstrated by the position of their branch points (see scale). Although the trees are based on quantitative data, they show only the relative branching orders (see text). Designations: Eco, *Escherichia coli*; Bsu, *Bacillus subtilis*; Bst, *Bacillus stearothermophilus*; Mpo, *Marchantia polymorpha* chloroplasts; Nta, *Nicotiana tabacum* chloroplasts; Sol, *Spinacia oleracea*; Hma, *Halobacterium marismortui*; Hcu, *Halobacterium cutirubrum*; Mva, *Methanococcus vannielii*; Sac, *Sulfolobus acidocaldarius*; Sce, *Saccharomyces cerevisiae*; Spo, *Schizosaccharomyces pombe*; Asa, *Artemia salina*; Dme, *Drosophila melanogaster*; Rno, *Rattus norvegicus*; Cha, Chinese hamster; Hsa, *Homo sapiens*.

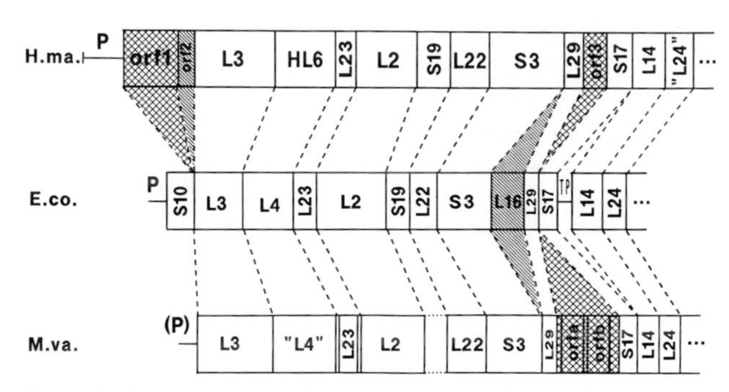

Figure 9. Organization of the S10 operons and the adjacent genes of the *spc* operons in *H. marismortui* (H.ma.), *E. coli* (E.co.), and *M. vannielii* (M.va.). T, Terminator; P, promoter.

related to each other, although at different extents for the various proteins.

(ii) Proteins of chloroplast ribosomes are closely related to proteins of eubacterial ribosomes.

(iii) The archaebacterial ribosomal proteins form a monophyletic group that differs from that of eubacteria and eucaryotes. *M. vannielii* has ribosomal proteins that are usually more closely related to eucaryotic than to eubacterial proteins. On the other hand, *H. marismortui* ribosomes contain several proteins that are as much related to eubacterial as to eucaryotic proteins. In addition, there are some archaebacterial ribosomal proteins that have no counterparts in the eubacterial organelle but might be related to some eucaryotic sequences.

(iv) Ribosomal proteins from lower and higher eucaryotes, e.g., yeasts and rat, are very similar to each other.

The results of comparisons of the archaebacterial ribosomal sequences underline the heterogeneous nature of archaebacteria. They establish that a gradual increase in sequence relationship toward the eucaryotic sequences occurred when passing from the halobacteria (most eubacterial like) through the methanogens to the thermoacidophiles (less eubacterial like).

On the other hand, the arrangement of the ribosomal protein genes in the archaebacteria *H. marismortui*, *H. halobium*, *M. vannielii*, and *S. acidocaldarius* follows the general organization of the ribosomal protein genes in transcriptional units as in eubacteria. No introns were found within these genes. In contrast, ribosomal proteins in eucaryotes are monocistronic. It remains to be seen why the gene organization of ribosomal proteins on the chromosome is similar in the eubacterial and archaebacterial kingdoms, whereas the amino acid sequences of many archaebacterial proteins are more sequence related to their eucaryotic counterparts.

REFERENCES

Abraham, P. R., J. van 't Riet, and H. A. Raué. 1989. A putative mitochondrial ribosomal protein gene cluster on chromosome XI of *Saccharomyces cerevisiae*, p. 47. *In Yeast Genetics and Molecular Biology*. Meeting Atlanta, Ga.

Arndt, E., G. Breithaupt, and M. Kimura. 1986. The complete amino acid sequence of ribosomal protein H-S11 from the archaebacterium *Halobacterium marismortui*. *FEBS Lett.* **194:** 227–234.

Arndt, E., W. Krömer, and T. Hatakeyama. 1990. Organization and nucleotide sequence of a gene cluster coding for eight ribosomal proteins in the archaebacterium *Halobacterium marismortui*. *J. Biol. Chem.* **265:**3034–3039.

Arndt, E., and C. Weigel. 1990. Nucleotide sequence of the genes encoding the L11, L1, L10 and L12 equivalent ribosomal proteins from the archaebacterium *Halobacterium marismortui*. *Nucleic Acids Res.* **18:**1285.

Auer, J., K. Lechner, and A. Böck. 1989. Gene organization and structure of two transcriptional units from *Methanococcus* coding for ribosomal proteins and elongation factors. *Can. J. Microbiol.* **35:**200–204.

Baier, G., O. Hohenwarter, C. Hofbauer, H. Hummel, M. Stöffler-Meilicke, and G. Stöffler. *J. Biol. Chem.*, in press.

Bland, M. M., C. S. Levings III, and D. F. Matzinger. 1986. The tobacco mitochondrial ATPase subunit ζ gene is clearly limited to an open reading frame for a ribosomal protein. *Mol. Gen. Genet.* **204:**8–16.

Bonen, L. 1987. The mitochondrial S13 ribosomal protein gene is silent in wheat embryos and seedlings. *Nucleic Acids Res.* **15:**10393–10404.

Brimacombe, R., J. Atmadja, A. Kyriatsoulis, and W. Stiege. 1986. RNA structure and RNA-protein neighborhoods in the ribosome, p. 184–202. *In* B. Hardesty and G. Kramer (ed.), *Structure, Function, and Genetics of Ribosomes*. Springer-Verlag, New York.

Brockmöller, J., and R. M. Kamp. 1988. The crosslinked amino acids in the protein pair S13-S19 and sequence analysis of protein S13 of *Bacillus stearothermophilus* ribosomes. *Biochemistry* **27:**3372–3381.

Burke, J. M., and U. L. RajBhandary. 1982. Intron within the large rRNA gene of *N. crassa* mitochondria: a long open reading frame and a consensus sequence possibly important in splicing. *Cell* **31:**509–520.

Fearon, K., and T. L. Mason. 1988. Structure and regulation of a nuclear gene in *Saccharomyces cerevisiae* that specifies MRP7, a protein of the large subunit of the mitochondrial ribosome. *Mol. Cell. Biol.* **8:**3636–3646.

Garland, W. G., K. A. Louie, A. T. Matheson, and A. Liljas. 1987. The complete amino acid sequence of the ribosomal 'A' protein (L12) from *Bacillus stearothermophilus*. *FEBS Lett.* **220:**43–46.

Gatermann, K. B., C. Teletski, T. Gross, and N. F. Käufer. 1989. A ribosomal protein family from *Schizosaccharomyces pombe* consisting of three active members. *Curr. Genet.* **16:**361–367.

Giri, L., W. E. Hill, H. G. Wittmann, and B. Wittmann-Liebold. 1984. Ribosomal proteins: their structure and spatial arrangement in prokaryotic ribosomes. *Adv. Protein Chem.* **36:**1–77.

Graack, H.-R., L. Grohmann, and T. Choli. 1988. Mitochondrial ribosomes of yeast: isolation of individual proteins and N-terminal sequencing. *FEBS Lett.* **242:**4–8.

Grohmann, L., H.-R. Graack, and M. Kitakawa. 1989. Molecular cloning of the nuclear gene for mitochondrial ribosomal protein YmL31 from *Saccharomyces cerevisiae*. *Eur. J. Biochem.* **183:** 155–160.

Grohmann, L., H.-R. Graack, and A. R. Subramanian. Ribosomal RNAs, and proteins and their genes. *In* I. K. Vasil (ed.), *Cell*

Culture and Somatic Cell Genetics of Plants, vol. 8. *The Molecular Biology of Mitochondria*. Academic Press, Inc., New York, in press.

Gualberto, J. M., H. Wintz, J.-H. Weil, and J.-M. Grienenberger. 1988. The genes coding for subunit 3 of NADH dehydrogenase and for ribosomal protein S12 are present in the wheat and maize mitochondrial genomes and are co-transcribed. *Mol. Gen. Genet.* **215**:118–127.

Hatakeyama, T., T. Hatakeyama, and M. Kimura. 1988. The primary structures of ribosomal proteins L16, L23 and L33 from the archaebacterium *Halobacterium marismortui*. *FEBS Lett.* **240**:21–28.

Hatakeyama, T., F. Kaufmann, B. Schroeter, and T. Hatakeyama. 1989. Primary structures of five ribosomal proteins from the archaebacterium *Halobacterium marismortui* and their structural relationships to eubacterial and eukaryotic ribosomal proteins. *Eur. J. Biochem.* **185**:685–693.

Hatakeyama, T., and M. Kimura. 1988. Complete amino acid sequences of the ribosomal proteins L25, L29 and L31 from the archaebacterium *Halobacterium marismortui*. *Eur. J. Biochem.* **172**:703–711.

Hirano, H., K. Eckart, M. Kimura, and B. Wittmann-Liebold. 1987. Semi-preparative HPLC purification of ribosomal proteins from *Bacillus stearothermophilus* and sequence determination of the highly conserved protein L19. *Eur. J. Biochem.* **170**:149–157.

Hudspeth, M. E. S., W. M. Ainley, D. S. Shumard, K. A. Butow, and L. C. Grossman. 1982. Location and structure of the *var1* gene on yeast mitochondrial DNA: nucleotide sequence of the 40.0 allele. *Cell* **30**:617–626.

Isono, K., and S. Isono. 1976. Lack of ribosomal protein S1 in *Bacillus stearothermophilus*. *Proc. Natl. Acad. Sci. USA* **73**: 767–770.

Itoh, T. 1988. Complete nucleotide sequence of the ribosomal 'A' protein operon from the archaebacterium, *Halobacterium halobium*. *Eur. J. Biochem.* **176**:297–303.

Kamp, R. M., and B. Wittmann-Liebold. 1988. Ribosomal proteins from archaebacteria: high performance liquid chromatographic purification for microsequence analysis. *Methods Enzymol.* **164**:542–571.

Kimura, J., E. Arndt, and M. Kimura. 1987. Primary structures of three highly acidic ribosomal proteins S6, S12 and S15 from the archaebacterium *Halobacterium marismortui*. *FEBS Lett.* **224**: 65–70.

Kimura, J., and M. Kimura. 1987a. The primary structures of ribosomal proteins S14 and S16 from the archaebacterium *Halobacterium marismortui*. *J. Biol. Chem.* **262**:12150–12157.

Kimura, J., and M. Kimura. 1987b. The complete amino acid sequences of the 5S rRNA binding proteins L5 and L18 from the moderate thermophile *Bacillus stearothermophilus* ribosome. *FEBS Lett.* **210**:85–90.

Kimura, M. 1984. Proteins of the *Bacillus stearothermophilus* ribosome. *J. Biol. Chem.* **259**:1051–1055.

Kimura, M., E. Arndt, T. Hatakeyama, T. Hatakeyama, and J. Kimura. 1989. Ribosomal proteins in halobacteria. *Can. J. Microbiol.* **35**:195–199.

Kimura, M., and C. K. Chow. 1984. The complete amino acid sequences of ribosomal proteins L17, L27, and S9 from *Bacillus stearothermophilus*. *Eur. J. Biochem.* **139**:225–234.

Kimura, M., J. Dijk, and I. Heiland. 1980. The primary structure of protein BL17 isolated from the large subunit of the *Bacillus stearothermophilus* ribosome. *FEBS Lett.* **121**:323–326.

Kimura, M., and J. Kimura. 1987c. The complete amino acid sequence of ribosomal protein S12 from *Bacillus stearothermophilus*. *FEBS Lett.* **210**:91–96.

Kimura, M., J. Kimura, and K. Ashman. 1985a. The complete primary structure of ribosomal proteins L1, L14, L15, L23, L24 and L29 from *Bacillus stearothermophilus*. *Eur. J. Biochem.* **150**:491–497.

Kimura, M., J. Kimura, and T. Hatakeyama. 1988. Amino acid sequences of ribosomal proteins S11 from *Bacillus stearothermophilus* and S19 from *Halobacterium marismortui*. *FEBS Lett.* **240**:15–20.

Kimura, M., J. Kimura, and K. Watanabe. 1985b. The primary structure of ribosomal protein L2 from *Bacillus stearothermophilus*. *Eur. J. Biochem.* **153**:289–297.

Kimura, M., and G. Langner. 1984. The complete amino acid sequence of the ribosomal protein HS3 from *Halobacterium marismortui*, an archaebacterium. *FEBS Lett.* **175**:213–218.

Kimura, M., N. Rawlings, and K. Appelt. 1981. The amino acid sequence of protein BL10 from the 50S subunit of the *Bacillus stearothermophilus* ribosome. *FEBS Lett.* **136**:58–64.

Kitakawa, M., L. Grohmann, H.-R. Graack, and K. Isono. 1990. Cloning and characterization of nuclear genes for two mitochondrial ribosomal proteins in *Saccharomyces cerevisiae*. *Nucleic Acids Res.* **18**:1521–1529.

Köpke, A. K. E., G. Baier, and B. Wittmann-Liebold. 1989. An archaebacterial gene from *Methanococcus vannielii* encoding a protein homologous to the ribosomal protein L10 family. *FEBS Lett.* **247**:167–172.

Köpke, A. K. E., C. Paulke, and H. Gewitz. 1990. Overexpression of the methanococcal ribosomal protein L12 in *Escherichia coli* and its incorporation into halobacterial 50S subunits yielding active ribosomes. *J. Biol. Chem.* **265**:6436–6440.

Köpke, A. K. E., and B. Wittmann-Liebold. 1988. Sequence of the gene for ribosomal protein L23 from the archaebacterium *Methanococcus vannielii*. *FEBS Lett.* **239**:313–318.

Köpke, A. K. E., and B. Wittmann-Liebold. 1989. Comparative studies of ribosomal proteins and their genes from *Methanococcus vannielii* and other organisms. *Can. J. Microbiol.* **35**:11–20.

Kreader, C. A., C. S. Langer, and J. E. Heckman. 1989. A mitochondrial protein from *Neurospora crassa* detected both on ribosomes and in membrane fractions. *J. Biol. Chem.* **264**: 317–327.

Krömer, W., T. Hatakeyama, and N. Kimura. Nucleotide sequence of *Bacillus stearothermophilus* ribosomal protein genes: part of the ribosomal S10-operon. *Biol. Chem. Hoppe-Seyler*, in press.

Kuiper, M. T. R., R. A. Akins, M. Holtrop, H. de Vries, and A. M. Lambowitz. 1988. Isolation and analysis of the *Neurospora crassa* Cyt-21 gene. *J. Biol. Chem.* **263**:2840–2847.

Lechner, K., G. Heller, and A. Böck. 1989. Organization and nucleotide sequence of a transcriptional unit of *Methanococcus vannielii* comprising genes for protein synthesis elongation factors and ribosomal proteins. *J. Mol. Evol.* **29**:20–27.

Mankin, A. S. 1989. The nucleotide sequence of the genes coding for the S19 and L22 equivalent ribosomal proteins from *Halobacterium halobium*. *FEBS Lett.* **246**:13–16.

Matsushita, Y., M. Kitakawa, and K. Isono. 1989. Cloning and analysis of the nuclear genes for two mitochondrial ribosomal proteins in yeast. *Mol. Gen. Genet.* **219**:119–124.

McDougall, J., T. Choli, U. Kapp, V. Kruft, and B. Wittmann-Liebold. 1989. The complete amino acid sequence of ribosomal protein S18 from the moderate thermophile *Bacillus stearothermophilus*. *FEBS Lett.* **245**:253–260.

Moore, P. B., M. Capel, M. Kjeldgaard, and D. M. Engelman. 1986. A 19 protein map of the 30S ribosomal subunit of *Escherichia coli*, p. 87–100. *In* B. Hardesty and G. Kramer (ed.), *Structure, Function, and Genetics of Ribosomes*. Springer-Verlag, New York.

Myers, A. M., M. D. Crivellone, and A. Tzagoloff. 1987. Assembly of the mitochondrial membrane system. *J. Biol. Chem.* **262**:

3388–3397.

Noller, H. F., M. Asire, A. Barta, S. Douthwaite, T. Goldstein, R. R. Gutell, D. Moazed, J. Normanly, J. B. Prince, S. Stern, K. Triman, S. Turner, B. Van Stolk, V. Wheaton, B. Weiser, and C. R. Woese. 1986. Studies on the structure and function of ribosomal RNA, p. 143–163. *In* B. Hardesty and G. Kramer (ed.), *Structure, Function, and Genetics of Ribosomes.* Springer-Verlag, New York.

Nowotny, V., R. P. May, and K. H. Nierhaus. 1986. Neutron-scattering analysis of structural and functional aspects of the ribosome: the strategy of the glassy ribosome, p. 101–111. *In* B. Hardesty and G. Kramer (ed.), *Structure, Function, and Genetics of Ribosomes.* Springer-Verlag, New York.

Partaledis, J. A., and T. L. Mason. 1988. Structure and regulation of a nuclear gene in *Saccharomyces cerevisiae* that specifies MRP13, a protein of the small subunit of the mitochondrial ribosome. *Mol. Cell. Biol.* 8:3647–3660.

Planta, R. J., W. H. Mager, R. J. Leer, L. P. Woudt, H. A. Raué, and T. T. A. L. El-Baradi. 1986. Structure and expression of ribosomal protein genes in yeast, p. 699–718. *In* B. Hardesty and G. Kramer (ed.), *Structure, Function, and Genetics of Ribosomes.* Springer-Verlag, New York.

Pohl, T., and B. Wittmann-Liebold. 1988. Identification of a cross-link in the *Escherichia coli* ribosomal protein pair S13-S19 at the amino acid level. *J. Biol. Chem.* **263**:4293–4301.

Pon, C. L., M. Brombach, S. Thamm, and C. O. Gualerzi. 1989. Cloning and characterization of a gene cluster from *Bacillus stearothermophilus* comprising *infC*, *rpmI* and *rplT*. *Mol. Gen. Genet.* **218**:355–357.

Pritchard, A. E., S. E. Venuti, M. A. Ghalambor, C. L. Sable, and D. J. Cummings. 1989. An unusual region of *Paramecium* mitochondrial DNA containing chloroplast-like genes. *Gene* 78:121–134.

Ramirez, C., L. C. Shimmin, C. H. Newton, A. T. Matheson, and P. P. Dennis. 1989. Structure and evolution of the L11, L1, L10, and L12 equivalent ribosomal proteins in eubacteria, archaebacteria, and eukaryotes. *Can. J. Microbiol.* **35**:234–244.

Rich, B. E., and J. A. Steitz. 1987. Human acidic ribosomal phosphoproteins P0, P1, and P2: analysis of cDNA clones, in vitro synthesis, and assembly. *Mol. Cell. Biol.* 7:4065–4074.

Royden, C. S., V. Pirrotta, and L. Y. Jan. 1987. The *tko* locus, site of a behavioral mutation in *D. melanogaster*, codes for a protein homologous to prokaryotic ribosomal protein S12. *Cell* **51**: 165–173.

Schüler, D., and R. Brimacombe. 1988. The *Escherichia coli* 30S ribosomal subunit; an optimised three-dimensional fit between the ribosomal proteins and the 16S RNA. *EMBO J.* 7:1509–1513.

Schuster, W., and A. Brennicke. 1987. Plastid, nuclear and reverse transcriptase sequences in the mitochondrial genome of *Oenothera*: is genetic information transferred between organelles via RNA? *EMBO J.* 6:2857–2863.

Shimmin, L. C., C. H. Newton, C. Ramirez, J. Yee, W. L. Downing, A. Louie, A. T. Matheson, and P. Dennis. 1989. Organization of genes encoding the L11, L1, L10, and L12 equivalent ribosomal proteins in eubacteria, archaebacteria, and eukaryotes. *Can. J. Microbiol.* **35**:164–170.

Spiridonova, V. A., A. S. Akhmanova, V. K. Kagramanova, A. K. E. Köpke, and A. S. Mankin. 1989. Ribosomal protein gene cluster of *Halobacterium halobium*: nucleotide sequence of the genes coding for S3 and L29 equivalent ribosomal proteins. *Can. J. Microbiol.* **35**:153–159.

Stöffler, G., and M. Stöffler-Meilicke. 1986. Immuno electron microscopy on *Escherichia coli* ribosomes, p. 28–46. *In* B. Hardesty and G. Kramer (ed.), *Structure, Function, and Genetics of Ribosomes.* Springer-Verlag, New York.

Strobel, O., A. K. E. Köpke, R. M. Kamp, A. Böck, and B. Wittmann-Liebold. 1988. Primary structure of the archaebacterial *Methanococcus vannielii* ribosomal protein L12. *J. Biol. Chem.* **263**:6538–6546.

Suryanarayana, T., and A. R. Subramanian. 1983. An essential function of ribosomal protein S1 in messenger ribonucleic acid translation. *Biochemistry* **22**:2715–2719.

Suyama, Y., and F. Jenney. 1989. The $tRNA^{Glu}$ (anticodon TTU) gene and its upstream sequence coding for a homolog of the *E. coli* large ribosome-subunit protein L14 in the *Tetrahymena* mitochondrial genome. *Nucleic Acids Res.* **17**:803.

Tanaka, I., M. Kimura, J. Kimura, and J. Dijk. 1984. The amino acid sequence of two small ribosomal proteins from *Bacillus stearothermophilus*. *FEBS Lett.* **166**:343–346.

Thomm, M., G. Wich, J. W. Brown, G. Frey, B. A. Sherf, and G. S. Beckler. 1989. An archaebacterial promoter sequence assigned by RNA polymerase binding experiments. *Can. J. Microbiol.* 35:30–35.

Wada, A., and T. Sako. 1987. Primary structures of and gene for new ribosomal proteins A and B in *Escherichia coli*. *J. Biochem.* **101**:817–820.

Wahleithner, J. A., and D. R. Wolstenholme. 1988. Ribosomal protein S14 genes in broad bean mitochondrial DNA. *Nucleic Acids Res.* **16**:6897–6912.

Walleczek, J., D. Schüler, M. Stöffler-Meilicke, R. Brimacombe, and G. Stöffler. 1988. A model for the spatial arrangement of the proteins in the large subunit of the *Escherichia coli* ribosome. *EMBO J.* 7:3571–3576.

Warner, J. R. 1989. Synthesis of ribosomes in *Saccharomyces cerevisiae*. *Microbiol. Rev.* **53**:256–271.

Wittmann, H. G. 1986. Structure of ribosomes, p. 1–27. *In* B. Hardesty and G. Kramer (ed.), *Structure, Function, and Genetics of Ribosomes.* Springer-Verlag, New York.

Wittmann-Liebold, B. 1986. Ribosomal proteins: their structure and evolution, p. 326–361. *In* B. Hardesty and G. Kramer (ed.), *Structure, Function, and Genetics of Ribosomes.* Springer-Verlag, New York.

Wittmann-Liebold, B. 1988. Structure and evolution of ribosomal proteins and their genes. *Endocytobiosis Cell Res.* 5:259–285.

Wittmann-Liebold, B., and K. Ashman. 1985. On-line detection of amino acid derivatives released by automatic Edman degradation of polypeptides, p. 303–327. *In* H. Tschesche (ed.), *Modern Methods in Protein Chemistry.* Walter de Gruyter, Berlin.

Wittmann-Liebold, B., K. Ashman, and M. Dzionara. 1984. On the statistical significance of homologous structures among the *Escherichia coli* ribosomal proteins. *Mol. Gen. Genet.* **196**: 439–448.

Woese, C. R., and G. E. Fox. 1977. The concept of cellular evolution. *Mol. Evol.* **10**:1–6.

Woese, C. R., and G. J. Olsen. 1986. Archaebacterial phylogeny: perspectives on the urkingdoms. *Syst. Appl. Microbiol.* 7:161–177.

Yonath, A., M. A. Saper, and H. G. Wittmann. 1986. Studies on crystals of intact bacterial ribosomal particles, p. 112–127. *In* B. Hardesty and G. Kramer (ed.), *Structure, Function, and Genetics of Ribosomes.* Springer-Verlag, New York.

Yonath, A., and H. G. Wittmann. 1989. Challenging the three-dimensional structure of ribosomes. *Trends Biochem. Sci.* **14**: 329–335.

Chapter 54

Structure and Evolution of Archaebacterial Ribosomal Proteins

A. T. MATHESON, J. AUER, C. RAMÍREZ, and A. BÖCK

During the past decade, the translational apparatus has proven to be a valuable phylogenetic probe in the study of molecular evolution. A comparative study of rRNA structure (Woese and Fox, 1977) led to the concept of three evolutionary lines of descent, the eubacteria, the archaebacteria, and the nuclear-cytoplasmic components of the eucaryotes. It is now generally accepted that the archaebacteria are unique and form a third kingdom which consists of three distinct phenotypes: the methanogens, the extreme halophiles, and the sulfur-dependent extreme thermophiles.

The archaebacterial ribosomes show several distinct features. Electron microscopy studies (Lake et al., 1982; Stöffler and Stöffler-Meilicke, 1986) indicate that the ribosomal subunits show structural features more complex than those exhibited by the 70S eubacterial ribosome and approaching those shown by the 80S eucaryotic counterpart. Physical chemical studies have indicated that the archaebacteria contain two classes of ribosomes on the basis of size and protein content (Cammarano et al., 1986). The extreme halophiles (Strøm and Visentin, 1973) and most of the methanogens (Schmid and Böck, 1982b) contain ribosomes similar in composition (number of rRNA and ribosomal protein [r-protein] components) to those of the eubacteria. The ribosomes of the sulfur-dependent extreme thermophiles (Londei et al., 1983; Schmid and Böck, 1982a) and *Methanococcus* spp. (Schmid and Böck, 1982b), however, contain an increased number of r-proteins, and the average molecular weight of these proteins is larger than that of the r-proteins in the eubacteria.

Compared with the large amount of data now available on the structure of the rRNA, information on the structure of the r-proteins in archaebacteria has been slower to appear. Early data on the structure of the r-protein L12 from the extreme halophile *Halobacterium cutirubrum* and the methanogen *Methanobacterium thermoautotrophicum* (Matheson et al., 1980) strongly supported the concept of the archaebacteria as a monophyletic kingdom. These data also indicated that the archaebacterial r-proteins showed structural features closer to those of the equivalent proteins in eucaryotes than to their counterparts in eubacteria.

More recently, the availability of highly improved protein microsequencing techniques (Walsh et al., 1988), as well as the use of gene cloning to obtain the primary structure of the archaebacterial r-proteins from the DNA sequence of their genes, has resulted in a substantial increase in the amount of structural data available on these proteins. The latter approach has also provided valuable information on the structure and organization of the archaebacterial r-protein genes. Not only are these genes arranged in operons as in eubacteria, but the order of the genes within the operons is also similar to that found in the eubacteria (Auer et al., 1989a; Auer et al., 1989b; Itoh, 1988; Köpke et al., 1989; Lechner et al., 1989; Shimmin et al., 1989a). This contrasts with the eucaryotic r-protein genes, which are scattered throughout the genome and are present as monocistronic units (Mager, 1988).

In this chapter, we review the structural data available on the archaebacterial r-proteins, mainly from the operons equivalent to the *rif*, *spc*, and *str* regions in *Escherichia coli*, and compare the structures of these r-proteins with those of the equivalent proteins in eucaryotes and eubacteria. From these data, we speculate on the evolution of the archaebacterial ribosome and, in more general terms, on the evolution of the archaebacterial cells.

A. T. Matheson and C. Ramírez ■ Department of Biochemistry and Microbiology, University of Victoria, Victoria, British Columbia V8W 3P6, Canada. **J. Auer and A. Böck** ■ Lehrstuhl für Mikrobiologie der Universität, Maria-Ward-Strasse 1a, D-8000 Munich 19, Federal Republic of Germany.

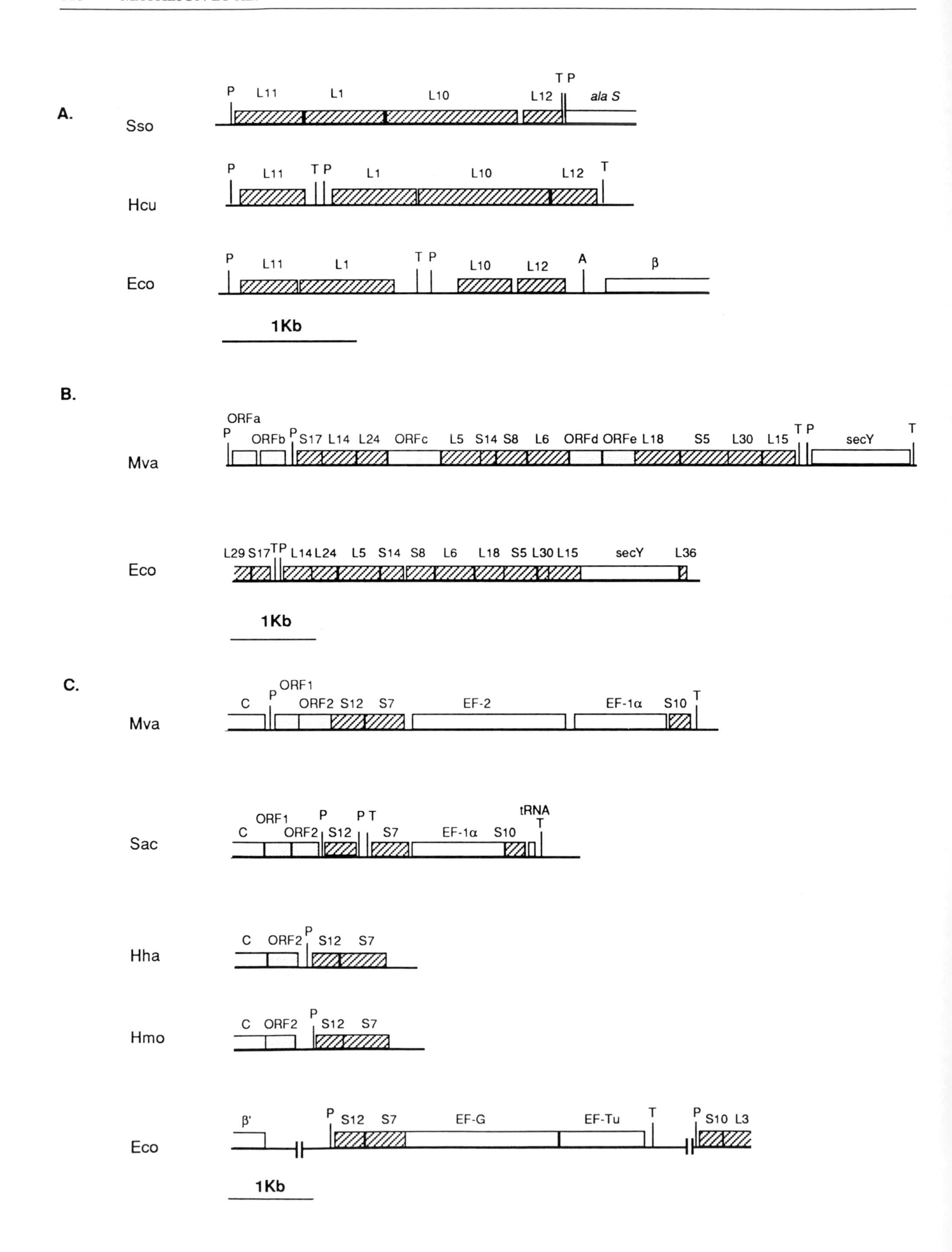

A.
Sso
P L11 L1 L10 L12 T P ala S
Hcu
P L11 T P L1 L10 L12 T
Eco
P L11 L1 T P L10 L12 A β
1Kb
B.
Mva
ORFa
P ORFb P S17 L14 L24 ORFc L5 S14 S8 L6 ORFd ORFe L18 S5 L30 L15 T P secY T
Eco
L29 S17 T P L14 L24 L5 S14 S8 L6 L18 S5 L30 L15 secY L36
1Kb
C.
Mva
C P ORF1 ORF2 S12 S7 EF-2 EF-1α S10 T
Sac
C ORF1 ORF2 P S12 P T S7 EF-1α S10 tRNA T
Hha
C ORF2 P S12 S7
Hmo
C ORF2 P S12 S7
Eco
β' P S12 S7 EF-G EF-Tu T P S10 L3
1Kb

STRUCTURES AND EVOLUTION OF THE *rif* OPERON PROTEINS

Organization of the r-Protein Genes

In *E. coli*, the *rif* region contains the genes that code for r-proteins L11, L1, L10, and L12 as well as the β and β' subunits of RNA polymerase (Post et al., 1979). These genes are organized as shown in Fig. 1.

In *E. coli*, the L11, L10, and L12 r-proteins form the L12 stalk in the large ribosomal subunit, a region involved with factor binding and with the GTPase center of the ribosome (Möller and Maassen, 1986). The L1 r-protein is located on a ridge in the lateral protuberance opposite the L12 stalk (Lake and Strycharz, 1981). This protein is involved in the interaction between the peptidyl-tRNA and the ribosome at the P and E sites and indirectly with the GTPase center (Moazed and Noller, 1989; Sander, 1983).

These proteins are located in similar regions in the archaebacterial 50S ribosomal subunit (Stöffler and Stöffler-Meilicke, 1986). In addition, the genes that code for these proteins in the genera *Sulfolobus* and *Halobacterium* are arranged in the same order as in *E. coli* (Fig. 1) (Post et al., 1979; Shimmin and Dennis, 1989; Shimmin et al., 1989a; C. Ramírez, L. C. Shimmin, P. A. Leggatt, and A. T. Matheson, unpublished data). However, in the archaebacteria the genes that code for the β and β' subunits of the RNA polymerase are not part of this cluster but are located instead close to the *str* operon (Lechner et al., 1989; Leffers et al., 1989; Pühler et al., 1989). The organization of these genes in transcriptional units is different from that found in *E. coli* (Fig. 1), and it varies between the genera *Sulfolobus* and *Halobacterium* (Shimmin and Dennis, 1989; Ramírez et al., unpublished data).

Structures of the r-Proteins

Table 1 gives the isoelectric points, molecular weights, and number of amino acid residues of the *rif* operon proteins from *Sulfolobus solfataricus*. The complete sequences of all of these proteins can be found in the Appendix (see p. 634). Figure 2 shows a summary of the percentage of identical amino acid residues between the *Sulfolobus* proteins and the homologous r-proteins from other archaebacteria, eubacteria, and eucaryotes.

In the remainder of this chapter, the following designations are used to indicate the sources of r-proteins: Ani, *Anacystis nidulans*; Asa, *Artemia salina*; Bst, *Bacillus stearothermophilus*; Egrch, *Euglena gracilis*, chloroplast; Eco, *Escherichia coli*; Hcu, *Halobacterium cutirubrum*; Hha, *Halobacterium halobium*; Hma, *Halobacterium marismortui*; Hmo, *Halococcus morrhuae*; Hsa, *Homo sapiens*; Mau, *Mesocricetus auratus*; Mca, *Mycoplasma capricolum*; Mlu, *Micrococcus luteus*; Mmu, *Mus musculus*; Mpoch, *Marchantia polymorpha*, chloroplast; Mra, *Mucor racemosus*; Mva, *Methanococcus vannielii*; Ntach, *Nicotiana tabacum*, chloroplast; Rno, *Rattus norvegicus*, liver; Sac, *Sulfolobus acidocaldarius*; Sce, *Saccharomyces cerevisiae*; Scemt, *Saccharomyces cerevisiae*, mitochondrion; Sso, *Sulfolobus solfataricus*; Tma, *Thermotoga maritima*; Tth, *Thermus thermophilus*; Zmach, *Zea mays*, chloroplast.

L11

Only the sequences of the Sso L11 and Hcu L11 proteins are currently available from archaebacterial sources (Ramírez et al., 1989; Shimmin and Dennis, 1989). The two archaebacterial proteins are slightly larger than the Eco L11 protein (Table 1) and show a higher sequence similarity to each other than to the Eco L11 protein (Fig. 2). Only the N-terminal sequence of the eucaryotic L11 protein is currently available (Otaka et al., 1984).

The L11 proteins are relatively rich in proline and glycine, and a considerable number of these residues are conserved between the two kingdoms. Since these residues play an important role in protein secondary structure, this conservation suggests that the secondary structure of the L11 proteins may be conserved between the two kingdoms.

L1

Three L1 proteins have been sequenced from the archaebacteria: Sso L1 (Ramírez et al., 1989), Hcu L1 (Shimmin and Dennis, 1989), and Hha L1 (Itoh,

Figure 1. Comparison of gene organization of the *rif*, *spc*, and *str* operons in archaebacteria with the related transcriptional units in *E. coli*. Gene sizes are drawn to scale. P and T, Sites for transcription initiation and termination (processing), respectively; A, transcriptional attenuation. (A) *rif* operon. *ala S*, Ala-tRNA synthetase gene; β, gene for the β subunit of RNA polymerase. Sso data are from Ramírez et al. (unpublished data); Hcu data are from Shimmin and Dennis (1989); Eco data are from Post et al. (1979). (B) *spc* operon. ORFa to -e, Genes that are not present in the *E. coli* operon. Mva data are from Auer et al. (1989a); Eco data are from Cerretti et al. (1983). (C) *str* operon. C, Gene for the C subunit of the archaebacterial RNA polymerase; β', gene for the β' subunit of the *E. coli* RNA polymerase; ORF1 and -2, genes that are not present in the eubacterial operon. Data for archaebacteria are from Auer et al. (1989a), Lechner et al. (1989), Leffers et al. (1989), Pühler et al. (1989), and Auer et al. (unpublished data); Eco data are from Post and Nomura (1980).

Table 1. Isoelectric points, molecular weights, and number of amino acid residues of the *rif* proteins from *S. solfataricus*

Gene product	pI[a]	M_r[a]	No. of amino acids		
			S. solfataricus	*E. coli*[b]	Eucaryotes
rif proteins					
L11	9.87	18,184	170	141	—[c]
L1	10.81	24,895	221	233	—[d]
L10	9.56	36,531	335	165	317 (Hsa P0)[e]
L12	4.74	11,139	105	120	115 (Hsa P2)[e]
Proteins upstream of the *rif* operon region					
L46	12.9	6,046	50		50 (Sce L46)[f]
LX	12.02	8,312	71		—[g]

[a] Calculated by using the PEP program from Bionet.
[b] Wittmann-Liebold, 1986.
[c] Equivalent protein exists but has not been sequenced (Juan-Vidales et al., 1983).
[d] Equivalent protein exists but has not been sequenced (Zimmermann et al., 1980).
[e] Rich and Steitz, 1987.
[f] Leer et al., 1985.
[g] Existence of an equivalent protein in eucaryotes remains to be shown.

1988). The proteins from the two extreme halophiles are identical except that Hha L1 has an Ala instead of Val in position 116. The archaebacterial proteins in this case are smaller than their eubacterial counterpart (Table 1). The three archaebacterial sequences show a higher sequence similarity to each other than to the equivalent proteins in eubacteria (Fig. 2). As was the case with the L11 r-proteins, considerable conservation of Pro and Gly residues has occurred in the L1 proteins between the two kingdoms. No sequence data are currently available on the eucaryotic L1.

Both L1 and L11 bind directly to the 23S rRNA (L1 to nucleotides 2100 to 2200 and L11 to nucleotides 1052 to 1112) (Gourse et al., 1981; Schmidt et al., 1981). The binding sites of these two proteins in the 23S rRNA have been conserved in the three kingdoms (Beauclerk et al., 1985; El-Baradi et al., 1987; Zimmermann et al., 1980). However, the rRNA-binding site on the proteins has yet to be determined. In this respect, it is interesting to note that there is a region in the L11 proteins that shows sequence similarity to a region in the L1 proteins (Fig. 3). Whether these regions might be involved with binding of the rRNA remains to be shown.

L10 and L12

A great deal of comparative structural information is available on L10 and L12. Comparison of the

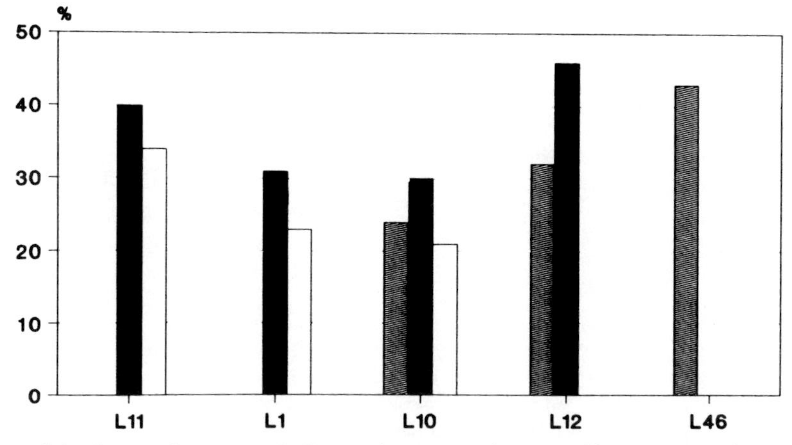

Figure 2. Comparison of the degree of sequence similarity of r-proteins from *S. solfataricus* (*E. coli* nomenclature except for proteins that have no eubacterial counterpart) to homologous r-proteins from eucaryotes (hatched columns), other archaebacteria (solid columns), and eubacteria (open columns). The degree of sequence similarity is given as the percentage of identical amino acid residues in the aligned sequences. References for the sequences used are given by Ramírez et al. (1989). Sequences were aligned by using the GENALIGN program from Bionet.

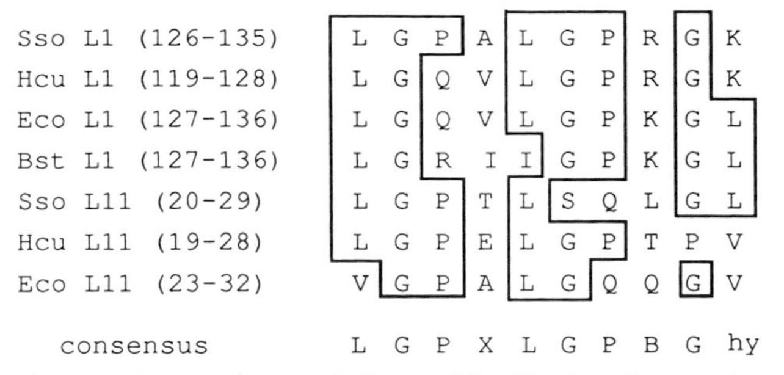

Figure 3. Conserved region in L11 and L1. Numbers in parentheses refer to the positions of these residues in each protein. Identical residues are boxed. A consensus sequence is shown at the bottom. X, Any amino acid; B, basic residue (R, K); Hy, hydrophobic.

structures of these two proteins within the three kingdoms reveals that they can be divided into two major subgroups: one formed by the archaebacteria and eucaryotes and the other formed by the eubacterial proteins (Ramírez et al., 1989; Shimmin et al., 1989b).

The L10 proteins from archaebacteria and eucaryotes are of similar size and are much larger than the *E. coli* protein (Table 1), which is missing several structural features present in the protein from the other two kingdoms (Ramírez et al., 1989; Shimmin et al., 1989b). The sequence similarity among the L10 proteins from the various kingdoms is shown in Fig. 2.

The L12 protein varies in size: from 99 to 114 residues in archaebacteria, from 105 to 116 residues in eucaryotes, and from 118 to 128 residues in eubacteria (Wittmann-Liebold, 1986). Considerable structural similarity is evident between the L12 proteins from archaebacteria and eucaryotes (Fig. 2). However, rearrangement of the eubacterial L12 protein is required in order to obtain maximum sequence similarity with the archaebacterial or eucaryotic proteins (Lin et al., 1982; Matheson, 1986). In addition, certain structural features common to the archaebacterial and eucaryotic proteins are absent in the eubacterial L12 (Ramírez et al., 1989; Shimmin et al., 1989b).

It is also evident that L10 and L12 share certain structural features. The L10 protein contains a module that is repeated three times in the archaebacteria and twice in the eucaryotes (Fig. 4). There is also evidence that the eubacterial L10 protein and the various L12 proteins may contain a single copy of this module (Ramírez et al., 1989; Shimmin et al., 1989b). In addition, there is a highly charged C-terminal region common to L10 and L12 in both archaebacteria and eucaryotes which is absent in the equivalent eubacterial proteins. It appears that about 75% of the L12 gene is duplicated in the L10 gene (Fig. 4).

The amino acid sequence of the common C-terminal region of archaebacterial and eucaryotic L10 and L12 proteins is shown in Fig. 5. Table 2 summarizes the amount of sequence similarity found in this C-terminal region. Of special note is the complete conservation of the last 31 amino acids in the C-terminal region of the protein and of the nucleotide sequence in the corresponding region of the genes for Sso L10 and L12 proteins. In contrast, while the human L10 and L12 proteins show almost complete conservation of the amino acid sequence, there is less conservation at the nucleotide level and a considerable number of the codons show changes in the wobble position. There is no obvious explanation for the lack of similar changes in the wobble position of the codons in these regions in *S. solfataricus*. A similar complete conservation of nucleotide structure has been noted (Reiter et al., 1987) in two viral proteins from the SSV1 virus that infects *Sulfolobus*

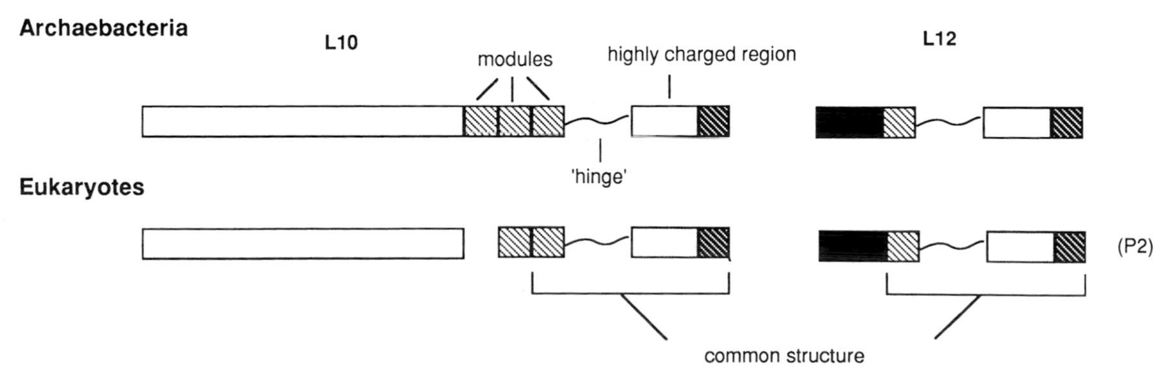

Figure 4. Structural summary of L10 and L12 of archaebacteria and eucaryotes. The archaebacterial L10 protein contains three copies of a 26-amino-acid-long module, whereas the equivalent eucaryotic protein contains only two copies. The C-terminal region of L10 contains a hinge region, a highly charged region followed by a neutral region as shown in Fig. 5. The L12 proteins appear to contain a copy of the module as well as the hinge region, the highly charged region, and the C-terminal neutral region. About 75% of the L12 protein is duplicated in the C-terminal portion of L10. (See Ramírez et al., 1989; Shimmin et al., 1989b).

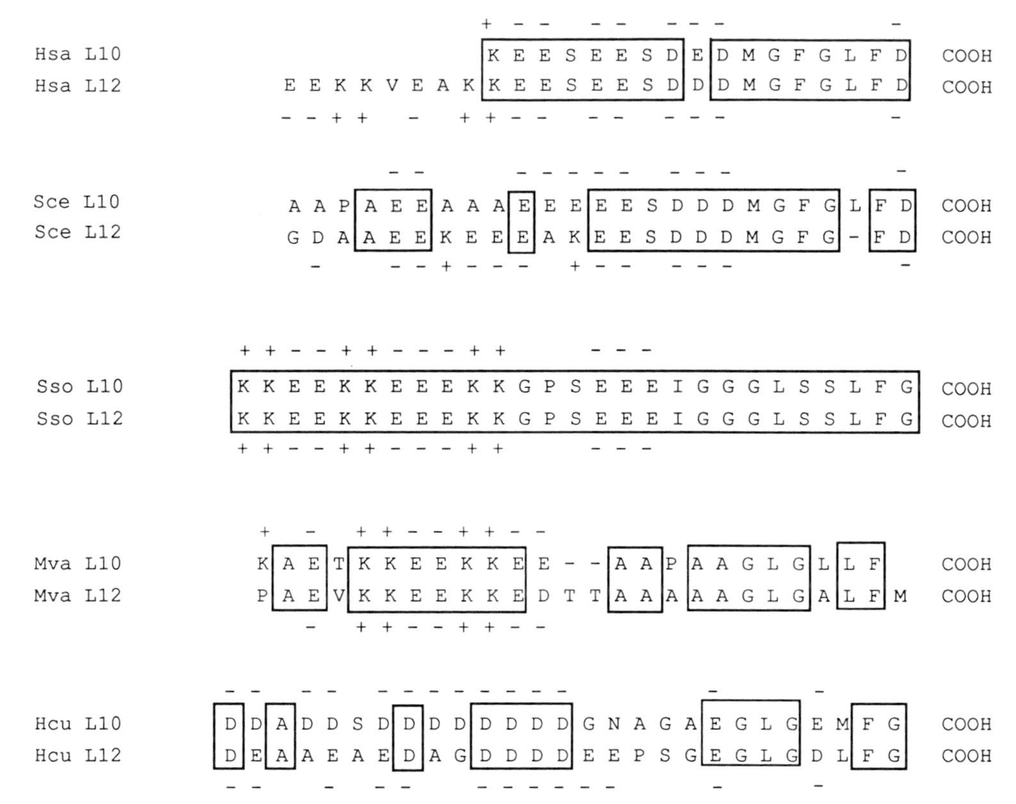

Figure 5. Comparison of the sequences of the C-terminal region of L10 and L12 from the archaebacteria and eucaryotes (see Köpke et al., 1989; Mitsui and Tsurugi, 1988a, 1988b; Ramírez et al., 1989; Rich and Steitz, 1987). Identical residues are boxed; acidic and basic residues are marked − and +, respectively.

sp. strain B12, in which 20 amino acids and the corresponding codons in the gene have been completely conserved.

In the L10 and L12 proteins from *M. vannielii* and from the extreme halophile, the amount of amino acid sequence conservation between the two proteins in the C-terminal region is less impressive, especially in *H. cutirubrum* (Köpke et al., 1989; Ramírez et al., 1989).

In *E. coli*, the N-terminal region of the L12 proteins binds to the single copy of the L10 protein (Gudkov et al., 1980; Schop and Maassen, 1982). It was postulated earlier (Liljas et al., 1986), on the basis of sequence and secondary structure predictions, that the C-terminal domain in archaebacteria might be equivalent to the N-terminal domain in *E. coli* and might therefore be the binding site for L10 in the archaebacteria. However, recent studies by M. Remacha, T. Nacanda, S. Zinker, M. D. Vilella, and J. P. G. Ballesta (personal communication) have shown that the N-terminal region rather than the C-terminal region of the eucaryotic L12 is involved

Table 2. Conservation of amino acid residues in the C-terminal regions of L10 and L12 in archaebacteria and eucaryotes, and the corresponding conservation of nucleotide structure in the equivalent region of the genes for these proteins

Source	Length (amino acids) of region	% Amino acid conservation	% Nucleotide conservation in codons where the amino acid has been conserved[a]	Reference
Archaebacteria				
S. solfataricus	31	100	100	Ramírez et al., 1989
M. vannielii	27 (1 gap)	70	88 (7)	Köpke et al., 1989
H. cutirubrum	43	40	90 (5)	Shimmin and Dennis, 1989
Eucaryotes				
H. sapiens	17	94	82 (9)	Rich and Steitz, 1987
S. cerevisiae	31 (1 gap)	92	87 (5)	Mitsui and Tsurugi, 1988a, 1988b

[a] Numbers in parentheses indicate number of changes observed in the wobble position.

Table 3. Isoelectric points, molecular weights, and number of amino acid residues of the *spc* proteins from *M. vannielii*

Gene product	pI	M_r	No. of amino acids		
			M. vannielii	*E. coli*[a]	Eucaryotes[b]
S17	7.8	12,082	109	83	158 (Rno S11)
L14	10.9	14,205	133	123	137 (Sce L17A)
L24	10.7	13,423	119	103	—[c]
ORFc	10.0	27,345	244		360[d] (Sce S6)
L5	10.2	10,293	180	178	174 (Sce L16)
S14	10.6	6,129	53	98	—
S8	9.7	14,335	130	129	130 (Sce S24)
L6	9.6	20,162	182	176	—
ORFd	12.3	15,524	135		135 (Mmu L32)
ORFe	11.2	17,286	149		196 (Rno L19)
L18	9.1	21,820	195	117	296 (Rno L5)
S5	10.0	24,359	225	166	—
L30	10.4	17,224	154	58	258 (Rno L7)
L15	10.4	15,951	143	144	149 (Sce L29)

[a] Wittmann-Liebold, 1986.
[b] The identity of the equivalent eucaryotic protein, shown in parentheses, was determined by computer analysis of the amino acid sequence data as described in Auer et al. (1989b).
[c] —, Existence of an equivalent protein in eucaryotes remains to be shown.
[d] The exact number of amino acid residues has not been determined.

with L10 binding. A more detailed study of the specific regions in the L12 proteins that interact with L10 is required to define the specific binding sites.

STRUCTURES OF THE *spc* OPERON PROTEINS

Organization of the r-Protein Genes

At min 72 on the *E. coli* chromosomal map, four r-protein operons are closely linked: the so-called *str*, S10, *spc*, and α operons (Nomura et al., 1984). The *spc* operon was so named because a mutation in the gene coding for r-protein S5 was found to confer resistance to the antibiotic spectinomycin (Nomura et al., 1977).

In *E. coli*, the *spc* region contains genes coding for r-proteins L14, L24, L5, S14, S8, L6, L18, S5, L30, L15, and L36 (X) and for SecY (PrlA) (Cerretti et al., 1983). The organization of these genes within the transcriptional unit is shown in Fig. 1.

On the chromosome of the archaebacterium *M. vannielii*, the genes homologous to those organized in the *spc* operon of *E. coli* were recently identified (Auer et al., 1989a; Auer et al., 1989b). The transcriptional units of *M. vannielii* and *E. coli* exhibit a striking similarity in both gene composition and gene order (Fig. 1). There are only three deviations: (i) three open reading frames (ORFc, -d, and -e) are quasi-"inserted" into the eubacterial-type gene organization; (ii) the transcriptional pattern is different in that it starts with the gene for protein S17, which in *E. coli* is part of the S10 operon, and that it terminates downstream of the gene for L15, whereas in *E. coli* the genes for SecY and L36 are part of this operon; and (iii) the gene for L36, which was previously called X (Markmann-Mulisch et al., 1987), is not present in this chromosomal region of *M. vannielii*.

Structures of the r-Proteins

The amino acid sequences of the r-proteins deduced from the DNA sequences are shown in the appendix. The proteins are small (their molecular

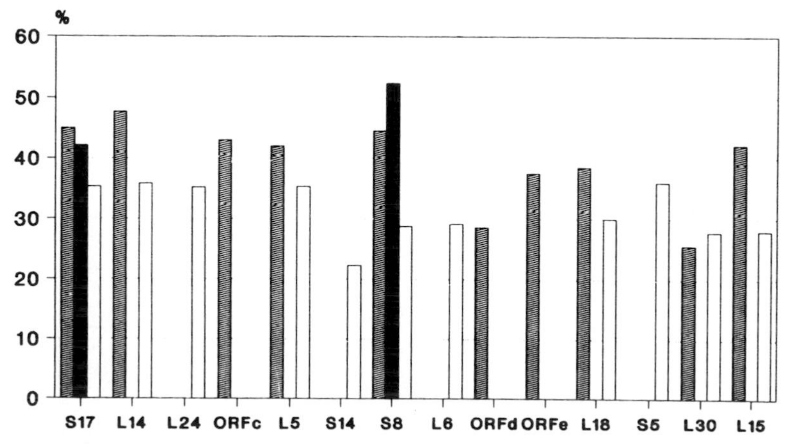

Figure 6. Comparison of the degree of sequence similarity of *spc* r-proteins from *M. vannielii* (*E. coli* nomenclature except for proteins that have no eubacterial counterpart) to homologous proteins from eucaryotes (hatched columns), *H. marismortui* (archaebacteria) (solid columns), and eubacteria (open columns), including chloroplasts. The degree of sequence similarity is given as the percentage of identical amino acid residues in the aligned sequences. The heights of the open columns represent the averages of sequence similarity of different eubacterial-type sequences. The sequences of homologous proteins were aligned pairwise, using the standard programs of the UWGCG package (Devereux et al., 1984). References for the sequences used are given by Auer et al. (1989b).

```
RnoL5    1  MGFVKVVKNK AYFKRYQVRF RRRREGKTDY YARKRLVIQD 40
SceL1a   1  .AFQKDAKSS AYSSRFQYPF RRRREGKTDY Y......... 30
MvaL18   1  .....(MI)M AQNAKHTVPF RRRREGKTDF ..RQRLGLLL 31
HcuL13   1  .......... ATGPRYKVPM RRRREVRTDY ..H....... 21
EcoL18   1  .......... ....MDKKSA RIRRATR... .ARRKLQELG 22

SceL16   1  .......... .......MST KAQNPMRDLK IEKLVLNISV GES 26
MvaL5    1  .......... ....MSFQEV WEKEPMKKPR IQKVTVNFGV GEA 29
HcuL19   1  .......... .....SETDS TDFHEMREPR IEKVVVHMGV GQG 28
EcoL5    1  MAKLHDYYKD EVVKKLMTEF NYNSVMQVPR VEKITLNMGV GEA 43
```

Figure 7. Alignment of amino acid sequences of *M. vannielii* proteins homologous to 5S rRNA-binding proteins from *E. coli*, *H. cutirubrum*, and eucaryotes.

weights vary within a range from 6,000 [Mva S14] to 27,000 [Mva ORFc]), and they are rich in basic amino acid residues, resulting in isoelectric points from 7.8 (Mva S17) to 12.3 (Mva ORFd). These data are summarized in Table 3.

As an overview, the percentage of identical amino acid residues between *M. vannielii* proteins and homologous r-proteins from eucaryotes, *H. marismortui*, and eubacteria is given in Fig. 6. In those instances where *M. vannielii* proteins could be compared with several r-proteins from eubacteria, the given value of similarity is the arithmetic average of all comparisons.

There is a considerable difference in the degree of similarity among homologous r-proteins. Some proteins, such as members of the L14 family, are well conserved, whereas others, such as members of the S14 family, are weakly conserved. This result might reflect the importance of a particular r-protein in the functioning of the ribosome. Comparative analyses of homologous r-proteins from *E. coli* and *B. stearothermophilus* showed that essential r-proteins are more conserved than those that are functionally less important (Wittmann-Liebold, 1986).

The average similarity of the *M. vannielii* proteins to all of their homologous proteins is 35%. The degree of similarity between *M. vannielii* proteins and eucaryotic proteins varies between 25.6% (Mva L30-Rno L7) and 47.7% (Mva L14-Sce L17A). The corresponding values for the comparison with eubacterial proteins range from 22.4% (Mva L30-Eco L30) to 42.7% (Mva L24-Eco L24).

A direct comparison of homologous r-proteins from the three kingdoms, which could be performed for proteins S17, L14, L5, S8, L18, L30, and L15, showed that except for L30, *M. vannielii* proteins are more closely related to their eucaryotic than to their eubacterial counterparts. From the available archaebacterial sequences, Hma S14 and S15 (Kimura and Kimura, 1987) showed a rather high degree of sequence similarity to Mva S17 (42%) and Mva S8 (52%), respectively. Interestingly, Mva S17 showed a higher degree of similarity to Rno S11 (45%) than to Hma S14 (42%).

The number of r-proteins constituting the 5S rRNA-protein complex is different in the three kingdoms: eucaryotic 5S complexes appear to contain only one r-protein; this was shown for yeast cells (Sce L1A [Nazar et al., 1979]) and for rat liver (Rno L5 [Chan et al., 1987]). Archaebacterial 5S complexes, as exemplified by *H. cutirubrum*, contain two proteins, L13 and L19 (Smith et al., 1978). Finally, eubacterial 5S complexes, as exemplified by *E. coli*, contain three r-proteins, L18, L5, and L25 (Erdmann, 1976; Garrett et al., 1981).

The N-terminal amino acid sequences of r-proteins Hcu L13 and Eco L18 and fragment CN1 of Sce L1a align very well with each other, as do Hcu L19 and Eco L5 sequences (Fig. 7). However, a specific relationship of fragment CN2 of Sce L1a with Hcu L19, Eco L5, or Eco L25 is less clear (Nazar et al., 1982). The complete amino acid sequences of Mva L18 and Mva L5 also align very well with the completely sequenced homologs from *E. coli* and the N-terminal sequences of Hcu L13 and L19, respectively (Fig. 7). Like its counterparts from *E. coli* and *H. cutirubrum*, Mva L18 shows a specific relationship with fragment CN1 of Sce L1a. However, Mva L5 exhibits a high degree of similarity (42%) to Sce L16. Thus, as discussed by Nazar et al. (1982), this result raises the possibility that some other r-pro-

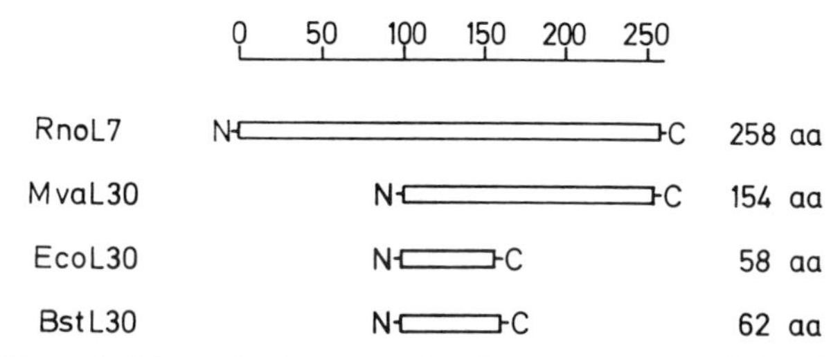

Figure 8. Schematic alignment showing an example of the size divergence of L30 homologs (*E. coli* nomenclature). Protein sizes are given as number of amino acid residues (aa). N and C represent the N and C termini, respectively.

tein(s) may interact with the 5S rRNA within the yeast ribosome.

Table 3 gives the size (i.e., number of amino acid residues) of the *M. vannielii spc* operon r-proteins in comparison with those from the *E. coli* and the eucaryotic homolog. In general, the *Methanococcus* proteins are larger than their homologs from eubacteria and smaller than their counterparts from the 80S ribosome. The size differences can be traced to internal deletions (insertions) or, quite frequently, to C- or N-terminal truncations (additions). A prominent example is r-protein L30, whose schematic alignment is presented in Fig. 8. The eubacterial proteins lack the 100 N-terminal amino acids as well as the 100 C-terminal amino acids, whereas the archaebacterial homologous protein only lacks the N-terminal domain. Interestingly, the N-terminal domain of the eucaryotic homolog (Rno L7) includes a fivefold repeated sequence of 12 amino acids in length. This domain is thought to be involved in the binding of the eucaryotic ribosome to the endoplasmic reticulum (Lin et al., 1987b).

STRUCTURES OF THE *str* OPERON PROTEINS

Organization of the r-Protein and Elongation Factor Genes

In *E. coli*, the genes for r-proteins S12 and S7 and for translation elongation factors EF-G and EF-Tu are organized in the so-called *str* operon (Fig. 1) (Post and Nomura, 1980). With few exceptions, this gene arrangement is well conserved among eubacteria (reviewed in Buttarelli et al., 1989; Lindahl and Zengel, 1986; Nomura et al., 1984; Ohama et al., 1987; Ohkubo et al., 1987).

Figure 1 gives a comparison of the *str* operon from *E. coli* with the equivalent chromosomal regions from *M. vannielii* (Lechner et al., 1989), from *S. acidocaldarius* (Pühler et al., 1989; Auer et al., unpublished results), and from two halophilic archaebacteria, *H. halobium* and *Halococcus morrhuae* (Leffers et al., 1989). The genes for r-proteins S12 and S7 and for elongation factors EF-G (EF-2) and EF-Tu (EF-1α) are present in the same order. However, there are some differences in the gene composition of the respective operons and in the transcriptional pattern.

In the *str* operon of *M. vannielii* there are two ORFs, ORF1 and -2, in front of the gene for r-protein S12, and at the 3′ end of the operon there is a gene coding for the archaebacterial counterpart to S10. These three genes and the genes for S12, S7, EF-2, and EF-1α are cotranscribed (Lechner et al., 1989). In the corresponding chromosomal region of *S. acidocaldarius*, the gene for EF-2 is not present and a gene for tRNA^{Ser}(GGA) is located just downstream of the gene for S10. In agreement with the situation in *M. vannielii*, there are also two ORFs upstream of the gene for r-protein S12, and the gene for r-protein S10 is located downstream of the gene for EF-1α. However, in contrast to the methanogenic transcriptional pattern, the *Sulfolobus* genes for ORF1 and ORF2 and for r-protein S12 are weakly cotranscribed with the genes for the subunits of the RNA polymerase; also, these genes have their own promoters (Pühler et al., 1989). In addition, the genes for S7, EF-1α, S10, and tRNA^{Ser} are cotranscribed in a different unit (Auer et al., unpublished results). Finally, in the extreme halophiles, the information on the *str* operon is limited to the genes for S12, S7, and one of the archaebacterial ORFs (ORF2) (Fig. 1). Again, the gene order is very similar to that of the other archaebacteria and eubacteria. In contrast to *M. vannielii* and *S. acidocaldarius*, the gene for ORF1 in the extreme halophile is missing in this region. ORF2 from extreme halophiles is cotranscribed with the genes for the subunits of the RNA polymerase (genes β'', β', A, and C), whereas the genes for r-proteins S12 and S7 are cotranscribed from a single promoter (Leffers et al., 1989). The overall high similarity of gene organization in this region from different archaebacteria raises the possibility that in the extreme halophiles, the genes for elongation factors lie downstream of the gene for r-protein S7.

There are two intriguing features of the archaebacterial gene organization when it is compared with that of a eubacterium, i.e., *E. coli*: (i) the archaebacterial genes homologous to those of the eubacterial *str* operon are closely linked to the genes encoding the subunits of the RNA polymerase, whereas in *E. coli* the genes for RNA polymerase subunits β and β' lie at a distance from the genes harbored by the *str* operon and are organized in the so-called *rif* operon together with the genes for r-proteins L10 and L12 (Nomura et al., 1977; Post et al., 1979); (ii) in the two archaebacteria in which the complete *str* operon has been studied, the gene for r-protein S10 is part of this operon, whereas in *E. coli* the gene for S10 is the promoter-proximal gene of the S10 operon, which is located about 15,000 base pairs downstream of the *str* operon.

Structures of the r-Proteins and Elongation Factors

The amino acid sequences of the r-proteins and elongation factors as deduced from the DNA sequence are given in the Appendix. The r-proteins are small, basic proteins except for the acidic homologs of S7 from the extreme halophiles (Table 4).

Table 4. Isoelectric points, molecular weights, and number of amino acid residues of the *str* operon proteins from *M. vannielii*, *S. acidocaldarius*, *H. halobium*, and *Halococcus morrhuae*

Gene product	pI	M_r	No. of amino acids		
			Archaebacterium	*E. coli*	Eucaryotes[a]
Mva ORF1	10.1	11,482	105	—[b]	114 (Rno L30)
Sac ORF1	9.9	11,536	104		
Mva S12	11.1	16,068	147	123[c]	—
Sac S12	10.0	12,665	118		
Hha S12	10.6	15,489	142		
Hmo S12	10.6	15,441	142		
Mva S7	10.4	20,741	188	153[c]	—
Sac S7	10.4	22,056	195		
Hha S7	5.5	22,984	210		
Hmo S7	4.8	22,424	203		
Mva EF-2	5.4	80,161	727	703[d]	858 (Mau EF-2)
Mva EF-1α	7.0	46,467	428	393[d]	460 (Asa EF-1α)
Sac EF-1α	9.7	48,200	435		
Mva S10	10.5	10,236	91	103[c]	—
Sac S10	10.7	11,984	102		

[a] Lechner et al., 1989.
[b] —, Sequence of the homologous protein is not yet available.
[c] Wittmann-Liebold, 1986.
[d] Post and Nomura, 1980.

The average similarity values of the *str* proteins from *M. vannielii* compared with the homologous proteins from eucaryotes, other archaebacteria, and eubacteria are schematically presented in Fig. 9; the degree of similarity among archaebacterial r-proteins varies from 33% (Mva ORF1-Sac ORF1) to 92% (Hha S12-Hmo S12). The corresponding values for the comparison of archaebacterial proteins with their eubacterial counterparts range from 20 to 36%. Thus, the archaebacterial sequences exhibit a closer relationship to each other than to any eubacterial homolog (Fig. 9). As mentioned above, the archae-

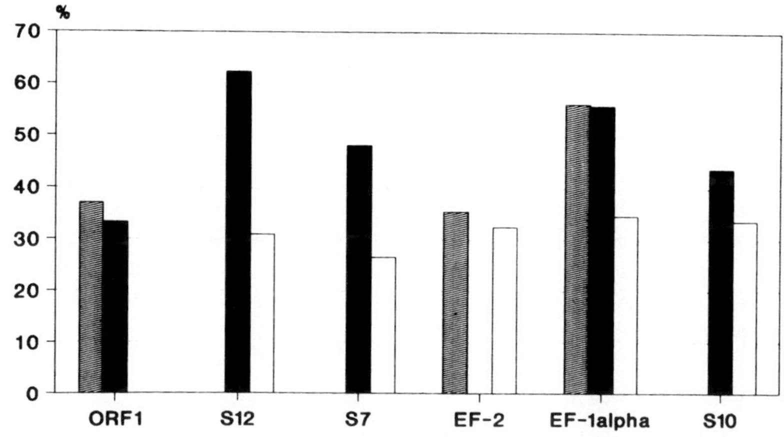

Figure 9. Comparison of the degree of sequence similarity of *M. vannielii str* r-proteins (*E. coli* nomenclature except for proteins that have no eubacterial counterpart) and elongation factors (eucaryotic nomenclature) to homologous proteins from eucaryotes (hatched columns), other archaebacteria (solid columns), and eubacteria (open columns), including chloroplasts (Auer et al., unpublished data). The degree of sequence similarity is given as the percentage of identical amino acid residues in the aligned sequences. The heights of the columns represent the average of the sequence similarity of several sequences. The sequences of homologous proteins were aligned pairwise, using the programs of the UWGCG package (Devereux et al., 1984).

Table 5. Other r-protein genes located in the genomic fragments under study (see Fig. 1)

Location	Gene	Equivalent protein		% Similarity
		Eucaryote	Eubacterium	
Upstream of *rif* region	Sso L46	Sce L46[a]	Absent	40
		Rno L39[b]	Absent	46
	Sso LX	?	Absent	
spc region[c]	C	Sce S6[d]	Absent	43
	D	Mmu L32	Absent	29
	E	Rno L19	Absent	38
str region[c]	1	Rno L30	Absent	37

[a] Leer et al., 1985.
[b] Lin et al., 1982.
[c] See Lechner et al. (1989) and Auer et al. (1989b) for complete lists of references.
[d] Only the N-terminal sequence is available.

bacterial r-proteins again generally have more amino acid residues than do their eubacterial homologs (Table 4).

The comparison of elongation factor sequences revealed a high degree of conservation of the primary structure, especially at functionally important regions such as the binding domains for phosphate or the tRNA (Kohno et al., 1986). As expected from this result, the overall degree of similarity of EF-1α (EF-Tu) from different kingdoms (Table 7) is relatively high; it varies within a range of 34 to 56% identity. The archaebacterial proteins share 56% identical amino acids with each other and have about the same degree of similarity to the eucaryotic factors. In contrast, the degree of similarity between the archaebacterial and the eubacterial proteins is on average only 34%, with a maximal deviation of 30 to 37%. A closer relationship between archaebacterial and eucaryotic sequences also was obtained from comparisons of EF-2 (EF-G) sequences, although the differences of similarity are less pronounced (Lechner et al., 1988) (Fig. 9).

There is also a size gradient when elongation factor sequences are compared: the archaebacterial proteins are smaller than their eucaryotic and larger than their eubacterial counterparts (Table 4).

OTHER r-PROTEINS PRESENT IN THE GENOMIC FRAGMENTS UNDER STUDY

During the sequencing of the r-protein genes reported above, a substantial number of ORFs were

Table 6. Codon usage of r-protein genes in *S. solfataricus* and *M. vannielii*

Codon	Amino acid	Usage		Codon	Amino acid	Usage		Codon	Amino acid	Usage		Codon	Amino acid	Usage	
		Sso	Mva			Sso	Mva			Sso	Mva			Sso	Mva
UUU	Phe	21	34	UCU	Ser	8	16	UAU	Tyr	1	12	UGU	Cys	0	11
UUC	Phe	10	42	UCC	Ser	2	11	UAC	Tyr	12	56	UGC	Cys	0	18
UUA	Leu	51	119	UCA	Ser	20	64	UAA	Term	3	13	UGA	Term	5	1
UUG	Leu	12	29	UCG	Ser	1	2	UAG	Term	0	1	UGG	Trp	9	13
CUU	Leu	15	23	CCU	Pro	22	46	CAU	His	11	17	CGU	Arg	1	1
CUC	Leu	3	21	CCC	Pro	13	2	CAC	His	3	35	CGC	Arg	0	0
CUA	Leu	19	7	CCA	Pro	26	53	CAA	Gln	31	45	CGA	Arg	1	3
CUG	Leu	3	2	CCG	Pro	1	4	CAG	Gln	16	23	CGG	Arg	0	1
AUU	Ile	28	95	ACU	Thr	30	38	AAU	Asn	45	35	AGU	Ser	19	17
AUC	Ile	7	56	ACC	Thr	6	22	AAC	Asn	16	70	AGC	Ser	9	16
AUA	Ile	74	12	ACA	Thr	31	44	AAA	Lys	96	272	AGA	Arg	45	136
AUG	Met	28	69	ACG	Thr	2	10	AAG	Lys	48	32	AGG	Arg	11	24
GUU	Val	35	121	GCU	Ala	32	66	GAU	Asp	34	44	GGU	Gly	13	90
GUC	Val	8	16	GCC	Ala	15	8	GAC	Asp	15	50	GGC	Gly	7	29
GUA	Val	42	86	GCA	Ala	57	90	GAA	Glu	63	178	GGA	Gly	41	96
GUG	Val	10	8	GCG	Ala	9	9	GAG	Glu	33	13	GGG	Gly	2	16

Table 7. Wobble base usage of r-protein genes in *S. solfataricus* and *M. vannielii*

Wobble base	Usage			
	Sso		Mva	
	No.	%	No.	%
A	605	48.2	1219	47.0
U	335	27.0	664	25.6
C	126	10.0	452	17.4
G	185	14.8	256	10.0

found. Although some of these ORFs coded for unknown proteins or for nonribosomal proteins, others coded for r-proteins that are equivalent to eucaryotic r-proteins or corresponded to r-proteins isolated from the archaebacterial ribosomes but having no equivalence, as yet, in either the eucaryotic or eubacterial ribosome. A list of these genes is given in Table 5.

Although most of the r-proteins listed in Table 5 show homology to eucaryotic r-proteins, protein Sso LX, for example, appears to be unique to the archaebacteria. However, although all of the eubacterial r-proteins have been sequenced (Giri et al., 1984), the same is true for only a portion of the eucaryotic r-proteins. This protein, therefore, may yet be found to have a eucaryotic counterpart.

CODON USAGE

Codon usage of the r-protein genes in *S. solfataricus* and *M. vannielii* is shown in Table 6. There is a strong bias to A or U in the wobble position (75% A or U in *S. solfataricus* and 73% in *M. vannielii*), which likely reflects the low C+G content of the DNA in these organisms (Table 7). The r-protein genes in these organisms rarely use the CGN codons for Arg. There are, however, some interesting differences in the codon usage of *S. solfataricus* and *M. vannielii*. When the use of C or G in the wobble position is considered, *S. solfataricus* shows a bias toward G, whereas *M. vannielii* prefers C in the third position. In addition, *S. solfataricus* shows a bias toward UGA over UAA as a chain terminator, whereas the opposite is true in *M. vannielii*. In *M. vannielii*, the chain initiator codon is almost always AUG, whereas in *S. solfataricus*, GUG and UUG are also used for initiation in the r-protein genes.

In the extreme halophiles, with the high C+G content in the DNA, the r-protein genes show a

Table 8. Sequence similarities between pairs of aligned r-protein S8 (top) and elongation factor EF-1α (EF-Tu) (bottom) sequences from archaebacteria, eucaryotes, and eubacteria[a]

r-Protein S8 from:	Sce	Hma	Mva	Eco	Mca	Bst	Mpoch	Ntach	Zmach
Sce									
Hma	40.0								
Mva	45.0	51.7							
Eco	25.8	27.5	32.5						
Mca	28.3	25.0	24.2	49.2					
Bst	30.0	30.0	31.7	50.8	57.5				
Mpoch	26.7	26.7	27.5	44.2	50.8	52.5			
Ntach	23.3	19.2	18.3	40.8	42.5	44.2	60.0		
Zmach	23.3	20.8	20.0	43.3	55.8	45.0	62.5	76.5	

EF-1α from:	Msa	Asa	Mra	Sce	Sac	Mva	Eco	Tth	Mlu	Tma	Ani	Egrch	Scemt
Hsa													
Asa	85.8												
Mra	85.3	80.2											
Sce	83.4	42.6	87.1										
Sac	55.2	52.8	54.2	54.2									
Mva	65.0	56.8	56.0	56.6	56.0								
Eco	34.3	34.0	34.6	35.1	32.7	34.9							
Tth	36.5	37.0	37.3	37.3	33.8	36.7	72.9						
Mlu	35.4	34.0	34.6	34.3	32.4	34.3	70.5	69.4					
Tma	35.7	36.2	35.7	36.2	34.0	35.4	70.2	74.0	67.3				
Ani	32.7	32.4	32.7	32.7	33.2	34.3	74.3	74.8	66.8	66.0			
Egrch	34.9	34.9	33.5	34.0	34.0	33.8	70.2	70.5	64.3	64.9	80.7		
Scemt	33.0	36.3	32.4	31.9	30.0	33.8	65.4	67.3	63.3	64.3	63.5	63.3	

[a] Values were calculated from scores obtained with the aid of distance matrices of program FITCHPRO (PHYLIP 3.0 program package [Felsenstein, 1984]).

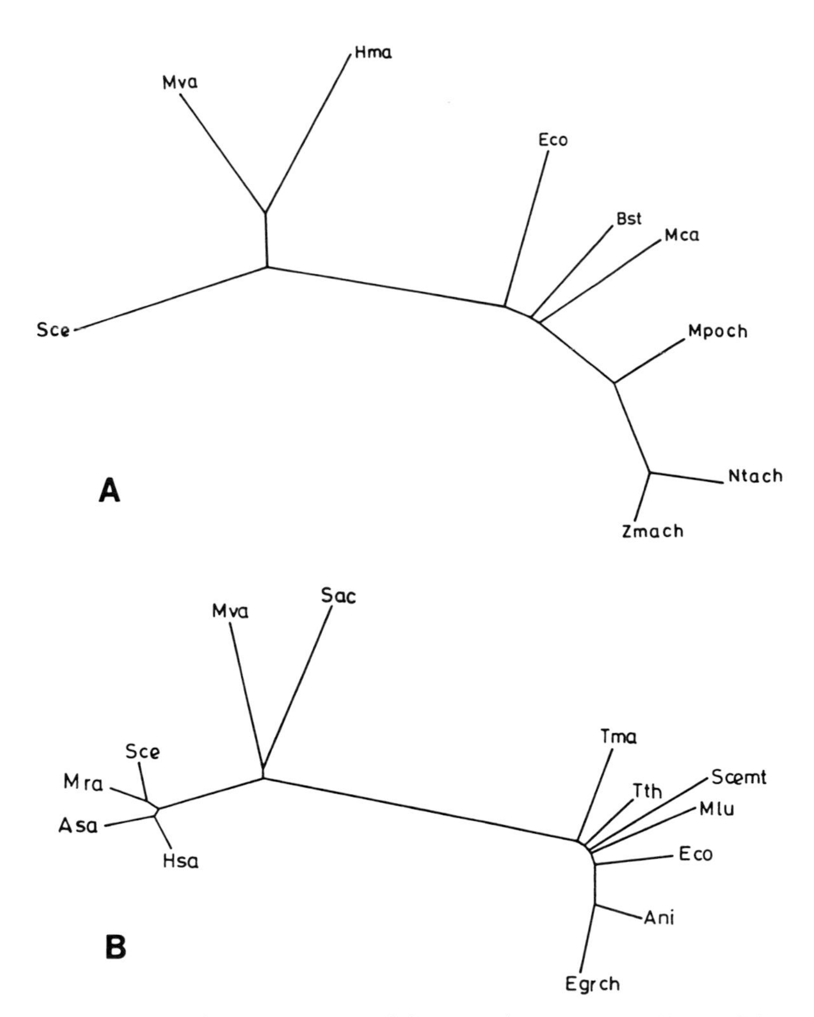

Figure 10. Phylogenetic dendrograms of r-protein S8 (A) and elongation factor EF-1α (EF-Tu) (B). Distance matrix trees (Fitch and Margoliash, 1967) were calculated from identity matrices (see Table 7); only sequence positions without gaps were considered. The distance values were corrected for multiple mutations according to Feng et al. (1985). The lengths of the lines are proportional to evolutionary distances.

strong bias toward C and G in the wobble position of the individual codons (Arndt and Kimura, 1988; Shimmin and Dennis, 1989). In addition, the CGN codons for Arg are extensively used, whereas the AGA and AGG codons are rarely used, unlike the case for *S. solfataricus* and *M. vannielii.*

PHYLOGENETIC IMPLICATIONS

Since sequences for r-protein S8 and elongation factor EF-1α are now available from all three lines of descent, phylogenetic trees were constructed for these molecules.

The amino acid sequences of homologous proteins from various organisms were aligned, using programs GAP, PROFILE, PROFILEGAP, and LINEUP (University of Wisconsin Genetics Computer Group [UWGCG] program package [Devereux et al., 1984]). These multiple alignments then were used to calculate sequence similarities between pairs of aligned molecules (Table 8) and to construct phylogenetic trees (Fig. 10).

Archaebacteria appear to be closer to each other than to any organism of either of the other two primary kingdoms. Therefore, these phylogenetic trees would support the monophyletic, holophyletic origin of the archaebacterial kingdom. The topology of these trees seems to be reasonable, since it reflects the branching order already deduced from the analysis of other macromolecules (Leffers et al., 1987; Woese, 1987). In the case of the EF-1α tree, the yeast sequence shows a closer relationship to EF-1α of the fungus than to those of the brine shrimp or of humans. Similarly, the assumed natural topology can also be seen on the eubacterial branch: *T. maritima*, which is considered to be a primitive eubacterium (Achenbach-Richter et al., 1987; Woese, 1987), branches off first, and the chloroplast sequence is located in the vicinity of the cyanobacterium *A. nidulans* (Woese, 1987). There are, however, two features of the EF-1α tree that should critically be discussed: (i) the archaebacterial branch has a low significance, and (ii) the branching order is not stable; i.e., use of the matrix of Dayhoff et al. (1978) instead of the distance matrix results in a different branching order within the eubacterial lineage. We therefore reevaluated the topology of this tree by using parsimony analysis (PROTPARSPRO [Felsenstein, 1984]) and the maximum likelihood method (DNAML [Felsenstein, 1984]). Application of the two methods resulted in the same overall topology (Fig. 10): eubacteria, eucaryotes, and archaebacteria are separated from each other, and there is a closer relationship between archaebacteria and eucaryotes. The parsimony analysis showed that the probability of a common archaebacterial branch is not significantly higher than that of two separate branching points for *M. vannielii* and *S. solfataricus*. With the maximum likelihood method, the branching order of the eubacterial kingdom shown in Fig. 10 could be obtained only when the first two bases of each codon were used instead of the total sequence. This result is expected, since bacteria differ greatly in chromosomal G+C content. However, this restriction results in two separate archaebacterial branches. Therefore, it has to be questioned whether the phylogenetic dendrograms shown in Fig. 10 reflect the natural relationship of the organisms.

It is evident from the trees shown in Fig. 10 that the branching point of the archaebacterial sequences is shifted to the side of the eucaryotic sequences in comparison with that of the 16S or 23S rRNA molecule (Leffers et al., 1987; Woese, 1987). A similar shift was reported for the RNA polymerase subunit sequences (Leffers et al., 1989; Zillig et al., 1988) and 5S rRNA sequences (Hori and Osawa, 1987). However, it should be noted that when the sequences of glyceraldehyde-3-phosphate dehydrogenase are compared, there is a much closer relationship between eucaryotes and eubacteria in relation to the archaebacteria (Hensel et al., 1989). Thus, different evolutionary relationships of primary lineages are obtained when different macromolecules are used. This result could mean that different genes have evolved with greatly different relative evolution rates in the three kingdoms. As an alternative for this explanation, Zillig and co-workers (Zillig et al., 1988; W. Zillig, H.-P. Klenk, P. Palm, W. Leffers, G. Pühler, F. Gropp, and R. A. Garrett, *Endocytobiosis Cell Res.*, in press) recently proposed that the eucaryote might be a chimera of ancestral archaebacterial and ancestral eubacterial origin. Their assumption implies that the eucaryote was generated by some sort of fusion event. This could explain why some macromolecules of the modern eucaryote, such as the RNA polymerases II and III or r-proteins, display a higher sequence similarity to their counterparts of modern archaebacteria whereas other macromolecules, such as glyceraldehyde-3-phosphate dehydrogenase, show a closer relationship to their homologs of modern eubacteria. However, this result could also reflect the different values of different macromolecules for evolutionary investigations. At present, we cannot exclude that elongation factors and r-proteins contain insufficient evolutionary information or have a level of overall conservation that is too low to allow one to determine the exact topology of deep branch points of primary lineages. Further analyses are necessary to evaluate these results.

EVOLUTION OF THE RIBOSOME

Although we are far from having a complete picture of the evolution of the ribosome, some of the possible stages in the evolution of this complex are beginning to be defined. (i) The primitive ribosome is thought to consist mainly of RNA (Woese, 1980). This proposal has received considerable support from the discovery of RNA catalysis (Cech and Bass, 1986) and the observation that rRNA plays a major functional role in protein synthesis (Noller, 1984). (ii) The first r-proteins were likely small peptides that interacted with rRNA. It has been proposed that the initial function of these molecules may have been to stabilize the structure of rRNA (Maizels and Weiner, 1987). Thus, rRNA-binding proteins such as L11, L1, and L10 probably contain domains that have derived from these ancient peptides. (iii) Modern r-proteins probably evolved by gene duplication and by joining of functional domains (Jue et al., 1980; Lin et al., 1987a). Evidence of this can be seen in the modules of the archaebacterial L10 proteins and in the conserved regions in the L1 and L11 proteins discussed above. (iv) The ancestral ribosome probably contained more proteins than do the eubacterial and archaebacterial ribosomes (Auer et al., 1989a; Auer et al., 1989b; Cammarano et al., 1986; Wool, 1980). As mentioned earlier, there are two different types of ribosomes within the archaebacteria (Cammarano et al., 1986). The observation that *Thermococcus celer* belongs to a branch within the archaebacterial tree that is believed to be very close to the root of the tree has led Cammarano et al. (1986) to propose that the larger ribosome of the sulfur-dependent extreme thermophiles and members of the order Methanococcales more closely resembles the ancestral ribosome.

The eubacterial and halophile-methanogen ribosomes, which contain fewer r-proteins, would have been the result of further streamlining (Auer et al., 1989a; Wool, 1980). During this process, the functions of different r-proteins were combined and the number of r-proteins in the ribosome was reduced. It is also likely that the length of many of the r-proteins was also reduced. It should be noted that archaebacterial r-proteins generally are larger than their eubacterial counterparts. The L12 and L10 eubacterial proteins seem to be good examples of this streamlining process, since whole domains appear to have been eliminated or modified (Fig. 4).

As indicated above, the halophilic and most of the methanogenic ribosomes are similar in number of proteins to the ribosomes from eubacteria. However, the presence of proteins in the ribosomes from the extreme halophiles that have no equivalent in the eubacterial ribosome (Kimura et al., 1989) would seem to indicate that this streamlining process has occurred in the eubacterial line and separately in the archaebacterial line. Since the streamlining processes that gave rise to the eubacterial and archaebacterial ribosomes appear to be different, it is possible that we will find proteins common to the ribosomes in all three kingdoms, proteins common to the ribosomes in two of the three kingdoms, and proteins that are unique to the ribosomes of a particular kingdom. Since structural data are available only for some of the archaebacterial and eucaryotic r-proteins, we must wait until all of the r-proteins in each kingdom have been sequenced before we will be able to determine whether all three possibilities have taken place during the evolution of the ribosome.

We thank Jennifer Duggan for typing the manuscript. The invaluable help of W. Zillig and P. Palm in the construction of phylogenetic trees is greatly acknowledged.

The work of A.T.M. and C.R. was supported by the Natural Sciences and Engineering Research Council of Canada; the work of A.B. and J.A. was supported by the Deutsche Forschungsgemeinschaft.

REFERENCES

Achenbach-Richter, L., R. Gupta, K. O. Stetter, and C. R. Woese. 1987. Were the original eubacteria thermophiles? *Syst. Appl. Microbiol.* **9**:34–39.

Arndt, E., and M. Kimura. 1988. Molecular cloning and nucleotide sequence of the gene for the ribosomal protein S11 from the archaebacterium *Halobacterium marismortui*. *J. Biol. Chem.* **263**:16063–16068.

Auer, J., K. Lechner, and A. Böck. 1989a. Gene organization and structure of two transcriptional units from *Methanococcus* coding for ribosomal proteins and elongation factors. *Can. J. Microbiol.* **35**:200–204.

Auer, J., G. Spicker, and A. Böck. 1989b. Organization and structure of the *Methanococcus* transcriptional unit homologous to the *E. coli* 'spectinomycin operon': implications for the evolutionary relationship of 70S and 80S ribosomes. *J. Mol. Biol.* **209**:21–26.

Beauclerk, A. A. D., H. Hummel, D. J. Holmes, A. Böck, and E. Cundliffe. 1985. Studies of the GTPase domain of archaebacterial ribosomes. *Eur. J. Biochem.* **151**:245–255.

Buttarelli, F. R., R. A. Calogero, O. Tiboni, C. O. Gualerzi, and C. L. Pon. 1989. Characterization of the *str* operon genes from *Spirulina platensis* and their evolutionary relationship to those of other prokaryotes. *Mol. Gen. Genet.* **217**:97–104.

Cammarano, P., A. Teichner, and P. Londei. 1986. Intralineage heterogeneity of archaebacterial ribosomes, evidence for two physicochemically distinct ribosome classes within the third kingdom. *Syst. Appl. Microbiol.* **7**:137–146.

Cech, T. R., and B. L. Bass. 1986. Biological catalysis by RNA. *Annu. Rev. Biochem.* **55**:599–626.

Cerretti, D. F., D. Dean, G. R. Davis, D. M. Bedwell, and M. Nomura. 1983. The *spc* ribosomal protein operon of *Escherichia coli*: sequence and cotranscription of the ribosomal genes and a protein export gene. *Nucleic Acids Res.* **11**:2599–2616.

Chan, Y.-L., A. Lin, J. McNally, and I. G. Wool. 1987. The primary structure of rat ribosomal protein L19. A determination from the sequence of nucleotides in a cDNA and from the sequence of amino acids in the protein. *J. Biol. Chem.* **262**: 12879–12886.

Dayhoff, M. O., R. M. Schwartz, and B. C. Orcutt. 1978. A model of evolutionary change in proteins, p. 345–358. *In* M. O. Dayhoff (ed.), *Atlas of Protein Sequence and Structure*, vol. 5, suppl. 3. National Biomedical Research Foundation, Washington, D.C.

Devereux, J., F. Haeberli, and O. Smithies. 1984. A comprehensive set of sequence analysis programs for the VAX. *Nucleic Acids Res.* **12**:387–395.

El-Baradi, T. T. A. L., V. C. H. F. de Regt, S. W. C. Einerhand, J. Teixido, R. J. Planta, J. P. G. Ballesta, and H. A. Raué. 1987. Ribosomal proteins EL11 from *Escherichia coli* and L15 from *Saccharomyces cerevisiae* bind to the same site on both yeast 26S and mouse 28S rRNA. *J. Mol. Biol.* **195**:909–917.

Erdmann, V. A. 1976. Structure and functions of 5S and 5.8S RNA. *Prog. Nucleic Acid Res. Mol. Biol.* **18**:45–90.

Felsenstein, J. 1984. The statistical approach to inferring evolutionary trees and what it tells us about parsimony and compatibility, p. 169–191. *In* T. Ducan and T. F. Stuessy (ed.), *Cladistics: Perspectives in the Reconstruction of Evolutionary History*. Columbia University Press, New York.

Feng, D. F., M. S. Johnson, and R. F. Doolittle. 1985. Aligning amino acid sequences: comparison of commonly used methods. *J. Mol. Evol.* **21**:112–125.

Fitch, W. M., and E. Margoliash. 1967. Construction of phylogenetic trees. A method based on mutation distances as estimated from cytochrome c sequences is of general applicability. *Science* **155**:279–284.

Garrett, R. A., S. Douthwaite, and H. F. Noller. 1981. Structure and role of 5S RNA-protein complexes in protein biosynthesis. *Trends Biochem. Sci.* **6**:137–141.

Giri, L., W. E. Hill, and H. G. Wittmann. 1984. Ribosomal proteins: their structure and spacial arrangement in prokaryotic ribosomes. *Adv. Protein Chem.* **36**:1–78.

Gourse, R. L., D. L. Thurlow, S. A. Gerbi, and R. A. Zimmermann. 1981. Specific binding of a prokaryotic ribosomal protein to a eukaryotic ribosomal RNA: implications for evolution and autoregulation. *Proc. Natl. Acad. Sci. USA* **78**:2722–2726.

Gudkov, A. T., L. G. Tumanova, G. M. Gongadze, and V. N. Bushuev. 1980. Role of different regions of ribosomal proteins L7 and L10 in their complex formation and in the interaction with the ribosomal 50S subunit. *FEBS Lett.* **109**:34–38.

Hensel, R., P. Zwickl, S. Fabry, J. Lang, and P. Palm. 1989. Sequence comparison of glyceraldehyde-3-phosphate dehydrogenases from the three urkingdoms: evolutionary implication. *Can. J. Microbiol.* **35**:81–85.

Hori, H., and S. Osawa. 1987. Origin and evolution of organisms as deduced from 5S ribosomal RNA sequences. *Mol. Biol. Evol.* **4**:445–472.

Itoh, T. 1988. Complete nucleotide sequence of the ribosomal 'A' protein operon from the archaebacterium, *Halobacterium halobium. Eur. J. Biochem.* **176**:297–303.

Juan-Vidales, F., F. Sanchez-Madrid, M. T. Saenz-Robles, and J. P. G. Ballesta. 1983. Purification and characterization of two ribosomal proteins of *Saccharomyces cerevisiae*: homologies with proteins from eukaryotic species and with bacterial protein ECL11. *Eur. J. Biochem.* **136**:275–281.

Jue, R. A., N. W. Woodbury, and R. F. Doolittle. 1980. Sequence homologies among *Escherichia coli* ribosomal proteins: evidence for evolutionarily related groupings and internal duplications. *J. Mol. Evol.* **15**:129–148.

Kimura, J., and M. Kimura. 1987. The primary structures of ribosomal proteins S14 and S16 from the archaebacterium *Halobacterium marismortui*. Comparison with eubacterial and eukaryotic ribosomal proteins. *J. Biol. Chem.* **262**:12150–12157.

Kimura, M., E. Arndt, T. Hatakeyama, T. Hatakeyama, and J. Kimura. 1989. Ribosomal proteins in halobacteria. *Can. J. Microbiol.* **35**:195–199.

Kohno, K., T. Uchida, H. Ohkubo, S. Nakanishi, T. Nakanishi, T. Fukui, E. Ohtsuka, M. Ikehara, and Y. Okada. 1986. Amino acid sequence of mammalian elongation factor 2 deduced from the cDNA sequence: homology with GTP-binding proteins. *Proc. Natl. Acad. Sci. USA* **83**:4978–4983.

Köpke, A. E. K., G. Baier, and B. Wittmann-Liebold. 1989. An archaebacterial gene from *Methanococcus vannielii* encoding a protein homologous to the ribosomal protein L10 family. *FEBS Lett.* **247**:167–172.

Lake, J. A., E. Henderson, M. W. Clark, and A. T. Matheson. 1982. Mapping evolution with ribosome structure: intralineage constancy and interlineage variation. *Proc. Natl. Acad. Sci. USA* **79**:5948–5952.

Lake, J. A., and W. A. Strycharz. 1981. Ribosomal proteins L1, L17, L27 from *Escherichia coli* localized at single sites on the large ribosomal subunit by immune electron microscopy. *J. Mol. Biol.* **153**:979–992.

Lechner, K., G. Heller, and A. Böck. 1988. Gene for diphtheria toxin-susceptible elongation factor 2 from *Methanococcus vannielii. Nucleic Acids Res.* **16**:7817–7826.

Lechner, K., G. Heller, and A. Böck. 1989. Organization and nucleotide sequence of a transcriptional unit of *Methanococcus vannielii* comprising genes for protein synthesis elongation factors and ribosomal proteins. *J. Mol. Evol.* **29**:20–27.

Leer, R. J., M. C. van Raamsdonk-Duin, P. Kraakman, W. H. Mager, and R. J. Planta. 1985. The genes for yeast ribosomal proteins S24 and L46 are adjacent and divergently transcribed. *Nucleic Acids Res.* **13**:701–709.

Leffers, H., F. Gropp, F. Lottspeich, W. Zillig, and R. A. Garrett. 1989. Sequence, organization, transcription and evolution of RNA polymerase subunit genes from the archaebacterial extreme halophiles *Halobacterium halobium* and *Halococcus morrhuae. J. Mol. Biol.* **206**:1–17.

Leffers, H., J. Kjems, L. Ostergaard, N. Larsen, and R. A. Garrett. 1987. Evolutionary relationships amongst archaebacteria. A comparative study of 23S ribosomal RNAs of a sulfur-dependent extreme thermophile, an extreme halophile and a thermophilic methanogen. *J. Mol. Biol.* **195**:43–61.

Liljas, A., S. Thirup, and A. T. Matheson. 1986. Evolutionary aspects of ribosome-factor interactions. *Chem. Scripta* **26B**: 109–119.

Lin, A., Y.-L. Chan, R. Jones, and I. G. Wool. 1987a. The primary structure of rat ribosomal protein S12: the relationship of rat S12 to other ribosomal proteins and a correlation of the amino acid sequences of rat and yeast ribosomal proteins. *J. Biol. Chem.* **262**:14343–14351.

Lin, A., Y.-L. Chan, J. McNally, D. Peleg, O. Meyuhas, and I. G. Wool. 1987b. The primary structure of rat ribosomal protein L7. The presence near the amino terminus of L7 of five tandem repeats of a sequence of 12 amino acids. *J. Mol. Biol.* **262**: 12665–12671.

Lin, A., J. McNally, and I. G. Wool. 1984. The primary structure of rat liver ribosomal protein L39. *J. Biol. Chem.* **259**:487–490.

Lin, A., B. Wittmann-Liebold, J. McNally, and I. G. Wool. 1982. The primary structure of the acidic phosphoprotein P2 from rat liver 60S ribosomal subunits. *J. Biol. Chem.* **257**:9189–9197.

Lindahl, L., and J. M. Zengel. 1986. Ribosomal genes in *Escherichia coli. Annu. Rev. Genet.* **20**:297–326.

Londei, P., A. Teichner, and P. Cammarano. 1983. Particle weights and protein composition of the ribosomal subunits of the extremely thermoacidophilic archaebacterium *Caldariella acidophila. Biochem. J.* **209**:461–470.

Mager, W. H. 1988. Control of ribosomal protein gene expression. *Biochim. Biophys. Acta* **949**:1–15.

Maizels, N., and A. M. Weiner. 1987. Peptide-specific ribosomes, genomic tags, and the origin of the genetic code. *Cold Spring Harbor Symp. Quant. Biol.* **52**:743–749.

Markmann-Mulisch, U., K. von Knoblauch, A. Lehmann, and A. R. Subramanian. 1987. Nucleotide sequence and linkage map position of the *secX* gene in maize chloroplast and evidence that it encodes a protein belonging to the 50S ribosomal subunit. *Biochem. Int.* **15**:1057–1067.

Matheson, A. T. 1986. Ribosomes of archaebacteria, p. 345–377. *In* C. R. Woese and R. S. Wolfe (ed.), *The Bacteria*, vol. 8. *Archaebacteria*. Academic Press, Inc., New York.

Matheson, A. T., M. Yaguchi, W. E. Balch, and R. S. Wolfe. 1980. Sequence homologies in the N-terminal region of the ribosomal 'A' proteins from *Methanobacterium thermoautotrophicum* and *Halobacterium cutirubrum. Biochim. Biophys. Acta* **626**:162–169.

Mitsui, K., and K. Tsurugi. 1988a. cDNA and deduced amino acid sequence of 38 kDa-type acidic ribosomal protein AO from *Saccharomyces cerevisiae. Nucleic Acids Res.* **16**:3573.

Mitsui, K., and K. Tsurugi. 1988b. cDNA and deduced amino acid sequence of acidic ribosomal protein A2 from *Saccharomyces cerevisiae. Nucleic Acids Res.* **16**:3575.

Moazed, D., and H. F. Noller. 1989. Interaction of tRNA with 23S rRNA in the ribosomal A, P and E sites. *Cell* **57**:585–597.

Möller, W., and J. A. Maassen. 1986. On the structure, function, and dynamics of L7/L12 from *Escherichia coli* ribosomes, p. 309–325. *In* B. Hardesty and G. Kramer (ed.), *Structure, Function, and Genetics of Ribosomes*. Springer-Verlag, New York.

Nazar, R. N., M. Yaguchi, and G. E. Willick. 1982. The 5S RNA-protein complex from yeast: a model for the evolution and structure of the eukaryotic ribosome. *Can. J. Biochem.* **60**: 490–496.

Nazar, R. N., M. Yaguchi, G. E. Willick, C. F. Rollin, and C. Roy. 1979. The 5-S RNA binding protein from yeast (*Saccharomyces cerevisiae*) ribosomes. Evolution of the eukaryotic 5-S RNA binding protein. *Eur. J. Biochem.* **102**:573–582.

Noller, H. F. 1984. Structure of ribosomal RNA. *Annu. Rev. Biochem.* **53**:119–162.

Nomura, M., R. Gourse, and G. Baughman. 1984. Regulation of the synthesis of ribosomes and ribosomal components in *Escherichia coli*: translational regulation and feedback loops. *Annu. Rev. Biochem.* **53**:75–117.

Nomura, M., E. A. Morgan, and S. R. Jaskunas. 1977. Genetics of bacterial ribosomes. *Annu. Rev. Genet.* **11:**297–347.

Ohama, T., F. Yamao, A. Muto, and S. Osawa. 1987. Organization and codon usage of the streptomycin operon in *Micrococcus luteus*, a bacterium with a high genomic G+C content. *J. Bacteriol.* **169:**4770–4777.

Ohkubo, S., A. Muto, Y. Kawauchi, F. Yamao, and S. Osawa. 1987. The ribosomal protein gene cluster of *Mycoplasma capricolum. Mol. Gen. Genet.* **210:**314–322.

Otaka, E., K. Higo, and T. Itoh. 1984. Yeast ribosomal proteins. *Mol. Gen. Genet.* **195:**544–546.

Post, L. E., and M. Nomura. 1980. DNA sequences from the *str* operon of *Escherichia coli. J. Biol. Chem.* **255:**4660–4666.

Post, L. E., G. D. Strycharz, M. Nomura, H. Lewis, and P. P. Dennis. 1979. Nucleotide sequence of the ribosomal protein gene cluster adjacent to the gene for RNA polymerase subunit in *Escherichia coli. Proc. Natl. Acad. Sci. USA* **76:**1697–1701.

Pühler, G., F. Lottspeich, and W. Zillig. 1989. Organization and nucleotide sequence of the genes encoding the large subunits A, B and C of the DNA-dependent RNA polymerase of the archaebacterium *Sulfolobus acidocaldarius. Nucleic Acids Res.* **17:**4517–4534.

Ramírez, C., L. C. Shimmin, C. H. Newton, A. T. Matheson, and P. P. Dennis. 1989. Structure and evolution of the L11, L1, L10 and L12 equivalent ribosomal proteins in eubacteria, archaebacteria and eukaryotes. *Can. J. Microbiol.* **35:**234–244.

Reiter, W. D., P. Palm, A. Henschen, F. Lottspeich, W. Zillig, and B. Grampp. 1987. Identification and characterization of the genes encoding three structural proteins of the *Sulfolobus* virus-like SSV1 particle. *Mol. Gen. Genet.* **206:**144–153.

Rich, B. E., and J. A. Steitz. 1987. Human acidic ribosomal phosphoproteins P0, P1, and P2: analysis of cDNA clones, in vitro synthesis, and assembly. *Mol. Cell. Biol.* **7:**4065–4074.

Sander, G. 1983. Ribosomal protein L1 from *Escherichia coli*: its role in the binding of tRNA to the ribosome and in elongation factor G-dependent GTP hydrolysis. *J. Biol. Chem.* **258:**10098–10102.

Schmid, G., and A. Böck. 1982a. The ribosomal protein composition of the archaebacterium *Sulfolobus. Mol. Gen. Genet.* **185:**498–501.

Schmid, G., and A. Böck. 1982b. The ribosomal protein composition of five methanogenic bacteria. *Zentralbl. Bakteriol. Parasitenkd. Infektionskr. Hyg. Abt. 1 Orig. Reihe C* **3:**347–353.

Schmidt, F. S., J. Thompson, K. Lee, J. Dijk, and E. Cundliffe. 1981. The binding site for ribosomal protein L11 within 23S ribosomal RNA of *Escherichia coli. J. Biol. Chem.* **256:**12301–12305.

Schop, E. N., and J. A. Maassen. 1982. Characterization of the region on protein L7/L12 involved in binding to ribosomal particles. *Eur. J. Biochem.* **128:**371–375.

Shimmin, L. C., and P. P. Dennis. 1989. Characterization of the L11, L1, L10 and L12 equivalent ribosomal protein gene cluster of the halophilic archaebacterium *Halobacterium cutirubrum. EMBO J.* **8:**1225–1235.

Shimmin, L. C., C. H. Newton, C. Ramírez, J. Yee, W. L. Downing, A. Louie, A. T. Matheson, and P. P. Dennis. 1989a. Organization of genes encoding the L11, L1, L10 and L12 equivalent proteins in eubacteria, archaebacteria and eukaryotes. *Can. J. Microbiol.* **35:**164–170.

Shimmin, L. C., C. Ramírez, A. T. Matheson, and P. P. Dennis. 1989b. Sequence alignment and evolutionary comparison of the L10 equivalent and L12 equivalent ribosomal proteins from Archaebacteria, Eubacteria and eucaryotes. *J. Mol. Evol.* **29:** 448–462.

Smith, N., A. T. Matheson, M. Yaguchi, E. Willick, and R. N. Nazar. 1978. The 5S RNA-protein complex from an extreme halophile, *Halobacterium cutirubrum*. Purification and characterization. *Eur. J. Biochem.* **89:**501–509.

Stöffler, G., and M. Stöffler-Meilicke. 1986. Electron microscopy of archaebacterial ribosomes. *Syst. Appl. Microbiol.* **7:**123–130.

Strøm, A. R., and L. P. Visentin. 1973. Acidic ribosomal proteins from the extreme halophile, *Halobacterium cutirubrum*: the simultaneous separation, identification and molecular weight determination. *FEBS Lett.* **37:**274–280.

Walsh, M. J., J. McDougall, and B. Wittmann-Liebold. 1988. Extended N-terminal sequencing of proteins of archaebacterial ribosomes blotted from two-dimensional gels onto glass fiber and polyvinylidene difluoride membrane. *Biochemistry* **27:** 6867–6876.

Wittmann-Liebold, B. 1986. Ribosomal proteins: their structure and evolution, p. 326–361. *In* B. Hardesty and G. Kramer (ed.), *Structure, Function, and Genetics of Ribosomes*. Springer-Verlag, New York.

Woese, C. R. 1980. Just so stories and Rube Goldberg machines: speculations on the origin of the protein synthetic machinery, p. 357–373. *In* G. Chambliss, G. R. Craven, J. Davies, K. Davis, L. Kahan, and M. Nomura (ed.), *Ribosomes. Structure, Function, and Genetics*. University Park Press, Baltimore.

Woese, C. R. 1987. Bacterial evolution. *Microbiol. Rev.* **51:** 221–271.

Woese, C. R., and G. E. Fox. 1977. Phylogenetic structure of the prokaryotic domain: the primary kingdoms. *Proc. Natl. Acad. Sci. USA* **74:**5088–5090.

Wool, I. G. 1980. The structure and function of eukaryotic ribosomes, p. 797–824. *In* G. Chambliss, G. R. Craven, J. Davies, K. Davis, L. Kahan, and M. Nomura (ed.), *Ribosomes. Structure, Function, and Genetics*. University Park Press, Baltimore.

Zillig, W., P. Palm, W.-D. Reiter, F. Gropp, G. Pühler, and H.-P. Klenk. 1988. Comparative evaluation of gene expression in archaebacteria. *Eur. J. Biochem.* **173:**473–482.

Zimmermann, R. A., D. L. Thurlow, R. S. Finn, T. L. Marshand, and L. K. Ferrett. 1980. Conservation of specific protein-RNA interactions in ribosome evolution, p. 569–584. *In* S. Osawa, H. Ozeki, H. Uchida, and T. Yura (ed.), *Genetics and Evolution of RNA Polymerase, tRNA, and Ribosomes*. University of Tokyo Press, Tokyo.

(Appendix begins next page.)

APPENDIX

Complete amino acid sequences of the archaebacterial r-proteins discussed in this chapter (see Fig. 1).

A SULFOLOBUS SOLFATARICUS: RIF REGION

L11
1 MPTKTIKIMV EGGSAKPGPP LGPTLSQLGL NVQEVVKKIN DVTAQFKGMS
51 VPVTIEIDSS TKKYDIKVGV PTTTSLLLKA INAQEPSGDP AHKKIGNLDL
101 EQIADIAIKK KPQLSAKTLT AAIKSLLGTA RSIGITVEGK DPKDVIKEID
151 QGKYNDLLTN YEQKWNEAEG

L1
1 MKKVLADKES LIEALKLALS TEYNVKRNFT QSVEIILTFK GIDMKKGDLK
51 LREIVPLPKQ PSKAKRVLVV PSFEQLEYAK KASPNVVITR EELQKLQGQK
101 RPVKKLAIQN EWFLINQESM ALAGRILGPA LGPRGKFPTP LPNTADISEY
151 INRFKRSVIV KTKDQPQVQV FIGTEDMKPE DLAENAIAVL NAIENKAKVE
201 TNLRNIYVKT TMGKAVKVKR A

L10
1 MIGLAVTTTK KIAKWKVDEV AELTEKLKTH KTIIIANIEG FPADKLHEIR
51 KKLRGKADIK VTKNNLFNIA LKNAGYDTKL FESYLTGPNA FIFTDTNPFE
101 LQLFLSKFKL KRYALPGDKA DEEVVVPAGD TGIAAGPMLS VFGKLKIKTK
151 VQDGKIHILQ DTTVAKPGDE IPADIVPILQ KLGIMPVYVK LNIKIAYDNG
201 VIIPGDKLSI NLDDYTNEIR KAHINAFAVA TEIAYPEPKV LEFTATKAMR
251 NALALASEIG YITQETAQAV FTKAVMKAYA VASSISGKVD LGVQIQAQPQ
301 VSEQAAEKKE EKKEEEKKGP SEEEIGGGLS SLFGG

L12
1 MEYIYASLLL HAAKKEISEE NIKNVLSAAG ITVDEVRLKA VAAALKEVNI
51 DEILKTATAM PVAAVAAPAG QQTQQAAEKK EEKKEEEKKG PSEEEIGGGL
101 SSLFG

L46
1 MSKHKSLGKK LRLGKALKRN SPIPAWVIIK TQAEIRFNPL RRNWRRNNLK
51 V

Lx
1 MAEVKIFMVR GTAIFSASRF PTSQKYVRAL NEKQAIEYIY SQLGGKNKIN
51 DTTYTYKRSK KLRKMKSQTR Q

B METHANOCOCCUS VANNIELII: STR REGION

S12 1 MSGSKSPKGE FAGRKLLLKR KATRWQPYKY VNRELGLKVK ADPLGGAPMG
51 RGIVVEKVGL EAKQPNSAIR KCVKVQLIKN GRVVTAFAPG NHAINFIDEH
101 DEVVIEGIGG PSGQAKGDIP GVRYKVLMVG KNSIRELVRG RQEKVKR

S7 1 LEIKLFGKWD STSVTVKDPS LKSHISLNPV LIPHTAGRNS KKMFDKNKMH
51 VVERLANKLM ATQVNTGKKN EVLSIIEEAL TIVENRTKEN PIQVVVDALE
101 NSGPREETTR ISYGGIAFLQ SVDVSPSRRL DTAFRNISLG ASQGAHKSKK
151 SIAQCLADEL VAASKADMQK SFAVKKKEEK ERVAQSAR

EF2 1 MGRRAKMVEK VKSLMETHDQ IRNMGICAHI DHGKTTLSDN LLAGAGMISK
51 DLAGDQLALD FDEEEAARGI TIYAANVSMV HEYNGKEYLI NLIDTPGHVD
101 FGGDVTRAMR AIDGAVVVCC AVEGVMPQTE TVLRQALKEK VKPVLFINKV
151 DRLINELKLT PEELQGRFMK IIAEVNKLIE KMAPEEFKKE WLCDVVTGKV
201 AFGSAYNNWA ISVPYMQKSG ISFKDIIDYC EQEKQSELAD KAPLHEVILD
251 MAIKHLPNPL QAQKYRIPNI WKGDAESEVG KSMAMCDPNG PLAGVVTKII
301 VDKHAGSISA CRLFSGRIKQ GDELYLVGSK QKARAQQVAI FMGAERVQVP
351 SISAGNICAL TGLREATAGE TVCSPSKILE PGFESLTHTS EPVITVAIEA
401 KNTKDLPKLI EILRQIGRED NTVRIEINEE TGEHLISGMG ELHIEVITDT
451 KIGRDGGIEV DVGEPIIVYR ETITGTSPEI EGKSPNKHNK LYMIAEPMEE
501 SVYAAYVEGK IHDEDFKKKT NVDAETRLIE AGLEREQAKK VMSIYNGNMI
551 VNMTKGIVQL DEARELIIEG FKEGVKGGPL ASERAQGVKI KLIDATFHED
601 AIHRGPSQII PAIRFGVRDA VSSAKPILLE PMQKIYINTP QDYMGDAIRE
651 INNRRGQIVD MEQEGDMAII KGSVPVAEMF GFAGAIRGAT QGRCLWSVEF
701 SGFERVPNEI QTKVVAQIRD RKGLKSE

EF1A 1 MAKTKPILNV AFIGHVDAGK STTVGRLLLD GGAIDPQLIV RLRKEAEEKG
51 KAGFEFAYVM DGLKEERERG VTIDVAHKKF PTAKYEVTIV DCPGHRDFIK
101 NMITGASQAD AAVLVVNVDD AKSGIQPQTR EHVFLIRTLG VRQLAVAVNK
151 MDTVNFSEAD YNELKKMIGD QLLKMIGFNP EQINFVPVAS LHGDNVFKKS
201 ERNPWYKGPT IAEVIDGFQP PEKPTNLPLR LPIQDVYTIT GVGTVPVGRV
251 ETGIIKPGDK VVFEPAGAIG EIKTVEMHHE QLPSAEPGDN IGFNVRGVGK
301 KDIKRGDVLG HTTNPPTVAT DFTAQIVVLQ HPSVLTDGYT PVFHTHTAQI
351 ACTFAEIQKK LNPATGEVLE ENPDFLKAGD AAIVKLIPTK PMVIESVKEI
401 PQLGRFAIRD MGMTVAAGMA IQVTAKNK

S10 1 MQKARIKLSS TKHEELDSVC NQIKAIAEKT GVDMAGPIPL PTKSLKITTR
51 KSTDGEGSSS FDRWTMRVHK RVIDIEADER TMKHIMKVKN S

C SULFOLOBUS ACIDOCALDARIUS: STR REGION

S12 1 MLALKEKFDP LEGAPMARGI VLEKVGIESR QPNSAVRKAV RVQLVKNGRI
51 VTAFVPGDGG VNFIDEHDEV VIAGIGGTLG RSMGDLPGVR YKVVMVNGVS
101 LDALYKGKEA ETSKINFR

S7 1 MSIYLPHTGG RHEHRRFGKS RIPIVERLIN NLMRPGRNKG KKMLAYNIVK
51 TTFDIIAVKT GQNPIQVLVR AIENAAPREE VTRIMYGGIV YYVAVDVAPQ
101 RRVDLALRHL VTGASEASFN NPKPIEEALA EEIIAAANND NKSVAIRKKE
151 EIERIALSSR

EF1A 1 MSQKPHLNLI VIGHVDHGKS TLIGRLLMDR GFIDEKTVKE AEEAAKKLGK
51 DSEKYAFLMD RLKEERERGV TINLSFMRFE TRKYFFTVID APGHRDFVKN
101 MITGASQADA AILVVSAKKG EYEAGMSAEG QTREHIILSK TMGINQVIVA
151 INKMDLADTP YDEKRFKEIV DTVSKFMKSF GFDMNKVKFV PVVAPDGDNV
201 THKSTKMPWY NGPTLEELLD QLEIPPKPVD KPLRIPIQEV YSISGVGVVP
251 VGRIESGVLK VGDKIVFMPV GKIGEVRSIE THHTKIDKAE PGDNIGFNVR
301 GVEKKDVKRG DVAGSVQNPP TVADEFTAQV IVIWHPTAVG VGYTPVLHVH
351 TASIACRVSE ITSRIDPKTG KEAEKNPQFI KAGDSAIVKF KPIKELVAEK
401 FREFPALGRF AMRDMGKTVG VGVIIDVKPR KVEVK

S10 1 MPTKARIRLW SSNIDSLNFV VNQIRNMAQK TGIQVSGPIP LPTTRMEVPV
51 MRLPHGEGKK KWEHWEMKVH KRIIDIAADE RVMRQLMRVR VPDDVYIEIE
101 LI

D METHANOCOCCUS VANNIELII: SPC REGION

S17 1 MSNIGIDVKA PENVCEDVNC PFHGTLSVRG QIFEGVVSGD KGHNTIVIKR
51 EVTGYISKYE RYEKRTTSLV AHNPPCINAK TGDVVKVMEC RPVSKTKSFV
101 VIEKTENLE

L14 1 MKGLGSTIVR SLPNGARLVC ADNTGAKELE VIAVKNYSGT VRRLPSAGVG
51 QIVFVSVKKG TPEMRKQVLP AIIIRQKKEY KRADGTRVKF EDNAAVIVTP
101 EGTPKGSDIK GPVSKEAAER WPGVSRLAKI IH

L24 1 MVLTNSKQPR KQRKALYNAP LHLRNSVMSA MLSKALKEKY GKNALPVKKG
51 DTVKVLRGIF KGIEGEVSKV NYSGYKIIVE GVVNKKQDGK ETPYPIHPSN
101 VMITKMEDSD EKRFKTSNK

L5 1 MSFQEVWEKE PMKKPRIQKV TVNFGVGEAG DRLTIGAKVI ETLTGQAPVR
51 TLAKQTNPAF GIRKKLPIGL KVTLRGKNAE EFLENAFVAF KVSGKVLYAS
101 SFDKVGNFSF GVPEHIDFPG QKYDPTVGIY GMDICVTFEK PGYRVKSRKL
151 KRSHIPAKHL VKKEEAIEYI EKKFGAEVVM E

S14 1 MTKEPFKTKY GQGSKVCKRC GRKGPGIIRK YGLDLCRQCF RELAPKLGFK
51 KYD

```
S8    1  MSLMDPLANA LNHVSNCEGV GKNVAYLKPA SKLIGRVLKV MQDQGYIGNF
     51  EYIEDGKAGV YKVDLIGQIN KCGAVKPRYA VKYQEFEKFE KRYLPAKGFG
    101  LLIVSTPKGL MTHDEARTAG VGGRLISYVY

L6    1  MPVAALIREE IEIPGNVSVE VNGSEVVVKS GAKVLKRELA FPGIEIKMEN
     51  EKVVVESTFP KKNQTAMVGT YRSHIQNMIK GVSEGFEYKL VIRYAHFPMK
    101  VTFKGNTVII DNFLGEKYPR TAKVMEGVTV KVNGEEVIVS GTNKEFVGQT
    151  AANIEQATKV KGRDTRIFQD GIYIVEKAGK VL

L18   1  MIMAQNAKHR VPFRRRREGK TDFRQRLGLL LSGKPRLVAR KSLNNIIAQL
     51  MAYDEKGDIV LVSAHSRELV KMGYKGHCGN LPAAYLTGLL LGKKAVEEGL
    101  EEAILDKGLH RATKGAAIFA VLKGALDAGM DIPCGEEIIA DEERLNGTHI
    151  KQYAELLKED EEAYKKQFSK YLEKGLNPED LPEHFEELKG KILNL

S5    1  MAEKRAEKRK FNTDSWEPKT QVGRMVKEGT ISDISYIMDK GLPLLEPEIV
     51  DVLLPDLEEQ VLDVKLVQRM HKSGRRARYR ATVVVGNKNG YVGVGMGKSK
    101  EVGPAIRKAI AQAKLSLIKV RVGCGSWECG CGSPHSIPFT AKGTCGSVKV
    151  ELLPAPRGVG LVAGNVAKAV LGLAGVKDAW TTTYGDTRTT YNFAEATFDA
    201  LNNLNFVRCL PEQKAKLGLT EGRVL

L30   1  MAYAVVRVRG SVGVRGDIAD TMKMLRLHRV NHCVVIPEND HYTGMIKKVK
     51  DYVTYGEIDK ETLVSLILKR GRLAGNKRLS EELLKELVEL PVDALAEKVL
    101  AGEIKLKDTP IKPVFRLHPP RRGYDRGGIK KGFSIGGALG YRAGKINDLL
    151  NKMM

L15   1  MIRKSKKITK QRGSRTCGYG EAKKHRGAGH RGGRGNAGHQ KHKWLSVCKF
     51  NPEYFGKYGF NRNPCLIKKL ETINVGELEE YVLKYKDAFK LKDGKVVVNA
    101  TEIGFEKILG KGRISTAMVV KAVEFSEGAK EKIEAAGGEF VEL
```

E METHANOCOCCUS VANNIELII: OPEN READING FRAMES IN STR AND SPC REGIONS

```
ORF1   1  MRRRYSMDIN RAIRVAVDTG NVVLGTKQAI KNIKHGEGKL VIIAGNCAKD
      51  VKEDIFYYTK LSETPVYTHQ VTSIELGAIC GKPFPVSALL VLEPGNSAIL
     101  NINNE

ORFc   1  MAIKGPRKHL KRLAAPANWQ LPRKERTFTV RPSPGPHSMD KSLPLLLIVR
      51  DTLKCADNAR EAKKIIQMGK ILIDGVKRKE YKHPVGLMDV LSIPELNENY
     101  LVLFDENGRI SLKKTEKTGV KLCKIVNKTV IKGGHIQLNL HDGRNQIVKV
     151  ANALKAEEDI YKTGDSVLVS LPEQAVVGHV EFNEGKLAYI TGGKHVGEFA
     201  KVVEVEKRTL YSDIVTLENK DGEKFKTIKP YVFIVGQDEP VISM

ORFd   1  MSEFKRLMRL KLKMKQKRPE FKRQDSHRFQ RIGTMWRRPT GHHSGQRIQV
      51  TYRLSPVKIG FRGPALVRGL HPSGLEDIIV NNVKQLAALN PKTQGARIAS
     101  AVGTRKRIEI VKKANELGIR VFNVSKQKQG EFLSL

ORFe   1  MDVSTQRRIA AAVLDCGIDR VWVDPENLEK VKMAITKDDI RLLINDGIIV
      51  KKQEKGISSA RKKEVQEQKR KGKRKGPGSR RGAKGARTPK KEKWMNTIRP
     101  LRTLLKELRE NEKIERSSYR KLYRMAKGGA FRSRNHMKLY MKEHGILAE
```

F SULFOLOBUS ACIDOCALDARIUS: OPEN READING FRAME IN STR REGION

```
ORF1   1  MSQSFEGELK TLLRSGKVIL GTRKTLKLLK TGKVKGVVVS STLRQDLKDD
      51  IMTFSKFSDI PIYLYKGSGY ELGTLCGKPF MVSVIGIVDE GESKILEFIK
     101  EVKQ
```

Chapter 55

Synthesis of Ribosomal Proteins as Ubiquitin Fusions

DANIEL FINLEY, BONNIE BARTEL, and ALEXANDER VARSHAVSKY

In this paper, we show that two ribosomal proteins of the yeast *Saccharomyces cerevisiae* are synthesized as fusions to ubiquitin, a small protein that is otherwise used to modify proteins posttranslationally. The mature forms of these ribosomal proteins, which we refer to as ubiquitin tails, are generated through rapid proteolytic processing reactions; the ubiquitin-tail hybrids are present at low levels in the ribosome, if at all. In at least one case, the transient covalent association between ubiquitin and the tail protein facilitates ribosome biogenesis, apparently by increasing the efficiency of incorporation of the tail into nascent ribosomes. These results suggest a novel, chaperone function for ubiquitin in which its covalent associations with other proteins can directly or indirectly promote the formation of specific cellular structures.

The posttranslational ligation of ubiquitin to various acceptor proteins in eucaryotic cells has been shown to be involved in a number of cellular processes, such as selective protein degradation (Hershko, 1988; Varshavsky et al., 1988), DNA repair (Jentsch et al., 1987), progression through the cell cycle (Finley and Varshavsky, 1985; Goebl et al., 1988), and a variety of stress responses (Finley et al., 1987; Goff et al., 1988). One major function of ubiquitin is to mark proteins destined for selective elimination. An alternative role for ubiquitin, in which its reversible joining to an acceptor protein modulates protein function without metabolically destabilizing the acceptor protein, has also been suggested (Pickart, 1988). In vivo acceptors of ubiquitin include various nuclear, cytoplasmic, and integral membrane proteins.

The posttranslational conjugation of ubiquitin to other proteins involves the formation of an isopeptide bond between the carboxyl-terminal glycine residue of ubiquitin and the ϵ-amino group of a lysine residue in the acceptor protein (Hershko, 1988). In the yeast *S. cerevisiae*, the six or more enzymes that catalyze such reactions include the products of the genes *RAD6*, which is required for DNA repair (Jentsch et al., 1987), and *CDC34*, which is required for the transition from the G1 to the S phase of the cell cycle (Goebl et al., 1988). Ubiquitinated proteins often contain single ubiquitin moieties linked to one or more of the lysine residues of the acceptor protein (Chau et al., 1989). Alternatively, multiple ubiquitin moieties can be attached sequentially to an acceptor protein to form a chain of branched ubiquitin-ubiquitin conjugates in which the carboxyl-terminal Gly-76 of one ubiquitin is joined to the internal Lys-48 of an adjacent ubiquitin. The attachment of such a multiubiquitin chain to the proteolytic substrate is apparently essential for the degradation of a variety of proteins (Chau et al., 1989).

Unlike the posttranslationally formed, branched ubiquitin-protein conjugates, linear ubiquitin adducts are formed as the translational products of natural gene fusions (Ozkaynak et al., 1987; Finley et al., 1988). Thus, in *S. cerevisiae* for example, ubiquitin is generated exclusively through proteolytic processing of precursors in which ubiquitin is joined either to itself, as in the (linear) polyubiquitin protein (UB14), or to unrelated (tail) amino acid sequences, as in the hybrid proteins UBI1, UBI2, and UBI3 (Fig. 1). In growing yeast cells, ubiquitin is generated largely from the UBI1–UBI3 hybrid proteins (Finley et al., 1987). The polyubiquitin (*UBI4*) gene is dispensable in growing cells but becomes essential (as the major supplier of ubiquitin) during stress (Finley et al., 1987).

The tail amino acid sequences of the ubiquitin precursors are highly conserved between yeasts and mammals (Finley et al., 1988), suggesting that they function similarly in all eucaryotes. The yeast *UBI1* and *UBI2* genes encode identical 52-residue tails,

Daniel Finley ■ Department of Cellular and Molecular Physiology, Harvard Medical School, Boston, Massachusetts 02115. **Bonnie Bartel and Alexander Varshavsky** ■ Department of Biology, Massachusetts Institute of Technology, Cambridge, Massachusetts 02139.

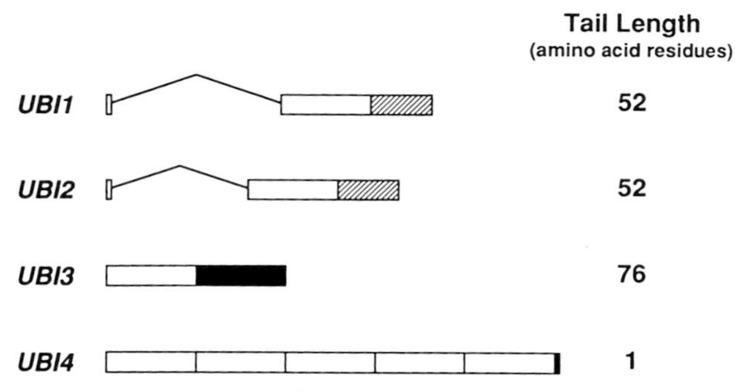

Figure 1. Organization of yeast ubiquitin genes. Open blocks represent the 228-base-pair ubiquitin-coding repeats; striped and shaded blocks represent tail-coding elements. *UBI1* and *UBI2* contain identically positioned, largely nonhomologous introns within their ubiquitin-coding elements (Ozkaynak et al., 1987).

whereas *UBI3* encodes a 76-residue tail with scant amino acid sequence similarity to the tails of the UBI1 and UBI2 proteins (Fig. 1). However, both the UBI1/UBI2 and UBI3 tails are highly basic and contain cysteine-rich, putative zinc-binding domains, suggesting that these proteins may bind nucleic acids (Ozkaynak et al., 1987).

DELETION ANALYSIS OF THE *UBI1–UBI3* GENES

To investigate the functions of the ubiquitin hybrid genes, we first precisely deleted the coding sequences of these genes from the *S. cerevisiae* genome, using previously described methods (Finley et al., 1987). Unlike the deletion of *UBI4* (the polyubiquitin gene) (Finley et al., 1987), deletions of the ubiquitin hybrid genes resulted in slow-growth phenotypes, with the effect of the *ubi3* deletion being much more severe than that of either the *ubi1* or *ubi2* deletion. The phenotypes of the *ubi1* and *ubi2* single mutants can be viewed essentially as gene dosage effects, because *UBI1* and *UBI2* encode identical proteins (Ozkaynak et al., 1987) and because the *UBI2* gene, carried on a centromere-containing (CEN) plasmid, fully complemented the growth defect of the *ubi1* mutant. The *ubi1 ubi2* double mutant is inviable, as indicated by our inability to recover *ubi1 ubi2* meiotic segregants upon sporulation of a *ubi1*/+ *ubi2*/+ diploid and by the fact that a yeast plasmid carrying the *UBI2* gene and an origin of replication, but lacking a CEN element, was fully stabilized against mitotic loss in a *ubi1 ubi2* genetic background (data not shown). Thus, *UBI1* and *UBI2* encode an essential protein and appear to be functionally interchangeable genes.

There is no correlation between growth rates of the various deletion mutants and their levels of free ubiquitin (Finley et al., 1989). The growth defects of the *ubi1*, *ubi2*, and *ubi3* deletion mutants must therefore be due at least in part to either the absence or decrease in concentration of specific tail components of the UBI1–UBI3 proteins.

To dissect the contributions of ubiquitin and the tails to the functions of the ubiquitin hybrid genes, genetic elements encoding the tail and ubiquitin moieties were separated from each other in plasmid constructs, introduced into the mutant strains described above, and tested for their ability to complement the cognate deletion mutations. Both UBI1 and UBI3 tails, when overexpressed in a ubiquitin-free form, were capable of complete complementation (Finley et al., 1989). In contrast, free ubiquitin, when expressed at a rate comparable to that of UBI1 to UBI4 combined, also failed to complement *ubi1* and *ubi3*. These results indicated that the tails of ubiquitin hybrid proteins have growth-related functions and that location of the tails within ubiquitin hybrid proteins is not strictly essential for tail function (Finley et al., 1989).

It has been proposed that the ubiquitin-tail fusion arrangement may instead be required for proper ubiquitin function (Lund et al., 1985), specifically that the tails, by virtue of their apparent nuclear targeting signals (Lund et al., 1985; Ozkaynak et al., 1987), mediate transport of ubiquitin to the nucleus. To test this possibility, a strain entirely lacking ubiquitin hybrid proteins was generated. We constructed a plasmid that expressed both UBI1 and UBI3 tails in a ubiquitin-free form and transformed it into a *ubi1*/+, *ubi2*/+, *ubi3*/+ diploid, which was then sporulated and subjected to tetrad analysis. In the presence of the free tail-expressing plasmid, *ubi1 ubi2 ubi3* meiotic products could be recovered as viable cells, indicating that linkage of the tails to ubiquitin within the ubiquitin hybrid proteins does not support an essential function of ubiquitin. The polyubiquitin gene, *UBI4*, the only ubiquitin gene that remained in the *ubi1 ubi2 ubi3* mutants, was strongly induced in this genetic background (unpublished data).

DEFICIENCIES OF RIBOSOMAL SUBUNITS IN *ubi1* AND *ubi3* MUTANTS

In the course of a Northern (RNA) hybridization experiment, RNA preparations from the various *ubi1–ubi3* mutants were analyzed electrophoretically. In doing so, we discovered a strikingly substoichiometric ratio of 18S to 25S rRNA in the *ubi3* mutants (Fig. 2a). The 18S and 25S rRNAs reside within the small (40S) and large (60S) ribosomal

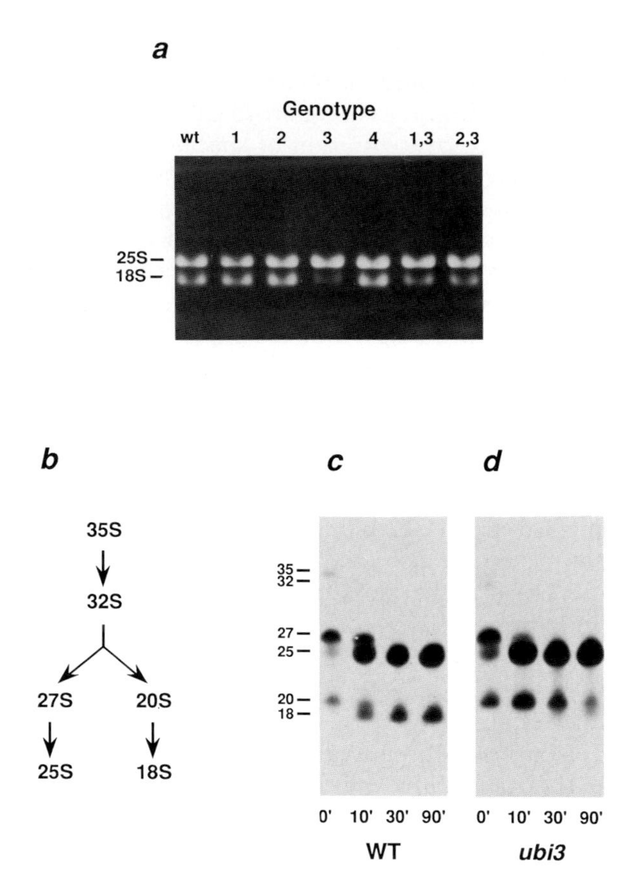

Figure 2. Defective pre-rRNA processing and reduced levels of 18S rRNA in the *ubi3* mutant. (a) RNA was prepared from mutant and wild-type cells grown in yeast extract-peptone-dextrose, electrophoresed on an agarose gel in the presence of ethidium bromide, and photographed under UV light. Lanes are numbered according to the ubiquitin gene or genes deleted in the corresponding strain (e.g., lane 1, *ubi1* strain). (b) Schematic representation of the pathway of pre-rRNA processing in *S. cerevisiae*. 35S RNA is the primary transcript. (c and d) Pulse-chase analysis of pre-rRNA processing in wild-type and *ubi3* deletion strains. rRNA precursors in exponentially growing yeast cells were labeled with [*methyl*-^{3}H]methionine for 6 min at 23°C. ([*methyl*-^{3}H]methionine donates methyl groups to newly synthesized pre-rRNA [Warner, 1981] via the general methyl donor *S*-adenosylmethionine.) Unlabeled methionine was added to initiate the chase incubation, and RNA was prepared from the cells 0, 10, 30, and 90 min later, as indicated. RNA was electrophoresed on a 1% agarose gel, followed by fluorography. For details, see Finley et al. (1989).

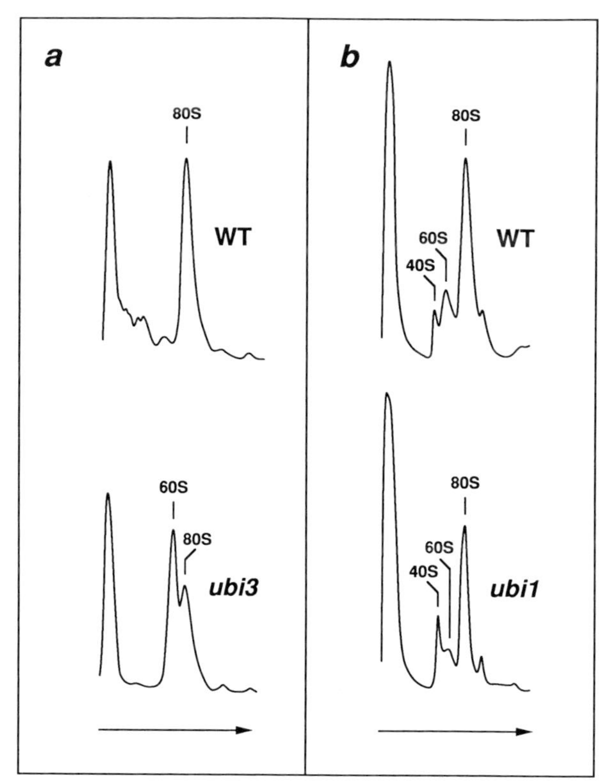

Figure 3. Aberrant stoichiometries of ribosomal subunits in *ubi1* and *ubi3* mutants. Sedimentation analysis of subunits present in cytoplasmic extracts from spheroplasts of either wild-type and *ubi3* strains (a) or wild-type and *ubi1* strains (b). Direction of sedimentation is indicated by an arrow. For details, see Finley et al. (1989).

subunits, respectively, and their normal 1:1 stoichiometry is maintained in part by their derivation from a single precursor transcript (Planta and Raué, 1988; Warner, 1981).

To test whether the deficiency of 18S rRNA in the *ubi3* mutant was accompanied by a deficiency of small ribosomal subunits, extracts from *ubi3* and wild-type cells were prepared under conditions that dissociate translating polyribosomes (Carter et al., 1980). The extracts were sedimented through 15 to 30% sucrose gradients at low ionic strength, which promotes the association of large and small ribosomal subunits into 80S ribosomes (monosomes). The major sedimentation peak in the wild-type extract was that of monosomes (Fig. 3a). In contrast, the major peak in the *ubi3* extract was composed of free large (60S) subunits, whereas the monosome (80S) peak was greatly reduced in magnitude. These results indicated that the *ubi3* mutant contains substoichiometric quantities of small ribosomal subunits.

To determine whether the *ubi3* mutant was defective in the formation (rather than metabolic stability) of 40S ribosomal subunits, we examined the processing of pre-rRNA and wild-type cells. Pre-rRNA processing was monitored by using a pulse-chase procedure, followed by agarose gel electrophoresis and fluorography. The results are shown in Fig. 2c and d, together with a simplified representation of the pre-rRNA-processing pathway (Fig. 2b). After a 6-min pulse, the predominant labeled species are the 27S and 20S RNAs, which are immediate precursors of the mature 25S and 18S rRNAs, respectively. In wild-type cells at 30°C, these intermediate species are processed to completion within 30 min; in the *ubi3* cells, however, the rate at which the 18S rRNA is generated from the 20S pre-rRNA is dramatically reduced, and the bulk of the 20S pre-rRNA is ultimately degraded rather than processed. In contrast, conversion of the 27S precursor to the 25S rRNA (residing in the large ribosomal subunit) is normal in the *ubi3* mutant (Fig. 2d). Thus, a specific pre-rRNA-processing defect explains the aberrant stoichiometry of rRNA in the *ubi3* mutant and accounts at least in part for its growth defect.

Remarkably, both the *ubi1 ubi3* and *ubi2 ubi3* double mutants, despite their relatively low free ubiquitin levels, grow faster than the *ubi3* single mutant. This is particularly striking since the *ubi1* and *ubi2* single mutants grow slower than wild-type cells. Thus, *ubi1* and *ubi2* are both weak suppressors of *ubi3*. *ubi1* and *ubi2* also partially restore the stoichiometry of 18S rRNA to 25S rRNA in the *ubi3* genetic background (Fig. 2a). This result indicates that *UBI1* and *UBI2* are, like *UBI3*, involved in some aspect of ribosome assembly or turnover. Although the ratio of 18S rRNA to 25S rRNA, as estimated from gel electrophoretic patterns (Fig. 2a), is not noticeably abnormal in either *ubi1* or *ubi2* mutants, a more sensitive assay, sucrose gradient sedimentation, revealed a deficiency of 60S subunits in the *ubi1* mutant, in contrast to the *ubi3* mutant, which was deficient in 40S subunits (Fig. 3b). A slight reduction in the monosome level (relative to total RNA in the extract) was also apparent in the *ubi1* cells (Fig. 3b; compare the 80S peak with the peak at the top of the gradient). Although the 60S subunit deficiency of *ubi1* cells is moderate, it is presumably sufficient to account for the mild growth defect of these cells, which still express (from the *UBI2* gene) a protein identical to UBI1.

The partial suppression of the *ubi3* growth defect by the *ubi1* and *ubi2* mutations may result from their lowering of 60S subunit levels, thus partially alleviating the relative deficiency of 40S subunits, perhaps without significantly affecting the absolute levels of 40S subunits. Although the coupling of 40S and 60S subunits within the ribosome occurs only after the 40S subunit associates with mRNA, the two subunits have significant residual affinity for each other in the absence of mRNA as well (van Holde and Hill, 1974). As a result, the abnormally high level of free 60S subunits in the *ubi3* mutant (Fig. 3a) may result in partial sequestration of 40S subunits, thereby further decreasing the concentration of free 40S subunits available for translation in *ubi3* cells. This effect, analogous to the effect of squelching described for transcriptional regulators (Gill and Ptashne, 1988), should weaken upon decrease in the level of 60S subunits, thus possibly accounting for the suppressive effect of the *ubi1* and *ubi2* mutations in the *ubi3* genetic background.

THE TAILS ARE RIBOSOMAL PROTEINS

The findings discussed above, which indicated that the UBI3 tail participates in ribosome biogenesis, suggested that this tail was either a component of the pre-rRNA-processing apparatus (which appears to include a variety of small nuclear ribonucleoproteins [Tollervey, 1987]) or a ribosomal protein whose incorporation within the nascent ribosome facilitates recognition of the 20S pre-rRNA by specific processing factors. To distinguish between these possibilities, we first added an immunological marker to the carboxyl terminus of the UBI3 protein by fusing the *UBI3* gene with a sequence encoding a 12-residue epitope from the human c-Myc protein (Munro and Pelham, 1986). The resulting *UBI3-myc* gene was found to fully complement the growth defect of the chromosomal *ubi3* deletion, indicating that the carboxyl-terminal Myc peptide did not obstruct the activity of the UBI3 protein. Whole-cell extracts of wild-type (*UBI3*) cells and of *UBI3-myc* cells carrying a chromosomal deletion of *ubi3* were fractionated by sodium dodecyl sulfate-polyacrylamide gel electrophoresis, transferred onto filters, and probed with a monoclonal antibody to the Myc peptide. A single band was observed in immunoblots of *UBI3-myc* cells, whereas extracts from wild-type (*UBI3*) cells were essentially free of antibody-binding species (Fig. 4a, lane W). The immunoreactive protein is apparently the deubiquitinated form of UBI3-Myc, since it is not recognized by antibodies to ubiquitin, and its apparent molecular weight is considerably less than that of *Escherichia coli*-produced UBI3-Myc (Fig. 4a). Thus, the UBI3 hybrid protein is efficiently processed (deubiquitinated) in vivo. Rapid in vivo deubiquitination in eucaryotic cells (but not in bacteria) is also characteristic of engineered ubiquitin fusions such as ubiquitin–β-galactosidase (Bachmair et al., 1986).

Extracts from wild-type and *UBI3-myc* cells were analyzed on sucrose gradients under conditions optimized for the resolution of polyribosomes (Fig. 5a). The proteins of the monosome and polyribosome peaks were then fractionated by sodium dodecyl sulfate-polyacrylamide gel electrophoresis, transferred onto filters, and probed with the anti-*myc* antibody. The *myc*-tagged UBI3 tail protein was found primarily within rapidly sedimenting structures and cosedimented with both 80S monosomes and the broadly distributed polyribosomes (Fig. 5a). A second extract was prepared from *UBI3-myc* cells

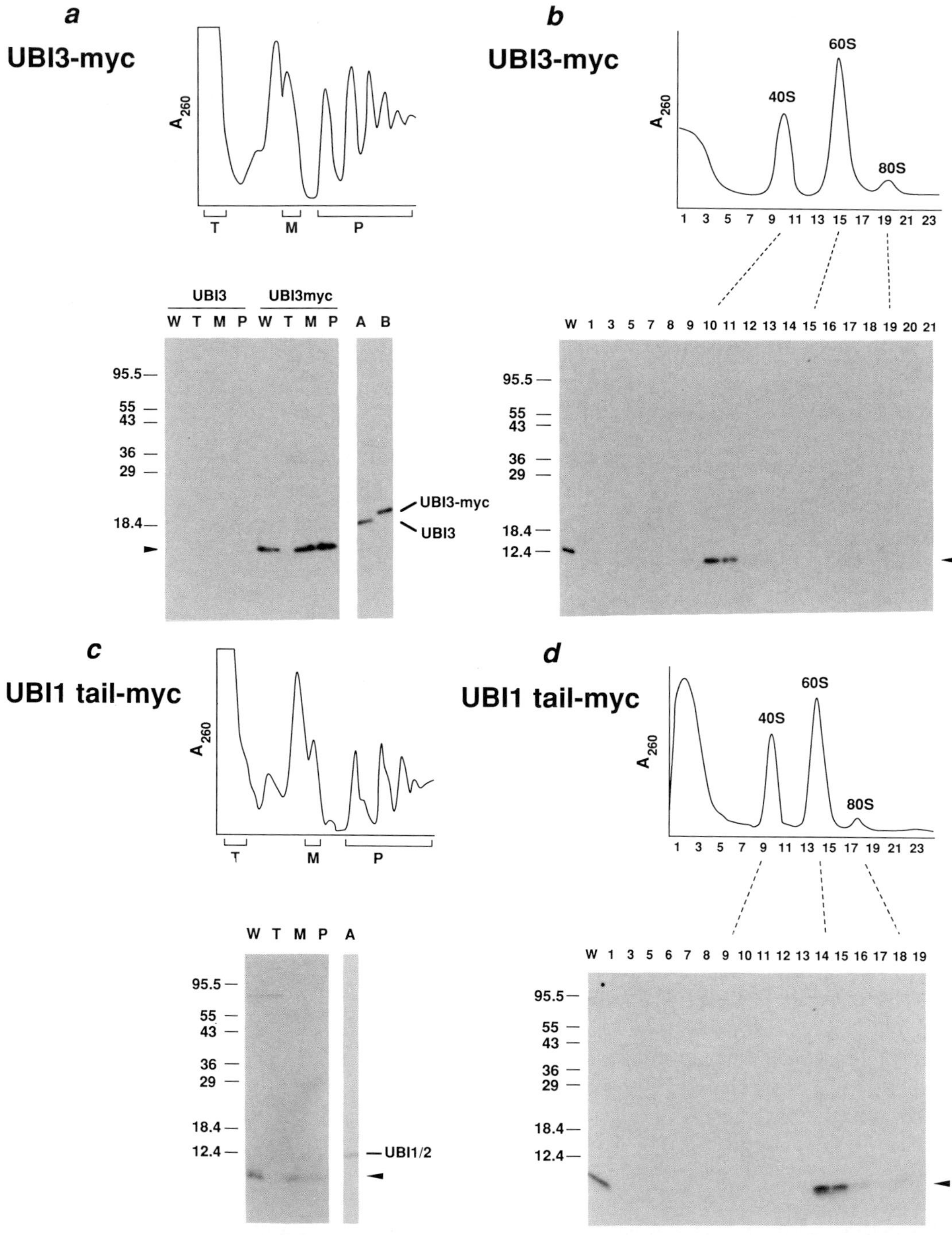

Figure 4. Immunoblot analysis of *S. cerevisiae* extracts fractionated in sucrose gradients to resolve either polyribosomes (a and c) or dissociated 40S and 60S subunits (b and d). In panels a and c, fractions were pooled as shown: W, whole-cell extract; T, top of gradient; M, monosomes; P, polyribosomes. For details, see Finley et al. (1989). (a) Polyribosomal fractions from *ubi3* cells carrying either *UBI3* (lanes 1 to 4) or *UBI3-myc* (lanes 5 to 8) on a plasmid, probed with anti-*myc* antibody. Lanes A and B show the unprocessed (nondeubiquitinated) UBI3 and UBI3-Myc proteins produced in *E. coli*, probed with antiubiquitin antibody. (b) Ribosomal subunit fractions from *ubi3* cells carrying *UBI3-myc* on a CEN plasmid, probed with anti-*myc* antibody. (c) Polyribosomal fractions from *ubi1 ubi2* cells carrying *UBI1* tail-*myc* on a 2μm plasmid, probed with anti-*myc* antibody. Lane A shows unprocessed UBI1 produced in *E. coli*, probed with antiubiquitin antibody. (d) Ribosomal subunit fractions from *ubi1 ubi2* cells carrying *UBI1 tail-myc* on a 2μm plasmid, probed with anti-*myc* antibody. In panel c, the monosome and polyribosome peaks each show a shoulder of more rapidly sedimenting material. Similar results have been obtained with unrelated mutants deficient in large ribosomal subunits (Rotenberg et al., 1988), as is the *ubi1* strain (Fig. 3). The shoulder appears to represent mRNAs that have bound the 40S ribosomal subunit to form a preinitiation complex but have not proceeded to bind a 60S subunit because the level of free 60S subunits is low (Rotenberg et al., 1988).

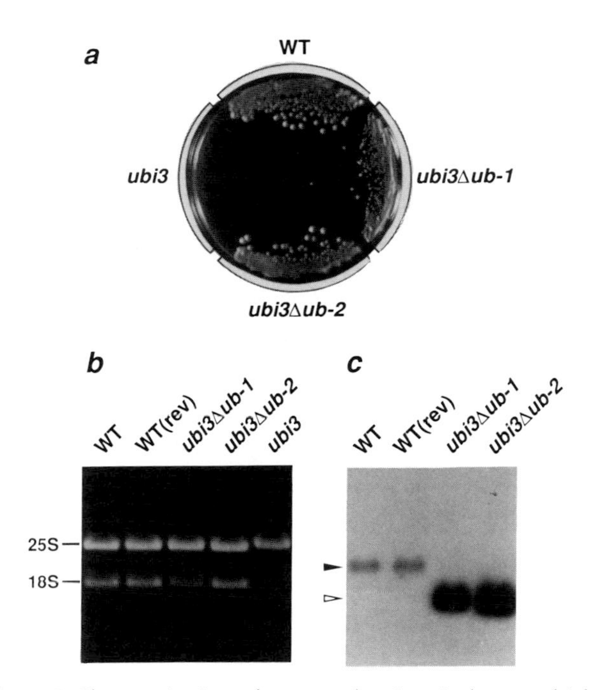

Figure 5. Characterization of mutants bearing single or multiple copies of the *ubi3Δub* allele. (a) Growth defect of the *ubi3Δub-1* mutant. Cultures were streaked onto yeast extract-peptone-dextrose plates and incubated for 3 days at 30°C. Quadrant designations are as follows: WT, wild type; *ubi3Δub-1*, a growth revertant obtained through integrative transformation of a *ubi3* strain with a DNA fragment carrying the *ubi3Δub* deletion mutation (single-copy integration); *ubi3Δub-2*, as *ubi3Δub-1* except that three to four copies of the transforming DNA fragment integrated tandemly at the *UBI3* locus; *ubi3*, which forms sizable colonies only after 1 week. (b) 18S rRNA deficiency in the *ubi3Δub-1* mutant. RNA was resolved on an agarose gel in the presence of ethidium bromide and photographed under UV light. Designations are as in panel a except for lane WT(rev), which contained RNA from a phenotypically wild-type revertant, obtained through integrative transformation of a *ubi3* strain with a DNA fragment carrying the wild-type *UBI3* gene. (c) Northern hybridization analysis of *UBI3* RNA levels in mutant strains. ►, *UBI3* RNA; ▷, *ubi3Δub* RNA. For details, see Finley et al. (1989).

under conditions that dissociate polyribosomes and centrifuged in the presence of a high-ionic-strength buffer, which dissociates monosomes into free subunits. Under these conditions, the UBI3 tail cosedimented with the 40S ribosomal subunit (Fig. 4b). The data of Fig. 4a and b show that the UBI3 tail is a component of the small subunit of the ribosome. The UBI3 tail does not appear to be a processing or assembly factor transiently associated with nascent 40S subunits, because it is also present within translationally active ribosomes. That the association of the UBI3 tail with the 40S subunit persists under high-salt conditions (0.5 M KCl) argues that the UBI3 tail is tightly associated with mature ribosomes.

To localize the UBI1 tail protein, we again used the *myc* tagging method. A *UBI1 tail-myc* gene in which the *myc* sequence was positioned upstream of the tail-coding sequence, in place of the ubiquitin-coding sequence, was found to rescue the viability of *ubi1 ubi2* mutants and to restore their growth rates to near wild-type levels. Extracts from *UBI1 tail-myc* cells were fractionated on sucrose gradients optimized for the analysis of either polyribosomes or ribosomal subunits, as described above for *UBI3-myc*. On the polyribosome gradient, the UBI1 tail cosedimented with 80S monosomes and polyribosomes (Fig. 4c), whereas on the ribosomal subunit gradient, the UBI1 tail cosedimented with the large (60S) subunit (Fig. 4d). Thus, like the UBI3 tail, the UBI1 tail is tightly associated with mature, translationally active ribosomes. We conclude that the UBI3 tail and the (identical) UBI1 and UBI2 tails are ribosomal proteins that reside in the small and large subunits, respectively. The major structural motifs of the UBI1–UBI3 proteins, their putative zinc fingers (Ozkaynak et al., 1987), are presumably involved in binding rRNA.

Although both *UBI1 tail-myc* and *UBI3-myc* were found in ribosomes, their localization to distinct ribosomal subunits precludes the possibility that these associations were mediated artifactually by the Myc peptide. Moreover, the human homolog of the UBI3 tail protein has independently been localized to the small ribosomal subunit and identified as the ribosomal protein S27a (Redman and Rechsteiner, 1989). The corresponding yeast protein is known as YS24 in the nomenclature of Otaka et al. (1984). Our data do not exclude the possibility that fractions of the UBI1 and UBI2 tails are present within yeast ribosomes in an unprocessed, ubiquitin-containing form. However, the presence of significant levels of free ubiquitin in *ubi3 ubi4* mutants indicates that the *UBI1* and *UBI2* gene products are subject to deubiquitination (Finley et al., 1989); furthermore, the full complementing activity of the free UBI1 tail indicates that ribosomes containing the free UBI1 tail are functionally indistinguishable from wild type.

A reexamination of the previously determined (Ozkaynak et al., 1987) nucleotide sequences upstream of the *UBI1–UBI3* genes revealed, in each case, near matches to the consensus sequence (Planta and Raué, 1988) of upstream activation sites within genes for ribosomal proteins. (Such sites have also been found within a few genes encoding nonribosomal proteins.) The *UBI1*, *UBI2*, and *UBI3* promoter regions also contain T-rich stretches that represent a second sequence motif characteristic of promoters of ribosomal protein genes (Planta and Raué, 1988). In *UBI1*, *UBI2*, and *UBI3*, the positioning of the upstream activation site and T-rich sequences with respect to each other and with respect to sites of translational initiation is similar to that of known ribosomal protein genes and consistent with the proposed function of these sequences in transcriptional regulation. With the exception of genes for ribosomal proteins, few yeast genes contain introns; their presence within the *UBI1* and *UBI2* genes further indicates that the sequence features of *UBI1*, *UBI2*, and *UBI3* are typical of ribosomal protein genes.

UBIQUITIN AND RIBOSOME BIOGENESIS

The tandem arrangement of ubiquitin and specific ribosomal proteins (tails) within proteolytically processed precursors has been conserved throughout the evolution of eucaryotes. Nevertheless, expression of the tails in ubiquitin-free form can fully complement chromosomal mutations in which the corresponding ubiquitin tail gene is entirely deleted (see above). To test further for a function of the cotranslational synthesis of ubiquitin and the tail proteins, we integrated the free tail-expressing derivative of the *ubi3* gene (*ubi3Δub*) into its natural chromosomal locus. The complete deletion mutant, *ubi3*, was transformed with linear DNA fragments carrying either the ubiquitin deletion mutation *ubi3Δub* or the wild-type *UBI3* gene, and growth revertants were selected. Growth revertants resulting from transformation with *ubi3Δub* DNA were predominantly of two classes: those growing at wild-type rates (1.6-h doubling time) and those in which the *ubi3* growth defect had been only partially reverted (3-h doubling time, as compared with the 6.8-h doubling time of the *ubi3* mutant). Southern hybridization analysis indicated that in all revertants with wild-type growth rates, the *ubi3* mutation had been replaced by multiple tandem insertions of *ubi3Δub* DNA, whereas in all partial growth revertants, *ubi3* had been replaced with a single, properly integrated *ubi3Δub* sequence (see Fig. 5a for a comparison of growth rates; we refer to the integrated, single-copy *ubi3Δub* allele as *ubi3Δub-1* and to the integrated multiple-copy allele taken for analysis as *ubi3Δub-2*). In contrast, single-copy integration of wild-type (*UBI3*) DNA resulted in complete reversion of the *ubi3* growth defect (Fig. 5a). Thus, full complementation of the *ubi3* phenotype requires approximately three- to fourfold amplification of the *ubi3Δub* copy number.

To determine the stoichiometry of 18S to 25S rRNA in the *ubi3Δub-1* mutant (which contained the integrated, single-copy *ubi3Δub* allele), RNA from this strain and from controls was electrophoresed on an agarose gel and visualized by ethidium staining (Fig. 5b). A deficiency of 18S rRNA was clearly apparent in the *ubi3Δub-1* mutant, although it was not as pronounced as in the *ubi3* mutant (Fig. 5b). A wild-type (1:1) stoichiometry of 18S to 25S rRNA was seen in the *ubi3Δub-2* mutant, which contained multiple copies of *ubi3Δub* (Fig. 5b). Thus, deletion of the ubiquitin-coding element of *UBI3* leads to a deficiency of 18S rRNA which can be suppressed by amplifying the copy number of the free tail-expressing allele of the *UBI3* gene. This result indicates that the ubiquitin-coding portion of the *UBI3* gene, although not strictly required for the processing of 20S pre-rRNA, facilitates this processing reaction, apparently by increasing in some manner the efficiency of incorporation of the UBI3 tail into the nascent 40S ribosomal subunit.

To test whether the effect of deleting the ubiquitin-coding sequence of *UBI3* on tail function was achieved through a posttranslational mechanism, *UBI3* mRNA levels were determined in *ubi3Δub-1* and control strains. Northern hybridization analysis (Fig. 5c) showed that the effect of the *ubi3Δub-1* mutation on UBI3 tail function is not achieved through decreasing the level of *UBI3* mRNA; instead, the *ubi3Δub-1* mRNA is considerably more abundant than that of *UBI3*. Thus, the primary functional defect of the *ubi3Δub-1* mutant lies in either a cotranslational or posttranslational process. We conclude that the effective levels of newly synthesized UBI3 tail protein, as reflected by its capacity to accelerate the processing of 20S pre-rRNA (Fig. 3 and 5), are significantly enhanced by its transient covalent association with ubiquitin.

POSSIBLE MECHANISMS OF UBIQUITIN FUNCTION

Ubiquitin has previously been shown to function as a posttranslational modifying group which signals the degradation of acceptor proteins. Here we describe a second mechanism of ubiquitin function in which ubiquitin is joined to an acceptor protein cotranslationally and serves to promote the assembly of the acceptor protein into a specific cellular structure. The role that we have described for ubiquitin in the biogenesis of ribosomes in yeast cells is likely to apply to eucaryotes in general, since close homologs of the *UBI1–UBI3* genes exist in animals, plants, and lower eucaryotes (Finley et al., 1988). Moreover, sequences homologous to the tail components of *UBI1–UBI3* have invariably been found in the form of fusions to ubiquitin.

UBI1–UBI3 form a subset of a larger class of genes in which an element encoding either ubiquitin or an amino acid sequence similar to that of ubiquitin is fused to a downstream coding sequence. Although no systematic effort to identify such genes has yet been reported, four loci encoding carboxyl-terminally extended ubiquitinlike polypeptides have already been described: the *AN-1* gene, whose transcripts are segregated to the animal pole of *Xenopus laevis* embryos (D. Weeks and D. Melton, personal communication); a human gene that encodes an interferon-inducible 15-kilodalton protein (Blomstrom et al., 1986; Haas et al., 1987); the constitutively expressed human gene *GdX* (Toniolo et al., 1988); and the large open reading frame of the togavirus BVDV, which encodes a ubiquitinlike sequence embedded within the nonstructural viral polyprotein (Meyers et al., 1989). Because of their substantial divergence from the canonical, highly conserved ubiquitin sequence, the ubiquitinlike proteins, in contrast to the UBI1–UBI3 proteins, cannot serve as precursors to ubiquitin. Moreover, the processing proteases that deubiquitinate ubiquitin fusion proteins are expected to be virtually inactive upon the ubiquitinlike proteins, since in each of these proteins the bond that would otherwise be subject of cleavage is flanked by amino acid substitutions that block (or are expected to block) cleavage of engineered ubiquitin fusion proteins (Bachmair et al., 1986; Butt et al., 1988). We suggest that the function of ubiquitinlike protein domains may be to enhance the function of linked nonubiquitin domains in a way similar to that seen in the UBI3 protein.

It has recently been shown that certain proteins, known as molecular chaperones, associate transiently (and noncovalently) with newly synthesized polypeptide chains and thereby facilitate their incorporation into oligomeric structures (Ellis, 1987; Hemmingsen et al., 1988). By definition, molecular chaperones do not form part of the final oligomeric structure, nor do they necessarily possess steric information specifying the assembly of the structure (Ellis, 1987). By these criteria, it appears that the ubiquitin moiety of the UBI3 protein acts as a molecular chaperone of the UBI3 tail in facilitating its incorporation into ribosomes. The covalent, cotranslational association of a chaperone protein with its ligand presents a number of potential mechanistic advantages: the amino-terminal chaperone is bound to and can protect its ligand from the moment the ligand is synthesized; the chaperone can transiently block, or shelter, the amino terminus of its ligand (which may serve in some proteins as a signal for degradation [Bachmair et al., 1986]); and finally, the cotranslational chaperone can bind its ligand without the necessity of having a specific recognition site in the ligand.

It is probable that the ubiquitin tails, like other ribosomal proteins (Planta and Raué, 1988), are extremely short-lived if not assembled within ribosomes. Thus, the transient attachment of ubiquitin to the amino termini of the UBI1–UBI3 tails could increase the efficiency of their incorporation into nascent ribosomes via transient metabolic stabilization of the newly formed tails. Although ubiquitination of a protein can signal its degradation, this activity appears to require the formation of multiubiquitin chains (Chau et al., 1989). Rather than directly protecting the UBI1–UBI3 tails against degradation, ubiquitination could alternatively increase the rate of transport of the tail to or assembly within nascent ribosomes and in this way decrease the proportion of newly synthesized tails that are degraded in transit to the ribosome. Experiments to distinguish among these possibilities are under way.

We thank Mary Jalenak for skilled assistance in characterizing mutants, John McGrath for carrying out sequence searches, Joan Park and Andreas Bachmair for helpful discussions, and Daniel Weeks and Douglas Melton for permission to cite unpublished data. We also thank Rita Cronan for secretarial assistance.

This work was supported by grants to Alexander Varshavsky from the National Institutes of Health. Bonnie Bartel was supported by a predoctoral fellowship from the National Science Foundation.

REFERENCES

Bachmair, A., D. Finley, and A. Varshavsky. 1986. In vivo half-life of a protein is a function of its amino-terminal residue. *Science* **234:**179–186.

Blomstrom, D. C., D. Fahey, R. Kutney, B. D. Korant, and E. Knight, Jr. 1986. Molecular characterization of the interferon-induced 15-kDa protein. *J. Biol. Chem.* **261:**8811–8816.

Butt, T. R., M. I. Khan, J. Marsh, D. J. Ecker, and S. T. Crooke. 1988. Ubiquitin-metallothionein fusion protein expression in yeast. *J. Biol. Chem.* **263:**16364–16371.

Carter, C. J., M. Cannon, and A. Jimenez. 1980. A trichodermin-resistant mutant of Saccharomyces cerevisiae with an abnormal distribution of native ribosomal subunits. *Eur. J. Biochem.* **107:**173–183.

Chau, V., J. W. Tobias, A. Bachmair, D. Marriot, D. Ecker, D. K. Gonda, and A. Varshavsky. 1989. A multiubiquitin chain is confined to a specific lysine in a targeted short-lived protein. *Science* **243:**1576–1583.

Ellis, J. 1987. Proteins as molecular chaperones. *Nature* (London) **328:**378–379.

Finley, D., B. Bartel, and A. Varshavsky. 1989. The tails of ubiquitin precursors are ribosomal proteins whose fusion to ubiquitin facilitates ribosome biogenesis. *Nature* (London) **338:**394–401.

Finley, D., E. Ozkaynak, S. Jentsch, J. P. McGrath, B. Bartel, M. Pazin, R. M. Snapka, and A. Varshavsky. 1988. Molecular genetics of the ubiquitin system, p. 39–75. *In* M. Rechsteiner (ed.), *Ubiquitin*. Plenum Publishing Corp., New York.

Finley, D., E. Ozkaynak, and A. Varshavsky. 1987. The yeast polyubiquitin gene is essential for resistance to high temperature starvation and other stresses. *Cell* **48:**1035–1046.

Finley, D., and A. Varshavsky. 1985. The yeast ubiquitin system: functions and mechanisms. *Trends Biochem. Sci.* **10:**343–346.

Gill, G., and M. Ptashne. 1988. Negative effect of the transcriptional activator GAL4. *Nature* (London) **334:**721–724.

Goebl, M. G., J. Yochem, S. Jentsch, J. P. McGrath, A. Var-

shavsky, and B. Byers. 1988. The yeast cell cycle gene CDC34 encodes a ubiquitin-conjugating enzyme. *Science* **241**:1331–1335.

Goff, S. A., R. Voellmy, and A. Goldberg. 1988. Protein breakdown and the stress response, p. 207–238. *In* M. Rechsteiner (ed.), *Ubiquitin*. Plenum Publishing Corp., New York.

Haas, A. L., P. Ahrens, P. M. Bright, and J. Ankel. 1987. Interferon induces a 15-kilodalton protein exhibiting marked homology to ubiquitin. *J. Biol. Chem.* **262**:11315–11323.

Hemmingsen, S. M., C. Woolford, S. M. van der Vies, K. Tilly, D. T. Dennis, C. P. Georgopoulis, R. W. Hendrix, and R. J. Ellis. 1988. Homologous plant and bacterial proteins chaperone oligomeric protein assembly. *Nature* (London) **333**:330–334.

Hershko, A. 1988. Ubiquitin-mediated protein degradation. *J. Biol. Chem.* **263**:15237–15240.

Jentsch, S., J. P. McGrath, and A. Varshavsky. 1987. The yeast DNA repair gene *RAD6* encodes a ubiquitin-conjugating enzyme. *Nature* (London) **329**:131–134.

Lund, P. K., B. M. Moats-Staats, J. G. Simmons, E. Hoyt, J. D. E'Ercole, F. Martin, and J. J. Van Wyk. 1985. Nucleotide sequence analysis of a cDNA encoding human ubiquitin reveals that ubiquitin is synthesized as a precursor. *J. Biol. Chem.* **260**:7609–7613.

Meyers, G., T. Rumenapf, and H. J. Thiel. 1989. Ubiquitin in a togavirus. *Nature* (London) **341**:491.

Munro, S., and H. R. B. Pelham. 1986. An Hsp70-like protein in the ER: identity with the 78 kd glucose-regulated protein and immunoglobulin heavy chain binding protein. *Cell* **46**:291–300.

Otaka, E., K. Higo, and T. Itoh. 1984. Yeast ribosomal proteins. *Mol. Gen. Genet.* **195**:544–546.

Ozkaynak, E., D. Finley, M. J. Solomon, and A. Varshavsky. 1987. The yeast ubiquitin genes: a family of natural gene fusions. *EMBO J.* **6**:1429–1439.

Pickart, C. M. 1988. Ubiquitin activation and ligation, p. 77–100. *In* M. Rechsteiner (ed.), Plenum Publishing Corp., New York.

Planta, R. J., and H. A. Raué. 1988. Control of ribosome biogenesis in yeast. *Trends Genet.* **4**:64–68.

Redman, K., and M. Rechsteiner. 1989. Identification of the long ubiquitin extension as ribosomal protein S27a. *Nature* (London) **338**:438–440.

Rotenberg, M. O., M. Moritz, and J. L. Woolford, Jr. 1988. Depletion of Saccharomyces cerevisiae ribosomal protein C16 causes a decrease in 60S ribosomal subunits and formation of half-mer polyribosomes. *Genes Dev.* **2**:160–172.

Tollervey, D. 1987. A yeast small nuclear RNA is required for normal processing of pre-ribosomal RNA. *EMBO J.* **6**:4169–4175.

Toniolo, D., M. Persico, and M. Alcalay. 1988. A "housekeeping" gene on the X chromosome encodes a protein similar to ubiquitin. *Proc. Natl. Acad. Sci. USA* **85**:851–855.

van Holde, K. E., and W. E. Hill. 1974. General physical properties of ribosomes, p. 53–91. *In* M. Nomura, A. Tissières, and P. Lengyel (ed.), *Ribosomes*. Cold Spring Harbor Laboratory, Cold Spring Harbor, N.Y.

Varshavsky, A., A. Bachmair, D. Finley, D. Gonda, and I. Wunning. 1988. The N-end rule of selective protein turnover: mechanistic aspects and functional implications, p. 207–238. *In* M. Rechsteiner (ed.), *Ubiquitin*. Plenum Publishing Corp., N.Y.

Warner, J. R. 1981. The yeast ribosome: structure, function and synthesis, p. 529–560. *In* J. Strathern, E. Jones, and J. Broach (ed.), *The Molecular Biology of the Yeast Saccharomyces cerevisiae: Metabolism and Gene Expression*. Cold Spring Harbor Laboratory, Cold Spring Harbor, N.Y.

Chapter 56

Phylogeny of Antibiotic Action

R. AMILS, L. RAMIREZ, J. L. SANZ, I. MARIN, A. G. PISABARRO, E. SANCHEZ, and D. UREÑA

Advances in our understanding of the complexity of the translational apparatus at the molecular level and the attractive characteristics of ribosomes, such as (i) the catalytic properties discovered in some RNAs (Kruger et al., 1982; Guerrier-Takada et al., 1983), which support the idea that these unusual macromolecules were involved in the design of the primordial living processes (Moore, 1988), and (ii) their utility as evolutionary clocks due to universal distribution, easy purification and analysis, constant function, and slow rate of mutation (Woese, 1982), have allowed interesting evolutionary issues to be explored.

After 30 years of accumulation of genetic, functional, and structural information, research on ribosomes can begin to address fundamental questions about function at the molecular level and the structure-function relationship of this complex cellular component, which involves the transformation of linear nucleic acid information into three-dimensional polypeptide structures of specific function. Furthermore, the wealth of comparative information on ribosomes has provided a unique data bank which facilitates the study of the evolution of ribosomal function.

In the past, living organisms had been classified according to the rules established by Linnaeus (1753), that is, on the basis of morphological traits. This kind of classification was feasible only when the organisms were of definite shape and their differences were identifiable. This methodology is not applicable to microorganisms because a wide range of organisms share features mainly as the result of convergent evolutionary processes. Whenever these macroscopic or phenotypic traits were the only source of taxonomic classification, misclassifications and "exceptional organisms" proliferated (Jones et al., 1987). Bearing this point in mind, it is important to consider that one of the classic methods of investigating phylogenetic relationships between organisms is study of their phenotypic relationships. This methodology is based on two assumptions: first, the greater the number of traits two organisms share, the closer their evolutionary history should be; and second, the hierarchy of taxonomic features must have a consistent evolutionary significance. Because these assumptions are not always true, as has been well documented in contemporary evolutionary studies, it was necessary to go a step further in the measurement of differences between organisms in order to understand their phylogenetic relationships.

The advent of molecular analysis techniques that allow rapid sequencing of nucleic acids and proteins partially resolved the problem. Direct comparison of DNA sequences allows the degree of genetic divergence between organisms to be measured and as a consequence gives a more accurate picture of their respective evolutionary paths. Nevertheless, the application and interpretation of sequence comparison data are difficult because of (i) our poor understanding of the rules governing selection, which are unavoidably related to the internal functions of gene products and to the external constraints of the ecological conditions in which the selection is produced, (ii) the recognition that horizontal transfer of information and efficient systems of recombination complicate this process, and (iii) the bias produced by the choice of chronometric genes and the systems of analysis (Sneath and Sokal, 1973; Gouy and Li, 1989).

Despite these problems, DNA sequence comparison is the most potent tool for the study of evolutionary relationships now available. However, this type of information can be obtained by using other macromolecules, such as RNA or proteins, provided

R. Amils, L. Ramirez, J. L. Sanz, I. Marin, A. G. Pisabarro, E. Sanchez, and D. Ureña ■ Centro de Biología Molecular, U.A.M.-C.S.I.C., Universidad Autonoma de Madrid, Canto Blanco, Madrid 28049, Spain.

that the general requirements of chronometric quality are fulfilled (Woese, 1982; Amils and Sanz, 1986).

In this respect, different chronometers have been used by several authors to study phylogenetic relationships between organisms; among these, rRNA genes are considered the best (Woese, 1987). These macromolecules are the core of the translational apparatus and fulfill all of the chronometric requirements mentioned above. Although 5S rRNA is much easier to sequence because of its smaller size, 16S-18S rRNA is more widely used to determine phylogenetic relationships because minor changes produce more disturbing noise in the 5S rRNA than in the 16S-18S rRNA (Woese, 1982). Recently 23S-28S rRNA analysis has been used to provide comparative validation of the phylogenetic trees obtained with the rRNA of the smaller subunit. Since important functional structures have been correlated with specific parts of the sequence of the 23S-28S rRNA, comparative information on the evolution of functional sites has been obtained (Leffers et al., 1987).

Our group has been involved in generating a comparative functional data bank based on studies of the inhibitory actions of antibiotics with diverse specificity (kingdom, functional, chemical, mode of action, etc.). Cell-free systems are used to avoid the interference produced in in vivo experiments by transport, inactivation, and regulation. Our inhibitory curves are studied in cell-free systems operating at optimal ionic conditions, using poly(U) as a messenger to avoid additional complication in analysis of the results by the radically different conditions in which the selected ribosomal species perform, e.g., 30 to 90°C or 0 to 4,000 mM monovalent cation concentration.

Our original goal was to ascertain the eubacterial or eucaryotic nature of the archaebacterial ribosomes (Amils and Sanz, 1986). In addition, we evaluated the phylogenetic value of the functional (phenotypic) analysis of the antibiotics. At present, we are exploring the possibilities that this system of analysis offers when it is used to complement genotypic or structural information. Our ultimate goal is to generate information that might contribute to an understanding of the evolution of the translational apparatus. This chapter reviews the development of the basic concepts in the field and the approaches used to generate information on the evolution of ribosomal function.

USE OF PROTEIN SYNTHESIS INHIBITORS AS PHYLOGENETIC MARKERS

The inhibitors of protein synthesis can be classified into three groups: inhibitors of eubacterial systems (group I), inhibitors of eucaryotic systems (group II), and inhibitors of both eubacterial and eucaryotic systems (group III) (Vazquez, 1979). This empirical classification, with obvious pharmaceutical applications, was established before archaebacteria were belatedly granted kingdom status because of their obvious genotypic differences from the eubacteria and eucaryotes (Woese and Fox, 1977).

In this context and because of the reported archaebacterial differences (reviewed in Amils and Sanz [1986]), the question of the sensitivity of archaebacteria to ribosomal antibiotics was pertinent. The sensitivities of archaebacteria to inhibitors of protein synthesis in in vivo systems were first reported by Hammes et al. (1979), Hilpert et al. (1981), and Pecher and Böck (1981) in the early 1980s.

To overcome the problems that might complicate interpretation of the results when in vivo experiments are performed (membrane permeability, inactivation of the antibiotic, existence of a specific target site for the antibiotic in the ribosome, or possible pleiotropic effects of the antibiotic) (reviewed in Böck and Kandler, 1985), it was necessary to develop cell-free systems that tested sensitivity in experimental conditions in which problems of permeability, inactivation, or side effects could be eliminated (Amils and Sanz, 1986; Eldhardt and Böck, 1982; Cammarano et al., 1985; Sanz et al., 1987; Altamura et al., 1988; Sanz et al., 1988). Our group, in collaboration with Böck's group in Munich and Cammarano's group in Rome, has accumulated data on the sensitivity of the translational apparatus of different organisms belonging to the main archaebacterial groups with the aim of characterizing the eubacterial or eucaryotic features of the archaebacterial ribosomes (Amils et al., 1989).

In our first attempts to determine the phylogeny of different organisms, we used 13 cell-free systems belonging to the three primary kingdoms. The sensitivity of a given system to an antibiotic was recorded as 1 or 0, depending on the corresponding inhibition curve compared with that of the eubacterial (*Escherichia coli*) or eucaryotic (*Saccharomyces cerevisiae*) reference system (Amils and Sanz, 1986). This approach corroborated initial observations (Eldhardt and Böck, 1982) that archaebacterial ribosomes exhibited a mosaic pattern straddling the eubacterial and eucaryotic patterns. More importantly, statistical analyses of the data (multivariate factor analysis by principal components, cluster analysis and compatibility, and parsimony phylogenetic trees) demonstrated, despite the differences observed between archaebacterial organisms, significant statistical segregation of all of the archaebacterial ribosomes from the eubacterial and eucaryotic ones, offering additional and independent functional evidence to support the conclusion based on genotypic analysis that archaebacteria should be considered a new primary kingdom (Woese and Fox, 1977; Oliver et al., 1987).

Further analysis of the antibiotic sensitivity data required two considerations: (i) that controls were necessary to eliminate the possibility that reported insensitivities were produced by lack of a target site and not by secondary effects produced by the extreme conditions in which some archaebacterial ribosomes carry out protein synthesis (e.g., temperature and high salt concentration) and (ii) that the transformation of an inhibition curve to a 1 or 0 value in the matrix data was an extreme oversimplification that had to be overcome by trying to preserve the wealth of structural and functional data found in the inhibition curves (Amils et al., 1989).

Thermophilic eubacteria were used as controls to show that the lack of sensitivity of some thermophilic archaebacteria (e.g., *Sulfolobus solfataricus*) was related to the archaebacterial nature of the ribosomes and not to the temperature at which the microorganisms develop. The eubacteria *Thermus thermophilus* and *Bacillus stearothermophilus* exhibit eubacterial inhibition behavior (similar to that of the mesophilic *E. coli*) at 75°C, proving that the lack of inhibition of thermophilic archaebacteria cannot be related to an ecological adaptation to high temperatures or to an inactivation of the effectors at this temperature (Cammarano et al., 1985; Amils et al., 1989).

For the extreme halophilic archaebacteria, more elaborate controls were needed to determine whether the lack of inhibition exhibited by some antibiotics was due to the lack of an interaction site for the antibiotic or to competition for the binding site between the cationic antibiotics and the high concentration of cations required for the maintenance of functional structure. Using ribosomes for a halotolerant eubacterium, *Vibrio costicola*, which function both at low (similar to *E. coli*) and at high ionic strength (same order of magnitude as halophilic cell-free systems [Sanz et al., 1988]), we have demonstrated that the apparent insensitivity of halophilic ribosomes to aminoglycosides is due to competition with the high concentration of monovalent cations (up to 4 M) present in the assay rather than to the lack of binding sites for those effectors. We cannot exclude the possibility that a structural change in *V. costicola* ribosomes produced by the change in cation concentration is responsible for the loss of aminoglycoside-binding sites, but we believe that this possibility is unlikely because polymerization activity is not changed drastically. Thus, the functional structures must be resistant to changes in the ionic strength; consequently, the structures responsible for binding the inhibitors of those functional sites must also be resistant. On the basis of these results, the aminoglycoside antibiotics were removed from the data bank when halobacteria were analyzed (Amils et al., 1989).

To use most of the structural and functional data provided by the inhibition curves, an algorithm capable of transforming the inhibition at different concentrations of antibiotic into a quantitative value that increases the inhibition values at lower effector concentration was used (Amils et al., 1989). The data matrix constructed (Table 1) was used to calculate Euclidian distances between organisms. These distances were then used to construct a tree by use of both a simple and a complete linkage clustering method (Fig. 1) (Dixon et al., 1983; Sneath and Sokal, 1973). The clusters obtained are basically in agreement with those obtained by using total sequence comparison of 16S-18S rRNA (Woese, 1987) and 23S rRNA (Pecher and Böck, 1981). As discussed previously (Amils et al., 1989), when functional analysis is used, *Thermoplasma acidophilum* and *Thermococcus celer* cluster together with the sulfur-dependent archaebacteria, in contrast to their proximity to the halophilic-methanogenic branch when total sequences of 16S-18S rRNA are analyzed (Woese, 1987; Achenbach-Richter and Woese, 1988; Achenbach-Richter et al., 1988). We cannot explain this difference, but it is worth mentioning that from a functional point of view, the ribosomes of both *Thermoplasma acidophilum* (Sanz et al., 1987) and *T. celer* (Altamura et al., 1988) are similar to those of sulfur-dependent archaebacteria. We have analyzed 10 sulfur-dependent archaebacteria and found that although they display differences in antibiotic sensitivity, they cluster together regardless of the statistical system used for the analysis (J. L. Sanz et al., unpublished results).

Despite these rather small differences, the strong correlation obtained between the phenetic classification of ribosomes on the basis of functional properties and the phylogenetic relationships generated by primary sequence (genotypic) comparison allows us to postulate that antibiotic sensitivity analysis can be used to investigate functional phylogenetic relationships, i.e., the evolution of the functional sites in ribosomes.

FUNCTIONAL PHYLOGENY

Many questions are posed by the remarkable coherence between phenotypic clustering and the

Table 1. Quantitative matrix data of the sensitivity of 10 cell-free systems to 28 protein synthesis inhibitors[a]

Antibiotic	Code	Antibiotic specificity group	Score									
			Methanobacterium formicicum	*Sulfolobus solfataricus*	*Desulfurococcus mobilis*	*Thermococcus celer*	*Thermoplasma acidophilum*	*Halobacterium salinarium*	*Escherichia coli*	*Bacillus stearothermophilus*	*Vibrio costicola*	*Saccharomyces cerevisiae*
Althiomycin	ALT	I	278	0	0	0	30	466	1142	1342	1147	74
Carbomycin	CAR	I	616	0	0	0	42	624	1280	1505	1312	72
Griseoviridin	GRI	I	209	0	112	0	0	649	1938	1020	1872	248
Thiostrepton	TIO	I	1098	192	616	1780	486	1316	1834	1644	1158	64
Tylosin	TIL	I	513	0	0	96	169	114	1603	1922	2048	6
Virginiamycin M	VIR	I	281	244	436	284	255	83	1668	1738	2370	206
Anisomycin	ANI	II	485	0	0	0	0	776	0	0	228	1025
Bruceantin	BRU	II	694	0	0	90	0	74	7	0	0	2025
Streptimidone	ESE	II	0	0	0	0	84	81	48	218	96	766
Streptovitacin A	ESA	II	0	0	0	0	0	105	45	0	60	824
Haemanthamine	HAE	II	0	0	0	0	0	31	306	0	0	888
Harringtonine	HAR	II	0	0	0	0	0	130	188	115	150	762
Mitogillin	MIT	II	120	211	544	225	614	100	584	296	168	1428
Narciclasine	NAR	II	1496	0	240	60	129	186	0	70	0	1691
Pretazetine	PRE	II	60	0	0	0	0	0	170	390	25	812
Restrictocin	RES	II	300	158	533	200	702	150	570	184	512	1394
α-Sarcin	SAR	II	736	409	1066	195	548	90	579	136	190	1871
Tylophorine	TYL	II	130	0	0	0	70	0	138	350	100	1142
Tubulosine	TUB	II	66	0	0	0	15	0	72	434	0	1091
Amicetin	AMI	III	0	0	0	0	69	537	1078	1078	968	156
Anthelmycin	ANT	III	1138	228	0	92	230	900	968	1035	1000	1078
Blasticidin S	BLA	III	2238	0	66	0	0	578	1608	467	1954	1992
Edein	EDE	III	645	755	90	1229	686	98	1292	892	2172	1668
Sparsomycin	ESP	III	1167	547	30	0	348	1502	998	1378	1394	1026
Higromycin B	HIG	III	1438	57	0	0	216	36	1545	474	1458	1354
Puromycin	PUR	III	448	105	298	456	187	1792	589	546	808	350
Tetracycline	TET	III	507	114	79	224	213	206	926	1115	1139	782
Fusidic acid	FUS	III	878	0	0	30	60	189	557	1056	908	1055

[a] The inhibition data were transformed into a quantitative matrix by using the algorithm described by Amils et al. (1989). Data on the various organisms are from the references indicated: *Sulfolobus solfataricus* (Cammarano et al., 1985), *Thermoplasma acidophilum* (Sanz et al., 1987), *Desulfurococcus mobilis* and *Thermococcus celer* (Altamura et al., 1988), *Escherichia coli*, *Bacillus stearothermophilus*, and *Saccharomyces cerevisiae* (Cammarano et al., 1985), and *Methanobacterium formicicum*, *Halobacterium salinarium*, and *Vibrio costicola* (J. L. Sanz, H. Hummel, A. Böck, I. Marin, and R. Amils, unpublished results).

phylogenetic trees generated with genotypic information. Several clarifications are needed before we can address some of these questions. The primary sequence of rRNA carries not only the cumulative phylogenetic history of the corresponding genes but also the intrinsic spatial structure information responsible for functional properties. When rapid sequencing techniques were developed, this vast quantity of genotypic information allowed highly conserved sequences to be distinguished from variable ones. This wealth of comparative information on otherwise conservative functional structures, the ribosomes, made it possible for a consensus to be reached rapidly regarding the secondary structure of the 16S-18S and 23S-28S rRNAs. In comparison, no consensus has been reached regarding the secondary structure of the 5S rRNA molecule despite its smaller size and the number of sequences available for analysis. This is probably due to its rather ambiguous structure necessarily related to its fundamental functional role in the ribosome. The availability of secondary structures for the large rRNAs allowed (i) permanent refinement of the models by addition of new sequences, (ii) correlation of the models with structural and functional data through complementary techniques (genetics, photoaffinity labeling, in-

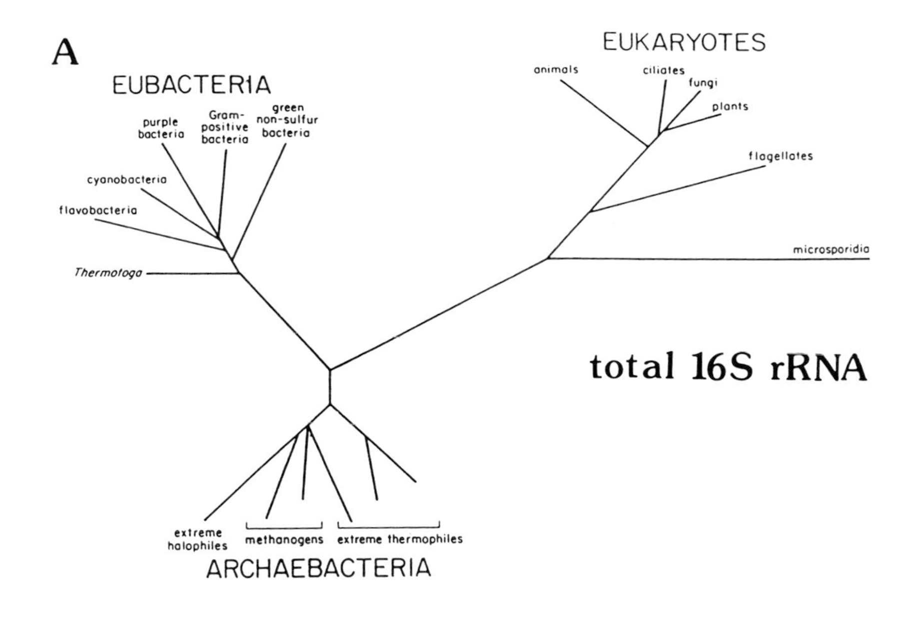

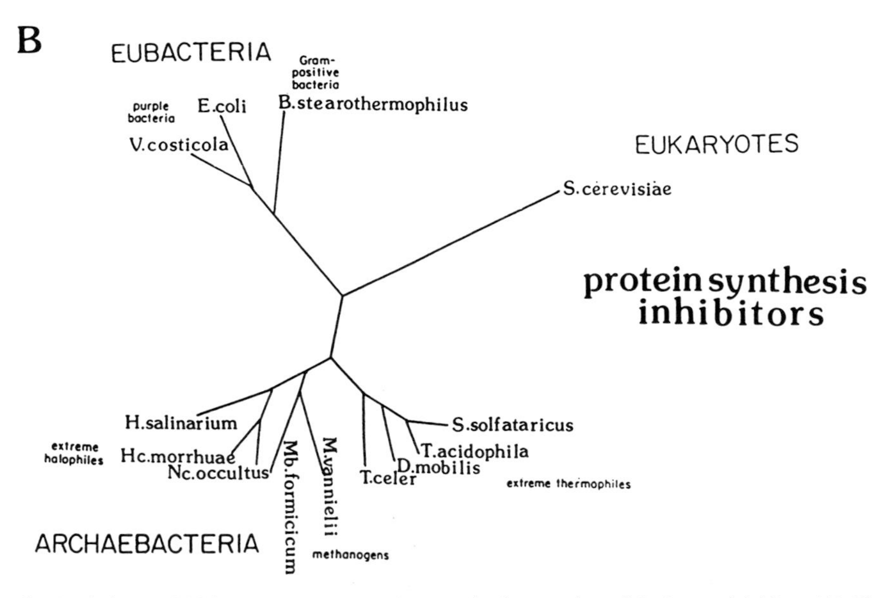

Figure 1. Trees obtained from rRNA sequence comparison and ribosomal antibiotic sensitivities. (A) Phylogenetic tree obtained with 16S rRNA sequences, adapted from Woese (1987). (B) Phenetic tree obtained by using the quantitative matrix data and the P2M subprogram described by Amils et al. (1989).

munoelectron microscopy, cross-linking, etc.), and (iii) use of this information to generate a three-dimensional model of the ribosome.

Taking into account the outstanding comparative work done by Woese's group with 16S rRNA sequences, which led to the identification of extremely conserved sequences of rRNA, it seems reasonable to postulate that the consistency between the functional and sequence data is in the fraction of primary sequence committed to functional space. Using our function-structure probe, we are able to detect specific configurations of the ribosomal functional spaces, which must be related to structures that depend on specific positions of bases on the primary sequence.

A simplified model of the correspondence between genotypic information of the rRNA and the phenotypic property detected by a functional probe is shown in Fig. 2. In this figure, the functional space detected by the antibiotic (functional probe) depends on specific positions of bases in the genotype (in our model, 16S rRNA). This structure allows the functional probe to interact specifically with a definite binding constant, making it possible to measure the corresponding phenotypic property of inhibition. The ribosome with this phenotype is considered the

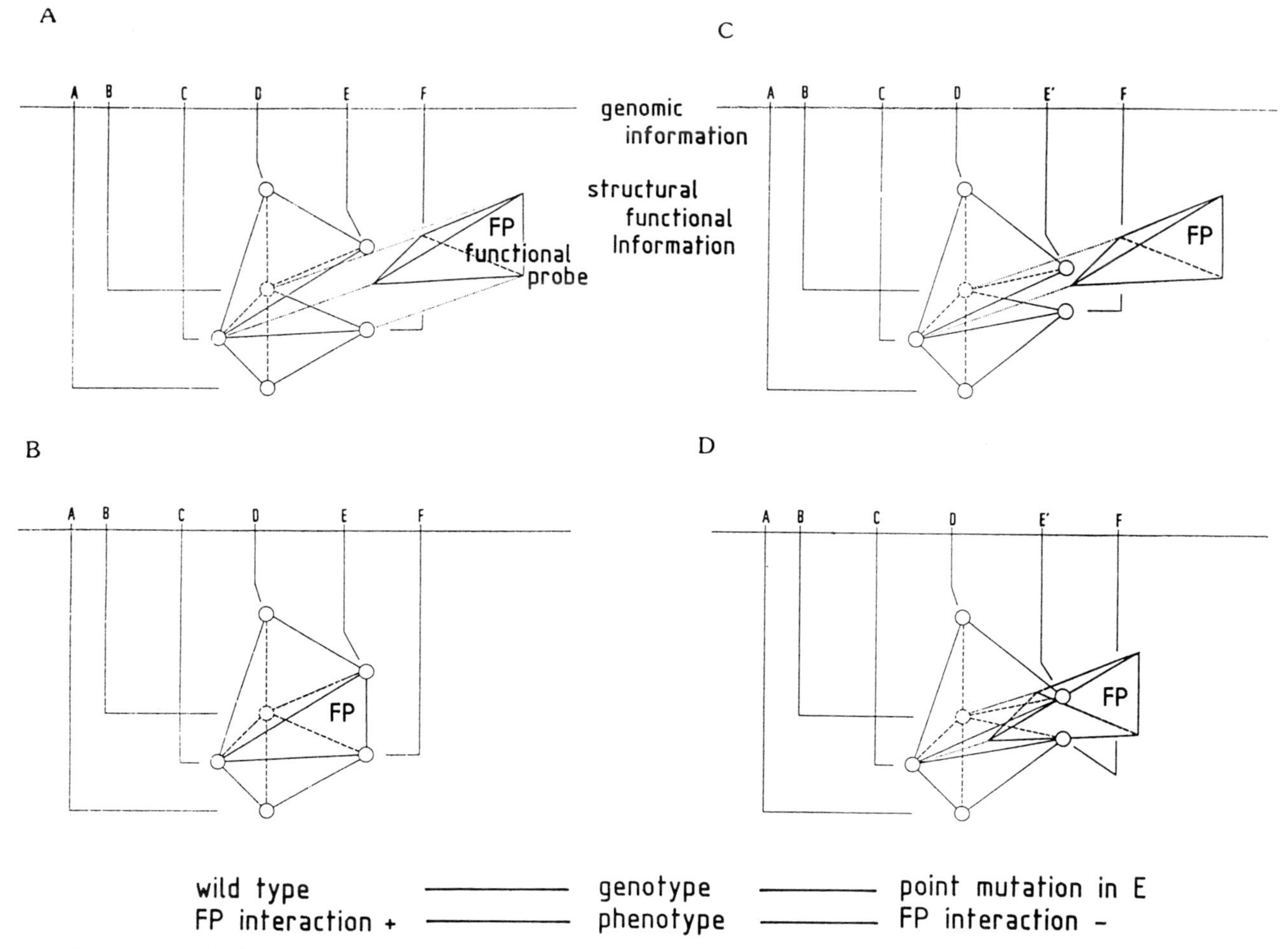

Figure 2. Model of the correlation between ribosomal genotypic information and functional phenotypic information obtained from antibiotic sensitivity spectra of ribosomes. (A and B) Simplified model showing the correspondence between specific positions of the 16S rRNA gene sequence and the functional structure tested by interaction with the functional probe (FP). (C and D) Effect of a single mutation in the 16S rRNA gene (E to E′), which produces a modification of the functional space that impedes the interaction with the functional probe without affecting the basic function of the ribosome. (After Amils et al., 1989.)

wild type, and the interaction with the functional probe is positive. The effect of a specific mutation in section E of the gene produces a small change in the structure of the functional space, which consequently impedes the binding of the functional probe. The genotype change from E to E′ is responsible for the negative phenotypic property. It is obvious that the model is too simple and does not take into consideration the cooperation existing in the quaternary structure of the ribosome. Similar models could be built in which several genes are responsible for the structure of a functional space. The results in terms of functional testing would be the same, but what would be much more complicated is the interpretation of their correlation with the structural features of the particle.

One critical aspect of this type of analysis is the lack of differential information introduced by a negative sensitivity result. The lack of affinity for an antibiotic can be produced by many different means, such as a mutation in a specific rRNA base or in an amino acid of a protein, deletions in ribosomal proteins, or modifications in rRNA bases. Two considerations are worth mentioning in this context: (i) most negative results have a taxonomic or phylogenetic consistency, and (ii) there is an increasing amount of evidence that a specific base change in the rRNA is sufficient to dramatically alter the affinity for the antibiotic without any apparent functional consequences. Such antibiotics include erythromycin (Douthwaite et al., 1985; Sor and Fukuhara, 1984; Ettayebi et al., 1985), macrolides, lincosamides, and streptogramin B (Pernodet et al., 1988), streptomycin (Montandon et al., 1985; Montandon et al., 1986), and kanamycin and gentamicin (De Stasio et al., 1989). Moreover, it has been shown that several ribosomal inhibitors protect specific sets of highly conserved nucleotides in the rRNA when bound to

helix
A B

U UA
1056 GAUG UGGCU G
A
1082 UUAC ACCGA A
U CG

E.coli

ribosomal types	thiostrepton sensitivity	helix B stability
E.coli	++	5 b.p.
M.vannielli	++	5 b.p.
H.halobium	++	4+1* b.p.
T.tenax	-	3+1* b.p.
S.solfataricus	+	3+1* b.p.
S.cerevisiae	-	2+2* b.p.

Figure 3. Correlation between the stability of a double helix of the 23S rRNA secondary structure corresponding to the GTPase center and the sensitivity to thiostrepton. The stability of helix B of the 23S rRNA segment related to GTPase activity (Egebjerg et al., 1989) is plotted against the sensitivity of the corresponding ribosomes to thiostrepton. Sensitivities are rated ++, +, ±, and − according the inhibition produced by the antibiotic. *, Noncanonical base pair (b.p.).

ribosomes (Moazed and Noller, 1987). These two arguments indicate that many negative sensitivity results have evolutionary significance.

The great diversity in antibiotic sensitivity displayed by ribosomes belonging to the three primary kingdoms makes it possible to use the functional data bank to select specific ribosomal systems that might contain differential comparative information that could be used to generate models of functional structures for testing by different techniques. For example, there is the striking correlation between the stability of a double helix within the 23S rRNA segment involved in GTPase activity by protection experiments and the sensitivity of the corresponding ribosomes to thiostrepton, a powerful and extremely specific inhibitor of this function (Fig. 3). This functional site has been studied by many groups using complementary techniques, but so far a molecular model of the functional interference produced by the antibiotic is lacking (Egebjerg et al., 1989). Obviously, this correlation should be tested by topological techniques such as footprinting and supported by the addition of sequences and sensitivities to the data bank.

This comparative phylogenetic sensitivity data bank can also be used to generate models for testing by selecting resistant ribosomes or by site-directed mutagenesis techniques. The obvious advantage of resistance over insensitivity mediated by phylogenetic selection is the specificity of the change, which in the case of insensitivity could depend on complementary structural features encoded by different genes, as mentioned above, that might be difficult to analyze with the current level of knowledge of the ribosomal quaternary structure. Although these techniques complement each other, the phylogenetic functional data have the additional value of representing the naturally selected functional mechanism. In theory, this mechanism must be understood at the molecular level in order to generate models that might lead to a greater understanding of the principles of functional selection, the ultimate goal in evolutionary studies in general and in research on ribosomes in particular.

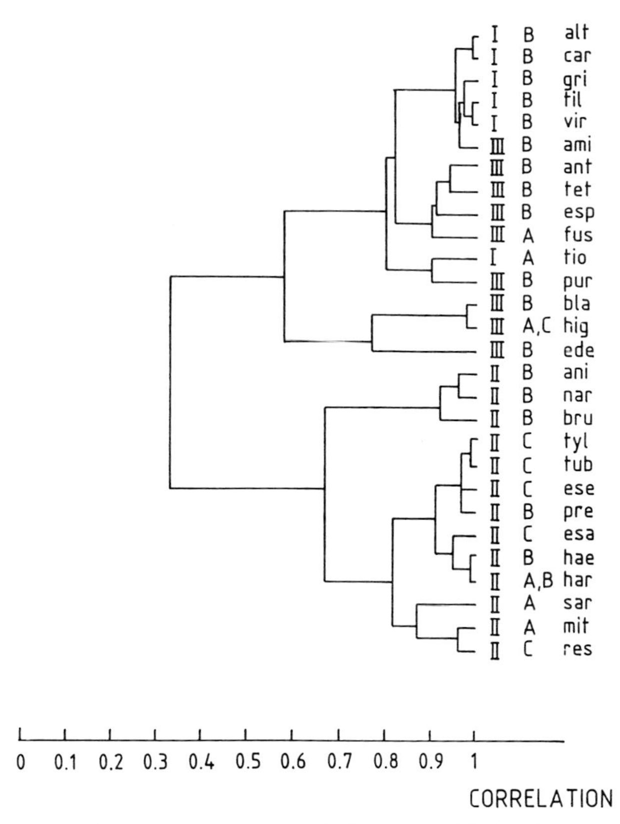

Figure 4. Functional correlation of different antibiotics. The cluster was obtained by using the sensitivity of 16 cell-free systems to 28 functional probes with different kingdom specificities (I, antibiotics with eubacterial specificity; II, antibiotics with eucaryotic specificity; III, antibiotics that inhibit both eubacterial and eucaryotic systems) and functional specificities (A, inhibitors of binding of the ternary complex; B, inhibitors of peptidyltransferase; C, inhibitors of translocation). The sensitivity matrix was analyzed according to the statistical methods described by Amils et al. (1989). The three-letter code for the antibiotics is given (capitalized) in Table 1.

PHYLOGENETIC CLASSIFICATION OF ANTIBIOTICS

Having addressed the phylogenetic value of the functional data bank, we now turn to the structural consistency of the antibiotic classification, using the functional matrix after incorporation of the archaebacterial sensitivity data. The results obtained by using 16 cell-free systems and 28 antibiotics are shown in Fig. 4. In this example, aminoglycoside antibiotics and halophilic archaebacteria were not considered in order to avoid the problems stated above. Analogous results are obtained when either one of these variables is considered, which proves that the data bank is sufficiently consistent to allow the removal of data without any dramatic distortion of the clustering obtained. The same result is also obtained with random elimination of data (Amils et al., 1989).

The main conclusion that can be drawn from this type of analysis is that the antibiotics currently used in our studies, which were selected to represent the principal structural and functional families of protein synthesis inhibitors, cluster in two groups. The first cluster is formed by the antibiotics belonging to group I and group III (nomenclature of Vazquez [1979]); this group is well separated from the antibiotics belonging to group II (specific for eucaryotic systems). In the first cluster there is a strong functional correlation of peptidyltransferase inhibitors, supporting the idea that this type of analysis has phylogenetic consistency (L. Ramirez et al., unpublished results). When the aminoglycosides are incorporated into the analysis, they are integrated into this cluster in two cohesive groups.

In the cluster formed by the eucaryote-targeted antibiotics, we also find interesting functional correlations such as the group formed by anisomycin, narciclasine, and bruceantine, which gives phylogenetic consistency to a long series of obscure antibiotic sensitivity data (reviewed in Vazquez, 1979).

The clustering of antibiotics does not change dramatically when the extremely variable sensitivities of archaebacterial ribosomes are incorporated. A careful statistical study using discriminant analysis can prove that there is internal consistency in the clusters of antibiotics, especially those that interfere with peptidyltransferase activity, an extremely well-defined functional space and one of the most important functions of the ribosome, preserved throughout evolution with only small modifications (Ramirez et al., unpublished results). To continue with this type of analysis, we are now dissecting the peptidyltransferase functional space in different domains, which we hope to analyze at the structural level with other techniques (sequence comparison, footprinting, heterologous reconstitution, photoaffinity labeling, etc.).

It is important to stress that on average, the inhibitory values obtained for archaebacterial ribosomes are significantly lower than those obtained for either eubacteria or eucaryotes. This could be explained by the fact that the antibiotics have been selected for their specificity in one of the reference kingdoms, eubacteria and eucaryotes. Although to date there is no report of specific antiarchaebacterial inhibitors of protein synthesis, preliminary results from our laboratory show that these inhibitors might be produced by different types of streptomycetes (L. Gravalos et al., unpublished results). Specific inhibitors for archaebacteria will add an interesting perspective to this project and have an important practical impact on the study of archaebacteria.

The analysis of individual antibiotics shows interesting patterns that can be studied by using comparative phylogenetic analysis, such as the coherent hypersensitivity of halobacterial ribosomes to puromycin and sparsomycin or to thiostrepton, which is hyperactive in eubacteria and in the methanogenic-halophilic branch of archaebacteria. In this context, it is noteworthy that the sulfur-dependent archaebacterium *T. celer* shows a significantly high value of inhibition by thiostrepton, similar to that of the methanogenic-halophilic archaebacteria, in contrast to the low inhibition values scored by other members of the sulfur-dependent group. This correlates with the results obtained with rRNA sequence comparison mentioned above (Sanz et al., 1987; Altamura et al., 1988). Another example of peculiar behavior is found for anisomycin. This antibiotic scores high values of inhibition for eucaryotes and halophiles but lower values for eubacteria, methanogens, and sulfur-dependent archaebacteria. This situation is opposite that for edeine, which inhibits all systems analyzed except halophilic ribosomes.

MODELS FOR FUNCTIONAL EVOLUTION OF RIBOSOMES

The finding that archaebacterial ribosomes exhibited a mosaic pattern of sensitivities to protein synthesis inhibitors indicated that this type of functional information could generate ribosomal evolutionary models. Assuming that the differences in antibiotic sensitivity are the result of structural differences related to the functional shaping of the particle throughout evolution, we have proposed a model of the gradual evolution of ribosomal structure (Amils and Sanz, 1986). With this model, it was difficult to ascertain the functional characteristics of the prototype ribosome and the corresponding progression of the sensitivity patterns that gave rise to the functional properties of modern-day ribosomes.

The existence of antibiotic-binding sites maintained in all systems strongly supported the idea that the basic components of the translational machinery existed much as in their present form before the transitions responsible for the creation of modern kingdoms. In principle, phylogenetically shared sensitivities should antedate the radiation of the three lineages of descent or should have come into existence by convergent evolution. At present, it seems more reasonable to assume that phylogenetically shared sensitivities represent primeval features of a common ancestral progenote that were retained by any two or by all three lines of descent. In this context, we can argue that antibiotic sensitivity differences displayed by archaebacteria, in contrast to the uniform pattern exhibited by eubacteria and eucaryotes, indicate that archaebacteria lost homology with one another by virtue of the differential loss of their sensitivities, whereas eubacteria and eucaryotes did not.

This conclusion implies that the early evolution of archaebacteria must have been fundamentally different from that of the other two lines of descent. One possibility is that the common archaebacterial ancestor split into different sublines very early, before stringent genetic integration imposed severe restrictions on the possibility of changes. Others are that archaebacterial ribosomes represent either alternate solutions for the same function or diverse states of evolution of ribosomal structures not yet optimized because of the lack of competition in the ecologically extreme habitats in which they develop.

Once the analysis of the functional data bank is refined by preserving the information provided by the inhibition curves, the homogeneous patterns of sensitivities described for the eubacterial and eucaryotic ribosomes disappear. The full spectrum of eubacterial and eucaryotic diversity has not yet been explored. In addition, the possibility of finding eubacterial and eucaryotic organisms with heterogeneous patterns of sensitivity is illustrated by the recently described extremely thermophilic, anaerobic eubacterium *Thermotoga maritima*, which on the basis of the 16S rRNA sequence homology is the eubacterium closest to the diversification point of the phylogenetic tree (Woese, 1987) and correspondingly exhibits an abnormal pattern of sensitivity to aminoglycoside antibiotics (Londei et al., 1988).

We still do not have a clear picture of the ribosomal prototype that gave rise to the existing ribosomal systems, although if we compare the current situation with the ideas discussed a few years ago (Amils and Sanz, 1986), we see substantial promising advances.

Once the phylogenetic consistency of the functional inhibitory data bank is established, it can be used to generate and discard functional models. Refinements in the statistical analysis will give us some insight into how sequence data and functional space data interact. An obvious direction to take is to expand our data bank by acquiring sensitivity data from species of phylogenetic interest.

At any rate, the current state of the art indicates that relevant information on the functional structures of the protoribosome can be obtained from the antibiotic sensitivity spectra of contemporary ribosomes. This may allow some of the fundamental questions about the evolution of the translational apparatus to be addressed.

REFERENCES

Achenbach-Richter, L., R. Gupta, W. Zillig, and C. R. Woese. 1988. Rooting the archaebacterial tree; the pivotal role of *Thermococcus celer* in archaebacterial evolution. *Syst. Appl. Microbiol.* **10:**231–240.

Achenbach-Richter, L., and C. R. Woese. 1988. The ribosomal gene spacer region in Archaebacteria. *Syst. Appl. Microbiol.* **10:**211–214.

Altamura, S., J. L. Sanz, R. Amils, P. Cammarano, and P. Londei. 1988. The antibiotic sensitivity spectra of ribosomes from the Thermoproteales: phylogenetic depth and distribution of antibiotic binding sites. *Syst. Appl. Microbiol.* **10:**218–225.

Amils, R., L. Ramirez, J. L. Sanz, I. Marín, A. G. Pisabarro, and D. Ureña. 1989. The use of functional analysis of the ribosome as a tool to determine archaebacterial phylogeny. *Can. J. Microbiol.* **35:**141–147.

Amils, R., and J. L. Sanz. 1986. Inhibitors of protein synthesis as phylogenetic markers, p. 605–620. *In* B. Hardesty and G. Kramer (ed.), *Structure, Function, and Genetics of Ribosomes.* Springer-Verlag, New York.

Böck, A., and O. Kandler. 1985. Antibiotic sensitivity of archaebacteria, p. 525–554. *In* C. R. Woese and R. S. Wolfe (ed.), *The Bacteria. A Treatise of Structure and Function*, vol. 8. *Archaebacteria.* Academic Press, Inc., New York.

Cammarano, P., A. Teichner, P. Londei, M. Acca, B. Nicolau, J. L. Sanz, and R. Amils. 1985. Insensitivity of archaebacterial ribosomes to protein synthesis inhibitors. Evolutionary implications. *EMBO J.* **4:**811–816.

De Stasio, E. A., D. Moazed, H. F. Noller, and A. E. Dahlberg. 1989. Mutations in 16S ribosomal RNA disrupt antibiotic-RNA interactions. *EMBO J.* **8:**1213–1216.

Dixon, W. J., M. B. Brown, L. Engelman, J. W. Frane, M. A. Hill, R. I. Jennrich, and J. D. Toporek. 1983. *BMDP Statistical Software.* University of California Press, Berkeley.

Douthwaite, S., J. B. Prince, and H. F. Noller. 1985. Evidence for functional interaction between domains II and V of 23S ribosomal rRNA from an erythromycin resistant mutant. *Proc. Natl. Acad. Sci. USA* **82:**8330–8334.

Egebjerg, J., S. Douthwaite, and R. Garrett. 1989. Antibiotic sensitivity at the GTPase associated centre within *Escherichia coli* 23S rRNA. *EMBO J.* **8:**607–611.

Eldhardt, D., and A. Böck. 1982. An "in vitro" polypeptide synthesizing system for methanogenic bacteria: sensitivity to antibiotics. *Mol. Gen. Genet.* **188:**128–134.

Ettayebi, M., S. M. Pasad, and E. A. Morgan. 1985. Chloramphenicol-erythromycin resistance mutations in a 23S rRNA gene of *Escherichia coli. J. Bacteriol.* **162:**551–557.

Gouy, M., and W.-H. Li. 1989. Phylogenetic analysis based on rRNA sequences supports the archaebacterial tree rather than the eocyte tree. *Nature* (London) **339:**145–147.

Guerrier-Takada, C., K. Gardiner, T. Marsh, N. Pace, and S. Altman. 1983. The RNA moiety of ribonuclease P is the catalytic subunit of the enzyme. *Cell* **35:**849–857.

Hammes, W. P., J. Winter, and O. Kandler. 1979. The sensitivity of pseudomureine-containing genus Methanobacterium to inhibitors of murein synthesis. *Arch. Microbiol.* **123:**275–279.

Hilpert, R., J. Winter, W. Hammes, and O. Kandler. 1981. The sensitivity of archaebacteria to antibiotics. *Zentralbl. Bakteriol. Hyg. Abt. 1 Orig. C* **2:**11–20.

Jones, W. J., D. P. Nagl, Jr., and W. B. Whitman. 1987. Methanogens and the diversity of archaebacteria. *Microbiol. Rev.* **51:**135–177.

Kruger, K., P. J. Grabowski, A. J. Zaug, J. Sands, D. E. Gottschling, and T. R. Cech. 1982. Self-splicing RNA: autoexcission and autocatalyzation of the ribosomal RNA intervening sequence of Tetrahymena. *Cell* **31:**147–157.

Leffers, H., J. Kjems, L. Ostergaard, N. Larsen, and R. A. Garrett. 1987. Evolutionary relationships amongst archaebacteria. A comparative study of 23S rRNA of a sulfur dependent extreme thermophile, an extreme halophile and a thermophilic methanogen. *J. Mol. Biol.* **195:**43–61.

Linnaeus, C. 1753. *Species Plantarum.*

Londei, P., S. Altamura, R. Huber, K. O. Stetter, and P. Cammarano. 1988. Ribosomes of the extreme thermophilic eubacterium *Thermotoga maritima* are uniquely insensitive to the miscoding-inducing action of aminoglycoside antibiotics. *J. Bacteriol.* **170:**4353–4360.

Moazed, D., and H. F. Noller. 1987. Interaction of antibiotics with functional sites in 16S ribosomal RNA. *Nature* (London) **327:** 389–394.

Montandon, P. E., P. Nicolas, P. Schürmann, and E. Stutz. 1985. Streptomycin-resistance of *Euglena gracilis* chloroplasts: identification of a point mutation in the 16S rRNA gene in an invariant position. *Nucleic Acids Res.* **13:**4299–4311.

Montandon, P. E., R. Wagner, and E. Stutz. 1986. *E. coli* ribosomes with a C_{912} to U base change in the 16S rRNA are streptomycin resistant. *EMBO J.* **5:**3705–3708.

Moore, P. B. 1988. The ribosome returns. *Nature* (London) **331:**223–227.

Oliver, J. L., J. L. Sanz, R. Amils, and A. Marín. 1987. Inferring the phylogeny of archaebacteria: the use of ribosomal sensitivity to protein synthesis inhibitors. *J. Mol. Evol.* **24:**281–288.

Pecher, R., and A. Böck. 1981. In vivo susceptibility of halophilic and methanogenic organisms to protein synthesis inhibitors. *FEMS Microbiol. Lett.* **10:**295–297.

Pernodet, J. L., F. Boccard, M. T. Alegre, M. H. Blondelet-Rouault, and M. Guérineau. 1988. Resistance to macrolides, lincosamides and streptogramin type B antibiotics due to a mutation in an rRNA operon of *Streptomyces ambofaciens. EMBO J.* **7:**277–282.

Sanz, J. L., S. Altamura, I. Mazzotti, R. Amils, P. Cammarano, and P. Londei. 1987. Unique antibiotic sensitivity of an "in vitro" polypeptide synthesis system from the archaebacterium *Thermoplasma acidophilum.* Phylogenetic implications. *Mol. Gen. Genet.* **207:**385–394.

Sanz, J. L., I. Marín, M. A. Balboa, D. Ureña, and R. Amils. 1988. An NH_4^+ dependent protein synthesis cell-free system for halobacteria. Biochemistry **27:**8194–8199.

Sneath, P. H. A., and R. R. Sokal. 1973. *Numerical Taxonomy.* W. H. Freeman & Co., San Francisco.

Sor, F., and H. Fukuhara. 1984. Erythromycin and spiramycin resistant mutations of yeast mithocondria: nature of the *rib* 2 locus in the large ribosomal RNA gene. *Nucleic Acids Res.* **12:**8313–8318.

Vazquez, D. 1979. *Inhibitors of Protein Biosynthesis.* Springer-Verlag KG, Berlin.

Woese, C. R. 1982. Archaebacterial and cellular origins: an overview. *Zentralbl. Bakteriol. Hyg. Abt. 1 Orig. C* **3:**1–17.

Woese, C. R. 1987. Bacterial evolution. *Microbiol. Rev.* **51:** 221–271.

Woese, C. R., and G. E. Fox. 1977. Phylogenetic structure of the prokaryotic domain: the primary kingdoms. *Proc. Natl. Acad. Sci. USA* **74:**5088–5090.

Chapter 57

Chloroplast Ribosomal Proteins and Their Genes

ALAP R. SUBRAMANIAN, PETER M. SMOOKER, and KLAUS GIESE

In this chapter, we present a summary of recent work on chloroplast ribosomal protein genes. We first describe those genes which are now known to be located in the circular DNA molecules (110 to 160 kilobase pairs [kb]) that constitute the chloroplast genome. The operon organization of these genes and the similarities to and differences from that of the corresponding genes of *Escherichia coli* are discussed. Finally, the most recent work on the isolation and characterization of cDNA clones for several of the nuclear-located ribosomal protein (r-protein) genes are summarized. We have restricted our discussion mainly to higher-plant chloroplasts and, for illustration, have concentrated on our results with maize and spinach.

r-PROTEIN GENES IN THE ORGANELLE GENOME

Chloroplast ribosomes (70S) are of the eubacterial type (Boynton et al., 1980) and contain three rRNA molecules (23S, 16S, and 5S). A 4.5S rRNA species found specifically in the chloroplast ribosomes of higher plants has been shown to correspond to the 3′-end region of 23S rRNA; it arises from an extra cleavage step in the processing of the rRNA precursor. The chloroplast rRNA operon has the same structural motif (including *trnA* and *trnI* genes in the 16S–23S spacer) as the *rrnA* operon of *E. coli*. It is located in the chloroplast genome (Schwarz and Kössel, 1980) either in two identical copies (most plants) or in a single copy (certain legumes and conifers).

The number of proteins in higher-plant chloroplast ribosomes has been estimated to be 56 to 60 by two-dimensional gel electrophoresis (Mache et al., 1980; Capel and Bourque, 1982; Udalova et al., 1984; Subramanian, 1985). Twenty of the chloroplast r-proteins are encoded in the organelle genome of higher plants. The chloroplast genome of a nonflowering lower plant (liverwort) has also been completely sequenced and contains the same number of r-protein genes (20), but two of the genes (*rps16* and *rpl21*) are found in the chloroplast of one or the other but not both plants. There are also differences in gene copy numbers between these two very distantly related plant species.

We and two other research groups have mapped all of the chloroplast-located r-protein genes in maize (*Zea mays*), a cereal plant of the monocot group of angiosperms, and sequenced most of them. The r-proteins encoded in chloroplast DNA are listed in Table 1, and their physical map positions on the circular maize chloroplast DNA (139 kb) are shown in Fig. 1. The list includes 12 proteins of the small ribosomal subunit (S2, S3, S4, S7, S8, S11, S12, S14, S15, S16, S18, and S19) and 8 of the large subunit (L2, L14, L16, L20, L22, L23, L33, and L36).

Protein Sequence Homology

The derived amino acid sequence of each of the chloroplast proteins showed significant identity to that of a corresponding *E. coli* protein (Table 1). The highest value was ~65% (S12 and L36) and the lowest was ~25% (L22 and L23), the average being 44%. Chain lengths of the chloroplast proteins are generally similar to those of their *E. coli* counterparts, exceptions being S11, S18, L20, L22, and L33, which are significantly longer in the chloroplast. Table 1 also lists the computed ALIGN scores (Dayhoff, 1978) of each of the proteins against those of their *E. coli* homologs. Small sequences of high identity (e.g., L36; 37 residues) may score lower

Alap R. Subramanian, Peter M. Smooker, and Klaus Giese ▪ Max-Planck-Institut für Molekulare Genetik, Abteilung Wittmann, Ihnestrasse 73, D-1000 Berlin-Dahlem, Federal Republic of Germany.

Table 1. r-Proteins encoded in the chloroplast DNA of higher plants

Protein	Chain length[a]		Δ Length	Identity to *E. coli* (%)	ALIGN score	Reference
	Chloroplast[b]	*E. coli*				
S2	236	240	−4	39	53	Stahl and Subramanian, unpublished data
S3	224	232	−8	38	47	McLaughlin and Larrinua, 1987c
S4	201	203	−2	39	26	Subramanian et al., 1983
S7	156	153[c]	+3	43	27	Giese et al., 1987
S8	136	129	+7	38	39	Markmann-Mulisch and Subramanian, 1988b
S11	143	128	+15	48	30	Markmann-Mulisch and Subramanian, 1988a
S12	124	123	+1	66	48	Giese et al., 1987
S14	103	98	+5	39	15	Srinivasa and Subramanian, 1987
S15	87	88	−1	34	22	Ohto et al., 1988
S16	85	82	+3	32	13	Ohto et al., 1988
S18	101	74	+27	38	12	Ohto et al., 1988
S19	93	91	+2	42	26	McLaughlin and Larrinua, 1987a
L2	274	272	+2	49	46	Ohto et al., 1988
L14	123	123	0	52	35	Markmann-Mulisch and Subramanian, 1988b
L16	136	136	0	53	40	Gold et al., 1987; McLaughlin and Larrinua, 1987d
L20	128	117	+11	38	18	Ohto et al., 1988
L22	126	110	+16	27	3	McLaughlin and Larrinua, 1987b
L23	93	100	−7	25	7	McLaughlin and Larrinua, 1988
L33	66	54	+12	38	9	Ohto et al., 1988
L36	37	38	−1	62	14	Markmann-Mulisch et al., 1987

[a] Chloroplast data are for maize except S15, S16, S18, L2, L20, L33, which are for tobacco (Ohto et al., 1988); *E. coli* data are from Wittmann-Liebold (1986).
[b] fMet is included.
[c] *E. coli* B.

numerical values than larger sequences of lower identity (e.g., S2; 236 residues).

The higher-plant chloroplast is known to be in many respects an evolutionarily conservative organelle. A comparison of the derived amino acid sequences of 14 of the 20 chloroplast-encoded r-proteins of maize and tobacco (for which such data are available) shows 70 to 95% sequence identity and little variation in peptide chain length between the homologous proteins of these two plants, even though the monocot (maize) and dicot (tobacco) groups of the angiosperm have been separated for over 110 million years (Friis et al., 1987).

Antibody Cross-Reactions and Protein-Chemical Data

Antibodies raised against purified *E. coli* r-proteins corresponding to 11 of the 20 listed in Table 1 have been tested for cross-reaction with spinach chloroplast proteins (Bartsch, 1985). Antisera against five (S7, S11, S12, S19, and L2) gave clear cross-reactions on Western immunoblots of both one- and two-dimensional gels. Among these five, sequence identity is very high only for S12; in the other cases the values are near average, indicating that these sequences may contain regions of highly conserved structures that are also strongly immunogenic. Protein L2, an immunologically conserved protein (Schmid et al., 1984), is a case in point.

Structural data from isolated proteins are available for six of the chloroplast-encoded proteins, S2, S4, S19, L2, L22, and L36 (Kamp et al., 1987; Zhou et al., 1989; A. R. Subramanian, unpublished data). The most extensive data are from L2, from which CNBr fragments of the high-performance liquid chromatography-purified protein were isolated and sequenced to yield 32 sequence positions (Kamp et al., 1987). These agreed perfectly with the corresponding nucleotide sequence-derived published data (Zurawski et al., 1984). Chloroplast L2 is also noteworthy because its N terminus is methylated (*N*-methylalanine), showing the occurrence of methyl modification in chloroplast r-proteins. The agreement between nucleotide sequence and protein-chemical data supports the view that these organelle r-protein genes are functionally expressed.

Gene Structure: Introns

The coding sequences in four of the chloroplast r-protein genes (*rps12*, *rps16*, *rpl2*, and *rpl16*) are interrupted by introns of the group II type, first observed in yeast mitochondrial genes (Michel and Dujon, 1983). They have conserved nucleotide sequences at the intron-exon boundaries, and each of the intron sequences can be folded into a characteristic secondary structure motif (Ohto et al., 1988). The length of the introns ranges from 500 to 1,000 base pairs. The exons are of varying length, the

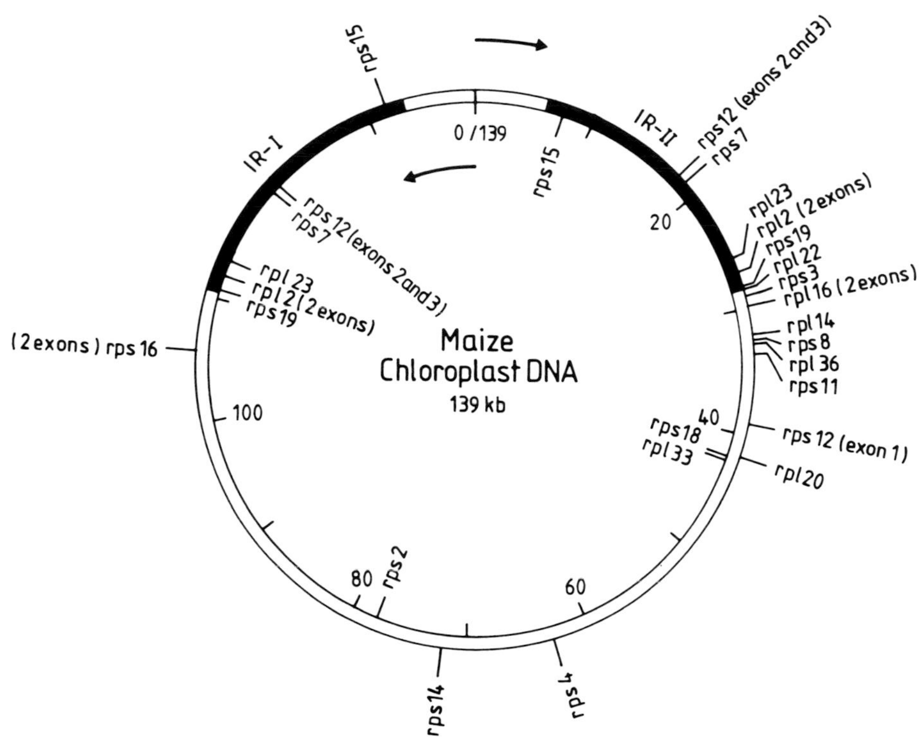

Figure 1. Physical map of the chloroplast genome of maize (*Z. mays*) showing the experimentally determined loci of the r-protein genes and directions of transcription. The darkened segments (IR-I and IR-II) show positions of the inverted repeats. The loci depicted outside the circle are transcribed clockwise; those inside the circle are transcribed counterclockwise.

shortest being exon 1 of *rpl16*, which in maize encodes only three residues, Met, Leu, and Ser.

The intron-exon structure of *rps12*, which contains three exons, is most unusual. Two of its exons are located near the *rps7* locus (as part of an operon including *rps7*), but exon 1 (encoding the N-terminal 38 amino acid residues) is located at a distant (~25-kb) site near the *rpl20* locus (Fig. 1). This unique organization of *rps12* has been found in liverwort, tobacco, maize, and rice. The mature mRNA is formed by the *cis* splicing of exons 2 and 3 and the unusual *trans* splicing involving exon 1 (e.g., Koller et al., 1987).

Gene Copy Number

Unlike in *E. coli*, where every r-protein gene is present in one copy per genome, six of the chloroplast r-protein genes in maize are present in two copies. These (the genes for S7, S12 [exons 2 and 3], S15, S19, L2, and L23) are located in a domain of the chloroplast genome that is repeated in the opposite orientation and therefore called an inverted repeat. There are several interesting rearrangements and transpositions at the boundaries between the inverted repeats and the single-copy regions (Palmer, 1985) often involving r-protein genes. Thus, *rps15* (Prombona and Subramanian, 1989) and *rps19* (McLaughlin and Larrinua, 1987b) occur in two copies in rye and maize but not in tobacco (Ohto et al., 1988), whereas no r-protein genes are repeated in liverwort (Ohyama et al., 1988). A third copy of *rpl23* has been found in wheat, but as a pseudogene (Bowman et al., 1988).

Promoter Elements

The regulatory structural elements of transcription have been identified in some of the chloroplast r-protein genes (Table 2). In these cases, −35 and −10 hexanucleotide elements and the nucleotide distance between them have been determined (Giese et al., 1987; Russel and Bogorad, 1987; D. Stahl and A. R. Subramanian, unpublished data). The promoter sequences of *rps2*, *rps4*, and *rps12* diverge only slightly from the canonical procaryotic consensus sequence of TTGACA–17 ± 1 nucleotides (nt)–TATAAT. Functional evidence for the *rps2* promoter was obtained by primer extension analysis; the transcripts were identified to start 2 nt downstream from the −10 sequence (Stahl and Subramanian, unpublished data).

Shine-Dalgarno Sequences

The 3′-end region of higher-plant chloroplast 16S rRNA is highly conserved and is similar to that

Table 2. Regulatory structural elements identified in some of the chloroplast-located r-protein genes of maize

Gene	Promoter sequence			Shine-Dalgarno sequence[a]		
	−35	Distance (nt)	−10	3′	Distance (nt)	5′
rps2	TTGTAC	17	TTTAAT	AAAGGA	12	AUG
rps4	TAGATA	17	TAATAT	AAGGAG	5	AUG
rps7	TTGATA	18	TATTAT[b]	GGA	22	AUG
rps8				AGGAGG	19	AUG
rps11				GAG	10	AUG
rps12	TTGATA	18	TATTAT[b]	AAG	5	AUG
rps14				AGGA	10	AUG
rps19				AAAGGAG	9	AUG
rpl14				GGA	12	AUG
rpl22				AAAGG	22	AUG
rpl23				GGA	10	AUG
rpl36				AAAGGA	10	AUG

[a] The 3′ nucleotide sequence of maize chloroplast 16S rRNA (1,491 nt) is 3′-UUUCCUCCACUAGG . . . −5′. For further analysis, see Ruf and Kössel (1988) and Bonham-Smith and Bourque (1989).
[b] Promoter of the operon *rps12* (exons 2 and 3)–160 nt–*rps7*. By inserting it upstream of a reporter gene, this promoter has been shown to be functional in *E. coli* (Giese et al., 1987).

of *E. coli* (footnote *a*, Table 2). Shine-Dalgarno-type ribosome-binding sequence regions found in 12 maize chloroplast r-protein genes are listed in Table 2. In a few cases (e.g., *rps3* and *rpl16* of maize), no such sequence stretches have been found within a reasonable nucleotide distance from the initiating AUG codon.

A canonical Shine-Dalgarno element does not occur with every r-protein (or other protein)-coding sequence in the chloroplast genome. Whether there are as yet unknown ribosome recognition signals in chloroplast mRNA (or, conversely, as yet unknown mRNA recognition elements in the chloroplast ribosome) remains to be seen. In this context, the architectural features of the spinach chloroplast 30S subunit derived from electron microscopy studies (B. Tesche and A. R. Subramanian, unpublished data) show some distinct differences from that generally accepted for the *E. coli* 30S subunit. Also, amino acid sequence data from purified spinach chloroplast r-proteins (Subramanian, unpublished data) suggest the existence of at least two to three chloroplast r-proteins with no *E. coli* counterparts. These may perform additional functions at initiation.

Gene Organization

Northern (RNA) blot analyses have revealed that all r-protein genes in the chloroplast genome are transcribed (e.g., Ohto et al., 1988; Neuhaus et al., 1989; D. Stahl, A. Prombona, and A. R. Subramanian, unpublished data). Multiple transcripts have been identified for each individual r-protein gene, suggesting processing of primary polycistronic transcripts. Several of the r-protein genes are cotranscribed with other genes not encoding components of the chloroplast translation apparatus (Hudson et al., 1987; Stahl and Subramanian, unpublished data). The small *rpl36* gene is cotranscribed with eight other upstream r-protein genes (Ohto et al., 1988).

Comparison of the clustering patterns of the r-protein genes in the chloroplast and the patterns of the corresponding genes in *E. coli* is shown in Fig. 2. A similar clustering pattern, but with several differences, is evident. Four clusters carry two or more r-protein genes in single transcription units: the L23, S12, L20, and L33 operons. The largest is the L23 operon, consisting of 10 r-protein genes and *rpoA*.

The 10 genes of the chloroplast L23 operon belong to three operons of the main gene cluster in *E. coli*. Evidence for its existence as a single transcription unit in the chloroplast is derived from Northern blot studies (Ohto et al., 1988), but further work is necessary to ascertain whether it is also driven by additional internal promoters. Genes *rpsD* (S4) and *rpsN* (S14), which in *E. coli* occur in the α and *spc* operons, are transposed to distant locations in the chloroplast genome.

The chloroplast counterpart of the S12 operon of *E. coli* shows many extraordinary features. The two elongation factor (EF-Tu and EF-G) genes are absent in the higher-plant chloroplast genome, and the S12 gene is split into three exons, two of which form an operon with the S7 gene while the third is located at a great distance in a transcription unit with the L20 gene. Interestingly, algal chloroplast genomes (e.g., *Euglena gracilis*) have a different organization: here the S12 gene is intronless and occurs clustered with both S7 and EF-Tu genes (Montandon and Stutz, 1984). A similarly altered gene organization has been found in the cyanelle (a chloroplastlike organelle with murein succulus) genome, in which

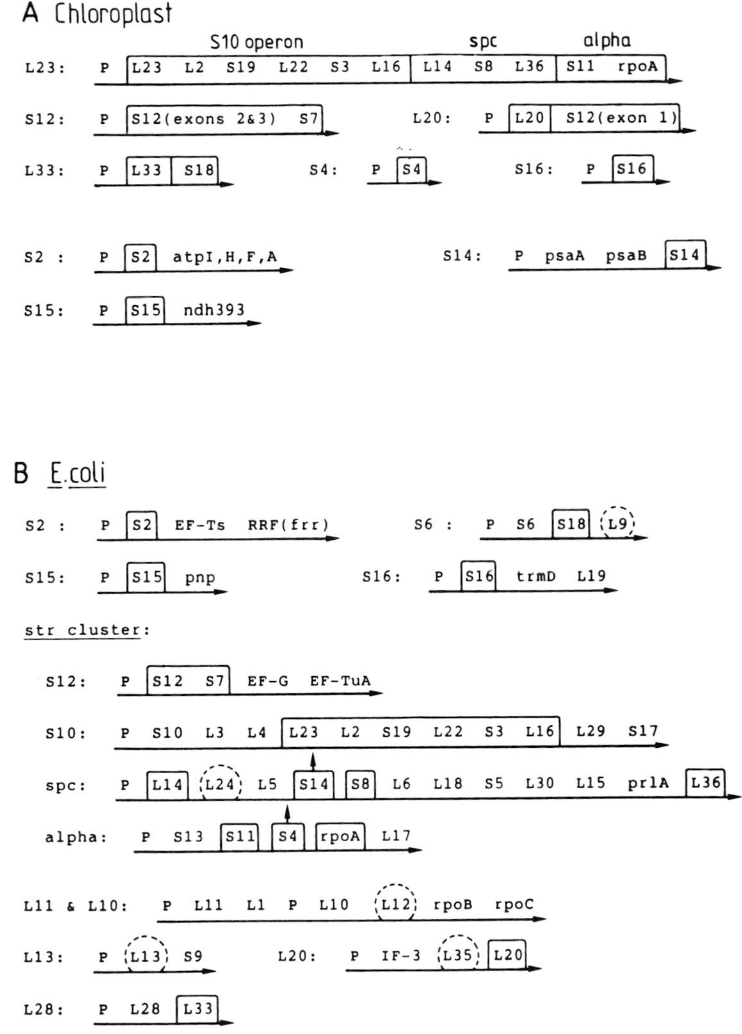

Figure 2. Operon organization of the chloroplast-located r-protein genes in maize and the homologous genes in *E. coli*. The chloroplast-located genes common to both are boxed. Arrows indicate two genes of *E. coli spc* and α operons which in the chloroplast genome occur unclustered with other r-protein genes. Broken circles enclose five nuclear-located chloroplast r-protein genes discussed in the text. P, Promoter.

the genes for EF-Tu and two additional r-proteins have been identified (W. Löffelhardt and M. Kuntz, personal communication).

The genes for S2, S14, and S15 also have unusual organizations: they occur in transcription units with the genes for components of the photosynthetic apparatus. The cotranscribed nonribosomal genes in the S2 and S14 operons have been shown to be photogenes (Rodermel and Bogorad, 1985; Hudson et al., 1987; Stahl and Subramanian, unpublished data); i.e., their expression is light inducible. The S15 gene (Prombona and Subramanian, 1989) occurs together with a gene for a component of the mitochondrial-type NADH-ubiquinone reductase. The occurrence of some r-protein genes in operons with light-inducible genes would suggest additional translational regulatory mechanisms for their expression.

Organelle Allocation of Chloroplast r-Protein Genes

Chloroplast genomes of higher plants carry 20 r-protein genes, encoding 12 proteins of the small ribosomal subunit and 8 proteins of the large subunit (Table 1). Taking 21 and 33, respectively, as the numbers of *E. coli* r-proteins in the two subunits (Wittmann, 1982), 57% of the small-subunit proteins and 24% of the large-subunit proteins are organelle encoded. There is thus an asymmetry in the organelle allocation of the r-protein genes for the two ribosomal subunits.

Dabbs (1986) has isolated *E. coli* mutants that lack any 1 of 15 r-proteins. It is a striking fact that only 1 of these 15, L33, is encoded in the chloroplast genome; all others are nuclear encoded (Table 3).

Table 3. Gene locations of the chloroplast r-proteins corresponding to those lacking in *E. coli* mutants[a]

Protein	Gene location	Protein	Gene location
L1	Nucleus	L30	Nucleus
L11	Nucleus	L33	Chloroplast
L15	Nucleus		
L19	Nucleus	S6	Nucleus
L24[b]	Nucleus	S9	Nucleus
L27	Nucleus	S13	Nucleus
L28	Nucleus	S17	Nucleus
L29	Nucleus	S20	Nucleus

[a] Data are taken from Dabbs (1986).
[b] Location experimentally confirmed (see Table 4).

Does this correlation imply that nearly all the organelle-encoded r-proteins belong to a fundamental, absolutely essential group of r-proteins? The known facts on these 20 proteins do not seem to support such a view. Dabbs (1986) had also included S1 in his table, but noted that in this case the protein was missing only from the ribosome, being present in the supernatant. Protein S1 is known to be an essential component of the *E. coli* ribosome (reviewed in Subramanian, 1983). Two types of evidence indicate that an S1 homolog is present in the chloroplast ribosome: (i) affinity binding experiments using a matrix-bound poly(U) column (Subramanian, unpublished data) and (ii) Western blot experiments using monoclonal antisera to *E. coli* S1 (Hahn et al., 1988).

The homologs of all organelle-encoded 50S r-proteins (with the exception of L36, for which assembly data are lacking) are known components of the *E. coli* RI_{50} particle. The organelle-encoded 30S subunit r-proteins include the two assembly initiator proteins of the *E. coli* 30S subunit (S4 and S7) as well as seven other proteins (S8, S11, S12, S15, S16, S18, and S19) of the RI_{30} particle. It therefore appears that some aspects of the assembly pathway may have played a key role in the evolutionary allocation of chloroplast r-protein genes between the organelle genome and the nuclear genome.

r-PROTEINS ENCODED IN THE NUCLEAR GENOME

Most chloroplast r-proteins are encoded in the nuclear DNA, and this raises some interesting points. First, the gene dosage between the chloroplast and nuclear genomes is dramatically different. Each mature leaf cell contains many chloroplasts (>100 in wheat, for example), and each chloroplast contains up to 1,000 copies of its genome (Hoober, 1984); nuclear genes are generally present in one or a few copies each. Hence, the coordinate regulation of expression of the r-protein genes in the two genetic systems to achieve stoichiometric levels probably includes mechanisms nonexistent in bacteria. Second, since these nuclear-encoded proteins are synthesized in the cytoplasm, they must all be equipped with a specific targeting mechanism for import into the chloroplast.

The work on nuclear genes encoding chloroplast r-proteins is still in its infancy. To date, cDNA clones for a total of five such proteins have been identified by virtue of their homology to *E. coli* proteins (Gantt, 1988; Phua et al., 1989; Giese and Subramanian, 1989; P. M. Smooker and A. R. Subramanian, unpublished data). We shall conclude this chapter with a brief summary of these results.

Figure 3 shows a schematic representation of the cDNA clones for these five r-proteins. They all encode proteins of the large subunit of the chloroplast ribosome: L9, L12, L13, L24, and L35. Protein L35 is the name given to the "A-protein" of Wada and Sako (1987), whose gene locus and nucleotide sequence in *E. coli* were already known (Fayat et al., 1983). We purified a small protein (8 kilodaltons)

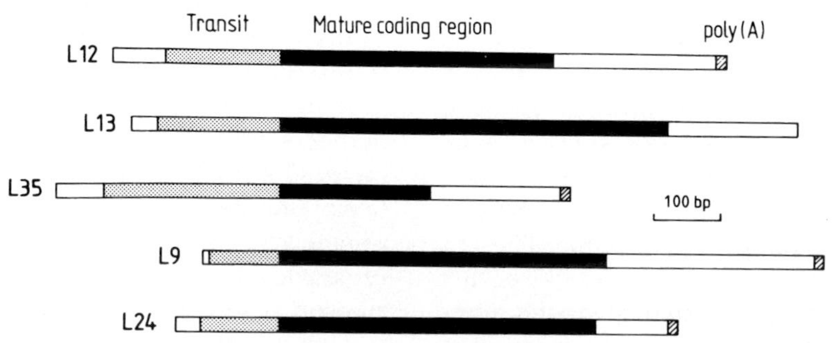

Figure 3. Schematic representation of the identified cDNA clones for five chloroplast r-proteins, encoded in nuclear DNA. The 5′ and 3′ noncoding regions of the mRNA are shown unshaded. There is evidence for the existence of two species of L13 mRNA, each polyadenylated at a separate site (Phua et al., 1989). The clones for L12, L13, and L35 are from spinach (*S. oleracea*); those for L9 and L24 are from pea (*P. sativum*). bp, Base pairs.

Table 4. Characteristics of nuclear-encoded chloroplast r-proteins identified at the cDNA level

Protein	Chain length		Δ Length	Identity to *E. coli* (%)	Protein-chemical sequence	Cross-reaction to *E. coli* antisera[a]
	Chloroplast[b]	*E. coli*				
L12[c]	133	120	+13	48	Complete	Positive
L13[d]	190	142	+48	54	No	Positive
L35[e]	73	65	+8	41	58 N terminal	Not done
L9[f]	160	148	+12	30	No	Negative
L24[f]	149	103	+46	35	No	Negative

[a] Bartsch, 1985.
[b] Transit peptide is not included.
[c] Giese and Subramanian, 1989.
[d] Phua et al., 1989.
[e] Smooker and Subramanian, unpublished data.
[f] Gantt, 1988.

from spinach chloroplast ribosomes and determined its N-terminal sequence to 58 residues, revealing 37% identity to the *E. coli* "A-protein" sequence. A synthetic oligonucleotide mixture was designed from these data and used to isolate the L35 cDNA clone (Smooker and Subramanian, unpublished data). Information on the mature proteins encoded by these five cDNA clones is summarized in Table 4.

Three of the clones (L12, L13, and L35) were isolated from a spinach (*Spinacia oleracea*) cDNA library, and two (L9 and L24) were isolated from a pea (*Pisum sativum*) cDNA library. Both libraries were constructed in the expression vector λgt11.

Each of these five cDNA clones carries the coding region for a presequence that may function as a transit sequence. The exact lengths of the transit sequences of L12 and L35 are known, since in these two cases the N-terminal sequences of the mature protein have been determined (Bartsch et al., 1982; Subramanian, unpublished data). The transit sequences of L12 and L35 are, respectively, 56 and 86 amino acid residues long. In the case of L35, the transit sequence is longer than the r-protein whose transport it directs. In the three other cases, the lengths of the transit sequences have only been estimated; the values are 60, 34, and 39 amino acid residues, respectively, for those of L13, L9, and L24 (Phua et al., 1989; Gantt, 1988). The transit sequences of pea r-proteins (34 and 39 residues) appear significantly smaller than those of spinach r-proteins (56, 60, and 86 residues), but whether this is a general feature remains to be seen.

CONCLUDING REMARKS

We have summarized the current state of knowledge of higher-plant chloroplast r-proteins and the genes encoding them. Following nucleotide sequencing of the complete genomes of three plant chloroplasts (tobacco, a liverwort, and rice), all r-proteins encoded by them have now been identified. This work demonstrated that the majority of genes encoding chloroplast r-proteins have been transferred to a nuclear location. Two major challenges now await researchers in this field.

The first is characterization of the full complement of nuclear-located r-protein genes and a study of the adaptations which the genes and their products have undergone. The most obvious adaptation is the addition of a coding sequence for a transit peptide. No primary structure identity has been found between the five putative transit peptides. However, they share a similar, rather unusual amino acid composition: a high percentage of serine and threonine (between 30 and 40 mol% of the total residues) and an absence of tryptophan and tyrosine. As more data are accumulated, the general features of these peptides, their evolutionary origins, and the mechanisms of import that they direct may be elucidated. Another feature of the nuclear-encoded r-proteins is that they are generally larger than the *E. coli* homologs. This is particularly marked for L13 and L24 (Table 4) and also for L1 (unpublished results from our laboratory), which all have large N-terminal extensions. The percent identity of the nuclear-encoded and the chloroplast-encoded r-proteins to their *E. coli* homologs is of the same order, indicating that the location of the genes in the nucleus has neither increased nor decreased protein divergence.

The second problem to be addressed is that of the coordination of assembly of the chloroplast ribosome. Organellar ribosome assembly faces the particular problem of coordinating the expression of genes located in two different cellular compartments. The organization of r-protein genes in the chloroplast genome is in operon structures with similarities to that in *E. coli* (Fig. 2), and it will be interesting to determine whether the nuclear genes are located only on certain chromosomes, arranged in a transcriptionally linked pattern. An understanding of the control mechanisms that have evolved to ensure efficient assembly of the chloroplast ribosome should yield insights into the evolution of the organelles themselves.

ADDENDUM IN PROOF

cDNA clones for three other chloroplast r-proteins have since been isolated and characterized. These are for L11, L21, and Psrp-1. The last two are particularly interesting with respect to the endosymbiont theory for the origin of chloroplasts: L21 is encoded in the organelle genome in a lower plant (liverwort) but is nuclear encoded in spinach. Psrp-1 is a chloroplast r-protein with no homolog in the *E. coli* ribosome (P. M. Smooker, V. Kruft, and A. R. Subramanian, *J. Biol. Chem.*, in press; C. H. Johnson, V. Kruft, and A. R. Subramanian, *J. Biol. Chem.*, in press).

REFERENCES

Bartsch, M. 1985. Correlation of chloroplast and bacterial ribosomal proteins by cross-reactions of antibodies specific to purified *Escherichia coli* ribosomal proteins. *J. Biol. Chem.* **260:** 237–241.

Bartsch, M., M. Kimura, and A. R. Subramanian. 1982. Purification, primary structure, and homology relationships of a chloroplast ribosomal protein. *Proc. Natl. Acad. Sci. USA* **79:** 6871–6875.

Bonham-Smith, P. C., and D. P. Bourque. 1989. Translation of chloroplast-encoded mRNA: potential initiation and termination signals. *Nucleic Acids Res.* **17:**2057–2080.

Bowman, C. M., R. F. Barker, and T. A. Dyer. 1988. In wheat ctDNA, segments of ribosomal protein genes are dispersed repeats, probably conserved by nonreciprocal recombination. *Curr. Genet.* **14:**127–136.

Boynton, J. E., N. W. Gillham, and A. M. Lambowitz. 1980. Biogenesis of chloroplast and mitochondrial ribosomes, p. 903–950. *In* G. Chambliss, G. R. Craven, J. Davies, K. Davis, L. Kahan, and M. Nomura (ed.), *Ribosomes. Structure, Function, and Genetics.* University Park Press, Baltimore.

Capel, M. S., and D. P. Bourque. 1982. Characterization of *Nicotiana tabacum* chloroplast and cytoplasmic ribosomal proteins. *J. Biol. Chem.* **257:**7746–7755.

Dabbs, E. R. 1986. Mutant studies on the prokaryotic ribosome, p. 733–748. *In* B. Hardesty and G. Kramer (ed.), *Structure, Function, and Genetics of Ribosomes.* Springer-Verlag, New York.

Dayhoff, M. D. 1978. Survey of new data and computer methods of analysis, p. 1–8. *In Atlas of Protein Sequence and Structure,* vol. 5. National Biomedical Research Foundation, Washington, D.C.

Fayat, G., J.-F. Mayaux, C. Sacerdot, M. Fromant, M. Springer, M. Grunberg-Manago, and S. Blanquet. 1983. *Escherichia coli* phenylalanyl-tRNA synthetase operon region: evidence for an attenuation mechanism, identification of the gene for the ribosomal protein L20. *J. Mol. Biol.* **171:**239–261.

Friis, E. L., W. G. Chaloner, and P. R. Crane. 1987. *The Origins of Angiosperms and Their Biological Consequences.* Cambridge University Press, Cambridge.

Gantt, J. S. 1988. Nucleotide sequences of cDNAs encoding four complete nuclear-encoded plastid ribosomal proteins. *Curr. Genet.* **14:**519–528.

Giese, K., and A. R. Subramanian. 1989. Chloroplast ribosomal protein L12 is encoded in the nucleus: construction and identification of its cDNA clones and nucleotide sequence including the transit peptide. *Biochemistry* **28:**3525–3529.

Giese, K., A. R. Subramanian, I. M. Larrinua, and L. Bogorad. 1987. Nucleotide sequence, promoter analysis, and linkage mapping of the unusually organized operon encoding ribosomal proteins S7 and S12 in maize chloroplast. *J. Biol. Chem.* **262:**15251–15255.

Gold, B., N. Carrillo, K. K. Tewari, and L. Bogorad. 1987. Nucleotide sequence of a preferred maize chloroplast genome template for *in vitro* DNA synthesis. *Proc. Natl. Acad. Sci. USA* **84:**194–198.

Hahn, V., A.-M. Dorne, R. Mache, J.-P. Ebel, and P. Stiegler. 1988. Identification of an *Escherichia coli* S1-like protein in the spinach chloroplast ribosome. *Plant Mol. Biol.* **10:**459–464.

Hoober, J. K. 1984. *Chloroplasts.* Plenum Publishing Corp., New York.

Hudson, G. S., J. G. Mason, T. A. Holton, B. Koller, G. B. Cox, P. R. Whitfeld, and W. Bottomley. 1987. A gene cluster in the spinach and pea chloroplast genomes encoding one CF_1 and three CF_0 subunits of the H^+-ATP synthase complex and the ribosomal protein S2. *J. Mol. Biol.* **196:**283–298.

Kamp, R. M., B. R. Srinivasa, K. von Knoblauch, and A. R. Subramanian. 1987. Occurrence of a methylated protein in chloroplast ribosomes. *Biochemistry* **26:**5866–5870.

Koller, B., H. Fromm, E. Galun, and M. Edelman. 1987. Evidence for in vivo *trans*-splicing of pre-mRNAs in tobacco chloroplasts. *Cell* **48:**111–119.

Mache, R., A.-M. Dorne, and R. M. Batlle. 1980. Characterization of spinach plastid ribosomal proteins by two-dimensional gel electrophoresis. *Mol. Gen. Genet.* **117:**333–338.

Markmann-Mulisch, U., and A. R. Subramanian. 1988a. Nucleotide sequence of maize chloroplast *rps*11 with conserved amino acid sequence between eukaryotes, bacteria and plastids. *Biochem. Int.* **17:**655–664.

Markmann-Mulisch, U., and A. R. Subramanian. 1988b. Nucleotide sequence and linkage map position of the genes for ribosomal proteins L14 and S8 in the maize chloroplast genome. *Eur. J. Biochem.* **170:**507–514.

Markmann-Mulisch, U., K. von Knoblauch, A. Lehmann, and A. R. Subramanian. 1987. Nucleotide sequence and linkage map position of the *sec*X gene in maize chloroplast and evidence that it encodes a protein belonging to the 50S ribosomal subunit. *Biochem. Int.* **15:**1057–1067.

McLaughlin, W. E., and I. M. Larrinua. 1987a. The sequence of the maize *rps*19 locus and the inverted repeat/unique region junctions. *Nucleic Acids Res.* **15:**3932.

McLaughlin, W. E., and I. M. Larrinua. 1987b. The sequence of the maize plastid encoded *rpl*22 locus. *Nucleic Acids Res.* **15:**4356.

McLaughlin, W. E., and I. M. Larrinua. 1987c. The sequence of the maize plastid encoded *rps*3 locus. *Nucleic Acids Res.* **15:** 4689.

McLaughlin, W. E., and I. M. Larrinua. 1987d. The sequence of the first exon and part of the intron of the maize plastid encoded *rpl*16 locus. *Nucleic Acids Res.* **15:**5896.

McLaughlin, W. E., and I. M. Larrinua. 1988. The sequence of the maize plastid encoded *rpl*23 locus. *Nucleic Acids Res.* **16:**8183.

Michel, F., and B. Dujon. 1983. Conservations of RNA secondary structures in two intron families including mitochondrial-, chloroplast-, and nuclear-encoded members. *EMBO J.* **2:**33–38.

Montandon, P.-E., and E. Stutz. 1984. The genes for the ribosomal proteins S12 and S7 are clustered with the gene for the EF-Tu protein on the chloroplast genome of *Euglena gracilis. Nucleic Acids Res.* **12:**2851–2859.

Neuhaus, H., A. Scholz, and G. Link. 1989. Structure and expression of a split chloroplast gene from mustard (*Sinapsis alba*): ribosomal protein gene *rps*16 reveals unusual transcription features and complex RNA maturation. *Curr. Genet.* **15:**63–70.

Ohto, C., K. Torazawa, M. Tanaka, K. Shinozaki, and M. Sigiura. 1988. Transcription of ten ribosomal protein genes from tobacco chloroplasts: a compilation of ribosomal protein genes found in the tobacco chloroplast genome. *Plant Mol. Biol.* **11:**589–600.

Ohyama, K., H. Fukuzawa, T. Kohchi, T. Sano, S. Sano, H. Shirai, K. Umesono, T. Shiki, M. Takeuchi, Z. Chang, S. Aota, H. Inokuchi, and H. Ozeki. 1988. Structure and organization of *Marchantia polymorpha* chloroplast genome. *J. Mol. Biol.* **203:** 281–298.

Palmer, J. D. 1985. Comparative organization of chloroplast genomes. *Annu. Rev. Genet.* **19:**325–354.

Phua, S. H., B. R. Srinivasa, and A. R. Subramanian. 1989. Chloroplast ribosomal protein L13 is encoded in the nucleus and is considerably larger than its bacterial homologue. *J. Biol. Chem.* **264:**1968–1971.

Prombona, A., and A. R. Subramanian. 1989. A new rearrangement of angiosperm chloroplast DNA in rye (*Secale cerale*) involving translocation and duplication of the ribosomal *rps*15 gene. *J. Biol. Chem.* **264:**19060–19065.

Rodermel, S. R., and L. Bogorad. 1985. Maize plastid photogenes: mapping and photoregulation of transcript levels during light-induced development. *J. Cell Biol.* **100:**463–476.

Ruf, M., and H. Kössel. 1988. Occurrence and spacing of ribosome recognition sites in mRNAs of chloroplasts from higher plants. *FEBS Lett.* **240:**41–44.

Russel, D., and L. Bogorad. 1987. Transcription analysis of the maize chloroplast gene for the ribosomal protein S4. *Nucleic Acids Res.* **15:**1853–1867.

Schmid, G., O. Strobel, M. Stöffler-Meilicke, G. Stöffler, and A. Böck. 1984. A ribosomal protein that is immunologically conserved in archaebacteria, eubacteria and eukaryotes. *FEBS Lett.* **177:**189–194.

Schwarz, Z., and H. Kössel. 1980. The primary structure of 16S rDNA from *Zea mays* chloroplast is homologous to *E. coli* 16S rRNA. *Nature* (London) **283:**739–742.

Srinivasa, B. R., and A. R. Subramanian. 1987. Nucleotide sequence and linkage map position of the gene for maize chloroplast ribosomal protein S14. *Biochemistry* **26:**3188–3192.

Subramanian, A. R. 1983. Structure and function of ribosomal protein S1. *Prog. Nucleic Acid Res. Mol. Biol.* **28:**101–142.

Subramanian, A. R. 1985. The ribosome: its evolutionary diversity and the functional role of one of its components. *Essays Biochem.* **21:**45–85.

Subramanian, A. R., A. Steinmetz, and L. Bogorad. 1983. Maize chloroplast DNA encodes a protein sequence homologous to the bacterial assembly protein S4. *Nucleic Acids Res.* **11:**5277–5286.

Udalova, G. V., N. P. Yurina, and M. S. Odinstova. 1984. Ribosomes of the chloroplasts: stoichiometry of proteins. *Dokl. Akad. Nauk SSSR* **278:**1258–1262.

Wada, A., and T. Sako. 1987. Primary structure of and genes for new ribosomal proteins A and B in *Escherichia coli*. *J. Biochem.* (Tokyo) **101:**817–820.

Wittmann, H. G. 1982. Components of bacterial ribosomes. *Annu. Rev. Biochem.* **51:**155–183.

Wittmann-Liebold, B. 1986. Ribosomal proteins: their structure and evolution, p. 326–361. *In* B. Hardestry and G. Kramer (ed.), *Structure, Function, and Genetics of Ribosomes*. Springer-Verlag, New York.

Zhou, D.-X., F. Quigley, O. Massenet, and R. Mache. 1989. Cotranscription of S10- and *spc*-like operons in spinach chloroplasts and identification of three of their gene products. *Mol. Gen. Genet.* **216:**439–445.

Zurawski, G., W. Bottomley, and P. R. Whitfeld. 1984. Junctions of the large single copy region and the inverted repeats in *Spinacia oleracea* and *Nicotiana debneyi* chloroplast DNA: sequence of the genes for $tRNA^{His}$ and the ribosomal proteins S19 and L2. *Nucleic Acids Res.* **12:**6547–6558.

INDEX